General College Chemistry

General College Chemistry
Sixth Edition

Charles W. Keenan

Professor of Chemistry

THE UNIVERSITY OF TENNESSEE, KNOXVILLE

Donald C. Kleinfelter

Professor of Chemistry

THE UNIVERSITY OF TENNESSEE, KNOXVILLE

Jesse H. Wood

Professor Emeritus of Chemistry

THE UNIVERSITY OF TENNESSEE, KNOXVILLE

HARPER & ROW,
PUBLISHERS

San Francisco

Cambridge
Hagerstown
New York
Philadelphia

London
Mexico City
São Paulo
Sydney

1817

Sponsoring Editor:	*Malvina Wasserman*
Special Projects Editor:	*Ann Ludwig*
Project Editor:	*Karen Judd*
Copyeditor:	*Carol Reitz*
Summary and Glossary Author:	*Fred Raab*
Production Manager:	*Laura Argento*
Designer and Cover Artist:	*Dare Porter*
Illustrator:	*J & R Technical Services Inc.*
Color Section Separator:	*Color Tech Corp.*
Cover Color Separator:	*Focus 4*
Cover Photographer:	*David Tise*
Compositor:	*York Graphic Services, Inc.*
Printer and Binder:	*Kingsport Press*

Our cover photograph is of the kaleidocycle—a three-dimensional ring formed by a chain of linked tetrahedra—designed especially for the Sixth Edition of *General College Chemistry*. Its six tetrahedra are decorated with a photomicrograph of magnesium nitrate under polarized light with a magnification of $40\times$ (Manfred P. Kage; Peter Arnold, Inc.).

Library of Congress Cataloging in Publication Data

Keenan, Charles William.
    General college chemistry.

    Includes index.
    1. Chemistry.   I. Kleinfelter, Donald C., joint author.   II. Wood, Jesse Hermon, joint author.
III. Title.
QD31.2.W66 1980   540   79-19311
ISBN 0-06-043615-8

Basic Organization of the Text

Preface xii
Chapter 1: Preview of Descriptive and Theoretical Chemistry 1
Chapter 2: Chemical Equations; Stoichiometry 34
Chapter 3: Structures of Atoms; Periodic Relationships 59
Chapter 4: Atomic Spectroscopy; Modern Electronic Theory 92
Chapter 5: Chemical Compounds 120
Chapter 6: Theories of the Covalent Bond 163
Chapter 7: The Gaseous State 197
Chapter 8: Changes of State: Liquids and Solids 229
Chapter 9: The Chemical Behavior of Hydrogen, Oxygen, and Water 261
Chapter 10: Solutions; Electrolytes and Nonelectrolytes 292
Chapter 11: Acids and Bases 321
Chapter 12: Properties of Solutions; The Colloidal State 342
Chapter 13: Thermochemistry; Thermodynamics 371
Chapter 14: Chemical Kinetics 403
Chapter 15: Chemical Equilibria 440
Chapter 16: Ionic Equilibria I: Solutions of Acids and Bases 472
Chapter 17: Ionic Equilibria II: Solubility and the Solubility Product 506
Chapter 18: Ionic Equilibria III: Redox and Electrochemistry 530
Chapter 19: Nuclear Chemistry 564
Chapter 20: Molecular Spectroscopy and Molecular Structure 606
Chapter 21: Metals I: Properties and Production 623
Chapter 22: Metals II: More About Transition Metals 655
Chapter 23: Groups VIIA and VIIIA: The Halogens and Noble Gases 679
Chapter 24: Groups VIA and VA: The Sulfur Family and the Nitrogen Family 706
Chapter 25: Carbon, Silicon, and Boron 740
Chapter 26: Organic Chemistry 764
Chapter 27: Complex Derivatives of Hydrocarbons; Biochemistry 798
Appendix 1: Arithmetic Procedures 827
Appendix 2: Derivation of Integrated Rate Equation 835
Appendix 3: Tables 836
Glossary 846
Answers to Exercises 860
Index 868

Contents

Preface *xii*

Chapter 1: Preview of Descriptive and Theoretical Chemistry 1

1•1 Our Chemical World 2   1•2 SI Units of Measurement 4   Special Topic ★ **Exact, Precise, and Accurate** 6   Special Topic ★ **The Arbitrary Character of Units of Measure** 12   1•3 Properties of Matter 14   1•4 Changes in Matter and Energy 15   1•5 Laws About Matter and Energy 19   1•6 Hallmarks of a Chemical Reaction 22   1•7 Dalton's Atomic Theory 24   1•8 Mendeleev's Periodic Table 26   **Chapter Review** 30

Chapter 2: Chemical Equations; Stoichiometry 34

2•1 Writing Formulas 35   2•2 Practice in Writing Balanced Equations 37   2•3 Atomic Weights 40   2•4 Molecular Weights 40   2•5 Weight Relationships in Chemical Reactions 41   2•6 Calculation of Percentage Composition from Formulas 49   2•7 Calculation of Formulas from Experimental Data 49   **Chapter Review** 55

Chapter 3: Structures of Atoms; Periodic Relationships 59

3•1 Nature of Charged Bodies 60   3•2 Experimental Evidence for the Election 61   3•3 Nucleus of the Atom 62   3•4 Calculation of Atomic Weights 70   3•5 Sizes of Atoms 73   3•6 Arrangements of Electrons in Atoms 73   3•7 A Modern Periodic Table 80   3•8 Energy Sublevels 81   3•9 Usefulness of the Periodic Table 88   **Chapter Review** 88

Chapter 4: Atomic Spectroscopy; Modern Electronic Theory 92

4•1 Electromagnetic Radiation 93   4•2 Atomic Spectra 94   4•3 The Wave Nature of Matter 104   4•4 The Heisenberg Uncertainty Principle 105   4•5 The Quantum Mechanical Description of Electrons in Atoms 106   Special Topic ★ **Computers—Tools of the Chemical Theorist** 114   4•6 Photoelectron Spectra of Atoms 115   **Chapter Review** 117

Chapter 5: Chemical Compounds 120

5•1 How Atoms Combine 121   5•2 Transfer of Electrons 123   5•3 Sharing of Electrons 128   5•4 Diagramming Lewis Formulas for Covalent Molecules 132   5•5 Polar Covalent Bonds 138   5•6 Sizes of Atoms, Molecules, and Ions 144   5•7 Oxidation Numbers 150   5•8 The Systematic Naming of Compounds 155   **Chapter Review** 158

Chapter 6: *Theories of the Covalent Bond* 163
6.1 Introduction to Valence Bond Theory 164   6.2 Sigma Bonds 165   6.3 Pi Bonds 167
6.4 Hybridized Orbitals 167   6.5 Valence Shell Electron Pair Repulsion (VSEPR) Model
for Molecular Geometry 178   6.6 Resonance 182   6.7 Introduction to Molecular Orbital
Theory 188   Special Topic ★ **The Use of Molecular Models** 188   6.8 Sigma and Pi Molecular
Orbitals 190   **Chapter Review** 194

Chapter 7: *The Gaseous State* 197
7.1 The Three States of Matter 198   7.2 The Pressure of Gases 199   7.3 Boyle's Law 201
7.4 Temperature Effects 204   7.5 Avogadro's Law 209   7.6 A General Gas Equation 211
7.7 Molar Relationships with Gaseous Reactants 216   7.8 Dalton's Law of Partial
Pressures 218   7.9 Graham's Law of Diffusion (Effusion) 220   7.10 Deviations from the
Gas Laws 221   7.11 The Kinetic Molecular Theory 222   **Chapter Review** 224

Chapter 8: *Changes of State: Liquids and Solids* 229
8.1 General Characteristics of Changes of State 230   8.2 Intermolecular Attractive
Forces 235   8.3 Liquefaction of Gases 239   8.4 Vaporization of Liquids 240
8.5 Solidification of Liquids 244   8.6 Crystalline Solids 245   8.7 Ionic Compounds 248
8.8 Solid Covalent Substances 251   8.9 Metallic Solids 253   8.10 Packing of Particles 256
Chapter Review 257

Chapter 9: *The Chemical Behavior of Hydrogen, Oxygen, and Water* 261
9.1 The Phlogiston Theory 262   9.2 Oxygen 263   Special Topic ★ **A Matter of Priority: The
Discovery of Oxygen** 263   9.3 Ozone 272   Special Topic ★ **Threatened Depletion of the Earth's
Ozone Shield** 276   9.4 Hydrogen 278   9.5 Water 286   9.6 Hydrogen Peroxide 288
Chapter Review 288

Chapter 10: *Solutions; Electrolytes and Nonelectrolytes* 292
10.1 Nature of Solutions 293   10.2 Why Substances Dissolve 293   10.3 Solubility
Relationships 296   10.4 Effect of Temperature on Solubility 299   10.5 Effect of Pressure
on Solubility 301   10.6 Expressing Concentrations 302   10.7 Electrolytes and
Nonelectrolytes 307   10.8 Ionic Equations 312   **Chapter Review** 317

Chapter 11: *Acids and Bases* 321
11.1 Arrhenius Acids and Bases 322   11.2 Brønsted–Lowry Acids and Bases 322
11.3 Naming Inorganic Acids 329   11.4 Structures of Hydroxy Compounds 330
11.5 Neutralization 331   11.6 Lewis Acids and Bases 337   **Chapter Review** 339

Chapter 12: Properties of Solutions; The Colloidal State 342

12.1 Properties of Solutions of Nonelectrolytes 343   12.2 Properties of Solutions of Electrolytes 350   12.3 Distillation 353   12.4 Osmosis 355   Special Topic ★ **Desalination of Seawater** 358   12.5 Particle Size and the Colloidal State 360   12.6 Importance of Colloid Chemistry 361   12.7 Types of Colloidal Systems 362   12.8 Properties of Colloidal Systems 362   12.9 Selective Separations 364   12.10 Stability of Colloidal Systems 366   **Chapter Review** 367

Chapter 13: Thermochemistry; Thermodynamics 371

13.1 Introduction 372   13.2 Experimental Determination of Heats of Reaction 373   13.3 Thermochemical Equations 374   13.4 Standard Enthalpies 376   13.5 Bond Dissociation Energies 383   13.6 Enthalpies of Electron Loss or Gain 385   13.7 Enthalpies of Ionic Crystals 386   13.8 The First Law of Thermodynamics 389   13.9 The Criteria for Spontaneous Chemical Processes 390   Special Topic ★ **Entropy and Natural Processes** 392   **Chapter Review** 399

Chapter 14: Chemical Kinetics 403

14.1 Reaction Mechanisms 404   14.2 Reaction Rate 407   14.3 Determination of Rate Laws or Rate Equations 418   14.4 Order of a Chemical Reaction 421   14.5 Determination of Reaction Order from Experimental Data 423   14.6 Order and Reaction Mechanism 431   14.7 Chain Reactions 432   **Chapter Review** 434

Chapter 15: Chemical Equilibria 440

15.1 Establishing Chemical Equilibria 441   15.2 Influence of the Nature of Reactants 442   15.3 Influence of Concentration 443   15.4 Influence of Pressure Changes 456   15.5 Influence of Temperature 465   15.6 Influence of a Catalyst 466   **Chapter Review** 467

Chapter 16: Ionic Equilibria I: Solutions of Acids and Bases 472

16.1 Ionization Constants of Weak Acids 473   16.2 Ionization Constants of Weak Bases 475   16.3 Determination of and Calculations Involving K_a and K_b 476   16.4 Ionization of Water 482   16.5 Hydrolysis Constants of Acidic Cations and Basic Anions 484   16.6 Expressing Hydrogen Ion Concentration 489   16.7 The Common Ion Effect 491   16.8 Buffered Solutions 493   16.9 Indicators 498   **Chapter Review** 502

Chapter 17: Ionic Equilibria II: Solubility and the Solubility Product 506

17.1 Solubility Product Constants 507   17.2 Calculation of Solubility Product Constants 508   17.3 Calculation of Solubilities from K_{sp} Values 510   17.4 The Common Ion Effect 512   17.5 Separation by Selective Precipitation 514   17.6 Dissolution of Precipitates 518   17.7 The Qualitative Analysis Scheme 523   **Chapter Review** 527

Chapter 18: Ionic Equilibria III: Redox and Electrochemistry 530

18.1 Voltaic Cells 531   18.2 Standard Electrode Potentials 535   18.3 Potentials of Voltaic Cells 539   18.4 Importance of Standard Electrode Potentials 544   18.5 Equilibrium Constants and Free Energies from Electrode Potentials 544   Special Topic ★ **The Glass Electrode** 546   18.6 Balancing Redox Equations 547   18.7 Driving Force of Equilibrium Constants 551   18.8 Electrolysis 552   18.9 Batteries and Fuel Cells 556   **Chapter Review** 559

Chapter 19: Nuclear Chemistry 564

19.1 Radioactive Elements 565   19.2 Detection of Radiations 567   19.3 Radioactive Series 569   19.4 Background Radiation 573   19.5 Half-Life 574   19.6 Applications of Radioactivity 576   19.7 Bombardment Reactions 579   19.8 Acceleration of Charged Particles 580   19.9 Mass Loss and Binding Energy 583   19.10 Nuclear Stability 589   Special Topic ★ **Possible Superheavy Nuclides** 590   19.11 Nuclear Fission 591   19.12 Nuclear Reactors 595   Special Topic ★ **Our Energy Dilemma: Nuclear or Coal or ?** 598   19.13 Fusion Reactions 600   **Chapter Review** 602

Chapter 20: Molecular Spectroscopy and Molecular Structure 606

20.1 Some Types of Spectroscopy 607   20.2 Ultraviolet and Visible Spectra 608   20.3 Infrared Spectra 611   Special Topic ★ **Color and Color Photography** 612   20.4 Nuclear Magnetic Resonance Spectra 615   20.5 Mass Spectra 617   20.6 Miscellaneous Uses of Spectroscopy 618   **Chapter Review** 620

Chapter 21: Metals I: Properties and Production 623

21.1 General Divisions of the Periodic Table 624   21.2 Physical Properties of IA and IIA Elements 626   21.3 Chemical Properties of IA and IIA Elements 627   21.4 Compounds of IA and IIA Elements 629   21.5 Classification 634   21.6 Physical Properties 635   21.7 Chemical Properties 637   Special Topic ★ **When Your Car Rusts Out** 639   21.8 Metallurgy 641   21.9 Industrial and Environmental Aspects of Metal Production 645   **Chapter Review** 652

Chapter 22: Metals II: More About Transition Metals 655

22.1 Emf Diagrams 657   22.2 Stability of Redox Species in Solution 658   22.3 Effect of pH on Stabilization of Redox Species in Solution 661   22.4 Effect of Disproportionation on Stabilities of Redox Species in Solution 661   22.5 Effect of Atmosphere on Stabilities of Redox Species in Solution 663   22.6 Coordination Compounds 663   22.7 Ligand-Field Theory 669   22.8 Uses of Coordination Compounds 673   **Chapter Review** 676

Chapter 23: Groups VIIA and VIIIA: The Halogens and Noble Gases 679

23.1 Properties of the Halogen Family 680   23.2 Characteristic Reactions 683
23.3 Production of the Halogens 687   23.4 Important Uses of the Halogens 691
Special Topic ★ **Mercury and Chlorine in the Environment** 692   23.5 Metal Halides 693
23.6 Nonmetal and Metalloid Halides 694   23.7 Halogen Oxy-Acids and Oxy-Salts 696
23.8 Isolation of the VIIIA Elements 697   23.9 Physical Properties 698   23.10 Chemical
Properties 698   Special Topic ★ **Facts, Theories, and Scientific Discovery** 698   23.11 Compounds
of the Noble Gases 699   23.12 Uses of the Noble Gases 701   **Chapter Review** 702

Chapter 24: Groups VIA and VA: The Sulfur Family and the Nitrogen Family 706

24.1 Properties of the Sulfur Family 707   24.2 Characteristic Reactions 712   24.3 Production
of the Sulfur Family Elements 712   24.4 Properties of the Binary Hydrogen Compounds 713
24.5 Sulfides, Selenides, and Tellurides 714   24.6 Metal Polysulfides 714   24.7 Oxides and
Oxy-Acids of Sulfur 715   Special Topic ★ **Acid Rain** 716   24.8 Sulfates, Selenates, and
Tellurates 720   24.9 Properties of the Nitrogen Family 721   24.10 Characteristic
Reactions 725   24.11 Production of the VA Elements 727   24.12 Properties of the Binary
Hydrogen Compounds 728   24.13 Oxy-Acids and Bases of the VA Elements 732
Special Topic ★ **Protection of Teeth by Fluoride Ions** 734   24.14 Miscellaneous Compounds of
the VA Elements 735   **Chapter Review** 736

Chapter 25: Carbon, Silicon, and Boron 740

25.1 Properties of Carbon, Silicon, and Boron 741   Special Topic ★ **Transistors and Solar
Cells** 744   25.2 Occurrence 747   25.3 Compounds of Carbon 747   25.4 Compounds of
Silicon 754   Special Topic ★ **Extraterrestrial Chemistry** 758   25.5 Compounds of Boron 760
Chapter Review 761

Chapter 26: Organic Chemistry 764

26.1 Alkanes 765   Special Topic ★ **Sources of Natural Gas** 768   26.2 Alkenes 769
26.3 Alkynes 771   26.4 Benzene Hydrocarbons 772   26.5 Cycloalkanes 774   26.6 Sources
of Hydrocarbons 775   26.7 Properties of Hydrocarbons 778   26.8 Alcohols 781   Special
Topic ★ **A Fuel Economy Based on Methanol** 784   26.9 Phenols 788   26.10 Carboxylic
Acids 788   26.11 Esters 790   **Chapter Review** 794

Chapter 27: Complex Derivatives of Hydrocarbons; Biochemistry 798

27.1 Stereoisomers and Chirality 799   27.2 Carbohydrates 803   27.3 Proteins 809
27.4 Oxidation–Reduction in Cells 815   27.5 Hormones 817   27.6 Viruses 818
27.7 Drugs 818   Special Topic ★ **Biochemical Benefits and Risks of Synthetic Organic
Compounds** 820   27.8 Polymers 822   **Chapter Review** 824

Appendix 1: Arithmetic Procedures 827

1.1 A Few Useful Formulas 827   1.2 Significant Figures 827   1.3 Expressing Numbers 829
1.4 Logarithms 830   1.5 More on Taking Roots 831   1.6 Factor-Units Method 832

Appendix 2: Derivation of Integrated Rate Equation 835

Appendix 3: Tables 836

A.1 SI Base Units 836   A.2 Certain SI Derived Units with Special Names 837   A.3 Prefixes
for Fractions and Multiples of SI Units 837   A.4 General Physical Constants 837   A.5 SI
Equivalents for Units to Be Abandoned Eventually 838   A.6 Equivalents (Conversion
Factors) 838   A.7 Vapor Pressure of Water at Different Temperatures 839   A.8 Four-Place
Logarithms 840   A.9 Standard Reduction Potentials, $\mathscr{E}^{\circ}_{red}$, in Acid Solutions 842
A.10 Standard Reduction Potentials, $\mathscr{E}^{\circ}_{B}$, in Basic Solutions 845

Glossary 846

Answers to Exercises 860

Index 868

Preface

This new edition of *General College Chemistry*, like its predecessors, presents an introduction to chemistry for the student who seeks a firm foundation in the subject. The coverage is broad, but it is also deep in certain areas. The subjects that are emphasized—such as atomic and molecular structure, chemical bonding, acids and bases, equilibrium, and the periodic relationships of elements—are fundamental to many disciplines other than chemistry. This text is designed not only for the science major, but also for the person who wants a cultural understanding of the chemical way of looking at the world around us.

Chemistry is divided more evenly than most sciences between theoretical and practical knowledge. On the one hand, the theories of the behavior of the electron seem to have little to do with our everyday life. On the other hand, chemists help create and produce most of the materials enjoyed by modern society.

The study and practice of chemistry play a major role in shaping our civilization. Our philosophical concepts of the nature of life itself, our production of food, medicines, and weapons, our effect on the environment that nourishes and sustains us—all these and practically everything else of interest to us have been affected by the human development of chemistry. The philosopher Garry Wills remarks that our environmental concerns are now, as ever, with earth, air, fuel, and water. In this provocative allusion to the four elements recognized by the ancients, we have a hint of the enormous influence of the chemical perspective on our culture.

The study of general chemistry is quite demanding. For most students the first-year college course, as it has developed in this country, requires the mastering of many new concepts. These are often abstract and commonly involve understanding numerical relationships. In this revision of *General College Chemistry*, great pains have been taken to make clear the objectives of the course, to show how topics are related to one another, to illustrate how typical numerical problems are solved, to provide exercises in developing problem-solving skills, and to guide the student in reviewing the large amount of subject matter. The text itself is organized to help the student develop a systematic plan of study.

Organization of the Text

● *Chapter outlines* are given at the beginning of each chapter.

● An *introduction* to each chapter provides in a few paragraphs an overview of its objectives.

● *Important terms* are set in boldface the first time they are used, and they are defined at that point.

● *Examples/solutions* are given to illustrate the most important types of problems. With each example there is a reference to one or more similar exercises at the end of the chapter.

● A *Summary* of each chapter focuses on the essential concepts. It recalls the terms that one should be able to define and the relationships one should be able to apply in solving problems.

● The *Exercises* at the end of each chapter, arranged under general topic headings, are designed to give practice in applying both verbal and numerical concepts.

● *Arithmetic procedures* essential for solving problems in general chemistry are reviewed in the appendix.

● *Tables* of units, equivalents, logarithms, and selected data are included in the appendix.

● A *Glossary*, following the appendix, collects the definitions of important terms for easy reference.

● *Answers* to numerical exercises are listed following the glossary.

- Finally, there is a detailed *Index* to guide one quickly to any topic in the text.

Changes and Features in This Edition

Some of the changes made for this revision were suggested by our students, colleagues, and editors. Many were suggested by peer educators who reviewed thoroughly the fifth edition and the preliminary manuscript for this revision with the aim of making the presentation even more effective in the classroom.

Previous users of *General College Chemistry* will recognize some of the features of the organization as new in this edition. The listing of *topic headings* for each chapter not only gives students a view of the forest before they become engrossed with the trees, but it is an aid to the teacher in making assignments and indicating any changes in coverage to a class. Both the *Summaries* and the *Glossary,* new in this edition, will be of great help to the teacher as well as to the student. In the main, the organization and coverage of material follow the pattern that has been so well received in previous editions, but there are a number of noteworthy changes in content and emphasis.

- *SI units* are stressed throughout the text, although they are not used exclusively. We view the present period as one of transition, during which students must be familiar with English, cgs, and SI solutions to problems. We therefore have many examples and exercises involving conversions between systems. There is a thoroughgoing use of the joule rather than the calorie as the unit of energy. One of the great benefits of this use of an SI unit is the consistency it brings to energy calculations throughout the course. A case in point is that the student finds it necessary to become expert in using only one value of the gas constant R, not three different values for gas law, electric energy, and thermodynamic calculations. In many instances non-SI units are used without slavish conversion to SI units before solving simple problems. For example, our usage reflects the opinion of the international commission that the units of angstroms and atmospheres can be phased out gradually.

- *Chemical equations and stoichiometry* are covered earlier in this edition, in Chapter 2 rather than mainly

in Chapter 6 as in the fifth edition. This helps students correlate their first lecture and laboratory work.

- *Electronic theory and theories of the covalent bond,* discussed in Chapters 4 and 6, are treated in less detail than in the previous edition. With some classes, teachers may find it desirable to delay covering these chapters until just prior to Chapter 20 on molecular structure. Chapters 3 and 5 are designed to work well with this optional change in order.

- *Solution relationships and acids and bases* are given increased attention by the new organization in Chapters 10 and 11.

- *Thermochemistry* is now placed before the work on kinetics and equilibria because of the importance to these two subjects of concepts of energy.

- *Ionic equilibria* are developed in greater depth in three correlated chapters, including new material on the solubility of ionic substances in Chapter 17.

- An overview of *qualitative analysis* is presented to illustrate solubility product relationships. In our companion text, *General College Chemistry*, Sixth Edition, *with Qualitative Analysis,* this material is correlated with laboratory instruction in semimicro analysis.

- *Simple electromotive force diagrams* of the classical Latimer type, used in most American texts, have replaced the graphs in the fifth edition that correlated both voltages and free energy changes.

- The *chemistry of the sulfur and nitrogen families* is coordinated in Chapter 24, rather than being presented in separate chapters.

- *Organic chemistry* is presented in one chapter rather than in three chapters as in the fifth edition.

- Deletions and reorganizations have made possible more illustrative *Examples*—178 in this edition versus 96 in the fifth—and more *Exercises*—1,270 in this edition versus 1,027 in the fifth. Most of the exercises are new.

- A complete list of *Answers* is now given for numerical exercises. A majority of present reviewers and users favored this approach over the previous consensus to omit about a third of the answers.

- The *reading level* of the text has received careful attention. Sentences have been shortened and written more often as simple declarative statements. The definitions of boldface terms are stated with particular care.

• Lists of *supplementary readings* for each chapter are now included in the Instructor's Manual rather than in the text.

• *Key Chemicals* that are produced in exceptional amounts by industry are featured throughout the text, but especially in later chapters. We are indebted to *Chemical and Engineering News* for permission to use its term for these substances and the latest available data on production and costs.

• There are new *Special Topics* on units of measurement, precision, the use of models, priority in scientific discovery, our energy dilemma, acid rain, transistors and solar cells, extraterrestrial chemistry, and the biochemical effects of synthetic organic chemicals.

• There are new and modified illustrations, including nine new ones in full color. The care devoted by the artist and the editors to the illustrations maintains the high standard that continues to distinguish the artwork in *General College Chemistry.*

• The completely new design of the book complements the other changes in increasing the pedagogical effectiveness of the text. There is a smooth integration of the text discussion with the illustrative examples, special topics, and figures.

Accompanying Studying/Teaching Aids

• The *Study Guide/Workbook* by Donald C. Kleinfelter and Joseph R. Peterson, which was widely adopted by users of the fifth edition, has been revised to accompany this edition. Arranged systematically with the sections in the text, it includes clearly stated *learning objectives* to help the student focus on essential material. It guides the student in identifying *key words* and *concepts* and in practicing their expression and use. Worked-out examples of *exercises* are presented to develop the skills necessary to analyze, set up, and solve numerical relationships. In each chapter there is a *self-test* to help students evaluate their success in mastering the learning goals. *Answers* are given for all exercises and self-test questions in the *Study Guide/Workbook.*

• The *Laboratory Manual for College Chemistry,* Sixth Edition, by William E. Bull, William T. Smith, Jr., and Jesse H. Wood, has been revised to accompany this edition of the text. The stress on *SI units* parallels that now in the text. Each laboratory exercise has been critically examined with the aim of using *nonhazardous chemicals* as well as *low-cost chemicals.* The result has been to eliminate or minimize the use of substances identified as dangerous in recent studies sponsored by the Department of Health, Education and Welfare. The laboratory program produced in this way continues to have an outstanding selection of exercises with full attention to health and financial considerations. Exercises in inorganic synthesis and in applied biochemistry have been added in this revision.

• An *Instructor's Manual* for the sixth edition by John T. Cone and the authors of this text is available to teachers. It includes *suggested course outlines* for courses with different objectives. The wealth of material in *General College Chemistry* makes it possible for the teacher to make selective omissions in designing somewhat different courses. The *Manual* also includes *solutions* to all numerical and most verbal exercises and lists of *supplementary readings* for each chapter.

Acknowledgments

We owe much to others for the form and content of this sixth edition. Of course, we have been stimulated by the exciting technical discoveries of the last five years and by many of the teaching developments reported in such places as the *Journal of Chemical Education* and *Chemistry* (now *SciQuest*). But our chief debt is to those who brought fresh ideas and devoted long hours to this revision.

The next to last version of the manuscript was critically reviewed, and in some areas reorganized, by Patricia A. Cunniff, Prince George's Community College, Robert Desiderato, North Texas State University, Gary L. Gard, Portland State University, Cecil N. Hammonds, Penn Valley Community College, Forrest C. Hentz, Jr., North Carolina State University, William K. Plucknett, University of Kentucky, and Vincent J. Sollimo, Burlington County College. Weldon Burnham, Richland College, reviewed the exercises for content and clarity.

Fred Raab, Mill Valley, California, did a superlative job of writing summaries that are both faithful to the content of the chapters and creative in the way concepts are related to one another. He also compiled the *Glossary,* and in the process made a number of perceptive suggestions for improving the clarity and precision of the text. Edward M. Jones, Copperhill, Tennessee, supplied detailed information on the history of the chemical operations there. John T. Cone, The University of Tennessee, edited and worked all the exercises, compiled the answers, and drafted the solutions section of the *Instructor's Manual.*

Working with Mal Wasserman and the staff at Harper & Row has been an exhilarating experience. Mal conceived and arranged for the latest and most thorough review process we have ever had. Ann Ludwig sharpened our definitions and directed the pedagogical reorganization, and Dare Porter brought forth a design that unifies the different teaching ele-

ments. Carol Reitz edited the manuscript not only with an eye for language but also with technical expertise. When the chips were down, it was Karen Judd who made the deadlines elastic and helped us most in implementing all the creative suggestions.

As always, the help at home base has been wonderfully generous and encouraging. The advice of our colleagues in the department continues to be our chief strength. For this edition, William E. Bull, James Q. Chambers, Charles A. Lane, and Torsti P. Salo reviewed and greatly improved certain sections. Joyce Wallace again brought order out of chaos with her typing and loyal support. Libba Keenan and Lynn Kleinfelter rounded out the team with devoted editing of the first efforts, proofreading of the final version, and preparation of the index, while shouldering family responsibilities abandoned by the scribes.

Knoxville, Tennessee
October, 1979

Charles W. Keenan
Donald C. Kleinfelter
Jesse H. Wood

*Bear in mind that the wonderful things you learn
in your schools are the work of many generations,
produced by enthusiastic effort and infinite labor in every
country of the world. All this is put into your hands as your
inheritance in order that you may receive it, honor it,
add to it, and one day faithfully hand it on to your
children. Thus do we mortals achieve immortality in the
permanent things which we create in common.*

Albert Einstein

Preview
of Descriptive and
Theoretical Chemistry

1•1   **Our Chemical World**

1•1•1   Chemistry and Our
Way of Life

1•1•2   Chemistry As a Science

Scientific Measurements

1•2   **SI Units of Measurement**

1•2•1   The Seven SI Base Units

1•2•2   Units of Length, Area,
and Volume

**Special Topic: Exact,
Precise, and Accurate**

1•2•3   Units of Mass and Density

1•2•4   Units of Temperature
and Energy

**Special Topic:
The Arbitrary Character
of Units of Measure**

**Fundamentals of Descriptive
Chemistry**

1•3   **Properties of Matter**

1•3•1   Chemical Properties

1•3•2   Physical Properties

1•4   **Changes in Matter
and Energy**

1•4•1   Changes in Matter

1•4•2   Changes in Energy

1•4•3   Classes of Matter

1•5   **Laws About Matter
and Energy**

1•5•1   Law of Conservation
of Energy

1•5•2   Law of Conservation
of Mass

1•5•3   Law of Definite
Composition

1•5•4   Law of Multiple Proportions

1•6   **Hallmarks of a Chemical
Reaction**

Early Theoretical Chemistry

1•7   **Dalton's Atomic Theory**

1•7•1   Relative Atomic Weights

1•7•2   Symbols of Elements

1•8   **Mendeleev's Periodic Table**

Chemistry is the study of the structure of matter and the changes matter undergoes in natural processes and in planned experiments. Through chemistry we become familiar with the composition and uses of inanimate materials, both natural and artificial, and with the vital processes of living things, including our own bodies. The chemical perspective of the world around us is fascinating. We can develop this perspective by our own observations and experiments, firmly based on our human desire for understanding and our search for order.

In this chapter, units for scientific measurements are described, and two broad areas of chemical study are introduced, the descriptive and the theoretical. When a space rocket is launched, the solid propellants erupt as a violent discharge of incandescent gases. Or consider the mixing of the liquid vinegar with the solid baking soda: a lot of gas bubbles are produced and the solid may completely disappear. **Descriptive chemistry** records the characteristics of substances that distinguish them from one another, describes the conditions under which they interact, and summarizes the properties and uses of any new substances produced. **Theoretical chemistry** seeks to explain why the changes take place. These explanations focus on the behavior of the tiniest particles of matter, particles too small to be seen but about which much has been learned from indirect measurements.

1•1   Our Chemical World

1•1•1   Chemistry and Our Way of Life   Although chemistry has been part of a liberal education for generations, today it is assuming an even greater role because of the growing awareness of the value of a chemical viewpoint by all educated people. This public recognition of the impact of chemistry is appropriate. Within the past 45 years the population of the world has doubled, and it is expected to double again during the next 25 years. One of the main reasons for this accelerating rate of growth has been the application of chemical knowledge in the areas of medicine and agriculture. The development of health-giving drugs and antibiotics to control disease and infection has been based in large part upon chemical research. The fertilizers and insecticides that have increased food production to feed the growing population have been discovered through the application of chemical knowledge.

Currently the attention that is being given to ecology, the environment, and the energy crisis is bringing about a wider concern for chemical questions by the general public. The increased number of persons living on our planet plus the increased use of material resources to raise standards of living have resulted in dangerous modifications in the environment. The poisoning of water supplies by weed-killer and insecticide residues, atmospheric pollution from chemical reactions taking place during the combustion of fuels, the killing of plant and animal life by the waste products of industrial processes—these are examples of modern problems with chemical overtones.

The economic growth of the great industrial nations is rapidly depleting our energy resources and producing a huge volume of waste products that threatens to upset the chemical balances of nature. The balances that are of most concern to us are those that operate in the zone near the surface of the earth, called the **biosphere,** the zone in which life exists (see Figure 1.1). As shown in the left-hand section of Figure 1.1, the biosphere is a relatively thin section close to the surface of the earth. From analyses of living plants and animals, chemists have identified millions of different substances. Samples of substances not associated with living things have been obtained from the atmosphere, the oceans, and the thin crust of the earth that can be sampled by the drilling of wells and other direct techniques. The total number of unique substances known today, including both those found in nature and those made synthetically in chemical laboratories, is about 4 million.

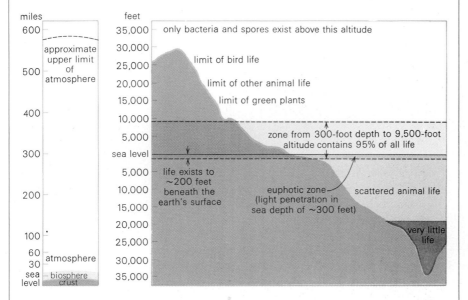

Figure 1•1 The biosphere, zone of life. The diagram at left shows the region in which life on earth exists relative to sea level, the crust of the earth, and the atmosphere. At the right some details of the biosphere are indicated schematically.
Source: Adapted by permission from "How Man Pollutes His World," National Geographic Society, Washington, D.C., 1970.

It is a formidable challenge to understand and control changes taking place on a huge scale in a system as complex as the earth's biosphere. Yet at the present time most scientists have an optimistic rather than a defeatist view of protecting the environment and developing safer ways to generate energy. For example, an understanding of the chemistry involved is the key to designing devices that minimize the amounts of undesirable substances emitted by automobile engines. Chemists are playing prominent roles in finding ways to make waste products less damaging and in developing new techniques for using nuclear, solar, and green-plant sources of energy.

1•1•2   **Chemistry As a Science**   In chemistry, as in all natural sciences, we constantly make observations and gather facts. As in all sciences except perhaps astronomy and geology, we focus our attention on *reproducible facts,* that is, on events or occurrences that take place in the same way under the same conditions. Most of the phenomena and materials of nature appear to us to be strictly reproducible. For example, the burning of 1 pound of carbon produces a given amount of energy, water freezes at a temperature of 32 °F, and carbohydrates are converted to carbon dioxide and water in the tissues of animals.

Whenever possible, as we seek to bring order to a growing collection of observations, we summarize facts in a concise statement that is called a **law.** We also seek to explain facts and laws by means of a **hypothesis** or a **theory** designed to suggest why or how something happens as it does. In the language of science, a hypothesis is similar to a theory but is less formal and is based on a less thorough study.

It is not possible, indeed not desirable, to separate the facts, laws, and theories of chemistry from those of other sciences. The arbitrary divisions of natural studies into physics, astronomy, geology, botany, zoology, chemistry, and other sciences are classifications imposed not by nature but by us. Chemistry draws heavily from all the sciences for helpful ideas, and it contributes to many areas in the other sciences. Chemical facts and theories are of great importance for anyone interested in medicine, meteorology, home economics, agriculture, or engineering. For both scientists and non-scientists, a study of chemistry enriches the appreciation of both the world of nature and the somewhat unnatural world of new materials that the chemical industry provides for all of us.

Scientific Measurements

Before looking at any specific chemical changes, we shall review the ways in which scientific observations are described. In all precise work, careful measurements of materials and conditions are necessary. Particularly in the sciences there must be standard methods for describing the size, mass, temperature, and other characteristics of any material that is studied. For a century the metric system, with its decimal structure, was the basis for scientific measurements and, in most countries, for everyday use as well. Until recently, however, England, the United States, and a few other countries retained the English system of weights and measures for all purposes except scientific work. Whereas the metric system is purposefully designed to provide for clear definitions and easy calculations, the English system is simply a collection of measurements and units that grew up over many years.

1•2   SI Units of Measurement

In 1960, a major step was taken to consolidate and simplify measurements when the General Conference of Weights and Measures adopted the International System of Units (SI). This system, which is a logical extension of the metric system, relates all units of measurement to the fewest possible base units.

1•2•1   The Seven SI Base Units

The seven SI base units are listed in Table 1.1. The measurements of all quantities can be expressed in terms of these seven units or units derived from them. Several derived SI units are described in Table A.2.

Because of their fundamental importance, the SI base units must be defined as precisely as possible. The official definitions of the SI base units are given in Table A.1. While these technical definitions are useful in laboratories with sophisticated equipment, they are of little help in giving us a practical idea of the size of any unit. It adds nothing to our sense of the length of one second of time to read that a second is "the duration of 9,192,631,770 periods" of a certain type of radiation from a certain type of cesium atom.[1]

Table 1•1 SI base units

Physical quantity	Name of unit	Symbol
length	meter[a]	m
mass	kilogram	kg
time	second	s
electric current	ampere	A
thermodynamic temperature	kelvin	K
amount of substance	mole	mol
luminous intensity	candela	cd

[a]In most countries the spelling "metre" is preferred.

To know that a second is one sixtieth of a minute and to have a long experience with second hands on watches make us comfortable with the second as a unit of time. However, the other SI base units are less familiar to us. For this reason, when each SI base unit is introduced in the text we will relate it to other, more familiar units of measurement. In Figure 1.2 are shown the relative sizes of three English units of measure and the corresponding SI units.

There is a growing commitment in the United States to adopt SI units for everyday use because of the great advantages of the system itself and because of the help it will be to education, science, industrial efficiency, and international trade and commerce. For these reasons, SI measurements generally will be used in this book, although a familiarity with the centimeter-gram-second (cgs) system and with the English system is assumed also. In Table A.6 the relationships between SI units and many other commonly used units are given. We will use the term **SI units** to refer collectively to the base units in Table 1.1, to derived units such as those in Table A.2, and to the decimal fractions and multiples of these units.

To express decimal fractions or multiples of SI units a system of prefixes is used. A list of these is given in Table A.3; the fractional prefixes are derived from Latin roots and the multiple prefixes are derived from Greek roots. The most commonly used prefixes are *centi-* for $\frac{1}{100}$, *milli-* for $\frac{1}{1,000}$, and *kilo-* for 1,000; they are abbreviated c, m, and k, respectively.

1•2•2   Units of Length, Area, and Volume   The SI base unit of length, the **meter,** m, is equivalent to about 39.37 in., or about 1.1 yd. Distances that in this country have been expressed in miles are expressed in the interna-

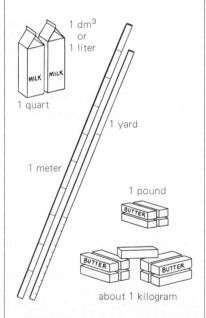

Figure 1•2 Sizes of three English units relative to SI units.
Source: **Adapted by permission from D. A. Mackay,** *National Geographic,* **152 (2), 293 (1977).**

---

[1]To read such a definition, however, does indicate to us that the unit of time can be measured with great precision. It also impresses on us that differences in lengths of time as short as one nine-billionth of a second can be measured.

tional community in **kilometers,** km. One kilometer is equivalent to about 0.62 mile. For objects in the laboratory convenient units of length are the **centimeter,** cm, and the **millimeter,** mm. One centimeter is equivalent to about 0.39 in. Two hundred leaves of this book have a thickness of about 15 mm, that is, about 1 mm thickness per 13 leaves.

Units of area and volume are derived from units of length. The fundamental SI unit of area is the **square meter,** m^2, and that of volume is the **cubic meter,** m^3. These two units are too large to be useful for laboratory

★   Special Topic ★ Special Topic ★ Special Topic ★ Special Topic ★ Special Topic ★ Special Topic ★ Special Topic ★ Special Topic   ★

Exact, Precise, and Accurate The discussion of units of measurement is an appropriate place for us to consider the meanings of three adjectives that are often applied in calculations: exact, precise, and accurate. The term *exact* is used for quantities that have an infinite number of significant figures. For example, by definition there are exactly 1,000 cubic centimeters in 1 liter. We could write that there are 1,000.0000000. . . cubic centimeters in 1.0000000. . . liter, with no limit to the number of zeros. For another example, the inch is defined as exactly 2.54×10^{-2} meter, or 2.54 centimeters. Exact quantities either are defined arbitrarily or result from counting objects one by one; *no measured quantities, other than by simple counting, are exact.*

The term *precise* refers to how reproducible measurements of the same quantity are. Suppose that two calibrated 250-cm^3 beakers are each tested three times to determine their volumes. Beaker A is found to have volumes of 248, 249, and 247 cm^3; beaker B is found to have volumes of 251, 247, and 249 cm^3. The set of values for beaker A is more precise, varying only ±1 from the average value of 248 cm^3.

The term *accurate* refers to how closely a measured value is to a true or accepted value. In the case of the two beakers, beaker B, with an average volume of 249 cm^3, is closer to the calibrated value of 250 cm^3. For most experimental measurements a true value is not known, so the accuracy cannot be determined. The relationship between precision and accuracy is illustrated in the drawings to the right.

As pointed out in Appendix 1, the number of significant figures in a quantity shows how precisely the quantity is known. The number of significant figures in a calculated answer is not limited by exact conversion factors. In Example 1.1 the factors 1 quart/2 pints and 1,000 cm^3/1 liter are exact by definition. However, the factor 1 liter/1.0567 quarts is a precise measured relation between two ways of expressing volume. The number of significant figures in the answer is limited to five by the five figures, 1.0567, in the measured conversion factor. If we had wished to calculate even more precisely the number of cubic centimeters in a pint, we could have found in a reference book the factor 1 liter/1.056718

quarts. The number of significant figures in very precisely measured quantities is rarely more than seven or eight.

The accuracy of a measurement is commonly expressed in terms of the error or the percent error. *Error* is defined as the difference between a measured value and the true (or most probable) value. In the case of beaker A, the error between the average of the three measurements and the calibrated value is

$$250 \text{ cm}^3 - 248 \text{ cm}^3 = 2 \text{ cm}^3$$

The error is stated as a positive quantity whether the measured value is higher or lower than the true (or most probable) value. The percent error is

$$\% \text{ error} = \frac{\text{error}}{\text{true (or most probable) value}} \times 100$$

$$= \frac{2 \text{ cm}^3}{250 \text{ cm}^3} \times 100 = 0.8\%$$

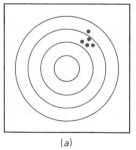

(a)

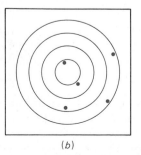

(b)

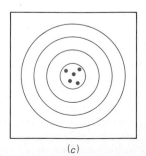

(c)

Rifle targets showing precision versus accuracy. (*a*) Shooting is precise, but not accurate. (*b*) Neither precise nor accurate. (*c*) Both precise and accurate.
Source: From G. H. Ayres, *Quantitative Chemical Analysis,* 2nd ed., Harper & Row, New York, 1968.

work. A standard outdoor telephone booth, for example, has a floor area of about 0.75 m² and a volume of about 1.5 m³.

In the laboratory, convenient units of area are the **square centimeter,** cm², and the **square millimeter,** mm². The convenient laboratory units of volume are the **cubic centimeter,** cm³, and the **cubic millimeter,** mm³. For working with liquid solutions, the chemist often finds it convenient to use 1,000 cm³ as a standard volume; this volume is called a **liter,** L. Laboratory glassware is commonly calibrated in terms of **milliliters,** mL, or cubic centimeters: 1 mL = 1 cm³. A liter is equivalent to 1.0567 quarts.

• Example 1•1 •

How many cubic centimeters are there in 1 pint? Express the answer to five, three, and two significant figures. (*Read Appendix 1 if you are unfamiliar with the use of significant figures, the use of exponents, or the factor-units method of solving problems.*)

• Solution • We begin with the volume given and select conversion factors by which we can multiply so that units cancel to yield the desired units, cubic centimeters. It is convenient to use abbreviations in calculations: 2 pt = 1 qt; 1.0567 qt = 1 L; and 1 L = 1,000 cm³.

$$\text{volume} = 1\ \text{pt}\left(\frac{1\ \text{qt}}{2\ \text{pt}}\right)\left(\frac{1\ \text{L}}{1.0567\ \text{qt}}\right)\left(\frac{1{,}000\ \text{cm}^3}{1\ \text{L}}\right)$$

$$= 473.17\ \text{cm}^3 \quad \text{(five significant figures)}$$

To three significant figures the volume is 473 cm³. To two significant figures the volume is 470 cm³. To show more clearly only two significant figures, we can write this expression as 4.7×10^2 cm³.

See also Exercises 9–16 at the end of the chapter.

• Example 1•2 •

Express the liter in terms of cubic meters, m³, and in terms of cubic decimeters, dm³.

• Solution •

$$1\ \text{L} = 1{,}000\ \text{cm}^3\left(\frac{1\ \text{m}}{100\ \text{cm}}\right)^3$$

$$= 1 \times 10^3\ \text{cm}^3\left(\frac{1\ \text{m}^3}{1 \times 10^6\ \text{cm}^3}\right)$$

$$= 1 \times 10^{-3}\ \text{m}^3$$

$$1\ \text{L} = 1{,}000\ \text{cm}^3\left(\frac{1\ \text{dm}}{10\ \text{cm}}\right)^3$$

$$= 1 \times 10^3\ \text{cm}^3\left(\frac{1\ \text{dm}^3}{1 \times 10^3\ \text{cm}^3}\right)$$

$$= 1\ \text{dm}^3$$

(Note that each of the relationships used in these solutions is exact by definition. Therefore, 1 L is exactly 1×10^{-3} m³ or 1 dm³, and the number of significant figures is unlimited.)

See also Exercises 9–16.

1•2•3   **Units of Mass and Density**   The SI base unit of mass, the **kilogram,** kg, is equivalent to about 2.2 pounds.[2] It is the only base unit still defined in terms of an artificial object (see Table A.1). The international standard kilogram is a piece of platinum–iridium alloy that is kept at Sèvres, France. A precise copy of it that is used as a standard in this country is shown in Figure 1.3. The one-thousandth part of the kilogram is the **gram,** g. In laboratory work, the most commonly used units are grams and **milligrams,** mg.

The **density** of a substance is defined as *mass per unit volume.* The units commonly used for chemical laboratory work are grams per cubic centimeter, g/cm^3. As shown in Table 1.2, the densities of substances vary widely. The densities of solids differ by a factor of more than 100. Gases differ greatly among themselves, but as a class, gases under ordinary conditions can be thought of as having densities about one-thousandth that of liquids and solids. The densities of gases are often expressed in grams per liter. Hydrogen gas is the least dense substance, and the solids osmium and iridium (see Table 21.7) are the most dense.

Figure 1•3 The United States Prototype Kilogram No. 20, shown here to actual size. The diameter is 39 mm; the height is 39 mm. *Source:* From the U.S. Department of Commerce, National Bureau of Standards, Washington, D.C.

━━━━━ • Example 1•3 • ━━━━━

From the data given in the caption of Figure 1.3, calculate the approximate density of the platinum–iridium alloy in grams per cubic centimeter and in SI base units.

━━━━━ • Solution • The volume, *V*, of a cylinder is calculated by

$$V = \pi r^2 h$$

For the platinum cylinder,

$$h = 39 \text{ mm} = 3.9 \text{ cm}$$

$$r = \frac{39 \text{ mm}}{2} = 19.5 \text{ mm} = 1.95 \text{ cm}$$

$$V = (3.14)(1.95 \text{ cm})^2(3.9 \text{ cm})$$

$$= 46.6 = 47 \text{ cm}^3$$

$$\text{density} = \frac{\text{mass}}{\text{volume}} = \frac{1 \text{ kg}\left(\dfrac{1{,}000 \text{ g}}{1 \text{ kg}}\right)}{47 \text{ cm}^3}$$

$$= \frac{1{,}000 \text{ g}}{47 \text{ cm}^3} = 21 \text{ g/cm}^3 \quad \text{or} \quad 21 \text{ g} \cdot \text{cm}^{-3}$$

---

[2]Often the term *weight* is used when the term *mass* should be. **Mass** refers to the quantity of matter in a body, whereas **weight** refers to the gravitational attraction of the earth for a body. However, it is a matter of common practice in this country to use the term *weight* for the term *mass.* Two bodies with the same mass have the same weight at a given location on the earth's surface.

In SI base units,

$$\text{density} = \frac{1 \text{ kg}}{47 \text{ cm}^3 \left(\dfrac{1 \text{ m}}{100 \text{ cm}}\right)^3}$$

$$= \frac{1 \text{ kg}}{47 \text{ cm}^3 \left(\dfrac{1 \text{ m}^3}{10^6 \text{ cm}^3}\right)} = 2.1 \times 10^4 \text{ kg/m}^3$$

(We see that the density in SI base units is numerically 1,000 times the density in grams per cubic centimeter.)

See also Exercises 19 and 21–23.

Closely related to the property of density is specific gravity. The **specific gravity** of a substance is the ratio of its mass to the mass of an equal volume of water at a specified temperature. To determine the specific gravity of a liquid, a container with a precise volume called a *volumetric flask* can be used (see Figure 1.4). Suppose that a 5.00-cm³ volumetric flask holds 4.99 g of water at 25 °C. When filled with a sample of gasoline, it holds 3.58 g. The specific gravity of the gasoline is

$$\text{specific gravity} = \frac{\text{mass of a sample of the gasoline}}{\text{mass of an equal volume of water}} = \frac{3.58 \text{ g}}{4.99 \text{ g}} = 0.717$$

Note that specific gravity has no units; it is a dimensionless ratio.

Table 1•2 Densities (at room temperature)

Substance	Density, g/cm³
hydrogen (gas)	0.000084
carbon dioxide (gas)	0.0018
balsa wood	0.16
cork wood	0.21
oak wood	0.71
ethyl alcohol	0.79
water	1.00
eucalyptus wood	1.06
magnesium	1.74
table salt	2.16
sand	2.32
aluminum	2.70
iron	7.9
silver	10.5
lead	11.3
mercury	13.6
gold	19.3

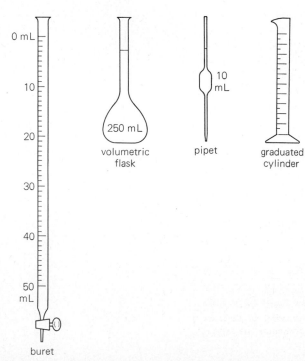

volumetric flask   pipet   graduated cylinder   calibrated beaker   buret

Figure 1•4 Some common laboratory vessels used for measuring the volumes of liquids.

When compared with water between 0 °C and 30 °C, the specific gravity of a substance is numerically equal to its density to two or three significant figures. This is because the density of water to three significant figures over this temperature range is 1.00 g/cm^3. The density of the gasoline can be calculated using the known volume of the volumetric flask:

$$\text{density} = \frac{\text{mass}}{\text{volume}} = \frac{3.58 \text{ g}}{5.00 \text{ cm}^3} = 0.716 \text{ g/cm}^3$$

• Example 1•4 •

The specific gravity of the sulfuric acid solution in an automobile battery is found to be 1.285 when measured with a hydrometer (see Figure 1.5). If the hydrometer has been calibrated against water at 20.0 °C, which has a density of 0.99823 g/cm^3, what is the density of the sulfuric acid solution?

• Solution • The mass of a unit volume of the sulfuric acid is 1.285 times the mass of an equal volume of water at 20 °C. Therefore,

$$\text{density of acid} = 1.285 \times 0.99823 \text{ g/cm}^3$$
$$= 1.283 \text{ g/cm}^3 \quad \text{or} \quad 1.283 \text{ g} \cdot \text{cm}^{-3}$$

See also Exercise 24.

• 1•2•4   **Units of Temperature and Energy**   For many years the scientific scale of temperature was based on the freezing and boiling points of water. The difference between these two points was divided into 100 parts, and so the scale was called the **centigrade scale** (Latin *centum*, a hundred). In 1948 the centigrade scale was designated as the **Celsius scale,** °C, in honor of the Swedish astronomer Anders Celsius, who developed the scale in 1742.

One of the disadvantages of the Celsius scale is that temperatures below the freezing point of water, 0 °C, are negative. This is awkward because temperature is an intensity factor that measures the relative hotness of a body. **Temperature** is the property that determines the direction of the spontaneous flow of heat. If two adjacent bodies are at different temperatures, the body at a lower temperature will receive heat energy from the body at a higher temperature. Temperature, therefore, should be measured on a scale on which the temperature of 0 means no heat energy and on which matter that contains any heat energy has a positive temperature.

A scale of temperature on which zero denotes the complete absence of heat energy is the **Kelvin scale.** The temperature of 0 K is absolute zero, the lowest possible temperature. In Section 7.4.1 we shall discuss a physical principle that led to the understanding that there is an absolute lowest possible temperature, −273.15 °C or 0 K. Relationships between the three temperature scales we use in this text are shown in Figure 1.6. The Celsius and Fahrenheit scales are compared in Figure 1.7.

The **heat energy** of a body is the type of energy that flows from a hotter body to a cooler one. The amount of heat in a body is determined not only by its temperature but also by the amount and kind of matter present. For example, 10 g of iron at 100 °C contains less heat than 100 g of iron at 100 °C; also, 10 g of iron at 100 °C contains less heat than 10 g of aluminum at 100 °C.

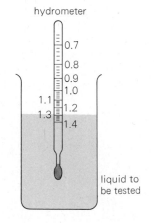

Figure 1•5 A hydrometer. The more dense the liquid, the higher a hydrometer floats. The scale reading that coincides with the level of the liquid is the specific gravity of the liquid.

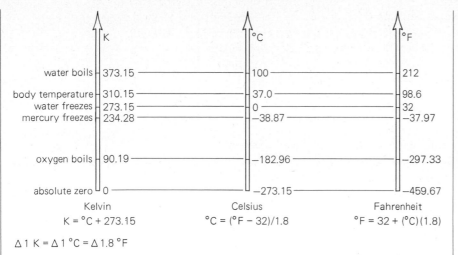

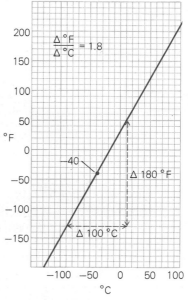

Figure 1•6 Relationships between three temperature scales. (The symbol Δ is read "a change in.")

Figure 1•7 A plot of temperatures in degrees Celsius versus degrees Fahrenheit shows that a change (Δ) of 1 °C is equivalent to a change of 1.8 °F. Note that at one point, −40, Celsius and Fahrenheit thermometers have the same reading.

For many years the scientific unit of heat energy was the **calorie,** cal, defined as the energy necessary to change the temperature of 1 g of water by 1 °C (from 14.5 °C to 15.5 °C for the precise definition).[3] The SI unit of energy is the **joule,** J, which is derived from three SI base units:

$$J = kg \cdot m^2 \cdot s^{-2}$$

One joule is the energy expended when the force needed to give 1 kg an acceleration of 1 m per second per second moves a distance of 1 m. Any mechanical quantity can be expressed in terms of mass, length, and time (see Table 1.3). The joule is named for James Prescott Joule, the British physicist who established the relationship between heat energy and mechanical energy in 1847. The calorie is now defined officially in terms of the joule:

$$1 \text{ cal} = 4.184 \text{ J} \quad \text{(exact)}$$

• *Specific Heat* • The amount of heat energy required to change the temperature of 1 g of a substance by 1 °C or 1 K is called the **specific heat** of the substance. The specific heat may be determined experimentally by measuring the heat required to change the temperature of a given weight of the substance from T_1 to T_2. Then,

$$\text{specific heat} = \frac{\text{no. of units of heat energy}}{(\text{no. of grams})(\Delta T)}$$

where the symbol Δ is read "the difference in" or "the change in." In this example, ΔT is $T_2 - T_1$. Note that 1 °C is exactly the same interval of temperature as 1 K. In any calculation, ΔT can be expressed as Δ °C or Δ K.

---

[3]Perhaps the most familiar use of the word *calorie* is in describing the energy content of foods. The popular "calorie" for measuring food energy is actually 1,000 calories and should be called a "kilocalorie."

Table 1•3 Dimensions of mechanical quantities in terms of mass (m), length (l), and time (t)

Mechanical quantity	Dimensions
area	l^2
volume	l^3
velocity	l/t
acceleration	l/t^2
density	m/l^3
momentum	ml/t
force	ml/t^2
energy	ml^2/t^2
power	ml^2/t^3

The specific heat of water is

$$\frac{1\ cal}{1\ g \times 1\ °C} \quad or \quad 1\ cal/(1\ g \times 1\ °C) \quad or \quad 1\ cal \cdot g^{-1} \cdot °C^{-1}$$

In SI units, the specific heat of water is

$$\frac{4.184\ J}{1\ g \times 1\ K} \quad or \quad 4.184\ J \cdot g^{-1} \cdot K^{-1}$$

As suggested by the data in Table 1.4, the specific heat of water is high compared with that of most substances. The fact that different amounts of heat are needed to change the temperatures of different substances is shown graphically in Figure 1.8.

A Pint's a Pound the World Around

*Then down with every "metric" scheme
Taught by the foreign school,
We'll worship still our Father's God!
And keep our Father's "rule"!
A perfect inch, a perfect pint,
The Anglo's honest pound,
Shall hold their place upon the earth,
Till Time's last trump shall sound!*

★ Special Topic ★ Special Topic ★ Special Topic ★ Special Topic ★ Special Topic ★ Special Topic ★ Special Topic ★ Special Topic ★

The Arbitrary Character of Units of Measure The units of measure are arbitrary. Each and every unit is unitary only because a person or a group of persons has declared it to be so. A later decision can change the size of any unit or abolish it altogether. The choice of a unit of measure is not forced upon us by nature's objects or processes.

Consider length. The earliest recorded unit of measurement, the cubit, was based on the length from a person's elbow to fingertip. In Genesis we read that Noah's ark was 300 cubits long. The great pyramids of Egypt were laid out in cubits. Both the length of the human foot and the width of the human hand have been used as units of length. In different places and times a unit called a foot represented the equivalent of about 10, 12, 13, 17, or even 27 modern inches! The hand, now defined as 4 inches, is still commonly used to measure the heights of horses. It might be possible to establish the precise average width of human hands, but the decision to use that length as a fundamental unit would be arbitrary.

The choices of the scientific units of measure are just as arbitrary as any of the foregoing examples, but in science an effort has been made to choose units that can be defined precisely. The present system of internationally adopted weights and measures had its origin in 1791 in the work of a committee of the French National Academy. In addition to carefully selecting the physical measurements on which to base units, the framers of the metric system provided for multiples and subdivisions on a decimal basis. Antoine Lavoisier, one of the most eminent scientists of that time, was moved to exclaim, "Never has anything more grand and simple, more coherent in all its parts, issued from the hand of man."

Of course, the early authors could not anticipate all needs. For example, consider their definition of the meter (Greek *metron,* a measure) as one ten-millionth of the earth's meridian from the equator to the North Pole.

When it became necessary to define length with greater precision, the original definition was replaced by a length marked on a carefully protected standard bar of platinum-iridium alloy. Later a certain number of wavelengths of light was used to define the meter.

Likewise the definition of the gram (Greek *gramma,* a small measure) has changed over time. Originally the gram was defined as the mass of 1 cm³ of water at the temperature of its maximum density. For convenience in handling, a mass of 1,000 times this, a kilogram, was fashioned out of a platinum-iridium alloy, and copies were made for laboratories around the world. Some time later, more precise measurements showed that the alloy object had a mass equal to that of 1,000.028 cm³ rather than 1,000 cm³ of water. It was then decided to abandon the original definition of the gram and to base all units of mass on the piece of platinum-iridium alloy that had been so carefully copied. This was designated the primary unit, and the name kilogram was retained.

Although the metric system has been legal in the United States since 1866 and "official" since 1893, for many years there was vocal and emotional resistance to it in nonscientific circles. Some of this resistance still remains. Early in this century an antimetric theme song included the refrain shown above. Like other such reactions to reasonable change, this ditty casts more joules than candelas on the subject. Today the United States appears to be moving toward full use of *Le Système International d'Unites,* SI. As in the past, the necessarily arbitrary definitions of units are subject to reexamination, and we can expect future scientists to recommend some changes.

Source: *Much of this information is from K. F. Weaver, "How Soon Will We Measure in Metric?"* National Geographic, *152,* 287 (1977).

Table 1•4 Specific heats

Substance	Specific heat	
	$cal \cdot g^{-1} \cdot {}^{\circ}C^{-1}$	$J \cdot g^{-1} \cdot K^{-1}$
water	1.000	4.184
ethyl alcohol	0.581	2.43
ice	0.478	2.00
aluminum	0.212	0.887
sand	0.188	0.787
table salt	0.185	0.774
carbon	0.127	0.531
iron	0.108	0.452
gold	0.0312	0.131
uranium	0.0280	0.117

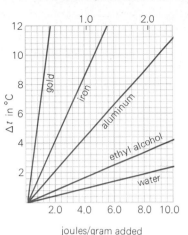

Figure 1•8 The change in temperature when heat is added to one gram of a substance. See also Table 1.4.

━━ • Example 1•5 • ━━

(a) How much heat in joules is required to raise the temperature of 35.2 g of aluminum from 17 °C to 85 °C?

(b) How much heat in joules is required to raise the temperature of the same weight of gold by the same amount?

━━ • Solution • ━━

(a) From Table 1.4, the specific heat of aluminum is found to be $0.887 \ J \cdot g^{-1} \cdot K^{-1}$. The ΔT of $(85 - 17) = 68 \ {}^{\circ}C = 68 \ K$. Therefore, the number of joules required is

$$(0.887 \ J \cdot g^{-1} \cdot K^{-1})(35.2 \ g)(68 \ K) = 2{,}123 \ J$$
$$= 2.1 \times 10^3 \ J \quad \text{(two significant figures)}$$

(b) From Table 1.4, the specific heat of gold is found to be $0.131 \ J \cdot g^{-1} \cdot K^{-1}$, so the problem can be solved just as in part (a). But, another approach is to note that, because the energy required to change the temperature of a material is directly proportional to its specific heat, the amount of heat required to change the temperature of gold is a fractional amount, 0.131/0.887, of the heat required for the same change in temperature of the same weight of aluminum. Therefore, for gold the number of joules required is

$$(2.1 \times 10^3 \ J)\left(\frac{0.131}{0.887}\right) = 3.1 \times 10^2 \ J$$

Only two significant figures are justified in answer (a) because the ΔT for the aluminum is 68 K. For the intermediate answer of 2,123 J for the heat gained by the aluminum, extra figures are retained, but the final answer is rounded to two significant figures. Because the value $2.1 \times 10^3 \ J$ is used in part (b), that answer also has only two significant figures, 3.1.

See also Exercises 32–34.

• Example 1·6 •

A piece of metal that weighs 42.0 g is heated to 97.0 °C and then dropped into 52.0 g of water at 21.0 °C in an insulated container. The temperature of the water rises to 22.9 °C. Calculate the specific heat of the metal.

• Solution • Because no other data are given, we will assume that all the heat lost by the metal is gained by the water. Also, we will assume that the final temperatures of the metal and the water are the same. For a heat change in any substance,

$$\text{heat} = (\text{specific heat})(\text{weight})(\Delta T)$$

So,

$$\text{heat lost by metal} = \text{heat gained by water}$$

$$\binom{\text{spec heat}}{\text{of metal}}\binom{\text{wt of}}{\text{metal}}\binom{\Delta T \text{ of}}{\text{metal}} = \binom{\text{spec heat}}{\text{of water}}\binom{\text{wt of}}{\text{water}}\binom{\Delta T \text{ of}}{\text{water}}$$

$$\binom{\text{spec heat}}{\text{of metal}}(42.0 \text{ g})(97.0 - 21.9 \text{ °C}) = \left(\frac{4.184 \text{ J}}{1 \text{ g} \times 1 \text{ °C}}\right)(52.0 \text{ g})(22.9 - 21.0 \text{ °C})$$

$$= 413 \text{ J}$$

$$\text{spec heat of metal} = \frac{413 \text{ J}}{42.0 \text{ g} \times 75.1 \text{ °C}}$$

$$= 0.13 \text{ J} \cdot \text{g}^{-1} \cdot \text{°C}^{-1} = 0.13 \text{ J} \cdot \text{g}^{-1} \cdot \text{K}^{-1}$$

See also Exercises 35–37.

Fundamentals of Descriptive Chemistry

In the period from about 1600 to 1900 in Western Europe, great advances were made in the study of the composition and classification of matter and of the changes that matter undergoes. Although modern research has brought about modification of some of these early ideas, an understanding of them is fundamental to the study of chemistry.

1·3   Properties of Matter

Every substance—for example, water, sugar, salt, silver, or copper—has a set of characteristics or properties that distinguish it from all other substances and give it a unique identity. Both sugar and salt are white, solid, crystalline, soluble in water, and odorless. But the former has a sweet taste, melts and turns brown on heating in a saucepan, and burns in air. The latter tastes salty, does not melt until heated above red heat, does not turn brown no matter how high it is heated, and does not burn in air, although it gives off a bright yellow light when heated directly in a flame. We have described both of these substances by listing some of their respective intrinsic properties.

Intrinsic properties are qualities that are characteristic of any sample of a substance, regardless of the shape or size of the sample. **Extrinsic**

properties are qualities that are not characteristic of the substance itself. Size, shape, length, weight, and temperature are extrinsic properties.

1•3•1 **Chemical Properties** **Chemical properties** are those qualities characteristic of a substance that cause it to change, either alone or by interaction with other substances, and in so doing to form different materials. Chemical properties are intrinsic properties. For example, it is characteristic of ethyl alcohol to burn, of iron to rust, and of wood to decay.

1•3•2 **Physical Properties** **Physical properties** are the characteristics of a substance that distinguish it from other substances, and that do not involve any change to another substance. Physical properties are intrinsic properties. Examples include melting point, boiling point, density, viscosity, specific heat, and hardness. The qualities in this group can be measured easily and expressed in definite numbers. The substance that we call ethyl alcohol melts at -117.3 °C (155.8 K), boils at 78.5 °C (351.6 K), has a density of 0.7893 g/cm^3, and has a specific heat of 2.43 J $\cdot$ g^{-1} $\cdot$ K^{-1}. No other substance has precisely this unique set of properties; such a substance is ethyl alcohol and nothing else.

1•4 **Changes in Matter
and Energy**

1•4•1 **Changes in Matter** The materials around us are subject to constant change. Plant and animal materials decay, metals corrode, gasoline burns, water changes to ice when the temperature drops sufficiently and changes back to the liquid form when the temperature rises, land areas erode, and lakes and seas evaporate. When we study these changes, we find we can classify them under two headings: chemical changes and physical changes. **Chemical changes** are those that result in the disappearance of substances and the formation of new ones. For example, when pieces of magnesium metal burn in a photoflash bulb, the magnesium and some oxygen from the bulb disappear. In their place, we find a powdery, incombustible solid, magnesium oxide, that has its own unique set of properties. Or, as another example, consider some of the changes in matter that occur as a stalk of corn matures. In this process, carbon dioxide and water disappear in the sense that they are converted to glucose in the growing plant. Much of this sugar accumulates in the ear of the corn, and as the ear matures the sugar is converted to starch. The glucose that appears has its own set of identifying properties that are completely different from those of the carbon dioxide and water from which it was made. The starch, in turn, has different properties from the sugar. Chemical changes are also referred to as *chemical reactions*.

     Physical changes are changes that do not result in the formation of new substances. For example, when ice melts to water or when sand is ground to a fine powder, no new substance is formed. However, it should be noted that in physical changes some properties do change and energy transformations do occur.

     It is not always clear whether a change should be classified as physical or chemical. At one time it was common to say that dissolving a solid in a liquid was a physical change, but today a better understanding of this process leads us to call it a chemical change in most cases. Taste and odor are often thought of as physical properties, but our detection of these properties probably involves chemical reactions.

1·4·2   Changes in Energy   The **energy** of a body or a system is that body's or system's ability to do work. Every change, chemical or physical, involves a change in energy. The flight of a bird, the breaking of the earth's crust by a new blade of grass, the burning of wood, the turning of a page in a book—all these actions involve energy.

We think of energy as having a number of forms. Heat energy, electric energy, radiant energy, chemical energy, and nuclear energy are some familiar examples. One form of energy can be converted into any other form.

Electric energy is the type of energy associated with the passage of a current of electricity. Electric energy is transformed into radiant energy in an electric light, into heat energy in a kitchen stove, or into mechanical energy in the starter of an automobile.

Radiant energy is the type of energy associated with ordinary light, X rays, radio waves, and infrared rays. Radiant energy is also called *electromagnetic radiation*. All such radiation travels through space with the speed of light, 3.00×10^8 meters per second, or 186,000 miles per second.

Chemical energy is the energy a substance has because of its chemical state. Chemical energy is transformed into other kinds of energy when matter undergoes the proper kind of change. For example, when coal and gasoline burn, or when the food we eat is "burned" in our cells, some chemical energy is converted to heat energy. Conversely, other kinds of energy can be transformed to chemical energy by the proper kind of change. Radiant energy from the sun is transformed in a growing corn plant to chemical energy that becomes associated with the substances making up the stalk, corn grains, and other parts of the plant. All life processes involve the change of chemical energy to other forms of energy, or vice versa. Chemical energy is the principal source of energy for our factories, our homes, and our transportation.

Nuclear energy or **atomic energy** is associated with the manner in which atoms are constructed. Methods for transforming this type of energy to heat, light, and other kinds of energy have been developed since 1942. Nuclear energy will be discussed in detail in Section 19.12.

Actually, all forms of energy fall into two general classes: potential energy and kinetic energy. **Potential energy** is the energy a body possesses because of its position or because of its existence in a state other than its normal state of lowest energy. The water held in a reservoir behind a dam is in a position to do work by turning a turbine or waterwheel and hence possesses potential energy. A tightly wound watch spring also possesses potential energy. Chemical energy is a form of potential energy.

Kinetic energy is the energy a body possesses because of its motion. As water flows through turbines, as a watch spring slowly uncoils, or as gasoline burns in an automobile engine, potential energy is transformed to kinetic energy. A moving automobile, a pitched baseball, an airplane in flight—all possess kinetic energy.

Kinetic energy depends on both mass and velocity. A change in the velocity of a moving body has a greater effect on its kinetic energy than does a proportionate change in its mass, because the kinetic energy is proportional to the square of the velocity. To illustrate, three identical automobiles, one moving at 20, one at 40, and one at 60 miles per hour, have kinetic energies in the ratio of $1^2:2^2:3^2$, that is, $1:4:9$. It requires nine times more work to bring the one moving at 60 miles per hour to a stop than is required to stop the one moving at 20 miles per hour. But if the mass is tripled and the velocity is not changed, the kinetic energy is increased by just three times. That is, the kinetic energy is directly proportional to the mass. The dependence of kinetic energy on the mass and velocity is given by the expression

$$\text{K.E.} = \tfrac{1}{2}mv^2$$

• Example 1•7 •

What will the increase in temperature, ΔT, be on impact for a raindrop weighing 0.22 g and falling at a speed of 41 m/s? (Assume that all kinetic energy is converted into heat energy.)

• Solution • If we convert the 0.22 g into kilograms, we can solve directly for the kinetic energy in joules:

$$\text{K.E.} = \tfrac{1}{2}mv^2 = \tfrac{1}{2}\left(0.22\ \cancel{g} \times \frac{1\ \text{kg}}{10^3\ \cancel{g}}\right)(41\ \text{m/s})^2$$

$$= \tfrac{1}{2}(2.2 \times 10^{-4})(41)^2\ \text{kg} \cdot \text{m}^2 \cdot \text{s}^{-2}$$

$$= 1.8 \times 10^{-1}\ \text{J}$$

For water, it requires 4.184 J to raise the temperature of 1 g by 1 K, so for the raindrop,

$$1.8 \times 10^{-1}\ \cancel{J} = (4.184\ \cancel{J} \cdot \cancel{g}^{-1} \cdot \text{K}^{-1})(0.22\ \cancel{g})(\Delta T)$$

$$\Delta T = \frac{1.8 \times 10^{-1}}{(4.184)(0.22)}\ \text{K}$$

$$= 0.20\ \text{K} \quad \text{or} \quad 0.20\ ^\circ\text{C}$$

See also Exercises 43 and 44.

• Exothermic and Endothermic Changes • If substances, or a single substance, change in such a way that energy is given to the surroundings, the change is said to be **exothermic** (heat comes out). The combining form -*thermic* originally referred to heat energy, but exothermic now refers to a change in which any type of energy is given off.

For example, when magnesium burns in oxygen to make magnesium oxide, chemical energy is converted to heat and radiant energy that is emitted to the surroundings. This is an exothermic chemical change. When a hot bowl loses heat to the table on which it has been placed, the cooling of the bowl is an exothermic physical change.

When carbon dioxide and water are changed to glucose in a growing plant, radiant energy from the sun is converted to chemical energy. Such a change in which materials take up energy from the surroundings is said to be

endothermic. This formation of glucose is an endothermic chemical reaction. In the example of the hot bowl on the table, the taking up of heat by the table is an endothermic physical change. For the bowl the change is exothermic, but for the table the change is endothermic.

The reaction of magnesium and oxygen to form magnesium oxide is exothermic, whereas the breaking down of magnesium oxide to yield magnesium and oxygen is an endothermic change. In fact, the amount of energy required to break down a given amount of magnesium oxide is equal to the amount of energy given off when that much magnesium oxide is formed. This example illustrates a fundamental principle: *A process that is exothermic in one direction is always endothermic in the opposite direction.*

1•4•3   **Classes of Matter** Pure substances are classified as either elements or compounds. **Elements** may be described as substances that cannot be decomposed by simple chemical change into two or more different substances. Some elements familiar to the early chemist were copper, silver, gold, sulfur, carbon, and phosphorus. **Compounds** may be described as substances of definite composition that can be decomposed by simple chemical change into two or more different substances. Common salt, sodium chloride, is an example of a compound. This white, crystalline substance may be decomposed into a shiny, active metal (sodium) and a poisonous, greenish-yellow gas (chlorine). The properties of the substances obtained by the decomposition of a compound are completely unrelated to the properties of the compound. Today, just over 100 elements are known, but there are over 4 million compounds. Some familiar compounds are water, sugar, alcohol, carbon dioxide, and ammonia.

Mixtures are materials that contain two or more different substances more or less intimately jumbled together. A mixture has no unique set of properties; rather, it possesses the properties of the substances of which it is composed. Air is an example of a gaseous mixture; it is composed principally of nitrogen, oxygen, argon, water vapor, and carbon dioxide, and each of these substances displays its own unique properties in the mixture. Further, it is usually possible to separate the components of a mixture by physical changes rather than chemical. For example, when the temperature of air is lowered, water vapor tends to separate as liquid or solid water, that is, as dew or frost. On extreme cooling, the carbon dioxide solidifies and then the remainder of the air liquefies. If the liquid air is carefully boiled, the mixture can then be separated, because each component tends to boil away over a particular temperature range, depending on its own boiling point. This method of separating the substances in a liquid mixture by boiling is called **distillation.** Figure 1.9 shows a simple apparatus used in laboratories to carry out distillations of common mixtures containing a low-boiling component (such as water, ethyl alcohol, or ethyl ether) and a much higher boiling component (such as sodium chloride, sugar, or glycerin). If the boiling points of the components are close together, a more complex apparatus must be used. (See Figure 12.6.)

In describing the appearance of materials, the chemist often finds two terms useful, homogeneous and heterogeneous. **Homogeneous** refers to material in which no differing parts can be distinguished even with a microscope, a solution of sugar in water, for example. **Heterogeneous** refers to material in which there are visible differing parts, a mixture of salt and pepper, for example.

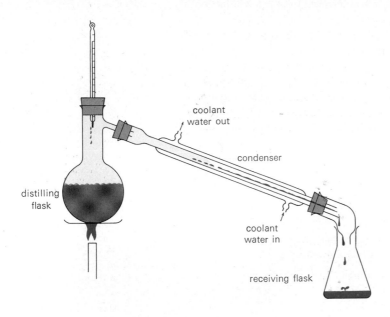

Figure 1•9 Apparatus for distilling certain types of mixtures.

1•5   Laws About Matter and Energy

1•5•1   Law of Conservation of Energy   *Energy is neither created nor destroyed in any transformation of matter.* This statement, the **law of conservation of energy,** seems to describe all natural phenomena accurately, whether in the field of chemistry, physics, biology, geology, astronomy, or any other. Consider as an example the reaction that takes place when a photoflash bulb is set off. We can ignore the tiny amount of heat due to the electric current that initiates the reaction and focus our attention on the chemical reaction and the energy it produces:

$$\text{magnesium} + \text{oxygen} \longrightarrow \text{magnesium oxide} + \text{heat and light}$$

$$\begin{array}{ccc} \text{chemical energy} & = & \text{chemical energy} & + & \text{energy} \\ \text{of the reactants} & & \text{of the product} & & \text{emitted} \end{array}$$

Because no energy is created or destroyed in the chemical reaction, the total energy of the system before the reaction equals the total energy after the reaction. In this exothermic reaction, therefore, the chemical energy of the product must be less than the chemical energy of the reactants, because some chemical energy is transformed to heat and light during the reaction. In an endothermic reaction, just the reverse would be true: the chemical energy of the products would be greater than the chemical energy of the reactants.

1•5•2   Law of Conservation of Mass   In the eighteenth century, experimental methods were developed for measuring volumes of gases; weighing gases, liquids, and solids; and conducting chemical reactions in a manner so that the weights of the reactants and the products of a reaction could be precisely measured (see Figure 1.10). Such experiments provided investigators with many facts and led to the statement of some fundamental laws describing chemical behavior. According to one of these laws, *mass is neither created nor destroyed in any transformation of matter.* This statement, which sums up the results of thousands of painstaking experiments, is the **law of conservation of mass.** The facts necessary to support this statement were correlated in 1756 by the Russian scientist M. V. Lomonosov. Perhaps because of translation difficulties, his work was not widely known in Western Europe. Antoine Lavoisier, in France, formulated the law independently in 1783. As an illustration of the law, consider the complete combustion of gasoline. The following relationship is true within the limits of our ability to determine the weights of the reacting substances and the products of the reaction:

$$\text{gasoline} + \text{oxygen} \longrightarrow \text{carbon dioxide} + \text{water vapor}$$
$$\text{weight of reactants} \quad = \quad \text{weight of products}$$

As we shall see in Section 19.9, the weights of reactants and products are not absolutely equal. In an exothermic chemical reaction, an insignificant amount of matter is converted to energy; in an endothermic reaction, the reverse is true. In nuclear reactions, these matter–energy changes are not insignificant.

Figure 1•10 When wood is burned in open air, matter appears to be destroyed. However, if wood is burned in a closed tube as shown, the products of combustion can be collected and weighed. It is found that the weight of products, as measured by the gain in weight of the second sodium oxide absorber, is actually more than the weight of the wood burned.

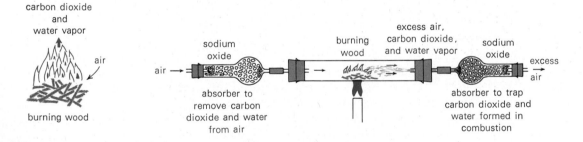

carbon dioxide and water vapor

air

burning wood

sodium oxide

air →

absorber to remove carbon dioxide and water from air

burning wood

excess air, carbon dioxide, and water vapor

sodium oxide

excess air

absorber to trap carbon dioxide and water formed in combustion

1•5•3   Law of Definite Composition   To determine the composition of a compound, one can decompose a weighed sample of the compound into its constituent elements and determine their individual weights. Or one can determine the weight of a compound formed by the chemical union of known weights of the elements. (Other methods are also available; some of them are discussed in Section 2.7.) The study of the composition of many compounds led, in the eighteenth century, to the statement of the following law: *A pure compound is always composed of the same elements combined in a definite proportion by weight.* This is the **law of definite composition** (also called the **law of definite proportions**).

By way of illustration consider water, whose composition has been determined by experiment many times. The same answer is always obtained. Water as it occurs in nature is composed of hydrogen and oxygen only, and these elements are always in the proportion of 11.19 percent hydrogen and 88.81 percent oxygen by weight. Or consider table sugar. The composition of sugar (no matter whether it comes from sugarcane, sugar beet, or maple syrup) as determined by analysis is carbon, 42.1 percent; hydrogen, 6.5 percent; and oxygen, 51.4 percent.

The fact that a compound has a certain, reproducible composition is useful in determining when a chemical change takes place. When pure oxygen gas is passed over powdered iron heated to just below 500 °C, a substance is produced that has the composition 72.36 percent iron and 27.64 percent oxygen. At a higher temperature, say, one above 600 °C, continued passing of oxygen results in the formation of a second substance that is 69.94 percent iron and 30.06 percent oxygen. If the supply of oxygen is then limited and the second substance is heated for a long time above 600 °C, still a third substance is formed that approaches the composition 77.73 percent iron and 22.27 percent oxygen. Although the second substance is usually reddish, each of these three oxides of iron can be black, so that they may not be distinguished visually. However, because of the different compositions of these substances, it is obvious that different chemical compounds are formed under conditions of different temperatures and amounts of oxygen present.

• *Limitations of the Law of Definite Composition* • Although the concept of definite composition was used by working chemists in the eighteenth century, later, from about 1802 to 1808, a famous controversy waged about its validity. The principal opponents were two French chemists, Joseph Proust and Claude Berthollet. Proust held that the law applied precisely to pure substances and supported his view with a series of elegant analyses. Berthollet, relying more on the laboratory work of others, argued that the composition of a compound depended upon the conditions of its preparation. Two factors were particularly important in deciding the issue in favor of Proust: (1) the experimental work cited by Berthollet was found in most cases to be based on impure substances or incomplete reactions, and (2) the influential new atomic theory (see Section 1.7) of John Dalton was compatible with Proust's experimental data.

For over a century after Proust's victory, the law of definite composition was accepted practically unchallenged. Today, however, we know that for many solid compounds the composition is not definite but can vary over a small range depending upon the method of preparation. For example, common iron sulfide can vary from about 63.5 percent iron to 60.1 percent; the iron can be missing from the solid owing to lattice vacancies (see Section 8.10). The vast majority of compounds do have definite compositions; they are called *daltonides*. Solid state compounds of somewhat variable composition are increasingly important; they are called *berthollides,* in memory of an able scientist who lost his most important argument.

1•5•4   **Law of Multiple Proportions**   As pointed out in the case of the oxides of iron, two elements may unite to form more than one compound. Analyses show that, although the compounds have different compositions by weight, these compositions are related in a simple way. If a given weight of the first element is considered, say, 1.00 g, it is found that the weights of the second element that will combine with this 1.00 g are related to each other in the ratio of small whole numbers. To take a specific example, let us consider the two common combinations of carbon with oxygen. Carbon burns in an excess of oxygen to form gas A, a dense, nonpoisonous, noncombustible gas. However, if not enough oxygen is present during the burning, gas B, a poisonous, combustible gas, is also formed. An analysis of these compounds reveals that each gas has its own definite composition. In the noncombustible gas, 1.00 g of carbon is always combined with 2.67 g of oxygen; whereas in the gas that will burn, 1.00 g of carbon is always combined with only 1.33 g of oxygen.

$$\text{Gas A} \qquad\qquad \text{Gas B}$$

$$\frac{2.67 \text{ g oxygen}}{1.00 \text{ g carbon}} \qquad\qquad \frac{1.33 \text{ g oxygen}}{1.00 \text{ g carbon}}$$

$$\text{per gram carbon,} \quad \frac{\text{gas A}}{\text{gas B}} = \frac{2.67 \text{ g oxygen}}{1.33 \text{ g oxygen}} = \frac{2}{1}$$

We see that the ratio of the weights of oxygen that combine with the same weight of carbon is 2:1.

Iron and chlorine also form two compounds, solid C and solid D. The amount of chlorine that combines with 1.00 g of iron in solid C is 1.26 g; the amount of chlorine that combines with 1.00 g of iron in solid D is 1.89 g.

$$\text{Solid C} \qquad\qquad \text{Solid D}$$

$$\frac{1.26 \text{ g chlorine}}{1.00 \text{ g iron}} \qquad\qquad \frac{1.89 \text{ g chlorine}}{1.00 \text{ g iron}}$$

$$\text{per gram iron,} \quad \frac{\text{solid C}}{\text{solid D}} = \frac{1.26 \text{ g chlorine}}{1.89 \text{ g chlorine}} = \frac{2}{3}$$

The weights of chlorine that combine with the same weight of iron are in the ratio of 2:3. A formal statement of these facts is known as the **law of multiple proportions:** *When two elements combine to form more than one compound, the different weights of one that combine with a fixed weight of the other are in the ratio of small whole numbers.*

1•6   **Hallmarks of a Chemical Reaction**

As mentioned in Section 1.4.1, it is not always obvious when a chemical change occurs. Definite criteria for recognizing a chemical change are based on the insights and information gained in the development of descriptive chemistry. Three kinds of changes always accompany chemical reaction. As a reaction progresses, the reactants change to products that have different (1) properties, (2) compositions, and (3) energy contents. In some cases, the changes are so dramatic that there is no doubt a chemical change has occurred.

Some examples of chemical changes are shown in Figure 1.11 on Plate 1. Part (*a*) of this figure shows the brilliant light that accompanies the burning of magnesium in air. A shiny strip of magnesium metal reacts with oxygen (and nitrogen) in air to give a white powdery substance, magnesium oxide (and magnesium nitride). The other changes illustrated by this figure, parts (*b*)–(*h*), are reactions that occur with water or in water solutions. In all but one case, a change in color or the formation of an insoluble solid substance is the indication that a chemical change has taken place.

When hydrochloric acid is mixed with sodium hydroxide (see part (*h*) of Figure 1.11), there is no change in appearance. However, there is a considerable rise in the temperature of the solutions when they interact. From this fact, it is clear that a change in energy occurs. The heat from this exothermic reaction is given off to the beaker holding the solution, to the surrounding air, and to the solution itself. If the proper amounts of hydrochloric acid and sodium hydroxide are mixed, and we allow all the water to evaporate, common salt, sodium chloride, remains. The sodium chloride has distinctly different properties and a different composition than the hydrochloric acid and sodium hydroxide used to produce it.

To illustrate further the three changes accompanying a chemical reaction, let us consider a different way of making sodium chloride, by the combustion of sodium in an atmosphere of chlorine:

$$\text{sodium} + \text{chlorine} \longrightarrow \text{sodium chloride} + \text{energy}$$
$$\underbrace{\phantom{\text{sodium} + \text{chlorine}}}_{\text{reactants}} \qquad \underbrace{\phantom{\text{sodium chloride}}}_{\text{product}}$$

• *Change in Properties* • Sodium is a soft, silvery metal that reacts violently with water, and chlorine is a poisonous, yellow-green gas. Sodium chloride is the white, crystalline solid we know as table salt. The differences in properties could not be more striking: two dangerous, harmful chemicals react to yield a substance that is essential in our diet.

• *Change in Compositions* • Sodium is 100 percent sodium; chlorine is 100 percent chlorine; sodium chloride is a compound composed of 39.34 percent sodium and 60.66 percent chlorine by weight.

• *Change in Energy* • During this reaction it is clear that a change in energy is occurring, because heat and light are given off. Some of this energy is taken up by the containing vessel and by the surrounding air. Because the reaction is exothermic, the sodium chloride must have a lower energy content than did the original sodium and chlorine.

Early Theoretical Chemistry

Simple chemical and physical changes have been studied since the dawn of history. Some of the earliest questions that occurred to the student of nature were: Of what is matter made? Is sand made of the same stuff that wood is? Is a piece of gold made of tiny pieces of gold, or is it made of tiny pieces of several substances mixed together to make gold? Could it be that all matter is composed merely of different proportions of four fundamental stuffs: earth, water, air, and fire?

Questions such as these were pondered by the ancient Greeks. The rational conclusion of some of them was: if a large piece of gold is cut (*tomos*) in half, both pieces are still gold. If these halves could be divided again and then again and again, one would finally get down to the smallest possible piece of gold. This tiny particle could not be cut (*a-tomos* is Greek for not cut), for it would be the unit particle of gold, a gold atom.

For more than 2,000 years after this early rational beginning, the concept of small particles called atoms had very little influence on the thinking of natural philosophers. It was not until the latter part of the seventeenth century that the birth of modern chemistry focused attention on the investigation of material things and led to the recognition of basic differences between elementary and complex substances. Indeed, it was not until early in the nineteenth century, about 1803, that the English chemist John Dalton stated his famous atomic theory of matter. Dalton's atomic theory was based directly on the ideas of elements and compounds and on the four empirical laws of chemical combination described in Section 1.5.

1•7   Dalton's Atomic Theory

John Dalton was an English schoolteacher who developed the first modern theory of *atoms* as the smallest particles of elements and *molecules* as the smallest particles of compounds. To explain the properties of elements, he developed the idea that an element contained only one kind of atom and that an atom was a simple, indestructible particle of matter. Elements, he said, could not be changed to simpler substances, because their atoms could not be broken down.

Dalton explained the constant composition of compounds by the theory that atoms of elements were joined to make more complex particles called molecules, which were the simplest units of compounds. According to Dalton, the favored atomic combination for just two elements was probably 1:1. Because all its molecules were identical, the compound would have a constant composition, having a greater percentage by weight of the element that had the heavier atom. Some of the early diagrams of atoms and molecules are shown in Figure 1.12.

The law of conservation of mass was easily explained also. The theory held that in any chemical reaction atoms could change partners, or molecules could be broken down into atoms, but the total number of atoms in the reactants and the products would be the same. If atoms were indeed indestructible, no mass could be gained or lost in a chemical reaction.

The law of multiple proportions is nicely accounted for if one assumes that under some conditions atoms of two types combine in a 1:1 combination and under other conditions they combine in a 1:2 or 1:3 or 2:3 or some other combination. If we go back to the example that we considered earlier of the two oxides of carbon, we will recall that the ratio of weights of oxygen that combined with a given amount of carbon under two different conditions was 2:1. With diagrams like Dalton's we can show clearly how the weight of oxygen per weight of carbon could be twice as great in one case as in the other, if we assume a 1:1 atomic combination in one case and a 1:2 in the other, as shown schematically in Figure 1.13. In modern symbols a molecule of the first oxide is given the formula CO and is named carbon monoxide. The formula for a molecule of the second oxide is CO_2, and this compound is named carbon dioxide.

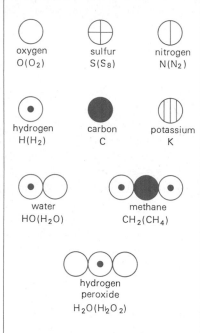

oxygen
$O(O_2)$

sulfur
$S(S_8)$

nitrogen
$N(N_2)$

hydrogen
$H(H_2)$

carbon
C

potassium
K

water
$HO(H_2O)$

methane
$CH_2(CH_4)$

hydrogen
peroxide
$H_2O(H_2O_2)$

Figure 1•12 Some of Dalton's symbols for elements and compounds. The correct formulas are given in parentheses when his were in error.

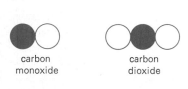

carbon
monoxide

carbon
dioxide

Figure 1•13 Dalton postulated that in one oxide of carbon, one atom of oxygen is combined with one atom of carbon; and that in the other oxide, two atoms of oxygen are combined with one atom of carbon.

Dalton's atomic theory can be summarized by listing the following assumptions:

1. All matter is made up of tiny, indestructible unit particles called atoms.
2. The atoms of a given element are all alike.
3. During chemical reactions atoms may combine, or combinations of atoms may break apart into separate atoms, but the atoms themselves are unchanged.
4. When atoms form molecules, they unite in small whole-numbered ratios, such as 1:1, 1:2, 1:3, 2:3.

Although some of these assumptions have been shown to be incorrect by later work, Dalton's theory was a guiding principle for a century of brilliant chemical discoveries.

1•7•1   **Relative Atomic Weights**   Dalton's atomic theory, coupled with the determination of the composition of many compounds, led to the development of *a scale of relative weights* of atoms.[4] Consider the example of carbon monoxide. We saw earlier that, to form this compound, 1.00 g of carbon combined with 1.33 g of oxygen. If we assume that the 2.33 g of carbon monoxide is a collection of billions upon billions of molecules all with the formula CO, it follows from the weight relationship of 1.00 g of C to 1.33 g of O that each oxygen atom must be one-third heavier than each carbon atom. If we could determine the weight of one of the atoms, the weight of the other atom could then be calculated.

It was not possible for Dalton and his contemporaries to determine the weight of a single atom or even to show beyond question that atoms existed at all. But they assumed that atoms did have definite weights and assigned *relative atomic weights* to them that agreed with the known compositions of compounds.

A few years after Dalton's initial work, oxygen atoms were chosen as the standard because this element combined readily with other elements, thereby allowing direct comparison of combining weights. Oxygen atoms were arbitrarily assigned a relative weight of 16 in order that the lightest atom, hydrogen, have a relative weight very close to 1. The weights of other atoms were compared with oxygen by analyzing as many compounds as possible and working out the most likely formulas for these compounds. In the case of the three elements hydrogen, carbon, and oxygen, the relative weights of H:C:O are 1:12:16. This early chemical method of determining relative atomic weights, however, is not used today, because of the discovery of a more precise, modern method that is described in Section 3.3.5.

1•7•2   **Symbols of Elements**   The medieval alchemists used symbols to stand for elements, such as a crescent ☽ for silver, symbolic of the silvery color of the moon, and a circle ○ for gold, symbolic of the golden sun and of

---

[4]Relative weights are weights that are compared with one another, even though the actual weights may be unknown. For example, the relative weights of three planets are Mars 0.1, Earth 1.0, and Venus 0.8.

perfection. Dalton made up other symbols, as shown in Figure 1.12. Our present system of using letters as symbols was begun by a contemporary of Dalton, the Swedish chemist J. J. Berzelius (1779–1848). He began by using the first letter of the name of the element as a symbol. Examples include those already used in this chapter: H for hydrogen, O for oxygen, and C for carbon. Because the names of several elements begin with the same letter, Berzelius found it convenient to use two letters in some symbols. Thus, carbon, calcium, chlorine, and cobalt are designated by the symbols C, Ca, Cl, and Co, respectively. Note that the first letter of the symbol is capitalized, the second is not. In some cases, the symbols we use today are related to the Latin names used centuries ago. For example, the symbols for silver (Ag), copper (Cu), and iron (Fe) are derived from the Latin names *argentum, cuprum,* and *ferrum,* respectively. The symbols for all known elements are listed with their names inside the back cover.

Possibly the most important use of the symbols of elements is in formulas to record the composition of the more than 4 million known compounds. The **formula** of a compound shows with symbols the kind and number of atoms that are chemically combined in the smallest unit of a compound. Consider the formula for glucose, $C_6H_{12}O_6$. The formula conveys the information that the smallest particle of this sugar contains a total of 24 atoms in chemical combination, 6 of which are carbon atoms, 12 are hydrogen, and 6 are oxygen. Figure 1.11 shows other examples of formulas.

1•8   Mendeleev's Periodic Table

At the time of Dalton's first statement of his atomic theory, there were about 40 known elements. One of the first attempts to group similar elements was made from 1817 to 1830 by Johann Döbereiner, when he grouped similar elements in threes, or *triads*. He noted that iron, cobalt, and nickel were alike in many ways, as were chlorine, bromine, and iodine.

By the early 1860s there were 63 known elements, and several chemists, including John Newlands in England, Lothar Meyer in Germany, and Dmitri Mendeleev in Russia, came to realize that an exciting principle applied to the elements they knew. These early chemists found that if they listed the elements in the order of increasing atomic weight, elements with similar properties appeared at regular intervals in the list.

Consider Table 1.5. In it we find two metals, sodium and potassium, that have properties very similar to those of lithium. Not only are these three elements similar in appearance, but they also form compounds with oxygen and chlorine that have similar formulas. In order of increasing atomic weight, sodium is the seventh element after lithium, and potassium is the seventh element following sodium:

$$1 \text{ lithium}$$
$$1 + 7 = 8 \text{ sodium}$$
$$8 + 7 = 15 \text{ potassium}$$

The elements beryllium, magnesium, and calcium are similar physically, and their compounds with oxygen have similar formulas. In order of

Table 1•5 Descriptions of some of the elements

Order of atomic weight	Atomic weight	Name	Symbol	Properties
1	7	lithium	Li	soft metal; low density; very active chemically; forms Li_2O, LiCl
2	9.4	beryllium	Be	much harder than Li; low density; less active than Li; forms BeO, $BeCl_2$
3	11	boron	B	very hard, nonmetallic; not very reactive; forms B_2O_3, BCl_3
4	12	carbon	C	brittle, nonmetallic; unreactive at room temperature; forms CO_2, CCl_4
5	14	nitrogen	N	gas; not very active; forms N_2O_5, NCl_3
6	16	oxygen	O	gas; moderately reactive; combines with most elements; forms Na_2O, BeO
7	19	fluorine	F	gas; extremely reactive; irritating to nose; forms NaF, BeF_2
8	23	sodium	Na	soft metal; low density; very active; forms Na_2O, NaCl (compare Li)
9	24	magnesium	Mg	much harder than Na; low density; less active than Na; forms MgO, $MgCl_2$ (compare Be)
10	27.4	aluminum	Al	as hard as Mg; not quite so active; forms Al_2O_3, $AlCl_3$ (compare B)
11	28	silicon	Si	brittle, nonmetallic; unreactive; forms SiO_2, $SiCl_4$ (compare C)
12	31	phosphorus	P	low melting point; solid; reactive; forms P_2O_5, PCl_3 (compare N)
13	32	sulfur	S	low melting point; solid; moderately reactive; combines with most elements; forms Na_2S, BeS (compare O)
14	35.5	chlorine	Cl	gas; extremely reactive; irritating to nose; forms NaCl, $BeCl_2$ (compare F)
15	39	potassium	K	soft metal; low density; very active; forms K_2O, KCl (compare Li and Na)
16	40	calcium	Ca	much harder than K; low density; less active than K; forms CaO, $CaCl_2$ (compare Be and Mg)

increasing atomic weight, these three also occur at intervals of every seventh element:

$$2 \text{ beryllium}$$
$$2 + 7 = 9 \text{ magnesium}$$
$$9 + 7 = 16 \text{ calcium}$$

Other examples of recurrent similarities are

$$4 \text{ carbon} \qquad\qquad 7 \text{ fluorine}$$
$$7 + 7 = 11 \text{ silicon} \qquad 7 + 7 = 14 \text{ chlorine}$$

We see that at every seventh entry in Table 1.5 there are elements with similar properties. An occurrence or phenomenon that is repeated in a regular way is said to be **periodic**. In Figure 1.14, there are plotted values of the boiling points for some elements along with the approximate modern atomic weights. The alternating rise and fall of the curve shows that the boiling point is roughly a periodic function of increasing atomic weight.

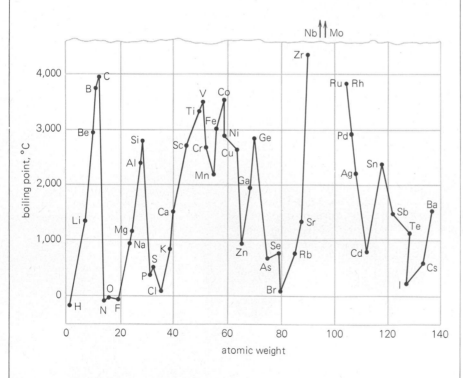

Figure 1•14 As the atomic weights increase, the boiling points of elements vary in a roughly periodic way. In this graph, modern data for boiling points are plotted against the atomic weights known in 1869. The boiling points of niobium and molybdenum are so high that they are off the graph.

Mendeleev saw most clearly how the periodicity of the properties of the elements could be shown in a **periodic table**. And he saw how such a table could be used to predict the existence and even the properties of the undiscovered elements. It was an idea that caught the imagination of the scientific world.

A modern version of a table based roughly on Mendeleev's original arrangement of 1869 is shown in Figure 1.15. Elements in the shaded boxes were not known at that time. Mendeleev left blank spaces for many of these, and wrote:[5]

> *With the periodic and atomic relations now shown to exist between all the atoms and the properties of their elements, we see the possibility not only of noting the absence of some of them but even of determining, and with great assurance and certainty, the properties of these as yet unknown elements; it is possible to predict their atomic weight, density in the free state or in the form of oxides, acidity or basicity, degree of oxidation, and . . . it is even possible also to describe the properties of some compounds of these unknown elements.*

He was confident that similar elements should fall in the same vertical groups and that properties should change smoothly from left to right in the horizontal periods. Within a few years, the discoveries of scandium, Sc, gallium, Ga, and germanium, Ge, proved most of Mendeleev's detailed predictions to be valid.

When the element argon, Ar, a chemically inactive gas, was found in 1894, its properties were unlike those of any other known element. With the periodic table as a guide, chemists assumed that argon must be a member of a family of elements that would form a new vertical column. An intensive search led to the discoveries, within just six years, of the five other family members from helium, He, to radon, Rn. All proved to be chemically inactive gases.

oxides	R_2O	RO	R_2O_3	RO_2	R_2O_5	RO_3	R_2O_7		RO_4	
period	group I	group II	group III	group IV	group V	group VI	group VII		group VIII	
1	H 1							He		
	A · B	A · B	A · B	A · B	A · B	A · B	A · B	A · B	A · B	A · B
2	Li 7	Be 9.4	B 11	C 12	N 14	O 16	F 19	Ne		
3	Na 23	Mg 24	Al 27.4	Si 28	P 31	S 32	Cl 35.5	Ar		
4	K 39	Ca 40	Sc ? 45	Ti 50	V 51	Cr 52	Mn 55	Fe 56	Co 59	Ni 59
	Cu 63.4	Zn 65.2	Ga ? 68	Ge ? 70	As 75	Se 79.4	Br 80	Kr		
5	Rb 85.4	Sr 87.6	Y (60)	Zr 90	Nb 94	Mo 96	Tc	Ru 104.4	Rh 104.4	Pd 106.6
	Ag 108	Cd 112	In (75.6)	Sn 118	Sb 122	Te 128	I 127	Xe		
6	Cs 133	Ba 137	La (94)	Hf ? 180	Ta 182	W 186	Re	Os 199	Ir 198	Pt 197.4
	Au 197	Hg 200	Tl 204	Pb 207	Bi 210	Po	At	Rn		

Figure 1•15 A portion of a modern periodic arrangement based on Mendeleev's original table. Elements in the shaded boxes were not known in 1869. The atomic weights shown are the ones tabulated by Lothar Meyer; those in parentheses were later found to be greatly in error. In the four cases with question marks, the atomic weights are those Mendeleev listed for elements he predicted would someday be found.

[5] From H. M. Leicester and H. S. Klickstein, *A Source Book in Chemistry 1400–1900*, McGraw-Hill, New York, 1952.

Chapter Review

Summary

The structure of matter and the changes that it undergoes are the domain of **chemistry,** which provides both a basis for our knowledge of the world and a means for making it a better place. Much chemical knowledge is summarized in well-established **laws** of nature; many explanations and speculations about nature are made as **hypotheses** and **theories.** Chemistry and all the other sciences are interwoven by the relatedness of everything in the world to everything else.

All chemical knowledge rests ultimately on measured quantities, which are best expressed in terms of the **SI units.** The **meter,** m, is the basis for measurements of length, area, and volume. The **kilogram,** kg, is the basis for measurements of **mass** (also often call **weight**). The mass per unit volume of a substance is its **density;** a closely related quantity is the **specific gravity,** which is dimensionless. The **temperature** of a substance is a measure of its relative hotness on any of several numerical scales, such as the **Celsius scale** and the **Kelvin scale.** Temperature is a manifestation of **heat energy,** two units of measure for which are the **calorie,** cal, and the SI unit **joule,** J. A measurable characteristic property of every substance is its **specific heat.**

The properties of substances and the changes that occur among substances are the province of **descriptive chemistry.** All properties are either **intrinsic** (characteristic of any sample of a given substance) or **extrinsic** (not characteristic of any particular substance). **Chemical properties** (intrinsic) are manifested in **chemical changes (chemical reactions),** in which substances are converted into other substances. **Physical properties** (also intrinsic) do not entail any change to a different substance. In a **physical change,** some properties of a substance may change, but the identity of the substance does not change.

All changes in matter entail a change in **energy.** Among the different forms of energy, all of them interconvertible, are **electric, radiant, chemical,** and **nuclear** (or **atomic**) energy. They can all be classified as either **potential energy** (energy by virtue of position or physical or chemical state) or **kinetic energy** (energy by virtue of motion). All processes entailing a change in heat energy are either **exothermic** (heat given to the surroundings) or **endothermic** (heat absorbed from the surroundings). A process that is exothermic in one direction is always endothermic in the opposite direction.

All pure substances are either **elements** or **compounds.** If two or more substances are brought together without undergoing chemical change, the result is a **mixture,** which may be either **homogeneous** or **heterogeneous.** If a chemical reaction does occur, three kinds of changes are observed: change in properties, in composition, and in energy. All chemical changes are governed by the **law of conservation of energy** and the **law of conservation of mass.** The compositions of chemical compounds are determined by the **law of definite composition** and the **law of multiple proportions.**

The fundamental principles underlying all chemical changes are the province of **theoretical chemistry.** A correlation of the concepts of elements and compounds with the four laws just mentioned was achieved in **Dalton's atomic theory,** the first modern theory of **atoms** and **molecules** as the fundamental particles of substances. An outgrowth of this theory was the scale of **relative atomic weights** of the elements. When elements were listed in order of atomic weight, the regular occurrence of elements with certain properties led Mendeleev to devise the **periodic table** of the elements and to predict the existence of several undiscovered elements. The numbers and hence the relative proportions of the different atoms in the smallest unit of a compound are given by its **formula,** in which the symbols of the chemical elements are used.

Exercises

At the end of each chapter there are both numerical and verbal exercises. These are arranged under general headings to help you review and strengthen your understanding of the major topics in the chapter. The more challenging exercises are marked with asterisks.

For working problems, your attention is again called to the sections in Appendix 1 on arithmetic procedures that are often encountered. In solving problems it is suggested that the data be used as given in entering values on a hand calculator. The answer then must be rounded to the proper number of significant figures. Answers to problems are given at the end of the book.

Arithmetic Relations

1. Convert each of the following numbers to the exponential form (see the appendix):
   (a) 381,000,000
   (b) 0.00001
   (c) 0.000763
   (d) $\sqrt{0.000001}$
   (e) 1/10,000

2. Perform the indicated calculations, first converting individual numbers to powers of ten if they are not expressed that way:
   (a) $12,000 \times 6.0 \times 10^{-4}$
   (b) $0.00001 \div 0.0000002$
   (c) $2.5 \times 10^3 \times 4.0 \times 10^{-4}$

(d) $\sqrt{3 \times 10^2 \times 12 \times 10^6}$

3. $A^x \times B^y = ?$ $A^x \div B^y = ?$ $A^0 = ?$ $A^1 = ?$
   $A^{1/n} = ?$

4. Solve, using the four-place logarithm table (Table A.8) in the appendix:

$$\frac{(3.60)(273)(760)(10^3)}{(10^{-2})(279)(388)(1,372)} = ?$$

5. Solve for x: $3x + 32 = 97 + 2x$

6. Solve the quadratic equation for x when $2x^2 + 3x = 2$. The formula for solving a quadratic equation is given in Appendix 1.

7. Suppose that the population of a town is reported to be 29,012.
   (a) Express this number in exponential form.
   (b) Express the number in exponential form to three significant figures.

Length, Area, Volume; Time

8. On a sheet or two of paper draw the following. If you can do this freehand without using a ruler, do so. If you are not quite familiar with the units, make the drawings with the aid of a ruler.
   (a) 1 cm, 1 dm, 1 mm, and 1 in.
   (b) 1 cm^2, 1 dm^2, 1 mm^2, and 1 in.^{2}
   (c) 1 cm^3, 1 dm^3, 1 mm^3, and 1 in.^{3}

9. On September 17, 1977, St. Olaf and Carleton Colleges played the first metric football game. Billed as the Liter Bowl, the field was 100 m long and 50 m wide (instead of the regulation 100 yd by 53.3 yd). An 86-kg back for St. Olaf, Tom Fiebiger, made the longest run of the day for 70 m.
   (a) Calculate the length and width of the Liter Bowl field in centimeters and in yards.
   (b) What was the length of the longest run in yards? In kilometers?
   (c) Calculate the area of the Liter Bowl field in square meters and in square yards. Which field, the regular one or the Liter Bowl, is greater in area and by what percent?

10. In a 1958 handbook, the inch is stated to be equivalent to 2.540005 cm. Today, it is stated as equivalent to exactly 2.54 cm. What is the probable reason for this change? Are there the same, more, or fewer significant figures in the 1958 equivalent as compared with the present one?

11. Calculate the volume of a quart in cubic meters (m^3).

*12. The average diameter of the earth is about 7,920 miles. Assuming the earth to be spherically shaped, calculate its volume in cubic kilometers (km^3).

13. In Canada a gallon is a larger volume (160 fluid oz) than the U.S. gallon (128 fluid oz). Calculate the volume of the Canadian gallon in milliliters, in cubic centimeters, and in liters.

14. For a period of six months, one of the authors of this text averaged 28.2 miles per gallon in his small automobile. Convert this gasoline consumption to kilometers per liter (km/L).

*15. A student calibrates two 500-mL beakers by filling them to the mark with water from a graduated cylinder. The volumes found in three tries are:

   beaker A: 502 mL, 502 mL, 503 mL
   beaker B: 498 mL, 501 mL, 502 mL

Which set of values is more precise? Explain. Which beaker, on the average, more accurately measures 500 mL? Explain.

16. Convert the speed of 20 miles per hour (mi/h) to:
   (a) feet per second
   (b) kilometers per hour

*17. Professor Jack Tomlison at California State University has suggested metric time. Each day would be a kilochron, composed of 1,000 chrons (from Greek *chronikos*, of time).
   (a) How many chrons are there in 6 hours and 36 minutes?
   (b) Convert a speed of 55 miles per hour to meters per chron.

Mass and Density

18. A 425-g package of natural breakfast food sells for 97 cents. What is the price per kilogram? The price per pound?

*19. A solution is made by dissolving 10.0 g of salt in 3.00×10^2 g of water. What is the composition of this solution in terms of percentage salt and percentage water by weight?

20. In Figure 1.2 the drawing indicates that nine quarter-pound sticks of butter weigh about 1 kg. What is the precise weight in kilograms, to five significant figures, of nine quarter-pound sticks? What is the percent error in considering nine quarter-pounds the equivalent of a kilogram?

*21. "A pint's a pound the world around." Assuming that the material is water at 4 °C and that 1 L weighs 1 kg, to five significant figures, calculate the error and the percent error in the quoted statement.

22. A rock that weighs 16.4 g has a volume of 7.1 cm^3 (or 7.1 mL). What is the density of the rock?

23. The precise dimensions of the standard shown in Figure 1.3 are: a mass of 1 kg minus 0.019 mg and a volume of 46.4030 cm^3 at 0 °C. Calculate the precise density to the proper number of significant figures. Compare the precise density to the approximate value found in Example 1.3.

*24. If the gasoline tank in your automobile holds 20 gal, what is the weight of the fuel when the tank is filled with gasoline that has a specific gravity of 0.68?

25. Express the following in SI base units: acceleration, force, and density.

Temperature and Heat

26. Some familiar temperatures in degrees Fahrenheit are: 90, a hot day; 98.6, normal body temperature; 72, a comfortable temperature; 30, could snow; −10, very cold. Convert each to the Celsius scale and to the Kelvin scale.

27. By reading Figure 1.7, make the following conversions approximately. Check each of your readings by an algebraic calculation. (a) 50 °F to °C; (b) −100 °F to °C; (c) −90 °C to °F; (d) 75 °C to °F.

28. Using an equation that relates degrees Celsius to degrees Fahrenheit, show algebraically that the two scales have the same reading at −40.

29. A temperature change of 32 degrees on the Fahrenheit scale is equal to what temperature change on the Celsius scale?

*30. Imagine that a temperature scale exists in degrees Peterson (°P) and that on this scale water boils at 165 °P and freezes at 25 °P.
(a) Convert a temperature of 45 °P to degrees Celsius.
(b) Convert a temperature of 32 °C to degrees Peterson.

*31. Plot a graph of the data in Figure 1.7 as follows. Label the vertical coordinate in Kelvins, making seven large divisions for 100-degree intervals from 0 to 700 K. Label the horizontal coordinate "°C or °F" in 12 large divisions from −500 to 700. Plot the Celsius versus the Kelvin temperatures and draw the best straight line you can near or through the points. Repeat for the Fahrenheit versus the Kelvin temperatures.
(a) At what temperatures in degrees Celsius and degrees Fahrenheit, respectively, do the two lines intersect the 0 K line?
(b) What is the ratio of the slope Celsius line/Fahrenheit line?
(c) What is the Kelvin, the Celsius, and the Fahrenheit temperature at which the two lines intersect?

Specific Heat

32. Draw a graph similar to Figure 1.8 that shows the data for water, ice, sand, carbon, and uranium.

33. (a) In Figure 1.8 read the approximate number of joules per gram needed to change the temperature by two degrees for ethyl alcohol, aluminum, and iron. Calculate from these readings the slopes of the lines, express them in the proper units, and compare them with the values in Table 1.4.
(b) In terms of the construction of a graph, what are the units for the slope of a line?

34. Calculate the amount of heat needed to raise the temperature of 10.0 g of gold from 20.0 °C to 100.0 °C. This amount of heat will raise the temperature of 10.0 g of water from 20.0 °C to what temperature?

35. A 54.0-g sample of metal at 98.0 °C is placed into 80.0 g of water at 24.0 °C. The final temperature of the water and the metal is 28.0 °C. Assuming no heat is lost to the surroundings, calculate the specific heat of the metal.

*36. A certain weight of a metal whose specific heat is $0.335 \text{ J} \cdot \text{g}^{-1} \cdot °\text{C}^{-1}$ is heated to 83.0 °C and is then placed in 200 g of water at 21.0 °C. The final temperature of the metal and water is 23.0 °C. Assuming no heat is lost to the surroundings, how much does the metal weigh?

37. What is the final temperature after 90.0 g of water at 90.0 °C is mixed with 50.0 g of water at 10.0 °C?

38. The British thermal unit (Btu) is defined as the heat required to raise the temperature of 1 lb of water by 1 °F. Calculate its value in joules.

39. The heat capacity of a calorimeter is known to be 10.0 J/°C. A 20.0-g piece of iron at 80.0 °C is added to the calorimeter when it contains a quantity of water initially at 25.0 °C. The final temperature of the system (water, iron, and calorimeter) is 30.0 °C. What is the weight of the water present in the calorimeter?

*40. One finds under *Nutrition Information* on the side of a box of popular breakfast cereal that a 1-oz serving provides 110 cal.
(a) Suggest a method for experimentally obtaining this information.
(b) Convert the value to kilojoules.
(c) This quantity of energy could raise the temperature of 1 gal of water at 40 °F to what temperature?

Changes in Matter and Energy

41. In the following description, which are extrinsic and which are intrinsic properties? Which are physical properties and which are chemical properties? Sucrose is a colorless solid that chars or blackens upon being heated or may even ignite and burn with a yellow flame. Its density is 1.6 g/mL. Usually it is in the form of small crystals (table sugar), although it can be in the form of a powder.

42. Pure substance A is a solid at room temperature. Upon being heated to about 250 °C, it gradually liquefies. Upon being cooled to room temperature, the liquid cannot be induced to solidify.
(a) Is substance A probably an element or a compound? Explain.
(b) Has a chemical change occurred? If so, can it be said that the change is endothermic, based on the information given?
(c) Can it be said that the resulting liquid is an element, based on the information given?

43. What is the mass of an object that has kinetic energy of 1 J when moving in uniform motion at a speed of 1 m/s?

44. (a) Calculate the kinetic energy in joules of an automobile weighing 2,000 lb (907 kg) traveling at a speed of 50 mi/h (80 km/h). What is its kinetic energy

when moving at 70 mi/h? When moving at 100 mi/h?

(b) Calculate the average kinetic energy of Tom Fiebiger if it took him 9.0 s to make his run (see Exercise 9).

Early Concepts of Chemical Reactions

45. When a candle weighing 1 g is burned in oxygen, the carbon dioxide and water vapor formed by the combustion weigh considerably more than 1 g. Is this in accord with the law of conservation of mass? Explain.

46. A 34.0-g sample of ammonia was shown to be composed of 6.0 g of combined hydrogen and 28.0 g of combined nitrogen; a 32.0-g sample of hydrazine was shown to be composed of 4.0 g of combined hydrogen and 28.0 g of combined nitrogen.

(a) Based on 1.0 g of hydrogen, calculate from these data the weight ratio of hydrogen to nitrogen for each compound.

(b) Taking a fixed weight of hydrogen, 1.00 g in each case, show that the ratio of the weights of nitrogen in the two compounds complies with the law of multiple proportions.

47. When carbon burns in a limited amount of oxygen, two gaseous compounds are formed. Suggest ways to distinguish these two substances from one another.

*48. In Section 1.5.4 the combining weight ratios for two oxides of carbon are given. Show that the relative atomic weights for carbon and oxygen listed in Section 1.7.1 agree with the combining weight ratios.

49. While Mendeleev was able to predict the existence of several undiscovered elements, he was unable to predict the existence of the noble gases. Why?

50. After Mendeleev devised his periodic table, he concluded that the atomic weights of certain elements had been incorrectly determined, and this was found to be true to a limited extent. How was Mendeleev able to predict that certain atomic weights were in error? Why was his prediction not valid in all cases?

Using the Index

51. To become familiar with how the Index can help you, find the following.

(a) the densities of lithium and of beryllium
(b) the melting points of silicon and sulfur
(c) the number of grams in an ounce
(d) an equation for the reaction of carbon and oxygen
(e) some major uses of chlorine
(f) information on carbon dioxide in the air
(g) references to the work of Marie Curie and Albert Einstein
(h) the chemistry of rain
(i) some causes of color of substances
(j) the compositions of aspirin, milk of magnesia, sucrose, and vinegar

Chemical Equations; Stoichiometry

2

Chemical Equations

2•1   **Writing Formulas**

2•1•1   Formulas of Elements

2•1•2   Formulas of Compounds

2•2   **Practice in Writing Balanced Equations**

2•2•1   Writing Balanced Equations for Some General Classes of Reactions

Stoichiometry

2•3   **Atomic Weights**

2•4   **Molecular Weights**

2•5   **Weight Relationships in Chemical Reactions**

2•5•1   The Mole

2•5•2   Molar Weights and Chemical Equations

2•5•3   Theoretical Versus Actual Yields

2•5•4   Limiting and Excess Reactants

2•6   **Calculation of Percentage Composition from Formulas**

2•7   **Calculation of Formulas from Experimental Data**

2•7•1   Laboratory Determination of Percentage Composition

2•7•2   Calculation of Empirical Formulas

2•7•3   Calculation of Molecular Formulas

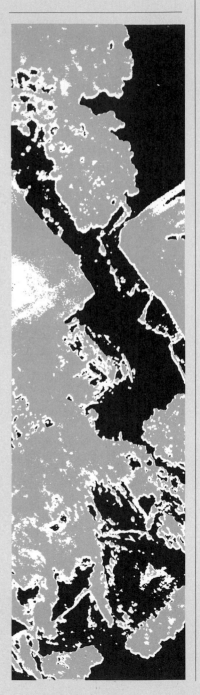

In the language of chemistry, each known pure substance, whether an element or a compound, has its own unique name and formula. The most concise way to describe a chemical reaction is to write the formulas for each substance involved in the form of a chemical equation. In this chapter we will lay the groundwork that will enable us to write and interpret chemical equations.

A chemical equation summarizes a great deal of information about the substances involved in reactions. It is not only a *qualitative* statement, describing what substances are involved, but also a *quantitative* statement, describing how much of each reactant or product is involved.

The process of making calculations based on formulas and balanced equations is referred to as **stoichiometry** (from the Greek *stoicheion*, element, and *-metria*, science of measurement). As the first step in developing our skills in stoichiometric calculations, we will discuss the writing of formulas for substances. Then we will work with balancing the equations for some general types of reaction. After practice in writing balanced equations, we proceed to stoichiometry. Our stoichiometric calculations will be devoted initially to determinations of the relative weights of reactants and products. At the conclusion of this chapter, we shall describe methods for determining the formulas of compounds from experimental data.

Chemical Equations

A **reactant** is any substance that is initially present and is changed during a chemical reaction. A **product** is any substance produced during a chemical reaction. A **chemical equation** (or **balanced chemical equation**) shows the formulas of reactants, then an arrow, and then the formulas of products, with equal numbers of atoms of each element to the left and right of the arrow. For example, the balanced equation for the reaction between hydrogen and oxygen to produce water is written as

$$2H_2 + O_2 \longrightarrow 2H_2O$$

The formula H_2 tells us that a hydrogen molecule is composed of two atoms. It is a *diatomic molecule*, as is the oxygen molecule, O_2. The water molecule, H_2O, is *triatomic* because it is composed of three atoms, two of hydrogen and one of oxygen. The equation states that two molecules of hydrogen react with one molecule of oxygen to form two molecules of water.

2•1   **Writing Formulas**

The formula of a substance expresses the kind and number of atoms that are chemically combined in a unit of the substance. There are several types of formulas, among them (1) molecular formulas and (2) empirical formulas.

A **molecular formula** states the actual number of atoms in a molecule or the smallest unit of a compound.[1] An **empirical formula** states the smallest whole-number ratio of atoms in a compound. The difference between a molecular formula and an empirical formula can be illustrated with the compound hydrogen peroxide, a common bleaching agent. The molecular formula of hydrogen peroxide is written as H_2O_2, which shows that each molecule is composed of four atoms. On the other hand, its empirical formula is HO, because the smallest whole-number ratio of atoms in a molecule is $1:1$.

2•1•1   Formulas of Elements   For the vast majority of elements, the formula is simply the symbol. Examples are sodium, Na, iron, Fe, silver, Ag, and tin, Sn. There are seven common elements, however, that almost always exist as diatomic molecules. These elements are hydrogen, nitrogen, oxygen, fluorine, chlorine, bromine, and iodine; their formulas are H_2, N_2, O_2, F_2, Cl_2, Br_2, and I_2. *It is essential in writing equations that the diatomic species, such as H_2, be indicated for any one of these seven elements when the ordinary form of the element is called for.* For example, the balanced equation for the reaction between hydrogen and oxygen to produce water is written as

$$2H_2 + O_2 \longrightarrow 2H_2O \qquad \text{(correct)}$$

not as

$$2H + O \longrightarrow H_2O \qquad \text{(incorrect)}$$

Under certain conditions, each of the aforementioned seven elements can exist as an individual atom. In addition, oxygen has a special triatomic form, O_3, ozone, which is described in Section 9.3. Elements other than the seven can also exist as molecules made of two or more atoms, called *polyatomic molecules.* Examples include the S_8 molecule of sulfur and the P_4 molecule of phosphorus. However, unless some special point is being made, these elements are commonly denoted in equations by their simple symbols, that is, as S and P, respectively.

2•1•2   Formulas of Compounds   In future chapters it will be shown why the chemist writes the formula for hydrogen peroxide as H_2O_2 instead of HO, and the formula for glucose as $C_6H_{12}O_6$ instead of CH_2O. To begin our use of formulas, we will take them as given in this text or in some other reference and gradually build up our own store of formulas for known substances.

In many cases the formula for a substance is the formula for one molecule of that substance. Examples are water, H_2O, each molecule of which is made from three atoms; hydrogen peroxide, H_2O_2, each molecule made from four atoms; and glucose, $C_6H_{12}O_6$, each molecule made from 24 atoms. In many other cases, the formula does not refer to a definite

[1]Uncombined elements are different from elements combined in a compound, but when referring to the composition of compounds, chemists are not precise in their use of the words *element* and *atom*. Chemists speak of water as being composed of "atoms of the elements" hydrogen and oxygen, even though no actual atoms or elements are present. The differences between the composition of combined and uncombined atoms will be taken up in Chapter 5.

molecule but merely shows the unit amount with the simplest ratio of atoms in the substance. Examples are magnesium oxide, MgO, which has one atom of magnesium per atom of oxygen, and sodium oxide, Na_2O, which has two atoms of sodium per atom of oxygen.

2•2   Practice in Writing Balanced Equations

To write a balanced equation, one may follow a three-step process: (1) write down the name(s) of the reactant(s), then an arrow, and then the name(s) of the product(s); (2) rewrite the expression using the formula for each substance; and (3) balance the equation by choosing the proper whole-number coefficients for each formula. This three-step process can be illustrated by looking at the burning of methane, the principal constituent of natural gas:

- *Step 1*

    methane + oxygen $\longrightarrow$ carbon dioxide + water

- *Step 2*

    $$CH_4 \ + \ O_2 \ \longrightarrow \ CO_2 \ + \ H_2O$$

- *Step 3*

    (a) $CH_4 \ + \ ?O_2 \ \longrightarrow \ CO_2 \ + \ 2H_2O$
    (b) $CH_4 \ + \ 2O_2 \ \longrightarrow \ CO_2 \ + \ 2H_2O$

        (the balanced equation)

In step 3(a), two molecules of water are shown on the right to account for the four H atoms of CH_4. At this point, four O atoms are shown to the right of the arrow. In step 3(b), the proper coefficient is chosen for O_2, so that four atoms of oxygen are also shown to the left of the arrow.

Under no circumstances are the subscript numbers of the correct formulas chosen in step 2 to be changed as one works to balance the equation in step 3.

As one begins the study of equation writing, it is frustrating to have to learn names and formulas while practicing the art of balancing equations. It is of some comfort to understand that even experienced chemists cannot write an equation unless they look up (or remember) the reactants, the products, and the formulas for each. Only an experienced person can be expected to predict what products might be formed from certain reactants, and even experience is no guarantee that the prediction will be correct. The beginner is expected to acquire gradually a background of known formulas for reactants and products as specific chemical reactions are studied.

2•2•1   Writing Balanced Equations for Some General Classes of Reactions

There are three general classes of reactions that provide examples of many useful equations.

1. A **direct combination reaction** is a reaction of two elements to produce a compound.

2.   A **single displacement reaction** is a reaction of an element with a compound to produce a different element and compound.

3.   A **double displacement reaction** is a reaction of two compounds to produce two different compounds by exchanging components.

These three classes of reactions are illustrated in Examples 2.1, 2.2, and 2.3.

• Example 2•1 •

Write the balanced equation for the reaction of the two elements aluminum, Al, and bromine, Br_2, to produce the compound aluminum bromide, $AlBr_3$.

• Solution •

- *Step 1*

  $$aluminum + bromine \longrightarrow aluminum\ bromide$$

- *Step 2*

  $$Al + Br_2 \longrightarrow AlBr_3$$

- *Step 3*

There are two bromine atoms per unit on one side of the arrow and three bromine atoms per unit on the other side. To balance the bromines, six atoms must be shown on each side. It takes three Br_2 molecules to supply the bromine atoms needed to make two $AlBr_3$ units:

$$Al + 3Br_2 \longrightarrow 2AlBr_3$$
$$(not\ balanced)$$

Now we must show two Al atoms on the left side to supply the aluminum atoms needed to make two $AlBr_3$ units:

$$2Al + 3Br_2 \longrightarrow 2AlBr_3$$
$$(balanced)$$

See also Exercises 8–11 at the end of the chapter.

• Example 2•2 •

Write the balanced equation for the reaction of magnesium, Mg, with ferric chloride, $FeCl_3$, to produce magnesium chloride, $MgCl_2$, and iron, Fe.

• Solution •

- *Step 1*

  $$magnesium + ferric\ chloride \longrightarrow magnesium\ chloride + iron$$

- *Step 2*

  $$Mg + FeCl_3 \longrightarrow MgCl_2 + Fe$$

- *Step 3*

It is good practice to start by trying to find the coefficients for the more complex formulas. Leave single atoms, such as Mg and Fe, to the last, because it is always possible to use any coefficient needed in the final steps. To balance the chlorine atoms, we must show six Cl atoms on each side of the arrow. As in Example 2.1, we use the coefficients 2 and 3:

$$Mg \quad + \quad 2FeCl_3 \quad \longrightarrow \quad 3MgCl_2 \quad + \quad Fe$$
(not balanced)

Now we need to show three Mg atoms as reactants and two Fe atoms as products to complete the balancing:

$$3Mg \quad + \quad 2FeCl_3 \quad \longrightarrow \quad 3MgCl_2 \quad + \quad 2Fe$$
(balanced)

See also Exercises 12 and 14.

- **Example 2•3**

Write a balanced equation for the production of silver chloride, AgCl, and calcium nitrate, $Ca(NO_3)_2$, from silver nitrate, $AgNO_3$, and calcium chloride, $CaCl_2$.

- **Solution**

- *Step 1*

silver nitrate + calcium chloride $\longrightarrow$ silver chloride + calcium nitrate

- *Step 2*

$$AgNO_3 \quad + \quad CaCl_2 \quad \longrightarrow \quad AgCl \quad + \quad Ca(NO_3)_2$$

The formulas $AgNO_3$ and $Ca(NO_3)_2$ show the polyatomic unit NO_3. Such a unit is treated just like a monatomic (one atom) unit in writing formulas and balancing equations. That is, each unit of $AgNO_3$ or AgCl may be considered as being composed of two units, an Ag unit and an NO_3 or a Cl unit. Each unit of $Ca(NO_3)_2$ or $CaCl_2$ is considered as being composed of three units, a Ca unit and two NO_3 or Cl units.

- *Step 3*

To balance the NO_3 units, we must show two NO_3 units on both sides of the arrow. We accomplish this by using the coefficient 2 in front of the $AgNO_3$:

$$2AgNO_3 \quad + \quad CaCl_2 \quad \longrightarrow \quad AgCl \quad + \quad Ca(NO_3)_2$$
(not balanced)

Now it can be seen that two units of AgCl are needed to complete the balancing:

$$2AgNO_3 \quad + \quad CaCl_2 \quad \longrightarrow \quad 2AgCl \quad + \quad Ca(NO_3)_2$$
(balanced)

See also Exercises 12 and 15.

Stoichiometry

2.3   Atomic Weights

Atoms are so minute that it is difficult for us to compare them to any familiar object. To three significant figures, the weight of an atom of hydrogen is 1.67×10^{-24} g, that of an atom of carbon is 1.99×10^{-23} g, and that of an atom of oxygen is 2.66×10^{-23} g. The use of such extremely small numbers to express the weights of these atoms is very cumbersome. As mentioned in Section 1.7.1, the relative weights of H:C:O atoms are 1:12:16. Because we usually are interested in comparing atoms with one another, we find it convenient to speak of their weights in units called *atomic mass units,* amu, rather than in units of grams.[2] Using atomic mass units, the relative atomic weights of hydrogen, carbon, and oxygen are 1.0079 amu, 12.011 amu, and 15.999 amu, respectively. A complete list of atomic weights of the elements is found on the inside back cover of this text. These relative atomic weights are proportional to the actual weights of the atoms. *One* **atomic mass unit** *(1 amu) is equivalent to about 1.661×10^{-24} g. It requires about 602,200,000,000,000,000,000,000 (6.022×10^{23}) amu to equal 1 g.*

2.4   Molecular Weights

The **molecular weight** of a substance is the sum of the weights of the atoms shown in its formula.[3] Example 2.4 illustrates the calculation of a molecular weight. The molecular weights of some common substances are listed in Table 2.1.

Table 2.1 **Molecular weights of some common substances**

Name	Formula	Molecular weight, amu[a]
hydrogen	H_2	2.016
oxygen	O_2	31.999
water	H_2O	18.015
hydrogen peroxide	H_2O_2	34.015
carbon dioxide	CO_2	44.010
sodium chloride	$NaCl$	58.443
calcium nitrate	$Ca(NO_3)_2$	164.090
glucose	$C_6H_{12}O_6$	180.157

[a] All values are rounded to three digits to the right of the decimal point, after using atomic weights as given on the inside back cover.

[2] The standard reference for this atomic weight scale will be described in Section 3.4.

[3] The use of the terminology "molecular weight of a substance" does not mean that a particular substance consists of molecules. As we will discuss fully in Chapter 7, the term *molecule* refers to an uncharged particle, but many substances are made of charged particles called *ions*. Some chemists use the term *formula weight* to refer to the sum of the weights of the atoms shown in the formula of a substance, and use the term *molecular weight* only to refer to substances that consist of molecules. Our more general definition of the term *molecular weight* is widely adopted because it allows the use of a familiar concept in all cases without forcing the user to find out first what sort of particles a particular substance contains.

━━━━ • Example 2•4 • ━━━━━━

Calculate the molecular weight of hydrogen sulfate, H_2SO_4, from the following atomic weights: H, 1.0079 amu; O, 15.999 amu; and S, 32.06 amu.

━━━━ • Solution • In one H_2SO_4 molecule,

weight of H = 2 × 1.0079 amu = 2.0158 amu
weight of O = 4 × 15.999 amu = 63.996 amu
weight of S = 1 × 32.06 amu = 32.06 amu
weight of one molecule of H_2SO_4 = 98.0718 = 98.07 amu

See also Exercise 17.

2•5   Weight Relationships in Chemical Reactions

A balanced chemical equation is the basis for calculating the weight relationships of reactants and products. In accord with the law of conservation of mass, the total weight of the reactants equals the total weight of the products in a balanced equation. Take the equation in Example 2.1, for instance:

$$2Al \quad + \quad 3Br_2 \quad \longrightarrow \quad 2AlBr_3$$

2 units   3 units   2 units

2(26.98) amu + 3[2(79.904)] amu = 2[26.98 + 3(79.904)] amu

53.96 amu + 479.42 amu = 533.38 amu

533.38 amu = 533.38 amu

Interpreted in this way, the equation shows the weight relationships when two atoms of aluminum react with three molecules of bromine.

Atoms are so small that it is not possible in ordinary laboratory work to study the reaction of only two atoms, or of even 2,000 atoms. We cannot weigh such small amounts on the most sensitive balances. In order to use a balanced equation to describe the amounts of substances with which one can actually work in the laboratory, chemists have devised the unit called the *mole*.

2•5•1   The Mole   The atomic weight of carbon to four significant figures is 12.01 amu. How many atoms of carbon are there in 12.01 g of carbon? Modern experimental methods show that this number of atoms is 6.022×10^{23}. This huge number is called the **Avogadro number,** named in honor of Amadeo Avogadro, the brilliant Italian contemporary of Dalton. The weight of 6.022×10^{23} atoms of oxygen is 16.00 g; the weight of 6.022×10^{23} molecules of carbon monoxide, CO, is 28.01 g; and the weight of 6.022×10^{23} molecules of carbon dioxide, CO_2, is 44.01 g. A **mole** of a substance is the amount that contains 6.022×10^{23} units of that substance.[4]

The weight of a mole of a substance is called its **molar weight.** *The molar weight in grams of a substance is numerically equal to its molecular weight in*

───────────────

[4]This definition is consistent with the SI definition in Table A.1. Mole is abbreviated mol.

atomic mass units. The molecular weight of sulfuric acid, H_2SO_4, is calculated in Example 2.4 as 98.07 amu. To express the molar weight of H_2SO_4, all we need to do is change the units from atomic mass units to grams. So the molar weight is 98.07 g. Additional examples are given in Table 2.2 and in Figure 2.1.

Table 2.2 **Mole relationships**

Name	Formula	Molecular weight, amu[a]	Molar weight, g
nitrogen	N_2	28.0	28.0
atomic nitrogen	N	14.0	14.0
silver	Ag	108	108
methanol	CH_3OH	32.0	32.0
sodium chloride	NaCl	58.4	58.4
barium chloride	$BaCl_2$	208	208
ammonium sulfate	$(NH_4)_2SO_4$	132	132

[a]See footnote 3 in this chapter.

Figure 2.1 Arranged from left to right around 1 mole of sugar ($C_{12}H_{22}O_{11}$, 342.3 g) are 1 mole of aluminum (Al, 27.0 g), 1 mole of water (H_2O, 18.0 g), 1 mole of copper (Cu, 63.6 g), and 1 mole of iron (Fe, 55.8 g). Each of these more or less pure samples contains approximately 6×10^{23} atoms if an element, or that number of molecules if a compound.

2.5.2 **Molar Weights and Chemical Equations** To interpret a chemical equation in terms of the amounts of substances that we can study in the laboratory, we first express all quantities in terms of moles. Using again the equation for the reaction of aluminum with bromine, we can write

$$2Al + 3Br_2 \longrightarrow 2AlBr_3$$

2 mol	3 mol		2 mol
2(26.98) g +	3[2(79.904)] g	=	2[26.98 + 3(79.904)] g
53.96 g +	479.42 g	=	533.38 g
	533.38 g	=	533.38 g

Example 2.5 shows the interpretation of three equations in terms of molar quantities. Example 2.6 illustrates one of the most fundamental types of chemical calculations, finding the weight of a product that is formed from a given weight of a reactant. Example 2.7 is similar, in that we are finding the weight of a reactant that is required to combine with a given weight of a second reactant.

• Example 2•5 •

In Section 1.5.3, it was pointed out that by varying the conditions under which iron and oxygen react, three different substances (each an iron oxide) can be prepared. The formulas for the iron oxides, in the order described in Section 1.5.3, are Fe_3O_4, Fe_2O_3, and FeO. Write balanced chemical equations for the reactions. Interpret the equations in terms of moles, and show the weights of all substances in grams.

• Solution •

(a) For Fe_3O_4:

oxygen	+	iron	$\xrightarrow{\text{below 500 °C}}$	the first iron oxide
$?O_2$	+	$?Fe$	$\longrightarrow$	$?Fe_3O_4$
$2O_2$	+	$3Fe$	$\longrightarrow$	Fe_3O_4
2 mol		3 mol		1 mol

$$2[2(16.00)] \text{ g} + 3(55.85) \text{ g} = [3(55.85) + 4(16.00)] \text{ g}$$
$$64.00 \text{ g } O_2 + 167.55 \text{ g Fe} = 231.55 \text{ g } Fe_3O_4$$

(b) For Fe_2O_3:

oxygen	+	iron	$\xrightarrow{\text{above 600 °C}}$	the second iron oxide
$?O_2$	+	$?Fe$	$\longrightarrow$	$?Fe_2O_3$
$3O_2$	+	$4Fe$	$\longrightarrow$	$2Fe_2O_3$
3 mol		4 mol		2 mol

$$3[2(16.00)] \text{ g} + 4(55.85) \text{ g} = 2[2(55.85) + 3(16.00)] \text{ g}$$
$$96.00 \text{ g } O_2 + 223.40 \text{ g Fe} = 319.40 \text{ g } Fe_2O_3$$

(c) For FeO:

the second iron oxide $\xrightarrow{\text{heat above 600 °C}}$ the third iron oxide + ?

$$Fe_2O_3 \longrightarrow FeO + oxygen$$

because the ratio of oxygen:iron is less in the product oxide, some oxygen also must be produced by the heating

$$?Fe_2O_3 \longrightarrow ?FeO + ?O_2$$
$$Fe_2O_3 \longrightarrow 2FeO + \tfrac{1}{2}O_2$$

To account for the two Fe atoms on the left, we place a 2 in front of the FeO. We then need a $\frac{1}{2}$ in front of the O_2 to give three O atoms on the right. In order to write a balanced equation in which all coefficients are whole numbers, the coefficients are doubled throughout:

$$2Fe_2O_3 \longrightarrow 4FeO + O_2$$

2 mol   4 mol   1 mol

$$2[2(55.85) + 3(16.00)]\,g = 4(55.85 + 16.00)\,g + 2(16.00)\,g$$

$$319.40\,g\ Fe_2O_3 = 287.40\,g\ FeO\ + 32.00\,g\ O_2$$

See also Exercises 21 and 22.

• Example 2.6 •

Using the equation in part (b) of Example 2.5, calculate the weight in grams of Fe_2O_3 produced when 14.0 g of oxygen reacts.

• Solution • Write the balanced equation for the reaction and interpret it in terms of moles of the substances mentioned in the problem, that is, oxygen and Fe_2O_3:

$$3O_2 + 4Fe \xrightarrow{} 2Fe_2O_3$$

3 mol of $\xrightarrow[\text{produce}]{\text{to}}$ 2 mol of this
oxygen is needed   iron oxide

First we calculate the number of moles of oxygen in the 14.0 g that react:

$$mol\ O_2\ available = 14.0\,g\,O_2 \times \frac{1\ mol\ O_2}{32.0\,g\,O_2} = 0.438\ mol\ O_2$$

From the balanced equation, we see that 2 moles of Fe_2O_3 are produced for each 3 moles of O_2 that react. So,

$$\frac{mol\ Fe_2O_3}{produced} = 0.438\ mol\,O_2 \times \frac{2\ mol\ Fe_2O_3}{3\ mol\,O_2} = 0.292\ mol\ Fe_2O_3$$

The weight of 1 mole of Fe_2O_3 is 160 g, so the weight of iron oxide produced is

$$wt\ Fe_2O_3 = 0.292\ mol\ Fe_2O_3 \times \frac{160\ g\ Fe_2O_3}{1\ mol\ Fe_2O_3} = 46.7\ g\ Fe_2O_3$$

• Check • A good way to check a yield calculation is to determine the weights of all reactants and make sure the sum equals the sum of the weights of all the products:

$$mol\ Fe\ required = 0.438\ mol\,O_2 \times \frac{4\ mol\ Fe}{3\ mol\,O_2} = 0.584\ mol$$

$$wt\ Fe\ required = 0.584\ mol\,Fe \times \frac{55.8\ g\ Fe}{1\ mol\ Fe} = 32.6\ g\ Fe$$

weight of reactants = weight of products

$$14.0 \text{ g} + 32.6 \text{ g} = 46.7 \text{ g}$$

$$46.6 \text{ g of reactants} = 46.7 \text{ g of products}$$

This is a satisfactory check for a mass balance calculation when data are rounded to three significant figures.

See also Exercises 23–27.

━━━ • Example 2•7 • ━━━

One of the suggested methods for removing dangerous mercury compounds from industrial wastes is to cause the compounds to react in such a way as to form elemental mercury, which can be trapped more easily. Consider the reaction of mercuric chloride, $HgCl_2$, with aluminum, Al, to produce mercury, Hg, and aluminum chloride, $AlCl_3$:

$$3HgCl_2 + 2Al \longrightarrow 3Hg + 2AlCl_3$$

What weight of aluminum would be required to react with 436 g of mercuric chloride?

• Solution • From the balanced equation, we see that the mole ratio of the reactants is

$$3 \text{ mol HgCl}_2 \text{ react with 2 mol Al} \longrightarrow$$

or

$$x \text{ mol HgCl}_2 \text{ react with } \tfrac{2}{3}x \text{ mol Al} \longrightarrow$$

The number of moles of $HgCl_2$ available is

$$\text{mol HgCl}_2 = 436 \text{ g HgCl}_2 \times \frac{1 \text{ mol HgCl}_2}{271 \text{ g HgCl}_2}$$

$$= 1.61 \text{ mol HgCl}_2$$

The number of moles of Al required for reaction is

$$\text{mol Al} = 1.61 \text{ mol HgCl}_2 \times \frac{2 \text{ mol Al}}{3 \text{ mol HgCl}_2}$$

$$= 1.07 \text{ mol Al}$$

$$\text{wt Al} = 1.07 \text{ mol Al} \times \frac{27.0 \text{ g Al}}{1 \text{ mol Al}}$$

$$= 28.9 \text{ g Al}$$

See also Exercises 23–27.

2•5•3   Theoretical Versus Actual Yields   In solving Example 2.6, it was assumed that the reaction between iron and oxygen proceeded so that both reactants were converted to the product shown in the equation. This seldom happens in actual practice, because many reactions do not "go to completion" but end in an equilibrium state with appreciable quantities of both reactants and products present. Also, the reactants may react to form two or more sets of products, so that a single equation does not tell the whole story about a chemical reaction. Frequently, too, in separating and purifying the product of a chemical reaction, some of the product is lost.

The amount of a product that we calculate will be obtained if the reaction goes to completion is called the **theoretical yield.** In actual practice, the recovery of a product is less than 100 percent, sometimes much less. The actual yield of a product divided by the theoretical yield times 100 equals the **percentage yield:**

$$\frac{\text{actual yield}}{\text{theoretical yield}} \times 100 = \text{percentage yield}$$

An illustration of a reaction in which the percentage yield is less than 100 percent is shown in Example 2.8.

• Example 2•8 •

A certain grade of coal contains 1.7 percent sulfur. Assume the burning of the sulfur compounds can be represented by the equation

$$S + O_2 \longrightarrow SO_2$$

Calculate the weight of the pollutant sulfur dioxide, SO_2, in the gases emitted to the atmosphere per metric ton (1,000 kg) of coal burned if the process is 79 percent efficient.

• Solution • The number of moles of sulfur per metric ton of coal is

$$\text{mol S} = 1{,}000 \text{ kg coal} \times \frac{1.7 \text{ kg S}}{100 \text{ kg coal}} \times \frac{1{,}000 \text{ g}}{1 \text{ kg}} \times \frac{1 \text{ mol S}}{32 \text{ g S}} = 530 \text{ mol S}$$

The balanced equation shows that 1 mole of SO_2 is formed per mole of S, assuming a 100-percent yield. Therefore, the theoretical or 100-percent yield of sulfur dioxide is 530 moles of SO_2. But, because the process is only 79 percent efficient,

$$\text{mol } SO_2 = 530 \text{ mol} \times 0.79 = 420 \text{ mol}$$

$$\text{wt } SO_2 = 420 \text{ mol} \times \frac{64 \text{ g}}{1 \text{ mol}} = 2.7 \times 10^4 \text{ g} = 27 \text{ kg}$$

See also Exercises 28 and 29.

2•5•4   Limiting and Excess Reactants   Calculations of the amounts of reactants needed or products yielded are made on the basis of the stoichiometric ratios shown in balanced equations. In practice, however, reaction conditions are almost always different. In the laboratory, in industry, or in nature, we do not expect that the amounts of reactants available will just happen to be precisely those required for the reaction.

Almost always there is less of one reactant than is needed to permit all the reactants to combine. The **limiting reactant** is the substance that reacts completely and thereby limits the possible extent of the reaction. The other reactant or reactants are said to be *in excess,* because some amount will be left over without reacting. Calculations made on the basis of the balanced equation must begin with the amount of the limiting reactant.

   To determine which of two reactants is the limiting one, we calculate the ratio of moles available for reaction and compare it with the stoichiometric ratio that is specified by the balanced equation. After we determine which is the limiting reactant, we can calculate the weight of the product of the reaction.

   Consider the reaction of calcium with hydrogen in a sealed vessel that contains 1.00 g of each reactant. From the table we see that the calcium is the limiting reactant; all of it, 1.00 g, reacts. The hydrogen is present in excess; only 0.050 g reacts, and 0.95 g of hydrogen is in excess (does not react). The weight of calcium hydride formed is 1.05 g.

	Ca	+	H_2	$\longrightarrow$	CaH_2
theoretical ratio, moles	1 mol		1 mol		1 mol
moles of reactants available	$(1.00 \text{ g})\left(\dfrac{1 \text{ mol}}{40.08 \text{ g}}\right)$ $= 0.0250 \text{ mol}$		$(1.00 \text{ g})\left(\dfrac{1 \text{ mol}}{2.016 \text{ g}}\right)$ $= 0.496 \text{ mol}$		
moles that can react and that can be formed	0.0250 mol Ca		0.0250 mol H_2		0.0250 mol CaH_2
moles in excess	none		0.471 mol		
weights that can react and that can be formed	$(0.0250 \text{ mol})\left(\dfrac{40.08 \text{ g}}{1 \text{ mol}}\right)$ $= 1.00 \text{ g Ca}$		$0.0250 \text{ mol}\left(\dfrac{2.016 \text{ g}}{1 \text{ mol}}\right)$ $= 0.0504 \text{ g } H_2$		$0.0250 \text{ mol}\left(\dfrac{42.1 \text{ g}}{1 \text{ mol}}\right)$ $= 1.05 \text{ g } CaH_2$

   Example 2.9 shows the method as applied to the reaction of gold with hot elemental chlorine.

• Example 2•9 •

Gold resists attack by most reactants, but hot chlorine gas is chemically active enough to react with it. At 150 °C the following reaction occurs:

$$2Au + 3Cl_2 \longrightarrow 2AuCl_3$$

Suppose that 10.0 g of gold and 10.0 g of chlorine are sealed in a container and heated until the reaction is complete. Which reactant is the limiting one? What weight of gold chloride is formed? What weight of the reactant in excess remains unreacted?

• Solution •

$$\text{stoichiometric ratio for reaction} = \frac{2 \text{ mol Au}}{3 \text{ mol Cl}_2} = \frac{0.67 \text{ mol Au}}{1 \text{ mol Cl}_2}$$

$$\text{available ratio for reaction} = \frac{(10.0 \text{ g Au})\left(\dfrac{1 \text{ mol Au}}{197 \text{ g Au}}\right)}{(10.0 \text{ g Cl}_2)\left(\dfrac{1 \text{ mol Cl}_2}{70.9 \text{ g Cl}_2}\right)}$$

$$= \frac{0.0508 \text{ mol Au}}{0.141 \text{ mol Cl}_2} = \frac{0.36 \text{ mol Au}}{1 \text{ mol Cl}_2}$$

A comparison of these two ratios shows that 0.67 mole of gold is required to completely react with 1 mole of chlorine, but that in the amounts available for reaction there is only 0.36 mole of gold per 1 mole of chlorine. Therefore, the gold is the limiting reactant; all the gold can react and there will be unreacted chlorine left in excess:

$$\text{mol AuCl}_3 \text{ formed} = 0.0508 \text{ mol Au} \times \frac{2 \text{ mol AuCl}_3}{2 \text{ mol Au}}$$

$$= 0.0508 \text{ mol AuCl}_3 \text{ formed}$$

$$\text{wt AuCl}_3 \text{ formed} = 0.0508 \text{ mol AuCl}_3 \times \frac{303 \text{ g AuCl}_3}{1 \text{ mol AuCl}_3}$$

$$= 15.4 \text{ g AuCl}_3 \text{ formed}$$

$$\text{mol Cl}_2 \text{ reacted} = \text{mol Au reacted} \times \frac{3 \text{ mol Cl}_2}{2 \text{ mol Au}}$$

$$= (0.0508)(1.5) \text{ mol Cl}_2$$

$$= 0.0762 \text{ mol Cl}_2 \text{ reacted}$$

$$\text{wt Cl}_2 \text{ reacted} = 0.0762 \text{ mol Cl}_2 \times \frac{70.9 \text{ g Cl}_2}{1 \text{ mol Cl}_2}$$

$$= 5.4 \text{ g Cl}_2 \text{ reacted}$$

$$\text{wt Cl}_2 \text{ in excess} = \text{wt Cl}_2 \text{ available} - \text{wt Cl}_2 \text{ reacted}$$

$$= 10.0 \text{ g} - 5.4 \text{ g} = 4.6 \text{ g Cl}_2 \text{ in excess}$$

There is an alternative way of calculating the weight of chlorine that reacted:

$$\text{wt Cl}_2 \text{ reacted} = \text{wt AuCl}_3 \text{ formed} - \text{wt Au reacted}$$

$$= 15.4 \text{ g AuCl}_3 - 10.0 \text{ g Au}$$

$$= 5.4 \text{ g Cl}_2 \text{ reacted}$$

See also Exercises 30–35.

2·6   Calculation of Percentage Composition from Formulas

The weights of the elements making up a mole of a compound are readily deduced from the formula of the compound and the atomic weights of the elements. The percentage composition by weight is then calculated from the weights of the elements and the weight of a mole of the compound.

• Example 2·10 •

Calculate to three significant figures the percentage composition by weight of each element in the anesthetic ethyl ether, $C_4H_{10}O$.

• Solution • In 1 mole of $C_4H_{10}O$, there are 4 moles of C atoms, 10 moles of H atoms, and 1 mole of O atoms. The weights of these atoms in 1 mole of the compound and the weight of 1 mole of the compound are found, then the percentages by weight are calculated as follows:

$$\text{wt of C} = \ \ 4 \text{ mol} \times 12.0 \text{ g/mol} = 48.0 \text{ g}$$
$$\text{wt of H} = 10 \text{ mol} \times 1.01 \text{ g/mol} = 10.1 \text{ g}$$
$$\text{wt of O} = \ \ 1 \text{ mol} \times 16.0 \text{ g/mol} = \underline{16.0 \text{ g}}$$
$$\text{wt of 1 mol of ethyl ether} = 74.1 \text{ g}$$

$$\% \text{ by weight of C} = \frac{48.0 \text{ g}}{74.1 \text{ g}} \times 100 = \ \ 64.8\%$$

$$\% \text{ by weight of H} = \frac{10.1 \text{ g}}{74.1 \text{ g}} \times 100 = \ \ 13.6\%$$

$$\% \text{ by weight of O} = \frac{16.0 \text{ g}}{74.1 \text{ g}} \times 100 = \ \ \underline{21.6\%}$$
$$\text{total} = 100.0\%$$

See also Exercises 36–38.

• Example 2•11 •

When added to toothpaste, stannous fluoride, SnF_2, has proven to be helpful in preventing tooth decay. Calculate to three significant figures the percentage by weight of each element in stannous fluoride.

• Solution • In 1 mole of SnF_2:

$$\text{wt of Sn} = \ \ 1 \text{ mol} \times 119 \text{ g/mol} = 119 \ \ \text{g}$$
$$\text{wt of F} = 2 \text{ mol} \times 19.0 \text{ g/mol} = \underline{\ \ 38.0 \text{ g}}$$
$$\text{wt of 1 mol of SnF}_2 = 157 \ \ \text{g}$$

$$\% \text{ by weight of Sn} = \frac{119 \text{ g}}{157 \text{ g}} \times 100 = 75.8\%$$

$$\% \text{ by weight of F} = \frac{38.0 \text{ g}}{157 \text{ g}} \times 100 = 24.2\%$$

See also Exercises 36–38.

2•7 Calculation of Formulas from Experimental Data

From the formula of a compound much quantitative information about the compound can be calculated, such as the molecular weight, the molar weight, and the percentage composition by weight. Logically, however, this is putting the cart before the horse, because the formula for a compound cannot be known with certainty until the molecular weight and the percentage composition by weight have been determined experimentally. Even for the simple compounds carbon dioxide and water, in order that we might have their useful formulas, someone had to go into the laboratory and determine experimentally (1) the elements that compose these compounds, (2) the proportions by weight of the elements in each, and (3) the relative molecular weight of each. In the remainder of this chapter, we shall discuss briefly how the composition of a compound is determined and how the formula is deduced from the experimental data.

2•7•1   Laboratory Determination of Percentage Composition It is estimated that over 300,000 new compounds are synthesized each year in the university and industrial research laboratories of the world. As these new compounds are obtained, their compositions must be experimentally determined. This is the first step in arriving at their formulas.

There are many methods for determining the percentage by weight of the different elements in a compound. These methods vary, depending on the nature of the compound and the elements in it. Two classical methods are precipitation analysis and combustion analysis.

Precipitation methods of analysis can be used when a slightly soluble compound is formed. For example, if a new compound contains silver, a weighed sample of the compound is dissolved in water and hydrochloric acid is added. This forms insoluble silver chloride, AgCl, which is removed by filtration, dried, and then carefully weighed on an analytical balance. The percentage of silver is calculated as follows:

$$\text{wt Ag} = \frac{\text{molar wt Ag}}{\text{molar wt AgCl}} \times \text{wt AgCl}$$

$$= \frac{107.9 \text{ g}}{143.4 \text{ g}} \times \text{wt AgCl}$$

$$\% \text{ Ag} = \frac{\text{wt Ag}}{\text{wt of sample}} \times 100$$

These calculations are applied in solving Exercises 41 and 43 at the end of the chapter.

Combustion methods of analysis are widely used. If a substance contains carbon and hydrogen, a weighed sample of the compound can be burned in a closed tube in a stream of oxygen to form carbon dioxide and water (Figure 2.2). The combustion products are swept from the tube by the stream of oxygen into two absorbing chemicals, one of which absorbs water vapor and the other, carbon dioxide.

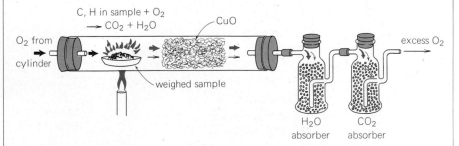

Figure 2•2 Apparatus for determining the percentages of carbon and hydrogen in a compound. Any traces of C or CO react to form CO_2 as they pass through the copper oxide, CuO. Traces of H_2 react to form H_2O as they pass through.

The gain in weight of one of the absorbers gives the weight of water formed and the other gives the weight of carbon dioxide. Because water, H_2O, is known to be $\frac{2}{18}$ hydrogen by weight, $\frac{2}{18}$ of the weight of the water is equal to the amount of hydrogen originally present in the compound. Similarly, CO_2 is $\frac{12}{44}$ carbon, and therefore $\frac{12}{44}$ of the weight gained by the CO_2 absorber is the weight of the carbon originally present in the sample. From these weights, the percentage composition is calculated as shown in Example 2.12.

———— • Example 2•12 • ————

Lactose, a white substance often used as a binder in the manufacture of pills and tablets, contains only carbon, hydrogen, and oxygen. To analyze a commercial preparation, a 0.5624-g sample was burned in a combustion train like that in Figure 2.2. The increase in weight of the water absorber was 0.3267 g, and the increase in the carbon dioxide absorber was 0.8632 g. Calculate directly the percentages of hydrogen and of carbon, and calculate by difference the percentage of oxygen in the sample.

• Solution •

$$\text{wt C} = \text{wt CO}_2 \times \frac{\text{molar wt C}}{\text{molar wt CO}_2}$$

$$= 0.8632 \text{ g CO}_2 \times \frac{12.011 \text{ g C}}{44.009 \text{ g CO}_2} = 0.2356 \text{ g C}$$

$$\% \text{ C} = \frac{0.2356 \text{ g C}}{0.5624 \text{ g sample}} \times 100 = 41.89\% \text{ C}$$

$$\text{wt H} = \text{wt H}_2\text{O} \times \frac{2 \text{ molar wt H}}{\text{molar wt H}_2\text{O}}$$

$$= 0.3267 \text{ g H}_2\text{O} \times \frac{2(1.008) \text{ g H}}{18.015 \text{ g H}_2\text{O}} = 0.03656 \text{ g H}$$

$$\% \text{ H} = \frac{0.03656 \text{ g H}}{0.5624 \text{ g sample}} \times 100 = 6.50\% \text{ H}$$

$$\% \text{ O} = 100 - \% \text{ C} - \% \text{ H} = 100 - 41.89 - 6.50 = 51.61\% \text{ O}$$

See also Exercises 39–43.

2•7•2   Calculation of Empirical Formulas   Once the composition of a compound is determined experimentally, the data, along with the known atomic weights, can be used then to calculate the simplest ratio of atoms in the compound and hence the simplest formula. As pointed out in Section 2.1, this simplest formula is called the empirical formula. It may or may not be the same as the molecular formula.

Examples 2.13 and 2.14 illustrate the calculation of empirical formulas.

———— • Example 2•13 • ————

Analysis of a certain compound composed of iron, Fe, and chlorine, Cl, reveals that 0.1396 g of iron is combined with 0.1773 g of chlorine. Calculate the empirical formula.

• Solution • These data show that the ratio by weight for iron and chlorine is 0.1396 : 0.1773. To obtain the empirical formula, we need to know the ratio by moles of atoms. First we calculate the moles of Fe and Cl atoms present in the compound:

$$0.1396 \text{ g Fe} \times \frac{1 \text{ mol Fe}}{55.85 \text{ g Fe}} = 0.002499 \text{ mol Fe}$$

$$0.1773 \text{ g Cl} \times \frac{1 \text{ mol Cl}}{35.45 \text{ g Cl}} = 0.005001 \text{ mol Cl}$$

The ratio by moles for iron and chlorine is 0.002499:0.005001. Because 1 mole of atoms of any element contains the same number of atoms (6.022×10^{23}), the ratio of moles must also be the ratio of atoms.

The calculated ratio of Fe atoms:Cl atoms, 0.002499:0.005001, is simplified by dividing each member by the smaller member:

$$\frac{0.002499 \text{ mol Fe}}{0.002499} = 1.000 \text{ mol Fe}$$

$$\frac{0.005001 \text{ mol Cl}}{0.002499} = 2.001 \text{ mol Cl}$$

This gives a ratio of Fe atoms:Cl atoms as 1:2. Because there are twice as many Cl atoms as Fe atoms, we conclude that the empirical formula for the compound is $FeCl_2$.

See also Exercises 45–47 and 50.

• Example 2•14 •

Analysis of a small amount of some apparently pure crystals recovered in the evaporation of a sample of municipal waste water showed the crystals to have 63.97 percent cadmium, Cd, 24.28 percent oxygen, O, and 11.75 percent phosphorus, P. Calculate the empirical formula.

• Solution • It is convenient to convert data for percentage composition into ratios by weight by taking 100 g of the substance as a basis for calculation. The number of moles of Cd, O, and P atoms in 100 g of the compound are:

$$63.97 \text{ g Cd} \times \frac{1 \text{ mol Cd}}{112.4 \text{ g Cd}} = 0.5691 \text{ mol Cd}$$

$$24.28 \text{ g O} \times \frac{1 \text{ mol O}}{16.00 \text{ g O}} = 1.518 \text{ mol O}$$

$$11.75 \text{ g P} \times \frac{1 \text{ mol P}}{30.97 \text{ g P}} = 0.3794 \text{ mol P}$$

The ratio of moles, 0.5691:1.518:0.3794, is also the ratio of atoms.

Our aim is to obtain the ratio of the atoms in small whole numbers. To simplify the ratio, we divide each member by the smallest member:

$$\frac{0.5691 \text{ mol Cd}}{0.3794} = 1.500 \text{ mol Cd} \simeq 1.5 \text{ mol Cd}$$

$$\frac{1.518 \text{ mol O}}{0.3794} = 4.001 \text{ mol O} \simeq 4.0 \text{ mol O}$$

$$\frac{0.3794 \text{ mol P}}{0.3794} = 1.000 \text{ mol P} = 1.0 \text{ mol P}$$

Frequently at this step, the ratio is not in the form of whole numbers but must be multiplied by a small number (usually 2, 3, or 4) to put it in that form. By inspection, we choose the multiplier 2 in this case to obtain the ratio Cd, $1.5 \times 2 = 3$ mol:O, $4.0 \times 2 = 8$ mol:P, $1.0 \times 2 = 2$ mol. The simplest, or empirical, formula is $Cd_3O_8P_2$.

As we shall learn in Chapter 5, oxygen and phosphorus are often present in compounds as the polyatomic unit, PO_4. We might guess that the crystals are of $Cd_3(PO_4)_2$, cadmium phosphate. Further tests would be necessary to prove this.

See also Exercises 45–47 and 50.

2•7•3   Calculation of Molecular Formulas   The molecular formula of a substance is a whole-number multiple of its empirical formula. The following are some examples that illustrate this relationship:

Name, Molecular Formula	Multiple (Empirical Formula)
ammonia, NH_3	$1 \times (NH_3)$
hydrazine, N_2H_4	$2 \times (NH_2)$
methane, CH_4	$1 \times (CH_4)$
acetylene, C_2H_2	$2 \times (CH)$
propene, C_3H_6	$3 \times (CH_2)$
glucose, $C_6H_{12}O_6$	$6 \times (CH_2O)$

In order to determine the molecular formula of a compound, the chemist must determine experimentally its molecular weight in addition to its empirical formula. Experimental methods for determining molecular weights will be discussed in later chapters, but Example 2.15 shows how a molecular weight is used with the empirical formula to calculate the molecular formula.

• Example 2•15 •

Compound X was found to be a minor constituent of a household liquefied fuel gas. By analysis, X was determined to be 85.69 percent carbon and 14.31 percent hydrogen by weight (see Example 2.12). The molecular weight was determined to be 55.9 amu (subject to ±2 percent error in the determination). Calculate the molecular formula of X.

• Solution •

• Step 1

Calculation of empirical formula:

$$\text{carbon} \quad \frac{85.69\ g}{12.01\ g/mol} = 7.13 \text{ mol of atoms}$$

$$\text{hydrogen} \quad \frac{14.31\ g}{1.008\ g/mol} = 14.20 \text{ mol of atoms}$$

The ratio of C atoms to H atoms, 7.13 : 14.20, is simplified as before to 1 : 2. Hence, the empirical formula is CH_2.

• Step 2

Calculation of molecular formula: As far as the weight relationship is concerned, the molecular formula for X can be any in which the atoms are in a 1 : 2 ratio:

$$CH_2, \ C_2H_4, \ C_3H_6, \ C_4H_8, \ C_5H_{10}, \ \ldots$$

The correct molecular formula is the one in such a list that agrees best with the experimental molecular weight:

formula	calculated mol wt = sum of atomic weights
CH_2	14.027 amu
C_2H_4	28.054 amu
C_3H_6	42.080 amu
C_4H_8	56.107 amu
C_5H_{10}	70.134 amu

The weight of 56.107 amu agrees most closely with the experimental value of 55.9 amu. Hence, the molecular formula for X is C_4H_8.

- *Alternate Step 2*

Finding the correct formula by writing a series of possible formulas and calculating the molecular weight of each is a trial-and-error method. There is a more efficient and systematic solution to this type of problem.

The molecular weight is a simple multiple of the empirical formula weight of CH_2; that is, it is $n(14.027 \ amu)$, where n is a whole number.

The experimental molecular weight is 55.9 amu. Therefore,

$$\frac{55.9 \text{ amu per molecule}}{14.027 \text{ amu per empirical formula}}$$

$$= 3.99 \simeq 4 \text{ empirical formulas per molecule}$$

The molecular formula is $4 \times (CH_2)$, or C_4H_8.

See also Exercises 48, 49, and 51.

• *Precise Molecular Weights* • Experimental methods for determining molecular weights give approximate values only. But as we have just seen, in the case of C_4H_8, only the approximate (experimental) molecular weight is needed to obtain the true formula.

Once the formula of a compound is obtained, its precise molecular weight can be calculated by adding the weights of the atoms that are combined in the molecule. The molecular weight thus calculated is correct to the same degree that the atomic weights are correct. In the case of C_4H_8, after the molecular formula has been obtained, the experimentally determined molecular weight of 55.9 amu is abandoned in favor of the precise molecular weight. The precise weight is calculated by adding the weight of four carbon atoms (4×12.011 amu) and eight hydrogen atoms (8×1.0079 amu) to obtain 56.107 amu.

Chapter Review

Summary

The most concise way to describe a chemical reaction is to write a **balanced chemical equation,** which is both a qualitative and a quantitative statement about the **reactants** and the **products** involved. Each substance is represented by its **molecular formula,** which states the actual numbers of atoms of each kind in a unit of the substance. The molecular formula is a whole-number multiple of the **empirical formula** of the substance, which states the minimum possible numbers, in the correct proportions, of atoms of each kind. The molecular formulas of many substances refer not to a definite molecule but to the simplest conceptual unit of the substance with the correct proportions of atoms. Three general classes of reactions that are found widely in chemistry are **direct combination reactions, simple displacement reactions,** and **double displacement reactions.**

The quantitative relations among the reactions and products in a balanced chemical equation provide the basis of **stoichiometry.** Stoichiometric calculations entail the use of the atomic weights of elements and the **molecular weights** of compounds. Atomic and molecular weights are usually given in terms of **atomic mass units,** amu. The number of atomic mass units in 1 gram of any substance is 6.022×10^{23}, which is the **Avogadro number.** The amount of any substance that contains this number of units (atoms, molecules, and so on) of the substance is 1 **mole.** The mass (or weight) of a mole of any substance is its **molar weight,** which is the same number in grams as the molecular weight in atomic mass units. Knowing the molar weights of substances is necessary for predicting or for interpreting the results of chemical changes.

The amount of a given product calculated to be obtained in a chemical reaction that goes to completion is the **theoretical yield.** This amount is seldom obtained in practice; a realistic figure is the **percentage yield,** which is calculated from the actual (measured) yield and the theoretical yield. In calculating the theoretical yield for a chemical reaction, it is essential to know which is the **limiting reactant**—the one that theoretically can react completely, leaving the other reactants in excess.

Stoichiometry enables the percentage composition (by weight) of a compound to be calculated from its empirical or molecular formula. More important in real life is that stoichiometry enables the empirical formula to be calculated from the percentage composition, which must be determined by experiment. Two of the classical methods for doing this are precipitation analysis and combustion analysis. Once the empirical formula is known, the molecular formula can be determined from an approximately measured molecular weight of the compound. Finally, from the known molecular formula, the *precise* molecular weight can be calculated.

Exercises

Writing Formulas and Naming Compounds

1. Why are oxygen and hydrogen represented by the formulas O_2 and H_2, whereas ozone is represented by O_3? Why not use something like Oz_2 for ozone?

2. (a) What is meant by the terms *diatomic* and *triatomic* when applied to molecules?
   (b) There are two common oxides of carbon. One is diatomic and the other is triatomic. Write the molecular formula and name for each oxide.

3. A molecule of neon is monatomic. What does this statement mean?

4. (a) What is meant by the term *tetratomic* when applied to a molecule?
   (b) White phosphorus is tetratomic. What is its molecular formula? What is its empirical formula?

5. Give the formula and name for a tetratomic molecule composed of hydrogen and oxygen.

6. What is the empirical formula for each of the following compounds? (a) hydrazine, N_2H_4; (b) styrene, C_8H_8; (c) diborane, B_2H_6; (d) naphthalene, $C_{10}H_8$; (e) pentane, C_5H_{12}; (f) oxalic acid, $H_2C_2O_4$; (g) ethylene glycol, $C_2H_6O_2$; (h) *para*-dichlorobenzene, $C_6H_4Cl_2$.

Writing Balanced Equations

7. Ammonia, NH_3, is produced by the combination of the elements nitrogen and hydrogen. Is the following equation an acceptable representation of this reaction? Explain.

$$3H_2 + 2N \longrightarrow 2NH_3$$

8. Balance the following:
   (a) $H_2 + O_2 \longrightarrow H_2O$
   (b) $Na_3PO_4 + CaCl_2 \longrightarrow NaCl + Ca_3(PO_4)_2$
   (c) $As_4 + O_2 \longrightarrow As_4O_6$
   (d) $Zn + AgNO_3 \longrightarrow Zn(NO_3)_2 + Ag$
   (e) $Ba(NO_3)_2 + (NH_4)_2SO_4 \longrightarrow BaSO_4 + NH_4NO_3$
   (f) $K_2SeO_4 + CaS \longrightarrow CaSeO_4 + K_2S$
   (g) $Al + O_2 \longrightarrow Al_2O_3$
   (h) $H_2SO_4 + Ga \longrightarrow H_2 + Ga_2(SO_4)_3$
   (i) $C_3H_8 + O_2 \longrightarrow CO_2 + H_2O$
   (j) $C_2H_6 + O_2 \longrightarrow CO_2 + H_2O$

9. Write balanced equations for the following reactions, given that the formulas for the elements are written as Na, Cl_2, H_2, O_2, and N_2:

(a) Sodium reacts with chlorine to form sodium chloride, NaCl.

(b) Hydrogen reacts with oxygen to form water, H_2O.

(c) Nitrogen reacts with hydrogen to form ammonia, NH_3.

(d) Hydrogen reacts with chlorine to form hydrogen chloride, HCl.

10. Write balanced equations for each of the following direct combination reactions:

(a) hydrogen + bromine $\longrightarrow$
hydrogen bromide, a diatomic unit

(b) silicon (Si) + oxygen $\longrightarrow$
silicon dioxide, a triatomic unit

(c) nitrogen + oxygen $\longrightarrow$
nitric oxide, a diatomic unit

(d) phosphorus (P) + oxygen $\longrightarrow$ P_4O_{10}

(e) phosphorus (P_4) + oxygen $\longrightarrow$ P_4O_{10}

(f) lithium (Li) + fluorine $\longrightarrow$
lithium fluoride, a diatomic unit

11. Elements in the same vertical group of the periodic table often form compounds with similar formulas. Write balanced equations for the following reactions, each of which has its analogy in Exercise 10 or 12:

(a) hydrogen + chlorine $\longrightarrow$

(b) carbon + oxygen $\longrightarrow$

(c) potassium + bromine $\longrightarrow$

(d) calcium + nitrogen $\longrightarrow$

12. Write balanced equations for the following reactions:

(a) barium reacts with hydrogen to yield barium hydride (BaH_2)

(b) magnesium reacts with nitrogen to yield magnesium nitride (Mg_3N_2)

(c) ammonium chloride (NH_4Cl) reacts with lead nitrate ($Pb(NO_3)_2$) to yield ammonium nitrate (NH_4NO_3) and lead chloride ($PbCl_2$)

(d) phosphorus tribromide (PBr_3) reacts with water (H_2O) to yield hydrogen bromide (HBr) and hydrogen phosphite (H_3PO_3)

(e) hydrogen chloride (HCl) reacts with aluminum hydroxide ($Al(OH)_3$) to yield aluminum chloride ($AlCl_3$) and water (H_2O)

13. When the reaction of oxygen with a hydrocarbon (a compound composed of H and C) or an oxygen derivative of a hydrocarbon (a compound composed of H, C, and O) is complete, the products are carbon dioxide and water. With this fact in mind, write balanced equations for each of the following:

(a) octane (C_8H_{18}) + oxygen $\longrightarrow$

(b) ethyl ether ($C_4H_{10}O$) + oxygen $\longrightarrow$

(c) glucose ($C_6H_{12}O_6$) + oxygen $\longrightarrow$

14. Write balanced equations for each of the following single displacement reactions:

(a) hydrogen + silver oxide (Ag_2O) $\longrightarrow$ water + ?

(b) magnesium + hydrogen chloride (HCl) $\longrightarrow$
magnesium chloride ($MgCl_2$) + ?

(c) zinc + ferric nitrate ($Fe(NO_3)_3$) $\longrightarrow$
zinc nitrate ($Zn(NO_3)_2$) + ?

(d) chlorine + sodium iodide (NaI) $\longrightarrow$ iodine + ?

15. Write balanced equations for each of the following double displacement reactions:

(a) sodium hydroxide (NaOH) reacts with hydrogen sulfate (H_2SO_4) to yield sodium sulfate (Na_2SO_4) and water (H_2O)

(b) zinc nitrate ($Zn(NO_3)_2$) reacts with hydrogen sulfide (H_2S) to yield zinc sulfide (ZnS) and hydrogen nitrate (HNO_3)

(c) magnesium chloride ($MgCl_2$) reacts with barium hydroxide ($Ba(OH)_2$) to yield magnesium hydroxide ($Mg(OH)_2$) and barium chloride ($BaCl_2$)

(d) zinc sulfate ($ZnSO_4$) reacts with hydrogen sulfide (H_2S) to yield zinc sulfide (ZnS) and hydrogen sulfate (H_2SO_4)

Simple Weight Relationships

16. Letting hydrogen be unity, calculate the whole-number weight ratios for the following four atoms, given these actual weights: H, 1.67×10^{-24} g; C, 1.99×10^{-23} g; O, 2.66×10^{-23} g; Au, 3.27×10^{-22} g.

17. Calculate the molecular weight of each of the following: Br_2 molecules; $C_{12}H_{22}O_{11}$ (sucrose); $Ga_3(PO_4)_2$.

18. The weight of 1.0 mole of oxygen molecules expressed in grams is 32 g. Would it be appropriate to say that the weight of 1.0 mole of oxygen molecules expressed in pounds is 32 lb? Why? Express the weight of 1.0 mole of oxygen in kilograms.

19. Calculate the molar weight of each of the following: neon, Ne; chlorine, Cl_2; sodium phosphate, Na_3PO_4; glycerine, $C_3H_8O_3$.

20. Calculate the number of moles in: 45.1 g of ammonium sulfate, $(NH_4)_2SO_4$; 11.8 g of chlorine, Cl_2; 11.8 g of chlorine, Cl; 150 g of neon, Ne.

21. (a) Balance the equation: $Sc + HBr \longrightarrow ScBr_3 + H_2$.

(b) Calculate the number of moles of H_2 produced when 0.50 mole of Sc reacts.

(c) How many atoms of Sc react?

22. Consider the reaction of hydrogen, H_2, with nitrogen, N_2, to form ammonia, NH_3.

(a) Write a balanced equation for the reaction.

(b) If hydrogen is present in excess and 42.0 g of nitrogen reacts, how many moles of ammonia can be formed?

(c) How many grams of ammonia can be formed?

(d) How many grams of hydrogen react with the 42.0 g of nitrogen?

23. One of the ways of preparing oxygen in the laboratory is by the reaction

$$2KClO_3 \xrightarrow{\text{heat}} 2KCl + 3O_2$$

How many grams of $KClO_3$ must be decomposed to produce 48.0 g of oxygen?

24. Aluminum, Al, reacts with hydrogen chloride, HCl, to produce aluminum chloride, $AlCl_3$, and hydrogen, H_2. How many grams of aluminum are required to produce 30 g of hydrogen?

25. Aluminum, Al, reacts with oxygen, O_2, to form aluminum oxide, Al_2O_3. If oxygen is present in excess and 5.40 g of aluminum reacts, what is the weight of aluminum oxide produced?

26. An important commercial process involves the use of an electric current (electrolysis) to bring about the reaction between common salt and water to produce lye, chlorine, and hydrogen. The equation is

$$2NaCl + 2H_2O \longrightarrow 2NaOH + Cl_2 + H_2$$

(a) What is the formula for lye?
(b) Assuming the theoretical yield is obtained, what weight of salt must be used to produce 100 g of lye?
(c) What weight of hydrogen is produced along with the 100 g of lye?
(d) What weight of water reacts?

27. The reaction involved with one kind of baking powder is

$$KHC_4H_4O_6 + NaHCO_3 \longrightarrow KNaC_4H_4O_6 + H_2O + CO_2$$
cream of   baking
tartar   soda

A recipe calls for two tablespoons (24.0 g) of cream of tartar. How much baking soda is needed to allow both substances to react completely?

Actual Yields; Limiting Reactants

28. A certain copper ore contains an average of 2.0 percent Cu_2S. The metallurgical process for recovering elemental copper from this ore gives 70 percent of the theoretical yield. What weight of copper is recovered from 2.0 metric tons of ore?

*29. A four-step synthesis of aspirin from benzene is outlined below:

$$C_6H_6 + Cl_2 \longrightarrow C_6H_5Cl + HCl \quad (80\%)$$
benzene

$$C_6H_5Cl + NaOH \longrightarrow C_6H_5OH + NaCl \quad (90\%)$$

$$C_6H_5OH + CO_2 \longrightarrow C_6H_4(OH)CO_2H \quad (70\%)$$

$$C_6H_4(OH)CO_2H + CH_3COCl \longrightarrow$$
$$C_6H_4(OCOCH_3)CO_2H + HCl \quad (90\%)$$
aspirin

Starting with 1.0 kg of benzene, calculate the weight of aspirin obtained. The percentage yield for each step is indicated after the equation for that step.

*30. One gram of magnesium is placed in a flask containing 1.00 g of nitrogen and the flask is heated. If we assume that the reaction goes to completion, what is the weight of magnesium nitride, Mg_3N_2, formed? Identify the reactant present in excess and state how much of it remains at the end of the reaction.

*31. In a reaction involving 3.0 g of ethane and 10.0 g of chlorine, 5.7 g of ethyl chloride is produced. Calculate the percentage yield of ethyl chloride. The equation for the reaction is

$$C_2H_6 + Cl_2 \longrightarrow C_2H_5Cl + HCl$$

*32. (a) Fluorine reacts with iron to produce iron fluoride, FeF_3. If 3.80 g of fluorine is added to 11.17 g of iron, what weight of iron fluoride can be produced?
(b) What weight of the reactant in excess remains at the end of the reaction?

33. A mixture of 56.0 g of carbon monoxide is heated with 64.0 g of oxygen. Calculate the theoretical yield of carbon dioxide that may be produced by this reaction.

*34. (a) Methane, CH_4, reacts with chlorine in the sunlight to form methyl chloride, CH_3Cl, and hydrogen chloride, HCl. If 640 g of CH_4 is mixed with 53.2 g of chlorine, what is the theoretical yield of methyl chloride that may be produced by the reaction?
(b) The actual yield of methyl chloride is 23.0 g. Calculate the percentage yield of methyl chloride.
(c) The actual yield of hydrogen chloride obtained in the experiment is 27.4 g. What is the percentage yield of hydrogen chloride for the reaction?
(d) Can you account for the difference between your answers to parts (b) and (c)? (Hint: CH_3Cl is known to react with chlorine to produce CH_2Cl_2.)

*35. A mixture of $KClO_3$ and KCl weighing 3.00 g was heated in an open container until the decomposition of the $KClO_3$ into KCl and O_2 was complete. After cooling, the residue of KCl weighed 2.30 g. What was the weight of the KCl in the initial mixture? The equation for the reaction is

$$2KClO_3 \longrightarrow 2KCl + 3O_2$$

Calculations Based on Formulas

36. In Example 2.5 the formulas for three oxides of iron are given. Show that these formulas are consistent with the data on their percentage compositions given in Section 1.5.3.

37. (a) Based on the formula for grain alcohol (ethanol, C_2H_6O), calculate to three significant figures its percentage composition by weight.

(b) Nitroglycerin, the explosive compound in dynamite, is an oily liquid with the formula $C_3H_5(NO_3)_3$. Calculate to four significant figures the percentage by weight of nitrogen.

38. In Section 1.5.4, two compounds of iron and chlorine were described as solid C and solid D. Which one of these is the compound discussed in Example 2.13?

39. It is suspected that a substance is the drug nicotine, $C_{10}H_{14}N_2$. If a 0.2500-g sample is combusted, what weight of carbon dioxide is expected as a product?

*40. A 0.2176-g sample of barium-containing mineral was dissolved in dilute HNO_3. Upon adding H_2SO_4 a white precipitate of $BaSO_4$ formed. This was removed by filtration, dried, and weighed. The weight of the precipitate was 0.0214 g. Calculate the percentage of barium in the mineral.

*41. A 1.000-g sample of silver-containing mineral was dissolved in concentrated nitric acid. The solution was then diluted with water to a volume of 400.0 mL. The addition of hydrochloric acid to a 50.00-mL portion of this solution gave a precipitate of AgCl that weighed 0.0081 g. Calculate the percentage of silver in the mineral.

*42. The formula for lactose is $C_{12}H_{22}O_{11}$. Compare the theoretical composition with the composition in the analysis of Example 2.12. What is the percent error between the analysis for carbon and the theoretical amount? What is a reasonable guess of a contaminant present that would account for the difference between the analytical and the theoretical values?

*43. A mixture of rubidium chloride (RbCl) and sodium chloride that weighed 0.2380 g was dissolved in water. Enough silver nitrate was then added to the solution to precipitate all the chlorine as silver chloride. After filtering and drying the silver chloride, it weighed 0.4302 g. Calculate the weights of RbCl and NaCl in the initial mixture. The equations for the precipitation reactions are

$$NaCl + AgNO_3 \longrightarrow AgCl\downarrow + NaNO_3$$
$$RbCl + AgNO_3 \longrightarrow AgCl\downarrow + RbNO_3$$

Calculation of Formulas
from Experimental Data

44. Example 2.12 describes a weight of lactose being burned in a stream of oxygen. Do the data obtained in this

quantitative experiment prove the validity of the law of conservation of mass? Explain.

45. A 0.9214-g sample of a hydrocarbon (a compound of carbon and hydrogen) is burned in an apparatus similar to that shown in Figure 2.2. The gains in weights of the CO_2 and H_2O absorbers are 3.0806 g and 0.7206 g, respectively.
(a) Calculate the percentage composition of the hydrocarbon.
(b) Calculate the empirical formula.

46. Calculate the empirical formula for the compounds with compositions described as follows:
(a) C_4H_8
(b) The number of C, H, and O atoms combined in a mole of the compound are 8.040×10^{23}, 1.608×10^{24}, and 2.680×10^{23}, respectively.

47. (a) A compound of hydrogen and nitrogen is 12.50 percent hydrogen and 87.50 percent nitrogen. What is the empirical formula?
(b) If the molecular weight is 50 ± 2 amu, what is the formula for a molecule?

48. A compound of nitrogen and oxygen is found to be 30.0 percent nitrogen and 70.0 percent oxygen. What is the molecular formula if the molecular weight is 90 ± 2 amu?

49. Suppose an examination of the compound described in Example 2.12 determined its molecular weight to be 341 ± 2 amu. What is a reasonable formula for the molecule?

50. Cysteine, a sulfur-containing compound found in the structural protein of human hair, has the following composition in weight percent: C, 29.8; N, 11.6; S, 26.4; O, 26.4; H, 5.8. Calculate the empirical formula of cysteine.

*51. (a) An oxide of phosphorus is composed of 56.34 percent phosphorus and 43.66 percent oxygen by weight. Given the atomic weights of phosphorus and oxygen as 30.97 amu and 16.00 amu, respectively, calculate the empirical formula for the compound.
(b) An experimental determination of the molar weight of the compound gives a value of 221 ± 1 g. What is the molecular formula of the compound?
(c) To six significant figures, what is the precise molecular weight of the compound?

Structures of Atoms; Periodic Relationships

3

Subatomic and Atomic Particles

3•1 Nature of Charged Bodies

3•1•1 Behavior of Charged Particles in Vacuum Tubes

3•2 Experimental Evidence for the Electron

3•3 Nucleus of the Atom

3•3•1 Radioactivity

3•3•2 Rutherford's Gold Foil Experiment

3•3•3 Atomic Number

3•3•4 Proton

3•3•5 Mass Spectrograph

3•3•6 Neutron

3•4 Calculation of Atomic Weights

3•4•1 Isotopes and the Law of Definite Composition

3•5 Sizes of Atoms

Periodic Relationships and Atomic Structure

3•6 Arrangements of Electrons in Atoms

3•6•1 Ionization Energies of Atoms

3•6•2 Periodic Behavior of the Elements

3•6•3 Periodic Electron Arrangements

3•7 A Modern Periodic Table

3•8 Energy Sublevels

3•8•1 Order of Filling Energy Sublevels

3•8•2 Summary of Energy Sublevel Filling and the Periodic Table

3•9 Usefulness of the Periodic Table

John Dalton wrote his papers on atomic theory in the first decade of the nineteenth century. Dalton's theory on atoms was accepted by most scientists because the idea of tiny particles was so successful in interpreting numerous chemical discoveries. For almost 100 years, his concept of an atom as the simple, indestructible unit of an element helped to stimulate and guide the experimental work of chemists throughout the world.

Toward the end of the 1800s, however, the easily understood world of Dalton's indestructible atoms was completely overturned by a series of astounding discoveries—X rays in 1895, radioactivity in 1896, the electron in 1897, and radium in 1898. The study of these phenomena revealed atoms to be complex structures made up of subatomic particles.

Ideas about how subatomic particles are arranged in the atoms of different elements are the basis for our present understanding of chemical behavior. In this chapter we shall describe some of the discoveries that led to our theories of atomic structure. We shall see how more knowledge about the structures of atoms makes a periodic table extremely helpful to students of chemistry.

Subatomic and Atomic Particles

The recognition by Mendeleev and others of the periodic behavior of the elements is considered by many historians to be the most important development in chemistry. One of the results was that even more attention was focused on the study of atoms. During the ensuing years, questions arose concerning the structure of atoms. Must not the atoms of elements in a given vertical group of the periodic table be similar to one another in some way? Must not the atoms of the elements in the same horizontal period differ from one another in some stepwise way? Answers to such questions were obtained from the study of the effects of electric discharges on small samples of gases. In order to understand the results of these studies, we must first know something about charged bodies in general.

3·1 Nature of Charged Bodies

In the experiments to be described, some familiar characteristics of charged bodies that should be kept in mind are:

1. A current of electricity is the movement of charged particles in a conductor.
2. Particles of opposite charge attract each other; those of the same charge repel each other.
3. Charged particles can move between charged wires or plates called **electrodes.** The positively charged (+) electrode is the **anode;** the negatively charged (−) one is the **cathode.** (See Figure 3.1.)

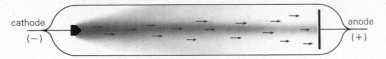

Figure 3•1 A cathode-ray tube.

3•1•1   Behavior of Charged Particles in Vacuum Tubes Much of the evidence for the structure of atoms is based on experiments done with vacuum tubes. Important points to note are:

1. A chamber in which a beam of charged particles is to travel efficiently must be evacuated. Otherwise, collisions of the particles in the beam with air molecules scatter the tiny charged particles so that they tend to travel in random directions.

2. A beam of charged particles can be obtained by using slits or holes to screen out all particles except those traveling in the desired direction. (See Figure 3.2 on Plate 3.)

3. Because the charged particles are not visible, their presence and their paths are revealed by the use of fluorescent compounds,[1] by photographic film, or by special electronic devices.

4. The path of a beam of charged particles is bent as the beam passes either between charged plates or near the poles of a magnet. From the direction of bending, the charges on the particles in the beam can be deduced. From the amount by which the beam is bent by a measured force, the masses of the particles can be calculated.

With this background material, we can now discuss the experimental evidence for the first subatomic particle that was clearly recognized, the electron.

3•2   Experimental Evidence for the Electron

If two wires are subjected to a high electric potential and then brought close together, a spark or arc jumps from one wire to the other. If the ends of the two wires are sealed in a glass tube (as shown in Figure 3.1) that is then highly evacuated, the discharge from one wire to the other is more gentle. This discharge, called a **cathode ray,** causes the glass tube to emit a faint yellowish-green glow.

Cathode rays, first studied intensively in 1858 by J. Plücker, have the following properties:

1. They travel in straight lines away from the cathode, unless acted upon by an outside force.

2. They are negatively charged. This is evident from the fact that they are attracted by a positively charged plate. Also, the path of the rays is bent

---

[1] Substances that absorb radiant energy that cannot be seen and then immediately emit visible radiant energy (light) are said to *fluoresce*.

by a magnetic field in the same direction as the path of particles known to be negative (Figure 3.2 on Plate 3). In the years 1909–1913, R. A. Millikan determined the absolute charge, e, on these tiny particles by an experiment involving the rise and fall of charged oil droplets in the presence and absence of an electric field. The present-day absolute value for e is 1.6022×10^{-19} coulomb (see Table A.2). For convenience,[2] this unit electric charge on the particle is assigned a relative value of -1.

3. They consist of particles of definite mass. The mass of a cathode-ray particle cannot be obtained directly. However, in 1897 J. J. Thomson determined the *ratio of charge to mass, e/m,* by an experiment involving the simultaneous deflection of cathode rays by an electric and a magnetic field. He was hailed as the discoverer of the **electron,** the subatomic particle with a unit negative charge. The absolute ratio of e/m for the electron as measured today is 1.7588×10^8 C/g. Using this ratio and the modern value for e, we can calculate the mass of the electron:

$$\frac{e}{m} = 1.7588 \times 10^8 \text{ C/g}$$

$$\frac{1.6022 \times 10^{-19} \text{ C}}{m} = 1.7588 \times 10^8 \text{ C/g}$$

$$m = \frac{1.6022 \times 10^{-19} \text{ C}}{1.7588 \times 10^8 \text{ C/g}} = 9.1096 \times 10^{-28} \text{ g}$$

The mass of the electron, 9.1096×10^{-28} g, is about $\frac{1}{1837}$ as much as the lightest atom, a hydrogen atom.

4. The nature of the cathode rays (electrons) is the same, irrespective of (a) the material of which the cathode is made, (b) the type of residual gas present in the evacuated tube, (c) the kind of metal wires used to conduct the current to the cathode, and (d) the materials used to produce the current.

All this evidence, especially the final item, indicates that electrons are fundamental particles found in all matter. The fact that an electron has a mass that is only a small fraction, $\frac{1}{1837}$, of that of the lightest known atom indicates that atoms can be made up of many subatomic particles.

3.3   Nucleus of the Atom

3.3.1   Radioactivity   The recognition that certain elements are *radioactive,* that is, they emit powerful, invisible radiations, is an example of the many scientific discoveries that have been made at least in part by accident. The French physicist Henri Becquerel, during an investigation of the light given off by various minerals, happened to store some samples of uranium ore side by side with carefully packaged photographic plates. When he later found the plates partially exposed, he realized that the ore samples had emitted a radiation so powerful that it penetrated the packaging materials. Discovery

---

[2]Recall that, also for convenience, weights of atoms are assigned relative values rather than absolute values.

by chance is sometimes attributed to serendipity, but, as Louis Pasteur observed, in science "chance favors the prepared mind." Physicists quickly applied many of the same techniques that they had applied to cathode rays to the study of the new invisible radiations. One of the most successful was young Ernest Rutherford, on his way to becoming one of the giants of twentieth-century physics.

Rutherford and others, using the method shown schematically in Figure 3.3, found that radioactive substances can produce three types of emissions. The whole apparatus is enclosed in a glass tube from which the air can be pumped so that the emissions are not deflected by collisions with air molecules. A sample of radioactive material emits radiations in every direction. But if the radioactive source is placed inside a lead block that has a hole in it, only the radiations that happen to be directed straight out the hole in the lead block escape and go through the hole in the metal shield. Unless acted upon by some outside force, the emissions then travel in a straight line toward a detection screen that is coated with zinc sulfide. In the absence of any outside force, only one fluorescent spot shows on the screen, and it is directly in line with the positions of the holes.

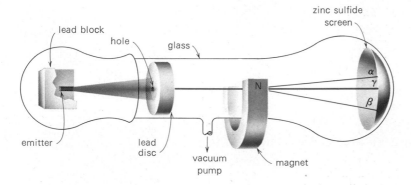

Figure 3•3 A schematic representation showing how the three kinds of natural radioactive emissions can be identified by determining the effect of a magnetic field on their paths of travel.

However, when the emissions have to travel through a magnetic field (the magnet is shown in place in Figure 3.3), three fluorescent spots appear on the screen; one spot is in line with the holes and the other two are displaced above and below this spot. The type of emission that passes through a magnetic field without being deflected by the field must be neutral in character. The other two spots must be the places that charged particles strike after being deflected by the magnetic field; that is, positive particles are deflected in one direction, and negative particles in the opposite direction.

This experiment reveals three kinds of emissions: one positive, one negative, and one neutral, named simply alpha (α), beta (β), and gamma (γ) after the first three letters of the Greek alphabet. Characteristics of these three types of emissions are summarized in Table 3.1. Although further research showed the beta particle to be identical with other electrons, it is still the convention to refer to electrons that are emitted by radioactive elements as "beta particles." Radioactivity will be taken up in more detail in Chapter 19.

With the revelation that elements could emit tiny particles, a fundamental assumption of the Dalton atomic theory had to be dropped. *Atoms cannot be indestructible if some of them spontaneously throw out parts of themselves.*

Table 3•1 Natural radioactive emissions

Name	Mass relative to H atom	Relative charge
alpha particle	~4	2+
beta particle	$\frac{1}{1837}$	1−
gamma ray	0	0

3•3•2   Rutherford's Gold Foil Experiment   As indicated in Table 3.1, an alpha particle has a mass about four times that of a hydrogen atom and a charge opposite in sign and twice the magnitude of the charge of the electron, that is, 2+. Because they have velocities of about 1.6×10^7 m/s (approximately 10,000 mi/s), alpha particles can be thought of as tiny, high-speed projectiles.

In 1908–1909, H. Geiger and E. Marsden, working with Rutherford, reported some of the most meaningful experiments of modern times. They constructed an apparatus (see Figure 3.4) that allowed them to study the effect of bombarding a thin sheet of gold with a beam of alpha particles. Gold was chosen as the target because it is a very malleable metal that can be beaten into extremely thin sheets, possibly only 100 atoms thick. Like other solids, however, gold can hardly be compressed at all, so one assumes that its atoms are packed tightly together. As shown in Figure 3.4, the experimenters made three principal observations:

1. Most of the alpha particles went straight through the supposedly closely packed gold atoms.
2. A few of the speeding alpha particles were deflected by something.
3. A very few of the alpha particles were bounced back from the gold.

Relative to the third observation, Rutherford remarked later, "It was almost as incredible as if you fired a 15-inch shell at a piece of tissue paper and it came back and hit you."

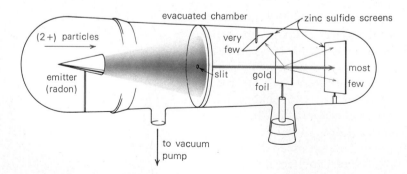

Figure 3•4 A schematic representation of several experiments in which gold foil was bombarded with a beam of alpha particles. Zinc sulfide emits a flash of light when struck by an alpha particle.

Even for the fertile mind of a genius, it takes time to correlate new facts and to frame theories to explain them. According to Geiger, Rutherford walked into the laboratory on a day in 1911 saying "I've got it." What he had was a clear mental picture of a relationship between the bombarding alpha particles and the gold atoms. He had worked out an explanation of all the observed facts in terms of a new concept of atomic structure.

Rutherford pictured a model of closely packed, spherical atoms similar to that shown in Figure 3.5. Each atom has a small center or nucleus. The nucleus is so small that only rarely does an alpha particle chance to pass near. Because the nucleus evidently repels alphas, it must be positive. The fact that an alpha particle occasionally bounces back shows that the nucleus of a gold atom must be considerably heavier than the high-speed alpha particle. In summary, the **nucleus** is the tiny, positively charged, massive center of an atom. Rutherford visualized the electrons in an atom as being far away from the nucleus, forming the outside surface of the atom. Between the nucleus and the outer electrons is space, empty except for other electrons.

Our present information, gleaned largely from X-ray investigations, indicates that the diameter of an atom is somewhat greater than 10^{-10} m. It has been estimated that a nucleus has a diameter of about 10^{-14} m, or about 1/50,000 that of an atom.

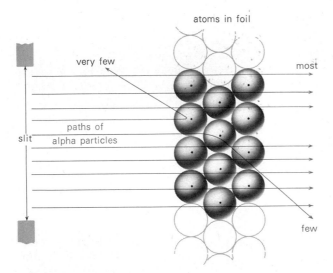

Figure 3•5 Rutherford's interpretation of the bombardment of gold atoms with alpha particles. The tiny dot in the center of each sphere represents the nucleus of the atom. In any drawing of this size, it is necessary to greatly exaggerate the sizes of the nuclei relative to the sizes of the atoms. The sizes of the atoms are also exaggerated relative to the size of the slit. The slit is perhaps 100 or more atomic diameters wide.

3•3•3   Atomic Number   When cathode rays strike matter, they may give rise to several effects. In 1895 W. C. Roentgen discovered that material that is struck by cathode rays may emit a penetrating radiation called X rays. **X rays** are a form of electromagnetic radiation, similar to light waves or radio waves, but with greater energy than either. X rays have no mass and no charge.

At the same time that Rutherford was developing his concept of the nuclear atom, a co-worker of his, H. G. J. Moseley, made an important discovery about the nuclei of atoms. He found that when cathode rays struck different elements used as anode targets in an X-ray tube (Figure 3.6), characteristic X rays were emitted. When Moseley tabulated his X-ray results, he found that in general the wavelengths of the X rays emitted decreased regularly as the relative atomic weights of the elements increased. Figure 3.7 is an artist's drawing of photographic records collected by Moseley. Note how the wavelengths of the X rays decrease in stepwise fashion

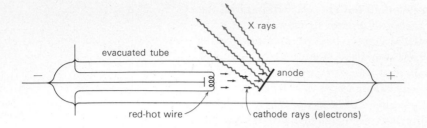

Figure 3•6 An X-ray tube.

from calcium, Ca, to zinc, Zn. (Brass, an alloy of copper, Cu, and zinc, shows lines characteristic of both these elements.)

As we will learn in Chapter 4 when atomic spectroscopy is discussed, the wavelength and energy of radiant energy are inversely proportional. Hence, the shorter the wavelength of an X ray, the greater is its energy. It is assumed that a high-energy cathode ray knocks a tightly held electron away from a bombarded atom. Immediately following this a replacement electron is attracted and falls in to fill the vacancy, and an X ray is emitted. Atoms that have high positive charges on their nuclei should attract replacement electrons strongly and emit high-energy (short-wavelength) X rays.

After studying his data, Moseley concluded that the number of positive charges on the nucleus "increases from atom to atom by a single electron unit. . . ." The number of positive charges on the nucleus is now called the **atomic number,** Z. Starting with the lightest element in the periodic table, Z is 1 for hydrogen 2 for helium, 3 for lithium, and so on. We can define an element in terms of its atomic number: *An element is a substance in which all the atoms have the same atomic number.*

Moseley calculated the charge on the nucleus of a calcium atom to be $20+$; of a titanium atom, $22+$; vanadium, $23+$; and so on to zinc, $30+$. The relative positions of cobalt and nickel in Figure 3.7 show that the regular pattern of decreasing wavelengths (increasing energies) is related to an increase in the atomic numbers of the elements rather than to an increase in the atomic weights. Even though atoms of nickel, Ni, weigh slightly less than atoms of cobalt, Co, nickel atoms have the higher atomic number. These determinations of atomic numbers were in excellent agreement with the order of elements assigned many years earlier by Mendeleev on the basis of chemical and physical similarities. For example, in Mendeleev's periodic table, tellurium, Te, comes before iodine, I, even though the atomic weight of iodine is the smaller. The fact that Te and I atoms were found to have atomic numbers of 52 and 53, respectively, showed in another way that the order should be Te–I.

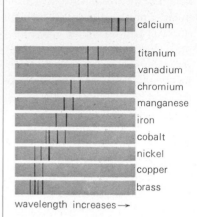

Figure 3•7 An artist's copy of some of Moseley's historic X-ray photographs. The X rays emitted by different elements were diffracted by a crystalline substance toward photographic plates. The positions of the lines on the plates are directly related to the wavelengths of the X rays. Of these elements, calcium emits the longest wavelength (least energetic), and zinc (which is mixed with copper to make brass) emits the shortest wavelength (most energetic) X ray.

3•3•4   Proton   Even before the electron was identified, E. Goldstein in 1886 noticed that a fluorescence appeared on the inner surface of a cathode-ray tube behind a cathode perforated with holes. This indicated that positive rays were moving in such a tube, and that some of them sped through the holes in the cathode and struck the end of the tube.

After the discovery of the electron, physicists devoted a great deal of effort to seeking a fundamental particle with a positive charge. Studies of cathode-ray tubes indicated that many different types of positive particles could exist, depending on the gas used to flush out the tube prior to evacuation. When hydrogen gas is used, the lightest of the positive particles is obtained. This particle has an e/m ratio of 9.5791×10^4 C/g. The absolute charge on the particle is the same as that on the electron, 1.6022×10^{-19} C. Because it is positively charged, the relative charge assigned is $+1$. From the e/m ratio and the known value of e, a mass of 1.6726×10^{-24} g is obtained, a mass that is about 1,836 times heavier than the electron. This subatomic particle with a mass of 1.6726×10^{-24} g and a unit positive charge is called a **proton.** Protons are formed when the high-speed cathode-ray electrons knock electrons off neutral (uncharged) hydrogen atoms, leaving the positive nuclei behind.

The fact that the proton was found to be about 1,836 times heavier than the electron supported Rutherford's theory of the nuclear atom very well. The light electrons could, at relatively great distances, surround a nucleus that might be made of the heavy, positive protons collected together. Additional support for this theory was afforded by Moseley's findings. His evidence indicated that an atom of hydrogen had one proton in its nucleus, a helium atom had two, a lithium atom three, and so on. The atomic number, then, must be the number of protons in the nucleus. It follows that a modern definition of an element may be stated in terms of protons as well as in terms of atomic numbers: *An element is a substance all of whose atoms contain the same number of protons.*

Because atoms are neutral particles, the number of protons must equal the number of electrons to provide for a balance of charges. In order to determine whether atoms could be thought of as being made solely of protons and electrons, it was necessary to determine the weights of specific atoms and to compare these weights with the total weights of the protons and electrons thought to be involved per atom.

3•3•5   **Mass Spectrograph**   Seeking to compare the weights of atomic-sized particles directly, J. J. Thomson and F. W. Aston, just prior to World War I, experimented with the deflection of streams of charged gaseous particles by means of known magnetic and electrostatic forces. After the war, Aston in England and other workers around the world perfected the **mass spectrograph,** an apparatus used to compare the weights of atoms precisely.

A schematic diagram of one of the early mass spectrographs developed by A. J. Dempster, of the University of Chicago, is shown in Figure 3.8. Atoms of an element are bombarded by electrons in the spark gap so that other electrons are knocked off the atoms, forming positively charged particles called **ions.**[3] The positive ions formed in the spark gap are accelerated by a high voltage, V, of 800 to 1,000 V in the region between the charged plates P_1 and P_2. A stream of ions speeds at high velocity through the slit in P_2 into the semicircular space A. Space A is between the poles of a powerful magnet. The magnet imposes a field of sufficient strength to cause the stream of ions to follow curved paths. If the stream contains ions of different masses,

---

[3]An **ion** is a positively or negatively charged particle in which the number of electrons does not equal the number of protons. If an atom or a molecule loses or gains one or more electrons, an ion is formed.

the paths of the lighter ions L are curved more than those of the heavier ions H. Also, the faster a particle is moving, the less its path is curved.

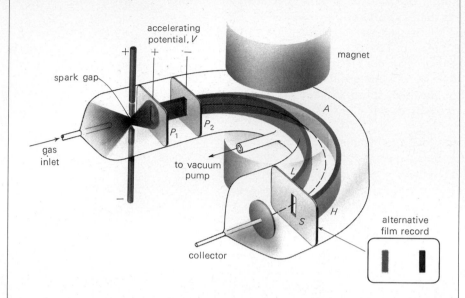

Figure 3.8 Schematic diagram of an early mass spectrograph.

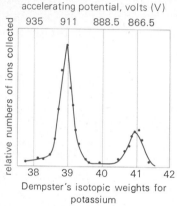

Dempster's isotopic weights for potassium

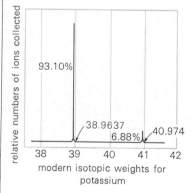

modern isotopic weights for potassium

Figure 3.9 Isotopic weights of potassium as obtained with an early and a modern mass spectrograph. The early apparatus of Dempster did not provide for sharp focusing of the two isotopes with mass numbers 39 and 41. An actual modern record is even more precise than the lower diagram indicates, and shows a trace of a rare isotope with weight 39.974 amu.

Dempster found that when he varied the accelerating voltage between plates P_1 and P_2 but kept the strength of the magnetic field constant, he could cause either light or heavy ions to go through a slit S and be counted by a negatively charged collector. The mass, m, of each of the different ions that go through slit S is inversely proportional to the original accelerating voltage, V. Letting m_1 stand for the mass of the lighter isotope, V_1 for the voltage required to collect it, m_2 for the mass of the heavier isotope, and V_2 for the voltage required to collect it, we can state mathematically that

$$m_1 \propto \frac{1}{V_1} \quad \text{and} \quad m_2 \propto \frac{1}{V_2} \qquad (1)$$

With the element potassium, K, Dempster found the results that are as shown in Figure 3.9. He knew from other measurements that most potassium atoms had a mass of about 39.0 amu. Ions of these atoms passed through the slit in the mass spectrograph at $V_1 = 911$ V. At the lower voltage, V_2, of 866.5 V, another stream of potassium ions was focused on slit S.

From the proportionalities between mass and voltage shown in Equation (1), we can write the following equations:

$$m_1 V_1 = \text{a constant} \quad \text{and} \quad m_2 V_2 = \text{a constant} \qquad (2)$$

Or, stated mathematically in an equivalent way,

$$m_1 V_1 = m_2 V_2 \qquad (3)$$

Using Equation (3), the mass, m_2, of each of the heavier ions may be calculated:

$$(39.0 \text{ amu})(911 \text{ V}) = (m_2)(866.5 \text{ V})$$

$$m_2 = 39.0 \text{ amu} \times \frac{911 \text{ V}}{866.5 \text{ V}} = 41.0 \text{ amu}$$

As indicated in Figure 3.9, the lighter potassium ions (39.0 amu) are more abundant than the heavier ones (41.0 amu).

Atoms of the same element that have different weights are called **isotopes.** The first isotopes discovered were those of neon, identified by Thomson and Aston in 1912–1913. Since that time isotopes for practically all the elements have been discovered. The mass spectrograph has been developed into a very precise instrument that can determine atomic masses to seven significant figures. Some instruments record results in photographic form, as shown in Figure 3.10.

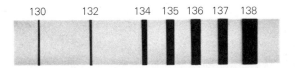

130   132   134 135 136 137   138

3•3•6   Neutron   By means of the mass spectrograph it has been determined that the most common isotope of hydrogen has a mass of 1.0078 amu. Also, it is known that the masses of a proton and an electron are 1.0073 amu and 0.00054859 amu, respectively. The mass of the most common hydrogen isotope can be accounted for by picturing a nucleus of one proton and one electron outside the nucleus:

$$\underset{\text{mass of proton}}{1.0073 \text{ amu}} + \underset{\text{mass of electron}}{0.00054859 \text{ amu}} \simeq \underset{\substack{\text{mass of common} \\ \text{H isotope}}}{1.0078 \text{ amu}}$$

But none of the other cases is so simple. Consider the common isotopes of nitrogen with masses of 14.00 amu and 15.00 amu. This element has an atomic number of seven, which indicates that nitrogen atoms contain seven protons and seven electrons. Atoms with this composition would weigh only a little more than 7 amu. However, nitrogen atoms weigh either 14 or 15 amu. All known atoms, except the most common hydrogen isotope, have masses that are much greater than the sum of the masses of their protons and electrons.

The fact that the mass of the protons and electrons does not account for the total mass of atoms led scientists to search for an uncharged particle whose presence would explain the additional mass of an atom but would not upset the balance of charges between protons and electrons. Because it had no charge, this elusive particle escaped detection for years, but in 1932 neutrons were identified by J. Chadwick.[4] A **neutron** is a subatomic particle that has no charge and has a mass of 1.0087 amu, that is, about the same as that of a proton.

---

[4]The experimental details of the identification will be presented in Section 19.7.

Figure 3•10 Artist's copy of a photographic record of barium isotopes obtained with a mass spectrograph. Precise measurement of the degree of film exposure shows that about 72 percent of the atoms have mass number 138, with other values ranging down to about 0.1 percent each for mass numbers 130 and 132. *Source:* From A. J. Dempster, *Phys. Rev.,* 49, 947 (1936).

With the discovery of the neutron, chemists were able to account for the main features of atoms in terms of the three fundamental particles listed in Table 3.2. An **atom** is defined as an extremely small, electrically neutral particle that has a tiny but massive positive core or nucleus and one or more electrons relatively far outside its nucleus:

atomic number (Z) = number of protons = number of electrons

The number of protons plus neutrons in the nucleus of an atom is called its **mass number,** A:

mass number (A) = number of protons (Z) + number of neutrons

For any isotope, the mass number is the whole number nearest its isotopic weight (mass in atomic mass units). From the mass number, the number of neutrons in the nucleus of an isotope is easily obtained. Consider the heaviest isotope of nitrogen with an isotopic weight of 15.00 amu:

mass number − atomic number = number of neutrons
(A)   (Z)
15   −   7   =   8

To refer to a specific isotope in words, we give the name of the element followed by the mass number, for example, nitrogen-15. To refer to a specific isotope with a symbol, we give the symbol for the element with the mass number, A, written to the upper left, for example, ^{15}N. Sometimes the symbol also includes the atomic number, Z, written to the lower left, for example, $^{15}_{7}N$.

Table 3.2 Fundamental particles of matter

Name	Symbol	Mass, g	Mass, amu	Relative charge
proton	p^+	1.6726×10^{-24}	1.0073	1+
neutron	n	1.6749×10^{-24}	1.0087	0
electron	e^-	9.1096×10^{-28}	5.4859×10^{-4}	1−

3.4   **Calculation of Atomic Weights**

Precise measurements with the mass spectrograph provide the best means of determining the atomic weights of elements. In 1961 the most common isotope of carbon, referred to as ^{12}C or as carbon-12, was designated by the International Bureau of Standards as the standard for atomic weights. The weight of this isotope is arbitrarily defined as exactly 12 amu, and all other atoms are compared to it. One **atomic mass unit** (amu) is defined as one-twelfth the weight of one carbon-12 atom.

The **atomic weight** of an element is the weighted average of the weights of its natural isotopes. In the case of potassium (Figure 3.9), precise measurements show that 93.10 percent of the atoms weigh 38.9637 amu and

6.88 percent weigh 40.974 amu. There is a third, very rare isotope (39.974 amu) with an occurrence of only 0.001 percent. In Table 3.3, the isotopic weights and the average atomic weights of several common elements are listed. Example 3.1 illustrates the calculation of an atomic weight from isotopic weights and percentages of isotopic abundances. Example 3.2 involves a different type of calculation.

Table 3•3 **Isotopic weights and average atomic weights**

Element	Mass number[a]	Symbol	Isotopic weight, amu	Isotopic abundance, percent	Average atomic weight, amu
hydrogen	1	^1H	1.0078	99.985	1.0079
	2	^2H	2.0141	0.015	
carbon	12	^{12}C	12 (exact)	98.892	12.011
	13	^{13}C	13.0034	1.108	
nitrogen	14	^{14}N	14.0031	99.635	14.007
	15	^{15}N	15.0001	0.365	
oxygen	16	^{16}O	15.9949	99.759	15.999
	17	^{17}O	16.9991	0.037	
	18	^{18}O	17.9992	0.204	
iron	54	^{54}Fe	53.940	5.84	55.85
	56	^{56}Fe	55.935	91.68	
	57	^{57}Fe	56.935	2.17	
	58	^{58}Fe	57.933	0.31	

[a]Only nonradioactive isotopes are listed.

The modern definition of a mole is also based on the carbon-12 isotope. A **mole** is defined as that quantity of a substance that contains the same number of ultimate particles (atoms, molecules, electrons, ions, or groups of ions) as are contained in precisely 12 g of carbon-12. (See also Section 2.5.1.)

— • Example 3•1 • —

The natural isotopes of sulfur have masses of 31.972, 32.971, 33.968, and 35.967 amu. Relative percent abundances in the same order are 95.01, 0.76, 4.22, and 0.01. Calculate the average atomic weight of sulfur.

— • Solution • To calculate the average atomic weight from the isotopic weights and their percent abundances, we multiply each isotopic weight by its fractional abundance and sum the proportional contributions of each isotope:

$$\frac{\text{isotopic}}{\text{weight}} \times \frac{\text{fractional}}{\text{abundance}} = \frac{\text{proportional}}{\text{contribution}}$$

31.972 × 0.9501 = 30.38
32.971 × 0.0076 = 0.25
33.968 × 0.0422 = 1.43
35.967 × 0.0001 = 0.00
average atomic weight = 32.06 amu

See also Exercise 20 at the end of the chapter.

• Example 3•2 •

The two naturally occurring isotopes of gallium, ^{69}Ga and ^{71}Ga, have masses of 68.9256 and 70.9247 amu, respectively. The atomic weight of gallium is 69.72 amu. Calculate the percentage of each isotope in a sample of gallium.

• Solution • Let x = the fractional abundance of mass 68.9256 amu; then $1 - x$ = the fractional abundance of mass 70.9247 amu:

$$68.9256(x) + 70.9247(1 - x) = 69.72$$
$$70.9247x - 68.9256x = 70.9247 - 69.72$$
$$1.9991x = 1.20$$
$$x = 0.600; \quad 1 - x = 0.400$$

Fractional abundance times 100 is equal to percent abundance, so there is 60.0 percent of mass 68.9256 amu and 40.0 percent of mass 70.9247 amu.

See also Exercises 21 and 23.

3•4•1   **Isotopes and the Law of Definite Composition**   Prior to the discovery of isotopes, it was believed that the law of definite composition applied precisely for any given formula. However, an example will show how the percentage composition of a compound varies depending on the isotopes present. Water as it occurs in nature has a composition of 88.81 percent oxygen and 11.19 percent hydrogen. Actually, water can have a range of compositions, depending on which isotopes of hydrogen are united with which isotope of oxygen. For example, it is possible to have water that is made up solely of hydrogen atoms of mass number 1 united with oxygen atoms of mass number 16, or to have water made up solely of hydrogen atoms of mass number 2 united with atoms of oxygen with mass number 16. Several possible percentage compositions are shown in Table 3.4. (See Table 3.3 for the weights of these isotopes.)

Table 3•4 **Percentage compositions of H_2O for various combinations of isotopes**

	Percentage of hydrogen by weight	Percentage of oxygen by weight
$^1H_2^{16}O$	11.19	88.81
$^2H_2^{16}O$	20.12	79.88
$^1H_2^{17}O$	10.60	89.40
$^2H_2^{17}O$	19.16	80.84

We see, then, that compounds may not always have their elements combined in a definite proportion by weight, although the proportion of atoms in a molecule is definite. In the case of water, the atomic ratio is always two atoms of hydrogen to one atom of oxygen.

It is common to find the various isotopes of an element so uniformly mixed in nature that any sample of the element will be composed of atoms whose average weight is the atomic weight listed in the table of elements inside the back cover. Unless the isotopes have been separated, compounds do have a definite composition by weight within the limits indicated in that table, and within the limits described in Section 1.5.3.

3•5   Sizes of Atoms

Just how big are particles of atomic dimensions? Although indirect images of atoms have been photographed, no atom has ever been seen, even through the most powerful optical microscopes. Physicists have been able, however, to measure the volume occupied by a known number of atoms and thus calculate the volume of an individual particle. If the atom is assumed to be spherical, its diameter can be calculated. Table 3.5 lists a number of common measurements of large and small objects. From the value of 2.8×10^{-10} m given for the diameter of the uranium atom, we can calculate that it would take a line of 6.8×10^7, or 68 million, uranium atoms placed side by side to span the diameter of a penny, a distance of 1.9×10^{-2} m. Measuring the vastness of the universe and the minuteness of the atom are two of the great experimental achievements of scientists.

Table 3•5 **Measurements, large and small**

distance to farthest observed galaxy	4.5×10^{25} m	5.0×10^9 light-years
distance to nearest star	4×10^{16} m	4.3 light-years
diameter of earth	1.3×10^7 m	8,000 miles
height of average person	1.7 m	68 in.
diameter of penny	1.9×10^{-2} m	0.75 in.
diameter of red blood cell	7.6×10^{-6} m	3×10^{-4} in.
diameter of smallest virus	1×10^{-8} m	4×10^{-7} in.
diameter of uranium atom	2.8×10^{-10} m	1×10^{-8} in.
diameter of hydrogen atom (smallest atom)	1×10^{-10} m	4×10^{-9} in.
diameter of proton	1.2×10^{-15} m	4.7×10^{-14} in.

Tiny fractions of meters are as inconvenient for expressing lengths of atomic size as are tiny fractions of grams for expressing weights. A common unit for describing the sizes of atoms is the **angstrom**, Å; $1 \text{ Å} = 1 \times 10^{-10}$ m. Atoms range in size from the hydrogen atom, with a diameter of about 1 Å, to the cesium atom, with a diameter of about 5 Å.

Periodic Relationships and Atomic Structure

3•6   Arrangements of Electrons in Atoms

The Rutherford concept of the massive, tiny, positive nucleus of an atom surrounded by negative electrons led to a most perplexing question. How are the electrons, which contribute hardly any mass, arranged so that they take up practically all the space occupied by an atom? This problem was attacked by many physicists. The breakthrough to understanding the arrangement of electrons is usually credited to Niels Bohr, whose studies of atomic spectra will be discussed in Chapter 4, but work done by others to test Bohr's ideas provided somewhat more direct evidence.

Following Rutherford, physicists made the fundamental assumption that the oppositely charged particles in atoms, the protons and electrons, attract each other according to the same law that describes the behavior of large, charged particles. Charles Coulomb had found in 1784 that the force between two charged points was directly proportional to the size of the charges and inversely proportional to the square of the distance between them:

$$\text{electrostatic force} \propto \frac{(\text{charge 1})(\text{charge 2})}{(\text{distance})^2}$$

If the charges are opposite in sign, the force is attractive; if the charges have the same sign, the force is repulsive.

As shown in Section 3.2, the symbol for the charge on an electron is e. The charge on a proton is opposite in sign but equal in magnitude, so its charge is e also. If the number of protons in a nucleus is Z, the charge on the nucleus is Ze. Researchers reasoned that if they could determine the forces of attraction between a nucleus and its electrons, they could use Coulomb's law to determine the relative distances, d, between the nucleus and the various electrons:

$$\text{force} \propto \frac{(\text{charge 1})(\text{charge 2})}{(\text{distance})^2} = \frac{(Ze)(e)}{d^2} = \frac{Ze^2}{d^2}$$

For an atom of a given element, Ze^2 is a constant, so we can say that the greater the force of attraction between the nucleus and an electron, the less is the distance between the nucleus and that electron.

3.6.1   Ionization Energies of Atoms

To measure the forces of attraction between electrons and a positive nucleus, one can determine the ionization energies of atoms. **Ionization energies** are the amounts of energy necessary to knock electrons off an atom to form ions. For some electrons very little energy is required, so these electrons must be relatively far from the positive nucleus; for others, great amounts of energy are needed, so these must be close to the nucleus.

One way to determine the ionization energy of an atom is to measure the minimum energy of speeding electrons (a cathode ray) required to knock electrons off gaseous atoms. Although simpler devices were used in the earliest investigations, the mass spectrograph is one of the instruments used today to determine ionization energies. A spark in the gap of the spectrograph (see Figure 3.8) is a beam of speeding electrons. The energies of these bombarding electrons can be controlled precisely by adjusting the difference in potential of the positive and negative spark gap electrodes. The energy of an electron is measured in electron volts; one **electron volt** (eV) is the energy acquired by a singly charged particle when it falls through a potential of 1 volt.[5]

With a sample of an element in the spark gap, there are no ions formed until the energy of a bombarding electron is raised enough to knock the most loosely bound electron off a gaseous atom. This amount of energy is called the **first ionization energy:**

$$\text{atom} + \text{bombarding electron} \longrightarrow \text{ion}^{1+} + \text{two electrons}$$

Only the positively charged ion travels through the spectrograph and is detected at the collector. For atoms with enough electrons, a second electron can be knocked off at a higher potential (the second ionization energy), a third at a still higher potential (the third ionization energy), and so on. The energy required for each of these steps can be determined with the mass spectrograph. The first three steps in the ionization of beryllium and magnesium atoms are represented schematically in Figure 3.11.

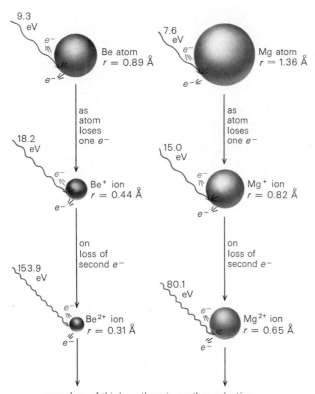

upon loss of third e^- there is another reduction in size and the formation of ions with 3+ charge, and so on

Figure 3•11 The ionization energy of an atom or an ion is related (a) to the distance of the outermost electron from the nucleus (the radius), (b) to the positive charge on the nucleus (the atomic number), and (c) to the effect of the other electrons. (Electrons repel one another and thus reduce the net attraction of the nucleus.) The relative sizes of atoms and ions will be discussed in Section 5.6.4. The bombarding electron (colored e^-) bounces away as it knocks off an electron (black e^-). Electrons can be thought of as having wavelengths, as indicated by the wavy lines (see Section 4.3). The smaller its wavelength, the greater is the energy of the bombarding electron.

In Table 3.6 some of the ionization energies of the first 22 elements are listed. Careful study of the relative magnitudes of these values has helped physicists to speculate about the arrangement of electrons around atomic

---

[5]The electron volt is a very small unit of energy. For example, it would require about 2×10^{22} eV to melt a 10-g ice cube at 0 °C.

Table 3.6 **Ionization energies, eV**[a]

Atomic number, Z	Symbol	1st e^-	2nd e^-	3rd e^-	4th e^-	5th e^-
1	H	13.6				
2	He	24.6	54.4			
3	Li	5.4	75.6	122.5		
4	Be	9.3	18.2	153.9	217.7	
5	B	8.3	25.1	37.9	259.3	340.2
6	C	11.3	24.4	47.9	64.5	392.1
7	N	14.5	29.6	47.4	77.5	97.9
8	O	13.6	35.1	54.9	77.4	113.9
9	F	17.4	35.0	62.7	87.1	114.2
10	Ne	21.6	41.0	63.4	97.1	126.2
11	Na	5.1	47.3	71.6	98.9	138.4
12	Mg	7.6	15.0	80.1	109.2	141.3
13	Al	6.0	18.8	28.4	120.0	153.8
14	Si	8.2	16.3	33.5	45.1	166.8
15	P	10.5	19.7	30.2	51.4	65.0
16	S	10.4	23.3	34.8	47.3	72.7
17	Cl	13.0	23.8	39.6	53.5	67.8
18	Ar	15.8	27.6	40.7	59.8	75.0
19	K	4.3	31.6	45.7	60.9	82.7
20	Ca	6.1	11.9	50.9	67.1	84.4
21	Sc	6.5	12.8	24.8	73.5	91.7
22	Ti	6.8	13.6	27.5	43.3	99.2

Source: *Data adapted with permission from J. E. Huheey,* Inorganic Chemistry, *2nd ed., Harper & Row, New York, 1978.*

[a]The horizontal colored lines mark sharp drops in ionization energies as atomic numbers increase. For each atom, the vertical colored line separates energies for electrons relatively easily knocked off (to the left) from electrons more tightly held (to the right).

nuclei. We summarize some of the data and the conclusions drawn from them.

1. Lithium, Li ($Z = 3$), sodium, Na ($Z = 11$), and potassium, K ($Z = 19$), atoms have the *smallest first ionization energies* of those elements listed. This suggests that each has one electron that is easily removed. The second ionization energy for each of these atoms is much larger than the first, which shows that their other electrons are more strongly held. These facts indicate that lithium, sodium, and potassium atoms each have one electron that is loosely held and therefore is relatively far from the positive nucleus. The other electrons that are more tightly held are presumed to be nearer to the nucleus.

2. Beryllium, Be ($Z = 4$), magnesium, Mg ($Z = 12$), and calcium, Ca ($Z = 20$), atoms each have first and second ionization energies that are

small compared with the third. This shows that each has two electrons that are removed relatively easily. Note that these elements have three of the *smallest second ionization energies* listed. These data indicate that beryllium, magnesium, and calcium atoms each have two electrons that are loosely held and are therefore relatively far from the positive nucleus. Once these electrons are removed, the remaining ones are more tightly held and therefore must be closer to the nucleus.

3. Atoms of helium, He ($Z = 2$), neon, Ne ($Z = 10$), and argon, Ar ($Z = 18$), each have *very large first ionization energies.* This means that these atoms hold on to all their electrons tightly.

Electrons in atoms are believed to be arranged around their nuclei in positions known as **energy levels.** Outer electrons that are considerably removed from the other electrons are said to be in *higher energy levels.* Inner electrons that are nearer the nucleus are in *lower energy levels.* As we imagine the building up of an atom from the one with the next lower atomic number, an electron is added to the lowest available energy level. As atoms become more complicated, energy levels become filled. Electrons add in turn to the next higher energy level, then to the next higher one, and so on. Energy levels, proceeding outward from the nucleus, are numbered 1, 2, 3, and so on.

In Table 3.6 in the 1st e^- column, notice that immediately after the large first ionization energies of atoms of helium, He ($Z = 2$), neon, Ne ($Z = 10$), and argon, Ar ($Z = 18$), there are the considerably smaller first ionization energies of atoms of lithium, Li ($Z = 3$), sodium, Na ($Z = 11$), and potassium, K ($Z = 19$). This indicates that Li, Na, and K atoms each have one electron in an outermost energy level. These single electrons must be in higher energy levels than the electrons in the outermost energy levels of He, Ne, and Ar atoms. The numbers of electrons remaining after these single electrons are removed are the same as in He, Ne, and Ar atoms: Li$^+$ and He, 2; Na$^+$ and Ne, 10; and K$^+$ and Ar, 18.

From the total number of electrons in the six atoms and their first ionization energies, we may conclude:

A helium atom has a single energy level, level 1, containing two electrons.

A lithium atom has two electrons in level 1, and one electron in level 2.

A neon atom has two electrons in level 1, and eight electrons in level 2.

A sodium atom has two electrons in level 1, eight electrons in level 2, and one electron in level 3.

An argon atom has two electrons in level 1, eight electrons in level 2, and eight electrons in level 3.

A potassium atom has two electrons in level 1, eight electrons in level 2, eight electrons in level 3, and one electron in level 4.

These conclusions are summarized in Table 3.7 along with the logical extrapolations for Be, Mg, and Ca atoms, which have atomic numbers greater by one than Li, Na, and K atoms. The plotted data in Figure 3.12 help confirm the idea that the twelve electrons of a magnesium atom are arranged in three energy levels.

Table 3.7 Electrons in energy levels for some elements[a]

Atomic number, Z	Symbol	Level 1	Level 2	Level 3	Level 4
2	He	2			
3	Li	2	1		
4	Be	2	2		
10	Ne	2	8		
11	Na	2	8	1	
12	Mg	2	8	2	
18	Ar	2	8	8	
19	K	2	8	8	1
20	Ca	2	8	8	2

[a]The numbers of electrons in the outermost energy levels are shown in color.

3.6.2   Periodic Behavior of the Elements   An examination of the first ionization energies, reading from top to bottom in column 3 of Table 3.6, reveals a pattern of rising values followed by abrupt drops. The positions of these abrupt discontinuities, marked by the horizontal colored lines, parallel precisely the positions of elements that begin new periods in the periodic tables shown in Figure 1.15 and inside the front cover of this book. A graph of first ionization energies versus atomic numbers (Figure 3.13) makes this alternating or periodic behavior even clearer. Each period begins with an element that has a relatively low first ionization energy and ends with one that has a high first ionization energy. The elements can be arranged in seven periods as follows:

period 1	H through He	2 elements
period 2	Li through Ne	8 elements
period 3	Na through Ar	8 elements
period 4	K through Kr	18 elements
period 5	Rb through Xe	18 elements
period 6	Cs through Rn	32 elements
period 7	Fr through ?	?

The seventh period is considered to be incomplete at present. The latest reported discovery (in 1977) was of element number 107, which is as yet unnamed.

Each of the six complete periods ends with an element whose atoms hold on to all their electrons very tightly. Except for helium, each of these atoms has eight electrons in its outermost energy level. Because these six gaseous elements are noted for their lack of chemical reactivity, they are called the **noble gases.** According to present theory, if enough elements are ever found, period 7 will have 32 members and will end with element 118, an element that should be a noble gas with the highest first ionization energy in that period.

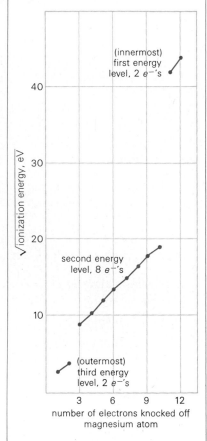

Figure 3.12 Plot of the ionization energies for the twelve electrons of a magnesium atom. The square root of eV is plotted, rather than eV, in order to emphasize the differences between levels. The twelve eV values are tabulated in Table 4.5.

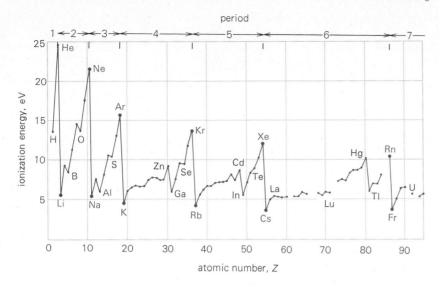

Figure 3•13 First ionization energies plotted against atomic numbers.

Comparison of the plotted regions in Figure 3.13 for periods 2 and 3 reveals certain similarities. In each period the third (B and Al) and sixth (O and S) elements have low energies that appear as interruptions in the otherwise regular energy increase. Comparison of periods 4 and 5 also reveals similarities: the pattern of rising, falling, and rising energies at the end of each of these periods is quite like the pattern in periods 2 and 3. Also, in periods 4 and in 5 there is a series of about ten elements that have nearly the same first ionization energies.

There appear to be correlations between ionization energy and chemical reactivity. The atoms with the lowest first ionization energies of each period, Li, Na, K, Rb, Cs, and Fr, are extremely reactive in chemical behavior, whereas the atoms with the highest first ionization energies of each period, He, Ne, Ar, Kr, Xe, and Rn, are the least reactive of all elements. Note that this correlation is not general. We cannot say that every element with a low first ionization energy will be more active than an element with a higher first ionization energy. The facts are much more complicated than that. For instance, the atoms F, Cl, Br, and I have rather high ionization energies but are very active chemically.

After the discovery of atomic numbers (see Section 3.3.3), scientists realized that the atomic number of an element, rather than its atomic weight, was the key to its placement in the periodic table. Also, the atomic number determines the number, and therefore the arrangement, of the electrons in an atom. A cardinal principle of modern chemical theory is that the properties of an element can be explained in terms of the electron arrangement of the atoms of that element. A statement of this principle is called the **periodic law:** *The chemical and physical properties of the elements are periodic functions of their atomic numbers.*

3•6•3   Periodic Electron Arrangements   Based on the study of the periodic repetition of properties, it is believed that electrons in atoms are arranged in as many as seven main regions or energy levels. Building from one element to the next in a given period, electrons add to the outermost energy level (along with certain inner levels that will be discussed in Section 3.8), until the electron arrangement of a noble-gas atom is attained. When the elements are placed in tabular form, with each period in a different row of the table, elements with the same numbers of outer electrons generally fall in the same vertical columns.

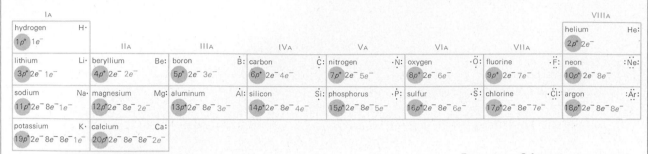

IA		IIA		IIIA		IVA		VA		VIA		VIIA		VIIIA	
hydrogen	H·													helium	He:
$1p^+$ $1e^-$														$2p^+$ $2e^-$	
lithium	Li·	beryllium	Be:	boron	Ḃ:	carbon	Ċ:	nitrogen	·N̈:	oxygen	·Ö:	fluorine	·F̈:	neon	:N̈e:
$3p^+$ $2e^-$ $1e^-$		$4p^+$ $2e^-$ $2e^-$		$5p^+$ $2e^-$ $3e^-$		$6p^+$ $2e^-$ $4e^-$		$7p^+$ $2e^-$ $5e^-$		$8p^+$ $2e^-$ $6e^-$		$9p^+$ $2e^-$ $7e^-$		$10p^+$ $2e^-$ $8e^-$	
sodium	Na·	magnesium	Mg:	aluminum	Äl:	silicon	Si̇:	phosphorus	·P̈:	sulfur	·S̈:	chlorine	·C̈l:	argon	:Är:
$11p^+$ $2e^-$ $8e^-$ $1e^-$		$12p^+$ $2e^-$ $8e^-$ $2e^-$		$13p^+$ $2e^-$ $8e^-$ $3e^-$		$14p^+$ $2e^-$ $8e^-$ $4e^-$		$15p^+$ $2e^-$ $8e^-$ $5e^-$		$16p^+$ $2e^-$ $8e^-$ $6e^-$		$17p^+$ $2e^-$ $8e^-$ $7e^-$		$18p^+$ $2e^-$ $8e^-$ $8e^-$	
potassium	K·	calcium	Ca:												
$19p^+$ $2e^-$ $8e^-$ $8e^-$ $1e^-$		$20p^+$ $2e^-$ $8e^-$ $8e^-$ $2e^-$													

Figure 3.14 Schematic representation of atoms of the first twenty elements, showing the periodic variation of electrons in the outer energy levels. The number of protons, p^+, in each nucleus is also shown. In the abbreviated version, dots are arranged around the symbol. Each dot represents an electron in the outermost energy level.

For the atoms of the first twenty elements, arrangements of the electrons are indicated schematically in Figure 3.14. With potassium, number 19, the fourth main energy level begins to be filled. The arrangement of electrons in energy levels becomes more complex as the atomic number increases above 20. In remembering the arrangements shown in Figure 3.14, it is helpful to note that, as the atomic number increases, (1) the first main energy level can have no more than two electrons, (2) the second main energy level can have no more than eight electrons, and (3) the third main energy level has no more than eight electrons unless an atom has 21 electrons or more.

As illustrated in Figure 3.15, the electron arrangements of isotopes are identical.

3.7   A Modern Periodic Table

A popular form of the periodic table is the *long form,* a copy of which appears inside the front cover. We shall refer to this table continually, so it is advantageous to be thoroughly familiar with it.

The elements in this periodic table are in vertical divisions, called **groups,** and horizontal rows, called **periods.** There are sixteen vertical divisions because there are eight groups, and each group has A and B **families.** There are seven periods, and each of the first six ends with a noble gas.

The key to the information provided for each element is given in Figure 3.16. For each element there are given (1) the symbol, (2) the atomic number in color in the upper left-hand corner, (3) the atomic weight below the symbol, and (4) the electron arrangement of the atom in main energy levels and energy sublevels (see Section 3.8). There are six complete periods; the seventh is incomplete. The first period has only two members; the next five have 8, 8, 18, 18, and 32 members, respectively. The eight groups are numbered with roman numerals I to VIII. An A family of a group always includes an element from period 2 and one from period 3, whereas a B family

^{35}Cl
$17p^+$
$18n$ $2e^-$ $8e^-$ $7e^-$

^{37}Cl
$17p^+$
$20n$ $2e^-$ $8e^-$ $7e^-$

Figure 3.15 Two isotopes of chlorine. Differing numbers of neutrons have no effect on electron arrangements.

of a group has none from these short periods. The following families illustrate these points:

IIA family: Be, Mg, Ca, Sr, Ba, Ra

IIB family: Zn, Cd, Hg

VIIA family: F, Cl, Br, I, At

VIIB family: Mn, Tc, Re

The table is arranged so that similar elements are in the same family. For example, family IB, the copper family, is made up of the metals copper, Cu, silver, Ag, and gold, Au. An element is usually more like a member of its own family than it is like any element in another family.

Usually the elements in the A family of a group are not physically similar to the elements in the B family of the same group. In most areas of the table the A and B families of a group are also chemically dissimilar. For example, the A family of group II is made up of very active metals (Mg, Ca, and so on), whereas the B family is made up of considerably less active metals (Zn, Cd, Hg). On the other hand, the formulas for the compounds of elements in the two families of a given group are often identical with respect to the ratio of atoms that are combined, for example, $MgCl_2$ and $CaCl_2$ versus $ZnCl_2$ and $CdCl_2$.

3•8   Energy Sublevels

To account for the existence of the A and B families and the great difference in the lengths of the periods, it is postulated that within a main energy level there must be **energy sublevels.** Much of the early evidence for electron arrangements came from spectroscopic studies (discussed in Chapter 4). The energy sublevels were given names suggested by the appearance or position of lines in the spectra of excited elements: *sharp, principal, diffuse,* and *fundamental.* Today we speak of these energy sublevels as the *s, p, d,* and *f* sublevels, respectively.

Within a given main energy level, electrons in an *s* sublevel on the average are closer to the nucleus than ones in a *p* sublevel; those in a *p* sublevel are closer than ones in a *d* sublevel; and those in a *d* sublevel are closer than ones in an *f* sublevel. As one progresses outward within a main level, the sublevels can accommodate greater numbers of electrons. The maximum numbers of electrons possible in the *s, p, d,* and *f* sublevels[6] are: *s,* 2; *p,* 6; *d,* 10; and *f,* 14.

The energy sublevels of atoms are divided further into regions of space called **orbitals.** Each orbital can accommodate a maximum of two electrons. An *s* sublevel is made up of one orbital called an *s* orbital; a *p* sublevel contains three orbitals called *p* orbitals; a *d* sublevel contains five orbitals called *d* orbitals; and an *f* sublevel contains seven orbitals called *f* orbitals. Note that the maximum number of electrons within a sublevel is two

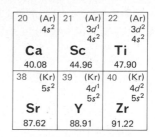

20 (Ar) $4s^2$	21 (Ar) $3d^1$ $4s^2$	22 (Ar) $3d^2$ $4s^2$
Ca 40.08	**Sc** 44.96	**Ti** 47.90
38 (Kr) $5s^2$	39 (Kr) $4d^1$ $5s^2$	40 (Kr) $4d^2$ $5s^2$
Sr 87.62	**Y** 88.91	**Zr** 91.22

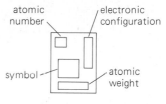

Figure 3•16 Portion of periodic table inside the front cover and key to the information given for each element. (Ar), for example, stands for $1s^2 2s^2 2p^6 3s^2 3p^6$. Each atomic weight is the average for the isotopic mixture found in nature.

---

[6]There are also *g, h, i,* and so on, sublevels, which may be occupied by excited electrons. However, unexcited electrons, with which we are mainly concerned, are not found in any of these sublevels.

times the number of orbitals. The theoretical foundation for the division of sublevels into orbitals will be discussed in Chapter 4.

Table 3•8 Subdivisions of main energy levels

main energy level (n)	1	2		3			4			
number of sublevels (n)	1	2		3			4			
number of orbitals (n^2)	1	4		9			16			
type of sublevel	s	s	p	s	p	d	s	p	d	f
number of orbitals per sublevel	1	1	3	1	3	5	1	3	5	7
maximum number of electrons per sublevel	2	2	6	2	6	10	2	6	10	14
maximum number of electrons per main level ($2n^2$)	2	8		18			32			

Table 3.8 gives the electron distribution for the sublevels of the first four main energy levels. A simple way of remembering the information in this table is to relate as much of it as possible to the particular main energy level, designated by the letter n:

1. For the main energy level closest to the nucleus, the first main energy level, $n = 1$, there is 1 energy sublevel, the $1s$ sublevel.

2. For the next higher main energy level, $n = 2$, there are 2 energy sublevels, the $2s$ and $2p$ sublevels. If the second main level is filled with its maximum of eight electrons, there are two electrons in the $2s$ sublevel and six electrons in the $2p$ sublevel.

3. Continuing outward from the nucleus, the next level is the third main level, $n = 3$; there are 3 energy sublevels, the $3s$, $3p$, and $3d$ sublevels. If the third main level is filled to capacity, it has eighteen electrons: two electrons in the $3s$ sublevel, six in the $3p$ sublevel, and ten in the $3d$ sublevel.

4. For the fourth main level, $n = 4$, there are 4 energy sublevels, the $4s$, $4p$, $4d$, and $4f$ sublevels. The fourth main level can hold a maximum of 32 electrons: two electrons in the $4s$ sublevel, six in the $4p$ sublevel, ten in the $4d$ sublevel, and fourteen in the $4f$ sublevel.

Note that the number of electrons required to fill a main energy level is $2n^2$. For example, if $n = 3$, the maximum number of electrons is $2(3)^2$, or 18.

3•8•1   **Order of Filling Energy Sublevels**   In Section 3.6.1, we described the electron arrangements in atoms of elements, based on measurements of ionization energies. From these measurements and others, the relative energies of electrons in various main levels and sublevels of atoms are known. Within a given main level, the order of the energies of sublevels, starting with the lowest, is s, p, d, f, and so on. Based on these relative energies of sublevels, one can determine an imaginary order of filling the sublevels as the atomic number increases.

The following convention is used to summarize the number of electrons in a given sublevel of an atom:

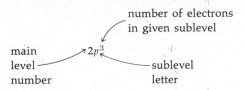

The **electronic configuration** of an atom is a listing of its sublevels that have electrons and the number of electrons in each. The electronic configurations of the first eighteen elements are given in Table 3.9.

Table 3•9 Electronic configurations

H	$1s^1$		
He	$1s^2$		
Li	$1s^2 2s^1$	Na	$1s^2 2s^2 2p^6 3s^1$
Be	$1s^2 2s^2$	Mg	$1s^2 2s^2 2p^6 3s^2$
B	$1s^2 2s^2 2p^1$	Al	$1s^2 2s^2 2p^6 3s^2 3p^1$
C	$1s^2 2s^2 2p^2$	Si	$1s^2 2s^2 2p^6 3s^2 3p^2$
N	$1s^2 2s^2 2p^3$	P	$1s^2 2s^2 2p^6 3s^2 3p^3$
O	$1s^2 2s^2 2p^4$	S	$1s^2 2s^2 2p^6 3s^2 3p^4$
F	$1s^2 2s^2 2p^5$	Cl	$1s^2 2s^2 2p^6 3s^2 3p^5$
Ne	$1s^2 2s^2 2p^6$	Ar	$1s^2 2s^2 2p^6 3s^2 3p^6$

With element 18, argon, the $3p$ sublevel is filled with six electrons. If we were to predict the electronic configurations for the two elements immediately after argon, that is, for potassium and calcium, we might guess that the next electrons occupy the $3d$ sublevel. However, based on the physical and chemical similarities of Li, Na, and K (see Table 1.5), it is established that potassium is in group IA (see Figure 1.15). This means that the potassium atom has only one electron in its outermost main energy level, and that electron must be in level 4 to begin a new period. Consequently the outermost electron of potassium is assigned to the $4s$ sublevel rather than to the $3d$ sublevel. The two outermost electrons of a calcium atom also are assigned to the $4s$ sublevel. The electronic configurations of potassium and calcium atoms are

$$K \qquad 1s^2 2s^2 2p^6 3s^2 3p^6 4s^1 \quad \text{or} \quad (Ar)4s^1$$
$$Ca \qquad 1s^2 2s^2 2p^6 3s^2 3p^6 4s^2 \quad \text{or} \quad (Ar)4s^2$$

In the abbreviated electronic configurations to the right, the term (Ar) stands for *argon core,* that is, for the electron arrangement of argon. It indicates that all sublevels through $3p^6$ are filled. In the periodic table inside the front cover, note that the electron arrangements for elements in period 2 include a helium core (He), in period 3 they include a neon core (Ne), and so on.

In period 4, after calcium, Ca ($Z = 20$), there follow the ten B family elements, scandium, Sc ($Z = 21$), through zinc, Zn ($Z = 30$). Recall that these elements have nearly the same first ionization energies (see Figure 3.13). After the $4s$ sublevel is filled in Ca atoms, electrons enter the $3d$ sublevel in these ten B family elements. In these elements, the number of electrons in the outer $4s$ sublevel remains fairly constant as $3d$ electrons are added one by one. When one of these ten elements is ionized, it is a $4s$ electron that is lost first.[7] It is for this reason that the first ionization energies of the ten elements are all nearly the same.

In period 5 we have another example of an outer energy sublevel taking on electrons before an inner sublevel begins to be filled. In rubidium, Rb ($Z = 37$), and strontium, Sr ($Z = 38$), electrons are added to the $5s$ sublevel before the $4d$ and $4f$ sublevels take on any electrons. After strontium there follow the ten B family elements yttrium ($Z = 39$) through cadmium ($Z = 48$) associated with the filling of the $4d$ sublevel.

Based on the ionization energies of the A and B family elements and their positions in the periodic table, we can estimate the order in which sublevels are filled with electrons. Although the relative positions of energy sublevels may change in different regions of the periodic table, the sublevels are filled roughly in order of increasing energy as follows: $1s$, $2s$, $2p$, $3s$, $3p$, $4s$, $3d$, $4p$, $5s$, $4d$, $5p$, $6s$, $4f$, $5d$, $6p$, $7s$, $5f$, $6d$, $7p$. This order is shown schematically in Figure 3.17. Figure 3.18 is useful in helping one remember this order. The filling of energy sublevels in the order of increasing energy is called the **aufbau** ("building up") **principle.** Although there are exceptions to the predicted order, this simple guideline applies in most cases.

After using the aufbau principle to determine the numbers of electrons in the various sublevels of an atom, we arrange the outer sublevels in order of their energies. Consider element 31, gallium. Using Figure 3.18 to determine the occupancy of sublevels, we find the following electronic configuration for gallium:

$$1s^2 2s^2 2p^6 3s^2 3p^6 4s^2 3d^{10} 4p^1$$

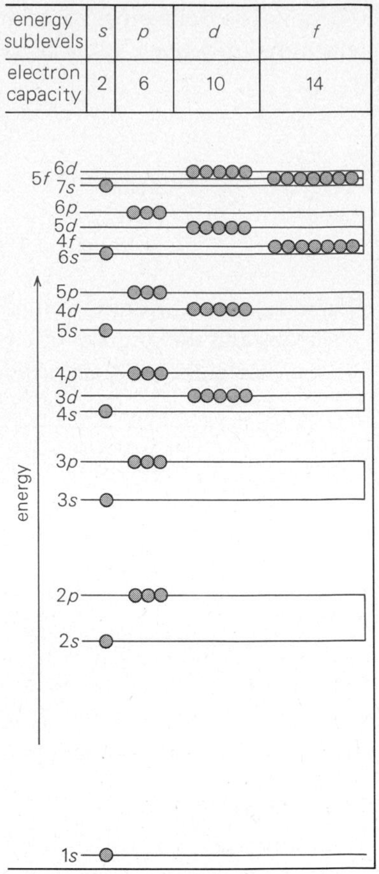

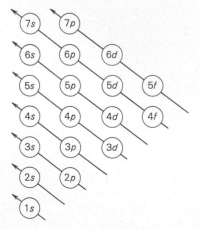

energy sublevels	s	p	d	f
electron capacity	2	6	10	14

Figure 3.17 Approximate relative energy ranking for filling sublevels of atoms with electrons. Each colored circle represents an orbital that can be occupied by either one or two electrons.

Figure 3.18 The approximate order in which sublevels are filled with increasing numbers of electrons. Follow through each slanting arrow in turn, starting with the lowest and continuing with the next higher. For example, after $3s$ is filled, there follow $3p$, $4s$, $3d$, $4p$, and so on. *Source:* Adapted from Therald Moeller, *Inorganic Chemistry,* Wiley, New York, 1952.

---

[7]The fact that the order of loss of electrons during ionization is not simply the reverse of the order of buildup will be discussed in Section 5.2.2.

Abbreviating the structure through $3p^6$ as the argon core, (Ar), we rearrange the outer sublevels so that the sublevels in main level 4 are outermost:

$$(\text{Ar})3d^{10}4s^24p^1$$

This is the electronic configuration for gallium that is given in the periodic table inside the front cover. It shows that the highest energy electrons are in the fourth main level, one in a p sublevel and two in an s sublevel.

For a more complicated case, consider element 73, tantalum. Following Figure 3.18, we write for tantalum:

$$1s^22s^22p^63s^23p^64s^23d^{10}4p^65s^24d^{10}5p^66s^24f^{14}5d^3$$

Abbreviating the structure through $5p^6$ as the xenon core, (Xe), we rearrange the outer three sublevels and write

$$(\text{Xe})4f^{14}5d^36s^2$$

This is the electronic configuration for tantalum shown inside the front cover. When a tantalum atom is ionized, the first electrons lost come from the $6s$ sublevel, then the $5d$, and so on. Summing up, the number of electrons per main energy level for tantalum is 2, 8, 18, 32, 11, 2.

The electronic configuration for an atom as determined experimentally does not always agree with that predicted by the use of the aufbau principle as shown in Figure 3.18. Chromium, Cr, for example, has the experimentally determined structure $(\text{Ar})3d^54s^1$. According to the regular prediction, however, its expected structure would be $(\text{Ar})3d^44s^2$. It is sufficient at this time to learn how to make regular predictions and to recognize that exceptions to the rule do exist.

• Example 3•3 •

Based on the order of increasing energies shown in Figures 3.18 and 3.19, write the complete and abbreviated electronic configurations for atoms of tin, Sn ($Z = 50$), and uranium, U ($Z = 92$). Check to see if the configurations written agree with those given inside the front cover of the text.

• Solution • For tin, the predicted electronic configuration is

$$1s^22s^22p^63s^23p^64s^23d^{10}4p^65s^24d^{10}5p^2 \quad \text{or} \quad (\text{Kr})4d^{10}5s^25p^2$$

For uranium, the predicted electronic configuration is

$$1s^22s^22p^63s^23p^64s^23d^{10}4p^65s^24d^{10}5p^66s^24f^{14}5d^{10}6p^67s^25f^4 \quad \text{or} \quad (\text{Rn})5f^47s^2$$

The predicted electronic configuration for tin agrees with that given on the inside front cover. However, that predicted for uranium does not agree. Its experimentally determined electronic configuration, using the abbreviated description, is $(\text{Rn})5f^36d^17s^2$.

See also Exercises 42–45.

3.8.2   Summary of Energy Sublevel Filling and the Periodic Table   The order in which electrons are added in building up atomic structures is clearly related to the periodic table, as summarized in Figure 3.19. There are several features to note in studying this figure.

1. With the first element of each period, a member of group IA, a new main level begins to fill with the addition of one electron to an *s* sublevel. The filling of an *s* sublevel with two electrons is completed in a member of group IIA.

2. With the third element of periods 2 and 3, a member of group IIIA, a *p* sublevel begins to fill with the addition of one electron after the *s* sublevel has filled. Although a IIIA element is not the third element in the other periods, a IIIA element is always associated with two electrons in the outer *s* sublevel and one in the outer *p* sublevel. Elements in groups IVA, VA, VIA, and VIIA involve the further successive addition of electrons to an outer *p* sublevel until that *p* sublevel is filled with six electrons in a member of group VIIIA.

Figure 3.19 The order in which sublevels are filled is related to the organization of the periodic table.

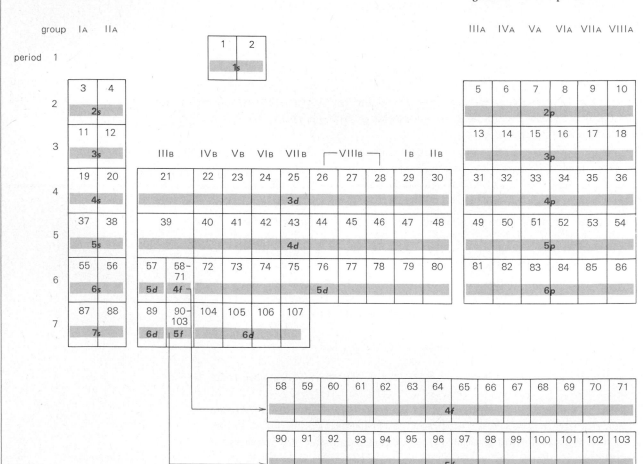

3.   Each period contains the number of elements corresponding to the filling of certain types of sublevels:

Period 1 has only an *s* sublevel and contains just 2 elements.
Periods 2 and 3 have *s* and *p* sublevels to fill and contain 8 elements each.

Periods 4 and 5 have *s*, *p*, and *d* sublevels to fill and contain 18 elements each.

Period 6 has *s*, *p*, *d*, and *f* sublevels to fill and contains 32 elements.

Period 7 contains the remaining known elements.

Presumably, in the event that enough elements are ever found or synthesized, period 7 will contain 32 elements also.

4.   Most of the B family elements in periods 4 through 7 involve the filling of an inner sublevel after an outer *s* sublevel has taken on electrons. In the 28 elements at the bottom of the periodic chart, electrons are being added to an inner *f* sublevel. For the other B family elements, an inner *d* sublevel is being filled with electrons.

5.   Each period except the first ends with the filling of a *p* sublevel in an atom of a noble gas. The electronic configuration that characterizes helium in period 1 is $1s^2$. Each of the noble gases ending periods 2 through 6 has an outer level electronic configuration of s^2p^6.

—— • Example 3.4 • ——

Using Figure 3.19 as a guide, answer each of the following questions about an atom of arsenic, As (Z = 33).
(a) How many electrons are there in the third main energy level?
(b) How many *p* sublevel electrons are there?
(c) What is its outer level electronic configuration?

—— • Solution • ——

(a) The position of arsenic in the periodic table shows that an arsenic atom has a partly filled $4p$ sublevel. Consequently, the inner $3s$, $3p$, and $3d$ sublevels are filled to capacity, with $2 + 6 + 10 = 18$ electrons in the third main level.
(b) The $2p$ and $3p$ sublevels are filled with electrons. There are three electrons in the $4p$ sublevel. The total number of *p* sublevel electrons is $6 + 6 + 3 = 15$ electrons.
(c) The 4 in front of the *p* indicates that main energy level 4 is the outer level being filled. Proceeding from left to right, we see that there are two electrons in the $4s$ sublevel and three in the $4p$ sublevel. The outer level electronic configuration is $4s^24p^3$. (This same answer could be obtained by noting that arsenic is in group VA of period 4. For A groups, the roman numeral is the number of electrons in the outermost main level (5 in this case), and the period number (4) is the number of the outermost main level.)

See also Exercises 46 and 47.

3.9 Usefulness of the Periodic Table

In subsequent chapters as we study the chemical and physical behavior of many of the elements, we shall find that the periodic table is a great help. First, the periodic table is an aid to remembering and understanding chemical data. A glance at the symbols in group IA helps us remember that there are a number of elements that in some ways resemble the well-known metal sodium, Na. These elements are lithium, Li, potassium, K, rubidium, Rb, cesium, Cs, and probably francium, Fr, although the last is so rare that it has not been studied a great deal.

Second, the table is a guide to chemical prediction and theory, because it enables us to determine which elements should be similar to one another. Not only do similar elements act alike, but their compounds may also be alike. For example, NaCl, sodium chloride or common table salt, has some properties that are similar to those of both KCl and RbCl.

Once confidence is gained in using the table, not only can we reason from a common compound, such as sulfuric acid, whose formula is H_2SO_4, and write the formula of a rarer one, H_2TeO_4, but we may successfully predict some of the ways in which the latter will differ from H_2SO_4.

The periodic table was marvelously useful to all those who studied atomic structure early in this century—Thomson, Aston, Rutherford, Moseley, Bohr, and hundreds of others. When Moseley was doing his X-ray studies, physicists were searching for some property of atoms that would increase regularly in the order in which the elements appeared in periodic tables. When studying ionization energies, scientists were searching for a feature of atomic structure that varied periodically as suggested by the tables. One way of beginning a scientific investigation is to design experiments to test a clearly stated prediction. The periodic table has its greatest usefulness in helping us clearly state our chemical expectations.

Chapter Review

Summary

The discovery of **subatomic particles** revolutionized chemistry and physics. Many studies were made of the behavior of charged particles in vacuum tubes, which contain two **electrodes:** an **anode** (positive charge) and a **cathode** (negative charge). The electric discharge in a vacuum tube, called a **cathode ray,** consists of negatively charged, light **electrons,** which are fundamental particles of all matter.

The **radioactivity** of certain elements means that atoms are not indestructible after all. The three kinds of radioactive emissions are called **alpha** (α), **beta** (β), and **gamma** (γ). Alpha particles were used in Rutherford's gold foil experiment, which demonstrated the existence of the atomic **nucleus.** The discovery of the emission of characteristic

X rays when different elements are bombarded by electrons led to a revised interpretation of the periodic table of the elements, in terms of **atomic numbers,** Z, instead of atomic weights. The atomic number of an element is the number of positively charged **protons** in the nucleus of its atom. In an electrically neutral atom, the number of protons equals the number of electrons. If an atom (or molecule) loses or gains one or more electrons relative to the total number of its protons, it becomes a positively or negatively charged **ion.** Positive ions analyzed in the **mass spectrograph** revealed the existence of **isotopes** of elements. The different atomic weights of the isotopes of a given element are due to the different numbers of uncharged **neutrons** in the atomic nuclei. The sum of the protons and neutrons in a given nucleus is the **mass number,** A, for that isotope of the element. These discoveries led to more precise definitions of the atom, the atomic mass unit, and the mole than had hitherto been possible. They also necessitated a refinement in the law of definite proportions.

The **ionization energies** of an atom are a measure of the **electrostatic force** of attraction between the positive nucleus and the various electrons. The **first ionization energy** is associated with the most loosely bound electron, the second ionization energy with the next most loosely bound electron, and so on. These energies are expressed in **electron volts,** eV. Certain periodic trends in the ionization energies of the elements indicate that the electrons in an atom are arranged in definite **energy levels,** which are filled systematically in increasing order of energy as the atomic number increases. The filling of energy levels correlates extremely well with the periodic properties of the elements, as exemplified by groups of elements with very similar properties, such as the **noble gases.**

These facts are embodied in the **periodic law,** which states that the intrinsic properties of the elements are periodic functions of their atomic numbers. The properties of the elements are related to the structures of their atoms, so electron arrangements are also periodic functions of atomic numbers. A modern form of the periodic table shows seven horizontal **periods** of elements, corresponding to the main energy levels of electrons, and eight vertical **groups** of elements, each of which is divided into two **families.** The existence of these families is accounted for by the postulate of **energy sublevels** within the main energy levels. The four sublevels commonly found in atoms are designated s, p, d, and f. The order in which electrons fill the main energy levels and the sublevels can be predicted from the **aufbau principle.** The listing of the number of electrons in each sublevel of an atom is called the **electron configuration** of that atom. It is a reliable indicator of many of the properties of the element in question because correlations can so readily be made with other elements in the periodic table.

Exercises

Nature of Charged Bodies

1. Discuss one type of experimental evidence that supports the belief that electrons are found in all kinds of matter.
2. When a neutral body gains electrons, what is its charge? Why?
3. In a television tube, cathode rays are focused so that when they collide with the fluorescent screen, a picture results. If the tube develops a small crack, will it continue to function? Explain.
4. The ratio of charge to mass, e/m, was determined for the electron before the value for m was determined. Why? Using the values of e/m and e from the text, calculate the mass of the proton in grams and in atomic mass units. Compare this mass with the mass of a hydrogen atom.

5. In Figure 3.3 the alpha particles are shown striking the zinc sulfide screen at a region above the area where the gamma rays strike. Could they be made to strike below the gamma ray region? Explain.
*6. Calculate in SI units the kinetic energy of an alpha particle (mass 4.00 amu) that has a velocity of 20,000 km/s. What is the velocity of the alpha particle in miles per hour?
*7. In 1977, a Stanford University research team used an adaptation of Millikan's oil drop experiment to try to find a smaller charge than that on one electron. Small balls of niobium, Nb, were used instead of oil drops. If the mass of an Nb atom is 92.9060 amu, and the e/m ratio is 3.4375×10^2 C/g, what is the value of the electric charge? How does this value compare with that given for the electron in Section 3.2? Assuming the result can be verified, what do you think is the importance of this research?

Evidence for Subatomic Particles

8. Two X-ray tubes are identical except that one has an anode of element D and the other of element E. Upon connecting each to the proper electric circuit, it is found that the E anode emits X rays with longer wavelengths than those from the D anode. Which element has the larger atomic number? Explain.
9. How did Moseley explain the fact that cobalt (at wt 58.93 amu) emitted a longer wavelength X ray than nickel (at wt 58.70 amu)?
10. Relative to Figure 3.4, what is the purpose of the small zinc sulfide screen that is placed in front of and at an angle to the gold foil? How was the information obtained interpreted in terms of atomic theory?
11. Why was gold chosen as the metal to be bombarded in the Rutherford experiment (see Figure 3.4) instead of an inexpensive metal?
12. When an atom loses an electron, what is the charge of the resulting particle? What are such particles called?
*13. The ratios of charge to mass for the electron and proton are 1.7588×10^8 C/g and 9.5791×10^4 C/g, respectively. Does this mean that the electron is more highly charged than the proton? Explain.

Mass Spectrograph; Isotopes

14. In determining the masses of atoms with a mass spectrograph, the atoms first have to be converted into ions. Why? How?
15. The carbon-12 isotope has a mass of 12 amu. How many significant figures would you attribute to this mass when using it in calculations? Explain.
16. (a) What is wrong with calculating the mass of a single chlorine atom as follows?

$$\frac{35.453 \text{ g/mol}}{6.022 \times 10^{23} \text{ atoms/mol}} = 5.887 \times 10^{-23} \text{ g/atom}$$

(b) The average weight of oxygen atoms is 15.999 amu. Calculate the average weight in grams and in pounds.

17. (a) A hydrogen molecule is usually represented by the formula H_2. In some instances it is represented by formulas such as $^1H^2H$, $^2H_2$, or $^1H_2$. What do these convey about a hydrogen molecule that is not conveyed by the formula H_2?

(b) Based on the data in Table 3.3, what is the largest possible weight of ammonia, NH_3, in atomic mass units and in grams of a molecule?

18. Write symbols for the following:
(a) an isotope of hydrogen (called tritium) with a mass number of 3
(b) an isotope of lead with a mass number of 210
(c) an isotope of lead with a mass number of 208
(d) an isotope of sulfur with a mass number of 34

19. For each of the isotopes in Exercise 18, list the numbers of protons, electrons, and neutrons in one atom.

20. The natural isotopes of boron have masses of 10.0129 and 11.0093 amu, with relative percent abundances of 19.7 and 80.3, respectively. Based on these data, calculate the atomic weight of boron.

*21. The two naturally occurring isotopes of copper, ^{63}Cu and ^{65}Cu, have masses of 62.9296 and 64.9278 amu, respectively. The atomic weight of copper is 63.546 amu. Calculate the percentage of each isotope in a piece of copper.

22. Show with symbols two likely isotopes of vanadium, V. State in words how the isotopes of an element differ from one another.

*23. Bromine, atomic weight 79.9 amu and atomic number 35, is very nearly a 50–50 mixture of two isotopes. Make a reasonable guess about the mass numbers of the two isotopes and describe each isotope in terms of the numbers of the three subatomic particles in one atom of each.

24. Since Dalton's time, there have been two discoveries that show that the law of definite composition may not apply precisely to two samples of the same compound. Take zinc sulfide, ZnS, as an example and discuss the two discoveries in terms of it.

*25. In a mass spectrograph of the type shown in Figure 3.8, positive chlorine ions of mass 34.97 amu are collected when the accelerating potential is 824.6 V. If the strength of the magnetic field is held constant, at what accelerating potential should the positive chlorine ions of mass 36.97 amu be collected?

*26. The value of e/m for a monatomic ion formed in the mass spectrograph is 6.8903×10^3 C/g. Assuming the ion has a charge of $1+$, what is the mass of the ion? From what kind of atom was this ion formed?

Weights and Sizes of Atoms

27. Calculate the weight in grams of the Avogadro number of atomic mass units (amu).

28. (a) The SI unit that is closest in length to the angstrom unit is the nanometer, nm. What is the range in size of atoms from hydrogen to cesium in nanometers?
(b) Calculate the diameters of the last four items listed in Table 3.5 in angstroms and in nanometers.

29. A helium atom has an isotopic weight of 4.0026 amu. What is the mass in grams of an alpha particle produced from this helium atom? Is the mass of the alpha particle different from the mass of the helium atom? Explain.

*30. If the gold foil with which Rutherford worked was 1/10,000 in. thick, what is the probable minimum number of gold atoms that an alpha particle passed through before hitting the fluorescent screen? (The radius of a gold atom is 1.5 Å.)

*31. Calculate the approximate density of the nucleus of a sodium atom in grams per milliliter. (The radius of an entire sodium atom is 1.57 Å or 0.157 nm.)

Periodicity and Atomic Structure

32. In Table 1.5, similar elements are found at every seventh entry for the periods beginning with Li and Na. In a modern periodic table, these same elements are found at every eighth entry. Why?

33. Diagram two likely isotopes of potassium, K, in the style of Figure 3.15.

34. How many elements are there in period 5? In period 6? How many electrons are there in the outermost energy level of the last member of each of these periods?

35. (a) Suppose that an element has values for its first through sixth ionization energies as follows: 6.3, 12.2, 20.5, 61.8, 77.0, 93.0 eV. What do these data indicate about the atom's electronic structure? In what family of the periodic table would the element probably belong?
(b) Repeat the questions in part (a) for an element with ionization energies of 4.2, 27.3, 40.0, 52.6, 71.0, 84.4 eV.

36. With the completion of a d sublevel, there is a peak in the ionization energy plot. Name three of the elements at these peaks in Figure 3.13.

*37. Sodium, Na, and potassium, K, are both in the A family of group I and have similar electronic configurations. Which has the lower ionization energy? Account for this difference in terms of atomic structure.

*38. The second ionization energy is always larger than the first. Why?

39. What is the electronic configuration of the outermost main level of the first element of each period of the periodic classification? Of the last element in each period?

40. Tell how many electrons:
    (a) can occupy the third main energy level of an atom
    (b) there can be in the d sublevel of the fourth main energy level
    (c) there can be in the s sublevel of the second main energy level

*41. Consider the following elements: beryllium, Be, oxygen, O, fluorine, F, sulfur, S, potassium, K, iron, Fe, and krypton, Kr. Based upon periodic trends, tell which one has:
    (a) the lowest ionization energy
    (b) eight electrons in the fourth main energy level
    (c) the highest third ionization energy
    (d) only four electrons in the p sublevel of the third main energy level
    (e) six electrons in the d sublevel of the third main energy level

Electronic Configurations

42. Write the complete electronic configuration for an atom of each of the following elements, based on the predicted order of filling sublevels.
    (a) phosphorus, P   (b) manganese, Mn
    (c) bromine, Br   (d) promethium, Pm

43. Repeat Exercise 42, but now write only the abbreviated electronic configuration that you would expect to find in the periodic table inside the front cover.

44. An atom of palladium has eighteen electrons in its fourth main energy level. With this information, write the abbreviated electronic configuration.

45. Using Figure 3.19 as a guide, answer each of the following questions about an atom of rhenium, Re ($Z = 75$).
    (a) How many electrons are there in the fourth main energy level?
    (b) How many d electrons are there?
    (c) What is its outer level electronic configuration?

46. Based on Figure 3.19, derive the number of sublevels and the number of electrons in the highest main level for the hypothetical noble gas with atomic number 118.

47. Using the noble-gas core convention, assign an abbreviated electronic configuration for the unknown element with an atomic number of 120. To what group in the periodic table would you assign such an element?

Use of the Periodic Table

48. Ammonia, NH_3, is a compound in which one atom of a Va element is combined with three hydrogen atoms. Write the molecular formulas for the four other binary compounds in which one atom of a Va element is combined with hydrogen atoms.

49. The boiling points of HCl, HBr, and HI are $-114\,°C$, $-87\,°C$, and $-51\,°C$, respectively. Which of the compounds H_2S, H_2Se, or H_2Te would you predict to have the highest boiling point? The lowest boiling point?

50. In the liquid state, NaF conducts an appreciable electric current, but HF does not. Which of the compounds NaCl, KBr, HCl, LiCl, HBr, CsF, and HI would you predict will conduct an appreciable electric current in the liquid state?

*51. (a) How many elements are there in group Va? In group VIIa (excluding astatine, At)?
    (b) The formula for nitrogen trichloride is NCl_3. How many other formulas can you write for compounds made up of one atom of a Va element and three atoms of a VIIa element?

52. There are eighteen elements in period 4, but the elements are divided into sixteen different groups. Account for the two "extra" elements.

53. One might make the statement that the properties of the elements are periodic functions of their atomic weights. Give as many examples as you can that show this statement is false.

Atomic Spectroscopy; Modern Electronic Theory

4

Atomic Spectroscopy

4•1 **Electromagnetic Radiation**

4•2 **Atomic Spectra**

4•2•1 Emission Spectra

4•2•2 Absorption Spectra

4•2•3 The Quantum Theory of Radiation

4•2•4 The Photoelectric Effect

4•2•5 Application of Quantum Theory to Atomic Spectra

Modern Electronic Theory

4•3 **The Wave Nature of Matter**

4•4 **The Heisenberg Uncertainty Principle**

4•5 **The Quantum Mechanical Description of Electrons in Atoms**

4•5•1 Quantum Numbers for Electrons

4•5•2 The Pauli Exclusion Principle

4•5•3 Order of Filling Orbitals

Special Topic: Computers— Tools of the Chemical Theorist

4•6 **Photoelectron Spectra of Atoms**

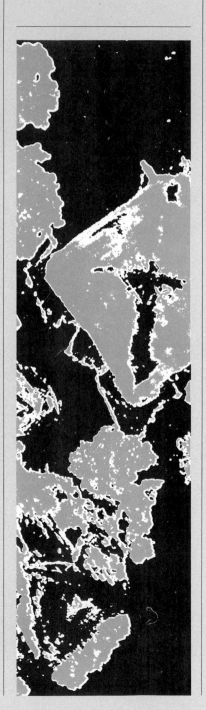

In Chapter 3 we defined an atom of an element as a neutral particle that has a massive, but relatively tiny, positive nucleus and one or more electrons far outside the nucleus. The arrangement of the electrons in an atom of any element was shown to be related to the position of that element in the periodic chart.

One of the problems that puzzled early twentieth-century scientists was what keeps the negative electrons at relatively great distances from the positive nucleus. Why does not the attraction of unlike charges cause the electrons to join the nucleus? The answer Rutherford and others proposed was that an atom is analogous to a miniature solar system, with the electrons revolving around the nucleus in well-defined orbits in much the same way as the planets move around the sun. The tendency for a speeding electron to fly off in a straight line tangent to its orbit must be just balanced by the electrostatic attractive force exerted on the electron by the positive nucleus. The planetary model of atomic structure had serious limitations, however. Certain fundamental laws that applied to large objects such as planets and baseballs did not seem to apply to atoms and their electrons. Because the model could not account for certain experimental observations, such as the radiation emitted by excited atoms of elements, it eventually was abandoned.

In this chapter we will present some more advanced ideas of atomic structure. We will follow the transition of ideas from the early work of Planck, Einstein, and Bohr to the later, more refined theories of de Broglie, Heisenberg, and Schrödinger. To lay the groundwork for an understanding of the contributions of these scientists, we begin by discussing the topics of atomic spectroscopy and radiant energy.

Atomic Spectroscopy

4•1 Electromagnetic Radiation

Radiant energy may be thought of as electric and magnetic fields that oscillate perpendicularly to the direction of travel. Visible light is one of several types of radiant energy. All types travel at the speed of light, c, 3.00×10^8 m/s (186,000 mi/s)[1] but differ in frequency and wavelength. **Frequency,** ν, is defined as the number of cycles of a wave that pass a given point in a unit of time. Usually the second is used as the interval of time, and the unit is expressed as cycles per second, or s^{-1} with cycles understood. The SI unit for frequency is the **hertz,** Hz; $1 \text{ Hz} = 1 \text{ s}^{-1}$. **Wavelength,** λ, is the distance between any two identical points on adjacent repeating cycles of the wave pattern, and it is expressed in units of length, for example, meters (m), centimeters (cm), or nanometers ($1 \text{ nm} = 10^{-9}$ m). Measurements are often made in terms of the wavelength rather than the frequency. The relationship

---

[1]More precise values of c are 2.997925×10^8 m/s and 186,282 mi/s.

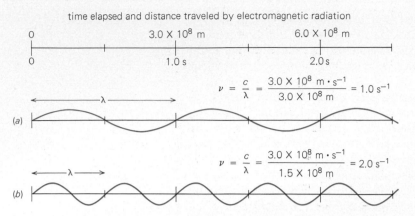

time elapsed and distance traveled by electromagnetic radiation

$$\nu = \frac{c}{\lambda} = \frac{3.0 \times 10^8 \text{ m} \cdot \text{s}^{-1}}{3.0 \times 10^8 \text{ m}} = 1.0 \text{ s}^{-1}$$

$$\nu = \frac{c}{\lambda} = \frac{3.0 \times 10^8 \text{ m} \cdot \text{s}^{-1}}{1.5 \times 10^8 \text{ m}} = 2.0 \text{ s}^{-1}$$

Figure 4•1 The relationship between wavelength and frequency of radiant energy. In (a) the wavelength λ is 3.0 × 10⁸ m, so the frequency ν is calculated to be 1.0 Hz; in (b) λ is half that in (a), or 1.5 × 10⁸ m, so the frequency is twice as great, and ν is calculated to be 2.0 Hz.

between the frequency and wavelength of radiant energy is illustrated by Figures 4.1 and 4.2. Their inverse proportionality is stated mathematically as

$$\nu \propto \frac{1}{\lambda} \quad \text{or} \quad \nu = \frac{c}{\lambda}$$

• Example 4•1 •

Calculate the wavelength in nanometers of electromagnetic radiation having a frequency of 6.26×10^{14} Hz, or 6.26×10^{14} s^{-1}.

• Solution •

$$\lambda = \frac{c}{\nu}$$

$$= \left(\frac{3.00 \times 10^8 \text{ m} \cdot \text{s}^{-1}}{6.26 \times 10^{14} \text{ s}^{-1}}\right)\left(\frac{1 \text{ nm}}{10^{-9} \text{ m}}\right)$$

$$= 4.79 \times 10^2, \text{ or } 479 \text{ nm}$$

See also Exercises 1–5 at the end of the chapter.

4•2 Atomic Spectra

A **spectrum** is the result produced when a beam of radiant energy is dispersed into its component wavelengths. If the dispersed radiation is from excited atoms, it is called an **atomic spectrum.** An optical instrument used for forming spectra is called a **spectroscope.** A diagram of a simple *prism spectroscope* is shown in Figure 4.3. The heart of the instrument is a glass prism that bends the path of light passing through it. The paths of different colors (wavelengths) are bent to different degrees and can be observed visually or recorded on film. Another device used to obtain spectra is the *grating spectroscope.* A grating is made by carefully scratching close, regular parallel lines on a polished surface. A beam of light is diffracted to produce a spectrum, as shown in Figure 4.4.

The field of study devoted to obtaining and analyzing spectra is called **spectroscopy.** Prism and grating spectroscopes are used to obtain spectra not only in the visible region but also over wide ranges on both sides of the visible region. Photographic film or electronic detectors are used to make permanent records of the spectral patterns so that they can be analyzed in detail.

Table 4•1 Some flame colors of elements

Element	Flame color
lithium	red
sodium	yellow
potassium	violet
rubidium	red
cesium	blue
calcium	orange-red
strontium	brick red
barium	green

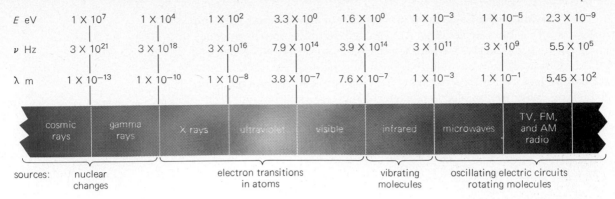

E eV	1×10^{7}	1×10^{4}	1×10^{2}	3.3×10^{0}	1.6×10^{0}	1×10^{-3}	1×10^{-5}	2.3×10^{-9}
ν Hz	3×10^{21}	3×10^{18}	3×10^{16}	7.9×10^{14}	3.9×10^{14}	3×10^{11}	3×10^{9}	5.5×10^{5}
λ m	1×10^{-13}	1×10^{-10}	1×10^{-8}	3.8×10^{-7}	7.6×10^{-7}	1×10^{-3}	1×10^{-1}	5.45×10^{2}

cosmic rays   gamma rays   X rays   ultraviolet   visible   infrared   microwaves   TV, FM, and AM radio

sources: nuclear changes   electron transitions in atoms   vibrating molecules   oscillating electric circuits rotating molecules

Figure 4•2 Regions of the electromagnetic spectrum. The approximate boundaries are stated at the top in frequencies with the corresponding wavelengths beneath. The frequencies and wavelengths are not shown to scale. The visible region for the average human eye is bracketed between 380 and 760 nm. Contrast the narrowness of this region with the infrared, which is bracketed between 760 and 1 million nm. The energies in electron volts corresponding to the given frequencies and wavelengths are also shown. The mathematical relationships of energy, frequency, and wavelength are presented in Section 4.2.3. *Source:* Adapted by permission from H. O. Hooper and P. Gwynne, *Physics and the Physical Perspective,* Harper & Row, New York, 1977.

After the spectroscope was invented in 1859, it was possible to study carefully the radiation emitted by excited atoms. It is through the study of atomic spectra that physicists and chemists have found how electron arrangements are related to the chemical properties of the elements.

4•2•1   Emission Spectra When an element absorbs sufficient energy, for example from a flame or an electric arc, it emits radiant energy. Although any element can be heated to incandescence, some elements have only to be heated in a Bunsen flame to vaporize them and make them emit a characteristic colored light. A list of some such elements and their characteristic flame colors is presented in Table 4.1. A convenient method of testing a substance for the presence of these elements is to dissolve a bit of the material in water and then dip a loop of platinum wire into the solution. If a droplet of the solution is carefully evaporated on the wire and the wire is then heated in the hot flame of a laboratory burner, the flame will have a color characteristic of the elements present.

An element can often be identified by visual observation of the flame and reference to a list such as the one in Table 4.1. For truly positive identification visual observation is insufficient; for example, many persons would have difficulty in distinguishing between a lithium and a strontium flame.

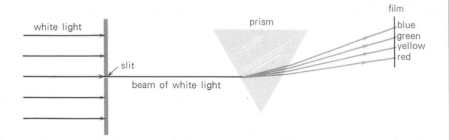

white light

slit

beam of white light

prism

film
blue
green
yellow
red

Figure 4•3 The separation of different wavelengths by a glass prism. See also Figure 4.5 on Plate 2.

Precise analysis of the color of a flame or other characteristic radiation emitted by an element can be made with a spectroscope. As the emitted radiation passes through a prism in a spectroscope, it is separated into component wavelengths to form an image called an **emission spectrum** (Figures 4.5 and 4.6 on Plate 2). There are two types of emission spectra: continuous and discontinuous.

A **continuous emission spectrum** consists of an uninterrupted band of colors of all visible wavelengths. At very high temperatures most solids become "white hot" and give a continuous emission spectrum. No absence of color (dark spaces) can be detected when the light is analyzed with a spectroscope (see Figure 4.6, spectrum 1). Elements and compounds with high melting points can be used as convenient sources of continuous spectra. Tungsten, which is commonly used as a filament material in incandescent lights, is an example of such an element.

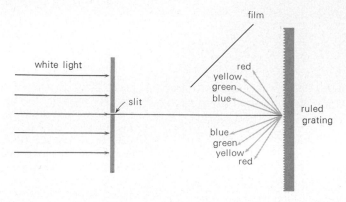

Figure 4•4 The separation of different wavelengths by a diffraction grating. See also Figure 4.6 on Plate 2.

When an electric current is passed through a hydrogen-filled tube that is similar in design to the common neon-sign tubes, radiation is emitted. If the emitted radiation is dispersed (spread out) by a prism or grating spectroscope, a discontinuous emission spectrum is obtained. A **discontinuous emission spectrum** of a substance consists of a pattern of bright lines on a dark background. Discontinuous emission spectra are also called **visible line spectra.** Such emission spectra have played important roles in scientific investigations, for the spectrum of a given element is as individual as a fingerprint. The elements rubidium, cesium, thallium, indium, gallium, and scandium were discovered (between 1860 and 1879) as a result of spectroscopic examination of minerals that revealed visible spectral lines unlike those of any previously known elements.

The element helium was discovered in the sun, 93 million miles away, before it was known to exist on earth. In 1868 European astronomers traveled to India to be in a good location to make observations during an eclipse of the sun. A French astronomer, Pierre Janssen, noted some unidentifiable lines in the spectrum of the sun's corona, whereupon one of his English contemporaries, J. N. Lockyer, suggested that these lines might belong to an element that existed in the sun but had not been found on earth. The unknown element was named *helium* from the Greek word *helios* (sun). It was not until 27 years later, in 1895, that Sir William Ramsay discovered that helium did exist on the earth in association with certain minerals. On heating these minerals he found that gases were evolved, one of which had a spectral pattern that matched the pattern discovered by Janssen and Lockyer.

Some discontinuous emission spectra of elements are shown in Figure 4.6 on Plate 2.

4•2•2 **Absorption Spectra** When continuous electromagnetic radiation, for example, white light, passes through a substance, certain wavelengths of radiation may be absorbed. These wavelengths are characteristic of the substance that absorbs the radiation, and the pattern of these dark lines is referred to as an **absorption spectrum.** When cool, a substance absorbs radiation of the same wavelengths that it emits when excited.

Part of the continuous spectrum emitted by the sun is absorbed by the gases in the sun's atmosphere. Because of this, there are narrow dark lines, called *Fraunhofer lines* after their discoverer, in the otherwise continuous spectrum that reaches us, and the position of these dark lines enables us to identify the gases that absorb this light (see Figure 4.6, spectrum 2).

Study of the absorption spectra of gases has led to the development of methods for analyzing unknown substances, regardless of whether they are gaseous, liquid, or solid. Transparent colored materials absorb certain wavelengths of visible light. Materials that do not absorb visible light may absorb characteristic wavelengths of ultraviolet or infrared radiation. Modern electronic devices can record the absorption spectra automatically (Figure 4.7). The pattern of absorption of radiant energy by a material gives an almost certain indication of the substances (elements or compounds) in the material and of the way in which the atoms are arranged in molecules.

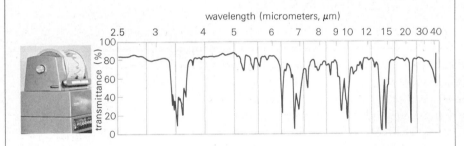

Figure 4•7 An electronic device automatically records an absorption spectrum (*left*). The recorded spectrum shown, unrolled and removed from the machine, is the absorption spectrum of the organic compound toluene, C_7H_8. The sharp dips in the curve indicate the wavelengths of radiant energy that have been absorbed most completely.
Source: Courtesy of Perkin-Elmer Corporation, Norwalk, Conn.

4•2•3   The Quantum Theory of Radiation   Prior to 1900 physicists had encountered problems in understanding the nature of radiant energy emitted by heated matter. According to the most influential theory at that time, radiant energy was emitted by tiny, oscillating charged particles whose frequencies of oscillation varied continuously as the temperature changed. However, the equations derived on the basis of this model did not fit the experimental data well. Then in 1900 a German physicist, Max Planck, resolved the conflict. Planck concluded that radiation can be neither emitted nor absorbed continuously, but that *radiant energy is discontinuous and consists of individual bundles of energy called* **quanta,** *or* **photons.** This concept, that radiant energy is *quantized,* is the fundamental postulate of the quantum theory that revolutionized twentieth-century physics.[2]

According to Planck, the energy of a photon is proportional to the frequency of the radiation, or

$$E = h\nu$$

[2]According to Webster, to *quantize* means to express as multiples of a definite quantity, or to state a property in a mathematical formula containing a factor *n*, where *n* may assume only integer values.

The constant h is called *Planck's constant of proportionality* and has dimensions of energy per photon multiplied by time. If E is in SI units of joules,

$$h = 6.626 \times 10^{-34} \, \text{J} \cdot \text{s} \cdot \text{photon}^{-1}$$

From the frequency one can calculate the energy associated with a single photon of radiant energy. Because $\nu = c/\lambda$, the energy for a given wavelength is easily calculated.

• Example 4.2 •

In the visible spectrum of excited hydrogen (see Figure 4.6), there is a red line with a wavelength of 656.3 nm. For this emission line, what is the frequency in hertz and the energy in joules per photon?

• Solution •

$$\nu = \frac{c}{\lambda} = \left(\frac{3.00 \times 10^8 \, \text{m} \cdot \text{s}^{-1}}{6.563 \times 10^2 \, \text{nm}} \right) \left(\frac{1 \, \text{nm}}{10^{-9} \, \text{m}} \right)$$

$$= 4.57 \times 10^{14} \, \text{s}^{-1} = 4.57 \times 10^{14} \, \text{Hz}$$

$$E = h\nu$$

$$= (6.626 \times 10^{-34} \, \text{J} \cdot \text{s} \cdot \text{photon}^{-1})(4.57 \times 10^{14} \, \text{s}^{-1})$$

$$= 3.03 \times 10^{-19} \, \text{J} \cdot \text{photon}^{-1}$$

See also Exercises 12 and 13.

In the electromagnetic spectrum diagrammed in Figure 4.2, the greater the frequency or the shorter the wavelength, the greater is the energy per photon. Changes in the energy of electrons in atoms are sources of radiation that range in frequency from visible light to X rays. One source of lower frequency (lower energy) electromagnetic radiation is an oscillating electric circuit; a source of higher frequency (higher energy) radiation is a change in the nucleus of an atom.

4.2.4   The Photoelectric Effect   One of the first applications of the quantum theory was made by Albert Einstein in 1905 to explain the **photoelectric effect,** the ejection of electrons from a metal surface by light. Suppose a metal such as cesium or potassium is exposed to light with a frequency that is increased gradually. No electrons are ejected from the surface of a particular metal until the frequency of the light reaches a certain minimum value called the **threshold frequency,** ν_0. Above the threshold frequency, however, electrons are emitted with kinetic energies that depend upon the frequency of the exciting light (Figure 4.8).

We must distinguish clearly between the *intensity* or brightness of a light, which is a measure of the number of photons per unit area per unit time, and the *energy per photon*. A light source may be very intense or bright, but if the energy of the individual photons is below the threshold energy, $h\nu_0$, no electrons will be ejected from the target metal. Another light source may have a low intensity (be very dim), but if the energy of the individual photons is above the threshold energy, electrons will be ejected from the target metal with a certain velocity.

In extending Planck's ideas, Einstein assumed that radiant energy is propagated through space as quanta or photons. Each photon has an energy

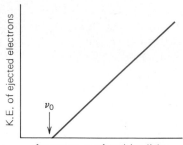

Figure 4.8 The photoelectric effect. When light has a frequency below the threshold frequency, ν_0, no electrons are ejected. Above ν_0, the kinetic energy of ejected electrons depends on ν.

that is determined by the frequency of the light; that is, $E = h\nu$. He further assumed that a definite amount of energy equal to $h\nu_0$ is needed to eject a single electron from a given metal surface. The amount of energy required to eject an electron from an atom bound to other atoms in the metal[3] is called the **work function,** W_0. Einstein reasoned that the energy of a photon, $h\nu$, would have to be equal to W_0 before an electron could be ejected. If the photon has more energy than this minimum threshold energy, the excess energy can be transformed into the kinetic energy, K.E., of the ejected electron:

$$E_{\text{photon}} = h\nu = W_0 + \text{K.E.}$$

Thus Einstein applied Planck's concept of quantized radiant energy to show how to account precisely for the energy of one quantum or one photon. For this explanation of the photoelectric effect, which gave support to a particle theory of light, Einstein was awarded the Nobel Prize in physics in 1921.

4•2•5   Application of Quantum Theory to Atomic Spectra It is the application of quantum ideas to explain atomic spectra that gives us an understanding of electron arrangements and their chemical importance. In 1885, J. J. Balmer, a Swiss mathematician, found that the wavelengths of a series of emission lines in the visible spectrum of atomic hydrogen were related to one another by a formula that may be written as

$$\frac{1}{\lambda} = R_H\left(\frac{1}{n_1{}^2} - \frac{1}{n_2{}^2}\right) \tag{1}$$

In this equation, R_H is a constant known as the *Rydberg constant* and has the value 1.09678×10^7 m^{-1}. For the series of lines called the *Balmer series*, n_1 has a value of 2 and n_2 has the values $2 + 1$, $2 + 2$, $2 + 3$, and so on. Four of these lines can be seen in the visible region of the hydrogen spectrum (see Figure 4.6).

Equation (1) is a general one that may be applied to other series of lines of the hydrogen spectrum. For longer wavelength lines appearing in the infrared, discovered by F. Paschen and known as the *Paschen series*, the value of n_1 is 3. In this series, n_2 is $3 + 1$, $3 + 2$, $3 + 3$, and so on. Although Equation (1) fits the experimental data quite well, there was no theoretical explanation for the formula when Balmer and Paschen applied it to their series of lines. In 1913, Niels Bohr, a Danish physicist, developed a theory that explained the positions of the known Balmer and Paschen lines. To account for the fact that excited samples of the same element always emit the same set of wavelengths, he proposed a new theory of atomic structure, known as the **Bohr theory.**

• *The Bohr Theory* • Bohr retained some features of the planetary model in that he visualized the atom as a positive nucleus surrounded by one

---

[3]The work function, W_0, of a metal is related to its ionization energy. However, the ionization energy (see Section 3.6.1) is the energy needed to eject an electron from an isolated gaseous atom. For a metal, the work function is always less than the ionization energy.

or more electrons traveling in definite circular orbits. However, he introduced two general assumptions which he applied to atoms *even though such behavior is unknown in large-scale systems.* These assumptions are summarized as follows:

1. *As long as an electron stays in a given orbit, or stationary state, it neither gains nor loses energy.*

2. *When an electron jumps from one orbit or stationary state to another, such a transition is accompanied by the absorption or emission of a definite amount of energy equal to the difference in energy between the two stationary states.*

The stationary states, or energy levels, normally occupied by electrons are those of relatively low energy, called **ground states.** When atoms are subjected to high temperature, or excited in other ways, as in an electric arc, the electrons, especially the outer ones in the excited atoms, absorb energy and are forced out to levels of higher energy, called **excited states.** When an excited electron falls back to a lower energy level, a certain amount of radiant energy is given off, and the amount of energy determines the wavelength of the emitted radiation.

By designating a higher energy level as E_H and a lower energy level as E_L, the difference in energy, $E_H - E_L$, between two given levels will be constant for all atoms of the same kind. Bohr assumed that Planck's quantization of radiant energy applied to atoms and that these constant energy differences, $E_H - E_L$, explained the fact that the radiation emitted by a given element always has the same set of frequencies (or wavelengths); that is,

$$E_H - E_L = \Delta E = h\nu$$

Using the planetary model, Bohr combined equations from classical physics in order to calculate the theoretical radii of electron orbits and the differences in energy of electrons in different orbits. He developed an equation, Equation (2), that related the frequency of a line in the hydrogen emission spectrum to the energy emitted when the electron falls from a higher energy level to a lower level:

$$\nu = \frac{2\pi^2 m e^4 Z^2}{h^3}\left(\frac{1}{n_L{}^2} - \frac{1}{n_H{}^2}\right) \tag{2}$$

In this expression, ν is the frequency of the emission; m and e are the mass and charge, respectively, of the electron; Z is the number of protons in the nucleus (1 for hydrogen); and h is Planck's constant. The symbol n stands for the **principal quantum number,** and it tells which main energy level of an atom an electron is in. The values of n are the integers 1, 2, 3, 4, and so on. For an electron transition, n_L is the number of the lower energy level into which the electron falls, and n_H is the number of the higher energy level from which the electron falls. In Bohr's words,[4]

We see that this expression accounts for the law connecting the lines in the spectrum of hydrogen. If we put $n_L = 2$ and let n_H vary, we get the ordinary

[4]N. Bohr, *Philosophical Magazine,* **26,** 1 (1913).

Balmer series. If we put $n_L = 3$, we get the series in the ultra-red observed by Paschen and previously suspected by Ritz. If we put $n_L = 1 \ldots$, we get [a] series . . . in the extreme ultraviolet . . . which [is] . . . not observed, but the existence of which may be expected.

Bohr's achievement met the test of a fruitful theory: it not only accounted for the known Balmer and Paschen lines, but it also enabled him to predict the existence of undiscovered spectral lines in the ultraviolet region. This series of lines was sought and found by F. Lyman at Princeton in 1915, thus crowning the theory with success.

In Figure 4.9, the spectral lines in the Paschen, Balmer, and Lyman series are related to theoretical transitions of electrons from one energy level to another. If we solve Equation (2) for the Balmer series, we find for the fall of an electron from $n_H = 3$ to $n_H = 2$ the calculated frequency of 4.57×10^{14} Hz. This corresponds to the red line at a wavelength of 656.3 nm (see Example 4.2 and Figure 4.6). For the fall from $n_H = 4$ to $n_L = 2$, the frequency corresponds to the green line at 486.1 nm. The fall from $n_H = 5$ to $n_L = 2$ corresponds to the blue line at 434.0 nm, and so on. With a fine spectroscope it is possible to observe twelve Balmer lines, all at frequencies consistent with Bohr's formula.

Of the three series shown in Figure 4.9, the Lyman radiations, arising from electrons falling into the energy level closest to the positive nucleus, have the shortest wavelengths (largest frequencies). The Paschen radiations, involving the fall of electrons into the third energy level, have the longest wavelengths (smallest frequencies). Within each series, the greater the difference between a high and low level, the shorter is the wavelength of the radiation emitted when an electron falls.

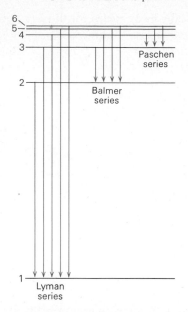

Figure 4•9 Schematic energy level diagram for an excited hydrogen atom. The energy levels are numbered 1 through 6. The different lines in the hydrogen emission spectrum are accounted for by jumps of excited electrons from higher to lower energy levels. Of the three series shown, the Lyman series has the largest energies and the Paschen series the smallest.

━━━━ • Example 4•3 • ━━━━

Use Equation (2) to calculate the frequencies in hertz for each of the five Lyman lines shown in Figure 4.9. (Given, $1 \text{ C} = 9.4805 \times 10^4 \text{ kg}^{1/2} \cdot \text{m}^{3/2} \cdot \text{s}^{-1}$, and other values in Appendix Table A.4.)

━━━━ • Solution • To simplify the calculations with Equation (2), we first evaluate the collection of constants in front of the parentheses:

$$\frac{2\pi^2 m e^4 Z^2}{h^3} =$$

$$\frac{2(3.1416)^2(9.1096 \times 10^{-31} \text{ kg})\left(1.6022 \times 10^{-19}\,\cancel{C} \times \dfrac{9.4805 \times 10^4 \text{ kg}^{1/2} \cdot \text{m}^{3/2} \cdot \text{s}^{-1}}{1\,\cancel{C}}\right)^4 (1)^2}{\left(6.626 \times 10^{-34}\,\cancel{J} \cdot \text{s} \cdot \cancel{\text{photon}^{-1}} \times \dfrac{1 \text{ kg} \cdot \text{m}^2 \cdot \text{s}^{-2}}{1\,\cancel{J} \cdot \cancel{\text{photon}^{-1}}}\right)^3}$$

$$= 3.291 \times 10^{15} \frac{\cancel{\text{kg}^3} \cdot \cancel{\text{m}^6} \cdot \text{s}^{-4}}{\cancel{\text{kg}^3} \cdot \cancel{\text{m}^6} \cdot \text{s}^{-3}} = 3.291 \times 10^{15} \text{ s}^{-1} = 3.291 \times 10^{15} \text{ Hz}$$

So, Equation (2) is simplified to:

$$\nu = 3.291 \times 10^{15} \text{ Hz}\left(\frac{1}{1^2} - \frac{1}{n_H^2}\right)$$

The frequencies for the five Lyman lines are calculated from this simplified equation. Summarizing,

n_H	$1 - \dfrac{1}{n_H{}^2}$	ν, Hz
2	3/4	2.468×10^{15}
3	8/9	2.925×10^{15}
4	15/16	3.085×10^{15}
5	24/25	3.159×10^{15}
6	35/36	3.200×10^{15}

See also Exercises 17–19.

• *Relationship of Equation (2) to Other Equations* • If we solve Equation (2) for the reciprocal of the wavelength, $1/\lambda$, we obtain

$$\frac{1}{\lambda} = \frac{\nu}{c} = \frac{2\pi^2 m e^4 Z^2}{h^3 c}\left(\frac{1}{n_L{}^2} - \frac{1}{n_H{}^2}\right) \tag{3}$$

For hydrogen, the value of the collection of constants preceding the parentheses is equal to 1.09737×10^7 m^{-1}, which agrees closely with the constant R_H that Rydberg calculated years earlier for hydrogen. Therefore, we can say

$$\frac{1}{\lambda} = R_H\left(\frac{1}{n_L{}^2} - \frac{1}{n_H{}^2}\right) \tag{4}$$

which has the same form as Equation (1), but with n_L and n_H substituted for n_1 and n_2, respectively. That is, Equation (4), which is based on a *theoretical concept* of electron energy levels, matches the equation worked out years earlier by Balmer to fit some *experimental facts*.

From Equation (4) we can obtain an equation that can be used to calculate the energy associated with an electron jump:

$$E = hcR_H\left(\frac{1}{n_L{}^2} - \frac{1}{n_H{}^2}\right) \tag{5}$$

The corresponding wavelength, λ, or frequency, ν, can also be calculated from Equation (4) or (5), using the equations $\nu = c/\lambda$ and $E = h\nu$.

• *Testing the Bohr Theory* • One way of testing a theory is to design an experiment, or experiments, with results that should either confirm or refute the theory. However, if the experimental results that you need are already available, then all you need to do is compare these results with those predicted from the theory.

Bohr reasoned that the energy given up by an electron falling from a higher level to a lower level must be equal to the energy required to excite an electron from that lower level to the higher one. There was one well-known excitation energy for hydrogen. Its ionization energy had been measured as 13.54 eV. Bohr assumed that this was the energy necessary to excite the electron from the lowest or ground-state level ($n_1 = 1$) to a position where there was essentially no attractive force between the electron and the nucleus. This, he said, would be mathematically equivalent to making the value of n_H equal to infinity, ∞.

Writing Equation (5) with $n_L = 1$ and $n_H = \infty$ gives

$$E = hcR_H\left(\frac{1}{1^2} - \frac{1}{\infty^2}\right) = hcR_H(1 - 0) = hcR_H \qquad (6)$$

Using such a formula and the value he had in 1913 for Planck's constant, Bohr calculated the ionization energy of hydrogen to be 13 eV. This close agreement with the experimental value gave many persons confidence in Bohr's theory. A modern value of the hydrogen atom's ionization energy is calculated in Example 4.4.

—— • Example 4.4 • ——

Use Equation (6) to calculate a modern value for the ionization energy of hydrogen. Compare this value with the modern experimental one of 13.60 eV. (Given, 1 eV = 1.602×10^{-19} J.)

—— • Solution •

$E = hcR_H$

$\quad = (6.626 \times 10^{-34} \text{ J} \cdot s \cdot \text{photon}^{-1})(2.998 \times 10^8 \text{ m} \cdot s^{-1})(1.097 \times 10^7 \text{ m}^{-1})$

$\quad = 2.179 \times 10^{-18} \text{ J} \cdot \text{photon}^{-1}$

$\quad = (2.179 \times 10^{-18} \text{ J} \cdot \text{photon}^{-1})\left(\dfrac{1 \text{ eV}}{1.602 \times 10^{-19} \text{ J}}\right) = 13.60 \text{ eV} \cdot \text{photon}^{-1}$

The calculated value agrees precisely with the experimentally determined one.

See also Exercises 20 and 21.

In Equation (3), the value of Z or Z^2 for the hydrogen atom is 1. If we separate Z^2 from the constants in this equation, the value of R_H is not changed. Separation of the Z^2 term gives Equation (7), which is similar to Equation (5) but can be used to calculate the ionization energies of multiply charged nuclei with only one electron:

$$E = Z^2 hcR_H\left(\frac{1}{n_L^2} - \frac{1}{n_H^2}\right) \qquad (7)$$

By letting $n_L = 1$ and $n_H = \infty$, as was done for the H atom, we can calculate the ionization energies for other species that have a nucleus and only one electron, such as He^+, Li^{2+}, and Be^{3+}. A comparison of values obtained by using Equation (7) with those obtained experimentally is presented in Table 4.2.

Although the Bohr model with its quantized energy levels, or quantum numbers, was successful in explaining most of the spectral properties of the hydrogen atom, it failed to apply to more complicated atoms. It could not be extended to the simple helium atom with its two electrons, even when numerous modifications were introduced into the theory to account for the interaction of the electrons. Continued study brought fundamental changes in the theory, such as the introduction of energy sublevels (see Section 3.8) and a corresponding second set of quantum numbers. Later, a third series of quantum numbers was introduced to account for the behavior of electrons in magnetic fields.

Table 4•2 Calculated versus experimental values for the ionization energy of some one-electron ions

Ion	Calculated value, eV[a]	Experimental value, eV[b]
He^+	54.39	54.42
Li^{2+}	122.38	122.45
Be^{3+}	217.57	217.71

[a]The calculated values are slightly less than the experimental values, with the difference increasing with increasing Z, due to the fact that the rotation of a nucleus about its own center of mass affects the electron energy. The greater the mass of the atom, the greater will be the deviation. By correcting for this mass effect, one can obtain Rydberg constants, R_c, that are slightly larger than that for hydrogen, and thereby obtain calculated values in nearly perfect agreement with the experimental ones.

[b]The experimental values are calculated by permission from J. E. Huheey, *Inorganic Chemistry*, 2nd ed., Harper & Row, New York, 1978. (Compare these values with those in Table 3.6.)

Bohr was a pioneer in the application of quantum theory to the electronic structure of atoms. Even though the theoretically accepted atom of today differs considerably from the early Bohr atom, Bohr is recognized with Planck, Einstein, and Rutherford as one of the creators of atomic physics.

Modern Electronic Theory

4•3 The Wave Nature of Matter

Prior to Planck's postulate of the quantization of radiant energy, it was generally accepted that a fundamental difference between matter and radiant energy was that the latter had no mass and was transmitted as a wavelike disturbance. However, in the years following Planck's discovery, the results of several experiments suggested that radiation could exhibit a particle-like character as well as a wavelike character, depending on the particular experiment carried out.

In 1924 the French physicist Prince Louis de Broglie extended to matter the concept that, like light, matter must act as both a particle (photon) and a wave. He postulated that all matter in motion has wave characteristics. He formulated the wave properties of a particle in motion as:

$$\lambda = \frac{h}{p}$$

where λ is the wavelength of the wave, h is Planck's constant, and p is the momentum. The momentum is the product of the mass times the velocity of the particle, $p = mv$. From the de Broglie equation, we see that the wavelength is inversely proportional to the momentum. For a relatively large particle in motion, such as a baseball, p is very large, so that λ is calculated to be extremely small, about 10^{-25} nm. This wavelength represents an insignificant property of the moving ball. For a moving electron, however, p is so small that values for λ are significant, often ranging between 1 and 100 nm. The calculated wavelength of an electron in motion is therefore of a convenient magnitude to be demonstrated experimentally, a feat that was accomplished within a short time after de Broglie proposed the dual theory of matter.[5]

4•4 The Heisenberg Uncertainty Principle

While the approach to atomic structure via the assumption that electrons move about the nucleus in well-defined orbits was invaluable in the early development of chemical theory, the futility of proceeding further in this direction was made apparent by the German physicist Werner Heisenberg in 1927. He recognized that any experimental method used to determine the position and momentum of an object in motion can cause changes in either or both the position and momentum and thus introduce an element of uncertainty into the measurement. Heisenberg developed mathematical equations to show that *no experimental method can be devised that will measure simultaneously the precise position as well as the momentum of an object*. For a particle as small as an electron, this lack of precision is critically important. **Heisenberg's uncertainty principle** states that the product of the uncertainty of the momentum Δp and that of the position Δx must be equal to or greater than Planck's constant, h:

$$(\Delta p)(\Delta x) \geq h$$

The more precisely one knows Δp, the less precisely Δx is known, and vice versa.

The uncertainty principle is one of the reasons classical theories of physics that apply to macroscopic systems are not applicable to phenomena on a submicroscopic atomic scale. From Newton's law of motion for a particle, $F = ma$, where F is the force, m the mass, and a the acceleration, one theoretically can measure precisely not only the present position and momentum of a particle, but also its position and momentum in the past and for any future time as well. Two centuries after Newton, it was found that the reason his law may be applied to our macroscopic world is that Planck's constant, h, is so very small relative to the size of the macroscopic quantities. The uncertainties associated with the position and momentum of a baseball dropped from the top of a building to the street below are much less than the possible accuracy of measurement. However, for an electron with its apparent high velocity and extremely small momentum, the uncertainties associated with its position and momentum are relatively large.

An electron is so small that it is disturbed or put in unpredictable motion by an attempt to examine it, as by irradiating it with light or X rays. Although we might attempt to determine where the electron is, we cannot measure its position exactly, and we cannot tell where it is going after we attempt to observe it. As a consequence of this limitation, the model of an

---

[5]Experiments performed in 1927 showed that electrons are diffracted by crystals in a manner similar to the diffraction of X rays (see Section 8.6.1).

atom that postulates electrons traveling in definite orbits and with definite momentums is not satisfactory. As we will learn in the next section, the best we can do is to find the electron's probable position in an atom.

4.5   **The Quantum Mechanical Description of Electrons in Atoms**

The dual nature of matter and radiation, the exhibition of both wavelike and particle-like character, confronts us with the task of finding a description for submicroscopic phenomena that is quite different from classical Newtonian descriptions. The goal is to present a unified description that applies to all possible experiments, both submicroscopic and macroscopic. The unified description that achieves this goal is called **quantum mechanics** or **wave mechanics.** This approach is based upon the assumption that there are waves associated with both matter and radiation and that the mathematical descriptions of such waves provide information about the energies of electrons and their positions.

Many theoreticians have contributed to the field of quantum mechanics, but the initial developments are credited to Heisenberg and Erwin Schrödinger. Building on the work of de Broglie, Schrödinger in 1926 developed an equation that related the wave properties associated with electrons to their energies. The **Schrödinger equation,**

$$\frac{\partial^2 \psi}{\partial x^2} + \frac{\partial^2 \psi}{\partial y^2} + \frac{\partial^2 \psi}{\partial z^2} + \frac{8\pi^2 m}{h^2}(E - V)\psi = 0$$

is a second-order differential equation that describes the total energy, E, and the potential energy, V, of a particle in terms of its mass, m, with respect to its position in three dimensions, x, y, and z. (The equation's derivation and application are presented in great detail in many advanced texts.)

Calculations based on the Schrödinger equation to determine the positions and energies of electrons in atoms are difficult and lengthy. They have been performed satisfactorily only for the hydrogen atom and one-electron ions. For atoms with large atomic numbers, that is, with many protons and electrons, the electrostatic interactions of the electrons with one another and with the nucleus cause the solutions to the equation to be much more difficult. After a number of reasonable approximations are introduced into the calculations, the results indicate that electrons in complex atoms occupy positions similar to those occupied by an electron in a hydrogen atom. Therefore, we use the ideas that apply to hydrogen to describe electrons in all atoms.

In this section, we will first describe some of the results found for the hydrogen atom. In quantum mechanics, the electron is treated as a three-dimensional wave. The impossibility of determining the precise position of an electron at a given instant is recognized from the uncertainty principle. However, the **wave function,** ψ, describes the region within which an electron is most likely to be found. This region, or the wave function itself, is called an **orbital.**

Although we cannot determine the precise position of an electron, the probability of the electron being at a definite location can be calculated from the Schrödinger equation. An electron "occupies" a whole orbital, although there is not an equal probability of the electron being at all positions within the atomic orbital. To help understand this, we can think of the electron as a particle that moves from place to place so rapidly that it behaves somewhat like an "electron cloud" whose density varies within the orbital. The square of the wave function, ψ^2, at a point is interpreted as the probability of finding the electron in a certain small volume, or the **probable electron density.**

When the electron in the ground state, or lowest energy orbital, of the hydrogen atom absorbs a definite quantum of energy, it is raised to a higher energy orbital. In quantum mechanics, the difference in the energies of any two orbitals is quantized. The quantized emission of energy by an excited H atom, which was explained in terms of the Bohr atom as an electron falling from a higher to a lower orbit, is accounted for in the quantum mechanical model as a change from a higher energy orbital to a lower energy one.

4•5•1   Quantum Numbers for Electrons   The energy states that an electron can occupy in a hydrogen atom are described by three quantum numbers. For other atoms, four quantum numbers are needed.

• *First or Principal Quantum Number, n* • The **first or principal quantum number,** n, of the Bohr atom is retained in the quantum mechanical model. It has integral values of 1, 2, 3, 4, and so on, and it tells which main energy level the electron is in. The larger the value of n, the greater is the size of the orbital the electron occupies.

The one orbital in main level 1 of the hydrogen atom is a spherically symmetrical region called the $1s$ orbital. A plot of the probable electron density, ψ^2, versus the distance, r, from the nucleus is shown in Figure 4.10. The maximum electron density is near the nucleus, with the density decreasing exponentially as r increases.

Because the probable electron density decreases exponentially with increasing distance from the nucleus, the density theoretically never falls to zero. A common limit to a schematic representation of an orbital is to show the region within which there is a 90 percent probability of finding the electron. The boundary for this 90-percent probability region is called the **90-percent contour.** Beyond this boundary the electron density becomes so small, that is, the electron cloud becomes so diffuse, that it is impractical to represent it in a drawing. Two ways of diagramming the electron cloud in the $1s$ orbital of a hydrogen atom are shown in Figure 4.11.

If the electron in the $1s$ orbital is represented as a cloud of negative charge whose density decreases with increasing distance from the nucleus, how can a radius be obtained for the H atom? One way in which this is accomplished in the quantum mechanical description is to imagine that the space represented by the $1s$ orbital is divided into a series of uniform, thin, concentric, spherical shells. The shells are spaced at a small distance, Δr, apart from one another (something like the layers of an onion). The probability of finding the electron between the shell with radius r and the shell

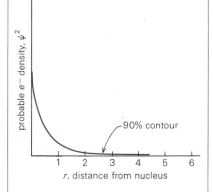

Figure 4•10 A plot of the probability of finding the $1s$ electron of hydrogen in a tiny unit volume versus the distance r from the nucleus. The volume within which there is a 90 percent probability of finding the electron has an r of about 2.66 bohrs, or 1.41 Å. (The units of r are bohrs; 1 bohr = 0.529 Å.)

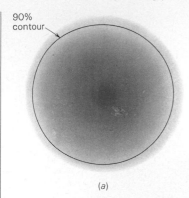

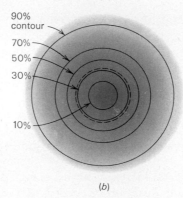

(a)

(b)

Figure 4•11 Schematic representations of the 1s orbital of hydrogen: (a) a cross section showing the probable density of the electron cloud; this cross section represents a slice through the middle of the cloud; (b) a diagram showing the contours of different regions within which there are certain percent probabilities of finding the electron. The dotted line of about 32.3 percent probability corresponds to the Bohr radius of 0.529 Å. The 90-percent contour is at 2.66 bohrs, or 1.41 Å.

with radius $r + \Delta r$ (comparable to the volume of each onion layer) is called the **radial probability.** The radial probability is a product of two variables, the volume, V, of the electron region between the shells and the electron density between the shells, ψ^2. At $r = 0$, the radial probability is zero because the volume is zero. As r increases, the volume between the thin shells increases. At the same time, the electron density decreases. Initially the volume increase is more than sufficient to compensate for the exponential decrease in ψ^2, so that the radial probability, $V \times \psi^2$, increases (see Figure 4.12). At large values of r, the decrease of ψ^2 is more important, and the radial probability approaches zero as ψ^2 approaches zero. The maximum value of the radial probability is found where $V \times \psi^2$ is a maximum. For the 1s orbital of hydrogen, this maximum is at 0.529 Å, the same radius that Bohr had calculated for his planetary model.[6]

• *Second Quantum Number, l* • The **second quantum number,** l, tells what kind of energy sublevel the electron is in and specifies the shape of the orbital(s) in a sublevel. The allowed values of l depend upon the value of n and range from 0 to $n - 1$.

If $n = 1$, l can only have a value of 0. This means there is only one type of orbital, the 1s orbital, for the first main level. When $n = 2$, l has values of 0 and 1. Consequently there are two sublevels and two types of orbitals associated with the second main level. The two sublevels are called the 2s and 2p sublevels, and the corresponding orbitals are the 2s and 2p orbitals. When $n = 3$, l has values of 0, 1, and 2; when $n = 4$, l has values of 0, 1, 2, and 3. Thus, the third main level has three kinds, and the fourth main level has four kinds of orbitals.

When $l = 0$, the orbital is an s orbital. Because l may equal 0 when $n = 1, 2, 3$, or any integer, there is a spherically shaped s orbital in every main energy level. Cross sections of the 2s and 3s orbitals, showing the relative electron densities within the 90-percent contours, are shown in Figure 4.13.

When $l = 1$, the orbital is designated as a p orbital. Because l may have a value of 1 for any value of n except $n = 1$, a p orbital is part of each main energy level except the first. The p orbitals are often described as

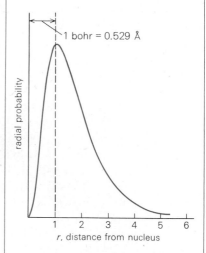

Figure 4•12 A plot of the probability of finding the 1s electron of hydrogen in a thin spherical shell (the radial probability) versus the distance r from the nucleus. The maximum in the radial probability is at 1 bohr (0.529 Å).

---

[6]This radius of 0.529 Å is called the **Bohr radius.** The distance 0.529 Å is referred to as 1 *bohr,* the unit of atomic distance.

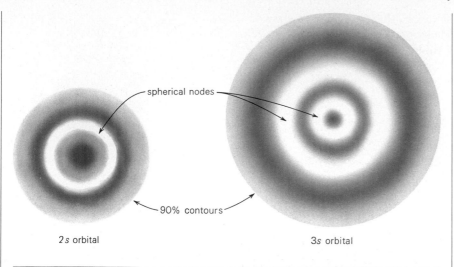

Figure 4•13 Cross sections of the $2s$ and $3s$ orbitals, showing relative electron densities. The regions where the electron density is zero (or practically zero) are called *nodes*. Note that the $2s$ orbital has one node and the $3s$ orbital has two nodes.

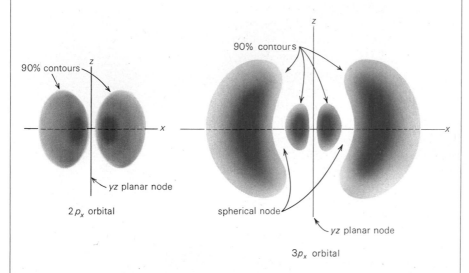

Figure 4•14 Cross sections of $2p$ and $3p$ orbitals centered on the x-axis. In the $2p$ orbital there is one node between the two "halves" of the orbital. In the $3p$ orbital there are two nodes.

having dumbbell shapes. Cross sections of a $2p$ and a $3p$ orbital, showing the relative electron densities within the 90-percent contours, are shown in Figure 4.14.

The letter designations for the orbitals described by $l = 2$ and $l = 3$ are d and f, respectively. The d sublevel is encountered first in the third main level and the f sublevel in the fourth main level. The corresponding orbitals are called d and f orbitals.

• *Third Quantum Number,* m_l • The **third quantum number,** m_l, tells which particular orbital the electron occupies within an energy sublevel. Also, it determines the specific orientation in space of the orbital relative to the nucleus.

In all main energy levels except the first, there is more than one energy sublevel and therefore more than one kind of orbital. The number of orbitals in a given kind of sublevel is equal to the number of possible m_l

values for that sublevel. For any value of l, there are $(2l + 1)$ allowed values of m_l. The quantum number m_l can have any integral value from $+l$ through $-l$, including 0, as illustrated by the following:

for s sublevel, $l = 0$; $m_l = 0$:   one s type

for p sublevel, $l = 1$; $m_l = +1, 0, -1$:   three p types

for d sublevel, $l = 2$; $m_l = +2, +1, 0, -1, -2$:   five d types

for f sublevel, $l = 3$; $m_l = +3, +2, +1, 0, -1, -2, -3$:

seven f types

The m_l values of $+1, 0$, and -1 for a p sublevel correspond to three p orbitals, each centered on a different coordinate axis, x, y, and z. These orbitals are called the p_x, p_y, and p_z orbitals. The approximate geometric shapes of the three $2p$ orbitals are shown in Figure 4.15. The m_l values of $+2$, $+1, 0, -1$, and -2 for a d sublevel correspond to five d orbitals. The geometric shapes and orientations of the five $3d$ orbitals are shown in Figure 4.16. For an f sublevel, the m_l values are $+3, +2, +1, 0, -1, -2$, and -3, corresponding to seven f orbitals. The geometric shapes of f orbitals are very complicated.[7]

• *Fourth Quantum Number, m_s* • The **fourth quantum number, m_s,** describes the two ways in which an electron may be aligned in a magnetic field: parallel to or opposed to the magnetic field. The electron can be thought of as rotating in either a clockwise or a counterclockwise direction. At one time, the electron was considered to be spinning like a top. Although this literal interpretation is no longer valid, the term *spin* is still used, and the alignments of two electrons relative to one another are still spoken of as "like

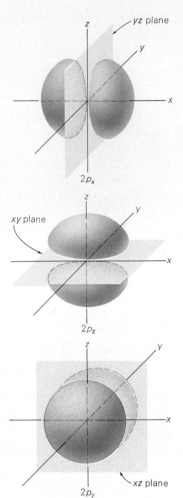

Figure 4.15 (*above*) Shapes and orientations of the three $2p$ orbitals. The plane dividing the two regions of each orbital is shown for perspective.

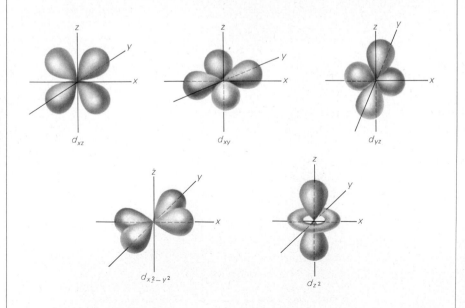

Figure 4.16 Shapes and orientations of the five $3d$ orbitals. Actually, $3d$ orbitals are larger than $2s$, $2p$, $3s$, or $3p$ orbitals, but this drawing is small relative to the previous three figures to save space.

[7] Interested students are referred to a fuller discussion of orbitals in Frank L. Lambert, *Chemistry*, **41** (2), 10 (1968).

spin" or "opposite spin." The so-called spin quantum number, m_s, can have only two values, $+\frac{1}{2}$ or $-\frac{1}{2}$.

• 4.5.2   **The Pauli Exclusion Principle**   The states of the electrons within atoms are described by four quantum numbers, n, l, m_l, and m_s. *No two electrons within the same atom may have the same four quantum numbers.* This is the famous **Pauli exclusion principle,** named for Wolfgang Pauli. Another statement of this principle is: *If two electrons occupy the same orbital, they must have different values of m_s.*

In a helium atom, two electrons occupy the 1s orbital in the ground state. These two electrons have opposite spins and are said to be *paired.* The quantum numbers for these two electrons are:

n	l	m_l	m_s
1	0	0	$+\frac{1}{2}$
1	0	0	$-\frac{1}{2}$

If the two electrons in a helium atom were to have like spins, they would be unpaired.

The electron distributions for the sublevels of the first four main energy levels were shown in Table 3.8. A summary of the allowed values of the four principal quantum numbers, in line with the Pauli exclusion principle, is given in Table 4.3. Examples 4.5, 4.6, and 4.7 illustrate the assignments of quantum numbers to main levels, sublevels, and specific atoms.

Table 4.3 Summary of the quantum numbers for the first four main energy levels

Main level n	Sublevel l	Orbital m_l	Electron spin for each value of m_l m_s
1	0 (1s)	0	$+\frac{1}{2}, -\frac{1}{2}$
2	0 (2s)	0	$+\frac{1}{2}, -\frac{1}{2}$
	1 (2p)	$+1, 0, -1$	$+\frac{1}{2}, -\frac{1}{2}$
3	0 (3s)	0	$+\frac{1}{2}, -\frac{1}{2}$
	1 (3p)	$+1, 0, -1$	$+\frac{1}{2}, -\frac{1}{2}$
	2 (3d)	$+2, +1, 0, -1, -2$	$+\frac{1}{2}, -\frac{1}{2}$
4	0 (4s)	0	$+\frac{1}{2}, -\frac{1}{2}$
	1 (4p)	$+1, 0, -1$	$+\frac{1}{2}, -\frac{1}{2}$
	2 (4d)	$+2, +1, 0, -1, -2$	$+\frac{1}{2}, -\frac{1}{2}$
	3 (4f)	$+3, +2, +1, 0, -1, -2, -3$	$+\frac{1}{2}, -\frac{1}{2}$

— • **Example 4.5** • ————————————————

Write all possible sets of quantum numbers for electrons in the second main energy level.

━━━ • Solution •

$n = 2$

$l = 0, 1$

$m_l = +1, 0, -1$

$m_s = +\frac{1}{2}, -\frac{1}{2}$

n	l	m_l	m_s	
2	0	0	$+\frac{1}{2}$	two e^-'s in the
2	0	0	$-\frac{1}{2}$	one 2s orbital
2	1	+1	$+\frac{1}{2}$	
2	1	+1	$-\frac{1}{2}$	
2	1	0	$+\frac{1}{2}$	six e^-'s in the
2	1	0	$-\frac{1}{2}$	three 2p orbitals
2	1	-1	$+\frac{1}{2}$	
2	1	-1	$-\frac{1}{2}$	

See also Exercises 33–35.

━━━ • Example 4•6 • ━━━

Write all possible sets of quantum numbers for electrons in the third sublevel of the fourth main level, that is, for the electrons in the 4d orbitals.

━━━ • Solution • For the 4d orbitals, the values of n, l, m_l, and m_s are as follows:

$n = 4$

$l = 2$

$m_l = +2, +1, 0, -1, -2$

$m_s = +\frac{1}{2}, -\frac{1}{2}$

n	l	m_l	m_s	
4	2	+2	$+\frac{1}{2}$	
4	2	+2	$-\frac{1}{2}$	
4	2	+1	$+\frac{1}{2}$	
4	2	+1	$-\frac{1}{2}$	
4	2	0	$+\frac{1}{2}$	ten e^-'s in the
4	2	0	$-\frac{1}{2}$	five 4d orbitals
4	2	-1	$+\frac{1}{2}$	
4	2	-1	$-\frac{1}{2}$	
4	2	-2	$+\frac{1}{2}$	
4	2	-2	$-\frac{1}{2}$	

See also Exercises 33–35.

━━━ • Example 4•7 • ━━━

Write the sets of quantum numbers for the electrons in an atom of boron.

━━━ • Solution • The electronic configuration for an atom of boron is $1s^2 2s^2 2p^1$. Five sets of quantum numbers account for the electrons:

n	l	m_l	m_s	
1	0	0	$+\frac{1}{2}$	two 1s orbital electrons
1	0	0	$-\frac{1}{2}$	
2	0	0	$+\frac{1}{2}$	two 2s orbital electrons
2	0	0	$-\frac{1}{2}$	
2	1	0	$+\frac{1}{2}$	one 2p orbitral electron

See also Exercises 33–35.

4.5.3 Order of Filling Orbitals Just as for energy sublevels, orbitals are filled in accord with the aufbau principle (see Section 3.8.1). The filling begins with the sublevel lowest in energy and continues upward (see Figures 3.17 and 3.18). Within a sublevel, electrons must have their spins unpaired as much as possible. This rule, known as **Hund's rule,** means that *each orbital is occupied by one electron before any orbital has two,* and that *the first electrons to occupy orbitals within a sublevel must have parallel spins.* As an example, each of the 2p electrons of a nitrogen atom occupies a different 2p orbital rather than two of them occupying one orbital (see Table 4.4). The next electron to be added to the 2p sublevel, in an oxygen atom, pairs with one of the electrons in one of the 2p orbitals.

Table 4.4 **Electron arrangements according to the aufbau principle, Hund's rule, and the Pauli exclusion principle**[a]

Main levels	1	2			3	Summary
Sublevels	s	s		p	s	
H	↑					$1s^1$
He	↑↓					$1s^2$
Li	↑↓	↑				$1s^2 2s^1$
Be	↑↓	↑↓				$1s^2 2s^2$
B	↑↓	↑↓	↑ ○ ○			$1s^2 2s^2 2p^1$
C	↑↓	↑↓	↑ ↑ ○			$1s^2 2s^2 2p^2$
N	↑↓	↑↓	↑ ↑ ↑			$1s^2 2s^2 2p^3$
O	↑↓	↑↓	↑↓ ↑ ↑			$1s^2 2s^2 2p^4$
F	↑↓	↑↓	↑↓ ↑↓ ↑			$1s^2 2s^2 2p^5$
Ne	↑↓	↑↓	↑↓ ↑↓ ↑↓			$1s^2 2s^2 2p^6$
Na	↑↓	↑↓	↑↓ ↑↓ ↑↓		↑	$1s^2 2s^2 2p^6 3s^1$
Mg	↑↓	↑↓	↑↓ ↑↓ ↑↓		↑↓	$1s^2 2s^2 2p^6 3s^2$

[a] Each single arrow within a circle represents an electron in an orbital. Two arrows pointing in opposite directions indicate that the two electrons within an orbital have opposite spins; that is, the electrons are paired.

The application of the aufbau principle and Hund's rule to the first twelve elements in the periodic chart is shown in Table 4.4. The rules are illustrated further in Example 4.8.

────── • Example 4.8 • ──────

Use the circle-and-arrow representation shown in Table 4.4 to describe the electron arrangements in the third and fourth main energy levels of an atom of iron, Fe ($Z = 26$).

• Solution • The electronic configuration for an atom of iron is $1s^2 2s^2 2p^6 3s^2 3p^6 4s^2 3d^6$. The $3s$, $3p$, and $4s$ orbitals are filled to capacity. For the six electrons distributed in the five $3d$ orbitals, two of these electrons will have paired spins and four will be unpaired.

$$
\begin{array}{ccc}
\text{⥮} & \text{⥮ ⥮ ⥮} & \text{⥮ ↑ ↑ ↑ ↑} \qquad \text{⥮} \\
3s & 3p & 3d \qquad\qquad 4s
\end{array}
$$

See also Exercise 37.

• Hund's Rule and Ionization Energies • In Section 3.6.2, the trend of increasing ionization energies from left to right across a period was described. It was pointed out that the VIA elements oxygen and sulfur have ionization energies that are somewhat lower than expected on the basis of the general trend. (See the graph in Figure 3.13.) These low values are apparently associated with the pairing of electrons in p orbitals. As shown in Table 4.4 and according to Hund's rule, when electrons are added within a sublevel, one electron is added to each orbital before a second is added to any orbital. In each element in family VIA, as for oxygen in Table 4.4, a fourth p electron is added to make a pair of electrons in one orbital. Probably owing to the repulsion between two electrons close to one another, when two electrons occupy an orbital the ionization energy needed to remove one of them is decreased. In Figure 3.13 the elements boron, carbon, and nitrogen lie on one smooth line, and the elements oxygen, fluorine, and neon on another. The members of the second set, in which p orbital electrons are being paired, have lower ionization energies than predicted on the basis of the first set.

★ Special Topic ★ Special Topic ★ Special Topic ★ Special Topic ★ Special Topic ★ Special Topic ★ Special Topic ★ Special Topic ★

Computers—Tools of the Chemical Theorist Because a computer can handle with fantastic speed information in the form of numbers, it can solve precisely problems that previously have had to be attacked by estimation and approximation. The equations that describe the theoretical behavior of electrons in atoms are so complex that theoreticians are forced to simplify them in order to state an approximate equation that can be solved. The use of computers has greatly reduced the need for simplification and has increased the precision with which theoretical descriptions can be calculated.

A. C. Wahl is one of the chemical theorists who have taken advantage of the rapid development of large electronic computers and the concurrent sophistication in their use to obtain computer-produced, accurate contour maps of electronic charge density derived from quantum mechanical computations. Contour diagrams of orbital and total charge density have been mapped out for a number of atoms.

A simple example of one such two-dimensional contour diagram is the probable electron density of the hydrogen atom shown in the accompanying figure. In this diagram, the innermost contour line corresponds to an electron density of 0.25 e^-/bohr^3, or 1.69 e^-/Å^3, and each successive outer line signifies a decrease by a factor of two, that is, 0.125, 0.0625, 0.0312, and so on, e^-/bohr^3. A more complex example, the fluorine atom's probable electron density, is also shown. Here, the innermost contour line corresponds to an electron density of 1.0 e^-/bohr^3, and each successive line signifies a decrease by a factor of two.

For atoms with more than one electron, the map of the total probable electron density is obtained by a merging of the contour maps of probable electron density for every occupied atomic orbital. For a fluorine atom, the electrons in the $1s$, $2s$, $2p_x$, $2p_y$, and $2p_z$ orbitals must be so merged. The size of the $1s$ orbital in the fluorine atom is much smaller than that of hydrogen or helium (not shown) due to the increased nuclear charge ($Z = 9$). This decrease in orbital size with increasing atomic number is a general trend for occupied atomic orbitals. It is an interesting fact that, just as the s orbitals

4•6 Photoelectron Spectra of Atoms

One possible way to determine experimentally the energy levels of electrons in atoms is to irradiate a sample with electromagnetic radiation of known energy and then measure the kinetic energies of the ejected electrons. This is an application of the photoelectric effect (see Section 4.2.4). It is only within about the past 25 years that instruments have been developed that are capable of determining the kinetic energies precisely enough to be useful. The fundamental relation has the same form as the equation for the photoelectric effect. We assume the total energy of the absorbed photon is used, first, in overcoming the attraction of the atom for the electron, called the **binding energy,** E_b. Second, any excess energy of the photon is used in imparting a certain kinetic energy to the ejected electron. The relation is

$$E_{\text{photon}} = h\nu = E_b + \text{K.E.}$$

From the known energy of the absorbed photon, $h\nu$, and the measured kinetic energies of the emitted electrons, K.E., we can calculate the binding energies of the electrons, E_b. If the energy of the radiation is sufficiently high, the energy of the absorbed photon is large enough to eject an electron from any one of the energy levels of the atom. Therefore, a series of binding energies, called a **photoelectron spectrum,** can be determined for an atom.

The photoelectron spectrum of neon, shown in Figure 4.17, reveals that electrons in a neon atom are held with three different binding energies. The largest is 667.0 eV:

★ Special Topic ★ Special Topic ★ Special Topic ★ Special Topic ★ Special Topic ★ Special Topic ★ Special Topic ★ Special Topic ★

are spherically symmetrical, the summation of a set of three *p* orbitals is also spherical. The overall shape of the complete fluorine atom does not have any of the characteristic dumbbell shape of the individual 2*p* orbitals.

The computer can solve many of our problems of atomic structure once the proper data are provided to it.

And this is apparently only the beginning of its sophisticated use. Computer programs have been extended to the description of the formation of molecules from atoms and the orbital structure of covalent diatomic molecules (see Section 6.2) and other species.

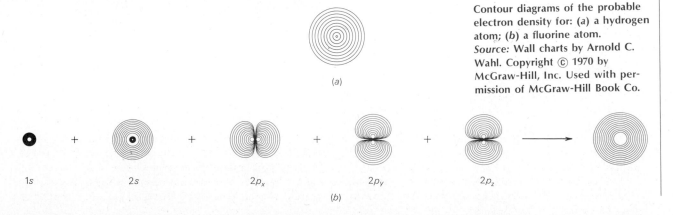

Contour diagrams of the probable electron density for: (*a*) a hydrogen atom; (*b*) a fluorine atom.
Source: Wall charts by Arnold C. Wahl. Copyright © 1970 by McGraw-Hill, Inc. Used with permission of McGraw-Hill Book Co.

(*a*)

1*s*   +   2*s*   +   2p_x +   2p_y +   2p_z ⟶

(*b*)

$$E_b = h\nu - \text{K.E.} = 1{,}253.6 - 586.6 = 667.0 \text{ eV}$$

The other two E_b values are calculated to be 45.0 and 21.6 eV. These three binding energies are associated with $1s$, $2s$, and $2p$ energy levels, respectively.

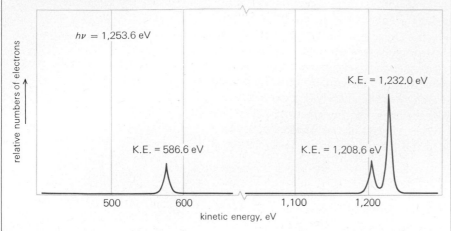

Figure 4•17 An idealized photoelectron spectrum of electrons from neon atoms irradiated by the K_α X ray of magnesium, which has an energy of 1253.6 eV.
Source: Courtesy of Dr. G. K. Schweitzer, The University of Tennessee, Knoxville.

Note that the binding energy of the most weakly held electron, 21.6 eV, is equal to the first ionization energy of neon (see Table 3.6). This is to be expected, for in both cases one outermost electron is knocked off a neutral gaseous atom. The second, third, and subsequent ionization energies, however, are not equal to the binding energies of electrons as determined by photoelectron spectroscopy. In measuring ionization energies, electrons are in turn knocked off particles of increasingly greater positive charge (see Figures 3.11 and 3.12). For example, the fourth ionization energy is that necessary to remove the outermost electron from an ion with a charge of $3+$. In measuring a binding energy, only one electron is knocked off, usually from a neutral particle, but it can be from any energy level. Table 4.5 gives the comparative data for values of the ionization energies and the binding energies for the magnesium atom.

Table 4•5 Comparison of ionization energies and binding energies for electrons in the magnesium atom, in electron volts

	1st	2nd	3rd	4th	5th	6th	7th	8th	9th	10th	11th	12th
E_{ion}	7.6	15	80	109	141	186	225	266	328	368	1,762	1,963
		$3s$			$2p$				$2s$		$1s$	
E_b	7.6			56					93		1,309	

When different orbitals have the same energies, they are said to be equivalent, or *degenerate*.[8] The binding energy data in Table 4.5 show that the three orbitals, $2p_x$, $2p_y$, and $2p_z$, are triply degenerate.

---

[8] The term *degenerate* is applied to different sets of quantum numbers that have the same energy. The term is used to point out that there is no way to distinguish between such states experimentally or even theoretically.

Chapter Review

Summary

All electromagnetic radiation has a **frequency,** ν, and a **wavelength,** λ, which are inversely proportional. The dispersion of radiation into its component wavelengths produces a **spectrum.** Most of our knowledge of the electronic structure of atoms comes from **atomic spectroscopy,** which entails the **emission** of radiation by hot atoms and the **absorption** of radiation by cool atoms. This radiation—indeed, *all* electromagnetic radiation—exists in the form of discrete **quanta,** or **photons,** of definite energy. *Thus, radiation has some particle-like properties.*

The energy of a photon is directly proportional to its frequency. This fact underlies the **photoelectric effect,** in which electrons are ejected by a metal surface only when the impinging photons are at or above the **threshold frequency,** ν_0. This frequency is determined by the **work function,** W_0, for that metal.

In the **Bohr theory** of atomic structure, the hydrogen emission spectrum was explained by proposing that (1) the electron can be in certain **stationary states,** that is, energy levels of discrete (quantized) values, and (2) radiation of a characteristic frequency is emitted or absorbed during a transition between the **ground state** and any one of various **excited states** of an electron. The energies of these states are specified by the **principal quantum number,** n, that tells which main level an electron is in.

The spectra and structures of more complex atoms are explained by **quantum mechanics,** or **wave mechanics,** in which the positions of electrons can be described only in terms of probable values—never definite values—as a consequence of the **Heisenberg uncertainty principle.** These probabilities are expressed mathematically in the **Schrödinger equation** as electron **wave functions,** ψ. *Thus, matter has some wavelike properties.* Wave functions are also called **atomic orbitals,** which describe regions in which a given electron is most likely to be found. The square of a wave function gives the **probable electron density** in a particular volume of space within the atom. The practical extent of an electron cloud around an atomic nucleus is given by the **90-percent contour.**

With the principal quantum number, three additional numbers are needed to describe the position (energy) of an electron. The **second quantum number,** l, specifies the energy sublevel; the **third quantum number,** m_l, specifies the orbital; and the **fourth quantum number,** m_s, specifies the alignment of the electron spin in the orbital. Each electron in an atom is uniquely characterized by its four quantum numbers, as required by the **Pauli exclusion principle.**

The order of filling orbitals (aufbau principle) is governed in part by **Hund's rule,** which is based on certain trends in ionization energies in the periodic table. From the photoelectron spectrum of an element we can determine the **binding energies,** E_b, of electrons in different energy sublevels.

The study of the interaction of electromagnetic radiation with matter has led to an understanding of the electronic structure of atoms. Experiment and theory have been fruitfully combined, each stimulating and guiding the other's development.

Exercises

Electromagnetic Radiation

1. A 120-V automatic turntable is set to operate at 60-Hz alternating current. Convert the frequency in hertz to a wavelength in meters.

2. A microwave oven produces electromagnetic radiation at a frequency of 2,450 MHz. Is this frequency in the microwave region of the electromagnetic spectrum? What is the wavelength in meters that corresponds to this frequency?

3. Calculate the frequency in hertz of an X ray that has a wavelength of 1.0×10^{-9} m.

4. Channels 2 through 13 (VHF) operate at frequencies between 59.5 and 215.75 MHz; channels 14 through 83 (UHF) operate at frequencies between 475.75 and 889.76 MHz. FM radio stations transmit signals between the wavelengths 2.78 and 3.41 m. Are FM radio signals received below channel 2, between channels 13 and 14, above channel 83, or within some other portion of the electromagnetic spectrum?

5. Standard AM radio reception is in the frequency range of 550 to 1,600 kHz. Does this correspond to a more narrow wavelength band than that of FM radio?

*6. A light-year is the distance that visible light and other forms of electromagnetic radiation travel in one year. The small red star known as *Proxima Centauri,* one of the three stars in the Alpha Centauri group, is the star closest to our sun. It is now about 4.3 light-years away. (It gets closer every day.) This star emits X rays, gamma rays, and visible light. Which of these types of radiation will reach the earth first? How long will it take each of these radiations to reach the earth? The earth is 93 million miles from the sun. Is this distance significant to the calculations?

Atomic Spectra

7. How could you determine whether the colored light from a neon sign is due to a mixture of colors (polychromatic) or to a single color (monochromatic)?

8. Does light from the sun have more than, less than, or the same number of wavelengths in the visible range as

white light from an incandescent solid? Cite information from a figure in this text to substantiate your choice.

9. Can you determine the color of toluene by a study of its absorption spectrum shown in Figure 4.7? Explain why or why not.

10. What is the relationship, if any, between the discontinuous emission spectrum and the absorption spectrum of sodium?

11. In Section 4.2.3, Planck's constant of proportionality, h, is given as 6.626×10^{-34} J $\cdot$ s $\cdot$ photon^{-1}. What is the value of this constant in units of eV $\cdot$ s $\cdot$ photon^{-1} and kcal $\cdot$ s $\cdot$ photon^{-1}?

12. In the visible spectrum of excited hydrogen there is a green line with a wavelength of 486.1 nm. Without calculating the frequency, calculate the energy for this emission line in joules per photon.

13. Calculate the energy in joules per photon for each of the frequencies given in Exercises 1, 2, and 5.

14. If your car's headlights are dim, does this mean that the energy per photon of the light emitted is less than when the headlights glow brightly? Explain.

*15. When ultraviolet radiation of frequency 1.6×10^{15} s^{-1} strikes the metal cesium, photoelectrons with a maximum kinetic energy of 4.5 eV are ejected. Calculate each of the following:
(a) the energy of the incident photons in electron volts
(b) the work function of the metal in electron volts
(c) the threshold frequency of the metal

16. The work function for potassium is 2.2 eV. Will photons with a wavelength of 3,500 Å be sufficient to dislodge electrons from the metal surface? If electrons are dislodged, what is the maximum kinetic energy of the emitted electrons?

17. The three Paschen lines in the hydrogen emission spectrum involve the fall of electrons to the third main level from higher energy levels. Use Equation (1) in Section 4.2.5 to calculate the wavelengths in meters that correspond to these emission lines. In what region of the electromagnetic spectrum are these emission lines found?

18. In addition to the Lyman, Balmer, and Paschen series of lines in the hydrogen emission spectrum, there are lines discovered by Brackett and Pfund. The Brackett lines are said to result from the fall of an electron from the sixth and fifth main levels to the fourth main level. The line discovered by Pfund is said to result from the fall of an electron from the sixth to the fifth main level. Calculate the frequencies in hertz and the wavelengths in nanometers that correspond to these emission lines. In what region of the electromagnetic spectrum is each of these lines found?

19. Use Equation (2) in Section 4.2.5 to calculate the frequency of the Pfund emission line in Exercise 18. (Given, e is 1.6022×10^{-19} C, m is 9.1096×10^{-31} kg, h is 6.626×10^{-34} J $\cdot$ s $\cdot$ photon^{-1}, and 1 C = $9.4805 \times$

10^4 kg$^{1/2}$ $\cdot$ m$^{3/2}$ $\cdot$ s^{-1}.) Compare your answer with that obtained in Exercise 18.

20. Prove by calculation that the Rydberg constant, R_H, of 1.097×10^7 m^{-1} is the reciprocal of the wavelength for the ionization energy of the hydrogen atom.

21. How much energy in joules is required for the ionization of an electron in the hydrogen atom from the third main level? How many joules would be required for such an excitation for a mole of hydrogen atoms?

22. What is the relationship between the ionization energy for a hydrogen atom and the following?
(a) the second ionization energy of a helium atom
(b) the third ionization energy of a lithium atom
(c) the fourth ionization energy of a beryllium atom
Predict a value in electron volts for the fifth ionization energy of a boron atom. Compare your prediction with the value listed in Table 3.6.

Wave Nature of Matter;
Uncertainty Principle

*23. Major league baseball pitcher James Rodney Richard has had his fastball clocked from the pitcher's mound to home plate at a speed of 98 mi/h. Calculate the de Broglie wavelength of the moving baseball, if the ball weighs 5.0 oz.

*24. Calculate the de Broglie wavelength of an alpha particle that has a kinetic energy of 27 MeV.

25. Calculate the uncertainty in the position, Δx, of an electron with an uncertainty in its velocity equal to half the speed of light.

26. (a) Would the uncertainty in the position of the baseball in Exercise 23 be more than or less than that of the electron described in Exercise 25? (Assume an uncertainty in the speed of the baseball of 1.0 mi/h.)
(b) Would your answer in part (a) be different if the baseball were moving with an uncertainty in its speed equal to half the speed of light?

*27. Show mathematically that one can measure with greater precision the position of a moving particle with a mass of 100 amu than one with a mass of 0.01 amu if both are moving with a velocity of 1×10^3 m $\cdot$ s^{-1}. (Assume that the measured momentum is in error by 100 percent in each case.)

28. Describe each of the following:
(a) a wave function
(b) an atomic orbital
(c) the probable electron density
(d) the radial probability

29. The radius of the second Bohr orbit for the hydrogen atom is 2.12 Å, that is, $(2)^2(0.529)$ Å.
(a) Is this radius within the 90-percent contour line (see Figure 4.11)?
(b) What do you think is the radius of the third Bohr orbit for the hydrogen atom? Of the fourth Bohr orbit?

Quantum Numbers; Electronic Configurations

30. (a) What are the values of the second quantum number, l, for the fourth main energy level? Which types of orbitals do these values of l indicate?
    (b) What are the values of m_l for the fourth main energy level? For each value of l, how many allowed values of m_l are there? What do these values of m_l tell us about the orbitals of an atom?

31. How many orbitals are there in the fourth main energy level of an atom? How many of these orbitals are p orbitals? d orbitals? f orbitals? How many electrons can each p orbital hold? Each d orbital?

32. One of the four quantum numbers is not required to describe the orbitals of a hydrogen atom. Which one is it? Why is this quantum number not required?

33. Write the sets of quantum numbers for the electrons in an atom of nitrogen. Repeat for the electrons in an atom of sulfur.

34. In an excited state of a carbon atom with all electrons in the first two main levels, there are no paired electrons in the second main level. With this information, write the sets of quantum numbers for the electrons in this excited carbon atom.

35. Write all possible sets of quantum numbers for electrons in the $4f$ orbitals of an atom.

36. Using subscripts x, y, and z where appropriate to distinguish between orbitals, write the core abbreviated electronic configuration for the atoms listed in Exercises 33 and 34.

37. Repeat Exercise 36, but now use the circle-and-arrow representations shown in Table 4.4 to describe the electron arrangements.

38. Consider Hund's rule and Pauli's exclusion principle. Did you have occasion to use either or both in answering Exercises 33 and 34? How?

39. According to the aufbau principle, the nineteenth electron goes into a $4s$ orbital rather than a $3d$ orbital. What experimental data seem to justify this order?

Photoelectron Spectra of Atoms

40. How many different values of the binding energy for electrons in an atom of phosphorus would you expect there to be?

*41. Given that the binding energy of a $2s$ electron in an atom of argon is 326 eV, calculate the frequency of a photon that can bring about the ejection of a $2s$ electron if the electron is ejected from the argon atom with a kinetic energy of 784 eV.

42. The binding energies for the electrons in an atom of argon are 16, 29.5, 249.5, 326.5, and 3,206 eV. Assign these binding energies to their respective orbitals.

43. Account for the differences in the binding energies of the outer main level electrons in atoms of neon and argon. (Refer to Figure 4.17 and Exercise 42 for data.)

Chemical Compounds

5

5•1 **How Atoms Combine**

5•1•1 Classes of Elements

5•1•2 Valence

5•1•3 Atomic Structure and Chemical Reactions

5•1•4 Inferences from Properties of Noble Gases

5•2 **Transfer of Electrons**

5•2•1 Rules of Eight and Two for Ions

5•2•2 Limitations of Rule of Eight for Ions

5•3 **Sharing of Electrons**

5•3•1 Limitations of Rule of Eight for Covalent Compounds

5•4 **Diagramming Lewis Formulas for Covalent Molecules**

5•4•1 Molecules with Single Bonds

5•4•2 Molecules with Multiple Bonds

5•4•3 Coordinate Covalent Bonds

5•5 **Polar Covalent Bonds**

5•5•1 Electronegativity

5•5•2 Dipole Moments

5•6 **Sizes of Atoms, Molecules, and Ions**

5•6•1 Covalent Radii and van der Waals Radii

5•6•2 Metallic Radii

5•6•3 Ionic Radii

5•6•4 Trends in Sizes of Atoms and Ions

5•7 **Oxidation Numbers**

5•7•1 The Periodic Table and Oxidation Numbers

5•7•2 Calculation of Oxidation Numbers

5•7•3 Polyatomic Ions

5•7•4 Formulas from Oxidation Numbers

5•8 **The Systematic Naming of Compounds**

5•8•1 Binary and Related Compounds

5•8•2 Ternary Compounds Containing Oxygen

5•8•3 Oxidation Numbers and Names of Compounds

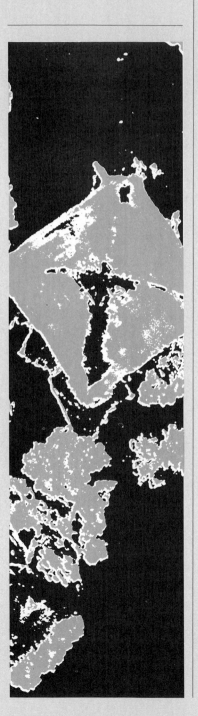

Since the discovery of the electronic structure of atoms, chemists and physicists have been able to investigate the ways in which atoms of one kind join with another kind. The guiding principle is that atoms act on one another by way of the electrons in their outer energy levels. This interaction of electrons leads to the strong forces of attraction, the *chemical bonds,* that hold atoms together in compounds.

A compound does not resemble the elements from which it is formed; it must be considered a unique substance. The characteristics of a compound depend in great measure on the types of chemical bonds that hold its atoms together. In this chapter we will begin to study the various types of chemical bonds that enable a few more than 100 different kinds of atoms to combine to form more than 4 million compounds.

In the compound sodium chloride, NaCl (ordinary table salt), produced by the reaction of sodium with chlorine, there is a different type of bonding than in the compound water, H_2O, produced by the reaction of hydrogen with oxygen. In this chapter, we will seek to understand the differences in such compounds by analyzing the processes of their formation in terms of the individual particles of the elements that are involved—the atoms. We will also learn how to deduce the formulas and names of some common types of compounds.

5•1 How Atoms Combine

5•1•1 Classes of Elements Based on physical properties, the elements are divided into three classes: metals, nonmetals, and metalloids. Elements that show a metallic luster when polished and are malleable (can be hammered into thin sheets), ductile (can be drawn into wires), and good conductors of heat and electricity are classed as **metals.** Elements that do not possess these characteristics to a considerable degree are **nonmetals.** Elements in a class between these, called the **metalloids,** or borderline elements, have some properties that resemble those of metals and some like those of nonmetals. The three classes are indicated in the periodic table inside the front cover. In the middle of the right-hand page there is a zigzag line that separates the metals (to the left) from the nonmetals (to the right). Near the line are the metalloids, elements such as boron, B, silicon, Si, germanium, Ge, arsenic, As, antimony, Sb, and tellurium, Te.

5•1•2 Valence From the formulas of such compounds as H_2O, H_2O_2, HCl, NaCl, $CaCl_2$, NH_3, CH_4, CO, CO_2, FeO, and Fe_2O_3, it is clear that atoms of different elements have different capacities for combining with each other. The combining capacity of an element is called its **valence.** At one time, valence was defined as the number of atoms of hydrogen that one atom of an element could combine with or take the place of in forming compounds. As indicated by the formulas, in H_2O, oxygen was assigned a valence of 2; in HCl, chlorine was assigned a valence of 1; and in NH_3 and CH_4, nitrogen and carbon were assigned valences of 3 and 4, respectively.

Today with the more recent knowledge about chemical reactions based on the electron structures of atoms, the early definition of valence is of little value. But the word *valence* is often used to refer to the combining power of an atom, and valence numbers are still assigned on the basis of formulas of compounds.

5•1•3   Atomic Structure and Chemical Reactions   The most important structural feature of atoms in determining chemical behavior is the number of electrons in their outermost energy levels. These outermost electrons are commonly referred to as **valence electrons.** When atoms of one element combine with those of another, there is always some change in the distribution of electrons in the outermost energy levels. The study of many elements and compounds has led to the idea that, in compound formation, atoms of certain elements tend to gain electrons and others tend to lose electrons. As a result of these tendencies, two atoms may either *transfer* or *share* electrons. Either process may provide for a stable arrangement of electrons that results in the formation of a compound.

By analyzing the properties of a compound composed of a metal and a nonmetal, sodium chloride (NaCl), for example, we can draw conclusions about what happens to the electrons as the compound is formed from the atoms. Consider the following behavior as illustrated by Figure 5.1. When charged electrodes are placed in molten sodium chloride, the metal portion, the sodium, tends to migrate toward the negatively charged ($-$) electrode (the cathode). The nonmetal portion, the chlorine, tends to migrate toward the positively charged ($+$) electrode (the anode). Many other metal–nonmetal compounds behave similarly. These observations lead to the following conclusions:

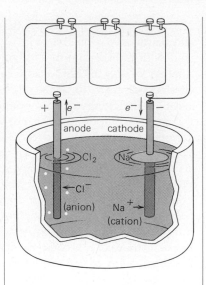

Figure 5•1 Schematic representation of what happens when charged electrodes are placed in molten sodium chloride, NaCl.

1.   A compound of a metal and a nonmetal is composed of positive and negative particles called ions. In the case of NaCl, they are sodium ions, Na^+, and chloride ions, Cl^-. Positively charged ions are attracted to the cathode and are called **cations.** Negatively charged ions are attracted to the anode and are called **anions.**

2.   The metal atoms lose electrons to form positive ions (cations) when they react with the nonmetal, and the nonmetal atoms form negative ions (anions) by gaining electrons from the metal; that is, metal and nonmetal atoms react chemically by the transfer of electrons.

On the other hand, a compound composed of only nonmetals, for example, ethyl alcohol (C_2H_6O), does not show the property of having its parts migrate toward charged electrodes. In this case the individual particles are neutral because electrons are shared between atoms.

5•1•4   Inferences from Properties of Noble Gases   Among the nonmetals there is a family of elements, the noble gases (group VIIIA), that do not combine readily with other elements. Prior to 1962 these elements were called *inert,* because they were thought to be completely unreactive. In that year their first compounds were made. The fact that there is a family of elements whose members form compounds rarely, if at all, suggests that the atoms of these elements have very stable electron arrangements. They have little tendency to gain, lose, or share electrons with other atoms. The arrangement of electrons in the noble gases is summarized in Table 5.1. All noble-gas atoms, except those of helium, have eight electrons in the outermost energy level. This is the number of electrons needed to fill the s and p sublevels of that level (only two electrons are needed for helium atoms, but they completely fill the first main level). Each of the noble gases, except

Table 5•1 Arrangements of electrons in noble gases

Noble gas	Symbol	Atomic number	Inner levels (noble-gas core)	Outermost main level
helium	He	2		$1s^2$
neon	Ne	10	(He)	$2s^2 2p^6$
argon	Ar	18	(Ne)	$3s^2 3p^6$
krypton	Kr	36	(Ar)$3d^{10}$	$4s^2 4p^6$
xenon	Xe	54	(Kr)$4d^{10}$	$5s^2 5p^6$
radon	Rn	86	(Xe)$5d^{10} 4f^{14}$	$6s^2 6p^6$

helium, has an outer level electronic configuration of $s^2 p^6$. Further, it can be seen from a study of the periodic table inside the front cover that, except for palladium, no atom contains more than eight electrons in its outer main level.

The electronic configurations of the unreactive noble-gas atoms help us to explain how atoms of reactive elements interact with one another. It is thought that by combining with one another, atoms of many of the other elements tend to achieve the electronic configurations of the noble gases. These configurations are attained in one of two ways: (1) *by the transfer of outer level electrons from the atoms of one element to those of another,* or (2) *by the sharing of electrons by two or more atoms.*

The new substances that result when two or more elements combine by sharing or transferring electrons are called **compounds.**

5•2   Transfer of Electrons

In general, when a metallic element combines with a nonmetallic element, electrons are lost by atoms of the metal and gained by atoms of the nonmetal.

Atoms of lithium, sodium, and potassium lose one electron easily; beryllium, magnesium, and calcium lose two. These six elements are metals. Fluorine and chlorine atoms each gain one electron; oxygen and sulfur atoms each gain two. These four elements are nonmetals. When a sodium atom loses an electron to a chlorine atom, we say that they have combined to make sodium chloride, NaCl (ordinary table salt). The reaction between the sodium and chlorine atoms can be indicated as in Figure 5.2, or it can be shown diagrammatically as follows, using dots to indicate only the electrons in the highest main energy levels, that is, the valence electrons:

$$\text{Na} \cdot + \cdot \overset{..}{\underset{..}{\text{Cl}}} : \longrightarrow \text{Na}^+ + : \overset{..}{\underset{..}{\text{Cl}}} : {}^-$$

The notation Na· refers to the uncharged sodium atom, which has one valence electron; the notation Na$^+$ refers to the sodium cation, which has a net charge of 1+ (11 p^+ in the nucleus, 10 e^- outside). The notation ·Cl: refers to the uncharged chlorine atom, which has seven valence electrons; the notation :Cl:$^-$ refers to a chloride anion, which has a net charge of 1− (17 p^+ in the nucleus, 18 e^- outside).

| sodium atom | chlorine atom | sodium ion | chloride ion |

Figure 5•2 Sodium and chlorine atoms react by the transfer of an electron to yield ions.

In arranging the dots around the atomic symbols, it is customary to indicate two electrons in an s sublevel with a pair of dots, for example, showing only electrons in the highest main level,

$$\text{Mg:} \qquad \text{for magnesium, } 3s^2$$

Further, as we learned in Chapter 4, within the p sublevel of an atom there are three atomic orbitals, which are designated as p_x, p_y, and p_z. Each of these orbitals can be occupied by a pair of electrons. However, one electron will go into each orbital before pairing occurs (see Section 4.5.3). Showing only electrons in the highest main levels of atoms, some examples are

$$\overset{\textstyle\cdot}{\text{Al}}: \qquad \text{for aluminum , } 3s^2 3p^1$$

$$\cdot\overset{\textstyle\cdot}{\underset{\textstyle\cdot}{\text{P}}}: \qquad \text{for phosphorus, } 3s^2 3p_x{}^1 3p_y{}^1 3p_z{}^1$$

$$\cdot\overset{\textstyle\cdot\cdot}{\underset{\textstyle\cdot\cdot}{\text{O}}}: \qquad \text{for oxygen, } 2s^2 2p_x{}^2 2p_y{}^1 2p_z{}^1$$

$$\cdot\overset{\textstyle\cdot\cdot}{\underset{\textstyle\cdot\cdot}{\text{Cl}}}: \qquad \text{for chlorine, } 3s^2 3p_x{}^2 3p_y{}^2 3p_z{}^1$$

Other examples are shown in Figure 3.14.

When electrons are transferred from one atom to another, ions are formed. Figure 5.3 illustrates that when magnesium atoms combine with fluorine atoms to produce magnesium fluoride, MgF_2, magnesium tends to lose its two third energy level electrons and attain a highest main level with eight electrons, while fluorine gains one electron in completing its highest level. This also can be indicated by the electron-dot notations shown below:

$$\text{Mg:} + 2\cdot\overset{\textstyle\cdot\cdot}{\underset{\textstyle\cdot\cdot}{\text{F}}}: \longrightarrow \text{Mg}^{2+} + 2:\overset{\textstyle\cdot\cdot}{\underset{\textstyle\cdot\cdot}{\text{F}}}:^{-}$$

The magnesium atom loses two electrons to form a magnesium cation with a charge of $2+$ (12 p^+ in the nucleus, 10 e^- outside). The particle formed when the fluorine atom gains one electron is a fluoride anion, which has a charge of $1-$ (9 p^+ in the nucleus, 10 e^- outside). The formation of *one* magnesium cation must result in the formation of *two* fluoride anions. The total cation charge must be balanced by the total anion charge. In this case, the charge of the magnesium cation $(2+)$ is balanced by that of the two fluoride anions, $(2)(1-) = (2-)$. Another way of looking at the balance of charges is to note that *the total electrons lost must equal the total electrons gained in any electron transfer reaction.*

magnesium   two fluorine   magnesium   two fluoride
atom   atoms   ion   ions

Figure 5.3 A magnesium atom gives up two electrons to form a stable Mg^{2+} ion; a fluorine atom gains one electron to form a stable F^- ion. Hence, magnesium and fluorine react in a $1:2$ ratio by atoms.

Using electron-dot notations for the combination of metal atoms with nonmetal atoms, we can account for the ionic composition of compounds such as potassium sulfide, K_2S, calcium oxide, CaO, and sodium nitride, Na_3N.

1. Combination of potassium and sulfur atoms:

$$2K\cdot + \cdot \ddot{\underset{..}{S}}: \longrightarrow 2K^+ + :\ddot{\underset{..}{S}}:^{2-}$$

2. Combination of calcium and oxygen atoms:

$$Ca: + \cdot \ddot{O}: \longrightarrow Ca^{2+} + :\ddot{\underset{..}{O}}:^{2-}$$

3. Combination of sodium and nitrogen atoms:

$$3Na\cdot + \cdot \ddot{N}: \longrightarrow 3Na^+ + :\ddot{\underset{..}{N}}:^{3-}$$

It should be emphasized that each of these atom combinations is merely a scheme to help us write the electron-dot formula for a known compound. For example, the electron-dot formula for calcium oxide was constructed from the combination of calcium and oxygen atoms. This is not the balanced equation that is written for the chemical reaction of calcium and oxygen. The equation for that direct combination reaction is written as

$$2Ca + O_2 \longrightarrow 2CaO$$

which shows oxygen in its normal elemental state, the diatomic molecule, O_2. We write combinations of diagrams of atoms only to help us construct the electron-dot structures of compounds, not to take the place of the standard balanced equations.

• Example 5•1 •

Show with electron-dot formulas how atoms react with one another to form the following ionic compounds: (a) lithium oxide, Li_2O; (b) barium bromide, $BaBr_2$; and (c) aluminum oxide, Al_2O_3.

• Solution •

(a) Electron-dot notations are $Li\cdot$ and $\cdot\ddot{O}:$ for lithium and oxygen atoms. A lithium atom loses its second main level electron to form a lithium cation. Because an oxygen atom gains two electrons to complete its main level and form an oxide anion, there must be two lithium atoms. Summarizing,

$$2Li\cdot + \cdot\ddot{O}: \longrightarrow 2Li^+ + :\ddot{\underset{..}{O}}:^{2-}$$

(b) Electron-dot notations are $Ba:$ and $\cdot\ddot{Br}:$ for barium and bromine atoms. A barium atom loses two electrons and a bromine atom gains one electron. Hence two bromine atoms are required for each barium atom:

$$Ba: + 2\cdot\ddot{Br}: \longrightarrow Ba^{2+} + 2:\ddot{Br}:^-$$

(c) Electron-dot notations are Al: and ·Ö: for aluminum and oxygen atoms. An aluminum atom loses three electrons and an oxygen atom gains two electrons. In order that the total number of electrons lost by Al atoms equals the total electrons gained by O atoms, a ratio of Al:O atoms of 2:3 is required. That is, $2 \times 3\,e^- = 6\,e^-$ lost by Al atoms are needed to balance $3 \times 2\,e^- = 6\,e^-$ gained by O atoms. Summarizing,

$$2\text{Al:} + 3\cdot\ddot{\text{O}}\text{:} \longrightarrow 2\text{Al}^{3+} + 3\text{:}\ddot{\text{O}}\text{:}^{2-}$$

See also Exercises 9 and 10 at the end of the chapter.

When the atoms of two elements combine with each other by transferring electrons, the substance formed does not resemble either of the original materials. The new substance is composed not of atoms but of ions. These ions, some positive and some negative, are strongly attracted to one another by the electrostatic attraction between unlike charges. A substance such as sodium chloride consists of positive and negative ions arranged in a well-ordered fashion in a crystal. Each positive ion is surrounded by negative ions and each negative ion by positive ones, as shown in Figure 5.4. The attraction that binds oppositely charged ions together is called an **ionic bond.** Compounds in which particles are held together by ionic bonds are called **ionic compounds.**[1] Both sodium chloride and magnesium fluoride are ionic compounds. The formula for sodium chloride is NaCl, because the Na^+ and Cl^- ions are present in a 1:1 ratio. For magnesium fluoride, the Mg^{2+} and F^- ions are present in a 1:2 ratio, and the formula is MgF_2.

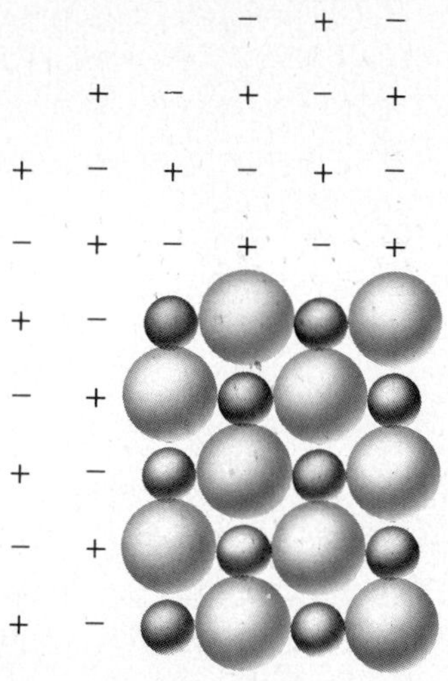

Figure 5.4 A two-dimensional layer of ions in one type of ionic crystal. In three dimensions, each ion (other than those on the crystal surface) will have six oppositely charged ions as its closest neighbors.

5.2.1   Rules of Eight and Two for Ions The remarkable stability of a highest main energy level that has its s and p sublevels filled is not only seen in atoms of the noble gases but is also revealed by the tendency of atoms of other elements to form ions. For example, we have shown in Figures 5.2 and 5.3 that sodium (atomic number 11) forms a unipositive ion, Na^+, whereas magnesium (atomic number 12) forms a dipositive ion, Mg^{2+}. Both fluorine (atomic number 9) and chlorine (atomic number 17) form uninegative ions, F^- and Cl^-. Each of these four ions has the outer level electronic configuration of a noble gas, s^2p^6. The arrangements of electrons of these atoms and ions are as follows:

Atoms		Ions	
Na	$1s^22s^22p^63s^1$	Na^+	$1s^22s^22p^6$
Mg	$1s^22s^22p^63s^2$	Mg^{2+}	$1s^22s^22p^6$
F	$1s^22s^22p^5$	F^-	$1s^22s^22p^6$
Cl	$1s^22s^22p^63s^23p^5$	Cl^-	$1s^22s^22p^63s^23p^6$

The sodium cation, the magnesium cation, and the fluoride anion (as well as the oxide, O^{2-}, and nitride, N^{3-}, anions) all have the electronic structure $1s^22s^22p^6$, identical with that of the neon atom. Atoms and ions that

[1] The term *electrovalent* is often used instead of *ionic* in referring to substances, as in electrovalent bonds or electrovalent compounds.

have the same number of electrons are said to be **isoelectronic.** The chloride ion, Cl^-, the potassium ion, K^+, the calcium ion, Ca^{2+}, and the sulfide ion, S^{2-}, are all isoelectronic with the argon atom, Ar. They all have the configuration $1s^2 2s^2 2p^6 3s^2 3p^6$.

There is a tendency for atoms with atomic numbers within about three units of that of a noble gas to gain or lose electrons so as to form ions that are isoelectronic with the noble gases. Such ions have eight electrons in their highest main energy level. This tendency was called the **rule of eight** or **octet rule** long before it was correlated with the $s^2 p^6$ structures of the highest main energy levels.

The elements H, Li, and Be, with atomic numbers 1, 3, and 4, tend to follow a **rule of two** when forming ions. The ions formed, H^-, Li^+, and Be^{2+}, are isoelectronic with helium; that is, they have a single energy level designated $1s^2$.

We can often predict the electronic configurations of ions from the structure of their atoms. Using the aufbau principle introduced in Chapter 3, the structure of an atom of yttrium ($Z = 39$) is written as

$$1s^2 2s^2 2p^6 3s^2 3p^6 4s^2 3d^{10} 4p^6 5s^2 4d^1 \quad \text{or} \quad (Kr)4d^1 5s^2$$

We predict that yttrium would lose three electrons, the two $5s$ electrons and the one $4d$ electron, to form a Y^{3+} ion that has a structure isoelectronic with krypton:

$$1s^2 2s^2 2p^6 3s^2 3p^6 4s^2 3d^{10} 4p^6 \quad \text{or} \quad (Ar)3d^{10} 4s^2 4p^6$$

Note the eight electrons (in color) in the fourth main energy level.

The structure of an arsenic atom is written as

$$1s^2 2s^2 2p^6 3s^2 3p^6 4s^2 3d^{10} 4p^3 \quad \text{or} \quad (Ar)3d^{10} 4s^2 4p^3$$

We predict that arsenic would gain three electrons to form As^{3-}. The completion of the $4p$ sublevel is shown by writing $4p^6$ as the last term, indicating that As^{3-} has a structure isoelectronic with krypton. Thus, As^{3-}, Kr, and Y^{3+} are isoelectronic.

5•2•2   Limitations of Rule of Eight for Ions The idea that ions are formed when atoms gain or lose electrons to attain the structure of a noble gas is a rule that is sometimes of value and sometimes not. In the examples just given, the rule of eight enables us to predict correctly the ionic structures of elements closely before and closely after family VIIIA, for example, Na^+, Cl^-, Mg^{2+}, F^-, Y^{3+}, and As^{3-}. However, consider gallium atoms, which commonly lose three electrons to form Ga^{3+} ions. The atom has the structure $1s^2 2s^2 2p^6 3s^2 3p^6 4s^2 3d^{10} 4p^1$. The ion has a similar structure, but without the $4s^2$ and $4p^1$ electrons: $1s^2 2s^2 2p^6 3s^2 3p^6 3d^{10}$. As a rule, electrons in the outermost energy level are the first to be lost in forming cations, although they may not be the last electrons added in following the aufbau principle. In the formation of the Ga^{3+} ions, the two $4s$ electrons are lost and the ten $3d$ electrons remain, although the $3d$ electrons were added after the $4s$ electrons in the theoretical buildup of the atom.

The ionization of iron atoms to Fe^{2+} and Fe^{3+} ions also illustrates the rule that the outer electrons are lost first in forming ions. The complete electronic configuration of an iron atom is $1s^2 2s^2 2p^6 3s^2 3p^6 4s^2 3d^6$. When iron forms an Fe^{2+} ion, it loses its two $4s$ electrons to attain the configuration $1s^2 2s^2 2p^6 3s^2 3p^6 3d^6$. When iron forms an Fe^{3+} ion, it loses the two $4s$ electrons and one of the $3d$ electrons and attains the configuration $1s^2 2s^2 2p^6 3s^2 3p^6 3d^5$.

The loss of electrons to form ions is not simply the reverse of the imaginary buildup of an atom according to its position in the periodic table. In the aufbau scheme, we consider the addition of a proton and an electron for each new atom; but in forming ions, only electrons are lost and the positive charge on the nucleus remains the same.

Note that the structures of Ga^{3+}, Fe^{2+}, and Fe^{3+} are not isoelectronic with any of the noble gases, so that the rule of eight is of no use in these cases. Another example to which the rule is not applicable is the common As^{3+} ion. Arsenic, As, is in group Vᴀ. Its outer level electronic configuration is $4s^2 4p^3$. When an atom of arsenic forms an As^{3+} ion, the three $4p$ electrons are lost, but the two $4s$ electrons remain.

In beginning our study of ions we shall limit our predictions to those ions that do achieve a structure with eight electrons in their highest main levels. However, we shall also encounter examples of other types of ions formed by many common elements, such as the ions of gallium, iron, and arsenic just mentioned. We shall memorize these ionic charges as we need them. As we develop skill in using sublevel structures, we shall be able to predict charges on some ions that do not follow the simple rule of eight.

5•3   Sharing of Electrons

Two nonmetal atoms, both of which tend to gain electrons, may combine with each other by sharing one or more pairs of electrons. For example, an atom of bromine and an atom of fluorine, each having seven electrons in its highest level (valence level), unite to form a molecule of bromine fluoride, BrF, by sharing two of these fourteen electrons between them. Using electron-dot notations, this may be diagrammed as follows:

$$: \overset{..}{\underset{..}{Br}} \cdot \ + \ \cdot \overset{..}{\underset{..}{F}} : \ \longrightarrow \ : \overset{..}{\underset{..}{Br}} : \overset{..}{\underset{..}{F}} :$$

As a result of sharing a pair of electrons, each atom can be considered isoelectronic with a noble gas. By including the pair of shared electrons common to both atoms, each atom has eight electrons in its valence level:

For many simple compounds, the rule of eight is a satisfactory guide for predicting the number of electrons to be shared in building up the highest levels.

In general, when a nonmetallic element combines with another nonmetallic element, electrons are neither gained nor lost by the atoms, but are shared.

The strong force that binds the bromine atom to the fluorine atom is the attraction of each for the electrons that are held jointly. A shared pair of electrons is called a **covalent bond.** Compounds whose atoms are joined by covalent bonds are called **covalent compounds.**

The hydrogen atom is noted for its tendency to form a covalent bond. The sharing of electrons between two hydrogen atoms (H· + ·H) leads to the formation of the covalent hydrogen molecule (H:H), H_2. In this molecule, each H atom follows the rule of two. Using electron-dot notations, we can account for the known compositions of hydrogen fluoride, HF, water, H_2O, ammonia, H_3N (generally written as NH_3), and methane, H_4C (generally written as CH_4), on the basis of the rules of eight and two:

$$H\cdot \ + \ \cdot \ddot{\underset{..}{F}}: \ \longrightarrow \ \ \ H:\ddot{\underset{..}{F}}: \qquad (H\ (:)\ \ddot{\underset{..}{F}}:)$$
hydrogen fluoride

$$2H\cdot \ + \ \cdot \underset{\cdot}{\ddot{O}}: \ \longrightarrow \ \underset{\underset{H}{\displaystyle |}}{H:\ddot{O}:} \qquad (H\ (:)\ \overset{..}{\underset{..}{O}}\ (:)\ H)$$
water

$$3H\cdot \ + \ \cdot \underset{\cdot}{\ddot{N}}\cdot \ \longrightarrow \ \underset{\underset{H}{\displaystyle |}}{H:\ddot{N}:H} \qquad (H\ (:)\ \overset{..}{N}\ (:)\ H)$$
ammonia

$$4H\cdot \ + \ \cdot \underset{\cdot}{\dot{C}}\cdot \ \longrightarrow \ \underset{\underset{H}{\displaystyle |}}{\overset{\overset{H}{\displaystyle |}}{H:\ddot{C}:H}} \qquad (H\ (:)\ C\ (:)\ H)$$
methane

Note that, in order to fill its highest energy level with eight electrons, fluorine (group VIIA) with *seven* valence electrons shares *one* pair with a hydrogen atom. Oxygen (group VIA) with *six* valence electrons shares pairs with each of *two* hydrogen atoms. Nitrogen (group VA) with *five* valence electrons shares pairs with each of *three* hydrogen atoms, and carbon (group IVA) with *four* valence electrons shares pairs with each of *four* hydrogen atoms.[2] For each compound, the number of H atoms plus the number of valence electrons in the other nonmetal atom equals eight.

As with hydrogen, each of the group VIIA elements can be isoelectronic with a noble gas by sharing just one electron pair. Therefore, these elements form compounds with many of the same elements that combine with hydrogen. For example, a series of covalent compounds can be formulated by using chlorine in place of hydrogen, as shown in the following cases

---

[2]In the formulation of methane, the expected diagram for the carbon atom might be $\dot{\ddot{C}}$: (see Figure 3.14). But because the known formula CH_4 indicates that four bonds are to be formed, the diagram $\cdot\dot{C}\cdot$ is used for carbon.

for chlorine fluoride, ClF, dichlorine oxide, Cl_2O, nitrogen trichloride, NCl_3, and carbon tetrachloride, CCl_4:

$$:\ddot{\underset{..}{Cl}}\cdot \;+\; \cdot\ddot{\underset{..}{F}}: \;\longrightarrow\; :\ddot{\underset{..}{Cl}}:\ddot{\underset{..}{F}}:$$

chlorine fluoride

$$2:\ddot{\underset{..}{Cl}}\cdot \;+\; \cdot\ddot{\underset{..}{O}}: \;\longrightarrow\; :\ddot{\underset{..}{Cl}}:\ddot{\underset{..}{O}}:$$
$$:\ddot{\underset{..}{Cl}}:$$

dichlorine oxide

$$3:\ddot{\underset{..}{Cl}}\cdot \;+\; \cdot\dot{N}\cdot \;\longrightarrow\; :\ddot{\underset{..}{Cl}}:\dot{N}:\ddot{\underset{..}{Cl}}:$$
$$:\ddot{\underset{..}{Cl}}:$$

nitrogen trichloride

$$:\ddot{\underset{..}{Cl}}:$$
$$4:\ddot{\underset{..}{Cl}}\cdot \;+\; \cdot\dot{C}\cdot \;\longrightarrow\; :\ddot{\underset{..}{Cl}}:C:\ddot{\underset{..}{Cl}}:$$
$$:\ddot{\underset{..}{Cl}}:$$

carbon tetrachloride

For each of these compounds, the number of Cl atoms and the number of valence electrons in the other nonmetal atom sum to eight.

As was stated in Section 5.2, it should be emphasized that each of these atom combinations is merely a scheme to help us write the electron-dot formula for a known compound. For example, the electron-dot formula for ammonia was constructed from the combination of three hydrogen atoms with one nitrogen atom. This is not the balanced equation that is written for the chemical reaction of hydrogen and nitrogen to produce ammonia. That equation is written as

$$3H_2 + N_2 \longrightarrow 2NH_3$$

which shows hydrogen and nitrogen in their normal elemental states as diatomic molecules.

Many covalent compounds are known in which more than one kind of atom is bonded to a single nonmetal atom. For example, beginning with methane, CH_4, one hydrogen atom at a time can be replaced with a chlorine atom to produce methyl chloride, CH_3Cl, methylene chloride, CH_2Cl_2, and chloroform, $CHCl_3$:

H   H   H

H : C : $\ddot{\underset{..}{Cl}}$:   H : C : $\ddot{\underset{..}{Cl}}$:   : $\ddot{\underset{..}{Cl}}$: C : $\ddot{\underset{..}{Cl}}$:

H   : $\ddot{\underset{..}{Cl}}$:   : $\ddot{\underset{..}{Cl}}$:

methyl chloride   methylene chloride   chloroform

These electron-dot notations that we have been using are called **Lewis structural formulas,** in honor of G. N. Lewis, who in 1916 proposed that a covalent bond consists of a pair of shared electrons. In these structural formulas, **unshared pairs** of electrons, that is, valence level electron pairs not involved in covalent bond formation, are customarily shown with pairs of dots. Unshared pairs are also referred to as **lone pairs.** It is also common when drawing the structures of covalent compounds to use dashes for pairs of shared electrons. Examples of such Lewis formulas are

$$
\text{H}{-}\overset{\displaystyle ..}{\underset{\displaystyle \text{H}}{\text{O}}}{:} \qquad :\!\overset{\displaystyle ..}{\text{Cl}}{-}\overset{\displaystyle ..}{\underset{\displaystyle :\overset{..}{\text{Cl}}:}{\text{N}}}{-}\overset{\displaystyle ..}{\text{Cl}}\!: \qquad \text{H}{-}\overset{\displaystyle \text{H}}{\underset{\displaystyle \text{H}}{\text{C}}}{-}\text{H} \qquad \text{H}{-}\overset{\displaystyle ..}{\underset{\displaystyle ..}{\text{F}}}\!:
$$

———— • Example 5•2 • ————————————————————

Show with electron-dot formulas how atoms react with one another to form the following covalent compounds: (a) hydrogen selenide, H_2Se; (b) phosphorus tribromide, PBr_3; and (c) silicon tetrafluoride, SiF_4. Write formulas for the compounds, first with electron dots and second with dashes for covalent bonds.

—— • Solution •

(a) $2\text{H}\cdot \; + \; \cdot\overset{..}{\text{Se}}: \; \longrightarrow \; \text{H}\!:\!\overset{..}{\underset{\displaystyle \text{H}}{\text{Se}}}\!: \;$ or $\; \text{H}{-}\overset{..}{\underset{\displaystyle \text{H}}{\text{Se}}}\!:$

(b) $3\!:\!\overset{..}{\underset{..}{\text{Br}}}\cdot \; + \; \cdot\overset{}{\underset{\displaystyle \cdot}{\text{P}}}\cdot \; \longrightarrow \; :\!\overset{..}{\underset{..}{\text{Br}}}\!:\!\overset{}{\underset{\displaystyle :\overset{..}{\underset{..}{\text{Br}}}:}{\text{P}}}\!:\!\overset{..}{\underset{..}{\text{Br}}}\!: \;$ or $\; :\!\overset{..}{\underset{..}{\text{Br}}}{-}\overset{}{\underset{\displaystyle :\overset{..}{\underset{..}{\text{Br}}}:}{\text{P}}}{-}\overset{..}{\underset{..}{\text{Br}}}\!:$

(c) $4\!:\!\overset{..}{\underset{..}{\text{F}}}\cdot \; + \; \cdot\overset{\displaystyle \cdot}{\text{Si}}\cdot \; \longrightarrow \; :\!\overset{..}{\underset{..}{\text{F}}}\!:\!\overset{\displaystyle :\overset{..}{\text{F}}:}{\underset{\displaystyle :\overset{..}{\underset{..}{\text{F}}}:}{\text{Si}}}\!:\!\overset{..}{\underset{..}{\text{F}}}\!: \;$ or $\; :\!\overset{..}{\underset{..}{\text{F}}}{-}\overset{\displaystyle :\overset{..}{\text{F}}:}{\underset{\displaystyle :\overset{..}{\underset{..}{\text{F}}}:}{\text{Si}}}{-}\overset{..}{\underset{..}{\text{F}}}\!:$

See also Exercises 16–19.

5•3•1   Limitations of Rule of Eight for Covalent Compounds   In the formation of covalent compounds the rule of eight, or octet rule, is strictly applicable to only the four second-period elements: carbon, nitrogen, oxygen, and fluorine. The reason the rule is so helpful is that one or more of the first three of these elements is found in the majority of the known covalent compounds. For many other elements the rule has a limited but valuable application in predicting and diagramming Lewis structures from known molecular formulas.

Two examples of molecules to which the rule of eight is not applicable are beryllium chloride, $BeCl_2$, and boron trifluoride, BF_3. An atom of each of the VIIA elements, chlorine and fluorine, has one electron it can share with other atoms. From the molecular formulas, $BeCl_2$ and BF_3, and the number of valence electrons in atoms of beryllium (two) and boron (three), we can see that in these compounds, beryllium has only four and boron six electrons in the valence level. The Lewis structures of these compounds are

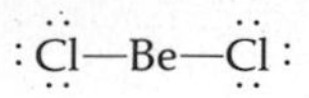

$$: \ddot{C}l—Be—\ddot{C}l : \quad \text{and}$$

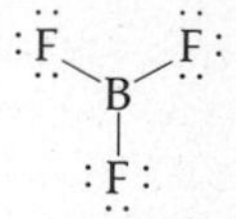

beryllium chloride
(four electrons in
valence level of Be)

boron trifluoride
(six electrons in
valence level of B)

There are many molecules in which the central atoms have more than eight electrons in their outermost energy levels. Two examples of such molecules are phosphorus pentachloride, PCl_5, and sulfur hexafluoride, SF_6. From these molecular formulas and the number of valence electrons per uncombined atom, five for phosphorus and six for sulfur, we see that the numbers of valence electrons around the central atoms are ten for PCl_5 and twelve for SF_6. The Lewis structures of these compounds are

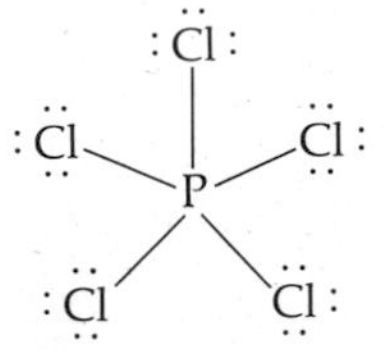

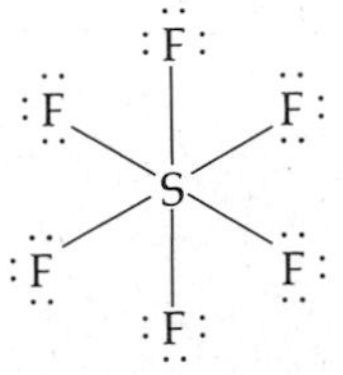

phosphorus pentachloride
(ten electrons in
valence level of P)

and

sulfur hexafluoride
(twelve electrons in
valence level of S)

The maximum number of electrons in the valence level of atoms of the elements carbon, nitrogen, oxygen, and fluorine in the second period is indeed limited to eight, because only four orbitals are available, one $2s$ and three $2p$. However, in molecules such as PCl_5 and SF_6, whose central atoms have more than eight electrons in their valence levels, more than just the s and p orbitals can be used in bonding. Because the elements phosphorus and sulfur are in the third period, the $3s$, $3p$, and $3d$ orbitals of their atoms may be involved in the covalent bonds they form.

5.4   Diagramming Lewis Formulas for Covalent Molecules

In addition to classifying compounds as ionic or covalent, compounds are classified as organic or inorganic. **Organic compounds** are the compounds of carbon.[3] Other organic compounds, whose structures were given in Section 5.3, are CH_4, CH_3Cl, CH_2Cl_2, $CHCl_3$, and CCl_4. Estimates of the number of known organic compounds range from 3 million upward. The

---

[3]The word *inorganic* means, literally, *not organic*. There are certain rocklike or earthy, carbon-containing substances that are usually classed as inorganic compounds, for example, the carbonates, bicarbonates, and cyanates. In addition, carbonic acid, H_2CO_3, carbon monoxide, CO, and carbon dioxide, CO_2, have been traditionally classified as inorganic compounds and thus are exceptions to the simple definition of organic compounds.

large number of organic compounds is explained on the basis of two characteristics of carbon atoms: (1) carbon atoms unite with one another by sharing one or more pairs of electrons to form chain or ring molecules; and (2) carbon atoms, with four valence electrons, can form four covalent bonds. This means that the carbon atoms are able to form rings and chains and still have valence electrons left over that can be used to form bonds with atoms of other elements.

In the preceding section, we showed Lewis formulas for a number of molecules. Many of the compounds in which we are interested have structures that are more involved than these. Because the Lewis structural formula of a molecule is often a great help to us in understanding its physical and chemical properties, it is necessary that we be able to diagram such formulas. In this section, the Lewis formulas for several covalent molecules will be diagrammed as examples. We need just two general guidelines for writing these diagrams. (1) *The molecular formula must be given (or remembered).* (2) *Wherever possible, the rule of eight (rule of two for hydrogen) is followed.*

5•4•1 Molecules with Single Bonds One way to construct a Lewis structural formula for a covalent molecule is to write first the electron-dot diagrams for each atom of the molecular formula. Then the atoms are redrawn closer together so that atoms share pairs of electrons to produce a structure in which all atoms follow the rule of eight or the rule of two. To illustrate this approach, we will diagram Lewis structures for hydrogen peroxide, H_2O_2, ethane, C_2H_6, and methyl alcohol, CH_3OH. In each of these compounds, hydrogen, with only one electron in the first main level, forms only one covalent bond with another atom by sharing a pair of electrons. Hence, the atoms other than hydrogen must be bonded to one another. With this information alone, we can diagram the Lewis structures of these molecules.[4]

Name, Formula	Formulation	Lewis Structure
hydrogen peroxide, H_2O_2	H· + ·Ö· + ·Ö· + ·H	H—Ö—Ö—H
ethane, C_2H_6	H· + ·C· + ·C· + ·H (with H above and below each C)	H—C—C—H (with H above and below each C)
methyl alcohol, CH_3OH	H· + ·C· + ·Ö· + ·H (with H above and below C)	H—C—Ö—H (with H above and below C)

---

[4]In drawing electron-dot structures, it is common to arrange them in two-dimensional form and to show as simply as possible that one has accounted for the outer electrons. Such facts as that CH_4 is tetrahedral in shape, that NH_3 is pyramidal, and that H_2O_2 is zigzag rather than linear are usually ignored. The common electron-dot diagram is used merely as a device to account for the numbers of electrons and the numbers of bonds.

In the CH_3OH Lewis structure, the total number of electrons shown is

$$\left(5 \text{ bonds} \times \frac{2 e^-}{\text{bond}}\right) + \left(2 \text{ lone pairs} \times \frac{2 e^-}{\text{lone pair}}\right) = 10 + 4 = 14\, e^-$$

This is also the total number of valence electrons in the six atoms used in the formulation: 4(1 per H) + 1(4 per C) + 1(6 per O) = 14 electrons. *The total number of electrons shown in the diagrams of the atoms that are brought together must equal the total electrons accounted for in the Lewis structure.*

For some compounds information in addition to the molecular formula must be provided in order to write the correct Lewis structure. This is especially true for organic compounds. For example, it is known that both ethyl alcohol and methyl ether have the same molecular formula, C_2H_6O. For this formula, we can diagram only two Lewis structures that comply with the rule of eight (and two). One has C—C—O bonds, and the other has C—O—C bonds. Because alcohols are known to have a hydrogen bonded to oxygen, the structure with C—C—O is assigned to ethyl alcohol:

Name, Formula	Formulation	Lewis Structure
ethyl alcohol, C_2H_5OH	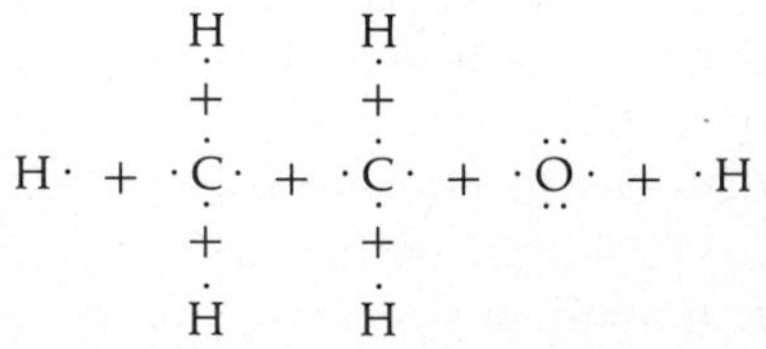	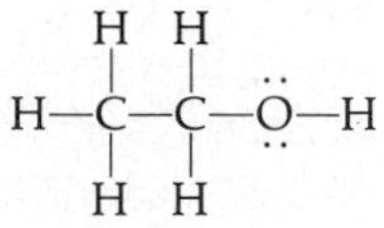
methyl ether, $(CH_3)_2O$	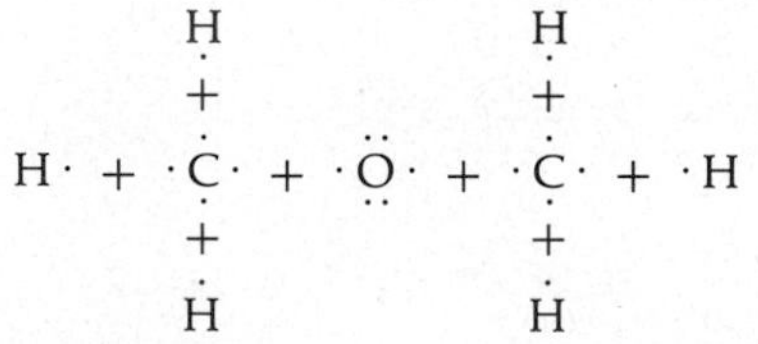	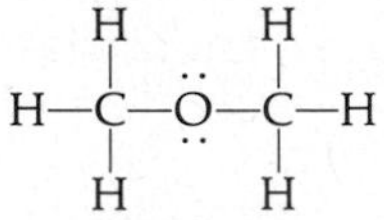

Up to this point, all of the atoms in the molecules for which we have written Lewis structures are connected by single bonds. A **single bond** is a covalent bond in which only one pair of electrons is shared between two atoms.

5·4·2   Molecules with Multiple Bonds   In addition to single bonds between atoms, multiple bonds are common. For example, in some compounds there are **double bonds,** in which two atoms share two pairs of electrons. There are also compounds with **triple bonds,** in which two atoms share three pairs of electrons.

• *Double Bonds* • Experimental data obtained for the two common compounds ethylene, C_2H_4, and carbon dioxide, CO_2, show that each molecule is symmetrical. Structural formulas involving *double bonds* are consistent with these data. When the atoms of these molecules are joined together so that each shares a pair of electrons, unpaired electrons remain. Pairing these two electrons in compliance with the rule of eight gives the other shared electron pair of the double bond:

Name, Formula	Formulation		Lewis Structure

ethylene, C_2H_4

$$\text{H} \cdot \; + \; \overset{+}{\cdot}\overset{\cdot}{\underset{\cdot}{C}}\cdot \; + \; \cdot\overset{\cdot}{\underset{\cdot}{C}}\overset{+}{\cdot} \; + \; \cdot\text{H} \longrightarrow$$

pairing of remaining two electrons

eight e^-'s around each C

carbon dioxide, CO_2

$$:\overset{\cdot\cdot}{\underset{\cdot\cdot}{O}}\cdot \; + \; \cdot\overset{\cdot}{\underset{\cdot}{C}}\cdot \; + \; \cdot\overset{\cdot\cdot}{\underset{\cdot\cdot}{O}}: \longrightarrow \; :\overset{\cdot\cdot}{\underset{\cdot\cdot}{O}} - C - \overset{\cdot\cdot}{\underset{\cdot\cdot}{O}}: \qquad :\overset{\cdot\cdot}{O}=C=\overset{\cdot\cdot}{O}:$$

pairing of remaining electrons

eight e^-'s around C and each O

• *Triple Bonds* • Two well-known molecules for which the experimental facts require *triple bonds* are acetylene, C_2H_2, and nitrogen, N_2. When the atoms of these molecules are joined together so that each shares a pair of electrons, two unpaired electrons remain on each carbon and nitrogen atom. Pairing these electrons in compliance with the rule of eight gives the two additional shared pairs of the triple bonds:

Name, Formula	Formulation		Lewis Structure

acetylene, C_2H_2

$$\text{H} \cdot \; + \; \cdot\overset{\cdot}{\underset{\cdot}{C}}\cdot \; + \; \cdot\overset{\cdot}{\underset{\cdot}{C}}\cdot \; + \; \cdot\text{H} \longrightarrow \; \text{H}-C-C-\text{H} \qquad \text{H}-C\equiv C-\text{H}$$

pairing of remaining electrons

eight e^-'s around each C

nitrogen, N_2

$$:\overset{\cdot}{\underset{\cdot}{N}}\cdot \; + \; \cdot\overset{\cdot}{\underset{\cdot}{N}}: \longrightarrow \; :N-N: \qquad :N\equiv N:$$

pairing of remaining electrons

eight e^-'s around each N

In working out the Lewis structure for a comparatively simple molecule, we can generally determine whether multiple bonds are required by first joining together the atoms other than hydrogen or a group VIIA element. If unpaired electrons remain, these are paired to produce the multiple bond. Consider the compounds formaldehyde, H_2CO, and hydrogen cyanide, HCN. Lewis structures for these compounds may be constructed as follows:

Name, Formula	Formulation	Lewis Structure

formaldehyde, H₂CO

$$H \cdot $$

pairing of remaining electrons   eight e^-'s around C and O

hydrogen cyanide, HCN H· + ·Ċ· + ·N̈· ⟶ H—C⦂⦂N: H—C≡N:

pairing of remaining electrons   eight e^-'s around C and N

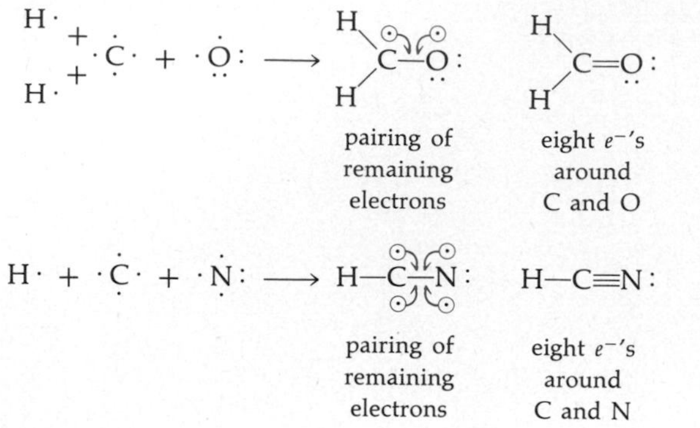

• **Example 5.3** •

Formulate Lewis structures for (a) hydrazine, N_2H_4; (b) methylamine, CH_3NH_2; (c) carbon disulfide, CS_2; and (d) formic acid, HCO_2H.

• **Solution** •

Formula	Formulation	Lewis Structure

(a) N_2H_4 H· + ·N̈· + ·N̈· + ·H   H—N̈—N̈—H

     +   +   H H

     H   H

(b) CH_3NH_2 H· + ·Ċ· + ·N̈· + ·H   H—C—N̈—H

     +   +   H H

     H   H

(c) CS_2 :Ṡ· + ·Ċ· + ·Ṡ: ⟶ :S⦂⦂C⦂⦂S: :S=C=S:

(d) HCO_2H H· + ·Ċ· + ·Ö: ⟶ H—C⦂⦂O: H—C=O:

     +   :Ö—H   :Ö—H

     :Ö· + ·H

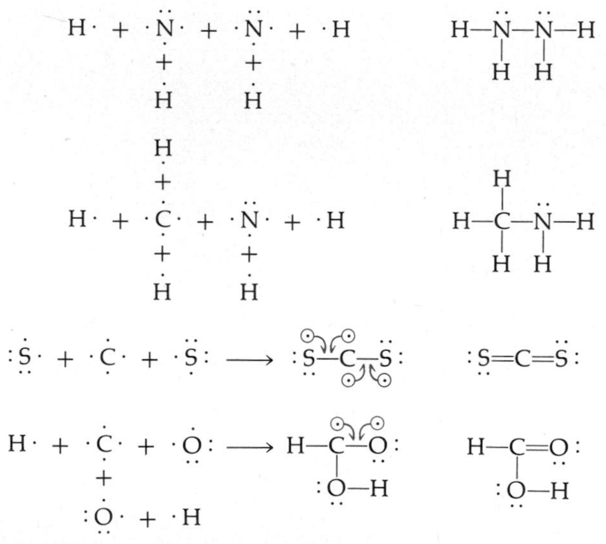

See also Exercises 21–24.

5.4.3 Coordinate Covalent Bonds A covalent bond in which both electrons are donated by one of the atoms is called a **coordinate covalent bond.** To illustrate such a case, let us first write the structural formula for hydrogen nitrite, HNO_2, known also as nitrous acid. For this compound, as well as for most acids containing hydrogen and one or more oxygen atoms, it is known that there is an O—H bond. With this information, and following our usual guidelines, we arrive at the following Lewis structure for hydrogen nitrite:

Name, Formula	Formulation	Lewis Structure
hydrogen nitrite, HNO_2	$:\ddot{O}\cdot + \cdot \ddot{N}\cdot + \cdot\ddot{O}: \longrightarrow :\ddot{O}-N-\ddot{O}:$ $+$ H H	$:\ddot{O}=N-\ddot{O}:$ H

Now let us consider a common compound with a molecular formula that differs from that of hydrogen nitrite by one oxygen atom. That compound is hydrogen nitrate, HNO_3, known also as nitric acid. In order to comply with the rule of eight for nitrogen and the additional oxygen atom, we must form a covalent bond to oxygen in which both electrons are donated by the nitrogen (a coordinate covalent bond). The Lewis structure for hydrogen nitrate is formulated from hydrogen nitrite as follows:

Name, Formula	Formulation	Lewis Structure
hydrogen nitrate, HNO_3	$:\ddot{O}:$ $+$ $:\ddot{O}=N-\ddot{O}:$ H	$:\ddot{O}:$ $:\ddot{O}=N-\ddot{O}:$ H

An additional example of a molecule in which a coordinate covalent bond is required to rationalize its Lewis structure is found in the interesting form of elemental oxygen known as ozone, O_3, an atmospheric pollutant that contributes to our daily smog buildup (Section 9.3.4). So that all oxygen atoms in a molecule of ozone can follow the octet rule, in one of the O-to-O bonds, the central oxygen must donate both electrons. The Lewis structure of ozone is formulated as follows:

Name, Formula	Formulation	Lewis Structure
ozone, O_3	$:\ddot{O}\cdot + \cdot\ddot{O}: + \ddot{O}: \longrightarrow :\ddot{O}-\ddot{O}-\ddot{O}:$	$:\ddot{O}=\ddot{O}-\ddot{O}:$

Coordinate covalent bonds can also be used to explain the structures of certain ions. The common ammonium ion, NH_4^+, is an example. This ion is present in a number of ionic compounds, for instance, ammonium chloride, NH_4Cl. Ammonium chloride can be formed from the reaction of ammonia with hydrogen chloride. To help us write its Lewis structure, we can imagine that the covalent bond joining H to Cl is broken in such a manner that both electrons of the bond remain with chlorine. This, of course, gives a chloride anion, Cl^-, and a hydrogen cation, H^+. Then the addition of H^+ to the neutral ammonia molecule produces an ammonium ion with a charge of

1+. These two steps are combined in one step in the following formulation:

Name, Formula	Formulation	Lewis Structure

ammonium chloride, NH_4Cl

$$H-\underset{\underset{H}{|}}{\overset{\overset{H}{|}}{N}}: + \text{(H)}-\ddot{\underset{..}{Cl}}: \qquad \left[H-\underset{\underset{H}{|}}{\overset{\overset{H}{|}}{N}}-H \right]^+ + :\ddot{\underset{..}{Cl}}:^-$$

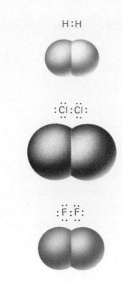

The bond between NH_3 and H^+ is formed as a coordinate covalent bond.

The concept of a coordinate covalent bond is useful to help us write structures, but the actual bond is not different from any other electron pair bond. In the NH_4^+ ion all four bonds are identical, and the positive charge on the ion is associated equally with all four hydrogen atoms.

In the compound ammonium chloride, there are both ionic bonds (between NH_4^+ and Cl^- ions) and covalent bonds (between N and H in the NH_4^+ ions). Such a compound is referred to as an ionic compound.

Figure 5.5 Some examples of non-polar diatomic molecules. These molecules are 100 percent covalent.

5•5   Polar Covalent Bonds

Two general types of chemical bonding have been described, the transfer of electrons that results in ionic bonding and the sharing of electrons that results in covalent bonding. While it is a temptation to classify things according to hard and fast rules, we must not generalize too quickly and definitely in our study of chemistry. Bonds can rarely be classified as either completely ionic or completely covalent but are actually somewhere between these outer limits.

A bond that is 100 percent ionic is one in which the attractive force between ions of opposite charge represents a *complete transfer* of an electron from a metal atom to a nonmetal atom. No substance has a bond that is 100 percent ionic.

A bond that is 100 percent covalent is one in which a pair of electrons is *shared equally* between two nonmetal atoms. This situation exists only when the atoms are the same, as is the case with diatomic molecules of elements (see Figure 5.5). On the other hand, if the two nonmetal atoms are different, the pair of electrons is attracted more toward one of the atoms. Such a pair of electrons constitutes a **polar covalent bond,** a covalent bond in which there is some electrostatic attraction between the two atoms. This electrostatic attraction is due to the fact that one of the atoms is partially negative and the other partially positive.

In the case of a diatomic molecule, a polar covalent bond results in a **polar molecule,** a molecule that has one end relatively positive and the other relatively negative. Examples are HCl and ClF, as shown in Figure 5.6. In these diagrams, the δ^+ or δ^- signifies a partial positive or partial negative charge that is less in magnitude than the charge associated with a free proton (1+) or electron (1−).

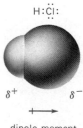

dipole moment

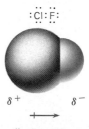

dipole moment

Figure 5•6 Polar covalent diatomic molecules. Note that the arrow indicating polarity is drawn from the partial positive toward the partial negative charge.

5•5•1   Electronegativity   From the energy necessary to break the polar bond between two atoms, we can calculate the relative attraction the atoms have for the bonding pair of electrons. The power that an element has for

attracting the pair of electrons in a covalent bond is called its **electronegativity.** This concept can be illustrated by considering the formation of hydrogen and fluorine atoms by two different paths. Let us compare the average energy of reactions (1) plus (2) with that of reaction (3):

Reaction	Energy Required, Joules
(1) H—H $\longrightarrow$ 2H·	7.24×10^{-19}
(2) :F̈—F̈: $\longrightarrow$ 2:F̈·	2.62×10^{-19}
(3) H—F̈: $\longrightarrow$ H· + ·F̈: $\delta^+\delta^-$	9.43×10^{-19}

The energy necessary to form one H atom and one F atom from nonpolar molecules by the pathway of reactions (1) and (2) is $(\frac{1}{2})(7.24 \times 10^{-19})$ + $(\frac{1}{2})(2.62 \times 10^{-19}) = 4.93 \times 10^{-19}$ J. The difference between this energy and the energy necessary to form one H atom and one F atom by breaking the polar bond in reaction (3) is

$$(9.43 \times 10^{-19}) - (4.93 \times 10^{-19}) = 4.50 \times 10^{-19} \text{ J}$$

This additional energy may be thought of as the energy required to overcome the attraction between the positive H end and the negative F end of the polar H—F bond. The additional energy is therefore related to the degree to which the electrons in the covalent bond are drawn toward the F atom and away from the H atom. From this type of calculation on a large number of diatomic molecules, Linus Pauling derived a scale of electronegativity values based on bond strengths. On Pauling's scale, the element fluorine, F, is assigned an arbitrary electronegativity of 4.0; all the other elements are compared with this standard.

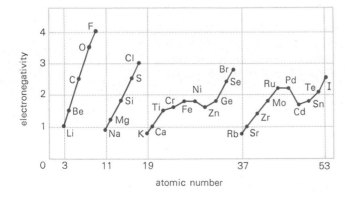

Figure 5•7 As the atomic numbers increase, the electronegativities of the elements vary in a periodic way, increasing from group IA to group VIIA in each period.

In general, the higher the first ionization energy of an element, the greater is its electronegativity. For example, potassium, K, which has a low ionization energy and therefore a relatively slight attraction for electrons, has an electronegativity of only 0.8. Silicon, Si, whose electron attraction is moderate, has a value of 1.8, and oxygen, O, with a strong electron attraction, has a value of 3.5, almost as high as that of fluorine. In Figure 5.7, the electronegativities are plotted versus the atomic numbers for four periods of elements, excluding the noble gases. The plot shows the regular increase from left to right through a period for A family members. As the atomic number increases from 3 to 9, the electronegativity rises; however, at 11, Na, it is low once more and then rises again until 17, Cl, is reached. In Figure 5.8, the electronegativities of most of the elements are listed. In contrast to the regular change in a single period from small to large values shown for A family members in these figures, electronegativities for B family members are much more nearly the same.

IA	IIA	IIIB	IVB	VB	VIB	VIIB	VIIIB			IB	IIB	IIIA	IVA	VA	VIA	VIIA	VIIIA
H 2.1																	He (2.7)
Li 1.0	Be 1.5											B 2.0	C 2.5	N 3.0	O 3.5	F 4.0	Ne (4.4)
Na 0.9	Mg 1.2											Al 1.5	Si 1.8	P 2.1	S 2.5	Cl 3.0	Ar (3.5)
K 0.8	Ca 1.0	Sc 1.3	Ti 1.5	V 1.6	Cr 1.6	Mn 1.5	Fe 1.8	Co 1.8	Ni 1.8	Cu 1.9	Zn 1.6	Ga 1.6	Ge 1.8	As 2.0	Se 2.4	Br 2.8	Kr (3.0)
Rb 0.8	Sr 1.0	Y 1.2	Zr 1.4	Nb 1.6	Mo 1.8	Tc 1.9	Ru 2.2	Rh 2.2	Pd 2.2	Ag 1.9	Cd 1.7	In 1.7	Sn 1.8	Sb 1.9	Te 2.1	I 2.5	Xe (2.6)
Cs 0.7	Ba 0.9	La-Lu 1.1-1.2	Hf 1.3	Ta 1.5	W 1.7	Re 1.9	Os 2.2	Ir 2.2	Pt 2.2	Au 2.4	Hg 1.9	Tl 1.8	Pb 1.8	Bi 1.9	Po 2.0	At 2.2	Rn (2.4)
Fr 0.7	Ra 0.9	Ac 1.1	Th 1.3	Pa 1.5	U 1.7	Np 1.3											

electronegativity scale

0.7–1.0   1.1–1.9   2.0–2.5   2.6–4.4

Figure 5•8 Electronegativities of the elements, Pauling scale. The black zigzag line separates metals from nonmetals.
Source: Most values are reprinted with permission from Linus Pauling, *The Nature of the Chemical Bond.* Copyright © 1939 and 1940, third edition copyright © 1960 by Cornell University. Used by permission of the publisher, Cornell University Press. All values in parentheses for the noble gases are based on indirect calculations instead of bond strengths and are taken from Bing-Man Fung, *J. Phys. Chem.*, 69, 596 (1965).

When two atoms of different electronegativities are joined by a covalent bond, the pair of electrons is drawn toward the atom of higher electronegativity. This results in one of the atoms being more negative than the other and imparts some polar or ionic character to the bond. Attempts have been made to correlate the electronegativity difference between two atoms with the ionic character of the bond between them. Pauling states that if atoms differ sufficiently (by about two units) in electronegativity, they will form bonds that are mainly ionic. If the electronegativity difference is less than this, the bonds are mainly covalent.

5•5•2   Dipole Moments   A polar diatomic molecule is a *dipole*, a body that has opposite charges at two points. For such a molecule we can determine a **dipole moment,** a measure of the degree of its polarity. The dipole moment of a molecule depends directly on the magnitude of the fractional charges and on the distance separating the positive and negative charges; that is,

$$\text{dipole moment} = \text{charge} \times \text{distance}$$

For gaseous molecules, dipole moments can be measured in a suitable electric field. In an electric field polar molecules tend to orient themselves, pointing their positive ends toward the negative plate and their negative ends toward the positive plate (see Figure 5.9). This orientation of the molecules tends to neutralize and thereby decrease the magnitude of an initially applied electric field. For a given applied field, two separated plates will have a maximum charge when separated by a vacuum. Any substance that is introduced between the plates reduces the charge. The ratio by which a substance reduces the charge is called its **dielectric constant** (see Table 10.1). From this property, and by the use of calculations initially developed by Nobel laureate Peter Debye, dipole moments of individual molecules, expressed in debye units, D, may be evaluated.[5]

   In Table 5.2 values are given for the dipole moments of some simple molecules. The two ionic cesium compounds, CsF and CsCl, are included for contrast with the other compounds that are largely covalent. Simple gaseous diatomic molecules of ionic compounds, such as CsF and CsCl, exist only at very high temperatures. Under ordinary conditions, as will be shown in Chapter 8, ionic compounds exist almost exclusively as solids at room temperature.

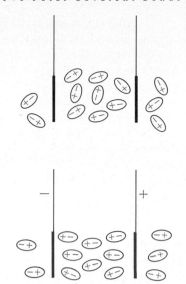

Figure 5•9 Polar molecules tend to become aligned in an electric field.

Table 5•2 Dipole moments[a]

AB	Dipole moment, D	Shape of molecule[b]	AB$_3$	Dipole moment, D	Shape of molecule
CsCl (gas)	10.44	linear	NH_3	1.47	pyramidal
CsF (gas)	7.89	linear	PF_3	1.03	pyramidal
HF	1.78	linear	PH_3	0.56	pyramidal
HCl	1.07	linear	NF_3	0.23	pyramidal
HBr	0.79	linear	BF_3	0	trigonal planar
HI	0.38	linear	SO_3	0	trigonal planar
CO	0.12	linear			

AB$_2$			AB$_4$		
H_2O	1.85	angular	CH_3Cl	1.92	tetrahedral
SO_2	1.62	angular	CH_2Cl_2	1.59	tetrahedral
H_2S	0.95	angular	$CHCl_3$	1.03	tetrahedral
CO_2	0	linear	CCl_4	0	tetrahedral
CS_2	0	linear	CH_4	0	tetrahedral

[a]Most of the values of the dipole moments are selected with permission from A. L. McClellan, *Tables of Experimental Dipole Moments*, Vol. II, Rahara Enterprises, El Cerrito, CA, 1974.

[b]Precise shapes of molecules are usually determined by X-ray measurements.

[5]The dipole moment unit, the debye, is defined on the basis of the magnitude in electrostatic units (esu) of the charges and the distance separating them. A unit negative charge (4.80×10^{-10} esu) and a unit positive charge (4.80×10^{-10} esu) separated by a distance of 1 Å is defined as a dipole with a moment of 4.80 debyes:

$$\text{dipole moment} = \text{charge} \times \text{distance}$$
$$= 4.80 \times 10^{-10} \text{ esu} \times 1 \text{ Å} = 4.80 \ D$$

Hence, 1 debye unit ($1 \ D$) $= 1 \times 10^{-10}$ esu $\cdot$ Å. In SI units $1 \ D = 3.34 \times 10^{-30}$ C $\cdot$ m.

All the diatomic molecules listed in Table 5.2 are polar. In the case of the four hydrogen halides, HF, HCl, HBr, and HI, the dipole moments are in accord with the decrease in electronegativities of the VIIA elements going down the group. Hydrogen fluoride has the largest dipole moment, and hydrogen iodide the smallest. The difference in the electronegativities of H and F is 1.9; the differences between H and the other VIIA elements are considerably less, for example, only 0.9 for H and Cl. Evidently this greater difference in H and F electronegativities more than compensates for the shorter distance between the H and F atoms in HF (see Table 5.4). (Both the magnitude of the charges *and* the distance between them are important in dipole moment determinations.)

• *Molecular Geometry and Dipole Moment* • Atoms in a molecule are at relatively fixed positions with respect to one another, so a molecule has a definite geometry or shape. The simplest molecules to describe are diatomic molecules. Because the two nuclei define a straight line, a diatomic molecule is *linear*. Some models of diatomic molecules are shown in Figures 5.5 and 5.6. Additional examples are given in Table 5.2.

Molecules containing more than two atoms may be linear or they may have more complex shapes, depending on the arrangement of the nuclei of their atoms. Some common molecular shapes, with the central atom designated A and attached atoms designated B, are shown in Figure 5.10.

The shape of a molecule can be related to whether or not it has a dipole moment. A triatomic or a more complex molecule may or may not have a dipole moment, depending upon the geometric arrangement of the bond moments in the molecule (see Figure 5.11). A **bond moment** is the polar moment for an individual bond. For a diatomic molecule, the dipole moment and the bond moment are the same; but for a polyatomic molecule, the dipole moment is equal to the vector sum of the various bond moments. A vector has a magnitude (equal to the bond moment) and a direction (arbitrarily pointing toward the more electronegative atom). The sum of the bond moment vectors is the resultant dipole moment of the molecule, as shown in Figure 5.11.[6]

Carbon dioxide, CO_2, is characteristic of any molecule in which the bond moments cancel one another, that is, in which the vector sum is zero. Although such molecules have polar bonds, the symmetrical arrangement of the bonds results in the bond moments canceling each other and leaves the molecule nonpolar as a whole. The linear geometry assigned to CO_2 is consistent with its having a dipole moment of zero.

From its formula alone, we might expect the sulfur dioxide molecule, SO_2, to be analogous to CO_2. However, SO_2 has an appreciable dipole moment of 1.62 D. Therefore, we know that the bond moments do not cancel and that the molecular geometry is angular. In a similar manner we conclude that a molecule of H_2O also has angular geometry, a fact consistent with its high dipole moment of 1.85 D. Note that in SO_2 and H_2O, the central atom has lone pairs of electrons and the molecules are angular. In CO_2, there are

---

[6]From the known dipole moment of water of 1.85 D and the bond angle of 104.5°, the O—H bond moment is calculated to be 1.51 D.

no lone pairs on C, and the molecule is linear. Lone pairs of electrons affect the shapes of molecules, which we will discuss in Chapter 6.

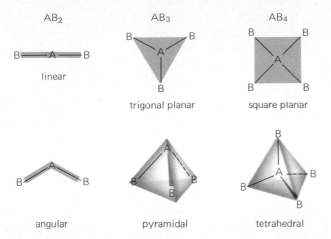

Figure 5•10 Some common shapes of simple molecules in which a central atom A is bonded to two or more attached atoms B. A solid line indicates a bond in the plane of the paper. A dashed line shows a bond that projects behind the plane of the paper. A wedge shows a bond that projects in front of the plane of the paper (the wide end of the wedge is toward the viewer).

For a molecule with the general formula AB_2, with A as the central atom, there are two possible shapes. If the molecule has a dipole moment, it has an angular shape; if its dipole moment is zero, it has a linear shape.

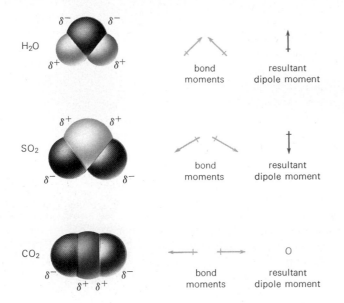

Figure 5•11 Polar and nonpolar triatomic molecules. The point of a vector is toward the more electronegative atom, the plus end toward the less electronegative atom.

Carbon disulfide, CS_2, boron trifluoride, BF_3, sulfur trioxide, SO_3, methane, CH_4, and carbon tetrachloride, CCl_4, are other examples of molecules that have zero dipole moments. From this fact we can deduce that in these molecules the bond moment vectors must cancel one another. Therefore, we predict that these molecules have symmetrical shapes. Their actual shapes are shown in Table 5.2.[7]

---

[7]For a molecule with the general formula AB_4, either tetrahedral or square planar would be a possible symmetrical, nonpolar shape. X-ray studies reveal CH_4 and CCl_4 to be tetrahedral.

The AB$_3$ molecules ammonia, NH$_3$, phosphine, PH$_3$, nitrogen trifluoride, NF$_3$, and phosphorus trifluoride, PF$_3$, all have appreciable dipole moments. From this, we can conclude that they are not trigonal planar molecules, but that they are probably T-shaped or pyramidal. Actually, each of them is pyramidal (see Table 5.2), as X-ray measurements have shown.

To account in detail for the dipole moment of a molecule, the lone pairs of electrons must be considered. A lone pair occupies approximately the same volume as a shared pair of electrons and affects both the size and the distribution of charge in the molecule (see Section 6.5).

5.6 Sizes of Atoms, Molecules, and Ions

In the preceding section we introduced the topic of molecular geometry. An understanding of molecular geometry is essential to the understanding of the properties of molecules. As we shall see in subsequent chapters, how molecules pack together in solids, how they are attracted to one another in liquids, and, in some cases, why the boiling point of one compound differs from another all depend upon molecular geometry. In order to understand molecular geometry and its application to chemical principles, the sizes of atoms, molecules, and ions must be determined. About 60 years ago, X-ray methods (see Section 8.6.1) were developed that enabled the distances between atomic nuclei to be measured. With this information the sizes of individual atoms could be calculated.

5.6.1 Covalent Radii and van der Waals Radii The volume occupied by a single unit of an element is difficult to represent. One complication is the size of the region occupied by the shared and unshared (lone) pairs of electrons. Another is whether the atom is held to a neighbor by a covalent bond, as in a Cl—Cl molecule, or is just touching a neighbor, as in solid neon. However, by using a few arbitrary rules and taking advantage of modern data on the distances between nuclei, the volume of an imaginary spherical atom can be calculated. The radius of an atom is determined not by a measurement made on an isolated single atom, but by a measurement of the distance between the nuclei of bonded atoms in a solid sample of an element. The volume of an atom is then calculated on the basis of its covalent radius if it is a nonmetal or its metallic radius if it is a metal.

The problem of assigning sizes to atoms can be illustrated by the case of elemental chlorine, as pictured in Figure 5.12. The distance between the nuclei of two atoms bonded together is the *interatomic bonded distance* or **bond length.** For chlorine this distance is 1.98 Å. The part of the bond length that is attributed to one of the two covalently bound atoms is called the **covalent radius** of that atom. For a single chlorine in the Cl—Cl molecule, this radius is 1.98 Å/2 = 0.99 Å. The distance between the nuclei of two atoms in adjacent, touching molecules is the *interatomic nonbonded distance*. For solid chlorine this distance is 3.50 Å. The part of the nonbonded distance between two touching atoms that is attributed to one of the atoms is called the **van der Waals radius** of that atom, named for the Dutch physicist J. H. van der Waals. For a Cl atom in solid Cl$_2$, this radius is 3.50 Å/2 = 1.75 Å.

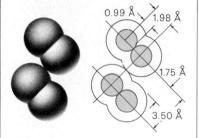

Figure 5.12 Schematic diagrams of chlorine molecules, Cl$_2$. The interatomic bonded distance (bond length) is 1.98 Å, and the covalent radius is 1.98/2 = 0.99 Å. The interatomic nonbonded distance is 3.50 Å, and the van der Waals radius is 3.50/2 = 1.75 Å. Each inner circle corresponds to the size of the chlorine atom represented in Figure 5.14.

Interatomic nonbonded distances may also be measured between atoms in the same molecule. In Figure 5.13 the distance of 2.89 Å is shown between two Cl atoms in a molecule of carbon tetrachloride, CCl_4. This distance is less than the sum of the two van der Waals radii (3.50 Å) because the two Cl atoms are bonded to the C atom. Each Cl atom is distorted as the nuclei are forced closer together than 3.50 Å.

Some single bond covalent and van der Waals radii for nonmetallic elements are listed in Table 5.3. Single bond covalent radii for atoms range from 0.37 Å for hydrogen to about 1.43 Å for antimony. Figure 5.14 shows the approximate relative volumes taken up by atoms bonded to other atoms. To visualize the approximate size of a complete molecule, one can consider the size of bonded atoms and then imagine expanded electron clouds reaching out as far as the van der Waals radii, as shown for chlorine in Figure 5.12. The sizes of the spheres representing atoms of the nonmetals in Figure 5.14 are drawn on the basis of covalent radii.

CCl₄

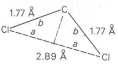

Figure 5•13 The interatomic non-bonded distance of 2.89 Å between the nuclei of two chlorine atoms in CCl_4 is less than the 3.50 Å between the nuclei of two chlorine atoms in adjacent molecules of Cl_2. (The distance of 2.89 Å was calculated from the law of sines, $\sin A/a = \sin B/b$; $a = (\sin 54.75°)(1.77 \text{ Å})/\sin 90° = 1.445$ Å. The 54.75° is half the tetrahedral angle of 109.5°.)

Table 5•3 **Single bond covalent and van der Waals radii**

Element	Covalent[a] radius, Å	van der Waals[b] radius, Å
H	0.37	1.20
C	0.77	1.70
N	0.75	1.55
O	0.73	1.52
F	0.71	1.47
Cl	0.99	1.75
Br	1.14	1.85
I	1.33	1.98
S	1.02	1.80
P	1.10	1.80

[a] Values of the covalent radii are reproduced with permission from L. E. Sutton, Ed., *Tables of Interatomic Distances and Configurations in Molecules and Ions, Spec. Publ. No. 11 and 18,* The Chemical Society, London, 1958, 1965 (except for nitrogen and oxygen). Values of the covalent radii of nitrogen and oxygen are reproduced with permission from J. E. Huheey, *Inorganic Chemistry,* 2nd ed., Harper & Row, New York, 1978.

[b] Mean values of the van der Waals radii are taken from A. Bondi, *J. Phys. Chem.,* **68**, 441 (1964).

• *Bond Distances Between Different Atoms* • When two different atoms are connected by a covalent single bond, we can estimate the bond distance by adding their covalent radii. For example, the estimated C—Cl bond distance in CCl_4 is the sum of the covalent radii for carbon and chlorine:

0.99 Å	(chlorine radius)
0.77 Å	(carbon radius)
1.76 Å	(sum)

The experimentally measured C—Cl bond distance is 1.77 Å (see Figure 5.13). The agreement between calculation and measurement is not

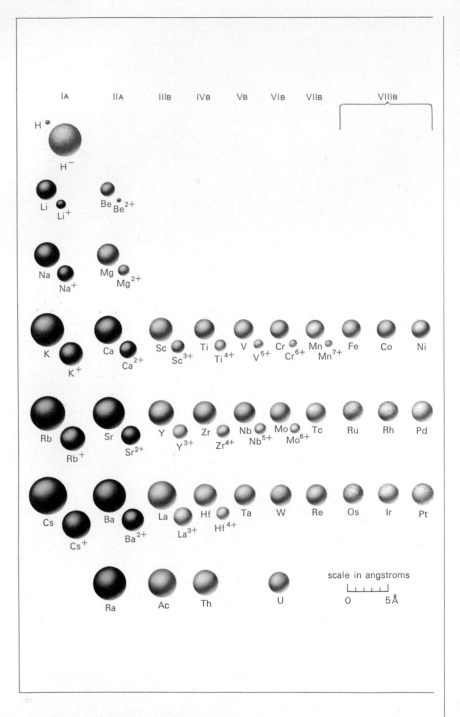

always so close as this. Some experimental single bond lengths for H and C bound to representative nonmetals are listed in Table 5.4 along with the values calculated from the addition of the covalent radii. The differences between the experimentally determined and calculated bond lengths are significant when hydrogen is bound to atoms with high electronegativities. The hydrogen atom has no inner core electrons to be repelled by the electron

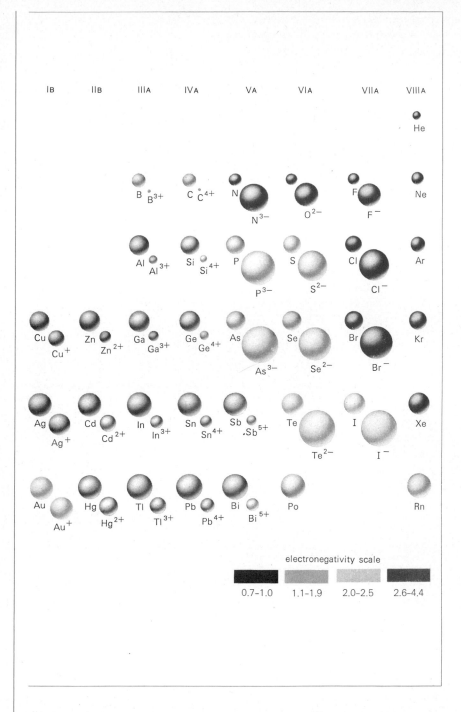

Figure 5•14 The relative sizes of the atoms and ions of most of the elements. Electronegativities are indicated by the color code in the lower right corner.
Source: Radii of noble-gas atoms are from Bing-Man Fung, *J. Phys. Chem.*, *69*, 596 (1965).

cloud of the larger, more electronegative atom. Therefore the hydrogen atom is drawn closely toward the other atom and is assigned a correspondingly smaller radius. An average or best value of 0.30 Å for the radius of hydrogen in covalent bonds to other atoms is often assigned to adjust for this effect. Differences are also observed when carbon forms bonds with highly electronegative atoms such as fluorine and oxygen. Such differences are common to covalent bonds in which the electronegativity difference between the atoms is about 1.0 or more. Otherwise, the experimental values compare rather well with those obtained by addition of the covalent radii.

Table 5•4 Single bond lengths, Å

Bond	Experimental	Calculated[a]	Bond	Experimental	Calculated
H—C	1.09	1.14 (1.07)	C—C	1.54	1.54
H—N	1.01	1.12 (1.05)	C—N	1.47	1.52
H—O	0.96	1.10 (1.03)	C—O	1.43	1.50
H—F	0.92	1.08 (1.01)	C—F	1.38	1.48
H—Cl	1.27	1.36 (1.29)	C—Cl	1.77	1.76
H—Br	1.41	1.51 (1.44)	C—Br	1.94	1.91
H—I	1.61	1.70 (1.63)	C—I	2.14	2.10
H—S	1.34	1.39 (1.32)	C—S	1.82	1.79
H—P	1.44	1.47 (1.40)	C—P	1.84	1.87

Source: *Reproduced with permission from L. E. Sutton, Ed.*, Tables of Interatomic Distances and Configurations in Molecules and Ions, Spec. Publ. No. 11 and 18, *The Chemical Society, London, 1958, 1965.*

[a] Values in parentheses are calculated by using the radius of hydrogen as 0.30 Å.

───── • Example 5•4 • ─────

(a) Using the values of the covalent radii listed in Table 5.3, calculate the interatomic bonded distances in molecules of BrF and BrCl.
(b) For one of these two compounds, the difference between the experimentally determined and the calculated bond distances is negligible, whereas in the other compound the difference is 0.09 Å. For which compound is the difference significant? What is the experimentally determined bond distance for this compound?

───── • Solution • ─────

(a) The calculated interatomic bond distances are the sum of the covalent radii of the atoms:

for BrF, this distance is 1.14 Å + 0.71 Å = 1.85 Å
for BrCl, this distance is 1.14 Å + 0.99 Å = 2.13 Å

(b) The electronegativity difference between F and Br is $4.0 - 2.8 = 1.2$, whereas the difference between Cl and Br is only $3.0 - 2.8 = 0.2$. Because the difference between F and Br is greater than 1.0, we might expect the actual (experimentally determined) bond distance in BrF to be less than the calculated one. The actual distance is then 1.85 Å − 0.09 Å = 1.76 Å.

See also Exercises 43 and 44.

5•6•2 Metallic Radii The structure of metals will be taken up in more detail in Section 8.9, but we should note here that in elemental metals, atoms are bonded together in a special way. We consider a pure crystal of a metal as a huge molecule consisting of millions of atoms held together by bonds that differ greatly from either ionic or covalent bonds. In metals, each atom, other than surface atoms, commonly has either eight or twelve nearest neighbors to which it is held by these bonds. Half of the internuclear distance between a metal atom and one of its nearest neighbors is called its **metallic radius.** The diagrams of metal atoms in Figure 5.14 are based on such measured metallic radii. Cesium, with a radius of 2.35 Å, is the largest atom shown.

     The covalent radius of a nonmetal atom or the metallic radius of a metal atom is often referred to as the **atomic radius.**

5•6•3   Ionic Radii   When metal atoms react with nonmetal atoms to form ions, the cations are always smaller than their atoms, and the anions are always larger. This is illustrated in Figure 5.15 by the reaction of atoms of sodium and chlorine to give sodium chloride. The sizes of ions relative to their respective atoms depend on the forces of electron–proton attraction and electron–electron repulsion. The outer electrons not only are attracted by the positive nucleus, but in addition are repelled by electrons in lower energy levels and in their own level. This repulsion, which has the net effect of reducing the *effective nuclear charge,* is called the **screening effect.** When an atom gains one or more electrons, the number of electrons in the highest energy level is increased, and each electron is more effectively screened from the nucleus. These electrons repel one another, and since the effective nuclear charge is less, they force one another farther from the nucleus. The result is that a negative ion is larger than its atom. In accounting for the relative sizes of positive ions, one needs to consider factors in addition to changes in the screening effect. Because all electrons in the highest energy level are often lost, the large volume due to these electrons disappears. The overall result is a positive ion that is smaller than the initial atom.

     The sizes of certain ions are shown to approximate scale in Figure 5.14. Simple monatomic ions vary in size from the beryllium ion, Be^{2+}, with a radius of 0.31 Å, to an anion eight times larger, the antimonide ion, Sb^{3-}, with a radius of 2.45 Å. Very small and very large ions may not exist separately, because such particles tend to form polar covalent bonds when they combine with ions of opposite charge.

5•6•4   Trends in Sizes of Atoms and Ions   In Figures 5.8 and 5.14 differences in color and shading indicate the approximate relative electronegativities of the elements. The following general trends are easily seen:

1. Atomic size generally decreases and electronegativity increases from left to right in a given period.

2. Atomic size increases and electronegativity decreases from top to bottom in the A families of the periodic table.

3. Atoms of low electronegativity (at the left of the periodic table) tend to lose electrons and form small positive ions.

4. Within a period, the larger the positive charge, the smaller is the ion.

5. Atoms of high electronegativity (at the right of the periodic table) tend to

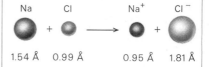

Na	Cl	Na$^+$	Cl$^-$
1.54 Å	0.99 Å	0.95 Å	1.81 Å

Figure 5•15 The reaction of an atom of sodium with an atom of chlorine to give a sodium cation and a chloride anion. The numbers beneath the spheres correspond to the atomic and ionic radii. The sizes of the ions are those determined for crystalline sodium chloride (compare Figure 5.14).

gain electrons and form large negative ions. The noble gases are exceptions.

6.   Within a period, the larger the negative charge, the larger is the ion.

Figure 5.16 shows trends 4 and 6 very clearly for a series of isoelectronic ions.

5•7   Oxidation Numbers

In using the rule of eight (and two) to write formulas for ionic compounds of A family elements (see Section 5.2.1), the group numbers are the key. We assume that a metal atom loses electrons to form an ion with a positive charge equal to its group number, and that a nonmetal gains electrons to form an ion with a negative charge equal to the group number minus 8. The formula is written with the requirement that the sum of the positive charges must equal the sum of the negative charges.

Consider the compounds lithium oxide and aluminum fluoride. For each compound, we can deduce the formula with the aid of two *half-reactions*, one for the metal and one for the nonmetal.

For lithium oxide,

$$Li \longrightarrow Li^+ + e^-$$

each atom / loses one e^- → in group IA, charge is $1+$

$$O + 2e^- \longrightarrow O^{2-}$$

each atom / gains two e^-'s → in group VIA, charge is $6 - 8 = 2-$

To make the sum of the positive charges equal to the sum of the negative charges, we need two Li^+ ions for each O^{2-} ion. The formula is Li_2O.

For aluminum fluoride,

$$Al \longrightarrow Al^{3+} + 3e^-$$

each atom / loses three e^-'s → in group IIIA, charge is $3+$

$$F + e^- \longrightarrow F^-$$

each atom / gains one e^- → in group VIIA, charge is $7 - 8 = 1-$

To make the sum of the positive charges equal to the sum of the negative charges, we need three F^- ions for each Al^{3+} ion. The formula is AlF_3.

The half-reactions in which electrons are lost are called *oxidations*; those in which electrons are gained are called *reductions*.

A set of small whole numbers called **oxidation numbers** or **oxidation states,** which are related to the combining ratios of elements, aid us in remembering formulas for compounds and in correlating certain chemical phenomena. In Chapters 9 and 18 we will find that the oxidation numbers are very useful in helping us balance oxidation–reduction equations.

In ionic compounds, the oxidation number of an ion is the same as the charge of the ion. In the compounds lithium oxide and aluminum fluoride, just described, the oxidation numbers of lithium, oxygen, alumi-

$2e^- 8e^- 8e^-$

S^{2-}

16+

1.84 Å

Cl^-

1.81 Å

K^+

1.33 Å

Ca^{2+}

0.99 Å

Figure 5•16 For isoelectronic ions, the size decreases as the positive charge on the nucleus increases.

num, and fluorine are $+1$, -2, $+3$, and -1, respectively. When oxidation numbers are designated in the formulas of compounds, they are written directly above the symbol, with a plus or minus sign in front of the number. When there is more than one atom in the formula, the oxidation number is placed within parentheses, and the number of atoms is written as a subscript to the right of the parentheses. Some examples are

$$\overset{+1-1}{\text{NaCl}} \quad \overset{(+1)_2-2}{\text{Li}_2\text{O}} \quad \overset{+3(-1)_3}{\text{AlF}_3} \quad \overset{(+2)_3(-3)_2}{\text{Ba}_3\text{N}_2}$$

If the magnitude of the charge on an ion is to be shown, rather than the oxidation number of the element in the compound, the number followed by a plus or minus sign is written above and to the right of the symbol. Examples are Li^+, O^{2-}, Al^{3+}, and F^-. (Note that if the charge is $1+$ or $1-$, the number 1 is omitted.)

For covalent compounds and for ionic compounds with covalent bonds in one or more ions, the assignment of oxidation numbers depends on arbitrary rules based on the electronegativities of the elements comprising the compounds. Three of these rules are:

1. The oxidation number of an uncombined element is zero. Examples are

$$\overset{0}{\text{Na}} \quad \overset{0}{\text{Mg}} \quad \overset{0}{\text{H}_2} \quad \overset{0}{\text{Cl}_2} \quad \overset{0}{\text{O}_2} \quad \overset{0}{\text{N}_2}$$

In a compound, the more electronegative elements are assigned negative oxidation numbers, and the less electronegative elements are assigned positive numbers. Examples are

$$\overset{+1-1}{\text{HCl}} \quad \overset{(+1)_2-2}{\text{H}_2\text{O}} \quad \overset{-3(+1)_3}{\text{NH}_3} \quad \overset{+6(-2)_3}{\text{SO}_3}$$

2. The element with a positive oxidation number usually comes first in the formula. Ammonia, NH_3, is an exception to this rule.

   It should be understood that the oxidation number of an atom in a covalent compound does not imply discrete charges. For example, in HCl the oxidation numbers of H and Cl are $+1$ and -1. But there is only a partial positive charge on hydrogen and a partial negative charge on chlorine, with no ions present. The bond between H and Cl is a polar covalent bond.

3. In the formula for a compound, the sum of the negative oxidation numbers and the positive oxidation numbers equals zero. For example, in SO_3, $+6 + 3(-2) = 0$.

5•7•1   The Periodic Table and Oxidation Numbers   Common oxidation states for several elements are shown in Figure 5.17. Examination of this figure reveals two important general observations:

1. A common positive oxidation number of all elements, except those of group VIII and oxygen and fluorine, is the group number. These numbers lie on the colored lines above oxidation number zero.

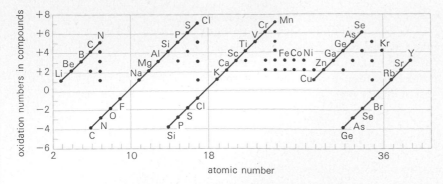

Figure 5•17 Some common oxidation states. From the periodic law we would expect oxidation numbers to be a periodic function of the atomic numbers. Study of this graph reveals that this is indeed the case, particularly for the maximum and minimum states (through which the lines have been drawn for emphasis). It is important to note that an element may have a number of oxidation states that are not evident from its position in the periodic table.

2. A common oxidation number of all nonmetals is the group number minus 8. For example, oxygen in H_2O has an oxidation number of $6 - 8 = -2$. These numbers lie on the colored lines below oxidation number zero. The data in Figure 5.17 for the A family elements are summarized with examples and pertinent comments in Table 5.5.

Table 5•5 Oxidation numbers for A family elements

Group	Common oxidation number(s)	Comments	Examples		
I	+1	H is −1 when combined with metals	$\overset{+1-1}{NaCl}$	$\overset{+1-1}{HCl}$	$\overset{+1-1}{LiH}$
II	+2	none	$\overset{+2(-1)_2}{MgCl_2}$	$\overset{+2-2}{CaO}$	
III	+3	none	$\overset{+3(-1)_3}{BCl_3}$	$\overset{+3(-1)_3}{AlF_3}$	
IV	+4, +2	Sn and Pb are +4 and +2; C is commonly +4 or −4, but has other oxidation numbers too	$\overset{+4\ (-1)_4}{SnF_4}$ $\overset{+4(-2)_2}{CO_2}$	$\overset{+2(-1)_2}{PbCl_2}$ $\overset{-4(+1)_4}{CH_4}$	$\overset{+2-2}{CO}$
V	+5, +3, −3	many other oxidation numbers are displayed by N and P; for example, N can be +4, +2, +1, −1, and −2	$\overset{+5(-1)_5}{SbF_5}$ $\overset{+4(-2)_2}{NO_2}$ $\overset{(+1)_2-2}{N_2O}$	$\overset{+3(-1)_3}{BiCl_3}$ $\overset{(+3)_2(-2)_3}{N_2O_3}$ $\overset{-1(+1)_2-2+1}{NH_2OH}$	$\overset{-3(+1)_3}{NH_3}$ $\overset{+2-2}{NO}$ $\overset{(-2)_2(+1)_4}{N_2H_4}$
VI	+6, +4, −2	O is almost always −2, −1 in peroxides, positive only with F compounds; other elements are commonly −2, +6, or +4	$\overset{(+1)_2-2}{H_2O}$ $\overset{(+1)_2-2}{H_2S}$	$\overset{(+1)_2(-1)_2}{H_2O_2}$ $\overset{+6(-2)_3}{SO_3}$	$\overset{+2(-1)_2}{OF_2}$ $\overset{+4(-2)_2}{SeO_2}$
VII	+7, +5, +3, +1, −1	all are −1 with H or metals, positive with O and other more electronegative elements; F is always −1	$\overset{+1-1}{NaF}$ $\overset{+1+3(-2)_2}{HClO_2}$	$\overset{+1+7(-2)_4}{HClO_4}$ $\overset{+1+1-2}{HClO}$	$\overset{+1+5(-2)_3}{HClO_3}$
VIII	+8, +6, +4, +2	no compounds of He, Ne, and Ar known; Kr combines only with F and is only +2; Xe is combined mainly with F and O	$\overset{(+1)_4\ +8\ (-2)_6}{Na_4XeO_6}$ $\overset{+4(-1)_4}{XeF_4}$	$\overset{+6(-1)_6}{XeF_6}$ $\overset{+2(-1)_2}{KrF_2}$	

The B family elements of group I may have oxidation numbers of $+1$, $+2$, and $+3$. Under ordinary laboratory conditions, the oxidation numbers usually exhibited are $+2$ for copper, $\overset{+2}{\text{Cu}}$, $+1$ for silver, $\overset{+1}{\text{Ag}}$, and $+3$ for gold, $\overset{+3}{\text{Au}}$.

The B family members of groups II and III commonly exhibit $+2$ and $+3$ oxidation numbers, respectively. However, mercury, Hg, in IIB also forms compounds in which its oxidation number is $+1$, for example, Hg_2Cl_2.

Each of the elements in groups IIIB to VIIB can exhibit a maximum positive oxidation number equal to its group number. Most exhibit a variety of positive oxidation states. As we see in Figure 5.17, Mn in group VIIB has oxidation states of $+2$, $+3$, $+4$, $+5$, $+6$, and $+7$. Only two of the nine group VIIIB elements have an oxidation state of $+8$. For the important metals iron, Fe, and cobalt, Co, the common oxidation states are $+2$ and $+3$, for example, $FeCl_2$ and $FeCl_3$.

5•7•2 Calculation of Oxidation Numbers Because the sum of the oxidation numbers for all the components of a compound is zero, we can calculate the oxidation number of an unfamiliar element in a compound, provided we know the formula for the compound and the oxidation numbers of all the other elements in it. Examples 5.5 and 5.6 illustrate the method.

━━━ • **Example 5•5** • ━━━

What is the oxidation number of rhenium in the chloride with the formula $ReCl_5$?

━━━ • **Solution** • Because chlorine is the more electronegative, we assign it a negative oxidation number; because it is in group VII, this number is $7 - 8 = -1$. We then let x be the oxidation number of rhenium:

$$\overset{x \quad (-1)_5}{ReCl_5} \qquad x + 5(-1) = 0 \qquad x = 5$$

The oxidation state of rhenium is $+5$:

$$\overset{+5(-1)_5}{ReCl_5}$$

See also Exercises 51–53.

━━━ • **Example 5•6** • ━━━

What is the oxidation number of chromium in potassium dichromate, $K_2Cr_2O_7$?

━━━ • **Solution** • Because potassium is in group IA, we assign it an oxidation number of $+1$. Of the other two elements, oxygen is the more electronegative so we assign it its expected negative oxidation number; in group VI this number is $6 - 8 = -2$. The number for chromium, x, is solved for to make the sum of the numbers zero:

$$\overset{(+1)_2 \quad (x)_2 (-2)_7}{K_2Cr_2\,O_7} \qquad +2 + 2x - 14 = 0 \qquad x = 6$$

The oxidation state of chromium is $+6$:

$$\overset{(+1)_2\,(+6)_2\,(-2)_7}{K_2Cr_2O_7}$$

See also Exercises 51–53.

5•7•3 Polyatomic Ions An ion that is made of two or more atoms joined by covalent bonds is called a **polyatomic ion.** Because these ions occur so often in common substances, students should memorize the name, formula, and charge on the polyatomic ions listed in Table 5.6. For polyatomic ions, a net oxidation number equal to the charge on the ion is assigned. Two examples are

phosphate $\overset{-3}{PO_4}$ or $[PO_4]^{3-}$ ammonium $\overset{+1}{NH_4}$ or $[NH_4]^+$

At times it may be necessary to know the oxidation numbers of individual elements in polyatomic ions. In such cases, the sum of the individual oxidation numbers must equal the charge on the ion, as shown in these examples:

$$\overset{+5(-2)_4}{[PO_4]^{3-}} \qquad \overset{-3(+1)_4}{[NH_4]^+}$$
$$+5 + 4(-2) = -3 \qquad -3 + 4 = +1$$

Table 5•6 Some common polyatomic ions

Name	Formula and charge
sulfate	$SO_4{}^{2-}$
sulfite	$SO_3{}^{2-}$
nitrate	$NO_3{}^-$
nitrite	$NO_2{}^-$
phosphate	$PO_4{}^{3-}$
phosphite	$PO_3{}^{3-}$
carbonate	$CO_3{}^{2-}$
cyanide	CN^-
acetate	$C_2H_3O_2{}^-$
hydroxide	OH^-
hydronium	H_3O^+
ammonium	$NH_4{}^+$

5•7•4 Formulas from Oxidation Numbers In using oxidation numbers to predict the formula of a compound, we choose the simplest ratio of atoms or ions that makes the plus and minus oxidation numbers add to zero.

• *Ratio 1:1* • The ratio is 1:1 when the oxidation state of one component equals that of the second. The term *component* may refer to a simple ion, such as Na^+, O^{2-}, or Cl^-, to a covalently bonded atom, such as H or Cl in HCl, or to a polyatomic ion, such as $NO_3{}^-$ or $SO_4{}^{2-}$. Examples:

$\overset{+1}{Na}$ with $\overset{-1}{Cl}$ $[1 \times (+1)] + [1 \times (-1)] = 0$, so: NaCl

$\overset{+2}{Mg}$ with $\overset{-2}{O}$ $[1 \times (+2)] + [1 \times (-2)] = 0$, so: MgO

$\overset{+3}{Al}$ with $\overset{-3}{PO_4}$ $[1 \times (+3)] + [1 \times (-3)] = 0$, so: $AlPO_4$

• *Ratio 1:2* • The ratio is 1:2 when the oxidation state of one component is twice that of the other. Examples:

$\overset{+2}{Ca}$ with $\overset{-1}{Cl}$ $[1 \times (+2)] + [2 \times (-1)] = 0$, so: $CaCl_2$

$\overset{+4}{Sn}$ with $\overset{-2}{S}$ $[1 \times (+4)] + [2 \times (-2)] = 0$, so: SnS_2

$\overset{+1}{NH_4}$ with $\overset{-2}{CO_3}$ $[2 \times (+1)] + [1 \times (-2)] = 0$, so: $(NH_4)_2CO_3$

• *Ratio 1:3* • The ratio is 1:3 when the oxidation state of one component is three times that of the other. Examples:

$$\overset{+3}{\text{Fe}} \text{ with } \overset{-1}{\text{NO}_3} \quad [1 \times (+3)] + [3 \times (-1)] = 0, \text{ so:} \quad \text{Fe(NO}_3)_3$$

$$\overset{+1}{\text{H}} \text{ with } \overset{-3}{\text{PO}_4} \quad [3 \times (+1)] + [1 \times (-3)] = 0; \text{ so:} \quad \text{H}_3\text{PO}_4$$

• *Ratio 2:3* • The ratio is 2:3 when one component has an oxidation state of 3 and the other of 2. Examples:

$$\overset{+3}{\text{Al}} \text{ with } \overset{-2}{\text{O}} \quad [2 \times (+3)] + [3 \times (-2)] = 0, \text{ so:} \quad \text{Al}_2\text{O}_3$$

$$\overset{+2}{\text{Ca}} \text{ with } \overset{-3}{\text{PO}_4} \quad [3 \times (+2)] + [2 \times (-3)] = 0, \text{ so:} \quad \text{Ca}_3(\text{PO}_4)_2$$

• *Other Ratios* • Other ratios, such as 1:4, 1:5, 2:5, and 2:7, are not uncommon. For example, in a certain oxide of chlorine, the oxidation numbers of chlorine and oxygen are +7 and −2, respectively. Accordingly, the simplest formula is Cl_2O_7, the simplest ratio in which the oxidation numbers add to zero: $2(+7) + 7(-2) = 0$.

5•8 The Systematic Naming of Compounds

Every substance has a name that differentiates it from all other substances. In the early days of the science, names were chosen for a variety of reasons: *water* because the compound was well known by this name even before it was recognized as a chemical compound; *ammonia* because it was made from sal ammoniac, a salt originally obtained from camel dung near a shrine to Jupiter Ammon in Egypt; and *oil of vitriol* because it was distilled at high temperatures from minerals called vitriols from ancient times. As the science of chemistry matured, it became necessary to develop a systematic nomenclature in order to give unambiguous names to the huge and growing list of substances being discovered.

Today the vast majority of substances have **systematic names** that indicate their compositions; pure oil of vitriol is now called hydrogen sulfate. There are still a few compounds that retain their unsystematic or **trivial names;** water and ammonia are examples. We will introduce systematic nomenclature for inorganic substances by considering two classes of compounds: (1) binary and related compounds and (2) ternary compounds containing oxygen as one of the three elements.

5•8•1 Binary and Related Compounds
A compound composed of just two elements is called a **binary compound.** The formula of a binary compound is usually written with the symbol for the less electronegative element shown first. Therefore, the symbol for the element with a positive oxidation number is followed by the symbol for the element with the negative oxidation number. For binary compounds, the name is composed of the name of

the first element and a word ending in *-ide* derived from the name of the second element. Examples:

HCl	hydrogen chloride	$AlBr_3$	aluminum bromide
H_2S	hydrogen sulfide	Mg_3N_2	magnesium nitride
K_2O	potassium oxide	LiH	lithium hydride

Although the systematic names of all binary compounds end in *-ide*, it is not true that all names ending in *-ide* refer to binary compounds. There are related compounds that contain polyatomic anions whose systematic names end in *-ide*, such as hydroxide, OH^-, and cyanide, CN^-. Examples:

$Ca(OH)_2$	calcium hydroxide	(not binary)
NaCN	sodium cyanide	(not binary)

Also, the names of negative ions are retained when these anions are found in more complex compounds. Example:

NH_4Cl	ammonium chloride	(not binary)

• *Using Prefixes to Indicate Proportions* • The stoichiometric proportions in formulas are often indicated in the name with Greek numerical prefixes. Examples:

N_2O	dinitrogen oxide	BF_3	boron trifluoride
NO_2	nitrogen dioxide	PCl_5	phosphorus pentachloride

Such use of prefixes should be limited to cases where it helps avoid confusion, as with the many oxides of nitrogen. We will not be expected to choose those instances where prefixes should be used. If a name that includes a prefix is presented to us, it should help us write the proper formula.

5•8•2   **Ternary Compounds Containing Oxygen**   A compound composed of three elements is called a **ternary compound.** There are several classes of ternary compounds that contain oxygen and two other elements. As for binary compounds, the elements are placed in order of increasing electronegativity in both the formula and the name. And again, the first part of the name is simply the name of the first element. The second part of the name is based on the name of an **oxy-ion,** an ion that contains oxygen and one other element.

• *The -ate and -ite Classes* • Two common classes of compounds whose names are based on oxy-ion names are the *-ate* class and the *-ite* class. Names for compounds in these classes are obtained as follows:

1. The root of the name of the oxy-ion indicates the parent element, the element other than oxygen. For example, *sulf-* indicates sulfur.

2. Some common oxy-ion is arbitrarily called the *-ate* ion. In the case of sulfur, SO_4^{2-} is designated as the sulfate ion.

3.   An anion whose parent element has an oxidation number two units less (one oxygen atom less) than the *-ate* ion is called the *-ite* ion. In the case of sulfur, SO_3^{2-} is the sulfite ion.

The formulas and names of some representative compounds in the *-ate* and *-ite* classes are

H_2SO_4	hydrogen sulfate	Na_2SO_4	sodium sulfate
H_2SO_3	hydrogen sulfite	Na_2SO_3	sodium sulfite
HNO_3	hydrogen nitrate	KNO_3	potassium nitrate
HNO_2	hydrogen nitrite	KNO_2	potassium nitrite
$HClO_3$	hydrogen chlorate	$NaClO_3$	sodium chlorate
$HClO_2$	hydrogen chlorite	$NaClO_2$	sodium chlorite

The *-ate* and *-ite* names are retained even though the anions are found in combination with polyatomic cations. An example is

NH_4ClO_3 ammonium chlorate   (not ternary)

• *The Per——ate Class* • There are some oxy-ions that do not fall into the *-ate* or *-ite* classifications. If the oxidation number of the parent element in an oxy-ion is higher than in the *-ate* ion (generally, there is one more oxygen atom), the anion is called the *per——ate* anion. This usage is restricted to oxy-ions of the group VIIA and VIIB elements. Examples are

$HClO_4$ hydrogen perchlorate $LiClO_4$ lithium perchlorate

• *The Hypo——ite Class* • If the oxidation number of the parent element in an oxy-ion is lower than in the *-ite* ion (generally, there is one less oxygen atom), the anion is called the *hypo——ite* anion. A familiar example of this type of anion is the active ingredient of certain household bleaches, ClO^-, the hypochlorite ion. Examples of compound names are

$HClO$ hydrogen hypochlorite $NaClO$ sodium hypochlorite

In this section we have written a number of formulas for ternary compounds containing oxygen. When such compounds are ionic, the formulas for the probable ions can be deduced as follows: the first element in the formula forms a simple positive ion, and the second element is combined with oxygen in a negative oxy-ion. Applying this scheme to $NaClO_3$, we predict the ions to be Na^+ and ClO_3^-; or for Na_2SO_3, we predict two Na^+ ions and one SO_3^{2-} ion. The overall sum of positive and negative charges must be zero. Another example is $Ca_3(PO_4)_2$. We predict three Ca^{2+} ions; to balance these six positive charges, there must be a 3− charge on each of the two PO_4^{3-} ions.

5•8•3   **Oxidation Numbers and Names of Compounds**   Generally, the difference between the oxidation states of the parent element in a series of oxy-ions is two units. This difference is consistent with a difference of one oxygen atom in each formula, as illustrated by the following series of compounds:

$\overset{+7}{\text{HClO}_4}$ hydrogen perchlorate $\overset{+7}{\text{KClO}_4}$ potassium perchlorate

$\overset{+5}{\text{HClO}_3}$ hydrogen chlorate $\overset{+5}{\text{KClO}_3}$ potassium chlorate

$\overset{+3}{\text{HClO}_2}$ hydrogen chlorite $\overset{+3}{\text{KClO}_2}$ potassium chlorite

$\overset{+1}{\text{HClO}}$ hydrogen hypochlorite $\overset{+1}{\text{KClO}}$ potassium hypochlorite

For elements that have two or more familiar oxidation states, it is common to show the state in a given compound with a roman numeral. Iron forms two series of compounds, one in which its oxidation number is +2 and another in which it is +3. To give these compounds unambiguous names, the oxidation number is given in parentheses. Examples:

$FeCl_2$ iron(II) chloride $FeCl_3$ iron(III) chloride
$FeSO_4$ iron(II) sulfate $Fe_2(SO_4)_3$ iron(III) sulfate

Copper also has two oxidation states in compounds, namely, +1 and +2. Examples:

Cu_2O copper(I) oxide CuO copper(II) oxide

Because the common state for copper is +2, it is customary to omit the oxidation number in the name of copper(II) compounds. Therefore, if no oxidation state is shown, it is assumed that the copper(II) compound is meant.

Unfortunately for the student of chemistry, the method of naming just described was developed after many common compounds were already named in accordance with other systems. Examples:

$FeCl_2$ ferrous chloride $FeCl_3$ ferric chloride
Cu_2O cuprous oxide CuO cupric oxide

These older names persist, particularly in industry, which means that the student is sometimes faced with the necessity of knowing two or more names for a single compound.

Chapter Review

Summary

The three main classes of elements are **metals, non-metals,** and **metalloids.** Elements typically bear no resemblance to the compounds they form in which atoms of different elements are held together by **chemical bonds.**

The combining capacity of an element with other elements is its **valence,** which is determined by the **valence electrons** in the outermost energy levels.

The *transfer* of a valence electron from a metal atom to a nonmetal atom produces a positive **cation** and a negative **anion.** The electrostatic force that binds these two kinds of ions together is called an **ionic bond;** the resulting substance is an **ionic compound.** Atoms and ions that have the same number of electrons are **isoelectronic.** The tendency for the elements near the ends of the periods in the periodic table to form ions that are isoelectronic with one of the noble gases is called the **rule of eight,** or **octet rule.** The very lightest of elements tend to form ions that follow the **rule of two.**

The *sharing* of a pair of valence electrons between two nonmetal atoms is called a **covalent bond;** the resulting substance is a **covalent compound.** Molecules of covalent compounds can contain many more than two atoms. A convenient way to account for all the valence electrons in a compound is to write the **Lewis structural formula,** in which a **shared pair** of electrons is shown by a line and an **unshared pair** of electrons (a **lone pair**) is shown by a pair of dots. In most covalent molecules, all the atoms follow the rule of eight or the rule of two.

All compounds are either **organic** or **inorganic.** The enormous number and diversity of organic compounds are due to the ability of carbon atoms to form molecular chains and rings and to form four covalent bonds simultaneously. Most of these bonds are **single bonds** (one pair of shared electrons), but **double bonds** (two pairs) and **triple bonds** (three pairs) are common. A kind of bond found in many inorganic compounds is the **coordinate covalent bond,** in which both electrons are donated by one atom.

A covalent bond between any two unlike atoms is a **polar covalent bond.** One atom—the one with the greater **electronegativity**—attracts the electron pair more strongly than the other and is hence partially negative; the other atom is partially positive. If the molecule is diatomic, the result is always a **polar molecule,** or **dipole.** The **dipole moment** of such a molecule is a measure of the strength of its polarity. This is determined by measuring the **dielectric constant** of the substance. More complex molecules are also polar and have dipole moments, unless the **bond moments** for all of their individual bonds cancel each other, as is often true for highly symmetric molecules.

The properties of a compound depend in part on its molecular geometry and **bond lengths.** The length of a covalent bond is the sum of the **covalent radii** of the two atoms that are bound together. The distance between neighboring atoms in adjacent molecules of a covalent compound is determined by the **van der Waals radii** of the two atoms. The radius of a metal atom in a metal is called the **metallic radius. Ionic radii** differ from the corresponding atomic radii by amounts determined in part by the electron **screening effect.**

In writing chemical formulas and balancing chemical equations, it is useful to assign **oxidation numbers,** or **oxidation states,** to the atoms or ions involved. These numbers are related to the combining ratios of elements and the group numbers in the periodic table. The **systematic names** of inorganic compounds, whether the compounds are **binary** (two kinds of atoms), **ternary** (three kinds), or more complex, depend in part on the oxidation numbers of the constituent atoms and ions..

Exercises

Ideas About Atoms
and Their Combinations

1. Based on their location in the periodic table, classify the following elements as metals, nonmetals, or metalloids: hafnium, palladium, radon, sulfur, antimony, silicon, arsenic, bromine, nitrogen.
2. Given that the valence of chlorine is one, and reasoning from the first compound to the last, what are the valences of the other elements in the compounds $BiCl_3$, Bi_2S_3, Ag_2S, Ag_2O, and SnO_2?
3. For one of the compounds in Exercise 2, could you predict some of its physical properties from a knowledge of the properties of the elements from which it is made?
4. Explain how a current of electricity can be used to show that the atoms in potassium bromide, KBr, are united in a way that differs from the way the atoms in wood alcohol (methanol), CH_3OH, are united.
5. Showing only electrons in the highest main level, write the electron-dot notation for each of the following atoms: boron, nitrogen, sulfur, calcium, cobalt, zinc, arsenic, iodine, xenon.
6. What is an outstanding chemical property of the elements whose valence level electronic configuration is: (a) s^2p^6; (b) s^1; (c) s^2p^5?
7. Locate in the periodic table the elements that are composed of atoms in which the highest sublevel that is partly or completely filled is an s sublevel.
8. Locate in the periodic table the elements that are composed of atoms in which the highest p sublevel lacks only one or two electrons.

Transfer of Electrons; Ions

9. When an element described in Exercise 7 reacts with an element described in Exercise 8, what type or class of compound is likely to form? What change takes place within the atoms to bring about the formation of the compound?
10. Show, in a manner comparable to that in Figures 5.2 and 5.3, how the following atoms react to form ions:
    (a) Ca + F $\longrightarrow$ (b) Na + O $\longrightarrow$
    (c) Al + S $\longrightarrow$ (d) Ga + Cl $\longrightarrow$

11. Which of the following formulas represent cations and which represent anions: NH_2^-, H_2Se, VO^{2+}, Co^{3+}, NaCl, ClO^-?

12. Write the electronic configuration for the antimony ion, Sb^{3+}, and state whether or not its structure follows the rule of eight.

13. (a) Based upon information from the periodic table, write the electronic configuration for the following ions: Rb^+, Se^{2-}, Ga^{3+}, Zn^{2+}.
    (b) Do any of the ions follow the rule of eight? If so, which one(s)?
    (c) Are any of the ions isoelectronic? If so, which one(s)?

14. Which of the following are probably not stable ions: Rb^{2+}, Ba^{3+}, Se^{2-}, Sn^{4+}, S^{3-}? Why?

15. How many electrons are present in each of the following ions: I^-, NH_4^+, SO_3^{2-}, Sr^{2+}?

Sharing of Electrons; Molecules

16. Write electron-dot formulas for each of the following atoms, and for each pair formulate the Lewis structure of the molecule produced by their combination. Use the rule of eight and the rule of two as guides.
    (a) P + Cl $\longrightarrow$ (b) H + S $\longrightarrow$
    (c) I + F $\longrightarrow$ (d) H + Si $\longrightarrow$

17. Using the electron-dot notation, rewrite the following equations:
    (a) Br + Cl $\longrightarrow$ BrCl
    (b) $Br_2 + Cl_2 \longrightarrow$ 2BrCl
    (c) Do these two equations represent the same chemical reaction? Explain.

18. Which of the following formulas probably do not represent stable covalent compounds: KCl, SiH_3, PH_3, $GeCl_4$, H_2S, $HeCl_2$, CsF?

19. Use electron-dot notations to formulate the combination of the following atoms. Show clearly whether ionic or covalent compounds are formed. Use the rule of eight and the rule of two as guides.
    (a) nitrogen + fluorine   (b) calcium + bromine
    (c) hydrogen + arsenic   (d) nitrogen + aluminum
    (e) chlorine + sulfur   (f) sulfur + sodium

20. Phosphorus pentachloride, PCl_5, is a well-known compound, but nitrogen pentachloride, NCl_5, is as yet unknown. Explain.

More on Lewis Formulas

21. Using the rules of eight and two as guides (but there are exceptions), write Lewis formulas for each of the following compounds (use a dashed line to indicate a pair of shared valence electrons; unshared valence electrons are to be shown with dots): dichlorine oxide, Cl_2O; chloroform, $CHCl_3$; phosphine, PH_3; hydrogen sulfide, H_2S; boron trifluoride, BF_3.

22. Write Lewis structural formulas for the following compounds:
    (a) methyl mercaptan, CH_3SH
    (b) ethylamine, $C_2H_5NH_2$
    (c) dimethylamine, $(CH_3)_2NH$
    (d) ethyl ether, $(C_2H_5)_2O$
    (e) hydroxylamine, NH_2OH

23. Write Lewis structural formulas for the following compounds, each of which is known to have one or more multiple bonds per molecule:
    (a) carbon disulfide, CS_2
    (b) acetonitrile, CH_3CN
    (c) cyanogen, $(CN)_2$
    (d) propene, C_3H_6

*24. Write Lewis structural formulas for the following compounds, each of which is known to contain a carbon-to-oxygen double bond:
    (a) urea, $(NH_2)_2CO$
    (b) acetaldehyde, C_2H_4O
    (c) acetone, $(CH_3)_2CO$
    (d) formic acid, HCO_2H
    (e) acetic acid, CH_3CO_2H
    (f) phosgene, $COCl_2$

25. The Lewis structural formulas for sulfur dioxide, SO_2, and ozone, O_3, are very much alike. Write the Lewis structural formula for a molecule of SO_2.

26. Formulate a Lewis structural formula for sulfur trioxide, SO_3, from that drawn for sulfur dioxide in Exercise 25.

27. When one writes a Lewis structure for sulfuric acid, H_2SO_4, in compliance with the octet rule for sulfur and oxygen, two coordinate covalent bonds are required. With the additional knowledge that there are two O—H bonds in a molecule of H_2SO_4, write its Lewis structural formula.

28. A common acid containing sulfur is sulfurous acid, H_2SO_3. Note that its formula differs from that of sulfuric acid by one oxygen atom. Write a Lewis structural formula for sulfurous acid.

29. When water reacts with hydrogen chloride, hydronium and chloride ions are formed. By analogy with the production of ammonium and chloride ions from ammonia and hydrogen chloride in Section 5.4.3, formulate the production of the hydronium and chloride ions.

Electronegativity; Polarity

30. Using the periodic table as a guide, list the elements in each of the following sets in order of increasing electronegativity (lowest to highest):
    (a) O, S, Cl, Se
    (b) Mg, Ca, Rb, Cs
    (c) S, Cl, Ga, Ge, As

31. Based upon atomic structure, give a logical explanation for the following facts:
    (a) Electronegativity decreases from element to element

in an A family as the atomic number increases.

(b) Electronegativity increases for A families in a given period as the atomic number increases.

(c) There are only relatively small changes in electronegativity for B family members as compared with the changes described in part (b).

32. (a) Which is the more electronegative, H or N? H or P?

(b) What is the difference in the electronegativities between H and N? Between H and P?

(c) The dipole moment of NH_3 is listed in Table 5.2 as 1.47 D. If PH_3 had the same geometry as NH_3, would you predict that PH_3 has a dipole moment larger than, smaller than, or about the same as that of NH_3? Explain.

33. Classify the bonds in the following as more than 50 percent ionic or more than 50 percent covalent: Cl_2, HI, RbF, Br_2, H_2S, SO_2, Cl_2O.

*34. Consider the following reactions and the energies required in joules per molecule as experimentally determined:

Reaction	Energy Required, Joules
(a) H—H $\longrightarrow$ 2H $\cdot$	7.24 $\times$ 10^{-19}
(b) $:\!\ddot{C}l\!-\!\ddot{C}l\!:\ \longrightarrow\ 2:\!\ddot{C}l\cdot$	4.03 $\times$ 10^{-19}
(c) $:\!\ddot{B}r\!-\!\ddot{B}r\!:\ \longrightarrow\ 2:\!\ddot{B}r\cdot$	3.22 $\times$ 10^{-19}
(d) $H\!-\!\ddot{C}l\!:\ \longrightarrow\ H\cdot\ +\ \cdot\ddot{C}l\!:$ $\ \ \delta^+\ \ \delta^-$	7.17 $\times$ 10^{-19}
(e) $H\!-\!\ddot{B}r\!:\ \longrightarrow\ H\cdot\ +\ \cdot\ddot{B}r\!:$ $\ \ \delta^+\ \ \delta^-$	6.08 $\times$ 10^{-19}

Calculate the energy required to overcome the attraction between the positive and the negative ends of each of the polar H—Cl and H—Br bonds. Compare these two values with that calculated for the polar H—F bond in Section 5.5.1. How do the calculated bond strengths relate to the electronegativities of F, Cl, and Br?

*35. (a) Which is the more electronegative, N or H? N or F?

(b) What is the difference in the electronegativities between N and H? Between N and F?

(c) The dipole moment of NH_3 is listed in Table 5.2 as 1.47 D. Assuming that the molecular shapes of NH_3 and NF_3 are about the same, would you predict that NF_3 has a dipole moment larger than, smaller than, or about the same as that of NH_3? Explain.

(d) Actually, the dipole moment is found to be 0.23 D. What does this tell you about the contribution of

the unshared pair of electrons to the dipole moments of these molecules?

36. The dipole moment of carbon tetrachloride, CCl_4, is 0 D, and its shape is tetrahedral. Would the value for the dipole moment be different if the shape of the molecule were square planar? The dipole moment of methylene chloride, CH_2Cl_2, is 1.59 D. It too has a tetrahedral shape. Would the value of its dipole moment be different if the shape of the molecule were square planar?

37. (a) Select from the following list the molecules that contain one or more polar bonds: HI, O_2, SO_2, H_2Se, Cl_2, ICl, SiH_4.

(b) Using the symbols δ^+ and δ^-, indicate the partial charges of the atoms joined by the polar bonds.

(c) Should each of those molecules with polar bonds have a dipole moment? Explain.

38. Which of the following statements are true?

(a) All diatomic molecules are linear.

(b) All diatomic molecules are polar.

(c) Ammonia, NH_3, and boron trichloride, BCl_3, are both polar molecules.

(d) All polar diatomic molecules are composed of two different elements.

(e) All molecules of the type AB_4 are polar.

*39. Write a Lewis structural formula for each of the following compounds. Which of these compounds violates the rule of eight? For parts (a), (b), and (f), write a Lewis structure that depicts the approximate geometry, based on the facts that (a) has no dipole moment but that (b) and (f) do.

(a) BCl_3; (b) NCl_3; (c) PCl_5; (d) SF_4; (e) SF_6; (f) ClF_3

40. Match each of the following compounds with its dipole moment.

Compound	Dipole Moment
ClF	0.57 D
BrF	0.62 D
ICl	0.88 D
BrCl	1.29 D

*41. Given that the charge of the electron in SI units is 1.602 $\times$ 10^{-19} C, show that 1 debye unit, 1 D, is equal to 3.34 $\times$ 10^{-30} C $\cdot$ m. (See footnote 5, Section 5.5.2.)

*42. (a) Show how the O—H bond moment of 1.51 D is calculated from the data given in footnote 6 in this chapter. (The solution to this exercise requires vector analysis and a table of or calculator with trigonometric functions.)

(b) Calculate the S—H bond moment in H_2S. The dipole moment of H_2S is 0.95 D, and the bond angle is 93.3°.

Sizes of Atoms,
Molecules, and Ions

43. (a) Using the values of the covalent radii listed in Table 5.3, calculate the interatomic bond distances in molecules of ClF and IBr.

   (b) For one of these two compounds, there is no difference between the experimentally determined and the calculated bond distances, whereas in the other compound the difference is 0.07 Å. For which compound is there this difference? What is the experimentally determined bond distance for this compound?

*44. (a) Using data from Table 5.4, calculate the percentage by which the calculated bond length differs from the experimentally determined bond length in HF, HCl, HBr, and HI.

   (b) Account for the very large percentage for HF as compared with the other three.

45. The experimentally determined interatomic bond distances in the molecules HCl and ClF are 1.27 Å and 1.63 Å, respectively. If one of each of these molecules were to move about and collide without reacting, what is the closest distance in angstroms that a hydrogen nucleus could come to a fluorine nucleus? In nanometers? (Use data in Table 5.3.)

*46. Each B—Cl single bond length in a molecule of BCl_3 is 1.75 Å. Calculate the interatomic nonbonded distance between the nuclei of two chlorine atoms in BCl_3.

*47. The H—Te—H bond angle in hydrogen telluride, H_2Te, is 90°. The H—Te single bond length is 1.7 Å. Calculate the interatomic nonbonded distance between the nuclei of the two hydrogen atoms in H_2Te.

48. (a) What is the radius of the sphere that circumscribes a CF_4 molecule?

   (b) What would be the volume of the sphere in cubic angstroms, $Å^3$, and in cubic nanometers, nm^3?

49. Explain how a metallic bond differs from an ionic bond. From a covalent bond.

50. With only the periodic table inside the front cover as a guide, arrange each of the following sets of atoms and ions in order of increasing size:
   (a) K, Ca, Rb, Ti, Br, Cl
   (b) Na^+, H^+, Cl^-, Br^-
   (c) Rb, Sr, Cs, Rb^+, Sr^{2+}, Cs^+
   Now check your arrangements with Figure 5.14.

Oxidation Numbers

51. Assign oxidation numbers to the compounds whose formulas are given in Exercises 2, 21, 37, and 39.

52. Assign oxidation numbers to the individual elements in the polyatomic ions listed in Table 5.6.

53. Assign oxidation numbers to the compounds whose formulas are given in Exercises 54, 58, and 59 below.

Naming Compounds

54. Name each of the following compounds: H_3PO_4, H_3PO_3, H_3PO_2, Na_3PO_4, Na_3PO_3, Na_3PO_2.

55. For K_2MnO_4 the systematic name is potassium manganate. With this information, name each of the following compounds: K_2MnO_3, $KMnO_4$, and $HMnO_4$.

56. Write the formulas for four different ternary compounds of oxygen in which the other two elements are sodium and bromine. Repeat for calcium and chlorine as the other two elements.

57. Name each compound in Exercise 2.

58. Name each of the following compounds by using a roman numeral to identify the oxidation number of the first element: $CuSO_4$, $Fe(NO_3)_2$, N_2O, NO, NO_2, SO_3, MnO_2, OF_2.

*59. Suggest systematic names for the following compounds: Na_4XeO_6, Na_2XeO_4, Na_3AsO_3, Na_3AsO_4.

60. The name of $NaNO_3$ is sodium nitrate. Is there a compound known by the name sodium pernitrate? Explain.

Theories of the Covalent Bond

6

Valence Bond Theory

6•1 **Introduction to Valence Bond Theory**

6•2 **Sigma Bonds**

6•3 **Pi Bonds**

6•4 **Hybridized Orbitals**

6•4•1 In Molecules with Single Bonds

6•4•2 In Molecules with Multiple Bonds

6•4•3 Hybridization Involving *d* Orbitals

6•5 **Valence Shell Electron Pair Repulsion (VSEPR) Model for Molecular Geometry**

6•5•1 Limitations of VSEPR Theory

6•6 **Resonance**

6•6•1 An Orbital Interpretation of Resonance

Molecular Orbital Theory

6•7 **Introduction to Molecular Orbital Theory**

Special Topic: The Use of Molecular Models

6•8 **Sigma and Pi Molecular Orbitals**

6•8•1 Filling Molecular Orbitals for Homonuclear Diatomic Molecules

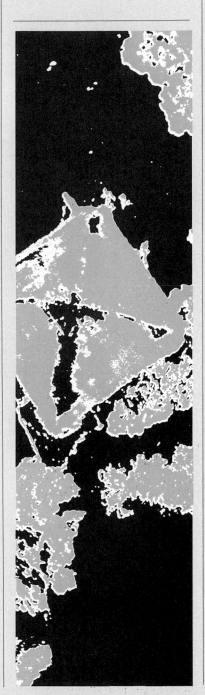

After the pioneering suggestion of Lewis that a covalent bond consists of a pair of shared electrons, the study of covalent bonding became more and more intense and sophisticated. The methods of wave functions and orbital descriptions presented in Chapter 4 to describe atoms were used to describe the electronic structures of molecules. In the 1920s two competing approaches began to be developed: (1) the valence bond (VB) theory, and (2) the molecular orbital (MO) theory.

The VB theory initially was the more successful, reaching its greatest influence in the 1940s. Recently, the MO theory has received more attention. For a detailed study of these two theories of covalent bond formation and their applications, a working knowledge of quantum mechanics is required. Such a mathematical treatment is beyond the scope of this text; however, there are many important concepts derived from the two theories that enrich our understanding of chemical bonds.

We shall devote some attention to both theories because both have proved successful in accounting for the structures and properties of a wide variety of covalent molecules. The two theories are being changed continually as new discoveries are made. Often, some concept developed in one theory has its counterpart in the other, and frequently the two theories are found to give identical results. The VB theory follows more directly from the electron pairing idea of Lewis. In the first part of this chapter, applications of this approach to covalent bonding will be presented. The latter part of the chapter will be devoted to a brief discussion of the MO theory.

Valence Bond Theory

6•1   Introduction to Valence Bond Theory

Consider the formation of a diatomic molecule. With the valence bond approach, we first describe the atomic orbitals in two isolated atoms that are separated sufficiently so that they do not interact. Each of these atoms is assumed to have one or more half-filled orbitals. Next, we imagine the two atoms moving toward one another until two of the half-filled orbitals, one from each atom, overlap. The result is that two electrons, one from each atom, occupy the overlapping orbitals of both atoms. *This overlap of atomic orbitals is a covalent bond that joins the two atoms to form a molecule.*

As a consequence of the Pauli exclusion principle, it is a basic requirement for a covalent bond that the two electrons have opposite spins ($\uparrow\downarrow$), because they occupy the same orbital. The electrons are said to be paired. If this is the only orbital overlap between the two atoms, a single bond is formed. If additional orbitals of the two atoms overlap, then multiple bonds result. For the sake of simplicity, any electrons that are not in overlapping orbitals are considered to be in atomic orbitals attracted to their respective nuclei. One can readily see the similarity between valence bond theory and the earlier electron pairing idea of Lewis.

Three examples of the formation of diatomic molecules by the valence bond approach are diagrammed in Figure 6.1. In part (a) of the figure, a pair of 1s atomic orbitals of hydrogen atoms overlap to form the molecular orbital of the H_2 molecule.[1] The resulting molecular orbital includes both s orbital electrons and both nuclei. The formation of the molecular orbital of the F_2 molecule is shown in part (b). The electronic configuration of an isolated fluorine atom, $1s^2 2s^2 2p_x^2 2p_y^2 2p_z^1$, indicates the presence of one half-filled p orbital. The half-filled p orbitals of two F atoms are diagrammed, with each half-filled orbital arbitrarily centered on the x-axis and lying in a head-to-head direction. (Each of the other atomic orbitals of the two atoms is filled with two electrons of opposite spins.) As a consequence of the interaction of these two orbitals, a single molecular orbital is formed that includes both p orbital electrons and both nuclei. Part (c) is a representation of the formation of the molecular orbital of the HF molecule from the overlap of a

Figure 6•1 As two half-filled atomic orbitals approach each other, the orbitals begin to overlap. Diagrams are shown for the formation of (a) H_2, (b) F_2, and (c) HF molecules. In each case, a bonding orbital is formed in which the electron density of the shared pair is greatest between the two nuclei. The three-dimensional diagrams show the regions occupied before and after bond formation by the two electrons that pair to form the molecular orbital. Two equivalent ways of drawing the molecular orbitals are shown. All the valence electrons are indicated in the circle-and-arrow diagrams, but those that do not form bonding orbitals are lightly shaded.

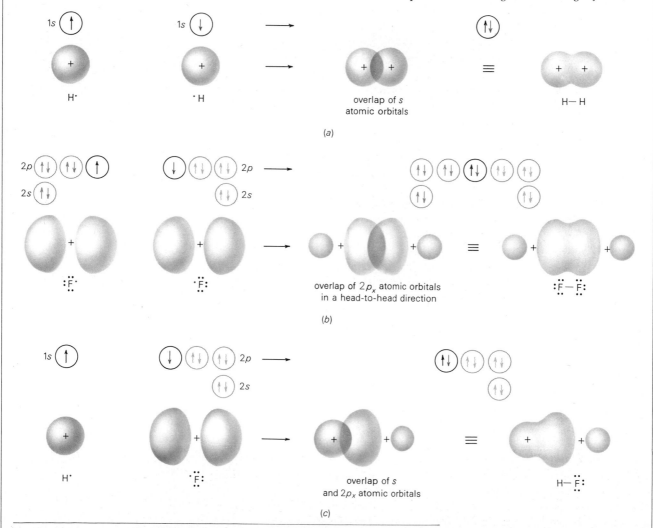

[1]The term *molecular orbital* is used in both the VB and the MO theories.

half-filled 1s atomic orbital of a hydrogen atom and a half-filled 2p orbital of a fluorine atom.

In H_2, F_2, and HF, the molecular orbitals that are formed are called *sigma orbitals*. A **sigma (σ) orbital** is a molecular orbital that is symmetrical around a line that passes through the two nuclei. When the electron density in this orbital is concentrated in the bonding region between the two nuclei, the bond is called a **sigma bond.**

The diagrams of the F_2 and HF molecules shown in Figure 6.1 do not include the filled 2s and 2p orbitals of the fluorine atoms. These orbitals have been omitted because it is assumed that they are not markedly involved in the bonding process. These nonbonding electron pairs do, of course, contribute to the overall size of the molecules. To describe molecular structure properly, we must consider the probable electron densities due to all electrons of the bonded atoms. As for an atom, we can set a certain boundary limit, for example, a 90-percent contour line. However, because of the rigorous mathematical treatment involved, it is difficult to construct an adequate model of molecular structure that describes the systems of electrons and nuclei. As discussed in the special topic in Chapter 4, "Computers—Tools of the Chemical Theorist," Wahl has expanded the computer's function to produce diagrams that pictorially depict the formation of molecules from atoms.

Figure 6.2 presents Wahl diagrams showing several stages in the formation of an H_2 molecule. Stage 1 shows two isolated hydrogen atoms. The points plotted for internuclear distances on the graph correspond to the various positions of the atom at the right as it approaches the atom at the left. Because there are no attractive or repulsive forces between the two atoms when they are isolated, the potential energy of the system at stage 1 is essentially zero. At stage 2 each hydrogen nucleus is beginning to attract the other's electron, and the process of covalent bond formation has begun. Because the attractive force between the two atoms exceeds the coulombic repulsive forces of electron–electron and proton–proton interactions at this point, the potential energy of the system at stage 2 is lower than that at stage 1. Each successive stage represents an overall contraction of the electron density cloud and a gradual buildup of charge density between the two nuclei as the bond formation process continues.

At stage 7 the bonding distance of lowest potential energy is reached. This optimum bonding distance is the internuclear distance of 0.74 Å (see Section 5.6.1). The minimum energy corresponds to the point where the attractive force exceeds the proton–proton and electron–electron repulsive forces by the greatest amount. After stage 7, there would be a rapid increase in energy through stages 8 and 9 due to the increased repulsion between the two positively charged nuclei at the shorter internuclear distances. These contour diagrams show that the volume occupied by the electrons decreases when the molecule forms. Obviously, the Lewis representation of two electrons between two hydrogen nuclei is oversimplified; the charge concentrated between the two nuclei is only a small percentage of the total of two electrons in the hydrogen molecule.

Two stages in the formation of an F_2 molecule from two fluorine atoms are shown in Figure 6.3. Only the beginning stage where the fluorine atoms are isolated and the final stage of optimum bonding distance and

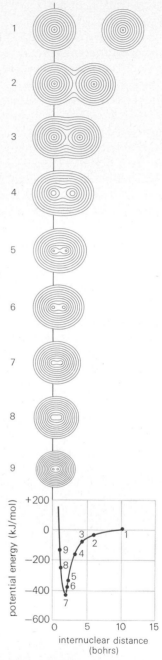

Figure 6.2 Formation of the hydrogen molecule, H_2. The internuclear distances for each stage, 1 through 9, are keyed to the plot below. These distances are in bohr units; 1 bohr = 0.529 Å.

Source: Wall chart by Arnold C. Wahl. Copyright © 1970 by McGraw-Hill, Inc. Used with permission of McGraw-Hill Book Co.

lowest energy are diagrammed. The intermediate stages in this molecular formation are more involved since considerably more than just the two $1s$ orbitals used in the case of the H_2 molecule are involved. The orbital structure of the fluorine atom obtained by merging of the $1s$, $2s$, $2p_x$, $2p_y$, and $2p_z$ orbital electronic charge densities was shown in Chapter 4. The contour diagram for the F_2 molecule displays an optimum bonding distance of 1.42 Å at the minimum in potential energy.

In the contour diagram of the H_2 molecule in Figure 6.2, the eighth contour line from the nuclei approximately corresponds to the van der Waals radius of 1.20 Å (see Table 5.3). The tenth line from the nuclei of the F_2 molecule in Figure 6.3 approximates its van der Waals radius of 1.47 Å. Beyond these contour lines, there is only a small percentage of the total electronic charge. The van der Waals radius of an atom is approximately the radius of the 98-percent contour line.

Diagrams of the type shown in Figure 6.1 generally will be sufficient for our use in depicting sigma bonds and their formation. We can often avoid the complication of trying to picture electron orbital clouds by using simple space-filling or scale models to help us visualize molecular structure. In Figure 6.4, models of four molecules are shown (also see similar figures in Chapter 5). Each atom is represented by a colored ball of appropriate size that is attached to another ball at approximately the proper angle.

6•3 Pi Bonds

In the formation of the bonding orbital between two fluorine atoms, the $2p$ orbitals overlap in a head-to-head fashion (see Figure 6.1). There is a second way in which half-filled p orbitals of two different atoms may overlap to form a bonding orbital. If the two p orbitals are situated perpendicular to the line passing through the two nuclei, then the lobes of the p orbitals will overlap extensively sideways to form an electron cloud that lies above and below the two nuclei. The molecular orbital resulting from this sideways or lateral overlap is called a **pi** (π) **orbital.** This covalent bond is called a **pi bond.** The basic features of a pi bond are shown in Figure 6.5. Pi bonds are present in molecules having two atoms connected by double or triple bonds such as those found in ethylene, C_2H_4, carbon dioxide, CO_2, acetylene, C_2H_2, and nitrogen, N_2 (see Section 5.4.2). In these compounds the two atoms are also joined by a sigma bond. The sigma bond has greater orbital overlap and is generally the stronger bond; a pi bond, with less overlap, is generally weaker.

6•4 Hybridized Orbitals

6•4•1 In Molecules with Single Bonds The ground-state electronic configurations for beryllium ($1s^2 2s^2$), boron ($1s^2 2s^2 2p_x^1$), and carbon ($1s^2 2s^2 2p_x^1 2p_y^1$) suggest that atoms of these elements would form zero, one, and two covalent bonds, respectively. The configuration for the beryllium atom ($1s^2 2s^2$), for example, indicates that all its electrons are paired in spin and therefore are not able to join with electrons of other atoms to form bonds. However, the experimental facts concerning covalent bond formation by these elements may be summarized as follows:

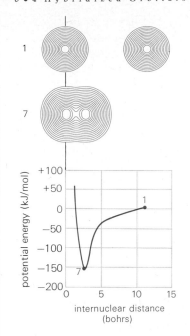

Figure 6•3 Formation of the fluorine molecule, F_2.
Source: Wall chart by Arnold C. Wahl. Copyright © 1970 by McGraw-Hill, Inc. Used with permission of McGraw-Hill Book Co.

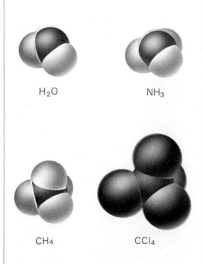

H_2O NH_3

CH_4 CCl_4

Figure 6•4 Scale models of molecules of four covalent compounds.

Beryllium, Be

1. Forms two covalent bonds, for example, $BeCl_2$.
2. In $BeCl_2$, or similar compounds, both bonds are identical.
3. The bond angle between the two bonds is 180°; the molecular geometry is linear.

Boron, B

1. Forms three covalent bonds, for example, BCl_3.
2. In BCl_3, or similar compounds, all bonds are identical.
3. The bond angle between any two bonds is 120°; the molecular geometry is planar triangular.

Carbon, C

1. Forms four covalent bonds, for example, CH_4.
2. In CH_4, or similar compounds, all bonds are identical.
3. The bond angle between any two bonds is 109°28', or about 109.5°; the molecular geometry is tetrahedral.

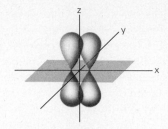

Figure 6•5 (above) A representation of a pi bond formed by the lateral overlap of two half-filled p_z orbitals. These p_z orbitals overlap above and below the xy plane. In a similar fashion, another pi bond can be formed by lateral overlap of two half-filled p_y orbitals on either side of the xz plane. The p_z orbitals have been exaggerated for clarity.

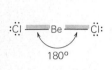

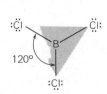

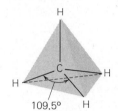

Figure 6•6 (below) Schematic representations of the formation of two equivalent Be—Cl bonds in $BeCl_2$. (Nonbonding $1s$ orbitals are not shown.) The p orbitals of Cl and the hybridized orbitals are simplified for clarity. Only atomic orbital overlaps are pictured. The ↓ in ↑↓ denotes the electron donated by the Cl atom.

In order to accommodate these experimental facts in our theories of covalent bond formation, it is necessary to recognize that the electronic structures of these elements in molecules are not the same as their structures

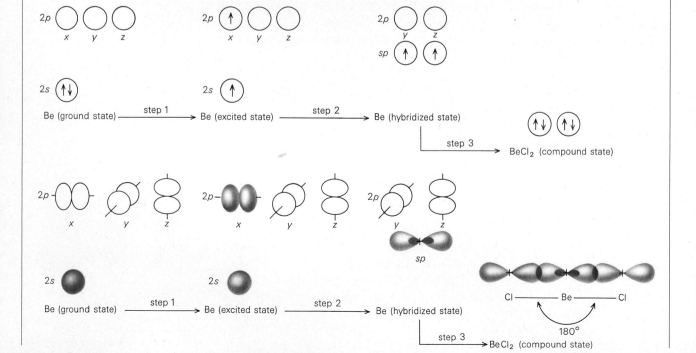

in separate atoms. First, in these compounds beryllium must have two, boron three, and carbon four bonds, with each bond containing only one electron donated by the central atom. Second, for each compound the bonding orbitals that result from the overlap of the atomic orbitals from two different atoms, Be and Cl, B and Cl, or C and H, must be identical. These bonding orbitals are sigma bonds just like those described in Figure 6.1 for F_2, H_2, and HF.

The diagrams in Figures 6.6, 6.7, and 6.8 schematically represent imaginary processes of bond formation in these compounds. Each process illustrated by these diagrams has as its foundation three basic steps. *Step 1 involves electron promotion from a filled 2s orbital to an unoccupied 2p orbital.* This electron promotion occurs from the lower energy ground state of the atom to the higher energy excited state and therefore requires an addition of energy. *Step 2 involves the mixing of the 2s with all of the 2p orbitals that now contain one electron.* The numbers of 2p orbitals involved in the mixing process are one for beryllium, two for boron, and three for carbon. After the orbitals are mixed, they no longer have the spherical shape of *s* orbitals or the dumbbell shape of *p* orbitals, but have shapes that are part *s* and part *p* in character. The process of mixing different orbitals of the same atom to form equivalent orbitals is termed **hybridization.** The orbitals formed are called **hybrid orbitals.**

In the case of beryllium, the hybrid orbitals are a mixture of one 2s orbital and one 2p orbital and are called *sp hybrid orbitals*. Because two orbitals mixed to form them, there must be two *sp* hybrid orbitals. A representation of the shape of the two *sp* orbitals is shown in Figure 6.6.

Figure 6•7 Schematic representations of the formation of three equivalent B—Cl bonds in BCl_3. (Compare with Figure 6.6.) In step 3, the boron sp^2 orbitals have been rotated 90° and then overlapped with orbitals of chlorine atoms, so that the B and Cl nuclei are in the plane of the paper. The smaller portions of the lobes of the sp^2 hybrid orbitals have been omitted for clarity. The ↓ in ↑↓ denotes the electron donated by the Cl atom.

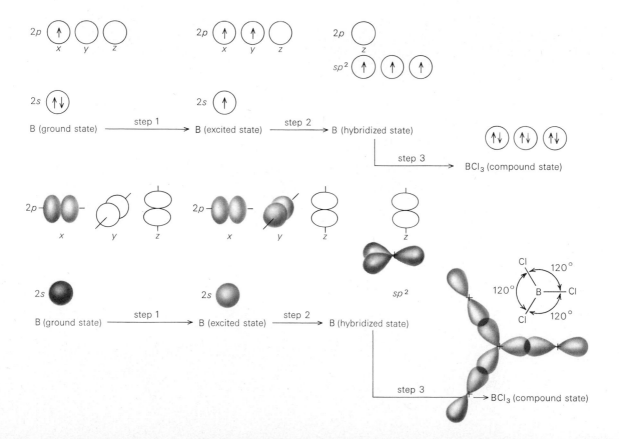

In the case of boron, the hybrid orbitals are a mixture of a $2s$ orbital and two $2p$ orbitals and are called sp^2 (pronounced s-p-two) *hybrid orbitals*. Three such sp^2 hybrid orbitals are formed. Because these orbitals have more p character than s character in their composition, they differ somewhat from the sp hybrids. A representation of the shape of the three sp^2 orbitals is shown in Figure 6.7.

In carbon, the hybrid orbitals are a mixture of one $2s$ and three $2p$ orbitals and are called sp^3 (pronounced s-p-three) *hybrid orbitals*. Four of these orbitals are formed. These orbitals possess still more p character than sp^2 hybrid orbitals and therefore have a geometry that differs from both sp and sp^2 hybrids (see Figure 6.8).

Step 3 involves overlap of the hybrid orbitals of Be, B, and C with atomic orbitals from Cl, Cl, and H, respectively. Beryllium has each of its two sp hybrid orbitals overlapping with the half-filled $3p$ orbital of a chlorine atom to form two sigma molecular orbitals. Each of the three sp^2 hybrid orbitals of boron overlaps with the half-filled $3p$ orbital of a chlorine atom to form three sigma bonds. The four sigma bonds in CH_4 result from the overlap of each of the four sp^3 hybrid orbitals of carbon with the $1s$ orbital of a hydrogen atom.

The energy gained from the formation of the two additional bonds in step 3 more than compensates for the energy required for the promotion of the $2s$ electrons. Although we have considered step 2 to be separate from step 3 in these imaginary processes of bond formation, they are not independent

Figure 6·8 Schematic representations of the formation of four equivalent C—H bonds in CH_4. (Compare with Figures 6.6 and 6.7.) The smaller portions of the lobes of the sp^3 hybrid orbitals have been omitted for clarity. The ↓ in ↑↓ denotes the electron donated by the H atom. The diagram beside the final orbital representation depicts schematically the three-dimensional character of the bonds. Solid lines indicate bonds joining atoms in the plane of the paper, dashed lines stand for bonds projected behind the plane of the paper, and wedges (◿) stand for bonds projected in front of the plane of the paper.

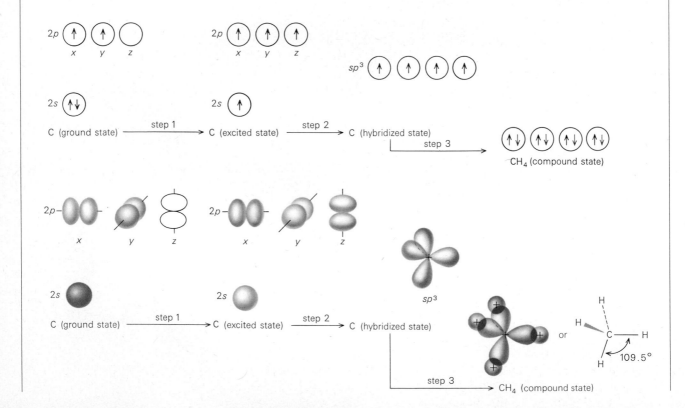

of one another. Hybridization and bond formation should be thought of as being simultaneous processes. In a sense, the orbitals hybridize upon the approach of the half-filled atomic orbitals of the other atoms. In these compounds, hybridization of the central atom occurs because the hybrid orbitals are able to overlap more effectively with the atomic orbitals of the other atoms. Also, the molecular geometry resulting from the equivalent hybrid orbitals is that in which the electron pairs of the bonds are as far away from each other as possible. The combination of the more effective overlap and the minimization of repulsion energies between bonding electron pairs results in the molecule having the lowest energy when the orbitals of the central atom have the hybridized arrangement.

—— • Example 6•1 • ——

When hydrogen fluoride reacts with glass, the gaseous silicon compound, SiF_4, is formed. Using circle-and-arrow diagrams similar to those used in Figures 6.6, 6.7, and 6.8, diagram the imaginary processes of bond formation in a molecule of SiF_4.

—— • Solution • The ground-state electronic configuration of silicon is $(Ne)3s^2 3p_x^1 3p_y^1$. Because its outer energy level electronic configuration, $s^2 p^2$, is identical with that of carbon, we would expect silicon to form structurally similar compounds. We write a schematic representation similar to that shown for CH_4 in Figure 6.8:

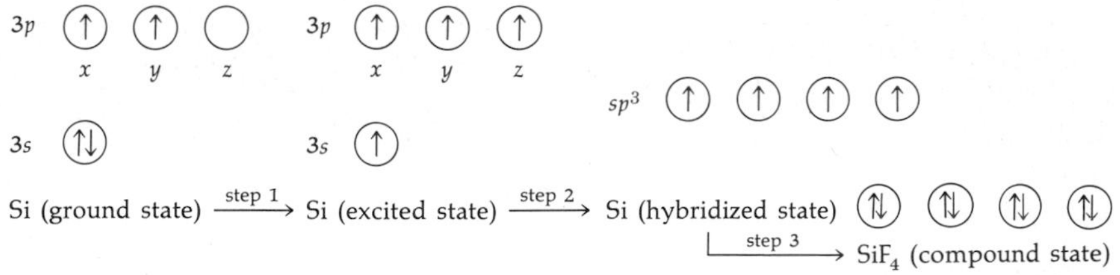

In the representation of the compound state, the colored arrows represent the electrons donated by the F atoms.

See also Exercises 7, 8, and 11 at the end of the chapter.

Other combinations of these hybrid orbitals with atomic orbitals are possible. For example, overlap of each of the four sp^3 hybrid orbitals of a carbon atom with the half-filled 3p orbital of a chlorine atom gives the four sigma bonds in CCl_4. Combination of two hybrid orbitals is also known. For example, the sigma bond between the two carbon atoms of ethane, C_2H_6, results from overlap of two sp^3 hybrid orbitals, as shown in Figure 6.9.

The shape or geometry of a simple molecule depends upon the number and arrangement of the bonding orbitals of the central atom. Thus far in our descriptions of hybridized orbitals, three types of molecular geometry have been introduced. For any molecule in which the central atom is sp hybridized, such as $BeCl_2$, the geometry is linear about the central atom. If the central atom is sp^2 hybridized, as in BCl_3, the three orbitals are directed from the central atom toward the corners of an equilateral triangle with all

four atoms in the same plane. That is, the geometry is planar triangular.[2] If
the central atom is sp^3 hybridized, as in CH_4 and CCl_4, the four orbitals are
directed toward the corners of a tetrahedron; that is, the geometry is tetrahedral. The bond angles associated with these molecules in which the central
atoms are sp, sp^2, and sp^3 hybridized are 180°, 120°, and 109.5°, respectively.

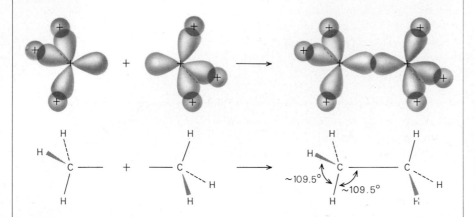

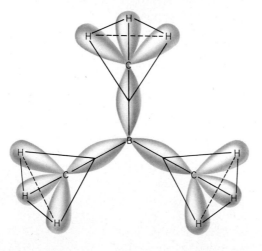

Figure 6•9 A representation of the
formation of the sigma bond between carbon atoms in ethane,
C_2H_6.

     In ethane, C_2H_6, both the H—C—H and H—C—C bond angles are
approximately 109°. Complex molecules possess a combination of geometries. For example, in trimethylboron (Figure 6.10), the C—B—C bond angles
are 120°, whereas the H—C—B and H—C—H bond angles are approximately 109°. The three hybrid orbitals of boron are directed toward the
corners of an equilateral triangle (carbons at the corners), and the four hybrid
orbitals of each carbon are directed toward the corners of a tetrahedron
(hydrogens at three corners and boron at the fourth).

Figure 6•10 A schematic representation of the geometry and the general location of the bonding orbitals
in a trimethylboron molecule,
$B(CH_3)_3$.

[2] The geometry of sp^2 hybridized orbitals is often referred to as planar trigonal geometry; the
terms planar triangular and planar trigonal are synonymous.

• sp^3 *Hybridization in* NH_3 *and* H_2O *Molecules* • The fact that beryllium, boron, and carbon form more covalent bonds than would be predicted from their atomic ground-state electronic configurations illustrates one of the fundamental principles of the scientific method. When one makes predictions based on existing chemical theory about properties of molecules, for example, molecular structure, which later are found not to be supported by the experimental facts, then it becomes necessary to alter the theory to accommodate these facts. It is difficult to argue against the reality of experimental data. The cases of the molecular structure of ammonia, NH_3, and water, H_2O, illustrate this principle once again.

The known ground-state electronic configuration for an atom of nitrogen $(1s^2 2s^2 2p_x^1 2p_y^1 2p_z^1)$ suggests that in NH_3 the three sigma bonds joining the three hydrogen atoms to the nitrogen atom result from the overlap of the three half-filled $2p$ orbitals of nitrogen with $1s$ orbitals of the hydrogen atoms. On the basis that p orbitals are oriented at right angles to one another, the prediction is that the bond angles between N—H bonds should be 90°. Also, from the ground-state electronic configuration for an oxygen atom $(1s^2 2s^2 2p_x^2 2p_y^1 2p_z^1)$, it is predicted that in H_2O there should be an angle of 90° between the two O—H sigma bonds that result from the overlap of the two half-filled p orbitals of an oxygen atom with $1s$ orbitals of two hydrogen atoms.

Figure 6•11 Schematic representations of the formation of sp^3 hybrid orbitals in ammonia, NH_3, and the three sigma bonds. The figure legend is similar to that shown for CH_4 in Figure 6.8.

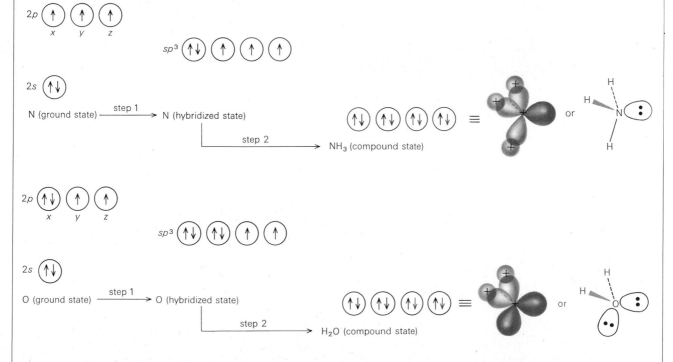

Figure 6•12 Schematic representations of the formation of sp^3 hybrid orbitals in water, H_2O, and the two sigma bonds. The figure legend is similar to that shown for CH_4 in Figure 6.8. (Also see Figure 6.11.)

Experimentally, however, it is found that the bond angle is 107.3° in ammonia and 104.5° in water. One method of accounting for these larger angles is to assume that the repulsion between positively charged hydrogen nuclei tends to increase the distance between them, thereby giving a larger bonding angle. A more widely accepted explanation is based on sp^3 hybridization, with the resultant formation of four orbitals by the central atom directed toward the four corners of a regular tetrahedron. As we will see in Section 6.5, the regular tetrahedral angles of 109.5° are distorted due to the presence of orbitals that are not involved in bonding. Schematic representations to account for sp^3 orbitals in ammonia and water molecules are shown in Figures 6.11 and 6.12. These representations are similar to that shown for methane, CH_4, with the exception that no promotion of an electron from a ground state to an excited state is required.

6.4.2   **In Molecules with Multiple Bonds**   As pointed out in Section 6.3, sigma and pi bonds are present in molecules having two atoms connected by double or triple bonds. The theory of hybridized orbitals assists us in rationalizing the bonding in such molecules. Diagrams similar to those used for representing the imaginary process of bond formation in $BeCl_2$, BCl_3, and CH_4 can be used to explain the bonding in ethylene, C_2H_4, acetylene, C_2H_2, and other molecules with multiple bonds.

• *Ethylene, C_2H_4* • The formation of ethylene is diagrammed in Figure 6.13. Step 1, the electron promotion step, is exactly like the first step in the bond formations of methane except that two carbon atoms are involved rather than one. Step 2 differs somewhat. For each carbon atom, one $2s$ and two $2p$ orbitals are hybridized (as is done with boron in the formation of BCl_3 in Figure 6.7) to give three sp^2 hybrid orbitals of equal energy. The remaining electron on each carbon atom stays in a $2p$ orbital. Step 3 involves overlap of

Figure 6.13 Schematic representation of the formation of the sigma bonds and the pi bond in ethylene. The *p* orbitals that make up the pi bond are shaded gray.

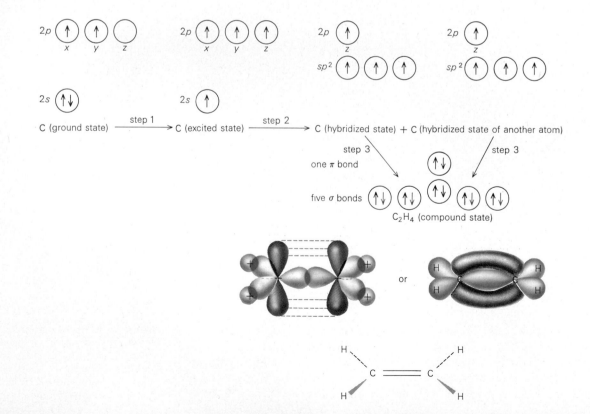

an sp^2 hybrid atomic orbital on each carbon to form a C—C sigma orbital.

The remaining sp^2 orbitals of each carbon overlap with 1s orbitals of hydrogen atoms to form two additional sigma bonds on each carbon. The two p orbitals overlap sideways, or laterally, to form a pi orbital. Thus, there are six bonding molecular orbitals in ethylene, five sigma and one pi. The five sigma bonds all lie in the same plane; the p orbitals that overlap to give the pi bond are perpendicular to this plane.

• *Acetylene, C_2H_2* • In the formation of acetylene, C_2H_2, diagrammed in Figure 6.14, step 1 is identical with that shown for ethylene in Figure 6.13. Step 2 involves for each carbon atom the hybridization of one 2s and only one 2p orbital (as is shown with beryllium in the formation of $BeCl_2$ in Figure 6.6) to give two sp hybrid orbitals of equal energy. The remaining electrons, two per carbon atom, stay in the 2p orbitals. Step 3 involves overlap of an sp hybrid atomic orbital on each carbon to form a C—C sigma orbital.

The remaining sp orbitals of each carbon overlap with 1s orbitals of hydrogen atoms to form two additional C—H sigma orbitals. The linear sigma bonding framework of the acetylene molecule is shown centered on the x-axis at the lower left of Figure 6.14. The p orbitals on each carbon atom are centered on the y- and z-axes. One pair of p orbitals overlaps laterally, as was shown for ethylene in Figure 6.13, to form a pi orbital. In the same manner, the second pair of p orbitals overlaps laterally to form a second pi orbital. Thus there are five bonding molecular orbitals in acetylene, three sigma and two pi. The four atoms of the three sigma bonds lie in a straight line; the two sets of p orbitals that overlap to give the pi bonds are oriented at right angles to each other. These two pi orbitals are identical in all respects.

Figure 6•14 Schematic representation of the formation of the sigma and pi bonds in acetylene. The p orbitals that make up the pi bonds are shaded gray.

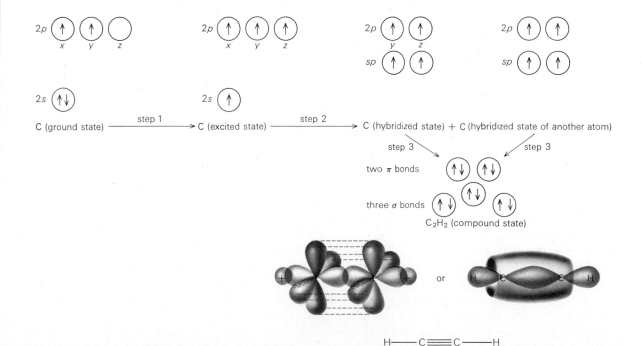

The electron clouds of the two pi orbitals are visualized as merging to form a hollow, thick-walled, cylindrical electron cloud wrapped around the sigma orbital joining the two carbon atoms.

The carbon-to-carbon distance is less for a double bond than for a single bond, and the distance is still less for a triple bond. Also, as shown in Table 6.1, there is a slight shortening of the C—H bond length when it is adjacent to a double or triple bond.

Table 6•1 Bond lengths in ethane, ethylene, and acetylene

Molecule	Carbon–carbon bond type	Carbon–carbon bond length, Å	Carbon–hydrogen bond length, Å
C_2H_6	C—C	1.54	1.09
C_2H_4	C=C	1.34	1.07
C_2H_2	C≡C	1.20	1.06

• *Other Molecules* • Elemental nitrogen, $:N{\equiv}N:$, like acetylene, is a molecule with one sigma and two pi bonds. As in acetylene, the sigma bond results from atomic orbital overlap in a head-to-head fashion. And the pi bonds are the result of lateral overlap of the half-filled $2p_y$ and $2p_z$ orbitals of each nitrogen atom.

Another molecule with both sigma and pi bonds is formaldehyde, H_2CO. There are three sigma bonds and one pi bond in each molecule. Two of the three sigma bonds result from overlap of two of the three sp^2 hybrid orbitals of carbon with $1s$ orbitals of hydrogen atoms; the third sigma bond involves overlap of the remaining sp^2 hybrid orbital of carbon with a half-filled atomic orbital of oxygen. The pi bond results from lateral overlap of p orbitals of the carbon and oxygen atoms.

━━━ • Example 6•2 • ━━━

Consider a molecule of carbon dioxide, CO_2.
(a) How many sigma and pi bonds are there?
(b) The geometry is linear. What is the most likely hybridized state of carbon?
(c) Show the sigma bonds between the atoms with straight lines; then add the overlapping p orbitals.

━━ • Solution •

(a) The Lewis structure is $:\overset{..}{O}{=}C{=}\overset{..}{O}:$, and each double bond is composed of a sigma bond and a pi bond. Hence there are two sigma and two pi bonds.
(b) Because the molecule is linear about the central carbon atom, we presume the carbon atom is sp hybridized.
(c) If carbon is sp hybridized, it has two half-filled p orbitals at right angles to one another. Therefore, the lateral overlaps of the p orbitals between carbon and oxygen are also at right angles to one another. We can trace outlines from Figure 6.14 to diagram these orbitals (see margin).

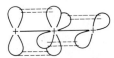

See also Exercises 11–15.

6•4•3   **Hybridization Involving d Orbitals**

In Section 5.3.1, some exceptions to the rule of eight for covalent compounds were presented. Among the exceptions are the compounds PCl_5 and SF_6, molecules of which have ten and twelve electrons, respectively, in the outermost energy level of the central atoms. In molecules that have central atoms with more than eight electrons in their valence levels, s, p, and d orbitals of the central atoms are utilized in bonding.

The diagrams in Figures 6.15 and 6.16 schematically represent imaginary processes of bond formation in PCl_5 and SF_6, respectively. In both compounds, the first step involves the formation of the number of half-filled orbitals that is equal to the number of bonds the central atom must form. The second step involves hybridization of these half-filled orbitals of the central atom. In the case of PCl_5, in which phosphorus forms five bonds to chlorine, the one $3s$ and the three $3p$ orbitals of phosphorus are hybridized with one $3d$ orbital to form five sp^3d hybrid bonding orbitals. In the case of SF_6, in which sulfur forms six bonds, the $3s$ and $3p$ orbitals are hybridized with two $3d$ orbitals to form six sp^3d^2 hybrid bonding orbitals.

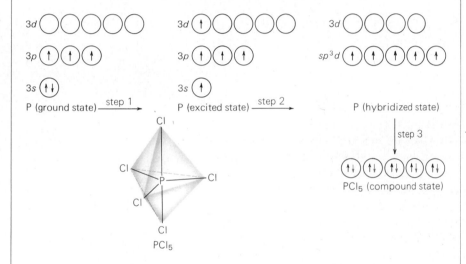

Figure 6•15 Schematic representation of the formation of sp^3d hybrid orbitals in PCl_5 and the five sigma bonds. The shape of a PCl_5 molecule is also shown.

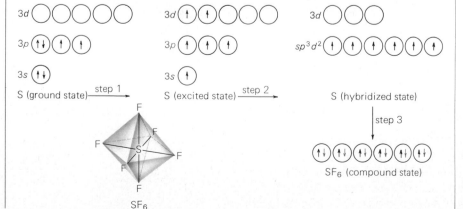

Figure 6•16 Schematic representation of the formation of sp^3d^2 hybrid orbitals in SF_6 and the six sigma bonds. The shape of an SF_6 molecule is also shown.

The hybrid bond structures of PCl$_5$ and SF$_6$ are shown in Figures 6.15 and 6.16, respectively. Five hybridized orbitals (sp^3d) are directed toward the five corners of a trigonal bipyramid. Six hybridized orbitals (sp^3d^2) are directed toward the six corners of an octahedron. Other examples of molecules for which hybridization of s, p, and d orbitals is used to explain molecular structure are presented in Chapters 22–24.

6•5   Valence Shell Electron Pair Repulsion (VSEPR) Model for Molecular Geometry

The arrangements of the bonds around the central atom in the simple molecules BeCl$_2$, BCl$_3$, and CH$_4$ result in geometrically symmetrical molecules. The shapes of these molecules are shown in Figures 6.6, 6.7, and 6.8. In these preferred arrangements, the pair of electrons in each bonding orbital is as far as possible from a pair in another orbital. In this way, the net repulsion of the pairs of electrons for one another is minimized. This approach to explaining the geometry of molecules is called the **valence shell electron pair repulsion (VSEPR) model.**

The principle of minimizing the repulsion between valence shell electron pairs in bonds also applies to the repulsion between pairs of electrons in bonding orbitals and pairs in nonbonding orbitals. Consider the shapes of the three molecules shown in Figure 6.17. In the CH$_4$ molecule, the four bonding orbitals are identical, so the repulsions between the electron pairs in them are precisely the same. Because these electron pairs tend to be as far from one another as possible (for the minimum net repulsion), the bond angles are the regular tetrahedral angles of 109.5°. In the NH$_3$ molecule, one of the four sp^3 orbitals contains a lone pair (unshared pair) of electrons. The charge cloud in the lone-pair orbital is under the influence of only one nucleus, so this orbital is larger than a bonding orbital that is between two positive nuclei. Because the repulsion between a lone pair of electrons in a nonbonding orbital and a shared pair of electrons in a bonding orbital is greater than the repulsion between shared pairs in two bonding orbitals, the N—H bonds are pushed together. The H—N—H bond angles are 107.3°, which is smaller than the regular tetrahedral angle that results when all the repulsions are equal.

In the H$_2$O molecule, there are two electron pairs in bonding orbitals and two lone pairs in nonbonding orbitals. The repulsion between these lone-pair, occupied orbitals increases the angle between them to more than 109.5°. At the same time, their repulsions toward the electron pairs in the bonding orbitals push the bonding orbitals closer together, thereby reducing the H—O—H angle from the tetrahedral value to 104.5°. The effect of the two lone pairs in H$_2$O is greater than that of the single lone pair in NH$_3$.

From the bond angles in the molecules of CH$_4$, NH$_3$, and H$_2$O, we can see how repulsion energies vary between valence shell electron pairs. Letting LP stand for lone pair and BP for bonding pair, the order of repulsion energies is LP–LP > LP–BP > BP–BP.

Relative to Figure 6.17, it should be understood that the purpose of these diagrams is to make clear the geometry of the orbitals in the valence shell of the central atom. In referring to the shape of the molecules them-

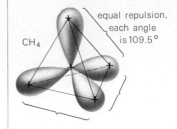

CH$_4$ equal repulsion, each angle is 109.5°

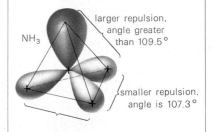

NH$_3$ larger repulsion, angle greater than 109.5°

smaller repulsion, angle is 107.3°

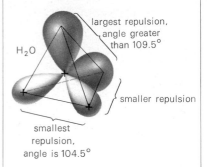

H$_2$O   largest repulsion, angle greater than 109.5°

smaller repulsion

smallest repulsion, angle is 104.5°

Figure 6•17 The fact that bond angles in H$_2$O and NH$_3$ molecules are smaller than the regular tetrahedral angle of 109.5° is accounted for in terms of unequal repulsions between bonding orbitals (smaller) and orbitals with lone pairs of electrons (larger). The plus signs represent atomic nuclei.

selves, the lone pairs are usually ignored and only the positions of the atoms are indicated. The NH_3 molecule is described as pyramidal, with the nitrogen atom at the apex and the three hydrogen atoms at the corners of the triangular base of the pyramid. The H_2O molecule is simply described as angular, with the oxygen atom at the apex.

Based on the VSEPR model, we can predict the geometric arrangement of the electron pairs in the valence shell of the central atom A for any molecule AB_n from the number of the bonding pairs and the lone pairs around A. Some hypothetical examples are given in Figure 6.18. Examples of actual molecules are shown in Figures 6.15, 6.16, 23.4, 23.5, 23.9, and 23.11,

Figure 6•18 Some geometric arrangements of valence shell electron pairs and predictions of molecular shapes based on VSEPR theory. In the figures for the geometric arrangement of valence shell electron pairs, a heavy line shows a lone pair of electrons.

	Number of bonding pairs	Number of lone pairs	Number of electron pairs in valence shell	Geometric arrangement of valence shell electron pairs		General shape of molecule	
AB_3	3	0	3	triangular	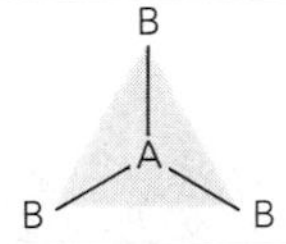	triangular	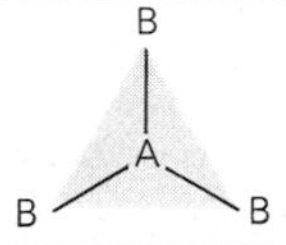
AB_2	2	1	3	triangular	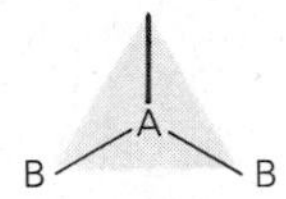	angular	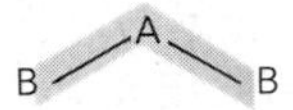
AB_4	4	0	4	tetrahedral	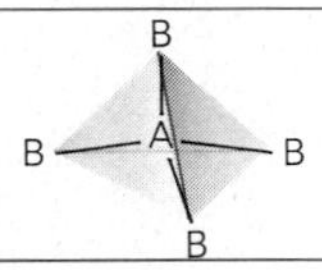	tetrahedral	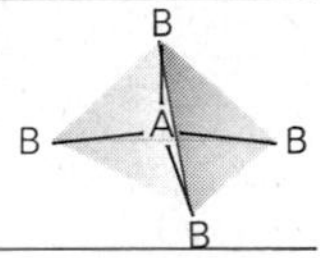
AB_3	3	1	4	tetrahedral	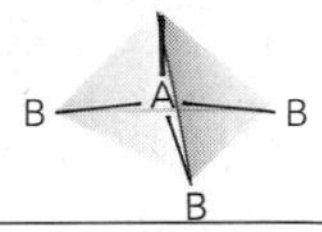	trigonal pyramidal	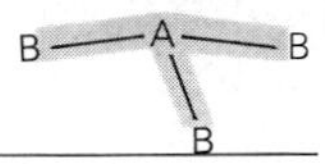
AB_2	2	2	4	tetrahedral	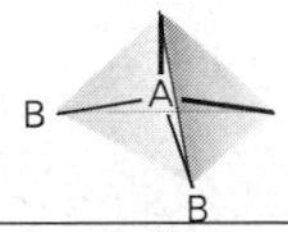	angular	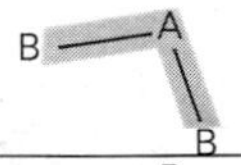
AB_6	6	0	6	octahedral	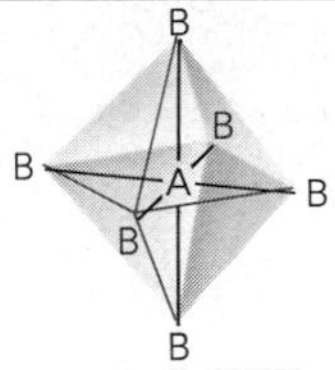	octahedral	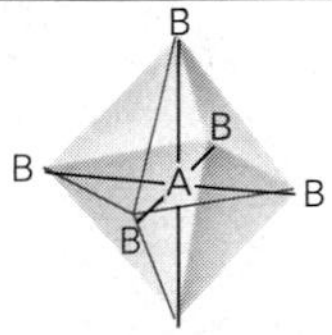
AB_5	5	1	6	octahedral	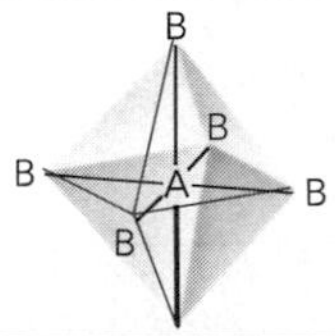	square pyramidal	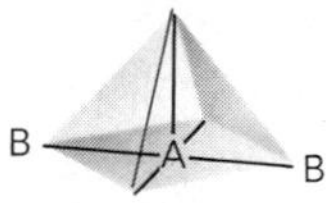
AB_4	4	2	6	octahedral	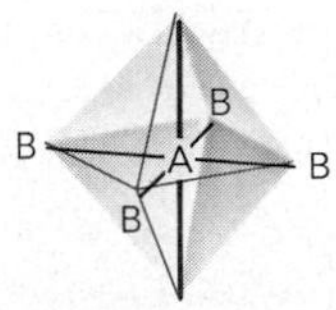	square planar	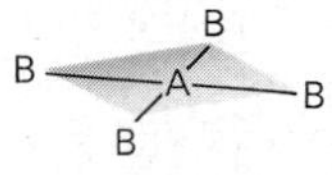

and in similar drawings elsewhere in the text. Predictions of structures are based on the following assumptions:

1. When there are no lone pairs in the valence shell of the central atom, the geometric arrangement of the valence shell electron pairs and the geometry of the molecule are identical.

2. When there are lone pairs, it is generally possible to predict the correct shape of a molecule by taking into account the differences in repulsion energies between bonding pairs and lone pairs.

3. If there are more than eight electrons in the valence shell of the central atom, and there are one or more lone pairs, the lone pairs always adopt positions that minimize 90° interactions.

The application of the third assumption is illustrated in Example 6.3. Similar applications of the VSEPR model will help us account for properties of compounds in later chapters.

• Example 6.3 •

Consider the molecule sulfur tetrafluoride, SF_4.
(a) How many bonding pairs and lone pairs of electrons are there in the valence shell of the sulfur atom? Based on the VSEPR model, predict the approximate geometric arrangement of the valence shell electron pairs.
(b) Draw all possible arrangements of bonding and lone pairs that correspond to the geometric arrangement you predicted in part (a). Tabulate the different kinds of repulsions where angles are approximately 90°, and then predict the correct molecular shape. (Assume that repulsions at angles greater than 90° have much less effect than the repulsions at 90°.)

• Solution •

(a) An atom of sulfur has six valence electrons. Of these electrons, four are associated with the covalent bonds to fluorine atoms and two are unshared. Hence, in SF_4 there are four bonding pairs and one lone pair in the valence shell of sulfur. We predict that the geometric arrangement of the five valence shell electron pairs will be similar to the molecular shape of PCl_5, in which there are also five pairs around the central atom. Therefore, the geometric arrangement is a trigonal bipyramid.
(b) There are two possible arrangements for the five orbitals pointing toward the corners of a trigonal bipyramid. Below the drawing of each are tabulated the repulsions at 90°.

(1)   four BP–BP
       two LP–BP

(2)   three BP–BP
       three LP–BP

In configuration (2) there are three LP–BP repulsions, whereas in (1) there are only two such repulsions. Because the energy of an LP–BP interaction is greater than that of a BP–BP interaction, we predict that the molecular shape corresponds to that shown in configuration (1).

See also Exercises 16, 25, 28, and 29.

The VSEPR theory can be applied to ions as well as to molecules. Conversion of NH_3 to the ammonium ion, NH_4^+, gives a species in which there are no lone pairs and all bonding orbitals are identical. Consequently, the geometry of NH_4^+ is like that of CH_4 with the regular tetrahedral angles of 109.5°.

In the cases of NH_3 and H_2O the bonds from the central atom are all projected toward hydrogen atoms. If one compares atoms other than hydrogen bound to the central atom, different values of bond angles may be obtained. For example, the bond angles in nitrogen trifluoride, NF_3, are only 102.1°, compared with those in NH_3 of 107.3°. In NF_3, the more electronegative fluorine atoms attract electrons so strongly that the N—F bonding orbital takes up less space than the N—H bonding orbital. Therefore, the repulsion of the lone-pair orbital pushes the N—F bonding orbitals closer together.

A multiple bond, such as C=C or C=O, contains more than one pair of electrons and therefore occupies more space than a single electron pair. The result of the larger sizes of multiple bonding orbitals can be seen in the list of bond angles in Table 6.2. The X—C—X (X is H, F, or Cl) bond angles for the compounds listed are less than the 120° value predicted for equivalent sp^2 hybridized orbitals, and the angles involving the double bond in X—C=C and X—C=O are greater than 120°. Also, the X—C—X bond angles grow smaller as the electronegativity of X increases, thereby further illustrating the trend observed when NH_3 and NF_3 were compared.

Table 6•2 **Bond angles in some molecules containing a double bond**

Source: *Reproduced with permission from L. E. Sutton, Ed.,* Tables of Interatomic Distances and Configurations in Molecules and Ions, *Spec. Publ. No. 11 and 18,* The Chemical Society, London, *1958, 1965.*

6·5·1   Limitations of VSEPR Theory   It is informative to compare the bond angles in families of compounds. Such a comparison for the hydrogen compounds of the VIA and VA elements is given in Table 6.3. For the compounds other than water and ammonia in this table, sp^3 hybridization is presumably not involved. The fact that the bond angles are very near to 90° indicates that the larger atoms use p orbitals for bonding. In PH_3, for example, it appears that the three sigma bonds are formed largely by the overlap of s orbitals of three hydrogen atoms with the $3p_x$, $3p_y$, and $3p_z$ orbitals of a phosphorus atom, rather than by the overlap of s orbitals with sp^3 hybridized orbitals of phosphorus. Evidently, the VSEPR model applies only to the hydrogen compounds of the smaller nitrogen and oxygen atoms.[3] There are other compounds in which the simple VSEPR theories described here do not adequately account for the observed bond angles. In fact, many factors influence the geometry of molecules; the repulsion between valence shell electron pairs is only one of them.

Table 6·3 Bond angles in some hydrogen compounds of VIA and VA elements

VIA compound	Bond angle	VA compound	Bond angle
H_2O	104.5°	NH_3	107.3°
H_2S	93.3°	PH_3	93.3°
H_2Se	91.0°	AsH_3	91.8°
H_2Te	89.5°	SbH_3	91.3°

Source: *Reprinted with permission from* Handbook of Chemistry and Physics, 1978–1979, *Robert C. Weast, Ed. Copyright The Chemical Rubber Co., CRC Press, Inc.*

6·6   Resonance

With the development of experimental methods for determining the shapes of molecules and the distances between atoms in molecules, it became increasingly apparent that in many cases Lewis formulas based on completed shells of eight electrons were not satisfactory. A few examples will illustrate how Lewis structures are not consistent with the determined properties of certain substances such as sulfur dioxide, SO_2, ozone, O_3, the nitrite anion, NO_2^-, the nitrate anion, NO_3^-, the carbonate anion, CO_3^{2-}, and sulfur trioxide, SO_3.

The Lewis structures of three representative substances listed in Table 6.4 indicate that in each example there are both single and double bonds. In Table 6.1 it is shown that single bonds are longer than double bonds. However, in each of the three cases listed in Table 6.4, the molecules or ions are symmetrical; all the bonds are identical in length as far as can be determined experimentally. Because the formulas indicate incorrectly that

[3] An indication that VSEPR theory does not apply to PH_3 is seen in the comparison of its bond angles (93.3°) with those of PF_3 (97.8°). Recall the values for NH_3 (107.3°) and NF_3 (102.1°).

there are bonds of different kinds, they do not adequately describe the structures of the molecules. It is a striking fact that there is more than one logical way of writing Lewis formulas for all these substances. For example, in Table 6.4, the double bond in sulfur dioxide is arbitrarily drawn toward the oxygen on the left; it could just as well have been drawn toward the right.

Table 6•4 Lewis formulas and some properties of molecules

Unit	One possible Lewis formula	Determined properties
sulfur dioxide molecule	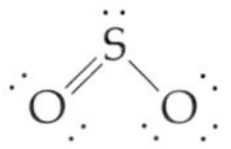	molecule is symmetrical; both S—O bonds are same length, 1.43 Å, a little shorter than a single bond but longer than a double bond
ozone molecule	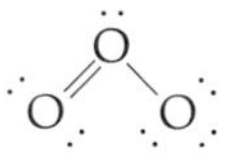	molecule is symmetrical; both O—O bonds are same length
nitrate anion	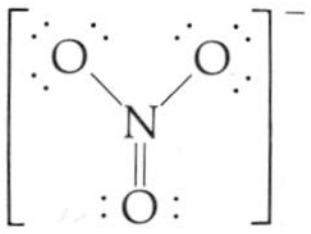	ion is symmetrical; all N—O bonds are same length, and all bond angles are equal

It is believed that when a molecule or ion can be represented by two or more arbitrary Lewis structures that differ only in the arrangement of electrons, no one of these structures accurately represents the molecule or ion. The actual structure is a hybrid that is intermediate between the two or more structures that can be written. Each of the written structures is called a **resonance structure;** the intermediate structure is said to be a **resonance hybrid.**

The concept of resonance is often indicated by drawing all the probable Lewis structures and indicating with double-headed arrows that the actual structure is intermediate between these. In the case of sulfur dioxide, we write

(a)   (b)

This representation is somewhat misleading because it suggests that the molecule has two forms, (a) and (b), and switches back and forth between them. The double-headed arrow does not have any dynamic significance; it merely implies that the molecule or ion is better represented by both structures than by one of them alone.[4] One can arrive at an even better representation of the hybrid structure by superimposing one structure upon

---

[4] The designation $\leftrightarrow$ must be carefully distinguished from the designation $\rightleftharpoons$, which indicates a reversible chemical reaction (see Section 10.7.3 and Chapter 15).

the other. The result of this superimposition is shown below; the colored dashes in the hybrid structure represent four electrons:

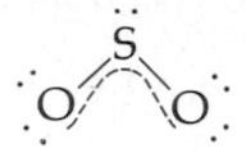

This representation indicates that there is one stable hybrid form, that this form is symmetrical, that the two S—O bonds are the same length, and that the hybrid has an electron arrangement intermediate between the two resonance structures drawn previously.

As a further example, the three resonance structures (*a*), (*b*), and (*c*) and the superimposed hybrid structure for the nitrate anion, NO_3^-, are depicted as follows:

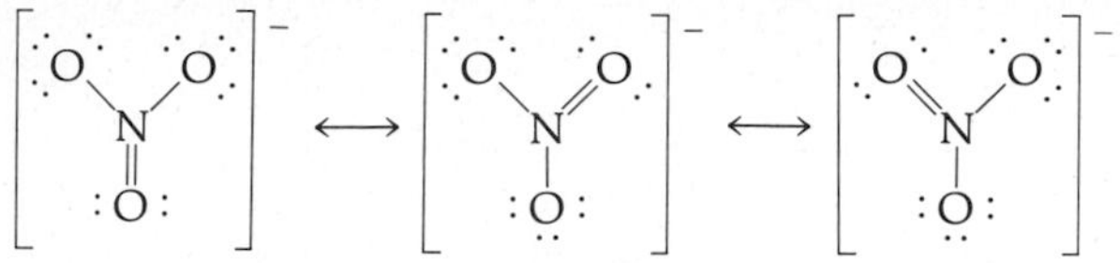

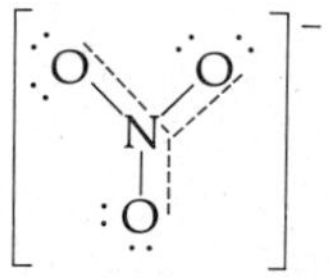

The colored dashes in the hybrid structure represent six electrons.

Another example of resonance is that shown by the benzene molecule, C_6H_6. In a benzene molecule there are six carbon atoms linked together in the form of a hexagonal ring. A hydrogen atom is bonded to each carbon atom. These ideas were first incorporated into a structural formula by the German chemist F. A. Kekulé in 1865. In order to allow for the four bonds to each carbon, he proposed alternate double and single bonds between the carbon atoms:

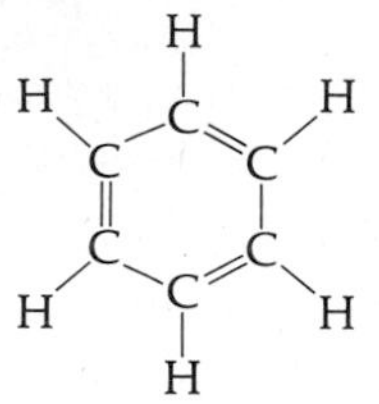

benzene

The three double bonds in the Kekulé formula suggest that benzene hydrocarbons have chemical properties very much like the alkene hydrocarbons. They do not. Also, the Kekulé formula suggests that the distances from carbon to carbon in the ring are alternately the distance between two adjacent carbon atoms in alkane molecules (1.54 Å) and the distance between two carbon atoms joined by a double bond in alkene molecules (1.34 Å). However, the distances, measured by X-ray diffraction, are uniform around the ring and all six are 1.39 Å. Today the benzene molecule is considered to be planar, with all carbon–carbon bonds the same length and all bond angles 120°. These data are consistent with benzene having a hybrid structure of two resonance structures, (*a*) and (*b*):

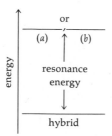

hybrid structure

The colored dashes in the hybrid structure represent six electrons.

A significant feature of resonance theory is that a molecule that has resonance forms has a lower energy than a molecule with similar bonds that does not have resonance forms. Using benzene as the model, the following diagram illustrates this concept:

In benzene, the hybrid structure has a lower energy than either of the (*a*) or (*b*) resonance structures considered separately. The calculated amount by which the hybrid structure is lower in energy than one of the resonance forms is called the **resonance energy** of the molecule. The hybrid structure is said to be stabilized by this resonance energy.

A corollary of resonance theory is that the greater the number of equivalent resonance structures that can be drawn for a molecule, the greater is its resonance energy. For example, sulfur trioxide, SO_3, which has the same valence shell Lewis structure as the nitrate anion, has three resonance structures, as compared with two for sulfur dioxide. Hence, the SO_3 molecule presumably possesses greater resonance energy than the SO_2 molecule.

━━━ • Example 6•4 • ━━━━━━━━━━━━━━━━━━━━━━━━━

Write Lewis structural formulas to depict the resonance structures of the bicarbonate ion, HCO_3^-, and the carbonate ion, CO_3^{2-}. For which of these two ions would you predict a greater resonance energy?

━━━ • Solution • For the bicarbonate ion, there are two resonance structures:

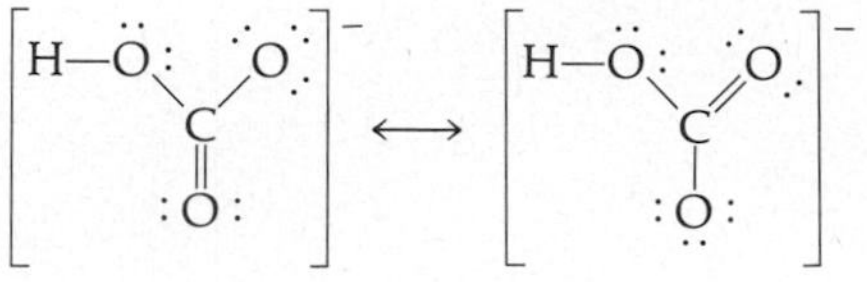

And for the carbonate ion, there are three:

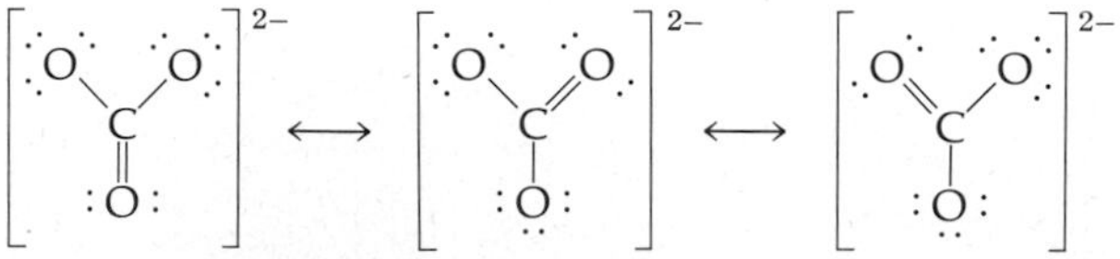

Because there is one more resonance structure for the carbonate ion, it presumably possesses greater resonance energy.

See also Exercises 35 and 36.

6•6•1   **An Orbital Interpretation of Resonance**   By using the more modern concepts of sigma and pi orbitals, the concept of resonance can be put in another perspective. Consider once more the benzene molecule. The resonance structures (a) and (b) show the double bonds as fixed, or *localized,* between two carbon atoms. However, in our discussion of the structure of benzene, it was pointed out that we do not think there actually are double bonds or single bonds between carbon atoms, but rather that all the bonds are alike. In Figure 6.19, the symmetrical, equivalent bonding in benzene is depicted in terms of hybridized orbitals. The 120° bond angles in Figure 6.19(a) suggest sp^2 hybridization for carbon of the type described for ethylene (see Section 6.4.2). The three bonds to each carbon shown in Figure 6.19(b) are sigma bonds. They result from the overlap of the three sp^2 orbitals of one carbon atom with the sp^2 orbitals of adjacent carbon atoms and with the s orbital of a hydrogen atom.

As in ethylene, the fourth valence electron of each carbon in benzene may be thought of as initially occupying a p orbital that lies above and below the plane containing the carbon nuclei, as shown in part (c) of Figure 6.19. Because there are six carbon atoms in the ring, there are six of these orbitals, all parallel to one another and close enough for lateral overlap, that is, pi bond formation. However, each p orbital overlaps two neighboring p orbitals, one on either side. These overlaps result in two doughnut-shaped electron

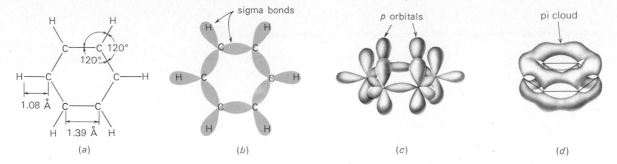

clouds. One such doughnut lies above and one lies below the plane through the carbon nuclei [see Figure 6.19(d)].

  The *p* orbital electrons in benzene interact with more than two atomic nuclei and are therefore involved in the formation of more than one bond. Such electrons are said to be *delocalized*. As a result of this delocalization, the energy of the system is lower than it would be if the electrons were localized. The amount by which a system's energy is lowered due to the delocalization of electrons is called its **delocalization energy.** Resonance energy and delocalization energy are equivalent terms used to describe the same phenomenon: *a structure possesses a lower energy when it has electrons in molecular orbitals that are associated with more than two atomic nuclei.*

  Orbital diagrams that are equivalent to the hybrid structures for sulfur dioxide and some other molecules and ions are shown in Figure 6.20. In each of these structures there are sigma bonds and overlapping *p* orbitals joining the atoms. The resulting pi bond system involves electron delocalization over more than two atoms.

Figure 6•19 Schematic representation of a benzene molecule: (*a*) bond angles and positions in space of carbon and hydrogen nuclei; (*b*) sigma bonds involving sp^2 and *s* orbitals; (*c*) *p* orbitals of carbon not involved in sigma bond formation (note that the plane of the benzene ring is now shown almost perpendicular to the plane of the paper); (*d*) the overlap of the *p* orbitals, forming the pi electron clouds above and below the benzene ring. (Centers of sigma bonds are indicated by the colored hexagon.)

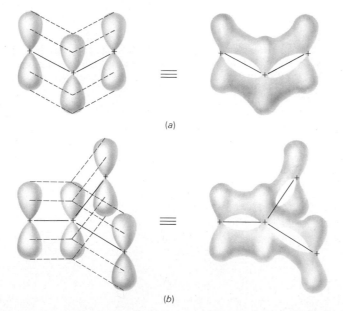

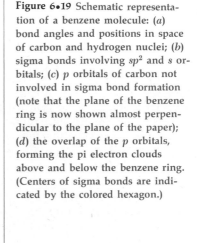

Figure 6•20 Delocalized pi electron structures for representative molecules and ions. Examples include (*a*) SO_2, O_3, and NO_2^-; (*b*) NO_3^-, CO_3^{2-}, and SO_3. Distances between *p* orbitals have been exaggerated for clarity. The centers of sigma bonds are indicated by the solid black lines.

Molecular Orbital Theory

6·7   Introduction to Molecular Orbital Theory

In describing the bonding between two atoms in terms of the valence bond (VB) theory, the method is to begin with the electronic structures of the two separated atoms. The two atoms are then thought of as moving toward one another until their half-filled atomic orbitals interact to form the bonding molecular orbital(s).

In molecular orbital (MO) theory, the method is to begin with a given arrangement of the atoms in the molecule. The distances between

---

The Use of Molecular Models What "truth" can we show with models of atoms and molecules? This question should never be far from us as we use models to help explain and predict chemical changes. A model cannot represent fully the thing modeled, and it may even mislead us if we imagine that the thing modeled *must* behave as predicted with the model.

In the accompanying photographs are shown four different types of molecular models for three organic compounds, ethyl alcohol (grain alcohol), ethylene, and *ortho*-cresol (a chemical component of creosote, which is one of the most effective wood preservatives). We can recognize some advantages and disadvantages of each of the four model types.

Ball-and-Stick or "Tinkertoy" Models. These models provide good three-dimensional representations. The number of bonds to each atom is correctly shown, and the bond angles can be correct if the models are carefully made. These models give an unfortunate idea of the "openness" of a molecule, however. Also, all the

balls are the same size, so differences between atoms are not indicated.

In the model of ethylene, both bonds (springs) of the double bond are shown as being the same. This, of course, is a poor representation of our theory that the sigma and pi bonds are distinctly different from one another. In the model of *ortho*-cresol, the carbon–carbon bonds of the benzene ring are depicted as the alternating single and double bonds of a Kekulé formula. The OH and CH_3 groups of this model appear to be attached to carbon atoms joined by a double bond. This model suggests that there is another *ortho*-cresol in which the OH and CH_3 groups are attached to carbon atoms joined by a single bond. We would prefer a model that showed all carbon–carbon bonds in the benzene ring to be identical, and that allowed us to construct only one molecular model for *ortho*-cresol.

Scale Models. The basic units of these models are wooden spheres from which parts have been cut off. When they are fitted together, they roughly represent

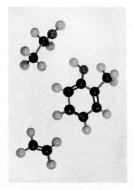
ball-and-stick models

Dreiding models

scale models

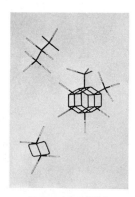
framework models

nuclei and the positions of nuclei are known from spectroscopic and other experimental measurements. For a diatomic molecule, we begin with two bare nuclei separated by the bond length, and calculate the sets of wave functions or molecular orbitals that electrons must occupy if they are attracted to both the nuclei. A number of electrons equal to the total number originally associated with the individual atoms is assigned to these sets of wave functions or molecular orbitals that encompass the entire molecule. The rules for adding electrons to the various molecular orbitals are similar to those for adding electrons to atomic orbitals, that is, the aufbau principle, Hund's rule, and the Pauli exclusion principle are all involved.

Ideally, the MO approach pictures all the electrons of a diatomic molecule as moving in the fields of both nuclei, rather than being restricted to one or the other nucleus. In practice, however, only the electrons in the

★   Special Topic ★ Special Topic ★ Special Topic ★ Special Topic ★ Special Topic ★ Special Topic ★ Special Topic ★ Special Topic   ★

van der Waals and covalent radii. The scale of the actual models is 1.00 cm = 1.00 Å. The assembled models show rather well both bond angles and the relative sizes of the atoms. The model of *ortho*-cresol does show that the six carbon atoms of the benzene ring are all bonded to one another in an equivalent fashion. However, as with a ball-and-stick model, the difference between sigma and pi bonds in multiple bond systems is not shown.

Dreiding Models. These models are constructed of metal rods and tubes joined together at a point that represents an atomic nucleus. Each unit represents a single molecule, for example, CH_4, NH_3, H_2O, C_2H_4, and C_2H_2. A benzene ring is made as one unit and shows the carbon atoms bonded to one another in an equivalent manner. The units may be assembled to give more complicated models by inserting the rod of one unit into the tube of another. For a tube or rod corresponding to a C—H or O—H bond, a colored tip represents the hydrogen nucleus.

These models represent bond angles well, and they show relative internuclear distances precisely. The scale of the models is 1.00 cm = 0.40 Å. When two tetrahedral carbon units are fastened together as in the ethyl alcohol model, the measured C—C distance of 3.85 cm corresponds to 1.54 Å. The models do not show the three-dimensional nature of atoms, and multiple bonds are not represented. Dreiding models are often the favorites of research chemists because they represent so precisely the geometric relationships in molecules.

Framework Models. These models are constructed from two simple components: (1) metal valence clusters that look like a child's jacks, and (2) plastic tubing, which can be cut to the proper scale of 1.00 in. = 1.00 Å. On a tube representing a C—H bond, the covalent radius of carbon (black) extends from its attachment at the

center of the valence cluster to where black meets white. The covalent radius of hydrogen extends from this point out to the little black circle. The van der Waals radius of hydrogen extends from the little black circle out to the tip of the tubing. Similar tubes are used to represent O—H bonds.

Lone pairs of electrons are represented by short sections of tubing, such as the two sections on the oxygen in the ethyl alcohol model. The differences in carbon-carbon multiple bonds are clearly shown. The representation of *p* orbital overlap is a special feature of the framework models that makes them attractive for instructional use. Note how the model of *ortho*-cresol shows the equivalent *p* orbital overlap above and below the plane of the benzene ring. All the carbon-carbon bonds in benzene rings are shown as identical.

A large variety of molecular structures can be depicted with these models. One drawback is that the three-dimensional nature of the bonded atoms is not shown. These models are relatively inexpensive, and students may wish to purchase a set for personal use.

So which type of model best represents the "true" structures of the three compounds? No model can exhibit all the characteristics of the object it is meant to represent. The remarkable fact is that models have been so successfully used by scientists as they think about the unseen and unseeable tiny molecules that make up all matter. Linus Pauling, in bed ill, made models by sketching diagrams on pieces of paper and folding them as he thought about possible structures for protein molecules. Watson and Crick arranged ball-and-stick models and cardboard cutouts for planar rings as they worked out the double helix structure for DNA. From Nobel Prize–winning discoveries to the studies of beginners, models play a key role in our understanding of chemistry.

higher energy levels interact with both nuclei to any great extent. Electrons in very low energy levels are called *nonbonding electrons* and are placed in atomic orbitals that are associated with only one of the nuclei.

6•8   Sigma and Pi Molecular Orbitals

In contrast to VB theory, in which *one* molecular orbital is said to form as a result of the interaction of two atomic orbitals, in MO theory *two* molecular orbitals are said to result from the combination of the two atomic orbitals. In mathematical terms, the two molecular orbitals result from addition and subtraction of the atomic orbitals that overlap. The molecular orbital that results from the *addition* of the atomic orbitals that overlap is called a **bonding orbital.** (See σs and $\sigma 2p_x$ in Figure 6.21 and $\pi 2p_z$ in Figure 6.22.) In a bonding molecular orbital, because the electron density is concentrated between the two nuclei, the energy of the system is lowered compared with that of the isolated atomic orbitals. The molecular orbital that results from the *subtraction* of the atomic orbitals that overlap is called an **antibonding orbital.** (See $\sigma^* s$ and $\sigma^* 2p_x$ in Figure 6.21 and $\pi^* 2p_z$ in Figure 6.22.) In an antibonding molecular orbital, the electron density is concentrated away from the region between the two nuclei. The net effect of having a low electron density between the nuclei is that the two nuclei repel each other. Therefore, the energy of an antibonding orbital is increased compared with that of the two isolated atomic orbitals.

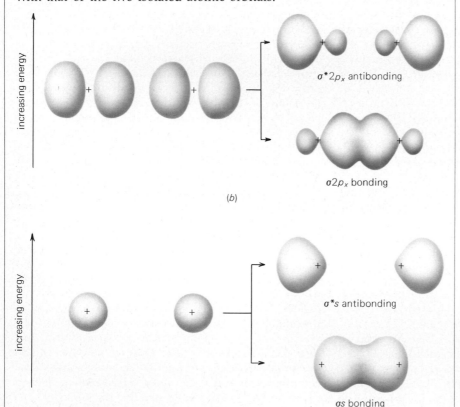

$\sigma^* 2p_x$ antibonding

$\sigma 2p_x$ bonding

(b)

$\sigma^* s$ antibonding

σs bonding

(a)

Figure 6•21 Formation of sigma bonding and antibonding orbitals: (*a*) combination of two *s* atomic orbitals; (*b*) combination of two $2p_x$ atomic orbitals in a head-to-head fashion.

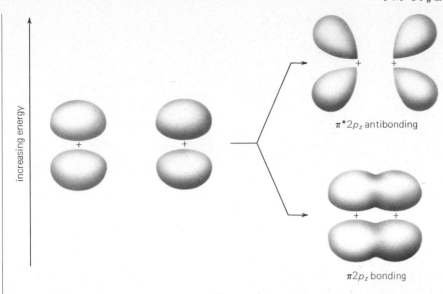

increasing energy

π^*2p_z antibonding

$\pi 2p_z$ bonding

Figure 6•22 Formation of pi bonding and antibonding orbitals by combination of two $2p_z$ atomic orbitals.

When two s orbitals are combined to form two molecular orbitals [Figure 6.21(a)], the resulting molecular orbitals are called **sigma (σ) molecular orbitals.** As in VB theory, each σ orbital is symmetrical about the line that passes through the two nuclei. The bonding sigma orbital is designated σs and the antibonding sigma orbital is designated $\sigma^* s$. Two sigma molecular orbitals also result when two p orbitals are combined in a head-to-head fashion along the x-axis [Figure 6.21(b)]. Here the bonding and antibonding orbitals are designated σp_x and $\sigma^* p_x$, respectively.

The diagrams in Figure 6.21 help make clear physical effects of bonding and antibonding orbitals. For a bonding orbital, each positive nucleus is attracted to the region of concentrated electron density; the net attractive effect is a chemical bond between the nuclei. For an antibonding orbital, the low electron density between the nuclei results in increased repulsion between the positive nuclei; the net repulsive effect is to act against the chemical bonding of the nuclei. In a possible molecule, if the antibonding effects neutralize the bonding effects, the molecule is so unstable that it cannot exist.

In addition to the sigma molecular orbitals that result from overlap of p orbitals in a head-to-head fashion on the x-axis, two p orbitals can be combined in a sideways or lateral arrangement to give **pi (π) molecular orbitals** (see Figure 6.22). When there are two sets of pi orbitals oriented at right angles, the bonding orbitals are designated πp_y and πp_z and are said to be **degenerate;** that is, they have the same energy. The antibonding orbitals, designated $\pi^* p_y$ and $\pi^* p_z$, are also degenerate. Again, as in the case with sigma orbitals, the bonding orbitals are lower in energy than the antibonding ones.

6•8•1   Filling Molecular Orbitals for Homonuclear Diatomic Molecules

The simplest molecules to which MO theory can be applied are **homonuclear diatomic molecules,** molecules made of two atoms of the same element. As mentioned in Section 6.7, electrons are added to molecular orbitals in order

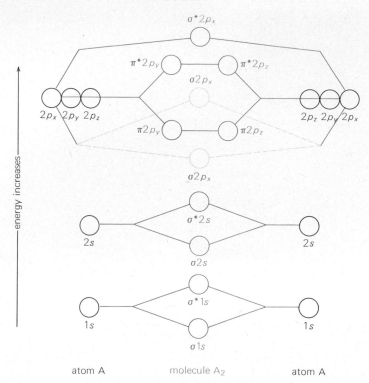

energy increases →

atom A   molecule A_2 atom A

Figure 6•23 Energy order for filling molecular orbitals for homonuclear diatomic molecules is shown in color; black circles show, for comparison, atomic orbitals. For some molecules, $\sigma 2p$ is above $\pi 2p$; for others, it is below. Asterisks denote antibonding orbitals. Note that the $2p_y$ and $2p_z$ orbitals are degenerate.

of increasing energy. Figure 6.23 shows how the electrons that originally were in atomic orbitals of two homonuclear, separated atoms, A and A, are assigned to molecular orbitals in the molecule A_2.

Table 6.5 lists the MO assignments of electrons to molecular orbitals for six possible homonuclear diatomic molecules. The net number of bonds, called the **bond order,** is calculated as

bond order $= \frac{1}{2}$(number of bonding e^- − number of antibonding e^-)

When the bond order is zero, the possible molecule is too unstable to exist, as in the cases of the hypothetical molecules He_2 and Ne_2.

Table 6•5 Filling of molecular orbitals

Molecule	Number of e^-	MO configuration[a]	Bond order
H_2	2	$(\sigma 1s)^2$	1
He_2	4	$(\sigma 1s)^2(\sigma^* 1s)^2$	0
N_2	14	$KK(\sigma 2s)^2(\sigma^* 2s)^2(\pi 2p_y)^2(\pi 2p_z)^2(\sigma 2p_x)^2$	3
O_2	16	$KK(\sigma 2s)^2(\sigma^* 2s)^2(\sigma 2p_x)^2(\pi 2p_y)^2(\pi 2p_z)^2(\pi^* 2p_y)(\pi^* 2p_z)$	2
F_2	18	$KK(\sigma 2s)^2(\sigma^* 2s)^2(\sigma 2p_x)^2(\pi 2p_y)^2(\pi 2p_z)^2(\pi^* 2p_y)^2(\pi^* 2p_z)^2$	1
Ne_2	20	$KK(\sigma 2s)^2(\sigma^* 2s)^2(\sigma 2p_x)^2(\pi 2p_y)^2(\pi 2p_z)^2(\pi^* 2p_y)^2(\pi^* 2p_z)^2(\pi^* 2p_x)^2$	0

[a] K refers to main level 1; KK represents the electrons in both atoms which occupy the two $1s$ levels, a total of four electrons. These are treated as nonbonding electrons and are assigned to atomic orbitals associated with their respective nuclei.

In Section 4.6 we described the determination of the binding energies of electrons in atoms by photoelectron spectroscopy. Probably the most exciting present use of photoelectron spectroscopy is as a tool to study the chemical bond. The spectrum of a molecule gives a precise picture of the binding energy for each electron. From these energies and with the use of the energy order in Table 6.5, we can deduce which electrons are in bonding orbitals, which are in antibonding orbitals, and which are in nonbonding orbitals. The photoelectron spectrum of molecular nitrogen, N_2, has five peaks, from which the following binding energies are determined: 409.9, 37.3, 18.6, 16.8, and 15.5 eV. They are assigned as follows:

$$\underbrace{(\sigma1s)^2(\sigma^*1s)^2}_{409.9} \quad \underbrace{(\sigma2s)^2}_{37.3} \quad \underbrace{(\sigma^*2s)^2}_{18.6} \quad \underbrace{(\pi2p_y)^2(\pi2p_z)^2}_{16.8} \quad \underbrace{(\sigma2p_x)^2}_{15.5}$$

It would be expected from Figure 6.23 that there would be a difference in energy between the $\sigma1s$ and σ^*1s orbitals. However, the data for the N_2 molecule indicate that the electrons in these orbitals have the same energies to four significant figures. The comparatively large value of 409.9 eV shows that these four inner level electrons are much more tightly held than the other electrons. It is reasonable to think of them as nonbonding because they are attracted so strongly by their respective nuclei.

The data show that the $\pi2p_y$ and $\pi2p_z$ orbitals are degenerate; that is, they have the same energy. This finding is consistent with the scheme shown in Figure 6.23.

• *The Case of the O_2 Molecule* • When the VB and the MO theories were in their early stages of development, one of the pressing puzzles was the electronic structure of the O_2 molecule. The Lewis formula was usually written as $:\ddot{O}=\ddot{O}:$, and, according to approximate VB theory, the molecule had a normal double bond consisting of one sigma bond and one pi bond. Both the Lewis formula and the approximate VB approach described all the electrons in O_2 as paired. However, oxygen has an easily measured physical property that indicates that there are unpaired electrons in its molecules. A sample of oxygen is slightly attracted by a magnetic field; that is, oxygen is paramagnetic. It is well established that in molecules of a paramagnetic substance there are unpaired electrons (see Section 22.7.1).

The solving of the O_2 puzzle was one of the first cases in which the MO theory proved superior to the VB theory. In filling molecular orbitals in the order of increasing energy according to the MO scheme in Figure 6.23, one electron is placed in each of the two π^* orbitals of equal energy before either of the orbitals is filled with an electron of opposite spin (Hund's rule). The application of these procedures to O_2 is shown in Figure 6.24. The unpaired electrons in the π^*2p_y and π^*2p_z orbitals account for the paramagnetism of O_2. For diamagnetic substances such as H_2, N_2, and F_2, the MO theory shows all electrons paired.

Interpreting the MO configuration given in Table 6.5 for the O_2 molecule in terms of the diagram in Figure 6.24, we would expect to find seven peaks in the photoelectron spectrum of O_2, or six if the $\sigma1s$ and σ^*1s overlapped as they do in the case of the N_2 molecule. The actual O_2 spectrum

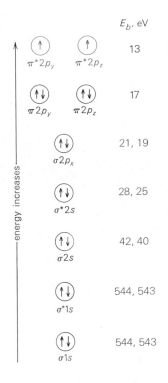

Figure 6•24 Molecular orbitals in an O_2 molecule, showing the assignment of the ten values of binding energy found in the photoelectron spectrum for molecular oxygen. For the electrons in five orbitals, there are two slightly different values for binding energies; these differences are attributed to the difference in the shielding effect of the two π^* electrons on an electron of like spin compared with one of unlike spin.

shows ten peaks. The extra lines have been attributed to the effect on other electrons of the same spin by the two unpaired electrons in the two π^* orbitals of the molecule. The binding energies of electrons in other orbitals depend in part on whether they have the same or different spin from the two π^* electrons. For electrons in all the orbitals except the other π orbitals (17 eV), the difference in binding energy due to the spin of an electron is great enough to be detected. The ten binding energies found are shown in Figure 6.24.

Chapter Review

Summary

The **valence bond theory** is one of the two major theories of the chemical bond. In VB theory, the half-filled atomic orbitals of two atoms are thought of as overlapping to form one or more covalent bonds between them. This overlapping of atomic orbitals is sometimes called a molecular orbital. Electrons not in overlapping orbitals are called **nonbonding electron pairs**. A **sigma orbital**, σ, is one that results from the head-to-head overlap of half-filled atomic orbitals along the line between the two atoms; this is called a **sigma bond**. A **pi orbital**, π, is one that results from the side-by-side overlap of half-filled p orbitals that are perpendicular to the sigma bond; this weaker bond is called a **pi bond.**

The promotion of electrons to higher energy levels and the mixing of these levels with others to form a new set of equivalent orbitals is called **hybridization.** The formation of these **hybrid orbitals** accounts for many major features of molecular structure and bonding, including many examples that do not follow the rule of eight. A useful theoretical concept for explaining the geometries of many simple molecules and ions is the **valence shell electron pair repulsion model** (VSEPR model). The basis for this model is the assumption that pairs of valence shell electrons, both bonding and nonbonding, keep as far from each other as is geometrically possible.

Many molecules and ions can be depicted by two or more hypothetical Lewis formulas called **resonance structures,** none of which accurately describes the known structure of the species. Such a molecule or ion is a **resonance hybrid,** a species whose actual structure is intermediate among the written resonance structures. The energy of the resonance hybrid is lower than those of its hypothetical resonance structures by an amount called the **resonance energy,** which stabilizes the hybrid form. All such forms have **delocalized electrons:** p-orbital electrons that participate in the formation of more than one bond. The resonance energy is thus also called the **delocalization energy.**

Molecular orbital theory is the other major theory of the chemical bond. In MO theory, two atomic nuclei are

thought of as being separated by the distance of their bond length, and the **molecular orbitals** of all the electrons that belong to the molecule are calculated in sequence. These orbitals are called **bonding orbitals** or **antibonding orbitals** depending on whether they promote or hinder bond formation. As with VB theory, there are **sigma molecular orbitals** and **pi molecular orbitals;** both of these can be either bonding or antibonding. A measure of the stability of possible **homonuclear diatomic molecules** is the **bond order** calculated from MO theory.

The two major theories, VB and MO, are complementary: each has its strengths and weaknesses. Together they explain virtually all known aspects of chemical bonding.

Exercises

Sigma and Pi Bonds

1. Show with a diagram similar to those in Figure 6.1 the half-filled orbitals in an atom of oxygen. How many sigma bonds could an oxygen atom form with other atoms?

2. Show with a diagram similar to those in Figure 6.1 what presumably happens when an atom of Cl combines with an atom of I to form a molecule of ICl. If you wished to describe the overall size of an ICl molecule, what would you need to add to the diagram?

3. If the internuclear distance in bohrs is plotted against the potential energy in kilojoules per mole for the process indicated in Exercise 2, there would be a minimum potential energy at some point. To what distance does this minimum correspond?

4. Repeat Exercises 2 and 3 for the combination of hydrogen and sulfur atoms to form hydrogen sulfide, H_2S. Assume that the molecule forms in one step.

5. Write the Lewis structural formulas for the following:
   (a) four molecules with pi bonds, other than those named in Section 6.3
   (b) two molecules, other than C_2H_2 and CO_2, that have two pi bonds per molecule

Hybridized Orbitals

6. Consider the atoms silicon, nitrogen, and boron.
   (a) Based solely on the ground-state electronic configuration of these atoms, how many single covalent bonds could each form?
   (b) How many single bonds do each of these atoms form in molecules that are familiar to you?
   (c) What experimental evidence suggests that hybridization in the nitrogen atom might be involved in an ammonia molecule?

7. Show with circle-and-arrow diagrams similar to those in Figures 6.6, 6.7, 6.8, 6.11, and 6.12 the imaginary processes of bond formation in molecules of BH_3, CH_2F_2, and NCl_3.

8. The H—O—C bond angle in a molecule of methyl alcohol, CH_3OH, is about 109°. What type of hybridization, if any, would you propose for the oxygen atom? For the carbon atom? Show the overlapping sigma orbitals in a molecule of methanol. Also show the orbitals containing the unshared (lone) pairs of the oxygen atom.

9. What is the hybridization of phosphorus in PH_4^+? Predict the logical orbital and molecular geometries for a unit of PH_4^+.

10. (a) Based solely on analogies with the compounds CH_4, NH_3, and H_2O, predict the orbital geometries and molecular shapes of SiH_4, AsH_3, and H_2Se.
    (b) The H—E—H bond angles (E is Si, As, or Se) in these three compounds are 109.5°, 91.8°, and 91.0°, respectively. With this additional information, answer part (a) again.

11. How many sigma and how many pi bonds are there in a molecule of each of the following compounds?
    (a) ethylene, C_2H_4
    (b) disilane, Si_2H_6
    (c) acetylene, C_2H_2
    (d) hydrazine, N_2H_4
    (e) hydrogen cyanide, HCN
    (f) acetaldehyde, CH_3CHO

12. What is the hybridization state of each atom, other than hydrogen, in each compound listed in Exercise 11? [Ignore any possible hybridization of nitrogen and oxygen in parts (e) and (f).]

13. Refer again to Exercise 11. Using symbols for atoms of elements and solid lines, dashed lines, and wedges for bonds, show a three-dimensional drawing of the sigma bonds for each compound. For the compounds with multiple bonds, show the p orbital overlap of the pi bond system(s).

14. A molecule of allene, C_3H_4, has two double bonds. What simple inorganic molecule also has two double bonds? What is the geometry of the three carbon atoms in allene? How are the C—H bonds arranged with respect to each other? Propose the most likely hybrid-

ized state of each carbon in allene.

15. Based on your answers to Exercise 14, show with diagrams similar to those you used in Exercise 13 the sigma and pi bonds in a molecule of allene. How are the overlapping p orbitals aligned with respect to each other?

16. Write a Lewis structural formula for a molecule of each of the following: (a) ClF_3; (b) SF_4; (c) PF_5; (d) BrF_5.

17. State in words the meaning of the symbol sp^3d^3.

*18. What type of hybridization might you expect for the central atom in each of the molecules listed in Exercise 16? Use circle-and-arrow diagrams of the type shown in Figures 6.15 and 6.16 to represent the formation of the hybrid orbitals and sigma bonds for each molecule.

VSEPR Models
for Molecular Geometry

19. Write a Lewis structural formula for the amide ion, NH_2^-. Predict a value for the H—N—H bond angle in an NH_2^- ion. Explain the reasoning for your choice.

20. Write a Lewis structural formula for the hydronium ion, H_3O^+. Predict a value for each bond angle in the H_3O^+ ion. Explain the reasoning for your choice.

21. Consider the following compounds and the listed values of bond angles:

Compound	Angle	Value
H_2O	H—O—H	104.5°
CH_3OH	H—O—C	109°
$(CH_3)_2O$	C—O—C	111.5°

How do you explain the differences in the listed values?

22. Boron trifluoride, BF_3, forms a coordinate covalent bond with ammonia, NH_3, to produce the compound H_3NBF_3. Diagram a molecule of the product and predict the geometric arrangement of the F atoms around the B atom and the H atoms around the N atom.

23. Which would you expect to be larger, the C—O—C bond angle in methyl ether, $(CH_3)_2O$, or the C—S—C bond angle in methyl sulfide, $(CH_3)_2S$? Explain your choice.

24. Explain the meaning of the notation LP–LP > LP–BP > BP–BP.

*25. Which molecules in Exercise 16 have a similar geometric arrangement of the electron pairs in the valence shells of their central atoms? Do any of the molecules have the same molecular shape? Based on the differences in repulsion energies of lone pairs (LP) and bonding pairs (BP), and on the assumption that lone pairs always adopt positions that minimize 90° interactions, predict the shape of each molecule.

*26. The ionic compound K_2SiF_6 is known. What is the most probable shape of the SiF_6^{2-} ion? Which orbitals of silicon would you suppose hybridize in the imaginary process of bond formation for the SiF_6^{2-} ion?

27. The F—C—F and Cl—C—Cl bond angles are 110° and 113° in molecules of C_2F_4 and C_2Cl_4, respectively. Explain the difference in the bond angles from the standpoint of VSEPR theory.

*28. When ICl reacts with Cl^-, the ion ICl_2^- is formed. Write a Lewis structure for ICl_2^-. Based on the criteria listed in Exercise 25, predict the shape of the ICl_2^- ion. What type of hybridization, if any, would you propose for the I atom in ICl_2^-?

*29. When BrF_3 reacts with F^-, the BrF_4^- ion is formed. Write a Lewis structure for BrF_4^-. Based on the criteria listed in Exercise 25, predict the shape of the BrF_4^- ion. What type of hybridization, if any, would you propose for the Br atom in BrF_4^-?

30. An extra column with the heading **Hybridized Orbitals of Central Atom** might be added to Figure 6.18. Based on the data in this chapter and the known geometric arrangement of the electron pairs in the valence shell of the central atom, assign the most probable hybridization state associated with each of the following geometries:

geometric arrangement of valence shell electron pairs	hybridized orbitals of central atom
linear	
trigonal planar	
tetrahedral	
trigonal bipyramidal	
octahedral	

31. Predict the O—N—O bond angles in the nitrate ion, NO_3^-. What are two other ions or molecules containing three atoms bonded to a fourth central atom in which the B—A—B (A is the central atom) bond angles are the same as the O—N—O angles in NO_3^-?

Resonance

32. Would you expect the O—N—O bond angle in the nitrite ion, NO_2^-, to be more than, less than, or the same as each O—N—O bond angle in NO_3^-? Explain.

33. Write Lewis structural formulas for ozone, O_3, and sulfur dioxide, SO_2, that depict their approximate molecular shape. The O—S—O bond angle in SO_2 is 120°. Would you expect the O—O—O bond angle in O_3 to be more than, less than, or the same as the O—S—O

angle in SO_2? Check your prediction with the value given in Section 9.3, and explain any differences in the angles for the two molecules. Why do you think the O—S—O bond angle is 120° and not more or less than that value?

34. What type of hybridization would you predict for (a) N in NO_2^-; (b) N in NO_3^-; (c) the central O in O_3; (d) C in CO_3^{2-}?

*35. The electron-dot structures of nitrogen dioxide, NO_2, and the nitronium ion, NO_2^+, can be formulated by the successive removal of an electron from the nitrite ion, NO_2^-. For the nitronium ion, three resonance structures may be written, two of which are essentially equivalent.
(a) Write the three resonance forms for the nitronium ion.
(b) Do these structures suggest anything to you about the geometry of the NO_2^+ ion?
(c) What do you think is the hybridization of the nitrogen atom in NO_2^+?
(d) Draw an orbital representation of NO_2^+, showing the sigma bonds with lines and any pi bonds as in Figure 6.20.

36. Write Lewis structural formulas to depict the resonance forms of the acetate ion, $CH_3CO_2^-$. Draw an orbital picture that is equivalent to the hybrid structure for $CH_3CO_2^-$.

Molecular Orbital Theory

37. Distinguish between bonding and antibonding orbitals.

38. Show how the bond order of a hypothetical Ne_2 molecule is calculated from the bond assignment given in Table 6.5. What does the bond order indicate about the existence of a diatomic neon molecule?

39. How does the description of the bonding in a molecule of F_2 by VB theory differ from that by MO theory?

40. Explain why the Lewis formula for oxygen, $:\ddot{O}=\ddot{O}:$, does not account for its observed magnetic property. How does the MO representation of the bonding in an O_2 molecule account for this property? It appears that an acceptable Lewis structural formula for O_2 does not follow the rule of eight. Write what you believe to be an acceptable Lewis formula.

41. The lowest binding energy value in N_2 is 15.5 eV, whereas that for O_2 is 13 eV. Are the same type of electrons being removed from both molecules? These values represent the ionization energies of the molecules. Were the same magnitudes and relative order of values shown for the N and O atoms? (Consult Table 3.6.)

The Gaseous State

7

7•1 **The Three States of Matter**

7•1•1 Individual Particles in Gases

7•2 **The Pressure of Gases**

7•2•1 Measurement of Atmospheric Pressure

7•2•2 Variations in Atmospheric Pressure

7•2•3 Measurement of Gas Pressures

The Gas Laws

7•3 **Boyle's Law**

7•3•1 Working Gas Law Problems

7•4 **Temperature Effects**

7•4•1 Absolute Scale of Temperature

7•4•2 Charles's Law

7•4•3 Relation of Pressure and Temperature

7•5 **Avogadro's Law**

7•6 **A General Gas Equation**

7•6•1 Evaluating the Gas Constant R

7•6•2 Solving Problems with the Ideal Gas Law Equation

7•6•3 Calculation of Gas Densities

7•7 **Molar Relationships with Gaseous Reactants**

7•7•1 Weight–Volume Relationships

7•7•2 Volume–Volume Relationships

7•8 **Dalton's Law of Partial Pressures**

7•9 **Graham's Law of Diffusion (Effusion)**

7•10 **Deviations from the Gas Laws**

7•11 **The Kinetic Molecular Theory**

7•11•1 Gaseous State: Facts and Theories

7•11•2 A Molecular Model for Gas Behavior

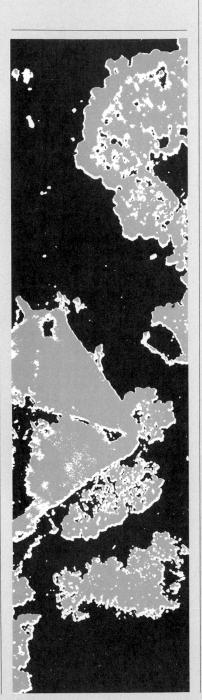

An understanding of the gaseous state of matter is an essential part of the study of chemistry in the laboratory. Usually the amount of a gaseous substance is determined by measuring its volume. However, because the volume of a gas varies with the pressure and temperature, these two conditions must be measured also. One of the chief aims of this chapter is to show how we use such measurements to calculate amounts of gases.

Gases of all sorts act in remarkably similar ways when subjected to changes in pressure and temperature. We can describe the behavior of gases in terms of a few rather broad and simple laws of nature called *the gas laws*. Our study of these laws has a twofold purpose: first, to show how to make a number of the calculations that are necessary in dealing with gaseous reactants and, second, to relate the behavior of gases to the molecular theory of chemical reactions.

The study of gases is also fundamental to our understanding of the ways in which the particles of reactants come together to interact with each other. One might think that because we can more easily handle samples of liquids and solids that the study of their behavior would be simpler than the study of gases. Such is not the case. It is through working with gases that we can best become familiar with one of the most powerful and useful ways of looking at matter—the moving particle theory or the kinetic molecular theory.

7·1   The Three States of Matter

Although the three states of matter—gaseous, liquid, and solid—are familiar to us all, it is useful to state clearly the characteristics that distinguish them.

A gas has no shape of its own; rather it takes the shape of its container. It has no fixed volume but is compressed or expanded as its container changes in size. The volume of the container is the volume of the gas.

A liquid has no specific shape; it takes the shape of its container as it seeks its own level under the influence of gravity. But it does have a specific volume. Although it is not absolutely incompressible, it is compressed only a negligible amount even by moderately high pressures.

A solid has a fixed shape and a fixed volume. Like a liquid, it is not compressed appreciably by moderately high pressures.

A given pure substance may be able to exist in each of the three states, depending on its temperature. It is simply a result of our usual surrounding temperatures that we think of water, H_2O, as a liquid and of ammonia, NH_3, as a gas. At somewhat lower temperatures, the former can be a solid and the latter a liquid. And at a low enough temperature, any substance exists as a solid.

Each of the elements can exist as a gas, liquid, or solid. Many compounds, however, decompose when heated. Such compounds can exist only in the solid state or, at the most, in the solid and liquid states. Ordinary sugar, for example, decomposes instead of melts when heated; potassium chlorate melts to a clear liquid that decomposes when heated further. Water is a familiar example of a compound that can exist in all three states.

7•1•1   Individual Particles in Gases   Measuring the properties of gases is one of the main sources of information that enables us to determine the composition of molecules. In this chapter we will show how the interpretations of the measurements of gases lead to definite formulas for molecules. For example, in the case of the noble gases we find that the individual particles are single atoms or *monatomic* molecules. In the cases of a few other elements that are gases at ordinary room temperature and pressure, the data show that the particles are usually *diatomic* molecules. (See Section 2.1.1 for seven common examples.) The first evidence for the *triatomic* formula for ozone, O_3, came from measurements of the masses of samples of this form of elemental oxygen.

Compounds that are normally gases at room temperature or that exist as gases at higher temperatures may have molecules of two or more atoms. Examples include the oxides of carbon, CO and CO_2, ammonia, NH_3, methane, CH_4, propane, C_3H_8, and benzene, C_6H_6.

Air contains molecules of both elements and compounds. The principal constituents are indicated in Figure 7.1. Also, in air there are minute amounts of all sorts of other molecules, many produced by natural biochemical processes or by domestic and industrial activities.

7•2   The Pressure of Gases

A striking property of the gaseous state is its *compressibility* or the opposite, its *expansibility*. The modern world rolls on the gases compressed in the pneumatic tires of its vehicles. Air is a mixture of gases that behaves physically in the same manner as pure oxygen, pure nitrogen, or any other gaseous substance. A volume of air that is initially two or three times the volume of a tire can be forced into the tire under pressure. If the tire is punctured, the extra air will rush out. Such behavior is characteristic of all gases.

7•2•1   Measurement of Atmospheric Pressure   In the case of the punctured tire just described, instead of saying that the "extra" air rushes out, it would be preferable to say that the air rushes out until the pressure inside the tire drops to become equal to the pressure outside. The pressure outside is that of the atmosphere. In about 1643, Torricelli, a student of Galileo, made the first **barometer,** an instrument for measuring the pressure of the atmosphere. One procedure that he used was to fill a glass tube with mercury and then invert it in a dish of mercury (Figure 7.2). If the tube were long enough, he found that the mercury would fall away from the upper end of the tube. In this simple barometer, the height of the mercury column is directly proportional to the pressure of the atmosphere.

Mercury is nearly an ideal liquid for use in a barometer. Because of its great density (13.6 g/cm^3 at 0 °C), a relatively short column of mercury weighs the same as a column of the atmosphere that is many miles high. Also, the space above the liquid mercury in a barometer is a very good vacuum, containing only a few gaseous mercury molecules. The pressure in this evacuated space can be considered as practically zero under most conditions.

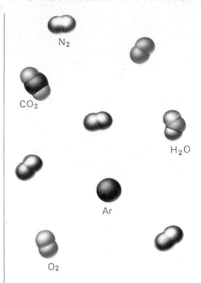

Figure 7•1 In a sample of air, the abundant molecules are particles of nitrogen (diatomic molecules, N_2) and oxygen (diatomic molecules, O_2), with lesser amounts of argon (monatomic molecules, Ar), water (triatomic molecules, H_2O), and carbon dioxide (triatomic molecules, CO_2).

Like Torricelli, we can think of the atmosphere as a "sea of gas" that presses on objects just as a sea of water presses on objects beneath its surface. At any point beneath the surface of a fluid, pressure is exerted in all directions, up as well as down.

7•2•2   **Variations in Atmospheric Pressure**   Three of the main reasons for the variation in the pressure of the atmosphere from place to place are changes in the weather, the height of the atmosphere, and the attraction of gravity. The last of these has the smallest effect.[1] At a given place, the height of the barometer changes with the weather. High readings tend to be associated with fair weather and low readings with foul weather. There are also important differences at different altitudes. At a low altitude, we are deeper in the sea of air and find that its pressure is greater.

The influences of weather and altitude on atmospheric pressure can be seen in the recorded barometric pressure readings in Miami and Denver. In Miami, the U.S. Weather Bureau barometer, which is 3 m above sea level, had an average reading of 763 mm of mercury in 1976. Over the period 1871–1970, the maximum reading was 775 mm and the minimum was 701 mm. In Denver, where the station barometer is 1,625 m above sea level, the average reading was 628 mmHg in 1976. From 1871 to 1970, the maximum reading was 655 mm and the minimum was 605 mm.

The pressure of the atmosphere often has an important effect on the volumes of gases that we measure. In order to compare volumes measured under different pressures, a standard for pressure measurements is needed. Because the average pressure of the atmosphere at sea level supports a column of mercury 760 mm high, this pressure is referred to as **one atmosphere** (atm) or **standard atmospheric pressure.** Standard atmospheric pressure is also expressed in other units, as in the following equivalents:

$$1 \text{ atm} = 760 \text{ mmHg (or 760 mm)} = 760 \text{ torr} = 1.01325 \times 10^5 \text{ pascals (Pa)}$$
$$= 14.7 \text{ pounds per square inch (psi)}$$

For some years there were efforts to replace the term *millimeters of mercury* with the term *torr* (named for Torricelli). The reason was to avoid using a unit of length as a unit of pressure (pressure = force per unit area). However, before the torr could win acceptance, the major shift to SI units had begun. Now it is anticipated that the commonly used unit will be the pascal. One **pascal,** Pa, is the pressure exerted by a force of 1 newton per square meter. (See Tables A.2 and A.5.) The pascal is named for the French scientist Blaise Pascal, who carried out many studies of the barometer in the years just after its discovery.[2]

Because most laboratory barometers and other pressure gauges now in use are calibrated in millimeters, we will often use pressure units of

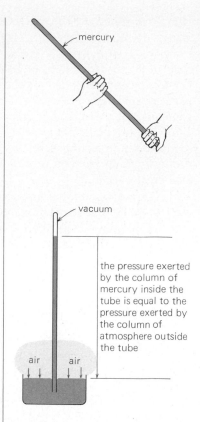

Figure 7•2 A Torricellian barometer is made by completely filling a long glass tube with mercury, inverting it while taking care not to spill any, and then releasing the open end under the surface of a pool of mercury.

---

[1] The attraction of gravity is greater at the poles than at the equator. It is greater at low altitudes than at high. It also varies depending on differences in the density of the earth from place to place.

[2] The U.S. Weather Bureau plans to use the kilopascal, kPa, as the unit of pressure in its regular reports of weather conditions. Atmospheric pressures in the United States range between about 80 and 110 kPa, with readings between 96 and 102 kPa being most common. It is felt that the public will find such kilopascal values more convenient than 80,000 to 110,000 Pa.

millimeters of mercury. We also will often use the unit atmosphere. However, in any SI calculations, we will use the unit pascal.

7•2•3   Measurement of Gas Pressures   A **manometer** is a U-tube filled with a liquid and is a convenient device for measuring small differences in pressure. The manometer shown in Figure 7.3 has one arm open to the atmosphere and measures the difference between the pressure of the atmosphere on one side and the pressure in a closed container attached to the other side. With a mercury-filled manometer, the difference in the levels in the two limbs can be directly added to or subtracted from the barometric pressure.

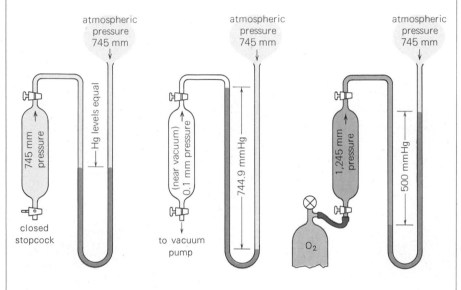

Figure 7•3 The use of a manometer to measure gas pressures. In this illustration it is assumed that a barometer in the laboratory shows the pressure of the atmosphere to be 745 mmHg. The gas pressure in the flask at the left equals that of the atmosphere. For the flask in the center, the position of the mercury shows that almost all of the gas has been pumped out of the flask. At the right, the flask has been filled with oxygen to a pressure of 1,245 mm. This is shown by the sum of the pressure indicated by the height of the mercury column (500 mm) and the air pressure (745 mm).

A manometer is not suitable for measuring great differences in pressure. One type of pressure gauge for measuring high pressures has a metal bellows that expands as the pressure increases. The movement of the bellows is transmitted to a pointer that indicates the pressure on a dial.

The Gas Laws

7•3   Boyle's Law

The discovery that the pressure of the atmosphere could be measured in terms of the height of a column of fluid soon led to the precise study of the changes in volume of trapped samples of gas due to changes in pressure. The behavior revealed by experiments similar to those shown schematically in Figure 7.4 is typical of all gases. At any constant temperature, the greater the pressure on a sample of gas, the less is the volume. Because all gases act in this way, this behavior is called *a law of nature*. First demonstrated in about 1660 by Robert Boyle, it is known as **Boyle's law.** *If* *the temperature remains constant, the volume of a given mass of gas varies inversely with the pressure.* Stated mathematically,

$$V \propto \frac{1}{P}$$

Using data from the specific example in Figure 7.4, it can be seen that the product of the pressure and the volume is a constant:

$$1{,}480 \text{ mm} \times 50 \text{ mL} = 74{,}000 \text{ mm} \cdot \text{mL}$$
$$740 \text{ mm} \times 100 \text{ mL} = 74{,}000 \text{ mm} \cdot \text{mL}$$

That is, $PV = $ a constant. Stated mathematically in an equivalent way,

$$P_1V_1 = P_2V_2 \quad \text{or} \quad \frac{V_1}{V_2} = \frac{P_2}{P_1}$$

The symbols V_1 and P_1 refer to the original volume and pressure, V_2 and P_2 to the volume and pressure under the new or changed conditions.

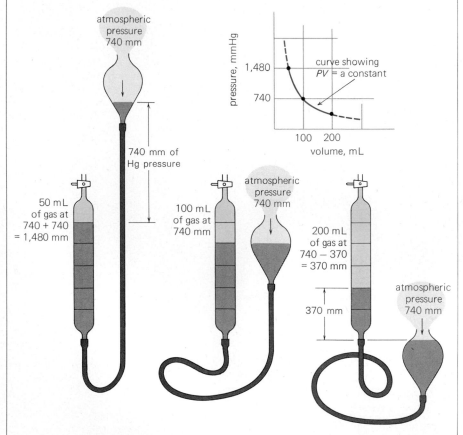

Figure 7.4 Illustration of Boyle's law. Initially, the apparatus is as shown in the middle view; the mercury levels are equal, so the trapped gas is at a pressure of 740 mm. When the mercury bulb is raised to a position so that the pressure is doubled (*left*), the volume of the trapped gas becomes half of the original value; when the mercury is lowered (*right*) so that the pressure is halved, the volume becomes twice the original value. These data are plotted on the graph at the upper right.

7·3·1   **Working Gas Law Problems** In this chapter we shall solve problems associated with several different gas laws. Many of them can be approached in the same systematic fashion. First, we must recognize that to describe completely a sample of a gas, four quantities must be known: the amount of matter present (expressed in terms of mass or number of moles), the volume, the pressure, and the temperature. Second, often we will find it helpful to list one set of conditions that describes the gas in its original state and another set that describes the gas in its changed state. Usually we can state the problem as one in which we must calculate one unknown quantity in the changed state.

Suppose we have a given mass of gas, m, occupying an original volume, V_1, at a specified pressure, P_1, and change the pressure on the gas to P_2. The problem is to calculate the volume in the changed state, V_2. Additional information is that the original temperature and the final temperature are the same; that is, the temperature, T, is held constant. We naturally assume (generally without stating it) that there are no leaks in the apparatus, so the mass of the gas is constant also. We can arrange this information in a table as follows:

	m	V	P	T
original	k	V_1 (known)	P_1 (known)	k
changed to	k	V_2 (?)	P_2 (known)	k

To indicate that a variable does not change, we write the symbol k, which represents a *constant*. Tabulating the information makes it very clear that only the pressure and volume change, and that therefore Boyle's law applies.

• Example 7•1 •

A sample of gas that weighs 0.312 g is trapped in a cylinder, such as that shown in Figure 7.4, at a volume of 183 cm³ and a pressure of 740 mmHg. Sometime later the volume is observed to be 116 cm³. Assuming there are no leaks in the apparatus and the temperature is held constant, what must the pressure be now?

• *Analysis* • First we tabulate the data:

	m	P	V	T
original	k	740 mm	183 cm³	k
changed to	k	P_2	116 cm³	k

Because the mass and temperature of the gas remain constant and only the pressure and volume change, we can use Boyle's law to solve the problem. We reason that because the volume has decreased, the pressure has increased. Hence P_2 must be greater than the original pressure, P_1, of 740 mm.

• Solution •

$$P_1V_1 = P_2V_2$$

$$P_2 = P_1 \times \frac{V_1}{V_2}$$

$$= 740 \text{ mm} \times \frac{183 \text{ cm}^3}{116 \text{ cm}^3}$$

factor is greater than
one, so $P_2 > 740$ mm

$$= 1{,}167 = 1{,}170 \text{ mm} \quad \text{(to three significant figures)}$$

See also Exercises 9–11 at the end of the chapter.

• Example 7•2 •

Solve the problem in Example 7.1 using SI units.

• Solution • First we convert all the data needed to work the problem to SI units. For the pressure,

$$P_1 = 740 \text{ mm} \left(\frac{1.01325 \times 10^5 \text{ Pa}}{760 \text{ mm}} \right) = 9.87 \times 10^4 \text{ Pa}$$

For the volumes,

$$V_1 = 183 \text{ cm}^3 \left(\frac{1 \text{ m}^3}{10^6 \text{ cm}^3} \right) = 1.83 \times 10^{-4} \text{ m}^3$$

$$V_2 = 116 \text{ cm}^3 \left(\frac{1 \text{ m}^3}{10^6 \text{ cm}^3} \right) = 1.16 \times 10^{-4} \text{ m}^3$$

Using Boyle's law,

$$P_2 = P_1 \times \frac{V_1}{V_2}$$

$$= 9.87 \times 10^4 \text{ Pa} \times \frac{1.83 \times 10^{-4} \text{ m}^3}{1.16 \times 10^{-4} \text{ m}^3} = 1.56 \times 10^5 \text{ Pa}$$

See also Exercises 9–11.

Since we have solved a problem by two methods, it is of interest to convert the answer we found in Example 7.2 to check the answer in Example 7.1. Multiplying 1.56×10^5 Pa by $(760 \text{ mm}/1.01325 \times 10^5 \text{ Pa})$ gives 1,170 mm.

7•4   **Temperature Effects**

If a certain quantity of gas is confined at constant pressure in a vessel, the volume of the gas changes with the temperature. Figure 7.5 is a diagram of an apparatus used to demonstrate this effect. Gas is trapped above liquid in a graduated tube encased in a jacket through which fluid at some temperature can be circulated. When the temperature increases, the volume of the gas increases, and vice versa. By raising and lowering the leveling bulb, the level of the liquid inside the tube is kept equal to the level in the bulb; in this way the pressure of the trapped gas can be kept constantly equal to the pressure of the atmosphere. (By properly positioning the leveling bulb, a chosen constant pressure above or below atmospheric pressure can be maintained.)

Suppose a tube like the one shown in Figure 7.5 contains 100 mL of dry air at 0 °C. Table 7.1 gives the volumes that the air would have at various

Table 7•1 **Change in volume of air with change in temperature, at constant pressure (see Figure 7.6)**

Temperature, °C	Volume, mL
273	200
200	173
150	155
100	137
50	118
0	100
−50	82
−100	63
−150	45

other temperatures. Some liquid other than mercury would have to be used to confine the air below −38.87 °C, the freezing point of mercury; also, above about 100 °C the evaporation of the mercury would begin to add appreciably to the volume of trapped gas.

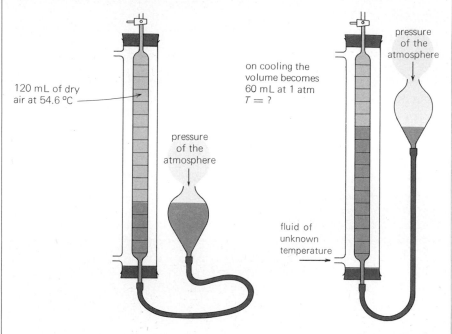

120 mL of dry air at 54.6 °C

on cooling the volume becomes 60 mL at 1 atm $T = ?$

pressure of the atmosphere

pressure of the atmosphere

fluid of unknown temperature

Figure 7•5 A gas thermometer. The volume of a quantity of dry gas is measured first at a known temperature and second at the temperature to be measured. The pressure must be kept constant during the measurements by raising or lowering a bulb of liquid; in this illustration, the pressure of the gas is always the pressure of the atmosphere. A similar thermometer could be calibrated by determining the volume over a wide range of temperatures (see Table 7.1).

The data from Table 7.1 are plotted on the graph in Figure 7.6. Over a wide range of temperature, there is a straight-line relationship between a change in temperature and a change in volume. At very low temperatures, air becomes liquid. The volume decreases suddenly to a small value when the liquid is formed. The straight-line relationship of temperature to volume

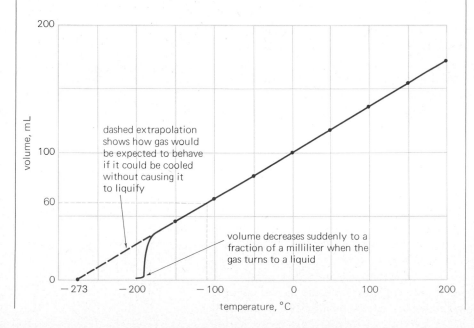

dashed extrapolation shows how gas would be expected to behave if it could be cooled without causing it to liquify

volume decreases suddenly to a fraction of a milliliter when the gas turns to a liquid

Figure 7•6 A graph of the data in Table 7.1, showing that, at constant pressure, the change in volume of a gas is directly proportional to the change in temperature. The dashed black lines are drawn for Example 7.3.

shows that the *changes* in the volume of a gas are directly proportional to the *changes* in temperature; that is,

$$\Delta V \propto \Delta T$$

This proportionality was first recognized in about 1787 by the French scientist Jacques Charles, and was stated in a general way by J. L. Gay-Lussac in 1802.

───── • Example 7•3 • ─────

Using the graph in Figure 7.6, estimate the unknown temperature in Figure 7.5 of the fluid that caused the volume of the gas to decrease to 60 mL.

───── • Solution • The reading of the graph for this problem is indicated by the black dashed lines in Figure 7.6. The first is drawn horizontally from the volume of 60 mL to intersect the straight colored line that is drawn through the data points. The second is then drawn from the point of intersection straight down to the temperature coordinate. We can estimate the temperature corresponding to a volume of 60 mL as about −110 °C.

See also Exercise 16.

7•4•1   Absolute Scale of Temperature   Extrapolation of the straight line in Figure 7.6 leads to the idea that, if the temperature were lowered enough, the air would occupy zero volume. Although it is inconceivable that matter can ever have a volume of zero, the temperature associated with "zero volume" on the graph has great significance. This temperature, calculated to be 273.15 degrees below 0° Celsius, is called **absolute zero.** Although a simple extrapolation like that shown in Figure 7.6 indicates that an absolute zero temperature exists, it was not until 1848 that Lord Kelvin convincingly demonstrated the validity of the absolute temperature scale. On the Kelvin scale, absolute zero is designed to be 0 K (see Section 1.2.4). A change of 1 K is equal in magnitude to a change of 1 °C, so that the freezing point of water, which is 273.15 degrees above absolute zero, has the value of 273.15 K on the Kelvin scale. *To convert °C to K, 273 (more precisely, 273.15) is added to the Celsius temperature.*

No highest temperature possible has been calculated because there is no theoretical upper limit for temperatures. The temperature inside the sun is estimated at 30,000,000 K; the temperature attained in a hydrogen bomb blast is estimated at 100,000,000 K.

7•4•2   Charles's Law   In Figure 7.6 the straight-line graph of the temperature of a gas versus its volume shows that changes in these quantities are directly proportional to each other. However, no direct ratio exists between volume and temperature if the temperature values are taken from the Celsius or Fahrenheit scales. The numbers on these scales are relative values only. Neither 0 °C nor 0 °F signifies a complete lack of temperature, because on both scales we can read temperatures "below zero."

Because it is only on the absolute scale that 0 means no temperature, any reference to a direct ratio of volume to temperature must mention that absolute values are being used. The statement of this relationship is known as **Charles's law.** *If the pressure does not change, the volume of a given mass of gas is directly proportional to the absolute temperature.* Stated mathematically,

$$V \propto T$$

Using data from Table 7.1 and converting to absolute temperatures, it is seen that the quotient of the volume divided by the absolute temperature is a constant:

$$\frac{155 \text{ mL}}{150 + 273 \text{ K}} = 0.366 \text{ mL/K}$$

$$\frac{100 \text{ mL}}{0 + 273 \text{ K}} = 0.366 \text{ mL/K}$$

$$\frac{82 \text{ mL}}{-50 + 273 \text{ K}} = 0.368 \text{ mL/K}$$

That is, V/T = a constant, or

$$\frac{V_1}{T_1} = \frac{V_2}{T_2} \quad \text{or} \quad \frac{V_1}{V_2} = \frac{T_1}{T_2}$$

— • **Example 7.4** • —

Calculate the Celsius temperature of the unknown fluid in Figure 7.5.

• *Analysis* • First we tabulate the data. Because absolute temperatures are always used in solving gas law problems, we tabulate all temperatures in absolute units. We see by the figure that the pressure is kept constant. We also must assume that there are no leaks in the apparatus, so the amount of gas is taken to be constant.

	m	V	T	P
original	k	120 mL	(55 °C + 273) K	k
changed to	k	60 mL	T_2	k

Because only the volume and the temperature change, we see that we can use Charles's law to solve this problem. Also, we reason that, because the volume has decreased, the temperature must have decreased.

• **Solution** •

$$\frac{V_1}{T_1} = \frac{V_2}{T_2}$$

$$T_2 = T_1 \times \frac{V_2}{V_1}$$

$$= 328 \text{ K} \times \frac{60 \text{ mL}}{120 \text{ mL}} = 164 \text{ K}$$

factor is less than one, so $T_2 < 328$ K

$$°C = 164 \text{ K} - 273 = -109 \text{ °C}$$

This calculated value agrees well with the $-110\,°C$ we read off the graph in Example 7.3.

See also Exercises 15–18.

7.4.3   Relation of Pressure and Temperature   After being checked in the cool morning hours, the air pressure in a tire on a hot summer day can increase markedly after several hours of driving. There may be no noticeable increase in the tire's volume. The relationship between pressure and temperature at constant volume is not so commonly referred to by a discoverer's name, probably because it was recognized gradually by several investigators. It is sometimes named after Joseph Gay-Lussac and sometimes after Guillaume Amontons, who related the pressure of a gas to its temperature and constructed a gas thermometer on this basis in 1703. We will recognize the contributions of both scientists by calling the relationship **Gay-Lussac's and Amontons's law.** *The pressure of a given mass of gas is directly proportional to the absolute temperature if the volume does not change.* Stated mathematically,

$$P \propto T$$

or $P/T =$ a constant. Equivalent statements are

$$\frac{P_1}{T_1} = \frac{P_2}{T_2} \quad \text{or} \quad \frac{P_1}{P_2} = \frac{T_1}{T_2}$$

This relationship is applied to a practical question in Example 7.5.

━━━━ • Example 7.5 • ━━━━

Consider a small steel tank of methane gas at 20 °C that has a pressure of 30.0 psi measured on a gauge. What will the gauge pressure be if the tank is placed near a steam radiator and the temperature is increased to 85 °C?

• *Analysis* • First we must recognize that an ordinary gauge measures the *difference* between the pressure inside a container and the pressure of the atmosphere outside (see Figure 7.7). If the pressure in the tank were just equal to the outside pressure of the atmosphere, the gauge would show a reading of zero. If we assume the pressure of the atmosphere to be 1 atm or 14.7 psi, the actual pressure of the gas in the tank at 20 °C is $30.0 + 14.7 = 44.7$ psi.

A steel tank is so strong that we can assume the volume is practically constant in spite of the increase in pressure at the higher temperature.

Changing to absolute temperatures, we find $20\,°C + 273 = 293\,K$ and $85\,°C + 273 = 358\,K$.

The data are tabulated:

	m	V	T	P
original	k	k	293 K	44.7 psi
changed to	k	k	358 K	P_2

Assuming that the mass of the gas in the tank is constant and that the volume is constant, we know that the pressure is directly proportional to

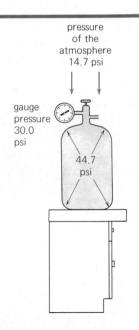

Figure 7.7 An ordinary gas gauge measures the difference between the pressure on one side and the pressure of the atmosphere.

the absolute temperature. We reason that the new pressure must be greater than the original pressure, because the temperature has increased.

• **Solution** •

$$P_2 = P_1 \times \frac{T_2}{T_1}$$

$$= 44.7 \text{ psi} \times \frac{358 \text{ K}}{293 \text{ K}} = 54.6 \text{ psi}$$

factor is greater than
one, so $P_2 > 44.7$ psi

The gauge pressure is the difference between the total tank pressure and the pressure of the atmosphere; therefore:

gauge pressure at 85 °C = 54.6 psi − 14.7 psi = 39.9 psi

See also Exercises 19–21.

7•5 Avogadro's Law

At a set pressure and temperature, a vessel of given volume contains a certain number of molecules of gas, regardless of the kind of gas. This relationship was formulated by Amadeo Avogadro in 1811 and is known as **Avogadro's law.** *Equal numbers of molecules are contained in equal volumes of different gases if the pressure and temperature are the same.*

For example, the number of molecules of nitrogen in 1 L at a pressure of 1.20 atm and a temperature of 25 °C is definite. According to Avogadro's law, at the same volume, pressure, and temperature there would be present the same number of hydrogen molecules, carbon dioxide molecules, or molecules of any other gas.

We use the concept of Avogadro's law to help us analyze situations in which volumes or pressures or temperatures are not equal. For example, suppose containers A and B have equal volumes and temperatures but the pressure in A is greater. This means that more molecules have been crowded into A than into B. Consider another example, and one that can be trickier to handle. Suppose containers C and D have equal volumes and pressures but the temperature in D is higher. This means that there are fewer molecules in D than in C. It takes fewer molecules at a higher temperature to exert the same pressure that is exerted by a larger number of molecules at a lower temperature.

A hot-air balloon (see Figure 7.8) operates at about a constant volume and a constant pressure. The lifting power of the balloon is approximately equal to the difference in the weight of the less dense hot air inside it and the weight of an equal volume of the more dense cooler air surrounding it. The density of a gas is inversely proportional to its absolute temperature. At 0 °C and 1 atm pressure, the density of dry air is 1.29 g/L.

For convenience in working with gases, **standard conditions** have been defined as 0 °C (273 K) and 1 atm (760 mmHg). These conditions are referred to as **standard temperature and pressure,** STP. If only the volume of

Figure 7•8 The lifting power of a hot-air balloon depends on the difference in the weight of the gases in the balloon compared with the greater weight of the equal volume of air that the balloon displaces.

a gas is specified, for example, 3.5 L of nitrogen, but no pressure and temperature, we assume that this volume is at STP.

The amount of a gas is conveniently measured in terms of the number of moles, designated n. At constant temperature and pressure, Avogadro's law is stated mathematically as

$$V \propto n$$

or $V/n =$ a constant.

At STP, with volume given in liters, the value of $V/n = 22.414$ L/mol. In SI base units, $V/n = 0.022414$ m^3/mol. The volume of 1 mole of a gas at standard temperature and pressure is called **Avogadro's volume** or the **molar gas volume**. (See also Figure 7.9.)

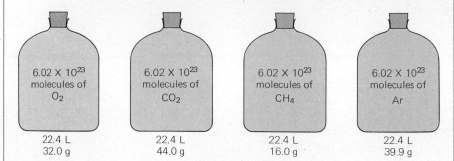

6.02 × 10^{23} molecules of O$_2$	6.02 × 10^{23} molecules of CO$_2$	6.02 × 10^{23} molecules of CH$_4$	6.02 × 10^{23} molecules of Ar
22.4 L 32.0 g	22.4 L 44.0 g	22.4 L 16.0 g	22.4 L 39.9 g

Figure 7.9 One mole of gas contains 6.02×10^{23} molecules and has the same volume as 1 mole of any other gas at the same temperature and pressure, 22.4 L or 0.0224 m^3 at 273 K and 1 atm. These identical volumes of different gases can be contrasted with the widely varying molar volumes of solids and liquids, examples of which are shown in Figure 2.1.

• Example 7.6 •

Compare the number of H$_2$ and N$_2$ molecules in two containers described as follows: a 2-L container of hydrogen filled at 127 °C and 5 atm; a 5-L container of nitrogen filled at 27 °C and 3 atm.

  • *Analysis* • The data are tabulated as follows:

	V	T	P
hydrogen	2 L	400 K	5 atm
nitrogen	5 L	300 K	3 atm

We can multiply the number of H$_2$ molecules by factors to see how many times larger or smaller the number of N$_2$ molecules is. First, the volume factor is set as 5 L/2 L, because a larger volume contains a larger number of molecules. Second, because a container filled at a lower temperature contains a larger number of molecules, the temperature factor is set as 400 K/300 K. Third, a container filled at a lower pressure contains fewer molecules, so that the pressure factor is set as 3 atm/5 atm.

• Solution •

$$\frac{\text{number}}{\text{of N}_2} = \frac{\text{number}}{\text{of H}_2} \times \frac{5 \text{ L}}{2 \text{ L}} \times \frac{400 \text{ K}}{300 \text{ K}} \times \frac{3 \text{ atm}}{5 \text{ atm}} = \frac{2(\text{number}}{\text{of H}_2)}$$

There are twice as many N$_2$ molecules as H$_2$ molecules.

See also Exercise 23.

• Example 7•7 •

A 1.40-L volume of a gas measured at a temperature of 27 °C and a pressure of 890 mm was found to weigh 2.273 g. Calculate the molecular weight of this gas.

• Analysis • We know that the weight of 22.4 L of a gas at STP is the weight of 1 mole (equal numerically to the weight of one molecule). Therefore, we wish to calculate the weight, m_2, that this gas would weigh if the original conditions were changed to 22.4 L at STP. So, the data are tabulated as follows:

	m	V	T	P
original	2.273 g	1.40 L	300 K	890 mm
changed to (as calculated)	m_2	22.4 L	273 K	760 mm

where m_2 is the weight of 1 mole of gas.

Using these data, we formulate an equation that can be solved to give the weight of 1 mole, m_2:

$$m_2 = 2.273 \text{ g} \times \frac{\text{volume}}{\text{volume}} \times \frac{\text{temperature}}{\text{temperature}} \times \frac{\text{pressure}}{\text{pressure}}$$

The analysis is undertaken as before. We consider separately how each of the three changes in conditions affects the weight of the gas.

Change in volume: There is an increase in volume; the weight of gas in 22.4 L is much greater than the weight in 1.40 L. The volume factor is 22.4 L/1.40 L.

Change in temperature: There is a decrease in temperature. Because a container filled at 273 K holds more molecules or a greater weight of gas than when filled at 300 K, the temperature factor is 300 K/273 K.

Change in pressure: There is a decrease in pressure; the weight of gas needed to exert 760 mm pressure is less than that needed to exert 890 mm pressure. The pressure factor is 760 mm/890 mm.

• Solution • We set up an equation using the factors worked out in our analysis, cancel units, and solve:

$$m_2 = 2.273 \text{ g} \times \frac{22.4 \text{ L}}{1.40 \text{ L}} \times \frac{300 \text{ K}}{273 \text{ K}} \times \frac{760 \text{ mm}}{890 \text{ mm}} = 34.1 \text{ g}$$

The weight of 1 mole (6.02×10^{23} molecules) in grams is equal numerically to the weight of one molecule in atomic mass units. Therefore, the weight of one molecule is 34.1 amu.

See also Exercises 22 and 26–28.

7•6   A General Gas Equation

As previously stated, there are four variable quantities that completely describe a given amount of a gas: m, V, T, and P. The amount of gas present can also be expressed in terms of the number of moles, n, instead of the mass, m. For a sample of a gas, these variables do not change independently of one another. Indeed, if three of them are fixed, the other is also fixed.

The volume of a gas is directly proportional to the number of moles present, n, and to the absolute temperature, T, and is inversely proportional to the pressure, P. This combination in a single statement of Boyle's, Charles's, Gay-Lussac's and Amontons's, and Avogadro's laws is called the **ideal gas law.** Stated mathematically,

$$V \propto nT\frac{1}{P} \quad \text{or} \quad V = RnT\frac{1}{P}$$

This equation is commonly written as

$$PV = nRT$$

and is called the **ideal gas law equation.** In this equation, the constant of proportionality, designated R, is called the **ideal gas constant.**

In the special case of the same sample of a gas at two conditions of pressure, volume, and temperature, n is constant, and

$$\frac{PV}{T} = nR = \text{a constant} \quad \text{or} \quad \frac{P_1V_1}{T_1} = \frac{P_2V_2}{T_2}$$

If any five of these quantities are known, the sixth can be calculated.

• Example 7•8 •

A quantity of methane has a volume of 100.0 L at 27 °C and 700 mmHg pressure. What will be its volume at 0 °C and 760 mm pressure?

• Solution • The data are tabulated:

	m	V	P	T
original	k	100.0 L	700 mm	$(27 + 273)$ K
changed to	k	V_2	760 mm	$(0 + 273)$ K

Because the mass, m, and consequently the number of moles of gas, n, are constant, and both the pressure and the temperature change, we can use the ideal gas law equation in the following form:

$$\frac{P_1V_1}{T_1} = \frac{P_2V_2}{T_2} \quad \text{or} \quad V_2 = V_1 \times \frac{P_1}{P_2} \times \frac{T_2}{T_1}$$

$$V_2 = 100.0 \text{ L} \times \frac{700 \text{ mm}}{760 \text{ mm}} \times \frac{273 \text{ K}}{300 \text{ K}} = 83.8 \text{ L}$$

See also Exercises 35 and 37.

7•6•1   Evaluating the Gas Constant R   The value of the constant, R, in the ideal gas law can be evaluated if we know the other four quantities for any sample of gas. In order to use the ideal gas law equation in solving problems involving gases, it is convenient to calculate a value of R for the case of 1 mole of gas at STP.

n	V	T	P
1 mol	22.414 L	273.15 K	1 atm

We can substitute these values in the equation and solve for R:

$$R = \frac{VP}{Tn} = \frac{(22.414 \text{ L})(1 \text{ atm})}{(273.15 \text{ K})(1 \text{ mol})}$$

$$= 0.082057 \text{ (L} \cdot \text{atm)/(K} \cdot \text{mol)} = 0.082057 \text{ L} \cdot \text{atm} \cdot \text{K}^{-1} \cdot \text{mol}^{-1}$$

Rounding this value to three significant figures gives

$$PV = RnT = 0.0821 \, nT$$

for units of liters, atmospheres, Kelvins, and moles.

Different values of R are calculated for different systems of units. For SI base units:

n	V	T	P
1 mol	$2.2414 \times 10^{-2} \text{ m}^3$	273.15 K	1.01325×10^5 Pa

$$R = \frac{(2.2414 \times 10^{-2} \text{ m}^3)(1.01325 \times 10^5 \text{ Pa})}{(273.15 \text{ K})(1 \text{ mol})}$$

$$= 8.314 \text{ (m}^3 \cdot \text{Pa)/(K} \cdot \text{mol)} = 8.314 \text{ m}^3 \cdot \text{Pa} \cdot \text{K}^{-1} \cdot \text{mol}^{-1}$$

Exercises 31 and 32 at the end of the chapter show a number of other ways in which R can be expressed.

7•6•2 **Solving Problems with the Ideal Gas Law Equation** The ideal gas law equation is particularly useful in finding the amount of a gas present if the volume, pressure, and temperature are known. To find the number of moles, we solve for n:

$$n = \frac{PV}{RT}$$

To find the molar weight, if we know the mass, m, of the gas, we note that $n = m/\text{molar wt}$. So we can write

$$\frac{m}{\text{molar wt}} = \frac{PV}{RT} \quad \text{or} \quad \text{molar wt} = \frac{mRT}{PV}$$

When using the ideal gas law equation, we must be careful to express all data in the same units used for the specific gas constant, R, that we choose. To illustrate the use of the ideal gas law equation, and to relate it to other ways of solving gas law problems, let us first apply it to Example 7.7, which was previously solved by considering, in turn, the separate effects of changing the volume, temperature, and pressure.

━━━ • Example 7•7 (Alternate approach) • ━━━

A 1.40-L volume of a gas measured at a temperature of 27 °C and a pressure of 890 mm was found to weigh 2.273 g. Calculate the molecular weight of this gas.

• *Analysis* • Because P, V, and T are known, we can use the ideal gas law to find n, the number of moles. Using the relationship $n = (m/\text{molar wt})$ enables us to solve directly for the molar weight, knowing that $m = 2.273$ g.

━━━ • Solution 1 • Express all quantities in appropriate units in order to use $R = 0.0821$ (L·atm)/(K·mol):

P	V	n	T
890 mm	1.40 L	$\dfrac{2.273 \text{ g}}{\text{molar wt}}$	27 °C
1.17 atm	1.40 L	$\dfrac{2.273 \text{ g}}{\text{molar wt}}$	300 K

$$\text{molar wt} = \frac{RmT}{PV}$$

$$= 0.0821 \frac{\text{L·atm}}{\text{K·mol}} \times \frac{(2.273 \text{ g})(300 \text{ K})}{(1.17 \text{ atm})(1.40 \text{ L})} = 34.2 \text{ g/mol}$$

molecular wt = 34.2 amu

This is a check, to three significant figures, of the previous answer, 34.1 amu.

━━━ • Solution 2 (SI base units) • Express all quantities in appropriate units in order to use $R = 8.314$ (m³·Pa)/(K·mol).

P	V	n	T
890 mm	1.40 L	$\dfrac{2.273 \text{ g}}{\text{molar wt}}$	27 °C
1.187×10^5 Pa	0.00140 m³	$\dfrac{0.002273 \text{ kg}}{\text{molar wt}}$	300 K

$$\text{molar wt} = \frac{RmT}{PV}$$

$$= 8.314 \frac{\text{m}^3 \cdot \text{Pa}}{\text{K·mol}} \times \frac{(0.002273 \text{ kg})(300 \text{ K})}{(1.187 \times 10^5 \text{ Pa})(0.00140 \text{ m}^3)}$$

$$= 0.0341 \text{ kg/mol} = 34.1 \text{ g/mol}$$

molecular wt = 34.1 amu

This is a check, to three significant figures, of the previous answer, 34.1 amu.

See also Exercises 38, 41, and 42.

• *Representative Gas Law Problems* • To demonstrate further the utility of the ideal gas law, we will apply it to two typical problems. The first step in the solution of many problems is to express the given data in the proper units. Examples 7.9 and 7.10 illustrate the method.

━━━ • Example 7•9 • ━━━

What is the pressure in pounds per square inch exerted by 100 g of nitrogen, N_2, which is confined at $-10\ °C$ in a cylinder that has a volume of 3.10 gallons?

━━━ • Solution • See Table A.6 for the gallon-to-liter conversion factor.

P	V	n	T
?	(3.10 gal)(3.79 L/gal)	$\dfrac{(100\ g)}{(28.0\ g/mol)}$	$-10\ °C + 273$
?	11.7 L	3.57 mol	263 K

$$P = \frac{nRT}{V} = R\,\frac{nT}{V}$$

$$= 0.0821\ \frac{\cancel{L}\cdot atm}{\cancel{K}\cdot \cancel{mol}} \times \frac{(3.57\ \cancel{mol})(263\ \cancel{K})}{11.7\ \cancel{L}} = 6.59\ atm$$

$$P = 6.59\ \cancel{atm} \times \frac{14.7\ psi}{1\ \cancel{atm}} = 96.9\ psi$$

See also Exercises 38, 39, and 42.

━━━ • Example 7•10 • ━━━

A 30-m^3 tank of CCl_2F_2, a refrigerant gas, is at a pressure of 0.54 atm and a temperature of 11 °C. What is the weight of the gas? Convert all data to SI base units before solving.

━━━ • Solution • ━━━

$$11\ °C + 273 = 284\ K$$
$$P = (0.54\ \cancel{atm})(1.01 \times 10^5\ Pa/\cancel{atm}) = 5.5 \times 10^4\ Pa$$
$$molar\ wt = 121\ g/mol = 0.121\ kg/mol$$

$$m = \frac{PV\ (molar\ wt)}{RT}$$

$$= \frac{(5.5 \times 10^4\ \cancel{Pa})(30\ \cancel{m^3})(0.121\ kg/\cancel{mol})}{[8.314(\cancel{m^3}\cdot \cancel{Pa})/(\cancel{K}\cdot \cancel{mol})](284\ \cancel{K})} = 85\ kg$$

See also Exercises 38, 39, and 42.

• **7•6•3   Calculation of Gas Densities**   Once the formula of a gaseous substance is known, the weight of 1 mole is also known, and the density at STP or at other conditions can be calculated. The densities of gases are usually expressed in grams per liter. These densities are numerically equal to those in the SI base units of kilograms per cubic meter.

━━━ • Example 7•11 • ━━━

Calculate the densities in grams per liter and in SI base units for nitrogen dioxide, NO_2, at STP, and at 100 °C and five times the standard pressure.

• **Solution 1 (grams per liter)** • At STP, the direct solution of this problem is to recall that 1 mole of gas occupies Avogadro's volume of 22.4 L. Because 1 mole of NO_2 weighs 46.0 g, its density is

$$d_{NO_2} = \frac{46.0 \text{ g/mol}}{22.4 \text{ L/mol}} = 2.05 \text{ g/L}$$

Or, we can rearrange the ideal gas law equation with n expressed as $m/$molar wt and solve for m/V:

$$\frac{m}{V} = P\frac{(\text{molar wt})}{0.0821T}$$

$$= \frac{(1 \text{ atm})(46.0 \text{ g/mol})}{[0.0821(\text{L} \cdot \text{atm})/(\text{K} \cdot \text{mol})](273 \text{ K})} = 2.05 \text{ g/L}$$

Similarly, at 100 °C and 5.00 atm,

$$\frac{m}{V} = \frac{(5.00)(46.0)}{(0.0821)(373)} \text{ g/L} = 7.51 \text{ g/L}$$

• **Solution 2 (SI base units)** • We recall that at STP, 1 mole of gas occupies Avogadro's volume of 0.0224 m³. Because 1 mole of NO_2 weighs 0.0460 kg, its density is:

$$d_{NO_2} = \frac{0.0460 \text{ kg/mol}}{0.0224 \text{ m}^3/\text{mol}} = 2.05 \text{ kg/m}^3$$

Or, we can rearrange the ideal gas law equation with n expressed as $m/$molar wt and solve for m/V:

$$\frac{m}{V} = P\frac{(\text{molar wt})}{8.314T}$$

$$= \frac{(1.013 \times 10^5 \text{ Pa})(0.0460 \text{ kg/mol})}{[8.314(\text{m}^3 \cdot \text{Pa})/(\text{K} \cdot \text{mol})](273 \text{ K})} = 2.05 \text{ kg/m}^3$$

Similarly, at 100 °C and $(5.00)(1.013 \times 10^5 \text{ Pa})$,

$$\frac{m}{V} = \frac{(5.00)(1.013 \times 10^5)(0.0460)}{(8.314)(373)} \text{ kg/m}^3 = 7.51 \text{ kg/m}^3$$

See also Exercises 29, 30, and 56.

7•7   Molar Relationships with Gaseous Reactants

The volume of a mole of any gas measured at STP is 22.4 L. This relationship is the basis for calculating the volumes of gaseous reactants and products involved in chemical reactions. If a volume is specified at some temperature and pressure other than STP, corrections are made by means of the gas laws.

Stoichiometric calculations in which gaseous substances are involved may be classified as weight–volume relationships or volume–volume relationships.

7•7•1 **Weight–Volume Relationships** Examples 7.12 and 7.13 illustrate weight–volume relationships.

• Example 7•12 •

Calculate the volume in liters at STP of dry hydrogen and oxygen theoretically obtainable by the decomposition of 100 g of water by electrolysis.

• Solution • To solve this problem, we first write a balanced chemical equation and calculate the number of moles of hydrogen and oxygen that are formed from the decomposition of 100 g of water. Multiplying the number of moles by the molar gas volume, 22.4 L/mol, yields the volume.

$$2H_2O \longrightarrow 2H_2 + O_2$$
$$2\ \text{mol} \qquad 2\ \text{mol} \quad 1\ \text{mol}$$

$$\text{mol } H_2O = 100\ \text{g } H_2O \times \frac{1\ \text{mol } H_2O}{18.0\ \text{g } H_2O} = 5.56\ \text{mol } H_2O$$

$$\text{mol } H_2 = 5.56\ \text{mol } H_2O \times \frac{2\ \text{mol } H_2}{2\ \text{mol } H_2O} = 5.56\ \text{mol } H_2$$

$$\text{mol } O_2 = 5.56\ \text{mol } H_2O \times \frac{1\ \text{mol } O_2}{2\ \text{mol } H_2O} = 2.78\ \text{mol } O_2$$

$$\text{vol } H_2 = 5.56\ \text{mol } H_2 \times \frac{22.4\ \text{L } H_2}{1\ \text{mol } H_2} = 125\ \text{L } H_2 \text{ at STP}$$

$$\text{vol } O_2 = 2.78\ \text{mol } O_2 \times \frac{22.4\ \text{L } O_2}{1\ \text{mol } O_2} = 62.3\ \text{L } O_2 \text{ at STP}$$

See also Exercises 43–45 and 47.

• Example 7•13 •

If 500 kg of water is decomposed (see Example 7.12), what volumes of dry hydrogen and oxygen are obtained if each gas is collected at 25.0 atm and 26 °C? Solve this problem using SI base units.

• Solution • From the solution to Example 7.12, we know that 5.56 moles of H_2 are obtained from 100 g of water. From 500 kg of water, there would be

$$\frac{5.56\ \text{mol } H_2}{100\ \text{g } H_2O} \times \left(500\ \text{kg } H_2O \times \frac{1{,}000\ \text{g } H_2O}{1\ \text{kg } H_2O}\right) = 2.78 \times 10^4\ \text{mol } H_2$$

At 25.0 atm, the pressure in SI units is $(25.0\ \text{atm})(1.013 \times 10^5\ \text{Pa/atm}) = 2.53 \times 10^6\ \text{Pa}$.

$$V = \frac{nRT}{P} = R\frac{nT}{P}$$

$$= 8.314\,\frac{m^3 \cdot Pa}{K \cdot mol} \times \frac{(2.78 \times 10^4\,mol\,H_2)(299\,K)}{(2.53 \times 10^6\,Pa)} = 27.3\,m^3\,H_2$$

$$vol\,O_2 = \tfrac{1}{2}(vol\,H_2) = 13.6\,m^3$$

See also Exercises 43–46.

7.7.2 Volume–Volume Relationships In 1805 and shortly thereafter, J. L. Gay-Lussac performed a series of experiments that established what today is called **Gay-Lussac's law of combining volumes.** *Gases react with one another in small, whole-numbered ratios by volume if the volumes are measured at the same temperature and pressure.* Gay-Lussac's empirical law was an important contribution in the development of chemistry, because it provided part of the background that enabled Avogadro, in 1811, to hypothesize that equal volumes of gases contain the same number of molecules. Gay-Lussac's law of combining volumes is illustrated in Example 7.14.

• Example 7.14 •

What is the theoretical volume of ammonia, NH_3, formed by the reaction of 10 m^3 of hydrogen, H_2, with nitrogen, N_2? What volume of nitrogen reacts? (All volumes are measured at the same temperature and pressure.)

• Solution • The equation for the reaction is interpreted in terms of moles and molar volumes:

$$N_2 \quad + \quad 3H_2 \quad \longrightarrow \quad 2NH_3$$

1 mol	3 mol	2 mol
1 molar volume	3 molar volumes	2 molar volumes

Because only the volume of hydrogen is given, we cannot calculate the number of moles (unless we assume some conditions, such as STP). However, we can take advantage of the fact that, at the same temperature and pressure, the volume of a gas is directly proportional to the number of moles or molecules present. Because equal volumes contain the same number of molecules (Avogadro's law), the volume of the nitrogen that reacts is one-third the volume of H_2. This follows from the balanced equation that states that 1 mole of N_2 reacts with 3 moles of H_2. From the same line of reasoning, we see that the volume of NH_3 produced is two-thirds that of H_2:

$$vol\,NH_3\,formed = 10\,m^3\,H_2 \times \frac{2\,vol\,NH_3}{3\,vol\,H_2} = 6.7\,m^3\,NH_3$$

$$vol\,N_2\,that\,reacts = 10\,m^3\,H_2 \times \frac{1\,vol\,N_2}{3\,vol\,H_2} = 3.3\,m^3\,N_2$$

See also Exercises 46 and 48.

7.8 Dalton's Law of Partial Pressures

In a mixture of different gases each gas exerts part of the pressure. This is the same pressure that it would exert if it alone occupied the volume containing the mixed gases. This relationship was formulated by John Dalton in about 1803 and is known as **Dalton's law.** *The total pressure in a mixture of gases is the sum of the individual partial pressures.* Stated mathematically,

$$P_{total} = p_1 + p_2 + p_3 + \cdots$$

The small p's refer to the partial pressures, the pressures exerted by each different gaseous substance in the mixture.

• Example 7•15 •

A sample of hydrogen is collected in a bottle over water. By carefully raising and lowering the bottle, the height of the water inside is adjusted so that it is just even with the water level outside (see Figure 7.10). The following measurements are taken: volume of gas, 425 cm³; atmospheric pressure, 753 mm; temperature of water (and of gas also), 34 °C. Calculate the volume the hydrogen would have if it were dry and at a pressure of 760 mm and a temperature of 0 °C.

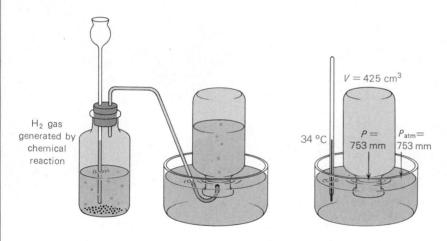

Figure 7•10 When a sample of gas is collected, its pressure, its volume, and its temperature are measured. It is often convenient to measure the latter two when the gas is held at the pressure of the atmosphere, that is, with the water level inside the bottle equal to that outside.

• *Analysis* • The hydrogen is mixed with water vapor. Therefore not all of the 753 mm pressure is due to hydrogen; a small part is due to the water vapor. (See Vapor Pressure, Section 8.4.1.) The vapor pressure of water varies with the temperature, so we find from Table A.7 in the appendix that its value at 34 °C is 40 mm. The pressure due to hydrogen can now be calculated by means of Dalton's law:

$$P_{total} = p_{H_2} + p_{H_2O} \quad \text{or} \quad p_{H_2} = P_{total} - p_{H_2O}$$

$$= 753 \text{ mm} - 40 \text{ mm} = 713 \text{ mm}$$

This latter pressure, which is the pressure exerted by the hydrogen itself, is the one that we must use as the original pressure in solving for the volume of the dry hydrogen at 0 °C and 760 mm:

	n	V	T	P
original	k	425 cm³	307 K	713 mm
changed to	k	V_2	273 K	760 mm

Because n is constant, we can use the equation

$$\frac{P_1 V_1}{T_1} = \frac{P_2 V_2}{T_2}$$

• Solution •

$$V_2 = V_1 \times \frac{P_1}{P_2} \times \frac{T_2}{T_1}$$

$$= 425 \text{ cm}^3 \times \frac{713 \text{ mm}}{760 \text{ mm}} \times \frac{273 \text{ K}}{307 \text{ K}} = 355 \text{ cm}^3$$

See also Exercises 50–52.

7.9   Graham's Law of Diffusion (Effusion)

A gas that has a high density diffuses more slowly than one with a lower density. Discovered in 1830 by Thomas Graham, the precise statement is called **Graham's law.** *The rates of diffusion of two gases are inversely proportional to the square roots of their densities.* Stated mathematically,

$$\frac{r_1}{r_2} = \frac{\sqrt{d_2}}{\sqrt{d_1}}$$

where r_1 and r_2 are the rates of diffusion of the two gases, and d_1 and d_2 are the respective densities.

• Example 7.16 •

At standard conditions 1 L of oxygen gas weighs almost 1.44 g, whereas 1 L of hydrogen weighs only 0.09 g. Which gas diffuses faster? Calculate how much faster.

• *Analysis* •

rate	density, g/L
r_{H_2}	$d_{H_2} = 0.09$
r_{O_2}	$d_{O_2} = 1.44$

According to Graham's law, the hydrogen will diffuse faster because it has the lower density. The factor is $\sqrt{1.44 \text{ g/L}}/\sqrt{0.09 \text{ g/L}}$.

• Solution •

$$r_{H_2} = r_{O_2} \times \frac{\sqrt{1.44}}{\sqrt{0.09}}$$

The equation is arranged in a logical way. We have decided that r_{H_2} is greater than r_{O_2}, so we know that the multiplier for r_{O_2} must be greater than one if the equation is to be correct:

$$r_{H_2} = r_{O_2} \times \frac{1.2}{0.3} = 4 r_{O_2}$$

See also Exercises 53 and 54.

Diffusion is the spontaneous equalization of physical states. When it involves different substances, it refers to the intermingling of particles as they move to become uniformly distributed among one another. An examination of Graham's original experiments shows that he was actually measuring rates of effusion more nearly than rates of diffusion. *Effusion* is the movement of gas particles through a small hole. The equation for Graham's law describes rates of effusion very precisely (see Figure 7.11).

According to Avogadro's law, the densities of gases are proportional to their molecular weights. We may therefore substitute molecular weight for density in the expression of Graham's law of diffusion (effusion):

$$\frac{r_1}{r_2} = \frac{\sqrt{d_2}}{\sqrt{d_1}} = \frac{\sqrt{\text{molecular wt}_2}}{\sqrt{\text{molecular wt}_1}}$$

We can determine the molecular weight of an unknown gas by measuring its rate of effusion and comparing this with the rate for a gas whose molecular weight is known.

7•10 Deviations from the Gas Laws

Thus far in our discussion of the gas laws, there has been no indication that the various laws do not hold equally well for all gases and that they are not absolutely exact. Actually, the gas laws are exact only when applied to what might be called an *ideal* gas. The molecules of such a gas would have no attraction for one another; nor would the molecules themselves occupy any space in the containing vessel (an impossible situation). However, real gases do not act in a completely ideal way; that is, their molecules do attract one another and they do take up some space. A given gas acts less ideally as the pressure is increased and/or as the temperature is decreased.

1. As pressure is increased, the molecules of a gas are forced closer together, thus increasing the effectiveness of the attractive forces and also increasing the proportion of the volume occupied by the molecules themselves.

2. As temperature is decreased, the molecules of a gas have less kinetic energy, thus increasing the effectiveness of the attractive forces.

The amount of deviation from ideal gas behavior depends also on the gas. Gases such as hydrogen, helium, nitrogen, and oxygen, whose molecules have relatively small attraction for one another, act more ideally at ordinary temperatures and pressures than do gases such as chlorine and ammonia, whose molecules have relatively great attraction for one another. The actual volumes per mole of fifteen gases are listed in Table 7.2. The differences between these values and 22.414 L is one measure of nonideal behavior. Deviations such as these introduce errors into the determination of molecular weights by either the gas density method or the effusion method. The more nearly ideal a gas, the more nearly it behaves as predicted by Boyle's law, Charles's law, and the other gas laws.

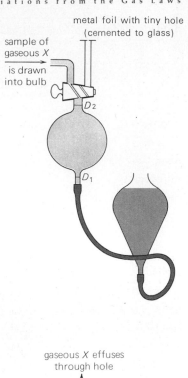

metal foil with tiny hole (cemented to glass)

sample of gaseous *X* is drawn into bulb

D_2

D_1

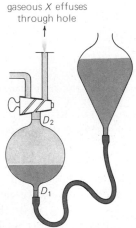

gaseous *X* effuses through hole

D_2

D_1

Figure 7•11 Molecular weight determined by effusion method. *Top:* The mercury bulb is lowered to allow the sample bulb to fill with gaseous X. *Bottom:* After the sample bulb is filled, the two-way stopcock is closed, and the mercury bulb is elevated. With the apparatus completely filled with X, the stopcock is opened to the right limb (as shown), and the time necessary for the mercury to rise from D_1 to D_2 is noted. The effusion time required for an equal volume of gas of known molecular weight is determined in a similar manner.

___

Table 7•2 Molar volume and gas constant, R, of some gases

Gas	Formula	Molar volume	$R = PV/nRT$ $\frac{(L \cdot atm)}{(mol \cdot K)}$	$\frac{(m^3 \cdot Pa)}{(mol \cdot K)}$
ideal[a]		22.414	0.082057	8.3144
hydrogen	H_2	22.428	0.082109	8.3197
helium	He	22.426	0.082101	8.3189
neon	Ne	22.425	0.082098	8.3186
nitrogen	N_2	22.404	0.082021	8.3108
carbon monoxide	CO	22.403	0.082017	8.3104
oxygen	O_2	22.394	0.081984	8.3071
argon	Ar	22.393	0.081981	8.3068
nitric oxide	NO	22.389	0.081966	8.3052
methane	CH_4	22.360	0.081860	8.2945
carbon dioxide	CO_2	22.256	0.081845	8.2930
hydrogen chloride	HCl	22.249	0.081453	8.2533
ethylene	C_2H_4	22.241	0.081424	8.2503
acetylene	C_2H_2	22.19	0.08124	8.232
ammonia	NH_3	22.094	0.08087	8.194
chlorine	Cl_2	22.063	0.08076	8.183

Source: *Adapted from* Chemical Systems: Energetics, Dynamics, Structure *by J. A. Campbell, Freeman, San Francisco. Copyright © 1970. Reproduced with permission.*

[a]All but the last two substances are "ideal" to ± 1 percent when $P \leq 1$ atm and $T \geq 273$ K.

7•11   The Kinetic Molecular Theory

The early discoveries of the behavior of gases were not easily explained, as we might expect when we recall that Torricelli and Boyle worked in the seventeenth century, whereas Dalton's theories of atoms were not forthcoming until the dawn of the nineteenth century. There were fanciful suggestions, such as the explanation of the action of the barometer on the basis of the limited strength of an invisible *funiculus* (Latin for *little cord*) that attached itself between the upper surface of the mercury and the top of the barometer tube. Boyle himself thoughtfully worked with the interesting notion that the individual particles of gases might be similar to the tangled, springy strands of sheep's wool. The intriguing gaseous property of expanding to fill uniformly all available space provoked arguments that revived an ancient Greek dilemma. Should we consider matter as continuous or particulate, that is, as infinitely subdivisible or as consisting of atoms and the void?

7•11•1   Gaseous State: Facts and Theories   From the characteristics of gases the early scientists built up a reasonable theoretical picture of the fundamental structure of a gas. Following are some experimental observations made over a century ago, paired with ideas that each supports.

Fact	Theory
A gas sample of any weight will uniformly fill a closed container. If the container is porous, the gas leaks out through openings that cannot be seen under a microscope.	Gases are made up of submicroscopic particles, called molecules, that are in rapid, random motion. A molecule moves in a straight line until it collides with another molecule or with the walls of a container. Being small, it can move through tiny pores and leave the container.
For a given weight of substance, the volume it occupies as a gas is much greater than that which it occupies as a liquid.	The molecules of a gas are widely separated at ordinary temperatures and pressures.
In a closed container a gas exerts uniform pressure (force per unit area) on every part of the wall of the container.	The pressure of the gas is the sum of the forces on the walls due to the random collisions of billions upon billions of moving molecules.
When the pressure on it is released, a gas expands.	The molecules move in random directions and have very little attraction for one another. As more space becomes available, they become more widely separated.
The volume of a certain quantity of gas can be decreased by compressing the gas.	The molecules of the gas are forced closer together by an increase in pressure.
In a closed container a gas exerts a certain pressure. As long as the volume remains the same and no heat is gained or lost, the pressure remains the same indefinitely.	The collisions between molecules are perfectly elastic; that is, there is no net change in kinetic energy. If energy were lost, the temperature and pressure would fall.
In a closed container of definite volume, a gas exerts a certain pressure as long as the temperature is not changed. If the gas is heated, the pressure rises. If the gas is cooled, the pressure falls.	As the temperature rises, the molecules move more rapidly and hit the walls harder and more often. The kinetic energy of the molecules is directly proportional to the absolute temperature.
If a gas is compressed sufficiently, and perhaps cooled at the same time, it liquefies.	The molecules of a gas do have some attraction for one another, enough to hold the particles together as a liquid under the proper conditions.
A moving body such as a golf ball or a hammer has an amount of kinetic energy that depends on two quantities: the mass of the body and its speed, or velocity. Expressed as an equation: K.E. $= \frac{1}{2}mv^2$.	The fact that the expression K.E. $= \frac{1}{2}mv^2$ holds for all known moving bodies is good reason for believing that it also holds for the moving molecules of gases.
Experiments show that dense gases diffuse more slowly than less dense gases at the same temperature.	At a given temperature, heavy molecules move more slowly than lighter ones; their average kinetic energies are equal.

7·11·2   A Molecular Model for Gas Behavior   The theoretical points just listed constitute the moving particle theory of gases, or, to use the language of the scientist, the kinetic molecular theory of gases. The theory may be summarized in terms of a model as follows:

1. Gases are composed of molecules that are widely separated from one another in otherwise empty space.

2. The molecules move about at high speeds, traveling in straight paths but in random directions.

3. The molecules collide with one another, but the collisions are perfectly elastic (result in no loss of energy).

4. The average velocity of the molecules increases as the temperature increases and decreases as the temperature decreases. Individual molecules in a sample of a pure gas are not all moving at the same speed, but for a given gas at a given temperature the average velocity is the same in all samples regardless of the pressure.

5. At a given temperature, the molecules of gases A and B have the same average kinetic energy. An increase in mass, m, is compensated for by a decrease in average velocity, v. That is, at a certain temperature, K.E. $= \frac{1}{2}m_A v_A{}^2 = \frac{1}{2}m_B v_B{}^2$. If m_A is greater than m_B, then v_A must be less than v_B.

Since these basic ideas were first developed, theories of molecules have been developed in greater and greater detail. It is now known that gross movement from place to place, *translational motion,* is just one of the possible motions molecules have. Molecules have *rotational motion,* and atoms in the same molecule continuously change their relative positions due to *vibrational motion.* Three types of motion for a diatomic molecule are diagrammed in Figure 7.12.

The kinetic molecular theory is firmly embedded in the practice and philosophy of chemistry because from its *theoretical* principles we can derive precisely each of the *experimental* gas laws. As we contrast the theories of early scientists with what we take for granted today, it becomes clear that the theories of one age may become the facts of the next. To paraphrase the physicist Percy Bridgman, today we are as sure of the existence of moving molecules as we are of our hands and feet.

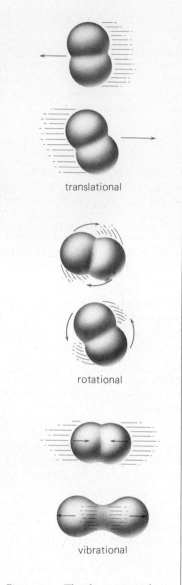

translational

rotational

vibrational

Figure 7·12 The three types of motion of diatomic molecules.

Chapter Review

Summary

All gases obey certain simple laws that provide a deep insight into the nature of matter. The compressibility of a gas is the basis for measurements of its pressure, which can be made with a **barometer.** Because atmospheric pressure is highly variable, it is useful to define a **standard atmospheric pressure** equal to 760 mmHg, or **one atmosphere,** atm. Pressures of about 1 atmosphere or less are conveniently measured with a **manometer.**

Four physical quantities—the amount of matter, the pressure, the volume, and the temperature—must be known in order to describe a gas completely. The relations among these quantities are expressed in the gas laws. For a given amount of gas, **Boyle's law** states the inverse relation between volume and pressure at constant temperature. **Charles's law** states the direct relation between volume and temperature at constant pressure. For a direct proportionality between volume and temperature to be shown numerically, the temperature must be expressed on an absolute scale such as the Kelvin scale. The lowest temperature that is physically possible is called **absolute zero**—the complete absence of temperature. Closely related to Charles's law is **Gay-Lussac's and Amontons's law,** which states the direct relation between pressure and temperature at constant volume. **Avogadro's law** equates the numbers of molecules in equal volumes of any two gases at constant temperature and pressure. The **standard conditions** most often chosen for convenience when working with gases are 273 K and 1 atm, called **standard temperature and pressure,** STP. At STP the volume of 1 mole of any gas is a constant, 22.4 L, called the **molar gas volume,** or **Avogadro's volume.**

The **ideal gas law** is a combined form of the four individual gas laws stated above. Its mathematical expression is the **ideal gas law equation,** which can be used to calculate any one of the four physical quantities mentioned above, when the other three are known. The proportionality constant in this equation is the **ideal gas constant,** R, one of the fundamental constants of nature. There are several other gas laws that are useful in doing stoichiometric calculations. **Gay-Lussac's law of combining volumes** deals with the volume ratios in which gases react with each other. **Dalton's law** deals with the partial pressures exerted by different gases in a mixture. **Graham's law** deals with the rates of diffusion of different gases.

All calculations based on the gas laws described above presuppose that the gases in question are **ideal gases,** that is, that they obey these laws perfectly. Real gases behave somewhat differently, however, particularly as the pressure is increased or the temperature is decreased. These deviations cause the real behavior to differ from the calculated ideal behavior.

The abundant facts about the properties of gases and the well-established laws describing their behavior provide the basis for one of the cornerstones of physical science, the **kinetic molecular theory,** which explains the gaseous state in terms of the motions of molecules.

Exercises

Note that variables that are not mentioned in the statement of a problem—temperature in the case of Exercise 10—are assumed to be unchanged. For Exercise 10(a), knowing that a gas tends to become warmer as it is compressed, we assume that heat energy is dissipated so that the final temperature is the same as the initial temperature.

The Three States of Matter

1. (a) With respect to the three states of matter, what is the natural state of nitrogen? Of copper? Of sugar?
   (b) Could nitrogen exist in either or both of the other two states? If not, why? If yes, what conditions are necessary?
   (c) Repeat part (b) for copper and sugar.
2. A sample of gas contains argon, hydrogen, nitrogen, carbon monoxide, water, ethane, and mercury. How many atoms are in each type of molecule?

The Pressure of Gases

3. (a) For the atmospheric pressure we usually observe in the laboratory, the barometer has a temperature correction of about 3 mm at 25 °C, because the density of mercury changes with temperature. Tell why you think the correction of 3 mm should be added to or subtracted from the barometric reading.
   (b) If the reading of the barometer shows a height of 745 mm, what percentage error is made by ignoring the correction in part (a) that should be made?
4. The density of mercury in grams per cubic centimeter is 13.5951 at 0 °C and 13.5340 at 25 °C. If a barometer is calibrated at 0 °C to read 760.00 mmHg, what will it read at 25 °C at the same atmospheric pressure?
5. (a) Suppose that a water barometer is constructed in the manner shown in Figure 7.2, except that the glass tube is very long. Based only on the densities of water and mercury (1.00 and 13.6 g/cm³, respectively), calculate the height of the water column in feet when the atmospheric pressure is 760 mmHg.
   (b) Could water be pumped from a well 35 ft deep with a suction pump? Explain.
*6. From the height of the barometer at standard pressure, show how to calculate the value and units of the pascal given in Tables A.2 and A.5 from the relationship pressure = force/area = mass × acceleration/area. The density of mercury is 1.35951×10^4 kg · m⁻³ and the acceleration of gravity is 9.80665 m · s⁻².
7. Show that the units of a pascal are kg · m⁻¹ · s⁻², based upon the definition of a pascal as being the pressure of a force of 1 newton per square meter, N/m².
8. What is the value for a standard atmosphere in kilopascals? What is the approximate average atmospheric pressure in kilopascals at Miami and at Denver (see Section 7.2.2)?

Boyle's Law

9. A quantity of hydrogen has a volume of 25 cm³ at 720 mm. What pressure will this hydrogen exert in a 75-cm³ container?
10. A 14-g sample of oxygen has a volume of 10.0 L at 740 mm. What will be its volume at each of the following pressures?
    (a) 820 mm
    (b) 600 torr
    (c) 2.00 atm
    (d) 50.0 psi
    (e) 1.00×10^6 Pa

11. A quantity of helium held in a 1-L flask has a pressure of 2.0 atm. The helium is allowed to flow through a valve into a 3-L flask that has been previously evacuated. What is the resulting pressure in the 4 L?

12. In Exercises 9–11, a fundamental assumption is made in order to solve the problems. What is that assumption?

Temperature Effects on Gases

13. Consider the manometer at the left in Figure 7.3. How will the mercury levels change under the following conditions?
    (a) The gas in the flask is warmed.
    (b) More gas is added to the flask.
    (c) The atmospheric pressure is changed to 740 mm.
    (d) The closed stopcock is opened.

14. Convert -10 °C to K; -10 °F to K; $1,000,000$ °C to K; $1,000,000$ °F to K.

15. Calculate V_2 for the following changes in temperature.
    (a) Ten liters of air is heated from 300 K to 600 K.
    (b) A volume of 250 mL of oxygen is cooled from 50 °C to 25 °C.
    (c) A balloon with a volume of 580 mL is cooled from 30 °C to 0 °C.
    (d) Five quarts of nitrogen at 0 °F is heated to 100 °F.

16. (a) In determining the temperature of a fluid in the gas thermometer (see Figure 7.5), why is it necessary to raise the bulb (right figure) so that the two mercury levels are the same?
    (b) Suppose that the resulting volume of the air is 100 mL instead of 60 mL. Using the graph of Figure 7.6, estimate the temperature of the fluid.
    (c) Calculate the temperature of the fluid, using Charles's law.

17. A 1.43-g sample of oxygen has a volume of 1.00 L at 0.00 °C and 1 atm. What will be its volume at 127 °C and 1 atm?

18. Consider Figure 7.5. Beginning with the apparatus shown on the left, the stopcock at the top is opened and air at 54.6 °C is forced in through it until the amount (weight) of air has been increased by 25.0 percent. The stopcock is then closed and a fluid of unknown temperature is circulated through the outer jacket until its temperature and that of the air in the inner jacket are the same. After adjusting to atmospheric pressure, the volume of the air is read as 110 mL. Calculate the temperature of the fluid.

19. Air confined in a household container has a pressure of 1.0 atm at 25 °C. What will be the pressure of the gas if the container is thrown into a trash fire and the temperature increases to 500 °C?

20. Two identical cylinders of nitrogen contain the same weight of nitrogen. The temperature of one cylinder is 20 °C, the other is 100 °C. If the pressure of the one at 20 °C is 3.5 atm, what is the pressure of the other one?

*21. If an automobile tire is filled to a gauge pressure of 28.0 psi in a heated garage at 70.0 °F, what is the gauge pressure after it is parked outside at 0.0 °F? The atmospheric pressure is steady at 740 mm during the period.

Avogadro's Law

22. Calculate the volume of 0.10 mole of chlorine at STP.

23. Which contains the larger number of molecules, 1 L of hydrogen at 740 mm and -13 °C, or 1 L of oxygen at 770 mm and 27 °C? How much more?

*24. (a) Calculate the approximate lifting power in kilograms of a hot-air balloon with a volume of 2,200 m³ when the air inside the balloon is at 60 °C and that outside is at 20 °C. The density of dry air at 60 °C is 1.060 g/L and at 20 °C is 1.205 g/L.
    (b) Repeat the calculation for a temperature inside the balloon of 120 °C.

25. After a hot-air balloon is full of hot gases, would it increase the lifting power of the balloon if more hot air could be forced into it so the balloon contained more hot air in the same volume? Explain.

26. Calculate the molecular weight of a gaseous compound with a density of 2.05 g/L at STP.

27. Calculate the molecular weight of an unknown gas if 250 mL of it weighs 0.500 g at STP.

28. What is the volume of 3.2 g of sulfur dioxide, SO_2, at STP?

29. Calculate the number of O_2 molecules in 1 L of air that weighs 1.29 g and is 21.0 percent oxygen by weight.

*30. The components that significantly determine the density of dry air are: 20.95 percent O_2, 78.09 percent N_2, and 0.93 percent Ar by volume. Based on these data, calculate the density of dry air at STP in grams per liter.

Ideal Gas Law Equation

31. Show by calculation that the gas constant, R, is properly expressed as
    (a) 82.1 cm³·atm·K⁻¹·mol⁻¹
    (b) 1.987 cal·K⁻¹·mol⁻¹

32. Calculate the value of the gas constant, R, in the following units:
    (a) L·mmHg·K⁻¹·mol⁻¹
    (b) J·K⁻¹·mol⁻¹
    (c) cm³·torr·K⁻¹·mol⁻¹

33. Calculate the density of the gas SF_6 in grams per liter and in SI base units for each of the following conditions:
    (a) STP
    (b) 150 °C and 1.00 atm
    (c) 0 °C and 2.00 atm
    (d) 150 °C and 2.00 atm

34. Which has the greater density at the same temperature and pressure, humid air or dry air? Justify your answer.

35. A sample of nitrogen gas occupies 80.0 mL at 25.0 °C and 1.00 atm. What will be its volume at 50.0 °C and 3.00 atm?

36. What is the volume of 0.500 mole of methane when measured at 273 °C and 1,520 torr?

37. A quantity of ethane has a volume of 200 cm^3 at 27.0 °C and 20.0 psi. Convert to SI base units and solve for the volume at STP.

38. Assuming ideal behavior, calculate the weight of hydrogen in a 75-L steel tank, if the pressure is 120 atm at 25 °C.

39. Given the following data about gases, use the ideal gas law equation to calculate the unknown quantity. Solve parts (a) and (d) by using SI base units.
    (a) Given, volume, 1.00 m^3; temperature, 273 K; and pressure, 1.013 $\times$ 10^5 Pa. How many moles of gas are present?
    (b) If 2.00 moles of gas occupy a volume of 11.2 L at a temperature of 27 °C, what is the pressure in atmospheres?
    (c) In outer space at a pressure of 1 $\times$ 10^{-6} mmHg and a temperature of 5 K, how many moles of gas are there per cubic meter?
    (d) If 0.50 mole of a gas in a volume of 5.6 L exerts a pressure of 1.013 $\times$ 10^7 Pa (100 atm), what is the temperature?
    (e) A balloon with 1.0 mole of gas is submerged in the sea at a pressure of 200 atm and a temperature of 5 °C. What is the volume?
    (f) Given, volume of scuba tank, 0.25 ft^3; weight of helium, 0.10 lb; and temperature, 25 °C. What is the pressure?

40. A 1.0-L container filled with hydrogen, H$_2$, at 20 °C and 740 mm is evacuated with a vacuum pump until the pressure gauge reads 1.0 $\times$ 10^{-4} mm (a rather poor vacuum). How many hydrogen molecules remain in the container?

*41. A 0.434-g sample of a gas, known to be 81.1 percent boron and the remainder hydrogen, is found to occupy a volume of 256 mL at a pressure of 730 mm and a temperature of 98 °C. What is the empirical formula? What is the molecular formula? What is the precise molecular weight?

42. The volume of a gas is 0.131 m^3 when measured at -73 °C and 0.500 atm. If the weight of the gas is 400 g and the empirical formula is CF$_2$, what is its molecular weight?

Molar Relationships
with Gaseous Reactants

43. (a) What volume of hydrogen at STP will combine with 12.0 g of oxygen to form water?
    (b) What is the volume of the oxygen?

(c) What weight of water is formed?
(d) What weight of hydrogen reacts?

44. How many liters of hydrogen can form at STP when 5.4 g of aluminum reacts completely with hydrochloric acid, HCl, in a single displacement reaction?

45. What volume of hydrogen at 0.80 atm and 25 °C is required to react with 0.50 g of copper(II) oxide to produce copper and water?

46. What volume of oxygen, measured at STP, is theoretically required for the complete combustion of 11.2 L of ethylene, C$_2$H$_4$, to produce carbon dioxide and water?

47. Consider the following reaction:

$$4Al + 3O_2 \longrightarrow 2Al_2O_3$$

    (a) What weight of aluminum oxide is formed per mole of aluminum that reacts?
    (b) How many cubic meters of oxygen are consumed per mole of aluminum that reacts?
    (c) If 10.0 kg of Al and 7.00 m^3 of O$_2$ at STP react, what weight of Al$_2$O$_3$ can be formed? How much of the excess reagent remains?

48. A mixture of hydrogen and oxygen in a mole ratio of 4 hydrogen to 3 oxygen is exploded in a steel bomb with an electric spark. The volume of the bomb is 125 mL and the initial total pressure of the two gases was 2.13 atm at 25 °C. Assuming that the reaction to form water proceeds so that one of the elements reacts completely, how many moles of the element in excess remain?

Dalton's Law

49. To a 2.0-L container of hydrogen at 3.0 atm is added 300 cm^3 of nitrogen measured originally at 28 psi, 0.10 mole of oxygen, and 2.0 L of sulfur dioxide measured originally at 200 kPa. Calculate the partial pressure of each gas in the mixture. Calculate the total pressure. (All measurements are at 0.0 °C.)

*50. Suppose 1 L of dry oxygen is bubbled through water and collected at the same temperature and pressure in a bottle over water. Will the volume of gas be less than, equal to, or greater than 1 L? Explain fully.

51. A quantity of helium, collected over water at 17.0 °C and 770 mm, has a volume of 0.85 L. Calculate the volume when dry at STP.

52. When a water solution containing sodium nitrite was heated with an excess of ammonium chloride, the following reaction occurred:

$$NH_4Cl + NaNO_2 \longrightarrow N_2 + 2H_2O + NaCl$$

The volume of gas evolved was 143 mL, measured over water at 22 °C and at a pressure of 740 mm. Calculate the weight of sodium nitrite in the original solution.

Graham's Law

53. Helium is found to diffuse two times faster than X. What is the molecular weight of X?

54. The ratio of the gas velocities of hydrogen and another diatomic element is $\sqrt{19}:1$. What is the formula of the element?

Deviations from the Gas Laws

55. The measured densities of ozone, O_3, and ethane, C_2H_6, at STP are 2.144 g/L and 1.357 g/L, respectively. Calculate the molar gas volume and the gas constant, R, for each. Calculate the percent deviation from the ideal values for each (see Table 7.2).

56. Calculate the ideal density of sulfur dioxide, SO_2, in SI base units when the gas is confined at 740 torr and 25 °C. Why might the actual density differ somewhat from what you calculated?

57. The expression $V/n = k$ is true with what limitations?

58. Explain the following in terms of the kinetic theory of gases:

(a) A liquid flows up to one's mouth through a straw.

(b) A quantity of gas exerts twice the pressure when it is compressed to half of its original volume.

(c) Hot air rises.

(d) When a gas is cooled sufficiently, it liquefies.

(e) When carbon dioxide is put under sufficient pressure at 20 °C, it liquefies; when oxygen at 20 °C is put under even greater pressure, it does not liquefy.

(f) A gas spreads more rapidly through a room when the temperature is 35 °C than when the temperature is 0 °C.

(g) A gas released into quiet air at Denver, Colorado, diffuses more rapidly than when released into quiet air at the same temperature at Miami, Florida.

(h) The air pressure in an automobile tire may increase considerably on driving some distance at a high speed.

(i) A bubble of air released at the bottom of a pool of water becomes larger as it approaches the surface.

Changes of State: Liquids and Solids

8

Changes of State

8•1 **General Characteristics of Changes of State**

8•1•1 Heat Capacity

8•1•2 Heat of Fusion

8•1•3 Heat of Vaporization

8•1•4 Temperature Changes and Changes of State

8•1•5 Attractive Forces and Changes of State

Liquid State

8•2 **Intermolecular Attractive Forces**

8•2•1 Instantaneous Induced Dipoles

8•2•2 Dipole–Dipole Attractions

8•2•3 Hydrogen Bonding

8•2•4 Summary of Attractive Forces

8•2•5 Other Physical Behavior of Liquids

8•3 **Liquefaction of Gases**

8•4 **Vaporization of Liquids**

8•4•1 Vapor Pressure

8•4•2 Boiling

Solid State

8•5 **Solidification of Liquids**

8•5•1 Phase Diagrams

8•6 **Crystalline Solids**

8•6•1 X-ray Studies

8•6•2 Characteristics of Crystals

8•7 **Ionic Compounds**

8•7•1 Sizes of Ions and the Radius Ratio

8•7•2 Polarization of Ions

8•8 **Solid Covalent Substances**

8•8•1 Structure of Ice and Water

8•9 **Metallic Solids**

8•9•1 Properties of Metals

8•9•2 Electrons in Metallic Bonds

8•10 **Packing of Particles**

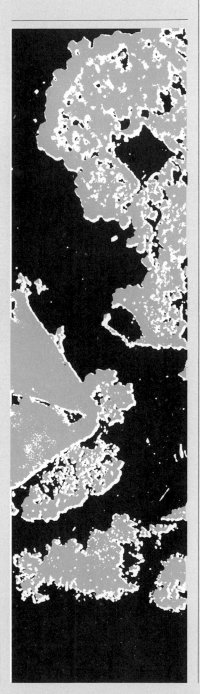

Changes in state are often encountered in chemical reactions. Sometimes a substance produced initially in the gaseous state quickly condenses to its liquid form. The energy change that accompanies a chemical reaction depends on the states of the reactants and products. Consider the burning of methane, the main constituent of natural gas, to produce carbon dioxide and water. A different amount of energy is released if the water produced is in the gaseous state than if it condenses to the liquid state. In this chapter we shall discuss the structures of liquids and solids and the changes in energy that occur when a substance changes its state.

Although reactions do take place in the gaseous and solid phases, most chemical reactions involve liquids. Both the planned reactions carried out in the laboratory or in industry and the reactions that happen in nature usually occur in the liquid phase. The great majority of the reactions that are important to living plants and animals take place in that most common liquid, water.

An understanding of the liquid and solid states of matter is essential to our understanding of most of the chemical reactions that interest us. Fortunately the kinetic molecular theory that applies to gases can be extended in a straightforward way to liquids and solids. The main differences are due to the fact that the particles in gases are widely separated from one another, whereas the particles in liquids and solids are touching.

Molecules in liquids move about more than in solids, but in both media, movements are much more restricted and slower than the movements of molecules in gases. Changes in pressure and temperature do have slight effects on the volume of a liquid or solid, but the changes are very small compared with those in gases. In fact, pressure has such a slight effect that we can consider liquids and solids to be "incompressible" by pressures up to several hundred atmospheres.

Changes of State

8•1 General Characteristics of Changes of State

For many substances the changes solid $\longrightarrow$ liquid $\longrightarrow$ gas do not involve a change in the nature of the molecules present. For example, in solid bromine, Br_2 molecules are rather effectively attracted to one another; in the liquid, Br_2 molecules roll over one another freely; and in the gas, Br_2 molecules are widely separated individual particles that bounce around at high velocities in random directions. This picture is generally applicable to covalent substances, although there may be complications. In the case of aluminum bromide, for example, the molecular form that first escapes as the gas is Al_2Br_6. At higher temperatures this molecule breaks into two $AlBr_3$ molecules.

At very high temperatures, ionic solids and metals do break up to form simple molecules, but these are very unusual states for such substances. For ionic compounds and metals, the molecular forms that exist in the gaseous state do not exist in the condensed phases. For example, when heated enough to become a gas, sodium chloride forms simple molecules such as Na_2Cl_2 or NaCl. But in solid NaCl, an ion is attracted equally to all its closest neighbors of unlike charge, with no molecular structures formed. Likewise, in the gaseous state, molecules of metals may consist of single atoms or a few atoms bound together. But in solid metals there are no definite molecules.

Although there may be changes in bonding and changes of molecular structure when a change of state occurs, a change of state is commonly referred to as a physical rather than a chemical change. Figure 8.1 summarizes the various changes of state and their names.

There are two essential concepts to keep in mind as we begin our study of the amounts of energy involved in various changes. First, the amount of energy required for an endothermic change is equal to the energy evolved in the reverse exothermic change, for example,

$$\text{solid} \underset{\text{exo-}}{\overset{\text{endo-}}{\rightleftharpoons}} \text{liquid}$$

This is the same fundamental principle that applies to chemical reactions, as discussed in Section 1.4.2. Second, the total energy for a given change in conditions is the same regardless of the path taken to achieve the change. As an example, consider the change of 1 gram of ice at −20 °C to 1 gram of gaseous water at 150 °C. Precisely the same amount of total energy would be required to cause the ice to sublime directly to the gas and then heat the gas to the final temperature as would be needed to heat the ice, melt it to liquid water, heat the liquid, vaporize it to gas, and heat the gas to the final temperature.

We speak informally of the melting point as the temperature at which a solid changes to a liquid, and the boiling point as the temperature at which a liquid bubbles consistently as it changes to a gas. Both of these important temperatures are defined more precisely later in this chapter. As we will see, under different pressures, the melting and boiling temperatures of a substance vary. The temperatures for these changes of a substance in contact with air at 1 atm are referred to as *normal melting or boiling points*. For water, they are 0 °C and 100 °C, respectively.

8•1•1   Heat Capacity   When heat is added to a solid, liquid, or gas at a temperature other than the melting point or boiling point, the temperature of the substance increases.

In order to compare the effect of heat on the temperature of different substances, we can measure the substance's **heat capacity,** the amount of heat needed to change the temperature of a given weight of a substance by 1 °C. The more a substance weighs, the larger is its heat capacity. If we have just 1 gram of a substance, the quantity of heat energy is called the *heat capacity per gram*. As defined in Section 1.2.4, this quantity is also called the *specific heat*. Table 1.4 lists the specific heats for several common substances. For example, the specific heat (heat capacity per gram) of ice is 2.00 J/(g · °C). The amount of heat needed to change the temperature of 1 mole of a substance by 1 °C is called the **molar heat capacity.** The molar heat capacity for ice is

$$[2.00 \text{ J}/(g \cdot °C)](18.0 \text{ g/mol}) = 36.0 \text{ J}/(\text{mol} \cdot °C)$$

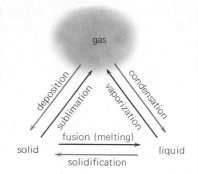

Figure 8•1 Names of changes of state. Endothermic changes are indicated by black arrows and exothermic changes by colored arrows.

8•1•2   **Heat of Fusion**   If 1.00 mole of ice at −10.0 °C is carefully heated, we find that to bring the temperature to 0 °C, it takes

$$(1.00\ \text{mol})(10.0\ °\text{C})[36.0\ \text{J}/(\text{mol} \cdot °\text{C})] = 360\ \text{J}$$

If more heat is added, the ice begins to melt but the temperature does not change. In fact, 6,010 J, or 6.01 kJ, is required to change 1 mole of ice at 0 °C to liquid water at 0 °C. This heat energy is used in counteracting part of the intermolecular attractive forces between molecules. The amount of heat needed to change 1 mole of a solid substance at its melting point to the liquid at the same temperature is called the **molar heat of fusion.** The molar heats of fusion for a number of substances are given in Table 8.1.

Table 8•1 **Molar heats of fusion and vaporization**

Substance	Molar heat of fusion, kJ/mol	Molar heat of vaporization, kJ/mol
ammonia	5.76	23.26
benzene	9.84	30.82
bromine	10.54	30.0
carbon dioxide	7.95	16.23[a]
hydrogen	0.12	0.904
hydrogen chloride	2.11	16.15
mercury	2.30	59.15
oxygen	0.44	6.82
water	6.01	40.66

[a]Subliming from solid carbon dioxide at −60 °C.

The amount of heat energy given up when a mole of a substance solidifies, called the **molar heat of solidification,** is the same as the amount taken up when a mole of solid melts. For example, 6.01 kJ is liberated to the surroundings when 1 mole of water freezes.

8•1•3   **Heat of Vaporization**   When 1 mole of water at 100 °C is vaporized, 40,660 J, or 40.66 kJ, of heat must be added to change this water to 1 mole of steam. Just as in the case of fusion (melting), the temperature is not increased during vaporization. The 40.66 kJ of heat energy is used in counteracting the attractive forces so that molecules of water can break away from each other as gas molecules (steam). The amount of heat required to change 1 mole of a liquid substance to a gas at the same temperature is called the **molar heat of vaporization.** The molar heats of vaporization of a number of substances are given in Table 8.1. The heat of vaporization is usually determined at the normal boiling point of the substance unless otherwise stated.

The amount of heat energy given up when a mole of a substance condenses, called the *molar heat of condensation*, is the same as the amount taken up when a mole of the substance vaporizes. For example, 40.66 kJ is liberated to the surroundings when 1 mole of steam condenses at 100 °C.

Figure 8.2 shows the result of continually adding heat energy to the element bromine over a wide range of temperatures. As heat energy is

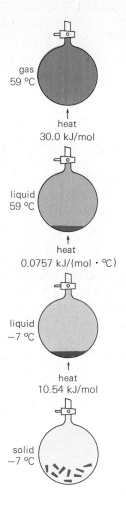

gas
59 °C

↑
heat
30.0 kJ/mol

liquid
59 °C

↑
heat
0.0757 kJ/(mol · °C)

liquid
−7 °C

↑
heat
10.54 kJ/mol

solid
−7 °C

Figure 8•2 Heat requirements to change solid bromine at its melting point to a gas at its boiling point. Enough air is in the bulb at all times to make a total pressure of 1 atm. More complex apparatus than that shown here is needed to measure separately the processes of melting, heating, and vaporizing.

progressively changed to kinetic energy, the intermolecular forces are overcome. Two abrupt changes in state are observed (1) when the solid bromine melts and (2) when the liquid bromine boils. Compare Figure 8.3 on Plate 3, which shows the deposition and sublimation of bromine in a sealed bulb at reduced pressure. Example 8.1 is a calculation of the amount of energy required to change solid bromine to vapor.

• Example 8•1 •

Calculate the amount of heat required to change 10.0 g of solid bromine at −7.0 °C to vapor at 59.0 °C. (See Figure 8.2 for data.)

• Solution • In 10.0 g of bromine, Br_2, there is

$$10.0 \text{ g} \times \frac{1 \text{ mol}}{159.8 \text{ g}} = 0.0626 \text{ mol}$$

To melt the bromine:

$$(10.54 \text{ kJ/mol})(0.0626 \text{ mol}) = 0.660 \text{ kJ}$$

To heat the bromine from −7.0 to 59.0 °C:

$$[(0.0757 \text{ kJ/(mol} \cdot °C)](0.0626 \text{ mol})(66.0 °C) = 0.313 \text{ kJ}$$

To vaporize the bromine:

$$(30.0 \text{ kJ/mol})(0.0626 \text{ mol}) = 1.88 \text{ kJ}$$

Total heat required:

See also Exercises 1 and 3–5 at the end of the chapter.

$$0.660 \text{ kJ} + 0.313 \text{ kJ} + 1.88 \text{ kJ} = 2.85 \text{ kJ}$$

8•1•4   Temperature Changes and Changes of State   According to the kinetic molecular theory, as heat energy is added to a substance, the energy is used to overcome the attractive forces holding the particles together. The higher the temperature, the greater is the kinetic energy of the particles. At the points of changes in state, large amounts of energy are involved although the temperature is constant. Also, if the change in state involves an increase in volume, energy must be used in pushing back the atmosphere. These facts are shown graphically in Figure 8.4 for the two substances water and bromine. The stair-step form of these plots is typical, with the horizontal portions showing the addition of heat at constant temperature at the melting and boiling points. Water has an exceptionally high heat capacity, heat of fusion, and heat of vaporization compared with most other substances of comparable molecular weight.

The slope of the line in Figure 8.4 that describes the change in temperature as heat energy is added to a single phase is the heat capacity per gram or the specific heat of that phase. For liquid water, the specific heat is 418 J/(1 g × 100 °C), or $4.18 \text{ J} \cdot \text{g}^{-1} \cdot °\text{C}^{-1}$ (compare with Table 1.4). The steeper slope for the liquid bromine line tells us that the specific heat of bromine is less than that of water. A calculation confirms this: specific heat of bromine is 31.3 J/(1 g × 66 °C), or 0.47 J · g^{-1} · °C^{-1}.

8•1•5   **Attractive Forces and Changes of State**   The attractive forces that influence melting and vaporization vary greatly for different classes of substances. Three broad classifications of substances are covalent, ionic, and metallic. Most covalent substances are composed of molecules attracted to one another by rather weak forces. The ions in ionic substances and the atoms in metals are usually attracted to each other by strong forces.

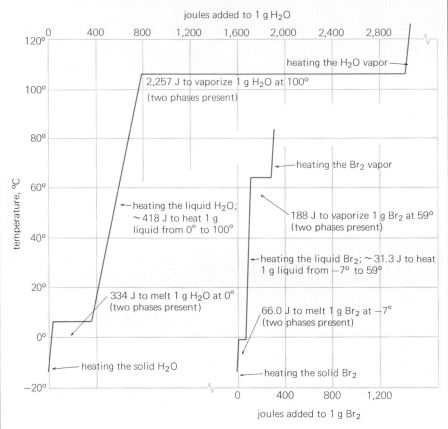

Figure 8•4 Plot of temperature versus joules added for 1 g of H_2O and 1 g of Br_2, both initially at −20 °C.

Because of these differences, covalent substances that form definite molecules tend to have low melting and boiling points. Ionic compounds have high melting and boiling points. The temperature ranges for metals are similar to those for ionic compounds, although there are a number of exceptions, such as mercury, gallium, and the easily melted metals in group IA of the periodic table.

Up to a few hundred degrees above room temperature, most ionic compounds and metals exist as solids. Near room temperature, covalent compounds may exist as solids, liquids, or gases, depending on the strength of the intermolecular forces. All substances that are liquids near room temperature are covalent substances, for example, water, alcohols, and petroleum products.

Liquid State

8•2 Intermolecular Attractive Forces

In discussing intermolecular forces in this chapter, we will limit most of our considerations to the attractions between two or more molecules of the same pure substance. However, molecules of different substances also attract each other. We will take up these cases again in Chapter 10, when we discuss solutions and other mixtures.

When a liquid covalent substance evaporates, molecules break away from their neighbors. The weak attractive forces between molecules are overcome, but the strong covalent bonds that join atoms within the molecule are not broken (see Figure 8.5). In this section we will consider three types of intermolecular attractive forces. Two of these are collectively called **van der Waals attractive forces.** The weaker of these two attractions is due to instantaneous induced dipoles and exists between all molecules, even nonpolar molecules. The stronger van der Waals attraction, called dipole–dipole, exists between molecules that have permanent dipole moments. The third type of attraction is stronger than either of the van der Waals forces. It exists only between certain molecules and is called hydrogen bonding.

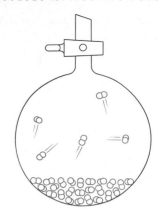

Figure 8•5 A schematic diagram of liquid and gaseous bromine. Compare Figure 8.2. The Br_2 molecules break away from each other as they go from the liquid to the gaseous state, but the Br—Br covalent bonds are not broken.

8•2•1 Instantaneous Induced Dipoles
There are attractions between the electrons of one molecule and the nuclei of adjacent molecules that are thought to be due to shifts in the positions or vibrations of the electrons and nuclei. A vibration in one molecule induces a shift in the electrons of a neighbor, as shown in Figure 8.6. When many molecules are together, as in the liquid state, these shifts are synchronized so there is a net attraction between many neighbors. The induced dipoles are said to be instantaneous because the vibrations occur billions of times a second. An instant later there may be no dipole present, or the direction of polarity may even be reversed.

The weak attractions due to instantaneous induced dipoles were described first in the 1930s by the German physicist Fritz London, so they are called **London forces.** It is these London forces that are responsible for attractions between nonpolar compounds. Large molecules are more effectively attracted to one another than are small ones. We can compare methane, CH_4, with propane, $CH_3CH_2CH_3$:

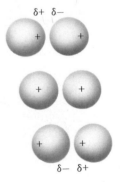

Figure 8•6 Schematic diagram of vibrations of electrons relative to the nuclei in two atoms of a noble gas. The symmetrical atoms (*center*) are nonpolar, but vibrations in electrons can induce instantaneous dipole attractions between these neighbors. Note that the positions of the nuclei do not change.

```
     H                H  H  H
     |                |  |  |
  H—C—H           H—C—C—C—H
     |                |  |  |
     H                H  H  H
  methane             propane
```

Two molecules of propane have more attractive interactions with one another than do two molecules of methane. Molecules with large and diffuse electron distributions are attracted to one another more than are molecules with more tightly held electrons. For example, compare iodine, I_2, with fluorine, F_2. The large iodine molecules are attracted to each other more strongly than are the smaller fluorine molecules.

Substances with molecules that are attracted to one another only by London forces have low boiling and low freezing points, compared with other substances of about the same molecular weight. If their molecules are small, they usually exist as gases at room temperature.

8.2.2 **Dipole–Dipole Attractions** Molecules that have permanent dipole moments are said to be *polar*. Examples are shown in Figure 5.6 and 5.11. The attractive force between two polar molecules is called a **dipole–dipole attraction.** This attraction is stronger than the attraction between nonpolar molecules. Hence, substances composed of polar molecules tend to have higher melting and boiling points than nonpolar molecules of about the same size.

8.2.3 **Hydrogen Bonding** Unusually strong intermolecular attractions can exist between molecules if one of the molecules has a hydrogen atom bound to an atom of high electronegativity, and the neighboring molecule contains an atom of high electronegativity that has an unshared pair of electrons. The hydrogen nucleus, a proton, is attracted by the neighboring electron pair and oscillates back and forth between the two atoms. The attraction between two molecules that share an oscillating proton is called **hydrogen bonding.** Although hydrogen bonding might be considered an extreme case of dipole-dipole interaction, it is sufficiently different to merit a separate discussion.

Strong hydrogen bonds are formed only by molecules that contain nitrogen, oxygen, or fluorine atoms. It appears that a lone pair of electrons in a small atom is more effective in attracting a neighboring hydrogen atom than is a lone pair in a larger atom. For example, although nitrogen and chlorine have the same electronegativity, the smaller nitrogen atom forms much stronger hydrogen bonds. Three pure substances with properties greatly influenced by hydrogen bonding are water, H_2O, ammonia, NH_3, and hydrogen fluoride, HF.

In the case of water, hydrogen bonding is very effective:

$$H\!-\!\overset{..}{\underset{H}{O}}\!: \;+\; H\!-\!\overset{\overset{\textstyle H}{|}}{\underset{..}{O}}\!: \;\longrightarrow\; H\!-\!\overset{..}{\underset{H}{O}}\!:\text{---}H\!-\!\overset{\overset{\textstyle H}{|}}{\underset{..}{O}}\!:$$

Because each simple water molecule contains two hydrogen atoms and two unshared pairs of electrons, this hydrogen bonding can continue in three dimensions until large aggregations are formed. For example, H_4O_2 can react with another molecule of H_2O to form H_6O_3:

$$H\!-\!\overset{..}{\underset{H}{O}}\!:\text{---}H\!-\!\overset{\overset{\textstyle H}{|}}{\underset{..}{O}}\!: \;+\; H\!-\!\overset{..}{\underset{H}{O}}\!: \;\longrightarrow\; H\!-\!\overset{..}{\underset{H}{O}}\!:\text{---}H\!-\!\overset{\overset{\textstyle H}{|}}{\underset{..}{O}}\!:\text{---}H\!-\!\overset{..}{\underset{H}{O}}\!:$$

For liquid water it is not presumed that definite H_4O_2 or H_6O_3 molecules exist; rather the picture is one of loose associations, with hydrogen bonds being made and broken as the molecules change partners and move about. (See also Section 8.8.1.) The degree of molecular association increases as the temperature decreases, so the formula $(H_2O)_x$ is probably better. Because x is indefinite, chemists continue to write the formula as H_2O.

Studies of liquid ammonia, including both X-ray and molecular weight determinations, indicate that it behaves similarly to water.

Hydrogen fluoride molecules are attracted to one another so strongly that they form zigzag chains of indefinite length that act like larger molecules. X-ray measurements reveal the following structure in solid HF, and it is thought that in the liquid there are chains with similar geometry:

$$1.57\text{Å} \qquad 0.92\text{Å}$$

$$F\text{---}F\text{---}F$$
$$\angle = 120°$$

8•2•4   **Summary of Attractive Forces**   Differences in the attractive forces that the molecules of pure substances have for one another are reflected in the melting and boiling points of these substances. In general, strong attractive forces and large sizes of molecules both lead to high melting and boiling points.

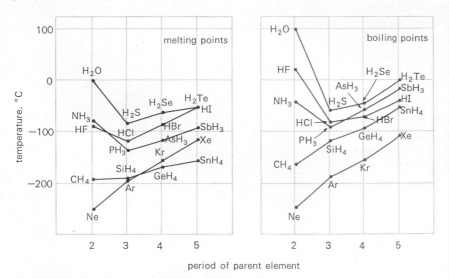

Figure 8•7 The melting and boiling points of the noble gases and the hydrogen compounds of groups IVₐ, Vₐ, VIₐ, and VIIₐ.
Source: Redrawn by permission from Linus Pauling, *The Nature of the Chemical Bond,* 3rd ed. Copyright 1939 and 1940, third edition © 1960, by Cornell University. Used by permission of Cornell University Press.

In Figure 8.7 the melting and boiling points for five series of substances are plotted. In two of these series, Ne–Xe and CH_4–SnH_4, nonpolar molecules are attracted to one another by instantaneous induced dipoles or London forces. In these two series it is interesting to compare the differences in the boiling points of pairs of molecules with about the same molecular weight. Consider Ne versus CH_4 and Kr versus GeH_4. The noble-gas molecules have simple spherical electron distributions, as shown in Figure 8.6, but the hydrogen compounds of the group IVA elements are lumpy tetrahedra (see CH_4 in Figure 6.4) that attract one another more strongly. Consequently, the boiling points for CH_4 and GeH_4 are higher than those for Ne and Kr, respectively.

Comparing molecules of different structure but roughly the same molecular weight, we can look at the group Xe, SnH_4, HI, SbH_3, and H_2Te or the group Kr, GeH_4, HBr, AsH_3, and H_2Se. In each group the boiling points of the three polar substances are higher, showing that these molecules require greater kinetic energies to break away from each other.

Finally, we note in Figure 8.7 the striking behavior of NH_3, HF, and H_2O. Their high boiling (and melting) points, compared with PH_3, HCl, and H_2S, are due to the strong hydrogen bonds between their molecules.

• Example 8•2 •

What would the boiling point of ammonia, NH_3, be if it followed the trend of the other group VA hydrogen compounds? Approximately how much difference is there between the actual boiling point and the predicted one?

• Solution • Extrapolation of the line for SbH_3, AsH_3, and PH_3 in Figure 8.7 intersects the vertical line for period 2 at about $-125\ °C$, so this would be the predicted boiling point. This is 90 °C lower than the approximate plotted value of $-35\ °C$. (The precise normal boiling point of NH_3 is $-33.4\ °C$.)

See also Exercise 11.

8•2•5   **Other Physical Behavior of Liquids**   Intermolecular attractions affect many properties of liquids other than melting and boiling points. The heat of fusion and the heat of vaporization are two such properties. Also related to intermolecular attractions is **surface tension,** the force that makes a liquid tend to form a spherical drop or to form a curved surface, or *meniscus,* where it comes in contact with a container (see Figures 8.8 and 8.9). In the case of both the drop and the meniscus, the curved surface that is formed has the smallest area possible under the circumstances and therefore minimizes the surface energy.

The intermolecular forces between molecules of the same substance or similar substances may be called **cohesive forces.** The forces between molecules of unlike substances, and particularly between liquids or gases on the one hand and solids on the other, may be called **adhesive forces.**

Water molecules are strongly attracted to bound —O—H groups in the surface of glass; thus water tends to creep up the sides of a glass container. Due to cohesive intermolecular attractions, the molecules that creep up the sides draw other water molecules up with them.

The rise of a column of liquid in a small tube is called **capillary rise.** The height of the column is related to the radius of the tube, the surface tension of the liquid, and the weight of the column of liquid that is drawn up. By measuring the height, the radius of the tube, and the weight of the column, the surface tension can be calculated. Because of strong cohesive intermolecular attractions, water has a high surface tension compared with other substances.

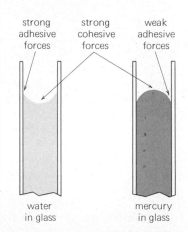

Figure 8•8 Schematic diagram of the concave meniscus formed by water and the convex meniscus formed by mercury in small glass tubes.

Mercury is an example of a substance with cohesive intermolecular forces that are greater than its adhesive forces for glass. The result is that, when it is in contact with glass, liquid mercury forms a convex meniscus (see Figure 8.8).

8•3   Liquefaction of Gases

A gas can be condensed or liquefied by the proper combination of decreasing the temperature and/or increasing the pressure. In the discussion of the absolute scale of temperature in Section 7.4.1, it was pointed out that the decrease in the volume of a gas with a decrease in temperature follows Charles's law until the temperature falls near the point at which the gas begins to condense to a liquid. According to kinetic theory, if the kinetic energy of the gas molecules is lowered by sufficiently decreasing the temperature, the intermolecular forces will be effective in holding the particles in the liquid state. Also, crowding gas molecules together by increasing the pressure makes intermolecular forces more effective. If the molecules are far apart, the attractions are weak, but as molecules come close to one another, the attractions increase. The gas liquefies if the attractive forces are great enough.

However, for each gaseous substance there is a temperature, called the **critical temperature,** above which the gas cannot be liquefied no matter how much pressure is applied. The pressure that must be applied to bring about liquefaction when a gas is at its critical temperature is called the **critical pressure.** The nonpolar molecules of such gases as hydrogen, oxygen, and nitrogen have relatively small attraction for one another. The kinetic energy of molecules of these gases must be decreased a great deal before the slight attractive forces (made more effective by the application of pressure) can hold them in the liquid form. Their critical temperatures are quite low (Table

Figure 8•9 If a droplet of liquid is small enough, it will assume a spherical shape on falling freely in still air. The attractive forces between molecules (indicated by the colored arrows) have the net effect of producing a surface tension that makes the total surface area as small as possible. The balancing of these forces results in the droplet's spherical shape.

Table 8•2 Critical temperatures and pressures

Gas	Normal boiling point, °C	Critical temperature, °C	Critical pressure, atm
hydrogen	−252.8	−239.9	12.8
nitrogen	−195.8	−146.9	33.5
oxygen	−183.0	−118.4	50.1
methane	−161.5	−82.6	45.4
carbon dioxide	(−78.5)[a]	31.0	72.9
propane	−42.1	96.8	42.0
ammonia	−33.4	132.4	111.3
freon-12	−29.8	111.5	39.6
sulfur dioxide	−10.0	157.5	77.9
butane	−0.5	152.0	37.5
water	100.0	374.2	218.3

[a] At 1 atm pressure, carbon dioxide does not exist as a liquid, so it does not have a normal boiling point. Rather than melting and boiling, it sublimes at 1 atm at −78.5 °C.

8.2). On the other hand, polar molecules of such gases as ammonia and sulfur dioxide have relatively great attraction for one another. These attractive forces hold the molecules in the liquid state at temperatures well above room temperature if sufficient pressure is applied.

The substances in Table 8.2 can be divided into two classes. The first four cannot be liquefied at room temperature, regardless of the pressure that is applied. The next seven have higher critical temperatures and so can be liquefied at average room temperatures by increasing the pressure alone.

8•4   Vaporization of Liquids

Liquids that evaporate readily consist of molecules that have weak intermolecular forces; they tend to scatter because of their own motion. As shown in Figure 8.10(a), some molecules escape from the body of the liquid (vaporize) if they happen to be headed upward with sufficient velocity to overcome the weak attractive forces. **Vapor** is the name given to the gaseous state of a substance at a temperature and pressure at which we ordinarily think of the substance as a liquid or solid. A liquid that evaporates readily is said to be **volatile.** Ethyl ether is a very volatile liquid; lubricating oil is only slightly volatile.

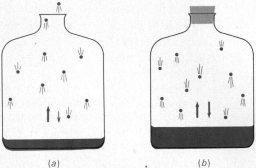

(a)   (b)

Figure 8•10 (a) Evaporation from an open container: more molecules of the substance are escaping than are returning. (b) Evaporation in a closed container at saturation or equilibrium: the number of molecules escaping from the liquid per unit time equals the number returning.

8•4•1   Vapor Pressure   Water escapes from a vessel by evaporation only if the vessel is open. If the vessel is closed, the process of evaporation continues, but the molecules that escape from the surface of the liquid are trapped in the container. As they collide with one another and the walls of the container, they may hit the surface of the liquid and rejoin it; that is, the molecules condense. The bottle of water shown in Figure 8.10(b) represents a system in which the liquid is evaporating and the vapor is condensing at the same rate. An **equilibrium** exists when an action in one direction is just balanced by an action in the reverse direction.[1]

The **vapor pressure** of a substance is defined as the pressure exerted by the gas of that substance when it is in equilibrium with the liquid or solid phase. The vapor pressures of liquids (or solids) increase as the temperature

---

[1] A system in which the rate of a process in one direction is equal to the rate of a reverse process can be referred to as a *dynamic equilibrium* to contrast it with the state of *static equilibrium.* Static equilibria are of interest in many mechanical systems, for example, a pendulum hanging at rest in its lowest position.

increases. Figure 8.11 shows one method of determining the vapor pressures of liquids.

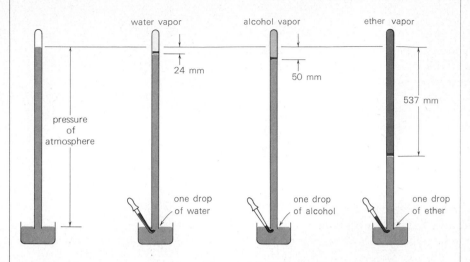

Figure 8•11 Determination of the vapor pressures of liquids at 25 °C. When a drop of a liquid is introduced into a barometer, it rises to the top of the mercury column and part of it evaporates. The space above the mercury quickly becomes saturated; that is, an equilibrium is established between the liquid and vapor phases of the sample. The pressure exerted by the vapor forces the mercury column to fall a distance that is a direct measure of the vapor pressure. The barometer at the left has a Torricellian vacuum above the mercury.

Figure 8.12 shows how the vapor pressure changes with temperature for four liquids. See Table A.7 in the appendix for data for water.

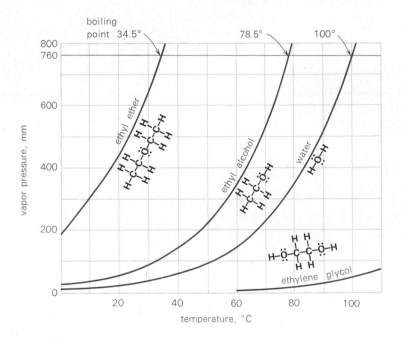

Figure 8•12 A plot showing the effect of temperature on the vapor pressures of four common liquids. The temperature at which the vapor pressure is 760 mm is the normal boiling point of the liquid.

The four compounds shown in Figure 8.12 illustrate the effect of hydrogen bonding on vapor pressures. Water, HOH, though of a lower molecular weight than ethyl alcohol, C_2H_5OH, has a lower vapor pressure at every temperature because of the effectiveness of its hydrogen bonding. Ethylene glycol, $HOCH_2CH_2OH$, with two —OH groups per molecule, forms hydrogen bonds much more effectively than ethyl alcohol and probably forms long chains. Hydrogen atoms bound to carbon, as in ethyl ether, $C_2H_5OC_2H_5$, generally are not positive enough to function effectively in hydrogen bonding. The high vapor pressure of ethyl ether shows how weak its attractive forces are.

• *Relative Humidity* • The moisture content of the air is commonly expressed in terms of relative humidity. The **relative humidity** is defined as the percent saturation of the air with water vapor. In Figure 8.10(*b*), for example, if the liquid is water, the relative humidity in the closed container is 100 percent.

Relative humidity is an important factor for our comfort. When the humidity is high, perspiration does not evaporate readily. On hot days this causes us to feel uncomfortable, because the body cannot cool itself efficiently by evaporation.

The relative humidity of a given sample of air decreases when the temperature increases. Therefore, in cold weather, the heated air inside buildings has a low relative humidity if the outside air is simply brought inside and warmed. This low relative humidity causes undesired heat loss due to the evaporation of water from the body, as well as excessive drying of the skin and mucous membranes.

When moist air is cooled in the absence of solid surfaces on which it can condense easily, the partial pressure of water vapor may exceed the vapor pressure of water at that temperature. Such air is said to be *supersaturated* with water vapor. The system is *metastable*,[2] and if it is disturbed a cloud or fog of tiny water droplets may suddenly condense. Tiny crystals of silver iodide, AgI, under the proper atmospheric conditions can cause rain clouds to form in supersaturated air.

• **Example 8.3** •

If the partial pressure of water vapor in the air is 12.8 mm and the temperature is 22.0 °C, what is the relative humidity?

• **Solution** • Referring to Table A.7, we find that the vapor pressure of water at 22 °C is 19.83 mm. This is the partial pressure of water vapor in air saturated with water vapor at that temperature.

$$\text{relative humidity} = \frac{\text{pressure of water vapor}}{\text{pressure of saturated water vapor}} \times 100$$

$$= \frac{12.8 \text{ mm}}{19.83 \text{ mm}} \times 100 = 64.5\%$$

See also Exercises 18–22.

• 8.4.2   **Boiling** The **boiling point** of a liquid is the temperature at which the pressure of vapor escaping from the liquid equals the external pressure. When the vapor pressure equals the external (or applied) pressure, bubbles of vapor begin to form in the liquid. Because the pressure of the vapor in the bubble equals the pressure of the atmosphere, the bubble can push through the surface and move into the gas phase above the liquid. The liquid boils.

---

[2] A **metastable system** is a system that is apparently stable but is not at equilibrium.

The boiling point of water (and other liquids) varies with the atmospheric pressure. As previously mentioned, the **normal boiling point** of a liquid is the temperature at which its vapor pressure is 1 atm. In mountainous regions the boiling point of water is considerably below 100 °C, because the atmospheric pressure is below 1 atm (see Table 8.3).

Table 8•3 **Changes in boiling point of water with altitude**

	Altitude, m	Average atmospheric pressure, mm	Boiling point of water, °C
sea level	0	760	100
Mt. Mitchell (NC)	2,037	589	93
Mt. Whitney (CA)	4,418	451	86
Mt. Everest (Asia)	8,882	244	71

At high altitudes, one must boil foods longer because of the low boiling temperatures. Conversely, foods cook more rapidly in pressure cookers, because they can be heated above the normal boiling point. For example, a pressure of about 0.70 atm (10 psi) more than the standard atmospheric pressure of 1 atm (14.7 psi) raises the boiling point of water to about 115 °C.

• Example 8•4 •

If the average atmospheric pressure in your hometown is 740 mm, what is the boiling point of water in an open pan? What is the boiling point of water in a pressure cooker operated at about twice the atmospheric pressure?

• Solution • From Table A.7 the boiling point of water at 740 mm can be calculated by interpolating between the boiling points at 738.5 mm and 743.9 mm:

$$\text{b.p. at 740 mm} = 99.2\ °C + (99.4 - 99.2)\ °C \times \frac{(740 - 738.5)\ \text{mm}}{(743.9 - 738.5)\ \text{mm}}$$

$$= 99.2 + 0.06 = 99.3\ °C$$

At twice 740 mm, or 1,480 mm, we see in the table:

$$\text{b.p. at 1,480 mm} \simeq \text{b.p. at 1,489.1 mm} = 120\ °C$$

See also Exercise 25.

• *Superheating* • When a liquid is heated in a vessel with a smooth inner surface, bubbles may not form easily when the boiling point is reached. If the liquid is heated to a temperature above its boiling point, it is said to be *superheated*. Such a liquid is in a metastable state, and it may suddenly and violently form bubbles or "bump." Bumping can be dangerous in the laboratory. One way to avoid it is to add to the liquid, before heating, some boiling chips made of small pieces of insoluble, unreactive solids, which provide irregular surfaces on which bubbles form easily.

Solid State

8•5 Solidification of Liquids

The **melting point** (or **freezing point**) of a substance is the temperature at which its solid and liquid phases are in equilibrium. If such an equilibrium is disturbed by the addition or withdrawal of heat energy, the system will change to form more liquid or more solid. However, the temperature will remain at the melting point as long as two phases are present (see Figure 8.4).

As pointed out in Section 8.4.2, the boiling point of a liquid changes markedly with changes in the opposing pressure. However, small differences in pressure, such as changes in the atmospheric pressure, have a negligible effect on the freezing point of a liquid. A large increase in pressure does favor the state with the smaller volume. For most substances, the solid state is more dense (smaller volume per given weight) than the liquid state. For a few substances, water and bismuth as examples, the liquid state is more dense.

8•5•1 Phase Diagrams

The relationships between the states or phases of a substance can be summarized in a *phase diagram* that shows what phases are present at different temperatures and pressures. Figure 8.13 is a phase diagram for H_2O over a limited range of temperatures and pressures. The following observations can be made on the basis of the figure:

1. Along the line AC, liquid and gaseous H_2O are in equilibrium.

2. Consider a point P on line AC. If the pressure is raised, say, to Q, gaseous H_2O will condense, and only liquid water will be present. If the pressure is lowered, say, to R, all the liquid will evaporate, and only gaseous H_2O will be present. We can generalize that for a temperature and pressure corresponding to any point within one of the three regions, only one phase can be present. For example, at the temperature and pressure of point X, only solid H_2O is present.

3. Along the line AB, gaseous and solid H_2O are in equilibrium, for example, at point W.

4. Line AD is the equilibrium line for the liquid and solid phases. The line is almost vertical, but its slant to the left is exaggerated in the figure to make it clear that the temperature for the normal freezing point of water, 0 °C, is not identical with the temperature for point A.

5. Point A is called a **triple point,** a condition of pressure and temperature at which all three phases are in equilibrium.

• *The Triple Point of Water* • Variations in air pressure affect the freezing point of water and water solutions slightly, both because of the change in the solubility of air in water and because of the influence of pressure on a change of state. Because water expands on freezing, an increase in pressure tends to prevent the freezing; that is, the freezing point is depressed by an increase in pressure (see Le Chatelier's principle, Section 10.4.1). Dissolved air also lowers the freezing point. With only pure H_2O

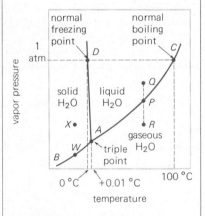

Figure 8•13 A phase diagram for H_2O. (This diagram is not drawn to scale.)

present the equilibrium between the three phases, solid, liquid, and vapor, is established at the **triple point of water,** a basic fixed point on the International Temperature Scale defined as 273.16 K. In the presence of air at 1 atm, the increase in pressure from the water vapor pressure alone of 4.58 mm to 760 mm causes a drop of 0.0075 °C in the freezing point, and the dissolved air lowers the freezing point an additional 0.0025 °C (0.0024 due to dissolved oxygen and nitrogen). The so-called normal freezing point of water is therefore 273.16 − (0.0075 + 0.0025) = 273.15 K = 0 °C. See Figure 8.14.

8•6   Crystalline Solids

Crystals are solid bodies bounded by planar surfaces. Because many solids such as rock salt, quartz, and snowflakes exist in forms that are obviously symmetrical, scientists long suspected that the atoms, ions, or molecules of these solids were also arranged symmetrically.

• *Appearance of Crystals* • We must not jump to conclusions about the arrangement of the particles inside a large crystal from a consideration of its outer appearance. When a substance in the liquid state or in solution crystallizes, the crystal may form by growing more in one direction than another. Figure 8.15 shows how a small cube may develop into one of three other possible shapes—a large cube, a flat plate, or a long needle-like structure. All three of these solids have the same cubic crystal structure but their overall shapes differ.

• *Isomorphism* • From the Greek word *morphe*, meaning form, we derive the word ending -*morphous*. Two substances that have the same crystal structure are said to be **isomorphous** (Greek *isos*, equal, and *morphe*, form). The formulas of a pair of such substances usually reveal that their atoms are in the same ratios. Examples are:

NaF and MgO	1:1	K_2SO_4 and K_2SeO_4	2:1:4
Cr_2O_3 and Fe_2O_3	2:3	$NaNO_3$ and $CaCO_3$	1:1:3

Isomorphous substances may or may not crystallize together in homogeneous mixtures. However, similarity of both formula and chemical properties is not enough to insure homogeneous crystallization. Two well-known similar substances that do not crystallize homogeneously are NaCl and KCl.

A single substance that crystallizes in two or more different forms under different conditions is said to be **polymorphous** (many forms). Calcium carbonate, $CaCO_3$, silicon dioxide, SiO_2, sulfur, S, and carbon, C, are examples of polymorphous substances. For example, graphite and diamond are polymorphic forms of carbon. In the case of elements, polymorphic forms are also called *allotropic forms* (see Section 9.2.1).

A substance that appears to be solid but has no well-developed crystal structure is said to be **amorphous** (without form). Tar and glass are such solids. Unlike crystalline solids, amorphous substances do not have sharp, definite melting points. Instead, they soften gradually as they are heated and they melt over a range of temperature.

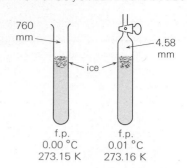

760 mm
4.58 mm
ice
f.p. 0.00 °C 273.15 K
f.p. 0.01 °C 273.16 K

Figure 8•14 The freezing point of water in air at 1 atm is 0.00 °C. In a closed container with only H_2O present, the freezing point is 0.01 °C and the pressure of the water vapor is 4.58 mm. (The air is removed from the right-hand container with a vacuum pump, the stopcock is closed, and the tube is cooled until ice crystals form.)

growth in three dimensions

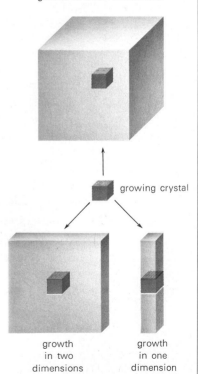

growing crystal

growth in two dimensions

growth in one dimension

Figure 8•15 Three shapes that may be assumed by a substance that crystallizes in a cubic system. *Source:* Based on A. F. Wells, *Structural Inorganic Chemistry,* The Clarendon Press, Oxford, 1945.

8•6•1   X-ray Studies   The regular three-dimensional array of similar points in a crystalline solid is called a **crystal lattice.** In 1912 the German physicist Max von Laue suggested that, if X rays were a type of electromagnetic radiation like light, the X rays should be of the proper wavelength to be diffracted by atoms or ions in a crystal lattice. Experimental tests of this hypothesis showed that an X ray did behave as electromagnetic radiation of short wavelength. This discovery opened a new field of research in measuring distances between planes of atoms or ions. These measurements made possible the calculation of the sizes and arrangements of these small particles.

The simplest case to consider is that in which the X radiation has a single wavelength; that is, the radiation is **monochromatic.** When a monochromatic beam of X rays falls on a crystalline substance, the radiation is strongly diffracted only when the crystal is turned at certain angles to the incident beam. Two English physicists, the father and son team of W. H. and W. L. Bragg, showed that the X ray can be thought of as being reflected at certain angles and that these angles are related to the wavelength of the X rays and to the distances between the parallel planes of atoms or ions in the crystal.

Figure 8.16 is a schematic representation of the relationship discovered by the Braggs. Waves whose maxima and minima occur together are said to be *in phase.* When the X rays are reflected, the waves that are in phase at *ABC* reach *A'B'C'* in phase. This is possible only if the differences in the distances *AEA'*, *BGB'* , and *CKC'* are whole-number multiples of the wavelength λ. In the figure, the difference between the distances represented by *AEA'* and *BGB'* is *FGH*, a distance of 1λ. The difference between *AEA'* and *CKC'* is *JKL*, a distance of 2λ.

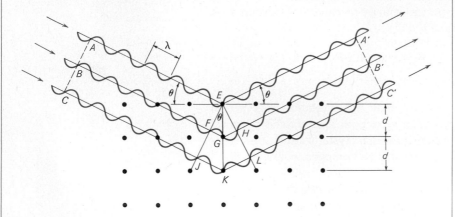

Figure 8•16 Schematic representation of the diffraction of a beam of X radiation of wavelength λ at the angle θ by a crystal made of planes of atoms whose centers (black dots) are separated by the distance *d*.

In Figure 8.16 we see that, for the right triangle *EFG*, the sine of angle θ is *FG/EG* (the length of the side opposite θ divided by the hypotenuse). But *EG* equals *d*, the distance between planes; and *FG* is $\frac{1}{2}λ$ in our drawing. So,

$$\sin\theta = \frac{FG}{EG} = \frac{\frac{1}{2}\lambda}{d} \quad \text{or} \quad \lambda = 2d\sin\theta$$

For the general case, it is only necessary that the distance *FGH* be a whole-

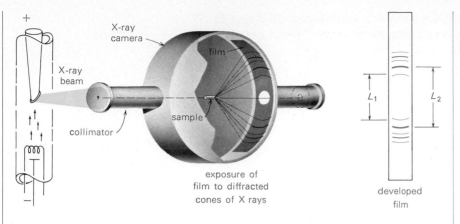

exposure of
film to diffracted
cones of X rays

developed
film

Figure 8•17 X-ray diffraction camera used with rotating powdered sample. The many minute crystals diffract the pinpoint X ray in conical envelopes at various angles; the developed film reveals a pattern of lines that show where the various cones strike the film. By measuring distances between pairs of lines on the film, for example, L_1 or L_2, and the diameter of the camera, the angles of diffraction (see Figure 8.16) can be determined. The angle between a pair of lines is 4θ, where θ is the angle used in the Bragg equation.

number multiple, n, of the wavelength. So the **Bragg equation** for the diffraction of X rays is

$$n\lambda = 2d \sin \theta$$

With the Bragg equation, values of d for a crystalline solid can be calculated. It is necessary that λ be known and that values of θ be precisely measured. One method of measuring the angles of reflection is with an X-ray camera such as that in Figure 8.17. A small sample of powdered crystals is slowly rotated and irradiated with a beam of X rays. The beam is reflected at certain angles and gives rise to a pattern of lines on a strip of photographic film. The X-ray powder pattern of a substance may consist of perhaps ten to more than 100 pairs of lines of various intensities and spacings. The pattern for a substance is practically unique for purposes of identification. Even mixtures of substances can be analyzed if they are not too complex, as suggested by the patterns in Figure 8.18. Patterns of thousands of compounds have been determined and recorded in the chemical literature.

8•6•2   Characteristics of Crystals   The smallest section of a crystal lattice that can be used to describe its structure is called a **unit cell.** Theoretically, the entire crystal can be reproduced by stacking together unit cells. Different crystal lattices can be classified according to the relative lengths of the unit cell's three axes and the relative sizes of the three angles betweeen the axes. Classified in this manner, there are six fundamentally different crystal arrangements. We will not describe these arrangements; they can be found in many texts on physical or inorganic chemistry. Two of the most common systems for simple crystals are the cubic and the hexagonal (see Figure 8.29).

A unit cell of the potassium bromide crystal is diagrammed in Figure 8.19. The arrangements and the sizes of the ions are calculated on the basis of X-ray data. Example 8.5 shows how the distance of 3.29 Å is determined.

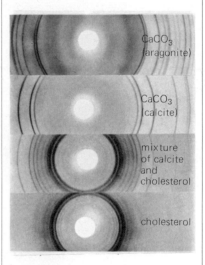

CaCO₃
(aragonite)

CaCO₃
(calcite)

mixture
of calcite
and
cholesterol

cholesterol

Figure 8•18 X-ray powder diffraction patterns obtained with known samples of crystalline calcium salts and cholesterol, made in order to compare with patterns obtained from samples of gallstones.
Source: Courtesy of the Eli Lilly Research Laboratories, Indianapolis, Indiana.

───── **Example 8•5** ─────

Suppose a powder diffraction camera such as that in Figure 8.17 has a film chamber with a diameter of 71.62 mm. When a sample of potassium bromide crystals is irradiated with X rays of $\lambda = 1.5418$ Å, the distance between the most intense pair of lines on the film is 33.85 mm (similar to the distance L_2 in Figure 8.17). Calculate the distance between the planes of ions in potassium bromide that cause this most intense pair of lines.

• Solution • The circumference of the camera is:

$$\pi \text{ (diameter)} = (3.1416)(71.62 \text{ mm}) = 225.0 \text{ mm}$$

The angle between the pair of lines is the angle defined by an arc of 33.85 mm:

$$\text{angle between lines} = \left(\frac{33.85 \text{ mm}}{225.0 \text{ mm}}\right)(360°) = 54.16°$$

Comparing Figures 8.16 and 8.17, we see that the Bragg angle, θ, is one-fourth the angle between a pair of lines. For this case, $54.16°/4 = 13.54°$. From a handbook (or hand calculator), $\sin 13.54°$ is found to be 0.2341.

Using the Bragg equation and letting $n = 1$:

$$d = \frac{n\lambda}{2 \sin \theta} = \frac{1.5418 \text{ Å}}{2(0.2341)} = 3.29 \text{ Å}$$

See also Exercise 36.

• *Calculation of the Avogadro Number* • X-ray measurements can be used to calculate the number of atoms (or molecules or pairs of ions) in a mole. Consider the unit cell of potassium bromide drawn in Figure 8.19. From the dimensions of the cell we calculate that 4 K^+ ions and 4 Br^- ions occupy a volume of 285 Å^3. From a handbook we find the density of KBr to be 2.75 g/cm^3 at 25 °C.

$$\frac{\text{number of } K^+, Br^-}{\text{pairs per mole}} = \left(\frac{4 \text{ pairs}}{285 \text{ Å}^3}\right)\left(\frac{1 \times 10^8 \text{ Å}}{1 \text{ cm}}\right)^3\left(\frac{1 \text{ cm}^3}{2.75 \text{ g}}\right)\left(\frac{119.0 \text{ g}}{1 \text{ mol}}\right)$$

$$= 6.07 \times 10^{23} \text{ pairs of ions per mole}$$

This basic method, with many refinements, has been used for the most precise determination of the Avogadro number to date. Scientists at the National Bureau of Standards have reported the value of the number to be $6.0220943 \times 10^{23} \pm 6.3 \times 10^{17}$.

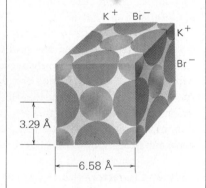

Figure 8•19 Unit cell of potassium bromide. (Bromide ions are shown in color.)

8•7   Ionic Compounds

The most familiar class of crystalline substances is that of ionic compounds. The tendency of ions to attract other ions of opposite charge and to repel ions of the same charge leads to well-ordered, three-dimensional arrangements of ions. Three of the main influences on the specific patterns formed by ionic compounds are (1) the charges on the ions, (2) the relative sizes of the two ions involved, and (3) the ease with which the ions, particularly the negative ones, can be distorted or polarized.

The electrostatic attractive and repulsive forces between ions act equally in all directions. In accord with Coulomb's law (see Section 3.6), these forces decrease in proportion to the square of the distance between charges; therefore, they are strongest between nearest neighbors. The melting and boiling points of ionic substances are high. A great deal of energy is needed to weaken the strong ionic bonds and give the particles enough energy to move as liquid particles or to evaporate as simple molecules.

8•7•1   Sizes of Ions and the Radius Ratio   The relative sizes of two ions are described by the **radius ratio:**

$$\frac{\text{radius of cation}}{\text{radius of anion}} = \frac{r_+}{r_-} = \text{radius ratio}$$

The radius ratio affects the ways in which spherical particles can be packed together. In ionic compounds, the radius ratio is an important factor in determining the structure of a crystal.

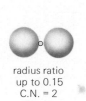

| radius ratio
up to 0.15
C.N. = 2
linear | radius ratio
0.15 to 0.22
C.N. = 3
trigonal planar | radius ratio
0.22 to 0.41
C.N. = 4
tetrahedral | radius ratio
0.41 to 0.73
C.N. = 6
octahedral | radius ratio
0.73 and up
C.N. = 8
cubic |

Figure 8•20 Arrangements favored by particles with different radius ratios.

The number of nearest neighbors that a particle has in a crystal is called its **coordination number,** C.N. In Figure 8.20 groups are shown in which the coordination numbers are 2, 3, 4, 6, and 8. If the radius ratio is less than 0.15, the smaller particle cannot touch three larger particles. As the relative size of the smaller particle increases, the number of larger particles that can touch it increases. The relationships of radius ratios to coordination numbers are summarized in Table 8.4.

There are many actual ionic crystals with structures that agree with predictions based on the radius ratio of the two ions. The most familiar ionic structure is the *sodium chloride structure* (see Figure 8.21). In this structure, the C.N. is 6. The Na^+/Cl^- radius ratio itself is 0.95 Å/1.81 Å = 0.53, which is within the range of 0.41 to 0.73 calculated for this structure. Potassium bromide, which has the sodium chloride structure, has the radius ratio K^+/Br^- of 1.33 Å/1.96 Å = 0.68. By examining the unit cell of KBr in Figure 8.19 or the drawing of NaCl in Figure 8.21, we can see that each ion (other than those on a surface) has six nearest neighbors of opposite charge.

Although all but three alkali metal halides crystallize at ordinary temperatures and pressures in the sodium chloride structure, those with the largest cations may have C.N. = 8. This crystalline structure is called the *cesium chloride structure.* It is formed by CsCl, CsBr, and CsI, as well as some nonalkali metal salts. The ratio for Cs^+/Cl^- is 1.69 Å/1.81 Å = 0.93.

A crystalline structure with C.N. = 4 is the *zinc blende structure,* named for the mineral form of zinc sulfide, ZnS. The radius ratio of Zn^{2+}/S^{2-} is 0.74 Å/1.82 Å = 0.41. Some other compounds that have the zinc blende structure are CdTe and SiC.

Table 8•4 Ranges of radius ratios versus coordination numbers

r_+/r_-	C.N.
<0.15	2
0.15–0.22	3
0.22–0.41	4
0.41–0.73	6
>0.73	8

By extending the diagrams in Figure 8.21, it can be shown that C.N. (cation) = C.N. (anion) in each of the three cases. For ionic compounds of more complex stoichiometry, for example, AB_2, AB_3, and so on, there may be more complex structures. But simple structures are often found also, as in the example of FeS_2, which crystallizes in the sodium chloride structure.

• *Steric Hindrance* • Often one ion can take the place of another in an ionic lattice if the ions are of the same charge and about the same size. For example, chloride, with a radius of 1.81 Å, might substitute for bromide, with a radius of 1.96 Å. Bromide might even be able to substitute for a chloride ion in some crystals. But iodide, with a radius of 2.16 Å, hardly could be expected to substitute for a fluoride ion, with a radius of 1.36 Å. When a particle is kept out of a place because of space limitations, the effect is called *steric hindrance*. This effect is not limited to ionic substances.

• *Assigning Sizes of Ions* • The size of an ion, as determined by X-ray measurements, may differ slightly from one compound to another. For example, the radius determined for the bromide ion is slightly different in the compounds lithium bromide and potassium bromide. This may be attributed to the differences in the charge densities of the two positive ions and to various packing effects. Due to packing, in different crystals the negative ions are at different distances from one another. If the radius ratio is small enough, there may be anion–anion contact, as shown in Figure 8.22.

By comparing hundreds of crystals, chemists and physicists have chosen the "best" values for the radii of individual ions. Typical values for many ions are given in the first tables in Chapters 21 and 23–25. Approximate sizes can be estimated from Figure 5.14.

8.7.2   Polarization of Ions   The ratio of the charge of an ion to its volume is called the **charge density.** Positive ions with high charges have small radii, so they have high charge densities (see Figure 8.23). Such ions have strong attraction for the electrons of adjacent ions and molecules.

Ions are considered to pack together as perfect spheres, like marbles, unless the shape of one of the ions is distorted by a strong attraction to a neighboring ion of opposite charge. An isolated simple ion—for example, a large negative ion—should be spherically symmetrical with its charge evenly distributed. But if a positive ion with a highly concentrated charge (high charge density) comes very close to it, the negative ion may be pulled out of shape.

As shown in Figure 8.24, the distorted ion is more negative on one side (toward the positive ion) and positive on the other, so it is said to be **polarized.** This polarization does not vary from one instant to the next, as in the case of London forces, but is constant as long as the ions are adjacent. Because of their greater sizes, negative ions tend to be polarized more easily than positive ones. The large I^- and Br^- ions are relatively easily distorted.

Among effective polarizing agents are the ions Mg^{2+}, Fe^{3+}, Zn^{2+}, and Al^{3+} (refer to Figure 5.14 to estimate their relative sizes). In a compound made of large negative ions and small positive ions, the polarization may be so marked that the bonds become quite covalent. Such *covalent-ionic* com-

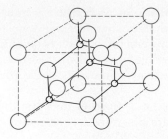

zinc blende
structure, C.N. = 4

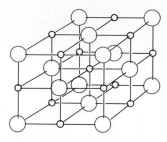

sodium chloride
structure, C.N. = 6

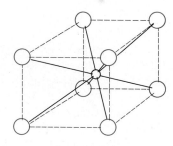

cesium chloride
structure, C.N. = 8

Figure 8·21 Unit cells for three well-known structures. The negative ions are shown in color. Solid lines show ionic bonds between nearest neighbors; dashed lines are drawn only to show perspective and cubic geometry. It would be more correct to draw these unit cells in the space-filling style of Figure 8.19. Can you see that these diagrams could be extended to show that in each structure the coordination number for negative ions equals the coordination number for positive ions?

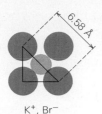

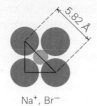

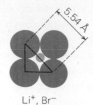

$a^2 + b^2 = c^2$

K⁺, Br⁻   Na⁺, Br⁻   Li⁺, Br⁻

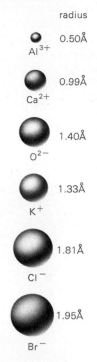

Figure 8•22 (*left*) In an ionic crystal, if the cation is small enough the anions touch each other. Bromide ions are in color, and positive ions are gray. In lithium bromide there is anion–anion contact.

radius

Al^{3+} 0.50Å

Ca^{2+} 0.99Å

O^{2-} 1.40Å

K^+ 1.33Å

Cl^- 1.81Å

Br^- 1.95Å

Figure 8•23 The charge density of the six ions decreases from top to bottom.

pounds include ZnS, $AlCl_3$, $FeCl_3$, and SnI_4. (See Table 23.3.)

It is difficult to generalize, but apparently one effect of polarization can be to lower the melting point. The data in Table 8.5 illustrate this for four different bases of comparison. A low melting point indicates that molecules, for example, $(HgCl_2)_x$ or $(AlBr_3)_y$, are probably present in the liquid rather than a simple ionic array. In the case of these two examples, the principal molecules in the vapor are $HgCl_2$ and Al_2Br_6.

Table 8•5 **Melting points of several ionic or covalent-ionic salts, °C**

Cations of different sizes	Cations of different sizes and charges	Anions of different sizes	Cations of same size but different e^- structure
$BeCl_2$, 405	NaBr, 755	LiF, 870	$HgCl_2$, 276
$CaCl_2$, 772	$MgBr_2$, 700	LiCl, 613	$CaCl_2$, 772
	$AlBr_3$, 97.5	LiBr, 547	
		LiI, 446	

Source: *Adapted with permission from J. E. Huheey,* Inorganic Chemistry, *2nd ed., Harper & Row, New York, 1978, p. 93.*

8•8   **Solid Covalent Substances**

Most substances that are largely covalent exist as discrete molecules. Bromine, Br_2, carbon dioxide, CO_2, hexane, C_6H_{14}, ammonia, NH_3, and ethyl alcohol, C_2H_5OH, are typical examples. As pointed out in Section 8.1.5, the melting and boiling points of covalent substances tend to be much lower than those of ionic substances. Actually ionic bonds and covalent bonds have about the same range of strengths. The differences in melting and boiling points are due to the fact that to melt a molecular solid or to vaporize a molecular liquid requires only enough energy to overcome the weak van der Waals attractive forces between molecules (see Figure 8.5). The intramolecular covalent bonds between atoms may be strong, but the intermolecular attractive forces between adjacent molecules are relatively weak.

A few nonmetallic elements, such as carbon and silicon, and a few compounds, such as boron nitride or silicon carbide, form crystals in which the atoms are tightly held in all directions by strong covalent bonds. These

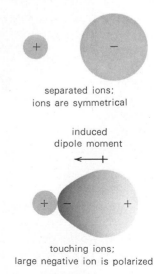

separated ions; ions are symmetrical

induced dipole moment

touching ions; large negative ion is polarized

Figure 8•24 Schematic representation of the polarization of an ion.

substances do not exist as discrete, small molecules. Indeed, a whole crystal can be considered to be one huge molecule. Such substances are extremely hard and brittle and have very high melting points.

In solid elements or compounds made up of covalent molecules, there are two general types of crystals. If the individual molecules are nonpolar or weakly polar, they pile together like marbles, taking up as little space as possible. If the individual molecules are polar, they tend to arrange themselves so that unlike charges are near one another and like charges are as far from one another as possible. A two-dimensional representation of the packing of strongly polar molecules is shown in Figure 8.25.

Substances with small nonpolar or weakly polar molecules form solids only at low temperatures. Such substances include the lighter halogens, CO, CO_2, and many organic compounds such as methane, CH_4, and other hydrocarbons.

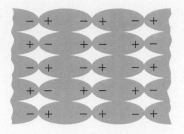

Figure 8·25 A schematic representation of the packing of a layer of polar molecules.

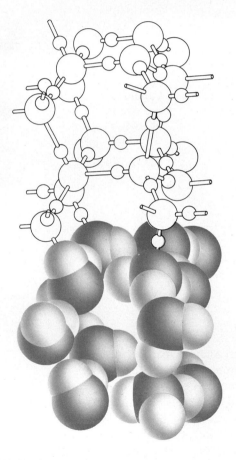

Figure 8·26 The orderly arrangement of water molecules in ice. Careful study of the drawing reveals that each oxygen atom is connected through four hydrogen atoms to four other oxygen atoms, and that each oxygen atom has two hydrogens close to it and two farther away (connected by hydrogen bonds). This pattern, in which the hydrogen bonds and the open spaces line up, forms planes through the ice. A block of ice is probably cleaved along such planes when struck with an ice pick.

● 8·8·1   **Structure of Ice and Water**   From the dimensions of a water molecule it is calculated that, if these molecules were close-packed, like marbles in a box, each would occupy about 15 Å³. For 1 mole of water, 6×10^{23} molecules or 18 g, the calculated volume would be about 9 cm³:

$$\left(\frac{15 \text{ Å}^3}{\text{molecule}}\right)(6 \times 10^{23} \text{ molecules})\left(\frac{1 \times 10^{-8} \text{ cm}}{1 \text{ Å}}\right)^3 = 9 \text{ cm}^3$$

But the measured volume of 1 mole of water is 18 cm³, or twice the calculated value. It appears, therefore, that about half the space in water is occupied by water molecules and half is empty.

X-ray studies of ice show that each H_2O molecule has four nearest neighbors, with the overall structure featuring repeating hexagonal groupings of six molecules arranged so that large empty spaces penetrate the ice crystal (see Figure 8.26). (Compare also the overall hexagonal structure of snow crystals, Figure 8.27.)

X-ray studies of liquid water indicate that its structure is somewhat like that of solid ice. But, on melting, the volume occupied by the liquid water is only about 90 percent of the volume of the ice. This indicates that the structure of the ice collapses partially and that there is less empty space in water than in ice. For a liquid to be more dense than its solid is unusual, because most substances expand on melting.

The unusual behavior of H_2O is related to the structures of ice and water. Many of the hydrogen bonds are broken when ice is melted. As the ice structure collapses, the molecules become more closely packed, and the density suddenly increases (see Figure 8.28). After the ice has melted, there is a further breaking up of the associated water molecules as the water is heated. This causes a further slight increase in density because of the closer packing of the molecules. The density of water is at a maximum at 4 °C, as shown in Figure 8.28. Above 4 °C, the expansion due to the increased motion of the molecules becomes greater than the contraction due to the breaking up of the hydrogen bonds, and the liquid expands; that is, the molecules occupy more space. In the vapor form, the molecules have too much kinetic energy for effective bonding to occur. The molecules in water vapor actually conform to the simple formula, H_2O.

8•9   Metallic Solids

The simplest solids are those that contain atoms of only one element—for example, the metals. The atoms in metals are usually arranged in the most efficient fashion possible, just as we might pack oranges into a box. With identical atoms the radius ratio is 1 and the maximum coordination number is 12. Three very common structures are drawn in Figure 8.29.

In the body-centered cubic structure, any atom (except those in the surface) has C.N. = 8; in the face-centered cubic, any atom has C.N. = 12; in the hexagonal close-packed structure, any atom has C.N. = 12. The latter two are examples of the closest packing possible for identical spheres, whether one is considering atoms or cannonballs. The face-centered cubic pattern can also be looked upon as a hexagonal type of packing, as shown in Figure 8.30, where it is compared with the hexagonal close-packed structure.

Elements such as chromium, iron, tungsten, sodium, and potassium crystallize in the body-centered pattern. Face-centered lattices are formed by aluminum, copper, silver, gold, platinum, lead, and many other metals. Among the hexagonal close-packed lattices are magnesium, zinc, and cadmium.

The most common metallic substances are elements, but there are many intermetallic compounds that have typical metallic properties. Examples of intermetallic compounds are Na_3Bi, $AgZn$, $FeAl$, and Cu_3Sn.

Figure 8•27 The beautiful symmetry of a snowflake never fails to amaze. We are filled with wonder when we focus on the fact that the growth on one spire of the flake cannot reasonably affect or be affected by the similar growth on the other five spires—yet they grow in unison. The six-fold symmetry of the snowflake can be correlated with the hexagonal symmetry of the ice crystal shown in Figure 8.26.
Source: Courtesy of the American Museum of Natural History, New York.

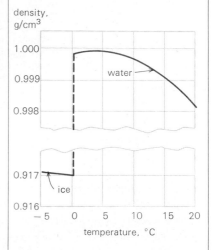

Figure 8•28 Densities of ice and water.

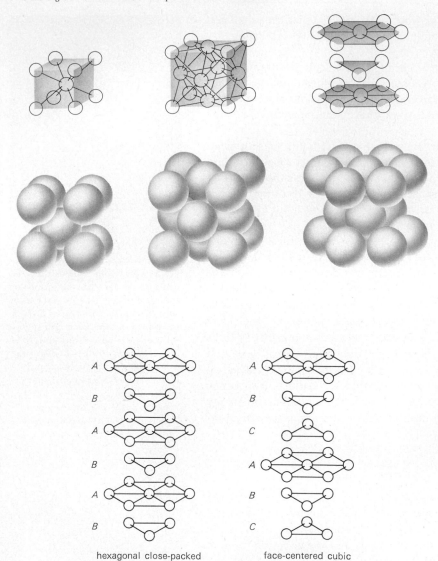

Figure 8•29 The three common types of metal structures. From left to right: body-centered cubic, face-centered cubic, and hexagonal close-packed. Bond lines between layers in hexagonal detail are omitted for clarity.

hexagonal close-packed   face-centered cubic

Figure 8•30 Two ways of stacking planes of hexagonally packed spheres. When planes touch, each sphere touches three spheres in each of the two adjacent planes; that is, C.N. = 6 + 3 + 3 = 12. Planes containing spheres aligned directly over one another are identified with a common letter, for example, A, A. (Note that the six spheres in a diagonal plane cutting three corners of the face-centered cubic model in Figure 8.29 are part of a layer of hexagonally packed atoms.)

Metals are similar to ionic structures in one important respect: the crystal is thought of as a single unit of indefinite size. Any single crystal of these solids, regardless of size, can be thought of as a molecule.

The field ion microscope has given the most striking direct experimental evidence for the symmetrical arrangement of atoms in metals. In this instrument there is a very highly charged, needle-pointed metal anode inside a tube similar to a television picture tube (Figure 8.31). Stray atoms of helium gas strike atoms on the anode, are ionized, and are sent streaming toward the negatively charged screen end (or picture end) of the tube. The highly symmetrical picture that appears on the screen has a pattern that is characteristic of the pattern in which atoms are crystallized in the metal anode.

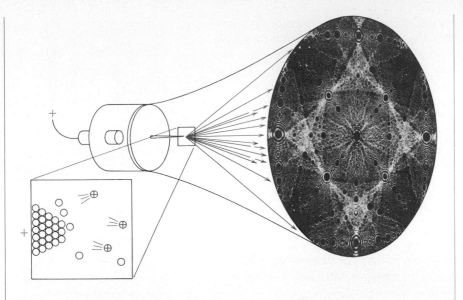

Figure 8•31 Schematic representation of the field ion microscope. The pattern shown here was produced by using an anode made from a needle-pointed crystal of platinum. Each small luminous dot is an image of an individual platinum atom in the surface of the crystal. *Source:* Platinum pattern courtesy of Professor E. W. Mueller, The Pennsylvania State University.

8•9•1   **Properties of Metals**   Of the hundred-odd elements known, approximately three fourths are classed as metals. Although they show a great diversity of properties, there are a few unifying characteristics, both chemical and physical, that serve to set them apart from the nonmetallic elements.

• *Chemical Properties* • It is a chemical property of most metals to act as *electron donors* in reactions. Metallic ions are usually positive ions. This is related to the low ionization energy of metal atoms and to the fact that there are usually fewer than four electrons in their outside energy levels.

• *Physical Properties* • The physical properties of most metals include the following characteristics:

a high degree of *electric conductivity*

a high degree of *thermal conductivity*

a definite *luster*. The freshly broken, uncorroded surface of all metal elements, except gold and copper, is lustrous gray or silvery in color.

a *close-packed arrangement of atoms*. Typical structures include the three described in Figure 8.29. As a result of this close packing, metals tend to have relatively high densities.

the ability under strain to *change shape without cracking*. Metals are malleable and ductile in spite of the fact that their atoms hold strongly to one another.

8•9•2   **Electrons in Metallic Bonds**   The physical properties of the metals indicate that the valence bonds that hold the atoms in a metal together are not ionic, nor are they simply covalent in nature. Although metal atoms commonly have only one to four outside electrons, they have eight or twelve nearest neighbors; that is, there is less than one electron available per bond. According to the present view, a metal is composed of a rigid lattice of positive ions around which there is a sea, or atmosphere, of valence electrons. These valence electrons are restricted to certain energy levels, but they have sufficient freedom so that they are not shared continuously by the same two ions. When energy is applied, these electrons are easily transferred from atom to atom. This bonding system, unique to the metals, is known as the **metallic bond.**

Unlike the ionic or covalent bond, the metallic bond provides for the strength and toughness of metals and at the same time permits deformation. Even though the positive ions occupy relatively stationary positions in the atmosphere of their electrons, they can glide over one another rather easily (see Figure 8.32). Therefore, they do not give way in a complete fracture when the metal is hammered or rolled into thin sheets or pulled into a wire.

The high electric conductivity of metals is attributed to the metallic bond. The fact that electrons are relayed (as electric current) through a crystal array of metal atoms with practically the speed of light indicates that some electrons are not held by individual atoms. Rather, the valence electrons in a metal belong to the solid as a whole. This helps explain high heat conductivity also. If one end of a piece of metal is heated, the electrons there move faster than electrons at the cold end. The rapid flow of heat in metals indicates that the energetic electrons at a hot spot move about rapidly and mingle with the rapidly moving, but less energetic, electrons in the cooler regions. This movement of valence electrons in metals and their interchange of energies is analogous to the random movement of molecules of a gas.

In conclusion, note that the metallic state is described primarily in terms of its physical properties. Elements are not the only materials that behave physically like metals. Two or more metallic elements can be mixed or even combined chemically; the material thus formed is different from its components and has many of the properties of a simple metal.

8.10   Packing of Particles

The individual particles that make up substances tend to arrange themselves in the most efficient way possible under any given set of conditions. In both molecules and crystals, nuclei and electrons fall into patterns in which:

Particles attracted to one another are as close together as possible.

Particles repelled by one another are as far from one another as possible.

All the particles pack together in symmetrical arrays so as to minimize the net energies of attractions and repulsions.

In describing crystal structures by means of diagrams, it is common to represent them as geometrically perfect, both in arrangement and in numbers of particles involved. However, things are seldom perfect, and even crystals are no exception to this "law of nature." Some common types of *lattice defect structures* are those that involve missing atoms or ions (that is, holes in the crystal), extra particles squeezed in here and there, or simply a disordered arrangement of the particles. In typical samples of sodium

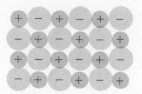

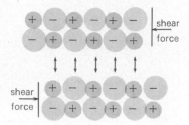

On being distorted, an ionic crystal fractures. When oppositely charged ions slide close to one another, the strong repulsions force the faces apart.

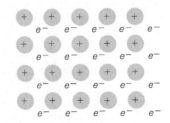

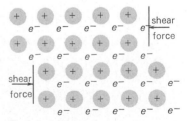

On being distorted, a metal can still be strong because new metallic bonds are formed.

Figure 8.32 A comparison of the behavior of ionic and metallic crystals when deformed by mechanical stress.

chloride, it has been found that there are perhaps 1×10^{18} holes per mole (58.5 g). Although this seems to be a huge number of imperfections, it is only one hole per million ions.

Defect structures may have special electric properties that are useful. Semiconductors, transistors, and insulators are made today from synthetic crystals that have carefully controlled defects.

Chapter Review

Summary

The effect of heat in increasing the temperature of a substance depends on the physical state of the substance and on its **heat capacity,** which can be expressed as the **molar heat capacity.** All changes of state among the gaseous, liquid, and solid phases of a substance involve the gain or loss of heat, although the temperature of the substance remains constant until the change is complete. The amount of heat needed to melt 1 mole of a given substance is the **molar heat of fusion;** an equal amount of heat is released as the **molar heat of solidification** when the substance freezes. Analogous to these two quantities are the **molar heat of vaporization** and the **molar heat of condensation.**

The three main types of substances—covalent, ionic, and metallic—vary greatly in their melting and boiling points, owing to the great differences in the nature and strength of the forces that enable their condensed phases to exist. The weak intermolecular forces in covalent substances called **van der Waals attractive forces** are of two kinds: **London forces** and **dipole–dipole attractions.** A somewhat stronger intermolecular force is **hydrogen bonding.** The latter causes the anomalous physical properties of water, including its high **surface tension.** In general, the intermolecular forces acting within a given substance or between like substances are **cohesive forces,** and those acting between unlike substances are **adhesive forces.** The **capillary rise** of a liquid is the result of a balance of such forces.

Gases condense to liquids when the intermolecular forces become strong enough to overcome the kinetic energy of the molecules. The temperature above which a gas cannot be liquefied, no matter how great the pressure, is the **critical temperature** for that gas; the pressure needed to liquefy the gas at that temperature is the **critical pressure.** The gaseous state of a substance under conditions at which it exists primarily as a liquid or solid is called a **vapor.** The **vapor pressure** of a substance is the pressure exerted by the vapor when it is in **equilibrium** with the liquid or solid phase; it increases with the temperature. When the vapor pressure of a liquid becomes equal to the pressure of the surrounding atmosphere, the liquid is at its **boiling point.** External pressure can have a great effect on the boiling point but it

has a negligible effect on the **freezing point** of a liquid (the **melting point** of the solid). The temperature–pressure relations among the three phases of a substance are shown by the **phase diagram** for that substance. The point at which all three phases can exist in equilibrium is the **triple point.**

The symmetrical arrangement of points in a crystal is called the **crystal lattice.** The smallest conceptual building block in a crystal lattice is the **unit cell,** the geometry of which provides a means for classifying the crystal. Physical measurements with **monochromatic** X rays reveal many details of crystal structure, such as the spacings between parallel planes of atoms or ions; these spacings are calculated from the **Bragg equation.**

The ways in which ions can be packed together in ionic crystals depend in part on the cation-to-anion **radius ratio.** This ratio is closely related to the **coordination number** of an ion in such a crystal. Another factor in the packing arrangement is the **charge density** of the ions, which determines the extent to which neighboring ions are **polarized** by the electrostatic interaction between them. The degree of polarity of covalent molecules also affects their packing arrangements in crystals. Metals are highly regular arrays of positive ions in a crystal lattice. The **metallic bond** can be envisioned as a loose "sea" of valence electrons that hold the ions together in what amounts to one giant molecule.

Exercises

Changes of State

1. The value for the heat of vaporization of water at 0 °C is 45.00 kJ/mol and at 100 °C it is 40.64 kJ/mol. Calculate the heat needed to evaporate 0.500 kg of water at 0 °C and 100 °C, with the temperature of the vapor being 0 °C and 100 °C, respectively.

2. How much heat is lost by your body due to the evaporation of 5.0 g of perspiration?

3. (a) How much heat is required to change 5.0 g of ice at −10 °C to water vapor at 110 °C?
   (b) How much heat is required to change 1 mole of

solid H_2O at its freezing point to 1 mole of gaseous H_2O at its boiling point? Repeat for 1 mole of Br_2.

*4. Assuming that the average molar heat capacity of liquid water is 75.3 J/(mol · °C) between 0 °C and 100 °C, calculate the average molar heat capacity of water vapor over this range of temperature. Use data in Exercise 1 as needed.

5. Calculate the heat of fusion and the heat of vaporization per gram for benzene, C_6H_6.

Intermolecular Attractive Forces

6. Write a Lewis structure for acetic acid, CH_3CO_2H, and one for a water molecule. Is it likely that hydrogen bonds would form between molecules of these two compounds? If so, indicate how the molecules might be positioned next to one another. Draw several possibilities.

7. Tell whether hydrogen bonding is expected between the two members of each of the following pairs of molecules:

(a) NH_3, NH_3 (b) CH_4, CH_4
(c) CH_3OH, CH_3OH (d) CH_3OCH_3, CH_3OCH_3
(e) NH_3, CH_4 (f) H_2O, CH_3OH
(g) NH_3, H_2O (h) CH_3OCH_3, H_2O
(i) CH_3OH, CH_3OCH_3

8. For each case in Exercise 7 where you expect hydrogen bonding, justify your expectation with electron-dot diagrams and show the hydrogen bond with a dashed line.

9. Years ago, methyl alcohol

$$H-\overset{\overset{\displaystyle H}{|}}{\underset{\underset{\displaystyle H}{|}}{C}}-\overset{..}{\underset{..}{O}}-H$$

was commonly used as an automobile antifreeze. Today ethylene glycol

$$H-\overset{..}{\underset{..}{O}}-\overset{\overset{\displaystyle H}{|}}{\underset{\underset{\displaystyle H}{|}}{C}}-\overset{\overset{\displaystyle H}{|}}{\underset{\underset{\displaystyle H}{|}}{C}}-\overset{..}{\underset{..}{O}}-H$$

is preferred. Give two reasons why the latter compound should form a mixture that boils at a higher temperature than does a mixture of methyl alcohol and water.

10. (a) With the help of a clearly labeled drawing, explain why a small drop of a liquid tends to become spherical in shape.
(b) Would the meniscus of benzene in a glass tube be more concave than that of ethyl alcohol? Why?

Liquefaction (Condensation)

11. Estimate the melting and boiling points for water and hydrogen fluoride if each followed the trend for the

hydrogen compounds in its group. By about how many degrees is the boiling point of each of these two compounds apparently increased due to hydrogen bonding?

12. (a) Methane, CH_4, is always in the gaseous state inside a pressurized cylinder at room temperature, but propane, C_3H_8, may be present as a liquid or a gas. Explain.
(b) Chlorine, Cl_2, and pentane, C_5H_{12}, have approximately the same molecular weight. However, chlorine is a gas and pentane is a liquid at 25 °C and 1 atm. Explain.

13. A bottle or a can of a cold beverage often becomes covered with drops of water.
(a) Describe clearly the process that leads to the formation of the drops. What is the name of this process?
(b) State whether the process is exothermic or endothermic, and explain why this is so on the basis of the kinetic molecular theory.

14. When the atmosphere is supersaturated with water vapor and is suddenly disturbed, a fog may form. Does the process of fog formation tend to cool, heat, or have no effect on the temperature of the air–fog system? Explain.

Vapor Pressure

15. By means of a clearly labeled graph, show how the vapor pressure of a liquid changes with temperature. Also show on the graph how the temperature of the normal boiling point is related to a certain pressure.

*16. Suppose a Torricellian barometer has 68 mm of liquid water on top of 710 mm of mercury. If the temperature of the barometer is 23 °C, what is the atmospheric pressure? Assume the density of mercury is 13.6 g/cm³.

*17. Look again at Exercise 5 at the end of Chapter 7. Did you take into consideration the vapor pressure of water when you first worked that exercise? Calculate the height of a water barometer if the pressure of the atmosphere is standard and the temperature of the water is 20 °C. Repeat for 0 °C. Assume the density of water is 1.00 g/cm³.

18. The relative humidity of a volume of air at 35 °C is 70.0 percent. What is the partial pressure of the water vapor?

19. Suppose a glass of iced tea has an outside surface temperature of 5 °C, and assume that air touching the glass is cooled to 5 °C. Determine whether or not moisture will condense on the glass under each of the following conditions of room temperature and relative humidity: (a) 20 °C, 50%; (b) 35 °C, 10%; (c) −5 °C, 100%; (d) 23 °C, 40%.

*20. An air-conditioning system may cool air below the temperature desired and then partially reheat it in order to provide cool air of the proper humidity.
(a) In a house with a volume of 465 m³, if the air has a relative humidity of 100 percent at 30 °C, what

weight of water will condense if the air is cooled to 10 °C?

(b) If the 100-percent relative humidity air at 10 °C produced in part (a) is heated to 20 °C, what will the relative humidity be?

21. Suppose a sample of air at 15 °C and 100 percent relative humidity is put in a closed container and heated. What is the relative humidity of the warmed air at 25 °C? At 50 °C? At 75 °C? At 100 °C?

22. If the temperature in a house is 22 °C and the relative humidity is 70 percent, what will the relative humidity be in the basement if the temperature is 10 °C and we assume all the basement air comes from the house above?

Freezing and Boiling

23. High-flying airplanes sometimes leave a long trail of fog behind them. Sometimes the trail remains visible for many minutes, but at other times it disappears quickly.

(a) Suggest some reasons for the differences in behavior.

(b) Write a hypothetical equation for the chemical reaction that produces the "fog substance," assuming that the airplane fuel is dodecane, $C_{12}H_{26}$, or similar compounds.

24. What are the two main molecular properties that determine the boiling point (or vapor pressure) of a liquid?

25. (a) If the average atmospheric pressure in Denver is 627 mm, what is the approximate average boiling point of water in an open beaker?

(b) What is the average atmospheric pressure in kilopascals?

26. Why must boiling chips be added *before* a liquid is heated rather than after it already may be superheated?

27. Predict whether the freezing points and boiling points of the following substances will be increased or decreased by an increase in pressure: mercury, lead, bismuth, ethyl alcohol.

28. One of the reasons that an ice skater glides so smoothly over the surface is that the ice melts to form liquid water just beneath the moving skate blade. What causes the ice to melt?

Phase Diagrams

29. In Figure 8.13, points R and X are at the same pressure. Describe the changes that will occur if gaseous H_2O at the pressure and temperature of point R is cooled at constant pressure until it attains the temperature at point X.

30. In Figure 8.13, points X and W are at the same temperature. Suppose a sample of H_2O exists at the conditions of point X, and the pressure is lowered below the pressure corresponding to point W. Describe the changes that will take place.

*31. At the conditions of point R in Figure 8.13, all the H_2O is in the gaseous phase. How could a sample of H_2O under these conditions be changed to solid ice at the conditions of point X without becoming a liquid at some point in the process?

Forms of Solids

32. Name two of the polymorphic forms of carbon. What is another general name that could be given to these forms?

33. We may say that sodium nitrate and calcium carbonate are isomorphous but that only calcium carbonate is polymorphous. Explain.

34. Atomic and ionic radii are commonly expressed in angstrom units. What is the nearest SI unit? Express dimensions of the unit cell and the radii of the two ions in Figure 8.19 in terms of this unit.

35. Name the equation $n\lambda = 2d \sin \theta$, and identify the four terms n, λ, d, and θ.

*36. On the X-ray film described in Example 8.5, the distance between a second pair of lines was found to be 48.30 mm.

(a) Calculate the distance between the planes of ions that are responsible for these two lines.

(b) Show that the distance calculated in part (a) and the 3.29 Å calculated in Example 8.5 are related by the Pythagorean formula, $a^2 + b^2 = c^2$.

(c) On a sketch similar to that for KBr in Figure 8.19, draw lines across one face of the unit cell to show planes of ions whose centers are 3.29 Å apart. Also draw lines to show planes separated by the distance found in part (a).

Ionic Compounds

37. What determines the specific pattern taken by ions of a given compound as they crystallize from solution or the molten liquid?

38. Calculate the radius ratio for calcium oxide. Radii for Ca^{2+} and O^{2-} are given in Tables 21.2 and 24.1.

39. Lithium bromide crystallizes in the sodium chloride structure. Would this be predicted on the basis of its radius ratio?

40. Lithium iodide crystallizes in the sodium chloride structure. Does it follow the rules of radius ratios or is it an exception? See Tables 21.1 and 23.1 for data.

41. For each of the unit cells shown in Figure 8.21, calculate the total number of positive and negative ions and show that the ratio is 1:1.

*42. Expand the diagrams for the three structures in Figure 8.21 to show that the anions have the same coordination numbers as the cations in these 1:1 compounds. (Suggestion: It is probably easiest to show C.N. = 6 for Cl⁻ in NaCl, next C.N. = 8 for Cl⁻ in CsCl, and hardest to show C.N. = 4 for S^{2-} in ZnS.)

*43. Prove by geometry that the minimum radius ratio for C.N. = 6 (octahedral geometry) is 0.414. (Hint: Note that the packing is square planar in the plane that includes the center atom and four neighbors, and use the Pythagorean formula.)

44. The density of sodium chloride is 2.16 g · cm^{-3}. How many pairs of Na^+, Cl^- ions are there in a grain of table salt that is a cube 0.10 mm on an edge?

*45. Using the X-ray data in Figure 8.22, calculate the radius of the sodium ion. Assume the bromide ion has the same radius in each of the three structures.

General Properties of Solids

46. Consider the following melting points: BeI_2, 510 °C; $BeBr_2$, 490 °C; $BeCl_2$, 405 °C; and BeF_2, sublimes at 800 °C. How can we account for the trend in the values for the first three? How can we explain the different behavior of BeF_2?

47. Ionic solids tend to be hard and brittle, whereas many covalent solids tend to be soft and easily distorted. How can this be explained?

48. Which of the following will float and which will not?
(a) solid lead in liquid lead
(b) solid water in liquid water
(c) solid bismuth in liquid bismuth
(d) solid ethyl alcohol in liquid ethyl alcohol

*49. Trace some of the field ion microscope pattern for platinum shown in Figure 8.31. Compare this with the structures in Figure 8.29 and show why you would classify platinum as body-centered cubic, face-centered cubic, or hexagonal close-packed.

50. In each of the following sets of three, choose the substance that has the highest boiling point.
(a) C_3H_8 CH_4 C_2H_6
(b) HBr   HCl   HF
(c) CH_4 HF   LiF

(d) Ne   He   Kr
(e) H_2O H_2S H_2Se

*51. In Figure 8.29 the corners of the unit cells for body-centered cubic and for face-centered cubic packing are at the centers of the eight corner atoms. Calculate the total number of atoms in each of these cells.

*52. Platinum crystallizes in the face-centered cubic lattice. The length of a side of the unit cell (see Figure 8.29) is 3.923 Å. The density is 21.4 g · cm^{-3}. Calculate a value for Avogadro's number based on these data. [Hint: Solve Exercise 51 to find the number of atoms per unit cell volume, and then calculate how many atoms it takes to have a volume of 1 mole (195.09 g) of platinum.]

53. Consider this list of substances and choose examples to illustrate each of the mentioned classes: Al, CH_4, CaO, Cl_2, Fe, NH_4Cl, H_2, CH_3CH_2OH, $NaNO_3$, $ScBr_3$.
(a) substances with only ionic bonds
(b) substances with only nonpolar covalent bonds
(c) substances with only polar covalent bonds
(d) substances with both ionic and covalent bonds

54. Which has the greater weight, 1 L of ice at 0 °C or 1 L of liquid water at the same temperature? One liter of water at 2 °C or one at 3 °C? One liter at 4 °C or one at 5 °C?

55. Consider again the ten substances listed in Exercise 53.
(a) List the substance(s) with polar bonds but with molecules that are nonpolar.
(b) List all the substances that are not bound by simple covalent or ionic bonds and name the kind of bonds in each.

56. (a) Thinking of the array of positive ions as interpenetrating the array of negative ions, could NaCl be thought of as interpenetrating body-centered cubic, interpenetrating face-centered cubic, interpenetrating hexagonal close-packed, or interpenetrating simple cubic?
(b) Repeat for CsCl.

The Chemical Behavior of Hydrogen, Oxygen, and Water

9

9•1 **The Phlogiston Theory**

9•2 **Oxygen**

9•2•1 Elemental Oxygen

Special Topic:
A Matter of Priority:
The Discovery of Oxygen

9•2•2 Compounds of Oxygen

9•2•3 Sources of Pure
Elemental Oxygen

9•2•4 Reactions of Oxygen

9•2•5 Oxidation–Reduction
(Redox) Reactions

9•2•6 Uses of Oxygen

9•3 **Ozone**

9•3•1 Preparation of Ozone

9•3•2 Reactivity of Ozone

9•3•3 Ozone in the Upper
Atmosphere

9•3•4 Ozone and Its Role
in Air Pollution
in the Lower Atmosphere

Special Topic:
Threatened Depletion of
the Earth's Ozone Shield

9•4 **Hydrogen**

9•4•1 Elemental Hydrogen

9•4•2 Compounds of Hydrogen

9•4•3 Laboratory Preparation
of Hydrogen

9•4•4 Commercial Production
of Hydrogen

9•4•5 Reactions of Hydrogen

9•4•6 Uses of Hydrogen

9•4•7 A Hydrogen Economy

9•5 **Water**

9•5•1 The Chemical Behavior
of Water

9•6 **Hydrogen Peroxide**

In this chapter attention will be focused on examples of descriptive chemistry to which we can apply a number of the ideas discussed in previous chapters. A study of oxygen and hydrogen, along with water and some of their other compounds, is well suited for this purpose. Many of the chemical changes these substances undergo are familar to us. The burning of a piece of paper, the decay of a log, the rusting of iron, the formation of starch in a kernel of corn, and the oxidation of sugar in body cells are common processes involving hydrogen, oxygen, or water.

Before developing these topics, it will be helpful to see how early experimental work with hydrogen, oxygen, and water established chemistry as a laboratory science and started it on a road of progress and service. Milestones along this road include (1) the discovery of the elements hydrogen and oxygen, (2) demonstration of the fact that water is a compound of hydrogen and oxygen, and (3) the unraveling of the role that oxygen plays in combustion. Experiments with oxygen led to a new theory of burning and to the overthrow of the phlogiston theory. These important events took place toward the end of the eighteenth century, from about 1765 to 1783.

9•1   The Phlogiston Theory

During the late seventeenth and most of the eighteenth centuries, combustion and associated reactions were explained in terms of the phlogiston theory. According to this theory, a combustible substance contained the mysterious material *phlogiston* plus some *calx*. When a substance burned, the phlogiston escaped into the air, leaving the calx behind as ashes. It was presumed that when a candle burned in a closed container, the air in the container finally became saturated with phlogiston. The candle ceased burning because the air could accept no more phlogiston.

Joseph Priestley, an English clergyman and scientist, in 1774 prepared a gas (Figure 9.1) in which things burned more readily than in air. A candle flamed up more brightly and burned for a longer time in this new gas than in ordinary air. Priestley prepared the gas by heating the compound that today is called mercury(II) oxide. He supposed his new gas was like ordinary air except that it contained no phlogiston to retard the burning of a candle. Hence, he called the gas "dephlogisticated air."

Earlier, in 1766, a countryman of Priestley, Sir Henry Cavendish, had found that a "flammable air" could be prepared by the reaction of metals such as iron, zinc, or tin with dilute acid solutions. When this combustible gas was burned in air or in Priestley's gas, water was formed. Cavendish thought that the combustible gas might be water combined with phlogiston.

The great French theoretician Antoine Lavoisier discarded the idea of phlogiston entirely. He showed that, when a substance burns, the products weigh more than the original substance. From such observations he concluded that a part of the air is used in the chemical reaction. This part of the air Lavoisier called *oxygen,* meaning "acid former," and he recognized that Priestley's gas was simply a pure form of oxygen.[1] Lavoisier also concluded that Cavendish's combustible gas was a new element that combined with

---

[1]Lavoisier's idea that all acids contain oxygen was soon proven wrong.

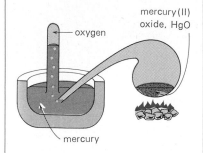

Figure 9•1 Priestley's preparation of oxygen. He frequently improvised his apparatus from household articles, using tall beer glasses for collecting gases over mercury or water, and even a gun barrel for heating solids in his fireplace.

oxygen to form water. He named this element *hydrogen,* from the Greek meaning "water former."

Today, using modern symbols, we can summarize this work of Priestley, Cavendish, and Lavoisier with the following equations:

$$2HgO \longrightarrow 2Hg + O_2$$ Priestley's preparation of oxygen

$$Zn + 2HCl \longrightarrow ZnCl_2 + H_2$$ Cavendish's preparation of hydrogen

$$2H_2 + O_2 \longrightarrow 2H_2O$$ Lavoisier's interpretation of the combustion of hydrogen

9•2   **Oxygen**

9•2•1   **Elemental Oxygen** The amount of elemental oxygen present in air is about 21 percent by volume (23 percent by weight). It exists in three gaseous forms, the molecules of which differ as follows: monatomic molecules, O, diatomic molecules, O_2, and triatomic molecules, O_3. Elements that

★ Special Topic ★ Special Topic ★ Special Topic ★ Special Topic ★ Special Topic ★ Special Topic ★ Special Topic ★ Special Topic ★

A Matter of Priority: The Discovery of Oxygen Scientists bestow their highest honors on those among them who discover new phenomena and new substances or who develop new theories. However, it is not uncommon for two or more persons to make the same or nearly the same discovery even though they work completely independently. In such cases the establishment of priority may be very difficult.

In determining who should get credit for a new finding, two dates are most important: the date the new observation or insight occurred and the date the new information was made known to other scientists. For the first date, one often can rely on records in notebooks. For establishing the second date, publication in a scientific journal is usually required, although a report at a scientific meeting or conference may also be acceptable. One of the main reasons for differences of opinion about priority for a discovery is that the original investigators themselves may have only a partial understanding of the nature of their findings. It may be only later that other investigators recognize the discovery to be of outstanding and permanent importance.

In the case of the discovery of oxygen, an event that today is said to mark the beginning of modern chemistry, historians are still evaluating old manuscripts and letters to answer questions of priority. The three principal researchers were Carl Wilhelm Scheele in Sweden, Joseph Priestley in England, and Antoine Lavoisier in France. Using modern formulas, we can reconstruct their work as follows.

In 1771 Scheele heated MnO_2 and H_2SO_4 (oil of vitriol) and collected an "air" that he called *vitriol air.* He recognized that it was associated with the part of the atmosphere that supports combustion, but this earliest work did not become widely known until long after his death. Although Priestley in 1771 also generated an air similar to Scheele's by heating KNO_3, he did not recognize clearly what he had prepared. It was not until August 1774 that he collected the same air by heating HgO and determined some of its properties. Both Scheele and Priestley explained the behavior of the new gaseous substance in terms of the phlogiston theory.

In September 1774 Scheele described his results to Lavoisier in a letter, and the next month Priestley visited Paris and discussed the new air with Lavoisier and others. It was Lavoisier, evidently, who had the flash of insight that the new substance was the key to explaining combustion. He generated samples of the gas and published a report of his discovery in May 1775. It was November 1775 when Priestley's first publication appeared, and it was mid-1776 when Scheele's results were published in a journal. The name *oxygine* was used first in a 1777 publication by Lavoisier.

It is the nature of science that persons in widely separated locations work on the same problems. Among the many famous questions of priority are those involving Newton versus Leibniz over the calculus, Mendeleev versus Meyer over the periodic table, and Darwin versus Wallace over the theory of natural selection. With the passage of time one of the researchers usually becomes popularly recognized as *the* discoverer. This is often due to the greater use that person made of the new knowledge to develop and stimulate more research. Sometimes it may be due merely to the accident of the country in which the person worked or the wide distribution of the journal in which the results were published.

exist in more than one molecular or crystalline form are said to be **allotropic** (Greek *allos tropos*, other way), and the forms are referred to as **allotropes.** The three allotropic forms of oxygen differ in their physical and chemical properties and are called atomic oxygen, ordinary oxygen, and ozone. In the atmosphere, almost all the oxygen is present as O_2 molecules, with the relative amounts of the other two forms increasing at high altitudes.

Three nonradioactive isotopes are found in nature: ^{16}O, ^{17}O, and ^{18}O. Precisely 99.76 percent of all the oxygen atoms existing in nature are of the ^{16}O variety, about 0.04 percent are ^{17}O, and 0.20 percent ^{18}O (see Table 3.3.)

9•2•2   Compounds of Oxygen

The earth's crust, consisting of the outer 4–20 miles of the planet's solid exterior surface, is composed almost exclusively of oxygen compounds. Thus oxygen is the most abundant element in the parts of our planet that we can sample and analyze (see Figure 9.2). Water is one of our most abundant compounds containing chemically combined oxygen. Other compounds are mostly complex ones called silicates, which are composed of oxygen, silicon, and various other elements. Lesser quantities of simpler oxygen compounds, for example, metallic sulfates, carbonates, phosphates, nitrates, and oxides, also occur in the earth's crust. There are small amounts of nonoxygen compounds such as metallic sulfides and chlorides, for example, ZnS and NaCl.

Oxygen is also a component of practically all organic compounds found in plants and animals. These include proteins, fats, carbohydrates, vitamins, hormones, enzymes, viruses, and bacteria.

9•2•3   Sources of Pure Elemental Oxygen

The atmosphere is the primary source of oxygen for commercial use. The oxygen is separated from the other gases present by first liquefying the air and then distilling this liquid. Nitrogen (boiling point, $-196\ °C$) and argon (boiling point, $-186\ °C$) tend to distill off first, because they have lower boiling points than oxygen (boiling point, $-183\ °C$). The relatively pure oxygen is then stored as a gas in steel cylinders under initial pressures that commonly are about 135 atm (2,000 psi).

In a modern air-separation plant where oxygen and nitrogen are produced at a rate of several tons per day, liquefaction of air and distillation proceed simultaneously. The equipment is arranged so as to permit the transfer of heat from the lines where liquefaction occurs to the columns where distillation takes place.

• *From Inorganic Compounds* • Most of the inorganic oxygen compounds found in nature are very stable. Many of these compounds probably have survived exposure to atmospheric conditions since the origin of our planet. Such compounds do not decompose when heated unless the temperature is very high. Some inorganic compounds, however, may be rather easily *pyrolyzed* (decomposed by heating) to give elemental oxygen as a reaction product. The following equations are illustrative:

$$2HgO \xrightarrow{\text{heat}} 2Hg + O_2$$
mercury(II)
oxide

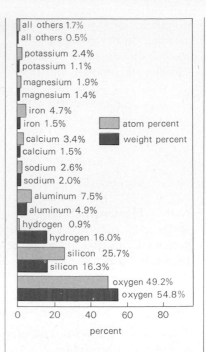

Figure 9•2 The relative abundance of elements in the earth's crust, the oceans, and the atmosphere. Elements occur in nature mainly in the form of compounds.

$$2Ag_2O \xrightarrow{\text{heat}} 4Ag + O_2$$

silver
oxide

$$2KClO_3 \xrightarrow{\text{heat}} 2KCl + 3O_2$$

potassium   potassium
chlorate   chloride

$$2NaNO_3 \xrightarrow{\text{heat}} 2NaNO_2 + O_2$$

sodium   sodium
nitrate   nitrite

Although oxygen is available inexpensively from commercial sources in cylinders of various sizes, it is frequently desirable in general chemistry laboratory exercises to use one of the thermally unstable oxygen compounds to prepare small amounts of the element. Potassium chlorate, $KClO_3$, is perhaps most often used for this purpose. It is pyrolyzed at a temperature slightly higher than its melting point of 368 °C. Because oxygen is not appreciably soluble in water, it may be collected over water, as shown in Figure 9.3. Although the reaction is exothermic, it tends to go slowly even after the decomposition temperature is reached.

The rate of decomposition is greatly increased if either manganese dioxide, MnO_2, or iron(III) oxide, Fe_2O_3, is mixed well with the potassium chlorate before heating. Neither the manganese dioxide nor the iron(III) oxide decomposes unless the temperature is excessively high; indeed, each can be recovered unchanged after the reaction is over. Substances that function in this manner are called catalysts. A **catalyst** is a substance that influences the speed of a reaction without itself being permanently changed. Just how a catalyst may do this will be discussed in Section 14.2.2. The presence of a catalyst is indicated by showing it above or below the arrow in a chemical equation, as in

$$2KClO_3 \xrightarrow[\text{heat}]{MnO_2} 2KCl + 3O_2$$

Many commercial products such as gasoline, rubber tires, polymers, and other organic synthetics that contribute to our present standard of living could not be produced without the use of catalysts. Moreover, many of the chemical changes that occur in the life processes of plants and animals are controlled by catalysts.

Oxygen also is obtained as a by-product in the preparation of hydrogen by the electrolysis of water (see Section 9.4.4) and by heating peroxides, for example, hydrogen peroxide, H_2O_2 (see Section 9.6). However, neither reaction is generally used as a laboratory or primary commercial method for producing oxygen.

9•2•4   **Reactions of Oxygen** The high electronegativity of oxygen atoms, 3.5, indicates a great tendency for oxygen to form compounds with ionic and polar covalent bonds. Generally, reactions with elemental oxygen give oxide products in which the oxidation state of oxygen is -2. Examples of oxide formation include reactions with certain metals, nonmetals, and organic compounds.

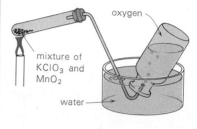

Figure 9•3 Laboratory preparation of oxygen.

With metals, ionic compounds are formed:

$$4Li + O_2 \longrightarrow 2Li_2O \qquad\qquad (1)$$
$$\text{lithium}$$
$$\text{oxide}$$

$$2Ca + O_2 \longrightarrow 2CaO \qquad\qquad (2)$$
$$\text{calcium}$$
$$\text{oxide}$$

$$2Zn + O_2 \longrightarrow 2ZnO \qquad\qquad (3)$$
$$\text{zinc}$$
$$\text{oxide}$$

With nonmetals, covalent compounds are formed:

$$C + O_2 \longrightarrow CO_2 \qquad\qquad (4)$$
$$\text{carbon}$$
$$\text{dioxide}$$

$$2H_2 + O_2 \longrightarrow 2H_2O \qquad\qquad (5)$$
$$\text{hydrogen oxide}$$
$$\text{(water)}$$

$$S + O_2 \longrightarrow SO_2 \qquad\qquad (6)$$
$$\text{sulfur}$$
$$\text{dioxide}$$

$$4P + 3O_2 \longrightarrow P_4O_6 \qquad\qquad (7)$$
$$\text{red} \qquad\qquad \text{phosphorus(III)}$$
$$\text{phosphorus}^2 \qquad \text{oxide}$$

Many organic compounds burn in oxygen to produce carbon dioxide and water. Some examples are

$$CH_4 + 2O_2 \longrightarrow CO_2 + 2H_2O \qquad\qquad (8)$$
$$\text{methane}$$
$$\text{(main component}$$
$$\text{of natural gas)}$$

$$C_2H_6O + 3O_2 \longrightarrow 2CO_2 + 3H_2O \qquad\qquad (9)$$
$$\text{ethyl}$$
$$\text{alcohol}$$

$$2C_8H_{18} + 25O_2 \longrightarrow 16CO_2 + 18H_2O \qquad\qquad (10)$$
$$\text{isooctane}$$
$$\text{(in gasoline)}$$

$$C_6H_{12}O_6 + 6O_2 \longrightarrow 6CO_2 + 6H_2O \qquad\qquad (11)$$
$$\text{glucose}$$
$$\text{(a sugar)}$$

All of these oxidation reactions, Equations (1) through (11), are exothermic. At a sufficiently high temperature, each reaction takes place

---

[2]Solid phosphorus exists in two allotropic forms, a white form with molecular formula P_4, and a red form involving large numbers of atoms crystallizing in layers. For simplicity, this red form is generally assigned the formula P.

rapidly and gives off both heat and light. When this occurs, the reaction is referred to as a **combustion.**

Oxygen in the atmosphere is in constant contact with such substances as wood, paper, gasoline, coal, zinc, and aluminum. However, at room temperature, these combustible materials react very slowly or practically not at all. This behavior illustrates an important principle. *There is no relationship between the speed of a chemical reaction and the energy change involved.*

• *Intermediate Compounds* • In certain cases, the products obtained depend upon the amount of oxygen available for the reaction. For example, when carbon reacts with a limited amount of oxygen, mostly carbon monoxide, CO, is formed:

$$2C + O_2 \longrightarrow 2CO \tag{12}$$

However, carbon monoxide is capable of undergoing further oxidation to carbon dioxide, CO_2:

$$2CO + O_2 \longrightarrow 2CO_2 \tag{13}$$

In the direct reaction of oxygen with carbon, if an excess of oxygen is available, the carbon reacts almost completely to form carbon dioxide.

$$C + O_2 \longrightarrow CO_2 \tag{4}$$

In this latter instance, carbon monoxide is probably an **intermediate compound,** a compound that is formed initially but reacts further under the reaction conditions. The equation for the overall reaction, Equation (4), is the algebraic sum of Equations (12) and (13):

$$\begin{array}{ll} 2C + O_2 \longrightarrow 2CO & (12) \\ \underline{2CO + O_2 \longrightarrow 2CO_2} & (13) \\ 2C + 2O_2 \longrightarrow 2CO_2 & \end{array}$$

or

$$C + O_2 \longrightarrow CO_2 \tag{4}$$

As another example, red phosphorus, P, reacts with a limited amount of oxygen to produce phosphorus(III) oxide, P_4O_6, or with excess oxygen to give phosphorus(V) oxide, P_4O_{10}. The production of P_4O_{10} can be carried out in two separate steps, in which case P_4O_6 is definitely an intermediate compound:

$$\begin{array}{ll} 4P + 3O_2 \longrightarrow P_4O_6 & (7) \\ \underline{P_4O_6 + 2O_2 \longrightarrow P_4O_{10}} & (14) \\ 4P + 5O_2 \longrightarrow P_4O_{10} & (15) \end{array}$$

The sum of the two equations, Equation (15), is the equation for the overall reaction.

Sulfur dioxide, produced as shown in Equation (6), does not react readily when heated with O_2 to form sulfur trioxide, SO_3. However, in the presence of a platinum catalyst, sulfur trioxide is obtained:

$$2SO_2 + O_2 \xrightarrow{\text{Pt}} 2SO_3 \tag{16}$$

When the reaction in Equation (6) is followed by (16), the SO_2 is an intermediate compound.

• *Air Pollution by Carbon Monoxide* • The products of the complete combustion of hydrocarbons or oxygen derivatives of hydrocarbons are carbon dioxide and water, Equations (8) through (11). Equation (10) is typical of the complete combustion of hydrocarbons, which are the major constituents of gasoline. Unfortunately, even in the best tuned internal combustion engines, combustion of gasoline is never complete. Some carbon monoxide is formed. This toxic substance is extremely dangerous because it readily combines with the hemoglobin in blood and thereby blocks the transportation of oxygen from the lungs to the tissues. We cannot detect the presence of carbon monoxide in the air, for it cannot be seen, smelled, or tasted. Although federal air quality standards recommend only 9 parts per million, or 9 ppm (9 parts of CO per million parts of air), as the maximum concentration permissible for an eight-hour exposure, levels of carbon monoxide may exceed 30 ppm for hours at a time on city streets or freeways packed with traffic. Other pollutants, such as aldehydes, ketones, and alkenes (see Chapter 26), are also products of incomplete combustion, but probably none is as harmful as carbon monoxide.

It has been estimated that during one year's operation, an average automobile emits 1,450 kg of carbon monoxide. Fortunately, carbon monoxide is oxidized slowly to carbon dioxide by natural processes; otherwise the concentration of the former would build up continually toward lethal concentrations.[3] Bacteria in the soil rather than atmospheric oxygen are believed to be responsible for transforming CO into CO_2. However, those who live in large paved cities cannot depend on this process to clear the air. It is in these locations during peak traffic hours where concentrations of the gas reach highly dangerous levels.

Many reactions involving oxygen take place much more readily if a catalyst is present. Certain catalysts cause practically complete oxidation of carbon monoxide to carbon dioxide. They are now used in automobile exhaust catalytic converters (see Figure 9.8).

• *Low-Temperature Oxidations* • We should not jump to the conclusion that oxygen participates in reactions only at elevated temperatures. If the other reactant is sufficiently active, a reaction will occur at room temperature, although perhaps slowly. The more active metals in the IA and IIA families react slowly with oxygen at low temperatures. Certain other elements, such as white phosphorus, ignite spontaneously in oxygen at room temperature. In this case, P_4 molecules react slowly at first, but heat produced by this exothermic reaction raises the temperature up to the combustion point.[4]

---

[3] Actually, human activities are the source of only 7 percent of atmospheric carbon monoxide; natural sources, including oxidation of methane from decaying organic matter, growth and decay of chlorophyll, and escape from the oceans, make up the remaining 93 percent.
[4] The reactions of white and red phosphorus with oxygen produce the same oxides.

• *Explosive Combustions* • Under normal conditions a combustible liquid or solid burns quietly because oxygen from the air is in contact with only its surface and the small amount of gases that volatilize from its surface. On the other hand, mixtures of air and combustible gases such as hydrogen may burn explosively. Once the combustion starts, it can flash through the gaseous mixture in a fraction of a second. The temperature rise is very rapid, so that an explosive expansion of hot gases occurs that can break windows or tear down walls. Mixtures of air with hydrogen, natural gas, or gasoline vapors, as well as air mixed with coal or wood dust, are examples of mixtures that can cause dangerous explosions when ignited. Figure 9.4 shows the use of potentially explosive but controlled combustion in rocket propulsion.

9•2•5   Oxidation–Reduction (Redox) Reactions   Each of the reactions represented by Equations (1) through (11) is an oxidation–reduction, or redox, reaction. A **redox reaction** is one in which there are changes in oxidation states. Oxidation and reduction always occur simultaneously, because an increase in oxidation state for one species is always accompanied by a decrease in oxidation state for another.

Consider, for example, the reaction of calcium with oxygen to produce calcium oxide, Equation (2). We can divide this reaction into the following half-reactions:

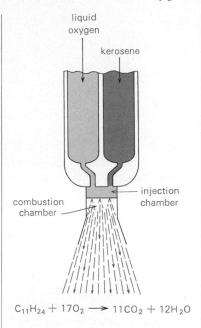

$C_{11}H_{24} + 17O_2 \longrightarrow 11CO_2 + 12H_2O$

Figure 9•4 The ejection of mass from the combustion of the fuel gives the thrust for rocket propulsion.

$$\text{oxidation:} \quad 2(\overset{0}{Ca} \longrightarrow \overset{+2}{Ca} + 2e^-)$$

$$\text{reduction:} \quad \overset{0}{O_2} + 4e^- \longrightarrow 2\overset{-2}{O}$$

As stated in Section 5.7, the half-reaction in which electrons are lost is an oxidation. The oxidation number of calcium increases from 0 in the uncombined element to +2 in the calcium ion. **Oxidation** is defined in a broad sense as a reaction in which atoms or ions undergo an *increase in oxidation state.*

The half-reaction in which electrons are gained is a reduction. The oxidation number of oxygen decreases from 0 in the uncombined element to −2 in the oxide ion. **Reduction** is defined in a broad sense as a reaction in which atoms or ions undergo a *decrease in oxidation state.*

A substance that brings about the oxidation of another substance is called an **oxidizing agent.** In Equation (2) oxygen is the oxidizing agent. It undergoes a decrease in its oxidation state. In the process of carrying out its role in the redox reaction, an oxidizing agent is reduced.

A substance that brings about the reduction of another substance is called a **reducing agent.** In Equation (2) calcium is the reducing agent. It undergoes an increase in its oxidation state. In the process of carrying out its role in the redox reaction, a reducing agent is oxidized.

In a reaction between a metal and a nonmetal, the metal is always the reducing agent, and the nonmetal is always the oxidizing agent.

In the reaction just used as an example, electrons are transferred as ions are formed. It is important to note that a complete transfer of electrons is not necessary in order to have changes in oxidation states. In the formation of molecules such as CO_2, H_2O, and SO_2, with polar covalent bonds, the pairs of electrons being shared are shifted more toward one of the atoms instead of being shared equally. For example, when sulfur burns to form SO_2,

$$S + O_2 \longrightarrow SO_2 \qquad\qquad (6)$$

atoms of sulfur and oxygen unite by sharing pairs of electrons. However, each electron pair is held closer to an oxygen. Each oxygen part of the molecule is assigned an oxidation state of -2 and the sulfur part, an oxidation state of $+4$. Sulfur is said to be oxidized, because its state changes from 0 to $+4$. Oxygen is reduced, with its state changing from 0 to -2.

Even though the electrons are shared in SO_2 rather than transferred from S to O, the half-reactions are written as though electrons are lost and gained:

$$\text{oxidation:} \quad \overset{0}{S} \longrightarrow \overset{+4}{S} + 4e^-$$

$$\text{reduction:} \quad \overset{0}{O_2} + 4e^- \longrightarrow 2\overset{-2}{O}$$

• *Balancing Redox Equations* • There are special methods that are helpful in balancing complex redox equations, but we will not consider these until we encounter such reactions in Chapter 18. The majority of redox equations can be balanced by the simple inspection technique illustrated in Section 2.2. However, important chemical details are revealed when one separates a balanced redox reaction into its component parts. To analyze reactions in this way, we must be able to (1) recognize a redox equation, (2) write the equations for the half-reactions for the oxidation and the reduction, and (3) show that the amount of oxidation is just balanced by the amount of reduction. These points are illustrated in Example 9.1.

━━━ • Example 9•1 • ━━━

Which of the following represent redox reactions? For these, write the equations for the half-reactions and balance the amounts of oxidation and reduction.
(a) $Al + Br_2 \longrightarrow AlBr_3$
(b) $Mg + Fe(NO_3)_3 \longrightarrow Mg(NO_3)_2 + Fe$
(c) $NH_4Cl + Fe_2(SO_4)_3 \longrightarrow (NH_4)_2SO_4 + FeCl_3$

━━━ • Solution • To identify a redox reaction, we assign oxidation numbers to all elements and determine whether any element undergoes a change in oxidation number.

(a)
$$\overset{0}{Al} + \overset{0}{Br_2} \longrightarrow \overset{+3\,(-1)_3}{AlBr_3}$$

Yes, this is a redox reaction because the oxidation states of aluminum and bromine change. The balanced equations for the half-reactions and the overall reaction are

$$\text{oxidation:} \quad 2(\overset{0}{Al} \longrightarrow \overset{+3}{Al} + 3e^-)$$

$$\text{reduction:} \quad 3(\overset{0}{Br_2} + 2e^- \longrightarrow 2\overset{-1}{Br})$$

$$\text{overall reaction:} \quad 2Al + 3Br_2 \longrightarrow 2AlBr_3$$

(b)
$$\overset{0}{Mg} + \overset{+3}{Fe}\overset{(-1)_3}{(NO_3)_3} \longrightarrow \overset{+2}{Mg}\overset{(-1)_2}{(NO_3)_2} + \overset{0}{Fe}$$

Yes, this is a redox reaction because the oxidation states of magnesium and iron change. Note that the net oxidation number of the nitrate ion is -1 (see Section 5.7.3). The balanced equations for the half-reactions and the overall equation are

oxidation: $3(\overset{0}{Mg} \longrightarrow \overset{+2}{Mg} + 2e^-)$

reduction: $2(\overset{+3}{Fe} + 3e^- \longrightarrow \overset{0}{Fe})$

overall reaction: $3Mg + 2Fe(NO_3)_3 \longrightarrow 3Mg(NO_3)_2 + 2Fe$

(c)
$$\overset{+1}{NH_4}\overset{-1}{Cl} + \overset{(+3)_2}{Fe_2}\overset{(-2)_3}{(SO_4)_3} \longrightarrow \overset{(+1)_2}{(NH_4)_2}\overset{-2}{SO_4} + \overset{+3}{Fe}\overset{(-1)_3}{Cl_3}$$

No, this is not a redox reaction because no element undergoes a change in oxidation state. Again, we have used a net oxidation number for a polyatomic ion that is not involved in any change of oxidation numbers.

See also Exercises 11, 14, 15, and 18 at the end of the chapter.

In writing half-reactions to show the balance of oxidation and reduction, we will usually use oxidation numbers above symbols, as in Example 9.1(a). For polyatomic ions, however, it is often convenient to use the charge on the ion as the net oxidation number (see Section 5.7.3). If a polyatomic ion does not change in the reaction, we know the oxidation numbers of its elements do not change [see Example 9.1(b) and (c)].

9•2•6   Uses of Oxygen   Oxygen is useful as it exists naturally in the air, as it exists in air enriched with oxygen, and as the pure element.

• *Oxygen–Carbon Dioxide Cycle in Nature* • In respiration in humans and other animals, inhaled air is brought into contact with blood in the lungs. There the dark red hemoglobin of the venous blood reacts with a certain amount of the oxygen to form oxyhemoglobin, the bright red component of arterial blood. This oxyhemoglobin is transported through the arteries to the cells, where the oxygen is used in the oxidation of food to carbon dioxide, water, and other products. The carbon dioxide is picked up by the blood, largely as the bicarbonate ion, and along with hemoglobin is transported through the veins to the lungs. The carbon dioxide is liberated in the lungs and exhaled, and the cycle starts anew. The oxidation reactions in the cells liberate the energy that keeps us warm and enables us to do work.

The energy to operate an automobile or a diesel engine or to heat homes with stoves or furnaces is derived in the same general way from the oxidation of gasoline, coal, and other fuels. In such activities humans are constantly removing oxygen from the air and converting it mainly to carbon dioxide and water. But we do not have to worry about the supply of oxygen becoming depleted so long as growing plants take up carbon dioxide and water, convert them into sugars, starch, and cellulose, and return oxygen to the air. This cycle in nature keeps the amount of elementary oxygen relatively constant.

The first balanced equation below represents the oxidation of sugar in the cells of animals; the second equation represents the formation of sugar in plants by photosynthesis:[5]

$$C_6H_{12}O_6 + 6O_2 \xrightarrow{\text{enzymes}} 6CO_2 + 6H_2O + \text{heat energy}$$

$$\text{radiant energy} + 6CO_2 + 6H_2O \xrightarrow{\text{chlorophyll}} C_6H_{12}O_6 + 6O_2$$

Today the amount of atmospheric carbon dioxide is increasing because of our intensive use of fossil fuel as an energy source. Although the accompanying reduction of the earth's elemental oxygen supply is insignificant, the increased concentration of CO_2 in the atmosphere may have long-range deleterious effects on our world's climate (see Section 25.3.2).

• *Commercial Uses* • Commonplace uses of oxygen-enriched air are in oxygen equipment for respiratory devices in hospitals, airplanes, mines, and underwater vessels. Pure oxygen is used in oxyacetylene and oxyhydrogen torches as well as in the propulsion of rockets. Large amounts of pure oxygen also are used to burn excess carbon out of molten iron in the production of steel. In fact, the steel industry uses by far the most commercial oxygen. The chemical industry also uses large quantities of oxygen in the production of oxygen-containing inorganic and organic compounds.

• *Oxygen, A Key Chemical* • The weekly magazine of the chemical profession, *Chemical and Engineering News*, identifies as *key chemicals* certain substances that have major industrial and commercial importance because of the huge amounts produced. These chemicals are useful not only in their own right, but in serving as raw materials for making other substances. A number of key chemicals are described at appropriate places in later chapters, as oxygen is here. Data are taken from the latest available report in *C & E News*.

Key Chemical: O_2
oxygen

How Made
Distillation of liquid air

U.S. Production; Cost
~430 billion ft³/year
~$0.31/100 ft³,
or $3.50/100 lb

Major Uses
65% steel making
7% fabrication of metals
15% chemical processing

9.3   Ozone

Ozone, O_3, is a colorless gas that condenses to form a blue liquid at $-111\ °C$; it freezes to a blue-black solid at $-192\ °C$ (see Figure 9.5 on Plate 3). The structure of ozone is represented as a resonance hybrid of two equivalent Lewis formulas:

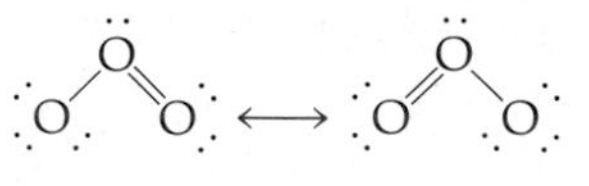

The O—O—O bond angle in ozone is 117 °.

The name ozone comes from the Greek *ozein*, to smell. Its sharp odor has been variously described as like new-mown hay, pungent, acrid, like

---

[5] These are overall equations that do not represent all the complex reactions necessary to convert the reactants into the products. In woody plants a large proportion of the end product is cellulose, $(C_6H_{10}O_5)_x$.

freshly ironed sheets, disagreeable, sweet, like chlorine bleach, and like something that gets you at the back of the nostrils. It is an extremely toxic substance, more toxic than cyanide (KCN or NaCN), strychnine, and carbon monoxide.

9•3•1   **Preparation of Ozone**   Ozone is commonly prepared by allowing a low-temperature electric discharge to pass through oxygen. (The sharp odor of ozone is frequently detected in the vicinity of electric motors, leaking power lines, and lightning.) The equation for the reaction is

$$3O_2(g) \longrightarrow 2O_3(g)$$

A simple ozonizing apparatus is shown in Figure 9.6. When pure dry oxygen at 20 °C is passed through the ozonizer, about 6 percent is converted into ozone. By allowing the oxygen to remain in the tube, at a low pressure and very low temperature, as much as 99.8 percent conversion to ozone has been obtained. Because the boiling point of ozone is 72 °C higher than that of oxygen, a liquefied mixture of the two can be separated by distillation. Such a distillation is hazardous, however, because of the unstable nature of the O_3 molecule.

9•3•2   **Reactivity of Ozone**   At room temperature, ozone tends to decompose on standing as follows:

$$2O_3(g) \longrightarrow 3O_2(g)$$

Ozone is a reactive molecule that can participate in oxidation reactions at temperatures at which ordinary oxygen is relatively inactive. For example:

$$6KI + O_3 \xrightarrow{20\ °C} 3K_2O + 3I_2$$

$$KI + O_2 \xrightarrow{20\ °C} \text{reaction extremely slow}$$

The oxidation of mercury and silver further illustrates this point. At normal temperatures these two metals are oxidized by ordinary oxygen at an imperceptible rate, if at all. When placed in ozone, they rapidly become tarnished with an oxide coating because of the oxidizing action of ozone. Also, moist sulfur is converted to sulfuric acid with O_3 (presumably via sulfur trioxide as an intermediate compound), whereas O_2 does not effect this conversion.

9•3•3   **Ozone in the Upper Atmosphere**   The various regions of the atmosphere are shown in Figure 9.7. Ozone is the most important trace constituent of the stratosphere, a stable, cloudless region with little vertical circulation. Although present only in small amounts, ozone provides a protective screen to shield living organisms from the sun's radiation, which is energetic enough to be quite harmful. The amount of ozone in the stratosphere is maintained by a delicate balance between the processes of formation and decomposition.

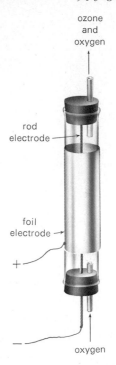

ozone and oxygen

rod electrode

foil electrode

+

−

oxygen

Figure 9•6 Laboratory ozonizer. The metal rod inside and the metal foil outside the glass tube are charged just enough so a "silent" electric discharge passes between the electrodes. Observable as a blue glow in a dark room, this discharge gives a greater yield of ozone than a high temperature spark.

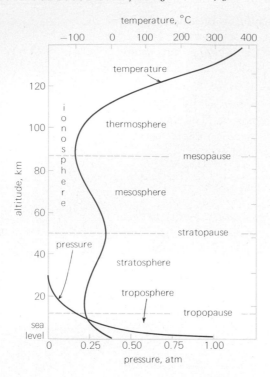

Figure 9.7 General trends of temperature and pressure in the atmosphere. Altitudes of the -pause layers vary with latitude. Erratic temperature inversions are not uncommon, particularly in the troposphere. At altitudes above 400 km, the temperature reaches a maximum of about 1,200 °C.

Ozone is formed predominantly in the upper stratosphere, at altitudes above 30 km. In this region, the short-wavelength, high-energy ultraviolet (uv) rays of the sun are absorbed by ordinary oxygen molecules, and oxygen atoms are produced:

$$O_2 \xrightarrow[\text{radiation}]{\text{uv}} 2O \tag{17}$$

Each oxygen atom then can combine with another molecule of oxygen to form a molecule of ozone:

$$O + O_2 + M \longrightarrow O_3 + M \tag{18}$$

In Equation (18), M is a third body such as an oxygen, nitrogen, or another molecule present in the atmosphere. This third molecule, M, must be present to take up part of the energy; otherwise the O_3 molecule, which is not very stable, would probably decompose rapidly. By doubling Equation (18) and adding it to Equation (17), we obtain Equation (19) for the overall reaction (see Example 9.2):

$$3O_2 \longrightarrow 2O_3 \tag{19}$$

Because this reaction is initiated by radiant energy, it is called a **photochemical reaction** (see also Figure 20.2).

• Example 9.2 •

Show that doubling Equation (18) and adding it to Equation (17) yields the overall equation, Equation (19).

• Solution • To double Equation (18), we multiply all parts by two:

$$2O + 2O_2 + 2M \longrightarrow 2O_3 + 2M \qquad 2(18)$$

Add (17)
$$\frac{O_2 \longrightarrow 2O}{2O + 3O_2 + 2M \longrightarrow 2O_3 + 2O + 2M} \qquad (17)$$

We cancel all species that occur on both sides of the arrow:

$$\cancel{2O} + 3O_2 + \cancel{2M} \longrightarrow 2O_3 + \cancel{2O} + \cancel{2M}$$

Finally, we write the simplified overall equation:

$$3O_2 \longrightarrow 2O_3 \qquad (19)$$

See also Exercise 19.

Most of the ozone is produced over the tropics at altitudes between 25 and 35 km. However, due to movement, the highest concentrations are found over the polar regions at altitudes of about 15 km. Ozone molecules absorb ultraviolet radiation according to the following equation:

$$O_3 \xrightarrow[\text{radiation}]{\text{uv}} O_2 + O \qquad (20)$$

This absorption process prevents most of the sun's lethal rays from reaching the earth's surface. Almost all of the oxygen atoms produced by this process rapidly combine to reform ozone by the reaction shown in Equation (18). Because reaction (18) is exothermic, the net result is the conversion of the sun's radiant energy into heat energy. This heat causes the temperature in the stratosphere to increase with altitude, as shown in Figure 9.7.

The ozone that is being formed continuously in the stratosphere is being destroyed continuously by natural processes. If the rates of the ozone-destroying processes were increased, ozone could be used up faster than it is formed (see the special topic, "Threatened Depletion of the Earth's Ozone Shield"). The net result would be a reduction in the concentration of the protective ozone shield, which would allow more harmful ultraviolet radiation to reach the earth's surface.

9.3.4   Ozone and Its Role in Air Pollution in the Lower Atmosphere
Whereas ozone in the stratosphere is essential to life on earth, its presence in the lower troposphere near ground level can be quite harmful. Continued breathing of air with as little as 0.1 ppm of ozone (0.1 part of ozone per million parts of air) produces headaches and other unpleasant symptoms. In some cities, the large amount of ozone in the air is a health hazard. It is toxic not only to humans but also to crops and other materials exposed to the atmosphere. It is destructive because it is such a strong oxidizing agent. Only fluorine and a few other substances are stronger oxidizing agents than ozone.

Effects on animals of exposure to ozone include structural changes in lung tissue, low resistance to respiratory bacterial infection, and disruption of cellular biochemistry. Ozone is particularly destructive to rubber and certain plastics. It is even found that in certain areas where the ozone concentration is higher than in other sections of the country, the lifetime of tires on automobiles is shorter.

Ozone is formed indirectly from one of the constituents of automobile exhaust. Although most of the nitrogen in the air drawn into an internal combustion engine does not react, a small amount does react with oxygen at the temperature of the combustion chamber to produce nitric oxide, NO:

$$N_2 + O_2 \longrightarrow 2NO \qquad (21)$$

Some of the nitric oxide, which is relatively harmless, is oxidized to brown,

★   Special Topic ★   Special Topic ★   Special Topic ★   Special Topic ★   Special Topic ★   Special Topic ★   Special Topic ★   Special Topic   ★

Threatened Depletion of the Earth's Ozone Shield For all living things on earth, the ozone in the lower stratosphere plays a crucial role in absorbing much of the sun's ultraviolet (uv) radiation, which otherwise would reach the earth's surface. The uv radiation has sufficient energy to break bonds in many molecules. This can be harmful to living organisms because the breaking of a bond can be the first step in an undesirable reaction. Also, it is thought that the warming of the stratosphere by the absorption of most of the uv radiation has an important effect on the climate at the earth's surface.

For some humans, overexposure to uv radiation can cause skin cancer. For some plants mutation rates can be increased, and for others the efficiency of photosynthesis can be decreased. Plankton, the basis for the ocean's food chain, is sensitive to uv radiation.

The possible depletion of the ozone shield by pollutants generated by human activities has become a great concern in the last decade or so. It is feared that the natural concentration of ozone could be changed dramatically by relatively small increases in the amounts of certain molecules. Oxides of nitrogen are such molecules, which have several artificial sources. A small amount of the nitrogen in fertilizers escapes to the atmosphere as N_2O. Some of this N_2O probably finds its way to the stratosphere where it is oxidized to NO, which reacts with ozone. Two sources that deliver oxides of nitrogen directly into the stratosphere are nuclear explosions and the high-flying supersonic transports (SSTs). At present most countries are observing the ban on atmospheric testing of nuclear bombs and only three countries are producing SSTs. The prediction that, in one year of full-scale operation, 500 SSTs could destroy 10–20 percent of the ozone layer was critical in the decision of the U.S. Congress in the 1970s not to support the building of supersonic commercial aircraft.

Other types of molecules that could attack and deplete the ozone shield are the chlorine compounds with low molecular weights that are used in refrigerators, air-conditioners, and some aerosol dispensers. Two such compounds are Freon-11, CCl_3F, and Freon-12, CCl_2F_2. Easily liquefied, they vaporize at room temperature when released from a pressurized container. These chlorofluorocarbons are so inactive chemically that they have long been considered almost ideal for uses around the home.

The chemical inactivity of these chlorine compounds toward other substances, however, is the very property that results in the threat that they pose for the ozone layer. Because they do not react with other constituents of the atmosphere, they diffuse around for years, and eventually many of the molecules can be expected to reach the upper stratosphere. Laboratory experiments show that CCl_3F and CCl_2F_2 are decomposed by uv radiation to form chlorine atoms that catalyze the destruction of ozone. Recognition in the 1970s of this threat to the ozone layer prompted a great reduction in the number of products packaged as aerosols and an intense search for other refrigerant gases.

Banning the production and use of freons has not been a simple matter. In the United States alone, freon-related industries have employed approximately a million persons in the manufacture and sales of products worth nearly $8 billion annually. Decisions that affect such huge undertakings rarely can take effect immediately and must be based on economic as well as scientific considerations. The threat from increased use of nitrogen fertilizers seems even more difficult to circumvent because, as population increases, requirements for food and the greater use of fertilizers presumably will increase.

pungent, toxic nitrogen dioxide, NO_2. Once it is formed, NO_2 undergoes photochemical decomposition by sunlight to yield nitric oxide and atomic oxygen:

$$NO_2 \xrightarrow[\text{radiation}]{\text{uv}} NO + O \qquad (22)$$

The atomic oxygen formed by reaction (22) is extremely reactive, and it can combine with oxygen in the air to produce ozone by reaction (18), shown previously:

$$O + O_2 + M \longrightarrow O_3 + M \qquad (18)$$

After ozone is formed it can react with more NO to produce NO_2:

$$NO + O_3 \longrightarrow NO_2 + O_2 \qquad (23)$$

Although reaction (23) removes some of the ozone built up in the air, it provides an additional source of NO_2 for the formation of more ozone via reactions (22) and (18). Hence, for every molecule of NO_2 that is delivered to the atmosphere (from an internal combustion engine or elsewhere), a molecule of O_3 may be formed. It is ironic that pollution decreases the amount of O_3 in the stratosphere where it is helpful but increases the amount in the troposphere where it is harmful.[6]

• *Smog* • Ozone and nitrogen oxides are some of the toxic pollutants that are added continuously to our air and contribute to the formation of smog. When smog results from photochemical reactions such as Equation (22), it is called *photochemical smog*. Table 9.1 lists some of the typical constituents of photochemical smog. Unburned and partially burned hydrocarbons emitted in automobile exhausts can react with ozone, atomic oxygen, or nitrogen oxides.

One way to reduce the amounts of chemically active pollutants produced directly or indirectly by automobile exhausts is to pass the exhausts through beds of catalysts. One such mixture of catalysts is known as a three-way catalyst because it will catalyze the reactions of exhaust hydrocarbons, carbon monoxide, and nitrogen oxides to produce carbon dioxide, water, and nitrogen. Three-way catalytic units may become standard equipment on many automobiles. The three-way catalyst is generally made of platinum and platinum–rhodium combinations supported on silica, SiO_2, alumina, Al_2O_3, or a combination of these two.

Catalysts for automobile exhaust gases must be chosen carefully to avoid the production of unwanted compounds. Nitrogen from the air and sulfur from impurities in gasoline may form such pollutants as hydrogen cyanide, HCN, hydrogen sulfide, H_2S, and the oxides of sulfur, SO_2 and SO_3. Unfortunately, some of the first catalysts used to control exhaust emissions caused sulfur compounds in fuels to form SO_3, which reacts with H_2O in the

---

Table 9•1 Concentrations of substances in photochemical smog

Constituent	Concentration, ppm
oxides of nitrogen	0.20
NH_3	0.02
H_2	0.50
H_2O	2×10^4
CO	4×10^1
CO_2	4×10^2
O_3	0.50
SO_2	0.20
CH_4	2.5
other alkanes	0.25
C_2H_4 (ethylene)	0.50
other alkenes and alkynes	0.50
C_6H_6 (benzene)	0.10
RCHO (aldehydes)	0.60

Source: *Data are reproduced with permission from R. D. Cadle and E. R. Allen,* Science, **167**, 246 (1970). *Copyright 1970 by the American Association for the Advancement of Science.*

---

[6]Only about 10 percent of all the NO and NO_2 in the atmosphere is produced artificially; the rest is formed in natural processes (see Section 24.10.1).

atmosphere to form the strong acid H_2SO_4. Figure 9.8 shows that 1976 automobiles equipped with some type of catalytic converter emit much less carbon monoxide than older models.

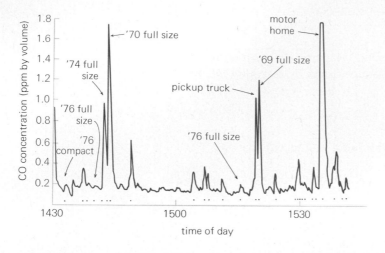

Figure 9.8 Record of CO concentrations inside an automobile traveling on an interstate highway as it was passed by other vehicles during a 75-minute period in April 1977. Note the low levels of CO in the exhausts from the 1976 models. *Source:* Reproduced by permission from L. W. Chaney, "Carbon Monoxide Automobile Emissions Measured from the Interior of a Traveling Automobile," *Science*, 199, 1203 (1978). © 1978 by the American Association for the Advancement of Science.

9.4 Hydrogen

9.4.1 Elemental Hydrogen

On our earth, hydrogen does not occur to any appreciable extent in the elemental form. Some volcanic gases contain elemental hydrogen, and there is a trace of it in the atmosphere, but probably less than 0.0001 percent. While elemental hydrogen is very rare in our planet's atmosphere, it is the most abundant element in the universe as a whole. For example, elemental hydrogen comprises about 75 percent of the sun's mass.

There are three known isotopes of hydrogen: ^1H, ^2H, and ^3H. Of all the hydrogen atoms existing in nature, 99.98 percent are of the ^1H variety, about 0.02 percent are ^2H, and only an infinitesimal percentage exist as ^3H. The isotopes of hydrogen have less resemblance to one another than do isotopes of other elements, due principally to the great percentage differences in their weights. For this reason, hydrogen isotopes are identified by individual names, whereas mass numbers suffice for all other isotopes. Table 9.2 gives the names, special symbols, and some of the properties of the hydrogen isotopes. Deuterium is also called *heavy hydrogen*.

Table 9.2 Isotopes of hydrogen

	^1H or H protium	^2H or D deuterium	^3H or T tritium
atomic mass, amu	1.0078	2.0141	3.0160
freezing point, °C	−259.1	−254.4	
boiling point, °C	−252.7	−249.6	
stability of nucleus	stable	stable	unstable; half of any sample decays to ^3He in 12.3 years

9●4●2   **Compounds of Hydrogen**   Compounds of hydrogen are widely distributed in nature. Practically all organic compounds contain hydrogen. Thus, sugars, starches, fats, cellulose, proteins, coal, natural gas, gasoline, fuel oils, kerosene, and lubricating oils all contain hydrogen. Water, an inorganic compound, is the most abundant compound of hydrogen on earth. The relative abundance of hydrogen (in compounds) in the earth's crust, the oceans, and the atmosphere is shown in Figure 9.2.

The common acids are a class of hydrogen compounds of special interest. Some acids (hydrochloric acid in gastric juice, acetic acid in vinegar, and citric acid in citrus fruits) are formed in natural processes. These and many other acids, including sulfuric, nitric, and phosphoric, are synthesized on a large scale by the chemical industry.

9●4●3   **Laboratory Preparation of Hydrogen**   Hydrogen is often prepared in the laboratory by the action of dilute solutions of strong acids with moderately active metals (see Figure 9.9). Dilute hydrochloric and sulfuric serve nicely as the acids; zinc, aluminum, iron, and magnesium are frequently used as the metals. The following equations are representative:

$$Zn + 2HCl \longrightarrow ZnCl_2 + H_2$$
$$2Al + 3H_2SO_4 \longrightarrow Al_2(SO_4)_3 + 3H_2$$

Metals in the IA family and the lower part of the IIA family are so reactive that they will react even with water to produce hydrogen. The following reactions are typical:

$$2Na + 2HOH \longrightarrow 2NaOH + H_2$$
$$\text{sodium hydroxide}$$
$$Ca + 2HOH \longrightarrow Ca(OH)_2 + H_2$$
$$\text{calcium hydroxide}$$

Sodium, potassium, rubidium, and cesium react so energetically with water at room temperature that the hydrogen may be ignited by the heat liberated. Such reactions with water are so violent that they must be carried out with considerable caution. Calcium and lithium react more slowly with water.

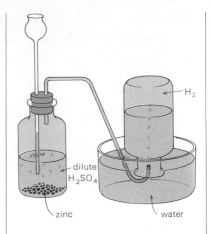

Figure 9●9 Laboratory preparation of hydrogen.

────── ● Example 9●3 ● ──────

For each of the following redox reactions carried out in water solution, write equations for the half-reaction and the overall reaction. Identify the oxidizing and reducing agents.
(a) magnesium + hydrochloric acid $\longrightarrow$
(b) cesium + water $\longrightarrow$

As the first step, write the formulas for reactants and products, and above each symbol write its oxidation number.

• Solution •

(a) The products of the first reaction are magnesium chloride, $MgCl_2$, and hydrogen, H_2:

$$\overset{0}{Mg} + \overset{+1\ -1}{HCl} \longrightarrow \overset{+2\ (-1)_2}{MgCl_2} + \overset{0}{H_2}$$

The equations for the half-reactions and overall reaction are

$$\begin{aligned}
\text{oxidation:} \quad & \overset{0}{Mg} \longrightarrow \overset{+2}{Mg} + 2e^- \\
\text{reduction:} \quad & \overset{+1}{2H} + 2e^- \longrightarrow \overset{0}{H_2} \\
\hline
\text{overall reaction:} \quad & \overset{+1}{Mg} + 2H \longrightarrow \overset{+2}{Mg} + H_2
\end{aligned}$$

or

$$Mg + 2HCl \longrightarrow MgCl_2 + H_2$$

Mg is the reducing agent and HCl is the oxidizing agent. (In HCl, only the hydrogen undergoes a change in oxidation state.)

(b) The products of the second reaction are cesium hydroxide, CsOH, and hydrogen, H_2:

$$\overset{0}{Cs} + \overset{(+1)_2\ -2}{H_2O} \longrightarrow \overset{+1\ -2+1}{CsOH} + \overset{0}{H_2}$$

The equations for the half-reactions and overall reaction are

$$\begin{aligned}
\text{oxidation:} \quad & 2(\overset{0}{Cs} \longrightarrow \overset{+1}{Cs} + e^-) \\
\text{reduction:} \quad & \overset{+1}{2H} + 2e^- \longrightarrow \overset{0}{H_2} \\
\hline
\text{overall reaction:} \quad & 2\overset{+1}{Cs} + 2H \longrightarrow 2\overset{+1}{Cs} + H_2
\end{aligned}$$

or

$$2Cs + 2H_2O \longrightarrow 2CsOH + H_2$$

Cs is the reducing agent and H_2O is the oxidizing agent. (One of the hydrogens in each H_2O molecule undergoes a change in oxidation state, but the other hydrogen does not.)

See also Exercises 28, 29, and 32.

9•4•4   Commercial Production of Hydrogen Hydrogen is one of the most important substances in the chemical spectrum. Domestic U.S. hydrogen production has increased about 15 percent annually since World War II. Of the present world production of hydrogen, estimated at about 10 trillion standard cubic feet, over one third is produced in the United States. It is estimated that in the year 2000, U.S. hydrogen production could be as much as 52 trillion standard cubic feet.

Three of the present methods of obtaining this increasingly valuable element are the water gas method, the steam–hydrocarbon method, and the electrolysis of water.

• *Water Gas Method* • When steam is passed over hot coke, carbon monoxide and hydrogen are produced:[7]

$$C + H_2O \longrightarrow CO + H_2$$

This mixture of carbon monoxide and hydrogen, called *water gas,* is of considerable value as a fuel, because both substances are combustible. If pure hydrogen is desired, the mixture is treated with steam in the presence of a catalyst to oxidize the carbon monoxide to carbon dioxide:

$$\underbrace{CO + H_2}_{\text{water gas}} + H_2O \longrightarrow CO_2 + 2H_2$$

The carbon dioxide is readily removed by passing the mixture of the two gases through water under pressure. The carbon dioxide dissolves; the hydrogen does not.

• *Steam–Hydrocarbon Method* • Large amounts of commercial hydrogen are made by passing mixtures of hydrocarbons and steam over a nickel catalyst at high temperatures. The equation for the reaction involving the simplest hydrocarbon, methane, is

$$CH_4 + 2H_2O \xrightarrow[\text{heat}]{\text{Ni}} CO_2 + 4H_2$$

The carbon dioxide and hydrogen can be separated as described previously, or the carbon dioxide can be removed by passing the mixture over lime (calcium oxide):

$$\underbrace{CO_2 + H_2}_{\text{gaseous mixture}} + \underset{\text{solid}}{CaO} \longrightarrow \underset{\text{solid}}{CaCO_3} + \underset{\text{gas}}{H_2}$$

• *From Water by Electrolysis* • Because our most abundant source of hydrogen is water, it would be ideal if water could be decomposed cheaply into hydrogen and oxygen. A satisfactory way of decomposing water is by passing a direct current through water to which a small amount of sulfuric acid has been added (Figure 9.10). The use of an electric current to bring about a redox reaction is called **electrolysis.** The equations for the half-reactions and the overall reaction for the electrolysis of water are

reduction: $2[2H^+ + 2e^- \longrightarrow H_2(g)]$

oxidation: $2H_2O(l) \longrightarrow O_2(g) + 4H^+ + 4e^-$

overall reaction: $2H_2O(l) \xrightarrow[\text{current}]{\text{direct}} 2H_2(g) + O_2(g)$

---

[7]Coke is the solid residue of carbon and ash that remains after coal is strongly heated in the absence of air.

When 36 g of water is decomposed by an electric current into hydrogen and oxygen, electric energy equivalent to about 572 kJ is used. Because of the high energy requirement, the preparation of oxygen and hydrogen by the electrolysis of water is too expensive for most commercial uses. However, the process is important in obtaining very pure hydrogen and oxygen.

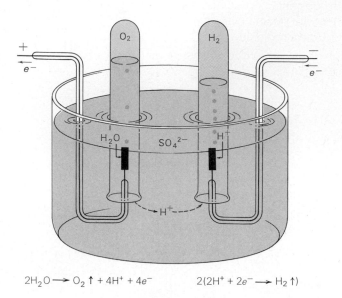

$$2H_2O \longrightarrow O_2 \uparrow + 4H^+ + 4e^- \qquad 2(2H^+ + 2e^- \longrightarrow H_2 \uparrow)$$

Figure 9.10 When a direct current is passed into water containing some H_2SO_4, water molecules give up electrons at the anode (+) to release O_2. Hydrogen ions acquire electrons at the cathode (−) and form H_2 molecules. Sulfuric acid is used to provide ions to conduct the current. H_2SO_4 reacts with water in an ionization reaction (see Section 9.5.1).

9.4.5   **Reactions of Hydrogen** With its intermediate electronegativity of 2.1, hydrogen has a tendency to form polar covalent bonds with nonmetals. Because most of the nonmetals with which it reacts have electronegativities greater than 2.1, hydrogen functions as a reducing agent and is oxidized in these reactions. Its change in oxidation state is from 0 to +1. Many such reactions are exothermic and proceed rapidly at elevated temperatures:

$$2H_2 + O_2 \longrightarrow 2H_2O$$
$$H_2 + S \longrightarrow H_2S$$
$$H_2 + Cl_2 \longrightarrow 2HCl$$
$$H_2 + F_2 \longrightarrow 2HF$$
$$3H_2 + N_2 \longrightarrow 2NH_3$$

Hydrogen also reacts with the more active metals. Ionic *metallic hydrides* containing the hydride ion, H^-, are formed:

$$2Na + H_2 \longrightarrow \quad 2NaH$$
$$\text{sodium hydride}$$

$$Ca + H_2 \longrightarrow \quad CaH_2$$
$$\text{calcium hydride}$$

Because the metals have low electronegativities, hydrogen functions in these cases as an oxidizing agent; its oxidation state changes from 0 to −1. Hydride ions, H^-, are isoelectric with helium atoms. The formation of such ions is predicted by the *rule of two*.

Elemental hydrogen is an excellent reducing agent, particularly in the reduction of metallic oxides. This may be demonstrated by passing hydrogen over hot copper oxide (Figure 9.11). The black copper oxide rapidly

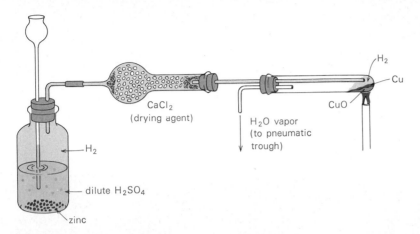

Figure 9•11 When hydrogen is passed over hot copper oxide, copper and water are formed.

disappears and is replaced by metallic copper; water meanwhile appears in the cooler part of the test tube. The following equation describes the reaction:

$$CuO + H_2 \longrightarrow Cu + H_2O$$

An interesting and important reaction of hydrogen is its tendency to add to a double or triple bond joining two carbon atoms. This reaction, called **hydrogenation,** is widely used in the production of solid cooking fats, petroleum products, and other synthetic organic chemicals. Ordinarily, a finely divided metal such as nickel or platinum must be present as a catalyst for the reaction. Hydrogenation is illustrated schematically in the following equation:

$$-\overset{|}{C}=\overset{|}{C}- \quad + H_2 \xrightarrow{\text{catalyst}} \quad -\overset{|}{\underset{H}{C}}-\overset{|}{\underset{H}{C}}-$$

portion of   hydrogenated
organic molecule   molecule

9•4•6   **Uses of Hydrogen**   About 42 percent of the hydrogen produced is used in making ammonia (see Section 24.12.1) and another 38 percent in refining of petroleum (see Section 26.6.1). The remaining 20 percent is divided among many industries, with metallurgical and food processing applications requiring the largest amounts. In the future, more petroleum refining, increased growth of plastics and elastomers (rubber-like substances), more desulfurization of fuel oils, more iron ore reduction, hydrogen–air fuel cells (see Table 18.4), and aerospace uses all will tend to increase hydrogen demand. When our fossil fuel sources become sufficiently depleted, hydrogen may also become the basis of our fuel economy.

9•4•7   A Hydrogen Economy Because our growing energy demands are depleting our reserves of fossil fuels at an alarming rate (Figure 9.12), attention is turning to long-term replacements for coal and petroleum. From the global viewpoint, total fossil fuel reserves are probably sufficient to maintain present living standards until the middle of the twenty-first century. However, examination of the data on the distribution of petroleum in Table 9.3 reveals that the United States may have to develop the technology for a new energy system sooner than might otherwise be expected.

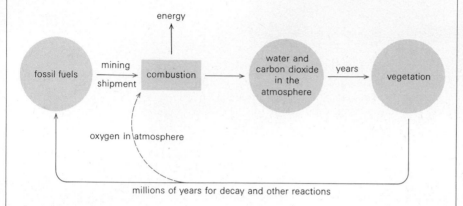

Figure 9•12 Fossil fuel cycle. The present worldwide rate of use of fossil fuels far exceeds the rate of renewal.
Source: Adapted by permission from *Chem. Eng. News,* 50 (*26*), 14 (1972).

Nuclear fission, and perhaps nuclear fusion, may become the prime energy source of the future. Nearly all proposals for utilizing nuclear power call for conversion to more electricity despite the fact that only about 20 percent of present and anticipated energy consumption is in the form of electric power. The remaining 80 percent is in the form of heat. The big problem is how to convert nuclear energy into chemical energy for storage and transportation. One proposed solution lies with a fuel economy based on hydrogen.

Electricity produced by nuclear reactors could be used to convert our most abundant compound, water, into hydrogen and oxygen. Unlike fossil fuels, the time required for the regeneration of hydrogen is relatively short (see Figure 9.13). One of the great advantages of hydrogen as fuel is its virtually pollution-free combustion to water. Standard gasoline engines have been converted to use hydrogen gas as fuel. Other than water vapor, the only

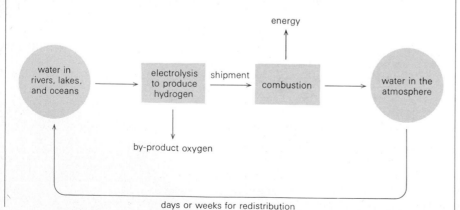

Figure 9•13 Hydrogen fuel cycle. Not only is the raw material for this cycle plentiful, but the rate of renewal is rapid. Oxygen is the most common by-product of electrolysis; if too much oxygen is produced to use, it can be released to the atmosphere.
Source: Adapted by permission from *Chem. Eng. News,* 50 (*26*), 14 (1972).

significant combustion product is a small amount of nitrogen oxides originating from the air used in combustion. A hydrogen-powered engine may be one type that will meet the federal emission requirements of the future.

Fuel cells based on hydrogen are another possibility for powering motor vehicles. Such fuel cell–battery combinations and electric motors would run trucks, cars, ships, and trains. Liquid hydrogen could be carried as fuel by aircraft. Iron ore could be economically reduced by hydrogen directly to iron:

$$Fe_3O_4 + 4H_2 \xrightarrow{\text{heat}} 3Fe + 4H_2O$$

It is expected that the exhaust gases from the reduction of iron ore by hydrogen would produce less atmospheric pollution than the present reduction by coke (see Section 21.9.2).

No great breakthroughs in science or technology are required to launch a hydrogen era. A hydrogen economy of the future can be achieved by solving a series of straightforward engineering problems. The phasing of hydrogen into our present fuel systems should not be difficult. Already hydrogen has been and is being used in significant amounts with natural gas and other fuels. Natural gas pipelines can be adapted for hydrogen transportation. One of the disadvantages of transporting large quantities of hydrogen in a pipeline is that hydrogen is lighter and has a lower heating value than natural gas. About three times the volume of hydrogen gas must be pumped for the same energy content as natural gas (mainly methane, CH_4), as shown by the following equations:

$$H_2(g) + \tfrac{1}{2}O_2(g) \longrightarrow H_2O(l) \quad \text{(about 286 kJ is liberated)}$$
22.4 L
(1 mol)

$$CH_4(g) + 2O_2(g) \longrightarrow CO_2(g) + 2H_2O(l) \quad \text{(about 890 kJ is liberated)}$$
22.4 L
(1 mol)

However, because hydrogen flows about three times faster, the pumps for a natural gas pipeline adapted for hydrogen use would be three times more efficient. For heating, hydrogen offers some advantages over natural gas because of its hotter flame. This would mean smaller appliances and simpler ventilating systems.

For aircraft and motor fuels, the requirement of producing hydrogen in refrigerated form at low temperatures may present some storage problems. However, trucks and ships are already transporting hydrogen in liquid form, and the technology of handling refrigerated materials at very low temperatures is well established, particularly in the aerospace industry.

One of the main drawbacks about hydrogen as a fuel concerns its safety. Unfortunately, hydrogen has a great ability to leak through small holes and even through some materials. Hydrogen was once used to fill balloons and dirigibles, but this practice was abandoned soon after many persons lost their lives when the hydrogen-filled dirigible *Hindenburg* was destroyed by fire in the spring of 1937 as it approached its mooring mast at Lakehurst, New Jersey, en route from Germany. This fear of the explosive

Table 9•3 The world's known extractable oil reserves

Geographic area	Percentage
Middle East	46
Mexico	16
Africa	13
Russia and other communist countries	12
United States	4
Canada	2
Europe	2
Indonesia	2
Venezuela; Caribbean and South American countries	3

Source: *Data are used with permission from F. J. Gardner,* Oil and Gas Journal, **70** (52), *79 (1972), but have been recalculated to reflect recent discoveries in Mexico.*

power of hydrogen–oxygen mixtures, which looms large in the minds of skeptics, has been termed by some the "Hindenburg syndrome." Certain hydrogen–air mixtures are explosive, but then so are methane–air mixtures.

Hydrogen already is being used in significant amounts with natural gas and other fuels. So-called town gas systems of the past routinely delivered gas containing 50 percent hydrogen. The town of Basilea, Italy, distributes a fuel containing 80 percent hydrogen, so far without incident. Dilution of hydrogen gas effectively reduces its wide flammability limits. One possible means of increasing the safety of hydrogen would be to dilute it with another gas at the point of distribution. Ideally the diluting gas would add to the fuel's heating value and would produce no toxic materials on combustion.

9.5　Water

9.5.1　The Chemical Behavior of Water

The reaction of water with active metals to form ionic hydroxides and hydrogen was discussed in Section 9.4.3. In Section 9.4.4, the reactions of steam with hot coke, with carbon monoxide, and with methane were presented. Obviously water will react with a large variety of substances. We will confine our discussion in this section to reactions with water at room temperature.

• *Reactions with Oxides* • Many metallic oxides react with water to form ionic hydroxides known as *bases*. Some examples are

$$K_2O + HOH \longrightarrow 2KOH$$

potassium oxide → potassium hydroxide

$$CaO + HOH \longrightarrow Ca(OH)_2$$

calcium oxide → calcium hydroxide

The bases, KOH and $Ca(OH)_2$ in these examples, are substances whose water solutions have a bitter taste, turn red litmus blue, and neutralize acids.

Many nonmetallic oxides react with water to form *acids*. Some examples are

$$CO_2 + H_2O \longrightarrow H_2CO_3$$

carbon dioxide → carbonic acid

$$SO_3 + H_2O \longrightarrow H_2SO_4$$

sulfur trioxide → sulfuric acid

$$P_4O_{10} + 6H_2O \longrightarrow 4H_3PO_4$$

phosphorus(V) oxide → phosphoric acid

The acids, carbonic, sulfuric, and phosphoric in these examples, are hydrogen-containing substances whose water solutions have a sour taste, turn blue litmus red, and neutralize bases. The chemistry of acids and bases will be presented in detail in Chapter 11.

• *Ionization Reactions* • Many polar covalent molecules react with water to produce ions. Reactions of compounds in which ions are produced are called **ionization reactions.** Some compounds that undergo ionization reactions when added to water are hydrogen nitrate (nitric acid), HNO_3, and hydrogen chloride (hydrochloric acid), HCl. The equations for these ionizations may be written as

$$HCl + H_2O \longrightarrow H_3O^+ + Cl^- \quad \text{or} \quad HCl \longrightarrow H^+ + Cl^-$$
$$HNO_3 + H_2O \longrightarrow H_3O^+ + NO_3^- \quad \text{or} \quad HNO_3 \longrightarrow H^+ + NO_3^-$$

These and other ionization reactions will be discussed further in Chapter 11.

• *Hydration* • The polar nature of water molecules is important when water is used as a solvent. Water readily dissolves many ionic compounds because of the *hydration* of ions. A **hydrated ion** is a cluster of the ion and one or more water molecules. In solution, the number of water molecules that cluster about many ions appears to be indefinite. Very often, however, when a water solution of a soluble salt is evaporated, the salt crystallizes with a precise number of water molecules called **water of crystallization.**

When copper chloride and magnesium chloride are crystallized from water solutions, the salts formed have the composition $CuCl_2 \cdot 4H_2O$ and $MgCl_2 \cdot 6H_2O$, respectively. In the former, the water molecules are thought to be at the corners of an imaginary square, with the Cu^{2+} ion at the center; in the latter, the water molecules are held in an octahedral structure with the Mg^{2+} ion at the center (see Figure 9.14). The hydrated $[Cu(H_2O)_4]^{2+}$ or $[Mg(H_2O)_6]^{2+}$ ions act as units with Cl^- ions to build up crystals of $CuCl_2 \cdot 4H_2O$ or $MgCl_2 \cdot 6H_2O$, respectively. It is found in most cases that the water of crystallization in salts is associated with the positive ions.

Often in naming salts or in writing formulas for them, the name or formula of the nonhydrated salt is used to represent the hydrated salt. For example, a water solution of copper sulfate might be represented in an equation by the formula $CuSO_4$, when in reality both the Cu^{2+} and SO_4^{2-} ions are hydrated in the solution. When it is important to emphasize the absence or presence of water of hydration, the terms *anhydrous* and *hydrate* are used in the name to distinguish between the two. Examples:

anhydrous copper sulfate, $CuSO_4$

copper sulfate pentahydrate, $CuSO_4 \cdot 5H_2O$

anhydrous zinc chloride, $ZnCl_2$

zinc chloride hexahydrate, $ZnCl_2 \cdot 6H_2O$

Note that *penta-* and *hexa-* denote the number of water molecules. In $CuSO_4 \cdot 5H_2O$, four water molecules are held close to each Cu^{2+} ion and one is held in the crystal between SO_4^{2-} ions.

It is to be noted that a pure hydrated salt, such as $CuSO_4 \cdot 5H_2O$, appears to be dry. There is no apparent moisture at all. However, often there are obvious differences between the anhydrous salt and the hydrate. For example, anhydrous copper sulfate, $CuSO_4$, is white, whereas the pentahydrate, $CuSO_4 \cdot 5H_2O$, is blue (see Figure 1.11 on Plate 1).

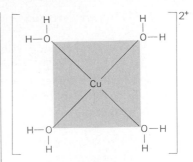

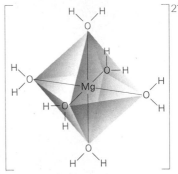

Figure 9•14 A schematic representation of the shape of a $[Cu(H_2O)_4]^{2+}$ ion (square planar) and an $[Mg(H_2O)_6]^{2+}$ ion (octahedral). The distances between the metal and the oxygen have been exaggerated to show better the spatial relationships.

9•6   Hydrogen Peroxide

Hydrogen peroxide, H_2O_2, is a colorless liquid with a melting point of -0.41 °C and a boiling point of 151 °C. The general structure of the hydrogen peroxide molecule is shown in Figure 9.15.

In concentrated form, H_2O_2 may explode violently if not handled properly. It decomposes into water and oxygen in an exothermic reaction:

$$2H_2O_2(l) \longrightarrow 2H_2O(l) + O_2(g)$$

This reaction, when carried out in water solution, may be used as a laboratory method to prepare oxygen. The common household variety of hydrogen peroxide is a 3-percent solution in water and is not dangerous at this dilution. Dilute water solutions of hydrogen peroxide can be obtained by treating barium peroxide or sodium peroxide with dilute acids:

$$BaO_2 + H_2SO_4 \longrightarrow BaSO_4 + H_2O_2$$
$$Na_2O_2 + H_2SO_4 \longrightarrow Na_2SO_4 + H_2O_2$$

Most of the hydrogen peroxide produced commercially is made by the reaction of oxygen with a complex organic compound that is fairly easily oxidized. Two H atoms are transferred from the organic compound to the O_2 molecule during the oxidation to form H_2O_2.

Like ozone, hydrogen peroxide can participate in oxidation reactions at temperatures at which oxygen is relatively inactive. For example:

$$PbS + 4H_2O_2 \xrightarrow{20\ °C} PbSO_4 + 4H_2O$$
$$PbS + O_2 \xrightarrow{20\ °C} \text{reaction extremely slow}$$

As shown by the equation, when hydrogen peroxide acts as an oxidizing agent, it gives up one atom of oxygen per H_2O_2 molecule. The oxidation number of each oxygen atom changes from -1 in hydrogen peroxide to -2 in the products of the reaction.

Hydrogen peroxide is used as an oxidizing and bleaching agent. In concentrated form, 90 percent or higher, it is used as an oxidizing agent in rockets and explosives.

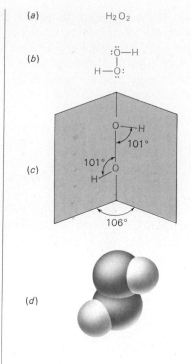

Figure 9•15 Different ways of representing the structure of a hydrogen peroxide molecule.

Chapter Review

Summary

Oxygen is the most abundant element in the earth's crust. Its three elemental forms are the **allotropes** atomic oxygen, ordinary oxygen, and ozone. Compounds of oxygen as gases, liquids, and solids are found almost everywhere. Most inorganic compounds of oxygen are very stable, but some can be pyrolyzed to yield gaseous oxygen. Many such reactions are facilitated by a **catalyst,** a substance that affects the rate of reaction without itself being permanently changed. Oxygen reacts exothermically with almost all the other elements and with many inorganic and most organic compounds to form a great variety of ionic and covalent compounds. Most such reactions, however, are imperceptibly slow except at high temperatures, at which rapid **combustion** often occurs. Most combustion processes involve the formation of **intermediate compounds.** The extremely rapid, violent combustion of a mixture is an explosion.

Virtually all reactions of oxygen fall in the broad category of oxidation–reduction reactions, or **redox reactions,** in which there is a change in the oxidation state of at least two elements. Such reactions can be visualized as two **half-reactions:** an **oxidation** half-reaction (increase in oxidation state, loss of electrons) and a **reduction** half-reaction (decrease in oxidation state, gain of electrons). Every redox reaction includes an **oxidizing agent,** which is reduced, and a **reducing agent,** which is oxidized.

Ozone, O_3, is a highly toxic, unstable, reactive gas that is found as a life-protecting ultraviolet radiation shield in the stratosphere and as a destructive air pollutant in the lower troposphere. In both regions it is produced by **photochemical reactions** that can be affected by human activities.

Hydrogen too is abundant in the earth's crust, but not in the elemental form. It is obtained for commercial use by the water gas and steam–hydrocarbon methods, and for special uses by the **electrolysis** of slightly acidified water. Hydrogen reacts, primarily as a reducing agent, with many different substances. An important reaction in organic chemistry is the **hydrogenation** of double or triple bonds in various compounds. Many attractive features of hydrogen as a fuel suggest the possibility of making it the basis of our fuel economy in the future.

Water reacts with certain metals and with many metal oxides to give ionic hydroxides (bases). It reacts with many nonmetal oxides to give hydrogen-containing acids. Many polar covalent compounds undergo **ionization reactions** in water, producing **hydrated ions,** which are ions with attached clusters of water molecules. Most ionic compounds dissolve in water, also producing hydrated ions. Water molecules can also be found in many kinds of salt crystals as water of crystallization.

The only other common compound of oxygen and hydrogen is hydrogen peroxide, H_2O_2, an unstable, highly reactive liquid. Like ozone, it is a very good oxidizing agent.

Exercises

Oxygen and Its Preparation

1. (a) How did Priestley explain the fact that more of a candle burned in a container filled with the "air" obtained from the decomposition of HgO than in the same container filled with common air?
   (b) How did he explain the fact that the candle gradually ceased to burn in either case?
2. An oxygen atom normally shares two electron pairs with other nonmetal atoms. However, the formula $:\ddot{O}=\ddot{O}:$ is inadequate for O_2 (see Section 6.8.1). What experimental evidence supports this conclusion? Write a Lewis structural formula that shows a molecule of O_2 has two unpaired valence electrons.
3. It is possible to have O_2 molecules with molecular weights of 32, 34, and 36 amu. Can such molecules be properly called allotropes of oxygen? Explain.

4. What is the average oxidation number of iron in Fe_3O_4? How can you account for the fractional number?
5. Oxygen generally has an oxidation number of -2 in its compounds. However, there are at least two other oxidation states that oxygen may have in compounds. Write the formulas for two compounds that exemplify these oxidation states. Name each compound.
6. What weight of oxygen can be formed by the decomposition of 50.0 kg of potassium chlorate? What will be the volume in liters of the oxygen when it is saturated with water vapor at 27 °C and under a pressure of 740 mm?
7. (a) Explain fully why the industrial method for obtaining oxygen from air requires the expenditure of much less energy than a method based upon obtaining oxygen from water.
   (b) Discuss the two preparations of oxygen from the standpoint of the value of the by-products.

Oxidation–Reduction Reactions

8. Write balanced equation(s) for each of the following:
   (a) Magnesium ribbon burns vigorously when heated to its ignition temperature.
   (b) An "empty" gasoline tank was being welded when a violent explosion of the tank occurred.
   (c) The combustion of carbon monoxide.
   (d) Gold(III) oxide was heated until only a yellow solid remained.
   (e) A combustion that gives only water as the product.
9. Priestley obtained oxygen by heating compounds other than HgO, one of which at that time was called *nitre*. Much later it was identified as KNO_3. Write a balanced equation for the decomposition of KNO_3 to yield KNO_2 and O_2. Which element is oxidized? Which element is reduced?
10. Scheele prepared oxygen by heating manganese dioxide, MnO_2, and sulfuric acid, H_2SO_4. In addition to oxygen, water and manganese(II) sulfate, $MnSO_4$, were produced. Write a balanced equation for this reaction. Which element is oxidized? Which element is reduced?
11. For each equation written in Exercise 8, which substance undergoes oxidation and which undergoes reduction? Identify the oxidizing and reducing agents. Indicate the changes in oxidation states involved.
12. Why does a brush fire burn more rapidly on a windy day than on a still day? Why does a match "blow out" on a windy day?
13. Write balanced equations for the combustion in oxygen of the following organic compounds: acetylene, C_2H_2; isopropyl alcohol, C_3H_8O; sucrose, $C_{12}H_{22}O_{11}$.
14. Several reactions are described below by unbalanced equations. Some are oxidation–reduction reactions and some are not. Choose those that are not and explain why they are not; then choose those that are and explain why they are.

(a) $S + K \longrightarrow K_2S$

(b) $NaCl + F_2 \longrightarrow NaF + Cl_2$

(c) $Mg_3N_2 + H_2O \longrightarrow Mg(OH)_2 + NH_3$

(d) $NaClO + H_2S \longrightarrow H_2SO_4 + NaCl$

(e) $HI + RbOH \longrightarrow RbI + H_2O$

(f) $SO_2 + H_2S \longrightarrow H_2O + S$

15. For each equation in Exercise 14 that describes a redox reaction, write the partial oxidation and reduction equations. Show in each case that the amount of oxidation equals the amount of reduction. Then write the balanced equation for the reaction.

16. (a) Write the formulas for two iron oxides that could be intermediate compounds in the reduction of iron(III) oxide, Fe_2O_3, to iron.

(b) Carbon monoxide is the principal reducing agent in the commercial production of iron from oxide ores. Write the equation for this reaction, assuming that the ore is iron(III) oxide.

Ozone

*17. The oxidation of H_2S to H_2SO_4 by ozone in a water solution can proceed stepwise as follows:

$$\overset{-2}{S} \longrightarrow \overset{0}{S} \longrightarrow \overset{+4}{S} \longrightarrow \overset{+6}{S}$$

Write an equation for each of the three steps (note that oxides of sulfur form acids with water). Then show that the three equations can be added to give a one-step equation for the oxidation of hydrogen sulfide to sulfuric acid.

18. Calcium bromide reacts with ozone to produce calcium oxide and bromine.

(a) Write the half-reactions that represent the oxidation and reduction.

(b) Write the balanced equation for the redox reaction.

19. Show how reactions (22), (18), and (23) in Section 9.3.4 provide for the presence of NO_2, O_3, or O in the atmosphere, but lead to no increase in the total amount of these active species. Assume that the reactions occur in sequence.

20. Write equations to account for the fact that nitrogen from the atmosphere may react in an automobile engine to initiate a series of reactions that produce ozone in photochemical smog. Would diesel engines initiate the same reactions?

Hydrogen and Its Preparation

21. Which has the greater density at STP, heavy hydrogen or tritium? What percentage greater?

22. In Section 9.4.7, the statement is made that hydrogen flows about three times faster than methane.

(a) Whose law supports this statement?

(b) Based on this gas law, calculate the relative rates of flow of the two gases.

*23. Four 1-L containers are filled at STP as follows: one with protium, one with deuterium, one with oxygen, and one with ozone.

(a) Calculate the number of molecules and their weight in each container.

(b) Assuming that the containers do not leak and are insulated so that no heat can flow in or out, which, if any, of the following will be changed after the containers and contents have stood for a long period of time (consider each container separately):

the number of molecules in a container
the weight of gas
temperature
pressure

In case of a change, state the magnitude of the change and why it occurred.

24. The reaction of potassium with water is frequently so exothermic that it is accompanied by an explosion. This explosion is the result of a second reaction involving one of the initial reaction products and air.

(a) Write an equation for the reaction of potassium with water.

(b) Write an equation for the explosive reaction.

*25. What volume of hydrogen, measured at STP, is theoretically obtainable by the reaction of hydrochloric acid and a cube of aluminum that is 2.0 cm on an edge? (See Table 1.2.)

26. Sodium hydride undergoes decomposition into sodium and hydrogen when an electric current is passed through the molten compound. For this electrolysis, write

(a) the oxidation half-reaction

(b) the reduction half-reaction

(c) the balanced equation for the reaction

Reactions of Hydrogen

27. Write a balanced equation for the complete combustion of water gas.

28. Calculate the amount of heat evolved when 1.00 lb of hydrogen is burned and the product is condensed to a liquid at 25.0 °C.

29. Hydrogen burns explosively when mixed with fluorine. Write equations for the half-reactions that represent the oxidation and the reduction. Identify the oxidizing and reducing agents. Write a balanced equation for this redox reaction.

30. When hydrogen reacts with the alkyne known as acetylene, C_2H_2, on a platinum catalyst, ethane, C_2H_6, is formed.

(a) Using Lewis structural formulas for all molecules, write a balanced equation for this reaction.

(b) What is a logical intermediate compound in the conversion of acetylene to ethane? Write its Lewis structural formula.

Reactions of Water

31. Write balanced equations for the following chemical reactions; in each case, indicate what needs to be done, if anything, beyond bringing the reactants together, for the reaction to occur:
    (a) strontium and water $\longrightarrow$
    (b) sodium oxide and water $\longrightarrow$
    (c) calcium and water $\longrightarrow$
    (d) nitrogen pentoxide (N_2O_5) and water $\longrightarrow$
    (e) $MgCl_2$ (anhydrous) and water $\longrightarrow$

32. When calcium hydride, CaH_2, is added to water, calcium hydroxide and hydrogen are formed.
    (a) Write an equation for the reaction.
    (b) Which element undergoes oxidation? Which element undergoes reduction?

33. Calculate in kilograms the weight of a mole of washing powder (sodium carbonate decahydrate). What is the weight in kilograms of a kilomole of this compound? What is the weight in milligrams of a millimole?

34. A 10.33-g sample of a certain hydrate is heated strongly until all the water of crystallization is expelled. The anhydrous salt weighs 6.73 g. Which one of the following is the hydrate: $MgCl_2 \cdot 6H_2O$, $CuCl_2 \cdot 4H_2O$, $ZnCl_2 \cdot 6H_2O$, $MgSO_4 \cdot 7H_2O$, or $Na_3PO_4 \cdot 10H_2O$? Give calculations to support your choice.

Hydrogen Peroxide

35. A water solution of hydrogen peroxide may be prepared by reaction of barium peroxide, BaO_2, with dilute sulfuric acid. How many kilograms of 30.0 percent hydrogen peroxide (70 percent water) by weight can be prepared from 75.0 kg of barium peroxide?

36. Write the formula and name for an example of a metal (a) hydride; (b) hydrate; (c) hydroxide; (d) peroxide; (e) hydroperoxide.

37. In Hershey, Pennsylvania, hydrogen peroxide is used to control hydrogen sulfide odors from waste treatment plants. Hydrogen sufide is oxidized to sulfuric acid by the peroxide. Write a balanced equation for this redox reaction.

38. One way by which hydrogen peroxide may be prepared is the hydrolysis of peroxydisulfuric acid, $H_2S_2O_8$, with steam. Sulfuric acid is obtained as a by-product. Write a balanced equation for this reaction. Is this a redox reaction? A molecule of peroxydisulfuric acid has an oxygen–oxygen bond. Write a Lewis structural formula for the acid.

*39. In 1970 the existence of two higher oxides of hydrogen, hydrogen trioxide, H_2O_3, and hydrogen tetroxide, H_2O_4, was confirmed.
    (a) Write Lewis structural formulas for these oxides.
    (b) What is the oxidation state of each oxygen?
    (c) Neither oxide is very stable. One decomposes at $-100\,°C$ into hydrogen peroxide and oxygen, the other at $-55\,°C$ into water and oxygen. Which oxide is less stable? Write equations for the decomposition reactions.

Solutions;
Electrolytes and Nonelectrolytes

10

10•1 **Nature of Solutions**

10•2 **Why Substances Dissolve**

10•2•1 Solvation

10•2•2 Insoluble Substances

10•3 **Solubility Relationships**

10•3•1 Saturated Solutions

10•3•2 Unsaturated and Supersaturated Solutions

10•3•3 Solvent Extraction

10•4 **Effect of Temperature on Solubility**

10•4•1 Solids in Liquids

10•4•2 Gases in Liquids

10•5 **Effect of Pressure on Solubility**

10•6 **Expressing Concentrations**

10•6•1 Percent by Weight

10•6•2 Percent by Volume

10•6•3 Mole Fraction

10•6•4 Molality

10•6•5 Molarity

10•6•6 Normality

10•7 **Electrolytes and Nonelectrolytes**

10•7•1 How Solutions Conduct Current

10•7•2 Sources of Ions

10•7•3 Ionic Equilibria

10•7•4 Strong and Weak Electrolytes

10•8 **Ionic Equations**

10•8•1 Oxidation–Reduction (Redox) Reactions

10•8•2 Precipitation Reactions

10•8•3 Formation of Covalent Compounds

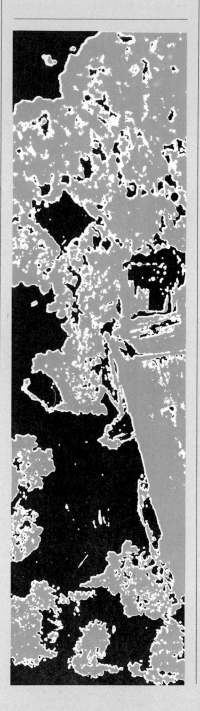

Chemical reactions usually take place between two mixtures of substances rather than between two pure substances. One common type of mixture is the solution. In nature most reactions take place in water solutions. The body fluids of both plants and animals are water solutions of numerous substances. Obviously the reactions in oceans, lakes, and rivers involve solutions. In the soil, the main reactions take place in thin layers of solution adsorbed on solids, even in desert regions.

The relative amount of a given substance in a solution is called its concentration. Concentration is an important factor in determining how rapidly a reaction occurs and, in some cases, in determining what products are formed. In this chapter we will describe several useful ways of expressing concentrations.

There are many different types of solutions. One way in which compounds and their solutions are classified is with respect to their electrical conductivities. In this chapter we will discuss the electrical properties of solutions and the classification of solutes as electrolytes and nonelectrolytes. Chemical reactions of electrolytes are often described by writing ionic equations, so we will look at a number of such equations.

10•1   Nature of Solutions

A **solution** is a homogeneous mixture of the molecules, atoms, or ions of two or more substances. A solution is called a mixture because its composition is variable. It is called *homogeneous* because its composition is so uniform that no differing parts can be detected, even with an optical microscope. In *heterogeneous* mixtures, definite surfaces can be detected between the separate parts or phases.

Although all gas phase mixtures are homogeneous and so may be called solutions, the molecules are too far apart to attract one another effectively. Solid phase solutions of metals are very useful and well known. Examples include brass (mainly copper and zinc), gold jewelry (usually gold and copper), and dental amalgam (mercury and silver).

We usually think of the liquid phase when we speak of solutions. Commonly one of the components of such a solution is a liquid before the mixture is made. This liquid is called the dissolving medium or the **solvent.** The other component, which may be a gas, liquid, or solid, is thought of as being dissolved in the first component. The dissolved substance is called the **solute.** In cases where there is a question, the substance present in the smaller amount is usually called the solute. As we might expect, there may be difficulties in applying this simple guideline. What is the solute in a 50–50 mixture of ethyl alcohol and water? Or in a syrup of 80 percent sucrose (table sugar) and 20 percent water? In the first case, either substance could be called the solute. In the second case, because water is the component that retains its physical state and sugar changes its physical state, most persons prefer to call the water the solvent.

10•2   Why Substances Dissolve

There is a strong tendency for nonpolar compounds to dissolve in nonpolar solvents and for ionic or polar covalent compounds to dissolve in polar solvents. In other words, like dissolves like. (See Figure 10.1 on Plate 3.)

Consider the two nonpolar liquids hexane, C_6H_{14}, and heptane, C_7H_{16}. These liquids have densities of 0.659 and 0.684 g/mL, respectively. We can carefully float the lighter liquid on top of the heavier one, as shown in Figure 10.2. The molecules diffuse randomly,[1] faster if the temperature is high, so that in time a uniform, homogeneous solution results. Two liquids that mix uniformly are said to be **miscible.**

There are several factors involved in the formation of a solution. One of these is the tendency of any system to attain maximum disorder, which we will take up in Chapter 13. The main factor we will consider in this chapter is the attraction between particles of solute and solvent that results in the formation of solvated particles.

10•2•1   Solvation   Solvation is the interaction of solvent molecules with solute particles to form aggregates. Some such aggregates have a definite number of solvent particles and some do not. When water is the solvent, the process is also called *hydration* or *aquation*.

When a small crystal of an ionic substance such as sodium chloride is placed in water, the polar water molecules orient themselves about the face of the crystal, as shown in Figure 10.3. The attractive force between the water molecules and the surface ions is great enough to cause the ions to leave their fixed positions in the crystal and to move to positions between the water molecules, as shown in Figure 10.4. Both ions are said to be solvated. In some cases the number of water molecules bound to an ion in solution may be the same as for that ion in a hydrated salt (see Section 9.5). In other cases the number is indefinite.

Any polar covalent solute may interact with polar solvents. Consider the dissolving of the sugar glucose in water.

$$H-\underset{\underset{H}{|}}{C}-\underset{\underset{H}{|}}{\overset{:\ddot{O}:}{C}}-\underset{\underset{H}{|}}{\overset{:\ddot{O}:}{C}}-\underset{\underset{H}{|}}{\overset{:\ddot{O}:}{C}}-\underset{\underset{H}{|}}{\overset{:\ddot{O}:}{C}}-\underset{\underset{H}{|}}{\overset{:\ddot{O}:}{C}}-C=\ddot{O}:$$

Each hydroxyl group (—O—H) and each aldehyde group $(H-\overset{|}{C}=O)$ is a region where relatively strong attractive forces exist for molecules of the solvent. When a small crystal of sugar is placed in water, water molecules tend to orient themselves about a surface sugar molecule so that the positive or negative parts of the water molecules attract the oppositely charged parts of the sugar molecule. The sugar molecule leaves the surface of the crystal and goes into solution as an aquated molecule. The water and sugar molecules are attracted to one another by hydrogen bonds:

$$H-\underset{|}{\overset{\delta-}{C}}=\overset{\delta+}{\ddot{O}}:---\overset{\delta+}{H}-\ddot{O}: \quad \text{and} \quad -\overset{|}{\underset{|}{C}}-O-\overset{\delta+}{H}---\overset{\delta-}{\ddot{O}}-H$$

---

[1] *Diffusion* must be distinguished from the mixing of masses of material by *convection,* a natural process due to density differences, or by *stirring,* an artificial mechanical process. Diffusion is the result of the constant motion of particles that occurs at any temperature above absolute zero.

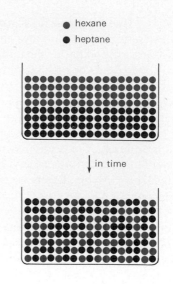

● hexane
● heptane

Figure 10•2 Two liquids made of very similar molecules are miscible; that is, they dissolve in one another to form a homogeneous mixture.

in time

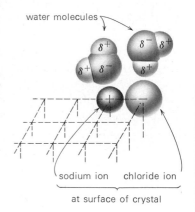

water molecules

sodium ion   chloride ion

at surface of crystal

Figure 10•3 Schematic representation of the forces that bring about the dissolving of salt in water.

The layer of solvent molecules held on the surface of solute particles helps to keep the ions or molecules separated in solution. This separation interferes with recrystallization and thereby aids in the solution process.

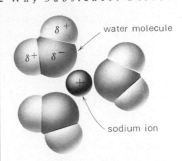
water molecule
sodium ion

• *Dielectric Constant* • In addition to the attraction between solute and solvent molecules, there is another important phenomenon to consider when ionic solutes dissolve. Solvents differ in their ability to reduce the attraction between positive and negative solute ions.

If two ions of opposite charge exist in a vacuum, there is a certain force of attraction, F, between them at a given distance. But if another substance—for example, a solvent—is in the space separating these ions, their attraction for each other is less. The attraction in a given solvent is F/ϵ, where ϵ is the dielectric constant of the solvent (see Section 5.5.2).

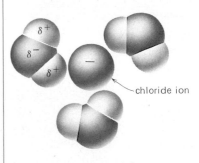
chloride ion

In general, polar liquids have high dielectric constants. For example, water, with a dipole moment of 1.85 debyes, has a high dielectric constant, 80 (Table 10.1). Ions that are widely separated in water attract one another with only $\frac{1}{80} F$. This property makes water an excellent solvent for ionic compounds.

Figure 10•4 Solvation of Na$^+$ and Cl$^-$ ions. The number of water molecules held by these simple ions is indefinite, but the upper limit is likely to be six.

Table 10•1 Dielectric constants

Solvent	Dielectric constant	Solvent	Dielectric constant
hydrogen sulfate, H_2SO_4	84 (20 °C)	liquid ammonia, NH_3	22 (−33°C)
water, H_2O	80 (20 °C)	ether, $(C_2H_5)_2O$	4.3 (20°C)
methyl alcohol, CH_3OH	33 (20 °C)	benzene, C_6H_6	2.3 (25°C)
ethyl alcohol, C_2H_5OH	24 (25 °C)	vacuum	1 (by definition)

• *Deliquescence and Efflorescence* • The attraction between solute and water molecules may be strong enough to cause a solid to attract water molecules from the air until a solution is formed. This process is called **deliquescence.** Calcium chloride, $CaCl_2$, is probably the most common deliquescent material in the laboratory. Crystals of it that are spilled and not swept up may become viscous droplets of solution within a few hours.

Whether a hydrate tends to pick up water from the air or lose water to the air depends on the humidity (see Section 8.4.1), as well as on the attraction between the substance and water. At a low humidity, a hydrate may lose water, a process called **efflorescence.** At water vapor pressures below about 8 mm, the beautiful blue crystals of $CuSO_4 \cdot 5H_2O$ crumble and become chalky as they effloresce, thus losing their water of crystallization.

Substances that attract water are said to be **hygroscopic.** If their attraction is strong enough, they may be used as **desiccants,** or drying agents. Calcium chloride and silica gel are moderate dessicants; phosphorus(V) oxide, P_4O_{10}, is one of the strongest known. Concentrated sulfuric acid is a powerful dessicant that often is used in bubble towers to dry gases.

10•2•2 Insoluble Substances If a substance is very slightly soluble, say, less than 0.1 g of solute in 1,000 g of solvent, we may describe it as **insoluble.**

Probably nothing is absolutely insoluble in a given solvent, but many things are insoluble for practical purposes, for example, glass in water.

Solids held together by strong bonds may be practically insoluble in any common liquid. Silicate rocks and minerals are such substances. Plastics made of huge molecules tend to be insoluble in most solvents. Metals are another example; they tend to dissolve only in other metals.

Guides to the solubilities of familiar ionic compounds in water are given in Table 10.2. Insoluble ionic compounds often are formed by mixing solutions of two soluble compounds (see Section 10.8.2).

Table 10.2 Solubilities of common metal compounds in water[a]

Compound	Solubility
nitrates	all soluble
nitrites	all soluble except Ag^+
acetates	all soluble except[b] Ag^+, Hg_2^{2+}, Bi^{3+}
chlorides	all soluble except Ag^+, Hg_2^{2+}, Pb^{2+}, Cu^+
bromides	all soluble except Ag^+, Hg_2^{2+}, Pb^{2+}
iodides	all soluble except Ag^+, Hg_2^{2+}, Pb^{2+}, Bi^{3+}
sulfates	all soluble except Pb^{2+}, Ba^{2+}, Sr^{2+}, $(Ca^{2+})^c$
sulfites	all insoluble except Na^+, K^+, NH_4^+
sulfides	all insoluble except Na^+, K^+, NH_4^+, Ba^{2+}, Sr^{2+}, Ca^{2+}
phosphates	all insoluble except Na^+, K^+, NH_4^+
carbonates	all insoluble except Na^+, K^+, NH_4^+
oxalates	all insoluble except Na^+, K^+, NH_4^+
oxides	all insoluble except Na^+, K^+, Ba^{2+}, Sr^{2+}, Ca^{2+}
hydroxides	all insoluble except Na^+, K^+, NH_4^+, Ba^{2+}, Sr^{2+}, $(Ca^{2+})^c$

[a] The compounds listed here include only those of the common metals of groups IA, IB, IIA, and IIB, and Mn, Fe, Co, Ni, Al, Sn, Pb, Sb, and Bi. The polyatomic ion NH_4^+ is included because of its importance.

[b] The mercury(I) ion is a diatomic ion, so its formula is Hg_2^{2+}.

[c] Both calcium sulfate and calcium hydroxide are slightly soluble substances.

When two liquids are mutually insoluble, they are said to be **immiscible** (see Figure 10.1 on Plate 3). Water molecules attract each other so strongly by hydrogen bonding that nonpolar molecules such as oils are squeezed out. Oil and water form separate layers, with the oil usually floating on top because of its lower density. Water tends not to dissolve molecules to which it is attracted only by London forces or weak dipole–dipole interactions.

10.3   Solubility Relationships

10.3.1   Saturated Solutions   When crystals of sugar are placed in water, molecules break away from the surface of the sugar and pass into the solvent, where they move about in the same manner as the water molecules. Because of this random motion, some of them collide with the surface of the sugar and are held there by the attractive forces of the other sugar molecules.

The sugar is constantly dissolving and crystallizing at the same time. When the sugar is first placed in the water, the rate of dissolving is very rapid as compared with the rate of crystallizing. As time goes on, the concentration of the dissolved sugar steadily increases, and the rate of crystallizing increases. When the rates of crystallizing and dissolving are equal, the processes are at equilibrium. The condition of equilibrium is indicated in an equation with double arrows, to show that two opposing processes are occurring simultaneously at equal rates:

$$\text{sugar} + H_2O \rightleftharpoons \text{sugar solution}$$

When these two processes are at equilibrium, the solution is said to be saturated.

A **saturated solution** is defined as one that contains the amount of dissolved solute necessary for the existence of an equilibrium between dissolved and undissolved solute. The formation of a saturated solution is hastened by vigorous stirring and an excess of solute. The amount of solute that dissolves in a given amount of solvent to produce a saturated solution is called the **solubility** of that solute. Solubility is commonly expressed in grams of solute per 100 cm^3 or per 100 g of solvent at a specified temperature.

Note that a saturated solution is not necessarily a concentrated solution.[2] For example, when limestone rock (calcium carbonate, $CaCO_3$) remains in contact with a quantity of water until equilibrium is reached between dissolved and undissolved calcium carbonate, the saturated solution is extremely dilute, because calcium carbonate is not very soluble.

10•3•2   **Unsaturated and Supersaturated Solutions**   An **unsaturated solution** is less concentrated (more dilute) than a saturated solution, and a **supersaturated solution** is more concentrated than a saturated solution. A supersaturated solution is usually prepared by first making a saturated solution at an elevated temperature. The solute must be more soluble in the warm than in the cool solvent. If any undissolved solute remains, it is removed. The hot solution is carefully cooled to avoid crystallization. This means that the solution must not be jarred or shaken and that dust and other foreign matter must be excluded. If no solute separates during cooling, the cool solution is supersaturated. Sucrose, sodium acetate, and sodium thiosulfate (hypo) readily form supersaturated solutions when treated in this way.

A supersaturated solution is a metastable system. It may be converted to a saturated solution by adding a small "seed" crystal (usually of the solute, although often a foreign substance works just as well). The crystal

[2] A solution with a relatively large amount of solute is called **concentrated.** One with a relatively small amount of solute is called **dilute.** These two terms are not defined precisely. A solution containing 10 g of NaCl per 100 g water might be called dilute in the laboratory but concentrated if you tasted it.

provides a nucleus about which the excess dissolved solute can crystallize (see Figure 10.5).

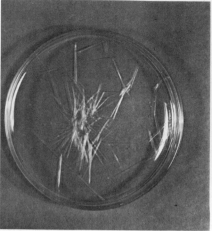

Figure 10•5 A supersaturated solution of sodium acetate is seeded (*left*). Excess solute precipitates as needle-like crystals (*middle*). The result is a saturated solution in equilibrium with excess solid (*right*).

• **10•3•3   Solvent Extraction**   When a solute dissolved in one solvent is extracted into another solvent, the process is called **solvent extraction.** In the laboratory, solvent extraction can be carried out in a separatory funnel, as shown in Figure 10.6. In industry, solvent extraction is often carried out in towers in which drops of a less dense solvent rise through a slowly descending stream of a more dense solvent. In Figure 10.7 this technique is shown applied to extracting the insecticide DDT from water into oil. Such a countercurrent extraction is very efficient because at the bottom end of the tower the solvent that has lost almost all its solute is extracted by fresh samples of the other solvent.

(a)
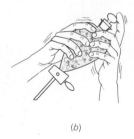
(b)

(c)

Figure 10•6 Solvent extraction in a separatory funnel. (*a*) Two immiscible liquids are placed in the funnel, and a solute is dissolved in the upper liquid. (*b*) The two liquids are thoroughly agitated together. (*c*) On standing, the liquid layers separate. In this case, the solute is now in the lower liquid, which is allowed to run out of the funnel.

DDT has a very low solubility in water but a higher solubility in fats and oils. In nature, the very dilute residues in lakes, streams, and oceans are extracted into the fats and oils in the bodies of plants and animals (see Figure 10.8). More than a decade ago the use of DDT was banned for most purposes in this country, because animals at the top end of the food chain had developed such dangerously high levels of it in their bodies.

10•4   Effect of Temperature on Solubility

In this section we shall consider the effect of temperature on two types of solutions: solids dissolved in liquids and gases dissolved in liquids.

10•4•1   Solids in Liquids

Most solids become more soluble in a liquid as the temperature rises; however, there are a few solids that become less soluble as the temperature increases. In Figure 10.9 the solubilities of several compounds, in grams of solute per 100 g of water, are plotted against temperature. With the exception of sodium sulfate and cerium sulfate, the solubilities increase as the temperature rises.

Consider the formation of an aqueous solution of potassium nitrate, $KNO_3(aq)$, at 20 °C. For the formation of this solution, we can write

$$KNO_3(s) + H_2O(l) \longrightarrow KNO_3(aq)$$

The solution process is endothermic; that is, heat is absorbed as the solid KNO_3 dissolves in water. For a saturated solution of KNO_3, the following equilibrium exists between the undissolved solid and the solution:

$$KNO_3(s) + H_2O(l) \rightleftharpoons KNO_3(aq)$$

or simply

$$KNO_3(s) \rightleftharpoons KNO_3(aq)$$

The process of forming solid crystals and water from a solution of KNO_3 is exothermic. *Any process that is endothermic in one direction is exothermic in the opposite direction.*

Because the processes of solution formation and crystallization are going on at the same rate at equilibrium, the net energy change is zero. However, if the temperature is raised, the process that absorbs heat, the formation of the solution, is favored. Immediately after the temperature is raised, the system is not at equilibrium as more solid dissolves.

Our interpretation of the effect of temperature changes on solubility is based on **Le Chatelier's principle,** stated by the French chemist Henri Louis Le Chatelier (1850–1936). *When a stress is brought to bear on a system at equilibrium, the system tends to change so as to relieve the stress.* The stress being brought to bear in the case of the saturated solution of KNO_3 is the addition of heat energy (temperature increase). A substance that absorbs heat when it dissolves tends to be more soluble at a higher temperature.

Consider now the formation of an aqueous solution of cerium sulfate, $Ce_2(SO_4)_3$:

$$Ce_2(SO_4)_3(s) + H_2O(l) \longrightarrow Ce_2(SO_4)_3(aq)$$

In this case, the solution process is exothermic; that is, heat is evolved as the solid $Ce_2(SO_4)_3$ dissolves in water. Hence the crystallization of $Ce_2(SO_4)_3$ from solution,

$$Ce_2(SO_4)_3(aq) \longrightarrow Ce_2(SO_4)_3(s) + H_2O(l)$$

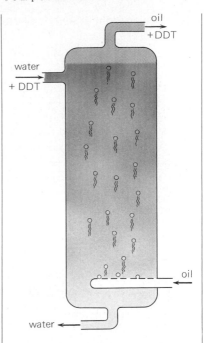

Figure 10•7 Solvent extraction as liquids move in opposite directions. Water containing a trace of DDT (indicated by color mixed with gray) is pumped into the top of the tower. Oil is pumped into the bottom, where it breaks up into drops. As the oil drops rise up through the water, they extract the DDT.

must be endothermic. For a saturated solution of $Ce_2(SO_4)_3$, we may write

$$Ce_2(SO_4)_3(s) + H_2O(l) \rightleftharpoons Ce_2(SO_4)_3(aq)$$

When the temperature is raised in this system, crystallization of solute is favored. A substance that evolves heat when it dissolves tends to be less soluble at a higher temperature.

Solids may be purified by taking advantage of differences in solubility at different temperatures. For most substances, when a hot concentrated solution is cooled, the excess solid crystallizes. The process can be facilitated by seeding the solution with a few tiny crystals of the pure solid. The overall process of dissolving the solute and crystallizing it again is known as **recrystallization.** This method is frequently used as an effective way of removing small amounts of impurities from solids, because the impurities tend to be left in solution. Unless the impurities have sizes, shapes, and polarities similar to those of the growing crystals of the recrystallized solid, very little of an impurity is likely to be incorporated, particularly if the crystals are grown slowly.

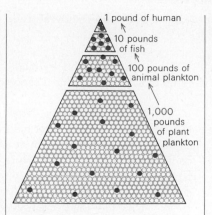

Figure 10•8 A part of nature's food chain. Large amounts of simpler forms of life are necessary to produce small amounts of more complex forms. Certain contaminants, represented by colored dots, are concentrated by the biochemical reactions involved. Therefore, in the more complex forms of life, harmful concentrations of mercury, DDT, and other pollutants may build up in body tissues.
Source: Adapted by permission from "How Man Pollutes His World," National Geographic Society, Washington, D.C., 1970.

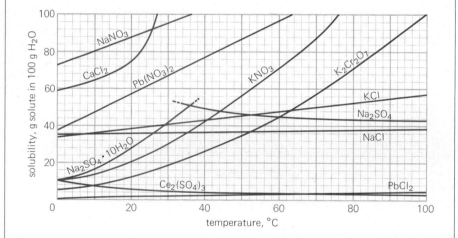

Figure 10•9 Solubilities of some salts at different temperatures.

━━━ • Example 10•1 • ━━━

The mineral sylvinite is approximately a 40:60 percent mixture of tiny potassium chloride and sodium chloride crystals. If water is saturated with the mineral at a cool temperature and then heated in contact with more mineral, a considerable amount of the more valuable KCl dissolves but only a small amount of NaCl dissolves.

Suppose that 2,000 kg of the crushed mineral is stirred with a saturated solution containing 1,000 kg of water at 100 °C, and then the solution is separated and cooled to 20 °C. According to Figure 10.9, what is the approximate weight of KCl crystals that will form? (Ignore the NaCl for purposes of this simplified calculation.)

• **Solution** • Referring to the figure, we see that about 57 g of KCl will dissolve in 100 g of water at 100 °C, but only about 38 g will dissolve at

20 °C. With 1,000 kg of water, 570 kg of KCl will dissolve at 100 °C but only 380 kg at 20 °C.

$$\text{wt KCl crystals} = 570 \text{ kg} - 380 \text{ kg} = 190 \text{ kg}$$

See also Exercises 14–16 at the end of the chapter.

10•4•2   Gases in Liquids   The solubility of a gas in a liquid usually decreases as the temperature increases. Carbon dioxide bubbles vigorously out of a carbonated drink if the liquid is warmed. When tap water is heated to lukewarm, the small amount of dissolved air begins to appear as small bubbles.

10•5   Effect of Pressure on Solubility

Pressure changes have little effect on solubility if the solute is a liquid or a solid. However, in the formation of a saturated solution of a gas in a liquid, the pressure of the gas plays an important part in determining how much of the gas dissolves. *The weight of a gas dissolved by a given amount of a liquid is directly proportional to the pressure exerted by the gas when in equilibrium with the solution.* This is **Henry's law,** stated by William Henry (1774–1836). Figure 10.10 is a graphic illustration of this law. Henry's law does not hold for gases that react with the solvent, for example, hydrogen chloride or ammonia in water.

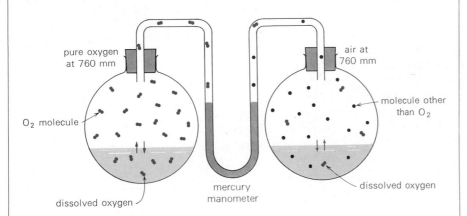

Figure 10•10 A diagrammatic representation of Henry's law. Note that the partial pressure of gaseous oxygen in the flask on the left is five times that on the right. Hence, five times as much dissolved oxygen is indicated in the liquid on the left.

Henry's law helps us understand how the concentrations of carbonic acid, H_2CO_3, and bicarbonate ion, HCO_3^-, are maintained in the blood (see Section 16.8.1). The amount of H_2CO_3 dissolved depends on the partial pressure of gaseous CO_2 over the system. If the CO_2 pressure is increased, say, as the result of cell metabolism, then more CO_2 gas dissolves in the blood to give more H_2CO_3. If the amount of H_2CO_3 decreases for some reason, then CO_2 from the large reservoir in the lungs dissolves in the blood to regenerate the supply.

Fear or excitement may make one breathe so rapidly that carbon dioxide is removed from the blood faster than the cells can replenish it. Convulsions or loss of consciousness results if the carbonic acid content of the blood is sufficiently altered. The condition may be alleviated by putting a paper bag over the patient's head for a short time, a triumph of medical science via Henry's law. As one breathes the same air over and over, the partial pressure of carbon dioxide increases, thereby increasing the amount dissolving in the blood and replenishing the supply of carbonic acid.

10•6 Expressing Concentrations

The **concentration** of a solution refers to the weight or volume of the solute present in a specified amount of the solvent or solution. There are several common methods of expressing these amounts.

10•6•1 Percent by Weight When expressing percent by weight, the percentage given refers to the solute; for example, a 5 percent aqueous NaCl solution contains 5 percent by weight of sodium chloride, the remaining 95 percent being water.

10•6•2 Percent by Volume The concentration of a solution of two liquids is frequently expressed as a volume percentage. The concentration of alcoholic beverages is usually expressed this way.[3] A wine that is 12 percent alcohol has 12 mL of alcohol per 100 mL of wine. It must be noted, however, that volumes of liquids are not additive; 89 mL of water must be added to 12 mL of alcohol to make 100 mL of solution. In chemical laboratory work, the use of the term "percent" by itself always means "percent by weight."

10•6•3 Mole Fraction Knowledge of the number of solute particles mixed with a known number of solvent particles is required in many laboratory operations. One way of expressing the number of particles present is in terms of the number of moles of solute and of solvent. The fractional part of the total number of moles that is due to the solute is the **mole fraction of the solute;** the fractional part of the total due to the solvent is the **mole fraction of the solvent.** The mole fraction multiplied by 100 is the **mole percent.**

1,000 mL
20 °C

Figure 10•11 Volumetric flasks are calibrated to hold a specified volume at a specified temperature when filled to the mark on the neck.

──── • Example 10•2 • ────

Calculate the mole fractions of ethyl alcohol, C_2H_5OH, and water in a solution made by dissolving 13.8 g of alcohol in 27.0 g of water.

──── • Solution •

$$\text{number of moles of } C_2H_5OH = \frac{13.8 \text{ g}}{46.1 \text{ g/mol}} = 0.300 \text{ mol}$$

$$\text{number of moles of } H_2O = \frac{27.0 \text{ g}}{18.0 \text{ g/mol}} = 1.50 \text{ mol}$$

$$\text{total number of moles} = 1.80 \text{ mol}$$

[3] The ethyl alcohol content of concentrated spirits, such as whiskey, is also expressed as the "proof" of the spirits. Liquor that is 100 proof is 50 percent ethyl alcohol by volume, 90 proof is 45 percent, and so on.

$$\text{mole fraction of } C_2H_5OH = \frac{\text{mol } C_2H_5OH}{\text{total mol}} = \frac{0.300}{1.80} = 0.167$$

$$\text{mole fraction of } H_2O = \frac{\text{mol } H_2O}{\text{total mol}} = \frac{1.50}{1.80} = 0.833$$

One sixth (0.167) of all the molecules in this solution are ethyl alcohol molecules. Note that the sum of the mole fractions of solute and solvent must be one.

$$\text{mole fraction of } H_2O = 1 - \text{mole fraction of } C_2H_5OH$$
$$= 1 - 0.167 = 0.833$$

See also Exercises 19 and 20.

10•6•4 Molality Another way to express concentration so that we know the number of solute particles in a given number of solvent particles is in units of molality. The **molality,** m, of a solution is the number of moles of solute per kilogram of solvent. No knowledge of the solution volume is involved in preparing molal solutions, just a knowledge of the weights of solute and solvent.

• Example 10•3 •

Calculate the molality of a solution made by dissolving 262 g of ethylene glycol, $C_2H_6O_2$, in 8,000 g of water.

• Solution • By definition,

$$\text{molality, } m = \frac{\text{mol of solute}}{\text{kg of solvent}} = \frac{\dfrac{\text{g solute}}{\text{g/mol}}}{\text{kg of solvent}}$$

$$= \frac{\dfrac{262 \text{ g } C_2H_6O_2}{62.1 \text{ g } C_2H_6O_2/\text{mol}}}{8{,}000 \text{ g} \times \dfrac{1 \text{ kg}}{1{,}000 \text{ g}}}$$

$$= \frac{0.527 \text{ mol}}{\text{kg of solvent}} = 0.527m$$

See also Exercise 26.

10•6•5 Molarity The **molarity,** M, of a solution is the number of moles of solute per liter of solution. To prepare 1 L of a one-molar ($1M$) solution of sucrose ($C_{12}H_{22}O_{11}$, mol wt 342 g), we place 342 g of sucrose in a 1-L volumetric flask (see Figure 10.11) and add water until the total volume is precisely 1 L. Similarly, to prepare 1 L of $1M$ sodium chloride (NaCl, mol wt 58.5 g), we place 58.5 g of the salt in a 1-L flask and add enough water to make the total volume precisely 1 L. In each case, we know the amount of solute in a given volume of solution. However, the amounts of solvent differ, and they are not known because the volumes of solute plus solvent are not additive.

• Example 10•4 •

Calculate the molarity of a solution made by dissolving 4.0 g of calcium bromide, $CaBr_2$, in enough water to give 200 mL of solution.

• Solution •

$$M = \frac{\text{mol of solute}}{\text{L of solution}} = \frac{\dfrac{\text{g solute}}{\text{g/mol}}}{\text{L of solution}}$$

$$= \frac{\dfrac{4.0 \text{ g } CaBr_2}{200 \text{ g } CaBr_2/\text{mol}}}{200 \text{ mL} \times \dfrac{1 \text{ L}}{1{,}000 \text{ mL}}} = 0.10 \text{ mol/L} = 0.10M$$

See also Exercises 22, 24, and 28.

• Example 10•5 •

What weight of baking soda (sodium bicarbonate), $NaHCO_3$, is needed to prepare 150 mL of a 0.350M solution?

• Solution •

$$M = \frac{\text{mol of solute}}{\text{L of solution}} = \frac{\dfrac{\text{g solute}}{\text{g/mol}}}{\text{L of solution}}$$

$$0.350M = \frac{\dfrac{\text{g } NaHCO_3}{84.0 \text{ g } NaHCO_3/\text{mol}}}{150 \text{ mL} \times \dfrac{1 \text{ L}}{1{,}000 \text{ mL}}}$$

$$\text{g } NaHCO_3 = \left(\frac{0.350 \text{ mol}}{L}\right)\left(150 \text{ mL} \times \frac{1 L}{1{,}000 \text{ mL}}\right)\left(\frac{84.0 \text{ g}}{\text{mol}}\right)$$

$$= 4.41 \text{ g } NaHCO_3$$

See also Exercises 22, 24, and 28.

For dilute aqueous solutions, the molarity and molality have practically the same numerical value. For example, 1 L of a 0.207M solution of sodium chloride has 12.1 g of NaCl dissolved in 994.7 g of water at 20 °C. The molality of this solution is 0.208m. An NaCl solution whose molarity is 0.414M has a molality of 0.417m.

• 10•6•6   **Normality**   The **normality**, N, of a solution is the number of equivalents of solute per liter of solution. Solution concentrations expressed in terms of normalities are used in oxidation–reduction and in acid–base reactions. The latter reactions will be discussed in Chapter 11. In order to express solution concentrations in terms of normalities, we must first explain what is meant by the terms equivalent and equivalent weight.

• *Equivalent Weights* • Stated simply, equivalent weights are the weights of substances that are equivalent to one another in chemical reactions. Equivalent weights are always obtained with reference to specific reactions.

Consider the following redox reaction: $2\overset{0}{Al} + 3\overset{0}{Cl_2} \longrightarrow 2\overset{+3(-1)_3}{AlCl_3}$

As in Section 9.2.5, we can divide this redox reaction into two half-reactions.

$$\text{oxidation:} \quad \overset{0}{\text{Al}} \longrightarrow \overset{+3}{\text{Al}} + 3e^-$$

$$\text{reduction:} \quad \overset{0}{\text{Cl}_2} + 2e^- \longrightarrow 2\overset{-1}{\text{Cl}}$$

For each mole of the reducing agent, aluminum, that undergoes oxidation to $\overset{+3}{\text{Al}}$, 3 moles of electrons are lost. The amount of aluminum required to lose 1 mole of electrons is 26.98 g/3 = 8.99 g:

$$\underset{\frac{1}{3}\text{ mol, 8.99 g}}{\tfrac{1}{3}\overset{0}{\text{Al}}} \longrightarrow \underset{\frac{1}{3}\text{ mol, 8.99 g}}{\tfrac{1}{3}\overset{+3}{\text{Al}}} + \underset{1\text{ mol}}{e^-}$$

For each mole of the oxidizing agent, chlorine, that undergoes reduction to $\overset{-1}{\text{Cl}}$, 2 moles of electrons are gained. The amount of chlorine required to gain 1 mole of electrons is 70.906 g/2 = 35.453 g:

$$\underset{\frac{1}{2}\text{ mol, 35.453 g}}{\tfrac{1}{2}\overset{0}{\text{Cl}_2}} + \underset{1\text{ mol}}{e^-} \longrightarrow \underset{\frac{1}{2}\text{ mol, 35.453 g}}{\overset{-1}{\text{Cl}}}$$

The equivalent weight of an oxidizing or reducing agent is the weight of the substance required to gain or lose 1 mole of electrons. In the previous example, the equivalent weight of Al is 8.99 g; the equivalent weight of Cl_2 is 35.453 g.

• Example 10•6 •

Calculate the equivalent weight of each reactant in the following redox reaction:

$$MnO_2 + 4HCl \longrightarrow MnCl_2 + Cl_2 + 2H_2O$$

• Solution • We write the oxidation and reduction half-reactions to show the loss or gain of 1 mole of electrons:

$$\text{oxidation:} \quad \underset{1\text{ mol}}{\overset{-1}{\text{Cl}}} \longrightarrow \underset{\frac{1}{2}\text{ mol}}{\tfrac{1}{2}\overset{0}{\text{Cl}_2}} + \underset{1\text{ mol}}{e^-}$$

$$\text{reduction:} \quad \underset{\frac{1}{2}\text{ mol}}{\tfrac{1}{2}\overset{+4}{\text{Mn}}} + \underset{1\text{ mol}}{e^-} \longrightarrow \underset{\frac{1}{2}\text{ mol}}{\tfrac{1}{2}\overset{+2}{\text{Mn}}}$$

The equivalent weight of $\overset{-1}{\text{Cl}}$ is the weight of 1 mole, or 35.5 g. We also can refer to the weight of HCl that contains this weight of chlorine as the equivalent weight of HCl, or 36.5 g.

The equivalent weight of $\overset{+4}{\text{Mn}}$ is the weight of $\tfrac{1}{2}$ mole, or 27.47 g. We also can refer to the weight of MnO_2 that contains this weight of manganese as the equivalent weight of MnO_2, or $\tfrac{1}{2}(86.94 \text{ g}) = 43.47$ g.

See also Exercises 30 and 31(a).

With respect to an oxidizing or reducing agent (and for acid–base reactions covered in the next chapter), a one-normal ($1N$) solution contains one equivalent weight, or simply one equivalent, per liter of solution. A $0.5N$ solution contains half an equivalent per liter, and so on. Expressed mathematically,

$$\text{normality, } N = \frac{\text{equiv of solute}}{\text{L of solution}} = \frac{\dfrac{\text{g solute}}{\text{g/equiv}}}{\text{L of solution}}$$

where g/equiv is the equivalent weight in grams of the substance.

In all cases that we will consider, the normality of a solution either is equal to its molarity or is a small-number multiple of it.

• Example 10•7 •

Calculate the weight of potassium permanganate, $KMnO_4$, present in 500 mL of a $0.100N$ solution. The redox reaction is

$$2KMnO_4 + 16HCl \longrightarrow 2KCl + 2MnCl_2 + 5Cl_2 + 8H_2O$$

• Solution • The weight of 1 mole of $KMnO_4$ is 158.0 g. Manganese, Mn, changes its oxidation number from $+7$ in $KMnO_4$ to $+2$ in $MnCl_2$. This represents a gain of 5 moles of electrons. For the gain of 1 mole of electrons, we need only $\frac{1}{5}$ mole of $KMnO_4$:

$$\overset{+7}{\tfrac{1}{5}Mn} + \ e^- \ \longrightarrow \ \overset{+2}{\tfrac{1}{5}Mn}$$

$$\tfrac{1}{5}\text{ mol} \quad 1\text{ mol} \qquad \tfrac{1}{5}\text{ mol}$$

Hence the equivalent weight of $KMnO_4$ is $158.0/5 = 31.6$ g.

$$N = \frac{\text{equiv of solute}}{\text{L of solution}} = \frac{\dfrac{\text{g solute}}{\text{g/equiv}}}{\text{L of solution}}$$

$$0.100N = \frac{0.100 \text{ equiv}}{L} = \frac{\dfrac{\text{g } KMnO_4}{31.6 \text{ g } KMnO_4/\text{equiv}}}{500 \text{ mL} \times \dfrac{1\text{ L}}{1{,}000\text{ mL}}}$$

$$\text{g } KMnO_4 = \left(\frac{0.100\text{ equiv}}{L}\right)\left(500\text{ mL} \times \frac{1\text{ L}}{1{,}000\text{ mL}}\right)\left(\frac{31.6\text{ g } KMnO_4}{\text{equiv}}\right)$$

$$= 1.58 \text{ g } KMnO_4$$

See also Exercise 31.

• Example 10•8 •

Calculate the molarity of the $KMnO_4$ solution described in Example 10.7.

• Solution • In a $0.100N$ solution, there is 0.100 equivalent per liter. As we can see from Example 10.7, there is $\frac{1}{5}$ or 0.200 mol/equiv. The molarity (moles per liter) of the $KMnO_4$ solution is

$$(0.100 \text{ equiv/L}) \left(\frac{0.200 \text{ mol KMnO}_4}{1 \text{ equiv KMnO}_4} \right) = 0.0200 \text{ mol/L} = 0.0200M$$

In this case, the normality of the solution is five times the molarity.

See also Exercise 31.

10•7 Electrolytes and Nonelectrolytes

To simplify the study of compounds, it is desirable to classify them into groups and subgroups so that compounds with similar properties can be studied together. In Chapter 5, one of the most useful classifications was developed in some detail, *ionic* versus *covalent* compounds. As pointed out there, if all the bonds are shared pairs of electrons, the compound is covalent. If one or more of the bonds is primarily due to attraction between ions, the compound is ionic.

Another method of classifying compounds in two broad divisions is based on whether the molten compound or a solution of it conducts a current of electricity. If a molten compound or its solution is a conductor of an electric current, the compound is termed an **electrolyte;** if not, the compound is a **nonelectrolyte.** The test is made quite easily. In the conductivity apparatus shown in Figure 10.12, either a light bulb or an ammeter is used to indicate conductance through the circuit. The metal strips or wires that dip into the liquid are the electrodes. Note that the two electrodes do not touch

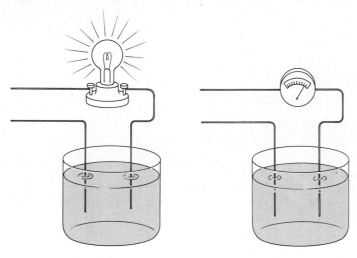

Figure 10•12 Two ways of testing a solution for electric conductivity.

each other, so that the liquid must be a conductor if the charge is to flow through the circuit. For example, if the light bulb is used, when the electrodes are immersed in pure water, the light will not glow. But when they are dipped into a water solution of sodium chloride, hydrogen chloride, hydrogen nitrate, or sodium hydroxide, the electric current is conducted by the solution and the light glows. Pure water is a nonelectrolyte, but sodium chloride, hydrogen chloride, hydrogen nitrate, and sodium hydroxide are classed as electrolytes. On the other hand, when the electrodes are immersed in a water solution of sugar, ethyl alcohol, or glycerin, the light does not glow. Sugar, ethyl alcohol, and glycerin are thus classed as nonelectrolytes.

It is important to note at this point that *all ionic compounds that dissolve in water form electrolyte solutions, whereas some covalent compounds that dissolve in water are electrolytes and some are not.*

10·7·1   How Solutions Conduct Current   Suppose we arrange the conductivity apparatus as shown in Figure 10.13. We shall use a direct current from a battery of dry cells,[4] and our electrolyte will be copper chloride, $CuCl_2$, dissolved in water. When the current is turned on, we are sure that charge is flowing through the circuit because of the deflection of the ammeter needle. We also notice that chlorine, a greenish-yellow gas, bubbles from the solution at the positive electrode (the anode) and that metallic copper begins to plate out on the negative electrode (the cathode).[5] When the battery is disconnected, the chemical changes cease. When the battery is reconnected, additional copper plates out and more chlorine bubbles appear. If the battery is left connected for a long time, with used cells being replaced as necessary, the deflection of the ammeter needle becomes less and less and eventually drops to zero. If we now examine the water solution, we find that no copper chloride remains. The principal reaction that has taken place is the electrolysis of copper chloride,

$$CuCl_2 \xrightarrow{\text{electric current}} Cu + Cl_2$$

Electrolysis is of great importance in the production of many elements and compounds and in covering articles with a thin layer of metal (electroplating).

The observations made during the electrolysis of the solution of $CuCl_2$ raise several questions:

1.   How did the charge pass through the solution?
2.   What caused the copper to be formed at the cathode?
3.   What caused the chlorine to be formed at the anode?
4.   What caused the current to stop eventually, even though worn-out batteries were replaced by fresh ones?

The answers to these questions begin with recognition of the fact that copper chloride, $CuCl_2$, is composed of ions. A more precise way of representing a unit quantity of copper chloride is as Cu^{2+}, Cl^-, Cl^-. When copper chloride is dissolved in water, the ions become separated and mix with the water, every ion being free to move around at random among the water molecules. When the current is turned on, there is a mass movement of

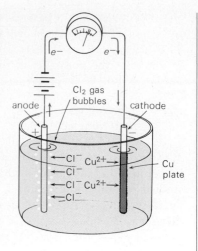

Figure 10·13 Electrolysis of copper chloride.

---

[4] There are two kinds of current: alternating and direct. Alternating current, the type commonly used in homes and industry for lighting, heating, and operating electric motors, is produced by rotating coils of copper wire in a magnetic field. In such current, the electrons oscillate rapidly, flowing first in one direction and then in the opposite direction. In direct current, produced by batteries and by the rectification of alternating current, the electrons flow in only one direction.
[5] For electric circuits associated with chemical reactions, by definition the **anode** is the electrode at which electrons are given up by a reactant (oxidation occurs) and the **cathode** is the electrode at which electrons are gained by a reactant (reduction occurs).

Cu^{2+} ions toward the negative pole, or cathode; there is also a mass movement of Cl^- ions toward the positive pole, or anode. At the cathode, electrons are picked up by the copper ions. In this manner, the copper ions are changed to copper atoms. Because the Cu^{2+} ions undergo a decrease in oxidation number, the half-reaction that takes place at the cathode is reduction:

$$Cu^{2+} + 2e^- \longrightarrow Cu$$

At the anode, electrons are given up by the chloride ions. Because the Cl^- ions undergo an increase in oxidation number, the half-reaction that occurs at the anode is oxidation:

$$2Cl^- \longrightarrow Cl_2 + 2e^-$$

The equation representing the overall reaction is

$$CuCl_2 \longrightarrow Cu + Cl_2$$

A more precise way of describing a reaction involving ions that are separated by water molecules is by means of an *ionic equation*. For the electrolysis of a water solution of copper chloride, the cathode and anode reactions may be summed to give the following ionic equation:

$$Cu^{2+} + 2Cl^- \longrightarrow Cu + Cl_2$$

We can summarize the answers to our four questions:

1. A solution of an electrolyte contains ions that take up or give up electrons.

2. The current (a flow of electrons) enters the solution at the cathode. The entering electrons are taken up by the positive ions (cations). Reduction takes place at the cathode.

3. Electrons leave the solution at the anode. Negative ions (anions) give up these electrons. Oxidation occurs at the anode.

4. When all the ions originally present have been changed to neutral particles, there are no longer any positive or negative particles to take up or give up electrons. No current can flow.

10•7•2 **Sources of Ions** In order to understand the properties of electrolytes, it is important to recognize the different ways in which ionic solutions are formed. Ionic solutions arise from two sources: ionic compounds and polar covalent compounds.

• *Ionic Compounds* • Ionic compounds are composed of ions even when in the dry, solid form. However, it is only when such substances are melted or dissolved in a solvent that the ions are free to migrate to an anode or cathode. All ionic compounds are electrolytes.

It should be emphasized that in the case of ionic compounds such as NaCl and $CuCl_2$, the water plays no special part in conducting the current beyond furnishing a medium in which the ions can move about. When pure sodium chloride is melted, at about 800 °C, this pure liquid is an excellent conductor of an electric current. X-ray studies show that sodium and chloride ions are present in even solid salt (see Section 8.6.1).

• *Polar Covalent Compounds* • It will be recalled that a polar covalent molecule as a whole is an electrically neutral particle (see Section 5.5). Pure liquid hydrogen chloride, HCl, pure liquid water, H_2O, pure liquid ammonia, NH_3, pure liquid acetic acid, CH_3CO_2H (also written as $HC_2H_3O_2$), and most organic compounds are very poor conductors of electric current. However, a solution of HCl in H_2O is a good conductor of electricity. On the other hand, if hydrogen chloride is dissolved in benzene, the solution does not conduct electricity. What is the reason for this difference? In order to account for the differences in conductivity, it is assumed that the covalent hydrogen chloride molecules are able to form ions in the water solution but are unable to do so in the benzene solution. Actually the ions result from a chemical reaction between hydrogen chloride molecules and water molecules, as shown in the following equation:

$$HCl + H_2O \rightleftharpoons H_3O^+ + Cl^- \qquad (1)$$

hydrogen   water   hydronium   chloride
chloride   ion   ion

As pointed out in Section 9.5.1, this type of reaction, in which two molecules react to form ions, is called an ionization reaction.

The formula H_3O^+ in Equation (1) is only one way of referring to the aquated H^+ ion in water solution.[6] Equations for ionization reactions in water are often written in a simplified form to show the formation of H^+ ion without showing that molecules of water are involved:

$$HCl \rightleftharpoons H^+ + Cl^-$$

hydrogen   hydrogen   chloride
chloride   ion   ion

Acetic acid, $HC_2H_3O_2$, when dissolved in water, forms a solution that conducts electricity. Acetic acid is a polar covalent compound that ionizes in water by a reaction similar to Equation (1), that is,

$$HC_2H_3O_2 + H_2O \rightleftharpoons H_3O^+ + C_2H_3O_2^- \qquad (2)$$

acetic   water   hydronium   acetate
acid   ion   ion

Ammonia, a covalent compound, also reacts with water to produce ions:

---

[6] Attempts to show conclusively that H_3O^+ is a separate ion in water solution have failed. Instead, evidence from spectroscopic and other studies indicates that the aquated proton is more complex. The actual species present may be a mixture of H_3O^+, $H_5O_2^+$, $H_7O_3^+$, and $H_9O_4^+$, depending upon the acid concentration and solution temperature.

$$NH_3 + HOH \rightleftharpoons NH_4^+ + OH^- \qquad (3)$$

ammonia   water   ammonium   hydroxide
ion   ion

Methylamine, CH_3NH_2, in which one of the hydrogens in ammonia is replaced with a methyl group, $-CH_3$, reacts like ammonia with water to form ions. Reactions (1) and (3) are shown by diagrams in Figure 10.14.

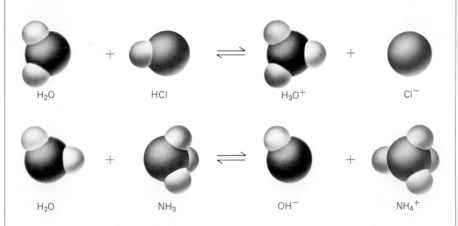

H_2O HCl H_3O^+ Cl^-

H_2O NH_3 OH^- NH_4^+

Figure 10•14 Pure hydrogen chloride and pure ammonia do not conduct electricity. Their water solutions are conductors because the pure substances react with water to form ions.

10•7•3   Ionic Equilibria   Note that double arrows are used in Equations (1) to (3). These arrows indicate that the ionization reactions are reversible and that the systems are at equilibrium. Consider what happens when ammonia is added to water, Equation (3). At the instant that the two are brought together, only NH_3 and H_2O molecules can react, because only these two are present. A very short time later some NH_4^+ and OH^- ions will have formed, and the reverse reaction starts, slowly at first, because very few of these ions are present to react. Both reactions continue to take place simultaneously. In time the concentrations of NH_3, H_2O, NH_4^+, and OH^- become adjusted so that both reactions are occurring at the same speed. At that point the system is at equilibrium.

The factors that determine the relative amounts of reactants and products at equilibrium will be discussed in detail in Chapter 15. Let us note here that when equilibrium is reached for reaction (1), the resulting solution is a very good conductor of electricity. This indicates that most of the hydrogen chloride has reacted to form ions; the concentration of HCl is relatively low and the concentrations of H_3O^+ and Cl^- are relatively high. We say that the equilibrium lies far to the right for this reaction. The opposite is true for reactions (2) and (3). At equilibrium these solutions are rather poor conductors of electricity, which indicates that the concentrations of $HC_2H_3O_2$ and NH_3 are relatively high and the concentrations of ions are relatively low. For these solutions, we say that the equilibrium lies far to the left.

10•7•4   Strong and Weak Electrolytes   Water solutions of sodium chloride and other ionic compounds as well as water solutions of certain covalent compounds are excellent conductors of electricity. Substances that exist in solution completely or almost completely in the form of ions are called **strong electrolytes.**

On the other hand, water solutions of many covalent compounds are poor conductors of electricity. Water solutions of ammonia and of acetic acid are examples. Substances in which a small percentage of the dissolved molecules react with water to form ions are called **weak electrolytes.** The great bulk of the dissolved substance is present as covalent molecules.[7]

Actually, the terms strong and weak electrolytes are not clear-cut classifications, for strong electrolytes may be weakly strong, moderately strong, strong, very strong, and so on. Weak electrolytes may be subclassified in the same manner. That is, there are all degrees of weak and strong, so the dividing line between the two is not always clear. This topic will be discussed in greater detail in Chapter 16.

In addition to being classified according to strength, electrolytes may be classified according to type. The three common types are *acids, bases,* and *salts.* For acids and bases there are both strong and weak electrolytes. Because **salts** are ionic compounds, they are all strong electrolytes.

10•8   Ionic Equations

In Section 10.7.1, a simple ionic equation was used to describe the reaction that occurs during the electrolysis of a water solution of copper chloride, $CuCl_2$:

$$Cu^{2+} + 2Cl^- \longrightarrow Cu + Cl_2$$

In such an equation, all reactants and products that are present as ions in the solution are written as separate ions in the equation.

Ionic equations are very clear and useful for several types of reactions, including (1) oxidation–reduction (redox) reactions, (2) formation of insoluble ionic compounds that separate, or precipitate, from solution, and (3) formation of covalent compounds.

10•8•1   Oxidation–Reduction (Redox) Reactions   The electrolysis of a water solution of copper chloride is an example of a redox reaction in which the Cu^{2+} and Cl^- ions undergo reduction and oxidation, respectively. Other examples of redox reactions for which we can write ionic equations are the reactions of zinc with the following iron(II) compounds in water solutions:

$$Zn + FeCl_2 \longrightarrow ZnCl_2 + Fe$$
$$Zn + FeBr_2 \longrightarrow ZnBr_2 + Fe$$
$$Zn + FeSO_4 \longrightarrow ZnSO_4 + Fe$$

---

[7] Water solutions of ammonia often are referred to as ammonium hydroxide and are represented by the formula NH_4OH. If one considers ionic compounds such as ammonium chloride, NH_4Cl, and sodium hydroxide, $NaOH$, then the name ammonium hydroxide and its formula NH_4OH imply that it, too, is an ionic compound. However, this is not the case because aqueous solutions of ammonia are only slightly ionized. A further complication is that no covalent compound with the formula NH_4OH exists in water solution, although the polar NH_3 molecules may attract several polar H_2O molecules and exist in solution as solvated molecules of varying composition. For these reasons, we will usually avoid the use of the name ammonium hydroxide and the formula NH_4OH.

These equations represent three seemingly different reactions. Because the iron and zinc compounds are present as ions, we can write the following *complete ionic equations* for the reactions:

$$Zn + Fe^{2+} + 2Cl^- \longrightarrow Zn^{2+} + 2Cl^- + Fe$$
$$Zn + Fe^{2+} + 2Br^- \longrightarrow Zn^{2+} + 2Br^- + Fe$$
$$Zn + Fe^{2+} + SO_4{}^{2-} \longrightarrow Zn^{2+} + SO_4{}^{2-} + Fe$$

In these three equations, Cl^-, Br^-, and $SO_4{}^{2-}$ ions are shown as both reactants and products. They do not undergo any chemical change. Such ions that are present but do not take part in the reaction are called **spectator ions.** In the formation of Zn^{2+} and Fe, each of these three reactions requires that zinc atoms undergo oxidation to Zn^{2+} ions and that Fe^{2+} ions be reduced to iron atoms. If we cross out the spectator ions by drawing lines through the symbols, we obtain

$$Zn + Fe^{2+} + \cancel{2Cl^-} \longrightarrow Zn^{2+} + \cancel{2Cl^-} + Fe$$
$$Zn + Fe^{2+} + \cancel{2Br^-} \longrightarrow Zn^{2+} + \cancel{2Br^-} + Fe$$
$$Zn + Fe^{2+} + \cancel{SO_4{}^{2-}} \longrightarrow Zn^{2+} + \cancel{SO_4{}^{2-}} + Fe$$

The single observable chemical change for each of the reactions is represented by the equation

$$Zn + Fe^{2+} \longrightarrow Zn^{2+} + Fe$$

Each of these four equations is a **net ionic equation,** an equation that shows only the species involved in the observable chemical change. If spectator ions are included, lines are drawn through their symbols.

• Example 10•9 •

The electrolysis of a concentrated aqueous solution of sodium chloride, NaCl, produces sodium hydroxide, NaOH, hydrogen, H_2, and chlorine, Cl_2.
(a) Write a complete ionic equation for this redox reaction.
(b) Write a net ionic equation for the reaction. Also write the net ionic equations for the electrolysis of concentrated aqueous solutions of lithium chloride, LiCl, and potassium chloride, KCl.

• Solution •

(a) Because NaOH and H_2 are produced along with Cl_2, both NaCl and HOH are reactants. The overall equation is

$$2NaCl + 2HOH \longrightarrow 2NaOH + H_2 + Cl_2$$

Sodium chloride and sodium hydroxide are soluble ionic compounds that exist as separate ions in water solution. The complete ionic equation is

$$2Na^+ + 2Cl^- + 2HOH \longrightarrow 2Na^+ + 2OH^- + H_2 + Cl_2$$

(b) We cross out the spectator ions, the Na^+ ions, or simply omit them to obtain the net ionic equation:

$$\cancel{2Na^+} + 2Cl^- + 2HOH \longrightarrow \cancel{2Na^+} + 2OH^- + H_2 + Cl_2$$

or

$$2Cl^- + 2HOH \longrightarrow 2OH^- + H_2 + Cl_2$$

Like NaCl, both LiCl and KCl are ionic IA metal chlorides. The Li$^+$ and K$^+$ ions would also be spectator ions, and the net ionic equation for the electrolysis of each of these chlorides is identical with that shown for NaCl, that is,

$$2Cl^- + 2HOH \longrightarrow 2OH^- + H_2 + Cl_2$$

See also Exercise 45.

Another example of a redox reaction for which we can write ionic equations takes place when a small piece of magnesium is dropped into dilute HCl. The magnesium disappears as bubbles of hydrogen gas come out of solution. A solution of magnesium chloride, $MgCl_2$, is formed. The equation for this single displacement reaction is

$$Mg + 2HCl \longrightarrow MgCl_2 + H_2$$

Because HCl and $MgCl_2$ are both strong electrolytes, we write their formulas in ionic form. For HCl the following equilibrium (see Section 10.7.3) lies far to the right:

$$HCl + H_2O \rightleftharpoons H_3O^+ + Cl^- \qquad (1)$$

The production of H_2 and Mg^{2+} ions occurs by the reaction of Mg with the H_3O^+ ions rather than with the un-ionized HCl molecules. The net ionic equation for this reaction may be written as

$$Mg + 2H_3O^+ + \cancel{2Cl^-} \longrightarrow Mg^{2+} + \cancel{2Cl^-} + 2H_2O + H_2$$

If the ionization of HCl is simplified by omitting the water, we write

$$HCl \rightleftharpoons H^+ + Cl^-$$

and the net ionic equation is written as

$$Mg + 2H^+ \longrightarrow Mg^{2+} + H_2$$

If a small piece of magnesium is dropped into dilute acetic acid, $HC_2H_3O_2$, the reaction is slow, but the magnesium gradually disappears as hydrogen gas bubbles are formed. A solution of magnesium acetate, $Mg(C_2H_3O_2)_2$, is produced. The equation for this single displacement reaction is

$$Mg + 2HC_2H_3O_2 \longrightarrow Mg(C_2H_3O_2)_2 + H_2$$

Because $Mg(C_2H_3O_2)_2$ is a strong electrolyte, we may write its formula in ionic form. The small equilibrium concentration of H_3O^+ ions present in the water solution of acetic acid (see Section 10.7.3),

$$HC_2H_3O_2 + H_2O \rightleftharpoons H_3O^+ + C_2H_3O_2^- \tag{2}$$

is sufficient to react with the magnesium metal. The equation representing this reaction is

$$Mg + 2H_3O^+ + 2C_2H_3O_2^- \longrightarrow Mg^{2+} + 2C_2H_3O_2^- + 2H_2O + H_2 \tag{4}$$

or, in the simplified form, using H^+ instead of H_3O^+,

$$Mg + 2H^+ \longrightarrow Mg^{2+} + H_2$$

These ionic equations are the same as those shown for the reaction of magnesium with dilute HCl. However, because acetic acid is a weak electrolyte and only a small percentage is present as ions, the net ionic equation is usually written with the un-ionized covalent molecule as a reactant. This equation is obtained by multiplying Equation (2) by 2, and adding the result to Equation (4):

$$2HC_2H_3O_2 + 2H_2O \longrightarrow 2H_3O^+ + 2C_2H_3O_2^-$$
$$\underline{Mg + 2H_3O^+ \longrightarrow Mg^{2+} + 2H_2O + H_2}$$

net ionic equation: $Mg + 2HC_2H_3O_2 \longrightarrow Mg^{2+} + 2C_2H_3O_2^- + H_2$

• **10•8•2   Precipitation Reactions**   As stated in Section 10.2.3, insoluble ionic compounds often are formed by mixing solutions of two soluble compounds. From Table 10.2 it can be deduced that barium sulfate, $BaSO_4$, is one of the ionic metal sulfates that is considered insoluble in water. Barium sulfate may be precipitated from water solution by reactions illustrated with the following three equations:

$$BaCl_2 + Na_2SO_4 \longrightarrow BaSO_4\downarrow + 2NaCl$$
$$Ba(NO_3)_2 + MgSO_4 \longrightarrow BaSO_4\downarrow + Mg(NO_3)_2$$
$$BaBr_2 + (NH_4)_2SO_4 \longrightarrow BaSO_4\downarrow + 2NH_4Br$$

The downward arrow indicates the formation of a water-insoluble precipitate. As also can be deduced from Table 10.2, the nine ionic compounds other than $BaSO_4$ are all water soluble, and their ions can be written separately. Complete ionic equations may be written for these double displacement reactions. But, when the spectator ions are crossed out or omitted, the resulting net ionic equation represents one observable chemical change. That change is the formation of insoluble barium sulfate, $BaSO_4$:

$$Ba^{2+} + 2Cl^- + 2Na^+ + SO_4^{2-} \longrightarrow BaSO_4\downarrow + 2Na^+ + 2Cl^-$$
$$Ba^{2+} + 2NO_3^- + Mg^{2+} + SO_4^{2-} \longrightarrow BaSO_4\downarrow + Mg^{2+} + 2NO_3^-$$
$$Ba^{2+} + 2Br^- + 2NH_4^+ + SO_4^{2-} \longrightarrow BaSO_4\downarrow + 2NH_4^+ + 2Br^-$$

or

$$Ba^{2+} + SO_4^{2-} \longrightarrow BaSO_4\downarrow$$

Such net ionic equations make it clear that the solutions brought together for a particular precipitation reaction need not be restricted to a specific pair of compounds. For example, any compound dissolving to give SO_4^{2-} ions in solution would serve for precipitating Ba^{2+} ions as $BaSO_4$; appropriate compounds include Na_2SO_4, K_2SO_4, $(NH_4)_2SO_4$, $MgSO_4$, H_2SO_4, and $Al_2(SO_4)_3$.

Magnesium hydroxide, $Mg(OH)_2$, may be precipitated by the reaction of aqueous magnesium nitrate, $Mg(NO_3)_2$, with a water solution of either sodium hydroxide, $NaOH$, or ammonia, NH_3. When sodium hydroxide is used, the equation for the double displacement reaction is

$$Mg(NO_3)_2 + 2NaOH \longrightarrow Mg(OH)_2{\downarrow} + 2NaNO_3$$

All compounds except $Mg(OH)_2$ are water soluble, and their ions can be written separately. The net ionic equation is

$$Mg^{2+} + \cancel{2NO_3^-} + \cancel{2Na^+} + 2OH^- \longrightarrow Mg(OH)_2{\downarrow} + \cancel{2Na^+} + \cancel{2NO_3^-}$$

When magnesium hydroxide is precipitated with aqueous ammonia, the equation is

$$Mg(NO_3)_2 + 2NH_3 + 2H_2O \longrightarrow Mg(OH)_2{\downarrow} + 2NH_4NO_3$$

Because NH_4NO_3 is a water-soluble, strong electrolyte, we may write its formula as separate ions. The small equilibrium concentration of hydroxide ions present in the water solution of ammonia (see Section 10.7.3),

$$NH_3 + H_2O \rightleftharpoons NH_4^+ + OH^- \tag{3}$$

is sufficient to combine with the Mg^{2+} ions to precipitate magnesium hydroxide, $Mg(OH)_2$:

$$Mg^{2+} + \cancel{2NO_3^-} + \cancel{2NH_4^+} + 2OH^- \longrightarrow Mg(OH)_2{\downarrow} + \cancel{2NH_4^+} + \cancel{2NO_3^-} \tag{5}$$

Because ammonia is a weak electrolyte, the net ionic equation is usually written with the un-ionized covalent molecule as a reactant. This equation is obtained by multiplying Equation (3) by 2, and adding the result to Equation (5):

$$2NH_3 + 2H_2O \longrightarrow 2NH_4^+ + 2OH^-$$
$$\underline{Mg^{2+} + 2OH^- \longrightarrow Mg(OH)_2{\downarrow}}$$

net ionic equation: $Mg^{2+} + 2NH_3 + 2H_2O \longrightarrow Mg(OH)_2{\downarrow} + 2NH_4^+$

— • Example 10.10 • —

Write net ionic equations for each of the following precipitation reactions as carried out in water solutions:
(a) reaction of iron(III) sulfate with sodium hydroxide
(b) reaction of silver nitrate with calcium chloride
(c) precipitation of silver cyanide by the reaction of silver nitrate with hydrogen cyanide, a weak electrolyte

• **Solution** • We first write a balanced equation for each reaction. In each of these cases, we predict that the reaction is a double displacement. Referring to Table 10.2, we identify the precipitate in the first two reactions as the less soluble of the products.

(a)
$$Fe_2(SO_4)_3 + 6NaOH \longrightarrow 2Fe(OH)_3\downarrow + 3Na_2SO_4$$
$$2Fe^{3+} + 3SO_4^{2-} + 6Na^+ + 6OH^- \longrightarrow 2Fe(OH)_3\downarrow + 6Na^+ + 3SO_4^{2-}$$

(b)
$$2AgNO_3 + CaCl_2 \longrightarrow 2AgCl\downarrow + Ca(NO_3)_2$$
$$2Ag^+ + 2NO_3^- + Ca^{2+} + 2Cl^- \longrightarrow 2AgCl\downarrow + Ca^{2+} + 2NO_3^-$$

(c)
$$AgNO_3 + HCN \longrightarrow AgCN\downarrow + HNO_3$$

Because HCN is a weak electrolyte, we write the net ionic equation with HCN in the un-ionized form:

$$Ag^+ + NO_3^- + HCN \longrightarrow AgCN\downarrow + H^+ + NO_3^-$$

See also Exercises 47 and 48.

10•8•3 Formation of Covalent Compounds The third type of reaction for which an ionic equation is useful is one in which ions react to form a covalent compound. The covalent compound may remain dissolved in solution as a weak electrolyte or nonelectrolyte, or it may escape as a gas, or both. The most common examples of this type of reaction are acid–base reactions, which will be discussed in Chapter 11.

Chapter Review

Summary

Most chemical reactions occur in **solutions.** A solution has a **solvent** (the major component) and a **solute** (the minor component). One of the main factors in the formation of liquid solutions is **solvation,** the interaction of solvent molecules with solute particles to form small aggregates. Water, a polar compound with a high dielectric constant, is an excellent solvent for most ionic and polar covalent substances (like dissolves like). The absorption of water vapor by certain kinds of crystals is called **deliquescence.** All such crystals are **hygroscopic,** and some are useful as **desiccants,** or drying agents. Crystals that release their water of crystallization to the atmosphere exhibit **efflorescence.**

Many substances are virtually **insoluble** in a given solvent (mutually insoluble liquids are **immiscible**). Depending on the relative amount of solute present, a solution can be described qualitatively as **dilute** or **concentrated.** The **solubility** of a substance that does dissolve is the amount of that substance that produces a **saturated solution** with a given amount of solvent. An **unsaturated solution** contains less solute than a saturated solution. A **supersaturated solution** contains more dissolved solute than is normally possible at the temperature in question; eventually the excess amount precipitates. A substance that is slightly soluble in one solvent may be very soluble in another solvent that is immiscible with the first; this is the basis for the **solvent extraction** process.

The response of any saturated solution to a change in temperature is always in accord with **Le Chatelier's principle,** which states that every system at equilibrium responds to an applied stress in such a way as to relieve the stress. In general, solids become more soluble with increasing temperature (the solution process is endothermic). This is the basis for the purification of a solid by **recrystallization** from solution. Most gases become less soluble with increasing temperature. In accord with **Henry's law,** a given gas becomes more soluble with increasing pressure exerted by that gas on the solution.

The best way to express the **concentration** of a solution depends on the nature of the solution, the way in which it was made, and the purpose for which the concentration must be known. The simplest ways are as the percent by weight and the percent by volume. It is often useful to know the **mole fractions** of the solute and solvent; the mole fraction times 100 is the **mole percent.** Three useful ways to express concentration are as the **molality** (*m*), **molarity** (*M*), and **normality** (*N*) of the solution. The latter is based on the concept of **equivalent weights** of substances. In a given reaction, one **equivalent** of one substance will react with one equivalent of another.

All compounds are either **electrolytes** (conductors of electricity) or **nonelectrolytes** (nonconductors) when in the pure liquid or dissolved state. Ionic solutions are solutions of ionic compounds, which are **strong electrolytes,** or polar covalent compounds, most of which are **weak electrolytes.** An ionic solution in which a redox reaction is caused by the passage of an electric current is said to undergo **electrolysis.** **Ionic equations** are useful in describing redox reactions, **precipitation reactions,** and reactions in which ions combine to form covalent compounds. In any such reaction in solution, those ions that remain unchanged throughout are called **spectator ions** and can be deleted from the equation. The resulting equation is the **net ionic equation.**

Exercises

Effect of Attractive Forces on Solubility

1. (a) The ions Na^+ and Cl^- are strongly solvated in water, but not in ether or benzene. Explain.
   (b) Compare the force of attraction between Na^+ and Cl^- in water and in benzene.
2. About 700 L (at STP) of gaseous ammonia, NH_3, dissolves in 1,000 g of water at room temperature.
   (a) How many ammonia molecules are there in solution for every ten water molecules?
   (b) Draw a likely hydrogen bonded structure linking several molecules of NH_3 and several of H_2O.
3. Based on dielectric constants (see Table 10.1), which is the better solvent for sodium chloride, methyl alcohol or common ether? Why?
4. The organic compound acetone

$$H-\underset{\underset{H}{|}}{\overset{\overset{H}{|}}{C}}-\underset{\underset{O:H}{\|}}{C}-\underset{\underset{H}{|}}{\overset{\overset{H}{|}}{C}}-H$$

   is very soluble in water. Explain.
5. What is meant by the statement *like dissolves like?* Illustrate it with three specific examples.
*6. Why is ethylene glycol, $C_2H_4(OH)_2$, soluble in water, but ethylene dibromide, $C_2H_4Br_2$, practically insoluble? Choose two solvents other than water in which ethylene glycol should be soluble. Choose two in which ethylene dibromide should be soluble.
7. Give a logical explanation for the fact that of the two ionic compounds Na_2SO_4 and $BaSO_4$, the former is water soluble and the latter is not.

8. Under what conditions could a saturated solution be referred to as a dilute solution? As a concentrated solution? Give examples of each.
9. For each of the following pairs of salts, consult Table 10.2 and tell what, if anything, will precipitate when their dilute solutions are mixed. Write an equation for each precipitation reaction and indicate the precipitate with a downward arrow ↓ to the right of its formula.
   (a) $CuCl_2 + H_2S$
   (b) $Ba(NO_3)_2 + K_2SO_4$
   (c) $NaCl + Ca(NO_3)_2$
   (d) $AgNO_3 + KI$
   (e) $CaBr_2 + (NH_4)_2CO_3$
*10. Mercury(II) chloride is 13 times more soluble in water than it is in benzene. If 50.0 mL of a benzene solution that contains 0.0200 g of $HgCl_2$ is extracted with 50.0 mL of water in a separatory funnel, how much $HgCl_2$ remains in the benzene? If the process is repeated, how much $HgCl_2$ remains?

Effect of Temperature and Pressure on Solubility

11. In a demonstration of a supersaturated solution similar to that shown in Figure 10.5, the temperature of the supersaturated solution just prior to the addition of the seed crystals was 20 °C. As the crystallization took place, the temperature rose, reaching a maximum of 28 °C. Relate this heat change to the heat changes that occurred in forming the supersaturated solution.
12. If 9.0 g of $K_2Cr_2O_7$ is heated with 10.0 g of water to 100 °C, will it all dissolve? If the solution is then cooled to 20 °C, what weight of solid $K_2Cr_2O_7$ will form?
13. How could you make a supersaturated solution of cerium sulfate, $Ce_2(SO_4)_3$? Examine Figure 10.9 before you answer.
14. The solubility of sodium selenate, Na_2SeO_4, is as follows: 84 g/100 mL of water at 35 °C; 72.8 g/100 mL of water at 100 °C. Would the dissolving of sodium selenate in water be endothermic or exothermic? On what do you base your conclusion?
15. Answer the following questions by referring to Figure 10.9.
   (a) A solution contains 60 g of $Pb(NO_3)_2$ in 100 g of water at 20 °C. Is it saturated, unsaturated, or supersaturated?
   (b) Suppose you have a solution that contains 60 g of KNO_3 in 100 g of water at 40 °C. Tell how you could get some crystals of pure KNO_3 and tell what weight you would expect to get.
16. Suppose that you have a sample of lead nitrate that is 99 percent pure; the contaminants are sodium chloride and potassium nitrate. Decribe how you would proceed to obtain some pure lead nitrate. Could you obtain 100 percent of the lead nitrate as the pure compound?

17. State Henry's law in the form of an algebraic equation.

Expressing Concentrations

18. A concentrated sucrose solution has a density of $1.438 \text{ g} \cdot \text{mL}^{-1}$ and contains 1,208 g of sucrose per liter of solution.
    (a) What are the percentages of water and of sucrose in the solution?
    (b) Is there any question about which of the substances should be called the solvent?
    (c) Calculate the mole fractions of sugar and of water.
19. What is the mole fraction of benzene, C_6H_6, in a solution prepared by dissolving 28.6 g of benzene in 72.0 g of carbon tetrachloride, CCl_4?
20. A solution contains 16.0 g of methyl alcohol, CH_3OH; 69.0 g of ethyl alcohol, C_2H_5OH; and 54.0 g of water. What are the mole fraction and the mole percent of methyl alcohol?
21. (a) Suppose someone asked you to make up 100 mL of a $3.00M$ solution of calcium carbonate, $CaCO_3$. What would your response be?
    (b) Repeat, for 100 mL of $3.00M$ calcium nitrate, $Ca(NO_3)_2$.
22. Calculate the molarity of each of the following aqueous solutions:
    (a) 36.0 g of glycerol, $C_3H_5(OH)_3$, in 1,000 mL of solution
    (b) 10.0 g of methanol, CH_3OH, in 100 mL of solution
    (c) 100 g of hydrogen sulfate, H_2SO_4, in 1.50 L of solution
*23. The ratio of solubilities of I_2 in CCl_4 versus H_2O is 86:1. A saturated water solution of iodine at 25 °C is $1.1 \times 10^{-3} M$. If 50 mL of this solution is extracted with 50 mL of carbon tetrachloride, what is the molarity of the aqueous I_2 solution?
24. How many milliliters of a $0.250M$ solution can be prepared from 14.8 g of calcium hydroxide?
25. Precisely 10.0 g of potassium chloride, KCl, is dissolved in 75.0 g of water.
    (a) Calculate the weight percent of chloride ion.
    (b) Calculate the mole percent of potassium chloride.
    (c) Calculate the molality, m, of potassium chloride.
    (d) If the volume of solution is 78.9 mL, what is the molarity, M?
26. Calculate the molality of the ethyl alcohol–water solution in Example 10.2 and the molality of the benzene–carbon tetrachloride solution in Exercise 19.
27. The density of a $0.00100M$ solution of calcium bromide, $CaBr_2$, is $1.00 \text{ g} \cdot \text{mL}^{-1}$. Show for this dilute solution that the molarity is very nearly the same as the molality. Explain why molarities and molalities are approximately equal for very dilute solutions.
28. A reagent bottle contains 5.00 g of sodium hydroxide,

NaOH. How many milliliters of a $2.00M$ solution can be prepared from this weight of NaOH?
29. The concentrated sulfuric acid commonly available in the laboratory is about $18M$. How many milliliters of this acid must be added to water to make each of the following?
    (a) 100 mL of $6M$ acid
    (b) 1 L of $0.05M$ acid
    (c) 250 mL of $0.2M$ acid
30. Calculate the equivalent weight of each reactant in the following redox reaction:

$$SnCl_2 + 2FeCl_3 \longrightarrow SnCl_4 + 2FeCl_2$$

31. (a) Calculate the equivalent weight of potassium dichromate, $K_2Cr_2O_7$, in the following redox reaction:

$$K_2Cr_2O_7 + 14HCl \longrightarrow$$
$$2KCl + 3Cl_2 + 2CrCl_3 + 7H_2O$$

    (b) How many equivalents of $K_2Cr_2O_7$ are present in 250 mL of a $0.250N$ solution? What weight of $K_2Cr_2O_7$ is required to prepare the 250 mL?
*32. What volume of a $0.100N$ HCl solution is required to react completely with the 250 mL of $0.250N$ $K_2Cr_2O_7$ solution described in Exercise 31(b)?
33. Calculate the molarity of the HCl and $K_2Cr_2O_7$ solutions described in Exercise 32.

How Substances Conduct Current

34. Explain how a current of electricity is conducted through a water solution of mercury(II) chloride.
35. Show with equations the cathode reaction, the anode reaction, and the overall reaction when each of the following is subjected to electrolysis to produce the elements:
    (a) a water solution of $NiCl_2$
    (b) molten KBr
36. When a current of electricity is passed through a water solution of cobalt bromide, $CoBr_2$, the elements are formed at the electrodes.
    (a) Just by looking at the substances as they are produced in the electrolysis cell, could you recognize which element is being formed at each electrode?
    (b) Write the equations for the two half-reactions, one for each electrode. Label each equation as an oxidation or a reduction reaction.
37. For the electrolysis of molten aluminum fluoride, write the equation for
    (a) the anode reaction
    (b) the cathode reaction
    (c) the overall reaction of the ions
38. In general, a polar covalent compound will form ions when dissolved in a polar solvent, but will not form ions

when dissolved in a nonpolar solvent. Explain why, and illustrate with a specific example.

39. How can one determine that hydrochloric acid is highly ionized in a water solution, whereas acetic acid is only slightly ionized, when each solution is made by dissolving 0.1 mole of acid in 1 L of water?

40. Give a logical explanation for each of the following:
    (a) A water solution of sulfuric acid that contains 0.05 mole of acid per liter is a better conductor of electricity than one of acetic acid that contains 0.1 mole of that acid per liter.
    (b) Pure molten potassium chloride is a much better conductor than solid potassium chloride.

41. Which of the following compounds would exist in water solution completely or almost completely in the form of ions: KCl, H_2S, $CaCl_2$, NH_3, CH_4, $NaNO_3$, CH_3OH, CsI, $RbOH$, $Mg(NO_3)_2$, HI, $C_{12}H_{22}O_{11}$ (sucrose), $CH_3CH_2CO_2H$, $NaHCO_3$? Which of these compounds would be classified as strong electrolytes? As weak electrolytes?

42. Write balanced equations for the ionization in water of formic acid, $HCHO_2$, and of methylamine, CH_3NH_2. These compounds are weak electrolytes that are similar to acetic acid and ammonia.

43. (a) A white, crystalline compound is found by analysis to be 35.0 percent N, 5.0 percent H, and 60.0 percent O. What is its empirical formula?
    (b) A solution of the compound in water conducts an electric current very well. What does this indicate about the compound?
    (c) Based on parts (a) and (b), what might be the name and formula of the compound?

Writing Ionic Equations

44. Write net ionic equations for the following redox reactions:

(a) A strip of aluminum is placed into a solution of nickel(II) nitrate and a coating of nickel forms on the aluminum strip.
(b) An iron nail is added to a hydrochloric acid solution; iron(II) chloride forms as hydrogen bubbles from the solution.

* 45. When a concentrated aqueous solution of calcium bromide is electrolyzed, the products are calcium hydroxide, bromine, and hydrogen.
    (a) Write the balanced equation for the reaction.
    (b) Write the half-reactions that occur at the anode and the cathode.
    (c) Write the net ionic equation for the reaction.

46. Write net ionic equations for each of the precipitation reactions in Exercise 9. [In Exercise 9(a), H_2S is a weak electrolyte.]

47. Write net ionic equations for the following precipitation reactions when carried out in water solutions:
    (a) the formation of insoluble strontium carbonate by the action of strontium iodide with ammonium carbonate (both reactants are strong electrolytes)
    (b) the formation of insoluble iron(III) hydroxide by the reaction of iron(III) nitrate, a strong electrolyte, with sodium hydroxide
    (c) the formation of iron(III) hydroxide by the reaction of iron(III) nitrate with ammonia, a weak electrolyte
    (d) the formation of insoluble mercury(II) sulfide by the reaction of mercury(II) bromide, a strong electrolyte, with hydrogen sulfide, a weak electrolyte
    (e) the formation of a precipitate of barium sulfate by the reaction of barium chloride with sodium sulfate

* 48. When a water solution of magnesium sulfate, $MgSO_4$, is added to a water solution of methylamine, a precipitate of magnesium hydroxide forms. Write a balanced net ionic equation for this reaction.

Acids and Bases

11

11.1 Arrhenius Acids and Bases

11.2 Brønsted–Lowry Acids and Bases

11.2.1 Aqueous Solutions of Acids and Bases

11.2.2 Relative Strengths of Acids and Bases

11.2.3 Acid–Base Reactions in Aqueous Salt Solutions

11.3 Naming Inorganic Acids

11.3.1 Binary and Related Acids

11.3.2 Ternary Oxy-acids

11.4 Structures of Hydroxy Compounds

11.5 Neutralization

11.5.1 Strong Acids and Strong Bases

11.5.2 Strong Acids and Weak Bases

11.5.3 Weak Acids and Strong Bases

11.5.4 Weak Acids and Weak Bases

11.5.5 Equivalent Weights and Normal Solutions in Neutralizations

11.5.6 Titrations of Acids and Bases

11.6 Lewis Acids and Bases

Most students are familiar with some acids and bases before they begin a formal course in chemistry. For example, it is common knowledge that hydrochloric acid is in the digestive juices of the stomach, acetic acid is the acid ingredient in vinegar, carbonic acid gives the "zip" to carbonated beverages, and citric acid is present in citrus fruits, such as oranges, grape-fruits, lemons, and limes. Many persons recognize the strong, pungent odor of the base ammonia, which is commonly used in water solution and in many cleansers as a disinfectant.

Acids and bases were defined by chemists centuries ago in terms of the properties of their water solutions. In these terms, an acid is a substance whose water solution has a sour taste, turns blue litmus red, reacts with active metals to form hydrogen, and neutralizes bases. Following a similar pattern, a base is defined as a substance whose water solution has a bitter taste, turns red litmus blue, feels soapy, and neutralizes acids.

Although these definitions of acids and bases have practical value, they greatly limit the scope of this field of chemistry. In this chapter, several concepts of acids and bases are presented that relate the properties of the solutions to the structures of the species present. Quantitative calculations that depend on the strengths of weak acids and bases will be covered in Chapter 16.

11•1 Arrhenius Acids and Bases

In 1887 Svante Arrhenius postulated that when molecules of elec-trolytes dissolve in water, both positive and negative ions are formed. Toward the end of the nineteenth century, definitions of acids and bases were stated in terms of the Arrhenius theory of ionization. **Arrhenius acids** are substances that dissolve in water to give H^+ ions, and **Arrhenius bases** are substances that dissolve to give OH^- ions. Examples:

Acids	*Bases*
hydrogen chloride, HCl	sodium hydroxide, NaOH
hydrogen nitrate, HNO_3	potassium hydroxide, KOH
hydrogen sulfate, H_2SO_4	calcium hydroxide, $Ca(OH)_2$
acetic acid, $HC_2H_3O_2$	ammonia, NH_3

The first three in each group are highly or completely ionized in water solution and are classed as strong acids or strong bases. On the other hand, acetic acid and ammonia are ionized only slightly in water solution and accordingly are classed as a weak acid and a weak base, respectively.

11•2 Brønsted–Lowry Acids and Bases

In 1923 J. N. Brønsted in Denmark and T. M. Lowry in England independently suggested a different way of describing acids and bases. According to this system, **Brønsted–Lowry acids** are *proton donors* and **Brøn-sted–Lowry bases** are *proton acceptors*. By this definition, a great variety of

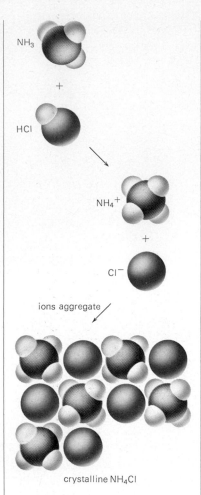

NH_3

+

HCl

NH_4^+

+

Cl^-

ions aggregate

crystalline NH_4Cl

Figure 11•1 Two invisible gases, HCl and NH_3, combine on contact to form a white solid, ammonium chloride, NH_4Cl. In this reaction, a proton is transferred from the hy-drogen chloride molecule (the acid) to an ammonia molecule (the base). The resulting chloride and ammo-nium ions then aggregate to form small crystals of the ionic com-pound, ammonium chloride.

chemical properties and chemical reactions can be correlated, including reactions that take place in solvents other than water or in no solvent at all, as depicted in Figures 11.1 and 11.2.

11•2•1 Aqueous Solutions of Acids and Bases

• *Monoprotic Acids* • Consider from the Brønsted–Lowry point of view what the proton-donating and proton-accepting species are in water solutions of acids such as HCl, HNO_3, and $HC_2H_3O_2$. If we use the symbol HA for the dissolved proton donor, then A^- represents the anion, as in, for example, Cl^-, NO_3^-, and $C_2H_3O_2^-$. We can write

$$HA + H-\overset{..}{\underset{H}{O}}: \rightleftharpoons \left[H-\overset{..}{\underset{H}{O}}-H \right]^+ + A^-$$

$$\text{acid} + \text{water} \rightleftharpoons \text{hydronium ion} + \text{anion}$$

Figure 11•2 Crystal growth found by a freshman student on the mouth of a test tube of concentrated hydrochloric acid left in a locker with an open ammonia solution. The white deposits on bottles, windows, and other objects in a chemistry laboratory are often ammonium salts. (Courtesy of Professor Lou Papenhagen, Mount Union College, Alliance, Ohio.)

Acids, such as HCl, HNO_3, and $HC_2H_3O_2$, with molecules that are capable of donating one proton to a water molecule are called **monoprotic acids.** Because donating protons is a reversible reaction, *every acid must form a base upon donating its proton.* Similarly, *every base must form an acid upon accepting a proton.* These relationships are said to be conjugate:

$$\underset{\text{acid}_1}{HA} + \underset{\text{base}_2}{H_2O} \rightleftharpoons \underset{\text{acid}_2}{H_3O^+} + \underset{\text{base}_1}{A^-}$$

conjugates
conjugates

The base that results when an acid donates its proton is called the **conjugate base** of the acid. Considering the above general reaction proceeding from left to right, A^- is the conjugate base of HA; for the reverse reaction, H_2O is the conjugate base of H_3O^+. The acid that results when a base accepts a proton is called the **conjugate acid** of the base. In the general reaction proceeding from left to right, H_3O^+ is the conjugate acid of H_2O; for the reverse reaction, HA is the conjugate acid of A^-. Hence, H_3O^+ and H_2O, and HA and A^-, are *conjugate acid–base pairs.* The following equations provide specific examples of this relationship.

$Acid_1$		$Base_2$		$Acid_2$		$Base_1$	
HCl	+	H_2O	$\rightleftharpoons$	H_3O^+	+	Cl^-	(1)
$HC_2H_3O_2$	+	H_2O	$\rightleftharpoons$	H_3O^+	+	$C_2H_3O_2^-$	(2)
NH_4^+	+	H_2O	$\rightleftharpoons$	H_3O^+	+	NH_3	(3)

$Base_1$ is the conjugate base of $acid_1$; $acid_1$ is the conjugate acid of $base_1$; $acid_2$ is the conjugate acid of $base_2$; and $base_2$ is the conjugate base of $acid_2$.

• *Polyprotic Acids* • Acids, such as H_2SO_4, H_3PO_4, and H_2CO_3, with molecules that are capable of donating more than one proton are called

polyprotic acids. Because molecules of H_2SO_4 and H_2CO_3 can donate two protons, they are also called **diprotic acids.** Acids with molecules that can donate three protons, such as H_3PO_4, are also called **triprotic acids.**

In water solution, sulfuric acid ionizes in two steps as shown in Equations (4) and (5):

$$H_2SO_4 + H_2O \rightleftharpoons H_3O^+ + HSO_4^- \tag{4}$$
$$\text{acid}_1 \quad \text{base}_2 \quad\quad \text{acid}_2 \quad \text{base}_1$$

$$HSO_4^- + H_2O \rightleftharpoons H_3O^+ + SO_4^{2-} \tag{5}$$
$$\text{acid}_1 \quad \text{base}_2 \quad\quad \text{acid}_2 \quad \text{base}_1$$

For the first step, as represented by Equation (4), the reaction of H_2SO_4 with water to produce H_3O^+ and HSO_4^- ions is essentially complete. However, in the second step, as represented by Equation (5), the reaction of HSO_4^- ions with water to produce H_3O^+ and SO_4^{2-} ions is far from complete. Hence, whereas H_2SO_4 is a strong acid, HSO_4^- is a relatively weak acid.

Note that in Equation (4) representing the reaction of HSO_4^- with H_3O^+, the HSO_4^- ion functions as a base. But in Equation (5), in its reaction with H_2O, the HSO_4^- ion functions as an acid. Ions or molecules that can either donate or accept a proton are said to be **amphiprotic.**[1]

• Example 11·1 •

Write equations for the ionization steps of the triprotic acid H_3PO_4 in water solution. For each equation, label the conjugate acid–base pairs as acid_1, base_1, acid_2, and base_2. Which species are amphiprotic?

• Solution •

$Acid_1$		$Base_2$		$Acid_2$		$Base_1$
H_3PO_4	+	H_2O	$\rightleftharpoons$	H_3O^+	+	$H_2PO_4^-$
$H_2PO_4^-$	+	H_2O	$\rightleftharpoons$	H_3O^+	+	HPO_4^{2-}
HPO_4^{2-}	+	H_2O	$\rightleftharpoons$	H_3O^+	+	PO_4^{3-}

Because the anions $H_2PO_4^-$ and HPO_4^{2-} function as both acids and bases in these reactions, they are amphiprotic species.

See also Exercises 2 and 3 at the end of the chapter.

• *Bases* • The properties that are common to aqueous solutions of bases are due to the hydroxide ion (OH^-), a Brønsted–Lowry base. The ionic hydroxides of the IA and IIA elements are strong bases. Because hydroxide ions are already present in these compounds, they simply dissolve in water and provide the hydroxide ions that are characteristic of a basic solution. The bases NaOH and KOH, which can furnish 1 mole of hydroxide ions per mole of compound, are called **monohydroxy bases.** Because 2 moles of hydroxide ions are furnished per mole of $Ca(OH)_2$ and $Ba(OH)_2$, these compounds are called **dihydroxy bases.**

---

[1]A substance that reacts as both an acid and a base is called an **amphoteric** substance. Examples are given in Section 24.13.1.

Covalent molecules that are Brønsted–Lowry bases can react in an ionization reaction with water to produce hydroxide ions. Ammonia is the most familiar such example:

$$H_2O + NH_3 \rightleftharpoons NH_4^+ + OH^- \qquad (6)$$
$$\text{acid}_1 \quad \text{base}_2 \qquad \text{acid}_2 \quad \text{base}_1$$

Water acts as an acid in this ionization reaction, whereas it reacts as a base in the ionization of acids [Equations (1)–(5)]. Water is thus amphiprotic.

A water solution of ammonia, NH_3, is ionized only slightly and so is a weak base. In an equilibrium mixture, covalent molecules of ammonia and water are present in abundance, and NH_4^+ and OH^- ions are present in relatively small amounts. The pungent odor of aqueous ammonia is due to some gaseous ammonia coming out of solution.

Many ions are Brønsted–Lowry bases that can react with water to produce hydroxide ions. For example, the acetate ion, $C_2H_3O_2^-$, and the cyanide ion, CN^-, react as follows:

$$H_2O + C_2H_3O_2^- \rightleftharpoons HC_2H_3O_2 + OH^- \qquad (7)$$
$$\text{acid}_1 \quad \text{base}_2 \qquad \text{acid}_2 \quad \text{base}_1$$

$$H_2O + \quad CN^- \rightleftharpoons \quad HCN \quad + OH^- \qquad (8)$$
$$\text{acid}_1 \quad \text{base}_2 \qquad \text{acid}_2 \quad \text{base}_1$$

11•2•2 Relative Strengths of Acids and Bases The strength of an acid, HA, in aqueous solution is a measure of its tendency to donate a proton to a water molecule:

$$HA + H_2O \longrightarrow H_3O^+ + A^- \qquad (9)$$

The extent to which this reaction proceeds from left to right is also a measure of the tendency of the conjugate base, A^-, to accept a proton from H_3O^+:

$$H_3O^+ + A^- \longrightarrow HA + H_2O \qquad (10)$$

If the reaction represented by Equation (9) predominates over that represented by Equation (10), then HA is a strong acid and A^- is a weak base. An example of a strong acid is HCl, so we may conclude that Cl^- is a relatively weak base. If the reaction represented by Equation (10) predominates over that represented by Equation (9), then A^- is a strong base and HA is a weak acid. Examples of weak acids are $HC_2H_3O_2$ and HCN, so we may conclude that $C_2H_3O_2^-$ and CN^- are relatively strong bases.

Acids and bases are ranked in order of their comparative strengths in Table 11.1. The fundamental principle upon which these rankings are based is, *the stronger the acid, the weaker is its conjugate base*. Note that among the three acids, HCl, $HC_2H_3O_2$, and HCN, the strongest acid is HCl and the weakest is HCN. Among the three bases, Cl^-, $C_2H_3O_2^-$, and CN^-, the strongest base is CN^- and the weakest is Cl^-.

• *The Leveling Effect* • In certain solvents it is easy to show that $HClO_4$ is a stronger acid than HNO_3. However, in water solutions the four acids listed in Table 11.1 as stronger than H_3O^+ form solutions of practically the same strength, because water is a strong enough base to take protons from each of the acids that are stronger than H_3O^+. For any very strong acid, HA, the ionization in water,

$$HA + H_2O \longrightarrow H_3O^+ + A^-$$

goes essentially to completion. This means that any acid stronger than H_3O^+ simply forms H_3O^+ in water solution, so H_3O^+ *is the strongest acid that can exist in water solution.*

Table 11.1 Relative strengths of acids and bases

Acid		Conjugate base	
$HClO_4$	strong	ClO_4^-	weak
HCl	acids	Cl^-	bases
H_2SO_4		HSO_4^-	
HNO_3		NO_3^-	
H_3O^+		H_2O	
H_2SO_3		HSO_3^-	
HSO_4^-		SO_4^{2-}	
H_3PO_4		$H_2PO_4^-$	
HF		F^-	
$HC_2H_3O_2$	decreasing	$C_2H_3O_2^-$	increasing
H_2CO_3	strength	HCO_3^-	strength
H_2S		HS^-	
HSO_3^-		SO_3^{2-}	
HCN		CN^-	
NH_4^+		NH_3	
HCO_3^-		CO_3^{2-}	
HS^-		S^{2-}	
H_2O		OH^-	
NH_3	weak	NH_2^-	strong
OH^-	acids	O^{2-}	bases

The reaction of a solvent to reduce different reagents to the same strength is called the **leveling effect.**

Water has a leveling effect on any base that has a basicity greater than OH^-. If sodium amide, $NaNH_2$, comes in contact with water, the following reaction takes place (as we would predict from Table 11.1):

$$NH_2^- + H_2O \longrightarrow NH_3 + OH^-$$

The very strong base NH_2^- cannot exist in water solution; neither can O^{2-}. Both these strong proton acceptors are leveled to form solutions of the OH^-

ion because OH⁻ *is the strongest base that can exist in water solution.*

Other solvents also level the strengths of potential acids and bases. In liquid ammonia, acids stronger than NH_4^+ are leveled to the strength of NH_4^+. The strongest base possible in ammonia is the amide ion, NH_2^-.

11•2•3   Acid–Base Reactions in Aqueous Salt Solutions

In pure water there is an extremely small concentration of H^+ (or H_3O^+) ions and an equal concentration of OH^- ions. When the concentration of H^+ equals the concentration of OH^-, an aqueous solution is **neutral.** If the concentration of H^+ is greater than that of OH^-, the solution is **acidic.** If the concentration of OH^- is greater than that of H^+, the solution is **basic.**

Water solutions of salts may be acidic, basic, or neutral, depending on the salt. An aqueous solution of ammonium chloride, NH_4Cl, turns blue litmus red; hence, this solution is acidic. A water solution of sodium acetate, $NaC_2H_3O_2$, has the opposite effect and turns red litmus blue; therefore, this solution is basic. An aqueous solution of sodium chloride, NaCl, or of ammonium acetate, $NH_4C_2H_3O_2$, has no effect on litmus and must be neutral.

In order to account for these differences, we must consider the reaction with water of the cation or the anion (or both) of a salt, a process known as hydrolysis. In a broad sense, **hydrolysis** is the reaction of any substance with water and is not limited to salt solutions. Examples of hydrolysis are the ionization of acetic acid in water

$$HC_2H_3O_2 + H_2O \rightleftharpoons H_3O^+ + C_2H_3O_2^-$$

and the ionization of ammonia in water

$$NH_3 + H_2O \rightleftharpoons NH_4^+ + OH^-$$

Reactions of ions of salts with water that change the acidity involve the transfer of protons and are hydrolysis reactions. We will consider the following four cases when a salt is dissolved in water.

1. *Neither the cation nor the anion of the salt acts appreciably as an acid or base.* Salts composed of cations (Li^+, Na^+, K^+, Ba^{2+}, Sr^{2+}) of strong bases and anions (Cl^-, NO_3^-, SO_4^{2-}) of strong acids form neutral solutions. Examples, in addition to NaCl, are potassium chloride, KCl; barium chloride, $BaCl_2$; strontium chloride, $SrCl_2$; sodium nitrate, $NaNO_3$; lithium sulfate, Li_2SO_4; and strontium nitrate, $Sr(NO_3)_2$.

    In water solutions of these salts, the cations do not react appreciably with water (hydrolyze) to form H^+ ions, and the anions do not react appreciably with water molecules to form OH^- ions.

2. *The cation of the salt acts as an acid, but the anion does not act appreciably as a base.* Salts composed of cations of weak bases and anions of strong acids form acidic solutions. Examples are ammonium chloride, NH_4Cl, and ammonium nitrate, NH_4NO_3. In water solutions of these salts, the NH_4^+ ion functions as an acid. Because the Cl^- and NO_3^- ions are very weak bases, they do not function appreciably as bases in water.

Other examples of cations that can donate protons to water molecules and therefore act as weak acids are certain hydrated metal cations. Examples are $Fe(H_2O)_6^{3+}$, $Al(H_2O)_6^{3+}$, and $Cu(H_2O)_4^{2+}$. A water solution of a salt that contains one of these ions is acidic, provided the anion of the salt does not have a strong tendency to react with protons. When salts such as iron(III) chloride, $FeCl_3$, aluminum nitrate, $Al(NO_3)_3$,

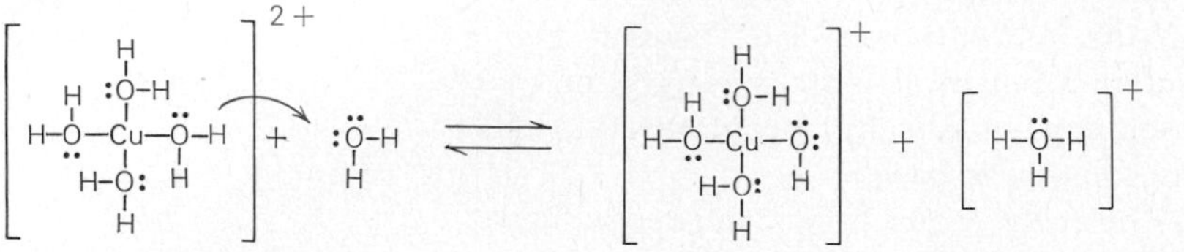

$$Cu(H_2O)_4^{2+} \;+\; H_2O \;\rightleftharpoons\; Cu(H_2O)_3OH^+ \;+\; H_3O^+$$

Figure 11•3 In solutions of copper ion, such as copper chloride, copper nitrate, and copper sulfate, protons are donated by hydrated copper ions to water molecules. When equilibrium is reached, the solutions are acidic, because they have a higher concentration of H_3O^+ ions than pure water has.

and copper(II) sulfate, $CuSO_4$, are dissolved in water, the metal cations are hydrated with water molecules. The hydrated ions hydrolyze to form acidic solutions (see Figure 11.3). The hydration and hydrolysis reactions for $FeCl_3$ and $CuSO_4$ are

$$FeCl_3 + 6H_2O \longrightarrow Fe(H_2O)_6^{3+} + 3Cl^-$$
$$Fe(H_2O)_6^{3+} + H_2O \rightleftharpoons H_3O^+ + [Fe(H_2O)_5(OH)]^{2+}$$

$$CuSO_4 + 4H_2O \longrightarrow Cu(H_2O)_4^{2+} + SO_4^{2-}$$
$$Cu(H_2O)_4^{2+} + H_2O \rightleftharpoons H_3O^+ + [Cu(H_2O)_3(OH)]^+$$

3.   *The anion of the salt acts as a base, but the cation does not act appreciably as an acid.* Salts composed of cations of strong bases and anions of weak acids form basic solutions. Examples are sodium acetate, $NaC_2H_3O_2$, and sodium cyanide, NaCN. Because the sodium ion is the cation of a strong base (see case 1), it does not function appreciably as an acid.[2] The acetate and cyanide ions function as bases. Because the cyanide ion is a stronger base than the acetate ion, a solution of NaCN is more basic than an equimolar solution of $NaC_2H_3O_2$.

Other examples of anions that can accept protons and therefore act as bases are the fluoride ion, F^-, and the carbonate ion, CO_3^{2-}. Salts that contain these ions tend to form basic solutions, assuming that the cations of the salts do not have a strong tendency to donate protons. Specific examples are sodium fluoride, NaF, and potassium carbonate, K_2CO_3. For a water solution of potassium carbonate, hydrolysis proceeds in two steps:

$$CO_3^{2-} + H_2O \rightleftharpoons HCO_3^- + OH^-$$
$$HCO_3^- + H_2O \rightleftharpoons H_2CO_3 + OH^-$$

Most of the OH^- ions are produced in the first step.

---

[2] Of course, the Na^+ ion by itself cannot function as an acid because it has no proton to donate. It is the hydrated sodium ion in solution that does not function appreciably as an acid.

4. *The cation of the salt acts as an acid, and the anion acts as a base.* With salts that are composed of cations of weak bases and anions of weak acids, both ions undergo hydrolysis. The resulting solution is neutral, acidic, or basic, depending upon the relative strengths of the acid cation and the basic anion.

    A solution of ammonium acetate, $NH_4C_2H_3O_2$, is essentially a neutral solution, because it turns out that the strength of the NH_4^+ ion as an acid is canceled by the strength of the $C_2H_3O_2^-$ ion as a base. However, a solution of ammonium carbonate, $(NH_4)_2CO_3$, is basic, because the carbonate ion is stronger as a base than the NH_4^+ ion is as an acid.

11•3 Naming Inorganic Acids

The two most common classes of inorganic acids are (1) binary and related acids and (2) ternary oxy-acids, which contain one or more oxygen atoms bound to an element other than hydrogen. The familiar acids often are referred to by both their compound names, in line with the rules given in Section 5.8, and their common or trivial names, according to custom established over the years. Because we have discussed the compound names earlier, our attention in this section will be centered primarily on the common names.

11•3•1 Binary and Related Acids This class includes the hydrogen halides, hydrogen sulfide, and hydrogen cyanide. The root of the common name of the acid indicates the parent element, for example, *chlor-* for chlorine. The name includes the root, the prefix *hydro-*, and the suffix *-ic*, as illustrated by the following examples:

Formula	Compound Name	Common Acid Name
HCl	hydrogen chloride	hydrochloric acid
HBr	hydrogen bromide	hydrobromic acid
H_2S	hydrogen sulfide	hydrosulfuric acid
HCN	hydrogen cyanide	hydrocyanic acid

One practice is to use the compound name for the anhydrous, pure substance and to use the common acid name for the water solution of the compound. However, this practice has not been adopted uniformly.[3]

11•3•2 Ternary Oxy-acids The systematic compound names of acids in this class, just as for the binary and related compounds, are based on the names of the anions that are formed after the removal of the protons (see Section 5.8.2). The common acid names of oxy-acids have the suffixes *-ic* and

---

[3]For the familiar inorganic acids, the internationally approved rules of nomenclature allow the use of either the systematic compound names or the common acid names to refer to the pure substances.

-ous that correspond to the *-ate* and *-ite* suffixes, respectively, of the compound names. Use of these suffixes is illustrated by the following examples:

Formula	Compound Name	Common Acid Name
H_2SO_4	hydrogen sulfate	sulfuric acid
H_2SO_3	hydrogen sulfite	sulfurous acid
HNO_3	hydrogen nitrate	nitric acid
HNO_2	hydrogen nitrite	nitrous acid
$HClO_4$	hydrogen perchlorate	perchloric acid
$HClO_3$	hydrogen chlorate	chloric acid
$HClO_2$	hydrogen chlorite	chlorous acid
$HClO$	hydrogen hypochlorite	hypochlorous acid

11•4 Structures of Hydroxy Compounds

In oxy-acids such as HNO_3, H_2SO_4, and $HClO$, and in oxy-bases such as NaOH and $Ca(OH)_2$, each hydrogen atom is covalently bonded to an oxygen atom. The question arises: why does one OH-containing compound, such as HOCl, act as an acid and another, such as NaOH, act as a base? One explanation is based on the differences in electronegativities of the elements attached to the OH groups. If the attached element has a low electronegativity, it has little attraction for a pair of electrons, and the compound is ionic. Recall that the hydroxides of the IA and IIA families are ionic compounds:

$$Na^+, \left[:\ddot{O}-H\right]^- \qquad Ca^{2+}, 2\left[:\ddot{O}-H\right]^-$$

The hydroxide ion is set free when these ionic hydroxides dissolve in water.

If the element attached to the OH group has a high electronegativity, it has a strong attraction for a pair of electrons, as does oxygen. As a result there is a covalent bond between the element and oxygen, rather than an ionic bond. The considerable attraction for electrons by the attached element and by oxygen leaves the bond between oxygen and hydrogen highly polar, so that the H nucleus is susceptible to removal by proton-seeking groups. Some hydroxy compounds of nonmetals that are good proton donors are shown in Figure 11.4.

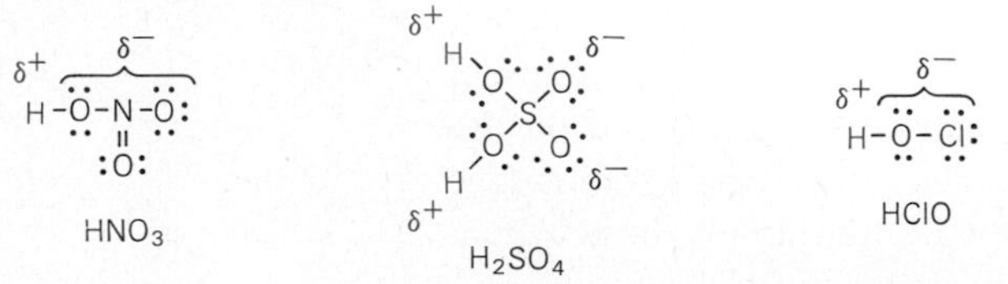

Figure 11•4 Polar —O—H groups are present in oxy-acids, as shown in these Lewis structural formulas.

In general, the greater the number of oxygen atoms bound to the other nonmetal element in a hydroxy compound, the stronger is the acid. The trend is illustrated by the following examples:

Weak Acids

$$H—\overset{..}{\underset{..}{O}}—N=\overset{..}{\underset{.}{O}}:$$

nitrous acid
(N has $+3$ oxidation number)

Strong Acids

$$H—\overset{..}{\underset{..}{O}}—N=\overset{..}{\underset{.}{O}}:$$
$$\qquad\qquad\quad |$$
$$\qquad\qquad\quad :\overset{..}{\underset{..}{O}}:$$

nitric acid
(N has $+5$ oxidation number)

$$H—\overset{..}{\underset{..}{O}}—\overset{}{\underset{|}{S}}—\overset{..}{\underset{..}{O}}—H$$
$$\qquad\qquad :\overset{..}{\underset{..}{O}}:$$

sulfurous acid
(S has $+4$ oxidation number)

$$\qquad\qquad :\overset{..}{\underset{.}{O}}:$$
$$\qquad\qquad\quad |$$
$$H—\overset{..}{\underset{..}{O}}—\overset{}{\underset{|}{S}}—\overset{..}{\underset{.}{O}}—H$$
$$\qquad\qquad :\overset{..}{\underset{..}{O}}:$$

sulfuric acid
(S has $+6$ oxidation number)

When the central element has a higher oxidation number and one more oxygen atom bonded to it, the O—H bond is weakened; the proton, H^+, is donated more easily, so the acid is stronger. This trend is shown by the oxy-acids of chlorine, $HClO$, $HClO_2$, $HClO_3$, and $HClO_4$. The range of strengths of these acids extends from the weak acid, $HClO$, to one of the strongest acids known, $HClO_4$. The steady increase in acidity observed corresponds to the increase in the number of oxygens bonded to chlorine.

Metals other than those with the lowest electronegativities follow the same trend as nonmetals and may form acidic hydroxy compounds if the metal has a high oxidation number. Compounds such as chromic acid, H_2CrO_4, and permanganic acid, $HMnO_4$, are strong acids in water solution. With their moderate electronegativities but very high positive oxidation states, the chromium and manganese in these two compounds attract electrons so strongly that the hydrogens are left with considerable positive character.

Two general rules describe the types of compounds just discussed: (1) hydroxy compounds of metals with low electronegativities yield the base OH^- in water solution; and (2) hydroxy compounds of nonmetals, or metals of relatively high electronegativities and high oxidation numbers, are H^+ donors in water solution.

11•5   Neutralization

11•5•1   Strong Acids and Strong Bases   When equimolar amounts of a strong acid such as hydrochloric acid, HCl, and a strong base such as sodium hydroxide, $NaOH$, are brought together in water solution, the hydronium ions of the acid and the hydroxide ions of the base combine to form water. This reaction is known as **neutralization.** The complete ionic equation is

$$H_3O^+ + Cl^- + Na^+ + OH^- \longrightarrow Na^+ + Cl^- + 2H_2O$$

or simply

$$H^+ + Cl^- + Na^+ + OH^- \longrightarrow Na^+ + Cl^- + H_2O$$

The net ionic equation is

$$H_3O^+ + OH^- \longrightarrow 2H_2O$$

$$H^+ + OH^- \longrightarrow H_2O$$

When the acid and the base species react, we say they **neutralize** one another.

After the reaction between hydrochloric acid and sodium hydroxide is complete, there remains a solution of Na^+ and Cl^- ions. Although these two spectator ions are not involved in the neutralization, we can say that the NaCl solution is formed as a result of the acid–base reaction.

An older representation of the neutralization reaction simply shows the initial acid and base that are brought together and the substances that are present when the reaction is completed, without regard to the solvents used, if any. The reaction between HCl and NaOH, either in the pure form or in water solution, is interpreted as

$$\underset{\text{acid}}{HCl} + \underset{\text{base}}{NaOH} \longrightarrow \underset{\text{salt}}{NaCl} + \underset{\text{water}}{HOH}$$

Overall neutralization equations still are often written in this fashion. Other examples are:

$$\underset{\substack{\text{nitric}\\\text{acid}}}{HNO_3} + \underset{\substack{\text{potassium}\\\text{hydroxide}}}{KOH} \longrightarrow \underset{\substack{\text{potassium}\\\text{nitrate}}}{KNO_3} + \underset{\text{water}}{HOH}$$

$$\underset{\substack{\text{sulfuric}\\\text{acid}}}{H_2SO_4} + \underset{\substack{\text{sodium}\\\text{hydroxide}}}{2NaOH} \longrightarrow \underset{\substack{\text{sodium}\\\text{sulfate}}}{Na_2SO_4} + \underset{\text{water}}{2HOH}$$

Note that in the neutralization of the diprotic acid sulfuric acid with sodium hydroxide, 2 moles of NaOH are required for each mole of H_2SO_4.

11.5.2 Strong Acids and Weak Bases Although the term *neutralization* is commonly applied to any reaction of an acid with a base, a strictly neutral solution may not always be obtained. In fact, a neutral solution is obtained only when the acid and base are of equal strengths.

Consider what happens when a strong acid, such as HCl, and the weak base ammonia, NH_3, are brought together in water solution. The following equations may be used to represent this reaction:[4]

older representation
$$\underset{\text{acid}}{HCl} + \underset{\text{base}}{NH_4OH} \longrightarrow \underset{\text{salt}}{NH_4Cl} + \underset{\text{water}}{HOH}$$

newer representation
$$H_3O^+ + \cancel{Cl^-} + NH_3 \longrightarrow NH_4^+ + \cancel{Cl^-} + HOH$$

The resulting solution of ammonium chloride is slightly acidic rather than

---

[4]Note that in order to satisfy the idea of a neutralization as the reaction

$$acid + base \longrightarrow salt + water$$

one must use the formula NH_4OH (see footnote 7, page 312) in the older representation.

neutral, because the NH_4^+ ion functions as an acid in water solution. (See case 2 in Section 11.2.3.)

11•5•3   Weak Acids and Strong Bases   The reaction in water solution of a weak acid such as acetic acid, $HC_2H_3O_2$, with the strong base NaOH may be represented by the following equations:

older
representation

$$\underset{\text{acid}}{HC_2H_3O_2} + \underset{\text{base}}{NaOH} \longrightarrow \underset{\text{salt}}{NaC_2H_3O_2} + \underset{\text{water}}{HOH}$$

newer
representation

$$HC_2H_3O_2 + \cancel{Na^+} + OH^- \longrightarrow \cancel{Na^+} + C_2H_3O_2^- + HOH$$

The resulting solution of sodium acetate is slightly basic rather than neutral, because the acetate ion functions as a base in water solution. (See case 3 in Section 11.2.3.)

11•5•4   Weak Acids and Weak Bases   As a final example of neutralization, consider the reaction in water solution of the weak acid acetic acid with the weak base ammonia. The following equations may be used to represent this neutralization:[4]

older
representation

$$\underset{\text{acid}}{HC_2H_3O_2} + \underset{\text{base}}{NH_4OH} \longrightarrow \underset{\text{salt}}{NH_4C_2H_3O_2} + \underset{\text{water}}{HOH}$$

newer
representation

$$HC_2H_3O_2 + NH_3 \longrightarrow NH_4^+ + C_2H_3O_2^-$$

The resulting solution of ammonium acetate, $NH_4C_2H_3O_2$, is practically neutral. This is because the acid strength of the NH_4^+ ion is just balanced by the base strength of the $C_2H_3O_2^-$ ion. (See case 4 in Section 11.2.3.)

In summary, reactions of acids and bases of the same strength produce neutral solutions. Both the combining acid and base may be strong or both weak. Reactions of acids and bases of different strengths produce either weakly acidic or weakly basic solutions, depending on the strengths of the conjugate acids and conjugate bases produced. If the acid produced is stronger than the base produced, a weakly acidic solution results. On the other hand, if the base produced is stronger than the acid produced, a weakly basic solution is obtained. Regardless of the relative strengths of the acids and bases involved, all such acid–base reactions are commonly referred to as neutralization reactions.

11•5•5   Equivalent Weights and Normal Solutions in Neutralizations   As mentioned in Section 10.6.6, equivalent weights and normal solutions are important in acid–base reactions. It follows from the equations below that 49.0 g of H_2SO_4 and 36.5 g of HCl are equivalent to one another in their neutralization reactions with NaOH:

$$\underset{\frac{1}{2}\text{ mol, 49.0 g}}{\tfrac{1}{2}H_2SO_4} + \underset{1 \text{ mol, 40.0 g}}{NaOH} \longrightarrow \underset{\frac{1}{2}\text{ mol, 71.0 g}}{\tfrac{1}{2}Na_2SO_4} + \underset{1 \text{ mol, 18.0 g}}{HOH}$$

$$\underset{1 \text{ mol, 36.5 g}}{HCl} + \underset{1 \text{ mol, 40.0 g}}{NaOH} \longrightarrow \underset{1 \text{ mol, 58.5 g}}{NaCl} + \underset{1 \text{ mol, 18.0 g}}{HOH}$$

Also, 56.1 g of KOH and 37.0 g of $Ca(OH)_2$ are equivalent in their neutralization reactions with HNO_3:

$$HNO_3 \quad + \quad KOH \quad \longrightarrow \quad KNO_3 \quad + \quad HOH$$
1 mol, 63.0 g   1 mol, 56.1 g   1 mol, 101.1 g   1 mol, 18.0 g

$$HNO_3 \quad + \quad \tfrac{1}{2}Ca(OH)_2 \quad \longrightarrow \quad \tfrac{1}{2}Ca(NO_3)_2 + \quad HOH$$
1 mol, 63.0 g $\tfrac{1}{2}$ mol, 37.0 g $\tfrac{1}{2}$ mol, 82.0 g   1 mol, 18.0 g

Specifically, the equivalent weight of an acid is the weight that supplies 1 mole of protons, that is, 6.022×10^{23} protons, to a base. The equivalent weight of a base is the weight that reacts with 1 mole of protons. The equivalent weights of sulfuric, hydrochloric, and nitric acids in these reactions are 49.0 g, 36.5 g, and 63.0 g, respectively. The equivalent weights of sodium hydroxide, potassium hydroxide, and calcium hydroxide are 40.0 g, 56.1 g, and 37.0 g, respectively.

Certain acids and bases can have more than one equivalent weight, depending on the reactions they undergo. In these cases balanced equations for the reactions that actually occur must always be considered in calculating equivalent weights. For example, if only one of the two acidic protons of H_2SO_4 is removed by reaction with NaOH, then the equivalent weight of H_2SO_4 is 98.1 g:

$$H_2SO_4 \quad + \quad NaOH \quad \longrightarrow \quad NaHSO_4 \quad + \quad HOH$$
1 mol, 98.1 g   1 mol, 40.0 g   1 mol, 120.1 g   1 mol, 18.0 g

Based upon the specific chemical reaction considered in each case, a **neutralization reaction** is defined as one in which equivalent amounts of an acid and a base react. Generally, when we speak of neutralizations we mean that all available protons from the acid and all available hydroxide ions from the base react to form water. For example, if we refer to the neutralization of H_2SO_4 by NaOH, we assume that the reaction produces Na_2SO_4 unless we have other information that tells us $NaHSO_4$ is the product in a particular case.

The normality of an acid or a base solution is defined as the number of equivalents of solute per liter of solution (see Section 10.6.6). A 1N solution of an acid or base contains one equivalent weight per liter of solution; a 0.5N solution contains half an equivalent weight per liter; and so on. Example 11.2 illustrates a calculation of the normality of an acid solution.

──── • Example 11.2 • ────

Calculate the normality of a solution made by dissolving 49.0 g of phosphoric acid, H_3PO_4, in enough water to prepare 600 mL of solution. Assume that all available protons are donated when the acid solution reacts with a base.

──── • Solution • The weight of 1 mole of H_3PO_4 is 98.0 g. Because 1 mole of this acid can supply 3 moles of protons to a base, the equivalent weight is one third of the molar weight:

$$\frac{\text{equiv wt}}{H_3PO_4} = \frac{98.0 \text{ g}}{\text{mol } H_3PO_4} \times \frac{1 \text{ mol } H_3PO_4}{3 \text{ equiv}} = 32.7 \text{ g/equiv}$$

$$N = \frac{\text{equiv of solute}}{\text{L of solution}} = \frac{\dfrac{\text{g solute}}{\text{g/equiv}}}{\text{L of solution}}$$

$$= \frac{\dfrac{49.0 \text{ g } H_3PO_4}{32.7 \text{ g } H_3PO_4/\text{equiv}}}{600 \text{ mL} \times \dfrac{1 \text{ L}}{1{,}000 \text{ mL}}}$$

$$= 2.50 \text{ equiv/L} = 2.50 \text{ } N$$

See also Exercise 32.

11•5•6   Titrations of Acids and Bases   Titration is the process of determining the amount of a solution of known concentration that is required to react completely with a certain amount of a sample that is being analyzed. The sample being analyzed is referred to as the **unknown.** Analytical procedures involving titration with solutions of known concentration are called **volumetric analyses.**

In the analysis of acidic and basic solutions, titration involves the careful measurement of the volumes of an acid and a base that just neutralize each other. Suppose that we wish to determine the concentration of a hydrochloric acid solution, and that we have on hand in the laboratory a base solution with a known concentration of 1.20N. The analysis is carried out as follows. Portions of the two solutions can be placed in separate burets (see Figure 11.5), and a convenient quantity of the acid, say, 15.0 mL, is measured from its buret into the flask. Alternatively, a known quantity of the acid may be taken from a beaker by a calibrated pipet fitted with a suction bulb, as shown on the right of Figure 11.5. An indicator, such as litmus or phenolphthalein, is added to the acid, and the flask is placed under the buret containing the base.

The base is permitted to run into the flask rather rapidly at first, then slowly, and finally drop by drop until a single, final drop causes the indicator to change color. The change in color is the signal that an amount of base has been added that is equivalent to the amount of acid in the 15.0 mL of the unknown solution. The total volume of base used is read from the buret. Suppose that this volume is 21.2 mL. This means that 21.2 mL of a 1.20N base is found to just neutralize 15.0 mL of hydrochloric acid of unknown concentration. From consideration of the experimental data,

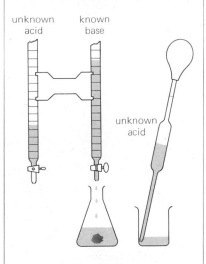

unknown acid   known base

unknown acid

Figure 11•5 Apparatus for titrations.

	base	acid
volume	21.2 mL	15.0 mL
normality	1.20N	?

we can see that the acid is more concentrated than the base, because a smaller volume of it is required. Even before we calculate it, we know that the normality of the acid is going to be greater than 1.20N.

For either solution, the product of the volume, V (in liters), times the normality, N, is the number of equivalents of the reacting species:

$$V_A \times N_A = \text{equiv}_A$$
$$V_B \times N_B = \text{equiv}_B$$

where A and B denote acid and base, respectively. At neutralization the number of equivalents of acid ($equiv_A$) is equal to the number of equivalents of base ($equiv_B$), and we can write:

$$equiv_A = equiv_B$$

and

$$V_A \times N_A = V_B \times N_B$$

Because a volume term appears on both sides of the equation, any volume unit can be used in this equation as long as both volumes are expressed in the same unit,[5] for example, both in liters (L) or both in milliliters (mL), that is:

$$L_A \times N_A = L_B \times N_B$$

or

$$mL_A \times N_A = mL_B \times N_B$$

With the latter relationship, the concentration of the unknown acid is calculated as follows:

$$15.0 \ mL_A \times N_A = 21.2 \ mL \times 1.20N$$

$$N_A = \frac{21.2 \ mL \times 1.20N}{15.0 \ mL} = 1.70N$$

The concentration of a basic solution of unknown concentration can be determined in a similar fashion if an acid solution of known concentration is available.

• Example 11.3 •

A laboratory technician prepared a calcium hydroxide solution by dissolving 1.48 g of $Ca(OH)_2$ in water. How many milliliters of $0.125N$ HCl solution would be required to neutralize this calcium hydroxide solution?

• Solution • At neutralization,

$$L_A \times N_A = L_B \times N_B$$

But we also know that

$$L_B \times N_B = equiv_B$$

So at neutralization, we can write

---

[5]When one uses mL, then the volume in mL times the normality (N) equals milliequivalents of acid or base.

$$L_A \times N_A = equiv_B$$

The number of equivalents of base, calcium hydroxide, is

$$equiv_B = \frac{g \ Ca(OH)_2}{g \ Ca(OH)_2/equiv} = \frac{1.48 \ g}{37.0 \ g/equiv} = 0.0400 \ equiv_B$$

And therefore, at neutralization,

$$L_A \times 0.125 N_A = 0.0400 \ equiv_B$$

$$L_A = \frac{0.0400 \ equiv}{0.125 \ equiv/L} = 0.320 \ L$$

$$mL_A = 0.320 \ L \times \frac{1,000 \ mL}{L} = 320 \ mL_A$$

See also Exercises 34, 38, and 39.

--- • Example 11•4 • ---

Quinine, a drug for the treatment of malaria, is an organic base that may be neutralized with HCl. If 0.675 g of quinine requires 42.0 mL of 0.100N HCl for neutralization, what is its equivalent weight?

• Solution • At neutralization, we can write

$$L_A \times N_A = equiv_B$$

and

$$L_A \times N_A = \frac{g \ base}{g/equiv}$$

$$0.0420 \ L_A \times 0.100 N_A = \frac{0.675 \ g}{g/equiv}$$

$$g/equiv = \frac{0.675 \ g}{0.0420 \ L \times 0.100 \ \frac{equiv}{L}} = 161 \ g/equiv$$

Thus, the equivalent weight of the organic base is 161 g; the molar weight will be an integral multiple of 161, that is, 161 g, 322 g, 483 g, and so on, depending on the number of basic units per molecular formula.

See also Exercises 35, 36, 40, and 42.

11•6   Lewis Acids and Bases

A very general theory of acid and base behavior was stated by G. N. Lewis. According to this concept, a **Lewis acid** is defined as any species that acts as an electron-pair acceptor in chemical reactions, and a **Lewis base** is an electron-pair donor. The Lewis definitions are consistent with the Brønsted–Lowry view, because the proton can be looked upon as an electron-pair acceptor. A substance that accepts a proton can be looked upon as an electron-pair donor. This is illustrated by the following reactions:

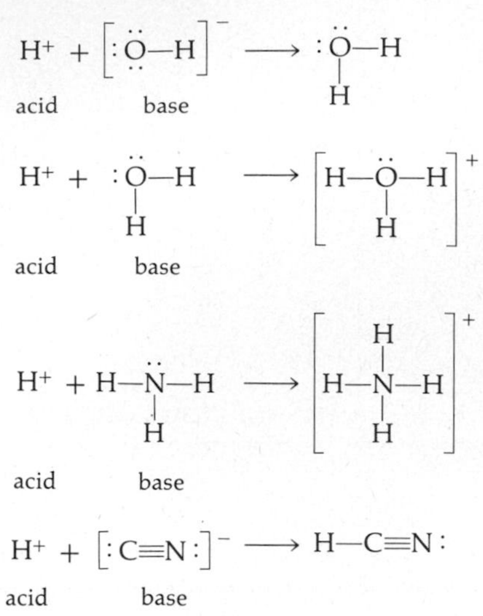

$$H^+ + \left[:\ddot{O}{-}H \right]^- \longrightarrow \ddot{O}{-}H \qquad \text{or} \quad H^+ + OH^- \longrightarrow H_2O$$

acid   base

$$H^+ + :\ddot{O}{-}H \longrightarrow \left[H{-}\ddot{O}{-}H \right]^+ \quad \text{or} \quad H^+ + H_2O \longrightarrow H_3O^+$$

acid   base

$$H^+ + H{-}\ddot{N}{-}H \longrightarrow \left[H{-}N{-}H \right]^+ \quad \text{or} \quad H^+ + NH_3 \longrightarrow NH_4^+$$

acid   base

$$H^+ + \left[:C{\equiv}N: \right]^- \longrightarrow H{-}C{\equiv}N: \quad \text{or} \quad H^+ + CN^- \longrightarrow HCN$$

acid   base

The Lewis definition extends the concept of the acid–base relationship to a number of reactions that do not involve proton transfer. For example, in the following reaction, boron trichloride acts as an acid, and ammonia acts as a base:

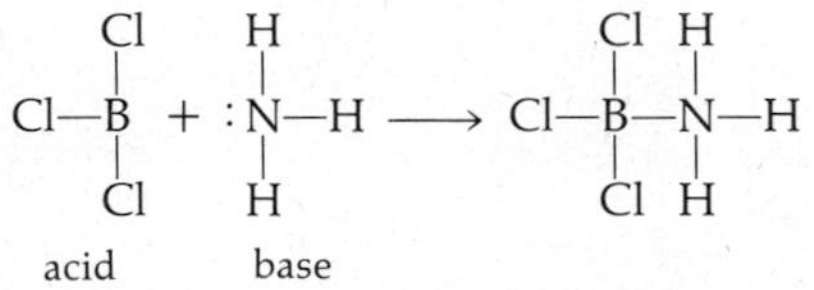

$$Cl{-}B + :N{-}H \longrightarrow Cl{-}B{-}N{-}H \quad \text{or} \quad BCl_3 + NH_3 \longrightarrow Cl_3BNH_3$$

acid   base

An advantage of the Lewis concept is that it identifies as acids certain substances that do not contain hydrogen but that have the same function as the common hydrogen-containing acids. Other examples of acid–base reactions that do not involve proton transfer but that are in accord with the Lewis definition include

$$SO_3 + O^{2-} \longrightarrow SO_4^{2-}$$
acid   base

$$SO_3 + OH^- \longrightarrow HSO_4^-$$
acid   base

$$AlCl_3 + Cl^- \longrightarrow AlCl_4^-$$
acid   base

$$Ag^+ + Cl^- \longrightarrow AgCl\downarrow$$
acid   base

Chapter Review

Summary

Acids and bases are defined at increasing levels of generality. An **Arrhenius acid** and an **Arrhenius base** are substances that dissolve in water to give H^+ (or H_3O^+) ions and OH^- ions, respectively. A more general theory, and the most useful for most purposes in inorganic chemistry, defines a **Brønsted–Lowry acid** as a proton donor and a **Brønsted–Lowry base** as a proton acceptor, in any solvent or in no solvent. When an acid donates a proton, it becomes its own **conjugate base,** which can accept a proton. When a base accepts a proton, it becomes its own **conjugate acid,** which can donate a proton. Such pairs of species are called **conjugate acid–base pairs.** Acids that can donate only one proton are **monoprotic acids;** those that can donate two and three protons are **diprotic acids** and **triprotic acids,** respectively, or simply **polyprotic acids.** A molecule or ion that can either donate or accept a proton, depending on the chemical conditions, is **amphiprotic.** Substances that have the capacity to react as either an acid or a base are called **amphoteric.** The OH^- ion in aqueous solutions is a Brønsted–Lowry base. Some compounds provide this ion directly when they dissolve in water. Those that can provide one OH^- ion are **monohydroxy bases;** those that provide two are **dihydroxy bases.**

The relative strengths of acids and bases in solution depend on their different affinities for protons. The stronger the acid, the weaker is its conjugate base. The action of a solvent in reducing a number of strong acids or strong bases to the same effective level of strength is called the **leveling effect.**

Pure water is **neutral.** The aqueous solution of a salt may be neutral or acidic or basic, depending on whether the cation or the anion (or both) undergoes **hydrolysis**—a chemical reaction with water. Some cations act like an acid and produce H_3O^+ ions; some anions act like a base and produce OH^- ions.

Most inorganic acids are either binary hydrogen compounds or ternary hydrogen compounds containing oxygen. The latter have one or more OH groups bound to the central atom. In general, the more O atoms bound to the central atom in a related series of acids, the stronger the acid.

When chemically equivalent amounts of an acid and a base are mixed, the result is a **neutralization reaction,** which produces an aqueous solution of a salt. This solution is strictly neutral only if the acid and the base are equal in strength; otherwise it is either weakly acidic or weakly basic. The concentration of an acid or base solution called an **unknown** can be determined by **titration** with a solution of known concentration. Such techniques are called **volumetric analyses.**

The most general theory of acids and bases is that of G. N. Lewis. A **Lewis acid** is an electron-pair acceptor and a **Lewis base** is an electron-pair donor. This theory encompasses the Brønsted–Lowry theory, and it goes one step further by ascribing acid–base behavior to many reactions that do not involve the transfer of a proton.

Exercises

Brønsted–Lowry Acids and Bases

1. Write equations to illustrate the ionization of each of the following compounds in water solution: formic acid, $HCHO_2$; methylamine, CH_3NH_2; hydrogen cyanide, HCN. Label the conjugate acid–base pairs for each by using $acid_1$, $base_1$, $acid_2$, and $base_2$.

2. A water solution of arsenic acid, H_3AsO_4, ionizes in three steps. Write an equation for each step, and label the conjugate acid–base pairs. Which species are amphiprotic?

3. Hydrazine, N_2H_4, is a weak base that ionizes in water solution in two steps. Write equations for these two steps. For which step is the equilibrium displaced farther to the right? Label the conjugate acid–base pairs. Which species are amphiprotic?

*4. Zinc reacts with dilute hydrochloric acid to produce zinc chloride and hydrogen. Write a net ionic equation for this reaction.

5. Write a balanced ionic equation, using Lewis formulas, for the reaction of dimethyl amine, $(CH_3)_2NH$, with acetic acid, $HC_2H_3O_2$.

6. Write equilibrium equations for the following reactions. Label the acids and bases present, according to the Brønsted–Lowry theory:
   (a) reaction of formate ion, CHO_2^-, with water
   (b) reaction of methylammonium ion, $CH_3NH_3^+$, with water

Strengths of Acids and Bases

7. Hydrogen nitrate, HNO_3, is a strong acid in water solution, whereas hydrogen fluoride, HF, is a weak acid. What can be said about the relative strengths of the anions, NO_3^- and F^-, in their abilities to function as Brønsted–Lowry bases?

8. Methylamine, CH_3NH_2, and ammonia, NH_3, are both weak bases in water solution; however, ammonia is slightly weaker than methylamine. What can be said about the relative strengths of the cations, $CH_3NH_3^+$ and NH_4^+, in their abilities to function as Brønsted–Lowry acids?

9. Explain why H_2SO_4 and $HClO_4$ are practically equal in acid strength in water, even though the latter is stronger in some other solvents.

*10. Sodium amide, $NaNH_2$, dissolves simply as a strong electrolyte in liquid ammonia but reacts quite rapidly with water. Explain. Write the equation for the reaction with water.

Acid–Base Reactions in Salt Solutions

11. Which of the following salts will hydrolyze to give acidic solutions and which will hydrolyze to give basic solutions? Assume that the solutions are all 0.50M. List the solutions that are acidic in order of decreasing acidity. List the basic solutions in order of decreasing basicity. (Refer to Table 11.1.)
    (a) sodium acetate, $NaC_2H_3O_2$
    (b) ammonium chloride, NH_4Cl
    (c) potassium cyanide, KCN
    (d) potassium fluoride, KF
    (e) sodium oxide, Na_2O
    (f) sodium bisulfate, $NaHSO_4$

12. Will a water solution of sodium sulfate, Na_2SO_4, be acidic, basic, or neutral? What about a water solution of potassium nitrate, KNO_3? Explain your answer for each salt solution.

13. A water solution of ammonium acetate, $NH_4C_2H_3O_2$, hydrolyzes to give a neutral solution. Explain. Based on the positions of $HC_2H_3O_2$ and $C_2H_3O_2^-$ in Table 11.1, explain whether a water solution of ammonium fluoride, NH_4F, will be weakly acidic, weakly basic, or neutral. Repeat for a water solution of ammonium sulfite, $(NH_4)_2SO_3$.

14. Write a chemical equation to explain each of the following statements. A water solution of:
    (a) sodium carbonate is basic
    (b) copper chloride is acidic
    (c) potassium chloride is neutral

15. The bicarbonate ion, HCO_3^-, is amphiprotic. Water, H_2O, also is amphiprotic. Write two equations for the hydrolysis of the HCO_3^- ion, one in which the ion acts as an acid and one in which it acts as a base.

*16. Refer to the statement in Exercise 13 concerning the ammonium acetate solution and to Exercise 15. Based on the positions of the ammonium and acetate ions in Table 11.1, explain whether a water solution of sodium bicarbonate, $NaHCO_3$, will be weakly acidic, weakly basic, or exactly neutral.

Naming Acids

17. Give the systematic compound name and the common acid name for each of the following acids: HBr, H_2Se, H_3AsO_4, H_3AsO_3, HI, H_2SeO_3, H_2SeO_4.

18. Give the systematic compound name and the common acid name for each of the following acids: $HMnO_4$, $HBrO_4$, $HBrO_3$, $HBrO_2$, $HBrO$, H_2CrO_4, H_2CrO_3.

Structures of Hydroxy Compounds

19. Compare the three hydroxy compounds of metals as to their acid–base properties: KOH, O_2AlOH, O_3MnOH.

*20. Consider the following acids: hypochlorous acid, $HClO$; hypobromous acid, $HBrO$; and hypoiodous acid, HIO. Which is the strongest acid in water solution? Which is the weakest? Explain your choices.

*21. List the following acids in order of increasing acidity, that is, weakest to strongest: acetic acid, CH_3CO_2H; chloroacetic acid, $ClCH_2CO_2H$; fluoroacetic acid, FCH_2CO_2H; trifluoroacetic acid, CF_3CO_2H. Explain your choices.

*22. Consider the following bases: ammonia, NH_3; hydrazine, N_2H_4; hydroxylamine, $HONH_2$. Which is the strongest base in water solution? Which is the weakest? Explain your choices.

23. In the series of acids $HClO$, $HClO_2$, $HClO_3$, and $HClO_4$, each oxygen atom is bonded to chlorine. With this information, write Lewis structural formulas for these four acids.

24. Which of the three acids H_3PO_4, $H_2PO_4^-$, and HPO_4^{2-} is the strongest acid? Why?

Neutralization

25. Consider the reaction of hydrochloric acid with ammonia to produce ammonium chloride (both reactants are in water solution).
    (a) Write equations for the several reactions that are possibly involved and the net ionic equation for the reaction.
    (b) Are the two solutions that result from the following both neutral: (1) the reaction in water of 1 mole of HCl with 1 mole of NH_3 and (2) the reaction in water of 1 mole of HCl with 1 mole of $NaOH$? Explain why or why not.

26. Consider the following neutralization reactions carried out in water solution. For each reaction, tell whether the so-called neutralization gives a solution that is neutral, weakly acidic, or weakly basic. Write the net ionic equation for each reaction.
    (a) sodium hydroxide + acetic acid
    (b) sodium hydroxide + nitric acid
    (c) ammonia + potassium bisulfate ($KHSO_4$)
    (d) ammonia + acetic acid
    (e) hydrocyanic acid + lithium hydroxide
    (f) hydrofluoric acid + potassium hydroxide
    (g) sodium bicarbonate + sodium hydroxide
    (h) ammonium chloride + lithium hydroxide

*27. Consider the neutralization of sulfuric acid, H_2SO_4, with sodium hydroxide, $NaOH$, in water solution. Does the net ionic equation $2H_3O^+ + 2OH^- \longrightarrow 4H_2O$ account completely for the neutralization of this diprotic acid? Explain.

Equivalent Weights of Acids and Bases

28. Calculate the weight of calcium hydroxide, $Ca(OH)_2$, required to neutralize a dilute nitric acid solution containing 21.0 g of HNO_3. Assume the products of the reaction are calcium nitrate, $Ca(NO_3)_2$, and water.
29. Determine the equivalent weight of phosphoric acid in each of the following reactions:
    (a) $H_3PO_4 + 3NaOH \longrightarrow Na_3PO_4 + 3H_2O$
    (b) $H_3PO_4 + Ca(OH)_2 \longrightarrow CaHPO_4 + 2H_2O$
    (c) $H_3PO_4 + NH_3 \longrightarrow NH_4H_2PO_4$
    (d) $2H_3PO_4 + 3Ca(OH)_2 \longrightarrow Ca_3(PO_4)_2 + 6H_2O$
30. How many grams of sodium hydroxide are required to convert 49.0 g of H_3PO_4 to Na_2HPO_4? To Na_3PO_4?
31. What weight of hydrogen bromide, HBr, is contained in 500 mL of a 1.00N solution?
32. Assume that sulfuric acid acts as a diprotic acid, and calculate the normality of a solution prepared by adding 24.5 g of H_2SO_4 to enough water to give 400 mL of solution. What is the molarity of this solution?
33. What is the normality of each of the following solutions?
    (a) 2.0M HCl   (b) 1.0M H_2SO_4
    (c) 0.50M H_3PO_4 (d) 0.50M $Ca(OH)_2$
34. Precisely 19.6 g of hydrogen phosphate, H_3PO_4, is dissolved in 327 mL of water. How many milliliters of 0.500N NaOH are required to neutralize completely the phosphoric acid solution?
35. Precisely 2.20 g of 2,6-lutidine, an organic base, is dissolved in 150 mL of water. To neutralize this basic solution requires 20.5 mL of 1.00N HCl. Calculate the equivalent weight of 2,6-lutidine.
36. If 200 mL of 0.250N sodium hydroxide, NaOH, is required to neutralize 2.95 g of succinic acid, calculate the equivalent weight of succinic acid.
37. (a) Write a balanced equation for the complete neutralization of phosphoric acid, H_3PO_4, with calcium hydroxide, $Ca(OH)_2$.
    (b) What volume of 0.200M $Ca(OH)_2$ is required to neutralize 25.0 mL of 0.150M H_3PO_4?
    (c) What volume of 0.600N $Ca(OH)_2$ is required to neutralize 25.0 mL of 0.450N H_3PO_4?
    (d) What, if any, is the relationship between the answers to parts (b) and (c)? Explain.
38. A 21.0-g sample of sodium oxide, Na_2O, is added to water to give 200 mL of sodium hydroxide solution. How many milliliters of 0.400N H_2SO_4 will be required

to neutralize this NaOH solution?
39. A laboratory technician prepares a phosphoric acid solution by dissolving 29.4 g of H_3PO_4 in 300 mL of water. How many milliliters of 0.250N NaOH will be required to neutralize this phosphoric acid solution?
40. An analytical chemist dissolved 9.00 g of oxalic acid in 225 mL of water and then neutralized the acid with 80.0 mL of 2.50N NaOH.
    (a) Calculate the equivalent weight of oxalic acid.
    (b) Oxalic acid is a diprotic acid. What is its molar weight?
*41. When barium nitrate, $Ba(NO_3)_2$, was added in excess to 50.0 mL of a sulfuric acid solution of unknown concentration, a precipitate of $BaSO_4$ formed that, when dried, was found to weigh 1.238 g. Calculate the molarity and normality of the sulfuric acid solution.
*42. A sample of a metal weighing 0.0730 g reacted completely when allowed to stand in 100 mL of 0.1000N HCl. The excess acid was neutralized by titration with 32.0 mL of 0.125N KOH.
    (a) Calculate the equivalent weight of the metal.
    (b) Based upon a comparison of your calculated equivalent weight with a list of atomic weights, choose two elements either of which could be the experimental metal. For each metal chosen, write the formula for the chloride that would have formed when the metal reacted with HCl. Discuss the reasons for your choices.

Lewis Acids and Bases

43. Gaseous ammonia reacts with gaseous hydrogen chloride to form the ionic compound ammonium chloride, NH_4Cl. Can this reaction be classified as an acid–base reaction, based upon the definition of acids and bases by Arrhenius? By Brønsted and Lowry? By Lewis? Explain in each case why or why not.
44. Identify the Lewis acids and Lewis bases in each of the following reactions. Which of these reactions also illustrate Brønsted–Lowry acid–base reactions? Rewrite equations (d), (f), and (g), using Lewis structural formulas.
    (a) $H^+ + CO_3^{2-} \longrightarrow HCO_3^-$
    (b) $2NH_3 + Ag^+ \longrightarrow [Ag(NH_3)_2]^+$
    (c) $H^+ + O^{2-} \longrightarrow OH^-$
    (d) $O^{2-} + SO_2 \longrightarrow SO_3^{2-}$
    (e) $CO_2 + OH^- \longrightarrow HCO_3^-$
    (f) $S^{2-} + SO_3 \longrightarrow S_2O_3^{2-}$
    (g) $ClF_3 + F^- \longrightarrow ClF_4^-$
    (h) $S^{2-} + Cu^{2+} \longrightarrow CuS$
45. Use Lewis structural formulas to illustrate the last four equations shown in Section 11.6.

Properties of Solutions; The Colloidal State

12

Properties of Solutions

12•1 **Properties of Solutions of Nonelectrolytes**

12•1•1 Vapor Pressure

12•1•2 Boiling Point

12•1•3 Freezing Point

12•1•4 Experimental Determination of Molecular Weights

12•1•5 Ideal Versus Real Solutions

12•2 **Properties of Solutions of Electrolytes**

12•2•1 Number of Particles in Solutions of Ionic Electrolytes

12•2•2 Number of Particles in Solutions of Covalent Electrolytes

12•3 **Distillation**

12•3•1 Vapor Pressures

12•3•2 Fractional Distillation

12•4 **Osmosis**

12•4•1 Osmotic Pressure
Special Topic:
Desalination of Seawater

The Colloidal State

12•5 **Particle Size and the Colloidal State**

12•6 **Importance of Colloid Chemistry**

12•7 **Types of Colloidal Systems**

12•8 **Properties of Colloidal Systems**

12•8•1 Tyndall Effect

12•8•2 Brownian Movement

12•8•3 Adsorption

12•9 **Selective Separations**

12•9•1 Efficient Solid Adsorbents

12•9•2 Precipitation of Aerosols

12•9•3 Chromatography

12•9•4 Dialysis

12•10 **Stability of Colloidal Systems**

Because solutions are such a common form of matter, understanding their behavior increases our knowledge of the world around us. Whereas a solution's chemical behavior depends on the specific solvent and solute involved, its physical behavior may be predicted on the basis of a few well-known laws.

In this chapter, we will focus attention on laws that affect such physical properties as freezing and boiling points, vapor pressures, and osmotic pressures. These properties are influenced by the way substances interact on the molecular level. We will discuss these interactions and look at a number of important practical applications related to them. Among these are the production of antifreeze solutions, the separation of solutions by distillation, and the role of osmosis in the cells of plants and animals.

Closely related to solutions is the study of the colloidal state, a state of subdivision that is intermediate between a solution and a suspension. Particles in a colloidal dispersion are too large to be considered truly dissolved, but they are so small that they do not settle out as do larger particles in a suspension. Some of the most interesting types of mixtures are classified as colloids—foams, fogs, gels, paints, smoke, and many foods, including milk. This chapter will conclude with a look at one of the most important concerns of colloidal chemistry, the adsorption of one or more substances on the surface of another.

Properties of Solutions

12•1   Properties of Solutions of Nonelectrolytes

When a solute is dissolved in a solvent, the properties of the solution differ from those of the pure solvent. There are four important physical properties that are affected in direct proportion to the number of solute particles present: (1) the vapor pressure, (2) the freezing point, (3) the boiling point, and (4) the osmotic pressure. Properties that depend on the number of solute particles and not on their kind are referred to as **colligative properties.** The kind of solute matters very little so long as it is a **nonvolatile nonelectrolyte,** a substance that does not have an appreciable vapor pressure and does not form ions. Examples of solutes that are very soluble in water are sugars, urea, ethylene glycol, and glycerin. We will first consider the behavior of such simple solutes, and later discuss the behavior of solutions with volatile solutes or electrolyte solutes.

The extent to which the properties of a solution are changed as compared with the properties of the pure solvent is stated by the **colligative property law.** *The vapor pressure, freezing point, and boiling point of a solution differ from those of the pure solvent by amounts that are directly proportional to the molal concentration of the solute.* A hypothetical solution that behaves exactly as described by the colligative property law is called an **ideal solution.** Most solutions approach ideal behavior only when extremely dilute.

12•1•1   Vapor Pressure   The fact that a mixture of a nonvolatile solute and a solvent has a lower vapor pressure than the pure solvent is easy to demonstrate. Figure 12.1 shows what happens when beakers of three liquids are exposed to one another in the same closed space. The pure solvent, water, evaporates away completely, because its vapor pressure is high. The non-volatile solute, sucrose, cannot move from whatever beaker it is in; a little of it is in the middle beaker, and a greater amount is in the beaker on the right. Because of the different vapor pressures, there is a net evaporation of water from the dilute solution and a net condensation in the concentrated solution. Eventually the two solutions attain equal concentrations and equal vapor pressures.

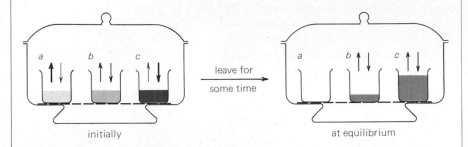

initially · at equilibrium

Figure 12•1 Three liquids—(a) pure water, (b) a dilute sucrose solution, and (c) a concentrated sucrose solution—are placed in a closed container. After enough time elapses, all the pure water evaporates and the two solutions attain equal concentrations.

The vapor pressure of a solution can be measured as shown in Figure 8.11 or in the more convenient apparatus pictured in Figure 12.2. After the bulb with a sample of solution is attached, the air in the connecting tube is removed with a vacuum pump. When the stopcock is closed, the pressure inside is due only to water vapor evaporating from the solution.

We can describe the number of solute molecules in a given number of solvent molecules in terms of molal concentrations or mole fractions. Both these ways of specifying concentration are useful in discussing colligative properties. The vapor pressure of an ideal solution is described by **Raoult's law.** *Each component in a solution exerts a pressure that is equal to its mole fraction times the vapor pressure of the pure component;* that is,

$$p_A = x_A p_A{}^\circ$$

where p_A is the vapor pressure exerted by component A in the solution, x_A is the mole fraction of component A, and $p_A{}^\circ$ is the vapor pressure of the pure substance A.

In a solution that has a nonvolatile solute, the vapor pressure of the solution is due only to the solvent. We can use p_A to refer to either the solvent or the solution, as in Example 12.1.

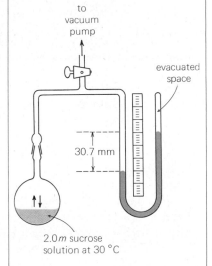

to vacuum pump

evacuated space

30.7 mm

2.0m sucrose solution at 30 °C

Figure 12•2 Use of a mercury manometer to measure the vapor pressure of an aqueous solution. The water vapor is in equilibrium with the solution.

• Example 12•1 •

Calculate the vapor pressure of a 2.0m aqueous solution of sucrose at 30 °C. (Because of the high concentration the solution will not act ideally, so our calculation will be approximate.)

• Solution • The vapor pressure of pure water at 30 °C is 31.82 mm (see Table A.7). In 1 L of a 2.0m solution,

$$\text{mol solute} = 2.0 \text{ mol}$$

$$\text{mol solvent} = 1{,}000 \text{ g}\left(\frac{1 \text{ mol}}{18.0 \text{ g}}\right) = 55.6 \text{ mol}$$

$$\text{mole fraction solvent} = \frac{55.6 \text{ mol}}{55.6 \text{ mol} + 2.0 \text{ mol}} = 0.965$$

$$\frac{\text{vapor pressure}}{\text{solvent}} = \frac{\text{vapor pressure}}{\text{solution}} = p_A = x_A p_A{}^{\circ}$$

$$p_A = (0.965)(31.82 \text{ mm}) = 30.7 \text{ mm}$$

See also Exercises 4–6 at the end of the chapter.

• *Relation Between Vapor Pressure and the Boiling and Freezing Points* • The way in which a nonvolatile, nonelectrolyte solute affects the vapor pressure of the solvent is shown graphically in Figure 12.3. This figure is a companion to Figure 8.13.

At every temperature, the vapor pressure of the solution is less than the vapor pressure of the pure solvent. At the normal boiling point, the vapor pressure of the solution is below 1 atm. To make the solution boil, it must be heated to a temperature that is above the normal boiling point. This change in the boiling point temperature is called Δt_{bp}.

At the normal freezing point, the vapor pressure of the solution is also less than the vapor pressure of the pure solid solvent. To make the solution freeze, it must be cooled to a temperature that is below the normal freezing point. At this temperature, the vapor pressure of the solution equals the vapor pressure of the pure solid solvent. This change in the freezing point temperature is called Δt_{fp}.

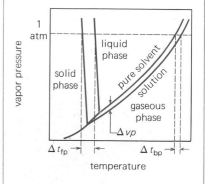

Figure 12•3 The effects on a solvent of a nonvolatile, nonelectrolyte solute are to lower the vapor pressure, Δvp, lower the freezing point, Δt_{fp}, and raise the boiling point, Δt_{bp}. Compare Figure 8.13.

12•1•2   Boiling Point   The boiling point of a solution may be higher or lower than that of the solvent, depending on the volatility of the solute as compared with that of the solvent. If the solute is nonvolatile—sugar is an example—its water solution boils at a temperature higher than water boils; if the solute is quite volatile—for example, ethyl alcohol—its water solution boils at a temperature below the boiling point of water.

In the case of an ethyl alcohol–water solution, the ethyl alcohol (boiling point, 78.3 °C) has a greater tendency to become a vapor than does water. The vapor pressure of the solution (the sum of the ethyl alcohol vapor pressure and the water vapor pressure) equals the atmospheric pressure at a temperature below 100 °C. That is, the boiling point of the solution is below that of pure water. The colligative property law is not applicable to solutions of volatile solutes, such as the ethyl alcohol–water solution.

The colligative property law is applicable in predicting the boiling points of solutions involving nonvolatile, nonelectrolyte solutes. It has been determined experimentally that 1.00 mole (6.02×10^{23} molecules) of any nonvolatile nonelectrolyte dissolved in 1.00 kg (1,000 g) of water raises the boiling point by approximately 0.512 °C. If 0.500 mole (3.01×10^{23} molecules) is dissolved in 1,000 g of water, the boiling point is raised to approximately 0.256 °C (that is, 0.512 °C × 0.500). If 2.00 moles are dissolved, the increase is approximately 1.024 °C (that is, 0.512 °C × 2.00), and so on. From these data, it can be seen that the *change* in the boiling point of the solution versus that of the pure solvent, that is, Δt_{bp}, is directly proportional to the molal concentration, m, of the solution:

$$\Delta t_{bp} \propto m \quad \text{or} \quad \Delta t_{bp} = k_b m$$

where k_b is a constant of proportionality called the **molal boiling point elevation constant** of the solvent. The values of this constant for several solvents are listed in Table 12.1.

Table 12.1 Molal freezing point and boiling point constants

Solvent	Freezing point, °C	k_f, °C/m	Boiling point, °C	k_b, °C/m
acetic acid	16.60	3.90	117.9	3.07
benzene	5.50	4.90	80.10	2.53
camphor	179.8	39.7	207.42	5.61
ethyl ether			34.51	2.02
nitrobenzene	5.7	7.00	210.8	5.24
phenol	40.90	7.40	181.75	3.56
water	0.00	1.86	100.00	0.512

• **Example 12.2** •

Calculate the boiling point of a solution that contains 1.5 g of glycerin, $C_3H_5(OH)_3$, in 30 g of water.

• **Solution** • Because the amount of change in the boiling point depends on the molality, we first calculate the molality of the solution. The molar weight of glycerin is 92 g.

$$\text{molality} = \frac{\text{mol of solute}}{\text{kg of solvent}}$$

$$= \frac{(1.5 \text{ g solute})\left(\dfrac{1 \text{ mol}}{92 \text{ g}}\right)}{30 \text{ g solvent} \times \dfrac{1 \text{ kg}}{1,000 \text{ g}}}$$

$$= 0.54 \frac{\text{mol solute}}{\text{kg solvent}} = 0.54m$$

$$\Delta t_{bp} = k_b m$$

$$= \frac{0.512 \text{ °C}}{1m} \times 0.54m = 0.28 \text{ °C}$$

$$\begin{array}{c}\text{boiling point} \\ \text{water}\end{array} + \Delta t_{bp} = \begin{array}{c}\text{boiling point} \\ \text{solution}\end{array}$$

$$100 \text{ °C} + 0.28 \text{ °C} = 100.28 \text{ °C}$$

See also Exercises 7 and 14.

12.1.3   Freezing Point In Figure 12.3, we see that the vapor pressure of the solution becomes equal to the vapor pressure of the solid at a temperature below the freezing point of the pure solvent. The depression of the freezing point, Δt_{fp}, is directly proportional to the molality (m) of the solution:

$$\Delta t_{fp} \propto m \quad \text{or} \quad \Delta t_{fp} = k_f m$$

where k_f is the **molal freezing point depression constant** of the solvent. Values of this constant are given in Table 12.1.

We should note that for changes in the freezing point, the colligative property law applies to volatile as well as nonvolatile solutes. At low temperatures the vapor pressures of both solvent and solute are low, and the effect of the vapor pressure of the solute is very slight.

• Example 12•3 •

Calculate the freezing point of a solution that contains 2.00 g of chloroform, $CHCl_3$, dissolved in 50.0 g of benzene.

• Solution • Because the amount of change in the freezing point is proportional to the molality, we first calculate the molality of the solution. The molar weight of chloroform is 119 g.

$$\text{molality} = \frac{\text{mol of solute}}{\text{kg of solvent}}$$

$$= \frac{(2.00 \text{ g solute})\left(\dfrac{1 \text{ mol}}{119 \text{ g}}\right)}{50.0 \text{ g solvent} \times \dfrac{1 \text{ kg}}{1,000 \text{ g}}}$$

$$= 0.336 \frac{\text{mol solute}}{\text{kg solvent}} = 0.336 m$$

$$\Delta t_{fp} = k_f m$$

$$= \frac{4.90 \text{ °C}}{1 m} \times 0.336 m = 1.65 \text{ °C}$$

$$\begin{array}{c}\text{freezing point} \\ \text{benzene}\end{array} - \Delta t_{fp} = \begin{array}{c}\text{freezing point} \\ \text{solution}\end{array}$$

$$5.50 \text{ °C} - 1.65 \text{ °C} = 3.85 \text{ °C}$$

See also Exercises 8, 9, 13, and 15.

12•1•4 Experimental Determination of Molecular Weights When a new compound is discovered—new ones are being discovered every day—a molecular weight determination is helpful in establishing the formula of the compound. One experimental method of obtaining molecular weights is based on Avogadro's law. The method was discussed in Chapter 7; it is based on determining the weight of 22.4 L of gas at standard conditions. This amount of gas contains 1 mole, and its weight in grams is numerically equal to the molecular weight. The method is limited to those substances that can be volatilized readily without decomposition.

A second method of determining molecular weights is based on the colligative property law. Experimentally, the problem is to find the weight of solute needed to lower the freezing point of 1 kg of solvent by the molal freezing point depression constant, k_f (or the weight needed to raise the boiling point of 1 kg of solvent by the molal boiling point elevation constant, k_b). This weight of solute contains 1 mole, and its weight in grams is equal numerically to the molecular weight. Example 12.4 illustrates a molecular weight determination for a water-soluble compound.

─── • Example 12•4 • ───

A solution made by dissolving 0.243 g of a new compound in 25.0 g of water has a freezing point of $-0.201\ °C$. Calculate the molecular weight of the new compound.

─── • Solution • Let us begin by calculating the molality of the solution. The change in the freezing point, Δt_{fp}, is $0\ °C - (-0.201\ °C) = 0.201\ °C$. With this value and the k_f for water, the molality of the solution can be calculated:

$$\Delta t_{fp} = k_f m \quad \text{or} \quad m = \frac{\Delta t_{fp}}{k_f}$$

$$m = \frac{0.201\ °\cancel{C}}{\frac{1.86\ °\cancel{C}}{1m}} = 0.108m$$

By definition,

$$\text{molality} = \frac{\text{mol solute}}{\text{kg solvent}} = \frac{\frac{\text{g solute}}{\text{g/mol}}}{\text{kg solvent}}$$

$$0.108m = \frac{0.108\ \text{mol}}{1\ \text{kg}} = \frac{\frac{0.243\ \text{g}}{\text{g/mol}}}{25.0\ \text{g} \times \frac{1\ \text{kg}}{1,000\ \text{g}}}$$

$$\text{g/mol} = \frac{0.243\ \text{g}}{\left(0.108\ \frac{\text{mol}}{\cancel{\text{kg}}}\right)\left(25.0\ \cancel{\text{g}} \times \frac{1\ \cancel{\text{kg}}}{1,000\ \cancel{\text{g}}}\right)}$$

$$= 90.0\ \text{g/mol}$$

Therefore, the molecular weight is 90.0 amu.

See also Exercises 18 and 19.

The majority of organic compounds are only slightly soluble or virtually insoluble in water. For molecular weight determinations of organic compounds, camphor is frequently used as a solvent. Because of its large molal freezing point depression constant of $39.7\ °C/m$ (Table 12.1), molecular weights can be determined by using only small amounts of valuable organic compounds. Example 12.5 illustrates this method.

─── • Example 12•5 • ───

A solution prepared by dissolving 0.115 g of quinine in 1.36 g of camphor has a freezing point of $169.6\ °C$. Calculate the molecular weight of quinine based on these experimental data.

• **Solution** • As in Example 12.4, we begin by calculating the molality of the solution. The change in the freezing point, Δt_{fp}, is 179.8 °C − 169.6 °C = 10.2 °C. The molality of the solution is

$$m = \frac{\Delta t_{fp}}{k_f} = \frac{10.2\ °C}{\dfrac{39.7\ °C}{1m}} = 0.257m$$

Proceeding as in Example 12.4,

$$g/mol = \frac{0.115\ g}{\left(0.257\ \dfrac{mol}{kg}\right)\left(1.36\ g \times \dfrac{1\ kg}{1,000\ g}\right)}$$

$$= 329\ g/mol$$

Therefore, the molecular weight of quinine is determined to be 329 amu.

See also Exercises 16, 18–20, and 22.

The molar weight for quinine of 329 g calculated in Example 12.5 can be compared with the equivalent weight of 161 g obtained in Example 11.4. Evidently, there are approximately 329/161 = 2.04, or two basic units per molecular formula; the molar weight is equal to twice the equivalent weight. In this case, one may wonder whether the molecular weight is 329 amu, as determined by the freezing point depression data, or 2(161) = 322 amu, as indicated by the titration data. Actually neither method provides a very accurate value. A more trustworthy determination may be one based on the percentage composition and the calculation of an empirical formula (see Examples 2.13 and 2.14).

Quinine contains C, H, N, and O. Combustion analysis gives the percentage of C and percentage of H; the percentage of N is determined by a standard analytical method; and percentage of O is obtained by difference [% O = 100% − (% C + % H + % N)]. For quinine, the empirical formula is $C_{10}H_{12}NO$, corresponding to a precise empirical weight of 162.211 amu. The precise molecular weight is 2(162.211) = 324.422 amu, so the molecular formula is $C_{20}H_{24}N_2O_2$. (Review Example 2.15.) Often the data from a colligative effect (for example, freezing point depression) or a volumetric method (for example, titration) are not as accurate as those based on combustion analysis.

Colligative effects (and molar gas volumes, see Example 7.7) are useful in determining approximate molecular weights, which can then be compared with precise empirical weights. Approximate methods are somewhat limited in accuracy because the systems do not behave ideally.

12•1•5 Ideal Versus Real Solutions The data in Table 12.1 are averages based on measurements of the effects of many different solutes. Actually, the changes in the colligative properties of a solvent may differ slightly from one solute to another. And, as mentioned previously, most real solutions approach ideal behavior only when extremely dilute.

The ways in which real solutions of moderate concentration behave are illustrated in Table 12.2. Two of the four solutes behave ideally to three significant figures. Each of the other two solutes has a freezing point depression for a 1*m* solution that differs appreciably from 1.86 °C. For this reason, we expect that calculations based on freezing point measurements will be only approximately correct. As shown for the case of quinine in Example 12.5, *the determination of an approximate molecular weight allows us to calculate a precise molecular weight* if a precise value for the empirical weight is known.

Table 12•2 Experimental molal freezing point depressions of nonelectrolytes

Solute	Formula	Molar weight, g	Observed freezing point depression for 1*m* water solutions, °C
methyl alcohol	CH_3OH	32.0	1.86
ethyl alcohol	C_2H_5OH	46.1	1.83
glycerin	$C_3H_5(OH)_3$	92.1	1.92
urea	$CO(NH_2)_2$	60.1	1.86

12•2   Properties of Solutions of Electrolytes

12•2•1   Number of Particles in Solutions of Ionic Electrolytes

It has been emphasized that the extent to which the freezing point, boiling point, and vapor pressure of a solution differ from those of the pure solvent *depends on the number of solute particles* (molecules, atoms, or ions) in a given weight of the solvent. With nonelectrolytes, 1 mole refers to the same number of particles, namely, 6.02×10^{23} molecules. But in the case of an electrolyte, 1 mole yields more than 1 mole of particles. Sodium chloride, NaCl, is not made up of molecules but of pairs of ions, Na^+ and Cl^-. This means that 58.5 g of NaCl contains not 6.02×10^{23} molecules, but 6.02×10^{23} Na^+ ions and 6.02×10^{23} Cl^- ions. The data in Table 12.3 show that for the ionic type of electrolyte, the number of particles in a mole is twice, three times, four times, and so on, the number in a mole of a covalent nonelectrolyte.

Table 12•3 Particles per mole for electrolytes

Formula	Particles represented by formula	Molar weight, g	No. of particles in 1 mole
NaCl	Na^+, Cl^-	58.5	$2 \times 6.02 \times 10^{23}$
KNO_3	K^+, NO_3^-	101.1	$2 \times 6.02 \times 10^{23}$
$CaCl_2$	Ca^{2+}, Cl^-, Cl^-	111.0	$3 \times 6.02 \times 10^{23}$
Na_2SO_4	Na^+, Na^+, SO_4^{2-}	142.0	$3 \times 6.02 \times 10^{23}$
AlF_3	Al^{3+}, F^-, F^-, F^-	84.0	$4 \times 6.02 \times 10^{23}$

When 1 mole of an ionic type of electrolyte is dissolved in 1 kg of water, the amounts by which the freezing and boiling points are changed are considerably more than 1.86 °C and 0.512 °C, respectively. Indeed, it appears at first glance that theoretically these constants would be exactly doubled for such compounds as sodium chloride, Na^+ and Cl^-, and potassium nitrate, K^+ and NO_3^-, and exactly tripled for such compounds as calcium chloride, Ca^{2+},

Cl^-, and Cl^-, and potassium sulfate, K^+, K^+, and SO_4^{2-}. In very dilute solutions, the observed values approach these multiples of k_f and k_b, but they are always somewhat less than the theoretical values. The explanation is believed to lie in the fact that each ion strongly attracts the oppositely charged ions in its vicinity, so no ion can act completely as an independent particle. These interionic attractions are reduced to a minimum when the solution is very dilute because the ions are separated by many water molecules. Consequently, the observed changes in the freezing and boiling points of very dilute solutions of KCl, NaCl, and $MgSO_4$ approach twice those of a solution of a nonelectrolyte of the same molal concentration, and the change for K_2SO_4 approaches three times the change for a nonelectrolyte.

We can illustrate the behavior of electrolytes by considering the depression of the freezing point of a $0.100m$ NaCl solution. The freezing point is not depressed by $2(0.100m)(1.86 \,°C/1m) = 0.372 \,°C$, but by only $0.348 \,°C$. We can describe this situation with the equation

$$\Delta t_{fp} = ik_f m$$

where i is a factor that shows how this NaCl solution compares with a nonelectrolyte solution of the same molality.[1] For a $0.100m$ NaCl solution,

$$\Delta t_{fp} = 0 \,°C - (-0.348 \,°C) = 0.348 \,°C$$

$$= i\left(\frac{1.86 \,°C}{1m}\right)(0.100m)$$

$$i = \frac{0.348 \,°C}{0.186 \,°C} = 1.87$$

This is the way the first value in Table 12.4 was determined.

Table 12•4 **Comparison of freezing point lowering by ionic electrolytes with lowering by nonelectrolytes of the same concentration**

Ionic type of electrolyte	Values of i for $\Delta t_{fp} = ik_f m$				Theoretical limit[a] for i
	$0.100m$	$0.0500m$	$0.0100m$	$0.00500m$	
NaCl	1.87	1.89	1.93	1.94	2
KCl	1.86	1.88	1.94	1.96	2
$MgSO_4$	1.42	1.43	1.62	1.69	2
K_2SO_4	2.46	2.57	2.77	2.86	3

[a] The theoretical limit of the ratio of the Δt_{fp} for an electrolyte to the Δt_{fp} for a nonelectrolyte is equal to the number of moles of ions that can form per mole of electrolyte.

The value of i is the ratio of the Δt_{fp} of the electrolyte to the Δt_{fp} of a nonelectrolyte solution of the same concentration. The theoretical limit of the value of i is the number of moles of ions that can form per mole of the electrolyte. The more dilute a solution is, the more closely the value of i approaches its theoretical limit. In Table 12.4, note how the values for each

---

[1] The factor i is known as the **van't Hoff factor**.

solution increase as the molality decreases. For the 0.00500m solutions, the values of i are approaching the theoretical limits of 2 or 3.

• Example 12•6 •

What must have been the experimentally determined value of the freezing point of the 0.100m solution of K_2SO_4 described in Table 12.4?

• Solution •

$$\Delta t_{fp} = ik_f m$$
$$= (2.46)(1.86\ °C/m)(0.100m)$$
$$= 0.458\ °C$$

$$\Delta t_{fp} = \frac{freezing\ point}{solvent} - \frac{freezing\ point}{solution}$$

$$\frac{freezing\ point}{solution} = 0.000\ °C - 0.458\ °C$$

$$= -0.458\ °C$$

See also Exercises 23–27, 29, and 30.

12•2•2 Number of Particles in Solutions of Covalent Electrolytes The i values for solutions of three covalent electrolytes are given in Table 12.5. We see that hydrochloric acid is a strong electrolyte because it has almost twice the effect on the freezing point that a nonelectrolyte has. Acetic acid is a weak electrolyte; it has only slightly more effect on the freezing point than a nonelectrolyte has. The i values for very dilute sulfuric acid solutions (for example, 0.00500m) show that this solute is a rather strong electrolyte.

Table 12•5 Comparison of freezing point lowering by covalent electrolytes with lowering by nonelectrolytes of the same concentration

Covalent type of electrolyte	Values of i for $\Delta t_{fp} = ik_f m$				Theoretical limit[a] for i
	0.100m	0.0500m	0.0100m	0.00500m	
HCl	1.91	1.92	1.97	1.99	2
$HC_2H_3O_2$	1.01	1.02	1.05	1.06	2
H_2SO_4	2.22	2.32	2.59	2.72	3

[a]See footnote to Table 12.4.

Example 12.7 shows how colligative data can be used to calculate the extent of ionization. We can see from this example and from Table 12.5 that *the percentage ionization of a covalent electrolyte is greater in a dilute solution than in a concentrated one.*

• Example 12•7 •

Calculate the percent ionization of 0.0100m acetic acid from its i value of 1.05 in Table 12.5.

• Solution • The i value of 1.05 shows that each mole of dissolved $HC_2H_3O_2$ forms 1.05 moles of particles in solution. It does this by reacting with water to form ions:

$$HC_2H_3O_2 + H_2O \rightleftharpoons H_3O^+ + C_2H_3O_2^-$$

or simply

$$HC_2H_3O_2 \rightleftharpoons H^+ + C_2H_3O_2^-$$

Let x be the number of moles of $HC_2H_3O_2$ that react to form ions. Then x is the number of moles of H^+ or $C_2H_3O_2^-$, and $0.0100 - x$ is the number of moles of un-ionized $HC_2H_3O_2$ at equilibrium:

$$0.0100 - x \rightleftharpoons x \ + \ x$$

The total number of moles of particles is $(0.0100 - x + x + x)$. So,

$$(1.05)(0.0100) = 0.0100 - x + x + x$$
$$0.0105 = 0.0100 + x$$
$$x = 0.0105 - 0.0100 = 0.0005$$

The percent ionization is

$$\frac{\text{mol } HC_2H_3O_2 \text{ that form ions}}{\text{total mol } HC_2H_3O_2}(100) = \frac{0.0005}{0.0100}(100) = 5\%$$

See also Exercises 27, 28, and 30.

12•3   Distillation

12•3•1   Vapor Pressures Consider two hypothetical liquids, A and B, that are completely miscible with each other. Let us imagine that we have measured the vapor pressure of each pure liquid at 20 °C and found it to be 100 mm for A and 200 mm for B. For any concentrations of A and B in a solution, we can calculate their ideal vapor pressures by using Raoult's law:

$$p_A = x_A p_A^\circ \quad \text{and} \quad p_B = x_B p_B^\circ$$

For the vapor pressure of the solution, assuming ideal behavior,

$$P_{\text{soln}} = p_A + p_B$$

For example, if 0.25 mole of A is mixed with 0.75 mole of B, p_A, p_B, and P_{soln} may be calculated as follows:

$$p_A = (0.25)(100 \text{ mm}) = 25 \text{ mm}$$
$$p_B = (0.75)(200 \text{ mm}) = 150 \text{ mm}$$
$$P_{\text{soln}} = 25 \text{ mm} + 150 \text{ mm} = 175 \text{ mm}$$

In another solution in which the mole fractions of A and B are both 0.50, we calculate p_A and p_B to be 50 and 100 mm, respectively, and P_{soln} to be 150 mm.
    The lines of Figure 12.4 show the total ideal vapor pressure, P_{soln}, and the partial pressures, p_A and p_B, of A and B for solutions of all concentrations of A and B, according to Raoult's law. Some examples of ideal solutions that most nearly follow Raoult's law are liquid mixtures of benzene and toluene, and of carbon tetrachloride and toluene.

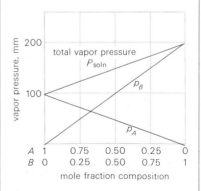

Figure 12•4 A typical vapor pressure versus mole fraction plot for two liquids, A and B, which follow Raoult's law.

Most liquid mixtures do not behave ideally, so that the measured vapor pressure of a liquid mixture may differ considerably from that calculated by Raoult's law. If the actual vapor pressure exceeds the ideal pressure, there is a *positive deviation* from Raoult's law. Examples of such liquid pairs are ethyl alcohol and water, carbon tetrachloride and ethyl alcohol, and acetone and methanol. If the actual vapor pressures are less than the ideal ones, there is a *negative deviation* from Raoult's law. Some liquid pairs that have a negative deviation are water and nitric acid, water and acetic acid, and water and hydrochloric acid.

12.3.2   Fractional Distillation   When a mixture of two miscible liquids is boiled, the vapor that escapes from the liquid usually has a different composition than that of the boiling liquid. The common behavior is for the vapor to be richer (more concentrated) in the more volatile component. By boiling away part of the liquid and condensing the vapor, the mixture can be separated into two parts. The condensed vapor is called the **distillate** and is richer than the original liquid in the more volatile component. The liquid left behind is called the **residue** and is richer in the less volatile component.

If we prepare liquid mixtures of known percentage composition, measure their initial boiling points and the compositions of the vapors that distill first, and then plot the resulting data, we obtain distillation curves of the type shown in Figure 12.5. Such curves allow us to determine the percentage composition of a liquid mixture and the composition of the distillate from the boiling point.

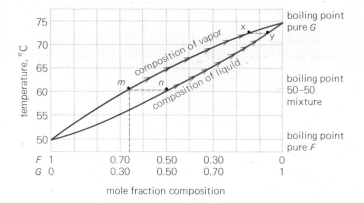

Figure 12•5 A graph showing how the boiling point and the composition of distilling vapor and liquid change (shown by arrows) as a solution of two liquids with approximately ideal behavior is distilled. Initially, there is a 50–50 mixture of the liquids.

The curves shown in Figure 12.5 are for mixtures of two hypothetical liquids F and G, with approximately ideal behavior. Pure F boils at 50 °C, and pure G boils at 75 °C. A horizontal line drawn at any boiling point between 50 and 75 °C intersects both the liquid composition curve and the vapor composition curve. The horizontal line at a boiling point of 60.5 °C intersects the liquid curve at mole fractions of 0.50F and 0.50G, as shown by point n; the mole fraction composition of the vapor at this boiling point is 0.66F and 0.34G, as shown by point m.

The effect of continually boiling away a liquid that has the original composition 0.50F and 0.50G is shown by the directional arrows in Figure 12.5. Because the escaping vapor is richer in F than in the liquid left behind, the residue becomes increasingly rich in G. As the mole fraction of G in the liquid residue increases, the boiling point steadily rises above 60.5 °C, as we can see by moving along the lower curve to the right and upward. In time, the mole fraction of F in the liquid remaining is reduced to some small value, say, 0.05; the mole fraction of G is then 0.95 (point y in the figure). The boiling point has climbed to about 73 °C, close to that of pure G.

It is apparent from a consideration of the composition of the vapor that comes off at the beginning of the distillation (point m) and that which comes off toward the end (point x) that F and G cannot be separated as pure liquids by a simple distillation, such as is illustrated in Figure 1.9.

If the vapor that is initially richer in F is condensed, and then that liquid is heated to boiling, the new vapor is further enriched in the more volatile component F. This may be repeated several times to give essentially pure F. The process of successive vaporization and condensation is called **fractional distillation.**

To bring about a separation of F and G as pure liquids, the distillate can be collected in batches. Each batch can be redistilled into smaller and smaller batches. Eventually, a first fraction will be obtained that approaches 100 percent F and a last fraction that approaches 100 percent G. This type of fractional distillation is called a **batch process,** because each part of the collected mixture is handled as a separate batch.

Fractional distillation by a batch process is relatively tedious and inefficient. A more efficient separation is achieved by a **continuous process,** which is a series of distillations, condensations, redistillations, and recondensations. A continuous distillation process is carried out in an expeditious manner in a **fractionating column** in which the vapor condenses and redistills many times before leaving the column. Fractionating columns, to be efficient, must provide good contact between the ascending vapor and the descending liquid formed by the condensation of vapor. With a well-designed column, essentially pure F is obtained as the distillate that comes off the top of the column, and pure G as the residue that is taken from the bottom. One design is shown in Figure 12.6.

• *Constant-Boiling Mixtures* • If the vapor pressure curve for a pair of substances deviates enough from Raoult's law, a **constant-boiling mixture** is formed.[2] The composition of the vapor of such a mixture is identical with that of the liquid. Therefore, the mixture cannot be separated by boiling. Instead it boils at a constant temperature with a constant composition. A mixture of hydrogen chloride and water has a constant maximum boiling point at 1 atm of 108.58 °C at a composition of 20.222 percent HCl. Another well-known constant-boiling mixture is ethyl alcohol and water. At 1 atm it boils at 78.2 °C with a composition of 95.6 percent ethyl alcohol.

12•4   **Osmosis**

A membrane that permits the passage of only certain types of molecules is called a **semipermeable membrane.** Consider two aqueous solutions separated by a semipermeable membrane through which only

---

[2]Such mixtures, also called **azeotropes,** are quite common. About 400 are described in *Lange's Handbook of Chemistry,* 12th ed., J. A. Dean, Ed., McGraw-Hill, New York, 1979.

water molecules can pass. Water passes through the membrane from the dilute solution into the concentrated one more rapidly than it passes in the opposite direction. This flow of fluid, which tends to equalize the concentration of water on both sides of the membrane, is called **osmosis.**

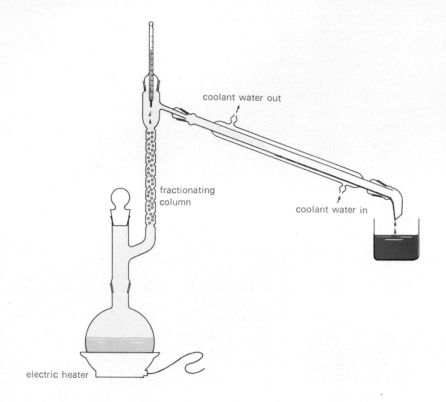

coolant water out

fractionating column

coolant water in

electric heater

Osmosis can be thought of as resulting from the unequal pressure of the water on the two sides of the membrane. There is a direct relationship between the vapor pressures of the solutions and osmosis. In Figure 12.1, we see that there is a net flow of water via the vapor phase from a dilute solution to a concentrated one until the concentrations become equal. In osmosis the net flow via the semipermeable membrane tends to accomplish the same result.

Osmosis may be readily demonstrated by fastening a piece of animal bladder or nonwaterproof cellophane over a thistle tube, as shown in Figure 12.7. An aqueous sugar solution is placed inside the thistle tube, which is then immersed in water. At the beginning, water molecules pass more rapidly from the pure water into the sugar solution than in the opposite direction. Because of this unequal rate, water accumulates in the thistle tube and the water level in it rises. As time goes on, two things happen to increase the rate of flow of water from the solution back into the pure water. First, the sugar solution becomes more dilute. Second, the height of the column of solution rises and exerts more mechanical pressure to drive water molecules back through the membrane. In time, the rate of flow in both directions becomes the same and the water level ceases to rise. Except for very dilute solutions, the equilibrium condition is quite difficult to achieve. As the water

Figure 12•6 In one type of fractionating column, the column is packed with glass beads or glass helices so that condensation and redistillation occur more efficiently over the entire length of the column. A fractionating column several meters tall may be needed to separate liquids whose boiling points are close together.

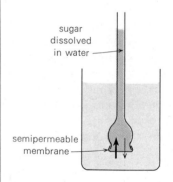

sugar dissolved in water

semipermeable membrane

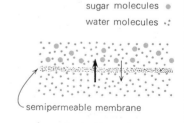

sugar molecules
water molecules

semipermeable membrane

Figure 12•7 Osmosis. The bottom figure is a schematic representation of the unequal rate of flow of water molecules through a small segment of the membrane.

level rises, the pressure of the solution on the membrane tends to increase until the membrane splits and spills the solution into the solvent.

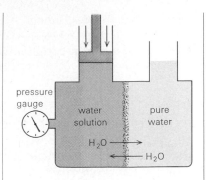

Figure 12•8 Apparatus for measuring osmotic pressure.

12•4•1   Osmotic Pressure   The **osmotic pressure** is defined as the pressure that must be applied to a solution to prevent any net transfer of pure solvent (at 1 atm pressure) into it through a semipermeable membrane. To measure the osmotic pressure in the laboratory, an apparatus of the type shown in Figure 12.8 is used. The solution is placed in a steel or other strong vessel that will not stretch or burst because of the additional pressure that is developed. A semipermeable membrane separates the solution reservoir from a reservoir of pure water. The semipermeable membrane is commonly a wall of an inorganic material that permits the passage of water molecules back and forth, but not solute molecules. The pressure necessary to maintain the volume of the solution constant is the osmotic pressure and can be read on the gauge.

For dilute solutions, the magnitude of the osmotic pressure, π, is proportional to the solute concentration and can be calculated from the equation

$$\pi V = nRT$$

where V (volume of solution), n (moles of solute), R (ideal gas constant), and T (absolute temperature) have the same units as in the ideal gas equation discussed in Chapter 7; and π is in atmospheres. Although this equation has the same form as the ideal gas equation, the basis for it is quite different. It is, in fact, an approximate form of a more complex relationship; this approximate form is not accurate for concentrated solutions.

If the mechanical pressure on a solution exceeds the osmotic pressure, pure solvent is driven out of the solution through a semipermeable membrane (Figure 12.9). This process is called **reverse osmosis** and is a method for recovering pure solvent from a solution. Two of the promising applications of reverse osmosis are the recovery of pure water from industrial wastes and the desalination of seawater.

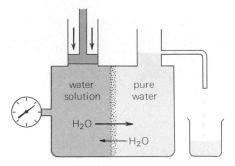

Figure 12•9 Reverse osmosis. Schematic representation showing that when the mechanical pressure is greater than the osmotic pressure, solvent is forced through a semipermeable membrane from a solution into a pure solvent. Compare Figure 12.8.

• *Biological Importance of Osmosis* • Osmosis is of great importance to plants and animals because it is the process by which water is distributed to all the cells of living organisms. Cell walls are semipermeable membranes through which water passes in both directions in accordance with the principles discussed in the preceding paragraphs. The membranes of living cells are also permeable to certain solutes, so nutrients and waste products are exchanged through them. The permeability of cell walls to solutes is frequently selective and to some extent independent of the size of the solute particles and their concentrations. For example, hydrated magnesium ions do

★ Special Topic ★ Special Topic ★ Special Topic ★ Special Topic ★ Special Topic ★ Special Topic ★ Special Topic ★ Special Topic ★

Desalination of Seawater "Water, water, everywhere," said the ancient mariner, "And all the boards did shrink. Water, water, everywhere, Nor any drop to drink." Around the world today, including many areas adjacent to the seas that surrounded the ancient mariner, people are threatened with poverty and even death because of lack of fresh water. The removal of dissolved salts from water, called *desalination,* is one of the most pressing technical problems facing humanity today. To keep pace with the growth of world population, fresh water must be provided for new cities, for their industries, and for the irrigation of arid croplands.

The cultivation of crops requires approximately one cubic meter of fresh water for every square meter of land per year. Industry requires enormous amounts of water. In the United States, the consumption for all needs is about 6.6 cubic meters per person every day. In many regions the only possible source of such quantities of water is the sea, but present methods of desalination are so expensive that their use is quite limited.

Much research and development are being devoted to five methods of desalination: (1) distillation, (2) freezing, (3) reverse osmosis, (4) electrodialysis, and (5) ion exchange.

Distillation processes require so much energy that one might assume they would be used only when small quantities of fresh water are absolutely required, as, for example, for drinking and washing on shipboard. However, most of the large-scale commercial desalination units now in operation throughout the world are boiling water *distillation* systems.

Water can also be desalinated by *freezing* the saline solution. The ice crystals formed in this way are separated from the salt and then melted to give fresh water. This method, on which research has been done for a number of years, has not been developed to a great extent because of its costs.

The *reverse osmosis* method of desalination, illustrated in Figure A, shows promise of becoming economically attractive. In this process, the salt is separated from the water by pressure applied across a semipermeable membrane separating the water source (saline feed) from the product (fresh water). The process possesses a high degree of structural and operational simplicity. Energy costs are quite low; the principal contribution to the cost comes from the initial and

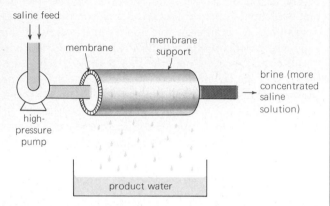

Figure A The reverse osmosis method of desalination. See also Figure 12.9. Although used to purify wellwater rather than seawater, a reverse osmosis plant designed for Sarasota, Florida, will process 4.5 million gallons of water daily.
Source: Reproduced by permission from Ronald F. Probstein, "Desalination," *American Scientist,* 61 (3), 280 (1973).

replacement costs of the membranes, which are generally made of cellulose acetate and hollow-fiber aromatic compounds that are structurally related to nylon. At the present time, reverse osmosis has been applied to the desalting of so-called brackish water, which is only moderately salty compared with seawater. [Standard seawater has a total dissolved solids content of 35,000 parts per million (ppm), whereas brackish water has a total dissolved solids content of 1,500 to 10,000 ppm.] Despite a large number of rosy predictions, the future of reverse osmosis for large-scale desalination of both brackish water and seawater rests upon the development of improved membranes.

The method of *electrodialysis* involves the forced movement of ions as well as the use of semipermeable membranes. Unlike reverse osmosis where the water is removed from the salt solution, in this process the salt is removed from the salt solution. Use is made of the fact that the dissolved salts exist in ionic form. A portion of an electrodialysis cell is shown in Figure B. The saline water is pumped into the top of the compartments. The membranes between the compartments are alternating positive-ion and negative-ion permeable. When a direct current is passed through the solution, the ions move through the membranes toward the electrodes of oppo-

not pass through the walls of the gastrointestinal tract to any great extent, whereas glucose molecules pass through at a rate too rapid to be accounted for by simple diffusion.

★

Special Topic ★ Special Topic ★ Special Topic ★ Special Topic ★ Special Topic ★ Special Topic ★ Special Topic ★ Special Topic

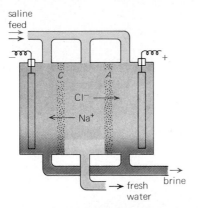

Figure B A portion of an electrodialysis cell. Upon application of the electric current, positive ions are attracted through the cation-permeable membrane C to the negative electrode; negative ions move in the opposite direction through the anion-permeable membrane A.
Source: Adapted by permission from *Du Pont Innovation*, 4 (2), 5 (1973).

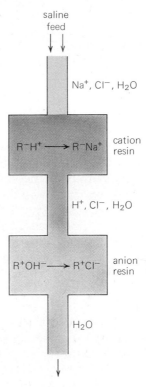

Figure C Flow diagram for desalination by ion exchange.

site charge, thereby depleting the salt in the middle compartment. Water undergoes reduction in the cathode compartment to give hydrogen and hydroxide ions, $2H_2O + 2e^- \longrightarrow H_2 + 2OH^-$, and a solution of sodium hydroxide is formed. In the anode compartment, water is oxidized to oxygen and hydrogen ions, $2H_2O \longrightarrow O_2 + 4H^+ + 4e^-$, and hydrochloric acid is formed. When the solutions in the two compartments mix in the exit pipe at the bottom of the cell, sodium chloride and water are formed once again. While this process is appealing due to its basic simplicity in design and operation, it suffers from the same limitations as does reverse osmosis.

In the method of *ion exchange*, a saline solution is allowed to flow through a bed of material made from a granulated zeolite (see Section 25.4.3) or an ion-exchange resin (see Section 21.6). The ions in solution become attached to the material in the bed and displace ions of the same sign. Ion exchange is used in water softening (see Section 25.4.3) and in water desalination. In one ion-exchange desalting scheme, illustrated in Figure C, the saline solution first flows through a resin bed where hydrogen ions of the resin exchange with cations, such as sodium ions, and then flows through a

second resin bed where the hydroxide ions of the resin exchange with anions, such as chloride ions. The ions of salt are removed, and the H^+ and OH^- ions that are released from the resins neutralize one another to form water. The resin materials (from synthetic polymers, see Section 27.8) are chemically similar to electrodialysis membranes. To date, the importance of ion exchange in desalination has been mainly in conjunction with one of the other processes—distillation, reverse osmosis, or electrodialysis—in which ion-exchange pretreatment of the water source has been used for the removal of scale-forming minerals.

At present, desalination on a moderate scale, for special needs, and with waters of low salinity is a reality. However, the technological advances required for economical large-scale seawater desalination have not yet been made.

When a solution to be injected into the bloodstream is formulated, the osmotic pressure of the solution must be considered. The average osmotic pressure of the blood is about 7.7 atm (it rises just after meals and then falls). If red blood cells are placed in a solution that has an osmotic pressure greater than that of normal blood, water passes out of the cells until they shrink and settle out of suspension. If the cells are put into a solution that has a lower osmotic pressure than blood, the cells may swell with water until the cell walls burst. Consequently, the osmotic pressure of solutions for injection is adjusted (chiefly with sodium chloride) until it is compatible with blood.[3]

The Colloidal State

The colloidal state is intermediate between a solution and a suspension. When matter is in this state of subdivision, it exhibits some interesting and important properties that are not characteristic of matter in larger aggregates. Before discussing these properties, we shall describe the particle size ranges associated with the colloidal state and how substances achieve this prescribed size.

12.5 Particle Size and the Colloidal State

The average diameter of atoms and simple ions is of the order of 2×10^{-10} m, or 2 Å, as we can see in Figure 5.14. The molecules of simple compounds, such as water, ammonia, and ethyl alcohol, are, of course, larger than atoms but usually not many times larger if we consider only the smallest dimension of the molecule. Only when several million such simple molecules are clustered together are particles formed that can be seen by the unaided eye.

Often particle clusters are formed that contain only a few hundred or a few thousand atoms, ions, or small molecules. The diameters of such particles may range from about 10 Å (10^{-9} m) to about 2,000 Å. Such particles cannot be seen clearly with the most powerful optical microscope.[4] Matter with at least one dimension in the range from approximately 10 Å to 2,000 Å is said to be in the **colloidal state.** The colloidal state is not characteristic of any particular substance; practically all substances, whether normally gaseous, liquid, or solid, can be put into the colloidal state. Biochemical entities often have colloidal dimensions, as the examples in Figure 12.10 indicate.

---

[3]In biological and medical terminology, solutions that have the same osmotic pressure as blood are called **isotonic** solutions; those that have lower osmotic pressures are called **hypotonic** solutions; and those that have higher osmotic pressures are **hypertonic** solutions.

[4]These minute particles are smaller than the wavelength of the visible light that they must reflect in order to be seen. The shortest wavelength of visible light is approximately 3,800 Å (3.8×10^{-7} m).

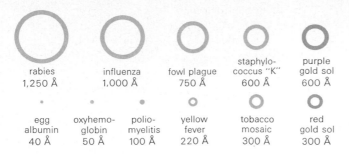

rabies
1,250 Å

influenza
1,000 Å

fowl plague
750 Å

staphylo-
coccus "K"
600 Å

purple
gold sol
600 Å

egg
albumin
40 Å

oxyhemo-
globin
50 Å

polio-
myelitis
100 Å

yellow
fever
220 Å

tobacco
mosaic
300 Å

red
gold sol
300 Å

Figure 12•10 Approximate sizes of some biochemical entities in the colloidal range, compared with particles in two gold sols (a sol is a colloidal system with solid dispersed in a liquid, see Table 12.6). Gold sols have been obtained in shades of red, blue, and violet, depending on the size of the dispersed gold particles.
Source: From W. J. Elford, *Trans. Faraday Soc.,* 33, 1103 (1937); and J. W. McBain, *Colloid Science,* Heath, New York, 1950.

Three idealized shapes of colloidal matter, *laminar, fibrillar,* and *corpuscular,* are diagrammed in Figure 12.11. For matter in the form of corpuscles, the diameter gives a measure of particle size. For laminar and fibrillar particles, the length, width, and thickness are all needed to indicate the particle size. However, only one of these dimensions has to be in the colloidal range for the material to be classed as colloidal. For example, soap in a soap bubble is classed as colloidal, because the soap film is only a few molecules thick.

colloidal systems

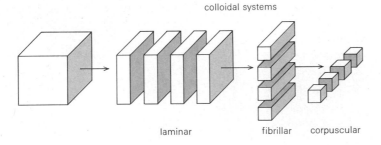

laminar   fibrillar   corpuscular

Figure 12•11 Schematic representation of the relation of bulk matter to three idealized shapes of colloidal matter. Note that, for a given amount of material in the bulk state, the amount of surface area exposed increases as the material is subdivided. Some of the new surfaces exposed at each step are shown in color.
Source: Adapted by permission from Egon Matijević, "Colloids—The World of Neglected Dimensions," *Chemical Technology,* 3 (*11*), 656 (1973).

12•6 Importance of Colloid Chemistry

Inasmuch as most substances can exist in the colloidal state, all fields of chemistry are concerned with colloid chemistry in some way or other. All living tissue is colloidal. Many of the complex chemical reactions that are necessary to life must be interpreted in terms of colloid chemistry. The portion of the earth's crust that is referred to as tillable soil is composed in part of colloidal material; therefore, soil science must include the application of colloid chemistry to soils. In industry, colloid science is important in the manufacture of paints, ceramics, plastics, textiles, photographic paper and films, glues, inks, cements, rubber, leather, salad dressings, butter, cheese, and other food products, lubricants, soaps, agricultural sprays and insecticides, detergents, gels and jellies, adhesives, and a host of other products. Such processes as bleaching, deodorizing, tanning, dyeing, and purification and flotation of minerals involve adsorption on the surface of colloidal matter and hence are concerned with colloid chemistry.

**12•7   Types of
Colloidal Systems**

Recall that in the stable, homogeneous mixture called a solution, individual molecules, atoms, or ions are dispersed in a second substance. In a somewhat similar manner, colloidal matter may be scattered or dispersed in a continuous medium to give a **colloidal dispersion** or **colloidal system.** Jelly, mayonnaise, India ink, milk, and fog are familiar examples. In such systems the colloidal particles are referred to as the *dispersed substance* and the continuous matter in which they are scattered as the *dispersing substance* or the *dispersing medium.*

Because either the dispersed or the dispersing substance may be gaseous, liquid, or solid (except that both may not be gases),[5] there are eight types of colloidal systems. A list of the eight types with examples is given in Table 12.6. The meanings of the terms foam, solid foam, liquid aerosol, solid aerosol, emulsion, solid emulsion, sol, and solid sol are evident from this table.

Table 12•6 Colloidal systems

Dispersed substance	Dispersing substance	Type name	Examples
gas	liquid	foam	whipped cream, beer froth, soap suds
gas	solid	solid foam	pumice, marshmallow, polyurethane foam
liquid	gas	liquid aerosol	fog, clouds
liquid	liquid	emulsion	mayonnaise, milk
liquid	solid	solid emulsion	cheese (butter fat dispersed in casein), butter
solid	gas	solid aerosol	smokes, dust
solid	liquid	sol	most paints, starch dispersed in water, jellies
solid	solid	solid sol	many alloys, black diamonds, ruby glass (gold in glass, a supercooled liquid)

**12•8   Properties of
Colloidal Systems**

12•8•1   Tyndall Effect All of us have observed the scattering of light by dust particles when a beam of sunlight enters a darkened room through a partly opened door or a slit in a curtain.[6] The dust particles, many of them

[5]Molecules in gaseous mixtures are not as large as colloidal particles.

[6]We can observe a very striking demonstration of light scattering by releasing some hair spray under these lighting conditions. We also get a good idea of what our lungs have to cope with when we breathe in the area where an aerosol spray has just been released.

too small to be seen, look like bright points in the beam of light. If, the particles are actually of colloidal size, we do not see the particles themselves; rather, we see the light that is scattered by them. This light scattering is called the **Tyndall effect.** It is due to the fact that small particles scatter light at all angles from the direction of the primary beam.

The Tyndall effect can be used to distinguish a colloidal dispersion from an ordinary solution, because the atoms, small molecules, or ions that are present in a solution do not scatter light noticeably in samples of little thickness (see Figure 12.12 on Plate 4). The Tyndall scattering of light accounts for the opacity of colloidal dispersions. For example, although both olive oil and water are transparent, a colloidal dispersion of the two has a milky appearance.

- **12•8•2   Brownian Movement**   If an optical microscope is focused on a colloidal dispersion at right angles to a light source and with the background in darkness, colloidal particles are observable, not as particles with definite outlines, but as small, sparkling specks. By following these points of reflected light, we can see that the dispersed colloidal particles are constantly moving in random, zigzag paths (see Figure 12.13). This random motion of colloidal particles in a dispersing medium is called **Brownian movement,** after the British botanist Robert Brown, who studied it in 1827.

The cause of the Brownian movement was a matter of mystery and conjecture until about 1905, when Albert Einstein published a mathematical analysis of it. Einstein showed that a microscopic particle suspended in a medium should exhibit a random motion because of the unequal number of collisions of molecules on different sides of the particle (see Figure 12.14). Even as late as 1905, a few leading scientists still regarded atoms and molecules as imaginary particles that were useful only in theoretical explanations. Also it was thought that the number of collisions at any one time would be approximately the same on different sides of a particle, so why should a particle move? Einstein's mathematical prediction of nonuniform, random collisions was verified experimentally by the French scientist Jean Perrin, who later received the Nobel Prize for his work. The prediction, followed by laboratory verification, overcame the last doubts that atoms and molecules are real entities and also gave conclusive support to the theory that molecules are in constant motion.

- **12•8•3   Adsorption**   Matter in the colloidal state has a tremendous amount of surface area. As illustrated in Figure 12.11, each time a solid is divided, two new surfaces are created. Consider a cube of nickel that is 1.00 cm on an edge. Such a cube weighs 8.90 g and has a surface area of 6.00 cm². If this cube is divided into a thousand billion billion small cubes, each of these tiny cubes will be 1×10^{-7} cm (10 Å) on an edge. The amount of surface area of the 8.90 g of nickel is increased by 10 million times, from 6.00 cm² to 60 million cm²—about 1.5 acres!

At the surface of a particle there are unsatisfied van der Waals forces or even valence forces that can attract and hold atoms (or molecules or ions) of foreign substances. This adhesion of foreign substances to the surface of a particle is called **adsorption.** Adsorbed substances are strongly held in layers that are usually no more than one or two molecules (or ions) thick. The

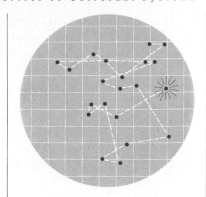

Figure 12•13 Brownian movement. A plot of the observed position of a colloidal particle at 30-second intervals. The particle is too small to be seen, even with a microscope, except by the sparkle of reflected light. *Source:* J. Perrin, *Les Atomes,* Presses Universitaires de France, 1948.

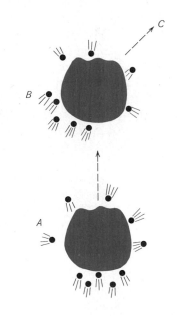

Figure 12•14 Dispersed colloidal particles move randomly because, at any one time, more molecules of the dispersing phase hit on one side of the particle than on another. The particle shown moves from position *A* to *B* to *C.* An increase in temperature increases the rate of Brownian movement, giving evidence that the kinetic energy of molecules is a function of temperature.

amount of foreign substance that can be adsorbed depends on the amount of exposed surface area. Although adsorption is a general phenomenon of solids, it is particularly efficient with colloidal matter due to the huge amount of surface area.

12•9   Selective Separations

Because different colloids adsorb different substances, the process of adsorption is widely used in the removal of undesirable colors and odors from certain materials, for the separation of mixtures, for the concentration of ores, and in various other purification processes.

12•9•1   Efficient Solid Adsorbents   Gas masks commonly contain activated charcoal or some other type of adsorbent. *Activated charcoal* is made by heating charcoal in air or steam so that more than half of the carbon is burned away. This leaves a very porous carbon "skeleton," so porous that a very large percentage of the remaining carbon atoms are surface atoms and hence capable of adsorbing. In addition to its present use in commercial gas masks, activated charcoal is widely used in industry and in laboratory work to remove impurities that have objectionable odors, flavors, colors, and other properties. It is also used in some cigarette filters. The so-called standard puff from a nonfilter cigarette has a volume of 35 mL and contains 2×10^{11} (200 billion) particles that vary in size from 1,500 Å to 10,000 Å in diameter. These particles are in addition to the gaseous molecules formed by the combustion, such as CO_2, CO, and H_2O. Over 1,100 compounds have been identified in the particles present in tobacco smoke, several of which are known to possess carcinogenic activity; that is, they induce certain types of cancer in the cells of experimental animals. Adsorption removes much, but by no means all, of the colloidal matter in tobacco smoke. To remove all would cause the disappearance of the smoke and probably the desire to puff away.

12•9•2   Precipitation of Aerosols   A commonly used process for destroying smoke and other types of aerosols is the Cottrell method of electric coagulation (see Figure 12.15). The smoke is led past a series of sharp points charged to a high potential (20,000 to 75,000 V). The points discharge high-velocity electrons that ionize molecules in the air. Smoke particles adsorb these positive ions and become so highly charged that they are attracted to, and held on, the oppositely charged electrodes. Cottrell precipitators are widely used in industry for two general purposes: to remove particulate pollutants from discharged flue gas and to recover valuable, finely divided solids that otherwise would be lost.

12•9•3   Chromatography   If a mixture of substances is passed over an adsorbing material, conditions may be controlled so that the components of the mixture are separated because of preferential adsorption. The first substances separated in this manner were colored, so the process was called **chromatographic separation** or **chromatography.** Applied in many areas, some of the greatest successes of chromatography have been in the separation of biological mixtures. Amino acids, proteins, fats, carbohydrates, vitamins, and hormones are examples of substances that have been discov-

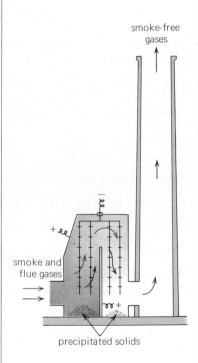

Figure 12•15 Schematic diagram showing a Cottrell precipitator attached to a smokestack. Positively charged smoke particles are attracted to the points of the hanging negative electrode, where they lose their charges, agglomerate, and fall off. Electrons are attracted to the positively charged walls of the precipitation chamber. The precipitated solids are removed periodically.

ered and studied. Adsorption can be a delicate treatment that does not destroy the fragile molecules of living systems. In recent years chromatography has played a key role in the work of many scientists who have won Nobel Prizes in chemistry or biochemistry.

Liquid chromatography is a process in which a moving liquid is used to carry substances over the surfaces of an adsorbent. In one form of liquid chromatography, a solution of two or more solutes creeps up through the pores of adsorbing paper by capillary attraction. In another form, a solution seeps down through a column of finely divided solid particles such as starch or aluminum oxide (see Figure 12.16). The most strongly adsorbed component moves most slowly along the paper or the column. If the adsorbed materials are colored, their locations are easily seen. Colorless materials can sometimes be located under ultraviolet light or by treatment with developing agents that react to form colored compounds.

Gas chromatography uses the gas phase to carry substances over the surfaces of an adsorbent. The method is especially valuable in separating and purifying small amounts of gases or liquids, amounts too small to handle by ordinary distillation methods. Mixtures of gases or vaporized liquids are passed through an adsorbing column (at a temperature above the condensation point of any liquid present). Some *carrier gas,* such as helium, is used to move the vapors through the column.

Because the heat conductivities of various gases differ greatly, the presence of a second gas mixed with the carrier gas is easily detected (see Figure 12.17). Commonly a heat conductivity detector is put at the end of a chromatographic column. As different samples of the separated gases come off the column, they can be collected in individual containers if desired.

Automated gas chromatographic units have been carried on interplanetary rockets to determine the composition of the atmospheres of other planets and to radio back the results. Used on the Mars *Viking Lander* and aboard the *Pioneer Venus,* there is one column to search for lower boiling, nonpolar gases, H_2 through Kr, and another column to look for higher boiling, polar gases, such as H_2O and HCl.

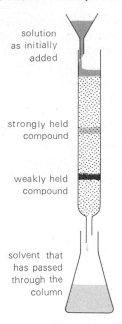

Figure 12•16 Column chromatography.

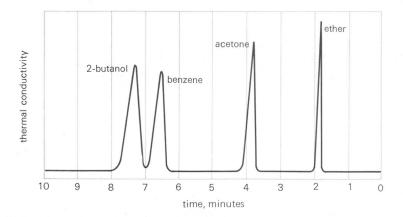

Figure 12•17 A gas chromatogram showing the separation of a mixture containing 0.004 mL of ethyl ether, 0.007 mL of acetone, 0.009 mL of benzene, and 0.012 mL of 2-butanol. Note that all of the ether came through the column within 2 minutes after the mixture was added. Acetone came through after 4 minutes, then came benzene, and finally 2-butanol, with the entire process requiring 8 minutes. The area under a peak is a measure of the relative amount of the substance causing the peak.
Source: Courtesy of Dr. N. S. Bowman, The University of Tennessee, Knoxville.

12•9•4   Dialysis The separation of ions from colloids by diffusion through the pores of a semipermeable membrane is called **dialysis.** The pores are commonly smaller than 10 Å in diameter and allow the passage of water molecules and small ions. Natural animal membranes, parchment paper, cellophane, and some synthetic plastics are suitable membrane materials. The particles that pass through the membrane probably do not do so simply by random diffusion. Rather they are adsorbed on the surface of the membrane and move from one adsorbent site to another as they move through the pores.

Dialysis is used to purify colloidal sols (see Table 12.6) and for some other specialized applications. Its most dramatic application is in the treatment of patients suffering from kidney failure. An essential function of the kidneys is to remove natural metabolic waste products, such as urea and creatine, from the blood. Failure to remove these waste products results in death. Artificial kidney units save the lives of thousands of persons each year (see Figure 12.18). Some persons use the dialysis treatment regularly over many years; others use dialysis to maintain life until they can have a kidney transplant operation.

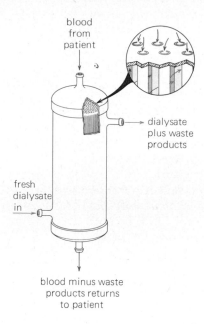

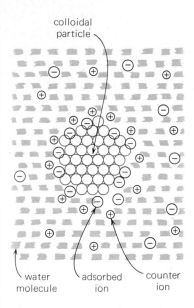

Figure 12•18 Schematic diagram of a blood dialysis unit.
Source: Adapted by permission from a diagram of the Travenol CF 1500 dialyzer, courtesy of Baxter Travenol Laboratories, Inc.

Figure 12•19 Stabilization of a colloidal particle by adsorbed ions.

12•10 Stability of Colloidal Systems

The carbon-in-water colloid shown in Figure 12.12 is so stable that it has not changed in appearance for about 30 years. In all gaseous and in most liquid colloids, the particles of the dispersed substance have a greater density than the dispersing medium. What keeps the small particles from clumping together and forming large particles that settle out? One of the most common ways in which colloidal particles become stabilized is by adsorbing a layer of ions (Figure 12.19). Particles that adsorb ions with like charges are stabilized because they repel one another rather than joining to become larger aggregates.

Another way in which a particle is stabilized is by the adsorption of a layer of molecules. Some substances that are adsorbed, such as soaps, are colloids themselves. Colloids that act as stabilizing agents are called **protective colloids.** Protective colloids are particularly effective in stabilizing colloids of liquids in liquids, called **emulsions** (see Figure 12.20). Milk is an emulsion of globules of butter fat in water, with the protein casein serving as the stabilizing agent. Mayonnaise is an emulsion of a liquid fat (such as olive oil or corn oil) in water, with egg yolk acting as the stabilizing agent. Gelatin is used as a protective colloid in making ice cream to prevent the formation of large particles of sugar or ice.

Sometimes we desire that a colloid not be stable. This is often the case in the laboratory when finely divided precipitates are formed. By controlling conditions, we try to cause the particles to clump together so they will settle or so we can trap them on a filter. If such a precipitate is washed incorrectly it may be **peptized;** that is, it may be dispersed as small colloidal particles. We try to prevent peptization by using the proper wash solution.

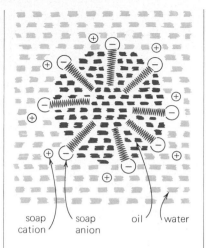

soap cation   soap anion   oil   water

Figure 12•20 Schematic representation of the stabilization with soap molecules of a particle in an oil–water emulsion.

Chapter Review

Summary

The vapor pressure, boiling point, freezing point, and osmotic pressure of a solution are **colligative properties,** which are useful in determining the molecular weights of compounds. The **colligative property law** states that the values of these properties differ from those of the pure solvent in direct proportion to the molal concentration of the solute. If the solute is a **nonvolatile nonelectroyte,** its identity is largely unimportant; if it is volatile or if it dissociates into ions, however, these factors must be taken into account. An **ideal solution** is one that obeys the colligative property law exactly. The lowering of the vapor pressure of a solution by a nonvolatile solute follows **Raoult's law.** Because the vapor pressure is lowered, the boiling point is elevated and the freezing point is depressed in direct proportion to the concentration of solute. The proportionality constants are called the **molal boiling point elevation constant** and the **molal freezing point depression constant,** respectively.

In solutions of strong electrolytes, the interaction between ions reduces the *effective* number of solute particles from the theoretical number; weak electrolytes dissociate only partially in solution. The effect on the solvent of a given electrolyte solute compared with that of nonelectrolytes is called the **van't Hoff factor** of the solute.

Solutions of liquids follow Raoult's law approximately. Their components can usually be separated by **fractional distillation,** using either the **batch process** or the **continuous process** (the latter in a **fractionating column**). The result is a **distillate** and a **residue.** In some cases a con-stant-boiling mixture, or **azeotrope,** is formed, which cannot be separated by distillation.

Osmosis is the flow of a solvent from a solution of lower concentration to one of higher concentration, through a **semipermeable membrane.** This tends to make the concentrations and hence the vapor pressures of the two solutions equal. The **osmotic pressure** of any solution is the pressure needed to counteract the flow of solvent (at 1 atm) into the solution through a semipermeable membrane. If this pressure is artificially exceeded in a real system, the solvent is driven *out* of the more concentrated solution; this is **reverse osmosis.** Osmosis is fundamental to the functioning of all living cells.

The **colloidal state** consists of aggregates, or clusters, of particles that are too large to be considered dissolved and too small to precipitate. Almost any substance—gas, liquid, or solid—can form a **colloidal dispersion,** or **colloidal system,** under the right conditions. Such dispersions are found virtually everywhere in nature and are of great importance in most fields of applied chemistry. Two natural phenomena that depend on the properties of colloids are the **Tyndall effect** and **Brownian movement.**

The enormous surface area presented by a given amount of colloidal matter makes possible the **adsorption** of large amounts of foreign substances by means of surface adhesion forces, as in a gas mask. An extremely useful scientific application of this property is in the **chromatographic separation** of a great variety of chemical compounds. Both **liquid chromatography** and **gas chromatography** are widely used for this purpose. Another application is in **dialysis,** a technique for the removal of ions from colloids, as in an artificial kidney machine. In most colloidal systems, the clusters of particles are stabilized (prevented from clustering further) by the adsorption of other substances. Some of these may themselves be colloids; they are called **protective colloids** and are especially important in the formation of **emulsions.**

Exercises

Vapor Pressure of Solutions

1. Consider Figure 12.1. If the relative humidity of the air in the container is 100 percent at the time the three beakers are inserted, will it be 100 percent at the conclusion? (The temperature remains unchanged.)

2. Suppose each beaker in Figure 12.1 has a capacity of 400 mL. Initially they contain: (*a*) 100 g of water, (*b*) 100 g of water plus 0.10 mole of sucrose, and (*c*) 100 g of water plus 0.30 mole of sucrose. What will be the volume in each beaker and the molal concentration of each solution at equilibrium? Assume that a negligible amount of water is in the form of water vapor in the container.

3. With respect to Figure 12.2, what would be the position of the mercury in the manometer tube if the sucrose solution were frozen and the stopcock were opened to the vacuum pump? Could the vacuum pump cause the mercury column to move around in the U-shaped tube to make it stand at a higher level in the left limb than in the right limb?

4. What is the mole fraction of each component in a solution that contains 10.0 g of glucose, $C_6H_{12}O_6$, in 100 g of water? What is the vapor pressure of this solution at 40 °C?

5. The vapor pressure of ethyl alcohol is 100 mm at 34.9 °C. What is the vapor pressure of a solution of 10.0 g of the nonvolatile, nonelectrolyte urea, $CO(NH_2)_2$, in 50.0 g of ethyl alcohol at 34.9 °C?

*6. Consider the following questions about an automobile radiator at 120 °C, assuming ideal behavior. The vapor pressure of ethylene glycol, $C_2H_4(OH)_2$, at that temperature is 40.0 mm.
   (a) What is the pressure if water alone is present (see Table A.7)?
   (b) What is the vapor pressure if 5.00 kg of ethylene glycol and 15.0 kg of water are present?
   (c) What is the vapor pressure if 10.0 kg of ethylene glycol and 10.0 kg of water are present?
   (d) What can happen when the radiator cap is quickly removed while the engine is hot? Why?

Boiling and Freezing Point Changes

7. A solution of a covalent compound in water is found to have a freezing point of −3.0 °C.

   (a) What is the molality of the solution?
   (b) What is the mole fraction of solute?
   (c) Assuming the solute is nonvolatile, what is the boiling point?

8. What is the freezing point of a 0.500*m* solution of a nonelectrolyte?

9. Calculate the freezing points for the radiator solutions described in Exercise 6, parts (b) and (c).

10. Given only the freezing point of water as 0.00 °C, the values of k_f and k_b (see Table 12.1), and the water vapor pressures (see Table A.7), construct a graph similar to that of Figure 12.3 for water and for a 2.0*M* solution of glycerin, $C_3H_5(OH)_3$. What is the boiling point of the solution as read from the graph?

11. (a) What is the vapor pressure at 0 °C of a sugar solution that freezes at −10 °C?
    (b) Repeat part (a) for sodium chloride.

12. A certain compound dissolved in acetic acid gives a solution that has a boiling point of 122.0 °C. What is the freezing point of this solution?

13. Comparing methyl alcohol, CH_3OH, with ethylene glycol, $C_2H_4(OH)_2$, which of the following mixtures will have the lowest freezing point?
    (a) 10 g methyl alcohol + 100 g water
    (b) 10 g ethylene glycol + 100g water
    (c) 10 g methyl alcohol + 1.0 kg water
    (d) 10 g ethylene glycol + 1.0 kg water
    (e) 1.0*m* methyl alcohol in water
    (f) 1.0*m* ethylene glycol in water

14. (a) Of the six solutions described in Exercise 13, which will have the highest vapor pressure at 50 °C?
    (b) Which will have the highest boiling point?

15. Suppose that a car radiator holds a volume of 30 L of water. Approximately what weight of ethylene glycol, $C_2H_4(OH)_2$, is required for addition to this quantity of water to give a solution that begins to freeze at −15 °C?

*16. Pure benzophenone, $C_{13}H_{10}O$, has a melting point of 48.0 °C. When 5.00 g of a nonelectrolyte with a molecular weight of 125 amu is mixed with 200 g of benzophenone, the mixture melts at 46.0 °C. Calculate k_f for benzophenone.

*17. What explanation can be given for the fact that each of the freezing point depressions in Table 12.2 is not precisely 1.86 °C?

Experimental Determination of Molecular Weights

18. Calculate the molecular weight of an organic compound if 1.01 g in 20.0 g of benzene yields a solution that begins to freeze at 3.18 °C.

19. A 0.680-g sample of an unknown compound with the empirical formula $C_3H_2NO_2$ is dissolved in 30.00 g of benzene. The boiling point of the solution is found to be 80.45 °C.

(a) What is the molality of the solution?

(b) What is the molecular weight of the unknown?

(c) What is the molecular formula of the unknown?

20. Pure acetic acid, $HC_2H_3O_2$, has a freezing point of 16.60 °C and a k_f of 3.90 °C/m.

 (a) If a certain weight of a nonelectrolyte dissolved in acetic acid causes the freezing point to be depressed to 14.82 °C, what is the molality of the solution?

 (b) If the weights of solute and solvent in part (a) are 4.00 g and 100 g, respectively, what is the molecular weight of the solute?

21. Suppose that the unknown compound in Example 12.4 is found to be 53.31 percent C and 11.18 percent H. Assuming that the remainder is oxygen, calculate an empirical formula and a precise molecular weight.

22. Suppose that a new compound whose molecular weight is to be determined by the freezing point method is soluble in either camphor or water to the extent of 0.10 g per 20 g of solvent. Which solvent would probably be better for the experiment? Why?

Properties of Solutions of Electrolytes

23. A 5.55-g sample of calcium chloride, an ionic electrolyte, is dissolved in 500 g of water.

 (a) Theoretically, how many moles of ions per liter of water would be formed if the solution behaved as an ideal electrolyte solution?

 (b) What would the freezing point of the solution be if the electrolyte solution behaved ideally?

 (c) The freezing point of the solution is found to be −0.483 °C. How many times the freezing point depression of an ideal nonelectrolyte solution of the same molal concentration is this?

24. The i values in Table 12.4 are based on certain experimentally measured freezing points. Calculate what these freezing points are for the following:

 (a) 0.100m $MgSO_4$ (b) 0.0500m KCl

 (c) 0.0100m K_2SO_4 (d) 0.00500m NaCl

25. Calculate the molalities and the i values for the following solutions, given the freezing points (percentages are by weight):

 (a) 1.00% NaCl, −0.593 °C

 (b) 0.500% KCl, −0.234 °C

 (c) 1.00% H_2SO_4, −0.423 °C

 (d) 0.500% $HC_2H_3O_2$, −0.159 °C

 (e) 1.00% $HC_2Cl_3O_2$, −0.211 °C

26. Suppose that you have experimentally determined the freezing point of a 0.10m $AlCl_3$ solution to be −0.93 °C. Using your data, calculate a value for i. Does your value for i indicate that the freezing point determination is correct to within a ±5 percent allowable error? Justify your conclusion with appropriate calculations.

27. A solution of hydrochloric acid, a covalent electrolyte, has 7.30 g of HCl dissolved in 2,000 g of water.

 (a) Theoretically, how many moles of ions would be formed if this solution behaved as an ideal electrolyte solution?

 (b) What would the freezing point of the electrolyte solution be if it behaved ideally?

 (c) The freezing point of this HCl solution is known to be depressed 1.91 times that of an ideal nonelectrolyte solution of the same molal concentration. What is the actual freezing point of the HCl solution?

28. What is the percent ionization of the weak acid HX in a 0.100m solution that has a freezing point of −0.205 °C?

*29. In 150 g of water, 9.0 g of a certain compound is dissolved. The solution has a boiling point of 100.45 °C.

 (a) Calculate the molecular weight of the compound as though it were a nonvolatile nonelectrolyte.

 (b) Suppose that you are reasonably sure that the dissolved compound is an ionic compound with the general formula AB_2. What do you calculate the molecular weight to be, assuming the compound is present completely as A^{2+} and B^- ions and acts ideally?

 (c) Suppose that you find the compound is magnesium nitrate, $Mg(NO_3)_2$. Assuming no experimental error, account for any difference between your answer to part (b) and the known molecular weight of magnesium nitrate.

 (d) Calculate the value of the i factor for this $Mg(NO_3)_2$ solution.

*30. Show that the i factor of 2.72 for 0.00500m H_2SO_4 indicates that H_2SO_4 ionizes in two steps, and that the first step may be 100 percent complete and the second one may be 72 percent complete.

Distillation

31. Describe the composition changes of the distillate and the residue as a 50–50 mixture of benzene (boiling point, 80 °C) and toluene (boiling point, 111 °C) is distilled.

*32. In Figure 12.5, the point n shows the boiling point of a 50–50 mixture.

 (a) What does point m show relative to point n?

 (b) If vapor with the composition of point m is condensed, what will be the boiling point of the resulting liquid?

 (c) If the resulting liquid of part (b) is boiled, what will be the composition of the vapor?

 (d) If the vapor in part (c) is condensed, what will be the composition of the resulting liquid?

 (e) Sketch a copy of Figure 12.5 and show on it, as a stair-step addition to the graph, the points described in parts (b), (c), and (d). Show the next steps, if the process is continued, by adding the next three points.

33. Can a 50–50 ethyl alcohol–water mixture be distilled in an apparatus such as that shown in Figure 12.6 so that some 100-percent alcohol is obtained? Explain.

Osmosis

34. In a system such as that diagrammed in Figure 12.7, what will be the height of the column above the level of the pure water if the concentration of the sugar solution at equilibrium is $0.10M$ and the temperature is $0\,°C$?

35. Too much fertilizer around the roots of a plant will cause it to wither and die. The plant is referred to as "burnt," and it has a dried appearance. Explain why this might occur.

36. Redwood trees grow to heights of more than 350 feet. Based simply on the height, calculate the osmotic pressure in their roots at ground level.

Particle Size

* 37. Consider a cube of wood painted blue that is 3 cm on an edge. Suppose you cut this cube at 1-cm intervals parallel to the six faces of the cube, so you make the maximum number of cubes 1 cm on an edge.
   (a) How many small cubes will you make?
   (b) How many small cubes will have all faces painted? All faces unpainted? One face painted? Two faces painted? Three faces painted?
   (c) What is the total area of the painted surfaces for all the small cubes? The total area of the unpainted surfaces?

38. In the United States, particulate concentrations in air are routinely measured at about 200 sites. An important property of this suspended matter is the size distribution. For example, a major fraction of particles with diameters of 0.5 micrometer (μm) or less are deposited in the lungs. Are such particles colloidal? What are the diameters in angstroms? In inches? (Given, $1\,\mu$m $= 10^{-6}$ m.)

39. A standard puff of cigarette smoke, 35 mL, is said to contain 2×10^{11} particles in the size range of 1,500 Å to 10,000 Å. Compare this number with the number of molecules in 35 mL of pure air at $0\,°C$ and 760 mm.

40. (a) Based on data given in Section 12.8.3, what is the weight of the tiny nickel cube that is 10 Å on an edge?
   (b) About how many nickel atoms are present in the cube?
   (c) Could you lift Avogadro's number of the tiny cubes?

*41. (a) Calculate the number of atoms in the tiny nickel cube described in Exercise 40, but this time based on the radius of the atom as 1.15 Å. Assume that the packing is simple cubic and that each atom occupies a cubical space that is 2(1.15) Å on an edge.
   (b) Compare your answer with that obtained in Exercise 40. Would the fact that nickel crystallizes in a face-centered cubic pattern account for the differences? Explain.

* 42. (a) A film of stearic acid, $C_{17}H_{35}CO_2H$ (molar wt 284.5 g), is adsorbed on the surface of water with the $-CO_2H$ end dissolved in the water and the hydrocarbon end sticking out. If it requires 4.5×10^{-3} g of stearic acid to cover a surface area of 2.0 m^2 with a film one molecule thick, what is the cross-sectional area of one molecule of stearic acid in the film? Give the answer in square meters and in square angstroms.
   (b) If the density of stearic acid is 0.90 g/cm^3, calculate the film thickness in meters and in angstroms.
   (c) Assuming a molecule of stearic acid to be cylindrical in shape, draw a cylinder and label it to show the length and diameter of the molecule in meters and in angstroms.

Properties of Colloidal Systems

43. Give some examples of water–oil and water–fat emulsions that are stabilized by protective colloids. In each example name the protective colloid and describe how stability is achieved.

44. (a) Separation by distillation is not a good method for separating a mixture of sugars, enzymes, and hormones. Why?
   (b) What method would you suggest for trying to separate mixtures of complex compounds such as sugars, proteins, and vitamins?

Thermochemistry; Thermodynamics

13

13•1 Introduction

Thermochemistry

13•2 Experimental Determination of Heats of Reaction

13•3 Thermochemical Equations

13•4 Standard Enthalpies

13•4•1 Standard States

13•4•2 Hess's Law

13•4•3 Standard Enthalpies of Formation, $\Delta H_f{}^\circ$

13•5 Bond Dissociation Energies

13•5•1 Average Bond Energies

13•5•2 Calculation of Heats of Formation of Radicals from Bond Dissociation Energies

13•6 Enthalpies of Electron Loss or Gain

13•6•1 Enthalpies of Ionization

13•6•2 Enthalpies of Electron Affinity

13•7 Enthalpies of Ionic Crystals

13•7•1 Born–Haber Cycle

Thermodynamics

13•8 The First Law of Thermodynamics

13•8•1 Internal Energy

13•8•2 Heat Changes at Constant Volume and Constant Pressure

13•9 The Criteria for Spontaneous Chemical Processes

13•9•1 Entropy

13•9•2 Free Energy

Special Topic: Entropy and Natural Processes

13•9•3 Free Energy and Reaction Rate

13•9•4 Influence of Temperature on Free Energy Changes

13•9•5 Summary of General Principles

Imagine that a chemist or a chemical engineer working for an industrial concern has been asked to develop an inexpensive and efficient preparation of methyl alcohol, CH_3OH. Of the numerous ways this valuable chemical might be produced, one that might be used is the following:

$$CO + 2H_2 \longrightarrow CH_3OH$$

One of the most important factors to consider is whether such a reaction will occur as written or whether the reverse reaction, the decomposition of methyl alcohol, predominates.

Energy requirements play a key role in determining the direction in which a chemical reaction will occur. Today, with attention focused on our energy needs, major efforts are devoted to searching for endothermic reactions powered by solar heat that yields substances that later can react exothermically to produce energy and the original reactants. Chemists hope to find a cycle of great practical value that is cleanly reversible under controlled conditions. The prediction of the direction and the extent of a reaction under different conditions is possible through the science of thermodynamics.

Thermodynamics, in the broad sense, is a study of the quantitative relationships between heat and other forms of energy, such as the energies associated with electromagnetic, surface, mechanical, and chemical phenomena. The concepts of thermodynamics are of fundamental importance to the engineer, the physicist, and the chemist. While the engineer may be interested primarily in problems of combustion and power and the physicist with radiation and electromagnetic problems, the chemist's prime objective is to determine the feasibility or spontaneity of a given chemical change.

In this chapter we will deal with the relationships between chemical energy and the laws of thermodynamics, laws that describe the behavior of all matter in the universe.

13•1   Introduction

Chemical thermodynamics may be defined as the branch of chemistry that is concerned with the relationship of heat, work, and other forms of energy to equilibria in chemical reactions and changes of state. Closely related to chemical thermodynamics[1] is the subject of **thermochemistry,** which is concerned with the measurement and interpretation of heat changes that accompany chemical reactions, changes of state, and formations of solutions.

Both subjects, thermodynamics and thermochemistry, are essential to chemical understanding. Not only do we often want to know how much energy can be obtained from reactions, but also the study of energy changes is fundamental to the theories of chemical bonding and structure. Because the measurement and calculation of heats of chemical reactions are basic to the understanding of thermodynamics, we will begin this chapter with the subject of thermochemistry.

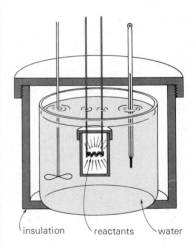

insulation   reactants   water

Figure 13•1 A bomb-type calorimeter.

---

[1]Henceforth, we will refer to chemical thermodynamics simply as thermodynamics, with the implied understanding that our main concern is with the chemical aspects of the subject.

Thermochemistry

13•2 Experimental Determination of Heats of Reaction

Because the standard unit of heat energy has been the calorie for so many years, the device used to measure the heat change during a chemical reaction is called a **calorimeter.** The techniques for its use were developed by Lavoisier and other early chemists and have been refined to great precision today in such laboratories as the National Bureau of Standards.

The two most common experimental methods of thermochemistry are called *combustion calorimetry* and *reaction calorimetry*. In the former method, an element or compound is burned, generally in oxygen, and one determines the energy or heat liberated in the reaction. Reaction calorimetry refers to the determination of the heat of any reaction other than a combustion reaction. This latter method has been used more commonly with inorganic compounds and their solutions.

As applied to organic compounds, combustion calorimetry involves the complete breakdown of the carbon skeleton when the compound burns in oxygen. The combustion method has had wide application with organic compounds that possess low reactivity with reagents other than oxygen, or that give more than one organic product with other reagents. Reaction calorimetry may be used with compounds that readily undergo fairly rapid reactions at moderate temperatures without the formation of unwanted side products.

There are many types of calorimeters that may be used efficiently by the thermochemist. A combustion-type calorimeter is shown in Figure 13.1. A calorimeter for the study of reactions in solution is shown in Figure 13.2. In a calorimeter of the type shown in Figure 13.1, the reaction takes place in a reaction chamber that is immersed in a weighed quantity of water inside an insulated vessel. One way of initiating the reaction inside the sealed chamber is by heating a nonreacting wire coil by the passage of a small amount of electricity. If the reaction is expected to be highly exothermic, the chamber (called a *bomb*) is made of heavy steel to withstand the pressure generated by any hot gases present.

The amount of heat evolved or absorbed is found by placing a weighed quantity of the reactants in the container, allowing the reaction to take place, and then noting the temperature change in the surrounding water. From the weights of the materials involved (water, products of the reaction, and calorimeter), the change in their temperature, and their heat capacities, the amount of heat change during the reaction can be calculated.

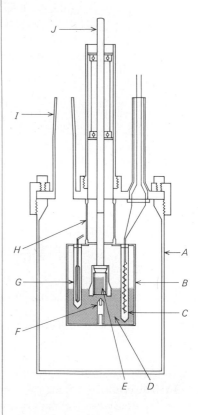

Figure 13•2 A calorimeter for study of reactions in solution: *A*, outer insulated jacket; *B*, calorimeter can; *C*, electric heater for calibration; *D*, one liquid reagent; *E*, other liquid reagent, held in fragile glass ampoule; *F*, spike on which ampoule is broken; *G*, thermistor for temperature measurement; *H*, glass tube for supporting calorimeter; *I*, connection to vacuum pump; *J*, stirrer and ampoule holder. *Source:* Adapted with permission from J. D. Cox and G. Pilcher, *Thermochemistry of Organic and Organometallic Compounds*, Academic Press Inc., London, 1970, p. 84.

• Example 13•1 •

The heat evolved in the reaction between powdered solid aluminum and gaseous oxygen to form solid aluminum oxide is determined in a calorimeter similar to the one shown in Figure 13.1. Suppose previous measurements showed that the materials of the calorimeter other than the water absorb 62.8 J for each 1.00 °C rise in temperature.

With 0.100 g of aluminum and excess oxygen sealed in the bomb and 500 g of water in the calorimeter, the temperature of the water and its contents rises from 20.58 to 22.02 °C after the reaction is complete. Calculate the heat evolved per mole of aluminum reacting.

■ • Solution • The total energy evolved when 0.100 g of aluminum reacts is:

joules evolved = joules absorbed by water + joules absorbed by calorimeter

$$= \left(\frac{4.184 \text{ J}}{g \times {}^{\circ}C}\right)(\text{wt g})[(t_2 - t_1)\ {}^{\circ}C] + \left(\frac{62.8 \text{ J}}{1.00\ {}^{\circ}C}\right)[(t_2 - t_1)\ {}^{\circ}C]$$

$$= [4.184 \times 500 \times (22.02 - 20.58) \text{ J}] + [62.8 \times (22.02 - 20.58) \text{ J}]$$

$$= 3,010 + 90 = 3,100 \text{ J}$$

For 1 mole of aluminum reacting,

$$\text{heat evolved} = \frac{3,100 \text{ J}}{0.100 \text{ g Al}} \times \frac{27.0 \text{ g Al}}{1 \text{ mol Al}}$$

$$= 837,000 \text{ J/mol Al} = 837 \text{ kJ/mol Al}$$

See also Exercise 1 at the end of the chapter.

13·3   **Thermochemical Equations**

When the heat change associated with a chemical reaction is indicated with an equation, the complete statement is referred to as a **thermochemical equation.** Because the physical state is important when energy changes are measured, the letters in parentheses, s, l, and g, specify the solid, liquid, and gaseous state, respectively:

$C(s) + O_2(g) \longrightarrow CO_2(g)$	393.52 kJ evolved	(1)
$N_2(g) + 2O_2(g) \longrightarrow 2NO_2(g)$	66.4 kJ absorbed	(2)
$H_2(g) + \frac{1}{2}O_2(g) \longrightarrow H_2O(l)$	285.83 kJ evolved	(3)

Such equations are interpreted in terms of molar quantities.

Equation (1) shows that when 1 mole (12.0 g) of solid carbon unites with 1 mole (32.0 g) of gaseous oxygen to form 1 mole (44.0 g) of gaseous carbon dioxide, 393.52 kJ of heat is liberated to the surroundings.

In Equation (2), when 1 mole (28.0 g) of gaseous nitrogen reacts with 2 moles (64.0 g) of gaseous oxygen to produce 2 moles (92.0 g) of gaseous nitrogen dioxide, 66.4 kJ of heat is absorbed from the surroundings.

Equation (3) illustrates the use of *fractional coefficients,* which are often used in equations that are to be interpreted in terms of moles rather than molecules: 1 mole (2.0 g) of gaseous hydrogen unites with $\frac{1}{2}$ mole (16.0 g) of gaseous oxygen to form 1 mole (18.0 g) of liquid water and to liberate 285.83 kJ.

Recall that a reaction that evolves heat is an exothermic reaction, and a reaction that absorbs heat is an endothermic reaction. In discussing energy changes during chemical reactions, the chemist finds it convenient to think of each substance as having a certain heat content, or **enthalpy,** H (Greek *enthalpein,* to warm in). The heat change in a chemical reaction is called the **enthalpy change,** ΔH. Strictly, the term *enthalpy change* refers to *the heat change during a process carried out at a constant pressure.* If the energy is to be precisely specified, the initial and final conditions of pressure and temperature must be known.

Reaction (1) is analyzed in the diagram at the left in Figure 13.3. Because heat is given off in this reaction, it is evident that the product, $CO_2(g)$, has a smaller heat content, or enthalpy, than the reactants, $C(s)$ plus $O_2(g)$. The decrease in enthalpy is shown by assigning a minus sign to the value of ΔH:

$$C(s) + O_2(g) \longrightarrow CO_2(g) \qquad \Delta H = -393.52 \text{ kJ}$$

Consider also reaction (3), the formation of water by the reaction of hydrogen with oxygen:

$$H_2(g) + \tfrac{1}{2}O_2(g) \longrightarrow H_2O(l) \qquad \Delta H = -285.83 \text{ kJ}$$

or

$$2H_2(g) + O_2(g) \longrightarrow 2H_2O(l) \qquad \Delta H = -571.66 \text{ kJ}$$

If we assume that each of the three substances $H_2(g)$, $O_2(g)$, and $H_2O(l)$ has its own characteristic heat content, or enthalpy, H, then for this exothermic reaction the sum of the enthalpy of hydrogen plus that of oxygen must be greater than the enthalpy of water (right side of Figure 13.3).

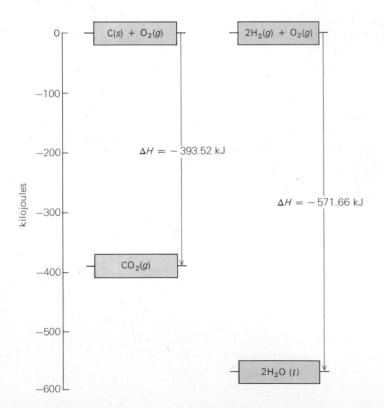

Figure 13•3 On the left is the schematic representation of the enthalpy of combustion of $C(s)$ to give $CO_2(g)$. On the right is the representation of the enthalpy of combustion of $H_2(g)$ to give $H_2O(l)$. The values shown are for the formation of 1 mole of $CO_2(g)$ and 2 moles of $H_2O(l)$.

Utilizing the sign Σ, meaning "the sum of all" or "the summation of," chemists have arbitrarily defined the change in enthalpy, ΔH, of a reaction as

$$\Delta H = \Sigma H \text{ products} - \Sigma H \text{ reactants}$$

When the enthalpy of the reactants is greater than that of the products, a reaction is exothermic. Conversely, if the enthalpy of the products is greater than that of the reactants, a reaction is endothermic. In reaction (2), the formation of nitrogen dioxide by the reaction of nitrogen with oxygen, the enthalpy of 2 moles of $NO_2(g)$ is greater than the enthalpy of 1 mole of $N_2(g)$ and 2 moles of $O_2(g)$. Therefore there is an overall increase in enthalpy and ΔH is positive:

$$N_2(g) + 2O_2(g) \longrightarrow 2NO_2(g) \qquad \Delta H = +66.4 \text{ kJ}$$

The decomposition of water into hydrogen and oxygen is also endothermic:

$$H_2O(l) \longrightarrow H_2(g) + \tfrac{1}{2}O_2(g) \qquad \Delta H = +285.83 \text{ kJ}$$

Comparing this endothermic reaction with the corresponding reverse reaction, reaction (3), we can make the following general statements. *If a reaction is exothermic, the reverse reaction is endothermic. The heat liberated in an exothermic reaction equals the heat absorbed in the reverse, endothermic reaction.*

13·4   Standard Enthalpies

13·4·1   Standard States   The enthalpy changes for chemical reactions depend upon the states of the substances involved. In the formation of carbon dioxide by the combustion of carbon, the ΔH value given is for the reactant solid carbon in the form of graphite. A different ΔH value is obtained if the solid carbon is in the form of diamond. In the formation of water, the ΔH of reaction of -285.83 kJ is for the product water in the liquid state. The ΔH value is -241.82 kJ if the water produced is in the gaseous state.

F. D. Rossini, an outstanding thermochemist who has worked at the National Bureau of Standards, has said that ". . . the aim of thermochemistry is to provide the experimental data for compiling a table of values from which may be calculated the heat of every possible chemical reaction." In the tabulation of values of ΔH for achieving this goal, it is the accepted practice to specify that the heats of reactions refer to reactions with the elements and compounds in certain *standard states*. For a solid or liquid, the standard state is the pure substance at 1 atm pressure; for a gas, it is the hypothetical ideal gas at 1 atm partial pressure.

The symbol ΔH_r° refers to the ΔH for a reaction in which each reactant and product is in its standard state at a specified reference temperature. Although ΔH_r° may be determined at any temperature, unless specified otherwise, the common reference temperature of 25 °C is assumed in this text. The values of $\Delta H_r°$ are for the balanced equation as written, assuming molar quantities are involved. For example, in the following reaction (at 25 °C),

$$H_2(g) + \tfrac{1}{2}O_2(g) \longrightarrow H_2O(l) \qquad \Delta H_r° = -285.83 \text{ kJ}$$

the $\Delta H_r°$ value is per 1 mole of $H_2(g)$ or $\tfrac{1}{2}$ mole of $O_2(g)$ reactants, or per 1 mole of $H_2O(l)$ product.

13•4•2   **Hess's Law** In 1840 the Swiss-Russian chemist G. H. Hess stated one of the most useful generalizations of thermochemistry. A modern version of **Hess's law** is, *for a given overall reaction, the change in enthalpy is always the same, whether the reaction is performed directly or whether it takes place indirectly and in different steps.*

As an example of Hess's law, consider the exothermic reaction between sulfur and oxygen to produce sulfur dioxide, followed by the exothermic reaction between sulfur dioxide and more oxygen to produce sulfur trioxide:

$$S(s) + O_2(g) \longrightarrow SO_2(g) \qquad \Delta H_r° = -296.83 \text{ kJ}$$
$$SO_2(g) + \tfrac{1}{2}O_2(g) \longrightarrow SO_3(g) \qquad \Delta H_r° = -98.9 \text{ kJ}$$

If these two steps are considered to take place as a simple one-step overall reaction, the heat evolved is the sum of the two steps:

$$S(s) + 1\tfrac{1}{2}O_2(g) \longrightarrow SO_3(g) \qquad \Delta H_r° = -395.73 \text{ kJ}$$

One useful consequence of Hess's law is that thermochemical equations can be added or subtracted to produce data that are difficult to determine experimentally. For example, carbon and carbon monoxide are important commercial fuels; therefore, it is of interest to compare the amount of heat liberated when carbon is burned to carbon dioxide with the amount of heat liberated when carbon is burned to carbon monoxide. The latter enthalpy change, or heat of reaction, is difficult to determine, because carbon monoxide is burned more readily than carbon. Consequently, when we burn carbon in the theoretical amount of oxygen needed to form carbon monoxide, we actually get a mixture of carbon dioxide, carbon monoxide, and unburned carbon. However, we can calculate the enthalpy change for the combustion of carbon to carbon monoxide from the enthalpy changes for two reactions that are easily carried out. These are the combustion of carbon to carbon dioxide and of carbon monoxide to carbon dioxide:

$$\Delta H_r°, \text{ kJ}$$
$$C(s) + O_2(g) \longrightarrow CO_2(g) \qquad -393.52 \qquad (4)$$
$$CO(g) + \tfrac{1}{2}O_2(g) \longrightarrow CO_2(g) \qquad -283.0 \qquad (5)$$

If we write the reverse reaction for Equation (5) with the corresponding change in sign for $\Delta H_r°$ and then add this equation to Equation (4), we obtain the desired information:

$$\Delta H_r°, \text{kJ}$$

$$C(s) + O_2(g) \longrightarrow CO_2(g) \qquad -393.52$$
$$CO_2(g) \longrightarrow CO(g) + \tfrac{1}{2}O_2(g) \qquad +283.0$$
$$\overline{C(s) + \tfrac{1}{2}O_2(g) \longrightarrow CO(g) \qquad -110.52}$$

Hence, $\Delta H_r°$ for the combustion of carbon to carbon monoxide is calculated to be -110.52 kJ per mole of $CO(g)$ formed.

• Example 13•2 •

Using the $\Delta H_r°$ values already given in this section for the combustion of carbon to carbon dioxide and of hydrogen to water, along with that for the combustion of methane, $CH_4(g)$,

$$CH_4(g) + 2O_2(g) \longrightarrow CO_2(g) + 2H_2O(l) \qquad \Delta H_r° = -890.37 \text{ kJ}$$

calculate the value of $\Delta H_r°$ for the following reaction:

$$C(s) + 2H_2(g) \longrightarrow CH_4(g)$$

• Solution • The overall aim is to use Hess's law by arranging known equations for addition in such a way that the resulting equation has $C(s)$ and $2H_2(g)$ as the reactants and $CH_4(g)$ as the product. Consider the following three equations:

$$\Delta H_r°, \text{kJ}$$

$$C(s) + O_2(g) \longrightarrow CO_2(g) \qquad -393.52$$
$$H_2(g) + \tfrac{1}{2}O_2(g) \longrightarrow H_2O(l) \qquad -285.83$$
$$CH_4(g) + 2O_2(g) \longrightarrow CO_2(g) + 2H_2O(l) \qquad -890.37$$

The first equation has the $C(s)$ we need as a reactant, so we write it down as is; to provide $2H_2(g)$, we multiply the second equation by 2; and to provide $CH_4(g)$ as a product, we reverse the third reaction. Finally, we add the three equations:

$$\Delta H_r°, \text{kJ}$$

$$C(s) + O_2(g) \longrightarrow CO_2(g) \qquad -393.52$$
$$2H_2(g) + O_2(g) \longrightarrow 2H_2O(l) \qquad -571.66$$
$$\underline{CO_2(g) + 2H_2O(l) \longrightarrow CH_4(g) + 2O_2(g) \qquad +890.37}$$
$$C(s) + 2H_2(g) \longrightarrow CH_4(g) \qquad -74.81$$

Hence, $\Delta H_r°$ for the formation of 1 mole of methane from carbon and hydrogen is -74.81 kJ.

See also Exercise 4.

13•4•3 **Standard Enthalpies of Formation, $\Delta H_f°$** The enthalpy of a substance is an example of a **state function,** that is, a function that depends only on the present state of the substance and not on the path by which the present state was attained. In the example of sulfur trioxide referred to previously, 1 mole of sulfur trioxide in its standard state has 395.7 kJ less enthalpy than the elements of which it is composed. This change in enthalpy is definite. It is the same if the sulfur trioxide is formed from the elements directly or in two steps.

The absolute enthalpies of substances are unknown, but relative values of enthalpies may be obtained by choosing a reference state for enthalpy and then measuring precisely the enthalpy changes of reactions. The universal reference state is the element in its standard state at 25 °C. By convention, the enthalpy of formation of an element in its standard state is assigned a value of *zero*.[2]

Consider again the reaction of carbon with oxygen to give carbon dioxide:

$$C(s) + O_2(g) \longrightarrow CO_2(g) \qquad \Delta H_r^\circ = -393.52 \text{ kJ}$$

The value of -393.52 kJ is not the absolute enthalpy of 1 mole of $CO_2(g)$ in its standard state, but it is the enthalpy of the compound relative to the enthalpies of the elements in their standard states. Only changes in enthalpy may be obtained in thermochemistry, but that is all we need to know to compare reactants and products. For the special case involving a reaction of elements, the heat of reaction is the heat of formation of the compound produced, that is, $\Delta H_r^\circ = \Delta H_f^\circ$. The **standard enthalpy of formation,** ΔH_f°, of a substance is defined as the change in enthalpy for the reaction in which 1 mole of the compound is formed from its elements at standard conditions. Figure 13.4 illustrates graphically how the ΔH_f° of

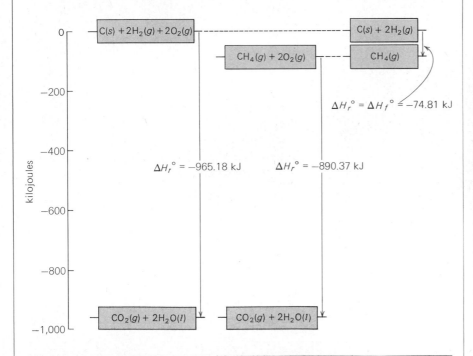

Figure 13•4 Schematic representation of how the enthalpy of formation, ΔH_f°, of $CH_4(g)$ is related to the ΔH_f°'s of five other substances. The ΔH_r° of -965.18 kJ is obtained by adding the ΔH_f° of $CO_2(g)$ to twice the ΔH_f° of $H_2O(l)$ (see also Figure 13.3 and Example 13.2). Because the products of the combustion of $CH_4(g)$ and of $C(s) + 2H_2(g)$ are the same, the difference in the heats of reaction must be due to the difference between the heats of formation of the reactants. Because the ΔH_f° of every element is assigned a relative value of zero by convention, this difference is attributed to the ΔH_f° of $CH_4(g)$, which has an enthalpy 74.81 kJ/mole less than that of 1 mole of $C(s)$ and 2 moles of $H_2(g)$.

---

[2] If the element exists in more than one form, zero enthalpy is assigned to the most stable allotrope. In the case of oxygen, for example, ordinary oxygen, O_2, is assigned zero enthalpy; ozone, O_3, has a relative heat content greater than zero.

methane is related to the standard enthalpies of formation of the elements (chosen as zero) and to the standard enthalpies of formation of carbon dioxide and water. Thousands of standard enthalpies of formation have been measured or calculated; some representative ones are listed in Table 13.1.

An examination of Table 13.1 reveals several items of interest. First, we see that when a compound is formed from its elements, the reaction may be either exothermic or endothermic. Compounds formed from their elements in exothermic reactions (ΔH_f° is negative) are called exothermic compounds; the others are called endothermic compounds. Endothermic

Table 13.1 Enthalpies of formation, ΔH_f°, in kJ/mole, at 25 °C

Substance	ΔH_f°	Substance	ΔH_f°
$AgCl(s)$	−127.07	$HBr(g)$	−36.4
$AgCN(s)$	146	$HCl(g)$	−92.31
$AlBr_3(s)$	−511.12	$HF(g)$	−271.1
$AlCl_3(s)$	−705.63	$HI(g)$	26.5
$AlF_3(s)$	−1,510	$HCN(g)$	135
$AlI_3(s)$	−310	$HN_3(g)$	294
$B_2H_6(g)$	36	$H_2O(l)$	−285.83
$BrCl(g)$	14.6	$H_2O(g)$	−241.82
$CH_4(g)$	−74.81	$H_2O_2(l)$	−187.8
$C_2H_2(g)$	226.7	$H_2O_2(g)$	−136.1
$C_2H_4(g)$	52.26	$H_2S(g)$	−20.2
$C_2H_6(g)$	−84.68	$H_2SO_4(l)$	−813.99
$C_6H_6(l)$	48.99	$ICl(g)$	17.8
$CH_3NH_2(g)$	−23.0	$LiF(s)$	−616.93
$CH_3OH(l)$	−239.0	$NH_3(g)$	−46.11
$CH_3OH(g)$	−201.1	$N_2H_4(l)$	50.63
$CO(g)$	−110.5	$NH_4Br(s)$	−270.8
$CO_2(g)$	−393.52	$NH_4Cl(s)$	−314.4
$CS_2(g)$	117.1	$NH_4F(s)$	−463.96
$ClF(g)$	−54.48	$NH_4I(s)$	−201.4
$CaO(s)$	−635.13	$NaCl(s)$	−411.0
$Ca(OH)_2(s)$	−986.17	$NaCl(aq)$	−407.1
$CaCO_3(s)$	−1,207	$NO(g)$	90.25
$CaCl_2(s)$	−795.8	$NO_2(g)$	33.2
$CaCl_2 \cdot H_2O(s)$	−1,109	$O_3(g)$	143
$CaCl_2 \cdot 2H_2O(s)$	−1,403	$SO_2(g)$	−296.83
$Fe_2O_3(s)$	−824.2	$SO_3(g)$	−395.7

Source: *Data are calculated mainly from* Circular of the National Bureau of Standards 500, *Washington, D.C., 1952, and from* Lange's Handbook of Chemistry, *12th ed., J. A. Dean, Ed., McGraw-Hill, New York, 1979. Used with permission of McGraw-Hill.*

compounds are often unstable. Hydrogen azide, HN_3, explodes violently when heated, decomposing into hydrogen and nitrogen with the evolution of 294 kJ/mole. Acetylene, C_2H_2, is also explosive at room temperature as a pure liquid. It is commercially available as a fuel gas in cylinders as a stable solution in acetone. An endothermic compound, such as acetylene, is a relatively efficient fuel. It gives off more heat in its combustion than it would if it were an exothermic compound.

A second generalization that is apparent in Table 13.1 is that an ionic compound usually has a more negative enthalpy of formation than a covalent compound has. Some covalent compounds, such as HF, are exceptions to this rule, but as a general guide it is sound. Of course, we have to make this comparison in terms of a similar number of atoms involved per mole.

A third item to note in Table 13.1 is that the physical state of a compound may make a considerable difference in its enthalpy. In the case of water, the difference between the gaseous and liquid states is 44.01 kJ/mole. In changing from a gas to a liquid, a substance loses enthalpy equivalent to its heat of vaporization and so has a lower enthalpy in the liquid state.

• *Calculation of Standard Enthalpies of Reaction from Standard Enthalpies of Formation* • The data in Table 13.1 are typical of those that can be used to calculate standard heats of reaction, $\Delta H_r°$, for all sorts of reactions, both actual and hypothetical. The equation used is similar to one stated previously:

$$\Delta H_r° = \Sigma \Delta H_f° \text{ products} - \Sigma \Delta H_f° \text{ reactants} \tag{6}$$

A few examples illustrate the method of calculation.

• **Example 13•3** •

Calculate $\Delta H_r°$ for the reaction between carbon monoxide, $CO(g)$, and hydrogen, $H_2(g)$, to give methyl alcohol, $CH_3OH(l)$.

• **Solution** • Write a balanced equation for the reaction and list the $\Delta H_f°$ value for each substance:

$$CO(g) + 2H_2(g) \longrightarrow CH_3OH(l)$$
$$-110.5 \qquad 0 \qquad\qquad -239.0$$
$$\Delta H_r° = (-239.0) - (-110.5 + 0)$$
$$= -128.5 \text{ kJ}$$

See also Exercises 7–12.

• **Example 13•4** •

Calculate $\Delta H_r°$ for the combustion of ammonia, $NH_3(g)$, to give nitric oxide, $NO(g)$, and water, $H_2O(l)$.

• **Solution** •

$$4NH_3(g) + 5O_2(g) \longrightarrow 4NO(g) + 6H_2O(l)$$
$$4(-46.11) \qquad 0 \qquad\quad 4(90.25) \quad 6(-285.83)$$
$$\Delta H_r° = [361.00 + (-1,714.98)] - (-184.44 + 0)$$
$$= -1,169.54 \text{ kJ}$$

This is the standard enthalpy change for the combustion of 4 moles of NH_3. For the combustion of 1 mole of NH_3,

$$\Delta H_r^\circ = \frac{-1,169.54}{4} = -292.38 \text{ kJ}$$

See also Exercises 7–12.

• *Calculation of Standard Enthalpies of Formation from Standard Enthalpies of Reaction* • By definition, the standard heat of formation of methane is expressed as follows:

$$C(s) + 2H_2(g) \longrightarrow CH_4(g) \qquad \Delta H_r^\circ = \Delta H_f^\circ = -74.81 \text{ kJ}$$

However, this reaction cannot be carried out at present under any known experimental conditions. The value of -74.81 kJ was obtained by the addition of known thermochemical equations according to Hess's law (see also Figure 13.4). Equation (6) is a convenient mathematical expression of Hess's law and may be used to calculate the ΔH_f° of a compound from the experimentally determined ΔH_r° and the heats of formation of all other participants in the reaction. The heats of formation of most hydrocarbons and oxygen derivatives of hydrocarbons (for example, alcohols and ethers) have been determined in this manner. Example 13.5 illustrates this approach.

• Example 13.5 •

Calculate ΔH_f° of octane, $C_8H_{18}(l)$, from the experimental ΔH_r° value of $-5,470.68$ kJ/mole for its combustion.

• Solution •

$$C_8H_{18}(l) + 12\tfrac{1}{2}O_2 \longrightarrow 8CO_2(g) + 9H_2O(l)$$
$$x \qquad\qquad 0 \qquad\qquad 8(-393.52) \quad 9(-285.83)$$

$$-5,470.68 = \Delta H_r^\circ$$
$$-5,470.68 = [(-3,148.16) + (-2,572.47)] - (x + 0)$$
$$x = [(-3,148.16) + (-2,572.47)] + 5,470.68$$
$$x = -249.95 \text{ kJ} = \Delta H_f^\circ, \; C_8H_{18}(l)$$

See also Exercises 13 and 14.

In all the foregoing examples, the values of ΔH_r° are for the conversion of reactants in their standard states at 25 °C to the products at the same conditions. In actual laboratory practice, the thermochemist may use a calorimeter similar to that shown schematically in Figure 13.1 to measure heats of reaction. This is especially true for combustions of organic compounds. With such equipment, the reaction is carried out not at constant pressure, but at constant volume in a small steel bomb. The final pressure in the bomb may be very high if most of the products tend to be gases, and the physical states may differ from those that would be obtained at constant pressure.

However, from the heat change measured under the condition of constant volume, we can calculate the enthalpy change for the reaction as though it had been carried out in an open container at constant pressure, with all substances in their standard states at 25 °C. To do this, we must consider the heat changes involved in the expansion or compression of gases, in the heating or cooling of substances, and in any solid–liquid–gas transformations. Extensive tables are available of heats of compression, heat capacities (or specific heats), heats of fusion (ΔH_{fus}), and heats of vaporization (ΔH_{vap}) for all kinds of substances under many different conditions. Using these data, the measured heat changes for thousands of reactions have been corrected to show what the standard enthalpy change, $\Delta H_r°$, would be if the reaction were carried out under the conditions of constant temperature and pressure at 25 °C and 1 atm. Conversely, changes in heat content for reactions at other conditions can be calculated from standard enthalpies by accounting for the differences in enthalpies of all substances at standard conditions compared with the specified reaction conditions.

13•5 Bond Dissociation Energies

A process that is of fundamental importance in interpreting chemical reactions is the dissociation of molecules into atoms and free radicals. For example, the reaction

$$CH_4(g) + Cl_2(g) \longrightarrow CH_3Cl(g) + HCl(g)$$

occurs in a series of steps, one of which involves the breaking of the bond in a molecule of chlorine to form two chlorine atoms:

$$Cl_2(g) \longrightarrow 2Cl(g)$$

The energies necessary for many such dissociation reactions have been measured. Examples for diatomic molecules are listed in Table 13.2. For diatomic molecules, the **bond dissociation energy,** $\Delta H_{dis}°$, is the amount of energy per mole required to break the bond and produce two atoms, with reactants and products being ideal gases in their standard states at 25 °C. For the dissociation of molecular chlorine into chlorine atoms, the $\Delta H_{dis}°$ is +242.6 kJ/mole. This value is a measure of the strength of the covalent bond joining the two atoms. The larger the value of $\Delta H_{dis}°$, the stronger is the bond joining the two atoms.

It is often convenient to speak of the **heat of formation of an atom,** which is defined as the amount of energy required to form 1 mole of gaseous atoms from the element in its common physical state at 25 °C and 1 atm pressure. In the case of diatomic gaseous molecules of elements, the $\Delta H_f°$ of an atom is equal to half the value of the bond dissociation energy; that is,

$$\Delta H_f° = \tfrac{1}{2}\Delta H_{dis}°$$

For the formation of chlorine atoms from chlorine molecules, $\Delta H_f° = +121.3$ kJ/mole of atoms. Some values for elements that exist as diatomic molecules in their standard states and those that do not are included in Table 13.3.

The bond dissociation energy for any chemical bond, A—B, may be calculated by use of Equation (7), which is a natural extension of Equation (6):

Table 13•2 Bond dissociation energies,[a] $\Delta H_{dis}°$, of diatomic molecules, in kJ/mole, at 25 °C

Molecule	$\Delta H_{dis}°$
H—H(g)	436.0
N≡N(g)	945.3
O—O(g)	498.3
F—F(g)	157
Cl—Cl(g)	242.6
Br—Br(g)	193.9
I—I(g)	152.6
H—F(g)	567.6
H—Cl(g)	431.6
H—Br(g)	366.3
H—I(g)	298.3
Cl—F(g)	254.3
Cl—Br(g)	218.6
Cl—I(g)	210.3

[a] Most of the values for the seven elements are calculated from those in the *Handbook of Chemistry and Physics,* R. C. Weast, Ed., © CRC Press, Inc., 1976. Used by permission of CRC Press, Inc. The values for the seven compounds are calculated by use of Equation (7).

$$AB(g) \longrightarrow A(g) + B(g)$$

$$\Delta H_{dis}^{\circ}\, A\text{—}B \;=\; \Delta H_f^{\circ}\, A(g) + \Delta H_f^{\circ}\, B(g) - \Delta H_f^{\circ}\, AB(g) \tag{7}$$

For hydrogen chloride, HCl, ΔH_{dis}° is calculated from data in Tables 13.1 and 13.3:

$$HCl(g) \longrightarrow H(g) + Cl(g)$$

$$
\begin{aligned}
\Delta H_{dis}^{\circ}\, H\text{—}Cl &= \Delta H_f^{\circ}\, H(g) + \Delta H_f^{\circ}\, Cl(g) - \Delta H_f^{\circ}\, HCl(g) \\
&= 218.0 + 121.3 - (-92.31) = 431.6 \text{ kJ}
\end{aligned}
$$

13.5.1 Average Bond Energies Consider the dissociation of 1 mole of ammonia into 3 moles of hydrogen atoms and 1 mole of nitrogen atoms:

$$NH_3(g) \longrightarrow 3H(g) + N(g)$$

The energy required to accomplish this process, ΔH_r°, is calculated as follows:

$$
\begin{aligned}
\Delta H_r^{\circ} &= 3\,\Delta H_f^{\circ}\, H(g) + \Delta H_f^{\circ}\, N(g) - \Delta H_f^{\circ}\, NH_3(g) \\
&= 3(218.0) + 472.6 - (-46.11) = 1{,}172.7 \text{ kJ}
\end{aligned}
$$

This value of 1,172.7 kJ is the total energy required to break the three N—H bonds in 1 mole of ammonia. One third of this value, $1{,}172.7/3 = 390.9$ kJ, is the average bond energy per mole of N—H bonds.

For polyatomic molecules, the **average bond energy,** $\Delta H_{dis\ avg}^{\circ}$ is the average energy per bond required to dissociate 1 mole of molecules into their constituent atoms. It is important to distinguish between the average bond energy, $\Delta H_{dis\ avg}^{\circ}$, and an individual bond dissociation energy, ΔH_{dis}°. In ammonia, there are three different individual bond dissociation energies for the stepwise dissociations:

	ΔH_{dis}°, kJ
$NH_3(g) \longrightarrow NH_2(g) + H(g)$	431
$NH_2(g) \longrightarrow NH(g) + H(g)$	381
$NH(g) \longrightarrow N(g) + H(g)$	360

The average of these three values, 391 kJ, is the average bond energy.

Consider also the case of methane, with its four bonds:

$$CH_4(g) \longrightarrow C(g) + 4H(g)$$

$$
\begin{aligned}
\Delta H_{dis\ avg}^{\circ} &= \tfrac{1}{4}[\Delta H_f^{\circ}\, C(g) + 4\,\Delta H_f^{\circ}\, H(g) - \Delta H_f^{\circ}\, CH_4(g)] \\
&= \tfrac{1}{4}[716.7 + 4(218.0) - (-74.81)] \\
&= \tfrac{1}{4}(1{,}663.5) = 415.9 \text{ kJ per mole of C—H bonds}
\end{aligned}
$$

Average bond energies are often useful for certain thermochemical calculations, but they have no real physical significance.

13.5.2 Calculation of Heats of Formation of Radicals from Bond Dissociation Energies The determination of bond dissociation energies and heats of formation of radicals are complementary problems. One can use Equation (7) to calculate the ΔH_f° of radicals from known values of ΔH_{dis}° and ΔH_f° of

Table 13.3 Heats of formation[a] of gaseous atoms, in kJ/mole, at 25 °C

Atom	ΔH_f°
H	218.0
N	472.6
O	249.2
F	78.5
Cl	121.3
Br	111.9
I	106.8
B	571.1
C	716.7
S	277.4

[a] Values are calculated from those in the *Handbook of Chemistry and Physics,* R. C. Weast, Ed., © CRC Press, Inc., 1976. Used by permission of CRC Press, Inc. (The values of ΔH_f° for the first five entries are half the corresponding ΔH_{dis}° values given in Table 13.2.) For the second five entries, heats of vaporization and sublimation are taken into account as well as dissociations of more complex molecules in the cases of boron, carbon, and sulfur; bromine is a liquid and the other four are solids at 25 °C.

other atoms and radicals. This principle is illustrated as follows for the calculation of $\Delta H_f°$ of the methyl radical, $CH_3(g)$:

$$CH_4(g) \longrightarrow CH_3(g) + H(g) \qquad \Delta H_{dis}° = 435.1 \text{ kJ}$$

$$\Delta H_{dis}° \text{ H—CH}_3 = \Delta H_f° \text{ CH}_3(g) + \Delta H_f° \text{ H}(g) - \Delta H_f° \text{ CH}_4(g)$$

$$435.1 = \Delta H_f° \text{ CH}_3(g) + 218.0 - (-74.81)$$

$$\Delta H_f° \text{ CH}_3(g) = 435.1 - 218.0 - 74.81 = 142.3 \text{ kJ}$$

Once the heats of formation of radicals are obtained, they can be used to calculate bond dissociation energies in polyatomic molecules. Example 13.6 illustrates such a calculation.

• Example 13•6 •

Calculate the carbon–carbon bond dissociation energy in ethane, C_2H_6.

• Solution • The reaction under consideration is

$$C_2H_6(g) \longrightarrow 2CH_3(g)$$

and the formula for the desired bond dissociation energy is

$$\Delta H_{dis}° \text{ CH}_3\text{—CH}_3 = 2 \Delta H_f° \text{ CH}_3(g) - \Delta H_f° \text{ CH}_3\text{CH}_3(g)$$

$$= 2(142.3) - (-84.68)$$

$$= 369.3 \text{ kJ}$$

See also Exercises 26–28.

13•6 Enthalpies of Electron Loss or Gain

13•6•1 Enthalpies of Ionization The removal of an electron from a neutral atom requires energy:

$$\text{Na} + \text{energy of ionization} \longrightarrow \text{Na}^+ + e^-$$

$$\text{Cl} + \text{energy of ionization} \longrightarrow \text{Cl}^+ + e^-$$

The ionization energies of a number of elements are listed in Table 3.6 in terms of electron volts per atom. To calculate the ionization energy in kilojoules per mole of atoms, that is, the **enthalpy of ionization,** ΔH_{ion}, from the ionization energy per atom, the conversion factor $1 \text{ eV} = 1.602 \times 10^{-22} \text{ kJ}$ is needed. The enthalpy of ionization for sodium is calculated from the ionization energy of 5.14 eV per atom in the following way:

$$\Delta H_{ion} = \left(\frac{5.14 \text{ eV}}{\text{atom}}\right)\left(\frac{1.602 \times 10^{-22} \text{ kJ}}{1 \text{ eV}}\right)\left(\frac{6.022 \times 10^{23} \text{ atoms}}{1 \text{ mol}}\right)$$

$$= \left(\frac{5.14 \text{ eV}}{\text{atom}}\right)\left(96.47 \frac{\text{kJ} \times \text{atom}}{\text{mol} \times \text{eV}}\right)$$

$$= 496 \text{ kJ/mol of sodium}$$

The sign of ΔH_{ion}, the enthalpy of ionization, is positive, indicating that energy has been absorbed in the process. The factor 96.47 can be used to multiply any value in Table 3.6 to convert it from electron volts per atom to kilocalories per mole.[3]

13•6•2   **Enthalpies of Electron Affinity**   Another fundamental process involving atoms and electrons is symbolized by the reaction $X + e^- \rightarrow X^-$ in which a neutral atom gains one electron to become a negative ion. The electron affinity is usually defined as the energy released when an atom gains one electron. However, this process is difficult to carry out experimentally, so the energy absorbed in the reverse reaction is actually measured to determine the electron affinity. We will define the **electron affinity**, ΔH_{ea}, as the enthalpy change for the loss of an electron by an ion with a charge of -1:

$$X^- \longrightarrow X + e^- \qquad \Delta H_{ea}$$

As we might expect, the electron affinity is positive and high for nonmetals, the elements that normally gain electrons in chemical reactions. But it is somewhat surprising to find that energy is also required to knock an electron off the negative ion of an active metal, such as Na^-. Negative metallic ions are formed only in certain environments such as in vacuum tubes; they cannot exist in ordinary environments where they are in contact with elements having higher electron affinities. Some examples of electron affinities are given in Table 13.4.

13•7   **Enthalpies of Ionic Crystals**

Enthalpies of dissociation, such as those we considered previously for such molecular substances as Cl_2, HCl, and NH_3, are of most interest in discussing covalent bond energies. For ionic compounds, the bonding energy is best approached by considering a large number of ions in a crystal. The **crystal energy** (or lattice energy), ΔH_{xtal}, of a substance is the enthalpy change per mole when widely separated gaseous ions come together to form a solid, crystalline substance. The enthalpy change in this process is a measure of how strongly the ions are bound together.

The crystal energy of an ionic compound can be calculated theoretically on the basis of the charges, sizes, and arrangements of the ions in the solid, or it can be calculated on the basis of the measured energies of other reactions. The latter technique, called the Born–Haber method, was developed by the eminent German scientists Max Born and Fritz Haber in 1919.

13•7•1   **Born–Haber Cycle**   The Born–Haber method makes it possible for us to use Hess's law to relate many of the thermochemical reactions we have been discussing. A **Born–Haber cycle** is a series of changes in a closed cycle in which the net change in energy must be zero.

Table 13•4 Some electron affinities at 0 K in kJ/mole of the anion[a]

Reaction	ΔH_{ea}
$F^- \longrightarrow F + e^-$	327.9
$Cl^- \longrightarrow Cl + e^-$	348.6
$Br^- \longrightarrow Br + e^-$	324.4
$I^- \longrightarrow I + e^-$	295.3
$N^- \longrightarrow N + e^-$	0
$O^- \longrightarrow O + e^-$	141.3
$S^- \longrightarrow S + e^-$	200.4
$Na^- \longrightarrow Na + e^-$	52.9
$K^- \longrightarrow K + e^-$	48.4

[a]Values are calculated from those given in electron volts by E. C. M. Chen and W. E. Wentworth, *J. Chem. Educ.*, **52**, 486 (1975).

---

[3]Calculations based on the data in Table 3.6 yield enthalpies of ionization at 0 K; values at 25 °C are about 6 kJ higher per mole of electrons removed.

A Born–Haber cycle for an alkali metal halide, MX, is represented in Figure 13.5. To distinguish the various molar enthalpy changes, ΔH, we use the following definitions and symbols:

Change in Enthalpy, as Defined	*Symbol*
formation, elements $\longrightarrow$ solid compound	ΔH_f
atomization, solid metal $\longrightarrow$ gaseous metal atoms	$\Delta H_{a,m}$
atomization, nonmetal $X_2 \longrightarrow$ gaseous nonmetal atoms	$\Delta H_{a,n}$
ionization, gaseous metal atom M $\longrightarrow$ ion $M^+ + e^-$	ΔH_{ion}
electron affinity, gaseous negative ion $X^- \longrightarrow$ atom $X + e^-$	ΔH_{ea}
crystal energy, gaseous ions $\longrightarrow$ crystal	ΔH_{xtal}

Note that $\Delta H_{a,m}$ is the same as the enthalpy of sublimation, ΔH_{sub}, for a metal that sublimes directly into monatomic gaseous atoms from its standard solid state. And $\Delta H_{a,n}$ is $\frac{1}{2}\Delta H_{dis}$ (see Table 13.2) for an element that is a diatomic gas in its standard state. Fluorine and chlorine are such elements, but bromine and iodine are not (see footnote b of Table 13.3).

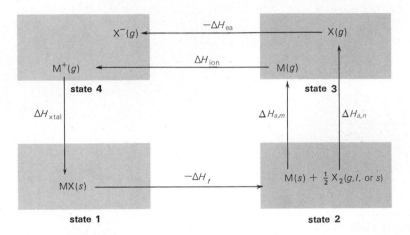

Figure 13•5 The Born–Haber cycle. The negative signs applied to ΔH_f and ΔH_{ea} show that the process occurring in the cycle as drawn is the reverse of the process as defined.

To illustrate the use of the Born–Haber cycle, we can consider different processes, using Figure 13.5 as a guide. First, suppose we start at state 1 with 1 mole of the solid compound at 25 °C and 1 atm, and pass successively through states 2, 3, 4, and back to state 1. Because the sum of all these enthalpy changes must be zero, we can write

$$-\Delta H_f + \Delta H_{a,m} + \Delta H_{a,n} + \Delta H_{ion} + (-\Delta H_{ea}) + \Delta H_{xtal} = 0$$

To assign the proper sign to each enthalpy quantity, we follow the code suggested by Figure 13.5. Note that the minus signs in $(-\Delta H_f)$ and $(-\Delta H_{ea})$ simply indicate the reverse of the defined processes, that is, decomposition rather than formation and electron gain rather than electron loss. Hence, when we need values of $(-\Delta H_f)$ and $(-\Delta H_{ea})$, we change the signs given in the tables for ΔH_f and ΔH_{ea}. For example, ΔH_f for NaCl is -411.0 kJ; hence $(-\Delta H_f)$, for the reverse process, is $+411.0$ kJ; or ΔH_{ea} for Cl^- is $+348.6$ kJ, so $(-\Delta H_{ea})$, for the reverse process, is -348.6 kJ.

Next, suppose we consider starting at state 1 and going to state 4 either directly ($1 \rightarrow 4$) or indirectly ($1 \rightarrow 2 \rightarrow 3 \rightarrow 4$). According to Hess's law, the enthalpy change by one path must equal that by any other, so that we can write

$$(1 \longrightarrow 4) = (1 \longrightarrow 2 \longrightarrow 3 \longrightarrow 4)$$

or

$$-\Delta H_{xtal} = -\Delta H_f + \Delta H_{a,m} + \Delta H_{a,n} + \Delta H_{ion} + (-\Delta H_{ea})$$

We see that this form of the Born–Haber relationship is algebraically the same as the first form we wrote.

At the time the Born–Haber cycle was developed, the most difficult quantity in the cycle to determine experimentally or to calculate theoretically was the electron affinity, ΔH_{ea}. The equation was used in the form

$$\Delta H_{ea} = -\Delta H_f + \Delta H_{a,m} + \Delta H_{a,n} + \Delta H_{ion} + \Delta H_{xtal}$$

to calculate electron affinities. In time, experimental determinations of ΔH_{ea} began to be made, and these were found to be in fairly good agreement with the previously calculated values. Today, the Born–Haber cycle is used to calculate thermochemical quantities not previously measured, especially crystal energies, ΔH_{xtal}. The approach has been standardized and proved valid by application to the alkali halides, a class of compounds that has been studied intensively for many years. For most other types of compounds, fewer data and less precise data are available, so that the correlations are not always successful.

Table 13.5 lists representative values of various enthalpy relations for the alkali halides. The values for the crystal energies, ΔH_{xtal}, have been adjusted to give constant values for ΔH_{ea} for each halogen and to make each Born–Haber cycle have a net enthalpy change of zero. To illustrate, the value

Table 13.5 Born–Haber data[a] for alkali halides, in kJ/mole

		Li	Na	K	Rb	Cs
	$\Delta H_{a,m} =$	161	107.7	89.1	85.8	76.6
	$\Delta H_{ion} =$	520.1	495.8	418.8	403.0	375.7
F	$\Delta H_{ea} = 327.9$	$\Delta H_f = -616.9$	-569.0	-562.6	-549.3	-530.9
	$\Delta H_{a,n} = 78.5$	$\Delta H_{xtal} = -1,049$	-923.1	-821.1	-788.7	-733.8
Cl	$\Delta H_{ea} = 348.6$	$\Delta H_f = -408.8$	-411.0	-435.9	-430.5	-433.0
	$\Delta H_{a,n} = 121.3$	$\Delta H_{xtal} = -863$	-787.2	-716.5	-692.0	-658.0
Br	$\Delta H_{ea} = 324.4$	$\Delta H_f = -350.3$	-359.9	-392.2	-389.2	-394.6
	$\Delta H_{a,n} = 111.9$	$\Delta H_{xtal} = -819$	-750.9	-687.6	-665.5	-634.4
I	$\Delta H_{ea} = 295.3$	$\Delta H_f = -297.9$	-288.0	-327.6	-328.4	-336.8
	$\Delta H_{a,n} = 106.8$	$\Delta H_{xtal} = -790$	-703.0	-647.0	-628.7	-600.6

[a] All values of ΔH_{ion} and ΔH_{ea} are for 0 K, that is, about 6 kJ less than the values at 25 °C. These two differences cancel one another in the Born–Haber cycle calculations.

of ΔH_{xtal} in the table for rubidium iodide, RbI, was calculated as follows after a value of ΔH_{ea} had been chosen:

$$\Delta H_{xtal} = \Delta H_f - \Delta H_{a,m} - \Delta H_{a,n} - \Delta H_{ion} + \Delta H_{ea}$$
$$= -328.4 - 85.8 - 106.8 - 403.0 + 295.3$$
$$= -628.7 \text{ kJ/mol}$$

The values for ΔH_{xtal} calculated in this way agree fairly well with those determined by other methods.

We can study the relationships between Figure 13.5 and the data in Table 13.5 to learn how to assign the proper sign to each enthalpy change on the basis of whether the change is exothermic ($-$) or endothermic ($+$). In a complete cycle, the heat given up in exothermic changes equals that taken up in endothermic changes.

Thermodynamics

13•8   The First Law of Thermodynamics

The success of the Born–Haber method is, of course, a check on the theory that all the important steps in the cyclical process have been considered. The concept that the net heat energy change of any cyclical process must be zero is based on one of the most fundamental of all scientific laws, the **first law of thermodynamics.** One statement of this law is: *energy can be converted from one form to another, but it cannot be created or destroyed.* Another way of stating the first law is: *the energy of the universe is constant.* The first law underlies the very idea of the Born–Haber cycle. If it were possible to go through the cycle with perfect efficiency, the net energy change would be zero.

13•8•1   Internal Energy

Internal Energy   The total energy of a system of chemical substances is referred to as its **internal energy,** E. This internal energy, which depends on the motion of the molecules, their arrangement, intermolecular attraction forces, and other factors, is a state function. Just as in the case of enthalpy, H, the absolute value of the internal energy, E, of a given state cannot be determined, but the change in internal energy can be determined and is called ΔE.

Chemical changes involve changes in the internal energy of the products relative to those of the reactants. If we let E_2 be the internal energy of the products and E_1 be the internal energy of the reactants, then the change in internal energy, ΔE, equals $E_2 - E_1$. A change in the internal energy of a system is brought about by the transfer of heat or the performance of work.

13•8•2   Heat Changes at Constant Volume and Constant Pressure

Heat Changes at Constant Volume and Constant Pressure   Although processes carried out at constant pressure, say, at atmospheric pressure, are of greater interest to the chemist than those carried out at constant volume, heat changes at constant volume are often more easily measured. As pointed out in Section 13.4.3, heats of combustion of organic compounds are commonly determined by using a bomb calorimeter. In this apparatus, the reaction is conducted in a tightly closed chamber, and the process occurs at constant volume. The heat change measured is equal to ΔE.

The heat change in a constant pressure process was defined in Section 13.3 as the change in enthalpy, ΔH. The desired ΔH is related to the easily measured ΔE by the following equation:

$$\Delta H = \Delta E + \Delta nRT \qquad (8)$$

where Δn is the number of moles of gaseous products minus the number of moles of gaseous reactants as indicated in the balanced equation for the reaction. The difference between ΔH and ΔE is small for most reactions at ordinary temperatures. Only when Δn or T is large will the difference be appreciable. (See Example 13.7.)

• Example 13.7 •

The heat of combustion of benzene, $C_6H_6(l)$, as determined in a bomb calorimeter is $-3,263.9$ kJ/mole at 25 °C and 1 atm pressure. Calculate the enthalpy change, $\Delta H_r°$ for this process.

• Solution • The value of $-3,263.9$ kJ/mole is the heat of combustion at constant volume, which is equal to the change in internal energy, ΔE, of the system. To calculate the enthalpy change, $\Delta H_r°$, using Equation (8), we must determine the value of Δn from the balanced equation for the combustion of 1 mole of benzene:

$$C_6H_6(l) + \tfrac{15}{2}O_2(g) \longrightarrow 6CO_2(g) + 3H_2O(l)$$

The value of Δn, the number of moles of gaseous products minus the number of moles of gaseous reactants, is

$$6CO_2(g) - \tfrac{15}{2}O_2(g)$$
$$6 \text{ mol} - 7.5 \text{ mol} = -1.5 \text{ mol}$$
$$\Delta H = \Delta E + \Delta nRT$$
$$\Delta H = -3,263.9 \text{ kJ} + \left[(-1.5 \text{ mol})\left(8.314 \frac{J}{K \times mol}\right)\right.$$
$$\left.\times (298 \text{ K})\left(\frac{1 \text{ kJ}}{1,000 \text{ J}}\right)\right]$$
$$\Delta H_r° = -3,263.9 \text{ kJ} + (-3.72 \text{ kJ}) = -3,267.6 \text{ kJ}$$

See also Exercise 35.

13.9   **The Criteria for Spontaneous Chemical Processes**

Some important goals of thermodynamics involve the answering of the following questions: (1) When two or more substances are mixed, do they tend to react spontaneously to form other substances? (2) If a reaction occurs, will it be accompanied by a release of energy to the surroundings? (3) If a reaction occurs, what will be the amounts of reactants and products present when chemical equilibrium is established?

We will attempt to answer the first two of these questions in this section. The answer to the third question will be taken up in Chapter 15 on chemical equilibria.

The first law of thermodynamics is useful in determining energy changes for various processes, for example, energy changes at constant volume (ΔE) or constant pressure (ΔH). However, it does not allow us to predict whether or not those processes will occur. Differences in the enthalpies of reactants and products have a great deal to do with the way in which reactions take place. Obviously they determine whether a reaction is exothermic or endothermic. But *enthalpy changes alone do not determine whether a chemical reaction is spontaneous or not.* To answer the questions posed above, we need to introduce another property of the system, the entropy.

13•9•1   Entropy   In addition to a change in enthalpy, a physical or chemical change involves a change in the relative disorder of the atoms, molecules, or ions involved. The disorder, or randomness, of a system is called its **entropy,** *S.* A few simple examples will serve to illustrate the concept of entropy further:

A quantity of gas confined in a 1-L flask has a higher entropy (greater disorder) than the same quantity of gas confined in a 10-mL flask at the same temperature. (See Figure 13.6.)

Sodium chloride in the form of its gaseous ions has a higher entropy than it does in the form of a solid crystal.

Liquid water at 0 °C has a higher entropy than ice at the same temperature.

Each of the preceding examples illustrates a general principle: *physical and chemical systems tend to attain the maximum state of disorder permitted by the energy of the system as balanced against the attractive forces between particles.* In each of these examples, we have referred to a change in disorder or entropy associated with a given system under examination, that is,

$$\Delta S = S_2 - S_1$$

• *Second Law of Thermodynamics* • Associated with the concept of changes in entropy is the **second law of thermodynamics.** *The total amount of entropy in the universe is increasing.* We can define the universe as consisting of a given chemical system under examination and the rest of the surrounding universe. The second law states that *when any spontaneous change takes place in a given system, there is an increase in the entropy of the universe.* This is another statement of the second law. To describe the spontaneity of a chemical process in which we might be interested in terms of its effect on the entropy of the universe is a bit grandiose. And this concept does not lead us to any practical way of predicting whether or not the process is spontaneous.

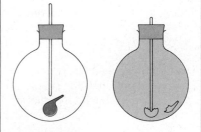

Figure 13•6 When a small bulb containing either gas or a volatile liquid is crushed, the molecules become more randomly separated in the larger container. The entropy of the substance increases.

However, the criterion for a spontaneous change can be stated in terms of the properties of a single chemical system, as is discussed in Section 13.9.2.

• *Third Law of Thermodynamics* • To imagine a completely ordered system, we can think of a perfect crystalline element or compound at the temperature of absolute zero. *A perfect crystal at absolute zero would have perfect order; hence its entropy would be zero.* This is a statement of the **third law of thermodynamics.** At any temperature other than absolute zero, there will be some amount of disorder due to thermal excitation. The entropy of a substance compared with its entropy in a perfectly crystalline form at absolute zero is called its **absolute entropy,** $S°$. The absolute entropy is, of course, always positive, because there is some disorder in any actual substance. The higher the temperature of a substance, the greater is the value of its absolute entropy.

13•9•2   Free Energy Overall changes in nature or other processes in which temperature and pressure changes occur are more complex than we can treat in detail here. However, the principles that apply to processes at constant

---

★ Special Topic ★ Special Topic ★ Special Topic ★ Special Topic ★ Special Topic ★ Special Topic ★ Special Topic ★ Special Topic   ★

Entropy and Natural Processes Care must be exercised in applying the great generalities of thermodynamics to processes and events in everyday life. Occasionally, a person may make the argument that processes involving living organisms apparently violate the second law, because the molecules in the body of the organism are more ordered than the simpler molecules in the food of the organism. Such an argument is based on misunderstanding.

The general statements of thermodynamic relationships are made with reference to one of several ideal conditions that scientists can approximate in their experimental testing of laws and theories. Examples of these conditions are *isothermal* (the system is maintained at the same temperature throughout an experiment, or at least the initial and final measurements are made at the same temperature), *isopiestic* (the system is at the same presssure), and *adiabatic* (the system is isolated so that no heat enters or leaves).

One statement of the second law is that in an isolated (adiabatic) system any spontaneous change is accompanied by an increase in the entropy of the system. But an adiabatic system is a very special system that can be approached in the laboratory only in well-insulated containers over short periods of time. In nature, parts of systems may approximate adiabatic changes. For example, great atmospheric turbulence may cause a large mass of air to be carried aloft so rapidly that part of it expands owing to the reduced pressure and becomes cooler under nearly adiabatic conditions.

The earth itself is not an adiabatic system. Far from being isolated, it receives high-energy radiation from the sun and emits longer wavelength, low-energy radiation to outer space (see the accompanying figure). The sun's radiation is the energy source that makes possible the reactions that lead to the ordered molecules in living organisms. These reactions are not spontaneous in the thermodynamic sense.

Philosophers have applied the second law on the grandest scale to that largest imaginable isolated system, the universe. If by definition the universe is an adiabatic system, applying the second law tells us that the net overall effect of all natural processes is to increase entropy. The end result will be maximum disorder, with all matter a few degrees Kelvin above absolute zero, a condition forebodingly referred to as the heat death of the universe.

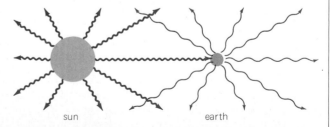

sun   earth

Photons of short wavelength have more available energy than photons of long wavelength. Even if the net heat gained by the earth equals the heat lost, there is a loss in the energy available to do work.

temperature and pressure are very important, and they can be used satisfactorily to illustrate the way in which changes in entropy affect the course of reactions.

For a process taking place at constant temperature and pressure, the total change in energy, ΔH, may be divided into two parts. One part is available for doing useful work and is called the change in the **Gibbs free energy,** ΔG. It is named for an American theoretician of the nineteenth century. The other part is not available for doing useful work because it is associated with the change in the disorder, that is, the change in the entropy, ΔS.

At constant temperature and pressure, the changes in enthalpy, free energy, and entropy for any process, physical or chemical, are related by the equation

$$\Delta H = \Delta G + T\,\Delta S \qquad (9)$$

$$\text{change in enthalpy} = \text{change in free energy} + T \times \text{change in entropy}$$

$$\text{change in total energy} = \text{change in available energy} + \text{change in unavailable energy}$$

From Equation (9), we can see that if ΔH is specified, an increase in ΔS must mean a decrease in ΔG.

Another way in which the second law is stated is that *in any spontaneous change, the amount of free energy available decreases.* The change in free energy (ΔG) rather than the change in enthalpy (ΔH) determines whether or not a process tends to take place. A process at constant temperature and pressure can take place spontaneously only if G decreases, that is, if ΔG is negative. A decrease in free energy is akin to a natural process going downhill. Heat passes from a hot to a cool body, water flows downhill, and a gas expands from a high to a low pressure (see Figure 13.6); but such spontaneous actions leave a system with less ability to do work, that is, less free energy.

Equation (9) is often written as

$$\Delta G = \Delta H - T\,\Delta S$$

For most chemical reactions at room temperature, $T\,\Delta S$ is small compared with ΔH; therefore, the sign of ΔG will usually be the same as that of ΔH. That is, if a change is highly exothermic, ΔH will have a large negative value and ΔG will be negative also. The process can take place spontaneously. A combustion reaction is an example of a process that can proceed spontaneously. On the other hand, if ΔH has a high positive value, ΔG will probably be positive also, and the process will not take place spontaneously. The decomposition of a stable compound, water, for example, into its elements hydrogen and oxygen at room temperature is an example of such a nonspontaneous reaction.

• *Standard Free Energies of Formation,* ΔG_f° • **The standard free energy of formation,** ΔG_f°, of a substance is defined as the change in free energy for the reaction in which 1 mole of a compound is formed from its elements, each in its standard state. As is the case for standard enthalpies of formation, the standard free energy of formation of an element is, by convention, assigned a value of zero. Values of $\Delta G_f°$ at 25 °C for compounds may be determined from the equation

$$\Delta G_f° = \Delta H_f° - T\,\Delta S° \tag{10}$$

where $\Delta H_f°$ is the standard enthalpy of formation (see Section 13.4.3 and Table 13.1), T is 298 K, and $\Delta S°$ is the sum of the absolute entropies of the products minus the sum of the absolute entropies of the reactants at 25 °C and 1 atm:

$$\Delta S° = \Sigma S° \text{ products} - \Sigma S° \text{ reactants} \tag{11}$$

Table 13.6 gives the values of $\Delta H_f°$, $\Delta G_f°$, and $S°$ for a number of elements and compounds. Note that the sign of $\Delta G_f°$ is usually the same as that for $\Delta H_f°$. Example 13.8 illustrates a calculation of $\Delta G_f°$ by use of Equation (10).

• Example 13.8 •

Calculate the standard free energy of formation, $\Delta G_f°$, for $NH_3(g)$ from the values of $\Delta H_f°$ and $S°$ given in Table 13.6.

• Solution • The balanced equation for the formation of ammonia from the elements is

$$N_2(g) + 3H_2(g) \longrightarrow 2NH_3(g)$$

For the formation of 2 moles of ammonia,

$$\Delta H_f° = 2(-46.11) = -92.22 \text{ kJ}$$
$$T\,\Delta S° = T(\Sigma S° \text{ products} - \Sigma S° \text{ reactants})$$
$$= 298\{2(0.1923) - [0.1915 + 3(0.1306)]\} \text{ kJ}$$
$$= 298(0.3846 - 0.5833) \text{ kJ}$$
$$= -59.2 \text{ kJ}$$
$$\Delta G_f° = \Delta H_f° - T\,\Delta S°$$
$$= -92.22 - (-59.2) = -33.0 \text{ kJ}$$

For the formation of 1 mole of ammonia,
$$\Delta G_f° = \frac{-33.0}{2} = -16.5 \text{ kJ}$$

See also Exercise 38.

For those cases that are only slightly exo- or endothermic, the change in entropy, $\Delta S°$, may determine that the sign of $\Delta G_f°$ will differ from that of $\Delta H_f°$. Such a case is the formation of iodine chloride, $ICl(g)$. Although this is an endothermic compound, $\Delta H_f° = 17.8$ kJ/mole, it has a favorable (that is, negative) standard free energy of formation, $\Delta G_f° = -5.44$ kJ/mole, owing to the increase in entropy as 1 mole of $I_2(s)$ and 1 mole of $Cl_2(g)$ react to form 2 moles of $ICl(g)$:

Table 13•6 Thermodynamic properties at 25 °C

Element or compound	$\Delta H_f°$ kJ/mol	$\Delta G_f°$ kJ/mol	$S°$ kJ/(mol $\times$ K)
$Br_2(l)$	0	0	0.1522
$Br_2(g)$	30.91	3.14	0.2454
$BrCl(g)$	14.6	−0.96	0.2400
$C(s)$	0	0	0.005694
$CH_4(g)$	−74.81	−50.84	0.1863
$C_2H_2(g)$	226.7	209.2	0.2008
$C_2H_4(g)$	52.26	68.24	0.2192
$C_2H_6(g)$	−84.68	−32.9	0.2295
$C_6H_6(l)$	48.99	124.3	0.1733
$CH_3NH_2(g)$	−23.0	32.3	0.2426
$CH_3OH(l)$	−239.0	−166.8	0.1272
$CH_3OH(g)$	−201.1	−162.4	0.2317
$CO(g)$	−110.52	−137.3	0.1979
$CO_2(g)$	−393.52	−394.40	0.2137
$CS_2(g)$	117.1	66.90	0.2378
$Cl_2(g)$	0	0	0.2230
$ClF(g)$	−54.48	−55.94	0.2178
$F_2(g)$	0	0	0.2027
$H_2(g)$	0	0	0.1306
$HBr(g)$	−36.4	−53.51	0.1986
$HCl(g)$	−92.31	−95.30	0.1868
$HF(g)$	−271.1	−273.2	0.1737
$HI(g)$	26.5	1.7	0.2065
$HCN(g)$	135	125	0.2017
$HN_3(g)$	294	328	0.2389
$H_2O(l)$	−285.83	−237.18	0.06991
$H_2O(g)$	−241.82	−228.59	0.1887
$H_2S(g)$	−20.2	−33.1	0.2058
$Hg(l)$	0	0	0.07602
$Hg(g)$	61.32	31.85	0.1748
$HgO(s)$	−90.83	−58.57	0.07029
$I_2(s)$	0	0	0.1161
$I_2(g)$	62.44	19.36	0.2606
$ICl(g)$	17.8	−5.44	0.2474
$N_2(g)$	0	0	0.1915
$NH_3(g)$	−46.11	−16.49	0.1923
$NO(g)$	90.25	86.57	0.2107
$NO_2(g)$	33.2	51.30	0.2400
$N_2O_4(g)$	9.16	97.82	0.3042
$NaCl(s)$	−411.0	−384.0	0.07238
$NaCl(aq)^a$	−407.1	−393.0	0.115
$O_2(g)$	0	0	0.2050
$O_3(g)$	143	163	0.2388
$S(s)$	0	0	0.0319
$S(g)$	277.4	236.9	0.1677
$SO_2(g)$	−296.83	−300.19	0.2481
$SO_3(g)$	−395.7	−371.1	0.2566

[a] This solution is ideal 1 molal.

Source: *Values are calculated mainly from* Circular of the National Bureau of Standards 500, *Washington, D.C., 1952, and from* Lange's Handbook of Chemistry, *12th ed., J. A. Dean, Ed., McGraw-Hill, New York, 1979. Used with permission of McGraw-Hill.*

$$I_2(s) + Cl_2(g) \longrightarrow 2ICl(g)$$

Methylamine, $CH_3NH_2(g)$, is an exothermic compound, $\Delta H_f^\circ = -23.0$ kJ/mole, yet it has a positive standard free energy of formation, $\Delta G_f^\circ = +32.3$ kJ/mole. This change in sign from a negative ΔH_f° to a positive ΔG_f° is due to the decrease in entropy associated with the following hypothetical reaction:

$$2C(s) + 5H_2(g) + N_2(g) \longrightarrow 2CH_3NH_2(g)$$

• *Calculation of Standard Free Energies of Reaction from Standard Free Energies of Formation* • The values of ΔG_f° in Table 13.6 may be used to calculate standard free energies of reaction, ΔG_r°, for all sorts of reactions, both actual and hypothetical. The equation used is similar to that used for calculations of standard enthalpies of reaction [see Equation (6) in Section 13.4.3]:

$$\Delta G_r^\circ = \Sigma \Delta G_f^\circ \text{ products} - \Sigma \Delta G_f^\circ \text{ reactants}$$

If ΔG_r° is negative, the reaction is a spontaneous one.

———— • Example 13.9 • ————

Calculate ΔG_r° for the reaction between carbon monoxide, $CO(g)$, and hydrogen, $H_2(g)$, to give methyl alcohol, $CH_3OH(l)$.

———— • Solution • Write a balanced equation for the reaction and list the ΔG_f° value for each substance:

$$\begin{array}{cccc} CO(g) & + & 2H_2(g) & \longrightarrow & CH_3OH(l) \\ -137.3 & & 0 & & -166.8 \end{array}$$

$$\Delta G_r^\circ = -166.8 - (-137.3 + 0)$$
$$= -29.5 \text{ kJ}$$

See also Exercises 40–42.

This value of ΔG_r° is less negative than the ΔH_r° of -128.5 kJ calculated in Example 13.3, due to the decrease in entropy associated with this reaction. Note that 1 mole of liquid product is formed from 3 moles of gaseous reactants.

13.9.3 Free Energy and Reaction Rate We must be careful not to confuse a spontaneous reaction with a reaction that occurs rapidly. A spontaneous reaction is one that has a natural tendency to occur and for which the change in free energy, ΔG, is negative. Such a reaction may proceed quite rapidly at room temperature, or it may proceed so slowly that no appreciable change can be observed.

All that thermodynamics can tell us about a chemical process is whether or not it has a natural tendency to occur, that is, whether or not it is spontaneous. *Thermodynamics tells us nothing about the speed (rate) of a chemical reaction.*

13.9.4 Influence of Temperature on Free Energy Changes If mercury is heated to just below its boiling point in oxygen (or air), it reacts with oxygen to form a red solid, mercury(II) oxide. But, if the reactants are heated to a higher temperature, they do not react appreciably. In fact, at high temperature, mercury(II) oxide decomposes (see Figure 9.1). Why should mercury and oxygen react only when heated, but not react if heated to too high a temperature? We can answer this question by considering the effect of temperature changes on the free energy change of a reaction.

With the data in Table 13.6, we can calculate the free energy change, ΔG, for the reaction

$$Hg(l) + \tfrac{1}{2}O_2(g) \longrightarrow HgO(s)$$

under different conditions. Our aim is to evaluate ΔG by estimating ΔH and ΔS.

We shall illustrate the calculation first at 25 °C, for which we can calculate $\Delta G°$ precisely. At 25 °C,

$$\Delta H_r° = \Sigma \Delta H_f° \text{ products} - \Sigma \Delta H_f° \text{ reactants}$$
$$= -90.83 - (0 + 0) = -90.83 \text{ kJ}$$
$$T \Delta S° = T(\Sigma S° \text{ products} - \Sigma S° \text{ reactants})$$
$$= 298\{0.07029 - [0.07602 + \tfrac{1}{2}(0.2050)]\}$$
$$= 298(0.07029 - 0.1785) = 298(-0.1082)$$
$$= -32.2 \text{ kJ}$$

and

$$\Delta G° = \Delta H° - T \Delta S°$$
$$= -90.83 \text{ kJ} - (-32.2 \text{ kJ})$$
$$= -58.6 \text{ kJ}$$

The moderately high negative value of $\Delta G°$ shows that the reaction between mercury and oxygen has a considerable tendency to occur at 25 °C. However, we know from experience that mercury does not react with oxygen at room temperature. This means that, although the reaction can take place as far as free energy change is concerned, the rate of the reaction is so slow that few if any molecules react at 25 °C. The reaction is too slow to be observed.

Values for ΔH, ΔG, and S differ at different temperatures. But we can use values like those in Table 13.6 for approximate calculations at various temperatures. One way of using the values is to assume that ΔH and S remain about the same and calculate ΔG. We shall use this method for calculating ΔG for the formation of HgO at two other temperatures. At 300 °C (573 K),

$$\Delta H \simeq -90.83 \simeq -91 \text{ kJ}$$
$$T \Delta S \simeq 573(-0.1082) \text{ kJ}$$
$$\simeq -62 \text{ kJ}$$

and

$$\Delta G = \Delta H - T \Delta S$$
$$\simeq -91 - (-62) = -29 \text{ kJ}$$

The negative value of ΔG indicates that the formation of mercury(II) oxide is spontaneous at about 300 °C. At this temperature the rate of the reaction is slow, but a long period of heating will yield some of the oxide.

At 500 °C (773 K), which is 143 degrees above the boiling point of mercury, we have to consider the mercury to be gaseous and calculate ΔG for the reaction

$$Hg(g) + \tfrac{1}{2}O_2(g) \longrightarrow HgO(s)$$

At 25 °C,

$$\Delta H_r^\circ = \Sigma \, \Delta H_f^\circ \text{ products} - \Sigma \, \Delta H_f^\circ \text{ reactants}$$
$$= -90.83 - (61.32 + 0) = -152.15 \text{ kJ}$$

At 500 °C (773 K),

$$\Delta H \simeq -152.15 \simeq -152 \text{ kJ}$$
$$T \, \Delta S^\circ = T(\Sigma S^\circ \text{ products} - \Sigma S^\circ \text{ reactants})$$
$$\simeq 773\{0.07029 - [0.1748 + \tfrac{1}{2}(0.2050)]\}$$
$$\simeq 733(0.07029 - 0.2773)$$
$$\simeq -160 \text{ kJ/mol}$$

and

$$\Delta G = \Delta H - T \, \Delta S$$
$$\simeq -152 - (-160) = +8 \text{ kJ/mol}$$

Even though using $Hg(g)$ as a reactant has caused ΔH to become more negative, $T \, \Delta S$ has changed to a greater extent with temperature, and ΔG is therefore positive.

The positive value of ΔG indicates that the reaction

$$Hg(g) + \tfrac{1}{2}O_2(g) \longrightarrow HgO(s)$$

has less tendency to take place at 500 °C than does the reaction

$$HgO(s) \longrightarrow Hg(g) + \tfrac{1}{2}O_2(g)$$

At 500 °C, mercury(II) oxide tends to decompose. If the oxide is heated in an open vessel and the oxygen allowed to escape, the decomposition goes to completion.

Let us conclude by repeating that these calculations are approximate, for the values of ΔH and S do change with temperature, and over a range of hundreds of degrees these changes may be quite important.

13.9.5 **Summary of General Principles**   The case of mercury(II) oxide serves to illustrate certain general principles which we have mentioned previously and which should be emphasized again.

1.   *A negative value of ΔG indicates that a reaction can take place spontaneously, but it does not tell us that the reaction will take place readily.* The reaction rate may be so low at a given temperature that it is zero for all practical purposes.

2.   *At ordinary temperatures, the value of ΔH as a rule determines whether ΔG is*

positive or negative. An exothermic reaction (negative ΔH) commonly has a negative ΔG and takes place spontaneously.

3. The entropy effect is usually small (a fraction of a kilojoule per mole per degree), and $T\,\Delta S$ at low temperatures is therefore small compared with ΔH (which usually is of the order of kilojoules per mole). *Only for those cases in which ΔH is small does the value of $T\,\Delta S$ affect the sign of ΔG at ordinary temperatures.*

4. *Any reaction that involves an increase in entropy will occur spontaneously at a high enough temperature.* If T is large enough, the product $T\,\Delta S$ subtracted from ΔH results in a negative ΔG, even if ΔH is positive. Some examples of changes that produce more disorder and therefore tend to be spontaneous at high temperatures are

$$\text{large molecules} \longrightarrow \text{small molecules} \longrightarrow \text{atoms}$$

and

$$\text{solids} \longrightarrow \text{liquids} \longrightarrow \text{gases}$$

For these changes, the entropy change, ΔS, is positive. As $-T\,\Delta S$ becomes greater, ΔG becomes more and more negative. The following case illustrates this well:

$$\text{HgO}(s) \xrightarrow[\text{temp.}]{\text{high}} \text{Hg}(g) + \tfrac{1}{2}\text{O}_2(g)$$

At a very high temperature, the oxygen molecules dissociate into atoms and increase the entropy even more:

$$\text{HgO}(s) \xrightarrow[\text{temp.}]{\text{very high}} \text{Hg}(g) + \text{O}(g)$$

5. *All compounds decompose if the temperature is high enough.* At extremely high temperatures, even atoms decompose. In the interior of the sun, nuclei and electrons are dissociated from one another.

Chapter Review

Summary

Thermodynamics, the study of energy relationships of all kinds, is fundamental to all the sciences. The province of **chemical thermodynamics** is the relationships of certain kinds of energy to chemical systems. **Thermochemistry** is concerned with the measurement and interpretation of the heat changes accompanying chemical processes. Most such measurements are made with a **calorimeter,** in which the heat evolved or absorbed in a chemical reaction is determined precisely. This heat change, in terms of molar quantities of reactants and products, is shown in the **thermochemical equation** for the reaction.

Every substance is regarded as having a certain **enthalpy,** or heat content, under specified conditions. The **enthalpy change,** ΔH, in any reaction is the heat change when that reaction occurs at constant pressure. The sign of ΔH indicates whether the reaction is exothermic or endothermic. Enthalpy is a **state function:** its value for a given substance depends only on the state of the substance, not on the process by which that state was attained. This fact is embodied in **Hess's law,** which enables many enthalpy changes to be calculated, rather than measured, by adding or subtracting known thermochemical equations. Most tabulated ΔH values are for 25 °C and refer to the elements and compounds in an arbitrarily chosen **standard state:** the most stable physical form of the substance at 1 atm. The **standard enthalpy of formation** of a substance is the enthalpy change for the reaction by which the substance is formed from its elements in the standard state. Standard enthalpies of formation can be used to calculate standard enthalpies of reaction, and vice versa, by applying Hess's law.

Other useful thermochemical quantities are the **bond dissociation energy** of a chemical bond, the **average bond energy** of a number of similar bonds in a compound, the **heat of formation of an atom** from the element in its standard state, the **enthalpy of ionization** of an atom, the **electron affinity** of an atom, and the **crystal energy** of an ionic solid. Many of these quantities are incorporated in the **Born–Haber cycle,** a conceptual closed cycle of thermochemical reactions in which the net enthalpy change is zero. This cycle is used for calculating thermochemical quantities whose direct measurement is difficult or impossible.

The **first law of thermodynamics** is a statement of the law of conservation of energy. The total energy of a chemical system is its **internal energy,** a state function. A change in the internal energy, ΔE, is brought about by the transfer of heat or the performance of work.

Whether or not a given chemical reaction can occur spontaneously depends not only on the change in enthalpy but also on the temperature and the change in **entropy,** ΔS, which is a measure of the change in the degree of disorder in a system. The tendency for entropy always to attain the maximum possible value permitted by the energy of the system is stated in the **second law of thermodynamics,** and the existence of positive values of the **absolute entropies** of all real substances is a consequence of the **third law of thermodynamics.** In a process occurring at constant temperature and pressure, the component of the total enthalpy change that is related to a change in the entropy of the system is regarded as energy that cannot do useful work. The remaining component, which is regarded as energy that *can* do useful work, is the change in the **Gibbs free energy,** ΔG, of the system. This change can be calculated from known values of the **standard free energies of formation** of the reactants and products. The sign of ΔG indicates whether or not the reaction *can* occur spontaneously at the temperature in question, but it does not indicate whether it *will* occur at an appreciable rate.

Exercises

Enthalpies of Reaction

1. Into a steel chamber of a calorimeter is placed 3.2 g of methanol, CH_3OH. The chamber is then filled with oxygen at a pressure of 10 atm. After the outer chamber is filled with water (3,000 g at 25.0 °C), the alcohol–oxygen mixture is ignited. Shortly thereafter, the combustion is complete. The final temperature of the water is measured as 30.2 °C. The steel combustion chamber weighs 2,100 g and has a specific heat of 0.46 J · g^{-1} · K^{-1}. Calculate the heat taken up by the combustion chamber and the water in which it is immersed. (See Figure 13.1.) Calculate the heat liberated in kilojoules for the combustion of 1 mole of methanol.

2. How are the states of reactants and products indicated in thermochemical equations? Why is it necessary to do this?

3. Describe fully the standard state for each of the following: hydrogen, mercury, sodium, carbon, oxygen.

4. By applying Hess's law to the equations below, calculate $\Delta H_r°$ at 25 °C for the combustion of 1 mole of acetylene, $C_2H_2(g)$, to form $H_2O(l)$ and $CO_2(g)$. Values of $\Delta H_r°$ at 25 °C for the three reactions are -393.52, -285.83, and $+226.7$ kJ, respectively.

$$C(s) + O_2(g) \longrightarrow CO_2(g)$$
$$H_2(g) + \tfrac{1}{2}O_2(g) \longrightarrow H_2O(l)$$
$$2C(s) + H_2(g) \longrightarrow C_2H_2(g)$$

5. Distinguish between ΔH_r and $\Delta H_r°$. Between $\Delta H_r°$ and $\Delta H_f°$.

6. Explain each of the following statements:
   (a) The enthalpy of a substance is a state function.
   (b) The absolute enthalpy of a substance is unknown, yet the enthalpy of an element is said to be zero.

7. Using data from Table 13.1, what is the value of $\Delta H_r°$ for each of the following reactions at 25 °C?
   (a) $\tfrac{1}{2}H_2(g) + \tfrac{1}{2}Br_2(g) \longrightarrow HBr(g)$
   (b) $2H_2(g) + \tfrac{1}{2}O_2(g) + C(s) \longrightarrow CH_3OH(l)$
   (c) $H_2(g) + 3N_2(g) \longrightarrow 2HN_3(g)$

8. Using Table 13.1 for data, calculate $\Delta H_r°$ for the reaction of $H_2(g)$ and acetylene, $C_2H_2(g)$, to give ethane, $C_2H_6(g)$.

9. Calculate $\Delta H_r°$ for the combustion of 1 mole of benzene, $C_6H_6(l)$, using data from Table 13.1.

10. Assuming reactants and products are in their standard states at 25 °C, which liberates the greater amount of heat per gram upon combustion, ethylene, $C_2H_4(g)$, or acetylene, $C_2H_2(g)$?

11. Using data in Table 13.1, calculate the standard enthalpy of reaction, $\Delta H_r°$, for the combustion of $CS_2(g)$ with oxygen, $O_2(g)$, to produce carbon dioxide, $CO_2(g)$, and sulfur dioxide, $SO_2(g)$.

12. Nitrogen tetroxide, $N_2O_4(g)$, was used as the oxidizer in the propulsion system of the first lunar landing craft. The fuel was a mixture of hydrazine, $N_2H_4(l)$, and dimethylhydrazine. The products of the reaction using hydrazine and nitrogen tetroxide were nitrogen, $N_2(g)$, and water vapor, $H_2O(g)$. Using data from Tables 13.1 and 13.6, calculate $\Delta H_r°$ for this reaction.

Enthalpies of Formation

13. The standard enthalpy of reaction, $\Delta H_r°$, at 25 °C for the combustion of 1 mole of liquid acetone, $C_3H_6O(l)$, with $O_2(g)$ to produce $H_2O(l)$ and $CO_2(g)$ is $-1{,}787.78$ kJ. Given the data in Table 13.1, calculate $\Delta H_f°$ for acetone.

14. The standard enthalpy of reaction is -73.55 kJ for

$$CH_3Br(g) + H_2(g) \longrightarrow CH_4(g) + HBr(g)$$

Using this information and the enthalpies of formation from Table 13.1, calculate $\Delta H_f°$ for methyl bromide.

Bond Dissociation Energies

15. Among the following gaseous molecules, in which one is the bond between the two atoms the strongest: H_2, O_2, F_2, N_2, HCl? In which one is the bond the weakest? On what do you base your conclusions?

16. (a) From data in Tables 13.1 and 13.3, calculate the bond dissociation energy, $\Delta H_{dis}°$, of NO(g) at 25 °C.
    (b) Why is it necessary that reactants and products be in the gaseous state to give data for the calculations of bond dissociation energies?

*17. From data in Tables 13.2 and 13.3, calculate:
    (a) the enthalpy of vaporization of bromine
    (b) the enthalpy of sublimation of iodine
    State any assumptions you make.

18. The standard heat of formation, $\Delta H_f°$, of a hydrogen atom, H(g), is half the standard bond dissociation energy, $\Delta H_{dis}°$, of a hydrogen molecule, $H_2(g)$. However, the standard heat of formation of the iodine atom, I(g), is not half the standard bond dissociation energy of an iodine molecule, $I_2(g)$. Why?

19. Using data in Tables 13.1 and 13.3, calculate the standard enthalpy of reaction for the decomposition of 1 mole of acetylene, $C_2H_2(g)$, into carbon, C(s), and hydrogen, $H_2(g)$. What would be the standard enthalpy of reaction if hydrogen atoms were produced instead of hydrogen molecules?

20. The nitrogen–nitrogen bond dissociation energy, $\Delta H_{dis}°$, in hydrazine, H_2NNH_2, is 297 kJ/mole, whereas it is 945.3 kJ/mole in nitrogen, N_2. Account for this difference.

21. From the following reactions and bond dissociation energies, calculate the average bond energy, $\Delta H_{dis\ avg}°$, in methane, CH_4.

reaction	$\Delta H_{dis}°$, kJ
$CH_4(g) \longrightarrow CH_3(g) + H(g)$	435
$CH_3(g) \longrightarrow CH_2(g) + H(g)$	444
$CH_2(g) \longrightarrow CH(g) + H(g)$	444
$CH(g) \longrightarrow C(g) + H(g)$	339

22. After checking the data for water in Table 13.1 and for hydrogen and oxygen in Tables 13.2 and 13.3, criticize the following calculation of the average O—H bond energy for water.

$$\Delta H_{dis\ avg}° = \frac{(2 \times 218.0) + 249.2 - (-285.83)}{2}$$

$$= 485.5 \text{ kJ/mol of O—H bonds}$$

23. Using data in Tables 13.1 and 13.3, calculate the average N—O bond energy in nitrogen dioxide, NO_2.

* 24. Consider the two compounds carbon monoxide and carbon dioxide, and use data in Tables 13.1 and 13.3 for the following.
    (a) Calculate the carbon–oxygen bond dissociation energy, $\Delta H_{dis}°$, in carbon monoxide.
    (b) Calculate the average bond energy, $\Delta H_{dis\ avg}°$ in carbon dioxide.
    (c) The correct Lewis structural formula for carbon dioxide is $:\ddot{O}=C=\ddot{O}:$ with two double bonds. Based on the answers to parts (a) and (b), does the Lewis formula for carbon monoxide as $:C=\ddot{O}$ with one double bond seem satisfactory? Explain.

25. What is the average C—F bond energy in CF_4, if its standard enthalpy of formation is -925 kJ/mole? See Table 13.3 for additional data.

26. Calculate the N—N bond dissociation energy in hydrazine, NH_2—NH_2, from the following experimental data: $\Delta H_f° NH_2(g) = 196$ kJ/mole; $\Delta H_f° N_2H_4(g) = 95.4$ kJ/mole.

27. Given the values of $\Delta H_f°$ for methyl ether, $CH_3OCH_3(g)$, the methyl radical, $CH_3(g)$, and atomic oxygen, O(g), as -183.7, $+142.3$, and $+249.2$ kJ/mole, respectively, what is the average C—O bond dissociation energy in methyl ether?

28. Given for hydrogen peroxide, H_2O_2:

$$HO-OH(g) \longrightarrow 2OH(g) \qquad \Delta H_{dis}° = 213 \text{ kJ/mol}$$

From this value and data in Table 13.1 and Section 13.5.2, calculate:
    (a) $\Delta H_f°$ of OH(g)
    (b) C—O bond energy, $\Delta H_{dis}°$, in CH_3OH

The Born–Haber Cycle

29. Calculate ΔH in kilojoules per mole for each of the following. See Table 3.6 for data.

$$K \longrightarrow K^+ + e^-$$
$$K \longrightarrow K^{2+} + 2e^-$$

30. Note in Table 13.4 that the electron affinity decreases regularly down the halogen family (group VIIA), except that the value for Cl⁻ appears to be out of line. What reasons can be given for the following?
    (a) expecting a regular decrease down the family
    (b) the seemingly large value for Cl⁻

*31. (a) The first ionization energy of silver is 7.576 eV; the crystal energy, ΔH_{xtal}, for silver chloride is -915.6 kJ/mole. Using appropriate data from tables in this chapter, calculate the enthalpy of sublimation of silver.

(b) From the enthalpy of sublimation of silver and data in tables in this chapter, calculate the crystal energy, ΔH_{xtal}, of silver fluoride, AgF, given $\Delta H_f°$ of AgF(s) $= -205$ kJ/mole.

32. What fundamental principle is the basis for both Hess's law and the Born–Haber cycle?

*33. The heat of combustion of n-octane, $C_8H_{18}(l)$, at 25 °C and 1 atm, as determined in a bomb calorimeter, is $-5,459.52$ kJ/mole. Calculate the standard enthalpy of reaction, $\Delta H_r°$, for the combustion of n-octane.

34. In which of the following is the entropy greater, given that equal quantities are compared and that the temperature is constant?

(a) liquid water at 0 °C or ice at 0 °C
(b) molten sodium chloride at its melting point or crystalline sodium chloride at its melting point
(c) hydrogen at 10 atm or hydrogen at 1 atm
(d) a mixture of hydrogen and oxygen in a 2:1 ratio or water

35. State the three laws of thermodynamics.

Standard Free Energies

36. For hydrogen peroxide, $H_2O_2(l)$, $\Delta H_f° = -187.78$ kJ/mole and $S° = 0.1096$ kJ $\cdot$ mol^{-1} $\cdot$ K^{-1}. From these and other data found in Table 13.6, calculate the standard free energy of formation, $\Delta G_f°$, of hydrogen peroxide.

*37. The standard enthalpies of formation, $\Delta H_f°$, at 25 °C of $COBr_2(g)$ and $COCl_2(g)$ are -96.2 and -221 kJ/mole, respectively. The standard free energies of formation, $\Delta G_f°$, are -111 and -207 kJ/mole. Why is the magnitude of $\Delta G_f°$ of $COBr_2(g)$ larger than its $\Delta H_f°$, whereas the reverse is true for the $\Delta G_f°$ and $\Delta H_f°$ of $COCl_2(g)$.

38. Calculate $\Delta G_r°$ at 25 °C for the combustion of 1 mole of ammonia, $NH_3(g)$, with oxygen, $O_2(g)$, to produce nitric oxide, $NO(g)$, and water, $H_2O(l)$, based on data in Table 13.6.

39. One of the ways carbon tetrachloride is produced is by the reaction of methane, $CH_4(g)$, with chlorine, $Cl_2(g)$. Hydrogen chloride, $HCl(g)$, is the other product of the reaction. Using data from Tables 13.1 and 13.6, and the values of $\Delta H_f°$ and $\Delta G_f°$ of $CCl_4(l)$ as -135.4 kJ/mole and -65.27 kJ/mole, respectively, calculate $\Delta H_r°$ and $\Delta G_r°$ for the reaction.

40. Calculate the change in standard free energy, $\Delta G_r°$, for each of the following reactions and then state whether or not each reaction will take place spontaneously (see Table 13.6 for data):

(a) $2HCl(g) + Br_2(g) \longrightarrow 2HBr(g) + Cl_2(g)$
(b) $4NH_3(g) + 7O_2(g) \longrightarrow 4NO_2(g) + 6H_2O(g)$
(c) $2NO(g) + 5H_2(g) \longrightarrow 2NH_3(g) + 2H_2O(g)$

Free Energy Change, Reaction Rate, and Temperature

41. Diamond is usually thought of as being chemically inactive, yet $\Delta G_r°$ for its reaction with $O_2(g)$ to form $CO_2(g)$ has a very high negative value. What does the negative value indicate about this reaction? What can be said about the rate of this reaction at 25 °C?

*42. From data given in Table 13.6, calculate the approximate Celsius temperature at which ΔG is zero for the reaction

$$C(s) + 2H_2(g) + \tfrac{1}{2}O_2(g) \longrightarrow CH_3OH(g)$$

43. Would 5,000 °C be a good temperature for forming water by direct union of the elements? Why?

Chemical Kinetics

14

14·1 **Reaction Mechanisms**

14·1·1 Transition State

14·2 **Reaction Rate**

14·2·1 Experimental Procedures

14·2·2 Factors Affecting Reaction Rates

14·3 **Determination of Rate Laws or Rate Equations**

14·4 **Order of a Chemical Reaction**

14·5 **Determination of Reaction Order from Experimental Data**

14·5·1 Determination of First-Order Reactions

14·5·2 Determination of Second-Order Reactions

14·6 **Order and Reaction Mechanism**

14·6·1 A Specific Example

14·7 **Chain Reactions**

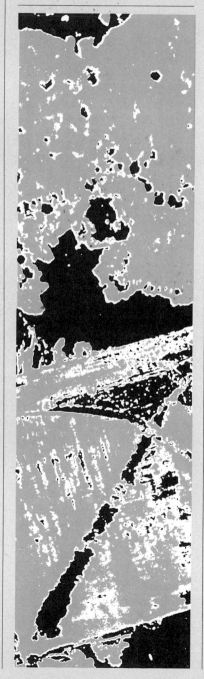

In the preceding chapter our attention was focused on what thermodynamics can tell us about a chemical system, namely, whether or not a reaction is spontaneous. In addition, as we will see in the following chapter, thermodynamics can tell us with precision the maximum possible extent of a chemical reaction. We can calculate to what extent reactants are converted to products when equilibrium is attained. Only the initial and final states are of concern in thermodynamics. How rapidly a chemical reaction occurs and how the system is converted from one state to another are not explained by this science. The answers to these two questions are the prime goals of the subject of chemical kinetics, which we will look at in this chapter.

Chemical kinetics is the study of the rates and mechanisms of chemical reactions. Iron rusts in moist air faster than in dry air; foods spoil faster when left unrefrigerated; and one's skin becomes tanned more readily in the summer than in the winter. These are three common examples of complex chemical changes with rates that vary according to the reaction conditions. Of more fundamental interest than the mere rate of a reaction is how a chemical change takes place. To understand this we examine the stepwise changes that atoms, molecules, radicals, and ions undergo as they are converted from reactants to products. For a given reaction, the sum of these steps is the **reaction mechanism.**

14·1   Reaction Mechanisms

Overall chemical equations indicate only the initial reactants and final products. They give no indication of how the reacting molecules change as the reaction progresses. As written, an equation such as

$$4NH_3 + 5O_2 \longrightarrow 4NO + 6H_2O$$

might lead one to think that nine molecules, four of ammonia and five of oxygen, must collide in order for the reaction to occur. Such a process, involving the simultaneous breaking of the bonds in nine reactant molecules and the making of the bonds in ten product molecules, could hardly occur in one step. It can be shown from a purely statistical viewpoint that it is extremely unlikely that such a large number of molecules would even bump together, let alone align themselves properly for the bond-breaking and bond-making processes.

No matter how complex the overall equation appears, reactions generally take place in a simple stepwise fashion, with each step usually involving only one, two, or three particles as reactants. The individual steps by which chemical changes occur are generally spoken of as **elementary processes** or **elementary reactions.** The familiar reaction of hydrogen and oxygen to form water is a good example of a reaction that follows a more complex pathway than is indicated by the overall reaction

$$2H_2 + O_2 \longrightarrow 2H_2O$$

This reaction, the subject of hundreds of research papers, is thought to involve several elementary reactions, some of which are the following:

$$H_2 + O_2 \longrightarrow H + HO_2 \qquad (1)$$
$$H + O_2 \longrightarrow O + OH \qquad (2)$$
$$O + H_2 \longrightarrow H + OH \qquad (3)$$
$$H_2 + HO_2 \longrightarrow H_2O + OH \qquad (4)$$
$$H_2 + OH \longrightarrow H_2O + H \qquad (5)$$
$$H + OH \longrightarrow H_2O \qquad (6)$$

Particles such as H, O, OH, and HO_2 are called radicals. A **radical** is an atom or group of atoms that has one or more unpaired electrons. Generally, radicals lead a transitory existence, probably existing for a tiny fraction of a second before colliding and reacting with other particles to form covalent bonds.

14•1•1 Transition State Often an elementary reaction between two substances proceeds in a simple manner that involves the collision of the two substances to form an activated species that directly gives rise to the products of the reaction. Consider the mechanism for the general elementary reaction

$$AB + AB \longrightarrow A_2 + B_2$$

as shown schematically in Figure 14.1. Not all collisions between two reactant AB molecules result in a chemical reaction, even though the molecules possess certain necessary attributes for the reaction to occur, including high energy and a natural tendency to react. In Figure 14.1(a), for example, collision between AB molecules is fruitless because the molecules are improperly oriented at the moment of impact, with the B portions turned toward one another. In (b), although the molecules are oriented properly, they do not collide with sufficient energy for a reaction to occur. In either case, (a) or (b), the molecules simply rebound without change. In (c), the colliding molecules are properly oriented and have enough energy for a reaction to occur. This condition of colliding molecules that is necessary for a reaction to occur is called the **transition state** or **activated complex.** This is the state that exists at the instant of a potentially efficient collision.

For the reverse of the reaction shown in Figure 14.1,

$$A_2 + B_2 \longrightarrow AB + AB$$

Figure 14•1 Mechanism for the reaction $2AB \rightarrow A_2 + B_2$. In (a), the colliding molecules are not properly oriented; and in (b), they do not collide with enough energy for a reaction to occur. In (c), the colliding molecules are properly oriented and have enough energy to form the activated complex, or transition state.

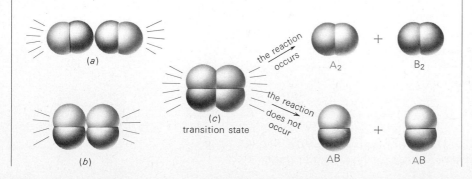

the colliding molecules, A_2 and B_2, also must be properly oriented and have the requisite energy for the breaking of certain existing bonds and the making of new ones. The transition state for this elementary reaction is also that depicted in Figure 14.1(c). In principle, regardless of whether we start with AB molecules or with A_2 and B_2 molecules, the transition complex can dissociate to form new molecules (the reaction occurs), or it can dissociate to re-form the original molecules (the reaction does not occur). Specific examples of reactions that are thought to proceed through cyclic transition states whose structures resemble Figure 14.1(c), in which four atomic centers collide, are the following:

$$2IBr \longrightarrow I_2 + Br_2$$
$$HCl + Br_2 \longrightarrow HBr + BrCl$$

For the reaction between two IBr molecules, the cyclic transition state is depicted as the middle structure in the following reaction scheme:

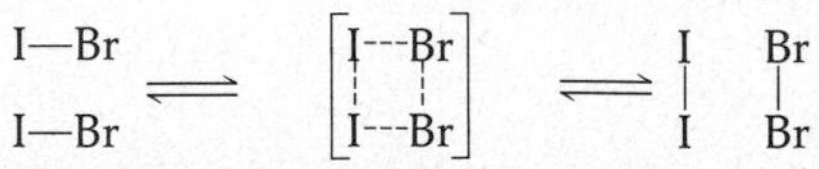

transition state

The dashed lines indicate that in the four-atom, cyclic transition state, the I—Br bonds are partially broken and the I—I and Br—Br bonds are partially formed.

Most elementary reactions between molecules do not proceed by way of such cyclic transition states involving simultaneous collisions of four atomic centers. For example, each of the following reactions presumably involves collisions between only two atomic centers:

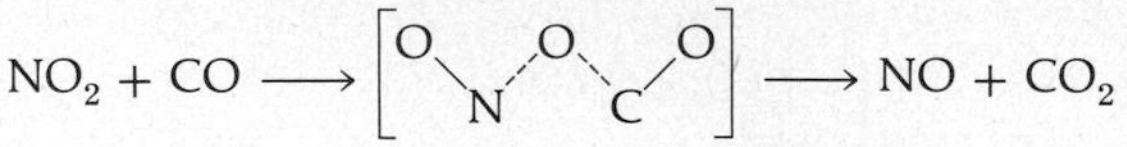

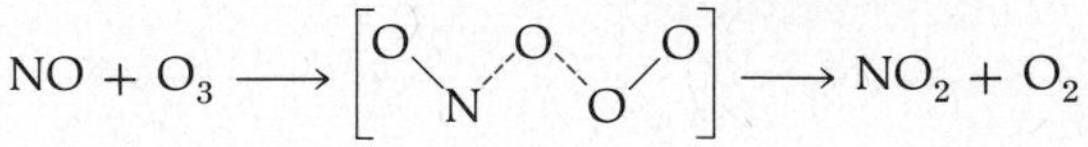

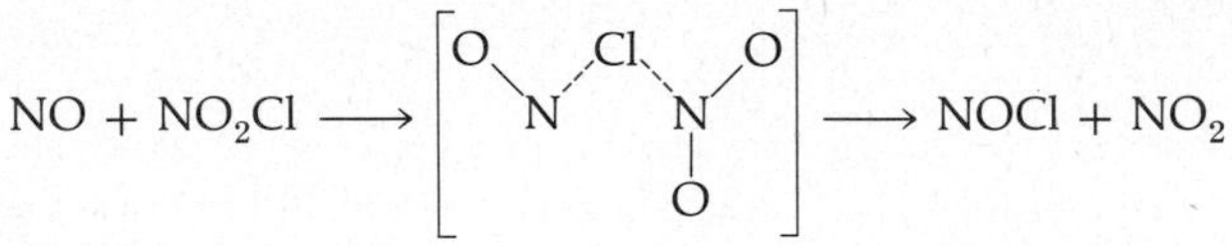

The transition states for these reactions (within the brackets) all involve the partial breakage of an old bond and the partial formation of a new one, as indicated by the dashed lines. Effective collisions between H_2 and O_2 molecules to give H and HO_2 radicals,

$$H_2 + O_2 \longrightarrow [H\text{-}\text{-}\text{-}H\text{-}\text{-}O\text{—}O] \longrightarrow H + HO_2$$

should also proceed by a similar pathway; in the transition state, the H—H bond is partially broken, while the H—O bond is partially formed.

Reactions in which products are formed in one elementary reaction step are not limited to inorganic molecules and radicals. Numerous reactions of organic compounds proceed in this way. The reaction between hydroxide ion and methyl bromide, CH_3Br, is an example, the course of which is shown schematically in Figure 14.2.

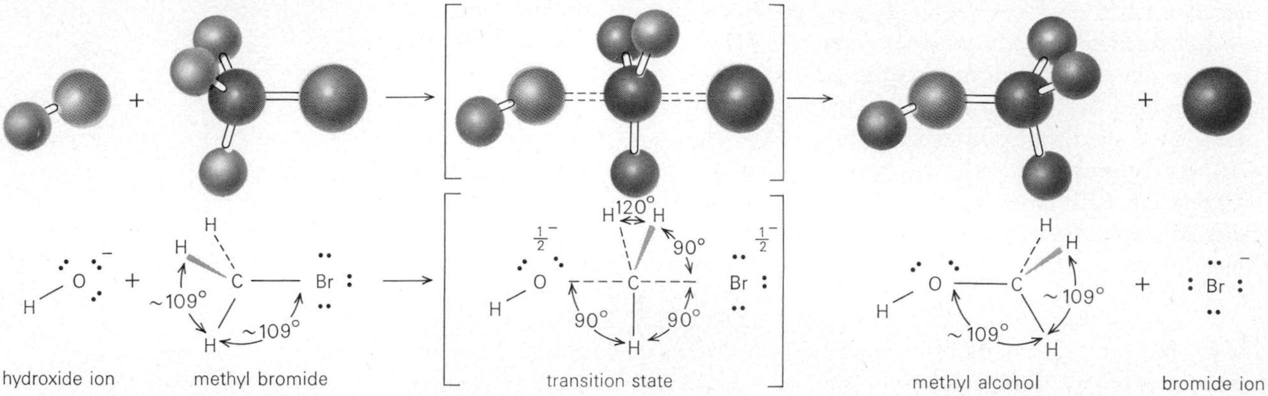

hydroxide ion   methyl bromide   transition state   methyl alcohol   bromide ion

In many instances, the final products of a reaction do not arise by means of one simple effective collision but by means of a series of effective collisions that involve a series of elementary reaction steps. For example, consider again the reaction of H_2 and O_2 molecules to give the H and HO_2 radicals:

$$H_2 + O_2 \longrightarrow H + HO_2 \qquad (1)$$

These products, H and HO_2, serve as reactants for the subsequent reactions:

$$H_2 + HO_2 \longrightarrow H_2O + OH \qquad (4)$$

$$H + OH \longrightarrow H_2O \qquad (6)$$

In these three elementary reactions, (1), (4), and (6), the H, HO_2, and OH radicals are initial products that are not isolated. Instead, they react further to give the final products of the reaction, H_2O molecules. Substances that act in this way are called **reaction intermediates.** Reaction intermediates may frequently be detected experimentally, and in some cases they may be isolated. They generally have only a transient existence, although for certain reactions and under certain conditions, they can be isolated and character-ized. A transition state, on the other hand, cannot be isolated. Its structure is only inferred from our suppositions of the probable reaction mechanism.

14•2   Reaction Rate

The **rate** or **speed of a reaction** is the change in concentration of reactants or products in a unit of time. The rate of a reaction may be expressed as the rate of decrease in the concentration of a reactant, or as the rate of increase in the concentration of a product. Concentrations are usually expressed in moles per liter, but for gas phase reactions, pressure units of

Figure 14•2 Mechanism for the re-action $HO^- + CH_3Br \rightarrow CH_3OH + Br^-$. In the transition state, the C—O bond is half formed and the C—Br bond is half broken. At this point, both the Br and OH groups bear a fraction ($\sim\frac{1}{2}$ each) of the unit negative charge. Note the bond an-gles between atoms attached to car-bon in the transition state; contrast these angles with those in the or-ganic reactant and product.

atmospheres, millimeters of mercury, or pascals may be used in place of concentrations. The unit of time may be a second, a minute, an hour, a day, or even a year, depending on whether the reaction is fast or slow.

14•2•1   Experimental Procedures   In order to measure the rates of chemical reactions, it is necessary to analyze directly or indirectly the amount of a product formed or the amount of a reactant left after suitable time intervals. Because rates of chemical reactions are markedly affected by temperature changes (see Section 14.2.2), it is necessary that the reaction mixture be maintained at a constant temperature. This requirement usually can be achieved by immersing the reaction vessel in a thermostatically regulated temperature bath of water or oil. Methods for determining the concentrations of reactants or products vary according to the type of reaction being investigated and the physical states of the reaction components.

• *Gas Phase Reactions* • For gas phase reactions, the composition of the gas mixture is often determined by chemical analysis. In recent years, the techniques of gas chromatography (see Section 12.9.3) and mass spectrometry have reduced considerably the labor involved in such analyses. A widely used indirect method for gaseous reactions involves the measurement of the pressure of the system. This method is applicable if there is an increase or decrease in pressure caused by an increase or decrease in the number of molecules produced by the reaction. Consider the decomposition of ammonia into nitrogen and hydrogen,

$$2NH_3 \longrightarrow N_2 + 3H_2$$

In a container of constant volume, the pressure will increase as the reaction progresses, because the pressure exerted by four product molecules, one of N_2 and three of H_2, is ideally twice that exerted by two reactant molecules. For example, if only NH_3 is present initially at a pressure of 2.00 atm, when 25 percent has reacted, the total pressure will be 2.50 atm. When 50 percent has reacted, the total pressure will be 3.00 atm:

percent reaction	p_{NH_3}, atm	p_{N_2}, atm	p_{H_2}, atm	P_{total}, atm
0	2.00	0	0	2.00
25	1.50	0.25	0.75	2.50
50	1.00	0.50	1.50	3.00

Conversely, in a container of constant volume, the formation of ammonia from nitrogen and hydrogen,

$$N_2 + 3H_2 \longrightarrow 2NH_3$$

causes a decrease in pressure as the reaction progresses. For example, suppose the initial pressures of N_2, H_2, and NH_3 are 1.00 atm, 3.00 atm, and 0 atm, respectively. When 25 percent of the N_2 and H_2 has reacted, the total pressure will be 3.50 atm. When 50 percent of each has reacted, the total pressure will be 3.00 atm:

percent reaction	p_{N_2}, atm	p_{H_2}, atm	p_{NH_3}, atm	P_{total}, atm
0	1.00	3.00	0	4.00
25	0.75	2.25	0.50	3.50
50	0.50	1.50	1.00	3.00

The pressure measurements can be made while the reaction is in progress by connecting a pressure-measuring device to the reaction vessel. If a gas phase reaction does not involve a change in pressure, that is, if the number of molecules remains constant as the reaction proceeds, analysis by pressure measurement still may be possible if one of the reaction components can be condensed to a liquid, thereby reducing the pressure of the system.

• *Reactions in Solution* • Analyses of reactant and product concentrations are generally simplest when a reaction is studied in solution. One of the traditional solution methods involves dividing the reactant solution among a number of sealed vials or ampoules, placing these in a constant-temperature bath, and then removing them one by one at appropriate time intervals. The reaction in an individual vial is essentially stopped by rapid chilling in an ice bath, after which the vial is opened and a precise volume of solution is withdrawn by means of a pipet. The contents of the solution are then determined by an appropriate analytical method.

The rate of the reaction of methyl bromide with water to produce methyl alcohol and hydrobromic acid,

$$CH_3Br + HOH \longrightarrow CH_3OH + HBr$$

may be followed by removing samples and titrating the hydrobromic acid with a standard NaOH solution. The quantity of moles of NaOH required to neutralize the HBr equals the quantity of moles of HBr formed. This is also the quantity of moles of CH_3Br that have reacted. (See Exercise 42 for an example of this experimental procedure.)

The rate of reaction of hydroxide ion with methyl bromide to produce methyl alcohol and bromide ion,

$$CH_3Br + OH^- \longrightarrow CH_3OH + Br^-$$

may also be followed by titration. In this case, the unreacted OH^- is titrated with a standard acid solution, such as hydrochloric acid. The moles of HCl required to neutralize the solution equal the moles of OH^- remaining. The moles of OH^- at the outset minus the moles of OH^- remaining equal the moles of OH^- that reacted. This quantity also is the moles of CH_3Br that have reacted. (See Exercise 42 for an example of this experimental procedure.)

Three other examples of reactions of organic compounds that are of interest[1] and whose rates are conveniently followed by titration are the

---

[1] Equation (7) is an example of the general type of organic reaction to form an ester called *esterification*. Equation (9) is an example of the general type of reaction called *saponification* (Latin *sapo,* soap), because the basic hydrolysis of certain complex natural esters has been used for centuries to make soap.

preparation of ethyl acetate, an ester, by the reaction of acetic acid with ethyl alcohol, Equation (7); the hydrolysis of ethyl acetate, Equation (8); and the basic hydrolysis of ethyl acetate, Equation (9):

$$CH_3CO_2H + C_2H_5OH \longrightarrow CH_3CO_2C_2H_5 + H_2O \qquad (7)$$
$$\text{acetic acid} \qquad \text{ethyl alcohol} \qquad \qquad \text{ethyl acetate}$$

$$CH_3CO_2C_2H_5 + H_2O \longrightarrow CH_3CO_2H + C_2H_5OH \qquad (8)$$

$$CH_3CO_2C_2H_5 + OH^- \longrightarrow CH_3CO_2^- + C_2H_5OH \qquad (9)$$

In the first two reactions, represented by Equations (7) and (8), the acetic acid is titrated with standard base, whereas in the reaction represented by Equation (9), the unreacted OH^- is titrated with standard acid.

   If a reaction produces a change in solution color owing to the disappearance of reactants or appearance of products, we can follow the progress of the reaction by measuring the change in the intensity of the color. Even if there is no color change during a reaction, rates still may be obtained if there is a change in the absorption spectrum of a reactant or product (see Section 20.6.2) as the reaction progresses.

   When ordinary analytical methods are used, it is difficult to measure the reaction rate for a rapid reaction, because the reaction may be practically over before even one analysis can be performed. However, by means of specialized instruments, chemists have been able to estimate the rates of extremely fast reactions that are complete within a few picoseconds. (A picosecond is 10^{-12} or a million-millionth of a second.) Some examples of such reactions are the detonation of explosives, the transfer of protons from acids to bases, the photochemical events of vision, and the primary events of photosynthesis.

14·2·2   Factors Affecting Reaction Rates   The rate at which a given chemical reaction takes place depends upon four factors: (1) the *nature* of the reactants, (2) the *temperature*, (3) the presence or absence of *catalytic substances*, and (4) the *concentration* of the reactants.

   • *Nature of Reactants* • Substances differ markedly in the rates at which they undergo chemical change. Hydrogen and fluorine molecules react explosively, even at room temperature, producing hydrogen fluoride molecules:

$$H_2 + F_2 \longrightarrow 2HF \qquad \text{(very fast at room temperature)}$$

Under similar conditions, hydrogen and oxygen molecules react so slowly that no chemical change is apparent:

$$2H_2 + O_2 \longrightarrow 2H_2O \qquad \text{(very slow at room temperature)}$$

Nickel and iron in the atmosphere corrode at different rates even when the temperature and concentrations are the same for both. In a relatively short time, iron oxide (rust) can be seen on the iron, but the surface of the nickel appears to be unchanged.

   Sodium reacts very rapidly with water at room temperature, but more slowly with methyl alcohol and ethyl alcohol.

Each of the preceding reactions is spontaneous; that is, the change in free energy, ΔG, is negative. The differences in reactivity may be attributed to the different structures of the atoms and molecules of the reacting materials. If a reaction involves two species of molecules with atoms that are already joined by strong covalent bonds, collisions between these molecules at ordinary temperatures may not provide enough energy to break these bonds. For example, in the reaction between H_2 and O_2 to produce H_2O, the bond dissociation energies for H_2 and O_2 are 436.0 and 498.3 kJ/mole (see Table 13.2). These high values indicate these covalent bonds are very strong. In contrast, the bond dissociation energy for F_2 of 157 kJ/mole is less than one-third that for O_2. This difference in bond dissociation energies helps explain why H_2 and F_2 react faster than H_2 and O_2 at room temperature.

• *Activation Energy* • During chemical changes, it is necessary for the reacting molecules to collide as they move about in a random way. However, for many spontaneous, exothermic reactions, at room temperature most of the molecules simply rebound from collisions without reacting, as shown in Figure 14.1(*a*) and (*b*). For example, in a mixture of hydrogen and oxygen at room temperature, the molecules collide repeatedly with each other and rebound without change. The situation is analogous to that depicted in Figure 14.3(*a*). A ball that is resting in a depression on a hillside will give up energy if it can roll down the hill. However, it will not do this unless its energy is first raised enough to get it out of the depression. Similarly, in a chemical system molecules cannot react unless they possess enough energy to form the transition state [see Figure 14.3(*b*)].

In the case of the hydrogen and oxygen molecules at room temperature, the molecules on the average do not have enough energy to form the transition state, similar to Figure 14.1(*c*). But if a lighted match is placed in the container, the molecules in the vicinity of the flame gain enough energy to react when they collide with each other. The overall energy release is transferred to nearby molecules, causing the reaction to spread very rapidly to all parts of the container.

The added energy that reacting substances must have to form the activated complex, or transition state, is called the **activation energy,** E_a, of the reaction. The activation energy for a specific reaction depends primarily on the nature of the reactants.

Suppose that a ball is resting at the bottom of a hill, and it is desired to roll the ball up to a depression on the hillside. In this case, as depicted by Figure 14.3(*c*), sufficient energy is required to push the ball up over the rim of the depression. This energy is analogous to the activation energy of the chemical system described in Figure 14.3(*d*).

To consider a system in which the products can react to form the original reactants, the plot in Figure 14.3(*e*) is useful; it is a combination of graphs similar to (*b*) and (*d*). Figure 14.3(*e*) shows clearly that the activation energy for an exothermic process is less than the activation energy for the reverse endothermic process.

In Section 13.8.1, the difference in internal energies between products and reactants was expressed as ΔE of the reaction. For most chemical reactions the difference between ΔE and ΔH, the change in enthalpy, is quite small, and we can say that $\Delta E \simeq \Delta H$. For the exothermic process depicted by Figure 14.3(*b*), energy is released and ΔE of the reaction is negative; for the endothermic process depicted by Figure 14.3(*d*), energy is absorbed and Δ*E* of the reaction is positive.

For a reversible reaction, the energy liberated in the exothermic reaction equals the energy absorbed in the endothermic reaction. The energy of the reaction, Δ*E*, is equal to the difference between the activation energies of the opposing reactions; that is, $\Delta E = E_a - E_a'$, as diagrammed in Figure 14.3(*e*).

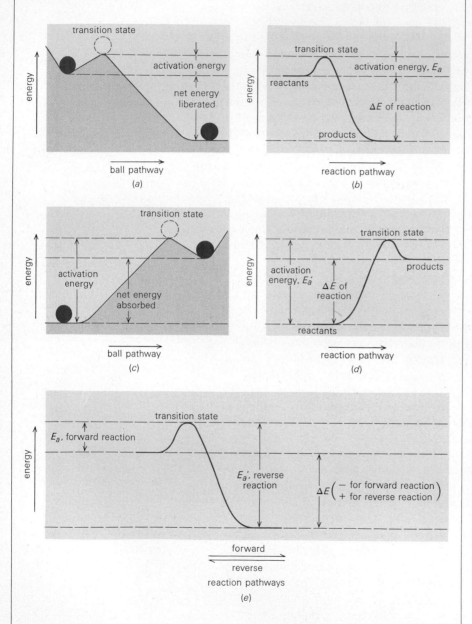

Figure 14•3 Mechanical analogies for exothermic and endothermic reactions. (*a*) A ball in a depression on a hillside must be raised (given additional energy) before it can roll down the hill spontaneously and release energy. (*b*) This graph describes a chemical system analogous to the mechanical system in part (*a*). (*c*) To roll a ball uphill to a depression requires enough energy to push it over the highest point. (*d*) This graph describes a chemical system analogous to the mechanical system in part (*c*). (*e*) For a reversible chemical reaction, $\Delta E = E_a - E_a'$.

• *Temperature* • The speed of a chemical reaction is increased by an increase in temperature. A rise of 10 °C in temperature usually doubles or triples the speed of a reaction between molecules. We can account for this increase in reaction rate partly on the grounds that molecules move about more rapidly at higher temperatures and consequently collide with each other more frequently. However, this does not tell the whole story, unless the activation energy for the reaction is essentially zero. As the temperature is increased, not only do the molecules collide more often, but they also collide with greater impact because they are moving more rapidly. At the elevated temperatures, larger percentages of the collisions result in a chemical reaction because larger percentages of the molecules have greater velocities and therefore sufficient energy to react (see Figure 14.4).

• *Presence of a Catalyst* • As stated in Chapter 9, a *catalyst* is a substance that increases the speed of a chemical reaction without itself undergoing a permanent chemical change. The process is called *catalysis*.

A catalyst is thought to influence the speed of a chemical reaction in one of two ways: (1) by the formation of intermediate compounds (homogeneous catalysis) or (2) by adsorption (heterogeneous catalysis).

• *Formation of Intermediate Compounds* • Let us again refer to the mechanical analogy depicted in Figure 14.3(*a*). The ball will roll down the hill if it is first pushed to the top of the hump. If the hump is very high relative to the depression, it may be desirable first to lower the hump or energy barrier by digging the top off. In chemical reactions that have high activation energies, we can "lift" the reactants over the energy barrier by raising the temperature. Frequently, however, it is not desirable to carry out a given reaction at a high temperature, because the products of the reaction may be unstable or because they may reform the reactants faster at the high temperature (see Section 15.5), thereby decreasing the yield of the desired reaction products. An alternate approach is to look for a way of lowering the energy barrier, that is, to provide a path with a lower activation energy so that molecules with lower energies can react.

How can the path of a reaction be changed? One way is to find a substance, that is, a catalyst, that can react with both energy-poor and energy-rich molecules to form an *intermediate compound*, which in turn reacts to form the desired substance. Consider the general reaction A + B → AB, with C representing the catalyst, as illustrated in Figure 14.5:

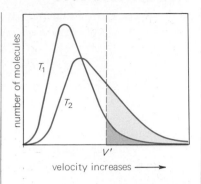

Figure 14•4 The number of molecules that have velocities equal to or greater than some value V' is much greater for the higher temperature, T_2, than for the lower temperature, T_1. Note that the majority of molecules at either temperature have velocities less than V'. A reaction that depends on the collision of two molecules having velocities V' or greater will proceed many times faster at T_2 than at T_1.

Figure 14•5 Graphic representations showing (1) that a catalyst lowers the activation energy of a reaction but does not change the ΔE of the reaction, and (2) that a catalyst lowers activation energies of forward and reverse reactions by the same amount. In this graph the designation $AB^{\ddagger}$ stands for the transition state (or activated complex) for the uncatalyzed reaction; $AC^{\ddagger}$ and $ABC^{\ddagger}$ stand for the transition states for the catalyzed reaction; and AC is the intermediate compound lying in the "valley" between the two transition states. The activation energy for either catalyzed reaction, A + B + C → AB + C or AB + C → A + B + C, always extends to the transition state of higher energy, $AC^{\ddagger}$ in this case.

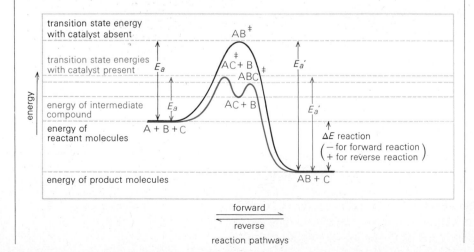

reaction pathways

$$A + B \longrightarrow AB \qquad \text{(very slow reaction)}$$

high activation energy; AB formed slowly

$$A + C \longrightarrow AC \qquad \text{(fast reaction)}$$
$$AC + B \longrightarrow AB + C \qquad \text{(fast reaction)}$$

lower activation energies; AB formed more rapidly

Note that C does not undergo a permanent chemical change; once separated from the products of the reaction, it can be used over and over.

Many chemical reactions are known to follow such a path when a catalyst is used. One such example is the gas phase reaction of sulfur dioxide, SO_2, with oxygen to give sulfur trioxide, SO_3:

$$2SO_2 + O_2 \longrightarrow 2SO_3$$

This oxidation is known to be very slow; it has a high activation energy. However, the rate of the reaction is significantly increased by nitric oxide, NO, which functions as a catalyst. The reactions in the presence of nitric oxide are

$$2NO + O_2 \longrightarrow 2NO_2$$
$$NO_2 + SO_2 \longrightarrow SO_3 + NO$$

These two faster reactions replace the one slow reaction. The nitrogen dioxide formed in the first reaction is the intermediate compound from which nitric oxide is regenerated in the second reaction. The intermediate compound, AC in the general case and NO_2 in the specific example, usually has only a temporary existence, being used up as rapidly as it is formed. However, sometimes the intermediate compound can be isolated and studied if desired.

There are numerous examples of homogeneous reactions in solution whose reaction rates are accelerated in the presence of catalytic substances. One example is the reaction of acetic acid with ethyl alcohol to produce ethyl acetate, a reaction that is catalyzed by strong acids such as H_2SO_4 and HCl:

$$\underset{\text{acetic acid}}{CH_3CO_2H} + \underset{\text{ethyl alcohol}}{C_2H_5OH} \xrightarrow{\;H^+\;} \underset{\text{ethyl acetate}}{CH_3CO_2C_2H_5} + \underset{\text{water}}{H_2O}$$

Without the catalyst present, it may take weeks to obtain the maximum yield of ethyl acetate. In the presence of the acid catalyst, the maximum yield is obtained within a few hours. Again, the catalyst does not increase the amount of ethyl acetate that can be obtained at equilibrium, because the rates of the forward and reverse reactions are increased by the same amount.

Practically all the chemical reactions that take place in living organisms require the presence of catalytic substances known as *enzymes*. The function of enzymes as specific catalytic agents will be taken up in Chapter 27.

• *Adsorption* • Many solid substances that act as catalysts can hold appreciable quantities of gases and liquids on their surfaces by adsorption (see Chapter 12). Finely divided nickel and platinum are well known for their

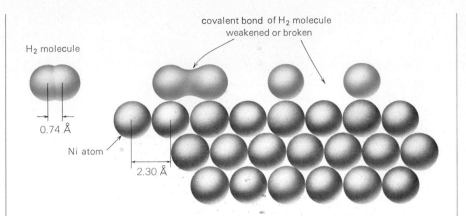

covalent bond of H₂ molecule
weakened or broken

H₂ molecule

0.74 Å

Ni atom

2.30 Å

Figure 14•6 Catalytic action by adsorption. Note that the interatomic dimensions are such that the bond in an H_2 molecule is probably stretched or broken when hydrogen is adsorbed on nickel. According to one theory, this accounts for the greater activity of hydrogen in catalyzed hydrogenation reactions.

ability to adsorb large amounts of various gases. The adsorbed molecules are frequently more reactive than the unadsorbed molecules. This increased reactivity can be attributed in some cases to the increased concentration of the adsorbed molecules; they are crowded close together on the surface of the solid, whereas in the gaseous state they are far apart. In other cases, the attractive forces between the molecules of the solid and those of the adsorbed liquid or gas make the adsorbed molecules more active chemically. This causes the reaction between molecules A and B to take place on the surface of the solid at a faster rate than if the catalyst were not present. (See Figures 14.6 and 14.7.) The catalyst must not strongly adsorb the product of the reaction. As the reaction proceeds, the product leaves the surface and more reactants are adsorbed. Thus the surface is used over and over.

  The hydrogenation of liquid fats to form solid fats is carried out on the surface of finely divided nickel. The oxidation of sulfur dioxide to sulfur trioxide on a platinum surface is another case in which a catalyst increases the speed of a reaction by adsorption. Impurities in a reacting mixture that are strongly adsorbed by the catalyst may act as inhibitors by decreasing the amount of surface. In time, the catalyst may become useless and is said to be poisoned.

  • *Concentration* • As mentioned earlier in this section, the rate of a reaction may be expressed as the rate of decrease in the concentration of a reactant, or as the rate of increase in the concentration of a product. Consider the general reaction

$$A \longrightarrow B + C$$

We express the concentration of A at time t_1 as $[A]_1$ and the concentration at t_2 as $[A]_2$, where the brackets mean concentrations in moles per liter. The average rate of decrease in the concentration of A is expressed as

$$\text{average rate of} \atop \text{decrease in [A]} = \frac{[A]_2 - [A]_1}{t_2 - t_1} = \frac{\Delta[A]}{\Delta t}$$

The average rates of increase in the concentrations of B and C are expressed as

Figure 14•7 Even though a catalyst does not undergo a permanent chemical change, it may change physically. High temperatures, coupled with forces of attraction to adsorbed molecules, can cause crystal changes. The upper photograph shows the polished surface of a copper catalyst before use (70,000 magnification). The lower shows the surface (3,000 magnification) of the copper as it appeared after use. This sample of copper was used continuously for seven days at 400 °C to catalyze the reaction of hydrogen and oxygen to form water.
Source: Courtesy of Professor Allan T. Gwathmey, University of Virginia, Charlottesville.

$$\text{average rate of increase in [B] or [C]} = \frac{[B]_2 - [B]_1}{t_2 - t_1} = \frac{[C]_2 - [C]_1}{t_2 - t_1} = \frac{\Delta[B]}{\Delta t} = \frac{\Delta[C]}{\Delta t}$$

In the expression for the average rate of decrease of [A], the quantity $(\Delta[A]/\Delta t)$ is negative because $[A]_1$ is greater than $[A]_2$. Because rates are conventionally expressed as positive values, we place a minus sign in front of this quantity so that $-(-) = +$. The three rate expressions are related to one another as follows:

$$-\left(\frac{\Delta[A]}{\Delta t}\right) = \frac{\Delta[B]}{\Delta t} = \frac{\Delta[C]}{\Delta t}$$

• Example 14.1 •

Consider the decomposition of nitrogen dioxide, NO_2, into nitric oxide, NO, and oxygen, O_2:

$$2NO_2 \longrightarrow 2NO + O_2$$

(a) Write the expressions for the average rate of decrease in the NO_2 concentration and the average rates of increase in the NO and O_2 concentrations.

(b) If an average rate of decrease in the NO_2 concentration is determined to be $4.0 \times 10^{-13} \text{ mol} \cdot L^{-1} \cdot s^{-1}$, what are the corresponding average rates of increase in the NO and O_2 concentrations?

• Solution •

(a) The average rate of decrease in the concentration of NO_2 is expressed as

$$-\frac{\Delta[NO_2]}{\Delta t}$$

The average rates of increase in the concentrations of NO and O_2 are expressed as

$$\frac{\Delta[NO]}{\Delta t} \quad \text{and} \quad \frac{\Delta[O_2]}{\Delta t}$$

(b) For every two molecules of NO_2 that react, two molecules of NO are produced. Hence, the decrease in the NO_2 concentration and the increase in the NO concentration are proceeding at the same rate:

$$-\frac{\Delta[NO_2]}{\Delta t} = \frac{\Delta[NO]}{\Delta t} = 4.0 \times 10^{-13} \text{ mol} \cdot L^{-1} \cdot s^{-1}$$

For every two molecules of NO that form, only one molecule of O_2 is obtained. Hence, the rate of formation of NO is twice the rate of formation of O_2; that is,

$$\frac{\Delta[NO]}{\Delta t} = 2\left(\frac{\Delta[O_2]}{\Delta t}\right) = 4.0 \times 10^{-13} \text{ mol} \cdot L^{-1} \cdot s^{-1}$$

$$\frac{\Delta[O_2]}{\Delta t} = \frac{4.0 \times 10^{-13} \text{ mol} \cdot L^{-1} \cdot s^{-1}}{2} = 2.0 \times 10^{-13} \text{ mol} \cdot L^{-1} \cdot s^{-1}$$

Note that the ratio of the average rates for the species NO_2, NO, and O_2 is $(4.0 \times 10^{-13} \, mol \cdot L^{-1} \cdot s^{-1}) : (4.0 \times 10^{-13} \, mol \cdot L^{-1} \cdot s^{-1}) : (2.0 \times 10^{-13} \, mol \cdot L^{-1} \cdot s^{-1})$, or $2:2:1$. This is the same ratio as for their coefficients in the balanced equation.

See also Exercises 15 and 17 at the end of the chapter.

One of the ways of obtaining information about the rate of a reaction is to plot the concentration of one of the reactants or products as a function of reaction time. Such a plot for the basic hydrolysis of ethyl acetate,

$$CH_3CO_2C_2H_5 + OH^- \longrightarrow CH_3CO_2^- + C_2H_5OH$$

is shown in Figure 14.8, in which the concentration of OH^- ions is initially 0.0500 mole/L, and the concentration is shown as determined at seven subsequent times (see the actual data given in Table 14.4). Drawn through the initial point, 0, and the seven experimental points is a smooth curve that shows how the concentration, c, of OH^- ions decreases with time.

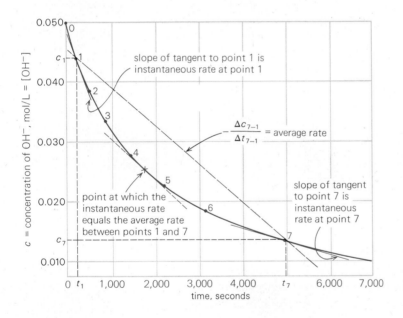

Figure 14•8 Plot of $[OH^-]$ versus time for the basic hydrolysis of ethyl acetate. The instantaneous rates of reaction at points 1 and 7 are described by the slopes of the colored lines, which are the tangents at these two points. The average rate between points 1 and 7 is equal to the slope of the dashed black line connecting these two points. A tangent whose slope is equal to the slope of this average rate line is shown by the dashed colored line; the point at which it is tangent is shown by the colored x.

Note the straight colored lines drawn through points 1 and 7. These are the tangents to the curve at these two points. The slope of the tangent at each point is equal to the rate of decrease in the concentration of OH^- ions at that point, or the **instantaneous rate.** At point 1, the steep slope of the tangent shows that the rate of decrease in the concentration of OH^- ions is very rapid at the beginning of the reaction. This contrasts with the lesser slope at point 7, which shows a much slower rate of reaction at that time. The slopes of the tangents to the curve decrease continually as $[OH^-]$ decreases. After a long enough time, the slope of the tangent would be zero; that is, the tangent would be horizontal, and the reaction would be complete. The instantaneous rate is continually decreasing as the reaction progresses from start to finish.

It is revealing to contrast the instantaneous rates at points 1 and 7 in Figure 14.8 with the average rate of decrease of $[OH^-]$ between these two points. Following the convention of expressing rates as positive values, we write

$$\text{average rate} = -\frac{c_7 - c_1}{t_7 - t_1} = -\frac{\Delta c_{7-1}}{\Delta t_{7-1}}$$

$$= 6.4 \times 10^{-6}\,\text{mol} \cdot \text{L}^{-1} \cdot \text{s}^{-1}$$

If the $[OH^-]$ decreased at this constant average rate between point 1 and point 7, the change in concentration would follow the dashed black line connecting the two points in Figure 14.8. The average rate is of little interest in the study of kinetics because it happens to be equal to the instantaneous rate at only one point.[2] When we refer to the rate of a reaction, *it is assumed that we are referring to an instantaneous rate* that is characteristic of the reaction at one instant under certain definite conditions.

14•3 Determination of Rate Laws or Rate Equations

The way that the rate of a reaction is affected by a change in the concentration of the reactants cannot be predicted from the overall chemical equation. Rather, it must be determined experimentally. For example, consider the following sets of experimental data for the decomposition of nitrogen pentoxide, N_2O_5, versus that of nitrogen dioxide, NO_2, at 25 °C:

$2N_2O_5 \longrightarrow 4NO_2 + O_2$		$2NO_2 \longrightarrow 2NO + O_2$	
$[N_2O_5]$	rate, mol · L^{-1} · s^{-1}	$[NO_2]$	rate, mol · L^{-1} · s^{-1}
0.020	0.70×10^{-6}	0.020	0.75×10^{-13}
0.040	1.4×10^{-6}	0.040	3.0×10^{-13}
0.080	2.8×10^{-6}	0.080	12.0×10^{-13}

Each time the concentration of N_2O_5 is doubled, the rate of the first decomposition is doubled. On the other hand, when the NO_2 concentration is doubled in the second reaction, the rate of its decomposition is increased by a factor of four. Therefore, the rate of the first reaction is directly proportional to the concentration of N_2O_5, whereas the rate of the second reaction is directly proportional to the square of the concentration of NO_2. These experimentally determined facts may be represented mathematically as

$$\text{rate} \propto [N_2O_5] \quad \text{and} \quad \text{rate} \propto [NO_2]^2$$

Each of these proportionalities can be converted to an equation by use of a constant, k:

---

[2] We can determine this point by finding where the tangent to the curve in Figure 14.8 is parallel to the dashed black line.

$$\text{rate} = k[N_2O_5] \quad \text{and} \quad \text{rate} = k[NO_2]^2$$

An equation that describes the relationship between the reaction rate and the concentration of reactant(s) is called a **rate equation** or a **rate law.** The proportionality constant, k, is referred to as the **rate constant** for a particular reaction. Because the concentration of the reactant, $[N_2O_5]$ or $[NO_2]$ in these examples, decreases as the reaction progresses, the rate must decrease with time. The rate constant, k, however, remains unchanged throughout the course of the reaction. Hence, the rate constant provides a convenient measure of reaction speed. The more rapid the reaction, the larger is the value of k; the slower the reaction, the smaller is its value.

━━━ • Example 14•2 • ━━━

Using the data given in the text for the N_2O_5 and NO_2 concentrations at $0.080M$, calculate the rate constants for the two decomposition reactions.

━━ • Solution • For N_2O_5, $\text{rate} = k[N_2O_5]$

$$2.8 \times 10^{-6}\,\text{mol}\cdot\text{L}^{-1}\cdot\text{s}^{-1} = k(8.0 \times 10^{-2}\,\text{mol}\cdot\text{L}^{-1})$$

$$k = \frac{2.8 \times 10^{-6}\,\text{mol}\cdot\text{L}^{-1}\cdot\text{s}^{-1}}{8.0 \times 10^{-2}\,\text{mol}\cdot\text{L}^{-1}} = 3.5 \times 10^{-5}\,\text{s}^{-1}$$

For NO_2, $\text{rate} = k[NO_2]^2$

$$12 \times 10^{-13}\,\text{mol}\cdot\text{L}^{-1}\cdot\text{s}^{-1} = k(8.0 \times 10^{-2}\,\text{mol}\cdot\text{L}^{-1})^2$$

$$k = \frac{12 \times 10^{-13}\,\text{mol}\cdot\text{L}^{-1}\cdot\text{s}^{-1}}{(8.0 \times 10^{-2}\,\text{mol}\cdot\text{L}^{-1})^2} = \frac{12 \times 10^{-13}\,\text{mol}\cdot\text{L}^{-1}\cdot\text{s}^{-1}}{6.4 \times 10^{-3}\,\text{mol}^2\cdot\text{L}^{-2}}$$

$$= 1.9 \times 10^{-10}\,\text{L}\cdot\text{mol}^{-1}\cdot\text{s}^{-1}$$

See also Exercise 16.

As another example of the influence of concentration, consider the union of hydrogen and iodine at a temperature at which all the substances are gases:

$$H_2 + I_2 \longrightarrow 2HI$$

This reaction has been thoroughly investigated by experiment, and the rate near 400 °C has been found to be directly proportional to the concentration of each reactant. The rate equation for this reaction is

$$\text{rate} = k[H_2][I_2]$$

Suppose the rate of this reaction is determined initially at a certain concentration of H_2 and I_2, and then in a second experiment the concentration of H_2 is doubled and the concentration of I_2 is tripled. The second reaction should proceed six times faster than the speed determined for the first. For the reverse reaction, the decomposition of hydrogen iodide,

$$2HI \longrightarrow H_2 + I_2$$

it has been determined that the rate is directly proportional to the square of the HI concentration:

$$rate = k[HI]^2$$

• **Example 14.3** •

In a container with a concentration of hydrogen iodide of $0.040M$, the rate of decomposition of HI is determined to be 8.0×10^{-6} mol $\cdot$ L^{-1} $\cdot$ s^{-1}. What will the reaction rate be, at the same temperature, when the concentration of HI has decreased to $0.010M$?

• **Solution** • The rate equation for the decomposition is

$$rate = k[HI]^2$$

For the first rate, at $0.040M$,

$$8.0 \times 10^{-6} \text{ mol} \cdot \text{L}^{-1} \cdot \text{s}^{-1} = k(0.040M)^2$$

and for the second rate, at $0.010M$, we can write

$$rate_2 = k(0.010M)^2$$

Solving each equation for k gives

$$k = \frac{8.0 \times 10^{-6} \text{ mol} \cdot \text{L}^{-1} \cdot \text{s}^{-1}}{(0.040M)^2} = \frac{rate_2}{(0.010M)^2}$$

$$rate_2 = \frac{(8.0 \times 10^{-6} \text{ mol} \cdot \text{L}^{-1} \cdot \text{s}^{-1})(0.010M)^2}{(0.040M)^2}$$

$$= 5.0 \times 10^{-7} \text{ mol} \cdot \text{L}^{-1} \cdot \text{s}^{-1}$$

Note that the second rate, 5.0×10^{-7} mol $\cdot$ L^{-1} $\cdot$ s^{-1}, is one sixteenth of the first rate:

$$\frac{5.0 \times 10^{-7} \text{ mol} \cdot \text{L}^{-1} \cdot \text{s}^{-1}}{8.0 \times 10^{-6} \text{ mol} \cdot \text{L}^{-1} \cdot \text{s}^{-1}} = \frac{1}{16}$$

The concentration decreased by a factor of four, $0.010/0.040 = 1/4$. Because the rate is proportional to the square of the HI concentration, a decrease in concentration by a factor of four causes the rate to be decreased by a factor of sixteen.

See also Exercise 20.

In summary, the rate of a reaction is proportional to the concentration of each reactant raised to some power. *The power to which the concentration is raised must be determined experimentally.* In many cases the power is 1 (as in the case of the reaction involving N_2O_5 and that involving H_2 and I_2); in other cases it is 2 (as for the decomposition of NO_2 and HI); and in still other cases it may be fractional or even zero.

14•4 Order of a Chemical Reaction

The **order** of a reaction is the sum of all the exponents of the concentration terms in the rate equation. If the rate of a chemical reaction is proportional to the first power of the concentration of just one reactant,

$$\text{rate} = k[A]$$

the reaction is said to be a **first-order reaction.**[3] The decomposition of N_2O_5 is an example of a first-order reaction. If the rate of a reaction is proportional to the second power of one reactant,

$$\text{rate} = k[A]^2$$

or if it is proportional to the first power of the concentration of two reactants,

$$\text{rate} = k[A][B]$$

the reaction is said to be a **second-order reaction.** We can also speak of the order with respect to each reactant. For example, in the latter rate equation, the reaction rate is first order in A and first order in B, or second order overall. A reaction can be of third order or possibly higher, but such cases are rare. In complicated reactions, the rate may be of fractional order, for example, first order in A and 0.5 order in B, or 1.5 order overall. Examples of first-order, second-order, third-order, and fractional-order reactions are listed in Table 14.1.

A reaction may be independent of the concentration of a reactant. Consider the general reaction $A + B \rightarrow AB$, which has been found to be first order in A. If increasing the concentration of B does not increase the rate of the reaction, then the reaction is said to be **zero order** with respect to B. This may be expressed as

$$\text{rate} = k[A][B]^0 = k[A]$$

The order of a reaction cannot be obtained from the coefficients of the reactants in the balanced equation. In the decompositions of N_2O_5 and NO_2, the coefficient for the reactant in each balanced equation is 2, but the former reaction is first order in N_2O_5 and the latter is second order in NO_2. As is illustrated by Example 14.4, a reactant may not even appear in the rate equation for a reaction. The order of a reaction is assigned only on the basis of experimental determinations and merely provides information about the way in which the rate depends on the concentrations of certain reactants. Theoretical predictions about orders of unfamiliar reactions are seldom successful. For instance, knowing that the reaction between H_2 and I_2 is second order, we might predict that the reaction between H_2 and Br_2 would

---

[3]First-order rate equations of the type rate $= k[A]^{1/2}[B]^{1/2}$ or $k = [A]^2[B]^{-1}$, and so on, are possible, but these rather unusual cases will not be treated in this text.

also be second order. It is not, but instead it has a more complex rate equation.

• Example 14•4 •

Consider the general reaction $2A + B_2 \rightarrow 2AB$ and the following experimental data:

experiment	[A]	[B$_2$]	rate, mol $\cdot$ L^{-1} $\cdot$ s^{-1}
1	0.50	0.50	1.6×10^{-4}
2	0.50	1.00	3.2×10^{-4}
3	1.00	1.00	3.2×10^{-4}

Write the most probable rate equation for this reaction.

• Solution • Comparing the data in experiment 2 with those in experiment 1, we note that when the concentration of B_2 is doubled, the rate is doubled. Hence the reaction is first order in B_2. Comparing the data in experiment 3 with those in experiment 2, we note that when the concentration of A is doubled, the rate is unchanged. Hence the reaction is zero order in A. The most probable rate equation is

$$\text{rate} = k[A]^0[B_2] \quad \text{or} \quad \text{rate} = k[B_2]$$

See also Exercises 18, 19, and 21.

Table 14•1 Examples of reactions of different orders

First-Order Reactions	Rate
$2N_2O_5 \longrightarrow 4NO_2 + O_2$	$k[N_2O_5]$
$2H_2O_2 \longrightarrow 2H_2O + O_2$	$k[H_2O_2]$
$SO_2Cl_2 \longrightarrow SO_2 + Cl_2$	$k[SO_2Cl_2]$
$C_2H_5Cl \longrightarrow C_2H_4 + HCl$	$k[C_2H_5Cl]$

Second-Order Reactions	
$NO + O_3 \longrightarrow NO_2 + O_2$	$k[NO][O_3]$
$2NO_2 \longrightarrow 2NO + O_2$	$k[NO_2]^2$
$NO_2 + CO \longrightarrow NO + CO_2$	$k[NO_2][CO]$
$H_2 + I_2 \longrightarrow 2HI$	$k[H_2][I_2]$
$C_2H_4Br_2 + 3KI \longrightarrow C_2H_4 + 2KBr + KI_3$	$k[C_2H_4Br_2][KI]$
$CH_3CO_2C_2H_5 + OH^- \longrightarrow CH_3CO_2^- + C_2H_5OH$	$k[CH_3CO_2C_2H_5][OH^-]$

Third-Order Reactions	
$2NO + O_2 \longrightarrow 2NO_2$	$k[NO]^2[O_2]$
$2NO + Cl_2 \longrightarrow 2NOCl$	$k[NO]^2[Cl_2]$
$2NO + Br_2 \longrightarrow 2NOBr$	$k[NO]^2[Br_2]$
$2NO + 2H_2 \longrightarrow N_2 + 2H_2O$	$k[NO]^2[H_2]$

Fractional-Order Reactions	
$CO + Cl_2 \longrightarrow COCl_2$	$k[CO][Cl_2]^{3/2}$
$COCl_2 \longrightarrow CO + Cl_2$	$k[COCl_2][Cl_2]^{1/2}$

14•5 Determination of Reaction Order from Experimental Data

One of the primary goals of the study of chemical kinetics is to determine each individual reaction step that is involved in the conversion of reactants to products. Once this mechanism of a reaction is known, it may be possible to alter the reaction conditions to increase the rate of formation and yield of a desired product. Or, it may be possible to alter the reaction conditions to decrease the rate of formation and yield of a product that is not desired.

One of the ways to study a reaction mechanism is to determine the order of the reaction. Once the reaction order is known, we can choose between those mechanisms that are highly probable and those that are not.

In Section 14.2.1, experimental procedures for following the rates of chemical reactions were described. Once the experimental data for a particular reaction have been collected, the data can be assembled in tabular form (see Table 14.3) or as a plot similar to that shown for the saponification of ethyl acetate in Figure 14.8. The order of a reaction can be found by determining how changes in the concentration affect the rate of the reaction.

In this discussion we will limit ourselves to methods used to determine first-order and second-order reactions. The customary procedure is to start with a rate equation that might apply to the reaction under consideration. Three such rate equations are listed in Table 14.2. The integrated forms of these rate equations are applied to the experimental kinetic data to see which equation best fits. The fit may be determined in a number of ways: (1) by calculations of the rate constant, k, (2) by a graphic plot, or (3) by determination of the half-life of the reaction. The equation that fits best reveals the order of the reaction.

Table 14•2 Rate equations[a]

Reaction	Order	Differential form[b]	Integrated form[b]	Units of k
A $\longrightarrow$ product(s)	1	rate $= k[A]$	$2.303 \log \dfrac{[A]_0}{[A]_t} = kt$	s^{-1}
2A $\longrightarrow$ product(s)	2	rate $= k[A]^2$	$\dfrac{1}{[A]_t} - \dfrac{1}{[A]_0} = kt$	$L \cdot mol^{-1} \cdot s^{-1}$
xA $+ y$B $\longrightarrow$ product(s)	2	rate $= k[A][B]$	$\dfrac{2.303}{[A]_0 - [B]_0} \log \dfrac{[B]_0[A]_t}{[A]_0[B]_t} = kt$	$L \cdot mol^{-1} \cdot s^{-1}$

[a]The units of the rate in every case are $mol \cdot L^{-1} \cdot s^{-1}$. $[A]_0$ is the concentration of A at time 0 (the initial concentration), and $[A]_t$ is the concentration after the passage of time t.

[b]The differential and integrated forms of a rate equation are simply two equivalent ways of mathematically relating the rate of the reaction to the changing concentrations of the reactant(s). To use these equations, it is not necessary that we be familiar with the operations of calculus that convert one form to the other.

14•5•1 Determination of First-Order Reactions

The simplest type of reaction to study is a first-order reaction in which there is just one reactant undergoing a chemical change.

• *Calculation of k for a First-Order Reaction* • As shown in Table 14.2, the rate equation for a first-order reaction, rate = $k[A]$, where [A] is the concentration of the reacting substance in moles per liter, can also be expressed in the form[4]

$$2.303 \left(\log \frac{[A]_0}{[A]_t} \right) = kt \qquad (10)$$

or

$$k = \frac{2.303}{t} \left(\log \frac{[A]_0}{[A]_t} \right) \qquad (11)$$

To illustrate the use of Equation (11) to calculate the rate constant, k, for a first-order reaction, we will look at a reaction between an alkali metal, M, and liquid ammonia as an example:

$$M + NH_3 \longrightarrow MNH_2 + \tfrac{1}{2}H_2$$

Many higher-order reactions appear to be first order if one of the reactants is present in such a large amount that its concentration does not change appreciably during the reaction. This is the case for the metal–ammonia reaction. A very small amount of a metal—potassium, for example—is dissolved in a relatively large volume of liquid ammonia (enough, say, to make about a $1 \times 10^{-3} M$ solution). Such a small fraction of the ammonia reacts that the amount of it present during the reaction is practically the same from start to finish. The ammonia is said to be present in a *swamping concentration*. The rate equation for the disappearance of the metal could be written as

$$\text{rate} = k'[A][NH_3]$$

where [A] stands for the concentration of the metal. But because [NH$_3$] is a constant, the product of $k'[NH_3]$ is also a constant, k, so that the rate equation is written

$$\text{rate} = k[A]$$

A reaction, like the one just discussed, that appears to be first order although it involves more than one reactant is called a **pseudo first-order reaction.** Irrespective of whether a reaction is truly first order or pseudo first order, the rate equation has the same form.

For the potassium–ammonia reaction, the rate is followed experimentally by determining the change in color of the solution with time. The reactant solution is an intense blue at the beginning, then slowly fades in color and becomes practically colorless as the reaction approaches completion. The amount of unreacted potassium remaining in solution at any time is determined from the depth of color left, which is measured with sensitive

---

[4]The solution to this integration for the first-order case is shown in Appendix 2.

photoelectric cells. (See Figure 24.15 on Plate 7.) Data for a typical experiment are listed in Table 14.3. From these data, the first-order rate constant can be calculated by the use of Equation (11). Example 14.5 illustrates one such calculation. *The fact that values of k calculated in this manner* (see the last column in Table 14.3) *are reasonably constant indicates that the reaction is indeed a first-order reaction.*

Table 14•3 **Change in molar concentration of potassium with time for the reaction** $K + NH_3 \longrightarrow K^+ + NH_2^- + \frac{1}{2}H_2$

| Time | | Intensity | Molar | Log molar | Rate constant |
h	min	of color	concentration	concentration	$k \times 10^2$, h^{-1}
0	0	1.702	1.13×10^{-3}	-2.947	—
2	33	1.476	9.80×10^{-4}	-3.009	5.59
4	34	1.322	8.77	-3.057	5.55
6	36	1.196	7.94	-3.100	5.35
9	38	0.996	6.62	-3.179	5.55
12	31	0.841	5.59	-3.253	5.62
21	26	0.506	3.36	-3.474	5.66

• Example 14•5 •

Use the initial potassium concentration in Table 14.3 and the concentration after 4 hours and 34 minutes to calculate the rate constant, k, for the potassium–ammonia reaction.

• Solution • We solve Equation (11) with $t = 4$ h, 34 min = 4.57 h; $[A]_0 = 1.13 \times 10^{-3} M$; and $[A]_t = 8.77 \times 10^{-4} M$:

$$k = \frac{2.303}{4.57 \text{ h}} \left(\log \frac{1.13 \times 10^{-3}}{8.77 \times 10^{-4}} \right) = \frac{2.303}{4.57 \text{ h}} (\log 1.288)$$

$$= \frac{2.303}{4.57 \text{ h}} (0.1099)$$

$$= 5.54 \times 10^{-2} \text{ h}^{-1}$$

This calculation has been made by using Table A.8 to look up the logarithm 0.1099. If your calculator has a log function, direct calculation gives $k = 5.55 \times 10^{-2} \text{ h}^{-1}$. To use the ln function on a calculator, we can write Equation (11) in the form

$$k = \frac{1}{t} \left(\ln \frac{[A]_0}{[A]_t} \right)$$

Again, $k = 5.55 \times 10^{-2} \text{ h}^{-1}$.

See also Exercises 39–42.

• *Graphic Plots for a First-Order Reaction* • The characteristic rate of decrease in concentration for first-order reactions is revealed by the plots shown in Figure 14.9. Because the first-order rate is proportional to [A], the rate of the reaction becomes slower as [A] becomes smaller. The rate appears to approach zero but never attain it, as shown by the curve in Figure 14.9(a). When changes in concentration can no longer be measured, the reaction is considered to be over. The system is then said to be at equilibrium.

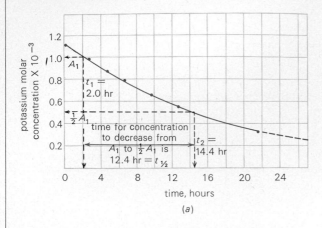

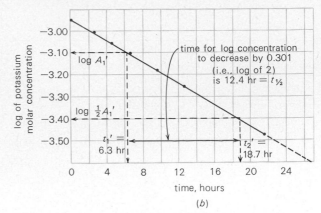

(a)   (b)

For a first-order reaction, a plot of the logarithm of the concentration, $\log [A]_t$, versus the time elapsed, t, is a straight line [see Figure 14.9(b)]. This type of plot is often used to evaluate the rate constant k. We write Equation (11) in the form

$$k = -\frac{2.303}{t}\left(\log\frac{[A]_t}{[A]_0}\right)$$

$$= -2.303\left(\frac{\log [A]_t - \log [A]_0}{t}\right) \qquad (12)$$

The quantity within the brackets is the slope of the straight line, so

$$k = -2.303(\text{slope}) \qquad (13)$$

Because the slope itself is negative, the value of k is positive. We can estimate the slope for the potassium–ammonia reaction from Figure 14.9(b). Using the data points at $t = 27$ and $t = 0$ hours, we obtain

$$\text{slope} = \frac{-3.60 - (-2.95)}{27\ \text{h}} = -\frac{0.65}{27}\ \text{h}^{-1}$$

Then k is calculated from Equation (13):

$$k = -2.303\left(-\frac{0.65}{27}\ \text{h}^{-1}\right) = 5.5 \times 10^{-2}\ \text{h}^{-1}$$

This graphic solution is in good agreement with the arithmetic solution in Example 14.5.

• *Determination of the Half-Life* • As illustrated by both plots shown in Figure 14.9, there is a certain time interval for a first-order reaction, called

Figure 14•9 Two ways of plotting the data in Table 14.3. In either case, after a smooth line is drawn passing through or close to the plotted points, some easily read place on the line is chosen as the "initial" concentration (A or A') at the "initial" time (t_1 or t_1'). Then the point on the curve is found that corresponds to half this initial concentration ($\frac{1}{2}A$ or $\frac{1}{2}A'$), and the time at which this half-initial concentration is reached is read (t_2 or t_2'). The time for half of the "initial" concentration to react is $t_2 - t_1 = t_{1/2}$.

the **half-life,** $t_{1/2}$, for any given concentration to fall to half of that concentration. For a rapid reaction, the half-life can be a small fraction of a second; for a slow reaction, it can be years.

The half-life for a first-order reaction can be estimated from a graphic plot of the data, as shown in Figure 14.9, or it can be calculated from the integrated form of the rate equation in Table 14.2. When $t = t_{1/2}$, the concentration $[A]_t$ is $\frac{1}{2}[A]_0$, so

$$k = \frac{2.303}{t_{1/2}} \left(\log \frac{[A]_0}{\frac{1}{2}[A]_0} \right) = \frac{2.303}{t_{1/2}} \log 2 = \frac{0.693}{t_{1/2}}$$

or

$$t_{1/2} = \frac{0.693}{k} \tag{14}$$

This simple equation is very useful. For example, k for the potassium–ammonia reaction is calculated from the average of the calculated values in Table 14.3 to be $5.55 \times 10^{-2}\,h^{-1}$. Therefore,

$$t_{1/2} = \frac{0.693}{5.55 \times 10^{-2}\,h^{-1}} = 12.5\,h$$

It should be noted that no concentration term appears in Equation (14). For a first-order reaction, *it takes the same amount of time to go from any known concentration to half that amount.* For example, if it takes 12.5 h for the concentration of potassium to decrease from $8.00 \times 10^{-4}\,M$ to $4.00 \times 10^{-4}\,M$, it will also take 12.5 h for the concentration to change from $4.00 \times 10^{-2}\,M$ to $2.00 \times 10^{-2}\,M$. *If the data for a reaction show that the half-life does not depend on the initial concentration, the reaction is shown to be first order.*

Time intervals other than $t_{1/2}$ can be calculated if we have enough information to calculate the rate constant. Example 14.6 illustrates such a calculation.

• Example 14•6 •

The thermal decomposition of acetone, $(CH_3)_2C{=}O$, at 600 °C is a first-order reaction with a half-life of 80 s.
(a) Calculate the rate constant, k.
(b) What time will be required for 25 percent of a given sample to decompose? For 85 percent?

• Solution •

(a) The rate constant, k, is

$$k = \frac{0.693}{t_{1/2}} = \frac{0.693}{80\,s} = 8.7 \times 10^{-3}\,s^{-1}$$

(b) When 25 percent of the initial molar concentration of acetone has decomposed, the molar concentration remaining is 75 percent. The ratio of $[(CH_3)_2C{=}O]_0$ to $[(CH_3)_2C{=}O]_t$ is then $1.0 : 0.75$, and t is calculated by use of Equation (10) as follows:

$$kt = 2.303 \left(\log \frac{[A]_0}{[A]_t} \right)$$

$$(8.7 \times 10^{-3} \text{ s}^{-1})t = 2.303 \left(\log \frac{1.0}{0.75} \right)$$

$$t = \frac{(2.303)(0.124)}{8.7 \times 10^{-3} \text{ s}^{-1}} = 33 \text{ s}$$

The ratio of $[(CH_3)_2C{=}O]_0$ to $[(CH_3)_2C{=}O]_t$ after 85 percent reaction is 1.0:0.15, and

$$t = \frac{2.303 \left(\log \dfrac{1.0}{0.15} \right)}{8.7 \times 10^{-3} \text{ s}^{-1}} = \frac{(2.303)(0.824)}{8.7 \times 10^{-3} \text{ s}^{-1}}$$

$$= 2.2 \times 10^2 \text{ s}$$

See also Exercises 28, 31, and 41.

14.5.2   Determination of Second-Order Reactions   As shown earlier, rate equations for second-order reactions are generally of two types:

1.   For 2A $\longrightarrow$ products,

$$\text{rate} = k[A]^2$$

2.   For A + B $\longrightarrow$ products,

$$\text{rate} = k[A][B]$$

With the integrated forms of these equations shown in Table 14.2, we can use experimental data to (1) calculate values of rate constants, k, (2) plot graphs, or (3) calculate values of half-lives, $t_{1/2}$. *If the data for a given reaction fit one of the second-order equations, that reaction is shown to be a second-order reaction.*

• *Calculation of k for Second-Order Reactions* • For the case of 2A $\rightarrow$ products, the integrated form in Table 14.2 can be written as

$$k = \frac{1}{t} \left(\frac{1}{[A]_t} - \frac{1}{[A]_0} \right) \tag{15}$$

For the case of A + B $\rightarrow$ products, in which the two initial concentrations are the same, that is, $[A]_0 = [B]_0$, Equation (15) also applies.[5]

The basic hydrolysis of ethyl acetate will be used to show how a second-order rate constant is determined. In Table 14.4 are listed some data for this reaction. Example 14.7 illustrates a calculation of k.

---

[5] If the reacting substances do not have equivalent initial concentrations, then the integrated form of the equation is the last one given in Table 14.2.

Table 14•4 Basic hydrolysis of ethyl acetate at 30° C

Time, s	$[OH^-]_t = [CH_3CO_2C_2H_5]_t$	Second-order rate constant $k \times 10^2$, $L \cdot mol^{-1} \cdot s^{-1}$	First-order rate constant $k \times 10^4$, s^{-1}
0	0.0500		
240	0.0441	1.12	5.23
540	0.0386	1.09	4.79
900	0.0337	1.07	4.38
1,440	0.0279	1.10	4.05
2,220	0.0228	1.07	3.54
3,180	0.0185	1.07	3.13
4,980	0.0136	1.07	2.61
8,580	0.00895	1.07	2.00

Source: H. A. Smith and H. S. Levenson, J. Amer. Chem. Soc., **61**, 1172 (1939).

• Example 14•7 •

Using the initial concentrations in Table 14.4 and the concentrations after 2,220 seconds, calculate a rate constant, k, for the basic hydrolysis of ethyl acetate.

• Solution • The initial concentrations of the reactants, ethyl acetate and sodium hydroxide, are the same, $0.0500M$, so Equation (15) is applicable. We solve the equation with $t = 2,220$ s, $[A]_0 = 0.0500M$, and $[A]_t = 0.0228M$:

$$k = \frac{1}{2.220 \times 10^3 \text{ s}} \left(\frac{1}{0.0228} - \frac{1}{0.0500} \right) L \cdot mol^{-1}$$

$$= \frac{1}{2.220 \times 10^3} (43.9 - 20.0) \, L \cdot mol^{-1} \cdot s^{-1}$$

$$= \frac{1}{2.220 \times 10^3} (23.9) \, L \cdot mol^{-1} \cdot s^{-1} = 1.07 \times 10^{-2} \, L \cdot mol^{-1} \cdot s^{-1}$$

See also Exercise 36.

Values of k, calculated as in Example 14.7, are shown in the third column of Table 14.4. Because the eight values are reasonably constant, we conclude that this reaction is second order. In contrast, the values of k in the fourth column were calculated using Equation (11) for a first-order reaction. The considerable variation in these values shows clearly that this reaction is not first order.

• Graphic Plots for a Second-Order Reaction • A plot of the molar concentration of OH^- versus time for the saponification of ethyl acetate is shown in Figure 14.8. For a second-order reaction to which Equation (15) applies, a plot of the reciprocal of the reactant concentration, $1/[A]_t$, versus the time elapsed, t, is a straight line (see Figure 14.10). This type of plot can be used to evaluate the rate constant k. Equation (15) can also be written as

$$k = \frac{\frac{1}{[A]_t} - \frac{1}{[A]_0}}{t} \tag{16}$$

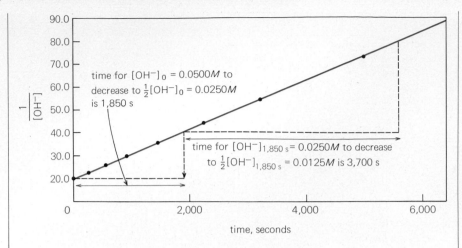

Figure 14•10 Graphic determination of two half-lives, $t_{1/2}$, for a second-order reaction. Note that the $t_{1/2}$ for the decrease from 0.250M is twice as long as the $t_{1/2}$ for the decrease from 0.500M. The data plotted are from Table 14.4.

The quantity on the right is the slope of the straight line, so

$$k = \text{slope} \tag{17}$$

The slope for the reaction of ethyl acetate with sodium hydroxide can be estimated from Figure 14.10. Choosing the points at $t = 6,000$ s and $t = 0$ s, we obtain

$$k = \text{slope} = \frac{(84.0 - 20.0)\,\text{L} \cdot \text{mol}^{-1}}{6,000\,\text{s}} = 1.07 \times 10^{-2}\,\text{L} \cdot \text{mol}^{-1} \cdot \text{s}^{-1}$$

• *Determination of the Half-Life* • The half-life for a second-order reaction of the type we have been discussing can be estimated from graphic plots such as Figures 14.8 and 14.10, or it can be calculated from the integrated second-order rate equation, Equation (15). Setting $[A]_t = \frac{1}{2}[A]_0$, we obtain

$$k = \frac{1}{t_{1/2}}\left(\frac{1}{\frac{1}{2}[A]_0} - \frac{1}{[A]_0}\right) = \frac{1}{t_{1/2}}\left(\frac{2}{[A]_0} - \frac{1}{[A]_0}\right) = \frac{1}{t_{1/2}} \times \frac{1}{[A]_0}$$

and

$$t_{1/2} = \frac{1}{k} \times \frac{1}{[A]_0} \tag{18}$$

For the saponification reaction, $t_{1/2}$ is calculated from the average of the calculated values of k in Table 14.4:

$$t_{1/2} = \frac{1}{1.08 \times 10^{-2}\,\text{L} \cdot \text{mol}^{-1} \cdot \text{s}^{-1}} \times \frac{1}{0.0500\,\text{mol} \cdot \text{L}^{-1}} = 1.85 \times 10^3\,\text{s}$$

Unlike a first-order reaction, $t_{1/2}$ for a second-order reaction is dependent upon the initial concentration of the reactants. The half-life is inversely proportional to the initial concentration. The larger the initial concentration, the smaller is the half-life. In the saponification reaction, for example, in which $[A]_0$ is 0.0500M, $t_{1/2}$ is 1,850 s; if $[A]_0$ were 0.0250M, $t_{1/2}$ would be 3,700 s.

14•6 Order and Reaction Mechanism

A knowledge of the order of the reaction allows one to postulate a reasonable reaction mechanism. To illustrate this point, let us consider the hypothetical reaction

$$A_2 + 2B \longrightarrow 2AB$$

Suppose we find by experiment that the reaction is first order in A_2 but zero order in B. The rate equation is

$$rate = k[A_2][B]^0 = k[A_2]$$

We might assume that the reaction proceeds through a slow step involving only A_2, followed by a relatively fast step involving B. Accordingly, we might postulate the following steps:

$$A_2 \longrightarrow 2A \quad \text{(slow)} \qquad (19)$$
$$A + B \longrightarrow AB \quad \text{(fast)} \qquad (20)$$

If a reaction takes place via a sequence of steps, there may be one step that is much slower than any other. This slower step is called the **rate-determining step** because it limits the overall rate of the reaction. In our hypothetical example, Equation (19) is the rate-determining step.

The indicated two-step mechanism is not the only possible mechanism. Another possibility consistent with the kinetic data would be the following three-step process:

$$A_2 \longrightarrow 2A \quad \text{(slow)}$$
$$2A + B \longrightarrow A_2B \quad \text{(fast)}$$
$$A_2B + B \longrightarrow 2AB \quad \text{(fast)}$$

A knowledge of the order of a reaction allows us to exclude certain reaction mechanisms. The two mechanisms proposed to explain first-order behavior are ruled out if the reaction is second order. For example, if the reaction is first order in A_2 and first order in B, the second-order overall rate equation is

$$rate = k[A_2][B]$$

Either of the following mechanisms might apply:

$$A_2 + B \longrightarrow A_2B \quad \text{(slow)}$$
$$A_2B + B \longrightarrow 2AB \quad \text{(fast)}$$

or

$$A_2 + B \longrightarrow AB + A \quad \text{(slow)}$$
$$A + B \longrightarrow AB$$

In each of these mechanisms, the slow step depends on the concentrations of A_2 and B.

14•6•1   A Specific Example In this section we will describe how kinetic data can help establish the most probable reaction mechanism. The reaction we will consider is

$$NO_2 + CO \longrightarrow NO + CO_2$$

In Section 14.1.1, this reaction is depicted as one that involves a single elementary process with a transition state in which only two atomic centers collide. However, this mechanism appears to operate only at temperatures of about 500 K and above. At lower temperatures, the reaction is second order in NO_2 and zero order in CO:

$$\text{rate} = k[NO_2]^2[CO]^0 = k[NO_2]^2$$

Although only one molecule of NO_2 is shown in the equation for the reaction, the rate equation requires that two molecules interact in a slow, rate-determining step. A number of mechanisms are consistent with the low-temperature kinetic data. One of these involves an initial, slow reaction between molecules of NO_2 to produce NO and O_2, followed by a fast reaction between CO and O_2 to produce CO_2:

$$\begin{array}{ll} 2NO_2 \longrightarrow 2NO + O_2 & \text{(slow)} \\ \underline{2CO + O_2 \longrightarrow 2CO_2} & \text{(fast)} \\ 2NO_2 + 2CO \longrightarrow 2NO + 2CO_2 & \\ NO_2 + CO \longrightarrow NO + CO_2 & \end{array}$$

or

This mechanism can be ruled out on the basis of additional data, however. First, the reaction between CO and O_2 to produce CO_2 is known to be very slow at temperatures below 500 K. Second, simultaneous collisions between three reactant molecules are highly improbable in gas phase reactions. Two consecutive two-body collisions are generally more likely to occur than one three-body collision.

Another mechanism that is consistent with the kinetic data is an initial, slow reaction between NO_2 molecules to produce NO and NO_3, followed by a fast reaction between NO_3 and CO to produce NO_2 and CO_2:

$$\begin{array}{ll} 2NO_2 \longrightarrow NO + NO_3 & \text{(slow)} \\ \underline{NO_3 + CO \longrightarrow NO_2 + CO_2} & \text{(fast)} \\ NO_2 + CO \longrightarrow NO + CO_2 & \end{array}$$

The NO_3 molecule appears to be a likely intermediate compound. It is an unstable, very reactive compound, but it has been made and studied.

14•7   Chain Reactions

Early in the development of chemical kinetics it was learned that certain chemical reactions that are normally very slow become extremely rapid in the presence of small amounts of a highly reactive species. Not only does the highly reactive species initiate the chemical reaction, but it also is regenerated in some subsequent step of the reaction. The overall reaction may be so exothermic and take place so rapidly that an explosion results. Such results are characteristic of a *chain mechanism*; the reaction itself is called a **chain reaction.**

An example of a chain reaction is the photochemical reaction of chlorine with hydrogen to produce hydrogen chloride:

$$Cl_2 + H_2 \longrightarrow 2HCl \quad \Delta H = -184.62 \text{ kJ}$$

A mixture of hydrogen and chlorine can be kept in the dark for a long time, but when the mixture is exposed to sunlight, a violent combustion occurs. The sunlight causes some of the chlorine molecules to dissociate into atoms, or free radicals, of chlorine:

$$Cl_2 \xrightarrow{\text{sunlight}} \cdot Cl + \cdot Cl \tag{21}$$

The very reactive chlorine atom (a free radical) can then combine with a hydrogen molecule to form a molecule of hydrogen chloride and a hydrogen atom (also a free radical):

$$\cdot Cl + H_2 \longrightarrow HCl + \cdot H \tag{22}$$

Next, the very reactive hydrogen atom can combine with a chlorine molecule. Note that a chlorine atom is again formed:

$$\cdot H + Cl_2 \longrightarrow HCl + \cdot Cl \tag{23}$$

In this chain reaction, reaction (21), which produces the first free radicals, is called the *initiation reaction*. Reactions (22) and (23) are *propagation reactions;* the chlorine radicals formed in reaction (23) replenish the supply to keep reaction (22) going.

Presumably, there are relatively few free radicals of hydrogen and chlorine in existence at any one time. But, as the reactions to form HCl proceed, free radicals are continually being used and being formed.

The final steps in the chain are the *termination reactions* shown in Equations (24), (25), and (26). A chain is broken or terminated by a reaction that uses a free radical but does not produce one. This occurs when atoms of chlorine react with atoms of hydrogen, Equation (24); when atoms of hydrogen react with one another, Equation (25); and when atoms of chlorine react with one another, Equation (26):

$$\cdot H + \cdot Cl \longrightarrow HCl \tag{24}$$
$$\cdot H + \cdot H \longrightarrow H_2 \tag{25}$$
$$\cdot Cl + \cdot Cl \longrightarrow Cl_2 \tag{26}$$

The breaking of the chain tends to stop the reaction between the hydrogen and chlorine. However, the overall reaction liberates a great deal of heat. If the temperature rises sufficiently, molecules of chlorine and hydrogen may become sufficiently activated to react directly.

Most of the processes by which ozone is destroyed in the upper atmosphere (see Section 9.3.3) involve chain mechanisms. Chain reactions initiated by oxides of nitrogen, NO and NO_2, by chlorine atoms, Cl, by hypochlorite radicals, ClO, and by hydroxy, OH, and hydroperoxy, HO_2, radicals are all known to contribute. The following example, starting with the decomposition of the chlorofluoromethane, CCl_2F_2, is illustrative:

$$CCl_2F_2 \xrightarrow[\text{radiation}]{\text{uv}} \cdot CClF_2 + \cdot Cl \qquad \text{(initiation)}$$

$$\cdot Cl + O_3 \longrightarrow \cdot ClO + O_2$$
$$\cdot ClO + \cdot O \longrightarrow \cdot Cl + O_2 \qquad \text{(propagation)}$$

$$\cdot Cl + \cdot Cl \longrightarrow Cl_2$$
$$\cdot ClO + \cdot ClO \longrightarrow Cl_2 + O_2 \qquad \text{(termination)}$$

Other reactions of chlorine atoms and ClO radicals may also terminate the chain. One example is with a hydroperoxy radical:

$$\cdot Cl + \cdot HO_2 \longrightarrow HCl + O_2$$

Other examples of chain reactions include the combustion of hydrogen with oxygen to produce water, the oxidation of gasoline hydrocarbons in an internal combustion engine, and certain reactions that produce smog. Reactions that proceed through chain mechanisms are of practical importance in the manufacture of plastics and in the production of halogen-substituted alkanes (see Section 26.7.2). Chain mechanisms are also important in nuclear reactions, as will be discussed in Chapter 19.

Chapter Review

Summary

The ultimate goal in the study of **chemical kinetics** is to understand in detail how chemical reactions occur. The stepwise changes that reactants undergo to become products add up to the **reaction mechanism.** Each individual step is called an **elementary process,** or **elementary reaction;** usually only one, two, or three particles are involved as reactants in each such step. When two particles collide in the proper orientation and with sufficient energy, a **transition state,** or **activated complex,** is formed for an instant. It immediately dissociates to give new particles (reaction) or the original particles (no reaction). The new particles may be the final products of the overall reaction or they may be **reaction intermediates,** which are usually unstable species that can be detected but only rarely isolated. Many such species are **radicals**—atoms or groups of atoms with one or more unpaired electrons.

Studying the **rate of a reaction** is the best way to unravel its mechanism. The rate is expressed as a change in the concentration of some species per unit time. It depends on four factors: (1) The infinitely variable *nature* of the reactants can account for enormous differences in reaction rates. The **activation energy** depends on the nature of the reactants and is the amount of energy needed to form the activated complex. (2) The *temperature* of the system is important because all chemical reaction rates increase markedly with increasing temperature. (3) A *catalyst* can increase the reaction rate by forming an **intermediate compound** with a relatively low activation energy for reaction, or by adsorbing reacting species on its surface. (4) The *concentrations* of the reactants continually decrease as the reaction progresses; in most cases this also means that the **instantaneous rate** of reaction decreases with time.

The effect of a changing concentration on the rate of a reaction cannot be predicted from the overall chemical equation; it must be determined experimentally. The equation that describes this effect is called a **rate equation,** or **rate law.** The proportionality constant between the rate and the concentration(s) is the **rate constant,** k, for that reaction.

Each concentration term in the rate equation is raised to some experimentally determined power; the sum of these exponents is the **order of the reaction.** If the sum is 1, the reaction is a **first-order reaction;** if the sum is 2, it is a **second-order reaction.** Some reactions are of fractional order overall, and some reactions are **zero order** with respect to a given reactant whose changing concentration has no effect on the rate. A reaction that has a first-order rate equation because the concentration of one or more of the reactants remains effectively constant throughout the reaction is a **pseudo first-order reaction.**

The order of a reaction is determined from the experimental data by calculations of the rate constant, by a graphic plot, or by determination of the **half-life** of the reaction. Knowing the reaction order, one can then propose and analyze possible reaction mechanisms, and eliminate impossible ones. A key factor is the **rate-determining step** because it limits the overall reaction rate. The mechanism of a **chain reaction,** which is typically very fast, depends on the formation of highly reactive species as reaction intermediates. These species are initiated, propagated, and terminated in separate elementary processes.

Exercises

Reaction Mechanisms

1. Write equations to show some elementary reactions that could occur when hydrogen chloride reacts with bromine to form hydrogen bromide and bromine chloride. Assume the reactions could occur by collisions between either two or four atomic centers.
2. For each elementary equation written for Exercise 1, propose a possible transition state.
3. The decomposition of nitrogen dioxide, NO_2, into nitric oxide, NO, and oxygen can be explained on the basis of one elementary equation:

$$2NO_2 \longrightarrow 2NO + O_2$$

   (a) For the decomposition to occur, how many atomic centers must collide simultaneously (NO_2 is an angular molecule)?
   (b) Diagram a possible structure for the transition state.
*4. The reaction between the ethoxide ion and ethyl iodide to form common ether proceeds in one elementary step:

$$CH_3CH_2O^- + CH_3CH_2I \longrightarrow$$
$$CH_3CH_2OCH_2CH_3 + I^-$$

   (a) How many atomic centers must collide simultaneously for the reaction to take place?
   (b) In a manner comparable to that of Figure 14.2, diagram the transition state.

Experimental Determination of Reaction Rates

5. A gas buret is filled to the 1.00-L mark with N_2O_4 at 1 atm. What is the total volume of gases at 1 atm after half the N_2O_4 has decomposed to NO_2? After 75.0 percent has decomposed to NO_2?
6. A reaction vessel is filled with N_2O_5 gas at a pressure of 1.01×10^5 Pa. If the N_2O_5 decomposes according to the equation

$$2N_2O_5 \longrightarrow 4NO_2 + O_2$$

   what is the partial pressure in pascals due to NO_2 when the total pressure rises to 1.51×10^5 Pa?
7. Describe in some detail a method for determining the reaction rate for each of the following (if no simple method can be used, explain why):
   (a) $CH_3Cl + KOH \longrightarrow CH_3OH + KCl$
   (b) $C_2H_5I + HOH \longrightarrow C_2H_5OH + HI$
   (c) $2NO + Cl_2 \longrightarrow 2NOCl$ (all are gases)
   (d) $NaOH + HCl \longrightarrow NaCl + HOH$

Factors Affecting Reaction Rates

8. The activation energy for the elementary reaction

$$NO + O_3 \longrightarrow NO_2 + O_2$$

   is 10.5 kJ/mole. Using data in Table 13.6, draw an energy versus reaction pathway diagram for this reaction. Label on this diagram each of the following:
   (a) E of reactants and E of products
   (b) ΔE of reaction
   (c) the transition state
   (d) the activation energies, E_a and E_a', for the forward and reverse reactions, respectively.
9. What is the value of the activation energy, E_a', for the reverse reaction in Exercise 8?
10. The decomposition of sulfur trioxide into sulfur dioxide and oxygen is catalyzed by nitric oxide, NO. Write the two equations representing the reactions involved in the catalyzed process. Show that these two equations may be summed to give the net reaction for the process.
11. List all the differences you can think of between a transition state and an intermediate compound.
12. Among the factors that influence reaction rates, which is involved in each of the following?
   (a) Powdered zinc reacts more rapidly with sulfuric acid than does a large piece of zinc of equal weight.
   (b) Sodium reacts more rapidly with water than does iron.
   (c) It is more dangerous to drop a lighted match into a gasoline tank that has just been emptied than into one that is completely full.
   (d) A brush fire burns more rapidly on a windy day than on a still day.

(e) The flame of a burning match is extinguished by the wind.

(f) The reaction between ethylene, C_2H_4, and hydrogen to form ethane, C_2H_6, proceeds much faster if platinum is present.

13. Give two reasons why a reaction with a high activation energy proceeds more rapidly when the temperature is raised.

14. Two substances that by themselves react slowly, react rapidly if a catalyst is present. The same products are formed in either case. How does the activation energy, E_a, differ for the two methods for carrying out the reaction? Explain. Does ΔE differ?

Writing Rate Expressions and Rate Equations

15. Consider the decomposition of ammonia, NH_3, into nitrogen, N_2, and hydrogen, H_2:

$$2NH_3 \longrightarrow N_2 + 3H_2$$

(a) Write the expressions for the average rate of decrease in the NH_3 concentration and the average rates of increase in the N_2 and H_2 concentrations.

(b) Indicate how these three rate expressions are related to one another.

16. (a) Given rate = $k[N_2O_5]$, calculate the rate constant for the decomposition of nitrogen(V) oxide,

$$2N_2O_5 \longrightarrow 4NO_2 + O_2$$

for a 0.040M solution if the instantaneous rate is 1.4×10^{-6} mol $\cdot L^{-1} \cdot s^{-1}$.

(b) Given rate = $k[NO_2]^2$, calculate k for the decomposition of nitrogen dioxide,

$$2NO_2 \longrightarrow 2NO + O_2$$

for a 0.040M solution if the instantaneous rate is 3.0×10^{-13} mol $\cdot L^{-1} \cdot s^{-1}$.

(c) In the respective rate equations, why is the concentration of NO_2 squared, whereas that for N_2O_5 is not?

17. Consider the first-order decomposition of hydrogen peroxide, H_2O_2, to produce water and oxygen.

(a) Write a balanced equation for this reaction.

(b) Assume that you are able to follow the rate of decomposition of H_2O_2 at a certain temperature by following the increase in the concentration of oxygen with the passage of time. The following data are collected:

time, seconds	$[O_2]$
20	0.15
140	0.45

Calculate $\Delta[O_2]/\Delta t$, $\Delta[H_2O]/\Delta t$, and $-(\Delta[H_2O_2]/\Delta t)$ for the 20-s to 140-s time interval.

18. Consider the general reaction

$$2A + B_2 \longrightarrow 2AB$$

and the following experimental data. Write the most probable rate equation for the reaction.

experiment	[A]	$[B_2]$	rate, mol $\cdot L^{-1} \cdot s^{-1}$
1	0.50	0.50	1.6×10^{-4}
2	0.50	1.00	3.2×10^{-4}
3	1.00	1.00	3.2×10^{-4}

19. The reaction of nitric oxide, NO, with hydrogen, H_2, produces nitrogen, N_2, and water, H_2O, according to the equation

$$2NO + 2H_2 \longrightarrow N_2 + 2H_2O$$

Consider the following data and then write the most probable rate equation for the reaction.

experiment	[NO]	$[H_2]$	rate, mol $\cdot L^{-1} \cdot s^{-1}$
1	0.20	0.20	3.20×10^{-3}
2	0.40	0.20	1.28×10^{-2}
3	0.20	0.40	6.40×10^{-3}

20. Consider the second-order reaction

$$2NO_2 \longrightarrow 2NO + O_2$$

The rate of decomposition of NO_2 at a certain temperature is determined to be 1.4×10^{-3} mol $\cdot L^{-1} \cdot min^{-1}$ at an NO_2 concentration of 0.500M. What will be the rate of decomposition when the concentration of NO_2 is 0.250M? When the concentration of NO_2 is 0.125M? (Assume constant temperature is maintained throughout the decomposition.)

21. Consider the reaction

$$2NO + Cl_2 \longrightarrow 2NOCl$$

and the following experimental data. Write the most probable rate equation for the reaction.

experiment	[NO]	$[Cl_2]$	rate, mol $\cdot L^{-1} \cdot s^{-1}$
1	0.380	0.380	5.0×10^{-3}
2	0.760	0.760	4.0×10^{-2}
3	0.380	0.760	1.0×10^{-2}

22. If a reaction involving reactants A, B, and C is known to be third order, what are the possible rate equations?

Half-Lives and Other Reaction Times

23. Write the general expression for the rate constant of a first-order reaction, and show how to derive from it the expression $k = 0.693/t_{1/2}$.

24. Write the general expression for the rate constant of a second-order reaction of the type $2A \rightarrow$ products, and show how to derive from it an equation for the half-life, $t_{1/2}$.

*25. The decomposition of ethyl chloride, C_2H_5Cl, into ethylene, C_2H_4, and hydrogen chloride, HCl,

$$C_2H_5Cl(g) \longrightarrow CH_2{=}CH_2(g) + HCl(g)$$

is a first-order reaction at 200 °C. Suppose you have determined the rate constant, k, the half-life, $t_{1/2}$, and the instantaneous rate when 25 percent of the ethyl chloride has decomposed at 200 °C and a constant pressure of 1.0 atm. What effect, if any, would a doubling of the pressure to 2.0 atm have on the following?
(a) the rate constant, k
(b) the half-life, $t_{1/2}$
(c) the instantaneous rate when 25 percent of the ethyl chloride has decomposed at 200 °C

26. Consider the reaction $A \rightarrow B + C$, and the tabulated experimental data given.

| experiment 1 | | experiment 2 | |
| $[A]_0 = 0.500M$ | | $[A]_0 = 1.00M$ | |
time, h	$[A]_t$	time, h	$[A]_t$
0.00	0.500	0.00	1.00
1.00	0.250	1.00	0.500
2.00	0.125	2.00	0.250
3.00	0.0625	3.00	0.125

(a) In experiment 1, how long does it take for the original concentration to be reduced to half that value? To one-fourth that value?
(b) Repeat part (a) for experiment 2.
(c) What is the order of the reaction?
(d) Calculate the rate constant, k, for the reaction.
(e) What are the instantaneous rates at 1.00, 2.00, and 3.00 h in the two experiments?
(f) How do the instantaneous rates at 3.00 h compare in the two experiments?

27. The reaction $A + B \rightarrow C + D$ is zero order with respect to B. Consider the tabulated experimental data for this reaction, and then answer the questions that follow.

| experiment 1 | | experiment 2 | |
| $[A]_0 = 1.00M$ | | $[A]_0 = 2.00M$ | |
time, h	$[A]_t$	time, h	$[A]_t$
0.00	1.00	0.00	2.00
1.00	0.800	1.00	1.33
4.00	0.500	2.00	1.00
12.00	0.250	6.00	0.500

(a) In experiment 1, how long does it take for the original concentration to be reduced to half that value? To one-fourth that value?

(b) Repeat part (a) for experiment 2.
(c) What is the order of the reaction with respect to A?
(d) Calculate the rate constant, k, for the reaction.
(e) What are the instantaneous rates at 4.00 h and 12.00 h in experiment 1? At 2.00 h and 6.00 h in experiment 2?
(f) Is there any relationship between the instantaneous rates in the two experiments?

28. The catalyzed decomposition of hydrogen peroxide, H_2O_2, into water and oxygen is known to be a first-order reaction. The rate constant at a certain temperature is $2.40 \times 10^{-4}\,s^{-1}$.
(a) Calculate the half-life, $t_{1/2}$, in seconds.
(b) How much time is required for 75.0 percent (three fourths) of a given quantity to decompose?
(c) How much time is required for 87.5 percent (seven eights) of a given quantity to decompose?
(d) What are the relationships between $t_{1/2}$, $t_{3/4}$, and $t_{7/8}$?

29. Consider the first-order reaction $A \rightarrow B + C$. If one starts with 10.0 g of A, how much remains after three half-lives? After six half-lives? After ten half-lives?

*30. Derive an equation for calculating the percentage of reactant remaining after n half-lives for a reaction.

31. The half-life, $t_{1/2}$, for the first-order decomposition of nitramide, NH_2NO_2, into nitrous oxide, N_2O, and water is 123 min at 15 °C. The equation is $NH_2NO_2 \longrightarrow N_2O + H_2O$.
(a) What is the value of the rate constant, k?
(b) How long will it take for 2.0 g of NH_2NO_2 to decompose until only 0.20 g remains? Until only 0.020 g remains?

32. The decomposition of SO_2Cl_2 into SO_2 and Cl_2 is a first-order reaction. The half-life at 320 °C is 8.75 h.
(a) Calculate the rate constant, k, for this reaction.
(b) If one starts with 2.50 g of SO_2Cl_2, how much will be left after 3.00 h?

33. Consider the first-order decomposition of ethyl chloride, C_2H_5Cl, into ethylene, C_2H_4, and hydrogen chloride, HCl:

$$C_2H_5Cl(g) \longrightarrow CH_2{=}CH_2(g) + HCl(g)$$

The half-life, $t_{1/2}$, for this reaction is 90.0 min at 200 °C.
(a) Calculate the rate constant, k.
(b) How many minutes will it take for 75.0 percent of the ethyl chloride to decompose?

34. Consider the second-order reaction, $2NO_2 \longrightarrow 2NO + O_2$. The rate constant, k, for this reaction at a certain temperature is $2.00 \times 10^{-8}\,L \cdot mol^{-1} \cdot s^{-1}$. If the initial concentration of NO_2 is $0.500M$, how many seconds are required for the NO_2 concentration to be reduced to $0.125M$?

*35. The reaction of hydrogen, H_2, with iodine, I_2, to produce hydrogen iodide, HI, follows the rate equation

$$rate = k[H_2][I_2]$$

The rate constant, k, for this decomposition at a certain temperature is $2.0 \times 10^{-6} \, L \cdot mol^{-1} \cdot s^{-1}$.

(a) If the initial concentrations of H_2 and I_2 are both $0.50M$, what is the half-life, $t_{1/2}$, for the reaction?

(b) How much time is required for 75 percent (three fourths) of the initial concentrations of H_2 and I_2 to react?

(c) What is the relationship between $t_{1/2}$ and $t_{3/4}$?

*36. For the saponification of ethyl acetate,

$$CH_3CO_2C_2H_5 + OH^- \longrightarrow CH_3CO_2^- + C_2H_5OH$$

the half-life, $t_{1/2}$, is 925 s at 30.0 °C when the initial concentrations of ethyl acetate and OH^- are both $0.100M$. How many seconds are required for 75.0 percent of the ethyl acetate to react?

*Determining Reaction
Order and Rate Constants*

37. The order of the reaction $2NO + O_2 \longrightarrow 2NO_2$ has been determined by experiment to be third order. The rate equation is

$$rate = k[NO]^2[O_2]$$

If the rate of the reaction is expressed in units of $mol \cdot L^{-1} \cdot s^{-1}$, what are the units of k?

38. Consider the following data for the reaction $A + B \rightarrow$ product. What is the order of the reaction with respect to [A]?

time, min	[A]
0	1.50
10.0	1.20
100.0	0.150
1,000.0	0.000

39. Calculate k for the reaction $A \rightarrow B + C$ from the following experimental data:

time, min	[A]
0	0.80
15.0	0.63
30.0	0.57
60.0	0.40
120.0	0.20

40. At 260 °C, trioxane(g) decomposes into formaldehyde(g) according to the equation

$$C_3H_6O_3(g) \longrightarrow 3H_2CO(g)$$

The following experimental data were obtained:

time, h	gas pressure, mm of Hg
0.00	100.0
1.00	173.0
2.00	218.0
3.00	248.0
4.00	266.0
infinity	300.0

(a) What is the order of the reaction?

(b) What is the rate constant, k, of the reaction?

(c) What is the half-life, $t_{1/2}$, of the reaction?

(d) What is the total gas pressure when half of the trioxane has reacted?

41. The thermal decomposition of acetone, C_3H_6O, to produce methane, CH_4, and ketene, C_2H_2O,

$$C_3H_6O(g) \longrightarrow CH_4(g) + C_2H_2O(g)$$

is a first-order reaction. Suppose that at a certain temperature the initial pressure of acetone is 0.100 atm. After 60 s has gone by, the gas pressure increases to 0.160 atm. (The volume is constant.)

(a) Based on these data, calculate the rate constant, k, for the reaction.

(b) How many seconds will elapse before the gas pressure becomes 0.180 atm? 0.190 atm?

(c) Write Lewis structural formulas for acetone, C_3H_6O, methane, CH_4, and ketene, C_2H_2O. (Hint: There is a carbon-to-carbon double bond in a molecule of ketene.)

42. The following experimental data were obtained by E. D. Hughes, *J. Chem. Soc.*, **255** (1936), for the hydrolysis of *tert*-butyl chloride, $(CH_3)_3CCl$, to *tert*-butyl alcohol.

experiment 1: $(CH_3)_3CCl + KOH \longrightarrow$
$$(CH_3)_3COH + KCl$$

experiment 2: $(CH_3)_3CCl + HOH \longrightarrow$
$$(CH_3)_3COH + HCl$$

In experiment 1, $0.0250M$ HCl was used to titrate the unreacted KOH in 5.00-mL samples of solution. Because $[(CH_3)_3CCl] = [KOH]$, the volume of HCl titrant required is directly proportional to the $[(CH_3)_3CCl]$ left unreacted. Hence the volume of HCl can be used in place of the $[(CH_3)_3CCl]$ in an integrated rate equation. In experiment 2, $0.0250M$ KOH was used to titrate the HCl formed.

experiment 1

$[(CH_3)_3CCl] = [KOH] = 0.0510M$

time, h	mL 0.0250M HCl
0.00	10.20
1.00	8.83
2.00	7.63
3.00	6.63
4.00	5.70
5.00	4.92
8.00	3.12
10.00	2.30
infinity	0.00

experiment 2

$[(CH_3)_3CCl] = 0.0465M$, $[KOH] = 0$

time, h	mL 0.0250M KOH
0.00	0.00
1.00	1.28
2.00	2.38
3.00	3.32
4.00	4.18
5.00	4.88
8.00	6.40
10.00	7.22
infinity	9.30

(a) Assume in experiment 1 that the reaction is first order in $(CH_3)_3CCl$ and zero order in KOH. Calculate the first-order rate constant, k, for each listed time interval, and compute an average value of k.

(b) Assume in experiment 2 that the reaction is first order in $(CH_3)_3CCl$. Calculate the first-order rate constant, k, for each listed time interval, and compute an average value of k.

(c) Compare the average values of k in the two experiments. What does this comparison tell you about the two reactions studied?

Order and Reaction Mechanism

*43. Consider the reactions studied in the two experiments of Exercise 42. Recall the order of these reactions. Identify the correct mechanism for either reaction among the following:

(a) $(CH_3)_3CCl + KOH \longrightarrow (CH_3)_3COH + KCl$
   (one elementary process)

(b) $(CH_3)_3CCl + KOH \longrightarrow$
   $(CH_3)_2C{=}CH_2 + HOH + KCl$ (slow)
   $(CH_3)_2C{=}CH_2 + HOH \longrightarrow$
   $(CH_3)_3COH$ (fast)

(c) $(CH_3)_3CCl \longrightarrow (CH_3)_3C^+ + Cl^-$ (fast)
   $(CH_3)_3C^+ + OH^- \longrightarrow (CH_3)_3COH$ (slow)

(d) $(CH_3)_3CCl \longrightarrow (CH_3)_3C^+ + Cl^-$ (slow)
   $(CH_3)_3C^+ + OH^- \longrightarrow (CH_3)_3COH$ (fast)

(e) $HOH \longrightarrow H^+ + OH^-$ (fast)
   $OH^- + (CH_3)_3CCl \longrightarrow$
   $(CH_3)_3COH + Cl^-$ (slow)

(f) $HOH \longrightarrow H^+ + OH^-$ (slow)
   $OH^- + (CH_3)_3CCl \longrightarrow$
   $(CH_3)_3COH + Cl^-$ (fast)

*44. The first-order decomposition of SO_2Cl_2 into SO_2 and Cl_2 presumably proceeds by a two-step mechanism.
(a) Write a reasonable mechanism for the reaction.
(b) Which step is the rate-determining step?

45. The reaction of methane, CH_4, with chlorine, Cl_2, in the presence of sunlight to produce methyl chloride, CH_3Cl, and hydrogen chloride, HCl,

$$CH_4 + Cl_2 \xrightarrow[\text{radiation}]{\text{uv}} CH_3Cl + HCl$$

is a chain reaction analogous to the reaction $H_2 + Cl_2 \rightarrow 2HCl$. Write equations to illustrate the initiation, propagation, and termination steps for this reaction.

46. In the reaction described in Exercise 45, a small amount of ethyl chloride, C_2H_5Cl, is produced. Write equations that explain how this compound may be obtained.

*47. Explain why each of the following is true or false.

(a) Particles such as O, H, OH, and CH_3 probably exist for only a fraction of a second as reaction intermediates.

(b) The activation energy for the forward reaction equals the activation energy for the reverse reaction.

(c) For a reversible reaction, an increase in temperature increases the reaction rate for both the forward and the reverse reactions.

(d) In a rate equation, the powers to which the concentrations are raised cannot be determined by inspection of the balanced equation.

(e) When Δt is infinitesimally small, the average rate equals the instantaneous rate.

(f) The reaction rate constant, k, for a given reaction varies directly as the rate varies.

(g) The half-life for a given first-order reaction depends upon the initial concentration.

(h) The speed of any chemical reaction is directly proportional to the molar concentrations.

(i) The larger the initial reactant concentrations for a second-order reaction, the shorter is its half-life.

(j) A chain reaction, once started, must continue until the initial reactants are consumed.

Chemical Equilibria

15

15•1 **Establishing Chemical Equilibria**

15•2 **Influence of the Nature of Reactants**

15•3 **Influence of Concentration**

15•3•1 Equilibrium Constants

15•3•2 Determination of Equilibrium Constants

15•3•3 Calculations Based on K_c

15•4 **Influence of Pressure Changes**

15•4•1 Case I, Relative Amounts of Reactants and Products Are Not Changed

15•4•2 Case II, Relative Amounts of Reactants and Products Are Changed

15•4•3 The Equilibrium Constant K_p

15•4•4 Relationship of $\Delta G°$ to K_p

15•5 **Influence of Temperature**

15•6 **Influence of a Catalyst**

In Chapter 13 we learned how thermodynamics can tell us whether or not a reaction is spontaneous. Then in our study of kinetics in Chapter 14, we looked at the factors that can influence the rate of conversion of reactants to products. This chapter will deal with the question of how much product is obtained at equilibrium under different experimental conditions.

In principle, the relationship between reaction rates and equilibria is relatively simple. When the products of a chemical system can react to form the original substances, the change is said to be **reversible.** In chemical reactions that are reversible, a condition of **chemical equilibrium** exists when the pair of opposing reactions, the forward and the reverse reactions, are proceeding at the same rate.

An understanding of the factors involved in establishing the state of equilibrium is important in many scientific pursuits. The chemist or chemical engineer who is concerned with the large-scale manufacture of a useful compound is interested in minimizing the influence of the reverse reaction. In a living organism there are many chemical equilibrium processes that are responsible for the well-being of the organism. For example, the acidity (or alkalinity) of the blood is maintained within very narrow limits by several opposing reactions. Numerous equilibria also are involved in the complex process of photosynthesis. Hence the medical researcher, the pharmacologist, the nutritionist, the biochemist, and the plant and soil chemist constantly study equilibrium processes in their efforts to solve the problems involved in making healthier plants and animals.

15•1 Establishing Chemical Equilibria

The following discussion will be concerned with a specific example of a reversible reaction and the establishment of a chemical equilibrium. Consider the union of hydrogen and nitrogen to form ammonia, Equation (1), and the reverse reaction, the decomposition of ammonia to form hydrogen and nitrogen, Equation (2):

$$3H_2(g) + N_2(g) \longrightarrow 2NH_3(g) \tag{1}$$

$$2NH_3(g) \longrightarrow 3H_2(g) + N_2(g) \tag{2}$$

When hydrogen and nitrogen are mixed in a $3:1$ ratio by volume at room temperature, a reaction does not occur at a detectable speed. However, at elevated temperatures in the presence of a catalyst, the reaction is rapid. At a temperature of 200 °C and a pressure of 30 atm, this mixture reacts rapidly until about 67.6 percent of the pressure exerted by the mixture is due to ammonia gas. No further apparent change occurs in the amounts of the three components present so long as the mixture is held at 200 °C and 30 atm.

Similarly, ammonia does not decompose at room temperature—Equation (2)—at a detectable rate. But if the temperature is raised and a catalyst is added, its decomposition into hydrogen and nitrogen occurs at a measurable speed. At 200 °C and 30 atm, the amount of ammonia diminishes until 32.4 percent of the pressure exerted by the mixture is due to hydrogen plus nitrogen; after this, there is no further apparent change.

Regardless of whether we start with pure ammonia or with pure hydrogen and nitrogen, neither reaction goes to completion. Each reaction appears to end by forming a mixture that contains 67.6 percent ammonia and 32.4 percent hydrogen and nitrogen. The percentage in terms of pressure exerted is the same as the percentage in terms of volume.

Here we are dealing with *two opposing reactions,* each of which takes place in such a way that the other can occur at the same time once the reaction has been started.

If we start with only hydrogen and nitrogen in the container, reaction (2) cannot occur at first because there is no ammonia. However, as reaction (1) proceeds, ammonia forms, and reaction (2) starts at a low rate initially because not much ammonia is present. Reaction (1) may be occurring at a very rapid rate initially, but as time goes on, the speed of reaction (1) steadily decreases, because hydrogen and nitrogen are being used up; and the speed of reaction (2) steadily increases, because the amount of ammonia is increasing. Eventually the speeds of the two opposing reactions become equal (Figure 15.1). If we start with ammonia in a closed container at 200 °C and 30 atm, the same reactions occur, but in the reverse order. Once the reaction rates are equal, the amounts of hydrogen, nitrogen, and ammonia do not change so long as the temperature and pressure do not change and nothing is added to or removed from the container.

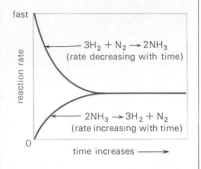

Figure 15•1 A graph showing how the reaction rates for the synthesis and decomposition of ammonia change with time. Initially, at time 0, only hydrogen and nitrogen are present in a sealed container.

15•2   Influence of the Nature of Reactants

The relative amounts of reactants and products at equilibrium vary greatly for different chemical reactions. Consider the general forward reaction

$$A_2 + B_2 \longrightarrow 2AB \qquad \Delta G_r° \text{ is negative} \qquad (3)$$

and its reverse opposing reaction

$$2AB \longrightarrow A_2 + B_2 \qquad \Delta G_r° \text{ is positive} \qquad (4)$$

Because the standard free energy change for the forward reaction is negative, reaction (3) has a greater tendency to occur than reaction (4). As mentioned in Chapter 13, the difference in tendencies to react is related in part to the strengths of the bonds in the reactants and products, and in part to the entropies of these substances.

The equilibrium system described by Equations (3) and (4) is represented as

$$A_2 + B_2 \rightleftharpoons 2AB$$

The greater tendency for reaction (3) to occur is reflected in the concentrations of molecules present at equilibrium. Whether one approaches equilibrium from the reactant side by starting with equal numbers of A_2 and B_2 molecules, or from the product side by starting with AB molecules alone, the predominant species at equilibrium will be AB.

In the specific case of the two opposing reactions,

$$3H_2(g) + N_2(g) \rightleftharpoons 2NH_3(g)$$

the nature of ammonia molecules is such that they have a moderate tendency to decompose at 200 °C and 30 atm. On the other hand, hydrogen and nitrogen molecules have a greater tendency to unite under these conditions and form ammonia. This is the main reason that in the equilibrium mixture, originating from stoichiometric amounts of N_2 and H_2, most of the material is in the form of ammonia (67.6 percent by volume). That is, the rate of decomposition can equal the rate of formation only when the concentration of the ammonia molecules is sufficiently greater than that of the hydrogen and nitrogen molecules to compensate for the smaller natural tendency of ammonia molecules to decompose.

In many reversible reactions, the tendency for one of the opposing reactions to occur is much, much greater than it is for the other. In these cases, we frequently ignore the other reaction because it has an insignificant effect on the amount of product obtainable from the reaction. Thus, in such chemical changes as

$$2H_2(g) + O_2(g) \rightleftharpoons 2H_2O(g)$$
$$C(s) + O_2(g) \rightleftharpoons CO_2(g)$$
$$2Na(s) + Cl_2(g) \rightleftharpoons 2NaCl(s)$$

the reverse reaction is not significant unless the temperature is abnormally high. For all practical purposes, we consider that the forward reaction *goes to completion*, and we indicate this by using a single arrow:

$$2H_2(g) + O_2(g) \longrightarrow 2H_2O(g)$$
$$C(s) + O_2(g) \longrightarrow CO_2(g)$$
$$2Na(s) + Cl_2(g) \longrightarrow 2NaCl(s)$$

15•3 Influence of Concentration

A balanced equation shows the ratio of the quantity of moles that react. For example, the equation

$$H_2 + Cl_2 \longrightarrow 2HCl$$

states that when 1 mole of H_2 reacts, 1 mole of Cl_2 also reacts, and 2 moles of HCl are formed. However, it is not necessary to bring the reactants together in the molar ratio indicated by the equation for a reaction to occur. The reaction between H_2 and Cl_2 will also take place in mixtures of the two in which the molar concentration of H_2 is small as compared with the concentration of Cl_2, or in mixtures in which it is large compared with the concentration of Cl_2.

To evaluate the influence of varying the initial concentrations on the amounts finally attained at equilibrium, we will first consider an equilibrium system that has been thoroughly studied. From experimental data it will be seen that a mathematical expression can be derived that defines the equilibrium concentrations in relation to each other.

15•3•1   Equilibrium Constants   The equilibrium system to be considered is

$$H_2(g) + I_2(g) \rightleftharpoons 2HI(g) \tag{5}$$

In one set of experiments, A. H. Taylor and R. H. Crist sealed mixtures of H_2 and I_2 in glass tubes and placed these tubes in a liquid bath maintained at 425.4 °C. After a suitable period of time to allow the system to come to equilibrium, the amounts of H_2, I_2, and HI were determined by analysis. In a second set of experiments, the procedure was repeated, except that the tubes were filled initially with HI instead of H_2 and I_2. Table 15.1 shows the amounts of H_2, I_2, and HI in each of the five tubes at equilibrium.

Table 15•1 **Study of the system $H_2 + I_2 \rightleftharpoons 2HI$ at 425.4° C**

Concentrations at beginning, mol/L			Concentrations at equilibrium, mol/L			K_c
$[H_2]$	$[I_2]$	$[HI]$	$[H_2]$	$[I_2]$	$[HI]$	
Combination						
0.010667	0.011965	none	0.001831	0.003129	0.01767	54.5
0.011354	0.009044	none	0.003560	0.001250	0.01559	54.6
0.011337	0.007510	none	0.004565	0.0007378	0.01354	54.4
Decomposition						
none	none	0.004489	0.0004789	0.0004789	0.003531	54.4
none	none	0.010692	0.001141	0.001141	0.008410	54.3

Source: *A. H. Taylor and R. H. Crist, J. Amer. Chem. Soc.,* **63**, *1377 (1941).*

A careful study of these results reveals that the quantities present at equilibrium are related to each other by a simple mathematical relationship: *when the concentration of HI is squared and this number is divided by the product of the H_2 and I_2 concentrations, the same number is obtained in each case.* This is illustrated using the concentrations in the first and last sets of data in Table 15.1:

$$\frac{[HI]^2}{[H_2][I_2]} = \frac{(0.01767)^2}{(0.001831)(0.003129)} = \frac{(0.008410)^2}{(0.001141)(0.001141)} = 54.4 \pm 0.1$$

We may express the mathematical relationship as

$$\frac{[HI]^2}{[H_2][I_2]} = K_c$$

in which K_c is known as the **equilibrium constant.** (The c stands for concentration expressed in moles per liter, as indicated by the use of square brackets, [   ].) K_c has a numerical value of 54.4 only if this equilibrium is established at 425.4 °C. If the equilibrium is established at another temperature, the numerical value of K_c will be different. But, for a given equilibrium system at a specific temperature, the value of K_c remains the same, no matter how the individual concentrations may be changed.

• *Writing K_c Expressions* • From the experimental study of a great number of chemical equilibrium systems, it has been found that similar mathematical expressions can be written that relate the concentrations at equilibrium. The general form of the equation depends solely on the balanced equation for the equilibrium. If the equilibrium is represented by

$$mA + nB \rightleftharpoons yC + zD$$

the general equilibrium constant expression is

$$K_c = \frac{[C]^y [D]^z}{[A]^m [B]^n}$$

where m, n, y, and z are the coefficients in the balanced equation, and the quantities in the brackets represent moles per liter of A, B, C, and D. Note that the products on the right of the equation are shown in the numerator and that the concentration of each substance is raised to the power equal to its coefficient in the equation. The following specific examples will serve to illustrate further the relationship:

$$2H_2(g) + O_2(g) \rightleftharpoons 2H_2O(g)$$

$$K_c = \frac{[H_2O]^2}{[H_2]^2[O_2]}$$

$$N_2(g) + 3H_2(g) \rightleftharpoons 2NH_3(g)$$

$$K_c = \frac{[NH_3]^2}{[N_2][H_2]^3}$$

$$\underset{\text{acetic acid}}{CH_3CO_2H(l)} + \underset{\text{ethyl alcohol}}{C_2H_5OH(l)} \rightleftharpoons \underset{\text{ethyl acetate}}{CH_3CO_2C_2H_5(l)} + \underset{\text{water}}{H_2O(l)}$$

$$K_c = \frac{[CH_3CO_2C_2H_5][H_2O]}{[CH_3CO_2H][C_2H_5OH]}$$

The value of K_c depends upon how the equilibrium equation is written. For example, for the H_2, I_2, HI system, if the equation is written

$$2HI(g) \rightleftharpoons H_2(g) + I_2(g) \tag{6}$$

then

$$K_c = \frac{[H_2][I_2]}{[HI]^2}$$

This equilibrium constant expression is the reciprocal of the expression for

Equation (5), that is,

$$K_c \text{ for Equation (6)} = \frac{1}{K_c \text{ for Equation (5)}}$$

From the K_c for Equation (5) of 54.4, we calculate K_c for Equation (6) to be

$$\frac{1}{54.4} = 1.84 \times 10^{-2}$$

Values of K_c can vary greatly among systems. For a system in which the concentrations of substances on the right of the double arrows are high compared with those on the left, the value of K_c will be large; if the concentrations of the substances on the left of the equilibrium equations are relatively high, the value of K_c will be small. To illustrate both possibilities, consider the following equilibria:

$$2H_2(g) + O_2(g) \rightleftharpoons 2H_2O(g) \qquad \text{at 1,000 °C}$$
$$N_2(g) + 3H_2(g) \rightleftharpoons 2NH_3(g) \qquad \text{at 1,000 °C}$$

At 1,000 °C hydrogen and oxygen have a great tendency to react to form H_2O, whereas water has little tendency to decompose into H_2 and O_2. Most of the equilibrium mixture will be H_2O; there will be only a trace of H_2 and O_2 present. The K_c for this system will be a very large number:

$$K_c = \frac{[H_2O]^2}{[H_2]^2[O_2]} = \frac{\left(\begin{array}{c}\text{relatively very}\\\text{large number}\end{array}\right)^2}{\left(\begin{array}{c}\text{very small}\\\text{number}\end{array}\right)^2 \times \left(\begin{array}{c}\text{very small}\\\text{number}\end{array}\right)} = \text{large number}$$

On the other hand, at 1,000 °C hydrogen and nitrogen have only a slight tendency to form NH_3, whereas ammonia has a great tendency to decompose into N_2 and H_2. Consequently, only a small fraction of the equilibrium mixture is composed of NH_3; most is in the form of N_2 and H_2. Hence K_c for this system is small:

$$K_c = \frac{[NH_3]^2}{[H_2]^3[N_2]} = \frac{\left(\begin{array}{c}\text{relatively small}\\\text{number}\end{array}\right)^2}{\left(\begin{array}{c}\text{large}\\\text{number}\end{array}\right)^3 \times \left(\begin{array}{c}\text{large}\\\text{number}\end{array}\right)} = \text{small number}$$

● **15·3·2   Determination of Equilibrium Constants**   In order to determine equilibrium constants, experiments similar to those described for the $H_2 + I_2 \rightleftharpoons 2HI$ system may be carried out. The equilibrium concentrations are generally approached by putting into a reaction vessel the reactants alone, H_2 and I_2 in this equilibrium equation, or the products alone, HI. One could start with a mixture of H_2 and HI, or a mixture of I_2 and HI, and still obtain the same value of K_c, but this approach is seldom used. If the initial concentrations of reactants are known, the calculation of K_c can be accomplished if the concentration of only one of the substances in the equilibrium mixture is determined. The following examples illustrate this method of obtaining K_c.

Captions to Color Section

PLATE 1

Figure 1•11 *Examples of chemical reactions. The chemicals are labeled as they usually are in the laboratory; for example, solid copper sulfate is labeled $CuSO_4$, and a solution of copper sulfate in water is also labeled $CuSO_4$, because the water in the solution is ignored. The balancing of the equations for these reactions is left for practice in later chapters. (a) Magnesium metal, Mg, burning in air combines mainly with oxygen, O_2, to form a gray powder that is mainly magnesium oxide, MgO. (b) When white solid copper sulfate, $CuSO_4$, is dissolved in water, H_2O, a blue solution of copper sulfate, $CuSO_4$, is formed. (Only a small part of the white solid shown is needed to form this pale blue solution.) (c) When a solution of ammonia, NH_3, in water is mixed with a solution of copper sulfate, $CuSO_4$, a deep blue solution of $Cu(NH_3)_4SO_4$ is formed. (d) When water solutions of copper sulfate, $CuSO_4$, and sodium hydroxide, NaOH, are mixed, the deep blue solid copper hydroxide, $Cu(OH)_2$, is formed as a precipitate, and sodium sulfate, Na_2SO_4, is dissolved in the colorless solution. (e) When water solutions of silver nitrate, $AgNO_3$, and hydrochloric acid, HCl, are mixed, the white solid silver chloride, AgCl, is formed, and nitric acid, HNO_3, is dissolved in the colorless solution. (f) A solution of potassium chromate, K_2CrO_4, combines with a solution of lead nitrate, $Pb(NO_3)_2$, to form the yellow solid lead chromate, $PbCrO_4$, and a colorless solution of potassium nitrate, KNO_3. The light yellow color of the solution on the right must be due to excess K_2CrO_4, because it is known that KNO_3 alone forms a colorless solution. (g) A solution of iron chloride, $FeCl_3$, combines with a solution of sodium hydroxide, NaOH, to form the brown solid iron hydroxide, $Fe(OH)_3$, and a solution of sodium chloride, NaCl, that is discolored by some iron hydroxide. (h) When a solution of hydrochloric acid, HCl, is mixed with a solution of sodium hydroxide, NaOH, no visible change takes place. However, with a thermometer we can observe that the temperature increases (in the case shown, from 24 °C to 57 °C). This indicates that a chemical reaction has taken place. (See page 22.)*
Source: *Photographs by Earl Walker.*

PLATE 2

Figure 4•5 *When a beam of white light (from the projector at the left) passes through a glass prism, the path of the light is bent from the original path. Because light of short wavelength (blue) is bent more than light of long wavelength (red), the beam of white light is spread out as a spectrum. In this case, the spectrum is continuous (there are no dark areas), showing that this white light is made up of all visible wavelengths (colors). (See page 95.)*

Figure 4•6 *(1) Artist's reproduction of a continuous spectrum from an incandescent solid, obtained with a diffraction grating. (2) The sun's absorption spectrum, showing several dark (Fraunhofer) lines. Below the sun's spectrum are shown the discontinuous emission spectra of sodium, hydrogen, calcium, mercury, and neon. (See pages 95, 96, and 97.)*

PLATE 3

PLATE 4

Figure 3•2 (a) *The spark of a 20,000-V Tesla coil at left generates a characteristic blue glow as it flashes through air to the left end of the tube. Inside the evacuated tube, the cathode "rays" or particles jump from the circular cathode toward the grounded anode at the far right, but are blocked by the aluminum plate except for a narrow open slit. Some of the cathode particles that fly through the slit strike a zinc sulfide coating, causing it to fluoresce with a greenish glow. The straight green line reveals that the particles are traveling in a straight line toward the anode.* (b) *When a magnet is placed near the beam of cathode particles, the path of the beam is bent. The direction of bending here (with the south pole of the magnet near the camera) shows that the particles are negative.* (See pages 61 and 62.)
Source: *Photographs by Joseph Carroll.*

Figure 8•3 *The cooling of bromine gas with an alcohol-dry ice bath.* (a) *A sealed, 2-L bulb of bromine gas at room temperature is placed in a bath of alcohol-dry ice which has a temperature of about −78 °C.* (b) *After several minutes, almost all the bromine has collected by deposition on the cold surface, forming crystalline platelets of solid bromine. At this point, because of the very low pressure inside the bulb and the considerable pressure of the atmosphere outside, there is danger of an implosion.* (c) *The bulb, just after removal from the cold bath, begins to return to room temperature. As the solid bromine warms, its vapor pressure rises and sublimation occurs; the gas again can be seen.* (See page 233.)
Source: *Photographs by Joseph Carroll.*

Figure 9•5 *Liquid ozone collected on the surface of silica gel granules in tubes that have been cooled in liquid air.* (See page 272.)
Source: *Photograph by Ross Chapple.*

Figure 10•1 *Selective solubility in different liquids (see Section 10.2) and immiscibility (see Section 10.2.2) are shown by this system of one polar and two nonpolar liquids. Left cylinder: Three immiscible liquids are shown—bottom, carbon tetrachloride; middle, water; top, ether. Ether and carbon tetrachloride are miscible with one another, but unless the container is shaken, the water layer effectively keeps them separated. Middle cylinder: The result is shown when a few crystals of iodine, I_2, are introduced; this nonpolar solute dissolves in the two nonpolar solvents. In the carbon tetrachloride, iodine has nearly the same color as pure I_2 vapor, indicating that there is little disturbance of the I—I bond; but in ether, the brown color shows that there is a definite chemical interaction which changes the absorption spectrum of the I_2 molecules. Right cylinder: Here the result is shown when a few crystals of the salt copper sulfate, $CuSO_4$, are introduced. The ions, Cu^{2+} and SO_4^{2-}, are quite soluble in the polar solvent but do not dissolve appreciably in the nonpolar solvents.* (See page 294.)
Source: *Photographs by Joseph Carroll.*

Figure 12•12 *The Tyndall effect.* (a) *The water on the left contains about 20 g of dissolved copper sulfate, whereas the water on the right contains only about 0.001 g of colloidal carbon (from India ink).* (b) *The blue light and the image of the bulb filament on the white card at left reveal that the spotlight shines through both liquids. The light is hardly reflected at all in the true solution, which contains many ions; but the light is brightly reflected (the Tyndall beam) by the relatively few particles in the colloid. The colloid shown was made almost 30 years ago and appears to be as stable now as when originally prepared.* (See page 363.)
Source: *Photographs by Joseph Carroll.*

Figure 18•6 *The electrolysis of an aqueous solution of potassium iodide containing some phenolphthalein.* (a) *The first view shows the solution and the platinum electrodes before the current is turned on. A strip of paper is used as a porous diaphragm to divide the solution into two compartments to retard mixing.* (b) *After the direct current is turned on, iodine forms at one electrode and bubbles of hydrogen at the other. The pink color of the phenolphthalein reveals that hydroxide ion is also formed.* (c) *The final view shows how mixing occurs unless prevented by removing the products and/or having a more effective diaphragm. The reactions at the electrodes are discussed in Section 18.8.3.*
Source: *Photographs by Joseph Carroll.*

Figure 19•23 *Polished cross sections of microspheres in fuel pellets for a converter reactor. On the left* (a) *are particles of thorium dioxide, ThO_2, encased in carbon, and on the right* (b) *are particles of uranium dioxide, UO_2, encased in carbon. See the discussion of the Fort St. Vrain reactor in Section 19.12.3.*
Source: *Courtesy of Union Carbide Corporation's Nuclear Division/Oak Ridge National Laboratory, Oak Ridge, Tenn.*

Ilene Cranston.　　　　£12

KEY

- Atomic number
- Electron arrangement
- Symbol
- Atomic weight

VIIIA

						2　1s² He 4.0026

	IIIA	IVA	VA	VIA	VIIA	
	5 (He) 2s² 2p¹ **B** 10.81	6 (He) 2s² 2p² **C** 12.011	7 (He) 2s² 2p³ **N** 14.007	8 (He) 2s² 2p⁴ **O** 15.999	9 (He) 2s² 2p⁵ **F** 18.998	10 (He) 2s² 2p⁶ **Ne** 20.179

IB	IIB	13 (Ne) 3s² 3p¹ **Al** 26.982	14 (Ne) 3s² 3p² **Si** 28.086	15 (Ne) 3s² 3p³ **P** 30.974	16 (Ne) 3s² 3p⁴ **S** 32.06	17 (Ne) 3s² 3p⁵ **Cl** 35.453	18 (Ne) 3s² 3p⁶ **Ar** 39.948

28 (Ar) 3d⁸ 4s² **Ni** 58.70	29 (Ar) 3d¹⁰ 4s¹ **Cu** 63.55	30 (Ar) 3d¹⁰ 4s² **Zn** 65.38	31 (Ar) 3d¹⁰ 4s² 4p¹ **Ga** 69.72	32 (Ar) 3d¹⁰ 4s² 4p² **Ge** 72.59	33 (Ar) 3d¹⁰ 4s² 4p³ **As** 74.92	34 (Ar) 3d¹⁰ 4s² 4p⁴ **Se** 78.96	35 (Ar) 3d¹⁰ 4s² 4p⁵ **Br** 79.904	36 (Ar) 3d¹⁰ 4s² 4p⁶ **Kr** 83.80
46 (Kr) 4d¹⁰ **Pd** 106.4	47 (Kr) 4d¹⁰ 5s¹ **Ag** 107.87	48 (Kr) 4d¹⁰ 5s² **Cd** 112.41	49 (Kr) 4d¹⁰ 5s² 5p¹ **In** 114.82	50 (Kr) 4d¹⁰ 5s² 5p² **Sn** 118.69	51 (Kr) 4d¹⁰ 5s² 5p³ **Sb** 121.75	52 (Kr) 4d¹⁰ 5s² 5p⁴ **Te** 127.60	53 (Kr) 4d¹⁰ 5s² 5p⁵ **I** 126.90	54 (Kr) 4d¹⁰ 5s² 5p⁶ **Xe** 131.30
78 (Xe) 4f¹⁴ 5d⁹ 6s¹ **Pt** 195.09	79 (Xe) 4f¹⁴ 5d¹⁰ 6s¹ **Au** 196.97	80 (Xe) 4f¹⁴ 5d¹⁰ 6s² **Hg** 200.59	81 (Xe) 4f¹⁴ 5d¹⁰ 6s² 6p¹ **Tl** 204.37	82 (Xe) 4f¹⁴ 5d¹⁰ 6s² 6p² **Pb** 207.2	83 (Xe) 4f¹⁴ 5d¹⁰ 6s² 6p³ **Bi** 208.98	84 (Xe) 4f¹⁴ 5d¹⁰ 6s² 6p⁴ **Po** (209)	85 (Xe) 4f¹⁴ 5d¹⁰ 6s² 6p⁵ **At** (210)	86 (Xe) 4f¹⁴ 5d¹⁰ 6s² 6p⁶ **Rn** (222)

In several cases, the atomic weight is rounded to fewer significant figures than given in the IPUAC Table of Atomic Weights 1977 inside the back cover. Electron configurations are from C. E. Moore, NSRDS-NBS 34, National Bureau of Standards, Washington, D.C., 1970, and are reproduced, with permission, from G. T. Seaborg, *Ann. Rev. Nucl. Sci.*, **18**, 53 (1968), except for elements 104–107, which are predicted by analogy. The state of the element at or near room temperature is indicated by the color of the symbol: black for solid, color for liquid, and gray for gas.

ᵃA value in parentheses denotes the mass number of the radioisotope of longest half-life.

ᵇIUPAC symbols and names for elements 104–106 are in the table inside the back cover; see unnilhexium, and so on. Only one laboratory has reported element 107, so its existence is not confirmed.

64 (Xe) 4f⁷ 5d¹ 6s² **Gd** 157.25	65 (Xe) 4f⁹ 6s² **Tb** 158.93	66 (Xe) 4f¹⁰ 6s² **Dy** 162.50	67 (Xe) 4f¹¹ 6s² **Ho** 164.93	68 (Xe) 4f¹² 6s² **Er** 167.26	69 (Xe) 4f¹³ 6s² **Tm** 168.93	70 (Xe) 4f¹⁴ 6s² **Yb** 173.04	71 (Xe) 4f¹⁴ 5d¹ 6s² **Lu** 174.97
96 (Rn) 5f⁷ 6d¹ 7s² **Cm** (247)	97 (Rn) 5f⁹ 7s² **Bk** (247)	98 (Rn) 5f¹⁰ 7s² **Cf** (251)	99 (Rn) 5f¹¹ 7s² **Es** (252)	100 (Rn) 5f¹² 7s² **Fm** (257)	101 (Rn) 5f¹³ 7s² **Md** (258)	102 (Rn) 5f¹⁴ 7s² **No** (259)	103 (Rn) 5f¹⁴ 6d¹ 7s² **Lr** (260)

PLATE 1

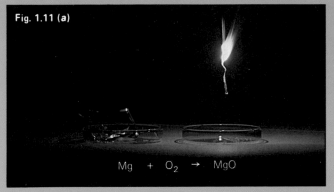

Fig. 1.11 (a)

$Mg + O_2 \rightarrow MgO$

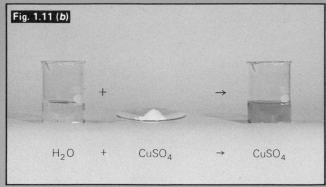

Fig. 1.11 (b)

$H_2O + CuSO_4 \rightarrow CuSO_4$

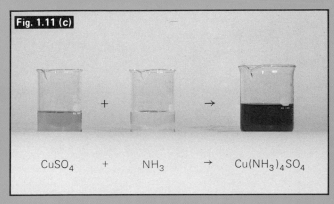

Fig. 1.11 (c)

$CuSO_4 + NH_3 \rightarrow Cu(NH_3)_4SO_4$

Fig. 1.11 (d)

$CuSO_4 + NaOH \rightarrow Cu(OH)_2 \downarrow + Na_2SO_4$

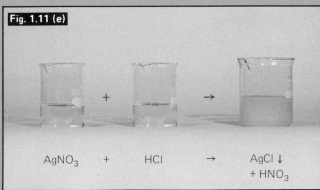

Fig. 1.11 (e)

$AgNO_3 + HCl \rightarrow AgCl \downarrow + HNO_3$

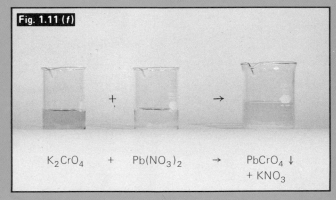

Fig. 1.11 (f)

$K_2CrO_4 + Pb(NO_3)_2 \rightarrow PbCrO_4 \downarrow + KNO_3$

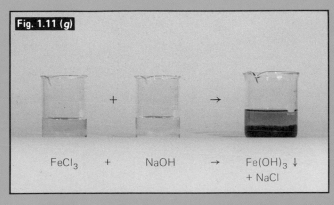

Fig. 1.11 (g)

$FeCl_3 + NaOH \rightarrow Fe(OH)_3 \downarrow + NaCl$

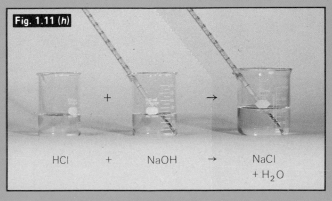

Fig. 1.11 (h)

$HCl + NaOH \rightarrow NaCl + H_2O$

PLATE 2

Fig. 4.5

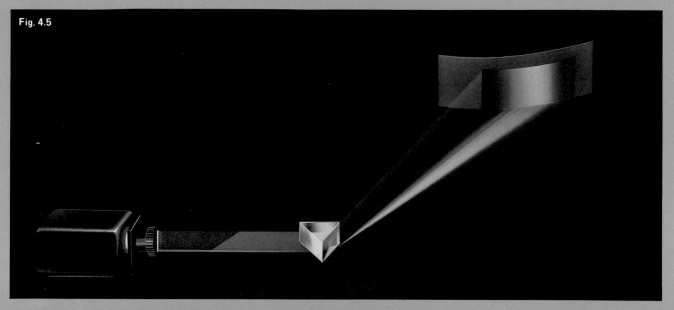

Fig. 4.6

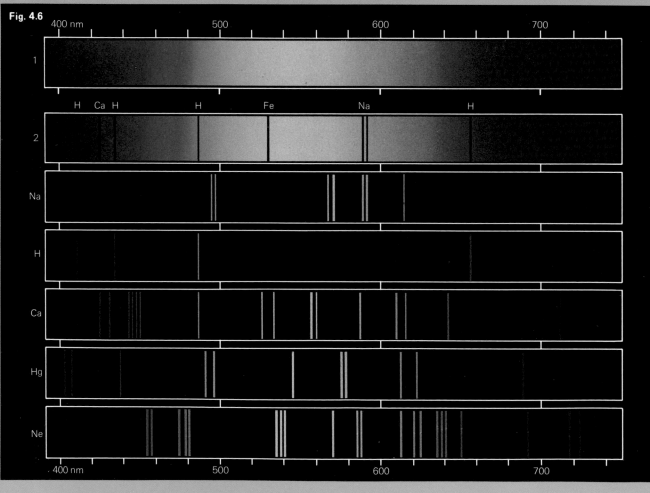

PLATE 3

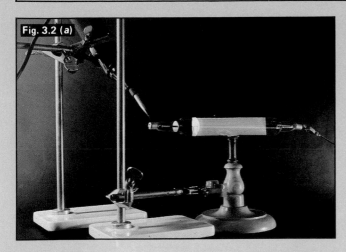
Fig. 3.2 (a)

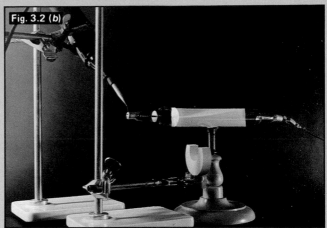
Fig. 3.2 (b)

Fig. 8.3 (a)

Fig. 8.3 (b)

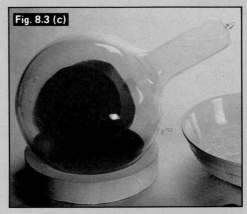
Fig. 8.3 (c)

Fig. 9.5

Fig. 10.1

PLATE 4

Fig. 12.12 (a)

Fig. 12.12 (b)

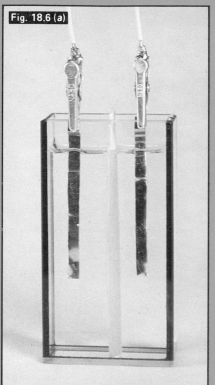

Fig. 18.6 (a)

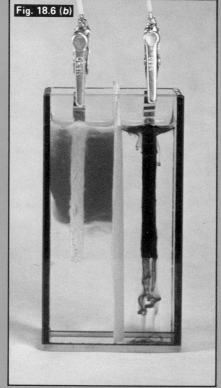

Fig. 18.6 (b)

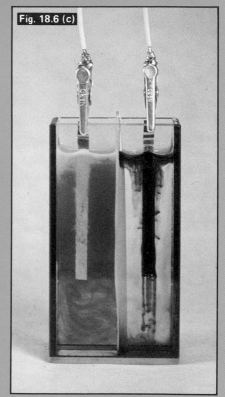

Fig. 18.6 (c)

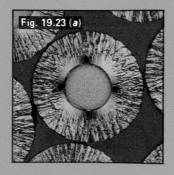

Fig. 19.23 (a)

Fig. 19.23 (b)

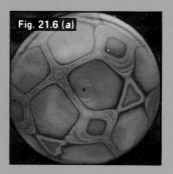

Fig. 21.6 (a)

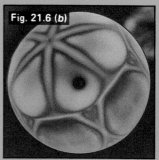

Fig. 21.6 (b)

PLATE 5

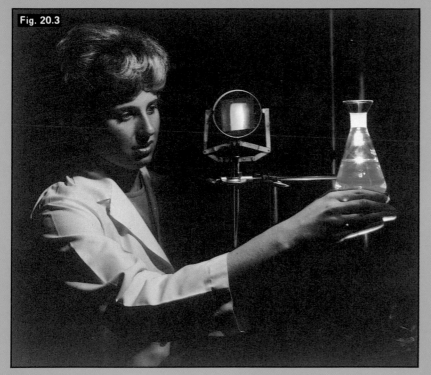

Fig. 20.3

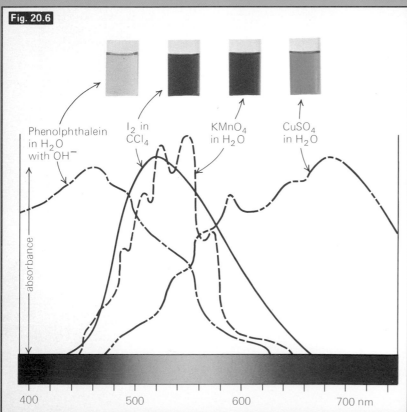

Fig. 20.6

Phenolphthalein in H_2O with OH^-

I_2 in CCl_4

$KMnO_4$ in H_2O

$CuSO_4$ in H_2O

absorbance

400   500   600   700 nm

Fig. 22.10

peridot

chrysoberyl

garnet

emerald

ruby

PLATE 6

Fig. 20.7

(a) Light transmitted by three primary subtractive filters

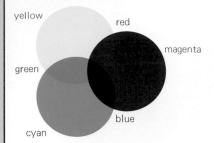

yellow

red

magenta

green

cyan

blue

(b) When white light is shone through a sandwich made of various combinations of three subtractive primaries,

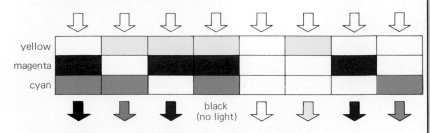

| yellow |
| magenta |
| cyan |

black (no light)

these examples and all intermediate hues can be transmitted.

(c) Three dyes which have colors of the three subtractive primaries:

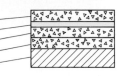
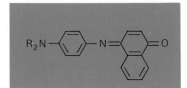

(d) <u>Kodachrome</u> film sandwich prior to exposure

blue-sensitive layer (will develop to yellow)
yellow filter (is bleached colorless in development)
green-sensitive layer (will develop to magenta)
red-sensitive layer (will develop to cyan)
transparent support of cellulose acetate

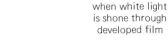

key to symbols used at left and below:

grains of AgX in film

light sensitized Ag⁺

developed (black) Ag

(e) Exposure, development, and final use of a <u>Kodachrome</u> slide

object being photographed

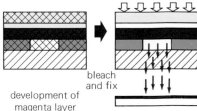

when white light is shone through developed film

original exposure: "taking a picture" of a red spot

development to form black silver deposit

exposure from below to red light and development of cyan dye

exposure from above to blue light and development of yellow dye

development of magenta layer

bleach and fix

the image reproduces the object

(f) Light transmitted by each of the three developed layers separately and by a composite sandwich of all three for a photograph of a geologic specimen

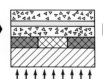

yellow layer

magenta layer

cyan layer

composite

PLATE 7

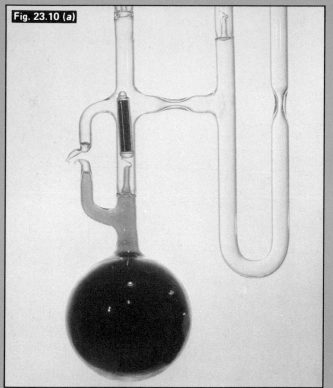

Fig. 23.10 (a)

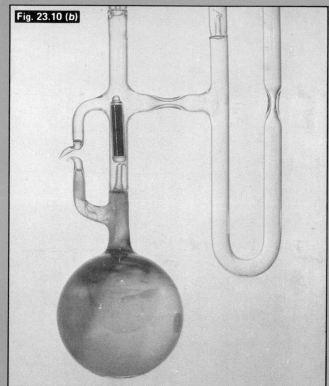

Fig. 23.10 (b)

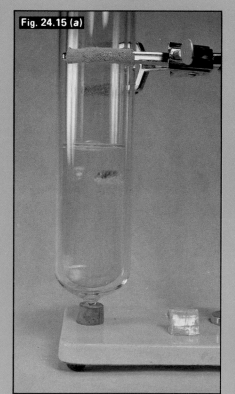

Fig. 24.15 (a)

Fig. 24.15 (b)

Fig. 24.15 (c)

PLATE 8

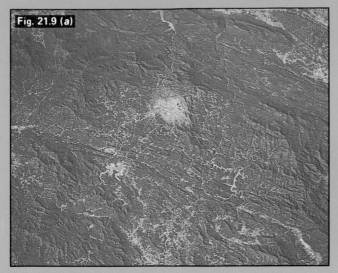

Fig. 21.9 (*a*)

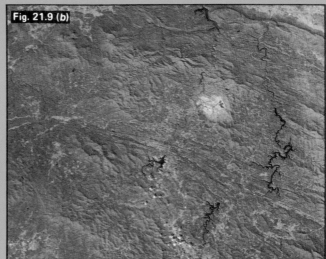

Fig. 21.9 (*b*)

Fig. 21.9 (*c*)

Fig. 21.9 (*d*)

Figure A

antenna

seismometer

GCMS processor

camera

meteorology
sensors

biology
processor

surface
sampler
boom

x-ray
fluorescence
tunnel

magnets

collector
head

Figure B

PLATE 4

Figure 21•6 (a) *The left sphere is a single crystal of copper, 1.3 cm in diameter, machined from a pure crystal formed by zone refining (see Figure 21.8). The surface was oxidized for 30 min at 250 °C, and the patterns are due to interference colors produced by thin copper oxide films of different thicknesses (similar to the color patterns in films of oil on wet streets). The rate at which the copper oxide forms is different on different exposed crystal faces, showing that the chemical activity of the copper depends on the arrangement of the atoms. Copper has a face-centered cubic structure (see Figure 8.29). The oxide film is thickest, about 4,000 Å, at the center of the cross in the pattern; the film is about 3,000 Å thick at the centers of the triangles; the film is thinnest, about 150 Å, at a point between the two thickest spots. (b) The right sphere is a single crystal of zirconium that was oxidized for 20 min at 360 °C. Zirconium has a hexagonal close-packed structure (see Figure 8.29). In the light blue areas, the oxide film is about 800 Å thick; thinner films are successively dark blue, purple, brown-yellow, and gold (350 Å). Both of these examples show that the susceptibility to chemical attack of a metal depends on which of its crystal faces is exposed. (In both photographs, the black circle is due to the reflection of the camera lens.) See also Figures 14.6 and 14.7.*
Source: *Courtesy of J. V. Cathcart and J. C. Wilson of the Oak Ridge National Laboratory, Oak Ridge, Tenn.*

PLATE 5

Figure 20•3 *Determining the absorption spectrum of water. A diffraction grating, shown here dispersing visible light, also can be used to separate photons of different energies in the ultraviolet region. Ionizing radiations excite water strongly at 21 eV. This collective excitation of electrons is thought to be the principal mechanism by which ionizing radiation initiates chemical reactions in water and living tissue. (See page 608.)*
Source: *Courtesy of Union Carbide Corporation's Nuclear Division/Oak Ridge National Laboratory and Dr. Linda Painter of ORNL and The University of Tennessee Department of Physics.*

Figure 20•6 *Four colored solutions and their absorbance curves. The phenolphthalein solution appears red to our eye because it absorbs light in the blue and green regions. The copper sulfate solution appears blue because it absorbs mainly in the red and orange regions. Purple is the color our eye reports when it receives a mixture of blue and red light such as is transmitted by the two middle solutions. Even though the spectral curves provide a certain identification of I_2 in CCl_4 versus $KMnO_4$ in H_2O, our eye has difficulty in distinguishing between them because they absorb light in the same region of the spectrum. (See page 612.)*
Source: *Photograph by Earl Walker.*

Figure 22•10 *The colors of gems and minerals are usually due to the presence of transition metals. In this figure (from top to bottom) are peridot, $(Fe,Mg)_2SiO_4$; yellow chrysoberyl, Al_2BeO_4, with some Al^{3+} ions replaced by Fe^{3+}; garnet, $(Fe,Mg)_3Al_2(SiO_4)_3$; emerald, $Be_3Al_2Si_6O_{18}$, with some Al^{3+} ions replaced by Cr^{3+}; and ruby, Al_2O_3, with some Al^{3+} ions replaced by Cr^{3+} (see also Figures 22.11 and 22.12).*
Source: *Used by permission from B. M. Loeffler and R. G. Burns, "Shedding Light on the Color of Gems and Minerals," American Scientist, 64, 636 (1976).*

PLATE 6

Figure 20•7 *Some of the chemical and mechanical details of the Kodachrome process for producing color slides.*
(a) *Any color can be produced by the transmittance of (white) light through filters containing the proper amounts of three primary dyes, yellow (Y), magenta (M), and cyan (C). (b) Schematic diagram showing how a three-layer sandwich of film can be arranged to make use of the phenomenon described in (a). (c) Examples of three dyes which can be synthesized in their respective film layers. (d) The Kodachrome film sandwich of three color-sensitive layers, a yellow filter, and a supporting base. In each color-sensitive layer, the amount of that color absorbed activates a proportional amount of Ag^+ ions. (e) A six-step sequence from film exposure, through development of the proper amount of primary dye in each layer, to the finished transparent color slide. Each development step, and the fix and bleach, involves immersing the film in a solution of the necessary reactants. (f) Photographs of a geologic specimen, showing the light transmitted by each of the three individual color filters in the developed film and the full color result of white light transmitted through all three layers. The white mineral is calcite, $CaCO_3$, the black is franklinite, $ZnFe_2O_4$, and the rose is willemite, Zn_2SiO_4. (See page 612.)*
Source: *From Infrared and Ultraviolet Photography and from A. Weissburger, "A Chemist's View of Color Photography," American Scientist, 58, 648 (1970). Reproduced with permission from a copyrighted Kodak publication and from American Scientist.*

PLATE 7

Figure 23•10 *The reaction of platinum hexafluoride and xenon. (a) At left is a section of a closed glass system in which a sample of the reddish gas PtF_6 is separated by a "break seal" from a high-pressure sample of the colorless gas xenon. Resting on the break seal is a glass-encased iron hammer. (b) When the hammer is raised by an outside magnet (not shown) and allowed to drop, the tip of the fragile break seal is broken off and the xenon rushes into the lower bulb. The solid yellow crystals, which form first as a smoke and then settle on the glass, are direct evidence of the reactivity of xenon. See Section 23.10.*
Source: *Photographed by Paul Weller at the Argonne National Laboratory.*

Figure 24•15 *(a) A sample of liquid ammonia at its boiling point, −33.4 °C, is shown in an unsilvered Dewar flask. At the base of the flask is a piece of sodium with a lustrous, freshly cut surface exposed to the air. (b) Dropping in a small piece of sodium cut from the large piece shown forms this characteristic blue-colored solution of an alkali metal in ammonia. Note: the surface of the sodium, now corroded, was photographed only a minute or two after the picture at the left. (c) After the original blue color fades, owing to reaction of the sodium, the faint yellow color characteristic of amide ion is seen in the solution. The blue color shown forming is due to the dissolution of a second small piece of sodium dropped in just before this picture was snapped. See also Figures 14.9 and 20.16 and Section 21.3.1.*
Source: *Photographs by Joseph Carroll.*

PLATE 8

Figure 21•9 *Copperhill, Tennessee. (a) A satellite photograph of the southern Appalachian forest showing treeless area of about 20 square miles. (b) Infrared photograph of the same region. This technique is preferred for routine mapping and monitoring because cloud cover does not interfere with it. (c) Ground level appearance of the area today.*
(d) Young trees growing in successful reforestation project. (See page 646.)
Source: *(a) and (b) Courtesy of NASA Goddard Space Flight Center, Greenbelt, Md. (c) and (d) Courtesy of Edward M. Jones, Copperhill, Tenn.*

Figure A. *Schematic drawing of a Viking Lander. Two of these spacecraft made successful landings on the surface of Mars. In addition to devices for measuring various meteorological and geological phenomena, they had chemical instruments for determining the composition of the molecules in the atmosphere, the major elements in the soil, and the possible presence of organic molecules associated with living organisms. (See Special Topic, page 758.)*
Source: *Courtesy of Martin Marietta Aerospace, Denver, Colo.*

Figure B. *Photograph, radioed from Mars, of the surface of the planet and the Viking I Lander's soil sampler. The sampler was able to dig trenches several inches deep in the loose soil. Soil samples were obtained for the gas chromatograph–mass spectrometer (GCMS) processor, for the X-ray fluorescence element analyzer, for three different types of biological tests for the presence of life, and for testing the magnetic and electrical properties of the soil. Because of the magnetic character and the high iron content of the surface (12 to 16 percent), it is thought that the red coloration is due to a form of reddish magnetic iron oxide not found on earth. (See Special Topic, page 758.)*
Source: *Courtesy of the National Aeronautics and Space Administration, Washington, D.C.*

━━━ • Example 15•1 • ━━━

Consider the equilibrium system $A(g) + B(g) \rightleftharpoons C(g) + D(g)$. The initial concentrations of A and B are 1.00M and 2.00M, respectively. After equilibrium is attained, the concentration of B is found to be 1.50M. Calculate K_c.

━━━ • Solution • The K_c expression for the equilibrium equation is

$$K_c = \frac{[C][D]}{[A][B]}$$

We need to know the concentrations in moles per liter of all substances at equilibrium to solve for the numerical value of K_c. Because the only concentration given at equilibrium is that of B, we must use the change in the concentration of B to calculate the other equilibrium concentrations.

Because nothing is said about the concentrations of C and D, we assume that neither is present initially. From the chemical equation, we see that for each mole of B that disappears, 1 mole of C and 1 mole of D are formed. At the outset, the concentration of B was 2.00M; at equilibrium the concentration of B is 1.50M. The concentrations of C and D at equilibrium equal the concentration of B that has disappeared, that is, 0.50M. The chemical equation also tells us that for each mole of B that disappears, 1 mole of A also disappears. The concentration of A present at equilibrium is $1.00 - 0.50 = 0.50M$. The solution can be set up in the following manner:

	A +	B $\rightleftharpoons$	C +	D
mol/L at the start	1.00	2.00	0	0
mol/L at equilibrium	0.50	1.50	0.50	0.50

$$K_c = \frac{(0.50)(0.50)}{(0.50)(1.50)} = 0.33$$

See also Exercises 5, 7, 8, and 12 at the end of the chapter.

━━━ • Example 15•2 • ━━━

G. B. Kistiakowsky, in *J. Amer. Chem. Soc.*, **50**, 2315 (1928), reported on the thermal equilibrium established at 321.4 °C for the decomposition of HI:

$$2HI(g) \rightleftharpoons H_2(g) + I_2(g)$$

In one experiment, the initial concentration of HI of 2.08 moles/L was reduced to 1.68 moles/L at equilibrium. Calculate the value of K_c.

━━━ • Solution • The K_c expression for the decomposition of HI is

$$K_c = \frac{[H_2][I_2]}{[HI]^2}$$

At the outset the concentration of HI was 2.08M; at equilibrium its concentration is 1.68M. Hence, 0.40 mole/L of HI has disappeared. From the chemical equation, we see that for every 2 moles of HI that disappear, 1 mole of H_2 and 1 mole of I_2 are produced. The concentrations of H_2 and I_2

present at equilibrium are each $0.40M/2 = 0.20M$.

Summarizing:

	2HI $\rightleftharpoons$	H_2 +	I_2
mol/L at the start	2.08	0	0
mol/L at equilibrium	1.68	0.20	0.20

$$K_c = \frac{(0.20)(0.20)}{(1.68)^2} = 0.014 = 1.4 \times 10^{-2}$$

See also Exercises 7, 8, and 12.

• Example 15.3 •

To study the decomposition of phosphorus(V) chloride, 35.7 g of PCl_5 is placed in a 5.00-L flask. The flask and contents are heated to 250 °C and then held at that temperature until the following equilibrium is established:

$$PCl_5(g) \rightleftharpoons PCl_3(g) + Cl_2(g)$$

It is then shown by analysis that 7.87 g of Cl_2 is present in the equilibrium mixture. Calculate K_c.

• Solution • Note that the quantities of PCl_5 and Cl_2 are expressed in grams rather than in moles, and that the volume of the container is 5.00 L. The expression for the equilibrium constant is

$$K_c = \frac{[PCl_3][Cl_2]}{[PCl_5]}$$

$$\text{mol/L of } PCl_5 \text{ at the start} = \frac{35.7 \text{ g}}{5.00 \text{ L}} \times \frac{1 \text{ mol}}{208 \text{ g}} = 0.0343M$$

$$\text{mol/L of } Cl_2 \text{ at equilibrium} = \frac{7.87 \text{ g}}{5.00 \text{ L}} \times \frac{1 \text{ mol}}{70.9 \text{ g}} = 0.0222M$$

From the chemical equation, we find these relationships at equilibrium:

$$[PCl_3] = [Cl_2] = 0.0222M$$
$$[PCl_5] = [PCl_5]_{start} - [Cl_2] = 0.0343 - 0.0222 = 0.0121M$$

Summarizing:

	PCl_5 $\rightleftharpoons$	PCl_3 +	Cl_2
mol/L at the start	0.0343	0	0
mol/L at equilibrium	0.0121	0.0222	0.0222

$$K_c = \frac{(0.0222)(0.0222)}{0.0121} = 0.0407 = 4.07 \times 10^{-2}$$

See also Exercise 11.

15.3.3 **Calculations Based on K_c** Once the value of the equilibrium constant is known for a specified reaction at a given temperature, the constant can be used to calculate the concentrations of substances in an equilibrium mixture. If the initial concentrations of one or more of the reacting substances is known, we can predict what all concentrations will be at equilibrium without performing any experimental determinations. The following examples illustrate this use of the equilibrium constant.

━━━ • Example 15•4 • ━━━━━━━━━━━━━━━━━━━━━━━━━━━━━━

The value of K_c at 25 °C is 4.00 for the following equilibrium:

$$CH_3CO_2H(l) + C_2H_5OH(l) \rightleftharpoons CH_3CO_2C_2H_5(l) + H_2O(l)$$

    acetic acid   ethyl alcohol   ethyl acetate   water

If 2.00 moles of acetic acid and 2.00 moles of ethyl alcohol are mixed and allowed to reach equilibrium at 25 °C, how many moles of ethyl acetate will be produced?

━━━ • Solution • Because the stoichiometric number of molecules on the right-hand side of the equilibrium equation equals the number of molecules on the left-hand side, the volume of the reaction container need not be specified. This can be proven in the following fashion: If V equals the volume of solution, then K_c may be expressed as

$$K_c = \frac{\dfrac{\text{mol } CH_3CO_2C_2H_5}{V} \times \dfrac{\text{mol } H_2O}{V}}{\dfrac{\text{mol } CH_3CO_2H}{V} \times \dfrac{\text{mol } C_2H_5OH}{V}}$$

The V's in the numerator and denominator cancel and we obtain

$$K_c = \frac{(\text{mol } CH_3CO_2C_2H_5)(\text{mol } H_2O)}{(\text{mol } CH_3CO_2H)(\text{mol } C_2H_5OH)}$$

From the equilibrium equation we can see that for each mole of $CH_3CO_2C_2H_5$ and H_2O that is formed, 1 mole each of CH_3CO_2H and C_2H_5OH disappears. Therefore, if we let x = moles of $CH_3CO_2C_2H_5$ = moles of H_2O that are present at equilibrium, then $(2.00 - x)$ = moles of CH_3CO_2H = moles of C_2H_5OH present at equilibrium. Summarizing:

	CH_3CO_2H	$+ C_2H_5OH$	$\rightleftharpoons CH_3CO_2C_2H_5$	$+ H_2O$
mol at the start	2.00	2.00	0	0
mol at equilibrium	$2.00 - x$	$2.00 - x$	x	x

$$K_c = 4.00 = \frac{(x)(x)}{(2.00 - x)(2.00 - x)} = \frac{x^2}{(2.00 - x)^2}$$

We could solve this equation by using the quadratic formula or by factoring, but it is easily solved by taking the square root of both sides:

$$2.00 = \frac{x}{2.00 - x}$$

$$x = 4.00 - 2.00x$$

$$x = \frac{4.00}{3.00} = 1.33$$

Therefore, 1.33 moles of $CH_3CO_2C_2H_5$ will be produced.

See also Exercises 15, 16, 19–21, and 24.

• Example 15.5 •

The value of K_c at 2,827 °C is 83.3 for the following equilibrium:

$$2NO(g) \rightleftharpoons N_2(g) + O_2(g)$$

A mixture of 0.0500 mole of N_2 and 0.0500 mole of O_2 is placed into a 2.00-L container and equilibrium is attained at 2,827 °C.
(a) How many moles of N_2, O_2, and NO will be present at equilibrium?
(b) What will be the concentrations in moles per liter of all substances at equilibrium?

• Solution •

(a) As in Example 15.4, the stoichiometric numbers of molecules of reactants and products are equal. Note that we are given the value of K_c for the decomposition of NO, but that in the experimental work pure N_2 and O_2 are the initial reactants.

Proceeding from right to left in the chemical equation, we see that for each mole of N_2 or O_2 that reacts, 2 moles of NO are produced. If we let x be the number of moles of N_2 or O_2 that react, then $2x$ is the number of moles of NO produced. Summarizing:

	2NO $\rightleftharpoons$	N_2	+	O_2
mol at the start	0	0.0500		0.0500
mol at equilibrium	$2x$	$0.0500 - x$		$0.0500 - x$

$$K_c = \frac{[N_2][O_2]}{[NO]^2}$$

$$83.3 = \frac{(0.0500 - x)(0.0500 - x)}{(2x)^2} = \frac{(0.0500 - x)^2}{(2x)^2}$$

We can solve this equation by taking the square root of both sides:

$$\sqrt{83.3} = 9.13 = \frac{0.0500 - x}{2x}$$

$$18.26x = 0.0500 - x$$

$$x = \frac{0.0500}{18.26} = 0.00260$$

$$\text{mol NO} = 2x = 0.00520$$

$$\text{mol } N_2 = \text{mol } O_2 = 0.0500 - 0.0026 = 0.0474$$

• Check •

$$K_c = \frac{(0.0474)(0.0474)}{(0.00520)^2} = 83.1$$

This is a good agreement with 83.3 when all intermediate calculations are to three significant figures.

(b) The reaction was carried out in a 2.00-L container. To obtain concentrations in moles per liter, we must divide the moles of each substance by 2.00 L:

$$[NO] = \frac{0.00520 \text{ mol}}{2.00 \text{ L}} = 0.00260M$$

$$[N_2] = [O_2] = \frac{0.0474 \text{ mol}}{2.00 \text{ L}} = 0.0237M$$

See also Exercises 15, 16, 19–21, and 24.

• Example 15•6 •

Refer to Example 15.4 and the equilibrium system

$$CH_3CO_2H + C_2H_5OH \rightleftharpoons CH_3CO_2C_2H_5 + H_2O$$

mol	0.67	0.67	1.33	1.33

If an additional 2.00 moles of CH_3CO_2H are added to the equilibrium mixture and then equilibrium is attained once again at 25 °C, what will be the moles of all substances at equilibrium?

• **Solution** • Because of the increased concentration of CH_3CO_2H, there will be more reaction of CH_3CO_2H and C_2H_5OH to form $CH_3CO_2C_2H_5$ and H_2O. If we let y be the number of moles of each of the first two that disappear, then y is also equal to the number of moles of each of the second two that are formed. Summarizing:

$$CH_3CO_2H + C_2H_5OH \rightleftharpoons CH_3CO_2C_2H_5 + H_2O$$

mol at the start	2.67	0.67	1.33	1.33
mol at equilibrium	$2.67 - y$	$0.67 - y$	$1.33 + y$	$1.33 + y$

$$K_c = 4.00 = \frac{(1.33 + y)(1.33 + y)}{(2.67 - y)(0.67 - y)} = \frac{1.77 + 2.66y + y^2}{1.79 - 3.34y + y^2}$$

$$3y^2 - 16.0y + 5.39 = 0$$

We can solve this equation with the quadratic formula (see Appendix 1):

$$y = \frac{-(-16.0) \pm \sqrt{(-16.0)^2 - 4(3)(5.39)}}{2(3)}$$

$$= 0.36 \text{ or } 4.97$$

The root $y = 4.97$ has no meaning because y cannot be greater than 2.67, the total number of moles of ethyl alcohol initially present. Therefore, we ignore the 4.97 and find $y = 0.36$.

$$1.33 + y = 1.33 + 0.36 = 1.69 = \text{mol of } CH_3CO_2C_2H_5 = \text{mol of } H_2O$$
$$0.67 - y = 0.67 - 0.36 = 0.31 = \text{mol of } C_2H_5OH$$
$$2.67 - y = 2.67 - 0.36 = 2.31 = \text{mol of } CH_3CO_2H$$

See also Exercise 22.

Example 15.6 helps to emphasize a principle stated earlier in Section 15.3.1: for a given equilibrium system at a specific temperature, the value of K_c remains the same, no matter how the individual concentrations are changed. The data from Examples 15.4 and 15.6 may be summarized as follows:

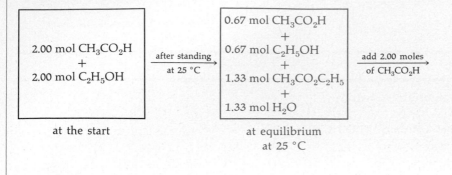

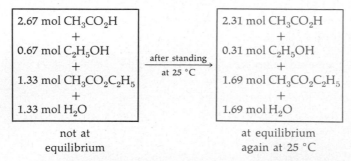

Immediately after the additional 2.00 moles of CH_3CO_2H are added to the initial equilibrium mixture, the rates of the forward reaction between CH_3CO_2H and C_2H_5OH and the reverse reaction between $CH_3CO_2C_2H_5$ and H_2O are not equal.

In time, the two rates become equal again and equilibrium is reestablished. Both the first equilibrium amounts and the new equilibrium amounts, when placed in the equilibrium constant expression, should give the same value of K_c:

$$K_c = \frac{(1.33)(1.33)}{(0.67)(0.67)} = \frac{(1.69)(1.69)}{(2.31)(0.31)}$$

$$3.9 = 4.0$$

This is agreement to two significant figures.

15•4   **Influence of Pressure Changes**

Pressure changes influence relative proportions at equilibrium only insofar as pressure changes bring about changes in concentration. Increasing the pressure on a liquid or solid does not appreciably increase the concentrations by crowding the molecules closer together and, therefore, has little effect on the equilibrium concentrations in reactions taking place in liquids or solids. However, increasing the pressure on a gaseous equilibrium system does increase the concentrations of the reacting species, and may or may not bring about a change in the relative amounts present in the equilibrium mixture.

15•4•1 Case I, Relative Amounts of Reactants and Products Are Not Changed In a gaseous system at equilibrium, if the stoichiometric number of reactant molecules on the left-hand side of the equilibrium equation equals the number of product molecules on the right-hand side, as in

$$H_2(g) + I_2(g) \rightleftharpoons 2HI(g)$$

$$\text{1 molecule} \quad \text{1 molecule} \quad \text{2 molecules}$$

the relative amounts are not changed by changes in pressure. For this system, the concentrations and relative amounts of reactants and products at different pressures are shown at the left in Figure 15.2. The relative amounts of $H_2 + I_2$ and HI are indicated by the mole fraction lines. The mole fractions remain constant at about 0.21 for H_2 and I_2 and at 0.79 for HI. However, the concentrations of H_2, I_2, and HI do increase as the pressure increases, as shown by the concentration lines. This increase in concentration is due to the fact that there are more molecules crowded into a unit of volume at the higher pressures.

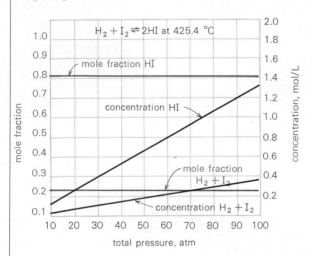

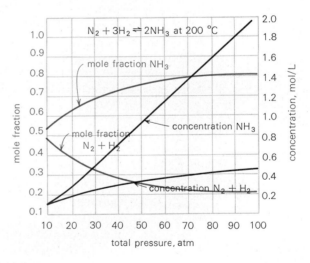

Let us take the data at 10.0 atm and 30.0 atm to illustrate these statements. At 10.0 atm, the equilibrium concentrations of H_2 and I_2 are each $0.0185M$ and that of HI is $0.137M$. Substituting these concentrations into the K_c expression gives

$$K_c = \frac{(0.137)^2}{(0.0185)(0.0185)} = 54.8$$

Now consider an increase of the pressure from 10.0 atm to 30.0 atm, a factor of 3. This increase can be accomplished by reducing the volume to one third of the volume at 10.0 atm. The concentrations of each substance

Figure 15•2 Changes in concentrations and in mole fractions of reactants versus products for two actual equilibrium systems as calculated for different pressures. In each system, substances are present in stoichiometric amounts.

will be three times that at 10.0 atm, that is, $[HI] = 3 \times 0.137M$, and $[H_2] = [I_2] = 3 \times 0.0185M$. Substituting these concentrations into the K_c expression gives the same value of K_c because all the 3's cancel:

$$K_c = \frac{(3 \times 0.137)(3 \times 0.137)}{(3 \times 0.0185)(3 \times 0.0185)} = 54.8$$

In this case, the concentrations of all substances are increased by a factor of 3 due to the pressure increase, *but the relative amounts of these substances are unchanged.*

15•4•2   Case II, Relative Amounts of Reactants and Products Are Changed
If the stoichiometric number of reactant molecules in the equilibrium equation does not equal the number of product molecules, as in

$$N_2(g) \quad + \quad 3H_2(g) \quad \rightleftharpoons \quad 2NH_3(g)$$
$$\text{1 molecule} \qquad \text{3 molecules} \qquad \text{2 molecules}$$

the relative amounts must change when changes in pressure occur. For this system, the concentrations and relative amounts of reactants and products at different pressures are shown at the right in Figure 15.2. In this case, we note that the relative amounts of NH_3 and $N_2 + H_2$ change with pressure, as shown graphically by the rise of the mole fraction curve for NH_3 and the fall of the curve for $N_2 + H_2$. However, a provocative fact about the concentrations of $N_2 + H_2$ is apparent from this graph. Even though the total amount of $N_2 + H_2$ in the equilibrium mixture is less at the higher pressure, their concentrations are actually greater.

We can illustrate these general observations by considering the experimental data at 10.0 atm and 30.0 atm. At 10.0 atm, the equilibrium concentrations of N_2 and H_2 are $0.0317M$ and $0.0953M$, respectively. The concentration of NH_3 is $0.131M$. Substituting these concentrations into the K_c expression gives

$$K_c = \frac{(0.131)^2}{(0.0317)(0.0953)^3} = 625$$

If the pressure is suddenly increased to 30.0 atm by decreasing the volume to one third of that at 10.0 atm, the concentration of each substance instantaneously becomes three times that at 10.0 atm. When these concentrations are substituted into the K_c expression, a value of K_c is obtained that is only one-ninth the value calculated for 10.0 atm:

$$K_c = \frac{(3 \times 0.131)(3 \times 0.131)}{(3 \times 0.0317)(3 \times 0.0953)(3 \times 0.0953)(3 \times 0.0953)} = 69.4$$

As emphasized in Section 15.3.3, the value of K_c remains the same at a specific temperature regardless of changes in concentrations. When compared with the value of 625, the small K_c value of 69.4 indicates that the system is not at equilibrium immediately after the pressure has been increased to 30 atm. For K_c again to be about 625, the two reactions must proceed so that the concentration of NH_3 increases and the concentrations of N_2 and H_2 decrease. In time, the amount of NH_3 increases sufficiently and the amounts of N_2 and H_2 decrease sufficiently to make the rates of the opposing reactions equal again. A new equilibrium is established. The concentrations of N_2, H_2, and NH_3 in the new equilibrium system have been determined experimentally to be $0.0626M$, $0.188M$, and $0.522M$, respectively. These data are summarized as follows:

$$\boxed{\begin{aligned} [N_2] &= 0.0317M \\ [H_2] &= 0.0953M \\ [NH_3] &= 0.131M \end{aligned}} \xrightarrow[\text{to 30.0 atm}]{\text{increase } P} \boxed{\begin{aligned} [N_2] &= 0.0951M \\ [H_2] &= 0.286M \\ [NH_3] &= 0.393M \end{aligned}} \xrightarrow[\substack{\text{but maintaining} \\ P = 30.0 \text{ atm}}]{\substack{\text{after} \\ \text{standing,}}}$$

at equilibrium
at 10.0 atm

not at
equilibrium

$$\boxed{\begin{aligned} [N_2] &= 0.0626M \\ [H_2] &= 0.188M \\ [NH_3] &= 0.522M \end{aligned}}$$

at equilibrium
at 30.0 atm

Substituting the equilibrium concentrations at 30.0 atm into the K_c expression gives

$$K_c = \frac{(0.522)^2}{(0.0626)(0.188)^3} = 655$$

This is close to the value of 625 calculated for 10.0 atm.[1]

For the two opposing reactions in an equilibrium system, the reaction that produces the smaller number of molecules is favored by an increase in pressure. In the $N_2 + 3H_2 \rightleftharpoons 2NH_3$ system, the favored reaction at high pressure is

$$\underset{\text{4 molecules}}{N_2 + 3H_2} \longrightarrow \underset{\text{2 molecules}}{2NH_3}$$

• *Pressure Changes and Le Chatelier's Principle* • The qualitative influence of pressure changes on the amounts in a gaseous equilibrium mixture can be predicted from a consideration of Le Chatelier's principle (see Section 10.4.1). According to this principle, if the pressure is changed on a system at equilibrium, the reaction that tends to diminish the change in pressure is favored. The reaction that produces the fewer number of molecules tends to lower the pressure, so this is the reaction favored at higher pressures. In the equilibrium system,

---

[1] Note that the value of the equilibrium constant, K_c, is slightly larger at the higher pressure. This is because the gases behave less like ideal gases as the molecules are crowded together at higher presssures (see Section 7.10).

$$N_2 \quad + \quad 3H_2 \quad \rightleftharpoons \quad 2NH_3$$

1 molecule   3 molecules   2 molecules

the forward reaction brings about a decrease in the total number of molecules in the container. Hence, it is favored at higher pressures. The experimental data tabulated in Table 15.2 show very nicely the effect of pressure changes on the concentrations for this equilibrium system. Also, the effect is shown schematically in Figure 15.3.

Table 15·2 Percentage yield by volume of ammonia

Temperature, °C	NH₃ yield, %						
	10.0 atm	30.0 atm	50.0 atm	100 atm	300 atm	600 atm	1,000 atm
200	50.7	67.6	74.4	81.5	90.0	95.4	98.3
300	14.7	30.3	39.4	52.0	71.0	84.2	92.6
400	3.9	10.2	15.3	25.1	47.0	65.2	79.8
500	1.2	3.5	5.6	10.6	26.4	42.2	57.5
600	0.5	1.4	2.3	4.5	13.8	23.1	31.4
700	0.2	0.7	1.1	2.2	7.3	12.6	12.9

Source: *A. T. Larson*, J. Amer. Chem. Soc., **46**, 371 (1924).

In the equilibrium

$$H_2 \quad + \quad Cl_2 \quad \rightleftharpoons \quad 2HCl$$

1 molecule   1 molecule   2 molecules

a change in pressure has no influence on the relative amounts in the equilibrium mixture. This is because neither the forward nor the reverse reaction tends to diminish the change in pressure. That is, the reactions produce no change in the total number of molecules present.

● **15·4·3 The Equilibrium Constant K_p** The equilibrium constant K_c is evaluated in terms of concentrations expressed in moles per liter. For an equilibrium system involving gases, measurements are commonly made of pressures rather than of concentrations. In such cases, the equilibrium constant may be calculated from the partial pressures of the gases. The constant evaluated in this way is called K_p. For the equilibrium system

$$2H_2O(g) \rightleftharpoons 2H_2(g) + O_2(g)$$

K_p is expressed as

$$K_p = \frac{(p_{H_2})^2(p_{O_2})}{(p_{H_2O})^2}$$

The total pressure equals the sum of the partial pressures,

$$P = p_{H_2O} + p_{H_2} + p_{O_2}$$

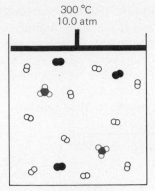

300 °C
10.0 atm

NH₃, 15%
H₂ and N₂, 85%

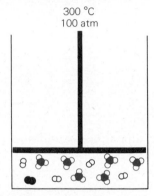

300 °C
100 atm

NH₃, 52%
H₂ and N₂, 48%

Figure 15·3 An illustration of Le Chatelier's principle, showing how an increase in pressure of ten times (top to bottom) can change the relative amounts of reactants. An increase in pressure favors the reaction that produces the fewer molecules. (See also Figure 15.2.)

From the ideal gas law equation (see Section 7.6), we see that the partial pressure of a gas is directly proportional to its concentration, c, in moles per liter:

$$pV = nRT \qquad c = \frac{n}{V} = \frac{p}{RT}$$

Hence, K_p is numerically related to K_c. For the general equilibrium equation,

$$wA + xB \rightleftharpoons yC + zD$$

the relationship between K_c and K_p is expressed by

$$K_c = K_p\left(\frac{1}{RT}\right)^{\Delta n}$$

where $\Delta n = (y + z) - (w + x)$, the number of gaseous product molecules minus the number of gaseous reactant molecules in the equilibrium equation. If the number of gaseous reactant molecules equals the number of gaseous product molecules, $\Delta n = 0$ and $K_p = K_c$.

• Example 15•7 •

Using the data in Table 15.2, calculate K_p and K_c for the equilibrium $N_2(g) + 3H_2(g) \rightleftharpoons 2NH_3(g)$ at 10.0 atm and 200 °C.

• Solution •

$$\text{total pressure} = 10.0 \text{ atm} = p_{NH_3} + (p_{N_2} + p_{H_2})$$
$$p_{NH_3} = 50.7\% \text{ of total pressure}$$
$$= 0.507 \times 10.0 \text{ atm} = 5.07 \text{ atm}$$
$$p_{H_2} + p_{N_2} = 10.0 \text{ atm} - 5.07 \text{ atm} = 4.93 \text{ atm}$$

Because the numbers of molecules of N_2 and H_2 are in a 1:3 ratio,

$$p_{N_2} = \tfrac{1}{4} \times 4.93 \text{ atm} = 1.23 \text{ atm}$$
$$p_{H_2} = \tfrac{3}{4} \times 4.93 \text{ atm} = 3.70 \text{ atm}$$
$$K_p = \frac{(p_{NH_3})^2}{(p_{N_2})(p_{H_2})^3} = \frac{(5.07)^2}{(1.23)(3.70)^3} = 4.13 \times 10^{-1}$$
$$K_c = K_p\left(\frac{1}{RT}\right)^{\Delta n}$$
$$= (4.13 \times 10^{-1})\left(\frac{1}{0.0821 \times 473}\right)^{-2}$$
$$= (4.13 \times 10^{-1})(0.0821 \times 473)^2 = 6.23 \times 10^2$$

See also Exercises 26, 29, 30, and 37.

• *Heterogeneous Equilibrium Systems* • Up to this point, all the equilibria we have considered are **homogeneous equilibria,** that is, equilibria with a single homogeneous phase. Equilibria involving two or more phases are known as **heterogeneous equilibria.** For example, the following equilibrium system involves both gaseous and solid phases:

$$C(s) + CO_2(g) \rightleftharpoons 2CO(g)$$

The equilibrium constant, K_c', for this system is

$$K_c' = \frac{[CO]^2}{[C][CO_2]}$$

In heterogeneous equilibria, if pure solids or pure immiscible liquids are reactants in a system with one or more gases, the equilibrium constant can be written in terms of only the partial pressures of the gases. Because the concentration of a pure solid or pure liquid is practically constant even though the pressure changes, we can write

$$[C] = k$$

so

$$K_c' = \frac{[CO]^2}{k[CO_2]}$$

$$(K_c')(k) = K_c = \frac{[CO]^2}{[CO_2]}$$

The equilibrium constant, K_p, for this system is

$$K_p = \frac{(p_{CO})^2}{(p_{CO_2})}$$

Expressions for equilibrium constants for systems that include pure liquid or pure solid substances do not include any terms for the concentrations or pressures of these substances. The constant effect of such substances is accounted for in the values of K_c or K_p.

• Example 15·8 •

At 1,000 °C and a total pressure of 20.0 atm, CO_2 makes up 12.5 molar percent of the gas in the system, $C(s) + CO_2(g) \rightleftharpoons 2CO(g)$.
(a) Calculate K_p.
(b) What percentage of the gas would be CO_2 if the total pressure were 40.0 atm?

• Solution •

(a) total pressure = 20.0 atm = $p_{CO_2} + p_{CO}$

$$p_{CO_2} = 0.125 \times 20.0 \text{ atm} = 2.50 \text{ atm}$$
$$p_{CO} = 20.0 \text{ atm} - 2.50 \text{ atm} = 17.5 \text{ atm}$$

$$K_p = \frac{(p_{CO})^2}{(p_{CO_2})} = \frac{(17.5)^2}{2.50} = 1.22 \times 10^2$$

(b) Let $x = p_{CO}$, then $(40.0 - x) = p_{CO_2}$

$$K_p = 122 = \frac{x^2}{40.0 - x}$$

$$x^2 + 122x - 4{,}880 = 0$$

Solving this equation with the quadratic formula, we obtain

$$x = 31.7 \text{ atm} = p_{CO}$$
$$40.0 - x = 8.3 \text{ atm} = p_{CO_2}$$
$$\text{percent } CO_2 = \frac{8.3 \text{ atm}}{40.0 \text{ atm}} \times 100 = 20.8 \text{ or } 21\%$$

See also Exercises 39–41.

15•4•4 **Relationship of $\Delta G°$ to K_p** The change in the standard free energy of a chemical reaction, $\Delta G°$ (see Section 13.9.2), is related to the equilibrium constant K_p by the equation

$$\Delta G° = -RT \ln K_p$$

or

$$\Delta G° = -2.303RT \log K_p \qquad (7)$$

The temperature is the absolute temperature. When R, the gas constant, is expressed as $8.314 \text{ J} \cdot \text{mol}^{-1} \cdot \text{K}^{-1}$, $\Delta G°$ is in units of joules.

The relationship between the magnitude of K_p and the sign of $\Delta G°$ is apparent from Equation (7). If K_p is greater than 1, $\log K_p$ is positive, $\Delta G°$ is negative, and the reaction from left to right is favored. If K_p is less than 1, $\log K_p$ is negative, $\Delta G°$ is positive, and the reaction from right to left is favored.

When an equilibrium system has appreciable quantities of all possible reactants, it is clear that neither the reaction toward the right nor the reaction to the left has an overwhelming tendency to occur. In such a case, $\Delta G°$ will be small, say, in the range of $+10$ to -10 kJ/mole. Example 15.9 illustrates an equilibrium system of this type.

• Example 15•9 •

F. H. Verhoek and F. Daniels, in *J. Amer. Chem. Soc.*, **53**, 1250 (1931), reported on the equilibrium system

$$N_2O_4(g) \rightleftharpoons 2NO_2(g)$$

At 25 °C and at a total pressure of 0.2118 atm, the partial pressure of NO_2 is 0.1168 atm. Calculate $\Delta G°$ for the conversion of $N_2O_4(g)$ to $NO_2(g)$.

• Solution •

$$p_{N_2O_4} = P_{\text{total}} - p_{NO_2}$$
$$= 0.2118 \text{ atm} - 0.1168 \text{ atm} = 0.0950 \text{ atm}$$

$$K_p = \frac{(p_{NO_2})^2}{p_{N_2O_4}} = \frac{(0.1168)^2}{0.0950} = 0.144$$

$$\Delta G° = -2.303RT \log K_p$$

$$= -2.303(8.314 \text{ J} \cdot \text{mol}^{-1} \cdot \text{K}^{-1})(298 \text{ K})(-0.842)$$

$$= 4.80 \times 10^3 \text{ J} \cdot \text{mol}^{-1} = 4.80 \text{ kJ} \cdot \text{mol}^{-1}$$

See also Exercise 44.

Perhaps the main use made of the relationship between $\Delta G°$ and K_p is the reverse of the type of calculation illustrated in Example 15.9. As shown in Section 13.9.2, using tabulated values of standard enthalpies, $\Delta H°$, and absolute entropies, $S°$, we can calculate $\Delta G°$ for a system from the equation

$$\Delta G° = \Delta H° - T \Delta S°$$

From $\Delta G°$, we can then calculate the equilibrium constant, K_p. (See Example 15.10.)

 • Example 15•10 •

At 25 °C, for the equilibrium system

$$N_2(g) + 3H_2(g) \rightleftharpoons 2NH_3(g)$$

the value of $\Delta G°$ for the reaction to the right is -32.98 kJ. Calculate K_p.

 • Solution •

$$\Delta G° = -2.303RT \log K_p$$

$$-32,980 = -2.303(8.314)(298) \log K_p = -5,706 \log K_p$$

$$\log K_p = \frac{32,980}{5,706} = 5.78 = 5.00 + 0.78$$

$$K_p = \text{antilog } (5.00 + 0.78)$$

$$= 6.0 \times 10^5$$

See also Exercise 42.

The large value of 6.0×10^5 for K_p for the equilibrium system $N_2(g) + 3H_2(g) \rightleftharpoons 2NH_3(g)$ at 25 °C shows that the reaction toward the right has a relatively great tendency to occur at that temperature. As a practical matter, as we shall see in Chapter 24, the reaction is too slow at this temperature to produce ammonia efficiently. The reaction is so slow at 25 °C that the equilibrium constant at that temperature cannot be measured by analyzing an equilibrium mixture. It must be calculated, as shown in Example 15.10, from other data.

 • *Free Energy at Equilibrium* • As a system approaches equilibrium, its free energy, G, approaches a minimum. Once a system is at equilibrium, it cannot change to any other state with a lower free energy. In a system at equilibrium, the tendency for a change to occur in one direction is just balanced by the tendency for change in the opposite direction. An example is a substance at its boiling point:

$$\text{liquid} \rightleftharpoons \text{vapor}$$

In this equilibrium system, the free energy used in vaporizing the liquid is equal to the free energy given up in condensing the vapor; the change in free energy, ΔG, at equilibrium is zero.

Because for any process at constant temperature and pressure,

$$\Delta G = \Delta H - T\,\Delta S$$

it follows that for a process at equilibrium,

$$0 = \Delta H - T\,\Delta S$$

or

$$\Delta S = \frac{\Delta H}{T} \qquad (8)$$

Equation (8) tells us that the change in entropy for a transition between two phases is equal to the heat necessary to accomplish the transition divided by the absolute temperature at which the transition takes place. For example, with this equation one can relate the change in entropy accompanying melting or boiling to the heat of fusion or vaporization, respectively. For many liquids, the value of ΔS found by dividing the heat of vaporization by the absolute temperature of the boiling point is near the value of $88\ \text{J} \cdot \text{mol}^{-1} \cdot \text{K}^{-1}$. This indicates that for many liquids there is about the same increase in disorder (increase in entropy) when the closely packed molecules of the liquid are vaporized into the randomly positioned, widely separated molecules of a gas.

15•5 Influence of Temperature

Changes in concentration or in pressure can change the relative amounts of reactants and products, but they do not change the value of the equilibrium constant. Changing the temperature, however, produces quite a different result. In a reversible reaction, the rate of one of the opposing reactions is generally increased more than that of the other by an increase in temperature. This difference in reaction tendencies is related to the individual nature of the product and reactant molecules. Because an increase in temperature affects the relative stabilities of reactants and products, an increase in temperature alters the value of the equilibrium constant itself.

How can we predict which reaction of an opposing pair is favored as the result of a change in temperature? One way is to apply Le Chatelier's principle. *If a system is at equilibrium, an increase in temperature causes the equilibrium to be displaced in the direction that absorbs heat.*[2] For the equilibrium system,

$$N_2(g) + 3H_2(g) \rightleftharpoons 2NH_3(g)$$

---

[2] This application of Le Chatelier's principle to enthalpies of reactions is known as **van't Hoff's law**.

the union of N_2 and H_2 to form NH_3 is exothermic, $\Delta H_r° = -92.22$ kJ. The decomposition of NH_3 into N_2 and H_2 is endothermic, $\Delta H_r° = +92.22$ kJ. Hence, the equilibrium system is displaced toward the left by an increase in temperature. Notice in Table 15.2 that the amount of NH_3 in the equilibrium mixtures is least at the highest temperature. Because the tendency for ammonia molecules to decompose is so much greater at the higher temperatures than is the tendency for nitrogen and hydrogen molecules to unite, very little ammonia is present in those equilibrium mixtures.

Now consider the following equilibrium system, which is important in air pollution (see Section 9.3.4):

$$N_2(g) + O_2(g) \rightleftharpoons 2NO(g)$$

For this system, the combination of N_2 with O_2 to produce NO is endothermic; $\Delta H_r° = +180.50$ kJ at 25 °C. Therefore, this equilibrium system is displaced more toward the right by an increase in temperature. However, the temperature must be increased to about 2,000 K before the volume percentage of NO at equilibrium is as high as one percent.

Although the K_p for the formation of NO does continue to increase with temperature, at very high temperatures (above 3,000 K), the following dissociation reactions become important:

$$O_2(g) \longrightarrow 2O(g)$$
$$N_2(g) \longrightarrow 2N(g)$$
$$NO(g) \longrightarrow N(g) + O(g)$$

The maximum percentage of NO obtained by heating air (78 percent N_2 and 21 percent O_2 by volume) is reached at about 3,500 K.

Actually all compounds tend to decompose if the temperature is high enough. The compounds may decompose into elemental molecules, which dissociate into atoms, or the molecules themselves may break down into atoms. In the case of the decomposition of water,

$$2H_2O(g) \rightleftharpoons 2H_2(g) + O_2(g) \qquad \Delta H_r° = +483.64 \text{ kJ at 25 °C}$$

the decomposition reaction is favored by an increase in temperature. At very high temperatures, in excess of 5,000 °C, the following dissociation reactions become important:

$$H_2(g) \longrightarrow 2H(g)$$
$$O_2(g) \longrightarrow 2O(g)$$
$$H_2O(g) \longrightarrow H(g) + OH(g)$$
$$OH(g) \longrightarrow H(g) + O(g)$$

15·6   Influence of a Catalyst

In an equilibrium system, a catalyst increases the speed of both forward and reverse reactions to the same extent. *A catalyst does not change the relative amounts present at equilibrium; the value of the equilibrium constant is not changed.* The catalyst does change the time required for reaching the equilibrium. Reactions that require days or weeks to come to equilibrium may reach it in a matter of minutes in the presence of a catalyst.

Furthermore, reactions that proceed at a suitable rate only at very high temperatures may proceed rapidly at much lower temperatures when a catalyst is used. This is especially important if high temperatures decrease the yield of the desired products. The synthesis of ammonia is a case in point. In the absence of a catalyst, the reaction between hydrogen and nitrogen is so slow, even at temperatures above 100 °C, that the reaction might take years to reach equilibrium.

Chapter Review

Summary

A **reversible** chemical change forms products that can act to produce the original reactants. A state of **chemical equilibrium** exists in a reversible system when the forward and reverse reactions proceed at the same rate. If the tendency for one of the two opposing reactions to occur is overwhelmingly dominant at a given temperature, the overall reaction is said to go to completion in that direction. The relative amounts of reactants and products at equilibrium depend on the nature of the substances involved and on the state of the system, but not on the direction from which equilibrium is approached.

The **equilibrium constant,** K_c, for a given reaction is usually written in terms of the concentrations of the species involved; at a given temperature it remains the same, regardless of how the individual concentrations are changed. The stronger the tendency for a reaction to proceed in a given direction, the larger the equilibrium constant for the reaction as written in that direction. The equilibrium constant for a reaction in which the stoichiometric numbers of moles of reactants and products are equal has no units, and the volume of the reaction vessel need not be known in order to calculate its value.

Pressure changes affect the *relative* concentrations of reactants and products at equilibrium only if the stoichiometric numbers of moles of gaseous reactants and products are *not* equal. In accord with Le Chatelier's principle, an increase in pressure shifts the equilibrium in the direction of the reaction that produces the smaller volume of gas. The equilibrium constant, however, does not change. For reactions involving gaseous reactants and products, it is customary to use an equilibrium constant, K_p, written in terms of the partial pressures of the gases.

Equilibria involving only one homogeneous phase (gaseous or liquid) are **homogeneous equilibria.** Those involving two or more phases are **heterogeneous equilibria.** Because the concentrations of any pure liquids or pure solids are essentially constant throughout a reaction, they are not included as separate terms in the equilibrium-constant expression but are "built into" the constant.

The equilibrium constant for a reaction is directly related to the standard free energy change for that reaction; either one can be calculated from the other, for a given temperature. At equilibrium the change in free energy of the system is zero. In accord with Le Chatelier's principle, an increase in temperature shifts the equilibrium in the direction of the reaction that absorbs heat (the endothermic reaction); this is known as **van't Hoff's law.** Changing the temperature also changes the value of the equilibrium constant itself, because the rates of the forward and reverse reactions tend to be affected differently. A catalyst changes both of these rates equally; hence it changes the overall rate at which equilibrium is reached, but it does not change the equilibrium constant.

Exercises

Writing and Calculating Equilibrium Constants, K_c

1. For the equilibrium A + B $\rightleftharpoons$ C, what determines the amounts of A, B, and C in the equilibrium mixture? Which of the two reactions would be referred to as the forward reaction?

2. Write the expression for the equilibrium constant, K_c, for each of the following gaseous equilibrium systems:

$$2ClO \rightleftharpoons Cl_2 + O_2$$
$$2NO + Cl_2 \rightleftharpoons 2NOCl$$
$$CH_4 + 4F_2 \rightleftharpoons CF_4 + 4HF$$
$$2NO_2 \rightleftharpoons 2NO + O_2$$
$$2O_3 \rightleftharpoons 3O_2$$
$$3O_2 \rightleftharpoons 2O_3$$

3. Show mathematically how the equilibrium constants of the last two equilibria of Exercise 2 are related.

*4. What is the molar concentration of the following?
(a) gaseous H_2 that is at a pressure of 10 atm at 0 °C
(b) 10 g of gaseous H_2 in a 1-L flask at 0 °C
(c) 10 g of gaseous NH_3 in a 1-L flask at 0 °C

5. The equilibrium mixture for the system

$$SO_2(g) + NO_2(g) \rightleftharpoons SO_3(g) + NO(g)$$

at a certain temperature was found to contain 0.60 mole of sulfur dioxide, 0.30 mole of nitrogen dioxide, 1.1 moles of sulfur trioxide, and 0.80 mole of nitric oxide, all in a 1.0-L container.
(a) Write the expression for the equilibrium constant, K_c.
(b) Calculate the value of K_c.
(c) Would K_c be larger, smaller, or unchanged if the same molar quantities had been in a 2.0-L vessel instead of the 1.0-L vessel? Why?

6. Consider the equilibrium

$$2H_2O(g) \rightleftharpoons 2H_2(g) + O_2(g)$$

A sample of 0.040 mole of water is put into a 1-L container at 2,500 °C. If 10 percent of the water decomposes, what is the concentration of oxygen in moles per liter?

7. Consider the homogeneous gaseous equilibrium system, $A + B \rightleftharpoons C + D$. If A and B are initially $1.00M$ and $2.00M$, respectively, and B is $1.50M$ at equilibrium, what is the value of K_c?

*8. Initially 0.20 mole of A, 1.0 mole of B, and 2.0 moles of C are mixed in a 1.0-L container, after which the homogeneous gaseous equilibrium $2A \rightleftharpoons 3B + C$ is established. Analysis shows 0.60 mole of A to be present at equilibrium. Calculate K_c.

9. Consider the equilibrium system

$$N_2(g) + O_2(g) \rightleftharpoons 2NO(g)$$

At 3,100 K, the value of K_c is only 1.20×10^{-2}. If the rates of the two opposing reactions must be equal at equilibrium, why is the number of moles of reactants not equal to the number of moles of products?

10. For the equilibrium system $2SO_2(g) + O_2(g) \rightleftharpoons 2SO_3(g)$, K_c at 600 °C is 6.8×10^2. What is the value of K_c at 600 °C for the decomposition of $SO_3(g)$ into $SO_2(g)$ and $O_2(g)$?

11. Nitrogen dioxide, a reddish-brown gas, tends to associate as the temperature is lowered, forming colorless nitrogen tetroxide:

$$2NO_2(g) \rightleftharpoons N_2O_4(g)$$

Into a sealed 100-mL flask is placed 0.235 g of NO_2, and equilibrium is established at 55 °C. At that temperature the flask is found to contain 0.115 g of NO_2. Calculate K_c.

12. Consider the gaseous equilibrium $2SO_2 + O_2 \rightleftharpoons 2SO_3$. Suppose at a temperature of 827 °C one introduced 2.00 moles of SO_2 and 1.00 mole of O_2 into a 500-mL container. At equilibrium 1.60 moles of SO_3 were obtained. Assuming no SO_3 was initially present, calculate K_c for this system.

13. Although we seldom express values of equilibrium constants with units, we may do so if we wish. What are the units of K_c for the following systems?
(a) $2SO_3(g) \rightleftharpoons 2SO_2(g) + O_2(g)$
(b) $N_2(g) + 3H_2(g) \rightleftharpoons 2NH_3(g)$
(c) $2NH_3(g) \rightleftharpoons N_2(g) + 3H_2(g)$
(d) $H_2(g) + Cl_2(g) \rightleftharpoons 2HCl(g)$
(e) $2NO_2(g) \rightleftharpoons N_2O_4(g)$

14. Write the chemical equations for the equilibria described by the following equilibrium constant expressions.

(a) $K_c = \dfrac{[N_2O_4]}{[NO_2]^2}$

(b) $K_c = \dfrac{[H_2O][NH_3]^2}{[N_2O][H_2]^4}$

(c) $K_c = \dfrac{[H_2]^4[CS_2]}{[H_2S]^2[CH_4]}$

Calculating Equilibrium Concentrations from K_c Values

15. Calculate the concentration of A at equilibrium for the homogeneous gaseous system $A + B \rightleftharpoons C + D$, if $K_c = 0.36$ and A and B are initially each $2.00M$.

16. Consider the following equilibrium system at 1,000 °C:

$$H_2(g) + I_2(g) \rightleftharpoons 2HI(g) \qquad K_c = 16.0$$

If 2.00 moles of H_2 and 2.00 moles of I_2 are placed into a 500-mL container and heated at 1,000 °C until equilibrium is attained, how many moles of each substance will be present?

17. Consider the equilibria described by the equations in Exercises 2 and 13. When making calculations involving the equilibrium constant expression, for which equilibria can the concentrations be expressed as either moles or moles per liter? For which must the concentrations be expressed as moles per liter? May the concentrations be expressed as grams per liter for any of the calculations?

18. For which of the following cases does the forward reaction of an equilibrium system go more nearly to completion: $K_c = 1$; $K_c = 100$; $K_c = 0.01$?

19. For the hydrolysis of ethyl acetate,

$$CH_3CO_2C_2H_5(l) + H_2O(l) \rightleftharpoons$$
$$CH_3CO_2H(l) + C_2H_5OH(l)$$

K_c is 0.25 at 25 °C.

(a) If 4.0 moles of ethyl acetate are mixed with 4.0 moles of water, how many moles of acetic acid, CH_3CO_2H, will be formed at equilibrium?

(b) Suppose the hydrolysis is repeated, but this time 40 moles of water are used instead of 4.0 moles. How many moles of acetic acid will be formed at equilibrium?

20. The value of K_c at 25 °C is 4.00 for the following equilibrium:

$$CH_3CO_2H(l) + C_2H_5OH(l) \rightleftharpoons CH_3CO_2C_2H_5(l) + H_2O(l)$$

(a) How is this value of K_c related to the value of K_c for the hydrolysis of ethyl acetate to produce acetic acid and ethyl alcohol?

(b) If one mixes 90.0 g of water with 440 g of ethyl acetate, how many grams of ethyl acetate should be left at equilibrium?

21. At 400 °C, K_c is 60.0 for the equilibrium

$$H_2(g) + I_2(g) \rightleftharpoons 2HI(g)$$

If the concentration of HI is 0.030M at equilibrium and the concentration of H_2 is 0.020M, what is the concentration of I_2 in moles per liter?

22. (a) The equilibrium constant, K_c, for the reaction

$$H_2(g) + CO_2(g) \rightleftharpoons H_2O(g) + CO(g)$$

at 750 °C is 0.771. If 0.50 mole of hydrogen and 1.0 mole of carbon dioxide are placed in a 2.0-L container that is held at a temperature of 750 °C, what is the concentration in moles per liter of each substance in the equilibrium mixture?

(b) If an additional mole of CO_2 is introduced into the equilibrium system described in part (a) and the system is then allowed to again reach equilibrium at 750 °C, what will be the resulting concentrations of reactants and products?

*23. At 298 °C, K_c is 1.8×10^{-56} for the system

$$3O_2(g) \rightleftharpoons 2O_3(g)$$

If only pure O_2 is present initially, and its concentration is 2.62M, what is the concentration of O_3 at equilibrium? What is the concentration of O_2 at equilibrium?

*24. For the system $2SO_2(g) + O_2(g) \rightleftharpoons 2SO_3(g)$, K_c is 3.00 at 1,000 °C. If 8.00 g of pure SO_3 is originally present in a 2.00-L container, what weights of SO_2, O_2, and SO_3 will be present at equilibrium?

Influence of
Pressure and K_p Calculations

25. Write the chemical equations for the equilibria described by the following equilibrium constant expressions.

(a) $K_p = \dfrac{(p_{NO})^4(p_{H_2O})^6}{(p_{NH_3})^4(p_{O_2})^5}$

(b) $K_p = \dfrac{p_{SO_3}}{(p_{SO_2})(p_{O_2})^{1/2}}$

(c) $K_p = \dfrac{p_{O_2}}{(p_{CO})^2}$

26. At equilibrium at 1,000 °C, hydrogen iodide is 33 percent dissociated into hydrogen and iodine. Calculate the K_p for the equilibrium:

$$2HI(g) \rightleftharpoons H_2(g) + I_2(g)$$

Is K_p numerically equal to K_c for this system? Why?

*27. For the decomposition of hydrogen iodide,

$$2HI(g) \rightleftharpoons H_2(g) + I_2(g)$$

K_c is 54.4 at 425.4 °C. If, prior to reaction, a 1:1 mixture (by volume) of H_2 and I_2 occupies a volume of 1.00 L at 425.4 °C and 1 atm, what are the concentrations of all substances at equilibrium? What are the relative mole fractions at equilibrium?

28. Refer to Exercise 8.
(a) Calculate the initial pressure in the container and the pressure at equilibrium, each at 0 °C.
(b) Calculate the mole percentage of A initially and at equilibrium.

29. Consider the equilibrium system

$$N_2(g) + 3H_2(g) \rightleftharpoons 2NH_3(g)$$

When a 1:3 molar mixture of $N_2:H_2$ is held at 100 atm and 300 °C until equilibrium is reached, the volume percentage of NH_3 is 52.0.
(a) What are the partial pressures of NH_3, N_2, and H_2 at equilibrium?
(b) Calculate K_p.
(c) Calculate K_c.

30. Consider again the equilibrium shown in Exercise 29. At 100 atm and 400 °C, the volume percentage of NH_3 at equilibrium is 25.1.
(a) Calculate K_p.
(b) How does this value of K_p compare with the value calculated in Exercise 29? How do you account for the different values of the "constant"?

*31. Consider the equilibrium system

$$2NH_3(g) \rightleftharpoons N_2(g) + 3H_2(g)$$

Pure ammonia, NH_3, is placed into a container which is then heated and brought to equilibrium at 300 atm and 400 °C. At equilibrium, p_{NH_3} is 47.0 percent of the total pressure.
(a) What is the value of K_p?
(b) The volume of the gases at equilibrium is 10.0 L. What volume would the original NH_3 have occupied at 300 atm and 400 °C?

*32. From Table 15.2, it is seen that the percentage yields of NH_3 at 50 atm and at 100 atm at 200 °C are 74.4 and 81.5, respectively. Assume that the molar ratio of $H_2 : N_2$ is 3:1 in answering the following.
  (a) Calculate the partial pressure and mole fraction of each component in each equilibrium mixture.
  (b) Calculate the concentration in moles per liter for each component.
  (c) Will the total mass of N_2, H_2, and NH_3 in 2.0 L of the equilibrium mixture at 50 atm equal the total mass in 1.0 L of the equilibrium mixture at 100 atm? Why?

33. A gas buret is filled to the 1.00-L mark with $N_2O_4(g)$ at 1.00 atm. What will be the value of K_p for the dissociation of $N_2O_4(g)$ to $NO_2(g)$ if the point where half of the N_2O_4 has dissociated is the point of equilibrium? (The pressure and the temperature are held constant.)

*34. At 677 °C, K_p is 6.0 for the system

$$COCl_2(g) \rightleftharpoons CO(g) + Cl_2(g)$$

A certain weight of $COCl_2$ was placed into a sealed container at 677 °C and 1.00 atm. The flask was heated for a short time at this temperature until the pressure increased to 1.25 atm.
  (a) What are the partial pressures of CO, Cl_2, and $COCl_2$?
  (b) Is the system at equilibrium? Explain.
  (c) If the system is not at equilibrium, how can you tell when equilibrium is attained?
  (d) What will be the partial pressure of $COCl_2$ at equilibrium?

*35. If, prior to reaction, a 3:1 mixture (by volume) of hydrogen and nitrogen occupies a volume of 1 L at 30 atm and 200 °C, what will be the volume occupied by the equilibrium mixture

$$3H_2 + N_2 \rightleftharpoons 2NH_3$$

at the same pressure and temperature if the ammonia exerts 67.6 percent of the pressure at equilibrium?

36. Repeat Exercise 13 but give units of K_p when the pressures are expressed in atmospheres.

37. Consider the equilibrium system

$$2NO_2(g) \rightleftharpoons 2NO(g) + O_2(g)$$

A certain weight of NO_2 is placed into a flask at 1.00 atm. When equilibrium is attained, the pressure has increased to 1.40 atm. Calculate K_p for the system.

38. Consider the gaseous equilibrium system $2SO_2 + O_2 \rightleftharpoons 2SO_3$. Starting with pure SO_3 in a vessel, it is found that heating to 727 °C, at a pressure of 300 atm, produces an equilibrium mixture in which SO_3 exerts 76.9 percent of the total pressure. What is the value of K_p for this system?

Heterogeneous Equilibria

39. Consider the following heterogeneous equilibrium:

$$C(s) + S_2(g) \rightleftharpoons CS_2(g)$$

At 1,000 °C and a total pressure of 10.0 atm, $S_2(g)$ makes up 14.5 mole percent of the gas in the system.
  (a) Calculate K_p.
  (b) What percentage would be $S_2(g)$ if the total pressure were 30.0 atm?

40. For the heterogeneous equilibrium system

$$2CO(g) \rightleftharpoons C(s) + CO_2(g)$$

K_p is 8.20×10^{-3} at 1,000 °C. The total pressure in the equilibrium system is calculated to be 16.0 atm. What is the mole percentage of CO in the system?

41. When an excess amount of ammonium carbamate, $NH_2CO_2NH_4$, is placed into an evacuated vessel, the following equilibrium is established:

$$NH_2CO_2NH_4(s) \rightleftharpoons 2NH_3(g) + CO_2(g)$$

For this equilibrium system at a certain temperature, the total pressure of the system is 150 atm.
  (a) What are the partial pressures of NH_3 and CO_2?
  (b) Calculate K_p.

Relationship Between $\Delta G°$ and K_p

*42. The standard free energies of formation, $\Delta G_f°$, of $NO_2(g)$ and $N_2O_4(g)$ at 25 °C are 51.30 and 97.82 kJ/mole, respectively. Calculate K_p for the following equilibrium system at 25 °C and 1 atm:

$$N_2O_4(g) \rightleftharpoons 2NO_2(g)$$

43. At 25 °C and at a total pressure of 0.298 atm, the partial pressure of N_2O_4 is 0.150 atm. Calculate K_p and $\Delta G°$ for the conversion of 2 moles of $NO_2(g)$ to $N_2O_4(g)$.

Influence of Temperature on Equilibria

44. Consider the following gaseous equilibrium system at 25 °C:

$$2SO_2(g) + O_2(g) \rightleftharpoons 2SO_3(g) \qquad \Delta H_r° = -197.7 \text{ kJ}$$

  (a) Write the K_c expression for this equilibrium system.
  (b) Assume that, in a certain experiment, the equilibrium concentrations of SO_2, O_2, and SO_3 were obtained at 700 °C. Suppose one then increased the pressure on the system from 5.0 atm to 50 atm and allowed equilibrium to be reestablished. What effect would this pressure increase have on each of the following (answer increase, decrease, or no change): the amount of O_2 present at equilibrium; the con-

centration in moles per liter of O_2 at equilibrium; the value of K_c?

(c) Suppose one increased the temperature to 900 °C and then allowed equilibrium to be reestablished. What effect would this temperature increase have on the amount of SO_3 present at equilibrium? The value of K_c?

(d) Suppose that one added 2.0 additional moles of SO_2 to the equilibrium mixture at 25 °C and then allowed equilibrium to be attained again. What effect would this additional SO_2 have on the amount of SO_3 present at equilibrium? The amount of O_2 present at equilibrium? The amount of SO_2 present at equilibrium? The value of K_c?

45. Consider the following equilibrium:

$$2NO_2(g) \rightleftharpoons 2NO(g) + O_2(g)$$

At 25 °C, K_p for this system is 4.3×10^{-13}; at 275 °C, K_p is 1.2×10^{-3}. Is the decomposition of NO_2 into NO and O_2 an exothermic or an endothermic reaction? Explain.

46. For the equilibrium system $2H_2O(g) \rightleftharpoons 2H_2(g)$ $+ O_2(g)$, the reaction to the right is endothermic. Will the value of K_p at 1,000 °C be greater than, less than, or the same as that at 750 °C? Explain.

47. For the gaseous system $Cl_2 + H_2 \rightleftharpoons 2HCl$, the reaction to the right is exothermic. Will the value for $\Delta G°$ at 1,000 °C be greater than, less than, or the same as that at 750 °C? Explain.

48. How is van't Hoff's law related to the effect of temperature on the solubility of solids in liquids? (See Section 10.4.1.)

49. (a) If the system

$$2NO_2(g) \rightleftharpoons N_2O_4(g) \qquad \Delta H_r° = -57 \text{ kJ}$$

is at equilibrium, what are two conditions that can be changed to shift the equilibrium in the direction of increasing the concentration of N_2O_4 and decreasing the concentration of NO_2?

(b) Do either or both of the changes you described in part (a) change the value of the equilibrium constant?

Ionic Equilibria I: Solutions of Acids and Bases

16

16•1 **Ionization Constants of Weak Acids**

16•1•1 Weak Monoprotic Acids

16•1•2 Weak Polyprotic Acids

16•2 **Ionization Constants of Weak Bases**

16•3 **Determination of and Calculations Involving K_a and K_b**

16•3•1 Determination of Ionization Constants

16•3•2 Calculations Involving Ionization Constants

16•4 **Ionization of Water**

16•5 **Hydrolysis Constants of Acidic Cations and Basic Anions**

16•5•1 Acidic Cations

16•5•2 Basic Anions

16•5•3 Acidic Cations with Basic Anions

16•6 **Expressing Hydrogen Ion Concentration**

16•7 **The Common Ion Effect**

16•8 **Buffered Solutions**

16•8•1 Buffered Systems in the Body

16•9 **Indicators**

16•9•1 Indicators and Titration

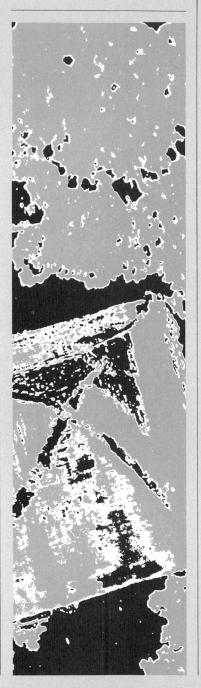

In the study of equilibria in Chapter 15, our attention was focused mainly on processes involving molecules. The concept of equilibrium is also very important in understanding reactions involving ions, particularly ions in solution. This first chapter on ionic equilibria will be devoted to quantitative determinations of the ion concentrations present at equilibrium for acids and bases.

As noted in Chapter 11, in water the ionization of some polar covalent acid molecules, such as HCl, HNO_3, and $HClO_4$, is essentially complete. Because there are so few un-ionized molecules present at equilibrium, equations for these ionization reactions generally are written with only a single arrow to the right. When we speak of a $0.50M$ solution of hydrochloric acid, we assume that the H^+ and Cl^- ion concentrations are both $0.50M$, and that the concentration of un-ionized HCl is practically zero.

On the other hand, for a weak acid such as acetic acid, $HC_2H_3O_2$, or a weak base such as ammonia, NH_3, proton transfer to or from water is far from complete. Equations for these ionization reactions are written with double arrows to emphasize that the equilibrium systems are reversible.

In this chapter we will deal with equilibrium constants for solutions of weak acids and weak bases. Much of our effort will be directed toward calculations of the H^+ ion concentrations of these solutions. No ion is more important to the chemistry of solutions than the H^+ ion. A convenient method of expressing the H^+ ion concentration on the pH scale will be presented in this chapter. Familiarity with this method of expression is useful to persons working in many fields, including medicine, chemistry, biology, nutrition, and agriculture.

16•1 Ionization Constants of Weak Acids

16•1•1 Weak Monoprotic Acids An equilibrium between ions and molecules can be treated mathematically in the same fashion as an equilibrium in which all species are molecules. Consider the ionization of any weak monoprotic acid, HA, in water solution:

$$HA + H_2O \rightleftharpoons H_3O^+ + A^- \tag{1}$$

The equilibrium constant, K_c, based on Equation (1) and following the conventions developed in Chapter 15, is

$$K_c = \frac{[H_3O^+][A^-]}{[HA][H_2O]} \tag{2}$$

For all dilute solutions, the molar concentration of water, $[H_2O]$, is practically the same, that is, about $55M$. With this knowledge, Equation (2) can be written as follows:

$$K_c = \frac{[H_3O^+][A^-]}{[HA](55)} \tag{3}$$

Because H_3O^+ and H^+ are merely different symbols for the proton in water solution, $[H_3O^+] = [H^+]$. After substituting $[H^+]$ for $[H_3O^+]$ in Equation (3) and rearranging the equation, we obtain

$$K_c \times 55 = \frac{[H^+][A^-]}{[HA]} = K_a \qquad (4)$$

The product of the two constants, $K_c \times 55$, is expressed as the constant K_a, called the **acid ionization constant.**

Ionization reactions are often written in a simplified form with the water omitted. For any weak monoprotic acid, HA, the simplified ionization equation is $HA \rightleftharpoons H^+ + A^-$. We may consider that the acid ionization constant expressed for HA is based on this simplified equation. However, we should remember that the constant, K_a, includes the molar concentration of water, $[H_2O]$, as a constant.

In the specific case of acetic acid, $HC_2H_3O_2$, the simplified ionization equation is

$$HC_2H_3O_2 \rightleftharpoons H^+ + C_2H_3O_2^-$$

and the expression for the acid ionization constant is

$$K_a = \frac{[H^+][C_2H_3O_2^-]}{[HC_2H_3O_2]} \qquad (5)$$

The strength of an acid is related to its degree of ionization, the magnitude of which is reflected by the value of its ionization constant (see Table 16.1). The smaller the value of the K_a, the weaker is the acid. The correlation of ranges of K_a values with descriptive adjectives such as "strong" and "very weak" is arbitrary. One scheme commonly used for describing ranges is shown in the lower part of Table 16.1. The same ranges are used to describe bases (see Section 16.2).

16•1•2   Weak Polyprotic Acids   The ionization of a polyprotic acid proceeds in a stepwise fashion. Each ionization step involves a different ionization constant expression. For the ionization of the diprotic acid, carbonic acid (see Figure 16.1), simplified equilibrium equations for the steps and their acid ionization constant expressions are:

- *Step 1*

$$H_2CO_3 \rightleftharpoons H^+ + HCO_3^- \qquad K_{a1} = \frac{[H^+][HCO_3^-]}{[H_2CO_3]}$$

- *Step 2*

$$HCO_3^- \rightleftharpoons H^+ + CO_3^{2-} \qquad K_{a2} = \frac{[H^+][CO_3^{2-}]}{[HCO_3^-]}$$

In a solution of a diprotic acid, the concentration of ions formed in the first step is much greater than that formed in the second step. From an

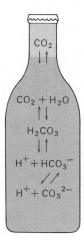

Figure 16•1 A carbonated beverage affords a good example of equilibria involving molecules and ions.

examination of Table 16.1, it can be seen that the numerical values of K_{a1} range from 10^4 to 10^7 times those of K_{a2}. Approximately the same ratio holds for each of the steps in the ionization of a triprotic acid such as phosphoric acid, H_3PO_4.

It is important to note that all steps in the ionization of polyprotic acids are taking place in the same solution. In a solution of H_2CO_3, there is only one H^+ concentration and only one HCO_3^- concentration. The same numerical values for these concentrations are used in calculations involving either K_{a1} or K_{a2}.

Table 16•1 Ionization constants[a] of acids in water solution at 25 °C

Name	Simplified ionization reaction	K_a
hydrochloric acid	$HCl \rightleftharpoons H^+ + Cl^-$	large
sulfuric acid	$H_2SO_4 \rightleftharpoons H^+ + HSO_4^-$	large
	$HSO_4^- \rightleftharpoons H^+ + SO_4^{2-}$	1.2×10^{-2}
sulfurous acid	$H_2SO_3 \rightleftharpoons H^+ + HSO_3^-$	1.3×10^{-2}
	$HSO_3^- \rightleftharpoons H^+ + SO_3^{2-}$	6.3×10^{-8}
chlorous acid	$HClO_2 \rightleftharpoons H^+ + ClO_2^-$	1.1×10^{-2}
phosphoric acid	$H_3PO_4 \rightleftharpoons H^+ + H_2PO_4^-$	7.5×10^{-3}
	$H_2PO_4^- \rightleftharpoons H^+ + HPO_4^{2-}$	6.2×10^{-8}
	$HPO_4^{2-} \rightleftharpoons H^+ + PO_4^{3-}$	4.4×10^{-13}
hydrofluoric acid	$HF \rightleftharpoons H^+ + F^-$	6.6×10^{-4}
nitrous acid	$HNO_2 \rightleftharpoons H^+ + NO_2^-$	5.1×10^{-4}
formic acid	$HCHO_2 \rightleftharpoons H^+ + CHO_2^-$	1.8×10^{-4}
acetic acid	$HC_2H_3O_2 \rightleftharpoons H^+ + C_2H_3O_2^-$	1.8×10^{-5}
carbonic acid	$H_2CO_3 \rightleftharpoons H^+ + HCO_3^-$	4.3×10^{-7}
	$HCO_3^- \rightleftharpoons H^+ + CO_3^{2-}$	5.6×10^{-11}
hydrosulfuric acid	$H_2S \rightleftharpoons H^+ + HS^-$	1.1×10^{-7}
	$HS^- \rightleftharpoons H^+ + S^{2-}$	1.0×10^{-14}
hypochlorous acid	$HClO \rightleftharpoons H^+ + ClO^-$	3.0×10^{-8}
hydrocyanic acid	$HCN \rightleftharpoons H^+ + CN^-$	6.2×10^{-10}

very strong	K_a greater than 1×10^3	
strong	K_a in range of 1×10^3 to 1×10^{-2}	
weak	K_a in range of 1×10^{-2} to 1×10^{-7}	
very weak	K_a less than 1×10^{-7}	

[a]Values of K_a in color are for the second or third steps in the ionization of polyprotic acids.

16•2 Ionization Constants of Weak Bases

Expressions for the equilibrium constants for dilute solutions of weak bases may be obtained in the same manner as was done for weak acids. Consider a dilute aqueous solution of any uncharged, weak Brønsted–Lowry base, designated by the symbol B. The equilibrium equation and the K_c expression are

$$B + H_2O \rightleftharpoons BH^+ + OH^- \tag{6}$$

$$K_c = \frac{[BH^+][OH^-]}{[B][H_2O]} \tag{7}$$

For dilute solutions, with an H_2O concentration of about $55M$,

$$K_c \times 55 = K_b = \frac{[BH^+][OH^-]}{[B]} \tag{8}$$

K_b is called the **base ionization constant.** For the specific case of an aqueous solution of ammonia, NH_3, in which B is NH_3 and BH^+ is NH_4^+,

$$K_b = \frac{[NH_4^+][OH^-]}{[NH_3]}$$

The ionization constants for several weak bases are listed in Table 16.2.

Table 16•2 Ionization constants[a] of bases in water solution at 25 °C

Name	Ionization reaction	K_b
dimethylamine	$(CH_3)_2NH + H_2O \rightleftharpoons (CH_3)_2NH_2^+ + OH^-$	5.9×10^{-4}
methylamine	$CH_3NH_2 + H_2O \rightleftharpoons CH_3NH_3^+ + OH^-$	4.2×10^{-4}
ethylenediamine	$H_2NCH_2CH_2NH_2 + H_2O \rightleftharpoons H_2NCH_2CH_2NH_3^+ + OH^-$	3.6×10^{-4}
	$H_2NCH_2CH_2NH_3^+ + H_2O \rightleftharpoons H_3NCH_2CH_2NH_3^{2+} + OH^-$	5.4×10^{-7}
trimethylamine	$(CH_3)_3N + H_2O \rightleftharpoons (CH_3)_3NH^+ + OH^-$	6.3×10^{-5}
ammonia	$NH_3 + H_2O \rightleftharpoons NH_4^+ + OH^-$	1.8×10^{-5}
hydrazine	$N_2H_4 + H_2O \rightleftharpoons N_2H_5^+ + OH^-$	9.8×10^{-7}
	$N_2H_5^+ + H_2O \rightleftharpoons N_2H_6^{2+} + OH^-$	1.3×10^{-15}
hydroxylamine	$HONH_2 + H_2O \rightleftharpoons HONH_3^+ + OH^-$	9.1×10^{-9}

[a]The two values of K_b in color are the second steps in forming polyprotic positive ions.

16•3   Determination of and Calculations Involving K_a and K_b

The numerical problems encountered in studying equilibria involving ions are no different from those presented in the study of the equilibria of un-ionized molecules in Chapter 15. If sufficient data are available from which the concentrations of the species on the right and left sides of the ionic equilibrium equation can be determined, then the value of the ionization constant can be calculated. Conversely, if the value of the ionization constant is known and the initial concentrations are given, one can calculate the concentrations of species present at equilibrium.

16•3•1 Determination of Ionization Constants In order to calculate the ionization constant of a weak electrolyte, we must in some way determine the number of ions present in solution and the number of molecules of the electrolyte that are not ionized. One method of determining the concentration of ions is by measuring changes in colligative properties (see acetic acid in Table 12.5 and in Example 12.7). Another way is by measuring the electric conductivity—the greater the degree of ionization of a dissolved electrolyte, the greater is the electric conductivity of its solutions.

   The amount of the electrolyte present as molecules is calculated by subtracting from the total amount of solute the amount that is determined to be present as ions. If our electric measurements indicate that 5.2 percent of the solute is present as ions, we assume that 94.8 percent of the solute molecules are not ionized.

   Measurement of the electric conductance of 0.100M acetic acid at 25 °C reveals that it is 1.34 percent ionized. In other words, when 0.100 mole of acetic acid is dissolved in enough water to make 1 L of solution, ionization proceeds rapidly to equilibrium. The result is that 1.34 percent of the acetic acid is in the form of ions and 98.66 percent is in the form of covalent molecules. The concentrations of ions and molecules remain constant indefinitely unless the temperature is changed or more water or acetic acid is added to the solution.

   Using these data, we can calculate K_a for the ionization of 0.100M acetic acid. The equilibrium concentrations in moles per liter are

$$HC_2H_3O_2 \qquad + H_2O \rightleftharpoons H_3O^+ \qquad + C_2H_3O_2^-$$

0.09866M	0.00134M	0.00134M
(98.66% of 0.100M)	(1.34% of 0.100M)	(1.34% of 0.100M)

Recalling that $[H_3O^+] = [H^+]$, we write

$$K_a = \frac{[H^+][C_2H_3O_2^-]}{[HC_2H_3O_2]}$$

$$= \frac{(0.00134)(0.00134)}{0.09866} = \frac{(1.34 \times 10^{-3})^2}{9.866 \times 10^{-2}} = 1.82 \times 10^{-5}$$

   Ionization constants for weak bases are obtained in a similar fashion. In the case of ammonia, NH_3, the ionization constant as calculated from conductance measurements is 1.8×10^{-5}. (The fact that it is numerically so close to that for acetic acid is just a coincidence.)

16•3•2 Calculations Involving Ionization Constants Once the ionization constant for a given equilibrium is known, it can be used in making calculations that involve different concentrations. Example 16.1 illustrates one such calculation. Several other examples will be given in later sections of this chapter.

─────── • Example 16•1 • ───────

Given the K_a for acetic acid as 1.8×10^{-5}, calculate the molar concentration of hydrogen ions, $[H^+]$, and the percent ionization of 0.50M acetic acid.

• Solution • Let $x =$ the moles of $HC_2H_3O_2$ ionized per liter; then $x =$ moles of H^+ and of $C_2H_3O_2^-$ ions produced per liter, and $(0.50 - x)$ = moles of un-ionized $HC_2H_3O_2$ per liter.

Summarizing,

	$HC_2H_3O_2 \rightleftharpoons H^+ + C_2H_3O_2^-$		
mol/L at the start	0.50	0	0
mol/L at equilibrium	$0.50 - x$	x	x

$$K_a = 1.8 \times 10^{-5} = \frac{[H^+][C_2H_3O_2^-]}{[HC_2H_3O_2]}$$

$$= \frac{(x)(x)}{0.50 - x}$$

For weak electrolytes, the degree of ionization is small. That is, the concentration of ionized solute, x, is very small compared with the concentration of un-ionized solute. In this example, $(0.50 - x)$ is probably very close to 0.50. For this reason, we can simplify the calculations and avoid solving a quadratic equation. Therefore,

$$1.8 \times 10^{-5} \simeq \frac{(x)(x)}{0.50}$$

$$x^2 \simeq 0.90 \times 10^{-5} = 9.0 \times 10^{-6}$$

$$x \simeq \sqrt{9.0} \times \sqrt{10^{-6}}$$

$$x \simeq 3.0 \times 10^{-3} M = [H^+]$$

which is also equal to the moles of acetic acid ionized per liter.

$$\frac{\text{percent}}{\text{ionization}} = \frac{\text{mol of solute ionized}}{\text{mol of solute at the start}} \times 100$$

$$= \frac{3.0 \times 10^{-3}}{0.50} \times 100$$

$$= 0.60\%$$

See also Exercise 9 at the end of the chapter.

• *Approximate Calculation Versus Precise Calculation* • When an expression such as $(0.50 - x)$ or $(0.50 + x)$ appears in a problem, it is often possible to simplify the calculation by neglecting x. This is possible when x is small compared with the figure from which it is subtracted or to which it is added. In the calculation in Example 16.1, x is 0.0030, which is quite small compared with 0.50. Actually in this case, for $(0.50 - x)$, x is not significant mathematically:

$$0.50 - 0.0030 = 0.497, \text{ which rounds to } 0.50$$

How can we decide whether we are justified in neglecting x, or whether we must use the quadratic formula? One way is first to carry out the simplified calculation and compare the value of x with the concentration of the weak acid. If x is greater than 5 percent of [HA], the calculation probably should be redone using the quadratic formula. Another way allows us to decide before doing the simplified calculation. If we divide K_a by [HA], and find that K_a/[HA] is less than 2.5×10^{-3}, we can use the simplified calculation.

• *Dilution of Solutions of Weak Electrolytes* • Comparing the result of Example 16.1 with data given in Section 16.3.1 for the 0.100M solution of acetic acid illustrates again a general concept of great importance that was discussed in connection with Table 12.5. *A dilute solution of a weak electrolyte is more completely ionized than a more concentrated solution.* The 0.50M acetic acid solution is only 0.60 percent ionized, whereas the much weaker 0.100M solution is 1.34 percent ionized. If one dilutes the 0.100M solution with water to twice its original volume, to 0.050M, the percent ionization increases to 1.9. The increase in the degree of ionization with dilution can be understood if we consider that the K_a for acetic acid of 1.8×10^{-5} is independent of the concentration of ions and un-ionized molecules.

If 100 mL of 0.50M acetic acid is diluted first with 400 mL of water to give 0.10M acetic acid, and then again with 500 mL of water to give 0.050M acetic acid, the amounts of the ions increase because of the reaction of acetic acid with water. The amount of un-ionized acetic acid decreases. Note that the *concentrations* of all species are *decreased* due to the diluting effect of the water. As shown by the following summary, the result for each solution is to establish by reaction the concentrations of ions and un-ionized acetic acid molecules that fit the expression

$$\frac{[H^+][C_2H_3O_2^-]}{[HC_2H_3O_2]} = 1.8 \times 10^{-5}$$

[HC$_2$H$_3$O$_2$]	[H$^+$] = [C$_2$H$_3$O$_2^-$]	moles of H$^+$ = moles of C$_2$H$_3$O$_2^-$	percent ionization	K_a
starting with 100 mL of 0.50M	0.0030M	0.00030 in 100 mL	0.60	$\frac{(0.0030)^2}{0.50} = 1.8 \times 10^{-5}$
add 400 mL water to make 0.10M	0.00134M	0.00067 in 500 mL	1.34	$\frac{(0.00134)^2}{0.10} = 1.8 \times 10^{-5}$
then 500 mL more to make 0.050M	0.00095M	0.00095 in 1,000 mL	1.9	$\frac{(0.00095)^2}{0.050} = 1.8 \times 10^{-5}$

In conclusion, note that the percent ionization does not depend on the volume of the solution. For example, 10 mL, 100 mL, and 1,000 mL of 0.050M acetic acid solutions are all 1.9 percent ionized.

• *Calculations Involving Polyprotic Acids* • Equilibrium problems involving polyprotic acids are solved by somewhat the same methods as those used for solving problems involving monoprotic acids. However, they are usually somewhat more complex due to the additional equilibria involved. The equilibria involved in the strong polyprotic acid H_2SO_4 are illustrated in Example 16.2. For weak polyprotic acids such as H_2SO_3, H_2CO_3, H_2S, and H_3PO_4, a few assumptions often can be made to simplify the calculations. These assumptions are presented in Example 16.3 for H_2S.

• Example 16.2 •

Calculate the concentrations of all ions present in 0.050M sulfuric acid, H_2SO_4.

• Solution • Simplified equations for the two steps involved in the ionization of H_2SO_4 are as follows:

• Step 1

$$H_2SO_4 \longrightarrow H^+ + HSO_4^-$$

• Step 2

$$HSO_4^- \rightleftharpoons H^+ + SO_4^{2-}$$

Step 1 proceeds essentially 100 percent to the right (indicated by the single arrow). Hence this step converts the 0.050M H_2SO_4 into 0.050M H^+ and 0.050M HSO_4^- ions:

	$H_2SO_4 \longrightarrow$	H^+	$+ HSO_4^-$
mol/L after first step of ionization	~0	0.050	0.050

The K_a for step 2 is 1.2×10^{-2} (see Table 16.1). Hence this step is far from complete. At equilibrium there will be SO_4^{2-}, H^+, and HSO_4^- ions.

Let x = moles per liter of HSO_4^- ions that are ionized; then x = moles per liter of H^+ and of SO_4^{2-} ions formed from the HSO_4^- ions. The molar concentration of HSO_4^- ions left at equilibrium equals $(0.050 - x)$.
Summarizing for step 2,

	$HSO_4^- \rightleftharpoons$	H^+	$+ SO_4^{2-}$
mol/L at the start	0.050	0	0
mol/L at equilibrium	$0.050 - x$	x	x

The ionization constant expression for HSO_4^- is

$$K_a = \frac{[H^+][SO_4^{2-}]}{[HSO_4^-]}$$

As emphasized in Section 16.1.2, in a solution of a polyprotic acid, there is only one H^+ concentration. Hence the value of $[H^+]$ used in the K_a expression equals the concentration furnished from step 1 plus that furnished from step 2; that is, $[H^+] = 0.050 + x$. Therefore,

$$K_a = 1.2 \times 10^{-2} = \frac{(0.050 + x)(x)}{0.050 - x}$$

The ratio $K_a/[\text{solute}] = (1.2 \times 10^{-2})/0.050 = 2.4 \times 10^{-1}$. This is greater than 2.5×10^{-3}, so we cannot simplify the calculation by ignoring the $+x$ and $-x$. We must use the quadratic formula to solve for x:

$$6.0 \times 10^{-4} - 1.2 \times 10^{-2}x = 5.0 \times 10^{-2}x + x^2$$

$$x^2 + 6.2 \times 10^{-2}x - 6.0 \times 10^{-4} = 0$$

$$x = \frac{-6.2 \times 10^{-2} \pm \sqrt{(6.2 \times 10^{-2})^2 - 4(1)(-6.0 \times 10^{-4})}}{2(1)}$$

$$x = 8.5 \times 10^{-3} \, M = [SO_4{}^{2-}]$$

$$0.050 + x = 5.8 \times 10^{-2} \, M = [H^+]$$

$$0.050 - x = 4.2 \times 10^{-2} \, M = [HSO_4{}^-]$$

See also Exercises 14–16.

• Example 16•3 •

A saturated aqueous solution of hydrogen sulfide (hydrosulfuric acid), H_2S, is about $0.10M$. Given that K_{a1} for H_2S is 1.1×10^{-7} and K_{a2} is 1.0×10^{-14}, calculate the concentrations of all ions present in a saturated H_2S solution.

• Solution • The simplified equilibrium equations and acid ionization constants involved are

$$H_2S \rightleftharpoons H^+ + HS^-$$

$$K_{a1} = 1.1 \times 10^{-7} = \frac{[H^+][HS^-]}{[H_2S]}$$

$$HS^- \rightleftharpoons H^+ + S^{2-}$$

$$K_{a2} = 1.0 \times 10^{-14} = \frac{[H^+][S^{2-}]}{[HS^-]}$$

The H^+ and HS^- concentrations are determined from the K_{a1} expression; these concentrations are not appreciably altered by the second ionization, because K_{a2} is very small compared with K_{a1}. Let $x = [H^+] = [HS^-]$ present at equilibrium; then $(0.10 - x) = [H_2S]$ at equilibrium.
Summarizing,

	H_2S	$\rightleftharpoons$	H^+	$+$	HS^-
mol/L at the start	0.10		0		0
mol/L at equilibrium	$0.10 - x$		x		x

We can use the simplified calculation because $K_a/[H_2S] = (1.1 \times 10^{-7})/0.10 = 1.1 \times 10^{-6}$, which is less than 2.5×10^{-3}.

$$K_{a1} = 1.1 \times 10^{-7} \simeq \frac{x^2}{0.10}$$

$$x^2 \simeq 1.1 \times 10^{-8}$$
$$x \simeq \sqrt{1.1} \times \sqrt{10^{-8}}$$
$$x \simeq 1.0 \times 10^{-4}\, M = [H^+] = [HS^-] \text{ from step 1}$$

The S^{2-} ions are formed from the second ionization step. Let y = moles of HS^- that are ionized per liter; then y = moles of H^+ and of S^{2-} ions per liter formed from the HS^- ions. The molar concentration of HS^- ions at equilibrium equals $1.0 \times 10^{-4} - y$.

Summarizing for this step,

	HS^-	$\rightleftharpoons$	H^+	$+ S^{2-}$
mol/L at the start	1.0×10^{-4}		1.0×10^{-4}	0
mol/L at equilibrium	$1.0 \times 10^{-4} - y$		$1.0 \times 10^{-4} + y$	y

Because K_{a2} is so small, we know y will be small. We can assume that the changes in $[H^+]$ and $[HS^-]$ due to ionization, $+y$ and $-y$, respectively, are not significant. Therefore, $[H^+] = [HS^-] \simeq 1.0 \times 10^{-4}\, M$.

$$K_{a2} = 1.0 \times 10^{-14} \simeq \frac{(1.0 \times 10^{-4})(y)}{1.0 \times 10^{-4}}$$

$$y \simeq 1.0 \times 10^{-14}\, M = [S^{2-}]$$

Hence, the $[S^{2-}]$ and K_{a2} have the same value of 1.0×10^{-14}. In fact, regardless of the molar concentration of H_2S, the concentration of S^{2-} is practically equal to the K_{a2} as long as no ions from other electrolytes are present.

See also Exercises 14–16.

16•4   Ionization of Water

Electric conductance measurements and other evidence show that water ionizes to a limited extent in accordance with the following equation:

$$2H_2O \rightleftharpoons H_3O^+ + OH^- \quad \text{or} \quad H_2O \rightleftharpoons H^+ + OH^-$$

Precise measurement of the electric conductance of very pure water at 25 °C reveals that its ionic composition is

$$[H_3O^+] = 1.0 \times 10^{-7}\, \text{mol/L}$$
$$[OH^-] = 1.0 \times 10^{-7}\, \text{mol/L}$$

The equilibrium constant expression for the ionization of water,

$$K_c = \frac{[H_3O^+][OH^-]}{[H_2O]^2}$$

can be modified, because the concentration of water molecules is considered to be constant at $55M$ for pure water and for dilute aqueous solutions. Therefore,

$$K_c = \frac{[H_3O^+][OH^-]}{(55)^2}$$

$$K_c \times (55)^2 = [H_3O^+][OH^-]$$

Following the same procedure of simplification used to obtain K_a for weak acids, the constant $K_c \times (55)^2$ is replaced with the ionization constant for water, K_w, to give

$$K_w = [H_3O^+][OH^-]$$

Because the ionization of water is often represented as $HOH \rightleftharpoons H^+ + OH^-$ and $[H_3O^+] = [H^+]$, the expression may be written simply as

$$K_w = [H^+][OH^-]$$

By substituting the hydrogen ion and hydroxide ion concentrations in this expression, we can calculate the value of the constant:

$$K_w = (1.0 \times 10^{-7})(1.0 \times 10^{-7}) = 1.0 \times 10^{-14}$$

This constant, K_w, is called the **ion product for water.** It indicates that in pure water or any water solution both hydrogen and hydroxide ions must be present and, moreover, that the product of their concentrations must be constant. It may seem odd, but acidic solutions contain a small concentration of hydroxide ions (base), and basic solutions contain a small concentration of hydrogen ions (acid). Further, if the concentration of one of these ions is known, the other can be easily calculated, because the product of the two must equal 1.0×10^{-14} (K_w at 25 °C).[1]

For example, in $1.0 \times 10^{-4} M$ HCl, the solution is $1.0 \times 10^{-4} M$ in H^+ ions, and the OH^- ion concentration is calculated as follows:

$$K_w = [H^+][OH^-]$$

$$1.0 \times 10^{-14} = (1.0 \times 10^{-4})[OH^-]$$

$$[OH^-] = \frac{1.0 \times 10^{-14}}{1.0 \times 10^{-4}} = 1.0 \times 10^{-10} M$$

Conversely, in $1.0 \times 10^{-3} M$ NaOH, the solution is $1.0 \times 10^{-3} M$ in OH^- ions, and the H^+ ion concentration can be calculated in this way:

$$1.0 \times 10^{-14} = [H^+](1.0 \times 10^{-3})$$

$$[H^+] = \frac{1.0 \times 10^{-14}}{1.0 \times 10^{-3}} = 1.0 \times 10^{-11} M$$

---

[1] The value of K_w increases slightly as the temperature rises. For ordinary purposes, the value is taken as 1.0×10^{-14} when the temperature is at or near room temperature.

━━━━ • Example 16•4 • ━━━━

Given that the K_b for ammonia, NH_3, is 1.8×10^{-5}, calculate the molar concentration of hydrogen ions in $0.50M$ ammonia.

━━━━ • Solution • Ammonia is a weak base. The equilibrium equation and base ionization constant expression involved are

$$NH_3 + H_2O \rightleftharpoons NH_4^+ + OH^-$$

$$K_b = \frac{[NH_4^+][OH^-]}{[NH_3]}$$

In order to calculate $[H^+]$, we must first calculate $[OH^-]$. Following the procedure of Example 16.1, we obtain

$$1.8 \times 10^{-5} = \frac{[NH_4^+][OH^-]}{[NH_3]} \simeq \frac{(x)(x)}{0.50}$$

$$x^2 \simeq 9.0 \times 10^{-6}$$

$$x \simeq 3.0 \times 10^{-3}\,M = [OH^-]$$

$$K_w = [H^+][OH^-] = 1.0 \times 10^{-14}$$

$$[H^+] \simeq \frac{1.0 \times 10^{-14}}{3.0 \times 10^{-3}} = 3.3 \times 10^{-12}\,M$$

See also Exercises 21 and 23.

In water solutions, common acidic properties are attributed to hydrogen ions (hydronium ions) and common basic properties to hydroxide ions. Because the product of the molar concentrations of these ions is constant at 1.0×10^{-14}, it is necessary to state the concentration of only one in order to describe the acidity or basicity of the solution:

If $[H^+]$ is greater than $1.0 \times 10^{-7}\,M$, the solution is acidic.

If $[H^+]$ is less than $1.0 \times 10^{-7}\,M$, the solution is basic.

If $[H^+]$ is equal to $1.0 \times 10^{-7}\,M$, the solution is neutral.

16•5 Hydrolysis Constants of Acidic Cations and Basic Anions

In Section 11.2.3, we learned that certain cations act as acids and certain anions act as bases in water solution. These ions hydrolyze to give weakly acidic, weakly basic, or neutral solutions. Some examples are given below:

1. A water solution of ammonium chloride, NH_4Cl, gives a weakly acidic solution, because the NH_4^+ ion acts as an acid but the Cl^- ion does not act appreciably as a base.

2. A water solution of sodium acetate, $NaC_2H_3O_2$, gives a weakly basic solution, because the $C_2H_3O_2^-$ ion acts as a base, but the Na^+ ion (hydrated) does not act appreciably as an acid.

3. A water solution of ammonium acetate, $NH_4C_2H_3O_2$, gives a neutral solution, because the acid strength of the NH_4^+ ion is just balanced by the basic strength of the $C_2H_3O_2^-$ ion.

16•5•1 Acidic Cations

The acidic cations and basic anions can be treated mathematically as polar covalent weak acids and weak bases. For example, consider the general case of a water solution of the conjugate acid, BH^+, of the weak base, B. The equilibrium equation for the reaction of this cation with water is

$$BH^+ + H_2O \rightleftharpoons H_3O^+ + B \tag{9}$$

The equilibrium constant, K_c, based on Equation (9) is expressed as

$$K_c = \frac{[H_3O^+][B]}{[BH^+][H_2O]} \tag{10}$$

By considering $[H_2O]$ to be constant and substituting $[H^+]$ for $[H_3O^+]$, as was done in Section 16.1.1 for the ionization of weak acids, we obtain

$$K_a = \frac{[H^+][B]}{[BH^+]} \tag{11}$$

By multiplying the numerator and denominator in Equation (11) by $[OH^-]$ and separating, the following relationship is found:

$$K_a = \frac{[H^+][B]}{[BH^+]} \times \frac{[OH^-]}{[OH^-]} = [H^+][OH^-] \times \frac{[B]}{[BH^+][OH^-]}$$

$$= K_w \times \frac{1}{K_b \text{ for B}}$$

That is, the value of the acid ionization or acid hydrolysis constant, K_a, for BH^+ is equal to the ionization constant for water, K_w, divided by the K_b for the conjugate base, B:

$$K_a = \frac{K_w}{K_b} \tag{12}$$

For the specific case of the ammonium ion, NH_4^+, we can write the following equilibrium equation:

$$NH_4^+ + H_2O \rightleftharpoons H_3O^+ + NH_3$$

Based on this equation and on the value in Table 16.2 for the K_b of NH_3, we can calculate the value of K_a for NH_4^+ in water solution:

$$K_a = \frac{[H^+][NH_3]}{[NH_4^+]} = \frac{K_w}{K_b} = \frac{1.0 \times 10^{-14}}{1.8 \times 10^{-5}} = 5.6 \times 10^{-10}$$

The values of K_a listed in Table 16.3 for the conjugate acids of weak bases have been calculated by using Equation (12) and the appropriate K_b values in Table 16.2. When we solve for the H^+ ion concentration in a water

Table 16•3 Acid hydrolysis[a] constants of cations in water solution at 25 °C

Cation	K_a
$N_2H_6^{2+}$	7.7
$N_2H_5^+$	1.0×10^{-8}
$HONH_2^+$	1.1×10^{-6}
NH_4^+	5.6×10^{-10}
$(CH_3)_3NH^+$	1.6×10^{-10}
$CH_3NH_3^+$	2.4×10^{-11}
$(CH_3)_2NH_2^+$	1.7×10^{-11}

[a] The general equation for the acid hydrolysis of a cation, CH^{n+}, is

$$CH^{n+} + H_2O \rightleftharpoons C^{(n-1)+} + H_3O^+$$

solution of a salt such as NH_4Cl, NH_4NO_3, or $(NH_4)_2SO_4$, in which the anion does not act appreciably as a base, the calculations are carried out in the same manner as is done for a water solution of a weak acid such as $HC_2H_3O_2$. Example 16.5 illustrates a calculation of $[H^+]$ for an ammonium chloride solution. This example should be compared with Example 16.1.

● Example 16·5 ●

Given that the K_a of the ammonium ion, NH_4^+, is 5.6×10^{-10}, calculate the H^+ ion concentration in $1.0M$ NH_4Cl.

● **Solution** ● First we write the simplified equilibrium equation for the hydrolysis reaction:

	$NH_4^+ \rightleftharpoons H^+ + NH_3$
mol/L at the start	1.0 0 0
mol/L at equilibrium	$1.0 - x$ x x

$$K_a = 5.6 \times 10^{-10} = \frac{[H^+][NH_3]}{[NH_4^+]}$$

$$= \frac{(x)(x)}{1.0 - x}$$

The ratio $K_a/[\text{solute}]$ is $(5.6 \times 10^{-10})/1.0 = 5.6 \times 10^{-10}$. This is less than 2.5×10^{-3}, so we will use the simplified calculation:

$$5.6 \times 10^{-10} \simeq \frac{x^2}{1.0}$$

$$x^2 \simeq 5.6 \times 10^{-10}$$

$$x \simeq \sqrt{5.6} \times \sqrt{10^{-10}}$$

$$x \simeq 2.4 \times 10^{-5} M = [H^+]$$

See also Exercise 26.

As mentioned in Section 11.2.3, certain hydrated metal cations can also donate protons to water molecules and therefore act as weak acids. One example is

$$Al(H_2O)_6^{3+} + H_2O \rightleftharpoons H_3O^+ + Al[(H_2O)_5OH]^{2+}$$

Some K_a values for such hydrated metal cations are listed in Table 16.4.

16·5·2 Basic Anions Consider the general case of the basic anion, A^-, of the weak acid, HA. The equilibrium equation for its reaction with water is

$$A^- + H_2O \rightleftharpoons HA + OH^- \tag{13}$$

and its base ionization or base hydrolysis constant, K_b, is expressed as

$$K_b = \frac{[HA][OH^-]}{[A^-]} \tag{14}$$

Table 16·4 Acid hydrolysis constants of hydrated metal cations in water solution at 25 °C

Cation	K_a
$Fe(H_2O)_6^{3+}$	6.5×10^{-3}
$Al(H_2O)_6^{3+}$	7.2×10^{-6}
$Ag(H_2O)_2^+$	1×10^{-7}
$Cu(H_2O)_4^{2+}$	3.0×10^{-8}
$Ni(H_2O)_6^{2+}$	4×10^{-10}
$Co(H_2O)_6^{2+}$	3×10^{-10}
$Fe(H_2O)_6^{2+}$	8×10^{-11}
$Mg(H_2O)_6^{2+}$	3.8×10^{-12}

Source: *Adapted with permission from J. E. Huheey,* Inorganic Chemistry, *2nd ed., Harper & Row, New York, 1978.*

Rearranging Equation (12), we can write Equation (15), which shows the relationship between the values of K_a and K_b for any conjugate acid–base pair:

$$K_a \times K_b = K_w \quad (15)$$

Solving this equation for K_b gives

$$K_b = \frac{K_w}{K_a}$$

Hence the value of K_b for A^- is equal to the ionization constant for water, K_w, divided by the K_a for the conjugate acid, HA.

For the specific case of the hydrolysis of the acetate ion, $C_2H_3O_2^-$, we write the following:

$$C_2H_3O_2^- + H_2O \rightleftharpoons HC_2H_3O_2 + OH^-$$

$$K_b = \frac{[HC_2H_3O_2][OH^-]}{[C_2H_3O_2^-]} = \frac{K_w}{K_a} = \frac{1.0 \times 10^{-14}}{1.8 \times 10^{-5}} = 5.6 \times 10^{-10}$$

The values of K_b listed in Table 16.5 for the conjugate bases of weak acids have been calculated in this fashion by using the K_a values in Table 16.1.

When we solve problems requiring the determination of the OH^- ion concentration in a water solution of a sodium or potassium salt of a weak acid, for example, $NaC_2H_3O_2$, $KC_2H_3O_3$, NaCN, or KCN, calculations are carried out in the same manner as is done for a water solution of a weak base, such as NH_3. Example 16.6 illustrates the calculation of OH^- for a sodium cyanide solution. Compare this calculation with that done for ammonia in Example 16.4.

Table 16•5 Base hydrolysis[a] constants of anions in water solution at 25 °C[b]

Anion	K_b
S^{2-}	1.0
HS^-	9.1×10^{-8}
PO_4^{3-}	2.3×10^{-2}
HPO_4^{2-}	1.6×10^{-7}
$H_2PO_4^-$	1.3×10^{-12}
CO_3^{2-}	1.8×10^{-4}
HCO_3^-	2.3×10^{-8}
CN^-	1.6×10^{-5}
$C_2H_3O_2^-$	5.6×10^{-10}
CHO_2^-	5.6×10^{-11}
NO_2^-	2.0×10^{-11}
F^-	1.5×10^{-11}
SO_4^{2-}	8.3×10^{-13}
HSO_4^-	very small
Cl^-	very small

[a] The general equation for the base hydrolysis of an anion, A^{n-}, is

$$A^{n-} + H_2O \rightleftharpoons HA^{(n-1)-} + OH^-$$

[b] The K_b values in color are for second or third steps in forming polyprotic hydrolysis products.

• **Example 16.6** •

Given that the K_b of the cyanide ion, CN^-, is 1.6×10^{-5}, calculate the OH^- ion concentration in 0.50M NaCN.

• **Solution** • First we write the equilibrium equation for the hydrolysis reaction:

$$H_2O + \quad CN^- \rightleftharpoons HCN + OH^-$$

mol/L at the start	0.50	0	0
mol/L at equilibrium	0.50 − x	x	x

$$K_b = 1.6 \times 10^{-5} = \frac{[HCN][OH^-]}{[CN^-]}$$

$$= \frac{(x)(x)}{0.50 - x}$$

The ratio $K_b/$[solute] is $(1.6 \times 10^{-5})/0.50 = 3.2 \times 10^{-5}$. This is less than 2.5×10^{-3}, so we will use the simplified calculation instead of using the quadratic formula:

$$1.6 \times 10^{-5} \simeq \frac{x^2}{0.50}$$

$$x^2 \simeq 8.0 \times 10^{-6}$$

$$x \simeq \sqrt{8.0} \times \sqrt{10^{-6}}$$

$$\simeq 2.8 \times 10^{-3} M \simeq [OH^-]$$

See also Exercise 29.

16.5.3   Acidic Cations with Basic Anions In some salts, the cation acts as an acid and the anion acts as a base. The resulting solution will be neutral, acidic, or basic, depending upon the relative strengths of the acid cation versus the basic anion. Consider the specific case of ammonium acetate, $NH_4C_2H_3O_2$, which exists in water solution largely as hydrated NH_4^+ and $C_2H_3O_2^-$ ions. Because the NH_4^+ ion ($K_a = 5.6 \times 10^{-10}$) is a stronger acid than water and the $C_2H_3O_2^-$ ion ($K_b = 5.6 \times 10^{-10}$) is a stronger base than water, these ions tend to react with each other much more than either reacts with water. The principal acid–base reaction taking place is between the acidic ammonium ions and the basic acetate ions:

$$NH_4^+ + C_2H_3O_2^- \rightleftharpoons NH_3 + HC_2H_3O_2$$

Because both the K_a for NH_4^+ and the K_b for $C_2H_3O_2^-$ have the value of 5.6×10^{-10}, their respective relative strengths as an acid and a base cancel one another. Therefore, a solution of ammonium acetate is essentially a neutral solution. A solution of ammonium carbonate, however, is basic because the value of the K_b for the carbonate ion, CO_3^{2-}, is greater than the value of the K_a for the ammonium ion; that is,

$$K_b \text{ for } CO_3^{2-} = 1.8 \times 10^{-4} > K_a \text{ for } NH_4^+ = 5.6 \times 10^{-10}$$

For salt solutions in which the cation and anion react as an acid and a base, respectively, the H^+ concentration may be calculated using the following formula:

$$[H^+] = \sqrt{\frac{K_w K_a}{K_b}} \tag{16}$$

From this relationship, the H^+ concentrations for ammonium acetate and ammonium carbonate solutions are $1.0 \times 10^{-7} M$ and $1.8 \times 10^{-10} M$, respectively. The much smaller value of $[H^+]$ in the ammonium carbonate solution shows that it is by far the more basic solution.

Equation (16) may be used to calculate the H^+ concentration in a water solution of a salt containing an amphiprotic cation or anion. For example, in a solution of sodium bicarbonate, $NaHCO_3$, the following equilibria are involved:

$$HCO_3^- + H_2O \rightleftharpoons H_3O^+ + CO_3^{2-} \qquad K_a = 5.6 \times 10^{-11}$$
$$\text{acid}$$

$$HCO_3^- + H_2O \rightleftharpoons OH^- + H_2CO_3 \qquad K_b = 2.3 \times 10^{-8}$$
$$\text{base}$$

Because the K_b of HCO_3^- is greater than the K_a, the solution is weakly basic. The H^+ concentration is

$$[H^+] = \sqrt{\frac{(1.0 \times 10^{-14})(5.6 \times 10^{-11})}{2.3 \times 10^{-8}}}$$

$$= \sqrt{2.4 \times 10^{-17}} = \sqrt{24 \times 10^{-18}} = 4.9 \times 10^{-9}\,M$$

16•6 Expressing Hydrogen Ion Concentration

It is rather cumbersome to express the hydrogen ion or hydroxide ion concentration in terms of some number times ten raised to a negative exponent. A widely used and more convenient method of stating the hydrogen ion concentration of dilute acids, bases, and neutral solutions is in terms of pH. The **pH** of a solution is defined as

$$pH = \log\frac{1}{[H^+]} \quad \text{or} \quad pH = -\log[H^+]$$

As an example of a simple pH calculation, consider pure water or a solution of sodium chloride, both of which are neutral. The H^+ ion concentration in each is $1.0 \times 10^{-7}\,M$. The pH is

$$pH = -\log[H^+]$$
$$= -\log(1.0 \times 10^{-7}) = -\log 1.0 - \log 10^{-7}$$
$$= -\log 1.0 - (-7) = 7 - \log 1.0 = 7 - 0.00 = 7.00$$

The extent of the acidity or basicity of a solution is completely and conveniently expressed by its pH value:

If the pH is 7.0, the solution is neutral.
If the pH is below 7.0, the solution is acidic.
If the pH is above 7.0, the solution is basic.

The smaller the pH value, the more acidic is the solution. A pH of 4.4 indicates a more acidic solution than a pH of 4.5. Conversely, the larger the pH value, the more basic is the solution. A pH of 10.7 indicates a more basic solution than a pH of 10.6. Example 16.7 illustrates two calculations of pH. Figure 16.2 shows the pH values for solutions of some common substances.

• Example 16•7 •

Calculate the pH of each of the following:
(a) 0.0063M hydrochloric acid, HCl
(b) 0.050M sodium hydroxide, NaOH

• Solution •

(a) Because HCl is a strong acid, we may assume that the 0.0063M HCl is 0.0063 or $6.3 \times 10^{-3}\,M$ in H^+ ions. The pH of the solution is

$$pH = -\log[H^+]$$
$$= -\log(6.3 \times 10^{-3}) = -\log 6.3 - \log 10^{-3}$$
$$= -\log 6.3 - (-3) = \underline{3 - \log 6.3} = 3 - 0.80 = 2.20$$

(b) Because NaOH is a strong base, we may assume that the 0.050M NaOH is 0.050 or 5.0×10^{-2} M in OH^- ions. Before calculating the pH, we first obtain the H^+ concentration:

$$K_w = [H^+][OH^-]$$
$$1.0 \times 10^{-14} = [H^+](5.0 \times 10^{-2})$$

$$[H^+] = \frac{1.0 \times 10^{-14}}{5.0 \times 10^{-2}} = 2.0 \times 10^{-13} M$$

We then calculate the pH in the usual manner:

$$pH = -\log[H^+]$$
$$= -\log(2.0 \times 10^{-13}) = -\log 2.0 - \log 10^{-13}$$
$$= -\log 2.0 - (-13) = \underline{13 - \log 2.0} = 13 - 0.30 = 12.70$$

See also Exercise 31.

Note the underlined portion of each part of the solution in Example 16.7. If we consider the general expression of $[H^+]$ as $a \times 10^{-x}$, the pH is equal to $x - \log a$. For example in part (a), $[H^+] = 6.3 \times 10^{-3}$, and pH = $3 - \overline{\log 6.3} = 2.20$.

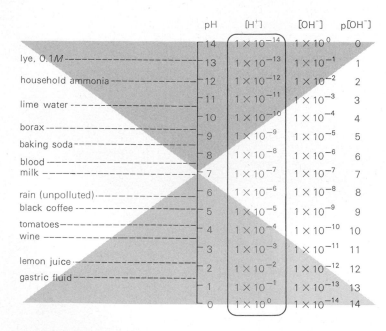

Figure 16•2 The pH values of some common substances.
Source: Courtesy of Beckman Instruments, Inc.

For some purposes it is more convenient to focus attention on the OH^- concentration and to speak of the **pOH**, the negative logarithm of $[OH^-]$. A useful relationship is obtained by taking the negative logarithms of both sides of the ion product expression for water:

$$[H^+][OH^-] = K_w$$
$$-\log[H^+] - \log[OH^-] = -\log K_w$$
$$pH + pOH = pK_w$$

The sum of the pH and pOH for the same solution is equal to 14, the value of pK_w, the negative logarithm of K_w. Thus, for the OH$^-$ concentration of $5.0 \times 10^{-2}\,M$ referred to in Example 16.7(b), the pOH of the solution is $2 - \log 5.0 = 2 - 0.70 = 1.30$; the pH is $14 - 1.30 = 12.70$.

In calculations of pH values for solutions of weak acids and weak bases, it is important to realize that one must first calculate the H$^+$ or OH$^-$ concentration. Problems to be solved are generally no different from those illustrated in Examples 16.1 through 16.6, except for the conversion of the H$^+$ or OH$^-$ concentration to pH or pOH. Example 16.8 illustrates this point.

───── • Example 16•8 • ─────

Given that the K_a for NH$_4^+$ is 5.6×10^{-10}, calculate the pH of a $1.0M$ NH$_4$Cl solution.

───── • Solution • The first step is to calculate the H$^+$ concentration. Because the data are identical with those in Example 16.5, we solve for the H$^+$ concentration as in that case and find

$$[H^+] = 2.4 \times 10^{-5}\,M$$

Then,

$$pH = -\log[H^+]$$
$$= -\log(2.4 \times 10^{-5}) = 5 - \log 2.4$$
$$= 5 - 0.38 = 4.62$$

See also Exercise 33.

Measurement of pH is one of the most important and frequently used analytical procedures in biochemistry. The precise determination of the pH of a solution can be made with an instrument called a pH meter (see the special topic on the Glass Electrode in Chapter 18). The H$^+$ ion concentration affects the structure and activity of biological macromolecules, and thus influences the behavior of cells and organisms. It is common practice to refer to the H$^+$ ion concentration in intracellular and extracellular fluids of living organisms in terms of pH.

16•7   **The Common Ion Effect**

The ionization of a weak electrolyte is markedly decreased by adding to the solution an ionic compound that has one ion in common with the weak electrolyte. This is an example of a **common ion effect,** the effect of adding an ion that is the same as one already involved in the equilibrium. Let us illustrate this effect by considering the following equilibrium in a $1.0M$ water solution of ammonia:

$$NH_3 + H_2O \rightleftharpoons NH_4^+ + OH^-$$

If solid ammonium chloride (NH_4^+, Cl^-) is added to this $1.0M$ solution until the solution is $1.0M$ in NH_4Cl, the concentration of the NH_4^+ is greatly increased. Because of this increased concentration of NH_4^+ ions, some of them react with OH^- ions to increase the amount of $NH_3 + H_2O$:

$$NH_4^+ + OH^- \longrightarrow NH_3 + H_2O$$
$$\text{increases} \quad \text{decreases} \quad \text{increases} \quad \text{increases}$$

When equilibrium is reestablished, the concentration of the NH_4^+ ions is greater than it was in the first equilibrium and the concentration of the OH^- ions is less. For all practical purposes we can say that the concentrations of ammonia and ammonium ion are each $1.0M$:

$$K_b = 1.8 \times 10^{-5} = \frac{[NH_4^+][OH^-]}{[NH_3]} = \frac{(1.0)[OH^-]}{1.0}$$

$$[OH^-] = 1.8 \times 10^{-5}\,M$$

So we see that the $[OH^-]$ is decreased from $3.0 \times 10^{-3}\,M$ in $1.0M$ NH_3 (see Example 16.4) to $1.8 \times 10^{-5}\,M$ in the NH_3–NH_4Cl solution.

Another way of looking at the effect on the OH^- concentration by the addition of NH_4Cl to the NH_3 solution is to consider the fact that the ammonium ion, NH_4^+, is a weak acid ($K_a = 5.6 \times 10^{-10}$). Addition of the weakly acidic NH_4^+ ions to the weakly basic ammonia solution causes a decrease in the OH^- concentration:

$$NH_4^+ + OH^- \longrightarrow NH_3 + H_2O$$

There is a corresponding increase in the H^+ concentration, because the product of the two, $[H^+][OH^-]$, is maintained at 1.0×10^{-14}, the value of K_w.

In the same manner, the ionization of weak acids is suppressed by the addition of the salts of these acids. For example, a dilute solution of acetic acid contains a lower H^+ concentration when sodium acetate is also present. Addition of the weakly basic $C_2H_3O_2^-$ ions ($K_b = 5.6 \times 10^{-10}$) to the weakly acidic acetic acid solution causes a decrease in the H^+ concentration:

$$C_2H_3O_2^- + H^+ \longrightarrow HC_2H_3O_2$$

A quantitative calculation illustrating the common ion effect in the acetic acid system is shown in Example 16.9.

• Example 16.9 •

Given that the K_a for acetic acid is 1.8×10^{-5}, calculate the pH of a $0.50M$ $HC_2H_3O_2$ solution that is also $0.20M$ in $NaC_2H_3O_2$.

• Solution • In Example 16.1, we found that both the $[H^+]$ and $[C_2H_3O_2^-]$ are $3.0 \times 10^{-3}\,M$ in $0.50M$ $HC_2H_3O_2$. If we make this solution $0.20M$ with respect to sodium acetate (by adding some solid solute), the $NaC_2H_3O_2$ dissociates to form $0.20M$ Na^+ and $0.20M$ $C_2H_3O_2^-$ ions. Some of the $C_2H_3O_2^-$ ions will react with the H^+ ions to form $HC_2H_3O_2$.

Let $y =$ moles per liter of $C_2H_3O_2^-$ ions that react with y moles per liter of H^+ ions to produce y moles per liter of $HC_2H_3O_2$ molecules. Then $x =$ moles per liter of H^+ ions present at equilibrium. And $(0.20 + 3.0 \times 10^{-3} - y) =$ moles per liter of $C_2H_3O_2^-$ and $(0.50 - 3.0 \times 10^{-3} + y) =$ moles per liter of $HC_2H_3O_2$ at equilibrium.

Summarizing,

	$HC_2H_3O_2$	$\rightleftharpoons$	H^+	$+$	$C_2H_3O_2^-$
mol/L at the instant $NaC_2H_3O_2$ is added	$0.50 - 3.0 \times 10^{-3}$		3.0×10^{-3}		$0.20 + 3.0 \times 10^{-3}$
mol/L at equilibrium	$0.50 - 3.0 \times 10^{-3} + y$		x		$0.20 + 3.0 \times 10^{-3} - y$

$$1.8 \times 10^{-5} = \frac{[H^+][C_2H_3O_2^-]}{[HC_2H_3O_2]} = \frac{(x)(0.20 + 3.0 \times 10^{-3} - y)}{0.50 - 3.0 \times 10^{-3} + y}$$

Fortunately, the determination of x can be greatly simplified. The value of y must be less than 3.0×10^{-3}. The effect of the $+y$ on $[HC_2H_3O_2]$ is to make $[HC_2H_3O_2]$ even closer to $0.50M$ than before. The effect of the $-y$ on $[C_2H_3O_2^-]$ is to make $[C_2H_3O_2^-]$ even closer to $0.20M$ than before the reaction with H^+ ions. Therefore, we can say that $[HC_2H_3O_2] \simeq 0.50M$ and $[C_2H_3O_2^-] \simeq 0.20M$, *which are the concentrations of acetic acid and sodium acetate given in the problem statement.* Substituting these concentrations into the K_a expression gives

$$1.8 \times 10^{-5} = \frac{[H^+][C_2H_3O_2^-]}{[HC_2H_3O_2]} \simeq \frac{(x)(0.20)}{0.50}$$

$$x \simeq \frac{(1.8 \times 10^{-5})(0.50)}{0.20} = 4.5 \times 10^{-5} M = [H^+]$$

$$pH = -\log[H^+]$$
$$= -\log(4.5 \times 10^{-5}) = 5 - \log 4.5$$
$$= 5 - 0.65 = 4.35$$

See also Exercises 35 and 36.

The foregoing examples of solutions of acetic acid and ammonia illustrate a general rule: *weak acids and bases become even weaker in solutions with their salts.*

The common ion effect also can be illustrated by solutions containing ions that hydrolyze. For example, in the reaction of cyanide ions with water,

$$CN^- + H_2O \rightleftharpoons HCN + OH^-$$

the extent of the reaction to the left is increased by the addition of more hydroxide ions. Consequently the extent of the reaction of cyanide ions with water is reduced.

16•8   Buffered Solutions

A solution that contains a weak acid plus a salt of that acid, or a weak base plus a salt of that base, has the ability to react with both strong acids and strong bases. Such a system is referred to as a **buffered solution,** because small additions of either strong acids or strong bases produce little change in the pH.

For example, if some hydrochloric acid is added to a solution containing acetic acid and sodium acetate, the basic acetate ions react with the added hydrogen ions to form more hydrogen acetate molecules:

$$H^+ + C_2H_3O_2^- \longrightarrow HC_2H_3O_2$$

The pH does not change appreciably.

On the other hand, if hydrogen ions are removed by addition of the base sodium hydroxide,

$$H^+ + OH^- \longrightarrow H_2O$$

the molecular hydrogen acetate ionizes to form more hydrogen ions:

$$HC_2H_3O_2 \longrightarrow H^+ + C_2H_3O_2^-$$

Again, the pH of the solution does not change significantly unless large quantities of the base are added.

As a further example, if some sodium hydroxide is added to a solution containing ammonia and ammonium chloride, the acidic ammonium ions react with the added hydroxide ions to form molecular ammonia:

$$OH^- + NH_4^+ \longrightarrow H_2O + NH_3$$

The pH does not change appreciably.

On the other hand, if hydroxide ions are removed by addition of the acid, hydrochloric acid, to form water, the ammonia ionizes to form more hydroxide ions:

$$NH_3 + H_2O \longrightarrow NH_4^+ + OH^-$$

Once more, the pH of the solution does not change significantly unless large quantities of the acid are added.

The difference between the changes in pH when small quantities of a strong acid are added to a sodium chloride solution (an unbuffered system) versus a buffered solution is illustrated in Examples 16.10 and 16.11.

─── • Example 16•10 • ───

Calculate the change in pH when 1.0 mL of 1.0*M* HCl is added to 50 mL of 1.0*M* NaCl.

• Solution • The addition of 1.0 mL of 1.0*M* HCl is equivalent to the addition of 0.0010 mole of H^+ ions (0.0010 L $\times$ 1.0 mol/L = 0.0010 mol). The addition of the 1.0 mL of solution changes the total volume to 51 mL, or 0.051 L. The H^+ ion concentration in the 0.051 L of solution is

$$\frac{0.0010 \text{ mol } H^+}{0.051 \text{ L}} = 2.0 \times 10^{-2}\, M = [H^+]$$

and the pH is

$$pH = -\log [H^+]$$
$$= -\log (2.0 \times 10^{-2}) = 2 - \log 2.0$$
$$= 2 - 0.30 = 1.70$$

For the addition of HCl to the NaCl solution, the change in pH is from 7.00 (the pH of the salt solution) to 1.70, a change of 5.30 units.

See also Exercises 42 and 43.

• Example 16•11 •

Calculate the change in pH when 1.0 mL of 1.0M HCl is added to 50 mL of a buffered solution that is initially 1.0M in $HC_2H_3O_2$ and 1.0M in $NaC_2H_3O_2$.

• Solution • For the $HC_2H_3O_2$–$NaC_2H_3O_2$ buffered solution, the H^+ concentration prior to the addition of the 1.0M HCl is calculated from the K_a for $HC_2H_3O_2$ in a manner similar to that shown in Example 16.9:

$$1.8 \times 10^{-5} = \frac{[H^+][C_2H_3O_2^-]}{[HC_2H_3O_2]} \simeq \frac{(x)(1.0)}{1.0}$$
$$x = [H^+] \simeq 1.8 \times 10^{-5} M$$
$$pH = -\log [H^+]$$
$$\simeq -\log (1.8 \times 10^{-5}) = 5 - \log 1.8$$
$$\simeq 5 - 0.26 = 4.74$$

The pH after addition of the 1.0M HCl is calculated as follows. In the 50 mL of solution, there is 0.050 mole of $HC_2H_3O_2$ and 0.050 mole of $C_2H_3O_2^-$ ions (0.050 L $\times$ 1.0 mol/L = 0.050 mol). The addition of 1.0 mL of 1.0M HCl is equivalent to the addition of 0.0010 mole of H^+ ions (see Example 16.10). This amount of H^+ ions reacts with 0.0010 mole of $C_2H_3O_2^-$ ions and forms an additional 0.0010 mole of $HC_2H_3O_2$ molecules:

	$HC_2H_3O_2 \rightleftharpoons$	H^+	$+ C_2H_3O_2^-$
mol at the instant HCl is added	0.050	~0.0010	0.050
mol at equilibrium	0.050 +0.0010	very small	0.050 −0.0010
	0.051		0.049

The addition of the 1.0 mL of solution changes the total to 51 mL, or 0.051 L. The concentrations of $HC_2H_3O_2$ and $C_2H_3O_2^-$ in moles per liter at equilibrium are

$$[HC_2H_3O_2] \simeq \frac{0.051 \text{ mol}}{0.051 \text{ L}} = 1.0M$$

$$[C_2H_3O_2^-] \simeq \frac{0.049 \text{ mol}}{0.051 \text{ L}} = 0.96M$$

Now we proceed with calculating $[H^+]$ and pH as before:

$$1.8 \times 10^{-5} = \frac{[H^+][C_2H_3O_2^-]}{[HC_2H_3O_2]} \simeq \frac{(x)(0.96)}{1.0}$$

$$x \simeq \frac{(1.8 \times 10^{-5})(1.0)}{0.96} = 1.9 \times 10^{-5}\,M = [H^+]$$

$$pH = -\log [H^+]$$
$$\simeq -\log (1.9 \times 10^{-5}) = 5 - \log 1.9$$
$$\simeq 5 - 0.28 = 4.72$$

The change in pH is from 4.74 to 4.72, a change of only 0.02 unit.

See also Exercises 40 and 45.

As these two examples show, for the addition of the HCl to the unbuffered solution (the NaCl solution) the change in pH is 5.30 units. In contrast, the addition of the same amount of HCl to the buffered solution causes the pH to change by only 0.02 unit.

Buffered solutions are used extensively in analytical chemistry, biochemistry, and bacteriology, as well as in photography and the leather and dye industries. In each of these areas, particularly in biochemistry and bacteriology, certain rather narrow pH ranges may be required for optimum results. If, during the course of a chemical reaction, the concentration of acids (or bases) is allowed to increase, an undesirable reaction may occur or the desired reaction may be inhibited. The action of enzymes, the growth of bacterial cultures, and other biochemical processes depend upon the control of pH by buffered systems.

Standard buffered solutions can be prepared from weak acids and the salts of weak acids. A convenient equation is available for calculating the pH of such solutions, or for calculating the ratio of acid to salt required to prepare a solution of a desired pH. The pH of a buffer containing the weak acid HA may be calculated as follows:

$$K_a = \frac{[H^+][A^-]}{[HA]}$$

$$[H^+] = K_a \times \frac{[HA]}{[A^-]}$$

$$-\log [H^+] = -\log K_a - \log \frac{[HA]}{[A^-]}$$

$$pH = pK_a - \log \frac{[HA]}{[A^-]}$$

or

$$pH = pK_a + \log \frac{[A^-]}{[HA]} \qquad (17)$$

For a solution in which the HA and A^- concentrations are the same,

$$pH = pK_a + \log 1.0 = pK_a + 0 = pK_a$$

For example, in a buffered solution of formic acid–formate ion, in which

$[HCHO_2] = [CHO_2^-],$

$$pH = pK_a = -\log K_a = -\log (1.8 \times 10^{-4}) = 3.74$$

16•8•1   Buffered Systems in the Body Intracellular and extracellular fluids in living organisms contain conjugate acid–base pairs that function as buffers at the pH of the fluids. The major intracellular buffer is the dihydrogen-phosphate–monohydrogenphosphate, $H_2PO_4^-$–HPO_4^{2-}, conjugate acid–base pair. The major extracellular buffer is the carbonic acid–bicarbonate, H_2CO_3–HCO_3^-, conjugate acid–base pair. This latter buffered system helps maintain the pH of the blood at a nearly constant value, close to 7.4, even though acidic and basic substances continually pass into the bloodstream. The buffering action of a solution containing carbonic acid and bicarbonate ions is based on the following reactions:

when an acid is added $HCO_3^- + H^+ \longrightarrow H_2CO_3$

when a base is added $H_2CO_3 + OH^- \longrightarrow H_2O + HCO_3^-$

• Example 16•12 •

What is the ratio of $[HCO_3^-]:[H_2CO_3]$ required to maintain a pH of 7.4 in the bloodstream, given that the K_a for H_2CO_3 in blood[2] is 8.0×10^{-7}?

• Solution • For a solution whose pH is 7.4, the H^+ concentration is calculated as follows:

$$pH = 7.4$$
$$-\log [H^+] = 7.4$$
$$\log [H^+] = -7.4$$
$$\log [H^+] = 0.6 - 8$$
$$[H^+] = \text{antilog } 0.6 \times \text{antilog } -8$$
$$[H^+] = 4 \times 10^{-8} M$$

Substitution of this $[H^+]$ value into the K_a expression for H_2CO_3 gives

$$8.0 \times 10^{-7} = \frac{[H^+][HCO_3^-]}{[H_2CO_3]}$$
$$= (4 \times 10^{-8})\frac{[HCO_3^-]}{[H_2CO_3]}$$
$$\frac{[HCO_3^-]}{[H_2CO_3]} = \frac{8.0 \times 10^{-7}}{4 \times 10^{-8}} = 20$$

Thus, the ratio of $[HCO_3^-]:[H_2CO_3]$ required to maintain a pH of 7.4 is $20:1$.

See also Exercises 46 and 47.

[2] The K_a and K_b values listed in Tables 16.1–16.5 are for water solutions. K_a and K_b values may be determined for systems other than water. In the case of the K_a value for carbonic acid given in this problem, the solvent system is blood.

The rather constant ratio of $[H_2CO_3]:[HCO_3^-]$ in the blood results from a balance between the rate of CO_2 production by oxidation in the cells and the rate of CO_2 loss by breathing. The same equilibria as shown in Figure 16.1 are presumably involved:

$$CO_2(g) + H_2O \rightleftharpoons CO_2(aq) + H_2O \rightleftharpoons H_2CO_3$$

As mentioned in Section 10.5, the partial pressure of $CO_2(g)$ in the lungs is very important in maintaining the concentration of H_2CO_3 in the buffered system. Metabolic processes continually form acidic substances such as the organic acid lactic acid, and the inorganic acids H_3PO_4 and H_2SO_4, which are liberated in body tissues. When these acids pass into the bloodstream, the bicarbonate ions react with them to produce more H_2CO_3. To keep the pH from falling too much, the excess H_2CO_3 decomposes into CO_2 and H_2O. The rate of respiration is increased and CO_2 is eliminated through the lungs. Under any condition in which the blood must absorb OH^- ions, H_2CO_3 is converted to HCO_3^-. Then more H_2CO_3 is quickly formed from the supply of CO_2 in the lungs.

If the pH-regulating mechanisms of the body fail, as may happen during illness, and if the pH of the blood falls below 7.0 or rises above 7.8, irreparable body damage may result. The catalytic activity of enzymes is extremely sensitive to small changes in pH. Their activity declines sharply on the high or low side of 7.4. A change in the H^+ ion concentration of as little as 2.5 times (say, from pH 7.4 to 7.0) can be fatal.

16.9   Indicators

An approximate determination of the pH of a solution can be made readily with acid–base indicators. **Acid–base indicators** are organic acids or bases that have one color if the hydrogen ion concentration is above a certain value and a different color if it is lower. We will use the general formula HIn for a weak acid indicator to illustrate the type of reaction involved. The equilibrium for its ionization is represented as

$$HIn \rightleftharpoons H^+ + In^-$$

If the indicator is the weak acid dinitrophenol, HIn is colorless and In^- is yellow. In solutions where the $[H^+]$ is moderately high, a pH of about 2.5 in this case, the presence of the common H^+ ions causes the reaction to the left to be favored. Consequently, the indicator is mainly in the colorless HIn form. Not enough of the colored anion is present to be visible. If the solution is made more basic by the addition of hydroxide ions, the OH^- ions react with the H^+ ions to form water. This causes the reaction to the right to be favored, and the mixture takes on a yellow color characteristic of In^-, the dinitrophenol anion.

The particular pH at which a visible color change occurs depends on the ionization constant of the indicator. Phenolphthalein is a much weaker organic acid than dinitrophenol. Consequently, its ionization is suppressed at a very low $[H^+]$, at a pH of about 8, so that the indicator is largely in the molecular form when $[H^+]$ in the solution is greater than 1×10^{-8}. When the $[H^+]$ is reduced to about 1×10^{-9}, a pH of 9, by adding a strong base, the colored ion concentration is increased a proportionate amount and the solution is a light pink color. In the two cases just discussed, the indicator molecule is colorless; for other indicators, litmus for example, the indicator molecule may have one color and the indicator ion a different color.

By using a variety of indicators and noting their colors in samples of a solution, one can arrive at a fair estimate of the acidity or basicity of soils, water, body fluids, and other types of solutions. Seven of these useful substances are listed in Table 16.6, but many others also are available.

Table 16•6 Indicators for acids and bases

Name	pH range in which color change occurs	Acid color	Base color
methyl yellow	2–3	red	yellow
dinitrophenol	2.4–4.0	colorless	yellow
methyl orange	3–4.5	red	yellow
methyl red	4.4–6.6	red	yellow
litmus	6–8	red	blue
phenolphthalein	8–10	colorless	red
thymolphthalein	10–12	yellow	violet
trinitrobenzene	12–13	colorless	orange

16•9•1 Indicators and Titration The selection of an indicator for a given acid–base titration (see Chapter 11) depends on the relative strengths of the acid and base to be used in the titration. Consider two titrations: (1) the titration of the strong acid HCl with the strong base NaOH, and (2) the titration of the weak acid $HC_2H_3O_2$ with the same base:

$$HCl + NaOH \longrightarrow NaCl + HOH \qquad (18)$$
$$HC_2H_3O_2 + NaOH \longrightarrow NaC_2H_3O_2 + HOH \qquad (19)$$

In such reactions, the **equivalence point** is the point at which just enough of one reactant has been added to react with a second reactant. An **endpoint** is the point at which a particular indicator changes color. Ideally, in a titration, the equivalence point and the endpoint of the chosen indicator should be identical. Figure 16.3 shows graphically how the pH changes as $0.10M$ NaOH is added (a) to 50.0 mL of $0.10M$ HCl and (b) to 50.0 mL of $0.10M$ $HC_2H_3O_2$. Because the solutions for both these titrations all have the same concentration, each equivalence point is reached when exactly 50 mL of base has been added. Several conclusions can be drawn from a study of these titration curves.

1. For the same amount of NaOH added, the pH of the HCl solution, Figure 16.3(a), is less than that of the $HC_2H_3O_2$ solution, Figure 16.3(b), until the equivalence point is passed. This is expected because, so long as any acid remains unreacted, the pH is determined mainly by the acid present, and HCl is a strong acid whereas $HC_2H_3O_2$ is weak. This conclusion is supported by the following sample calculations:

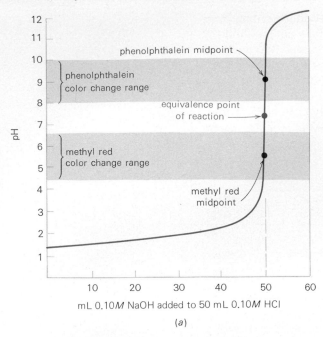

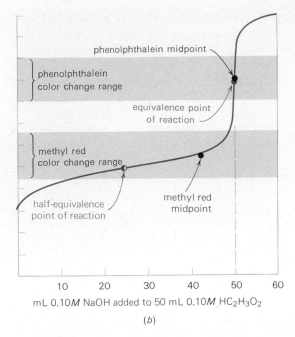

(a)

(b)

When 25.0 mL of 0.10M NaOH is added to 50.0 mL of 0.10M HCl, the moles of acid ($L_A \times M_A = moles_A$) and the pH of the resulting solution are calculated as follows:

	HCl	+	NaOH	⟶	NaCl	+ H$_2$O
before reaction	0.0050 mol$_A$		0.0025 mol$_B$		0 mol	
after reaction	0.0025 mol$_A$		0 mol$_B$		0.0025 mol	

Assuming that the volumes are additive, that is, assuming that the volume of the resulting solution is 75 mL, the moles of HCl per liter are

$$\frac{0.0025 \text{ mol}}{0.075 \text{ L}} = 3.3 \times 10^{-2} M$$

Because HCl is essentially 100 percent ionized in solution, the molarity of the HCl equals the molar concentration of H$^+$ ions, and [H$^+$] = $3.3 \times 10^{-2} M$. The pH $= -\log(3.3 \times 10^{-2}) = 2 - 0.52 = 1.48$.

For the addition of 25.0 mL of 0.10M NaOH to 50.0 mL of 0.10M HC$_2$H$_3$O$_2$, the pH of the resulting solution is calculated as follows:

	HC$_2$H$_3$O$_2$	+	NaOH	⟶	NaC$_2$H$_3$O$_2$ + H$_2$O
before reaction	0.0050 mol		0.0025 mol		0 mol
after reaction	0.0025 mol		0 mol		0.0025 mol

The moles of HC$_2$H$_3$O$_2$ and NaC$_2$H$_3$O$_2$ per liter are

$$\frac{0.0025 \text{ mol}}{0.075 \text{ L}} = 3.3 \times 10^{-2} M$$

Figure 16•3 The changes in pH as 0.10M NaOH is added to a strong acid, HCl, compared with a weak acid, HC$_2$H$_3$O$_2$.
Source: Data are taken from G. H. Ayres, *Quantitative Chemical Analysis*, 2nd ed., Harper & Row, New York, 1968.

For the titration of the weak acid, acetic acid, with the strong base, sodium hydroxide, the pH of the solution at any point may be calculated using Equation (17):[3]

$$pH = pK_a + \log \frac{[C_2H_3O_2^-]}{[HC_2H_3O_2]}$$

At the point when 25.0 mL of 0.10M NaOH has been added, $[HC_2H_3O_2] = [C_2H_3O_2^-] = 3.3 \times 10^{-2} M$.

$$pH = pK_a + \log \frac{3.3 \times 10^{-2}}{3.3 \times 10^{-2}} = 4.74 + \log 1.0 = 4.74$$

In summary, when half a solution of 0.10M HCl is neutralized by 0.10M NaOH, the pH is 1.48; when half a solution of 0.10M HC$_2$H$_3$O$_2$ is neutralized, the pH is 4.74.

2. The most striking fact illustrated in Figure 16.3 is the steepness of the titration curves in the region of the equivalence point. This almost vertical portion of the curve shows that near the equivalence point the addition of very little NaOH causes a great change in pH; only a drop or two of the base solution is needed to increase the pH by one or two units.

3. For acid–base titrations, the pH at the equivalence point depends upon the nature of the salt formed in the reaction. In the case of reaction (18), the solution at the equivalence point is simply a sodium chloride solution; the pH is 7.00 [see Figure 16.3(a)], because neither the Na$^+$ nor the Cl$^-$ ions react appreciably with water. For reaction (19), the solution at the equivalence point is a sodium acetate solution; the pH is 8.72 [see Figure 16.3(b)], because C$_2$H$_3$O$_2^-$ ions react with water to yield a basic solution:

$$C_2H_3O_2^- + HOH \rightleftharpoons HC_2H_3O_2 + OH^-$$

4. Ideally, in order to determine the equivalence point in a particular titration by observing the color change in an indicator added to the solution, one chooses an indicator whose endpoint occurs at the same pH as the equivalence point of the reaction. Practically, it is sufficient to choose an indicator whose color change occurs over a pH range that lies on the steepest part of the titration curve. In this region, the pH changes so abruptly with a drop or two of titrant that the indicator endpoint and the reaction equivalence point are probably achieved with the same drop of reagent.

5. For the titration of a strong acid with a strong base, shown in Figure 16.3(a), the abrupt pH change near the equivalence point encompasses a

---

[3]Because $(mL_B \times M_B)/1{,}000$ = moles of base used = moles of C$_2$H$_3$O$_2^-$ produced, and because $(50.00 - mL_B \times M_B)/1{,}000$ = moles of HC$_2$H$_3$O$_2$ remaining, the pH of the resulting solution for any point on the titration curve for the neutralization of HC$_2$H$_3$O$_2$ can be calculated from the formula

$$pH = pK_a + \log \frac{mL_B \text{ added}}{50.00 - mL_B \text{ added}}$$

wide range, from a pH of about 4 to 10. Any indicator that changes color within these limits will indicate when the equivalence point is reached. As shown, either methyl red or phenolphthalein will serve equally well.

6. For the $HC_2H_3O_2$ titration, Figure 16.3(b), the abrupt change in pH near the equivalence point ranges from about 7 to 10. The choice of an indicator is limited to one that changes color in this pH range, for example, phenolphthalein. Methyl red would not be satisfactory because its color change would occur when only about 90 percent of the acetic acid had reacted.

7. By analogy, one would expect the titration of a weak base with a strong acid to give a solution at the equivalence point that is acidic. For example, when 50 mL of 0.10M HCl is added to 50 mL of 0.10M NH_3, the 0.050M NH_4Cl that results has a pH of 5.28, because NH_4^+ ions react with water to form an acidic solution. Hence, phenolphthalein would not be a satisfactory indicator for this titration, but methyl red probably would serve adequately.

• *Ionization Constants from Titration Curves* • One method of determining the ionization constant of an acid is to titrate it and measure the values of pH with a pH meter. From graphic plots of the data, similar to those in Figure 16.3, it is relatively simple to obtain the value of K_a. From the titration curve, one locates the pH at the point where half of the required amount of NaOH has been added. (In Figure 16.3, this point is at 25.0 mL of NaOH.) At this point $[HA] = [A^-]$, and from Equation (17) we see that $pH = pK_a$. So, $K_a = -antilog\ pH$ *at the half-equivalence point.*

Chapter Review

Summary

Equilibria involving the ionization of weak acids and weak bases in aqueous solution are important in many fields of science. The equilibrium constants for these reactions are called the **acid ionization constant,** K_a, and the **base ionization constant,** K_b, respectively. Each of these constants incorporates the term for the concentration of water. The magnitude of the constant indicates the strength of the acid or the base. For polyprotic acids and polybasic bases, the first ionization constant is always much greater than the second, and so forth. In all solutions of weak electrolytes, the degree of ionization increases as the solution becomes more dilute. The ionization constant, however, does not change.

Pure water ionizes to a very slight degree. The equilibrium constant for this reaction is called the **ion product for water,** K_w; its value is the same in any aqueous solution at a given temperature and is equal to the product of the ionization constants for any conjugate acid–base pair.

The value of the H^+ ion concentration in any solution determines whether the solution is acidic, basic, or neutral.

In many solutions the H^+ concentration depends on the hydrolysis of acidic cations or basic anions. A numerically convenient measure of the acidity of a solution is its **pH,** which can be measured with a **pH meter.** For some purposes the **pOH** is a more convenient measure. The sum of the pH and the pOH of any solution is the constant pK_w.

The ionization of a weak electrolyte in solution is suppressed by the addition of a compound with which it has one ion in common. This is the **common ion effect,** the consequence of which is that weak acids and weak bases become even weaker in solutions with their salts. Such a solution is called a **buffered solution,** because its pH is only slightly affected by the addition of moderate amounts of strong acids or strong bases. In many natural and artificial systems, buffered solutions are important in maintaining a certain narrow range of pH values.

The pH of a solution can often be estimated with an **acid–base indicator,** which changes color at a certain pH value. In acid–base titrations, the **equivalence point** is the point at which chemically equivalent amounts of the two substances have been mixed. Typically, there is a large, abrupt change in the pH of the solution at this point. The **endpoint** of the titration is the point at which the indicator changes color; with careful choice of the indicator, the endpoint will fall precisely at the equivalence point. Much useful information about an acid–base reaction can be gained from its **titration curve,** which is a plot of the pH versus the volume of reactant added.

Exercises

*Writing Ionization
Reactions and Ionization Constants*

1. Unlike hydrochloric acid, hydrofluoric acid, HF, is a weak acid.
   (a) Write the equilibrium equation for the ionization of HF in water.
   (b) Write the expression for the acid ionization constant, K_a, including its numerical value (see Table 16.1).
2. Methylamine, CH_3NH_2, is a weak base in water solution.
   (a) Write the equilibrium equation for its ionization.
   (b) Write the K_c expression for its ionization.
   (c) Write the K_b expression for this weak base.
3. Write simplified equations to show all the steps in the ionization of the following:
   (a) H_2CrO_4 as a diprotic acid
   (b) $Zn(OH)_2$ as a dibasic base
   (c) $Zn(OH)_2$ as a diprotic acid
   (d) H_3AsO_4 as a triprotic acid
4. For each of the ionizations written for Exercise 3, write the expression for K_a or K_b.

Calculations Involving K_a and K_b

5. The acid ionization constant, K_a, for nitrous acid, HNO_2, is 5.1×10^{-4}. Calculate the value of the equilibrium constant, K_c, for the ionization of nitrous acid in water according to the equation

$$HNO_2 + H_2O \rightleftharpoons H_3O^+ + NO_2^-$$

6. Phenol, C_6H_5OH, is a typical weak monoprotic acid. A 0.50M solution is 1.7×10^{-3} percent ionized. Calculate the acid ionization constant, K_a, of phenol.
7. Using the value for the percent ionization of acetic acid found in Example 12.7, calculate a value for K_a. Is the calculated value a satisfactory check on the value listed in Table 16.1? Account for any discrepancy.
8. According to the guidelines in Section 16.3.2, for which of the following solutions should one use the quadratic formula to calculate $[H^+]$? (See Table 16.1 for K_a values.)
   (a) 1.0M $HCHO_2$ (b) 0.10M $HCHO_2$
   (c) 0.020M $HCHO_2$ (d) 0.10M H_2SO_4
   (e) 0.050M H_3PO_4 (f) 0.50M H_2S

9. Calculate the $[H^+]$ and the percent ionization in 1.5M HF. Would the 1.5M solution have a larger percent ionization than a 1.0M HF solution? Would the $[H^+]$ be larger or smaller in the 1.5M than in the 1.0M solution?
10. (a) Calculate the percent ionization for each of the following solutions: 0.50M NH_3; 0.10M NH_3; 0.050M NH_3.
    (b) Compare your answers to part (a) with the percent ionizations of the acetic acid solutions listed in Section 16.3.2. Explain any similarities or differences.
*11. Following the guidelines in Section 16.3.2, calculate $[H^+]$ for each solution listed in Exercise 8.
12. Hydrochloric acid is a very strong acid and hydrocyanic acid, HCN, is a very weak one, $K_a = 6.2 \times 10^{-10}$. What is the hydrogen ion concentration in 0.100M HCl? In 0.100M HCN?
13. For methylamine, CH_3NH_2, the value of K_b is 4.2×10^{-4}.
    (a) Calculate the $[OH^-]$ in 0.10M CH_3NH_2 by each of the following two methods:
        (i) Assume $[CH_3NH_2]$ at equilibrium is $0.10M - x \simeq 0.10M$.
        (ii) Let $[CH_3NH_2]$ be $0.10M - x$ and use the quadratic formula.
    (b) What is the percent ionization of 0.10M CH_3NH_2?

*Calculations with
Polyprotic Acids and Polybasic Bases*

14. Given that the K_a of HSO_4^- is 1.2×10^{-2}, calculate the $[H^+]$ in 0.250M H_2SO_4.
*15. Calculate the H^+ and SO_3^{2-} concentrations in a sulfurous acid solution made by dissolving 5.0 g of sulfur dioxide in enough water to make 1.0 L of solution.
16. Calculate the H^+, $H_2PO_4^-$, HPO_4^{2-}, and PO_4^{3-} concentrations in a 0.50M H_3PO_4 solution.
17. A solution of H_2SeO_4 is prepared by adding 0.100 mole of H_2SeO_4 to enough water to give 1 L of solution. The total concentration of ions in this solution is found to be 0.208M. Calculate the molar concentrations of the H^+, $HSeO_4^-$, and SeO_4^{2-} ions in the solution and the K_a for $HSeO_4^-$.
18. For ethylenediamine, $H_2NCH_2CH_2NH_2$, K_{b1} is 3.6×10^{-4} and K_{b2} is 5.4×10^{-7}. For hydrazine, H_2NNH_2, K_{b1} is 9.8×10^{-7} and K_{b2} is 1.3×10^{-15}.
    (a) Considering the values of K_{b1} for both compounds, which is the stronger base, ethylenediamine or hydrazine? By what magnitude is one base stronger than the other?
    (b) Explain the difference in the strengths of the two bases.
    (c) For each base, how many times larger is K_{b1} compared with K_{b2}? Explain the difference in magnitudes of K_{b1} versus K_{b2} for each compound.

Calculations Involving K_w

* 19. (a) Write two equilibrium equations that are used to indicate the self-ionization of water.
   (b) What is the K_a for water? The K_b?
   (c) Do the equilibrium constants (K_c) for the equations you wrote in part (a) have the same numerical value? If not, indicate how they differ and calculate the values.

20. (a) What is the hydrogen ion concentration in a sodium hydroxide solution that is $2.0 \times 10^{-3}\,M$?
   (b) What is the hydroxide ion concentration in a hydrochloric acid solution that is $2.0 \times 10^{-3}\,M$?

21. Calculate the molar concentrations at equilibrium of NH_4^+, OH^-, H^+, and NH_3 in $0.60M$ NH_3.

22. The $[H^+]$ values in $0.10M$ solutions of four substances are 2.0×10^{-5}, 3.6×10^{-11}, 4.0×10^{-9}, and 8.0×10^{-6}. For each weak acid, calculate the value of K_a; for each weak base, calculate the value of K_b.

* 23. For ethylenediamine, $H_2NCH_2CH_2NH_2$, K_{b1} is 3.6×10^{-4} and K_{b2} is 5.4×10^{-7}. Calculate $[OH^-]$ and $[H^+]$ in $0.40M$ ethylenediamine.

* 24. Like water, liquid ammonia, $NH_3(l)$, undergoes self-ionization; that is, molecules react with one another to produce ions.
   (a) Write the equilibrium equation for the self-ionization of ammonia.
   (b) The K_a of NH_3 is 1.0×10^{-33} at $-50\,°C$. What is the molar concentration of each ion in 500 mL of liquid NH_3 at $-50\,°C$?
   (c) If solid NH_4Cl is added to liquid NH_3 at $-50\,°C$ until the concentration of dissolved NH_4Cl is $0.001M$, what is the concentration of the amide ion, NH_2^-?
   (d) Pertaining to the liquid ammonia solvent, is the solution prepared in part (c) acidic, basic, or neutral? Justify your answer.

Calculations with Hydrolysis Constants

25. Can you say for certain that the four substances in Exercise 22 are weak electrolytes? Explain.

26. Calculate the $[H^+]$ in $0.50M$ methylamine hydrochloride, CH_3NH_3Cl. (K_a for $CH_3NH_3^+ = 2.4 \times 10^{-11}$.)

27. What are the molar concentrations of S^{2-}, HS^-, and H_2S in $1.0M$ Na_2S?

28. Cobalt(II) chloride dissolves in water to give the hexahydrated cobalt(II) cation.
   (a) Write an equation for this hydration reaction.
   (b) The hydrated cation undergoes hydrolysis to give a weakly acidic solution. Write the equilibrium equation for this hydrolysis reaction.
   (c) Write the K_a expression for the hydrolysis reaction.

29. The salt sodium fluoride, NaF, hydrolyzes to give a weakly basic solution.
   (a) Write the equilibrium equation for the important hydrolysis reaction.
   (b) What is the K_b expression for the fluoride ion?
   (c) Calculate the value of K_b for the fluoride ion. (K_a for HF $= 6.6 \times 10^{-4}$.)
   (d) What is the $[H^+]$ in $0.50M$ NaF?

30. (a) Will a $1.0M$ NH_4F solution be weakly acidic, weakly basic, or exactly neutral? Explain your answer in terms of the appropriate acid and base hydrolysis constants.
   (b) Repeat part (a) for a $1.0M$ $Cu(C_2H_3O_2)_2$ solution.

pH Calculations

31. Calculate the pH of each of the following solutions:
   (a) $0.0010M$ HCl
   (b) $0.0010M$ NaOH
   (c) $0.0050M$ HNO_3
   (d) $0.0050M$ $Ba(OH)_2$

* 32. The ions HCO_3^- and $H_2PO_4^-$ are amphiprotic. Calculate the $[H^+]$ and pH in $1.0M$ solutions of $NaHCO_3$ and NaH_2PO_4. (Note: For dilute solutions, the $[H^+]$ is independent of the salt concentration.)

33. Aluminum chloride is added to water to give a $0.50M$ solution. Given that the K_a for the hydrated Al^{3+} ion is 7.2×10^{-6}, calculate the pH of the solution.

34. A $1.0M$ solution of a certain salt has a pH of 12.50. Calculate the percent hydrolysis of the salt.

The Common Ion Effect

35. Explain why the pH of a solution that is $1.0M$ in both HF and NaF is higher than the pH in $1.0M$ HF. Support your explanation with balanced equations.

36. Calculate the pH of a solution that is $0.10M$ with respect to ammonia and $0.20M$ with respect to ammonium chloride. The K_b for ammonia is 1.8×10^{-5}. How much greater or smaller, as the case may be, is the hydrogen ion concentration of the same ammonia solution, but without the NH_4Cl?

37. The acid ionization constant, K_a, is 6.2×10^{-8} for the equilibrium equation

$$H_2PO_4^- + H_2O \rightleftharpoons H_3O^+ + HPO_4^{2-}$$

   Calculate the pH of the solution when the concentrations of $H_2PO_4^-$ and HPO_4^{2-} are equal.

38. Based upon your knowledge of weak and strong acids, weak and strong bases, and the meaning of pH, match each of the following pH values with its solution. Do not make calculations but consult tables for constants where desired.

Solutions	pH values
1. 0.050M NaOH	(a) 1.30
2. 0.050M HCl	(b) 2.52
3. 0.50M NH$_3$	(c) 5.05
4. 0.50M HC$_2$H$_3$O$_2$	(d) 7.00
5. 1.0M NaCl	(e) 8.95
6. 1.0M NaC$_2$H$_3$O$_2$ and	(f) 11.48
0.50M HC$_2$H$_3$O$_2$	(g) 12.70
7. 1.0M NH$_4$Cl and	
0.50M NH$_3$	

39. A solution of 0.100M HF is one-fourth neutralized with 0.100M NaOH. Calculate the pH. (See Table 16.1 for the K_a of HF.)

Buffered Solutions

*40. Consider a solution that is 0.90M in formic acid, HCHO$_2$, and 1.10M in sodium formate, NaCHO$_2$.
    (a) Calculate the pH of this formic acid–sodium formate solution. (K_a of HCHO$_2$ is 1.8×10^{-4}.)
    (b) To 100 mL of this formic acid–sodium formate solution is added 10.0 mL of 1.0M HCl. What is the pH of the resulting solution after equilibrium is reestablished?

41. Do the additions of common antacids, for example, MgO, NaHCO$_3$, and MgCO$_3$, to the stomach reduce the acidity of gastric juice by buffer action or by neutralization? Write equations to support your answer (the acid in gastric juice is HCl). Does the HCl serve a useful purpose in the digestive process?

42. If 1.0 mL of 5.0M NaOH is added to 99.0 mL of water, what is the pH of the resulting solution?

43. What is the pH of a solution prepared by mixing 100 mL of 0.0010M NaOH with 900 mL of 0.020M NaCl?

44. If 100 mL of 2.0M HCl is added to 100 mL of 2.0M NH$_3$, what is the pH of the resulting solution?

45. (a) What is the pH of a solution resulting from the addition of 1.0 mL of 2.0M NaOH to 50.0 mL of solution that is 1.0M in NH$_3$ and also 1.0M in NH$_4$Cl?
    (b) Repeat the calculation in part (a), but this time for the addition of 1.0 mL of 2.0M HCl instead of the NaOH.

46. What must be the ratio of [NH$_4^+$]:[NH$_3$] to maintain a pH of about 9.0, given that K_b for NH$_3$ is 1.8×10^{-5}?

47. The major intracellular buffer is H$_2$PO$_4^-$–HPO$_4^{2-}$. What is the ratio of [HPO$_4^{2-}$]:[H$_2$PO$_4^-$] if the pH is 7.4? (See Table 16.1 for data.)

Hydrazine, N$_2$H$_4$, ionizes in water according to the following equilibrium equation:

$$N_2H_4 + H_2O \rightleftharpoons N_2H_5^+ + OH^-$$

48. The pH of a solution of N$_2$H$_4$ is 8.30 when the N$_2$H$_4$ concentration is 2.0M and the N$_2$H$_5^+$ concentration is 1.0M. Calculate K_b for N$_2$H$_4$.

pH Changes During Titration

*49. Suppose you are going to titrate 50 mL of 0.10M NH$_3$ with 50 mL of 0.10M HCl.
    (a) Sketch a titration curve resulting from a plot of milliliters of HCl versus pH. [Refer to Figure 16.3(*b*) for comparison purposes.]
    (b) Show how a pH value can be obtained from this curve to which the pK_b for ammonia is related. What is this relation?
    (c) Explain how one can use a pH meter to measure the pH of a solution of a weak base as it is titrated with HCl, and can then plot a titration curve from which the pK_b is obtained.

Ionic Equilibria II: Solubility and the Solubility Product

17

17•1 Solubility Product Constants

17•2 Calculation of Solubility Product Constants

17•3 Calculation of Solubilities from K_{sp} Values

17•4 The Common Ion Effect

17•4•1 Effect of a Common Ion on Solubility

17•4•2 Use of a Common Ion in a Precipitation

17•5 Separation by Selective Precipitation

17•5•1 Use of the Common Ion Effect in Selective Precipitation

17•6 Dissolution of Precipitates

17•6•1 Dissolution with Formation of a Weak Electrolyte

17•6•2 Dissolution with Formation of a Complex Ion

17•7 The Qualitative Analysis Scheme

17•7•1 Group I, The Acid Chloride Group

17•7•2 Group II, The Acid Sulfide Group

17•7•3 Group III, The Basic Sulfide Group

17•7•4 Group IV, The Carbonate Group

17•7•5 Group V, The Soluble Group

This is the second of three chapters that deal with the application of equilibrium theory to ionic solutions. In this chapter we will discuss equilibria that exist when saturated solutions of slightly soluble salts are in contact with undissolved salt. For example, suppose we add several grams of a slightly soluble salt to a beaker of water. Because the ionic solid is only slightly soluble in water, only a very small amount dissolves to give ions in solution. In contrast to the very fast reaction of an acid with a base, the attainment of a solubility equilibrium is generally much slower. Nevertheless, just as in other chemical systems, the rates of the opposing reactions do become equal. When the rates of the forward reaction (dissolution) and reverse reaction (precipitation) are the same, a dynamic equilibrium is established between the ions in the solid phase of the undissolved salt and the ions in solution.

Learning about solubility equilibria is an essential part of the study of the chemistry of solutions. Many first-year college chemistry courses devote some time to laboratory experiments on inorganic **qualitative analysis,** the identification of the ions present in a compound or mixture. A knowledge of precipitation and the dissolution of precipitates is fundamental to the qualitative identification of cations and anions in solution as well as to the separation of one ion from another. In addition to solubility equilibria, classical schemes of metal cation analysis also use acid–base equilibria and equilibria with **complex ions,** ions in which neutral polar molecules or anions are attached to metal cations. In this chapter we will learn how these equilibrium systems interact to determine the solubility of ionic solutes.

17·1   Solubility Product Constants

Figure 17·1 Formation of a precipitate of silver chloride as hydrochloric acid is added to a solution of silver nitrate. In the final view, the precipitate is shown after coagulation and settling have occurred. (Photographs by Joseph Carroll.)

In Figure 17.1 we see what happens when a solution of hydrochloric acid is added to a solution of silver nitrate. At first, the tiny white particles of insoluble silver chloride form a milky suspension. On long standing, the solid precipitate settles to the bottom of the flask. The following equation can be written for the equilibrium between the solid phase of the undissolved salt and its ions in solution:

$$AgCl(s) \rightleftharpoons Ag^+ + Cl^-$$

The equilibrium constant for this dissolution reaction is

$$K_c = \frac{[Ag^+][Cl^-]}{[AgCl(s)]}$$

For a saturated solution of silver chloride, the effect of any undissolved solid solute, $AgCl(s)$, is constant, regardless of the amount of undissolved solute present at saturation:

$$[AgCl(s)] = k$$

Substituting k for $[AgCl(s)]$ in the K_c expression and rearranging gives

$$(K_c)(k) = K_{sp} = [Ag^+][Cl^-]$$

The product of the two constants, $(K_c)(k)$, is expressed as the constant K_{sp}, which is called the **solubility product constant.** For AgCl it equals the product of the concentration of the Ag^+ and Cl^- ions in moles per liter of saturated solution. For the general case, consider the slightly soluble ionic compound A_mB_n. The equation for the dissolution equilibrium is

$$A_mB_n(s) \rightleftharpoons mA^{n+} + nB^{m-}$$

and the K_{sp} expression is

$$K_{sp} = [A^{n+}]^m[B^{m-}]^n$$

For example, the dissolution equilibrium equation and the solubility product expression for lead chloride, $PbCl_2$, are

$$PbCl_2(s) \rightleftharpoons Pb^{2+} + 2Cl^-$$

and

$$K_{sp} = [Pb^{2+}][Cl^-]^2$$

For calcium phosphate, $Ca_3(PO_4)_2$, they are

$$Ca_3(PO_4)_2(s) \rightleftharpoons 3Ca^{2+} + 2PO_4^{3-}$$

and

$$K_{sp} = [Ca^{2+}]^3[PO_4^{3-}]^2$$

17·2   Calculation of Solubility Product Constants

Many numerical relationships that describe solutions of slightly soluble salts are similar to those we used in the study of the ionization constants of weak acids and bases. For example, as in the case of an ionization constant, we must know the concentrations of ions in solution in order to calculate the value of the solubility product constant. The number of moles per liter of ions present can be deduced from the number of moles of salt that are dissolved per liter of solution. For the three salts mentioned in the previous section,

$$x \text{ mol of AgCl dissolved per liter} \longrightarrow xM \text{ Ag}^+ \text{ and } xM \text{ Cl}^-$$
$$x \text{ mol of PbCl}_2 \text{ dissolved per liter} \longrightarrow xM \text{ Pb}^{2+} \text{ and } 2xM \text{ Cl}^-$$
$$x \text{ mol of Ca}_3(\text{PO}_4)_2 \text{ dissolved per liter} \longrightarrow 3xM \text{ Ca}^{2+} \text{ and } 2xM \text{ PO}_4^{3-}$$

Consider again the slightly soluble salt, silver chloride, AgCl. The solubility of silver chloride as experimentally determined is 0.00192 g/L at 25 °C. Because the molar weight of AgCl is 143 g, its solubility is

$$(1.92 \times 10^{-3} \text{ g/L})\left(\frac{1 \text{ mol}}{143 \text{ g}}\right) = 1.34 \times 10^{-5} \text{ mol/L}$$

The equation for the solubility equilibrium is

$$\text{AgCl}(s) \rightleftharpoons \text{Ag}^+ + \text{Cl}^-$$

If 1.34×10^{-5} mol/L of AgCl dissolves, the Ag$^+$ ion concentration is $1.34 \times 10^{-5} M$, and the Cl$^-$ ion concentration is also $1.34 \times 10^{-5} M$. Substituting these values in the solubility product expression gives

$$K_{sp} = [\text{Ag}^+][\text{Cl}^-]$$
$$= (1.34 \times 10^{-5})(1.34 \times 10^{-5}) = 1.80 \times 10^{-10}$$

• Example 17•1 •

The solubility of calcium fluoride, CaF$_2$, as experimentally determined is 0.015 g/L at 25 °C. Calculate its K_{sp}.

• Solution • First, we must calculate the solubility of CaF$_2$ in moles per liter. The molar weight of CaF$_2$ is 78.1 g. The solubility of CaF$_2$ is

$$(1.5 \times 10^{-2} \text{ g/L})\left(\frac{1 \text{ mol}}{78.1 \text{ g}}\right) = 1.9 \times 10^{-4} \text{ mol/L}$$

The equation for the solubility equilibrium is

$$\text{CaF}_2(s) \rightleftharpoons \text{Ca}^{2+} + 2\text{F}^-$$

Hence, if 1.9×10^{-4} mole/L of CaF$_2$ dissolves, the Ca^{2+} ion concentration is $1.9 \times 10^{-4} M$, and the F$^-$ ion concentration is $2 \times 1.9 \times 10^{-4} M$, or $3.8 \times 10^{-4} M$. Using these values in the solubility product expression, we obtain

$$K_{sp} = [\text{Ca}^{2+}][\text{F}^-]^2$$
$$= (1.9 \times 10^{-4})(3.8 \times 10^{-4})^2 = 2.7 \times 10^{-11}$$

See also Exercise 2 at the end of the chapter.

The solubility product constants for a number of salts are given in Table 17.1. Solubility products are calculated only for slightly soluble salts, because the K_{sp} relationship holds precisely only for dilute solutions. Most of these salts can be called *insoluble* according to the guideline stated in Section 10.2.2.

Table 17.1 Solubility product constants at 25 °C

Compound	K_{sp}	Compound	K_{sp}
aluminum hydroxide, $Al(OH)_3$	1.3×10^{-33}	lead chromate, $PbCrO_4$	2.8×10^{-13}
barium carbonate, $BaCO_3$	5.1×10^{-9}	lead sulfate, $PbSO_4$	1.6×10^{-8}
barium chromate, $BaCrO_4$	1.2×10^{-10}	lead sulfide, PbS	8.0×10^{-28}
barium sulfate, $BaSO_4$	1.1×10^{-10}	magnesium hydroxide, $Mg(OH)_2$	1.8×10^{-11}
bismuth hydroxide, $Bi(OH)_3$	4×10^{-31}	manganese hydroxide, $Mn(OH)_2$	1.9×10^{-13}
bismuth sulfide, Bi_2S_3	1×10^{-97}	manganese sulfide, MnS	2.5×10^{-13}
cadmium sulfide, CdS	8.0×10^{-27}	mercury(I) chloride, Hg_2Cl_2	1.3×10^{-18}
calcium carbonate, $CaCO_3$	4.8×10^{-9}	mercury(II) sulfide, HgS	1.6×10^{-52}
calcium chromate, $CaCrO_4$	7.1×10^{-4}	nickel hydroxide, $Ni(OH)_2$	2.0×10^{-15}
calcium sulfate, $CaSO_4$	9.1×10^{-6}	nickel sulfide, NiS	1.0×10^{-24}
cobalt sulfide, CoS	2.0×10^{-25}	silver bromide, $AgBr$	5.0×10^{-13}
copper(II) hydroxide, $Cu(OH)_2$	2.2×10^{-20}	silver chloride, $AgCl$	1.8×10^{-10}
copper(II) sulfide, CuS	6.3×10^{-36}	silver iodide, AgI	8.3×10^{-17}
iron(II) hydroxide, $Fe(OH)_2$	8.0×10^{-16}	strontium carbonate, $SrCO_3$	1.1×10^{-10}
iron(II) sulfide, FeS	6.3×10^{-18}	strontium chromate, $SrCrO_4$	2.2×10^{-5}
iron(III) hydroxide, $Fe(OH)_3$	4×10^{-38}	strontium sulfate, $SrSO_4$	3.2×10^{-7}
lead chloride, $PbCl_2$	1.6×10^{-5}	zinc sulfide, ZnS	1.6×10^{-23}

17.3 Calculation of Solubilities from K_{sp} Values

From the value of a solubility product, we can calculate the concentrations of ions in a solution of the pure electrolyte. Or we may wish to calculate the solubility of the salt directly. Consider the slightly soluble salt, barium sulfate, $BaSO_4$. The K_{sp} for $BaSO_4$ is listed in Table 17.1 as 1.1×10^{-10}. The solubility equilibrium is

$$BaSO_4(s) \rightleftharpoons Ba^{2+} + SO_4^{2-}$$

If we let x equal the moles per liter of $BaSO_4$ dissolved in pure water, then x is also the moles per liter of Ba^{2+} and the moles per liter of SO_4^{2-} in solution. Therefore,

$$K_{sp} = [Ba^{2+}][SO_4^{2-}]$$
$$1.1 \times 10^{-10} = (x)(x) = x^2$$
$$x = \sqrt{1.1} \times \sqrt{10^{-10}}$$
$$x = 1.0 \times 10^{-5} \text{ mol/L of } BaSO_4 \text{ dissolved}$$

The solubility of $BaSO_4$ in grams per liter is

$$(1.0 \times 10^{-5} \text{ mol/L})(233 \text{ g/mol}) = 2.3 \times 10^{-3} \text{ g/L}$$

• Example 17•2 •

Given that the K_{sp} for $Mg(OH)_2$ is 1.8×10^{-11}, calculate the solubility of this compound in grams per 100 mL of solution.

• Solution • The solubility equilibrium is

$$Mg(OH)_2(s) \rightleftharpoons Mg^{2+} + 2OH^-$$

Let x = moles per liter of $Mg(OH)_2$ dissolved; then x = moles per liter of Mg^{2+} ions in solution, and $2x$ = moles per liter of OH^- ions in solution. Therefore,

$$K_{sp} = [Mg^{2+}][OH^-]^2$$
$$1.8 \times 10^{-11} = (x)(2x)^2 = 4x^3$$

$$x^3 = \frac{1.8 \times 10^{-11}}{4} = 4.5 \times 10^{-12}$$

$$x = \sqrt[3]{4.5} \times \sqrt[3]{10^{-12}}$$
$$x = 1.7 \times 10^{-4} \text{ mol/L of } Mg(OH)_2 \text{ dissolved}$$

The solubility of $Mg(OH)_2$ per 100 mL of solution is

$$(1.7 \times 10^{-4} \text{ mol/L})(58.3 \text{ g/mol})\left(\frac{1 \text{ L}}{1{,}000 \text{ mL}}\right)(100 \text{ mL}) = 9.9 \times 10^{-4} \text{ g}$$

See also Exercise 3.

At the outset of this section, we calculated for $BaSO_4$, with a K_{sp} of 1.1×10^{-10}, a solubility of 1.0×10^{-5} mole/L. From Example 17.2 we learned that for $Mg(OH)_2$, with a K_{sp} of 1.8×10^{-11}, the solubility is 1.7×10^{-4} mole/L. If we repeat this type of calculation for manganese hydroxide, $Mn(OH)_2$, with a K_{sp} of 1.9×10^{-13}, we obtain a solubility of 3.5×10^{-5} mole/L. Let us summarize the results for these three slightly soluble salts:

compound	K_{sp}	solubility, mole/L
$BaSO_4$	1.1×10^{-10}	1.0×10^{-5}
$Mg(OH)_2$	1.8×10^{-11}	1.7×10^{-4}
$Mn(OH)_2$	1.9×10^{-13}	3.5×10^{-5}

Comparison of the solubility data for $Mg(OH)_2$ and $Mn(OH)_2$ leads us to a very important general relationship: *for two salts of the same type formula* (for example, both with an ionic ratio of 1:2), *the one with the smaller K_{sp} has the lower solubility in moles per liter*. On the other hand, if two salts of different type formulas are compared, there is no simple relationship between the values of K_{sp} and the solubilities in moles per liter. Note that the K_{sp} of $BaSO_4$ is the largest of the three salts listed, but it has the smallest solubility in moles per liter. The ionic ratio in $BaSO_4$ is $1:1$. A simple comparison can be made only when the ionic ratios of the salts are the same.

17.4   The Common Ion Effect

17.4.1   Effect of a Common Ion on Solubility   In the foregoing examples of solubility product calculations, we have considered cases of salts that are dissolved in *pure water*. Often, however, a solution contains an additional source of one of the ions of the insoluble salt. The influence that changing the concentration of one ion has on the concentration of the other ion is another instance of the common ion effect (see Section 16.7). In the case of barium sulfate, we calculated that 1 L of saturated $BaSO_4$ contains 1.0×10^{-5} mole of Ba^{2+} ions, or 1.0×10^{-5} mole of $BaSO_4$. However, if the SO_4^{2-} ion concentration is increased by adding a soluble salt containing the SO_4^{2-} ion—such as Na_2SO_4, K_2SO_4, or H_2SO_4—the concentration of Ba^{2+} must decrease in order to keep the product, $[Ba^{2+}][SO_4^{2-}]$, constant. Suppose solid Na_2SO_4 is added until the SO_4^{2-} concentration becomes $0.10M$. What must the Ba^{2+} ion concentration become? What is the solubility of $BaSO_4$ in this solution? From the K_{sp} expression and the concentration of SO_4^{2-}, we can calculate $[Ba^{2+}]$:

$$K_{sp} = 1.1 \times 10^{-10} = [Ba^{2+}][SO_4^{2-}]$$
$$1.1 \times 10^{-10} = [Ba^{2+}](0.10)$$

$$[Ba^{2+}] = \frac{1.1 \times 10^{-10}}{1.0 \times 10^{-1}} = 1.1 \times 10^{-9} M$$

The concentration of $BaSO_4$ that can dissolve in a $0.10M$ SO_4^{2-} solution is, therefore, $1.1 \times 10^{-9} M$. Compared with its solubility in water, $BaSO_4$ is less soluble in a $0.10M$ Na_2SO_4 solution, by a factor of about 10^4, due to the common ion effect.

• Example 17.3 •

In Example 17.2, the solubility of $Mg(OH)_2$ in 100 mL of solution was calculated to be 9.9×10^{-4} g. What is its solubility in grams per 100 mL of solution that is $0.050M$ in NaOH?

• Solution •   The $0.050M$ NaOH solution is $5.0 \times 10^{-2} M$ in Na^+ and $5.0 \times 10^{-2} M$ in OH^- ions. Using the K_{sp} expression, we can calculate $[Mg^{2+}]$:

$$K_{sp} = 1.8 \times 10^{-11} = [Mg^{2+}][OH^-]^2$$
$$1.8 \times 10^{-11} = [Mg^{2+}](5.0 \times 10^{-2})^2$$

$$[Mg^{2+}] = \frac{1.8 \times 10^{-11}}{2.5 \times 10^{-3}} = 7.2 \times 10^{-9} M$$

Or we can say that the concentration of $Mg(OH)_2$ that dissolves is $7.2 \times 10^{-9} M$. The solubility of $Mg(OH)_2$ per 100 mL of $0.050M$ NaOH is

$$(7.2 \times 10^{-9} \text{ mol/L})(58.3 \text{ g/mol})\left(\frac{1 L}{1,000 \text{ mL}}\right)(100 \text{ mL}) = 4.2 \times 10^{-8} \text{ g}$$

See also Exercises 5 and 6.

17•4•2   Use of a Common Ion in a Precipitation   Often there is a need to remove dissolved substances from solution. To prepare a compound, to remove undesirable ions from water, or to analyze an unknown solution, the chemist may choose to add a soluble salt that contains an ion that will combine with an ion in solution to form an insoluble precipitate. For example, consider the precipitation of calcium ions by carbonate ions when a solution of sodium carbonate, Na_2CO_3, is added to a solution of calcium chloride, $CaCl_2$. The complete ionic equation for this reaction is

$$Ca^{2+} + 2Cl^- + 2Na^+ + CO_3^{2-} \longrightarrow CaCO_3\downarrow + 2Na^+ + 2Cl^-$$

The net ionic equation is

$$Ca^{2+} + CO_3^{2-} \longrightarrow CaCO_3\downarrow$$

Suppose we add a solution of sodium carbonate to a solution that contains calcium chloride until the concentration of CO_3^{2-} ions in the resulting mixture is $2.0 \times 10^{-3}\,M$. What is the concentration of Ca^{2+} ions that can remain in solution? For $CaCO_3$, the solubility product constant at 25 °C is

$$K_{sp} = 4.8 \times 10^{-9} = [Ca^{2+}][CO_3^{2-}]$$

If $[CO_3^{2-}]$ is $2.0 \times 10^{-3}\,M$, $[Ca^{2+}]$ is easily calculated:

$$4.8 \times 10^{-9} = [Ca^{2+}](2.0 \times 10^{-3})$$

$$[Ca^{2+}] = \frac{4.8 \times 10^{-9}}{2.0 \times 10^{-3}} = 2.4 \times 10^{-6}\,M$$

We see that only 2.4×10^{-6} mole/L of Ca^{2+} ion can remain in solution at equilibrium when $[CO_3^{2-}]$ is $2.0 \times 10^{-3}\,M$.

Another conclusion can be drawn from this example. If the original calcium chloride solution is extremely dilute (less than $2.4 \times 10^{-6}\,M$), no calcium carbonate precipitate will form when only enough sodium carbonate is added to make $[CO_3^{2-}]$ equal to $2.0 \times 10^{-3}\,M$. Suppose $[Ca^{2+}]$ is only $1.0 \times 10^{-6}\,M$. Then,

$$[Ca^{2+}][CO_3^{2-}] = (1.0 \times 10^{-6})(2.0 \times 10^{-3}) = 2.0 \times 10^{-9}$$

which is less than 4.8×10^{-9}, the value of the K_{sp}. In this case, no precipitate of $CaCO_3$ will form. *If substitution of known ionic concentrations into the solubility product expression leads to a calculated value less than the K_{sp} for the salt, we may conclude that no precipitate forms.*

It is clear from a consideration of the solubility product relationship that an ion cannot be completely removed from solution by forming a so-called insoluble precipitate with another ion. However, the addition of a large excess of one ion may decrease the concentration of another ion to the vanishing point. In the precipitation of Ca^{2+} ion by the addition of CO_3^{2-} ion, if carbonate ion (as Na_2CO_3) is added to the solution until its concentration is $0.50M$, the Ca^{2+} ion concentration is reduced to 9.6×10^{-9} mole/L. (This is only 0.00000038 g of Ca^{2+} per liter.)

$$4.8 \times 10^{-9} = [Ca^{2+}][CO_3^{2-}]$$
$$4.8 \times 10^{-9} = [Ca^{2+}](0.50)$$

$$[Ca^{2+}] = \frac{4.8 \times 10^{-9}}{0.50} = 9.6 \times 10^{-9}\,M$$

— • Example 17•4 • —

To 50.0 mL of 0.050M sodium phosphate, Na_3PO_4, is added 50.0 mL of 0.0010M barium chloride, $BaCl_2$. The K_{sp} of barium phosphate, $Ba_3(PO_4)_2$, is 3.4×10^{-23}.
(a) Show by calculation whether or not $Ba_3(PO_4)_2$ precipitates.
(b) What is the concentration of Ba^{2+} ions in solution at equilibrium?
(c) What percentage of the Ba^{2+} ions remains in solution?

— • Solution • —

(a) Assuming the volumes are additive and that no precipitate forms, the resulting 100.0 mL of solution would be 0.025M in PO_4^{3-} ions and 0.00050M in Ba^{2+} ions. The halving of the concentrations is the result of the diluting of one solution by the other when they are mixed. We place these concentrations in the K_{sp} expression to see if a precipitate forms:

$$[Ba^{2+}]^3[PO_4^{3-}]^2 = (5.0 \times 10^{-4})^3(2.5 \times 10^{-2})^2 = 7.8 \times 10^{-14}$$

Because 7.8×10^{-14} is larger than the K_{sp} of 3.4×10^{-23}, we conclude that a precipitate of $Ba_3(PO_4)_2$ must form.
(b) Because the original concentration of the PO_4^{3-} ion is 50 times the original concentration of the Ba^{2+} ion, we can assume that the concentration of PO_4^{3-} in solution is changed very little when $Ba_3(PO_4)_2$ precipitates. For the purpose of making an approximate calculation, we use 0.025M for $[PO_4^{3-}]$ at equilibrium, and calculate $[Ba^{2+}]$ using the K_{sp} expression:

$$K_{sp} = 3.4 \times 10^{-23} = [Ba^{2+}]^3[PO_4^{3-}]^2$$
$$3.4 \times 10^{-23} \simeq [Ba^{2+}]^3(2.5 \times 10^{-2})^2$$

$$[Ba^{2+}]^3 \simeq \frac{3.4 \times 10^{-23}}{(2.5 \times 10^{-2})^2} = 5.4 \times 10^{-20} = 54 \times 10^{-21}$$

$$[Ba^{2+}] \simeq \sqrt[3]{54} \times \sqrt[3]{10^{-21}}$$
$$\simeq 3.8 \times 10^{-7}\,M$$

(c) The percentage of Ba^{2+} ions left in solution is calculated as follows:

$$\frac{[Ba^{2+}] \text{ left in solution}}{[Ba^{2+}] \text{ at the start}} \times 100 \simeq \frac{3.8 \times 10^{-7}\,M}{5.0 \times 10^{-4}\,M} \times 100 \simeq 0.076\%$$

See also Exercises 9–17.

17•5 Separation by Selective Precipitation

Consider a solution that contains two ionic species that can form similar insoluble compounds. If the solubilities of the potential precipitates differ sufficiently, one ion can be almost completely precipitated while the other is left in solution. This process is known as **selective precipitation** or **fractional precipitation.**

For example, consider a solution that is $0.10M$ in both Mn^{2+} and Mg^{2+} at a pH of 7.0. Because $Mn(OH)_2$, with $K_{sp} = 1.9 \times 10^{-13}$, is less soluble than $Mg(OH)_2$, with $K_{sp} = 1.8 \times 10^{-11}$, it should be possible to precipitate only the Mn^{2+} by adding the right concentration of OH^-. To precipitate $Mn(OH)_2$, the product of $[Mn^{2+}]$ and $[OH^-]^2$ must exceed the K_{sp} of 1.9×10^{-13}; to precipitate $Mg(OH)_2$, the product of $[Mg^{2+}]$ and $[OH^-]^2$ must exceed the K_{sp} of 1.8×10^{-11}. If we increase the OH^- ion concentration in solution to the point where the product of $[Mg^{2+}]$ and $[OH^-]^2$ equals the K_{sp} for $Mg(OH)_2$ of 1.8×10^{-11}, the solution is saturated with Mg^{2+} ions, but no precipitate of $Mg(OH)_2$ has yet formed. We can calculate what the OH^- ion concentration must be to achieve this result by using the K_{sp} expression for $Mg(OH)_2$:

$$K_{sp} = 1.8 \times 10^{-11} = [Mg^{2+}][OH^-]^2$$
$$1.8 \times 10^{-11} = (0.10)[OH^-]^2$$

$$[OH^-]^2 = \frac{1.8 \times 10^{-11}}{1.0 \times 10^{-1}} = 1.8 \times 10^{-10}$$

$$[OH^-] = 1.3 \times 10^{-5}\,M$$

If the OH^- ion concentration at equilibrium is $1.3 \times 10^{-5}\,M$, only $Mn(OH)_2$ will precipitate.

If the OH^- ion concentration at equilibrium is $1.3 \times 10^{-5}\,M$, what is the concentration of Mn^{2+} ions that remains in solution? This concentration is calculated using the K_{sp} expression for $Mn(OH)_2$:

$$K_{sp} = 1.9 \times 10^{-13} = [Mn^{2+}][OH^-]^2$$
$$1.9 \times 10^{-13} = [Mn^{2+}](1.3 \times 10^{-5})^2$$

$$[Mn^{2+}] = \frac{1.9 \times 10^{-13}}{(1.3 \times 10^{-5})^2} = 1.1 \times 10^{-3}\,M$$

The percentage of the Mn^{2+} ions that remains in solution is

$$\frac{1.1 \times 10^{-3}\,M}{1.0 \times 10^{-1}\,M} \times 100 = 1.1\%$$

That is, 98.9 percent of the available Mn^{2+} ions is precipitated at an OH^- ion concentration of $1.3 \times 10^{-5}\,M$. This is the maximum amount of Mn^{2+} ions that can be separated from the Mg^{2+} ions.

───── • Example 17•5 • ─────

Suppose you have a solution that is $0.10M$ in Zn^{2+} ions and $0.10M$ in Cd^{2+} ions.

(a) What S^{2-} ion concentration is needed to precipitate the maximum amount of one cation as the insoluble sulfide while leaving the other cation completely in solution?

(b) For the cation that is precipitated, calculate the percentage that will remain in solution.

• **Solution** • From Table 17.1 the values of K_{sp} for zinc sulfide, ZnS, and cadmium sulfide, CdS, are 1.6×10^{-23} and 8.0×10^{-27}, respectively.

(a) Because zinc sulfide has the larger K_{sp}, it is the more soluble sulfide. Using the K_{sp} expression, we can calculate how high the S^{2-} ion concentration can be before ZnS precipitates:

$$K_{sp} = 1.6 \times 10^{-23} = [Zn^{2+}][S^{2-}]$$
$$1.6 \times 10^{-23} = (0.10)[S^{2-}]$$

$$[S^{2-}] = \frac{1.6 \times 10^{-23}}{0.10} = 1.6 \times 10^{-22} M$$

(b) The S^{2-} concentration of $1.6 \times 10^{-22} M$ is the maximum concentration that can be used to precipitate the less soluble CdS and yet avoid the precipitation of ZnS. Using the K_{sp} expression for CdS, we can calculate the concentration of Cd^{2+} ions left in solution when this concentration of S^{2-} is present:

$$K_{sp} = 8.0 \times 10^{-27} = [Cd^{2+}][S^{2-}]$$
$$8.0 \times 10^{-27} = [Cd^{2+}](1.6 \times 10^{-22})$$

$$[Cd^{2+}] = \frac{8.0 \times 10^{-27}}{1.6 \times 10^{-22}} = 5.0 \times 10^{-5} M$$

The percentage of the Cd^{2+} ions that remains in solution is

$$\frac{5.0 \times 10^{-5} M}{1.0 \times 10^{-1} M} \times 100 = 0.050\%$$

See also Exercises 18, 19, and 23.

17.5.1   Use of the Common Ion Effect in Selective Precipitation   In Example 17.5 we concluded that the S^{2-} ion concentration should be $1.6 \times 10^{-22} M$ to achieve separation of $0.10M$ Cd^{2+} from $0.10M$ Zn^{2+}. How can we control the S^{2-} ion concentration to achieve the desired result? A common source of sulfide ion is the weak electrolyte hydrogen sulfide, H_2S. The overall simplified equilibrium equation for the ionization of aqueous H_2S is

$$H_2S \rightleftharpoons 2H^+ + S^{2-}$$

and

$$K_a = \frac{[H^+]^2[S^{2-}]}{[H_2S]} = 1.1 \times 10^{-21}$$

This value of K_a can be calculated from the first and second ionization constants for H_2S (see Example 16.3):

$$H_2S \rightleftharpoons H^+ + HS^- \qquad K_{a1} = \frac{[H^+][HS^-]}{[H_2S]} = 1.1 \times 10^{-7}$$

$$HS^- \rightleftharpoons H^+ + S^{2-} \qquad K_{a2} = \frac{[H^+][S^{2-}]}{[HS^-]} = 1.0 \times 10^{-14}$$

By multiplying the first ionization constant by the second, we obtain

$$(K_{a1})(K_{a2}) = \left(\frac{[H^+][\cancel{HS^-}]}{[H_2S]}\right)\left(\frac{[H^+][S^{2-}]}{[\cancel{HS^-}]}\right)$$

$$(1.1 \times 10^{-7})(1.0 \times 10^{-14}) = 1.1 \times 10^{-21} = \frac{[H^+]^2[S^{2-}]}{[H_2S]}$$

A saturated, aqueous solution of H_2S is about $0.10M$ in H_2S. For this solution,

$$1.1 \times 10^{-21} = \frac{[H^+]^2[S^{2-}]}{0.10}$$

$$1.1 \times 10^{-22} = [H^+]^2[S^{2-}]$$

By adjusting the H^+ ion concentration, we can control the concentration of the S^{2-} ion in the saturated solution of the weak electrolyte H_2S. If the solution is made $1.0M$ in H^+ by the addition of a strong acid, such as HCl, the S^{2-} ion concentration will be:

$$1.1 \times 10^{-22} = (1.0)^2[S^{2-}]$$

$$[S^{2-}] = 1.1 \times 10^{-22} M$$

If this S^{2-} ion concentration is used with $[Zn^{2+}] = 0.10M$ in the K_{sp} expression for ZnS, we find

$$[Zn^{2+}][S^{2-}] = (1.0 \times 10^{-1})(1.1 \times 10^{-22}) = 1.1 \times 10^{-23}$$

Because this value is less than the K_{sp} for ZnS of 1.6×10^{-23}, we see that ZnS will not precipitate from a solution that is $0.10M$ in Zn^{2+} ion, saturated with H_2S, and $1.0M$ in H^+ ion.
However, the product

$$[Cd^{2+}][S^{2-}] = (1.0 \times 10^{-1})(1.1 \times 10^{-22}) = 1.1 \times 10^{-23}$$

is greater than the K_{sp} for CdS of 8.0×10^{-27}. From this, we see that CdS will precipitate from the solution. This shows that by adding H^+ ions to a saturated solution of H_2S we can control the S^{2-} concentration so that Cd^{2+} ions can be separated from Zn^{2+} ions by selective precipitation. Example 17.6 illustrates how this type of application of the common ion effect may be used in the selective precipitation of Mn^{2+} ions from a solution of Mg^{2+} and Mn^{2+} ions.

• Example 17•6 •

The values of K_{sp} for $Mg(OH)_2$ and $Mn(OH)_2$ are 1.8×10^{-11} and 1.9×10^{-13}, respectively. If a solution is $0.10M$ with respect to both Mn^{2+} and Mg^{2+} ions at a pH of 7.0, which of the following will provide for the separation of one metal ion from the other?
(a) adding NH_3 until its concentration is $0.50M$
(b) adding NH_3 and NH_4Cl until they are $0.50M$ and $1.0M$, respectively

• Solution •

(a) In a solution that is $0.50M$ in NH_3, the OH^- ion concentration is calculated as follows:

$$K_b = 1.8 \times 10^{-5} = \frac{[NH_4^+][OH^-]}{[NH_3]}$$

$$1.8 \times 10^{-5} \simeq \frac{(x)(x)}{0.50}$$

$$x^2 \simeq 9.0 \times 10^{-6}$$

$$x \simeq 3.0 \times 10^{-3}\,M = [OH^-]$$

Using this $[OH^-]$ concentration in the K_{sp} expression for either of the metal hydroxides, we obtain

$$[M^{2+}][OH^-]^2 = (0.10)(3.0 \times 10^{-3})^2 = 9.0 \times 10^{-7}$$

Because this value of 9.0×10^{-7} is larger than either K_{sp}, both $Mg(OH)_2$ and $Mn(OH)_2$ will precipitate in $0.50M\ NH_3$.

(b) Compared with the $0.50M\ NH_3$ solution, a solution that is $0.50M$ in NH_3 and $1.0M$ in NH_4Cl has a considerably smaller OH^- ion concentration. From the equilibrium equation, $NH_3 + H_2O \rightleftharpoons NH_4^+ + OH^-$, we see that the addition of NH_4^+ ions will suppress the ionization of ammonia. To calculate the OH^- concentration, we can write

$$1.8 \times 10^{-5} = \frac{[NH_4^+][OH^-]}{[NH_3]}$$

$$1.8 \times 10^{-5} \simeq \frac{(1.0)[OH^-]}{0.50}$$

$$[OH^-] \simeq \frac{(0.50)(1.8 \times 10^{-5})}{1.0} = 9.0 \times 10^{-6}\,M$$

Using this value for $[OH^-]$ in either K_{sp} expression gives

$$[M^{2+}][OH^-]^2 = (0.10)(9.0 \times 10^{-6})^2 = 8.1 \times 10^{-12}$$

Because the value of 8.1×10^{-12} is less than the K_{sp} of $Mg(OH)_2$, magnesium hydroxide will not precipitate. However, this value exceeds the K_{sp} of $Mn(OH)_2$, so manganese hydroxide will precipitate.

See also Exercises 20–22 and 25.

17•6   Dissolution of Precipitates

If a solution contains a large number of ions, one group of ions may be separated from the others by precipitating a mixture of similar slightly soluble salts. After this mixture of precipitates is obtained, it is frequently necessary to dissolve one or more of them in order to determine which ions are present.

17•6•1   Dissolution with Formation of a Weak Electrolyte   Suppose we have a mixture of the slightly soluble hydroxides, $Mg(OH)_2$ and $Mn(OH)_2$.

Addition of the strong acid HCl dissolves both of them. For $Mn(OH)_2$, the dissolution equation is

$$Mn(OH)_2 + 2HCl \longrightarrow MnCl_2 + 2H_2O$$

and the net ionic equation is

$$Mn(OH)_2 + 2H^+ \longrightarrow Mn^{2+} + 2H_2O$$

Similarly, dissolution of a mixture of calcium carbonate, $CaCO_3$, barium carbonate, $BaCO_3$, and strontium carbonate, $SrCO_3$, occurs upon addition of HCl. For $CaCO_3$, the net ionic equation is

$$CaCO_3 + 2H^+ \longrightarrow Ca^{2+} + H_2CO_3 \longrightarrow Ca^{2+} + H_2O + CO_2\uparrow$$

Note that in the dissolution of the hydroxides and of the carbonates, the reaction with HCl produces a weak electrolyte. In solutions of these weak electrolytes (H_2O and H_2CO_3), the OH^- or CO_3^{2-} ion concentrations are reduced to such low levels that the products of the metal cation and anion concentrations do not exceed the values of K_{sp} for the slightly soluble salts.

By making use of the differences in solubilities based on known values of K_{sp}, it is often possible to carry out a process of selective dissolution. For example, consider again ZnS, with a K_{sp} of 1.6×10^{-23}, and CdS, with a K_{sp} of 8.0×10^{-27}. When a mixture of these sulfides is stirred in hydrochloric acid (approximately $3M$ HCl), the more soluble sulfide, ZnS, dissolves, whereas most of the CdS does not. The equation for the dissolution of ZnS is

$$ZnS + 2HCl \longrightarrow ZnCl_2 + H_2S$$

The concentration of S^{2-} ions in solution is lowered by the formation of the weak electrolyte H_2S. The ionization of H_2S to produce sulfide ions is suppressed by the strong acid solution (the common ion effect again). Hence, there are not sufficient S^{2-} ions present to keep the ZnS from dissolving. Because CdS is less soluble than ZnS, the S^{2-} ion concentration is still sufficiently large so that no appreciable CdS dissolves.

17•6•2   Dissolution with Formation of a Complex Ion   A metal ion in solution is usually present as a *complex ion* in which one or more ions or molecules are associated with the metal ion. The associated particles are called ligands. A **ligand** is a Lewis base that donates a pair of electrons that functions as a covalent bond between it and the central ion.

Complex ions will be discussed in detail in Chapter 22, but we have encountered examples in previous chapters. For example, the hydrated metal cations $Cu(H_2O)_4^{2+}$ and $Mg(H_2O)_6^{2+}$ were discussed in Section 9.5.1. Other hydrated ions were discussed in connection with hydrolysis in Section 16.5. In these examples, a neutral water molecule is the ligand. For Cu^{2+} we can write the equilibrium equation

$$Cu^{2+} + 4H_2O \rightleftharpoons Cu(H_2O)_4^{2+}$$

Another common neutral ligand is the ammonia molecule, NH_3. When ammonia is added to a water solution of Cu^{2+}, ammonia molecules replace the water molecules:

$$Cu(H_2O)_4{}^{2+} + 4NH_3 \rightleftharpoons Cu(NH_3)_4{}^{2+} + 4H_2O$$

light blue   deep blue

This reaction is easy to follow visually because the $Cu(NH_3)_4{}^{2+}$ ion has a deeper blue color than does the hydrated ion (see Figure 1.11 on Plate 1).

Anions that are common ligands include F^-, Cl^-, and CN^-. Familiar complex ions involving these ligands are $FeF_6{}^{3-}$, $CoCl_4{}^{2-}$, and $Zn(CN)_4{}^{2-}$. Note that when anions function as ligands, the charge on the resulting complex ion equals the algebraic sum of the charges on the metal ion and its associated anion ligands. For example, in the case of $FeF_6{}^{3-}$, the charge is $(3+) + 6(1-) = 3-$.

The replacement of multiple ligands by others in solution does not take place in one step but rather in a series of steps. For example, the reaction of the hydrated Cu^{2+} ion, $Cu(H_2O)_4{}^{2+}$, and NH_3 involves the following equilibria (each copper ion has four ligands, but the water molecules are omitted for simplicity):

$$Cu^{2+} + NH_3 \rightleftharpoons Cu(NH_3)^{2+} \tag{1}$$
$$Cu(NH_3)^{2+} + NH_3 \rightleftharpoons Cu(NH_3)_2{}^{2+} \tag{2}$$
$$Cu(NH_3)_2{}^{2+} + NH_3 \rightleftharpoons Cu(NH_3)_3{}^{2+} \tag{3}$$
$$Cu(NH_3)_3{}^{2+} + NH_3 \rightleftharpoons Cu(NH_3)_4{}^{2+} \tag{4}$$

Equilibrium constant expressions can be written for the formation of each of the complex ions shown in Equations (1) through (4). Quantitatively, the extent of complex ion formation can be expressed in terms of an equilibrium constant called the **stability constant** or **formation constant, K_f**. For reactions (1) through (4) each stability constant has been determined:

$$K_{f1} = \frac{[Cu(NH_3)^{2+}]}{[Cu^{2+}][NH_3]} = 2.0 \times 10^4$$

$$K_{f2} = \frac{[Cu(NH_3)_2{}^{2+}]}{[Cu(NH_3)^{2+}][NH_3]} = 4.7 \times 10^3$$

$$K_{f3} = \frac{[Cu(NH_3)_3{}^{2+}]}{[Cu(NH_3)_2{}^{2+}][NH_3]} = 1.1 \times 10^3$$

$$K_{f4} = \frac{[Cu(NH_3)_4{}^{2+}]}{[Cu(NH_3)_3{}^{2+}][NH_3]} = 2.0 \times 10^2$$

The overall stability constant for the formation of the $Cu(NH_3)_4{}^{2+}$ ion from Cu^{2+} and NH_3 is the product of the four stepwise stability constants:

$$K_f = (K_{f1})(K_{f2})(K_{f3})(K_{f4})$$

The conversion of the $Cu(H_2O)_4{}^{2+}$ ion to the $Cu(NH_3)_4{}^{2+}$ ion takes place because the $Cu(NH_3)_4{}^{2+}$ ion is a more stable complex ion; that is, K_f for $Cu(NH_3)_4{}^{2+}$ is greater than K_f for $Cu(H_2O)_4{}^{2+}$.

The type of complex formed in any situation depends on the relative stabilities and the concentrations of molecules and ions available. For the Cu^{2+} complexes just described, in solutions with a moderate concentration of NH_3, the most highly ammoniated species of the four shown will predominate, and the concentrations of the less ammoniated complexes often can be neglected.

Overall stability constants for some complex ions are listed in Table 17.2.

Table 17•2 **Stability constants at 25 °C**

Ligand	Metal ion	Complex ion	Stability constant, K_f
ammonia, NH_3	cadmium	$Cd(NH_3)_4{}^{2+}$	1.3×10^7
	cobalt(II)	$Co(NH_3)_6{}^{2+}$	1.3×10^5
	copper(II)	$Cu(NH_3)_4{}^{2+}$	2.1×10^{13}
	nickel	$Ni(NH_3)_6{}^{2+}$	5.5×10^8
	silver(I)	$Ag(NH_3)_2{}^{+}$	1.1×10^7
	zinc	$Zn(NH_3)_4{}^{2+}$	2.9×10^9
cyanide, CN^-	cadmium	$Cd(CN)_4{}^{2-}$	6.0×10^{18}
	copper(II)	$Cu(CN)_4{}^{2-}$	2.0×10^{30}
	iron(II)	$Fe(CN)_6{}^{4-}$	$1 \quad \times 10^{35}$
	iron(III)	$Fe(CN)_6{}^{3-}$	$1 \quad \times 10^{42}$
	nickel	$Ni(CN)_4{}^{2-}$	$2 \quad \times 10^{31}$
	silver	$Ag(CN)_2{}^{-}$	$1 \quad \times 10^{21}$
	zinc	$Zn(CN)_4{}^{2-}$	$5 \quad \times 10^{16}$
hydroxide, OH^-	aluminum	$Al(OH)_4{}^{-}$	1.1×10^{33}
	bismuth	$Bi(OH)_4{}^{-}$	$2 \quad \times 10^{35}$
	chromium	$Cr(OH)_4{}^{-}$	$8 \quad \times 10^{29}$
	copper	$Cu(OH)_4{}^{2-}$	$3 \quad \times 10^{18}$
	zinc	$Zn(OH)_4{}^{2-}$	4.6×10^{17}

• *Stability Constants in Analytical Chemistry* • The separation of a mixture of insoluble compounds is a familiar problem in analytical chemistry. Often it is possible to find a ligand that will react with one of the compounds to form a soluble complex. For example, chloride ion and bromide ion are so similar chemically that they tend to form similar precipitates with silver ion. However, a mixture of AgCl and AgBr can be separated by adding the right amount of aqueous ammonia. The NH_3 ligand tends to form the soluble complex ion $Ag(NH_3)_2{}^+$. Because AgCl ($K_{sp} = 1.8 \times 10^{-10}$) is more soluble than AgBr ($K_{sp} = 5.0 \times 10^{-13}$), the AgCl has a greater tendency to combine with NH_3 and dissolve. Example 17.7 shows the relationships between the various equilibria that are involved.

• Example 17•7 •

(a) For a mixture of 0.010 mole of solid AgCl and 0.010 mole of solid AgBr, what is the molar concentration of NH_3 needed to dissolve only the AgCl in a liter of solution?

(b) How many moles of NH_3 would be required to dissolve the AgBr? (See Tables 17.1 and 17.2 for equilibrium constants.)

• Solution •

(a) The equilibria and equilibrium constants that are important in describing the dissolution of AgCl are

$$AgCl \rightleftharpoons Ag^+ + Cl^-$$
$$K_{sp} = [Ag^+][Cl^-] = 1.8 \times 10^{-10}$$
$$Ag^+ + \cancel{Cl^-} + 2NH_3 \rightleftharpoons Ag(NH_3)_2^+ + \cancel{Cl^-}$$
$$K_f = \frac{[Ag(NH_3)_2^+]}{[Ag^+][NH_3]^2} = 1.1 \times 10^7$$

Because we assume that all of the AgCl dissolves, the Cl^- ion concentration is equal to the concentration of the dissolved AgCl, 0.010 mole/L. However, the value of $[Ag^+]$ will be very small, because most of the silver will be in the form of the complex ion $Ag(NH_3)_2^+$. The $[Ag^+]$ is calculated from the K_{sp} expression for AgCl:

$$[Ag^+][Cl^-] = [Ag^+](0.010) = 1.8 \times 10^{-10}$$

$$[Ag^+] = \frac{1.8 \times 10^{-10}}{1.0 \times 10^{-2}} = 1.8 \times 10^{-8}\,M$$

To calculate the concentration of $Ag(NH_3)_2^+$, we note that

$$[Ag(NH_3)_2^+] + [Ag^+] = [AgCl]_{dissolved} = 0.010M$$
$$[Ag(NH_3)_2^+] = 0.010M - 1.8 \times 10^{-8}\,M$$
$$\simeq 0.010M$$

The concentration of NH_3 in the equilibrium solution is found by using the values of $[Ag^+]$ and $[Ag(NH_3)_2^+]$ in the expression for K_f:

$$\frac{[Ag(NH_3)_2^+]}{[Ag^+][NH_3]^2} = \frac{0.010}{(1.8 \times 10^{-8})[NH_3]^2} = 1.1 \times 10^7$$

$$[NH_3]^2 = \frac{0.010}{(1.8 \times 10^{-8})(1.1 \times 10^7)} = 5.1 \times 10^{-2}$$

$$[NH_3] = \sqrt{5.1 \times 10^{-2}} = 2.3 \times 10^{-1} = 0.23M$$

The total moles per liter of NH_3 needed is the sum of that required to form the complex ions plus that which is dissolved in the solution at equilibrium:

$$\text{total } NH_3 = 2[Ag(NH_3)_2^+] + 0.23M$$
$$= 2(0.010M) + 0.23M$$
$$= 0.25M \text{ to dissolve the AgCl}$$

(b) The concentration of NH_3 required to dissolve the AgBr is calculated in a similar fashion:

$$[Ag^+][Br^-] = [Ag^+](0.010) = 5.0 \times 10^{-13}$$

$$[Ag^+] = \frac{5.0 \times 10^{-13}}{1.0 \times 10^{-2}} = 5.0 \times 10^{-11} M$$

$$\frac{[Ag(NH_3)_2^+]}{[Ag^+][NH_3]^2} = \frac{0.010}{(5.0 \times 10^{-11})[NH_3]^2} = 1.1 \times 10^7$$

$$[NH_3]^2 = \frac{0.010}{(5.0 \times 10^{-11})(1.1 \times 10^7)} = 18$$

$$[NH_3] = \sqrt{18} = 4.2M$$

$$\text{total } [NH_3] = 2(0.010M) + 4.2M$$

$$= 4.22M = 4.2M \text{ to dissolve the AgBr}$$

So, we see that $4.2M/0.25M = 17$, so 17 times more NH_3 is required to dissolve the AgBr than to dissolve the AgCl.

See also Exercises 33–37.

Another case in which aqueous ammonia provides the NH_3 ligands necessary to form a soluble complex is in the separation of Cu^{2+} ions from Bi^{3+} ions. Both these ions form hydroxide precipitates in aqueous ammonia:

$$Cu^{2+} + 2NH_3 + 2H_2O \longrightarrow Cu(OH)_2\downarrow + 2NH_4^+$$
$$Bi^{3+} + 3NH_3 + 3H_2O \longrightarrow Bi(OH)_3\downarrow + 3NH_4^+$$

The copper ion forms such a strong complex ion with ammonia that upon addition of excess ammonia, the precipitate of copper hydroxide dissolves:

$$Cu(OH)_2 + 4NH_3 \longrightarrow Cu(NH_3)_4^{2+} + 2OH^-$$

Consider the separation of Fe^{3+} ions from Al^{3+} ions. Reaction of a mixture of these ions with NaOH in solution initially gives precipitates of both $Fe(OH)_3$ and $Al(OH)_3$. Upon further addition of NaOH, the $Al(OH)_3$ dissolves to produce the soluble complex ion $Al(OH)_4^-$. Net ionic equations for these reactions are

$$Fe^{3+} + 3OH^- \longrightarrow Fe(OH)_3\downarrow$$

$$\underline{\begin{matrix} Al^{3+} + 3OH^- \longrightarrow Al(OH)_3\downarrow \\ Al(OH)_3 + OH^- \longrightarrow Al(OH)_4^- \end{matrix}}$$

$$Al^{3+} + 4OH^- \longrightarrow Al(OH)_4^-$$

17•7 The Qualitative Analysis Scheme

As mentioned at the beginning of this chapter, many first-year college chemistry courses include laboratory experiments on the qualitative analysis of salts and salt solutions. Although modern instrumental methods of analysis are used in industry and in other laboratories, studying the classical techniques for separating and identifying the cations and anions in a mixture still proves valuable to students. Qualitative analysis provides clear and striking examples of the principles that apply to precipitations, to the formation of complex ions, and to the control of ionic concentrations in solutions of weak electrolytes. An understanding of these important concepts is essential not only in the chemistry laboratory but also in most disciplines where some knowledge of solution chemistry is necessary—agriculture, biology, ecology, engineering, medicine, and other fields.

There are a number of systematic schemes for determining what cations and anions are present in a sample. Most of them involve separating the cations into several groups before identifying the individual ones. Some schemes separate the anions into groups, but the majority use individual tests for a limited number of anions.

The most popular schemes for the analysis of mixtures of cations use sulfide ions in precipitating two groups of ions. We can illustrate some of the essential concepts of one of these systematic sulfide outlines by describing the strategy of the stepwise separations of groups of ions. This method can be applied to a mixture of at least 25 different cations: Ag^+, Al^{3+}, As^{3+}, Ba^{2+}, Bi^{3+}, Ca^{2+}, Cd^{2+}, Co^{2+}, Cr^{3+}, Cu^{2+}, Fe^{2+}, Fe^{3+}, Hg_2^{2+}, Hg^{2+}, K^+, Mg^{2+}, Mn^{2+}, Na^+, Ni^{2+}, NH_4^+, Pb^{2+}, Sb^{3+}, Sn^{2+}, Sr^{2+}, and Zn^{2+}. A mixture containing all of these ions can be separated easily into five groups.

We will not describe how the mixture of ions within any one of the five groups is separated so that individual ions can be identified. Rather, we will limit this discussion to the principles involved in isolating the different groups. The strategy of the following scheme takes advantage of general rules of solubility, such as those presented in Table 10.2, and of differences in the solubility products of possible precipitates.

17.7.1 Group I, The Acid Chloride Group

Suppose we are told that a solution contains all 25 of the cations just listed. After referring to Table 10.2, we would conclude that the only anion of those listed that is present in appreciable concentration is the nitrate ion, NO_3^-. We could also conclude that the solution must be acidic; that is, the hydroxide ion concentration must be quite low, because most of the 25 cations form insoluble hydroxides.

To precipitate a few of the cations, we select an anion that does not form many insoluble salts. Chloride ion forms insoluble salts with only three of the 25 ions. Their K_{sp} values are listed in Table 17.1. It is convenient to add the chloride ion in the form of an HCl solution. This insures that the solution is acidic enough to prevent the precipitation of any of the hydroxides, and it avoids the addition of any cation that might interfere with a subsequent step in the analysis.

The three ions in the acid chloride group are Ag^+, Hg_2^{2+}, and Pb^{2+}. If all three are present in a solution, they precipitate as a mixture of AgCl, Hg_2Cl_2, and $PbCl_2$ when HCl is added. The net ionic equation for Hg_2^{2+} is

$$Hg_2^{2+} + 2Cl^- \longrightarrow Hg_2Cl_2\downarrow$$

The precipitation of group I ions must be at room temperature or colder, because lead chloride is too soluble in warm water. Also, care must be taken not to add too much hydrochloric acid. In concentrated HCl solutions, silver chloride and lead chloride dissolve, because Ag^+ and Pb^{2+} form soluble complexes by the reactions

$$AgCl + Cl^- \longrightarrow AgCl_2^-$$
$$PbCl_2 + 2Cl^- \longrightarrow PbCl_4^{2-}$$

17•7•2   **Group II, The Acid Sulfide Group**   After the group I precipitates are removed, the remaining solution is made less acidic and is saturated with hydrogen sulfide. The conditions in the solution are similar to those described in Section 17.5.1. There we saw that, in a saturated H_2S solution that is $1.0M$ in H^+ ions, the S^{2-} ion concentration is about $1.0 \times 10^{-22} M$.

With such a small value for $[S^{2-}]$, only salts with very small K_{sp} values form precipitates. In solutions in which the metal ions have concentrations of about $0.1M$, the sulfides of As^{3+}, Bi^{3+}, Cd^{2+}, Cu^{2+}, Hg^{2+}, Pb^{2+}, Sb^{3+}, and Sn^{2+} will precipitate. The K_{sp} values for five of these sulfides are listed in Table 17.1; the value of 8.0×10^{-27} for CdS is the largest of these very small constants. For the precipitation of arsenic sulfide, As_2S_3, we can write the following equations:

$$3H_2S \longrightarrow 6H^+ + 3S^{2-}$$
$$3S^{2-} + 2As^{3+} \longrightarrow As_2S_3\downarrow$$

The sum of these two equations gives the net ionic equation for the reaction:

$$2As^{3+} + 3H_2S \longrightarrow As_2S_3\downarrow + 6H^+$$

17•7•3   **Group III, The Basic Sulfide Group**   We see in Table 10.2 that relatively few of the commonly encountered cations do not form insoluble sulfides. After the group II sulfide precipitates are removed, the concentration of sulfide ion is increased. The solution is made basic by a buffer mixture of NH_3 and NH_4Cl, and the solution is again saturated with H_2S. In this solution the concentration of H^+ is about $1 \times 10^{-9} M$. From the equilibrium constant for $H_2S \rightleftharpoons 2H^+ + S^{2-}$ of 1.1×10^{-21}, we can calculate $[S^{2-}]$:

$$\frac{[H^+]^2[S^{2-}]}{[H_2S]} = \frac{(1 \times 10^{-9})^2[S^{2-}]}{0.10} = 1.1 \times 10^{-21}$$

$$[S^{2-}] = \frac{(1.1 \times 10^{-21})(0.10)}{1 \times 10^{-18}} = 1 \times 10^{-4} M$$

At such a high concentration of sulfide ion, the remaining insoluble metal sulfides tend to precipitate. However, the OH^- ion concentration is about $1 \times 10^{-5} M$ in this NH_3–NH_4^+ buffered solution. As we noted in discussing group I, there are many insoluble metal hydroxides. Seven ions precipitate in group III: Co^{2+}, Fe^{2+}, Mn^{2+}, Ni^{2+}, Zn^{2+}, Al^{3+}, and Cr^{3+}. The first five form insoluble sulfides and the last two form insoluble hydroxides. For the precipitation of aluminum hydroxide, $Al(OH)_3$, we can write the following equations:

$$3NH_3 + 3H_2O \longrightarrow 3NH_4^+ + 3OH^-$$
$$Al^{3+} + 3OH^- \longrightarrow Al(OH)_3\downarrow$$

The sum of these two equations gives the net ionic equation for the reaction:

$$Al^{3+} + 3NH_3 + 3H_2O \longrightarrow Al(OH)_3\downarrow + 3NH_4^+$$

The smallest K_{sp} value for any of the five group III sulfides is that for CoS, 2.0×10^{-25}. This is larger than the value of the largest K_{sp} for any group II sulfide.

Relative to the 25 ions for which the overall analytical scheme is designed, we might expect two additional ions to precipitate here as insoluble hydroxides. One is the Fe^{3+} ion, but this ion is reduced to Fe^{2+} by sulfide ion. The Fe^{2+} then combines with more S^{2-} to form the precipitate FeS. The second ion that can form an insoluble hydroxide but that does not precipitate as a hydroxide in group III is Mg^{2+}. As shown in Example 17.6(b), by controlling the OH^- ion concentration at about $1 \times 10^{-5}\,M$ with an NH_3–NH_4^+ buffer, the precipitation of $Mg(OH)_2$ can be prevented.

17•7•4 Group IV, The Carbonate Group The fourth group in the scheme is precipitated by adding CO_3^{2-} ion after the group III precipitates have been removed and the pH of the solution has been adjusted. Referring to Table 10.2, we see that practically all metal ions form insoluble carbonates. But after the ions in groups I, II, and III have been precipitated and removed, there are only seven of the original 25 ions left. These are Ba^{2+}, Ca^{2+}, K^+, Mg^{2+}, Na^+, NH_4^+, and Sr^{2+}.

Of the seven ions that remain, Ba^{2+}, Ca^{2+}, and Sr^{2+} form carbonates with the smallest K_{sp} values. These three constitute the group IV cations. The precipitation reaction for the strontium ion is

$$Sr^{2+} + CO_3^{2-} \longrightarrow SrCO_3\downarrow$$

The K_{sp} for $MgCO_3$ is 3.5×10^{-8}, which is slightly too large for this salt to precipitate at the CO_3^{2-} ion concentration that is provided.

17•7•5 Group V, The Soluble Group The four ions that remain in solution after the first four groups are precipitated constitute the group V cations. These are NH_4^+, Mg^{2+}, K^+, and Na^+. There is just one complication here. Because aqueous ammonia and ammonium chloride are added at several points in the analysis of groups II, III, and IV, the presence of NH_4^+ in an unknown must be tested for in a sample of the original solution.

• *Overview* • We have described how a mixture of many cations can be separated into five groups. Perhaps the most important concept relative to solubility equilibria is that there are two groups of sulfides precipitated. Sulfides with very small K_{sp} values are separated first with a very small concentration of S^{2-} ions in solution. Then another group of sulfides, which have larger K_{sp} values, are separated by using a much larger concentration of S^{2-} ions.

The identification of individual ions in each of the five groups is accomplished by separating each ion from the other members of its group and causing each to form a specific compound that has some unique characteristics.

Chapter Review

Summary

Many inorganic compounds are virtually insoluble in water, yet they all have a finite solubility, however small it may be. The **solubility product constant,** K_{sp}, is a measure of the solubility of such slightly soluble compounds. For any two salts with the same stoichiometric ratio of cations to anions, the one with the smaller solubility product constant is the less soluble in pure water. The relative solubilities of compounds can be greatly altered through the common ion effect, however. This effect is used to practical advantage in precipitating various ions from solution as insoluble salts, often to the point at which only a negligible concentration of an ion remains in solution.

Two dissolved cations that form similar precipitates with sufficiently different solubilities can be separated by **selective precipitation,** or **fractional precipitation.** This is done through careful selection of the requisite anion concentration, which can often be controlled by applying the common ion effect. Analogously, two similar precipitates with sufficiently different solubilities in a solution of a given pH range can be separated by **selective dissolution.** This is done by adjusting the pH appropriately, which results in the formation of a weak electrolyte involving the anion of the precipitate. Many precipitates can also be selectively dissolved through the formation of a **complex ion.** The **ligands** that surround the central metal cation are ions or molecules that behave like Lewis bases. The overall equilibrium constant for the addition of a series of identical ligands to the cation is called the **stability constant,** or **formation constant,** K_f, of the complex ion.

All the principles mentioned above are vital to the schemes of **qualitative analysis** by which the components of solutions and mixtures of inorganic compounds are separated and identified. One such scheme that is widely used divides the most common cations into five major groups: the acid chloride group, the acid sulfide group, the basic sulfide group, the carbonate group, and the soluble group.

Exercises

K_{sp} Expressions and Values

1. Write the expression for the solubility product constant, K_{sp}, for each of the compounds $SrSO_4$, $Fe(OH)_2$, and Sb_2S_3.

2. The solubilities of magnesium oxalate, MgC_2O_4, and magnesium fluoride, MgF_2, at 25 °C are 1.0 and 0.168 g/L, respectively. Calculate the solubility product constant for each compound. (The oxalate ion, $C_2O_4^{2-}$, is derived from oxalic acid, $H_2C_2O_4$.)

Calculation of Solubility from K_{sp}

3. Calculate in grams per 100 mL the solubilities at 25 °C of NiS and $Bi(OH)_3$ from the data in Table 17.1.

4. The values of K_{sp} for mercury(II) sulfide, HgS, and bismuth sulfide, Bi_2S_3, are 1.6×10^{-52} and 1×10^{-97}, respectively. Which sulfide has the greater solubility in water in moles per liter?

Effect of Common Ion on Solubility

5. (a) Calculate the solubility of calcium sulfate, $CaSO_4$, in water. Express the solubility in moles per liter and in grams per liter (see Table 17.1).
   (b) Calculate the solubility of calcium sulfate in a solution that is $0.050M$ in Na_2SO_4. Express the solubility in moles per liter and compare this solubility with that found in part (a). Account for any difference.

6. The solubility product constant, K_{sp}, of silver chromate, Ag_2CrO_4, is 1.1×10^{-12} at 25 °C.
   (a) Calculate its water solubility in grams per liter.
   (b) Calculate its solubility in grams per liter of solution that is $0.10M$ in sodium chromate, Na_2CrO_4.
   (c) Calculate its solubility in grams per liter of solution that is $0.10M$ in silver nitrate, $AgNO_3$.

7. The solubility of lead carbonate, $PbCO_3$, in $0.050M\ Na_2CO_3$ is 4.0×10^{-10} g/L. Calculate the K_{sp} for lead carbonate.

8. The solubility of silver carbonate, Ag_2CO_3, in $0.050M\ Na_2CO_3$ is 1.8×10^{-3} g/L. Calculate the solubility product constant for silver carbonate.

Precipitation of Insoluble Salts

9. To 10 mL of a $0.020M$ solution of $Ca(NO_3)_2$ is added 10 mL of $0.050M$ sodium chromate, Na_2CrO_4. Will calcium chromate precipitate? Explain (see Table 17.1).

10. To 100 mL of $0.020M\ BaCl_2$ is added 100 mL of $0.020M\ Na_2SO_4$. (Assume the volumes are additive.)
    (a) How many moles of barium sulfate, $BaSO_4$, will precipitate?
    (b) How many moles of Ba^{2+} and SO_4^{2-} ions will be left in solution? (The K_{sp} of $BaSO_4$ is 1.1×10^{-10}.)

11. To 100 mL of $0.020M\ BaCl_2$ is added 100 mL of $0.20M\ Na_2SO_4$. (Assume the volumes are additive.)
    (a) How many moles of $BaSO_4$ will precipitate?
    (b) How many moles of Ba^{2+} and SO_4^{2-} ions will be left in solution?

12. To 100 mL of $0.20M\ BaCl_2$ is added 100 mL of $0.020M\ Na_2SO_4$. (Assume the volumes are additive.)

(a) How many moles of $BaSO_4$ will precipitate?

(b) How many moles of Ba^{2+} and SO_4^{2-} ions will be left in solution?

*13. To 500 mL of $0.0010M$ barium chloride, $BaCl_2$, is added 500 mL of $0.050M$ sodium phosphate, Na_3PO_4.

(a) Show with appropriate calculations that a precipitate of barium phosphate, $Ba_3(PO_4)_2$, forms. (See Example 17.4.)

(b) When equilibrium is reached, will more than, less than, or exactly 99.9 percent of the barium ions, Ba^{2+}, have been precipitated?

(c) Repeat part (b) relative to 99.95 percent precipitation of barium ions. (Note that parts (b) and (c) are different versions of Example 17.4 and require a different method of solution.)

(d) Do your answers to parts (b) and (c) indicate that the approximation in Example 17.4 is valid? Why?

14. To a solution of sodium hydroxide is added iron(III) chloride until the Fe^{3+} ion concentration is $0.050M$. What is the molar concentration of OH^- ions that can remain in solution?

15. To a 100-mL solution of a calcium compound (pH about 7) known to contain 0.010 mole of Ca^{2+} ions is added 100 mL of $0.500M$ $(NH_4)_2CO_3$.

(a) What weight of $CaCO_3$ precipitates?

(b) What is the molar concentration of calcium ions left in the solution?

16. To 50.0 mL of $0.010M$ sodium phosphate, Na_3PO_4, is added 50.0 mL of $0.010M$ silver nitrate, $AgNO_3$. The K_{sp} of silver phosphate, Ag_3PO_4, is 1.4×10^{-16}.

(a) Will Ag_3PO_4 precipitate? If so, what weight of precipitate is formed?

(b) What is the concentration of Ag^+ ions in solution at equilibrium?

(c) What percentage of the Ag^+ ions remains in solution?

*17. The K_{sp} of silver carbonate, Ag_2CO_3, is 8.1×10^{-12}.

(a) Calculate $[H^+]$ and $[CO_3^{2-}]$ in a $0.10M$ H_2CO_3 solution.

(b) Would a solution maintained at $0.10M$ in H_2CO_3 produce sufficient CO_3^{2-} to precipitate Ag_2CO_3 if the solution is $0.10M$ in $AgNO_3$? Prove by calculations.

Selective Precipitation

*18. Repeat part (b) of Exercise 17 for a $0.10M$ solution of Hg_2^{2+} ions. (The K_{sp} of Hg_2CO_3 is 8.9×10^{-17}.) Do your calculations in this exercise and Exercise 17 suggest a possible way to separate the metal cations in a solution of Ag^+ and Hg_2^{2+}?

*19. The values of K_{sp} for silver chromate, Ag_2CrO_4, and mercury(I) chromate, Hg_2CrO_4, are 1.1×10^{-12} and 2.0×10^{-9}, respectively. Show by calculations whether $0.10M$ H_2CrO_4 can successfully separate the metal ions

from a solution that is $0.10M$ in Ag^+ and $0.10M$ in Hg_2^{2+} ions? (The K_{a1} for H_2CrO_4 is large; K_{a2} is 3.2×10^{-7}.)

The Common Ion Effect
in Selective Precipitation

*20. Repeat Exercise 19, but now have the $0.10M$ H_2CrO_4 solution $2.0M$ in H^+ due to the addition of HNO_3. If one ion forms a precipitate under these conditions, calculate the molarity of that ion left in solution.

21. A solution contains the following cations in $0.010M$ concentrations: Zn^{2+}, Ni^{2+}, Hg^{2+}, Cd^{2+}, Fe^{2+}, Co^{2+}, Cu^{2+}, Pb^{2+}, and Mn^{2+}. The solution is kept saturated with H_2S at an H^+ concentration that maintains the S^{2-} concentration at 2.0×10^{-23} M. Make an ion product calculation that enables you to select the cations that form sulfide precipitates by comparison with K_{sp} values in Table 17.1. List the cations that do form precipitates.

22. Suggest a way of separating the ions in a solution that is $0.10M$ in Ni^{2+} and $0.10M$ in Co^{2+} ions by use of $0.10M$ H_2CO_3. (The values of K_{sp} for $NiCO_3$ and $CoCO_3$ are 6.6×10^{-9} and 1.4×10^{-13}, respectively.)

23. The values of K_{sp} for manganese sulfide, MnS, and zinc sulfide, ZnS, are 2.5×10^{-13} and 1.6×10^{-23}, respectively.

(a) Which has the greater water solubility in grams per liter, MnS or ZnS?

(b) Can a saturated solution of H_2S be used to separate Mn^{2+} from Zn^{2+} ions in a solution that is $1.0M$ in both ions? If not, suggest a way by which they may be separated.

*24. To 100 mL of $0.20M$ magnesium nitrate, $Mg(NO_3)_2$, at a pH of 7.00, is added 100 mL of $1.0M$ NH_3.

(a) With a K_{sp} of 1.8×10^{-11}, almost all the $Mg(OH)_2$ precipitates. How many moles precipitate?

(b) What will be the pH of the solution after the $Mg(OH)_2$ has precipitated?

25. The values of K_{sp} for iron(II) sulfide, FeS, and zinc sulfide, ZnS, are 6.3×10^{-18} and 1.6×10^{-23}, respectively. Suggest a method by which the ions may be separated in a solution that is $0.10M$ in Fe^{2+} and $0.10M$ in Zn^{2+} ions.

Dissolution of Precipitates
with Formation of a Weak Electrolyte

* 26. A mixture of 0.10 mole of $Mg(OH)_2$ and 0.10 mole of $Mn(OH)_2$ is added to water maintained at a pH of 5.00 by the addition of small amounts of HCl. Assuming a volume of 1.0 L, what amounts of the hydroxides will dissolve? (The values of K_{sp} for $Mg(OH)_2$ and $Mn(OH)_2$ are 1.8×10^{-11} and 1.9×10^{-13}, respectively.)

27. Repeat Exercise 26, but use 0.10 mole of $Al(OH)_3$ in place of the $Mg(OH)_2$. (The K_{sp} of $Al(OH)_3$ is 1.3×10^{-33}.)

28. The K_{sp} of aluminum hydroxide is 1.3×10^{-33}.
    (a) What is the molarity of OH^- ions in a saturated solution of aluminum hydroxide?
    (b) Will a solution maintained at a pH of 7.00 precipitate aluminum hydroxide from a solution that is $0.10M$ in Al^{3+} ions?
    (c) Aluminum nitrate, $Al(NO_3)_3$, 0.10 mole, is added to water to give 1 L of solution. Should any precipitate of aluminum hydroxide form? Explain.

29. Which of the following sulfides should dissolve in $1.0M$ HCl: FeS, MnS, ZnS, CoS, NiS? Explain by showing the necessary calculations. [Assume that where any solution (reaction) takes place, the HCl concentration is maintained at $1.0M$.]

30. Explain why 1 mole of copper(II) carbonate, $CuCO_3$, will dissolve in 1 L of $2.0M$ HCl, but 1 mole of copper(II) sulfide, CuS, will not. (The values of K_{sp} for $CuCO_3$ and CuS are 1.4×10^{-10} and 6.3×10^{-36}, respectively.)

Use of Complex Ions in Analytical Chemistry

31. When $6.0M$ ammonia is added to a solution of aluminum nitrate, a precipitate of aluminum hydroxide $Al(OH)_3$, is produced. Although reaction of $6.0M$ sodium hydroxide with the aluminum nitrate solution initially gives a precipitate, a slight excess of sodium hydroxide causes the precipitate to dissolve. Explain.

32. Write net ionic equations for each of the following reactions. In each case, assume the reactant is in water solution and indicate all ligands of the complex ions.
    (a) copper(II) chloride + sodium cyanide (excess)
    (b) zinc nitrate + potassium hydroxide (excess)
    (c) cadmium sulfate + ammonia (excess)

*33. Imagine you are given a liter of solution that is $0.020M$ in Cl^-, $0.020M$ in Br^-, and $0.020M$ in I^- ions. Indicate qualitatively how you would separate the ions by each of the following methods. (The values of K_{sp} for AgCl, AgBr, and AgI are 1.8×10^{-10}, 5.0×10^{-13}, and 8.3×10^{-17}, respectively.)
    (a) selective dissolution of the silver salts
    (b) selective precipitation of the silver salts

34. Refer to Exercise 33. Show by calculation how you would separate AgBr from AgI by selective dissolution. If you wanted to convert the ions in the filtrate back to

the silver halide precipitate, what reaction would you use?

*35. The K_{sp} of zinc hydroxide, $Zn(OH)_2$, is 1.2×10^{-17}, and the K_f of $Zn(OH)_4{}^{2-}$ is 4.6×10^{17}. Calculate the concentrations of Zn^{2+} and $Zn(OH)_4{}^{2-}$ present in a solution that is initially $0.010M$ in zinc nitrate, $Zn(NO_3)_2$, at a pH maintained at 13.0. At a pH maintained at 14.0.

*36. How many moles of NH_3 per liter must be added to dissolve completely 0.010 mole of $Cu(OH)_2$ and produce $Cu(NH_3)_4{}^{2+}$? (The K_{sp} of $Cu(OH)_2$ is 2.2×10^{-20}, and the K_f of $Cu(NH_3)_4{}^{2+}$ is 2.1×10^{13}.)

37. (a) Given that $K_f = 1.1 \times 10^7$ for $Ag(NH_3)_2{}^+$, calculate the concentrations of Ag^+ and $Ag(NH_3)_2{}^+$ when a $0.010M$ solution of $AgNO_3$ is made $1.0M$ in NH_3.
    (b) Will AgCl precipitate if this solution is made $0.10M$ with respect to Cl^- ions? (The K_{sp} for AgCl is 1.8×10^{-10}.)

38. Write the four simplified equations, (1), (2), (3), and (4), in Section 17.6.2 in a more complete form to show how water molecules are involved.

39. Using the stability constants for the four stepwise ammoniation reactions for the hydrated copper ion, check the value of the overall stability constant given in Table 17.2 for $Cu(NH_3)_4{}^{2+}$.

Qualitative Analysis Scheme

40. In precipitating the group III cations, we might expect Mn^{2+} to form either $Mn(OH)_2$ or MnS. Assuming $[S^{2-}]$ is $1.0 \times 10^{-4}\ M$ and $[OH^-]$ is $1.0 \times 10^{-5}\ M$, calculate the Mn^{2+} concentration that would remain in solution if either precipitate formed by itself. State which precipitate you think will form, and justify your choice.

41. Suggest a method that you might use to separate the ions in each of the following solutions that are $0.10M$ with respect to each ion:
    (a) Ag^+ and Cu^{2+}
    (b) Cu^{2+} and Mn^{2+}
    (c) Al^{3+} and Ca^{2+}
    (d) Pb^{2+} and Cu^{2+}
    (e) Ca^{2+} and $NH_4{}^+$
    (f) Cu^{2+} and Bi^{3+}
    (g) Al^{3+} and Zn^{2+}
    (h) Cl^- and $SO_4{}^{2-}$

42. Consider the general case of a solution containing a cation M^{x+} from cation group II and a cation N^{y+} from group III. Describe in words and with chemical equations how we can precipitate M^{x+} as a sulfide, remove that precipitate, and then precipitate N^{y+} as a sulfide.

Ionic Equilibria III: Redox and Electrochemistry

18

Spontaneous Electrochemical Reactions

18•1 **Voltaic Cells**

18•1•1 Designation of Anode and Cathode

18•1•2 Gaseous Electrodes

18•2 **Standard Electrode Potentials**

18•2•1 Definitions of Units

18•2•2 Comparison of Electrodes

18•2•3 Assignment of Standard Reduction Potentials, $\mathscr{E}^{\circ}_{\text{red}}$

18•3 **Potentials of Voltaic Cells**

18•3•1 Representation of a Voltaic Cell

18•3•2 Cell Reaction and Cell Voltage

18•3•3 Spontaneity of Reaction

18•3•4 Effect of Concentration on Potential

18•4 **Importance of Standard Electrode Potentials**

18•5 **Equilibrium Constants and Free Energies from Electrode Potentials**

18•5•1 Equilibrium Constants

Special Topic: The Glass Electrode

18•5•2 Free Energies

Redox Reactions in Solution

18•6 **Balancing Redox Equations**

18•6•1 Oxidation-Number Method

18•6•2 Ion-Electron Method

18•7 **Driving Force and Equilibrium Constants**

Applications of Electrochemistry

18•8 **Electrolysis**

18•8•1 Faraday's Law

18•8•2 Decomposition Potentials

18•8•3 Electrode Products

18•8•4 Electroplating

18•9 **Batteries and Fuel Cells**

18•9•1 Small-Scale Power Sources

18•9•2 Lead Storage Cell

18•9•3 Sodium–Sulfur Cell

18•9•4 Fuel Cells

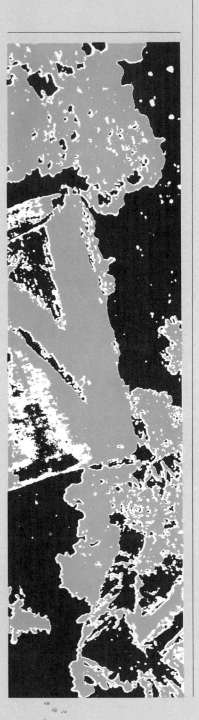

In the two preceding chapters, we looked closely at two of the major classes of reactions of ions in aqueous solution, acid–base reactions and precipitation reactions. In this chapter we will take up a third major class, oxidation–reduction reactions. All three types of reactions are important in solvents other than water, and even in the gaseous and solid phases. However, because reactions in aqueous solution are of the greatest importance, we will continue to devote almost all our attention to those systems.

Oxidation–reduction, or redox, reactions involve changes in the oxidation states of the reactants. In most simple examples there is an actual loss of electrons by one reactant and a corresponding gain by another. When the flow of electrons accompanying a reaction constitutes a current of electricity, the chemical change is referred to as **electrochemistry.**

To begin our study of redox equilibria, we will examine reactions that can be used to produce electric current. We will learn how the driving force for such a reaction can be measured and how it can be used to predict the chemical situation at equilibrium.

The principles of spontaneity and equilibria that apply to electrochemical reactions apply also to other redox reactions in solution. We will look at a number of typical reactions and learn how to balance redox equations that are difficult to balance by the ordinary trial-and-error method. Finally, some important applications of electrochemistry will be described.

Spontaneous Electrochemical Reactions

Electrochemical reactions can be subdivided into two classes: those that produce a current of electricity (the process occurring in a battery) and those that are produced by a current of electricity (electrolysis). The first type of reaction is spontaneous, and the free energy of the chemical system decreases; the system can do work, for example, run a motor. The second type must be forced to occur (by work done on the chemical system), and the free energy of the chemical system increases.

18•1   Voltaic Cells

How can we measure the driving force of an oxidation–reduction reaction, and how can we compare it with the driving force of another reaction? One direct way to make such determinations is to measure the potential (voltage) of a battery or voltaic cell.[1] A **voltaic cell** is an arrangement of chemicals and conductors that provides for the flow of electrons through an external circuit from a chemical that is being oxidized to a chemical that is being reduced.

In voltaic cells, *oxidation involves the loss of electrons by atoms, molecules, or ions;* and *reduction involves the gain of electrons by these particles.*

---

[1]The word *battery* was originally used to designate a series of voltaic cells, but it is now popularly used to denote any voltaic source of current, either a single dry-cell "battery" or the several cells of an automobile "battery."

As a simple example of spontaneous oxidation and reduction, let us consider the reaction that occurs when a strip of zinc is immersed in a solution of copper sulfate (Figure 18.1). There is a spontaneous reaction; metallic copper plates out on the zinc strip, the zinc strip is gradually dissolved, and heat energy is liberated. The overall reaction is

$$Zn + CuSO_4 \longrightarrow ZnSO_4 + Cu$$

Actually, the reaction occurs between zinc atoms and copper ions, as shown by the following net ionic equation:

$$Zn + Cu^{2+} \longrightarrow Zn^{2+} + Cu$$

Each atom of zinc loses two electrons to become a zinc ion, and each ion of copper gains two electrons to become a copper atom:

oxidation: $Zn \longrightarrow Zn^{2+} + 2e^-$

reduction: $Cu^{2+} + 2e^- \longrightarrow Cu$

Although this phenomenon is electric in nature, no flow of electrons can be detected when the zinc is in direct contact with the copper sulfate solution. The electrons are transferred directly from the atoms of zinc to the ions of copper.

When the zinc strip is not in direct contact with the copper sulfate solution, it is still possible to have a reaction between Zn and Cu^{2+} by using the arrangement shown in Figure 18.2. This provides for (1) the flow of electrons through a conductor from zinc atoms to copper ions and (2) diffusion of the positive and negative ions so that the solution, even in the vicinity of the metal surfaces, remains neutral. One of the characteristics of ionic solutions is that the number of positive charges must equal the number of negative charges in every portion of the solution. The reactions depicted in

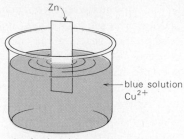

zinc is placed in copper sulfate solution

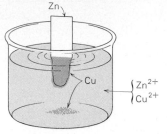

after short time

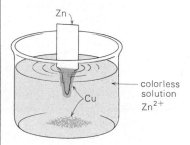

after reaction has gone to completion (zinc in excess)

Figure 18•1 When a piece of elemental zinc is in contact with a solution of copper sulfate, the zinc atoms change to zinc ions and the copper ions change to copper atoms.

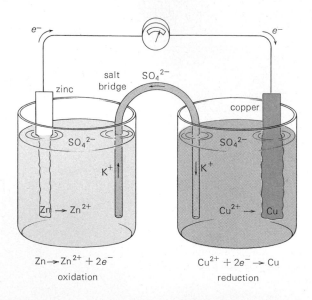

$Zn \longrightarrow Zn^{2+} + 2e^-$
oxidation

$Cu^{2+} + 2e^- \longrightarrow Cu$
reduction

Figure 18•2 Diagram of one type of a voltaic cell, the Daniell cell.

Figure 18.2 could not occur if the solution around the zinc electrode became crowded with positive ions and the solution around the copper electrode became depleted of positive ions.

There are several methods of providing for the diffusion of the ions. A common laboratory method is to immerse the zinc strip in a solution of a zinc salt, such as zinc sulfate, and to immerse a piece of copper in a solution of copper sulfate. The zinc sulfate solution is connected to the copper sulfate solution by a *salt bridge,* which provides for the diffusion of the ions. The salt bridge is filled with a solution of an electrolyte that does not change chemically in the process. Potassium sulfate ($2K^+$, SO_4^{2-}), serves nicely for this purpose. Such salts as NaCl, KCl, and KNO_3 are also satisfactory.

When the reaction

$$Zn + CuSO_4 \longrightarrow ZnSO_4 + Cu$$

proceeds as indicated in Figure 18.2, the action continues until one of the reactants, either the zinc atoms or the copper ions, is consumed so completely that the voltage falls to zero. Valence electrons flow from the zinc atoms into the conducting wire, and as the Zn^{2+} ions form, they enter the solution and diffuse away from the zinc strip:

$$Zn \longrightarrow Zn^{2+} + 2e^-$$

Negative ions diffuse through the salt bridge toward the zinc electrode. In time, it can be seen that the zinc strip is disappearing.

The electrons given up by the zinc atoms enter the connecting wire and cause electrons at the other end of the wire to collect on the surface of the copper electrode. These electrons react with copper ions to form copper atoms that adhere to the electrode as a copper plate:

$$Cu^{2+} + 2e^- \longrightarrow Cu$$

In time, the metallic copper electrode increases in size, and the blue color of the copper sulfate solution fades because of the decreasing concentration of the copper ions.

The SO_4^{2-} ions that are left behind by the copper ions diffuse away from the vicinity of the copper electrode. From the salt bridge, K^+ ions diffuse out toward the copper. Thus, while the reaction is in progress, there is an overall movement of negative ions toward the zinc electrode and an overall movement of positive ions toward the copper electrode. The pathway for this directional flow of ions through the solution can be thought of as the *internal circuit,* and the pathway for the flow of electrons through the conducting wire can be thought of as the *external circuit.*

A battery in which zinc, zinc sulfate, copper, and copper sulfate are used is known as a *Daniell cell,* after its inventor. Because the flow of electrons through the external circuit constitutes an electric current, the reaction between the zinc and copper sulfate can be harnessed to produce electric power. The energy that is given off as heat energy when these two react by direct contact (see Figure 18.1) can be evolved mostly as electric energy by the voltaic cell (see Figure 18.2).

18•1•1   Designation of Anode and Cathode We will designate which electrode is the cathode or anode on the basis of the type of chemical reaction taking place on the surface. *The electrode at which the oxidation reaction takes place is called the anode and the one at which the reduction reaction takes place is called the cathode.* In the Daniell cell in Figure 18.2, the zinc electrode is the anode and the copper electrode is the cathode.

18•1•2   Gaseous Electrodes An electrode easily can be made of one of the common metals; all that is necessary is to immerse a strip of the metal in a solution of its ions. However, building an electrode involving one of the gaseous elements and its ions presents more of a problem. Obviously, we cannot take a "piece" of gas, insert it in a solution of its ions, and connect a wire to it, thus making it part of a voltaic cell. Yet methods have been worked out that, in principle, accomplish just this.

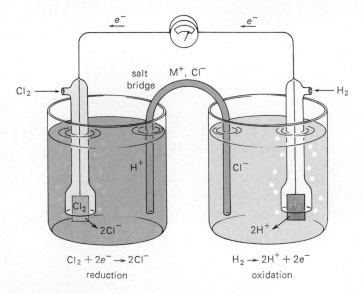

Figure 18•3 A voltaic cell that depends on the reactions of two gaseous elements.

Elemental hydrogen, maintained in contact with a solution of its ions, can be made one electrode of a voltaic cell by the method shown in Figure 18.3. The **hydrogen electrode** consists of a glass tube through which hydrogen gas can be continuously passed over a platinum foil (right-hand electrode in Figure 18.3). Coming down through the tube is a platinum wire that is connected to the platinum foil. The foil itself is covered with finely divided platinum to provide a large surface area. Platinum adsorbs hydrogen; in effect the electrode is a core of platinum with an adsorbed film of hydrogen exposed to a solution that contains H^+ ions. When hydrogen is removed by the withdrawal of current from the cell, the film of it is replenished from the hydrogen that passes through the glass tube. The platinum, being much less active than hydrogen, does not lose or gain electrons. The chlorine electrode, the other electrode in Figure 18.3, is constructed in a similar manner.

In the voltaic cell shown in Figure 18.3, the overall cell reaction is

$$H_2 + Cl_2 \longrightarrow 2HCl$$

Hydrogen molecules give up electrons to form hydrogen ions, and chlorine molecules take up electrons to form chloride ions. Individual electrode reactions can be written:

at the anode: $H_2 \longrightarrow 2H^+ + 2e^-$ (oxidation)

at the cathode: $Cl_2 + 2e^- \longrightarrow 2Cl^-$ (reduction)

18•2 Standard Electrode Potentials

18•2•1 Definitions of Units Because a current of electricity is the flow of electrons through a conductor (an electrolyte), a unit quantity of electricity is the number of electrons that pass through the circuit. The standard unit quantity, a derived SI unit, is the **coulomb,** C. One of the seven SI base units, the ampere, A, is a measure of the rate of flow of electrons. One **ampere** is the flow of one coulomb per second through a conductor:

$$A = C \cdot s^{-1} \quad \text{or} \quad C = A \cdot s$$

The unit of electric potential or voltage is the volt. The **volt,** V, is defined as the potential necessary to produce one joule of electric energy per second at a current flow of one ampere:

$$V = J \cdot A^{-1} \cdot s^{-1}$$

Because the coulomb has dimensions of amperes times seconds, we see that

$$V = J \cdot C^{-1}$$

18•2•2 Comparison of Electrodes We are not mainly interested in the flow of current or the production of energy by cells. Instead, our interest is in studying and understanding the forces that cause chemical reactions to occur. One way to do this is to determine electrode potentials at carefully specified concentrations and temperatures. The standard conditions for measuring potentials are to have electrodes in contact with solutions in which ions are $1m$ and gas pressures are maintained at 1 atm. Such electrodes are called **standard electrodes.**[2] The most common temperature for these measurements is 25 °C (298 K).

The potential of a voltaic cell is a measurement of the driving force of the redox reaction. In the most precise measurements, no current actually flows. The flow of electrons would cause reactions to take place at the electrodes and would change slightly the concentrations of the standard solutions.

As illustrated by the examples in Figures 18.2 and 18.3, a voltaic cell consists of two half-cells. It is not possible to measure the potential of a

---

[2]The standard ionic concentration is not precisely 1 mole per 1,000 g of solvent but is a concentration that behaves as an ideal $1m$ solution should behave; such a concentration is said to have an *activity* of unity.

single half-cell but only the potential between a pair of them. If we wish to compare the potential of one half-cell with another, we must measure the potential of each against a third comparison half-cell.

Because of the importance of the hydrogen ion in aqueous solutions, chemists selected the standard hydrogen electrode as the **standard comparison electrode** and arbitrarily assigned to it the potential of zero voltage. Consider the hypothetical voltaic cell shown in Figure 18.4. The voltage for this cell is taken as a measurement of the tendency of the substance M half-cell to undergo oxidation or reduction, as compared with the tendency of the hydrogen–hydrogen ion half-cell.

Ideally, standard electrode potentials would be measured directly, as pictured in Figure 18.4. In practice, however, a hydrogen electrode is so difficult to handle experimentally that other reference electrodes generally are used. Also, in actual laboratory measurements, two standard electrodes are not compared directly. The ionic concentrations cannot be adjusted independently, and $1m$ solutions are so concentrated that ion–ion interactions interfere with ideal independent ionic behavior. However, the potentials between two ideal standard electrodes can be calculated from measurements made on more dilute solutions. To simplify the discussion in the remainder of this section, the hypothetical behavior of ideal standard electrodes will be described in order to make clear the relationships between electrode potentials.

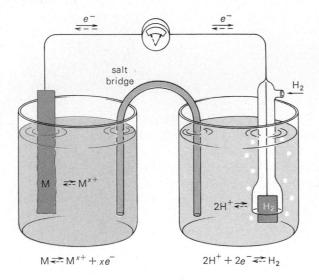

$$M \rightleftharpoons M^{x+} + xe^-$$ $$2H^+ + 2e^- \rightleftharpoons H_2$$

Figure 18.4 A hypothetical voltaic cell for determining electrode potentials. The hydrogen electrode is the comparison electrode. M stands for any substance and is shown here in contact with a solution of its positive ions. Some substances, though, form negative ions as shown in Figure 18.3. For those cases where M loses electrons more readily than H_2 does, the solid arrows show the direction of movement and the direction of the electrode reactions.

In the ideal comparison cell, the hydrogen electrode is always half of the cell, and the M electrode, the standard electrode of the substance being compared, is the other half (see Figure 18.4). If the M electrode were a standard copper electrode, the ideal voltage, as measured by the voltmeter, would be 0.34 V.

$$\text{at the anode:} \quad H_2 \longrightarrow 2H^+ + 2e^- \quad \text{(oxidation)}$$
$$\text{at the cathode:} \quad Cu^{2+} + 2e^- \longrightarrow Cu \quad \text{(reduction)}$$
$$\text{cell reaction:} \quad H_2 + Cu^{2+} \longrightarrow 2H^+ + Cu$$

If the M electrode were a standard silver electrode, the ideal voltage of the cell would be 0.80 V. The fact that this voltage is higher than the hydrogen–copper cell voltage indicates that silver ions have a greater tendency to undergo reduction (gain electrons) than copper ions do.

With an M electrode of magnesium, the deflection of the voltmeter is in the opposite direction; the ideal reading is 2.37 V. This opposite deflection means that magnesium atoms rather than hydrogen atoms are giving up electrons; that is, magnesium is oxidized.

$$\text{at the anode:} \quad Mg \longrightarrow Mg^{2+} + 2e^- \quad \text{(oxidation)}$$

$$\text{at the cathode:} \quad 2H^+ + 2e^- \longrightarrow H_2 \quad \text{(reduction)}$$

$$\text{cell reaction:} \quad Mg + 2H^+ \longrightarrow Mg^{2+} + H_2$$

If the M electrode is nickel, the deflection of the voltmeter is in the same direction as that obtained with a magnesium electrode; the ideal reading is 0.25 V. This smaller voltage indicates that nickel has less tendency than magnesium to give up electrons to hydrogen ions.

By comparing the voltage readings and the direction of current flow in the four cells described above, we can list the five elements—copper, silver, magnesium, nickel, and hydrogen—in a double column, with the most easily reduced ion at the bottom of one column and the most easily oxidized element at the top of the other column:

$$
\begin{array}{c c c c}
& Mg^{2+} & Mg \uparrow & \\
\text{increasing} & Ni^{2+} & Ni & \text{increasing} \\
\text{ease of} & H^+ & H_2 & \text{ease of} \\
\text{reduction} & Cu^{2+} & Cu & \text{oxidation} \\
\downarrow & Ag^+ & Ag &
\end{array}
$$

We can arrange the overall reactions that occur spontaneously in the comparison cells as follows:

$$Mg + 2H^+ \longrightarrow Mg^{2+} + H_2 \qquad \text{(oxidation of Mg, +2.37 V)}$$

$$Ni + 2H^+ \longrightarrow Ni^{2+} + H_2 \qquad \text{(oxidation of Ni, +0.25V)}$$

$$H_2 + Cu^{2+} \longrightarrow 2H^+ + Cu \qquad \text{(reduction of } Cu^{2+}\text{, +0.34 V)}$$

$$H_2 + 2Ag^+ \longrightarrow 2H^+ + 2Ag \qquad \text{(reduction of } Ag^+\text{, +0.80 V)}$$

Study of these equations indicates how metals or their ions react spontaneously:

1. Metals above hydrogen in the list undergo oxidation in the comparison cell. (The M electrode is the anode.)

2. Metal ions below hydrogen ion in the list undergo reduction in the comparison cell. (The M electrode is the cathode.)

18•2•3 **Assignment of Standard Reduction Potentials,** $\mathscr{E}^\circ_{red}$ Data such as the foregoing for Mg, Ni, Cu, and Ag have been determined for practically all of the elements and for many compounds. It is convenient to tabulate the results in an abbreviated form for each entry, as shown for some typical

Table 18•1 Standard reduction potentials, $\mathscr{E}^\circ_{red}$[a]

Couple[b] (ox/red)	Cathode reaction (reduction)	Reduction potential, volts (standard hydrogen electrode = 0)[c]
Li^+/Li	$Li^+ + e^- \rightarrow Li$	-3.04
K^+/K	$K^+ + e^- \rightarrow K$	-2.92
Ca^{2+}/Ca	$Ca^{2+} + 2e^- \rightarrow Ca$	-2.87
Na^+/Na	$Na^+ + e^- \rightarrow Na$	-2.71
Mg^{2+}/Mg	$Mg^{2+} + 2e^- \rightarrow Mg$	-2.37
Al^{3+}/Al	$Al^{3+} + 3e^- \rightarrow Al$	-1.66
Zn^{2+}/Zn	$Zn^{2+} + 2e^- \rightarrow Zn$	-0.76
Fe^{2+}/Fe	$Fe^{2+} + 2e^- \rightarrow Fe$	-0.44
$PbSO_4/Pb$	$PbSO_4 + 2e^- \rightarrow Pb + SO_4^{2-}$	-0.36
Co^{2+}/Co	$Co^{2+} + 2e^- \rightarrow Co$	-0.28
Ni^{2+}/Ni	$Ni^{2+} + 2e^- \rightarrow Ni$	-0.25
Sn^{2+}/Sn	$Sn^{2+} + 2e^- \rightarrow Sn$	-0.14
Pb^{2+}/Pb	$Pb^{2+} + 2e^- \rightarrow Pb$	-0.13
D^+/D_2	$2D^+ + 2e^- \rightarrow D_2$	-0.003
H^+/H_2	$2H^+ + 2e^- \rightarrow H_2$	0.000
Sn^{4+}/Sn^{2+}	$Sn^{4+} + 2e^- \rightarrow Sn^{2+}$	$+0.15$
Cu^{2+}/Cu	$Cu^{2+} + 2e^- \rightarrow Cu$	$+0.34$
I_2/I^-	$I_2 + 2e^- \rightarrow 2I^-$	$+0.54$
O_2/H_2O_2	$O_2 + 2H^+ + 2e^- \rightarrow H_2O_2$	$+0.68$
Fe^{3+}/Fe^{2+}	$Fe^{3+} + e^- \rightarrow Fe^{2+}$	$+0.77$
Hg_2^{2+}/Hg	$Hg_2^{2+} + 2e^- \rightarrow 2Hg$	$+0.79$
Ag^+/Ag	$Ag^+ + e^- \rightarrow Ag$	$+0.80$
NO_3^-/N_2O_4	$2NO_3^- + 4H^+ + 2e^- \rightarrow N_2O_4 + 2H_2O$	$+0.80$
NO_3^-/NO	$NO_3^- + 4H^+ + 3e^- \rightarrow NO + 2H_2O$	$+0.96$
Br_2/Br^-	$Br_2 + 2e^- \rightarrow 2Br^-$	$+1.07$
O_2/H_2O	$O_2 + 4H^+ + 4e^- \rightarrow 2H_2O$	$+1.23$
$Cr_2O_7^{2-}/Cr^{3+}$	$Cr_2O_7^{2-} + 14H^+ + 6e^- \rightarrow 2Cr^{3+} + 7H_2O$	$+1.33$
Cl_2/Cl^-	$Cl_2 + 2e^- \rightarrow 2Cl^-$	$+1.36$
PbO_2/Pb^{2+}	$PbO_2 + 4H^+ + 2e^- \rightarrow Pb^{2+} + 2H_2O$	$+1.46$
Au^{3+}/Au	$Au^{3+} + 3e^- \rightarrow Au$	$+1.50$
MnO_4^-/Mn^{2+}	$MnO_4^- + 8H^+ + 5e^- \rightarrow Mn^{2+} + 4H_2O$	$+1.51$
$HClO/Cl_2$	$2HClO + 2H^+ + 2e^- \rightarrow Cl_2 + 2H_2O$	$+1.63$
$PbO_2/PbSO_4$	$PbO_2 + SO_4^{2-} + 4H^+ + 2e^- \rightarrow PbSO_4 + 2H_2O$	$+1.68$
H_2O_2/H_2O	$H_2O_2 + 2H^+ + 2e^- \rightarrow 2H_2O$	$+1.78$
F_2/F^-	$F_2 + 2e^- \rightarrow 2F^-$	$+2.87$

Source: *Courtesy of Professor André J. de Béthune, Boston College. Taken by permission from A. J. de Béthune and N. A. Swendeman Loud,* Standard Aqueous Electrode Potentials and Temperature Coefficients, *C. A. Hampel, Publisher, Skokie, Ill., 1964.*

[a] Also called simply *standard electrode potentials, $\mathscr{E}^\circ$.*

[b] In the form used in this table, only the principal reacting species are shown for a couple. A more complete representation shows every reactant species. For example, instead of PbO_2/Pb^{2+}, we could write $PbO_2,H^+/Pb^{2+},H_2O$.

[c] Because the most active elements, such as lithium and fluorine, react with water, $\mathscr{E}^\circ$ in such cases is determined indirectly, for example, by way of the relationship $\Delta G^\circ = -nF\mathscr{E}_{cell}$, which is discussed later in this chapter.

substances in Table 18.1. (A more complete list is given in Table A.9 in the appendix.)

1. The voltage of the entire cell is assigned to the M electrode and is called the **standard reduction potential,** $\mathscr{E}^{\circ}_{red}$. Because the hydrogen electrode, $\mathscr{E}^{\circ}_{red} = 0$, is always half of each comparison cell, the effect of the hydrogen electrode on the voltage is constant. The different voltages assigned actually reflect the different tendencies of the M electrode to gain or lose electrons.

2. Half-reactions are sometimes referred to as *couples*, with the reactants indicated in the order of oxidized form/reduced form. Examples: the potential for the Ag^+/Ag couple is $+0.80$ V; for the $PbSO_4/Pb$ couple, -0.36 V.

3. The cathode, or reduction, reaction is shown in the table. When the electrode acts as the anode and undergoes oxidation, the reaction is the reverse of that given in the table.

4. Whether or not the cathode reaction will occur spontaneously when the electrode is connected to a hydrogen electrode can be inferred from the sign of the reduction potential in the table. If the sign is positive, the reaction to the right will occur as written; the electrode will act as the cathode and the hydrogen electrode as the anode. If the sign is negative, the reactions to the left will occur spontaneously, and the hydrogen electrode will act as the cathode (undergo reduction).

5. When a hydrogen electrode acts as the cathode, the reaction is

$$2H^+ + 2e^- \longrightarrow H_2 \quad \text{(reduction)}$$

When it acts as the anode, the reaction is

$$H_2 \longrightarrow 2H^+ + 2e^- \quad \text{(oxidation)}$$

6. The reduction potentials increase from -3.04 V for lithium to $+2.87$ V for fluorine. This means that there is an increasing tendency from top to bottom to gain electrons (undergo reduction) and a decreasing tendency to lose electrons (undergo oxidation).

18•3 Potentials of Voltaic Cells

18•3•1 Representation of a Voltaic Cell

Certain conventions are used to represent the reaction in a cell. These conventions are itemized below:

electrode; ions in solution ‖ ions in solution; electrode
    anode   cathode
    (oxidation)   (reduction)
    higher in Table 18.1   lower in Table 18.1

The two vertical parallel lines indicate the salt bridge separating the two electrodes. Specific examples are the Daniell cell in Figure 18.2,

$$Zn;Zn^{2+} \parallel Cu^{2+};Cu$$

and a cell made with platinum electrodes in which the overall reaction is $H_2 + Cl_2 \longrightarrow 2HCl$:

$$Pt,H_2;H^+ \parallel Cl^-;Cl_2,Pt$$

18·3·2 Cell Reaction and Cell Voltage From the reduction potentials in Table 18.1, we can predict the voltage of any voltaic cell that consists of two standard electrodes listed in the table. The reaction at an electrode is frequently referred to as a **half-reaction.** The **cell reaction** is the algebraic sum of the reactions that take place at the electrodes. For the cell shown in Figure 18.2, $Zn;Zn^{2+} \parallel Cu^{2+};Cu$, the half-reactions and cell reaction are

$$
\begin{aligned}
Zn &\longrightarrow Zn^{2+} + 2e^- & \text{(oxidation at anode)} \\
\underline{Cu^{2+} + 2e^- \longrightarrow Cu} & & \underline{\text{(reduction at cathode)}} \\
\text{sum: } Zn + Cu^{2+} &\longrightarrow Zn^{2+} + Cu & \text{(cell reaction)}
\end{aligned}
$$

The *cell voltage* is the algebraic sum of the oxidation potential and the reduction potential. (If we are dealing with standard electrodes, the cell voltage is designated by $\mathscr{E}^\circ_{\text{cell}}$.) The standard voltage for the cell above is as follows:

$$
\begin{aligned}
\mathscr{E}^\circ_{\text{cell}} &= \mathscr{E}^\circ_{\text{ox}} + \mathscr{E}^\circ_{\text{red}} \\
&= \mathscr{E}^\circ_{Zn;Zn^{2+}} + \mathscr{E}^\circ_{Cu^{2+};Cu} \\
&= 0.76 \text{ V} + 0.34 \text{ V} \\
&= 1.10 \text{ V}
\end{aligned}
$$

Note that the potential for the oxidation half-reaction has a sign opposite from that given in Table 18.1 and that the potential for the reduction half-reaction has the same sign as that in the table. Because Table 18.1 gives only reduction potentials, the sign must be changed if an oxidation potential is needed.

18·3·3 Spontaneity of Reaction *If the calculated voltage of the cell is positive, the cell reaction will take place spontaneously as written,* and the cell will provide current. Finding that $\mathscr{E}^\circ_{Zn;Zn^{2+}} + \mathscr{E}^\circ_{Cu^{2+};Cu}$ gives a positive 1.10 V reveals that $Zn + Cu^{2+} \rightarrow Zn^{2+} + Cu$ is a spontaneous process. Conversely, $\mathscr{E}^\circ_{Cu;Cu^{2+}} + \mathscr{E}^\circ_{Zn^{2+};Zn}$ gives a negative 1.10 V, showing that the chemical reaction $Cu + Zn^{2+} \rightarrow Zn + Cu^{2+}$ does not take place spontaneously. A mnemonic aid in using Table 18.1 to predict spontaneous reactions is shown in Figure 18.5.

One of the guides to predicting the activity of an element is its place in a table of standard reduction potentials, always keeping in mind two provisos. First, under other than standard conditions, relative activities may change. Second, relative voltages tell only what reactions *can* occur spontaneously, not that they *will* indeed occur (compare this with the first summary statement in Section 13.9.5).

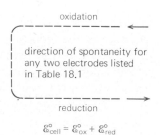

$$\mathscr{E}^\circ_{\text{cell}} = \mathscr{E}^\circ_{\text{ox}} + \mathscr{E}^\circ_{\text{red}}$$

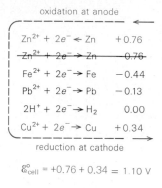

$$\mathscr{E}^\circ_{\text{cell}} = +0.76 + 0.34 = 1.10 \text{ V}$$

Figure 18·5 Mnemonic aid for predicting the directions of a pair of spontaneous electrode reactions used in a cell. The species appearing higher in Table 18.1 is oxidized, so this reaction is written with a direction and a potential opposite to that shown in the table. The species appearing lower in the table is reduced, so this reaction and its potential are taken directly from the table.

━━━ • Example 18•1 • ━━━

For each of the following voltaic cells, write the half-reactions, designating which is oxidation and which is reduction; then write the cell reaction. Next, calculate the voltage of the cell made from standard electrodes.

(a) Co;Co^{2+} ‖ Ni^{2+};Ni

(b) Cu;Cu^{2+} ‖ Ag$^+$;Ag

━━━ • Solution • ━━━

(a) Half-reactions and cell reaction:

$$\text{Co} \longrightarrow \text{Co}^{2+} + 2e^- \qquad \text{(oxidation)}$$
$$\underline{\text{Ni}^{2+} + 2e^- \longrightarrow \text{Ni} \qquad\qquad \text{(reduction)}}$$
$$\text{Co} + \text{Ni}^{2+} \longrightarrow \text{Co}^{2+} + \text{Ni} \qquad \text{(cell reaction)}$$

Cell voltage:

$$\mathscr{E}^{\circ}_{\text{cell}} = \mathscr{E}^{\circ}_{\text{ox}} + \mathscr{E}^{\circ}_{\text{red}}$$
$$= \mathscr{E}^{\circ}_{\text{Co;Co}^{2+}} + \mathscr{E}^{\circ}_{\text{Ni}^{2+};\text{Ni}}$$

(Now refer to Table 18.1 for reduction potentials; the oxidation potential is opposite in sign to the reduction potential.)

$$\mathscr{E}^{\circ}_{\text{cell}} = +0.28 \text{ V} + (-0.25 \text{ V}) = 0.03 \text{ V}$$

(b) Half-reactions and cell reaction:

$$\text{Cu} \longrightarrow \text{Cu}^{2+} + 2e^- \qquad \text{(oxidation)}$$
$$\underline{2\text{Ag}^+ + 2e^- \longrightarrow 2\text{Ag} \qquad\qquad \text{(reduction)}}$$
$$\text{Cu} + 2\text{Ag}^+ \longrightarrow 2\text{Ag} + \text{Cu}^{2+} \qquad \text{(cell reaction)}$$

Cell voltage:

$$\mathscr{E}^{\circ}_{\text{cell}} = -0.34 \text{ V} + 0.80 \text{ V} = 0.46 \text{ V}$$

See also Exercises 1 and 9–14 at the end of the chapter.

18•3•4   Effect of Concentration on Potential   All of the cell potentials discussed thus far have been $\mathscr{E}^{\circ}_{\text{cell}}$ values, that is, potentials for cells operating at standard conditions. For a cell at concentrations and conditions other than standard, a potential can be calculated using a relationship derived in 1889 by Walther Nernst, one of the early giants in physical chemistry. The Nernst equation, Equation (1), relates the experimental cell potential, $\mathscr{E}_{\text{cell}}$, to the standard cell potential, $\mathscr{E}^{\circ}_{\text{cell}}$:

$$\mathscr{E}_{\text{cell}} = \mathscr{E}^{\circ}_{\text{cell}} - \frac{RT}{nF} \ln Q \qquad (1)$$

In this equation, R is the familiar gas constant 8.314 J per mole per degree, T is the absolute temperature, n is the number of moles of electrons indicated in the balanced equation for the cell reaction, F is 96,500 coulombs per mole, and Q is a term that is similar in form to an equilibrium constant. At 25 °C

(298 K), and multiplying by 2.303 to convert to the log form, the equation becomes

$$\mathcal{E}_{cell} = \mathcal{E}^{\circ}_{cell} - \frac{0.0591}{n} \log Q \qquad (2)$$

For the generalized cell reaction, $wA + xB \rightarrow yC + zD$

$$Q = \frac{[C]^y[D]^z}{[A]^w[B]^x}$$

For precise calculations of Q, concentrations are expressed in terms of activities (see footnote 2 in this chapter). The following points should be noted:

1. In dilute solutions, good approximate calculations can be made using either molal or molar concentrations.
2. As in an equilibrium constant expression, the terms for solids are omitted.
3. If gases are involved, their pressures are expressed in atmospheres.
4. If water appears in the cell reaction (see Table 18.1), it does not appear in the calculation of Q. The concentration of water is treated as a constant that is automatically incorporated in all reactions in dilute solutions.

From the form of Q, it is clear that *the voltage of a cell will be increased by either (1) an increase in the concentration of the reactants, or (2) a decrease in the concentration of the products*. The form of Q and the use of Equation (2) to calculate $\mathcal{E}^{\circ}_{cell}$ are illustrated in Examples 18.2 and 18.3.

━━━ • Example 18.2 • ━━━

Calculate the voltage of the cell in Figure 18.2 if the copper half-cell is at standard conditions but the zinc ion concentration is only $0.0010M$. (Assume the temperature is 25 °C.)

━━━ • Solution • In the discussion of *cell voltage*, $\mathcal{E}^{\circ}_{cell}$ for this cell was shown to be 1.10 V. From the equation $Zn + Cu^{2+} \rightarrow Cu + Zn^{2+}$, the number of moles of electrons transferred, n, is seen to be 2. Therefore,

$$\mathcal{E}_{cell} = \mathcal{E}^{\circ}_{cell} - \frac{0.0591}{n} \log Q$$

$$= \mathcal{E}^{\circ}_{cell} - \frac{0.0591}{n} \log \frac{[Zn^{2+}]}{[Cu^{2+}]}$$

$$= 1.10 \text{ V} - \frac{0.0591}{2} \log \frac{0.0010}{1.0} \text{ V}$$

$$= 1.10 \text{ V} - (0.0296)(-3) \text{ V} = 1.19 \text{ V}$$

See also Exercises 17, 18, and 20–22.

━━━ • Example 18.3 • ━━━

What is the voltage of the $Pb;Pb^{2+} \parallel Ag^+$; Ag cell if the concentration of lead ion is $2.0M$ and that of silver ion is $0.0030M$?

• Solution • The cell reaction is $Pb + 2Ag^+ \rightarrow Pb^{2+} + 2Ag$:

$$\mathscr{E}_{cell} = \mathscr{E}^{\circ}_{cell} - \left(\frac{0.0591}{n} \log \frac{[Pb^{2+}]}{[Ag^+]^2} \right)$$

$$= [-(-0.13) + 0.80] V - \left(\frac{0.0591}{2} \log \frac{2.0}{(0.0030)^2} \right) V$$

$$= 0.93 \ V - 0.16 \ V = 0.77 \ V$$

See also Exercises 17, 18, and 20–22.

• Example 18•4 •

What must be the standard reduction potential of a couple in order for it to liberate hydrogen from a solution that is neutral?

• Solution • Let us assume that the couple is M^{2+}/M. For the reaction $M + 2H^+ \rightarrow M^{2+} + H_2$ to occur under standard conditions, $\mathscr{E}^{\circ}_{red}$ for the couple M^{2+}/M must be just less than 0.00, so that $\mathscr{E}^{\circ}_{cell}$ is therefore just greater than 0.00. If the hydrogen cell is in a neutral solution, pH = 7.00, our problem is to calculate the potential of $M + 2H^+ \rightarrow M^{2+} + H_2$ when H^+ is $1.0 \times 10^{-7} M$ and M^{2+} is standard $1.0m$. For an approximate calculation, we let $[M^{2+}] = 1.0$; $Q = [M^{2+}]/[H^+]^2$.

$$\mathscr{E}_{cell} = \mathscr{E}^{\circ}_{cell} - \frac{0.0591}{n} \log Q$$

$$= 0.00 \ V - \left(\frac{0.0591}{2} \log \frac{1.0}{(1.0 \times 10^{-7})^2} \right) V$$

$$= -\frac{0.0591}{2}(14) \ V = -0.41 \ V$$

Using more precise data, this potential is calculated as -0.414 V. For practical purposes, we would expect that metals with $\mathscr{E}^{\circ}_{red}$ values somewhat more negative than -0.414 V would react with pure water.

See also Exercise 23.

In Example 18.3, the potential of the cell is decreased by 0.16 V by decreasing the concentration of the reactant ion and increasing the concentration of the product ion. It is clear from this example that if the absolute value of $\mathscr{E}_{cell}$ is small enough, the change in the magnitude of Q could be great enough to reverse the direction of the reaction. Consider the cell reaction:

$$2Hg + 2Ag^+ \longrightarrow Hg_2^{2+} + 2Ag \qquad \mathscr{E}^{\circ}_{cell} = +0.01 \ V$$

From the voltage we see that the reaction has a slight tendency to go to the right at standard conditions. But suppose $[Hg_2^{2+}]$ were made $2.0M$, and $[Ag^+]$ were made $0.0030M$ (compare Example 18.3). Then

$$\mathscr{E}_{cell} = 0.01 \ V - 0.16 \ V = -0.15 \ V$$

The negative sign shows us that the cell reaction does not go as written. It is spontaneous in the reverse direction:

$$Hg_2^{2+} + 2Ag \longrightarrow 2Hg + 2Ag^+ \qquad \mathcal{E}_{cell} = +0.15 \text{ V}$$

This example shows why we must be careful in using the data in Table 18.1 to make predictions about the direction of reactions that involve small values of $\mathcal{E}_{cell}^{\circ}$ and/or that are far from standard conditions. Table 18.2 shows values of $\Delta\mathcal{E}_{cell}$ for various values of Q. For $n = 2$, the $\Delta\mathcal{E}_{cell}$ values would be only half those listed.

Table 18.2 Values of $\Delta\mathcal{E}_{cell}$ versus Q for $M + N^+ \rightarrow M^+ + N$ using the Nernst equation

Q	$\Delta\mathcal{E}_{cell}$, volt
10,000	−0.236
1,000	−0.177
100	−0.118
10	−0.0591
1 (standard)	0.000
0.1	+0.0591
0.01	+0.118
0.001	+0.177
0.0001	+0.236

18.4 Importance of Standard Electrode Potentials

Most of the voltaic cells that we could devise on the basis of standard electrode potentials would have no practical value. The real value of the data in Table 18.1 is to provide a better understanding of all oxidation–reduction reactions in solution, regardless of whether they take place in the compartments of a battery, by direct contact in a beaker in the laboratory, in the cells of living organisms, or in huge vats in industrial processes. In all cases electrons are being shifted about, and the standard potentials provide a comparison of the relative tendencies of different substances to acquire or to give up electrons.

From the relative positions of the metals in Table 18.1, we predict that when a strip of magnesium is placed in a nickel chloride solution, a rather rapid redox reaction will take place spontaneously:

$$Mg + Ni^{2+} + 2Cl^- \longrightarrow Ni + Mg^{2+} + 2Cl^- \qquad \mathcal{E}_{cell}^{\circ} \text{ is positive}$$

Similarly, we predict that when metallic nickel is placed in a solution of magnesium chloride, no reaction will occur. Nickel atoms will not spontaneously give electrons to magnesium ions:

$$Ni + Mg^{2+} + 2Cl^- \longrightarrow \text{no reaction} \qquad \mathcal{E}_{cell}^{\circ} \text{ is negative}$$

This does not mean that such a reaction could not be made to occur. However, we predict that it could occur only as a result of the expenditure of energy from an outside source.

18.5 Equilibrium Constants and Free Energies from Electrode Potentials

18.5.1 Equilibrium Constants
Note that Q in the Nernst equation is not an equilibrium constant, because the solutions it describes are initial concentrations and not those at equilibrium. *When a voltaic cell is completely discharged or run-down, then the system is at equilibrium.* Under this condition, $\mathcal{E}_{cell} = 0$ and the factor Q in the Nernst equation is equivalent to K, the equilibrium constant for the reaction. The standard electrode potential is, therefore, related to the equilibrium constant for the cell reaction by the expression

$$\mathcal{E}_{cell} = 0 = \mathcal{E}°_{cell} - \frac{RT}{nF} \ln K \qquad (3)$$

$$\mathcal{E}°_{cell} = \frac{RT}{nF} \ln K = \frac{0.0591}{n} \log K$$

• **Example 18•5** •

What will be the relative concentrations of Zn^{2+} and Cu^{2+} when the cell shown in Figure 18.1 has completely run down, that is, is at equilibrium? (Assume temperature is 25 °C.)

• **Solution** • At equilibrium, $[Zn^{2+}]/[Cu^{2+}]$ is the constant K.

$$\mathcal{E}°_{cell} = \left(\frac{0.0591}{n} \log K\right)$$

$$1.10 \text{ V} = \left(\frac{0.0591}{2} \log K\right) \text{V}$$

$$\left(\frac{0.0591}{2} \log K\right) \text{V} = 1.10 \text{ V}$$

$$\log K = \frac{1.10 \text{ V}}{\frac{0.0591}{2} \text{ V}} = 37.2$$

$$K = 2 \times 10^{37} = \frac{[Zn^{2+}]}{[Cu^{2+}]}$$

See also Exercise 28.

The ratio of concentrations found for the cell in Example 18.5 is also the ratio of concentrations in solution when excess solid zinc completely reacts with a solution of copper sulfate (see Figure 18.1). Two generalizations might be stated in connection with such systems. First, at equilibrium the calculated concentration of a species may be vanishingly small; however, it will not be zero. Second, for a reaction with a large enough driving force, the calculated amount of a reactant remaining may be so low that actual chemical determination of the concentration is impossible. This is true in the case of the final Cu^{2+} concentration in the zinc–copper ion reaction.

From the Nernst equation, it can be shown that for a simple one-electron reaction of the type $M + N^+ \rightarrow M^+ + N$, if the standard cell potential is about $+0.36$ V, the equilibrium concentration of M^+ will be over a million times greater than N^+. At equilibrium, such a reaction has indeed gone to completion for most purposes. The values of K for a range of $\mathcal{E}°_{cell}$ values are given in Table 18.3.

If the calculated cell voltage is negative, we conclude that the cell reaction as written is not spontaneous. Negative values of cell voltage are associated with values of K that are less than one. This is illustrated in a general way in Table 18.3, and a specific example is given in Example 18.6, part (b).

Table 18•3 Values of K versus $\mathcal{E}°_{cell}$ for $M + N^+ \rightarrow M^+ + N$

$\mathcal{E}°_{cell}$	K
$+2.00$	7×10^{33}
$+1.50$	2×10^{25}
$+1.00$	8×10^{16}
$+0.50$	3×10^{8}
$+0.25$	2×10^{4}
0.00	1
-0.25	6×10^{-5}
-0.50	3×10^{-9}
-1.00	1×10^{-17}
-1.50	4×10^{-26}
-2.00	1×10^{-34}

• **Example 18•6** •

Calculate the equilibrium constants for the following reactions.
(a) $3Cu + 2Au^{3+} \rightleftharpoons 3Cu^{2+} + 2Au$
(b) $Co^{2+} + Ni \rightleftharpoons Co + Ni^{2+}$

• Solution •

(a) From Table 18.1, $\mathscr{E}°_{cell} = -0.34 + 1.50 = 1.16$ V

$$\log K = \mathscr{E}°_{cell}\left(\frac{n}{0.0591}\right) = 1.16\left(\frac{6}{0.0591}\right) = 118$$

$$K = 1 \times 10^{118} = \frac{[Cu^{2+}]^3}{[Au^{3+}]^2}$$

(b) From Table 18.1, $\mathscr{E}°_{cell} = -(-0.25) + (-0.28) = -0.03$ V

$$\log K = -0.03\left(\frac{2}{0.0591}\right) = -1.0$$

$$K = 0.1 = \frac{[Ni^{2+}]}{[Co^{2+}]}$$

See also Exercises 25, 26, and 28.

★   Special Topic ★ Special Topic ★ Special Topic ★ Special Topic ★ Special Topic ★ Special Topic ★ Special Topic ★ Special Topic   ★

The Glass Electrode The most convenient way to determine the hydrogen ion concentration of a solution is to use a pH meter. Most of these instruments determine the [H+] in an unknown solution by measuring the difference in potential of two electrodes, one a standard reference electrode and the other an electrode made with a thin glass membrane that is sensitive to hydrogen ion. The membrane of the *glass electrode* (see the accompanying figure) is not penetrated by the hydrogen ions of the solution being tested, but different concentrations of ions cause differences in potential across the membrane. The region in which the potential difference arises can be described in terms of five parts:

| internal solution | hydrated gel layer | dry glass layer | hydrated gel layer | external solution |

The hydrated glass gel, H^+Gl^-, is mainly silicic acid, H_4SiO_4, and the equilibrium involved in the passage of current is

$$\begin{array}{cccc} H^+Gl^- & \rightleftharpoons & H^+ & + & Gl^- \\ \text{solid} & & \text{solution} & & \text{solid} \end{array}$$

At the boundaries with the dry glass, there is probably ion exchange between hydrogen ions and alkali metal ions in the glass. Current is conducted through the dry glass by the movement of alkali metal ions.

The amount of current conducted by a glass electrode is minute; it is the difference in potential that is measured and is determined by the [H+] concentration. Although with use there is a continued conversion of dry glass to hydrated glass gel, the useful life of commercially available electrodes in dilute solutions may be several years. The meter, which is calibrated by using external buffered solutions of known pH, gives a direct dial reading of the pH of an unknown solution. Meters commonly used measure values to ±0.01 pH units.

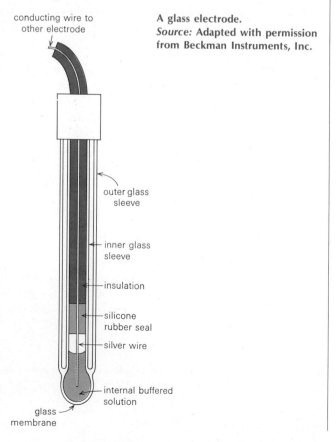

conducting wire to other electrode

A glass electrode.
Source: Adapted with permission from Beckman Instruments, Inc.

outer glass sleeve

inner glass sleeve

insulation

silicone rubber seal

silver wire

internal buffered solution

glass membrane

18•5•2   Free Energies In Section 15.4, the change in standard free energy of a reaction was related to the equilibrium constant by an expression similar to

$$\Delta G^\circ = -RT \ln K \tag{4}$$

Combining this equation with the equation relating K to $\mathscr{E}^\circ_{cell}$, the relation between the free energy change and the standard potential can be stated as

$$\Delta G^\circ = -nF\mathscr{E}^\circ_{cell} \tag{5}$$

The minus sign relates the two conventions that we have previously defined, in which a spontaneous reaction is denoted by a negative value of ΔG° or a positive value of $\mathscr{E}^\circ_{cell}$. If $\mathscr{E}^\circ_{cell}$ has the units of volts and F is 96,500 coulombs per mole, the value of ΔG° is given in joules of energy.[3] Expressing energy in kilojoules gives

$$\Delta G^\circ = -96.5n\mathscr{E}^\circ_{cell} \text{ kJ} \tag{6}$$

This equation correlates two important ideas. (1) For a standard voltaic cell, the change in free energy, ΔG°, is the maximum work that can be done by a process. (2) For such a voltaic cell, the maximum work available is calculated by multiplying the charge n that is moved times the voltage $\mathscr{E}^\circ_{cell}$ through which the charge is moved.

A major use of electrode potentials is in calculating free energy changes for chemical reactions. Consider the following cell reaction:

$$Mg;Mg^{2+} \parallel Zn^{2+};Zn$$

We calculate from Table 18.1 that $\mathscr{E}^\circ_{cell}$ is $2.37 - 0.76 = 1.61$ V. Therefore,

$$\Delta G^\circ = (-96.5)(2)(1.61) \text{ kJ} = -311 \text{ kJ}$$

Redox Reactions in Solution

18•6   Balancing Redox Equations

As mentioned in Section 9.2.5, simple redox equations can be balanced by inspection, but complex ones are best tackled in a systematic way. We shall describe two ways of balancing redox equations, the *oxidation-number method* and the *ion-electron method*. In both methods, it is necessary to assign oxidation numbers to each state of each element that appears in the equation. Rules for these assignments were given in Section 5.7. In both methods, partial equations are written for the oxidation and for the reduc-

---

[3]This important equation is often written with units of kilocalories: $\Delta G^\circ = -23.06n\mathscr{E}^\circ_{cell}$ kcal.

tion. Balancing the amount of oxidation with the amount of reduction reveals the coefficients that are needed to balance the desired overall equation.

● **18•6•1   Oxidation-Number Method**   We illustrate the oxidation-number method of balancing in Example 18.7. The reaction involves potassium permanganate, one of the most familiar strong oxidizing agents used in the laboratory. The first two steps are the same as for writing any equation. We must know what all the reactants and products are, and we must be able to write correct formulas for them.

━━━━━ **• Example 18•7 •** ━━━━━

Balance the equation for the reaction of potassium permanganate and sodium sulfite in the presence of sulfuric acid to form potassium sulfate, manganese(II) sulfate, sodium sulfate, and water.

━━━━━ **• Solution •**

• *Step 1*

$$\underset{\text{permanganate}}{\text{potassium}} + \underset{\text{sulfite}}{\text{sodium}} + \underset{\text{acid}}{\text{sulfuric}} \longrightarrow \underset{\text{sulfate}}{\text{potassium}} + \underset{\text{sulfate}}{\text{manganese(II)}} + \underset{\text{sulfate}}{\text{sodium}} + \text{water}$$

• *Step 2*

$$\mathrm{KMnO_4 + Na_2SO_3 + H_2SO_4 \longrightarrow K_2SO_4 + MnSO_4 + Na_2SO_4 + H_2O}$$
$$\text{(not balanced)}$$

• *Step 3*

Assign the oxidation number of each element in the equation:

$$\overset{+1}{\mathrm{K}}\overset{+7}{\mathrm{Mn}}\overset{(-2)_4}{\mathrm{O_4}} + \overset{(+1)_2}{\mathrm{Na_2}}\overset{+4(-2)_3}{\mathrm{SO_3}} + \overset{(+1)_2}{\mathrm{H_2}}\overset{+6(-2)_4}{\mathrm{SO_4}} \longrightarrow \overset{(+1)_2}{\mathrm{K_2}}\overset{+6(-2)_4}{\mathrm{SO_4}} + \overset{+2}{\mathrm{Mn}}\overset{+6(-2)_4}{\mathrm{SO_4}} + \overset{(+1)_2}{\mathrm{Na_2}}\overset{+6(-2)_4}{\mathrm{SO_4}} + \overset{(+1)_2}{\mathrm{H_2}}\overset{-2}{\mathrm{O}}$$

• *Step 4*

Select the elements that undergo a change in oxidation number, that is, undergo oxidation or reduction. Write the balanced equations for the half-reactions:

$$\text{oxidation:} \qquad \overset{+4}{\mathrm{S}} \longrightarrow \overset{+6}{\mathrm{S}} + 2e^-$$

$$\text{reduction:} \qquad \overset{+7}{\mathrm{Mn}} + 5e^- \longrightarrow \overset{+2}{\mathrm{Mn}}$$

• *Step 5*

Place coefficients in front of the reactants and products that contain these elements in the overall equation (step 2) so that the number of electrons lost in the oxidation equals the electrons gained in the reduction.

For this example, multiplying the oxidation half-reaction by 5 and the reduction half-reaction by 2 gives 10 e^-'s lost and 10 e^-'s gained:

$$\mathrm{2KMnO_4 + 5Na_2SO_3 + {?}H_2SO_4 \longrightarrow K_2SO_4 + 2MnSO_4 + 5Na_2SO_4 + {?}H_2O}$$

• *Step 6*

By inspection, determine the number of moles of the remaining substances, H_2SO_4 and H_2O in this case, required to balance the equation.

As the equation stands in step 5, 8 moles of sulfur are shown on the right (K_2SO_4, $2MnSO_4$, and $5Na_2SO_4$). To show the same amount on the left, 3 moles of H_2SO_4 must be shown:

$$2KMnO_4 + 5Na_2SO_3 + 3H_2SO_4 \longrightarrow K_2SO_4 + 2MnSO_4 + 5Na_2SO_4 + ?H_2O$$

The amount of water may be deduced in two ways:

(a) The total amount of oxygen atoms shown on the left of the last equation is 35 moles and on the right, not including H_2O, is 32 moles. Hence, 3 moles of water must be shown.

(b) The amount of hydrogen atoms shown on the left is 6 moles ($3H_2SO_4$). Hence, 3 moles of water must be shown.

The balanced equation is:

$$2KMnO_4 + 5Na_2SO_3 + 3H_2SO_4 \longrightarrow K_2SO_4 + 2MnSO_4 + 5Na_2SO_4 + 3H_2O$$

See also Exercises 31 and 32.

18·6·2   Ion-Electron Method In most redox reactions, the chemical action involves relatively few ions of the several that may be present. The ion-electron method focuses attention on these particles that we think actually exist in solution and that take part in the reaction. Symbols such as $\overset{+7}{Mn}$, $\overset{+4}{S}$, or $\overset{+6}{S}$ (see Example 18.7) are not used because such highly charged particles could not exist in aqueous solution.

A reaction involving a common oxidizing agent, the dichromate ion, is used in Example 18.8 to illustrate the ion-electron method of balancing redox equations.

• **Example 18·8** •

Balance the equation for the reaction of sodium dichromate, $Na_2Cr_2O_7$, and hydrochloric acid to yield sodium chloride, chromium(III) chloride, water, and chlorine.

• **Solution** •

• *Step 1*

$$\text{sodium dichromate} + \text{hydrochloric acid} \longrightarrow \text{sodium chloride} + \text{chromium(III) chloride} + \text{water} + \text{chlorine}$$

• *Step 2*

$$Na_2Cr_2O_7 + HCl \longrightarrow NaCl + CrCl_3 + H_2O + Cl_2$$
(not balanced)

• *Step 3*

Write the ionic form for each substance:

$$2Na^+ + Cr_2O_7^{2-} + H^+ + Cl^- \longrightarrow Na^+ + Cl^- + Cr^{3+} + 3Cl^- + H_2O + Cl_2$$

- Step 4

Assign the oxidation numbers:

$$\overset{+1}{2Na} + \overset{(+6)_2(-2)_7}{Cr_2O_7^{2-}} + \overset{+1}{H^+} + \overset{-1}{Cl^-} \longrightarrow \overset{+1}{Na^+} + \overset{-1}{Cl^-} + \overset{+3}{Cr^{3+}} + \overset{-1}{3Cl^-} + \overset{(+1)_2-2}{H_2O} + \overset{0}{Cl_2}$$

Write down only those items that involve an element that changes in oxidation number:

$$\overset{(+6)_2(-2)_7}{Cr_2O_7^{2-}} + \overset{-1}{Cl^-} \longrightarrow \overset{+3}{Cr^{3+}} + \overset{0}{Cl_2}$$

The chromium is reduced, because its oxidation number changes from +6 to +3. Some of the chlorine is oxidized (it changes from −1 to 0), but some of it is not changed.

- Step 5

For the reduction equation:

$$Cr_2O_7^{2-} \longrightarrow 2Cr^{3+} \qquad \text{(incomplete)}$$

Knowing that the oxygen goes to form water, we have

$$Cr_2O_7^{2-} \longrightarrow 2Cr^{3+} + 7H_2O \qquad \text{(incomplete)}$$

And, knowing that hydrogen ions must join the oxygen to form water, we have

$$Cr_2O_7^{2-} + 14H^+ \longrightarrow 2Cr^{3+} + 7H_2O$$
$$\text{(balanced as to atoms)}$$

Adding enough electrons to the left side to balance the equation electrically gives

$$Cr_2O_7^{2-} + 14H^+ + 6e^- \longrightarrow 2Cr^{3+} + 7H_2O$$
$$(-2) \;+ (+14) + (-6) \;=\; (+6) + \;(0)$$
$$\text{(balanced as to atoms and charges)}$$

For the oxidation equation:

$$2Cl^- \longrightarrow Cl_2$$
$$\text{(balanced as to atoms)}$$

To balance this equation electrically, $2e^-$ must be added to the right-hand side:

$$2Cl^- \longrightarrow Cl_2 + 2e^-$$
$$(-2) \;=\; (0) + (-2)$$
$$\text{(balanced as to atoms and charges)}$$

• *Step 6*

Next, we wish to add a balanced reduction equation to a balanced oxidation equation. We multiply the second equation by 3 so that the number of electrons lost in oxidation equals the number gained in reduction, and then add:

$$Cr_2O_7{}^{2-} + 14H^+ + 6e^- \longrightarrow 2Cr^{3+} + 7H_2O$$
$$\underline{3(2Cl^- \longrightarrow Cl_2 + 2e^-)}$$
$$Cr_2O_7{}^{2-} + 14H^+ + 6Cl^- + 6e^- \longrightarrow 2Cr^{3+} + 7H_2O + 3Cl_2 + 6e^-$$

The electrons are canceled to give the balanced ionic equation:

$$Cr_2O_7{}^{2-} + 14H^+ + 6Cl^- \longrightarrow 2Cr^{3+} + 7H_2O + 3Cl_2$$
$$(-2)\ \ + (+14) + (-6)\ \ =\ \ 2(+3) +\ \ \ 0\ \ \ +\ \ 0$$

• *Step 7*

In the ionic equation there is no mention of Na^+ ions; nor are enough negative ions shown to make the net charges on both sides of the equation equal zero. To write a balanced overall equation, both of these omissions must be taken care of by including two Na^+ ions for each $Cr_2O_7{}^{2-}$ and one Cl^- for each H^+. The final equation is

$$Na_2Cr_2O_7 + 14HCl \longrightarrow 2NaCl + 2CrCl_3 + 7H_2O + 3Cl_2$$

See also Exercise 32.

18•7 Driving Force and Equilibrium Constants

As pointed out in Section 18.4, the principles that apply to redox reactions at the electrodes of voltaic cells also apply to reactions that take place directly between substances. We can use the same methods to determine whether or not a reaction in solution is spontaneous and to calculate the equilibrium concentrations of the reactants and products.

In actual laboratory work, we hardly ever deal with solutions that are of standard concentrations. Therefore, calculations based on standard reduction potentials are of limited value. By taking into account the effect of concentration on the driving force of a reaction, we can usually make approximations and predictions that agree fairly well with experimental findings. We will illustrate these points with several practical examples.

— • Example 18•9 • —

Hydrogen peroxide is sometimes used to oxidize tin(II) ion to tin(IV) ion in laboratory analytical work. Does hydrogen peroxide oxidize practically all of the Sn^{2+} ions if equilibrium is approached?

— • Solution • From Table 18.1, we select two appropriate half-reactions. Writing the one higher in the table as an oxidation reaction and adding the two equations, we find

$$Sn^{2+} \longrightarrow Sn^{4+} + 2e^- \qquad -0.15 \text{ V}$$
$$\underline{H_2O_2 + 2H^+ + 2e^- \longrightarrow 2H_2O \qquad +1.78 \text{ V}}$$
$$H_2O_2 + 2H^+ + Sn^{2+} \longrightarrow Sn^{4+} + 2H_2O \qquad +1.63 \text{ V}$$

$$\log K = \mathscr{E}^\circ_{\text{cell}}\left(\frac{n}{0.0591}\right) = 1.63\left(\frac{2}{0.0591}\right) = 55.2$$

$$K = 2 \times 10^{55} = \frac{[Sn^{4+}]}{[H_2O_2][H^+]^2[Sn^{2+}]}$$

At any reasonable concentrations of H_2O_2 and H^+, $[Sn^{4+}]$ will be more than 10^{50} times larger than $[Sn^{2+}]$. So H_2O_2 can indeed oxidize practically all of the Sn^{2+} to Sn^{4+}.

See also Exercises 34 and 36.

• Example 18•10 •

Hydrogen sulfide, a weak electrolyte, is often encountered in laboratory work in solutions that contain the strong electrolyte, nitric acid. Is a reaction likely to occur? If so, what products are to be expected? That is, $H_2S + H^+ + NO_3^- \rightarrow$?

• Solution. • Looking at Table 18.1, we find

$$S + 2H^+ + 2e^- \longrightarrow H_2S \qquad +0.14 \text{ V}$$
$$NO_3^- + 4H^+ + 3e^- \longrightarrow NO + 2H_2O \qquad +0.96 \text{ V}$$

We reverse the first reaction so as to have H_2S on the left of the arrow as a reactant. Multiplying by the appropriate coefficients and adding, we find:

$$3(H_2S \longrightarrow S + 2H^+ + 2e^-) \qquad -0.14 \text{ V}$$
$$\underline{2(NO_3^- + 4H^+ + 3e^- \longrightarrow NO + 2H_2O) \qquad +0.96 \text{ V}}$$
$$3H_2S + 2NO_3^- + 2H^+ \longrightarrow S + 2NO + 4H_2O \qquad +0.82 \text{ V}$$

Note that 8 H^+'s on the left cancel the 6 H^+'s on the right, with 2 H^+'s in excess. Note, also, that *multiplying equations by coefficients to balance electrons lost and gained has no effect on the cell voltage.*

The cell voltage is positive, so we conclude that the reaction is likely to occur and that the products will be elemental sulfur, nitrogen(II) oxide, and water.

See also Exercises 34 and 36.

Applications of Electrochemistry

18•8   Electrolysis

18•8•1   Faraday's Law   The process whereby a current of electricity is used to bring about redox reactions that do not take place spontaneously is called **electrolysis.** The amount of chemical change produced by an electric current is directly proportional to the quantity of electricity passed. This fact was discovered by Michael Faraday in 1834 before the electron nature of an electric current was known. Our knowledge today enables us to see the basis for this relationship.

As mentioned in Section 18.2.1, the standard unit quantity of electricity that expresses the number of electrons that have passed through an electrolyte is the coulomb. On the basis of the charge of one electron, it is calculated that 96,500 C (more precisely, 96,487 C) corresponds to the passage of 1 mole of electrons. This quantity of electricity is called a **faraday:**

$$1 \text{ faraday} = 1 \text{ mol of electrons} = 9.65 \times 10^4 \text{ C}$$

During electrolysis, the passage of 1 faraday through the circuit brings about the oxidation of one equivalent weight of a substance at one electrode and the reduction of one equivalent weight at the other. This is a statement of **Faraday's law.**

Equivalent weights of oxidizing and reducing agents were discussed in Section 10.6.6. Examples 18.11 and 18.12 illustrate calculations based on Faraday's law.

• Example 18•11 •

What weights of gold and chlorine will be formed when 10,000 C of electricity is passed through a water solution of gold(III) chloride? The electrode reactions are

$$Au^{3+} + 3e^- \longrightarrow Au$$
$$2Cl^- \longrightarrow Cl_2 + 2e^-$$

• Solution • The equivalent weight of Au is $196.97/3 = 65.66$ g, and that of Cl_2 is $70.906/2 = 35.45$ g. Therefore, these amounts will be formed by the passage of 96,500 C. The amounts formed by the passage of 10,000 C will be

$$10,000 \text{ C} \times \frac{1 \text{ faraday}}{96,500 \text{ C}} \times \frac{65.66 \text{ g}}{1 \text{ faraday}} = 6.80 \text{ g Au}$$

$$10,000 \text{ C} \times \frac{1 \text{ faraday}}{96,500 \text{ C}} \times \frac{35.45 \text{ g}}{1 \text{ faraday}} = 3.67 \text{ g Cl}_2$$

See also Exercises 39 and 41–44.

• Example 18•12 •

How many coulombs will be required to plate out 5.60 g of iron from a solution of iron(III) chloride? The cathode reaction is

$$Fe^{3+} + 3e^- \longrightarrow Fe$$

• Solution • The equivalent weight of Fe is $55.85/3$, or 18.62 g. This amount will be formed by the passage of 1 faraday, or 96,500 C. The number of coulombs required to plate out 5.60 g will be

$$5.60 \text{ g} \times \frac{1 \text{ faraday}}{18.62 \text{ g}} \times \frac{96,500 \text{ C}}{1 \text{ faraday}} = 29,000 \text{ C}$$

See also Exercises 39 and 41–44.

18•8•2   **Decomposition Potentials**   The minimum voltage required to bring about electrolysis is called the **decomposition potential.** We can calculate decomposition potentials from tables of standard reduction potentials. The concepts can be illustrated with two familiar reactions:

$$Cu + Cl_2 \longrightarrow CuCl_2$$
$$CuCl_2 \longrightarrow Cu + Cl_2$$

The transfer of electrons from copper atoms to chlorine atoms (the first reaction) occurs spontaneously and could be utilized to make a voltaic cell. The reverse (second) reaction, that is, the transfer of electrons from chloride ions to copper ions, occurs only if energy is supplied; the electrons have to be "pumped" from the one ion to the other. This can be done by forcing a direct current to pass through an electrolytic cell in which the electrolyte is copper chloride:[4]

$Cu;Cu^{2+} \parallel Cl^-;Cl_2,Pt$
$\mathscr{E}^\circ_{cell} = -0.34 \text{ V} + 1.36 \text{ V} = +1.02 \text{ V}$ (voltaic cell)

$Pt,Cl_2;Cl^- \parallel Cu^{2+};Cu$
$\mathscr{E}^\circ_{cell} = -1.36 \text{ V} + 0.34 \text{ V} = -1.02 \text{ V}$ (electrolytic cell)

A voltaic cell made up of standard copper and chlorine electrodes has a positive voltage of 1.02 V. Now, if an outside current with a voltage slightly in excess of 1.02 V is connected so as to oppose the discharge of the cell, the electrons will flow in the opposite direction; that is, the reverse reaction will occur (electrolysis of copper chloride):

$$CuCl_2 \longrightarrow Cu + Cl_2$$

Voltages less than 1.02 will not bring about the electrolysis of copper chloride. It is apparent that the voltage required to decompose metallic salts of a given anion is related to the standard electrode potential of the metal, $\mathscr{E}^\circ_{red}$. *The lower a metal is on the table of potentials, the lower is the voltage required to decompose one of its compounds.*

Because electrode potentials vary with concentration, the voltage required to decompose a molten salt is not the same as that required to produce electrolysis of this salt in solution. Figure 5.1 illustrates one method of electrolytically decomposing salts of active metals.

18•8•3   **Electrode Products**   It is not always easy or even possible to predict what products will result when a direct current is passed through an aqueous solution of an electrolyte. In addition to the ions from the electrolyte, water molecules and the ions from water (H^+ and OH^-) are present. These also may participate in the electrochemical reactions. Furthermore, the electrode products obtained with concentrated solutions often differ from those ob-

---

[4] The cell used for electrolysis is called an electrolytic cell. Note, for example, that the lower unit in Figure 5.1 is an electrolytic cell and the three batteries at the top are voltaic cells.

tained with dilute solutions. To complicate the situation still more, one or both of the electrodes may react (the anode in Figure 18.7 is one of the reactants). To narrow the list of variables, we shall select inert electrodes, usually platinum, and then state a few rules that enable us to predict electrode products for a considerable number of reactants.

At the inert cathode, the following occurs:

1. If a metallic ion is more easily reduced than a hydrogen ion, the electrolysis of aqueous solutions of its salts causes the metal to form at the cathode. For example, if a water solution of $CuCl_2$, $AgNO_3$, or $HgCl_2$ is electrolyzed, Cu, Ag, or Hg, respectively, will form.
2. If a metallic ion is less easily reduced than a hydrogen ion, the electrolysis of aqueous solutions of its salts usually liberates hydrogen gas at the cathode. For example, if aqueous NaCl, KCl, or $MgCl_2$ is electrolyzed, hydrogen will be liberated at the cathode in each case. (See Figure 18.6 on Plate 4.)

At the inert anode:

1. Oxygen is liberated during the electrolysis of salts of oxyanions containing an element in a high state of oxidation, such as SO_4^{2-} and NO_3^-. That is, H_2O is more easily oxidized than such an anion. Example:

$$4AgNO_3 + 2H_2O \xrightarrow{\text{direct current}} 4Ag + O_2 + 4HNO_3$$
$$2H_2O \longrightarrow O_2 + 4H^+ + 4e^- \quad \text{(anode)}$$

2. Anions such as Cl^-, Br^-, and I^- (but not F^-) are more easily oxidized than water. Hence the free halogen is liberated. For example, during the electrolysis of concentrated sodium chloride, chlorine is liberated at the anode. (See Figure 18.6 on Plate 4, and also Section 23.3.2.) In the case of dilute solutions, oxygen as well as chlorine is formed, because the relative decomposition potentials for Cl^- and H_2O change as the concentration of chloride ion decreases.

18•8•4   Electroplating In the manufacture of metallic articles, an article that is fabricated from metal or an alloy of metals is frequently covered with a thin plate of some other metal. This is generally done to protect it against corrosion and to make it more attractive.

One method of plating is by electrolysis. The article to be plated is the cathode, and a block of the plating metal is the anode. Both electrodes are immersed in an aqueous solution of a salt of the plating metal and connected to a source of direct current.

The plating of pure silver on a fork made of a base metal is shown in Figure 18.7. Many factors are involved in getting a plate of uniform thickness that adheres strongly to the base metal. Among the important variables that must be controlled are the cleanness of the surface to be plated, the voltage, the temperature and purity of the solution, the concentration of the ion being plated out, and the total concentration of ions in the solution.

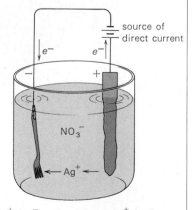

$$Ag^+ + e^- \rightarrow Ag \quad Ag \rightarrow Ag^+ + e^-$$

Figure 18•7 In silver plating base metals, the object to be plated is made the cathode; the anode is made of pure silver.

18•9   Batteries and Fuel Cells

18•9•1   Small-Scale Power Sources

The immediate experience most of us have with the generation of electric current is with battery-powered devices. Batteries for flashlights, calculators, watches, radios, and, of course, for starting the car, are all based on spontaneous redox chemical reactions.

The construction of batteries is as much an art as a science. Different substances and different physical forms provide voltaic cells with varying characteristics of size, power production, lifetime, recharging, and cost. Examples of electrochemical cells that are being used or seriously developed are listed in Table 18.4. A **primary cell** is one that is constructed with an anode and/or a cathode that is consumed chemically as the cell delivers current. A **secondary cell** is one that can be recharged by electrolysis, thus returning the electrodes to their initial condition. A **fuel cell** is a cell that uses continuously supplied reactants at unreactive electrodes and continuously discharges products.

Table 18•4 **Types of electrochemical cells**

System	Name
— Fuel Cells —	
$H_2 \mid KOH(aq) \mid O_2$	hydrogen–oxygen cell
$N_2H_4 \mid KOH(aq) \mid air$	hydrazine cell
— Primary Cells —	
$Zn \mid NH_4Cl(aq) \mid MnO_2$	Leclanché cell (dry cell)
$Zn \mid KOH(aq) \mid O_2$	zinc–oxygen cell
$Li \mid LiCl(l) \mid Cl_2$	lithium–chlorine cell
— Secondary Cells —	
$Pb \mid H_2SO_4(aq) \mid PbO_2$	lead–acid cell
$Cd \mid KOH(aq) \mid Ni(OH)_2$	nickel–cadmium cell
$Zn \mid KOH(aq) \mid Ag_2O$	silver–zinc cell
$Zn \mid KOH(aq) \mid air$	zinc–air cell
$Li \mid KPF_6(nonaq) \mid CuF_2$	lithium–copper fluoride cell

Source: *Mainly from E. J. Cairns and H. Shimotake,* Science, **164,** 1347 (1969).

• *The Leclanché Dry Cell* • The longest used small battery, and still the most common, is the so-called dry cell shown in Figure 18.8. Zinc metal acts as the negative electrode and is also a container for the other components of the battery. The positive electrode is an inert carbon rod located in the center of the can. This cell is called "dry" because the amount of water is relatively small; however, some moisture is essential to provide a solution for the diffusion of ions between the electrodes.

When the cell is delivering current, the reaction at the negative electrode involves the oxidation of zinc. The reaction at the positive electrode is complex, but under conditions of moderate use, hydrous manganous oxide, MnO(OH), appears to be the main reduction product. When the cell is

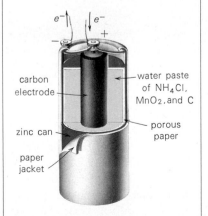

Figure 18•8 The Leclanché dry cell.

delivering moderate amounts of current, the reactions may be summarized as follows, although they are probably more complex.

$$\text{anode:}\quad Zn \longrightarrow Zn^{2+} + 2e^-$$
$$\text{cathode:}\quad 2MnO_2 + 2H_2O + 2e^- \longrightarrow 2MnO(OH) + 2OH^-$$

Non–oxidation–reduction reactions:

$$2NH_4^+ + 2OH^- \longrightarrow 2NH_3 + 2H_2O$$
$$Zn^{2+} + 2NH_3 + 2Cl^- \longrightarrow Zn(NH_3)_2Cl_2$$

Overall reaction:

$$Zn + 2MnO_2 + 2NH_4Cl \longrightarrow 2MnO(OH) + Zn(NH_3)_2Cl_2$$

It is a familiar fact that when a new flashlight battery is used continuously for an hour or so, the light dims and the battery appears to be run down. After standing idle for some time, the battery may again supply the normal amount of current. The idle period enables chemical reactions and diffusion processes to occur that remove reaction products from the immediate vicinity of the electrodes and permit the approach of fresh reactants.

A single dry cell has a voltage of 1.5 V. By connecting several in series, higher voltages are obtained.

• **Some Other Common Small Cells** • A cell that is often used instead of the Leclanché dry cell is the *alkaline battery*. This battery is composed of a zinc anode, a manganese dioxide cathode, and a potassium hydroxide electrolyte. The principal electrode reactions can be summarized as

anode:	$Zn + 2OH^- \longrightarrow Zn(OH)_2 + 2e^-$	+1.2 V
cathode:	$2MnO_2 + 2H_2O + 2e^- \longrightarrow 2MnO(OH) + 2OH^-$	+0.3 V
cell:	$2MnO_2 + 2H_2O + Zn \longrightarrow 2MnO(OH) + Zn(OH)_2$	+1.5 V

Under many conditions, especially when used heavily, this battery may deliver twice as much total energy as a Leclanché cell of the same size. The cell voltage is 1.5 V.

The *silver oxide–zinc battery* is often used as a miniature power source in such devices as hearing aids, watches, and small calculators. A schematic diagram of this cell is shown in Figure 18.9. The principal electrode reactions can be summarized as

cathode:	$Ag_2O + H_2O + 2e^- \longrightarrow 2Ag + 2OH^-$	+0.3 V
anode:	$Zn + 2OH^- \longrightarrow Zn(OH)_2 + 2e^-$	+1.2 V
cell:	$Ag_2O + H_2O + Zn \longrightarrow 2Ag + Zn(OH)_2$	+1.5 V

The *mercury(II) oxide–zinc battery* has a higher efficiency than other primary cells. From 80 to 90 percent utilization of the active materials is achieved. This battery, like the alkaline battery or the silver oxide–zinc

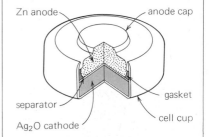

Figure 18•9 Silver oxide–zinc miniature cell.
Source: Courtesy of Battery Technical Services Division, Union Carbide Corporation.

battery, commonly has potassium hydroxide as an electrolyte. It has a voltage of about 1.4 V.

Although many anode materials have been tested—zinc, cadmium, indium, aluminum—zinc is the choice for the great majority of cells. The cathode is usually an easily reduced oxide or an inert electrode in contact with such an oxide.

An example of a battery that is not based on zinc and an oxide is the widely used *nickel–cadmium battery*. It is a common choice for rechargeable hand calculators. The overall reactions are

$$Cd + 2Ni(OH)_3 \; \underset{\text{recharge}}{\overset{\text{discharge}}{\rightleftharpoons}} \; CdO + 2Ni(OH)_2 + H_2O \qquad 1.35 \text{ V}$$

Again, the electrolyte is potassium hydroxide.

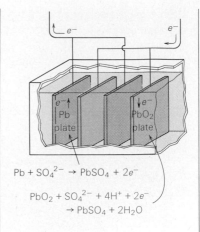

$$Pb + SO_4^{2-} \rightarrow PbSO_4 + 2e^-$$

$$PbO_2 + SO_4^{2-} + 4H^+ + 2e^- \rightarrow PbSO_4 + 2H_2O$$

Figure 18•10 Schematic representation of the lead storage battery, showing the changes that take place as the battery supplies current.

• **18•9•2   Lead Storage Cell**   The storage battery of an automobile can be recharged, so it acts as both a voltaic cell and an electrolytic cell. The battery is constructed of alternate plates of spongy lead and lead dioxide, separated by wood or glass fiber spacers, and immersed in an electrolyte, an aqueous solution of sulfuric acid (see Figure 18.10). When the battery supplies current, the lead plate, Pb, is the anode and the lead dioxide plate, PbO_2, is the cathode. The electrode reactions are shown in the figure. The complete reaction that occurs when current is drawn from the battery is

$$Pb + PbO_2 + 2H_2SO_4 \longrightarrow 2PbSO_4\downarrow + 2H_2O$$

Note that lead sulfate is formed at each electrode. Being insoluble, it is deposited on the electrode at which it forms, rather than dissolving in the solution. Note also that sulfuric acid is used up and water is formed. Because the diluted sulfuric acid is less dense than the original concentrated acid, the density of the electrolyte is commonly measured to determine the extent to which the battery is run down.

Recharging the battery consists of forcing the electrons through the battery in the opposite direction. In this electrolysis process all the above chemical changes are reversed. The lead sulfate and water are changed back to lead, lead dioxide, and sulfuric acid:

$$2PbSO_4 + 2H_2O \; \underset{\text{discharge}}{\overset{\text{recharge}}{\rightleftharpoons}} \; Pb + PbO_2 + 2H_2SO_4$$

When fully charged, a single cell of a lead storage battery has a potential of about 2.1 V. An automobile battery with six cells has a potential of about 12 V.

• **18•9•3   Sodium–Sulfur Cell**   A large-capacity battery to be used to propel vehicles or to store energy for electric utility peak load periods is being developed based on a sodium–sulfur cell (see Figure 18.11). Commercial models are already in experimental use in automobiles. The battery operates at temperatures above 300 °C in order to keep all the reactants in the liquid state.

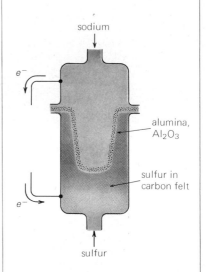

Figure 18•11 Diagram showing the principle of a sodium–sulfur cell.

As elemental sodium is oxidized at the ceramic alumina diaphragm, the Na^+ ions produced diffuse through the diaphragm into the sulfur side. Sulfur is reduced to S^{2-} ions, which react with sulfur atoms to form S_x^{2-} ions. The overall product is Na_2S_5 or Na_2S_3. The reactions are reversed when the battery is recharged.

18•9•4   Fuel Cells   *Fuel cells* commonly use oxygen at the cathode and usually use an oxidizable gas at the anode. The cell described in Figure 18.12 is a type that was used for electric power in the space ships in the *Apollo* moon program. The water vapor produced was condensed and added to the drinking water supply for the astronauts. The reactions are

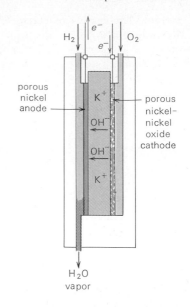

Figure 18•12 Schematic representation of a hydrogen–oxygen fuel cell. An operating temperature of about 200 °C is necessary to keep liquid the electrolyte that consists of approximately 75 percent KOH and 25 percent H_2O.

$$\text{cathode:} \quad O_2 + H_2O + 2e^- \longrightarrow HO_2^- + OH^-$$
$$HO_2^- \longrightarrow \tfrac{1}{2}O_2 + OH^-$$
$$\text{anode:} \quad H_2 + 2OH^- \longrightarrow 2H_2O + 2e^-$$
$$\text{overall:} \quad H_2 + \tfrac{1}{2}O_2 \longrightarrow H_2O$$

A less exotic use of fuel cells, but one of tremendous importance, is that proposed for major power plants and that uses coal to produce hydrogen or other oxidizable gases and air as the source of oxygen. The direct generation of electric current in such cells should not involve as much pollution of the air or thermal pollution of rivers as does the pollution from other types of power plants.

Chapter Review

Summary

Electrochemisty is the study of redox reactions carried out in such a way that a measurable electric potential exists in the system. In a **voltaic cell,** a spontaneous redox reaction generates an electric current that flows through an external circuit. In an **electrolytic cell,** a nonspontaneous redox reaction is "driven" by an electric current provided by an external source. All electrochemical cells must also have an internal circuit through which current, in the form of diffusing ions, can flow. Certain types of cells use a **salt bridge** for this purpose. In every cell, oxidation occurs at the anode and reduction occurs at the cathode. Electrons flow from the anode to the cathode in the external circuit. The units of measurement for the number of electrons, their rate of flow, and the electric potential difference that drives them are the **coulomb,** the **ampere,** and the **volt,** respectively.

For any given conditions, every electrode has an **electrode potential** determined by the half-reaction that occurs there. An electrode operating at standard conditions is called a **standard electrode.** Individual electrode potentials can be measured only indirectly; under standard conditions their values are determined relative to that of the **standard com-** parison electrode. The electrode used for this purpose is the standard **hydrogen electrode,** which has an arbitrarily assigned potential of zero. By convention, all electrode potentials are tabulated as **standard reduction potentials.**

Any two oxidation and reduction half-reactions can be added to give a hypothetical **cell reaction.** If the two standard reduction potentials in question are known, the **cell voltage** can be calculated directly, without the need for experimental measurement. The sign of the voltage indicates whether or not the cell reaction as written is spontaneous at standard conditions. Calculated or measured cell voltages provide an easy and accurate means for calculating both the equilibrium constant and the standard free energy change of the reaction. The Nernst equation is used to calculate the cell voltage under nonstandard conditions of temperature and species concentrations. All redox reactions can be balanced by either the oxidation-number method or the ion-electron method.

The **faraday** is a useful measure of the quantity of electricity corresponding to an electrochemical change in a given amount of substance. **Faraday's law** is the quantitative basis for the electrolysis of compounds by an electric current. The minimum voltage required to bring about electrolysis is called the **decomposition potential.** The opposite of electrolysis is the spontaneous production of electric current by batteries. A **primary cell** is irreversibly consumed by its own action, whereas a **secondary cell** is rechargeable by electrolysis. A **fuel cell** is a special kind of battery in which gaseous or liquid fuel flows continuously into the cell, where its free energy is converted directly to electricity by a redox reaction with an oxidizing agent, rather than to heat by a combustion reaction.

Exercises

Voltaic Cells

1. (a) Describe what you would observe after inserting a strip of clean magnesium into a solution of gold chloride, $AuCl_3$.
   (b) Write the equations for the half-reactions for the oxidation and for the reduction.
   (c) Write the equation for the overall reaction.

2. (a) Show with a labeled diagram how one could use beakers, glass tubing, copper wire, various chemicals, and strips of metals to make a voltaic cell based on the reaction described in Exercise 1.
   (b) Which part of the cell is the anode and which is the cathode? Why?
   (c) What provision did you make for the internal current flow? For the external flow?
   (d) What governed your choice of metals for the electrodes?

3. Some of the earliest batteries, called voltaic piles, were made by stacking together discs of metal and blotting paper soaked in sodium chloride solution, for example, zinc, silver, paper, zinc, silver, paper, zinc, silver, paper, etc. By touching the ends of a sufficiently long pile, a person could feel an electric shock. Diagram a pile, show the ends connected by a wire, and show the direction of electron flow in the wire.

4. In the voltaic pile described in Exercise 3, would chemical reactions occur? If so, write equations for them and state where they would occur.

5. Imagine that a voltaic cell like that shown in Figure 18.2 is constructed and set aside in a closed container that prevents the evaporation of the solutions. If the external circuit is never closed (that is, no electrons ever flow through the wire), will the cell remain ready for use indefinitely? Explain.

6. Electricity is stored in a battery. Do you agree or disagree with this statement? Explain fully, using an example with chemical equations to illustrate your viewpoint.

Standard Electrode Potentials

7. (a) What are the requirements for a standard electrode?
   (b) Explain why the standard hydrogen electrode has an assigned potential of 0.00 V.

*8. Given that $V = J \cdot A^{-1} \cdot s^{-1}$ and $1\,kWh = 3.6 \times 10^6\,J$, what does it cost to operate a window-type air-conditioner for one hour if the unit requires a 10-A current at 110 V and electric energy costs 5 cents per kilowatt-hour?

9. (a) Assuming that the concentrations of the ionic solutions used in making the voltaic cell in Exercise 2 are $1.0m$ and the temperature is 25 °C, what is the voltage of the cell?
   (b) What would be the voltage of a cell made by connecting a magnesium electrode to a standard hydrogen electrode? By connecting the gold electrode to a standard hydrogen electrode?

10. Write the half-reaction for the oxidation and for the reduction for each of the cells described in Exercise 9(b).

11. Represent in the conventional manner, that is, anode‖cathode, the voltaic cell in which each of the following reactions occurs spontaneously:
    (a) $Zn + Ni^{2+} \longrightarrow Zn^{2+} + Ni$
    (b) $2Al + 3Cl_2 \longrightarrow 2AlCl_3$
    (c) $H_2 + 2Ag^+ \longrightarrow 2H^+ + 2Ag$
    (d) $2H^+ + Pb \longrightarrow H_2 + Pb^{2+}$

12. Write the half-reactions occurring at the cathode and at the anode when each cell shown in Exercise 11 is delivering current.

13. Calculate $\mathscr{E}^\circ_{cell}$ for each cell shown in Exercise 11.

14. Represent the electrochemical cell based on each reaction shown below. Then calculate $\mathscr{E}^\circ_{cell}$ for each. Based on these calculations, which cells could be used to supply current?
    (a) $2Au + 3Cl_2 \longrightarrow 2AuCl_3$
    (b) $Fe + 2HCl \longrightarrow FeCl_2 + H_2$
    (c) $Pb + Sn(NO_3)_2 \longrightarrow Pb(NO_3)_2 + Sn$

15. Metal M is above hydrogen and silver is below hydrogen in the list of standard reduction potentials. If two standard half-cells are constructed to allow the reaction

    $$M + 2Ag^+ \longrightarrow 2Ag + M^{2+}$$

    to take place, the $\mathscr{E}^\circ_{cell}$ is found to be 1.08 V.
    (a) What is the standard potential of M?
    (b) Will metal M react with a $1m\,H^+$ solution? If so, write the ionic equation for the reaction.
    (c) Repeat (b), but for a $1m\,H^+$ solution substitute, first, a $1m\,Pb^{2+}$ solution and, second, a $1m\,Zn^{2+}$ solution.

Effect of Concentration on Potential

*16. In the Nernst equation, confirm by calculation the numerical value of the factor $0.0591/n$, and show that the unit associated with the factor is the volt.

17. Calculate the voltage of the cell in Figure 18.2 if the zinc half-cell is at standard conditions but the copper ion concentration is only $0.0010m$.

18. Calculate the voltage of the cell in Figure 18.4 under the following conditions:

(a) let M be Fe at standard conditions

(b) same as part (a), but let p_{H_2} be 100 atm

(c) let M be Ag at standard conditions

(d) same as part (c), but let p_{H_2} be 100 atm

19. Show that the $-(RT/nF)\ln Q$ term in the Nernst equation is zero for a cell operating at standard conditions.

20. (a) Write the cell reaction for Sn;Sn^{2+} ∥ Pb^{2+};Pb.

(b) Calculate the standard potential of the cell.

(c) Calculate the potential if $[Sn^{2+}] = 0.0010$ and $[Pb^{2+}] = 1.0$.

(d) Calculate the potential if $[Sn^{2+}] = 1.0$ and $[Pb^{2+}] = 0.0010$.

(e) What does the fact that the answers to parts (c) and (d) have different signs tell you about the value for the ratio of $[Sn^{2+}]/[Pb^{2+}]$ at equilibrium?

(f) Calculate $[Sn^{2+}]/[Pb^{2+}]$ at equilibrium.

21. Which of the following will produce a cell with the highest voltage?

(a) $1m\ Ag^+$, $1m\ Co^{2+}$

(b) $0.1m\ Ag^+$, $0.1m\ Co^{2+}$

(c) $2m\ Ag^+$, $2m\ Co^{2+}$

(d) $0.1m\ Ag^+$, $2m\ Co^{2+}$

(e) $2m\ Ag^+$, $0.1m\ Co^{2+}$

22. Calculate the voltage of the cell in Figure 18.3 under the following conditions:

(a) standard conditions

(b) $p_{H_2} = 1.0$ atm, $H^+ = 1.0m$, $Cl^- = 1.0m$, and $p_{Cl_2} = 0.10$ atm

(c) $p_{H_2} = 10$ atm, $H^+ = 1.0m$, $Cl^- = 1.0m$, and $p_{Cl_2} = 1.0$ atm

(d) $p_{H_2} = p_{Cl_2} = 5.0$ atm, $H^+ = Cl^- = 0.10m$

(e) $p_{H_2} = p_{Cl_2} = 0.10$ atm, $H^+ = Cl^- = 2.0m$

23. In Example 18.4, the $\mathscr{E}^\circ_{red}$ for an M^{2+}/M couple to liberate hydrogen from a neutral solution was calculated to be 0.414 V. Show that this value is the same for any couple, regardless of the charge on the ion M^{x+}.

24. When a piece of metal R is put into a solution of $T(NO_3)_2$, a spontaneous reaction occurs to form the metal T and a solution of $R(NO_3)_2$. The standard reduction potential for one of the metals is -0.30 V and for the other, -0.45 V.

(a) Write the half-cell equations for these two elements as they would appear in a table of standard reduction potentials. Show which one is -0.30 V and which is -0.45 V.

(b) Calculate the ratio of $[R^{2+}]/[T^{2+}]$ in solution when the reaction comes to equilibrium.

Equilibrium
Constants and Free Energies

25. (a) Calculate the equilibrium constants for the two reactions described in Example 18.1.

(b) How is K for the equilibrium in Example 18.1(a) related to the K for the reaction in Example 18.6(b)?

26. (a) Calculate the equilibrium constant for the reaction

$$2Au^{3+} + 3Ni \rightleftharpoons 2Au + 3Ni^{2+}$$

(b) Write the expression for K.

(c) Calculate the concentration of Au^{3+} at equilibrium, if the concentration of Ni^{2+} is 0.50M.

27. Construct a table similar to Table 18.3, with the same $\mathscr{E}^\circ_{cell}$ values but with $n = 2$.

28. (a) For a reaction of the type $M + N^+ \rightarrow M^+ + N$, what must the standard cell potential be in order to make the equilibrium concentration of $[M^+]$ a million times greater than $[N^+]$? Express your answer to three significant figures.

(b) Repeat part (a) for reactions of the following types:

$$M + N^{2+} \longrightarrow M^{2+} + N$$

$$3M + 2N^{3+} \longrightarrow 3M^{2+} + 2N$$

29. Consider two reactions with the same numerical value for $\mathscr{E}^\circ_{cell}$, but in the equation for one of the reactions $n = 1$, and in the equation for the other $n = 2$. What is the relation between the values of K for the two reactions?

30. (a) Calculate the free energy change in kilojoules and kilocalories for the reaction

$$Cu + NiCl_2 \longrightarrow CuCl_2 + Ni$$

Assume that the reaction is carried out in 1.0m solutions at 25 °C.

(b) Is the reaction spontaneous? Why?

Balancing Redox Equations

31. Balance the following equations by the oxidation-number method.

(a) $Fe_2O_3 + S \longrightarrow Fe + SO_2$

(b) $NH_3 + O_2 \longrightarrow NO + H_2O$

(c) $KMnO_4 + HCl \longrightarrow MnCl_2 + Cl_2 + KCl + H_2O$

(d) $N_2O + H_2 \longrightarrow H_2O + NH_3$

32. Balance the following equations. Use the ion-electron method for at least parts (a), (b), (d), (f), (i), and (j).

(a) $Cu + HNO_3 \longrightarrow Cu(NO_3)_2 + NO_2 + H_2O$

(b) $FeS + HNO_3 \longrightarrow Fe(NO_3)_3 + S + NO_2 + H_2O$

(c) $As_2S_3 + HNO_3 + H_2O \longrightarrow H_3AsO_4 + NO + S$

(d) $KMnO_4 + H_2SO_4 + FeSO_4 \longrightarrow$
$$K_2SO_4 + MnSO_4 + Fe_2(SO_4)_3 + H_2O$$

(e) $KMnO_4 + SnF_2 + HF \longrightarrow$
$$MnF_2 + SnF_4 + KF + H_2O$$

(f) $Cr_2O_3 + Na_2CO_3 + KNO_3 \longrightarrow$
$$Na_2CrO_4 + CO_2 + KNO_2$$

(g) $K_2S_5 + HCl \longrightarrow H_2S + S + KCl$

(h) $Na_3Fe(CN)_6 + Re + NaOH \longrightarrow$
$$Na_4Fe(CN)_6 + NaReO_4 + H_2O$$

(i) $Na_2SO_3 + H_2SO_4 + KMnO_4 \longrightarrow$
$$K_2SO_4 + MnSO_4 + Na_2SO_4 + H_2O$$

(j) $KBrO_3 + Fe(NO_3)_2 + HNO_3 \longrightarrow$
$$KBr + Fe(NO_3)_3 + H_2O$$

(k) $KMnO_4 + NaHSO_3 + H_2SO_4 \longrightarrow$
$$K_2SO_4 + MnSO_4 + Na_2SO_4 + H_2O$$

(l) $HgS + HNO_3 + HCl \longrightarrow$
$$HgCl_2 + NO + S + H_2O$$

*33. (a) Balance the following redox equation:

$$MnCl_2 + PbO_2 + HNO_3 \longrightarrow$$
$$HMnO_4 + Cl_2 + Pb(NO_3)_2 + H_2O$$

(b) If 10.8 g of $MnCl_2$ is converted to products, what weight of Cl_2 is formed?

(c) How many moles of HNO_3 react in the reaction involving 10.8 g of $MnCl_2$?

(d) If only 24.0 g of PbO_2 had been present, how much of the 10.8 g of $MnCl_2$ would have been left unreacted?

Driving Forces

*34. (a) Calculate $\mathscr{E}^\circ_{cell}$ and K for the reaction shown in Example 18.8.

(b) Is the reaction a spontaneous one?

(c) Write the expression for the equilibrium constant for the cell reaction.

(d) Under the proper laboratory conditions, the oxidation of chloride ion by dichromate ion does occur. Suggest what these conditions are.

35. In New Orleans, water from the Mississippi is thoroughly chlorinated to make it safe for drinking. To make fine coffee, a shop there holds the water at the boiling point in an open pot for two or three hours. How would this affect the amount of chlorine and its reaction products that remain dissolved in the water? Explain qualitatively on the basis of equilibrium considerations.

36. In Example 18.10, we picked for the nitrate couple NO_3^-/NO with $\mathscr{E}^\circ_{red} = +0.96$ V. We could have chosen NO_3^-/N_2O_4 with $\mathscr{E}^\circ_{red} = +0.80$ V.

(a) Is there any good reason for choosing one rather than the other?

(b) Using the NO_3^-/N_2O_4 couple, calculate the cell voltage and write the balanced net ionic equation for the reaction.

37. When gaseous chlorine is dissolved in water, the following equilibrium is involved:

$$Cl_2 + H_2O \rightleftharpoons HClO + H^+ + Cl^-$$

(a) Calculate the $\mathscr{E}_{cell}$ if the pressure of chlorine is 3.0 atm and the concentration of each of the products is 0.10m.

(b) Repeat part (a) for a chlorine pressure of 3.0 atm and 0.010m concentrations of each of the products.

(c) What do the calculations in parts (a) and (b) tell you about the spontaneity of the reaction at standard conditions? What do they tell you about equilibrium conditions for the reaction?

*38. Hydrogen peroxide solutions tend to decompose to form water and oxygen gas. At equilibrium, what is the relationship between the concentration of H_2O_2 in solution and the pressure of O_2 gas above the solution? Is it practical to prevent the decomposition from occurring by keeping the solution in a closed container so the pressure due to O_2 will be high? For purposes of estimating p_{O_2}, assume $[H_2O_2]$ at equilibrium is $1.0 \times 10^{-6} M$.

Faraday's Law

39. Review the definition in Section 10.6.6 of equivalent weights for reactants in redox reactions, and calculate the equivalent weight of each appropriate substance in the reactions in Exercise 31.

40. For the electrolysis of aluminum oxide write the equation for:

(a) the anode reaction

(b) the cathode reaction

(c) the balanced ionic equation

41. Refer to Exercise 40. What weight of aluminum would be produced if 24,125 C were passed through molten aluminum oxide? What weight of oxygen would be produced?

42. If the same amount of current is passed through separate solutions (or molten salts) of Li^+, Ni^{2+}, Au^{3+}, K^+, Cr^{3+}, and Hg^{2+}, in which case will the greatest weight of metal be plated out? In which case the least weight?

43. Calculate the number of electrons that must be gained at the cathode when 0.100 g of silver is electroplated on a metal serving tray. (Assume that the oxidation state of silver in the plating solution is +1.) How many coulombs are required to provide this number of electrons? How long must a 1.5-A current flow to supply this number of coulombs?

44. Repeat Exercise 43 for the electroplating of 0.100 g of chromium from a solution of chromium(III) chloride.

*45. A salt known to contain platinum and chloride ions only was electrolyzed in a water solution until 300.0 C had passed through the solution. The cathode was then removed, dried, and weighed. The gain in weight due to the deposit of elemental platinum was 0.152 g. Write the formula for the salt.

46. (a) A current from a supposedly constant source was passed through a solution of silver nitrate, $AgNO_3$, for 1 hour, depositing 6.58 g of silver. Calculate the number of faradays passed and the rate of flow in amperes.

(b) If the current actually fluctuated, how would the accuracy of the calculations in part (a) be affected?

Electrolysis and Battery Reactions

47. (a) Predict what is liberated at each electrode when aqueous solutions of the following are electrolyzed

between inert electrodes: $CuBr_2$, NaI, $Mg(NO_3)_2$, $HgSO_4$, HCl, KOH.

(b) Write equations for the chemical reactions that account for all the changes observed in Figure 18.6 on Plate 4.

48. Suppose that the standard electrode potential of a newly discovered metal is found to be -2.0 V. Would it be practical to attempt recovery of the metal from its chloride by electrolysis of a water solution of the chloride, assuming that this salt is water soluble? Explain. If the answer is no, suggest a method for obtaining the metal from its chloride.

*49. Write hypothetical electrode reactions for the following:
(a) the mercury(II) oxide–zinc battery
(b) the nickel–cadmium battery

(c) the sodium–sulfur battery

50. (a) Write equations for the anode and the cathode reactions when the lead storage battery is being used to supply current.

(b) What weight of lead sulfate is deposited on the plates that serve collectively as the anode when the battery supplies current to start an automobile engine, if 10 seconds with an average current of 30 A is required?

(c) What weight of lead sulfate is deposited on the plates that serve as the cathode?

(d) What weight of lead is formed when the battery is recharged to its initial strength? On which electrode would the lead form?

Nuclear Chemistry

19

19•1 **Radioactive Elements**

19•1•1 Radioactive Decay

19•1•2 Ionizing Effects of Radiation

19•2 **Detection of Radiations**

19•2•1 Photographic Methods

19•2•2 Fluorescent Methods

19•2•3 Cloud Chambers

19•2•4 Gas Ionization Counters

19•3 **Radioactive Series**

19•3•1 Alpha Emission

19•3•2 Beta Emission

19•3•3 Gamma Emission

19•4 **Background Radiation**

19•4•1 The Curie and the Rem

19•5 **Half-Life**

19•6 **Applications of Radioactivity**

19•6•1 Dating

19•6•2 Tracer Studies

19•6•3 Additional Uses of Isotopes

19•7 **Bombardment Reactions**

19•7•1 Neutron Bombardments

19•8 **Acceleration of Charged Particles**

19•8•1 Linear Accelerators

19•8•2 Circular Accelerators

19•8•3 Synthesis of Elements

19•9 **Mass Loss and Binding Energy**

19•9•1 Nucleons

19•9•2 Mass Loss

19•9•3 Mass-Energy Equivalence

19•9•4 Binding Energy

19•10 **Nuclear Stability**

Special Topic: Possible Superheavy Nuclides

19•10•1 The Transuranium Elements

19•11 **Nuclear Fission**

19•11•1 Fission Bombs

19•11•2 Sources of Fissionable Material

19•12 **Nuclear Reactors**

19•12•1 Controlling the Nuclear Reaction

19•12•2 Long-Term Reactor Operation

19•12•3 Breeder and Converter Reactors

Special Topic: Our Energy Dilemma: Nuclear or Coal or ?

19•13 **Fusion Reactions**

19•13•1 Stellar Energy

19•13•2 Thermonuclear Power

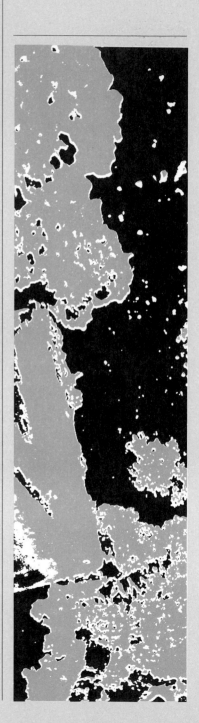

In all the chemical reactions we have studied in previous chapters, elements have maintained their identities in the reactants and products. This fundamental fact of chemical life has been reflected in our balanced equations. In this chapter, however, we will take up a special type of change called a nuclear reaction. A **nuclear reaction** involves the change of atoms of one or two elements into one or more atoms of a different element or elements.

Nuclear chemistry is concerned with the study of natural and artificial changes in the nuclei of atoms, and with the chemical reactions of radioactive substances. Natural radioactivity, which was mentioned in Section 3.3.1, is the most familiar example of nuclear chemistry. We will look at some of the uses of radioactive materials. Also, we will consider the effects of radioactive emissions (alphas, betas, and gammas) on substances, including living things.

The widespread use of nuclear reactors by public utilities for the production of electricity makes nuclear chemistry the concern of all citizens. This chapter will conclude with a review of some of the principles and practices of nuclear power production.

19•1   Radioactive Elements

An element that undergoes spontaneous emissions of radiations is said to be **radioactive.** It is widely known that elements such as radium and uranium are radioactive. It is not widely known that small amounts of radioactive isotopes of common elements such as carbon and potassium are present in living organisms, including our own bodies. Radioactive changes differ from ordinary chemical reactions in that the former involve transformations that originate in the nucleus of the atom, whereas the latter involve changes in only the arrangement of electrons.

We use the term *isotopes* to refer to the atoms of a given element that have different masses. To refer to a particular kind of atom of any element whatsoever, we use the term *nuclide*. A **nuclide** is a kind of atom that is distinguished from all others by the numbers of protons and neutrons it contains. Of the more than 3,000 known nuclides, about 280 are stable nuclides of the elements through $_{83}$Bi. The remainder are radioactive (unstable) nuclides. Most of the radioactive nuclides are not found in the earth's crust but are made synthetically as described in Sections 19.8.3 and 19.10.1.

19•1•2   Radioactive Decay

As the result of a radioactive emission, an unstable nuclide changes to a different nuclide. Such a change is spoken of as a **radioactive decay.** For example, when the radioactive nuclide $^{211}_{84}$Po emits an alpha particle, it changes to $^{207}_{82}$Pb.

The three emissions most characteristic of natural radioactivity, alpha (α), beta (β), and gamma (γ), were described in Table 3.1. Whenever a sample of an element emits alpha or beta particles, atoms of another element are formed. This observation leads to the inescapable conclusion that the alpha and beta particles come from the atomic nuclei of the radioactive substance. Beta radiation has been shown to be a stream of electrons; alpha radiation has been shown to be a stream of helium ions, He^{2+}. If a solid that emits alpha radiation is sealed in an evacuated glass tube, the tube gradually becomes filled with helium gas. This fact, coupled with the knowledge that an alpha particle has a charge of $2+$ and a mass of about 4 amu, shows that the alpha particle is identical with the nucleus of a helium atom. Evidently, when a speeding alpha particle slows down sufficiently, it takes two electrons from some atoms or ions it encounters and thereby becomes a helium atom:

$$_2^4He^{2+} + 2e^- \longrightarrow \, _2^4He$$

When an atom emits either an alpha or a beta particle, it often emits a gamma ray at the same time. Gamma rays are a type of electromagnetic radiation similar to visible light and X rays but of shorter wavelengths (see Figure 4.2).

Alpha particles emitted from a given nuclide have energies confined to a few discrete values, a fact that supports the energy level model of the nucleus. For instance, the nuclide ^{226}Ra emits alpha particles with either of two speeds, 1.517×10^7 or 1.488×10^7 m/s (approximately 9,300 mi/s). Overall, equal energy is emitted in each case, because the emission of the slower particle is followed by the emission of a gamma ray that removes the extra energy and leaves the nucleus in the same ground state as when a faster alpha particle is emitted. In a few cases, alpha particles are emitted at several speeds. ^{212}Bi emits alphas with five distinct speeds, ranging from 1.711×10^7 to 1.642×10^7 m/s. The emission of the slower alphas is also followed by the emission of gamma rays. On the other hand, some nuclides, ^{222}Rn, for example, emit alphas of one speed only. The energies of alpha particles emitted from an unknown material can sometimes be used as a method of identifying the nuclides present.

Beta particles, unlike alpha particles, are emitted from a given nuclide at all possible speeds from near zero to velocities of 99 percent of the speed of light. In other words, for two identical atoms, the beta particle from one can be of high energy and that from the other of low energy, yet the two atoms are at the same ground-state energy after the emissions. This apparent violation of the law of conservation of energy caused some early physicists to conclude that the conservation of energy did not apply to nuclear changes. To save the conservation law, Wolfgang Pauli in 1930 postulated that in beta emissions a neutral particle is emitted along with the electron and that the quantum of energy for the emission is shared between the two particles. This neutral particle, now called the **antineutrino,** has no rest mass or charge and travels at the speed of light.[1] It was not found by experimental work until 1956.

In the event that gamma radiation accompanies the emission of alpha and beta particles, the energies of the gamma radiation from a given nuclide are limited to definite values. Because gammas travel at the speed of light, the variation in energy means that the wavelengths of the emissions vary, as is the case with all radiant energy.

---

[1] In bombardment reactions (see Section 19.7), many particles are produced in artificial radioactive decay reactions. One of these particles is the **positron,** $_{+}^{0}e$, a subatomic particle with the same mass as an electron but with a unit positive charge. The emission of a positron is accompanied by the emission of a **neutrino,** a particle similar to an antineutrino. On the average, a positron exists for only 1×10^{-9} s before combining with an electron in a reaction that produces only radiant energy.

19•1•2   Ionizing Effects of Radiation   Radioactive emissions are called **ionizing radiations,** because they are able to cause the formation of ions in matter by knocking electrons off the atoms or molecules in their path. The chief effects of radioactivity on living plants and animals are traceable to these ionization reactions and the resulting chemical changes in cells. The ionizing power of alpha, beta, and gamma emissions is proportional to their energies.

Alpha particles produce from 50,000 to 100,000 ion pairs (a positive ion and an electron) per centimeter of path through air. Beta particles of similar energies produce only a few hundred ion pairs per centimeter of path. However, since the beta particle travels a greater distance before its energy is dissipated by collisions, the total number of ion pairs produced is about the same as that produced by an alpha particle of comparable energy. Gamma rays produce directly only a few ion pairs per centimeter as they pass through air, but they travel great distances.

The penetrating power of radioactive rays varies greatly. Depending on their energies, alpha rays have a range of about 2.8 to 8.5 cm in air. They can penetrate very thin metal foils but are stopped by a piece of ordinary paper. Beta rays penetrate thin sheets of metals. Gamma rays penetrate thick layers of metals (20 to 25 cm of lead). The relative penetrating power in aluminum is in the approximate ratio of $\alpha : \beta : \gamma = 1 : 100 : 10,000$.

19•2   Detection of Radiations

Because the emissions of radioactive substances are invisible, various indirect methods have been developed for detecting them. Four methods will be described. All are based on the fact that electrons are displaced to higher energy levels in the atoms and molecules affected by the emissions.

19•2•1   Photographic Methods   Photographic film and paper have long been used for the detection of radioactivity. The emissions affect the photographic emulsion in the same manner as ordinary light does. After exposure, the paper or film is developed in the usual way.

19•2•2   Fluorescent Methods   Many substances are capable of absorbing radiant energy of short wavelength (for example, gamma rays, X rays, ultraviolet rays) or kinetic energy from fast-moving particles (for example, beta and alpha particles) and transforming it to radiant energy of a wavelength that is in the region detectable by the eye. Substances that transform such energies into visible light are said to *fluoresce*. A common example of fluorescence is the luminous paint used on a watch dial. The paint ordinarily consists of one part of radium sulfate, $RaSO_4$, to 100,000 parts of zinc sulfide, ZnS. The invisible alpha emissions of the radioactive radium atoms bombard the zinc sulfide, with the result that some of the energy is transformed into visible radiant energy. The greenish glow is similar to that shown in Figure 3.2 on Plate 3.

19•2•3   Cloud Chambers   The *cloud chamber*, invented in 1911 by C. T. R. Wilson, an English physicist, makes it possible to see the path followed by a single ionizing radiation in its flight through a gas. A volume of air is saturated with water vapor and then cooled by expanding it rapidly. If there are no ions or other particles present to serve as nuclei for condensation to fog, the air becomes *supersaturated* with moisture. Substances other than water—for example, an alcohol—can be used to supersaturate the air with vapor, and other methods of cooling can be used.

Wilson found that when a radioactive substance was placed in the supersaturated air of the cloud chamber, thin lines of fog or cloud emanated from the radioactive material (see Figure 19.1).

Such cloud tracks result whenever an ionizing radiation flies through a gas that is properly supersaturated with a condensable vapor. The ion pairs formed serve as condensation centers for tiny droplets of liquid. The tracks give visible evidence of (1) the distances traveled by different kinds of particles, (2) the energies of particles (proportional to the number of ion pairs), (3) the collisions of particles, and (4) the variations in the paths of the particles affected by outside forces, for example, electrostatic attractions or magnetic fields.

The cloud chamber provides perhaps the most direct evidence for the existence of atoms. Not only can the tracks be observed with the unaided eye, but photographs also can be made and studied at leisure (Figure 19.2).

19•2•4   Gas Ionization Counters   In a *gas ionization counter*, an ionizing particle passes through a gas that is between two charged electrodes. The ions formed in the gas are attracted to the electrodes and cause a pulse of electric current to flow. Figure 19.3 shows a schematic diagram of a typical arrangement that takes advantage of the ionization phenomenon to count the number of charged particles that pass between two electrodes. The counting tube operates as follows:

1. A portion of a metal tube is cut out and replaced with a "window," a thin sheet of some material that can be penetrated by alpha or beta particles or gamma rays.[2]

2. A high potential is applied between an inside wire, *W*, and the outside of the metal tube.

3. A speeding particle or ray comes through the window and causes ionization of some of the gas molecules in the tube. The positive ions thus formed and the electrons knocked free conduct a pulse of current between the wire and the tube, and this current causes a detector light to flash or a clicker to sound. More elaborate models are provided with automatic recording devices, so that each surge of current is counted.

4. If the voltage between the electrodes is low, only the ions and electrons formed by the ionizing particle are available and the amount of current that flows is very slight; such a current would have to be amplified electronically in order to activate a light or clicker. If the voltage between the electrodes is high, the ions and electrons initially formed speed toward the electrodes with such force that they knock electrons off

---

[2]By using windows of different stopping power, radiations of different energies can be counted. Some detectors that are used for counting alpha particles are constructed so that there is no window between the source and the sensitive volume of the tube.

Figure 19•1 A cloud chamber. The emitter is glued onto a pin stuck into a stopper mounted on the chamber wall. The chamber has some methyl alcohol in the bottom and rests on dry ice. The cool air near the bottom becomes supersaturated with methyl alcohol vapor. When an emission speeds through this supersaturated vapor, ions are produced that serve as "seeds" about which the vapor condenses, forming tiny droplets, or fog.

Figure 19•2 A historic cloud chamber photograph of alpha tracks in nitrogen gas. The forked track was shown to be due to a speeding proton (going off to the left) and an isotope of oxygen (going off to the right). It is assumed that the alpha particle struck the nucleus of a nitrogen atom at the point where the track forks. (See Section 19.7.) Blackett made 20,000 such photographs (containing about 400,000 tracks) and found eight forked tracks.
Source: Photograph courtesy of P. M. S. Blackett at the Cavendish Laboratory.

molecules in their paths. The well-known Geiger counter tube operates at such a high voltage that a cascade of over a million electrons may be formed for each electron initially knocked free by a radioactive emission.

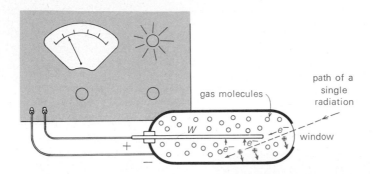

Figure 19•3 The principle of operation of a gas ionization counter.

19•3   Radioactive Series

A **radioactive series** is a collection of elements formed from a single radioactive nuclide by successive emissions of alpha or beta particles. Because each emission brings about the formation of an atom of a different element, the series begins with the radioactive decay of a *parent element* and continues from atom to atom until some nonradioactive atom is formed. Uranium-238 is the parent element for one naturally occurring series that contains 18 members (see Table 19.1), uranium-235 is the parent of a second series (15 members), and thorium-232 is the parent of a third decay series (13 members). A fourth series, beginning with plutonium-241 and ending with bismuth-209 and containing fifteen members, was discovered following the synthesis of certain elements not found in nature (see Section 19.8.3). Because the parent of this series, plutonium-241, does not occur in the earth's minerals to a measurable extent, the series is not considered a naturally occurring one.

There are certain nuclides in each of the series that can decay by emitting either an alpha or a beta particle. At these points, therefore, a series branches. In Table 19.1, the route taken by the fewer atoms in each branching is indicated by dashed arrows. It is interesting that immediately following a branching, the two different atoms formed in the first step decay to form identical atoms in the next step.

19•3•1   Alpha Emission

When an alpha or a beta particle is ejected from the nucleus of an atom, the resulting nucleus no longer has an atomic number characteristic of the original element. For example, a uranium-238 atom that emits an alpha particle suffers a loss of two neutrons and two protons. The atomic number is thereby reduced from 92 to 90, and the mass number is decreased from 238 to 234. In other words, the uranium-238 atom becomes a thorium-234 atom by the emission of an alpha particle. The alpha particle is identical with a helium nucleus and is indicated by the symbol $^4_2$He. The important generalization to be noted at this point is that *the emission of an alpha particle decreases the atomic number by 2 and the mass number by 4.*

It happens that atoms of the parent members of each of the three principal natural radioactive series decay by the emission of an alpha particle. With this information, one can write the equation for the nuclear change in

Table 19·1 Emissions and half-lives of members of natural radioactive series[a]

U-238 series	U-235 series	Th-232 series

U-238 series:

$$^{238}_{92}\text{U} \rightarrow \alpha \quad \downarrow 4.51 \times 10^9 \text{ y}$$

$$^{234}_{90}\text{Th} \rightarrow \beta \quad \downarrow 24.1 \text{ d}$$

$$^{234}_{91}\text{Pa} \rightarrow \beta \quad \downarrow 6.75 \text{ h}$$

$$^{234}_{92}\text{U} \rightarrow \alpha \quad \downarrow 2.47 \times 10^5 \text{ y}$$

$$^{230}_{90}\text{Th} \rightarrow \alpha \quad \downarrow 8.0 \times 10^4 \text{ y}$$

$$^{226}_{88}\text{Ra} \rightarrow \alpha \quad \downarrow 1.60 \times 10^3 \text{ y}$$

$$^{222}_{86}\text{Rn} \rightarrow \alpha \quad \downarrow 3.82 \text{ d}$$

$$\beta \leftarrow ^{218}_{84}\text{Po} \rightarrow \alpha$$
0.04% / \ 3.05 m

$$\alpha \leftarrow ^{218}_{85}\text{At} \quad ^{214}_{82}\text{Pb} \rightarrow \beta$$
2 s   26.8 m

$$\beta \leftarrow ^{214}_{83}\text{Bi} \rightarrow \alpha$$
99.96% / \ 19.7 m

$$\alpha \leftarrow ^{214}_{84}\text{Po} \quad ^{210}_{81}\text{Tl} \rightarrow \beta$$
1.6×10^{-4} s   1.32 m

$$^{210}_{82}\text{Pb} \rightarrow \beta \quad \downarrow 20.4 \text{ y}$$

$$\beta \leftarrow ^{210}_{83}\text{Bi} \rightarrow \alpha$$
~100% / \ 5.01 d

$$\alpha \leftarrow ^{210}_{84}\text{Po} \quad ^{206}_{81}\text{Tl} \rightarrow \beta$$
138 d   4.19 m

$$^{206}_{82}\text{Pb}$$

U-235 series:

$$^{235}_{92}\text{U} \rightarrow \alpha \quad \downarrow 7.1 \times 10^8 \text{ y}$$

$$^{231}_{90}\text{Th} \rightarrow \beta \quad \downarrow 25.5 \text{ h}$$

$$^{231}_{91}\text{Pa} \rightarrow \alpha \quad \downarrow 3.25 \times 10^4 \text{ y}$$

$$\beta \leftarrow ^{227}_{89}\text{Ac} \rightarrow \alpha$$
98.8% / \ 21.6 y

$$\alpha \leftarrow ^{227}_{90}\text{Th} \quad ^{223}_{87}\text{Fr} \rightarrow \beta$$
18.2 d   22 m

$$^{223}_{88}\text{Ra} \rightarrow \alpha \quad \downarrow 11.4 \text{ d}$$

$$^{219}_{86}\text{Rn} \rightarrow \alpha \quad \downarrow 4.00 \text{ s}$$

$$\beta \leftarrow ^{215}_{84}\text{Po} \rightarrow \alpha$$
5×10^{-4}% / \ 1.78×10^{-3} s

$$\alpha \leftarrow ^{215}_{85}\text{At} \quad ^{211}_{82}\text{Pb} \rightarrow \beta$$
10^{-4} s   36.1 m

$$\beta \leftarrow ^{211}_{83}\text{Bi} \rightarrow \alpha$$
99.7% / \ 2.16 m

$$\alpha \leftarrow ^{211}_{84}\text{Po} \quad ^{207}_{81}\text{Tl} \rightarrow \beta$$
0.52 s   4.79 m

$$^{207}_{82}\text{Pb}$$

Th-232 series:

$$^{232}_{90}\text{Th} \rightarrow \alpha \quad \downarrow 1.41 \times 10^{10} \text{ y}$$

$$^{228}_{88}\text{Ra} \rightarrow \beta \quad \downarrow 6.7 \text{ y}$$

$$^{228}_{89}\text{Ac} \rightarrow \beta \quad \downarrow 6.13 \text{ h}$$

$$^{228}_{90}\text{Th} \rightarrow \alpha \quad \downarrow 1.91 \text{ y}$$

$$^{224}_{88}\text{Ra} \rightarrow \alpha \quad \downarrow 3.64 \text{ d}$$

$$^{220}_{86}\text{Rn} \rightarrow \alpha \quad \downarrow 55.3 \text{ s}$$

$$\beta \leftarrow ^{216}_{84}\text{Po} \rightarrow \alpha$$
0.014% / \ 0.14 s

$$\alpha \leftarrow ^{216}_{85}\text{At} \quad ^{212}_{82}\text{Pb} \rightarrow \beta$$
3×10^{-4} s   10.6 h

$$\beta \leftarrow ^{212}_{83}\text{Bi} \rightarrow \alpha$$
66.3% / \ 60.6 m

$$\alpha \leftarrow ^{212}_{84}\text{Po} \quad ^{208}_{81}\text{Tl} \rightarrow \beta$$
3.0×10^{-7} s   3.10 m

$$^{208}_{82}\text{Pb}$$

Source: *From* Handbook of Chemistry and Physics, *R. C. Weast, Ed.,* © *CRC Press, Inc., 1978. Used by permission of CRC Press, Inc.; and from* Lange's Handbook of Chemistry, *12th ed., J. A. Dean, Ed., McGraw-Hill, New York, 1979.*

[a]The abbreviations are y, year; d, day; m, minute; and s, second.

the first step of each series:[3]

$$^{238}_{92}U \longrightarrow \,^{234}_{90}Th + \,^{4}_{2}He$$

$$^{232}_{90}Th \longrightarrow \,^{228}_{88}Ra + \,^{4}_{2}He$$

$$^{235}_{92}U \longrightarrow \,^{231}_{90}Th + \,^{4}_{2}He$$

The first equation is read: "Uranium-238 decays by alpha emission to yield thorium-234."

In these equations, note that the sum of the nuclear charges on the left side equals the sum of nuclear charges on the right side. Also the sum of the mass numbers on the left side of the arrow equals the sum of the mass numbers on the right side. As we shall see later, there is a slight change in mass (not in mass number), although this is not indicated by the equation. Because all alpha emitters have a neutron-proton ratio greater than 1, the emission of an alpha particle results in a slight increase in the neutron-proton ratio.

19•3•2   Beta Emission   The second step in each of the natural decay series involves the emission of an electron (beta particle) from the atoms formed in the first step. *The emission of a beta particle causes the atomic number to increase by 1, but the mass number remains unchanged.* It appears, therefore, that when an electron is emitted from the nucleus, a neutron changes to a proton. Consequently, beta emission results in a decrease in the neutron-proton ratio. The following equations show the second disintegration step for each of the three natural radioactive series:[4]

$$^{234}_{90}Th \longrightarrow \,^{234}_{91}Pa + \,^{0}_{-1}e$$

$$^{228}_{88}Ra \longrightarrow \,^{228}_{89}Ac + \,^{0}_{-1}e$$

$$^{231}_{90}Th \longrightarrow \,^{231}_{91}Pa + \,^{0}_{-1}e$$

The fourteen steps by which a radioactive ^{238}U atom changes to a stable ^{206}Pb atom are shown in detail in Table 19.1. In the ^{235}U and ^{232}Th

---

[3] In order to focus attention on the particles of greatest interest, nuclear chemical equations are purposely abbreviated. An example of a more complete equation is

$$^{238}_{92}U \longrightarrow \,^{234}_{90}Th^{2-} + \,^{4}_{2}He^{2+}$$

This equation takes into account the electrons around the atoms. The uranium atom had 92 protons and 92 electrons. The loss of an alpha particle leaves a particle with 90 protons and 92 electrons, that is, Th^{2-}. An isolated negative ion such as this would be unstable, and the two electrons would be lost immediately to nearby atoms or molecules, leaving the thorium atom $^{234}_{90}Th$. As the alpha particle flies away, it bumps into molecules and atoms that slow it down; eventually it gains two electrons and becomes a helium atom, $^{4}_{2}He$.

[4] Note that in these equations the negative charge on the electron is indicated by the -1 at the lower left of the e, so there is a balance of formal "atomic numbers" on each side of the equation. The antineutrino, without mass or charge, is not indicated in these equations. An example of a more complete equation is

$$^{234}_{90}Th \longrightarrow \,^{234}_{91}Pa + \,^{0}_{-1}\beta + \bar{\nu}_e$$

Here $^{0}_{-1}\beta$ is the symbol for an electron from the nucleus and $\bar{\nu}_e$ is the symbol for the antineutrino.

series, there are eleven and ten steps, respectively, before a stable isotope of lead is formed. The lead atoms that accumulate as the end products of the three natural series have different mass numbers, because the mass numbers of the parents, 238, 235, 232, are reduced by the same amount, 4, by each alpha emission (beta emissions do not change the mass numbers). The lead now present in uranium and thorium minerals was polonium, bismuth, radium, and other elements at various earlier times. Still earlier, the lead was uranium-238, uranium-235, or thorium-232.

The first two steps of the plutonium-241 decay follow the opposite order of emission of the three previously described series. That is, there is first a beta emission and then an alpha emission:

$$^{241}_{94}\text{Pu} \longrightarrow \,^{241}_{95}\text{Am} + \,^{0}_{-1}e$$
$$^{241}_{95}\text{Am} \longrightarrow \,^{237}_{93}\text{Np} + \,^{4}_{2}\text{He}$$

After ten more emissions, an atom of $^{209}_{82}\text{Pb}$ is formed. However, this isotope of lead is unstable and changes by beta emission to the stable end product, $^{209}_{83}\text{Bi}$.

──── • Example 19.1 • ────

Calculate to four significant figures the neutron-proton ratio, n/p^+, for each of the first three members of the uranium-235 series. Show how the magnitude of n/p^+ changes for alpha versus beta emission.

──── • Solution • ────

nuclide	n/p^+	$\Delta(n/p^+)$
$^{235}_{92}\text{U}$		
↓ $^{4}_{2}\text{He}$ emitted	$\dfrac{235-92}{92} = \dfrac{143}{92} = 1.554$	$\Delta = +0.013$ (for this α emission)
$^{231}_{90}\text{Th}$	$\dfrac{231-90}{90} = \dfrac{141}{90} = 1.567$	
↓ $^{0}_{-1}e$ emitted		
$^{231}_{91}\text{Pa}$	$\dfrac{231-91}{91} = \dfrac{140}{91} = 1.538$	$\Delta = -0.029$ (for this β emission)

See also Exercise 6 at the end of the chapter.

19.3.3 Gamma Emission Gamma radiation is a form of radiation that is similar in every respect to the longer known X radiation. After an alpha or a beta emission, for example, the nuclide produced may contain more energy than it does in its normal ground state. This excess energy may be emitted as gamma radiation, usually within about 1×10^{-12} s following the alpha or beta emission. The gamma emissions from a given nuclide have definite energies corresponding to the specific differences between the energy states of that nucleus (see Figure 19.4). One of the ways of determining the identity of a radioactive nuclide is to measure the energies of its gamma emissions; the pattern of emissions is different for each radioactive species, so the spectrum serves as a "fingerprint" identification. It is not uncommon for a given nuclide to have a dozen characteristic gamma emissions, and some synthetic nuclides have more than a hundred.

19•4   Background Radiation

There are radioactive species that occur naturally in practically all materials and that submit everything in our environment to a constant bombardment of subatomic particles and high-energy radiation. A major source of ionizing particles and radiation is **cosmic radiation,**[5] a rain of high-energy particles from outer space that strike the nuclei of atoms in the upper atmosphere and cause a cascade of secondary radiation that is similar in composition and effect to the radiations that arise from nuclear changes.

Together, natural radioactive substances and cosmic radiation contribute to an ever-present flux of *background radiation.* When a detection device is used, for example, a Geiger counter, the intensity of the background radiation must be measured in order that the reading due to it can be subtracted from the reading due to any sample that is to be tested.

19•4•1   The Curie and the Rem

The curie is a unit for specifying amounts of radioactivity. The unit is named for Marie Curie (1867–1934), the discoverer of polonium and radium and the first person to win two Nobel Prizes. In 1903 she shared the prize in physics with her husband, and in 1911 she was awarded the prize in chemistry.

Originally, the curie was defined in terms of a certain amount of radium. Today, a **curie** is defined as a source that decays at the rate of 3.700×10^{10} disintegrations per second. This is a rather large amount of radioactivity; one usually works in the laboratory with millicurie amounts.

Although the curie is a measure of the emission rate, it is not a satisfactory unit for setting up safety standards for handling radioactive materials. Different types of emissions have different penetrating and ionizing powers, so their biological effects are not the same. The biological effect of radiation is measured in units called rems. A **rem** is the amount of beta or gamma radiation that transfers 0.01 joule of energy to a kilogram of matter. A single exposure to 300 rems would result in the death within 30 days of 50 percent of the persons exposed. The minimum exposure that produces any immediately observable effect is about 50 rems. Background radiation is measured in millirems, as shown in Table 19.2.

It is interesting to consider some of the radioactive nuclides to which we are constantly exposed and which, to some extent, become a part of our bodies. Drinking water contains traces of compounds of uranium and uranium decay products such as radium in concentrations up to about 10^{-12} millicurie per mL. Milk contains about 6×10^{-11} millicurie of ^{40}K per mL, and living matter contains a small concentration of ^{14}C. It is estimated that the average human body contains enough radioactive nuclides to give rise to about 400,000 disintegrations per minute. It has been calculated that the disintegrations per minute arising from ^{40}K, ^{14}C, and ^{226}Ra indicate the presence in the body of about 10^{20}, 10^{14}, and 10^{11} atoms of ^{40}K, ^{14}C, and

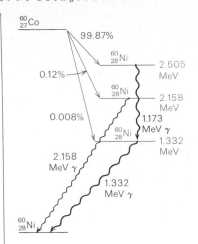

Figure 19•4 When cobalt-60 emits a beta particle, any one of three excited states of nickel-60 (shown in color) may be formed. The most common excited state, which is 2.505 MeV above the ground state, decays in two steps by emitting gamma radiations of 1.173 and 1.332 MeV. About 0.12 percent of the excited nickel nuclides formed are 2.158 MeV above the ground state; they decay in a single step.
Source: Courtesy of Dr. J. R. Peterson, The University of Tennessee, Knoxville.

---

[5]Cosmic rays are particles with very high kinetic energies, ranging upward from about 100 million electron volts (MeV). These particles are mainly hydrogen nuclei and, to a much lesser extent, helium, lithium, and other small nuclei. They move about through interstellar space at velocities approaching the speed of light. When they collide with components of the earth's atmosphere, a shower of secondary particles and rays is produced.

^{226}Ra, respectively. All of these, of course, are present in the body in compounds, not as elements.

19.5　Half-Life

Modern detection devices make it very convenient to measure the rate at which radioactive nuclides decay. How long does it take for a given sample of uranium to change to lead or, for that matter, how long does it take for any radioactive substance to change to other substances? The time required is independent of the amount of radioactive material present. For example, if it takes 1,600 years for half of a 1-g sample of radium-226 to change to radon, the same period of time (1,600 years) is required for half of a 2-g sample to change to radon, because the 2-g sample, with twice as many radium-226 atoms, will in all probability emit twice as many alpha particles as the 1-g sample does. That is, the rate of radioactive decay is proportional to the concentration or weight of just one reactant, and the reaction is first order.

The same equations that were discussed in Section 14.5.1 for first-order reactions are applicable here:

$$\text{rate} = k[A] \quad \text{and} \quad t_{1/2} = \frac{0.693}{k}$$

where k is the rate constant, $[A]$ is the concentration or weight of the radioactive nuclide, and $t_{1/2}$ is the time for half of a sample to decay. The half-life ($t_{1/2}$) of a radioactive substance is the length of time required for half of a given starting weight of a substance to change to other substances.

In Table 19.1, the half-lives of the members of the three natural radioactive series are given. The parent nuclide of each series has an extremely long half-life, of the order of billions of years.

Half-lives are determined experimentally by counting the number of emissions in a suitable period of time by a given weight of radioactive sample. Certain radioactive elements have very short half-lives and others have very long ones. For example, radon-219 is an alpha emitter that emits half of its total possible alpha particles in 4.00 s. Of a sample of pure ^{219}Rn weighing 0.008 g, only 0.004 g would be left after 4.00 s, because half of the original material would have turned into polonium-215:

$$^{219}_{88}\text{Rn} \xrightarrow[4.00\text{ s}]{t_{1/2}} {}^{215}_{86}\text{Po} + {}^{4}_{2}\text{He}$$

After 8.00 s, there would be only 0.002 g of ^{219}Rn; and after 12.00 s elapsed, only 0.001 g of the original 0.008 g would be left. *We cannot predict when any certain atom will decay, but we can predict the time required for half of a large number of them to do so.*

Two characteristic plots for a first-order reaction were given in Figure 14.9(a) and (b), and, as shown there, the half-life can be read directly. From a graph showing several half-lives (see Figure 19.5), it is clear that the activity of a radioactive substance theoretically never falls to zero. However,

Table 19.2 Background radiation in millirems per year

Average U.S. radiation per person from natural sources:	
from cosmic rays	45
from the earth	60
from own body	25
total	130
Variations due to certain activities:	
for a single round-trip, coast-to-coast flight at 40,000 ft	4
for an airline crew, per year	325
for average yearly medical exposure (chiefly due to X rays)	70
Variations due to living at different altitudes:	
sea level	35
5,000 ft	70
10,000 ft	140
Variations due to living in different structures:	
wooden	55
brick	80
granite	100
Variations due to geography (due mainly to altitude)	
Texas	100
California	115
Florida	120
Forty-one other states	120–180
South Dakota	200
New Mexico	210
Wyoming	245
Colorado	250

Source: "Nuclear Power in the Tennessee Valley," TVA, Knoxville, TN, 1978.

after a period of time equal to about ten half-lives, the radioactivity is so weak that it can hardly be measured. After x half-lives have elapsed, the number, n', remaining of a given original number, n, of radioactive atoms is[6]

$$n' = (\tfrac{1}{2})^x n$$

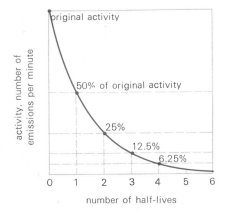
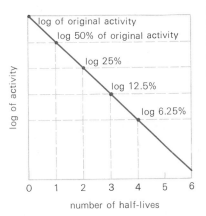

Figure 19•5 Two ways of plotting the loss in activity of a radioactive substance with time. Such first-order rate plots are characteristic of all radioactive substances; only the lengths of the half-lives vary. See also Figure 14.9.

• Example 19•2 •

The radioactivity of a 1.000-g sample of an alpha emitter was measured initially and again after 7.00 days; the second Geiger counter reading was only 27.9 percent of the initial reading. Calculate the half-life in days, assuming that all the radiation is emitted by one type of radioactive nuclide. Assume that, if any of the nuclides produced are radioactive, they are removed and not counted at the end of the seven-day period.

• Solution •

$$t_{1/2} = \frac{0.693}{k}$$

With the data given, we can use Equation (11) in Section 14.5.1 to calculate the value of the rate constant, k, and from this calculate the half-life, $t_{1/2}$. The weights of radioactive species are proportional to their concentrations and can be substituted for values of [A] in the equation:

---

[6] When n' is reduced to a very small number, say, a few thousand, the predictions based on the half-life become less sure. In the hypothetical case of two atoms remaining, it could not be said that one of them would surely emit its radiation before the passing of one half-life. It is an intriguing fact that, for a group of radioactive atoms, there is no known way of determining which atoms will be near the first or which will be near the last to emit their characteristic radiation.

$$k = \frac{2.303}{t}\left(\log \frac{[A]_0}{[A]_t}\right)$$

$$= \frac{2.303}{7.00 \text{ d}}\left(\log \frac{1.000}{0.279}\right)$$

$$= \frac{2.303}{7.00 \text{ d}}(\log 3.584)$$

$$= \frac{2.303}{7.00 \text{ d}}(0.5544) = 0.182 \text{ d}^{-1}$$

$$t_{1/2} = \frac{0.693}{0.182 \text{ d}^{-1}} = 3.81 \text{ d}$$

See also Exercises 12, 13, and 15.

• Example 19.3 •

If the emitter in Example 19.2 is radon-222, how many alpha particles are emitted per second by a 1.000-g sample?

• Solution • The product of the rate constant, k, times the number of atoms present is the number of disintegrations expected per unit time:

$$k = \left(\frac{0.182}{d}\right)\left(\frac{1 \text{ d}}{24 \text{ h}}\right)\left(\frac{1 \text{ h}}{3{,}600 \text{ s}}\right)$$

$$= 2.11 \times 10^{-6} \text{ s}^{-1}$$

$$\text{number of radon atoms} = 1.000 \text{ g}\left(\frac{6.02 \times 10^{23} \text{ atoms}}{222 \text{ g}}\right)$$

$$= 2.71 \times 10^{21} \text{ atoms}$$

$$\text{number of atoms decaying per second} = (2.71 \times 10^{21} \text{ atoms})(2.11 \times 10^{-6} \text{ s}^{-1})$$

$$= 5.72 \times 10^{15} \text{ atoms} \cdot \text{s}^{-1}$$

$$= 5.72 \times 10^{15} \text{ alphas} \cdot \text{s}^{-1}$$

See also Exercises 12, 13, and 15.

19.6 Applications of Radioactivity

19.6.1 Dating

• *Age of Minerals* • A fascinating calculation based on the concept of half-lives is the probable age of the earth. As shown in Table 19.1, the end product of the uranium-238 decay series is stable lead-206. In nature, where lead-206 and uranium-238 are found together in certain minerals, it is assumed that the lead has been formed as the result of radioactive decay. On the basis of the half-life of uranium, one can calculate the time required to establish the uranium-to-lead ratio found in the minerals. Because of the extremely long half-life of the parent element, the amounts in a mineral of the intermediate elements between uranium and lead are so tiny they can be ignored. Other pairs of parent–product substances used in dating minerals include ^{40}K–^{40}Ar and ^{232}Th–^{208}Pb.

Such calculations indicate that many rocks have existed in much their present state for billions of years. These and other studies support the theory that the earth is at least 4 to 6 billion years old. The ore described in Example 19.4 is about 1.2 billion years old, and may be sedimentary material formed long after the oldest igneous rocks.

—— • Example 19•4 • ——

Calculate the age of a uranium ore that contains 0.277 g of ^{206}Pb for every 1.667 g of ^{238}U.

—— • Solution • If we assume that all the ^{206}Pb came from the decay of ^{238}U, the weight of ^{238}U that has changed to ^{206}Pb is

$$\frac{238}{206} \times 0.277 \text{ g} = 0.320 \text{ g of } ^{238}\text{U changed to } 0.277 \text{ g of } ^{206}\text{Pb}$$

The weight of ^{238}U originally present was

$$1.667 \text{ g of } ^{238}\text{U} + 0.320 \text{ g of } ^{238}\text{U} = 1.987 \text{ g of } ^{238}\text{U}$$

To calculate the time for 1.987 g of ^{238}U to decay to 1.667 g of ^{238}U, we use the rate equations for first-order reactions, $k = 0.693/t_{1/2}$ and $kt = 2.303 \log [A]_0/[A]_t$. Using the first equation, the value of k is calculated from the known half-life of ^{238}U:

$$k = \frac{0.693}{4.5 \times 10^9 \text{ y}} = 1.5 \times 10^{-10} \text{ y}^{-1}$$

The value of k is then used in the second equation to calculate the time for the decay:

$$(1.5 \times 10^{-10} \text{ y}^{-1})t = 2.303 \log \frac{1.987}{1.667}$$

$$t = 1.2 \times 10^9 \text{ y}$$

See also Exercises 14 and 16–19.

Old rocks found in Finland and Canada have ages of about 3×10^9 years. Samples of some of the igneous volcanic rocks from the moon evidently solidified 3 to 4×10^9 years ago. An age of 4.6×10^9 years was calculated for a large meteorite that fell in Mexico in 1969 and from which about two tons of material was recovered. This is the oldest object yet found and indicates that the solar system is at least this old.

• *Age of Organic Materials* • An interesting method of dating ancient organic objects is based on the fact that the preserved object, if not too old, contains a measurable amount of radioactive carbon-14. In spite of its relatively short half-life of 5,730 years, a small amount of carbon-14 is present in the atmosphere (mainly as ^{14}CO$_2$). Although it is continually decaying to produce nitrogen, ^{14}C is also continually being produced by cosmic-ray activity that results in neutron capture by nitrogen atoms and the subsequent expulsion of protons:

$$^{14}_7\text{N} + ^1_0 n \longrightarrow ^{14}_6\text{C} + ^1_1\text{H}$$

Therefore, in the carbon dioxide in the atmosphere there is at all times a small quantity of $^{14}CO_2$ available to living plants. Once the life processes stop, carbon-14 is no longer taken up by the plant, and the amount of it in the plant tissues begins to diminish through radioactive decay.

We can measure the radioactivity due to carbon-14 per gram of carbon in living wood and compare this to the radioactivity in preserved wood or charcoal. From these values, we can calculate, using the known half-life of carbon-14, the time that must have elapsed to reduce the radioactivity to that of the preserved object. In most laboratories, the oldest objects that can be carbon-dated by their radioactivity are about 40,000 years old. Older samples have such low counting rates that background radiation interferes.

A newly developed technique for measuring small amounts of ^{14}C makes it possible to date organic objects 70,000 years and older. Instead of measuring the radioactivity of the sample, the concentration of ^{14}C is determined directly in an apparatus that operates on the basis of some of the same principles as does a mass spectrograph. Another advantage of the new method is that the sample size needed is only about $\frac{1}{100}$ g instead of about 1 g for the radioactivity measurement.

19.6.2   **Tracer Studies**   One of the chemical applications of radioactivity is in the study of reaction mechanisms. The photosynthesis equation,

$$6CO_2 + 6H_2O \xrightarrow[\text{chlorophyll}]{\text{sunlight}} C_6H_{12}O_6 + 6O_2$$

simply states that carbon dioxide and water react under the conditions shown to form glucose and oxygen. As pointed out in Chapter 14, complex reactions such as this proceed through a series of steps that usually involve one or two particles at a time. The use of radioisotopes to follow the course of such reactions has proved to be an invaluable tool. Isotopes used for this purpose are called **tracers.** In the case of photosynthesis, Melvin Calvin and his associates at the University of California used $^{14}CO_2$ to solve the very complicated course of this reaction. In this type of work, a detection device such as a Geiger counter is used to follow the radioactive atoms through the various intermediates to the final product.

Many chemical and biochemical processes are now studied, either in test tubes and beakers or in live animals and plants, via radioisotopes such as ^3H, ^{14}C, ^{24}Na, and ^{32}P.

Also, uncommon stable isotopes can be used as tracers. Samples of reactants and products are analyzed with a mass spectrograph to determine the reaction paths taken by certain isotopes during the reaction. Nonradioactive isotopes such as ^2H, ^{13}C, ^{15}N, and ^{17}O are particularly useful in following complex biochemical changes in living organisms, including human subjects. In tracer amounts, stable isotopes are considered completely safe because they are so nearly identical in chemical properties with the common isotopes. By labeling fat molecules with ^2H in place of ^1H, chemists have found that different persons digest and use fat in their bodies in quite different ways.

19•6•3 Additional Uses of Isotopes

In recent years all sorts of isotopes have become available. Most are produced by neutron bombardment in nuclear reactors (see Section 19.12) that are designed for that purpose. We shall mention only a few of the many uses of special isotopes.

Radioactivity is effectively used in the detection and treatment of illness. Technetium-99 ($t_{1/2} = 6.0$ h) can be incorporated in compounds that concentrate in malignancies. Its radiation then guides the surgeon in locating the diseased tissue. Thallium-201 ($t_{1/2} = 74$ h) is taken up quickly by normal heart muscle and blood vessel cells. Photographic images reveal the location of damaged or malfunctioning cells that do not absorb the isotope. Cancer of the thyroid is treated with iodine-131 ($t_{1/2} = 8.0$ d), which concentrates in that gland. Gamma-emitting units containing cobalt-60 are used in hundreds of hospitals in place of X-ray machines for the treatment of cancer, including leukemia.

Americium-241 (an alpha emitter, $t_{1/2} = 433$ y) is used in the ionization chambers of smoke-detecting devices that warn of possible fires (see Figure 19.6). Pellets so concentrated in strontium-90 ($t_{1/2} = 28$ y) or curium ($t_{1/2} = 18$ y) that they are red hot due to their intense energy emission are used as power sources for instruments sunk deep in the sea or on space probes.

Neutron bombardment (see Section 19.7.1) of natural materials has made possible one of the most effective ways to identify small samples of complex unknowns. When matter is bombarded by neutrons, many kinds of artificial radioactive nuclides are formed. By measuring the radiation spectrum of the bombarded sample, one obtains a *neutron activation analysis*. A few strands of hair found at the scene of a crime can be traced to the individual who lost them. Samples of oil illegally washed out of tankers at sea can be traced to the offending vessel.

19•7 Bombardment Reactions

We can observe natural radioactive changes and we can even concentrate and purify large quantities of radioactive materials, but we cannot control the nature of the particles emitted or the rate at which they are emitted. However, there is a second class of nuclear reactions, called bombardment reactions, that can be controlled. **Bombardment reactions** result when particles of atomic or subatomic size strike atoms of matter and permanently change these atoms. The idea that an atom of one element could be purposely changed to an atom of another violated a theory that had been universally accepted since Dalton's time. But in 1919 Blackett and Rutherford reported that this indeed happened when ordinary nitrogen was exposed to the alpha particles emitted by a small sample of radium. The equation for the reaction is

$$\ce{^{14}_{7}N + ^{4}_{2}He \longrightarrow ^{17}_{8}O + ^{1}_{1}H}$$

A high-speed alpha particle evidently meets a nitrogen nucleus head on with such force that it momentarily fuses with it. This fusion forms an unstable intermediate nuclear particle that ejects a proton, $^{1}_{1}H$, and leaves behind an atom of oxygen, $^{17}_{8}O$. The proton was identified by means of cloud chamber

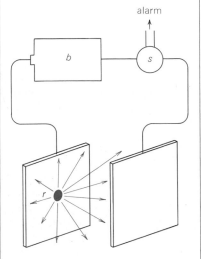

Figure 19•6 Ionization-type smoke detector. The battery, b, can force a small current to flow between the two plates, because the alpha emissions from the radioactive source, r, ionize the air between the plates. When smoke particles reduce the conductivity of the air, the drop in current triggers switch s, which turns on an alarm.

photographs (similar to that in Figure 19.2) that provided the data necessary to calculate its mass and charge.

Nuclear reactions of this kind generally do not give appreciable yields, for only rarely do the speeding alpha particles collide head on with an atomic nucleus. This is because the nucleus of an atom is so small compared with the total size of the atom. In the case of Blackett's work, most of the alpha particles emitted by the radium simply passed through the nitrogen gas without causing a reaction.

A most important bombardment reaction, historically speaking, was the reaction between beryllium atoms and alpha particles reported in 1930 by W. Bothe and H. Becker in Germany. This bombardment resulted in the appearance of a very penetrating radiation that had the characteristics of a stream of particles that was not deflected on passing through a magnetic field. In 1932 in England, James Chadwick showed that the particles had masses nearly equal to those of protons but were uncharged. Thus, the *neutron* was discovered:

$$^{9}_{4}Be + ^{4}_{2}He \longrightarrow ^{12}_{6}C + ^{1}_{0}n$$

19.7.1 Neutron Bombardments A neutron is a very effective "bullet" for bringing about a nuclear change. Unlike a positively charged proton, deuteron, or alpha particle, which is repelled by the positive nucleus of an atom, an uncharged neutron does not require a high velocity in order to strike a nucleus. Indeed, a slowly moving neutron passing close to a nucleus has a greater chance of being attracted into the nucleus than does a high-speed neutron, which might race on past.

Neutrons are easily generated by bombarding light elements such as lithium and beryllium with alpha particles or by the reactions occurring in a nuclear reactor (discussed in Section 19.12). The neutrons generated may cause additional reactions in materials placed in the target area. Two examples are

$$^{14}_{7}N + ^{1}_{0}n \longrightarrow ^{14}_{6}C + ^{1}_{1}H$$
$$^{35}_{17}Cl + ^{1}_{0}n \longrightarrow ^{35}_{16}S + ^{1}_{1}H$$

19.8 Acceleration of Charged Particles

Although atoms of low atomic numbers can be altered when struck by natural alpha particles, atoms of high atomic numbers are unaffected by these particles. These results indicate that atoms with highly charged positive nuclei repel low-energy, speeding positive alpha particles enough to prevent effective collisions with the nucleus. Calculations show, for example, that an alpha particle needs 27 MeV of energy in order to push in close enough to a radon nucleus so that short-range nuclear forces become effective and a radium nucleus is formed (see Figure 19.7):

$$^{4}_{2}He + ^{222}_{86}Rn \longrightarrow ^{226}_{88}Ra$$

Alpha particles from natural sources have energies ranging from about 5 to 8 MeV and are ineffective in producing such reactions. The energy of a particle can be increased in an **accelerator,** a device that increases the

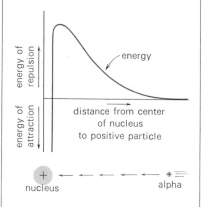

Figure 19.7 If an alpha particle has enough energy to overcome the coulombic repulsion of an atom's nucleus, it can penetrate close enough to the positive nucleus to come under the effective control of the powerful short-range nuclear attraction and so join the nucleus.

velocity of a charged particle and thereby increases its effect on an atom struck by it. Some of the modern particle accelerators now accelerate positive ions to energies in the 1,000-MeV range. Two types of these devices will be described.

19•8•1 Linear Accelerators A schematic diagram of an early model of a *linear accelerator* is shown in Figure 19.8. Suppose that an alpha emitter is placed at point zero. A (positive) alpha particle at 0 is attracted by cylinder 1 when 1 is negative. If the difference in potential is great enough, a particle that flies toward the middle of this cylinder tends to go right on through it. As the particle leaves the first cylinder, the charges on the alternating odd and even cylinders are reversed. Because of this reversal, the speeding particle is given a boost in speed between 1 and 2 (repelled by positive 1, attracted by negative 2). The automatic rapid alternation of charges on the odd and even cylinders causes the speed of the particles that are moving in phase with these changes to increase steadily. Protons, deuterons ($_1^2$H), alpha particles, and even heavy positive ions can be accelerated in this way.

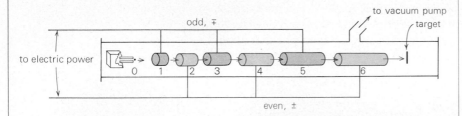

Figure 19•8 Diagram of an early type of linear accelerator. An alpha emitter is placed in the container at the left. Only those alphas can escape that happen to be emitted in line with the series of accelerating tubes.

When these high-speed projectiles strike the nuclei of the atoms in the target, nuclear reactions may occur. Examples are

$$_{19}^{39}K + _1^1H \longrightarrow _{18}^{36}Ar + _2^4He$$
$$_{83}^{209}Bi + _1^2H \longrightarrow _{84}^{210}Po + _0^1n$$
$$_3^7Li + _2^4He \longrightarrow _5^{10}B + _0^1n$$

Many nuclear changes are accompanied by the emission of gamma radiation, but it is not usually included in the equation unless particular importance is attached to it. The fact that a gamma ray is emitted in a reaction can be shown as in the equation that follows:

$$_6^{12}C + _1^1H \longrightarrow _7^{13}N + _0^0\gamma$$

More recently built linear accelerators (*linacs*) function in a somewhat different manner. Hollow tubes, in sections of equal length, serve as guides for the propagation of microwaves. When charged particles are injected into the instrument, they are carried forward by the traveling wave much as a surf rider is carried on the crest of a water wave. The largest linac built to date is at Stanford University. It consists of an almost 2-mile-long wave guide, made of 960 3-m sections of 10-cm copper tubing. It was designed to accelerate electrons to energies of about 2×10^4 MeV. Commercial linacs with a tube of only a meter in length are used by hospitals to supply high-energy electrons that bombard a gold target and generate X rays for medical radiation treatments.

19•8•2   Circular Accelerators

In circular accelerators, charged particles are accelerated by electric fields while being held in approximately circular orbits by magnetic fields. The operation of these devices is based on two well-known physical laws: (1) a charged particle is repelled by a like charge and attracted by an unlike charge, and (2) a charged particle that is moving in a magnetic field follows a curved path. Circular accelerators with a number of different designs have been invented. We will mention only two, the cyclotron and the synchrotron.

The *cyclotron* was the first circular accelerator to be developed. In the schematic diagram in Figure 19.9, the heart of the instrument is shown: a hollow disc split into halves (called *dees* because of their shapes). The electrostatic charge on the two dees is opposite in sign and can be rapidly alternated, + − + − + −, and so on. A positive particle is attracted back and forth between the dees, but because of the intense magnetic field the particle follows a curved path. As the particle goes faster and faster, the radius of its spiral path becomes larger and larger. At the end of its spiral, a speeding positive particle is deflected by the attraction of a negative electrode and flies out of the cyclotron to hit a target. The nuclear reactions that occur are similar to those carried out with a linear accelerator.

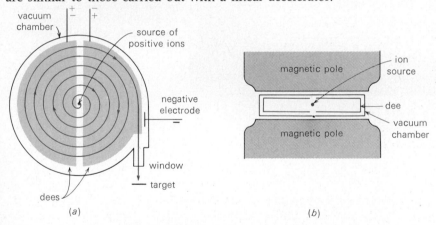

(a)   (b)

Figure 19•9 The cyclotron. (a) A horizontal view of the two dees, vacuum chamber, and path of the accelerated ions inside the dees. (b) A vertical cross section showing the placement of the dees between the poles of a magnet.

In the *synchrotron*, speeding particles are held in a path of constant radius, instead of a spiraling path, by a ring-shaped electromagnet that surrounds a circular tube through which the particles are orbiting. Because the magnet covers only the diameter of the tube, a few centimeters, it can be much smaller than the magnet of a cyclotron, whose field must cover the entire space through which the particles spiral. It is estimated that in a proton synchrotron, protons travel about 100,000 miles to attain an energy of 1.3×10^3 MeV. The journey takes about a second. Energies of over 1×10^5 MeV have been reached.

Over the past 50 years, the energies achieved in accelerators have increased by a hundred thousand times. Two of the most powerful systems today are at the Fermi National Accelerator Laboratory, Batavia, Illinois, and at the European Organization for Nuclear Research (CERN), Geneva, Switzerland. Both of these installations have huge magnetic storage rings. The

Fermi ring is over a mile in diameter. Within these rings, bunch after bunch of high-energy protons can be kept speeding in a circular path to build up an intense beam that can then be directed at a target. When nuclei are struck by such energetic particles, they shatter into a shower of subatomic particles. Almost 40 have been identified to date. Each time more powerful accelerators have been built, new subatomic particles have been discovered.

19•8•3   Synthesis of Elements   When it was found that atoms of one element could be changed into atoms of another by nuclear bombardment, there were still blank spaces in the periodic table between elements 1 and 92. Numbers 43, 61, 85, and 87 either had not been discovered or the evidence for their discovery was open to doubt. All of these were discovered shortly after 1937 by cyclotron bombardment or by nuclear fission. The last two have been found in minute amounts in nature.

   The synthesis of the unknown elements was guided by the principle that the nucleus of an atom that is bombarded with a small particle may absorb all or part of the particle and become a nucleus with greater mass and atomic number. In attempting to make the unknown element 43, it was logical to use a sample of molybdenum (number 42) as a target. The successful reaction was

$$\ce{^{98}_{42}Mo + ^{2}_{1}H \longrightarrow ^{99}_{43}Tc + ^{1}_{0}n}$$

The new element was named *technetium* (Greek *technetos,* artificial), because this was the first artificially made element. The synthesis of elements beyond atomic number 92, the transuranium elements, is discussed in Section 19.10.1.

19•9   Mass Loss and Binding Energy

19•9•1   Nucleons   Strictly speaking, it is incorrect to say that the nucleus of an atom contains protons and neutrons, just as it is incorrect to say that water contains hydrogen and oxygen atoms. In the latter case, atoms have interacted to form molecules and no longer have the properties of hydrogen and oxygen atoms. The term **nucleon** is often used to refer to a nuclear particle, irrespective of whether the particle is derived from a neutron or a proton. The statement, "a helium nucleus consists of four nucleons," is preferred to the statement, "a helium nucleus consists of two protons and two neutrons," for the former avoids the implication that protons and neutrons are packed in the nucleus like marbles in a box. However, as a matter of expediency, we shall continue to refer to nuclei as having definite numbers of neutrons and protons when it becomes necessary to point out differences in the composition of nuclei.

19•9•2   Mass Loss   An important discovery made during the 1930s was that the masses of atoms are always less than the sum of the masses of the individual electrons and nucleons that comprise them. To illustrate the difference, compare the mass of a $\ce{^{4}_{2}He}$ atom as determined in a mass spectrograph (see Figure 3.8) with the mass calculated by adding the masses of two electrons, two protons, and two neutrons. In this calculation, the mass of a hydrogen atom, ^1_1H, is taken as the mass of one electron and one proton, 1.007825 amu. The mass of one neutron is 1.008665 amu.[7] The calculated weight of a ^4_2He atom from these data is

$$2 \times 1.007825 = 2.015650 \text{ amu, mass of 2 protons and 2 electrons}$$
$$2 \times 1.008665 = \underline{2.017330 \text{ amu}}, \text{ mass of 2 neutrons}$$
$$4.032980 \text{ amu, calculated mass of } ^4_2He \text{ atom}$$

The **mass loss** of an atom is the difference between (1) the calculated sum of the masses of its total electrons plus nucleons, and (2) the actual mass of the atom as measured experimentally. Because the measured mass of a ^4_2He atom is 4.002603 amu, the mass loss is

$$4.032980 - 4.002603 = 0.030377 \text{ amu}$$

That is, a helium atom is about 0.8 percent lighter than we would expect from the masses of the electrons and nucleons that compose it.

As a second example, let us compare the mass of one of the isotopes of iron, $^{56}_{26}Fe$, with its calculated mass. This atom contains 26 protons, 26 electrons, and 30 neutrons:

$$26 \times 1.007825 = 26.20345 \text{ amu, weight of protons and electrons}$$
$$30 \times 1.008665 = \underline{30.25995 \text{ amu}}, \text{ weight of neutrons}$$
$$56.46340 \text{ amu, calculated weight of } ^{56}_{26}Fe$$

The measured mass of $^{56}_{26}Fe$ is 55.93494 amu. Hence, the mass loss is

$$56.46340 - 55.93494 = 0.52846 \text{ amu}$$

To compare the mass loss of a light nuclide with that of a heavy atom, we calculate for each the mass loss per nucleon. For a given nuclide, the **mass loss per nucleon** is the mass loss divided by the number of protons plus neutrons:

$$\text{for } ^4_2He, \frac{0.030377 \text{ amu}}{4} = 0.00759 \text{ amu per nucleon}$$

$$\text{for } ^{56}_{26}Fe, \frac{0.52846 \text{ amu}}{56} = 0.00944 \text{ amu per nucleon}$$

These calculations bring out an interesting and important relationship. Not only do neutrons and protons have slightly smaller masses when packed in the nuclei of atoms (as compared with their masses as isolated particles), but the amount of mass loss varies from one nuclide to another. For example, the mass of a neutron or proton is smaller by 0.00759 amu

---

[7] The mass of a neutron cannot be measured precisely by any existing method. It is calculated from the measured masses of a proton and a deuteron and from the energy required to effect the disintegration of a deuteron into a proton and a neutron.

when it is a part of a helium nucleus; the mass of the same particle is smaller by 0.00944 amu when it is a part of an iron nucleus.

If we make similar calculations for all the other elements, we find that the mass loss per nucleon is least for nuclides of very low mass numbers, for example, ^2_1H, ^6_3Li, ^7_3Li, ^9_4Be, and $^{10}_4Be$, and greatest for nuclides of mass numbers near $^{56}_{26}Fe$ and $^{58}_{28}Ni$. Figure 19.10 shows a plot of mass loss per nucleon, on the right ordinate, versus the mass number for various nuclides.

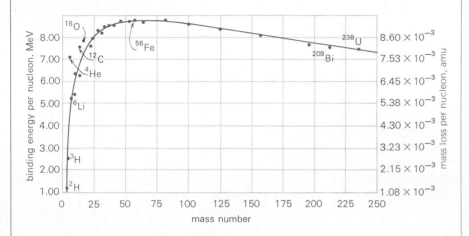

Figure 19•10 A graph showing binding energy per nuclear particle (left ordinate) and mass loss per nuclear particle (right ordinate), each plotted against the mass number of the nuclide.

19•9•3 Mass-Energy Equivalence

While the concepts *mass is a measure of the quantity of matter in a body* and *mass is neither created nor destroyed during transformations of matter* serve adequately for certain areas of science, they are not sufficiently precise for other areas. One use for which they are insufficient is to describe nuclear changes.

In 1905 Albert Einstein stated in his theory of relativity that the mass of a body is not necessarily constant. At the time this was a mind-boggling idea that seemed useless, if not absurd, to most scientists and nonscientists alike. The theory requires that for a particle in motion two masses must be recognized: (1) a rest mass, and (2) a relativistic mass that includes the rest mass and an added mass due to the kinetic energy (or other forms of energy) that the body possesses. According to Einstein's **law of the equivalence of mass and energy,** the mass to be ascribed to a certain amount of energy is given by the relationship

$$E = mc^2$$

where E is the energy in joules, m is the mass equivalent in kilograms, and c is the velocity of light in meters per second.

Einstein's relationship means, in effect, that mass must be recognized as another form of energy. A particle at rest possesses energy that is proportional to its rest mass, energy that can be thought of as *energy of matter*.

A particle in motion has a greater mass than when at rest because it possesses its rest mass as well as an added mass due to its energy of motion (kinetic energy). This added mass is not great unless the particle is moving at velocities that approach the speed of light. For example, a proton moving at

30,000 km/s (about 0.1 the speed of light) has a relativistic mass 1.005 times as great as its rest mass; at 0.5 the speed of light, a mass 1.155 times its rest mass; and at 0.9998 the speed of light, a mass 50.00 times its rest mass. It is seen from these figures that atoms, ions, and molecules that move with velocities of a few kilometers per second have relativistic masses so close to their rest masses that their masses can be treated in a nonrelativistic manner without introducing serious errors.[8]

A generation elapsed between Einstein's statement of his theory and its full acceptance as a law. The experimental verification was found in research on bombardment reactions. Measurements showed that the kinetic energy of the products formed was precisely equal to the energy calculated on the basis of $\Delta E = \Delta mc^2$, where Δm was the mass loss. An example of such a bombardment is

$$^7_3\text{Li} + ^1_1\text{H} \longrightarrow ^4_2\text{He} + ^4_2\text{He}$$

The mass loss is $8.0238 - 8.0052 = 0.0186$ amu, or 3.09×10^{-29} kg. The kinetic energy of the two alpha particles was determined, in a cloud chamber, to be 17.2 MeV. The agreement between this experimental value and the value predicted theoretically is good, as shown in Example 19.5. We may restate the laws of conservation of mass and energy as a single law. *In any process, the total mass-energy of an isolated system remains unchanged.*

• Example 19.5 •

For the reaction $^7_3\text{Li} + ^1_1\text{H} \rightarrow ^4_2\text{He} + ^4_2\text{He}$, show how well the measured total kinetic energy of 17.2 MeV for the two alpha particles produced accounts for the mass loss of 3.09×10^{-29} kg.

• Solution • Using the Einstein equation, we can predict theoretically the energy equivalent to the mass loss:

$$\Delta E = \Delta mc^2$$
$$= (3.09 \times 10^{-29}\text{ kg})(3.00 \times 10^8\text{ m} \cdot \text{s}^{-1})^2$$
$$= 27.8 \times 10^{-13}\text{ kg} \cdot \text{m}^2 \cdot \text{s}^{-2} = 27.8 \times 10^{-13}\text{ J}$$

From Table A.6 in the appendix, we find $1\text{ eV} = 1.6022 \times 10^{-19}$ J, so

$$\Delta E = (27.8 \times 10^{-13}\text{ J})\left(\frac{1\text{ eV}}{1.6022 \times 10^{-19}\text{ J}}\right)\left(\frac{1\text{ MeV}}{10^6\text{ eV}}\right)$$

$$= 17.4\text{ MeV predicted}$$

The experimental value of 17.2 MeV is in good agreement with this value of 17.4 MeV calculated theoretically.

See also Exercises 22, 40, and 42.

[8]Particles with kinetic energies approximately the same as gas molecules at room temperature are said to have **thermal energies.** Gaseous molecules such as H_2, O_2, N_2, and CO_2 have velocities of from a fraction to a few kilometers per second at 0 °C.

──── • Example 19•6 • ────

(a) Calculate the energy in kilojoules that would be produced by the nuclear reaction of 1 mole of lithium-7 with 1 mole of protons (total of 8.0238 g) to yield 2 moles of helium-4. See Example 19.5 for mass loss.
(b) Approximately what mass of gasoline and oxygen must react to yield the same amount of energy as that produced in part (a)?
(c) What would be the loss in mass of the gasoline and oxygen?
(d) Could the mass loss be measured by weighing the reactants and products in the combustion of gasoline?

──── • Solution • ────

(a) $\Delta E = \Delta mc^2 = \left(\dfrac{3.09 \times 10^{-29}\text{ kg}}{1\text{ atom }^7\text{Li}}\right)\left(\dfrac{6.022 \times 10^{23}\text{ atoms }^7\text{Li}}{1\text{ mol }^7\text{Li}}\right)(3.00 \times 10^8\text{m}\cdot\text{s}^{-1})^2$

$= 1.67 \times 10^{12}\text{ J}\cdot\text{mol}^{-1} = 1.67 \times 10^9\text{ kJ}\cdot\text{mol}^{-1}$

(b) We can use the compound octane as representative of the substances in gasoline. From Example 13.5,

$$C_8H_{18}(l) + 12\tfrac{1}{2}O_2(g) \longrightarrow 8CO_2(g) + 9H_2O(l) \qquad \Delta H = -5{,}470\text{ kJ}$$

$$\text{Mass gasoline + oxygen} = \left(\frac{114\text{ g }C_8H_{18} + 400\text{ g }O_2}{5{,}470\text{ kJ}}\right)(1.67 \times 10^9\text{ kJ})$$

$$= 1.57 \times 10^8\text{ g} = 1.57 \times 10^5\text{ kg}$$

$$= 1.57 \times 10^5\text{ kg}\left(\frac{1\text{ metric ton}}{1{,}000\text{ kg}}\right) = 157\text{ metric tons}$$

That is, it requires 157 tons(!) of gasoline and oxygen, reacting in an ordinary chemical reaction, to produce the same energy as 8.0238 g of lithium and hydrogen, reacting in a nuclear reaction.

(c) Because ΔE is the same for the nuclear reaction in part (a) as it is for the chemical reaction in part (b), the Δm is the same for both:

$$\Delta m = \frac{\Delta E}{c^2} = \frac{1.67 \times 10^{12}\text{ J}}{(3.00 \times 10^8\text{ m}\cdot\text{s}^{-1})^2}$$

$$= \frac{1.67 \times 10^{12}\text{ kg}\cdot\text{m}^2\cdot\text{s}^{-2}}{(3.00 \times 10^8\text{ m}\cdot\text{s}^{-1})^2}$$

$$= 1.86 \times 10^{-5}\text{ kg, mass loss}$$

(d)

mass $C_8H_{18} + O_2$	157,000.0000000 kg
mass loss	$-$ 0.0000186 kg
mass $CO_2 + H_2O$	156,999.9999814 kg

No balances ever made can determine weights to one part in 10^{12}. The calculation is made here with a ridiculous number of significant figures to emphasize that the tiny loss in mass could not be detected. By any method of weighing, the mass of the reactants is found to be precisely equal to the mass of the products in an ordinary chemical reaction:

$$\text{mass }(C_8H_{18} + O_2) = 1.57 \times 10^5\text{ kg} = \text{mass }(CO_2 + H_2O)$$

See also Exercises 22, 40, and 42.

19.9.4   Binding Energy   The energy that is equivalent to the mass loss for a given nuclide is called the **binding energy** of the nucleus. This energy may be thought of as the amount of energy that would be necessary to break an atom into its separated protons, neutrons, and electrons. Such an endothermic process is analogous to, though much more energetic than, the breaking of a molecule into the simpler molecules or atoms from which it is made.

It is convenient to interpret the mass loss curve in Figure 19.10 in terms of binding energies, which are plotted on the left ordinate in units of millions of electron volts per nucleon. As shown in Example 19.7, the binding energy in millions of electron volts per nucleon can be calculated by simply multiplying the mass loss in atomic mass units per nucleon by the mass–energy conversion factor of about 932 MeV/amu (or 931.479, to six significant figures). The graph in Figure 19.10 helps us understand several important concepts:

1. A number of irregularities occur for atoms of low mass numbers. In particular, ^4_2He, $^{12}_6C$, and $^{16}_8O$ have relatively large mass losses and binding energies per nucleon. These nuclei are especially stable for elements in this mass number range.

2. The mass losses and binding energies per nucleon are greatest for nuclei with mass numbers in the vicinity of 60 (iron and nickel). Neutrons and protons have their smallest masses in these nuclei. The remarkably high abundance of nickel and iron in the universe is thought to be associated with the great stability of nuclei of these elements.

3. When the nucleus of a very heavy atom splits into two or more nuclei of intermediate weight (of mass 70 to 160), there is a loss in mass, even though all protons, neutrons, and electrons are accounted for. This process is called **nuclear fission.**

    A well-known fission reaction is the splitting of a uranium-235 atom into two smaller atoms when struck by neutrons. Fission reactions are discussed in detail in Section 19.11.

4. When two nuclei of light atoms (mass less than 20) join to make one or more new nuclei, there is a loss in mass, even though the resulting atoms contain all the parts of the smaller atoms. This process is called **nuclear fusion.** Fusion reactions are also discussed in greater detail in Section 19.13.

5. Because a loss of mass must result in the appearance of an equivalent amount of energy, both nuclear fissions and nuclear fusions that proceed with the loss of mass are violently exothermic reactions. This is shown schematically in Figure 19.11.

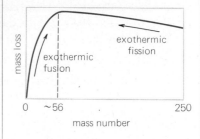

Figure 19.11 Regions in which exothermic fusions and exothermic fissions are theoretically possible. Compare with Figure 19.10.

• Example 19.7 •

Calculate the binding energy in millions of electron volts per nucleon for ^4_2He.

• Solution • From the Einstein equation, we calculate the energy associated with a mass loss of 1 kg:

$$\Delta E = \Delta mc^2 = (1 \text{ kg})(2.998 \times 10^8 \text{ m} \cdot \text{s}^{-1})^2$$

$$= 8.988 \times 10^{16} \text{ J}$$

Converting to MeV per amu:

$$\left(\frac{8.988 \times 10^{16}\,J}{1\,kg}\right)\left(\frac{1\,kg}{1,000\,g}\right)\left(\frac{1\,g}{6.022 \times 10^{23}\,amu}\right)\left(\frac{1\,eV}{1.602 \times 10^{-19}\,J}\right)\left(\frac{1\,MeV}{10^6\,eV}\right)$$

$$= 932\,MeV/amu$$

For $^4_2$He, the mass loss is 0.00759 amu per nucleon (see Section 19.9.2), so

$$E = \left(\frac{932\,MeV}{1\,amu}\right)\left(\frac{7.59 \times 10^{-3}\,amu}{nucleon}\right)$$

$$= 7.07\,MeV/nucleon\ for\ {}^4_2He$$

See also Exercise 23.

19•10 Nuclear Stability

All known elements have two or more nuclides. In a few cases (fluorine and aluminum are examples), only one nuclide occurs naturally, but synthetic nuclides of these elements have been produced by bombardment reactions. For elements with low and intermediate atomic numbers, most have both stable and unstable or radioactive nuclides. Consider hydrogen as an example. The nuclei of protium and deuterium atoms are stable, but those of tritium atoms are quite unstable. Tritium has such a short half-life (12.26 y) that none would be found in nature if it were not being continually formed in very small amounts by natural nuclear changes. Both stable and unstable nuclides are known for all the common elements, such as sodium, oxygen, nitrogen, chlorine, carbon, potassium, and silver. However, many of the unstable nuclides that have been synthesized do not occur appreciably, if at all, in nature.

For elements with high atomic numbers, beginning with $_{84}$Po, no stable nuclides are known, although some nuclides are much less stable than others. When we say nuclide A is less stable than B, we simply mean that A has the shorter half-life.

Apparently, one of the factors related to nuclear stability is a favorable neutron-proton ratio. Consider Figure 19.12, in which the number of protons is plotted against the number of neutrons for the stable nuclides. As the atomic number increases, the ratio of neutrons to protons increases; that is, neutrons become relatively more numerous in a given nucleus. For the nuclide $^{12}_6$C, the neutron-proton ratio is 6:6 or 1; for $^{200}_{80}$Hg, it is 120:80 or 1.5. That is, although the nuclei of stable carbon atoms contain an equal number of neutrons and protons, those of stable mercury atoms have 50 percent more neutrons than protons. If the neutron-proton ratio of a nuclide lies outside the so-called favorable belt for nuclear stability, the nuclide is radioactive. For example, both $^{14}_6$C and $^{205}_{80}$Hg are radioactive. To provide for a more favorable neutron-proton ratio, radioactive nuclides emit particles and, in so doing, achieve a neutron-proton ratio that provides for stability. As pointed out in Example 19.1, the n/p^+ ratio is increased by alpha emission and decreased by beta emission.

Except for the very light elements, there are more nuclides that have an even number of either neutrons or protons than those that have an odd number. From experimental observations and calculations of the binding

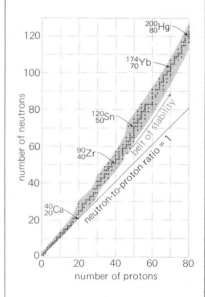

Figure 19•12 The increase of neutrons over protons as the atomic number increases is shown for stable nuclides through atomic number 80. Note that the nuclide $^{40}_{20}$Ca is the heaviest stable nuclide with a neutron-to-proton ratio of 1.

energies of nuclei, it has been found that there is a special stability associated with nuclei having so-called *magic numbers* of either protons or neutrons. These numbers are 2, 8, 20, 28, 50, 82, and 126. Nuclei that have magic numbers of *both* protons and neutrons are especially stable. Examples are $^{4}_{2}He$, $^{16}_{8}O$, $^{40}_{20}Ca$, and $^{208}_{82}Pb$. The word "magic" has been applied to these numbers to remind us that the mystery of their obvious importance has yet to be satisfactorily accounted for by any theory.

Several models have been proposed to describe the organization of nucleons in atomic nuclei. One such model, the *shell model*, is particularly interesting to the student of general chemistry. In this model, the principles of quantum mechanics (see Section 4.5) are used to describe the nucleus in a manner analogous to the description of electrons. The confined waves of nucleons are considered to produce vibrational patterns similar to electron-wave patterns, thus giving rise to unique energy levels.

An important aspect of the shell model from the chemist's point of view is that each energy level can accommodate a definite number of nucleons, just as each energy level of an atom can contain no more than a specified number of electrons. The magic numbers are believed to be the numbers of nucleons that fill nuclear shells or main energy levels.

★ Special Topic ★ Special Topic ★ Special Topic ★ Special Topic ★ Special Topic ★ Special Topic ★ Special Topic ★ Special Topic ★

Possible Superheavy Nuclides The magic numbers that are associated with known very stable nuclei have been correlated with calculations based on various models of the nucleus. The stability of a nucleus is a function of the number of neutrons, the number of protons, and the nuclear shape. When the outer shell is closed, the nucleus is spherical and is especially stable. The nuclei of uranium atoms are definitely oval, for example.

The accompanying figure is analogous to a topographic map, with "land areas" corresponding to regions of nuclear stability and regions below "sea level" representing possible unstable nuclei. The long peninsula shows the known nuclei. Note the peak of stability at ^{208}Pb at the intersection of the two magic numbers $p^{+} = 82$ and $n = 126$; lead-208 is an example of a *doubly magic* spherically symmetrical nucleus.

Theoretical calculations lead to the prediction that the doubly magic nucleus $^{298}114$ should be at the center of an "island of stability," but no nuclei as heavy as those on this island have as yet been discovered or synthesized. For a very heavy nucleus, the crucial question is whether it is stable toward spontaneous fission. If it is stable toward fission, the next question is, "What is its half-life with respect to alpha and/or beta decay?" If a combination of stability factors results in a nucleus that is stable for as long as a few billionths of a second, a small sample of the element should be detectable by modern methods.

Physicists hope to modify existing accelerators or construct new ones with which heavy, neutron-rich nuclei can be accelerated. The bombardment of uranium-238 with high-velocity uranium-238 nuclei might lead to the synthesis of a nucleus on the island of stability. For example:

$$^{238}_{92}U + {}^{238}_{92}U \longrightarrow {}^{298}114 + {}^{170}_{70}Yb + 8{}^{1}_{0}n$$

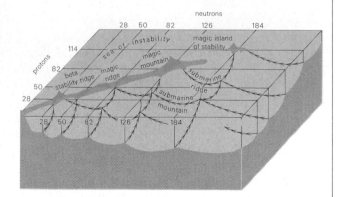

This figure shows the known and predicted regions of nuclear stability surrounded by a sea of instability.
Source: This special topic is adapted by permission from the article by G. T. Seaborg, "Prospects for Further Considerable Extension of the Periodic Table," *J. Chem. Educ.*, 46, 626 (1969).

19•10•1 The Transuranium Elements The decade of experimentation that led to nuclear fission, to the first elements with atomic numbers above 92 (transuranium elements), and to nuclear bombs began shortly after the discovery of neutrons in 1932. One aim of this early work (1934 to 1938) was to prepare elements beyond uranium, the last element in the periodic table at that time. Enrico Fermi and his associates in Rome predicted that ^{238}U, on neutron capture, could form unstable ^{239}U. They predicted that ^{239}U could decay by beta emission to form an element with atomic number 93, and that element 93 could, by beta decay, form element 94. In 1940 E. M. McMillan and P. H. Abelson, at the University of California, showed that element 93 was indeed formed from ^{238}U in the following reactions:

$$^{238}_{92}U + ^{1}_{0}n \longrightarrow ^{239}_{92}U$$
$$^{239}_{92}U \longrightarrow ^{239}_{93}Np + ^{0}_{-1}e$$

The name *neptunium* was given to element 93, because the planet Neptune is beyond Uranus, just as the new element was beyond uranium in the periodic classification. Because of the very small amount of neptunium produced in these experiments, the researchers were unable to show that element 94 was formed by the beta decay of neptunium.

The study of the transuranium elements at Berkeley was continued by G. T. Seaborg and his associates. To provide for larger scale experiments, they bombarded uranium with cyclotron-accelerated deuterons and were able to obtain enough neptunium to study it more fully. They showed that it does indeed decay to form element 94:

$$^{238}_{92}U + ^{2}_{1}H \longrightarrow ^{238}_{93}Np + 2^{1}_{0}n$$
$$^{238}_{93}Np \longrightarrow ^{238}_{94}Pu + ^{0}_{-1}e$$

The new element was named *plutonium* after the planet Pluto, which is just beyond Neptune in the solar system.

Seaborg and the group at the University of California at Berkeley, led by Albert Ghiorso in recent years, have played the major role in the development of transuranium-element chemistry. In addition to neptunium and plutonium, the Berkeley group is credited with the original syntheses of americium, curium, berkelium, californium, mendelevium, and lawrencium. Einsteinium and fermium were first identified in radioactive mixtures produced by nuclear bombs. Nobelium may also have been made first at Berkeley, but credit for the first synthesis was for some time a matter of dispute with a group that worked under the auspices of the Nobel Institute. The synthesis of element 106 was reported in 1974 both by a Russian group at the Laboratory for Nuclear Research, Dubna, U.S.S.R., and by the American group at Berkeley. In 1977 the Russian group claimed the synthesis of $^{261}107$.

19•11   Nuclear Fission

19•11•1 Fission Bombs Toward the close of World War II, nuclear bombs were developed that release the huge amounts of energy of exothermic nuclear fission reactions. Uranium-235 constituted the fissionable material in the first bomb used in warfare, and plutonium-239 was the fissionable material in the second and last one.

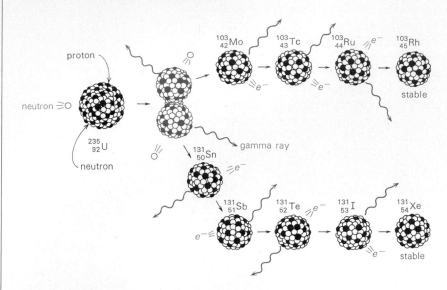

Figure 19•13 A uranium-235 nucleus, after capturing a neutron (*left*), fissions into two smaller nuclei with the emission of gamma rays (colored arrows) and two or three neutrons. The uranium-235 nuclei can fission in over 30 ways, producing a total of about 200 radioactive species, generally with atomic numbers 30 to 64 and masses 72 to 161.

Although all the elements heavier than bismuth are radioactive, only the nucleus of one natural nuclide, uranium-235, splits almost instantaneously into two fragments (undergoes fission) when struck by a slow-moving neutron. Plutonium-239 and uranium-233, which are synthetic nuclides, are also fissionable. Their production is discussed in the next section.

The fissioning process for uranium-235 nuclei is represented schematically in Figure 19.13. Typical equations for two of the 30 or more ways that uranium atoms may split are

$$\frac{1}{0}n + \frac{235}{92}U \longrightarrow \frac{103}{42}Mo + \frac{131}{50}Sn + 2\frac{1}{0}n$$

$$\frac{1}{0}n + \frac{235}{92}U \longrightarrow \frac{139}{56}Ba + \frac{94}{36}Kr + 3\frac{1}{0}n$$

The fissioning of uranium-235 by slow-moving (thermal) neutron bombardment produces a variety of primary nuclides, as shown in Figure 19.14. Primary fission products, for example, $^{103}_{42}Mo$, $^{131}_{50}Sn$, $^{139}_{56}Ba$, and $^{94}_{36}Kr$, have unstable, high n/p^+ ratios and decay by a series of beta (and gamma) emissions until stable nuclides are formed, as shown in Figure 19.13 for $^{103}_{42}Mo$.

• *Critical Mass* • Note in the two equations written above that each fission produces more neutrons. With the discovery that commonly two or three neutrons are released when a uranium-235 atom fissions, it was realized that neutrons could multiply. Nuclear fission can be a *chain reaction*, because the neutrons that are the products of one step can be reactants in the next step, and so on. The reaction kinetics are similar to those for the reactions described in Section 14.7.

Actually, under the proper conditions, billions of neutrons become available in a fraction of a second. Fission follows fission on a huge scale with the release of a tremendous amount of energy. To understand how the fission

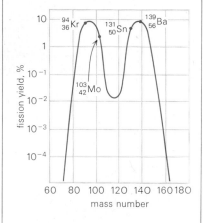

Figure 19•14 The percentage of nuclides of different mass numbers formed by the fission of uranium-235, induced by slow neutrons. The most common fissions produce one nucleus with a mass of about 95 amu and one of about 138 amu. A symmetrical fission, to produce two nuclei of about 117 amu each, occurs in only about 0.01 percent of the reactions.
Source: Adapted from the "Plutonium Project," J. M. Siegal, Ed., *J. Amer. Chem. Soc.*, **68**, 2437 (1946).

chain reaction can be controlled, first consider a natural uranium ore. Only a low concentration of uranium-235 is present in the ore, so that there is small probability of a neutron hitting a ^{235}U atom. Some fissions occur because of neutrons that are set in motion by cosmic radiation, but such natural fissions are relatively rare. If fissions are widely separated in space or in time, the energy is easily dissipated and the extra neutrons are absorbed by the nonfissionable atoms in the mineral. The chain reaction is not self-sustaining.[9]

Next, consider a sphere of pure uranium-235. If the sphere is small, about the size of a marble, most of the neutrons that are produced by occasional fissioning of uranium-235 atoms escape from the sphere. This occurs because the nucleus of an atom is extremely small compared with the total volume of the atom. It is quite improbable, therefore, that a neutron will accidentally hit a tiny nucleus as it passes through the atoms in a thin piece of uranium. Under these conditions, also, the chain reaction is not self-sustaining.

Now consider a larger sphere of pure uranium-235. Because the neutrons have to travel through a larger number of atoms before leaving this sphere, more of them will collide with uranium-235 nuclei than in a smaller sphere. As the sphere increases in size, a mass will be reached in which, on the average, one and only one neutron from each nucleus that undergoes fission will produce splitting in another nucleus. The remaining neutrons escape or are lost.[10] The chain reaction is now self-sustaining.

The amount and arrangement of fissionable material in which each fission produces only one new fission is called the **critical mass;** the neutron reproduction factor equals 1. Masses below this are said to be **subcritical;** the neutron reproduction factor is less than 1. Masses larger than the critical mass are said to be **supercritical;** the neutron reproduction factor is greater than 1 (see Figure 19.15). The actual amount of fissionable material needed to reach a critical mass depends on several factors, such as purity, the presence or absence of a casing to reflect neutrons, the shape of the fissioning material, and what isotope is fissioning. For example, the critical mass of a bare ball of 93 percent ^{235}U is 48.8 kg, but when the ball is surrounded by a casing of natural uranium, the critical mass is only 16.65 kg. The critical mass of a bare ^{239}Pu sphere is 16.28 kg, but when encased in natural uranium, it is only 5.91 kg.

In supercritical masses of fissionable substances, the reproduction of neutrons takes place at a fantastic rate, so that the supercritical mass becomes a beehive of neutrons within one millionth of a second. Under these conditions, fissioning is so rapid that the temperature rises to about 10,000,000 °C or more. Within a fraction of a second after becoming supercritical, the whole mass explodes into many subcritical masses.

---

[9]In a uranium mine in Africa, there is evidence of a prehistoric, natural, self-sustained fission reaction. The evidence was discovered during efforts to learn why uranium ore from one section of the mine has an unusually low ratio of ^{235}U to ^{238}U.

[10]The loss of neutrons may be due to a variety of factors in addition to the obvious one of neutrons missing all surrounding nuclei and speeding off into space. Bumping into the nuclei of impurities uses up some neutrons, and ineffective collisions with ^{235}U nuclei use up a considerable number. Approximately five sixths of the collisions with ^{235}U produce fission; one sixth produces ^{236}U nuclei.

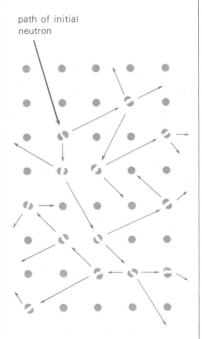

path of initial neutron

Figure 19•15 A schematic representation of how neutrons become very numerous in a supercritical mass of ^{235}U. (Only nuclei are represented; because the diameter of a uranium nucleus is so tiny relative to the diameter of the atom, the sizes of the nuclei are necessarily exaggerated here relative to the spaces between them.)

To cause an atomic bomb to explode, sufficient subcritical masses are suddenly brought together to make a supercritical mass. The neutrons and fissioning atoms must be held together long enough for the chain reaction to build up to a considerable rate. One way this is done is by enclosing the uranium (or plutonium) in a strong, dense case that reflects neutrons back into the fissionable material and also retards the bomb burst (see Figure 19.16).

The explosive power of fission and fusion bombs is usually described in terms of megatons. One **megaton** is equivalent to the energy released by the explosion of 1 million tons of TNT. The fissioning of about 50 kg of ^{235}U or ^{239}Pu releases this much energy. However, the fission bombs that were made were not this powerful. Those dropped in World War II were about 0.02-megaton bombs.

19.11.2 Sources of Fissionable Material Of the uranium present in a uranium ore, only 0.7 percent is the 235-isotope; the remainder is essentially uranium-238. Separation of the isotopes of an element is very difficult; because isotopes have practically identical chemical properties, ordinary chemical or physical methods of separation are not effective.

• *Gaseous Diffusion Separation*[11] • The rates of effusion of gases depend on their molecular weights, in accordance with the following expression:

$$\frac{r_1}{r_2} = \frac{\sqrt{m_2}}{\sqrt{m_1}}$$

where m_1 and m_2 are the molecular weights of two different molecules, and r_1 and r_2 are the rates of effusion (Graham's law; see Section 7.9). Above 57 °C, uranium hexafluoride, UF_6, is a gas whose molecules consist of $^{235}UF_6$ and $^{238}UF_6$. The weights of these molecules are 349 and 352 amu, respectively. Hence, if a portion of a sample of uranium hexafluoride diffuses through a porous barrier into a vacuum, the portion that diffuses will contain a slightly higher percentage of $^{235}UF_6$ than the undiffused portion does. If the process is repeated a sufficient number of times, practically pure $^{235}UF_6$ can be obtained.

Separation by gaseous diffusion is the way uranium isotopes are separated on a large scale. A gas centrifugation method, analogous to a cream separator, can also be used.

• *Plutonium-239 and Uranium-233* • The scarcity of uranium-235 in nature stimulated development of a way of synthesizing fissionable material. This has been accomplished by converting nonfissionable uranium-238 into fissionable plutonium-239 and converting thorium-232 into fissionable uranium-233 by neutron capture in nuclear reactors. The equations are

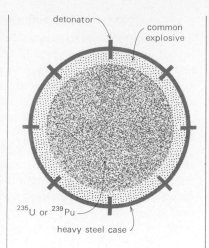

Figure 19.16 A method for producing an atomic explosion. When the common explosive, for example, TNT, in the outer portion of the sphere is detonated, the implosion wave compresses the loosely packed ^{235}U or ^{239}Pu into a compact, supercritical mass that explodes in a fraction of a second.

---

[11]As pointed out in the discussion of Graham's law in Chapter 7, the process described by the law is effusion rather than diffusion. However, the term *gaseous diffusion* is so widely used to refer to the separation of uranium isotopes that we will use it here.

for plutonium-239 production   for uranium-233 production

$$^{238}_{92}U + ^{1}_{0}n \longrightarrow ^{239}_{92}U \qquad\qquad ^{232}_{90}Th + ^{1}_{0}n \longrightarrow ^{233}_{90}Th$$

$$^{239}_{92}U \xrightarrow[24\ min]{t_{1/2}} ^{239}_{93}Np + ^{0}_{-1}e \qquad\qquad ^{233}_{90}Th \xrightarrow[22\ min]{t_{1/2}} ^{233}_{91}Pa + ^{0}_{-1}e$$

$$^{239}_{93}Np \xrightarrow[2.3\ d]{t_{1/2}} ^{239}_{94}Pu + ^{0}_{-1}e \qquad\qquad ^{233}_{91}Pa \xrightarrow[27\ d]{t_{1/2}} ^{233}_{92}U + ^{0}_{-1}e$$

19•12   Nuclear Reactors

The fissioning of uranium-235, uranium-233, or plutonium-239 can be controlled so that a chain reaction occurs without a disastrous explosion. Under the proper conditions, the energy that would be liberated in a fraction of a second in a nuclear explosion is liberated over a period of several days or weeks and is harnessed to produce useful work. An apparatus in which a controlled nuclear change takes place is called a **nuclear reactor.**

Over a hundred major nuclear reactors have been constructed in the United States and other countries. Although they have different designs, types of fuel, coolants, and so on, all operate in accord with certain fundamental physical principles. We can make these principles clear by discussing the construction and operation of the Browns Ferry Nuclear Plant, which is located in Alabama and operated by the Tennessee Valley Authority (TVA). Before taking up details, it is thought-provoking to realize that the single Browns Ferry plant has the capacity to produce 3,456,000 kW of electric power. The 29 dams and hydroelectric generators that TVA operates over several hundred miles of the Tennessee River and its tributaries have the capacity to produce 3,256,000 kW. TVA has under construction or in the final planning stages six more nuclear plants.

A nuclear power plant is similar to a coal-fired, gas-fired, or oil-fired power plant, in that each produces heat that is used to vaporize water into steam (see Figure 19.17). The gaseous steam rushes with great force from a hot high-pressure region to a cooler low-pressure area where it is condensed to liquid water. The force of the moving steam is used to turn huge turbines that are connected to electromagnetic generators of current.

Figure 19•17 In both coal-fired and nuclear electric power plants, energy from an exothermic process is used to produce steam. The steam, in turn, powers turbines that generate electricity.
Source: "Browns Ferry Nuclear Plant," TVA, Knoxville, Tenn., 1978.

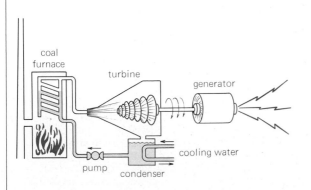

Coal-fired

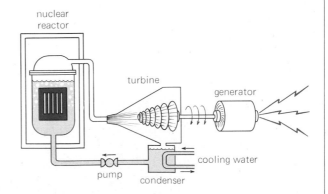

Nuclear

19•12•1 Controlling the Nuclear Reaction A controlled nuclear reaction is made possible by three essential components: a fissionable material, a moderator, and a neutron absorber (or control material). The fissionable material must be concentrated enough and arranged in such a way that a critical mass exists. The high-velocity neutrons produced by a fission are not as effective as low-velocity neutrons in causing additional fissions (see Figure 19.18). For this reason, a *moderator*, a material that slows down neutrons, is placed between small blocks or pellets of the fissionable material. To regulate precisely the flux of speeding neutrons, *control rods*, made of a material that completely absorbs neutrons, are moved into and out of the reactor as needed.

Graphite, heavy water (deuterium oxide), and light water (ordinary water) are the most common moderators. Boron and cadmium nuclei are the most common neutron absorbers. In the Browns Ferry reactor, uranium-235 is the fissionable material, ordinary water is the moderator, and boron carbide control rods are used. The uranium is in the form of small, bricklike pellets of uranium dioxide, UO_2, encased in metal (see Figure 19.19). The pellets are contained in metal tubes about 4 m (12 ft) long and about 1.2 cm (0.5 in.) in diameter. The tubes are bound together as a bundle of 196 tubes (see Figure 19.20), which is placed vertically in the water-filled reactor core. There are 191 such bundles in the complete core. Each bundle has in it a control rod, which is operated from the bottom of the reactor. Until these boron carbide absorbers of neutrons are withdrawn, no sustained chain reaction can occur.

The Browns Ferry reactor is a boiling-water reactor (see Figure 19.21). When the control rods are withdrawn to the point that the chain reaction is sustained, intense heat is produced, which causes the water around the core elements to boil. The reactor vessel acts as a giant pressure cooker that produces steam at about 285 °C and a pressure of 70 atm. This high-pressure steam rotates a turbine as it rushes toward the condenser region where it becomes a liquid at about 190 °C and 12 atm.

The rate of fissioning, and thereby the rate of power production, can be maintained automatically by the partial insertion and removal of the control rods. The rate of fissioning is also controlled by the amount of water flow. The water acts as the moderator. The more slowly the water flows past the hot tubes, the greater is the amount of boiling. The bubbles do not slow down neutrons effectively. So when the water boils too vigorously, there are fewer low-velocity neutrons and, therefore, fewer fissions. At an increased water speed, the amount of bubbling is decreased, the water is a better moderator, and the rate of fissioning increases.

19•12•2 Long-Term Reactor Operation As the ^{235}U atoms in the arrays of pellets undergo fissions, the radioactive fission-product nuclides are trapped inside the metal pellet containers. Some of these nuclides are effective neutron absorbers that interfere with the chain reaction. In a boiling-water reactor, from one fourth to one third of the fuel bundles must be replaced each year, both to replace the ^{235}U that has been used and to decrease the concentration of neutron-absorbing fission products.

The used fuel bundles are highly radioactive. Not only are the fission products radioactive, but the intense neutron bombardment produces "hot"

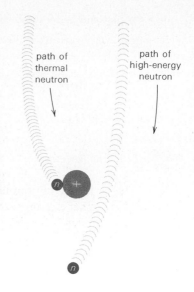

path of thermal neutron

path of high-energy neutron

n +

n

Figure 19•18 The attraction of a fast-moving neutron for a nucleus may not be great enough to change its direction to allow capture. On the other hand, a thermal neutron moving close to the nucleus may be attracted to and captured by the nucleus.

Figure 19•19 Fuel pellets containing UO_2 enriched in ^{235}U. Each of these metal-clad pellets can produce energy equivalent to that produced by burning 808 kg (1,780 lb) of coal. *Source:* Courtesy of the Atomic Industrial Forum, Washington, D.C.

nuclides of all sorts from all the materials in the assembly. Some of the ^{238}U is converted to ^{239}Pu and other nuclides. Originally it was planned to reprocess used fuel, especially to recover the unused ^{235}U and the ^{239}Pu, which is also a fissionable nuclide. At the present time, however, plants in the United States are simply storing used fuel rods.

It should be noted that it is extremely unlikely, experts say impossible, that any accident at a nuclear plant could result in a nuclear explosion. High concentrations of ^{235}U are needed for weapons-grade uranium for bombs. Reactor-grade uranium is only 1.5 to 3.5 percent ^{235}U, too low to support an explosive chain reaction. It is possible, however, for an accident or some malfunction to cause the liberation of dangerously radioactive materials. At the Three Mile Island plant near Harrisburg, Pennsylvania, in March 1979, the malfunction of equipment plus human error led to temporary loss of flow of cooling water and extreme overheating. Some radioactive nuclides were discharged to the surroundings and the core of the reactor was badly damaged. It will be years before repair and reuse of this billion-dollar plant will be possible.

To date, the safety record of the nuclear industry has been good in protecting the public and its own employees, although there are critics who feel the record is not good enough. The release of radioactivity to the environment from a well-run nuclear reactor is probably negligible. Barring an accident, the amount of radioactive nuclides released is less than that for a coal-burning power plant that releases in its stack gases much of the natural radioactive material found in coal.

19•12•3 Breeder and Converter Reactors

A reactor that uses ordinary uranium makes new fissionable material as it uses up its fissionable ^{235}U, a process called **breeding.** Breeding occurs because some of the neutrons produced by the fission are captured by ^{238}U nuclei, thereby setting up the series of reactions that produce the fissionable ^{239}Pu. Fissionable ^{233}U is produced in a similar manner from ^{232}Th. An ideal situation for reactor economy is for each fission to produce two effective neutrons: one causes another fission and one starts the change of ^{238}U or ^{232}Th to a new fissionable atom (see Figure 19.22). To accomplish this, the neutron yield per fission has to be more than 2, because an appreciable number of neutrons escape without resulting in either fission or breeding. The neutron yields per fission by thermal neutrons for ^{235}U and ^{239}Pu are 2.07 and 2.08, respectively. This is too few to keep the fissioning going and at the same time produce new fuel at the rate the original fuel is being used up. The solution to this problem has been the development of fast-breeder reactors in which moderators are not used to slow down high-energy neutrons. Compared with thermal neutrons, fast neutrons give a higher neutron yield per fission, 2.18 and 2.74, respectively, for ^{235}U and ^{239}Pu.

The Enrico Fermi reactor in Michigan, with a capacity power rating of 200,000 kW, is a fast-breeder, sodium-cooled reactor. The inner portion of the fuel component contains uranium enriched with 25 percent ^{235}U, and the outer part, or blanket, is made of uranium with a depleted ^{235}U content (0.4 percent instead of the 0.7 percent in natural uranium). The uranium assembly is immersed in a pool of circulating molten sodium. The molten metal, while being pumped through the fissioning fuel, is heated to a temperature of

Figure 19•20 An inspection of a bundle of 196 tubes of uranium dioxide pellets in a finished fuel assembly ready for use in a reactor. *Source:* Courtesy of the Atomic Industrial Forum, Washington, D.C.

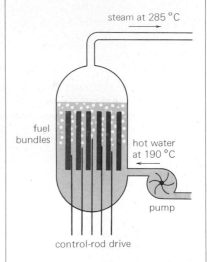

Figure 19•21 A schematic diagram of the core of a boiling-water power reactor. (Compare Figure 19.17.) Hot water, at about 190 °C under pressure, is pumped into the reactor near the bottom. As it rises between the tubes containing fissioning uranium, the water boils and becomes steam at about 285 °C and 70 atm.

Our Energy Dilemma: Nuclear or Coal or ? The average person living in an industrialized society today has a standard of living—food, clothing, housing, health, cultural opportunities, leisure activities, transportation—that the royalty of a few centuries ago could hardly have afforded. The principal basis for this standard is our lavish use of energy.

Oil and natural gas are being exploited to such an extent that it appears that in a few decades we will exhaust the world's supply of these two convenient, irreplaceable fossil fuels (see Table 9.3 and Figure 9.12).

The present pattern of supply of energy resources is shown in accompanying Table A. Figure A shows estimates of the times that different energy sources will last at their present rate of use. Figure B presents this information for fossil fuels, including coal, in a striking way. The era of fossil fuels promises to be just a blip of 200–400 years on the curve of human energy use.

Table A. Approximate energy resources for the United States for the early 1980s

oil	45%
gas	30%
coal	18%
hydroelectric	4%
nuclear	2%
wood	1%
solar	~0%
geothermal	~0%

Because of the rate at which oil and gas are being used, there seem to be only three immediate choices available: use more nuclear fission, use more coal, or make do with less energy. Solar energy, oil shale, and perhaps nuclear fusion are hopes for the future, and the increased use of wood and the incineration of garbage may help, but these possibilities will not be available in the next two to four decades on the huge scale needed.

The pros and cons of nuclear versus coal power production are complex, and they are so balanced that they face us with a dilemma. The chief drawback to nuclear fission is the problem of what to do with the radioactive fission products. If those nuclides that are easily incorporated by organisms—such as isotopes of strontium in bones, iodine in certain glands, hydrogen in all tissues—are ever set loose, their impact on all living things is likely to be devastating. One of the most ominous aspects of the development of nuclear power plants has been the failure to provide for the safe disposal of radioactive wastes before pressing ahead with the construction of huge plants. Large underground storage

vaults have been excavated, only to be left standing after geological studies pointed to the long-term possibility that the radioactive nuclides will get into the groundwater.

A major nuclear question is, To breed or not to breed? Most proponents of nuclear power see breeder reactors as the only way fission can tide humanity over the period between the exhaustion of our supplies of fossil fuels and the development of solar or fusion power (see Figure A). However, the plutonium-239 produced in the breeder is one of the most dangerous nuclides known. In the body, the alpha radiation from a 10-μg speck can cause cancer, and 12 mg will cause death within days or weeks. The large-scale production of something like plutonium-239, which has a half-life of 24,360 years, poses a problem of stewardship that humanity, with its changing political systems, has no reason to feel optimistic about solving.

Nuclear power has many positive characteristics to weigh in the balance, however. It will probably cost no more than oil or gas. A properly run power plant delivers a negligible amount of radioactivity to the environment, and overall produces much less pollution than a coal-fired power plant.

Figure A. Estimates, as of about 1980, of the years that energy sources will last at their present rate of use.

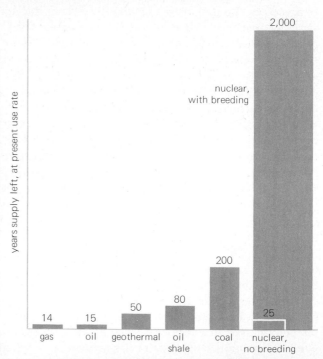

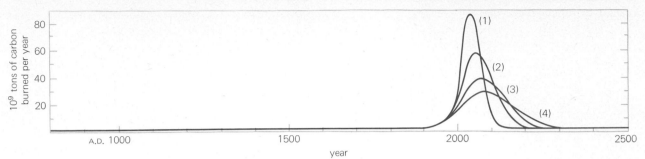

The chief advantages of coal are the fact that the technologies for its use and waste disposal are well known and that there is a large supply of coal in the United States. But coal has an alarming list of drawbacks, including atmospheric pollution by smoke, sulfur dioxide (see the special topic "Acid Rain" in Chapter 24), and carbon dioxide (see "The Greenhouse Effect" in Chapter 25). Ironically, the small traces of radioactive nuclides present in flue gases emitted from a coal-fired power plant are probably a health hazard equal to or greater than the nuclides present in gases discharged by a nuclear plant of the same capacity. The volumes of ash that will be produced if coal replaces oil and gas will be mountainous. Some of it can be used in making cement and for highway roadbed fill. Unfortunately, much of it may be simply piled up where rains can leach out a high pH runoff to pollute streams and lakes.

The mining of coal in deep mines is a dangerous, dirty, menial job that has a history of black lung, crippling injury, and death for miners. Strip mining is easier on miners but harder on the environment. Rains erode the bare scars on hillsides to discharge acids and sediments that ruin land and waterways downstream. Reclamation laws are beginning to eliminate the worst abuses, but they also have the effect of increasing the price of the coal. In Western states, coal seams close to the surface act as conduits for underground water circulation. Removal of these conduits may harm agricultural lands.

Nuclear and coal power plants share the problem of thermal pollution due to waste heat. In both plants, power is generated by vaporizing water to steam and then condensing the steam (see Figure 19.17). The water that cools the condenser is itself heated. When this warm water is discharged into a river, it may kill animal life because of its temperature or because it contains less oxygen than cool water (see Figure C). The TVA is developing uses for waste heat near one of its new nuclear plants. Waste hot water will heat industrial buildings, greenhouses, and fish-farms and be cooled before being discharged to the river.

Figure B. Past and projected rates of use of fossil fuel. All four projected rates assume an exponential demand due to increases in population and desired standards of living. Curves (1) through (4) vary only in the calculations of how rapidly the remaining available fuel is theoretically mined and used. These are "what if" calculations; costs and other fuel use will certainly prevent fossil fuel use at these rates.
Source: C. D. Keeling, "Impact of Industrial Gases on Climate," National Academy of Sciences, Washington, D.C., 1976; via C. F. Baes, Jr., et al., "The Global Carbon Dioxide problem," ORNL-5194, Oak Ridge National Laboratory, Oak Ridge, Tenn., 1976.

Figure C. Use of river water to cool a power plant condenser. Compare Figure 19.17. If the rate at which water is taken into the condenser is greater than the rate of flow of the river, the direction of flow in the river will be reversed between the points of intake and discharge. This happens in some places during dry weather.

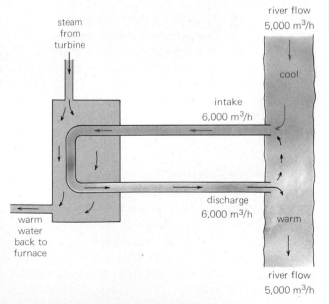

about 427 °C. Passing out of the reactor, the sodium gives up part of its heat in a heat exchanger to produce steam and is pumped back through the reactor to repeat the cycle.

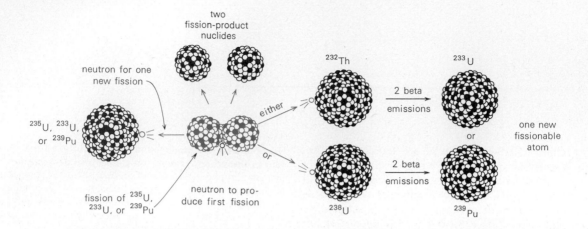

An advanced *converter reactor* is operated by the General Atomic Company at Fort St. Vrain, Colorado. The fissionable nuclide is ^{233}U. Particles of thorium oxide are mixed with particles of uranium carbide or uranium oxide so the thorium can be converted to additional fissionable material by the extra neutrons as power is being produced:

Figure 19•22 In a breeder reactor, one neutron from a fission (shown going to the left) produces a new fission to keep the chain reaction going, while a second neutron (shown going to the right) is captured by a nonfissionable atom to form a replacement fissionable atom.

$$^{232}_{90}\text{Th} + ^{1}_{0}n \longrightarrow ^{233}_{91}\text{Pa} + ^{0}_{-1}e$$

$$^{233}_{91}\text{Pa} \longrightarrow ^{233}_{92}\text{U} + ^{0}_{-1}e$$

The idea is to recover the ^{233}U produced and to use it to refuel the reactor. If as much ^{233}U could be produced and recovered as is consumed in the fission process, the result would be the continued production of nuclear power, so long as there is thorium available to convert.

The moderator in this reactor is graphite. The heat transfer fluid is a gas, so keeping the fission products in place in the fuel rods is quite important. Each tiny particle of UO_2 or ThO_2 is about the size of a grain of salt, and each is enclosed in a casing of charred carbon that traps the fission products. (See Figure 19.23 on Plate 4.) Masses of the tiny particles are molded with a binder material to form the fuel pellets.

19•13 Fusion Reactions

In Section 19.9.3, it was noted that matter is converted to energy when certain small atoms fuse to make new atoms. For example, in the fusion of four hydrogen atoms to form one helium atom, 0.8 percent of the matter is converted into energy. In the fission process, 0.1 percent of the mass of a uranium-235 atom is converted into energy. Examples of highly exothermic fusion reactions are

$$4^{1}_{1}\text{H} \longrightarrow {}^{4}_{2}\text{He} + 2_{+1}^{\;0}e + \text{energy} \qquad (1)$$

$$^{2}_{1}\text{H} + {}^{2}_{1}\text{H} \longrightarrow {}^{3}_{2}\text{He} + {}^{1}_{0}n + \text{energy} \qquad (2)$$

$$^{2}_{1}\text{H} + {}^{3}_{1}\text{H} \longrightarrow {}^{4}_{2}\text{He} + {}^{1}_{0}n + \text{energy} \qquad (3)$$

$$^{3}_{2}\text{He} + {}^{2}_{1}\text{H} \longrightarrow {}^{4}_{2}\text{He} + {}^{1}_{1}\text{H} + \text{energy} \qquad (4)$$

$$3^{4}_{2}\text{He} \longrightarrow {}^{12}_{6}\text{C} + \text{energy} \qquad (5)$$

$$^{12}_{6}\text{C} + {}^{4}_{2}\text{He} \longrightarrow {}^{16}_{8}\text{O} + \text{energy} \qquad (6)$$

Reactions such as these may take place spontaneously when the temperature is approximately 100 million degrees or more. At these high temperatures, atoms do not exist as such. Instead, there is a *plasma* of nuclei and of electrons (Figure 19.24) in which nuclei merge or interact. Fusion reactions that take place owing to high temperatures are often referred to as **thermonuclear reactions.** Equation (3) is an example of the type of reaction used in hydrogen bombs, bombs that are many times more powerful than fission bombs.

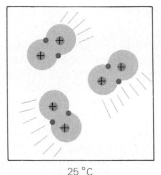 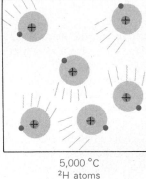 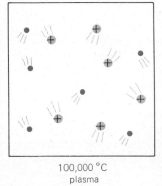 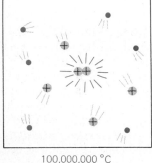

| 25 °C | 5,000 °C | 100,000 °C | 100,000,000 °C |
| $^{2}\text{H}_2$ molecules | ^{2}H atoms | plasma | fusion |

19•13•1 Stellar Energy Spectroscopic examination of the light from the sun indicates that large amounts of hydrogen and helium are present in its atmosphere. The fusion of hydrogen-1 to helium—reaction (1) above—is thought to be the overall reaction responsible for the tremendous amount of energy released by the sun and other stars. The energy emitted by the sun corresponds to a loss in weight of 5×10^{6} tons per second.

19•13•2 Thermonuclear Power Although the fission reaction can be carried out in nuclear reactors so the large amount of energy released can be used to produce power, so far researchers have been unable to control the fusion reaction in a similar way. However, scientists in the United States, Great Britain, Russia, and other countries to a lesser extent have already devoted a great deal of research effort to this end and continue to do so. The maintaining of the very high temperature necessary for the reaction is the chief stumbling block. Most of the research work has been based on the fusion of deuterium. The net energy yield of reactions shown in Equations (2), (3), and (4) are given in Table 19.3.

Various types of "magnetic bottles" have been designed to keep the hot deuterium plasma from coming in contact with the walls of the containing vessel. An experiment that has succeeded on a small scale uses an intense laser to excite a mixture of deuterium and tritium [Equation (3) above].

Figure 19•24 As the temperature is raised, $^{2}_{1}\text{H}_2$ molecules change to atoms, then to plasma (electrons and deuterium nuclei). At 100,000,000 °C, some nuclei fuse to form larger nuclei.

Table 19•3 Energy yield of fusion reactions (2), (3), and (4)

Reaction	Ignition KeV	Yield, KeV	Energy Ratio[a]
(2)	50	3,300	66
(3)	10	17,600	1760
(4)	100	18,300	183

Source: *Reproduced by permission from D. A. Dingee,* Chemical and Engineering News, **57**(14), 33 (1979).

[a]Ratio of fusion yield to energy required for ignition.

The laser beam strips electrons off the hydrogen atoms, and compresses the resulting plasma to a density 100 times that of solid hydrogen. Fusion occurs with much energy released, but the reaction is not yet self-sustaining.

Even though deuterium is present in natural waters to the extent of only one part in about 7,000 parts of hydrogen, the oceans contain enough deuterium to supply the world at its present energy consumption rate for a trillion years if a fusion process can be developed. One of the greatest advantages of fusion rather than fission power plants is that the former would not produce large amounts of dangerous radionuclides that must be stored for ages.

Chapter Review

Summary

Nuclear chemistry is concerned primarily with natural and artificially induced **nuclear reactions,** in which the compositions of atomic nuclei are changed. Each type of nucleus whose composition differs from that of every other type represents a distinct **nuclide.** Most nuclides are unstable; they undergo spontaneous **radioactive decay** by emitting alpha or beta particles, usually in combination with gamma rays or **antineutrinos.** The first three of these are called **ionizing radiations** because of their ability to ionize atoms or molecules in their path. This ability is exploited in such radiation-detecting devices as the **cloud chamber** and the **gas ionization counter.**

Three naturally occurring nuclides with extremely long half-lives are the starting points for **radioactive series** of nuclides that are formed by successive decay processes; each series ends when a stable nuclide is finally formed. The radiation from these and other natural radioactive nuclides, together with high-energy **cosmic radiation** from deep space, constitute the **background radiation** that is always present in our environment. Two units of measurement for radiation are the **curie** and the **rem.**

Among the more important applications of radioactivity are the dating of certain minerals and of ancient objects of organic origin, the use of radioisotopes as **tracers** in the study of chemical and biological processes, the diagnosis and treatment of disease, and the **neutron activation analysis** of complex materials.

Most radioactive nuclides are produced artificially by **bombardment reactions** that use neutrons or charged particles as projectiles. The kinetic energies of charged particles can be boosted to enormous values in **accelerators,** such as the **linac,** the **cyclotron,** and the **synchrotron.** Experiments with such machines have revealed a great variety of subatomic particles. Two of the most common particles, the proton and the neutron, are referred to collectively as **nucleons.** The calculated sum of the masses of the individual nucleons and electrons in an atom is always slightly greater than the actual mass of the whole atom. The difference is called the **mass loss,** which by Einstein's **law of the equivalence of mass and energy** is equivalent to the binding energy of the nucleus. A measure of the stability of a given nuclide is the **mass loss per nucleon,** which varies with atomic number. Another measure is the neutron-proton ratio, which is increased by alpha emission and decreased by beta emission. Certain **magic numbers** of protons or neutrons in a nucleus are associated with unusual stability. The **shell model** attempts to describe nuclear structure in terms of energy levels reminiscent of those of electrons in atomic structure.

A few very heavy nuclides undergo **nuclear fission,** yielding two lighter nuclei and several neutrons. The neutrons can cause the fission of neighboring nuclei, which release more neutrons, and so forth. The resulting chain reaction becomes self-sustaining only when there is a certain **critical mass** of the fissionable substance. The energy of the fission reaction can be released in an instant in an atomic bomb, whose yield is measured in terms of **megatons** of TNT, or in a controlled **nuclear reactor** for electric power generation. While generating power, **breeder reactors** and **converter reactors** produce more nuclear fuel for further power generation.

More promising in the long run is the prospect of unlimited energy from the **nuclear fusion** of very light nuclides. At extremely high temperatures, all matter exists in the **plasma** state, in which **thermonuclear reactions** like those that power the sun can occur. The energy can be released in a hydrogen bomb or, it is anticipated, in a controlled fusion reactor for power generation.

Exercises

Radioactive Decay

1. A sample of pure ozone kept in an inert chamber gradually changes so that in time very little is left. The same is true for radium, although a much longer time is required for the change. How do the two changes differ? What term is used to denote each type of change?

2. Consider the following atoms (X stands for the symbol that fits the data):

$$^{124}_{54}X, \ ^{124}_{52}X, \ ^{126}_{55}X, \ ^{124}_{53}X, \ ^{126}_{54}X$$

   (a) List by symbol all atoms that are isotopes of an element.
   (b) List all that can be referred to collectively as nuclides.

3. For two ^{226}Ra atoms, one may emit a lower energy alpha particle than is emitted by the other atom. What happens to the extra energy? Can this type of energy be used in the identification of different radioactive elements? Explain.

4. For two identical atoms that are beta emitters, the beta particle from one can be of high energy and that from the other, of low energy. Yet the two atoms are at the same ground-state energy after the emission. Is the law of conservation of energy violated? Explain.

5. (a) Describe the composition of a fog particle in a fog track that shows the path of an alpha particle.
   (b) Could the path of a gamma ray be shown in the fog track apparatus? Explain.

Radioactive Series

6. For a hypothetical atom with a neutron-proton ratio less than 1, show that an alpha emission would form a nuclide with an even smaller n/p^+ ratio.

7. The uranium-235 radioactive series starts with the emission of an alpha particle, followed by a beta particle, then another alpha particle emission. What nuclide is formed at this point? Write an equation for each step.

8. If radon-219 emits an alpha particle and the nuclide produced then emits a beta particle, what nuclide will be left? Write an equation for each step.

9. (a) What must be the end product of the natural radioactive series that begins with ^{235}U and involves seven alpha and four beta emissions?
   (b) If the ^{235}U series has fifteen members, how many branches must it have?

Units of Radioactivity

10. (a) Given that milk contains about 6×10^{-11} millicuries of ^{40}K per mL, calculate the number of radioactive

disintegrations per minute of the ^{40}K in a quart of milk.
   (b) Express the radioactivity of the radioactive nuclides in the average human body in terms of curies.

11. Why is the rem more satisfactory than the curie for setting up safety standards for working with radioactive materials?

Half-Life

12. Radiation from a purified isotope of bismuth is found to give 6,380 counts per minute. Five minutes later, the rate is 6,025 counts per minute.
   (a) Calculate the half-life of the isotope.
   (b) Locate the isotope in Table 19.1 and write the equation(s) for the decay reaction(s).

13. The specific rate constant for the radioactive decay of bismuth-210 is 1.60×10^{-6} s^{-1}.
   (a) Calculate the half-life of this isotope in seconds and in days.
   (b) If one starts working with a 1.600-g sample of this isotope, how many days will elapse until only 0.1000 g remains?
   (c) How many particles will be emitted by ^{210}Bi the next second after the decay has brought the amount of Bi down to 0.1000 g? (^{210}Bi is a beta emitter.)

14. At what time in the future will only 30 percent of nature's present supply of ^{238}U remain? (See Table 19.1.)

15. The half-life of radium-226 is 1,600 years. How long will it take for a 2.00-g sample to decay to the point where only 0.125 g of radium-226 remains?

16. A $^{14}_{6}$C assay of a human bone found by an archaeologist revealed that the $^{14}_{6}$C activity was one-tenth that of the $^{14}_{6}$C activity of living matter. What is the age of the bone?

17. (a) In dating with ^{14}C, it is assumed that there was as much ^{14}CO$_2$ in the atmosphere then as now. How can this be, since the age of the dated material may be as much as 40,000 years while the half-life of ^{14}C is only 5,730 years?
   (b) Why cannot the age of preserved organic matter as old as 80,000 years, or older, be determined by measuring the ^{14}C radioactivity?

18. Tritium is produced in the atmosphere by cosmic radiation and is carried to the earth's surface by rains. Hence, it is available to growing plants in very small amounts as ^{3}H$_2$O. Once growth stops, the amount of tritium in a plant steadily diminishes (compare with ^{14}C). An old wine from a wine cellar in Paris gave a tritium assay of only 10 percent of the tritium in living plants. Calculate the age of the wine. The half-life of tritium is 12.26 y.

19. Would a ^{14}C assay of the wine in Exercise 18 give more accurate data for the age calculation than the tritium assay? Why?

Bombardment Reactions

20. Balance the following nuclear equations by filling in data for the question marks; some are equations for natural radioactivity, but most are not:
    (a) $^{35}_{17}Cl + ^{1}_{0}n \longrightarrow ^{34}_{16}S + ?$
    (b) $^{96}_{42}Mo + ^{4}_{2}He \longrightarrow ^{100}_{43}Tc + ?$
    (c) $^{56}_{26}Fe + ^{2}_{1}H \longrightarrow ? + 2^{1}_{0}n$
    (d) $^{62}_{29}Cu + ? \longrightarrow ^{65}_{30}Zn$
    (e) $^{227}_{89}Ac \longrightarrow ^{4}_{2}He + ?$
    (f) $^{210}_{82}Pb \longrightarrow ^{0}_{-1}e + ?$
    (g) $^{23}_{11}Na + ? \longrightarrow ^{23}_{12}Mg + ^{1}_{0}n$
    (h) $? + ^{1}_{1}H \longrightarrow ^{28}_{14}Si + \gamma$
    (i) $^{238}_{92}U + ^{4}_{2}He \longrightarrow ? + ^{1}_{0}n$
    (j) $^{40}_{19}K \longrightarrow ^{0}_{-1}e + ?$
    (k) $^{6}_{3}Li + ^{1}_{1}H \longrightarrow ^{4}_{2}He + ?$
    (l) $? \longrightarrow ^{4}_{2}He + ^{230}_{90}Th$
    (m) $^{12}_{6}C + ? \longrightarrow ^{13}_{7}N + ^{0}_{0}\gamma$
    (n) $^{224}_{88}Ra \longrightarrow ? + ^{220}_{86}Rn$

21. We know that atoms of a radioactive element emit particles and change into atoms of a different element, but can nonradioactive atoms of a common element be changed to atoms of a different element? If so, how?

Mass Loss and Binding Energy

22. (a) Calculate the mass loss per nucleon for an atom of $^{58}_{28}Ni$ whose mass is 57.9353 amu.
    (b) Calculate the mass loss per nucleon for an atom of $^{28}_{14}Si$ whose mass is 27.97693 amu.
    (c) Based on your answers to parts (a) and (b), would the reaction $^{1}_{0}n + ^{58}_{28}Ni \longrightarrow 2^{28}_{14}Si + 3^{1}_{0}n$ take place spontaneously? Why?
    (d) Calculate the mass loss or gain for the reaction. What is the energy equivalence for the mass change if 1.00 mole of ^{58}Ni changes to ^{28}Si?
    (e) Approximately what weight of gasoline and oxygen must react to yield the same amount of energy as that calculated in part (d)?
    (f) Could the reaction described in part (c) be adapted for supplying energy to produce electricity? Why?

23. Based on data in Exercise 22, calculate the binding energy per nucleon for nickel-58 and for silicon-28.

*24. Using Figure 19.11, give hypothetical equations, with specific nuclides as examples, of the following:
    (a) a fission reaction that would be exothermic
    (b) a fission reaction that would be endothermic
    (c) a fusion reaction that would be exothermic
    (d) a fusion reaction that would be endothermic
    Which of the foregoing obviously could not be used to produce an explosion?

Nuclear Stability

25. The isotope of carbon with mass number 14 is radioactive.

(a) Calculate to four significant figures the neutron-proton ratio for this isotope.
(b) Based on an assumed most favorable neutron-proton ratio for elements of low atomic number, write a balanced nuclear equation for the most probable emission of ionizing radiation from this isotope. Justify your choice.

26. In Exercise 25, how is it possible to calculate the ratio n/p^+ to four significant figures when there is only one figure in the value for n or p^+?

27. The four most common isotopes of zirconium have the following atomic weights and percent abundances: 89.90 amu, 51.46%; 90.90 amu, 11.23%; 91.90 amu, 17.11%; and 93.90 amu, 17.40%.
    (a) What are the mass numbers of the four isotopes?
    (b) If there are only five naturally occurring isotopes, what is the mass number of the fifth one?

28. Tin has ten stable isotopes, the greatest number for any element. Is a magic number involved here? Explain.

Nuclear Fission

29. Balance the following fission reactions.
    (a) $^{1}_{0}n + ^{235}_{92}U \longrightarrow ^{124}_{49}In + 5^{1}_{0}n$
    (b) $^{1}_{0}n + ^{239}_{94}Pu \longrightarrow ^{144}_{58}Ce + 2^{1}_{0}n$
    (c) $^{1}_{0}n + ^{233}_{92}U \longrightarrow ^{133}_{51}Sb + ^{98}_{41}Nb$

30. Show with a balanced equation a possible nuclear fission reaction of ^{233}U that is brought about by neutron capture.

31. For the following statement, consider the five alternative conclusions and for each tell why the conclusion is true or false.
    The two large atoms that are initially formed by a nuclear fission reaction are likely to:
    (a) be alpha emitters
    (b) be beta emitters
    (c) be a beta emitter (for one of them) and an alpha emitter (for the other)
    (d) be nonradioactive
    (e) be higher in mass than isotopes commonly found in nature

32. Express answers to the following questions to two significant figures, but if the answer is less than 1×10^{-6} g, simply state this. Assume that among the fission products in a fuel assembly that is to be replaced, there are 0.50 g each of the following nuclides (half-lives are given in parentheses): ^{135}Xe (9.1 h); ^{131}I (8.0 d); ^{3}H (12.3 y); ^{90}Sr (27.7 y); ^{239}Pu (2.44×10^{4} yr). What weight of each will be present:
    (a) after one week
    (b) after one year
    (c) after 100 years
    (d) after 10,000 years

33. How much time would it take for the ^{239}Pu described in

Exercise 32 to decay to one-tenth its original radioactivity?

34. Only a small percentage of the ^{235}U or ^{239}Pu fissions when a fission bomb explodes. Why?

Nuclear Reactors

35. For a nuclear reactor for the production of electricity, explain
    (a) the function of rods such as boron carbide
    (b) the function of a moderator
    (c) how the reactor is made critical
    (d) how the reactor is shut down
    (e) how electricity is produced
36. Tell what sort of data would be needed to check the statement in the caption to Figure 19.19 that one uranium fuel pellet is equivalent to 808 kg of coal in producing energy. Show clearly how the data could be used in making the calculation.
37. Suppose an accident ruptured the containing vessel for the circulating water-steam in the core of a boiling-water nuclear reactor, and the water were suddenly lost. What effect would this have on the nuclear chain reaction?
38. In the advanced converter, ^{233}U–^{232}Th reactor, if as much ^{233}U could be produced and recovered as is used in fissioning, would this be an example of perpetual motion? Justify your answer.
39. Even though we desperately need to develop additional sources of energy, many people oppose the building of more nuclear reactors. There are several reasons for the opposition. Discuss as many as you can.

40. (a) With respect to Figure 19.13, calculate the mass loss per mole of ^{235}U for the overall process, that is, the fission and subsequent decay to $^{103}_{45}$Rh and $^{131}_{54}$Xe (note the emission of neutrons and electrons as part of the products). The masses in amu are ^{235}U, 235.0439; ^{103}Rh, 102.9055; ^{131}Xe, 130.9051; $^1_0 n$, 1.008665; $^{\;\;0}_{-1}e$, 0.00055.
    (b) Calculate the total amount of energy released by the fission of 1 mole of ^{235}U and the subsequent decay to ^{103}Rh and ^{131}Xe.
41. In most of the reactors now being used to produce electric power, natural uranium is enriched with ^{235}U. Why? What is the source of the ^{235}U?

Fusion Reactions

42. Equation (3) of Section 19.13 is an example of the type of reaction used in H bombs.
    (a) Calculate the mass loss when 1 lb of hydrogen forms helium according to Equation (3). Masses in amu: ^{2}H, 2.014102; ^{3}H, 3.016050; ^{4}He, 4.002603; and $^1_0 n$, 1.008665.
    (b) Calculate the energy equivalence of the mass loss calculated in part (a).
43. Why do we not build nuclear reactors based on fusion reactions instead of on fission reactions? What are the advantages and disadvantages of one type over the other type?

Molecular Spectroscopy and Molecular Structure

20

20•1 **Some Types of Spectroscopy**

20•2 **Ultraviolet and Visible Spectra**

20•2•1 Bond Energies

20•2•2 Electron Transitions and Color

20•3 **Infrared Spectra**

Special Topic: Color and Color Photography

20•4 **Nuclear Magnetic Resonance Spectra**

20•5 **Mass Spectra**

20•6 **Miscellaneous Uses of Spectroscopy**

20•6•1 Analyses

20•6•2 Rates of Reaction

In this chapter we present some of the details of molecular spectroscopy and their application to problems of molecular structure. Before beginning this subject, we should review the discussion of atomic spectroscopy at the beginning of Chapter 4, through the topics of emission and absorption spectra.

The great success of the Bohr theory (see Section 4.2.5) was that it accounted for the emission and absorption spectra of elements in terms of definite, *quantized* energy changes involving electron energy levels in atoms. Building on this theory, the absorption spectra of molecules are explained on the basis that each photon absorbed is used in exciting a molecule or a portion of a molecule in a definite way that requires a definite quantum of energy. Because radiant energy can be considered as waves or particles, there are different expressions for describing the effect of electromagnetic radiation on molecules. A molecule can absorb a certain frequency of radiation (1) if some natural periodic movement of the molecule is in phase with that certain radiant frequency or (2) if the energy associated with a photon of the radiation is precisely that needed to bring about a characteristic energy change of the molecule. Statement (1) uses the language of the wave theory of radiation, and statement (2) is made in terms of the particulate, or photon, theory. Both ways of looking at radiant energy are useful in our analysis of molecular spectra.

20•1 Some Types of Spectroscopy

When a substance is irradiated with electromagnetic radiation, it absorbs certain wavelengths of radiation and allows others to pass through (see Figure 20.1). As pointed out in Section 4.2.2, the pattern of wavelengths that a substance absorbs is called its *absorption spectrum*. The absorption spectra for molecular compounds are called *band spectra*, rather than line spectra, because the spectra usually consist of broad regions of absorption. Each region, or band, is found on close analysis to consist of many lines crowded closely together, so closely in many cases that only a spectrometer of great resolving power can separate them.

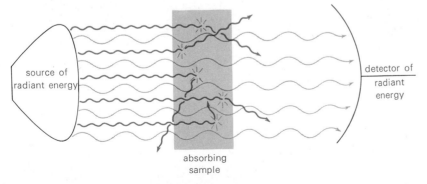

Figure 20•1 Depending on the wavelength, radiant energy may be absorbed by a substance or may pass through unaffected. Energy equivalent to that absorbed is radiated in random directions by the substance. Some substances when viewed by transmitted light have obviously different colors than when viewed by reflected light.

Even a simple molecule may have many more ways of absorbing energy than has an atom. Therefore, molecular absorption spectra generally are more complex than atomic absorption spectra. Consider a molecule of sulfur dioxide, SO_2. There are three nuclei with 32 electrons arranged such that some electrons are essentially held by a single atom, but other electrons are held jointly by two atoms. Each of the electrons can be excited to various higher atomic or molecular enegy levels; an electron can even be separated from the molecule in an ionization reaction to form SO_2^+; or the bonding electrons can be excited to such a degree that the bond is broken and the molecule is separated into two parts, SO and O (see Figure 20.2).

As is the case with atoms, the excitation of electrons in molecules usually requires photons of energy lying in the visible or ultraviolet regions of the electromagnetic spectrum (see Figure 4.2). There are other lower energy transitions a molecule can undergo that involve the bending and stretching of bonds in the molecule or the rotation of the molecule as a whole. These low-energy changes are brought about by photons in the infrared and microwave regions of the spectrum.

The study of the interaction of electromagnetic radiation with matter has been an inexhaustible mine of information during this century. The portions of the spectrum most useful in studying molecular structure are the ultraviolet, visible, and infrared regions, with some special applications of even lower fequency radiation. At higher frequencies, recent improvements in the measurement of the energies of electrons excited by X radiation have made possible the study of the electronic structure of atoms and molecules (see Section 4.6).

Although the term *spectrum* originally referred to a pattern of lines of radiant energy, there has been a proliferation of its use to refer to other patterns, for example, the properties of particles produced by some process. A *photoelectron spectrum* (see Figure 4.17) is a record of the various energies possessed by electrons that are knocked off an atom or a molecule by certain constant energy X radiation. A *mass spectrum* can be the record of the masses of the isotopes of an element (see Figure 3.9), or it can be the record of the masses of the various fragments produced when molecules of a single substance are decomposed (see Figure 20.15).

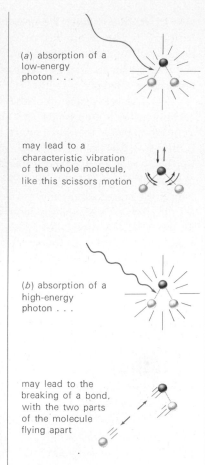

(a) absorption of a low-energy photon . . .

may lead to a characteristic vibration of the whole molecule, like this scissors motion

(b) absorption of a high-energy photon . . .

may lead to the breaking of a bond, with the two parts of the molecule flying apart

Figure 20•2 A low-energy photon may cause a molecule to vibrate (*a*), whereas a photon with a sufficiently high energy can cause a molecule to dissociate (*b*).

20•2 Ultraviolet and Visible Spectra

A high-energy photon in the ultraviolet range may pack a wallop sufficient to knock an outer electron away from an atom or a molecule. For the first ionization energies of the elements listed in Table 3.6, we see that 10 eV is about average. In Figure 4.2, we see that such energies are associated with photons in the ultraviolet region.

Energies sufficient to ionize atoms (that is, first ionization energies) are also sufficient to ionize molecules. Smaller energies suffice to excite electrons from lower to higher energy levels and give rise to *electronic absorption spectra*. In molecules as in atoms, electronic excitations require energies in the range of 1.5 to 8.0 eV. These energies are associated with photons in the visible region and the lower energy part of the ultraviolet region.

20•2•1 Bond Energies
If a photon has a high enough energy, it can cause a bond in a molecule to break. For example, ionizing radiations excite water strongly at about 21 eV (see Figure 20.3 on Plate 5). The ions formed can

cause chemical reactions to occur that involve the water and substances in contact with the water.

Such a gross chemical change as the decomposition of molecules is the main reason people must protect themselves from overexposure to high-energy radiation. The most serious damage by sunburn is due to the chemical reactions caused in tissues by the ultraviolet part of the sun's radiation. Ozone molecules in the earth's upper atmosphere help protect us by absorbing much of the sun's ultraviolet radiation, but they do not absorb it all. Suntan lotions may contain substances such as *p*-aminobenzoic acid (PABA) that absorb ultraviolet radiation and prevent it from harming the skin (see Figure 20.4).

The energies associated with radiant energy may be expressed in several ways. Two common ways are in electron volts per individual photon and in kilojoules per mole of photons. By studying the pattern of wavelengths absorbed by a substance, it is usually possible to determine the wavelength of the photon that has precisely the energy necessary to dissociate a bond. The energy of this photon is equal to the dissociation energy of the bond. By calculating the energy in kilojoules per mole of photons (or per mole of molecules dissociated), we can relate bond energies calculated from spectroscopic data to bond dissociation energies from thermochemical data.[1]

Radiations in the visible and ultraviolet ranges are energetic enough to break certain molecular bonds. For example, molecular hydrogen, H_2, has a spectroscopic bond energy of 432 kJ/mole, corresponding to ultraviolet radiation of 2.78×10^{-7} m (278 nm) wavelength. The Cl_2 molecule, with a spectroscopic bond energy of 240 kJ/mole, can have its bond broken by radiation of 5.00×10^{-7} m (500 nm) wavelength. This radiation is in the blue-green region of the visible spectrum, as we can see by referring to Figure 4.6 on Plate 2.

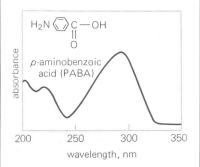

Figure 20•4 Absorbance curve for *para*-aminobenzoic acid, showing that it absorbs strongly in the near ultraviolet.

— • Example 20•1 • —

Examination of the absorption spectrum of bromine gas indicates that radiation of frequency 4.76×10^{14} Hz is required to break the bond. Calculate the spectroscopic bond energy in kilojoules per mole for the Br_2 molecule.

— • Solution • —

$$E = h\nu$$

$$= (6.626 \times 10^{-34} \text{ J} \cdot s \cdot \text{photon}^{-1})\left(4.76 \times 10^{14} \text{ Hz} \times \frac{1 s^{-1}}{1 \text{ Hz}}\right)$$

$$= 3.15 \times 10^{-19} \text{ J} \cdot \text{photon}^{-1}$$

Per mole,

$$E = (3.15 \times 10^{-19} \text{ J} \cdot \text{photon}^{-1})\left(\frac{6.02 \times 10^{23} \text{ photons}}{1 \text{ mol}}\right)\left(\frac{1 \text{ kJ}}{10^3 \text{ J}}\right)$$

$$= 190 \text{ kJ/mol}$$

See also Exercises 7–9 at the end of the chapter.

---

[1]Bond energies obtained from spectroscopic data are for 0 K; they are approximately 4 kJ/mole less than the values of $\Delta H^\circ_{\text{dis}}$ at 25 °C (298 K), such as those in Table 13.2.

Although spectroscopic analysis provides very precise data for the calculation of bond energies, the interpretation of spectra is often difficult. When a molecule is split into two parts, the parts themselves may be in excited states. The absorption line that corresponds precisely to the energy necessary to dissociate the molecule is surrounded in the spectrum by a great number of other lines, some at lower energies that are due to the excitation of electrons in the molecule and some at higher energies due to the breaking of the bond plus the excitation of one or both of the parts into which the molecule is split. The identification of the spectral line that corresponds to the simple bond-breaking energy can be made only after the consideration of many factors, often including the approximate determination of the bond energy by thermochemical methods. For example, iodine vapor, I_2, shows a continuous absorption of energy at wavelengths shorter than 4.99×10^{-7} m (499 nm) (see Figure 20.5). This indicates that with the shorter wavelengths the I_2 molecule is dissociated into excited atoms with variable amounts of kinetic energy. However, even the energy associated with photons at 4.99×10^{-7} m (499 nm) is equivalent to 240 kJ/mole, much more than the bond dissociation energy measured by thermochemical measurements of about 148 to 153 kJ/mole of iodine.

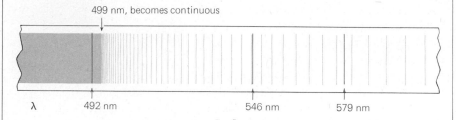

499 nm, becomes continuous

λ   492 nm   546 nm   579 nm

Figure 20•5 Artist's conception of the absorption spectrum of iodine vapor. On the actual film, the point at which the spectrum becomes continuous, 499 nm, cannot be detected visually; rather, it is calculated as the point at which $\Delta\lambda$ for a series of lines becomes zero. The three lines in color are superimposed from the spectrum of mercury as known reference points.
Source: Courtesy of Dr. W. H. Fletcher, The University of Tennessee, Knoxville.

Analysis of the iodine spectrum reveals that the dissociation of an I_2 molecule by a photon of 4.99×10^{-7} m (499 nm) wavelength produces one iodine atom in the unexcited (ground) state and one in an excited state:

$$I_2 \longrightarrow I + I^* \qquad \Delta E = 240 \text{ kJ/mol}$$

The excited iodine atom, I^*, emits energy equivalent to about 91 kJ/mole when it changes into a ground-state atom:

$$I^* \longrightarrow I \qquad \Delta E = -91 \text{ kJ/mol}$$

Addition of these two equations yields the energy required to dissociate 1 mole of iodine molecules into 2 moles of iodine atoms in the ground state, that is, the spectroscopic bond energy for I_2:

$$I_2 \longrightarrow I + I \qquad \Delta E = 149 \text{ kJ/mol}$$

Many other examples could be cited to show the difficulty of assigning specific frequencies to bond energies. Over two decades ago, the value of 266 kJ/mole was accepted as the bond dissociation energy of fluorine, F_2, on the basis of spectroscopic calculations. Later, thermochemical data indicated the value to be much less than this, about 151 to 155 kJ/mole.

The value of 157 kJ/mole is cited today, the precise calculation being based on a reinterpretation of the spectrum of fluorine.

The nitrogen molecule, N_2, has an absorption spectrum that shows many lines in the ultraviolet. Possible dissociation energies calculated spectroscopically have included 712.0, 826.8, 945.4, and 1,138 kJ/mole. The value of 945.4 is most consistent with thermochemical data. The measurement of distances between spectral lines can be done with great precision, but less precise thermal data are often needed to help the spectroscopist choose which of the energy changes revealed in the spectrum is probably due to the breaking of the bond.

The thermochemical bond dissociation energies of diatomic molecules cited in Table 13.2 are based mainly on spectroscopic calculations.

20•2•2 Electron Transitions and Color The absorption of visible wavelengths causes relatively low-energy reversible electron transitions in molecules. In general, colored substances have some electrons that are easily excited.

Certain organic compounds are particularly useful as sources of color for dyes. Molecules of organic compounds that have no double bonds or benzene rings do not selectively absorb in the visible portion of the spectrum; therefore, they are colorless. On the other hand, molecules with double bonds or benzene rings may absorb some visible wavelengths and transmit colored light. The electrons that are easily excited by visible light are usually found in a molecule where several atoms are joined by alternating double and single bonds. Three such groups of atoms, called **chromophores** (color bearers), are

The particular color a substance has is determined not only by the type of the chromophore present, but also by the structure of the molecule of which the chromophore is a part. Many different dyes can be made by introducing substituents, such as —OH, —NH_2, —$NHCH_3$, and —$N(CH_3)_2$, into the molecules that contain a given color-forming group. The groups that contribute to or change the color of the dye are referred to as **auxochromes** (auxiliary color producers). The auxochromes generally have the additional function of making the dye fast to cloth or other articles by salt formation.

20•3 Infrared Spectra

Photons in the infrared radiation region are not energetic enough to cause electronic transitions, but they can cause the bending and stretching of bonds; that is, they can cause vibrations in molecules in which atoms of a molecule change their relative positons. An atom has a large mass and is held in place in a molecule by a relatively weaker attractive force than is an electron; the atom, therefore, oscillates at lower frequencies. Vibrations of electrons take place fom 10^{14} to 10^{16} times per second (Hz), whereas vibrations of atoms in molecules occur at the slower but still respectable rate of 10^{13} to 10^{14} times per second (see frequencies in hertz in Figure 4.2).

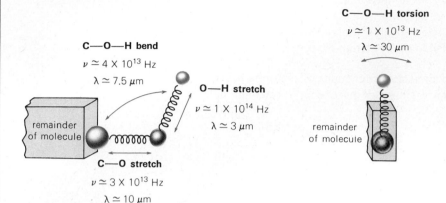

C—O—H torsion

$\nu \simeq 1 \times 10^{13}$ Hz

$\lambda \simeq 30\ \mu m$

C—O—H bend

$\nu \simeq 4 \times 10^{13}$ Hz

$\lambda \simeq 7.5\ \mu m$

O—H stretch

$\nu \simeq 1 \times 10^{14}$ Hz

$\lambda \simeq 3\ \mu m$

remainder of molecule

C—O stretch

$\nu \simeq 3 \times 10^{13}$ Hz

$\lambda \simeq 10\ \mu m$

remainder of molecule

Figure 20•8 Vibrations of the —C—O—H (alcohol) group, interpreted in terms of a model in which spheres are connected by elastic springs. Approximate frequencies and wavelengths are given; precise values will differ depending on the nature of the remainder of the molecule.
Source: Adapted from K. Nakanishi, *Infrared Absorption Spectroscopy,* Nankodo Company Limited, Tokyo, 1962.

★   Special Topic ★   Special Topic ★   Special Topic ★   Special Topic ★   Special Topic ★   Special Topic ★   Special Topic ★   Special Topic   ★

Color and Color Photography Color in substances is due to the selective absorption of part of the white light of the sun (or an artificial light) by molecules or atoms or ions. An object will appear to be blue, for instance, if it absorbs sufficient visible wavelengths in the green, yellow, orange, and red regions and reflects or transmits the radiation in the blue region (see Figure 20.6 on Plate 5). An object will also appear blue to the eye if it absorbs colors complementary to blue in the orange region (near 590 nm) and transmits or reflects the remaining wavelengths of white light. If the absorption is in the yellow region (near 550 nm), the object will appear dark blue.

Useful and highly developed applications of the chemistry of synthetic dyes have been made in the color photography industry. Different inventors and manufacturers use different techniques and processes, but all take advantage of the light sensitivity of silver halide emulsions. These emulsions are similar to those used in black-and-white photography and use the sensitized silver in reactions that result in color dyes being left behind to form a picture of the object photographed.

One type of color reproduction is based on the selective exposure and development of three superimposed layers, yellow, magenta, and cyan, in such a way that a transparent sandwich or color slide results. When white light shines through the slide, a full-color image of the object photographed is produced. In the *Kodachrome* process, each of the three dyes is formed in its respective layer by a reaction of the following type:

$$R_2N-\!\!\!\bigcirc\!\!\!-NH_2 + H_2C\begin{smallmatrix}X\\Y\end{smallmatrix} + 4Ag^+ \longrightarrow$$

developer   coupler

$$R_2N-\!\!\!\bigcirc\!\!\!-N=C\begin{smallmatrix}X\\Y\end{smallmatrix} + 4H^+ + 4Ag$$

dye

Examples of the dyes are given and the *Kodachrome* process is summarized in Figure 20.7 on Plate 6. When the picture is taken, the amount of Ag^+ that is sensitized in each of the three layers of the color film depends upon the wavelengths and intensity of the light reflected by the object photographed. During the first development step, the sensitized Ag^+ is reduced to black elemental Ag. This black deposit corresponds to a negative image of the object.

The Ag^+ ions that are not sensitized in taking the picture correspond to a positive image. These silver ions serve as the oxidizing agent to produce colored dyes during the next three steps in the development. All the black elemental silver formed during the development is removed in the final bleach-and-fix step.

Special films, sensitive to some specific part of the spectrum, for example, the infrared, ultraviolet, or X-ray region, are also of great use in scientific work.

Models of molecules in which small spheres (representing atoms) are attached to one another by steel springs (representing chemical bonds) can be used to help explain the vibrational frequencies obtained from spectral

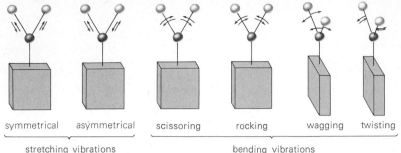

| symmetrical | asymmetrical | scissoring | rocking | wagging | twisting |

stretching vibrations   bending vibrations

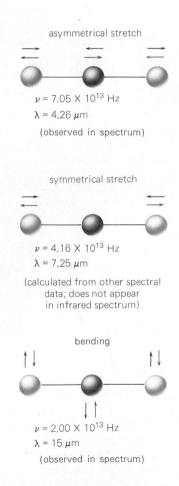

Figure 20•9 Six possible quantized motions, relative to the rest of a molecule, of two adjacent atoms. The motions alternate between those shown by the black arrows and those shown in color.

data. Such a model is pictured in Figure 20.8. The masses of the atoms, the geometry of a molecule, and the periods of the atomic vibrations can be related to one another and to the calculated stiffness of chemical bonds by use of the same equations of classical mechanics that apply to large vibrating masses or masses connected by stiff springs.

The terms that the infrared spectroscopist has given to the various modes of molecular vibrations—twisting, stretching, wagging, rocking, and so on—sound more like descriptions of a dance routine than like the staid language of science. Some of the possible motions in molecules are illustrated in Figure 20.9.

Only those stretching and bending motions that change the dipole moment of a molecule give rise to absorption of infrared radiation. Consider the three possible fundamental vibrations of the CO_2 molecule shown in Figure 20.10. Although we might expect bands at wavelengths equivalent to each of these frequencies, only two of the bands appear. During the symmetrical stretching of the C=O bonds, the dipole moment remains unchanged (zero), so that no absorption band is produced by this vibration. The frequency of this motion has been determined, however, by analysis of the electronic spectra and from other data.

The wavelength and frequency regions associated with the stretching and bending vibrations of some common bonds in organic molecules are shown in Figure 20.11. A few points can be emphasized in connection with these data. First, the frequencies for a given type of bond are approximately the same regardless of the kind of molecule in which the bond occurs. Second, the data in the figure show that double bonds have higher stretching frequencies than single bonds between similar atoms. This indicates that double bonds (two pairs of shared electrons) are stronger than single bonds, so that more energy is required to excite double bonds. Third, the data show the effect on stretching frequencies of the masses of atoms. Bonds involving hydrogen have high frequencies, stretching about 9×10^{13} times a second as compared with 2 to 5×10^{13} times a second for the other single bonds described in Figure 20.11. Note also that the substitution of the heavy hydrogen isotope, ^2H, for the common ^1H isotope lowers the frequency. The effect of a change in isotopic weight on stretching frequencies is observed in

asymmetrical stretch

$\nu = 7.05 \times 10^{13}$ Hz
$\lambda = 4.26\ \mu m$
(observed in spectrum)

symmetrical stretch

$\nu = 4.16 \times 10^{13}$ Hz
$\lambda = 7.25\ \mu m$
(calculated from other spectral data; does not appear in infrared spectrum)

bending

$\nu = 2.00 \times 10^{13}$ Hz
$\lambda = 15\ \mu m$
(observed in spectrum)

Figure 20•10 Three fundamental modes of vibration for the CO_2 molecule, an example of a linear, symmetrical, nonpolar molecule.

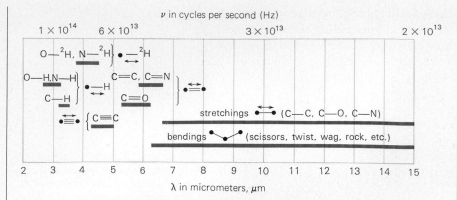

Figure 20•11 Regions in the spectrum where lines for certain bonds are usually found.
Source: Reproduced from K. Nakanishi, *Infrared Absorption Spectroscopy*, Nankodo Company Limited, Tokyo, 1962.

the case of other elements also, although the relative effect is less than for the change with hydrogen isotopes. The isotopes ^{18}O, ^{15}N, and ^{13}C were discovered by the spectroscopic detection of bonds involving them.

The infrared absorption spectrum of ethyl alcohol, C_2H_5OH, is shown in Figure 20.12. The more complex a molecule, the more ways it will be able to absorb energy and the more complicated its infrared absorption

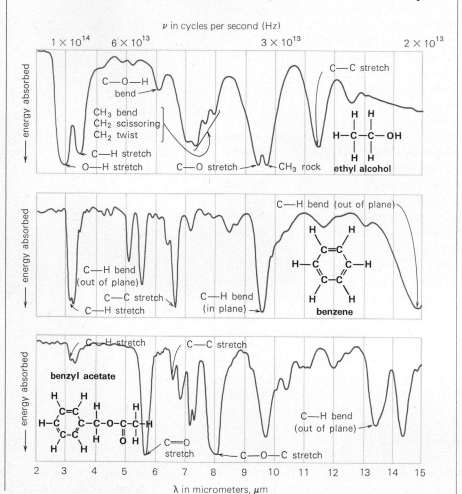

Figure 20•12 The infrared absorption spectra of three organic compounds.
Sources: Alcohol spectrum courtesy of Dr. W. H. Fletcher, The University of Tennessee, Knoxville; benzene courtesy of Prof. R. Mecke, University of Freiburg, Germany (via the DMS System); and benzyl acetate from the work of R. W. Silverstein and G. C. Bassler, *J. Chem. Educ.*, 39, 548 (1962).

spectrum will be. Spectra for benzene, C_6H_6, and benzyl acetate, $C_6H_5CH_2OCOCH_3$, are also shown in Figure 20.12.

From a study of infrared vibrational spectra, the spectroscopist is able to calculate the relative strengths of bonds, their stiffness (resistance to bending), bond angles, and, for very small molecules, the overall geometry. Rotational spectra, in the far infrared and microwave region, give especially important information on bond distances and bond angles. In spite of bending, stretching, and twisting, average bond distances and bond angles are quite definite, being measured often to three or four significant figures.

20•4 Nuclear Magnetic Resonance Spectra

In 1945 a powerful new spectroscopic technique was added to the arsenal of the student of molecular structures. Professors Purcell, Torrey, and Pound of Harvard University and Bloch, Hansen, and Packard of Stanford University independently developed a device for detecting the very low-energy interactions between certain atomic nuclei and magnetic fields, a phenomenon now called **nuclear magnetic resonance (nmr).**

The phenomenon of nmr depends on the possession of a magnetic moment by the nuclei of certain atoms. Nucleons can have either of two possible orientations with a magnetic field; that is, they behave as if they have either clockwise or counterclockwise spins. If the nuclear particles in a single nucleus do not have their spins paired, the nucleus as a whole will have a resultant magnetic moment. Such a nucleus tends to become aligned in an imposed magnetic field. The nucleus of the abundant isotope of hydogen, ^1H, has a magnetic moment, because it consists of but a single proton. The nucleus of a carbon-12 atom, $^{12}_6C$, has no resultant magnetic moment, because it consists of even numbers of protons and neutrons with paired spins.

In nmr spectroscopy, a sample of a compound is placed between the poles of a magnet powerful enough to align a portion of those nuclei that have magnetic moments (see Figure 20.13). The sample then is irradiated with electromagnetic radiation, usually in the radio-frequency range of 10^7 to 10^8 Hz. A spinning nucleus aligned with the magnetic field can be flipped into a different alignment by absorbing a photon of precisely the proper energy. Different nuclei, or similar nuclei held in different electron environments, absorb photons of different wavelengths.[2] The pattern of radio-frequencies absorbed is the nmr absorption spectrum of the compound.

The most widely used nmr spectrometers measure the energy absorbed by hydrogen nuclei, ^1H, only. This permits a broad application, however, because hydrogen is present in more compounds, including almost all organic compounds, than any other element. By determining the placement of the hydrogen atoms in the structure of a molecule, the placement of other atoms can often be deduced.

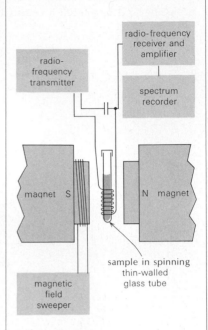

Figure 20•13 Schematic diagram of an nmr spectrometer.
Source: Adapted by permission from James Shoolery, *A Basic Guide to NMR*, Varian Associates, Palo Alto, Calif., 1972.

---

[2]In practice it is more difficult to vary the frequency of the radiation precisely than to vary the strength of the imposed magnetic field. Commonly nmr spectrometers operate at a fixed radio-frequency and measure the changes in absorption of photons as the magnetic field is slowly varied.

Three hydrogen nmr spectra are shown in Figure 20.14. The structures of the three compounds as drawn in the figure were well worked out by other techniques prior to the discovery of nmr, but their spectra will serve to illustrate the almost unbelievable power of the nmr spectrometer to "see" a molecule and to sketch a report for the chemist in graphic form.

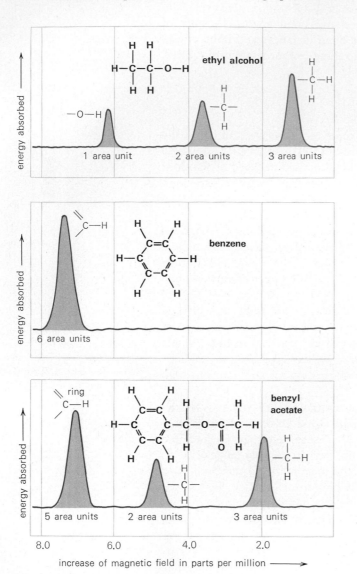

Figure 20•14 The low-resolution hydrogen nmr spectra of three organic compounds.

The hydrogen nmr spectrum of ethyl alcohol has three peaks, indicating that there are hydrogen atoms in this molecule held by three different kinds of bonds (or held in three different electron environments). The structure of ethyl alcohol as drawn shows three kinds of bonds: (1) an H—C bond involving a carbon atom on the end of the molecule, (2) an H—C bond to a carbon in the middle of the molecule, and (3) an H—O bond. The area under each peak in an nmr spectrum is proportional to the number of hydrogen atoms absorbing energy at that frequency. The relative areas under the peaks in the spectrum of ethyl alcohol show that in this molecule there are three hydrogen atoms held by the bond of type 1, two held by type 2, and one atom held by type 3.

The nmr spectrum of benzene, C_6H_6, is quite informative just because it is so simple. The single peak in the spectrum shows that each of the hydrogen atoms in the molecule is held by the same type of bond. This is strong evidence that the molecule is symmetrical, a property also indicated by other data (see Section 6.6).

The spectrum of benzyl acetate shows there are five hydrogen atoms held by one type of bond, two by another type, and three by a third. The structure drawn agrees with this assignment.

The nmr spectra shown in Figure 20.14 have the smooth, simple peaks characteristic of low-resolution spectra. With the precise measurements possible with a high-resolution spectrometer, some of these peaks can be shown to consist of several smaller peaks. The interpretations of high-resolution spectra reveal features of molecular structure that we shall not discuss.

20•5   Mass Spectra

The use of the mass spectrograph to determine the masses of atoms was discussed in Section 3.3.5. A wider use of this instrument today is in the study of the structure of molecules and their bond strengths. Also, mechanisms of reactions can be followed by using isotopes of known weight in certain positions in reactant molecules and tracing their positions in the products.

If a molecular substance is subjected to electron impact in the spark gap of a mass spectrograph (see Figure 3.8), many different sorts of positive ions can be formed. The spark can not only knock off electrons but can also decompose the molecule by breaking bonds. In the case of water, the major possibilities are

$$H_2O \longrightarrow H_2O^+ \longrightarrow OH^+ \longrightarrow O^+$$

Also, higher positive charges on each of these species are possible, and, of course, there will be protons, H^+, produced. In practical operation, a mass spectrograph can be adjusted so that singly charged positive ions of various sorts are formed and identified by their masses.

Figure 20.15 is a copy of the *mass spectrum* of a sample of CH_2Cl_2, dichloromethane, which contained some air and water vapor as impurities. Often air and/or water is purposely included with the sample to calibrate the spacings; the patterns of the three stair-step peaks for (O^+, OH^+, H_2O^+) and the high-low peaks for (N_2^+, O_2^+) serve to identify the points corresponding to masses of (16, 17, 18) and (28, 32), respectively. By counting from these known peaks, the mass numbers of other peaks can be found.

As indicated in Figure 20.15, many different singly charged positive ions are obtained by knocking off an electron and/or one H or one Cl atom from CH_2Cl_2. Also, the recombination of particles knocked off may form new particles, such as HCl^+. Because significant amounts of both chlorine-35 and chlorine-37 occur in natural chlorine (75.53%, 24.47%), there are peaks

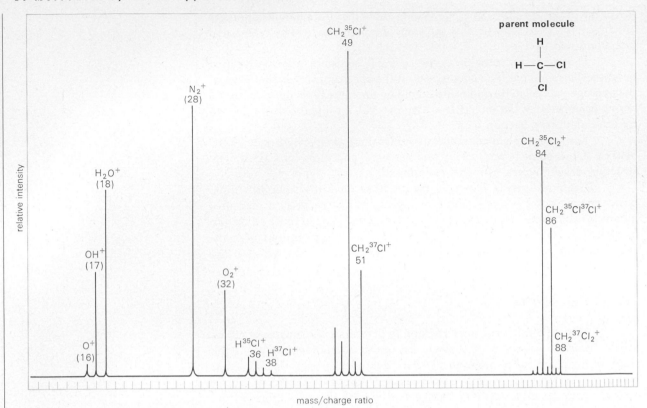

corresponding to the different weights of particles containing ^{35}Cl or ^{37}Cl. However, percentages of other isotopes are so small (e.g., ^{2}H or ^{13}C) that peaks for particles containing them are not visible in Figure 20.15. Also, note that no peaks for particles such as CH^+ or C^+ are found; the energy available in the spark gap is controlled so the molecule of CH_2Cl_2 is not broken up so completely.

20•6   Miscellaneous Uses of Spectroscopy

20•6•1   Analyses   The great utility of spectroscopy in determining the structures of molecules and the strengths of chemical bonds is obvious, but there are many other important uses. Once the spectrum of a compound is known, it can be used to identify the compound. By precise measurement of spectral intensities, the percentage of a compound in a mixture can sometimes be determined. One of the reasons that spectroscopic detection is so useful is that usually only a tiny sample of the unknown is needed. As might be expected, law enforcement officials often rely on spectral analyses. Spectroscopes are used in many industries to analyze raw materials and products. Spectroscopic standards have been established for colors of dyes and for brightness and colors of electric lights.

Infrared spectra such as those pictured in Figure 20.12 are useful in chemical analysis, particularly in the identification of complex organic

Figure 20•15 Mass spectrum of dichloromethane, CH_2Cl_2, in the presence of air and water. The peaks labeled in color are the known peaks due to water vapor and air that are used for calibration; by counting from these known peaks, the values for the unknown peaks are determined.
Source: Courtesy of Dr. C. A. Lane, The University of Tennessee, Knoxville.

chemicals. Spectra broadcast back from spectrometers mounted in rockets are one of the main sources of data about the atmospheres and surfaces of other planets.

Simple infrared spectrometers can be used to determine the temperature pattern of a surface surveyed. One application of this technique is the monitoring of the surfaces of dormant volcanos; the development of a hot area may give advance warning of volcanic activity. Spectrometers mounted in airplanes are used in scanning large areas to detect temperature differences that reveal plant diseases, water pollution, and patterns of ocean currents. Aerial infrared photography at night has been used to detect the warm spots that reveal the location of a moonshiner's fermenting mash.

Chemists studying air pollution are using practically every spectroscopic technique available both to determine the kinds and amounts of pollutants and to follow the chemical reactions in pollution-producing processes. For example, by the continuous analysis of engine exhausts as various pollution controls are tested, the chemist and automotive engineer can evaluate the controls. High-speed spectroscopic analysis is also a boon to the physician. In some cases, a mass spectroscopic analysis of blood can identify within a few minutes the drug in the system of an unconscious overdose victim.

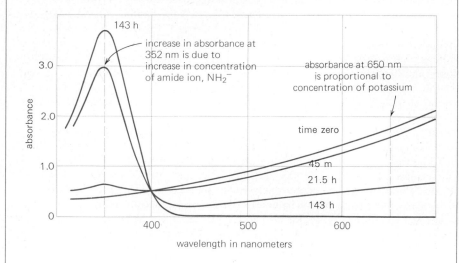

Figure 20•16 Absorption spectra at different times of a liquid ammonia solution of potassium during the $K + NH_3 \longrightarrow K^+ + NH_2^- + \frac{1}{2}H_2$ reaction. As the reaction takes place, potassium is used and the absorbance due to it decreases. (Compare values in Table 14.3 and Figure 14.9, and see Figure 24.15 on Plate 7.) As the reaction takes place, amide ion, NH_2^-, is produced, and the absorbance due to it increases. *Source:* Reproduced by permission from D. C. Jackman, "The Alkali Metal-Liquid Ammonia Reaction at Room Temperature," Ph.D. dissertation, The University of Tennessee, Knoxville, 1966.

20•6•2 **Rates of Reaction** Rates of reaction can be determined by following changes in concentration of reactants spectroscopically (see Section 14.2.1). Figure 20.16 shows spectral curves of the type used to determine the data in Table 14.3. The absorptions are in the visible and in the near ultraviolet regions. The colors of the solutions are similar to those shown in Figure 24.15 on Plate 7. At time zero, the solution is blue because of the greater absorption of light in the long-wavelength (red) region of the spectrum. As the reaction proceeds, the blue color gradually fades and a yellowish color takes its place because of the absorption in the short-wavelength (blue) region.

Chapter Review

Summary

Molecular spectroscopy, in its many forms, is the most versatile and powerful method for determining molecular structures. Because molecules can undergo rotational and vibrational excitation as well as electronic excitation, the absorption of photons by molecules produces complex band spectra rather than the simpler line spectra characteristic of atoms. Most molecular spectra are recorded in the ultraviolet, visible, and infrared regions of the electromagnetic spectrum.

Ultraviolet and visible spectra are the result of electronic transitions brought about by energetic photons. Photons of sufficiently high energy (typically in the ultraviolet) can ionize atoms and molecules. They can also break many chemical bonds; although this complicates the spectra, it allows very precise spectroscopic bond energies to be calculated. The absorption of visible and ultraviolet photons without bond breakage provides information about the compositions and electronic structures of molecules. Most organic compounds that are colored absorb visible photons as a result of their **chromophores;** a substituent group that modifies the color due to the chromophore is called an **auxochrome.**

Infrared spectra are the result of vibrational transitions brought about by lower-energy photons. They provide valuable information on the compositions and structures of molecules and on the nature of their chemical bonds. Many kinds of bonds have characteristic vibrational frequencies that are nearly independent of the kind of molecule in which the bond is found; often this allows such a bond to be "seen" in the infrared spectrum of a compound.

Nuclear magnetic resonance (nmr) spectroscopy is a powerful technique for studying molecules containing atomic nuclei that have magnetic moments. The interaction of these nuclei with a strong magnetic field and, simultaneously, a radio-frequency field reveals information about the chemical "environment" of the atoms in a molecule.

The **mass spectrum** of a substance is a record of the masses of the atomic and molecular ionic fragments produced by bombarding the molecules under study with electrons in a mass spectrograph. This information provides insight into the composition and structure of the molecule.

In addition to its central role in molecular structure determinations, molecular spectroscopy is also widely used for the qualitative and quantitative analysis of substances and for the study of the rates and mechanisms of reactions.

Exercises

Ultraviolet and Visible Spectra

1. Distinguish between electromagnetic spectra and mass spectra.

2. An atom may absorb a photon of energy or it may emit a photon of energy. Describe what takes place within the atom for each process. What conditons are necessary for each to occur?

*3. To excite electrons in an atom from lower to higher energy levels requires energies in the range of 1.5 to 8.0 eV. Show by calculation that such energies are provided by radiations that range from ultraviolet through the visible spectrum (see Figure 4.2).

4. Which would probably have a more complex absorption spectrum, a molecule of oxygen or a molecule of neon? Why? A molecule of neon or a molecule of argon? Why?

5. If a photon of radiation of wavelength 130 nm is the minimum energy required to remove the highest energy electron of atom X, can X absorb radiant energies of wavelength longer than 130 nm? Explain.

6. If a photon of 2.0 eV is absorbed in the excitation of an electron to a higher energy level, what is the wavelength of the photon emitted when the electron falls to its original level? Would it be possible for more than one wavelength to be emitted? Explain.

7. Calculate the spectroscopic bond energy in kilojoules per mole for the F_2 molecule, based on the absorption spectral line corresponding to a frequency of 3.94×10^{14} Hz. Repeat the calculation for a molecule of I_2 gas, based on the spectral line corresponding to a frequency of 3.8×10^{14} Hz.

*8. The spectroscopic bond energy of oxygen is 498.3 kJ/mole. Calculate the maximum wavelength of radiation that will bring about the dissociation of O_2 molecules into O atoms. Are O_2 molecules in the atmosphere likely to encounter this type of radiation?

*9. The bond dissociation energies in kilojoules per mole for H_2 and Cl_2 are 436.0 and 242.6, respectively.
   (a) Calculate the maximum wavelength of radiant energy that can dissociate H_2 molecules.
   (b) Repeat part (a) for Cl_2 molecules.
   (c) The violent reaction that occurs when a mixture of H_2 and Cl_2 is exposed to light is thought to proceed through a chain reaction with an initiating step of

$$X_2 \xrightarrow{h\nu} 2X$$

   Based on your calculations in parts (a) and (b), X_2 must be which one, H_2 or Cl_2? Why?

10. Calculate the wavelength of radiant energy that could produce Na^{2+} from Na^+. (See Table 3.6 for data.) Identify the type of radiant energy.

Color and Electron Transitions

11. When white light is passed through a colored solution,

the light that emerges is colored. What happens to produce the color?

12. Oxygen and sulfur are in the same family (VIA), yet oxygen is colorless and sulfur is yellow. Explain, based upon excitation energies.

13. Distinguish between auxochromes and chromophores.

14. If a film of the *Kodachrome* type that has not received any light, that is, no picture has been taken, is developed in the usual way, what color, if any, will result? Why?

Infrared Spectra

15. Which would require the higher energy photon:
    (a) the removal of an electron from H_2O or the exciting of the scissors motion in H_2O?
    (b) the rotation of an SO_2 molecule or the breaking of an S—O bond in SO_2?
    (c) the excitation of the Cs atom to produce its blue emission line or excitation of the Na atom to produce its yellow emission line?
    (d) the stretching of a C—H bond in C_6H_6 or the formation of the $C_6H_6^+$ ion?

16. What are some wavelengths at which infrared radiation with frequencies of 2×10^{13} to 1×10^{14} Hz will be most strongly absorbed when passing through ethyl alcohol? (See Figure 20.11.)

17. In what respect should the infrared absorption spectrum of ethylamine, $CH_3CH_2NH_2$, be similar to that of ethyl alcohol (see Figures 20.11 and 20.12). How might the two differ? Will the absorptions for both molecules be at the same frequencies for the same molecular vibrations, such as the C—C or C—H stretch?

18. Based on Figure 20.11, calculate in kilojoules per mole the average C≡C stretching energy in a molecule of acetylene, HC≡CH.

19. Which ones of the following pairs have the higher stretching energies:
    (a) $C—^1H$ or $C—^2H$?
    (b) C—C or C=C?
    (c) C—H or C≡N?

20. Based on Figures 4.7 and 20.12, state how the infrared absorption spectra for benzene and toluene (methylbenzene) differ.

Nuclear Magnetic Resonance

21. Of the following, which nuclei tend to become aligned in a magnetic field: ^1H, ^2H, ^3He, ^{12}C, ^{13}C, ^{16}O? If all atomic nuclei tended to become aligned in magnetic fields, how might nmr spectra for the three compounds shown in Figure 20.14 differ from what is the actual case?

22. The low resolution nmr spectrum for ethyl benzoate, $C_6H_5C—OCH_2CH_3$, should show how many peaks?
    ‖
    O

What should be the relative areas under the peaks?

23. Repeat Exercise 22 for $CH_3—\overset{\overset{\displaystyle H}{|}}{\underset{\underset{\displaystyle HO}{|}}{C}}—\overset{}{\underset{\underset{\displaystyle O}{\|}}{C}}—NH_2$.

*24. Consider two alcohols that have molecular formulas of C_3H_7OH. The nmr spectrum of one shows four peaks; the other shows three peaks. Write structural formulas for the alcohols that account for their nmr spectra.

Mass Spectra

25. Consider the following two compounds:

$$H—\overset{\overset{\displaystyle H}{|}}{\underset{\underset{\displaystyle H}{|}}{C}}—\overset{..}{\underset{..}{O}}—\overset{\overset{\displaystyle H}{|}}{\underset{\underset{\displaystyle H}{|}}{C}}—H \qquad H—\overset{\overset{\displaystyle H}{|}}{\underset{\underset{\displaystyle H}{|}}{C}}—\overset{\overset{\displaystyle H}{|}}{\underset{\underset{\displaystyle H}{|}}{C}}—\overset{..}{\underset{..}{O}}—H$$

methyl ether   ethyl alcohol

    (a) In a mass spectrographic analysis, both compounds might have lines at the same masses. Predict two such mass lines other than hydrogen. For each line give the formula of the fragment of the ether and of the alcohol.
    (b) Give the formula of a fragment that might be formed from the alcohol but not from the ether.
    (c) Give the formula of a fragment that might be formed from the ether but not from the alcohol.
    (d) How would the nmr spectra of the two compounds differ?

26. In Figure 20.15, peaks corresponding to several mass numbers are not labeled with formulas for probable ionic species. Give the formulas for species indicated by the peaks at 35, 37, 47, 48, 50, 82, 83, 85, and 87.

27. In Figure 20.15, might a peak at mass number 14 be expected? If there were a peak there, give formulas for two different species to which it could be attributed.

28. In Figure 20.17 (see page 622 top) is shown a copy of the mass spectrum[3] of $CHBr_2I$. Assign formulas to as many of the mass lines as you can. Begin by first assigning formulas to the lines at masses 79, 81, and 127.

Miscellaneous Uses of Spectroscopy

*29. Explain how one could use deuterium chloride, ^2HCl, to provide evidence that the ionization of an acid in water is more than the simple dissociation into ions, as represented by the equation $HCl \rightleftharpoons H^+ + Cl^-$.

---

[3]Source: *TRC Current Data News*, **6**, 1(1978), and the Center for Trace Characterization, Texas A & M University, College Station, Texas.

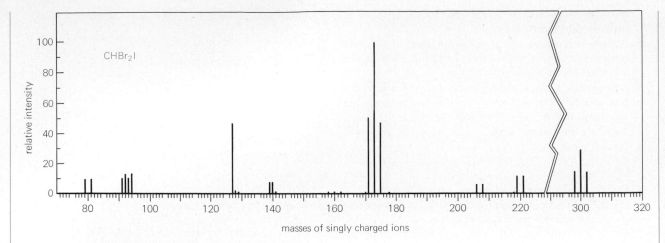

Figure 20•17 Mass spectrum of $CHBr_2I$. No species were tested for at masses less than 80 amu, and none was found between 222 and 297 amu. Each plotted mass is rounded to nearest whole number.

30. For the reaction of potassium and liquid ammonia (see Figure 20.16), what is measured to determine concentration changes? Describe how the measurements are made. Discuss how the measurements are treated to reveal the concentration changes.

31. The statement is made in Section 20.6.1 that a mass spectrographic analysis of blood might be made to identify a drug in an unconscious overdose victim. Because a mass spectrum could show many mass lines that could be attributed to the natural components of blood or to a variety of drugs, how could a specific drug be identified?

* 32. Use the data in Figure 20.15 to calculate the percentage abundance of the two chlorine isotopes.

* 33. It is said that one is more apt to sunburn on a hot, bright day if the wind is blowing strongly than if there is no wind. Can you argue either for or against this belief, based upon the fact that radiations of about 21 eV excite water strongly? Present your argument with calculations and reference to Figure 4.2.

Metals I: Properties and Production

21

21•1 **General Divisions of the Periodic Table**

21•1•1 Metals, Nonmetals, and Metalloids

21•1•2 Transition and Inner Transition Series Elements

Alkali and Alkaline Earth Metals

21•2 **Physical Properties of IA and IIA Elements**

21•2•1 Flame Spectra

21•3 **Chemical Properties of IA and IIA Elements**

21•3•1 Activity

21•3•2 Metallic Character

21•3•3 Characteristic Reactions

21•4 **Compounds of IA and IIA Elements**

21•4•1 General Considerations

21•4•2 Oxides

21•4•3 Hydroxides

21•4•4 Halides

21•4•5 Carbonates

21•4•6 Sulfates

21•4•7 Other Compounds

Transition Metals and Their Neighbors

21•5 **Classification**

21•5•1 Transition and Inner Transition Metals

21•5•2 Neighbors of Transition Metals

21•5•3 Metallo-Acid Elements

21•6 **Physical Properties**

21•7 **Chemical Properties**

21•7•1 Binary Compounds and Simple Salts

21•7•2 Oxidation States

21•7•3 General Chemical Activity

21•7•4 Corrosion

Special Topic: When Your Car Rusts Out

Production of Metals

21•8 **Metallurgy**

21•8•1 Concentration of Ore

21•8•2 Smelting

21•8•3 Refining

21•9 **Industrial and Environmental Aspects of Metal Production**

21•9•1 Copper

21•9•2 Iron

21•9•3 Aluminum

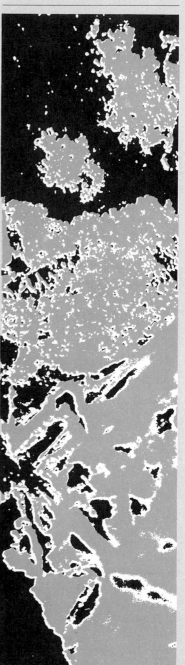

The next five chapters will be concerned with inorganic chemistry. The arrangement of the elements in the periodic table is a guide to family and group relationships that unify and clarify this broad field of chemistry. Fortunately, a number of the families are similar enough to be discussed together. This is particularly true of the metals, which comprise about three fourths of the elements.

In this chapter, we shall discuss the physical and chemical properties of the metals, some of their simpler compounds, and the production of metals. In Chapter 22, the oxidation–reduction reactions of transition metals and compounds will be discussed, and coordination compounds will be taken up. In Chapters 23–25, the nonmetals will be discussed, beginning with the most nonmetallic groups, VIIA and VIIIA, and working to the left in the periodic table. A number of *key chemicals* will be described that have great practical importance for our overall standard of living.

The periodic table was arranged more than a century ago on the basis of experimental, descriptive chemical relationships. More recent theoretical correlations with atomic structure have even increased the value of the table. For the student and for the practicing scientist, an understanding of periodic relationships is invaluable in organizing and extending one's knowledge of chemistry.

21•1   General Divisions of the Periodic Table

Before taking up the chemistry of particular families, we will review the general divisions of the periodic table.

21•1•1   Metals, Nonmetals, and Metalloids

In Section 5.1.1, the division of the elements in the periodic table into three major classes was described. These classes—the metals, nonmetals, and metalloids—each have important physical and chemical characteristics and occupy three reasonably distinct areas in the long form of the periodic table.

• *Metals* • About 80 elements are classified as metals, including some from every group except VIIIA, VIIA, and possibly VIA. The metals are at the left and in the center of the periodic table.

In chemical reactions with nonmetals, metal atoms tend to donate electrons and form cations. Their electronegativities are low, most of them being less than 2.0, as shown in Figure 5.8.

• *Nonmetals* • The nonmetals, consisting of about a dozen relatively common and important elements plus the noble gases, are to the right on the periodic table, with the exception of hydrogen. The atoms of the nonmetals tend to accept electrons and form anions in chemical reactions with metals. Nonmetals also readily react with one another by forming covalent bonds, for example, in SO_3, CO_2, and H_2O. Electronegativities of most nonmetals range from about 2.4 upward.

• *Metalloids* • The metalloids or borderline elements exhibit both metallic and nonmetallic properties to some extent; they usually act as electron donors with nonmetals and as electron acceptors with metals. These elements lie close to the zigzag line in the periodic table shown in Figure 21.1. Boron, B, silicon, Si, germanium, Ge, arsenic, As, antimony, Sb, and tellurium, Te, are included in this class. The electronegativities of the borderline elements range between 1.8 and 2.1.

IIIA	IVA	VA	VIA
B 2.0	C 2.5	N 3.0	O 3.5
Al 1.5	Si 1.8	P 2.1	S 2.5
Ga 1.6	Ge 1.8	As 2.0	Se 2.4
In 1.7	Sn 1.8	Sb 1.9	Te 2.1
Tl 1.8	Pb 1.8	Bi 1.9	Po 2.0

Figure 21•1 The metalloid elements lie near this zigzag line in the periodic table. The numerical values are the electronegativities. See also Figure 5.8.

21•1•2 Transition and Inner Transition Series Elements As discussed in Section 3.7, the long form of the periodic table shown on the inside front cover of the text is divided vertically into eight *groups* and horizontally into seven *periods*. Each of the eight groups is subdivided into A and B families. The elements in the A families often are referred to as the main group elements and those in the B families as the transition and inner transition series elements. All of the 58 elements in these latter two series are metals.

• *Transition Elements* • In the three periods with eighteen elements, periods 4, 5, and 6, the regular addition of electrons to the *d* sublevels begins with the third element in the period and continues until the eighth element from the end is reached (that is, family IB).[1] Beginning with scandium or yttrium, a *d* sublevel series is built up from atom to atom roughly as the atomic number increases, until it contains its maximum of ten electrons. In period 6, the *d* sublevel series begins with lanthanum and then skips to hafnium before continuing. There is some question among chemists about just which elements are to be grouped together in these series of elements. Most would include the IB family, but many would not extend the series through the IIB family. We will define the B family elements in periods 4, 5, 6, and 7 as the **transition elements.**

• *Inner Transition Elements* • The third elements in periods 6 and 7 (lanthanum, La, and actinium, Ac) have outer electronic configurations similar to those of scandium and yttrium. But with the fourth element in each of these last two periods, a series of fourteen elements begins in which the electronic configurations of the outer two main levels remain nearly constant. With cerium, Ce, a series begins in which electrons are added to the seven 4*f* orbitals of the fourth main energy level. A similar series is developed beginning with thorium, Th, in the seventh period, in which the 5*f* electrons are added in the fifth main level (thorium itself is an exception). These two series, the **lanthanide series** in the sixth period and the **actinide series** in the seventh period, are called **inner transition elements.**[2]

According to the most widely accepted classification, the actinide series ends with element 103, at which point fourteen 5*f* electrons have been added to the structure of actinium. Element 104 should be a transition element located in the periodic table just under hafnium, element number 72. Eight more elements would complete this transition series with element 112.

---

[1] Note in period 5 that the 4*d* sublevel is actually filled first in palladium, Pd.

[2] The elements of the lanthanide series were once referred to as the *rare earth elements*. Some years ago, the International Committee on Nomenclature recommended the names *lanthanoids* and *actinoids*, but most chemists have not adopted these names.

Alkali and
Alkaline Earth Metals

The alkali metals in family IA of the periodic table and the alkaline earth metals in family IIA are so named because most of their oxides and hydroxides are among the strongest bases (alkalis) known. In discussing these two families, particularly in describing trends and extremes in behavior, it is understood that francium and radium are not included, owing to their relative rarity and radioactivity.

21·2   Physical Properties of IA and IIA Elements

In Tables 21.1 and 21.2 are listed some of the important physical properties of the IA and IIA elements. The relative sizes of their atoms and ions are shown in Figure 5.14. The elements in both of these families have the silvery luster of typical metals on freshly cut surfaces. (But they tarnish rapidly on exposure to air.) They also have the high electric and thermal conductivities characteristic of metals. Pipes filled with sodium are used for

Table 21·1 Physical properties of alkali metals (excluding francium)

	Li	Na	K	Rb	Cs
melting point, °C	181	98	64	39	29
boiling point, °C	1,336	881	766	694	679
density, g/cm^3	0.54	0.97	0.87	1.53	1.88
electron distribution	2,1	2,8,1	2,8,8,1	2,8,18,8,1	2,8,18,18,8,1
ionization energy, eV	5.4	5.1	4.3	4.2	3.9
atomic radius, Å	1.34	1.54	1.96	2.16	2.35
ionic radius, Å	0.60	0.95	1.33	1.48	1.69
electronegativity	1.0	0.9	0.8	0.8	0.7
crystal structure	bcc	bcc	bcc	bcc	bcc

Table 21·2 Physical properties of alkaline earth metals (excluding radium)

	Be	Mg	Ca	Sr	Ba
melting point, °C	1,277	650	850	769	725
boiling point, °C	2,484	1,105	1,487	1,381	1,849
density, g/cm^3	1.86	1.74	1.55	2.6	3.59
electron distribution	2,2	2,8,2	2,8,8,2	2,8,18,8,2	2,8,18,18,8,2
ionization energy, eV	9.3	7.6	6.1	5.7	5.2
atomic radius, Å	1.25	1.45	1.74	1.92	1.98
ionic radius, Å	0.31	0.65	0.99	1.13	1.35
electronegativity	1.5	1.2	1.0	1.0	0.9
crystal structure	hex	hex	fcc	fcc	bcc

short, large-scale electric conductors. Molten sodium is used as the heat-transfer fluid in some nuclear reactors.

The crystal structures of the alkali and alkaline earth metals are the three most common for metals: body-centered cubic (bcc), face-centered cubic (fcc), and hexagonal close-packed (hex), as shown in Figure 8.29. Some of the properties of these elements not usually associated with metals are their relatively low melting points, their relatively low densities, and their softness. These three properties are especially typical of the alkali elements; one of these metals is a liquid at just above room temperature, and three have densities less than that of water. All of them, from lithium to cesium, can be deformed easily by squeezing them between thumb and forefinger (with proper protection for one's skin). The alkaline earth elements are somewhat harder, ranging from barium, which is about as hard as lead, to beryllium, which is hard enough to scratch most other metals. Beryllium is used in alloys to make springs with long-lived flexibility. Magnesium is used in lightweight alloys, particularly those in aircraft.

Familiarity with the data in Tables 21.1 and 21.2 is a prerequisite to understanding the chemical behavior of the two families. The relative simplicity of the chemical reactions of the IA and IIA elements is associated with the simple electronic structures of the members of these two families. Outside a stable core structure similar to that of a noble element, they have, respectively, one and two s electrons that are lost relatively easily.

The elements in groups IA and IIA have the lowest average ionization energies and electronegativities of all the families of elements. These properties are related to the sizes of the atoms (see Figure 5.14) and the relatively great distances from the nuclei of the outer s electrons.

21•2•1   **Flame Spectra**   As pointed out in Table 4.1, the IA and IIA elements impart characteristic colors to an ordinary flame. In analytical laboratory work, flame tests are often used to reveal the presence of various alkali and alkaline earth elements. The yellow flame test for sodium is one of the most sensitive; less than one part of sodium per billion parts of solvent (1 ppb) is detectable.

21•3   **Chemical Properties
of IA and IIA Elements**

21•3•1   **Activity**   The most striking characteristic of the alkali and alkaline earth metals is their extreme activity. The reason that most people are not familiar with the appearance of the very common metals sodium, potassium, and calcium is that these metals are so active that they do not exist as elements when in contact with air or water. (See the corrosion of sodium shown in Figure 24.15 on Plate 7.) None of the IA or IIA elements exists in nature in the elemental state. All the alkali elements exist in natural compounds as unipositive ions; all the alkaline earth elements exist as dipositive ions.

21•3•2   **Metallic Character**   Chemically, the metallic character of an element is associated with its tendency to lose electrons. The metallic character in A families tends to increase from top to bottom in the periodic table. This increase is probably less pronounced in the alkali family than in any other, because even lithium is quite metallic in character. In the great majority of chemical reactions, the elements from sodium to cesium act in the same way. Lithium is somewhat different, probably because its ion is so small that it has a very high charge density for a singly charged ion. Lithium is certainly a metal, but it is the least metallic of the elements in the IA family on the basis of its properties as an electron donor. Cesium is the most metallic.

In the alkaline earth family, there is also a great similarity in chemical properties. Calcium, strontium, and barium are notably alike, but magnesium and beryllium differ from these three by being somewhat less active. This can be related to the higher ionization energies of the latter two. All the alkaline earth elements are electron donors, beryllium being the least active and barium the most.

21•3•3 **Characteristic Reactions** The principal chemical reactions of the IA and IIA elements are listed in Tables 21.3 and 21.4. It is evident from these data that the elements in these two families have many common chemical characteristics.

The alkali and alkaline earth metals are powerful reducing agents because they lose electrons so readily. They combine readily with most nonmetallic elements, forming ionic compounds such as halides, hydrides,

Table 21•3 Reactions of group IA metals (M represents an alkali metal)

$4M + O_2 \rightarrow 2M_2O$	limited amount of O_2
$2M + O_2 \rightarrow M_2O_2$	heated in air
$2M + X_2 \rightarrow 2MX$	X = F, Cl, Br, or I
$2M + S \rightarrow M_2S$	Se and Te also react
$2M + 2H_2O \rightarrow 2MOH + H_2$	violent, except with Li
$2M + 2NH_3 \rightarrow 2MNH_2 + H_2$	with catalyst
$6M + N_2 \rightarrow 2M_3N$	only with Li
$2M + H_2 \rightarrow 2MH$	dry H_2 gas
$2M + 2H^+ \rightarrow 2M^+ + H_2$	violent

Table 21•4 Reactions of group IIA metals (M represents an alkaline earth metal)

$2M + O_2 \rightarrow 2MO$	Be and Mg must be heated
$M + O_2 \rightarrow MO_2$	Ba easily; Sr at high pressure; CaO_2 not formed directly
$M + X_2 \rightarrow MX_2$	X = F, Cl, Br, or I
$M + S \rightarrow MS$	Se and Te also react
$M + 2H_2O \rightarrow M(OH)_2 + H_2$	Mg and Be react only with steam to give oxides
$M + 2NH_3 \rightarrow M(NH_2)_2 + H_2$	with catalyst
$3M + N_2 \rightarrow M_3N_2$	heated
$M + H_2 \rightarrow MH_2$	heated; Be and Mg do not react
$M + 2H^+ \rightarrow M^{2+} + H_2$	rapid

oxides, and sulfides. Because lithium and the alkaline earth metals react directly with nitrogen at high temperatures, they continue to burn in air even after they have combined with all the oxygen available.

The alkali metals react spectacularly with water; calcium, strontium, and barium react less violently. Sample reactions are given in Section 9.4.3. In the case of potassium, rubidium, and cesium, the reactions are so rapid and so exothermic that the hydrogen evolved usually bursts into flame. Lithium reacts much more slowly than the other IA elements but still rapidly enough so that the lithium–water reaction has been studied for use as a propellant reaction for torpedoes. Of the elements in IIA, beryllium and magnesium do not react appreciably with water except at high temperatures.

All the elements except beryllium and magnesium corrode steadily in air until they are completely converted to oxides, hydroxides, or carbonates. Beryllium and magnesium react with oxygen readily, but the tough oxide film that is formed tends to protect the underlying metal from further attack at room temperature. When strongly heated, even these two metals burn violently. At high temperatures, magnesium burning in air reacts not only with oxygen but even with nitrogen and carbon dioxide.

Lithium is similar to magnesium in many chemical reactions, and beryllium is similar to aluminum. This diagonal relationship of similarities involving a period 2 element and a period 3 element in the next family is true also of boron and silicon, as we shall see in Chapter 25.

• Example 27•1 •

Write balanced equations for the following reactions:

(a) sodium + oxygen $\xrightarrow{\text{heat}}$
     (excess)

(b) calcium + water $\longrightarrow$
     (excess)

(c) barium + nitrogen $\xrightarrow{\text{heat}}$

(d) lithium + hydrogen $\longrightarrow$

• Solution •

(a) $2Na + O_2 \longrightarrow Na_2O_2$
(b) $Ca + 2H_2O \longrightarrow Ca(OH)_2 + H_2$
(c) $3Ba + N_2 \longrightarrow Ba_3N_2$
(d) $2Li + H_2 \longrightarrow 2LiH$

See also Exercise 12 at the end of the chapter.

21•4 Compounds of IA and IIA Elements

21•4•1 General Considerations Because of the striking chemical similarity of the alkali metals, their compounds are so alike that we need to discuss only the sodium and potassium compounds. Sodium compounds are used most widely because they are usually the cheapest.

The chemical differences between the various alkali ions are so slight that one can be substituted for another in most laboratory and industrial reactions. In some cases, however, substitution of one ion for another is not desirable. For instance, plants must secure potassium from the soil for proper growth, so fertilizers must contain potassium compounds. Sodium compounds cannot be used for this purpose.

All the alkali ions are colorless and quite inactive. Their simple salts—such as $LiCl$, KNO_3, Cs_2SO_4, and Rb_2CO_3—are usually very soluble in water. Solutions of these compounds are typically strong electrolytes. Lithium compounds resemble magnesium compounds. For example, the solubilities of their carbonates and phosphates are low.

Among the alkaline earth elements, calcium, strontium, and barium form compounds that are very similar to one another. Magnesium and, more particularly, beryllium form compounds that differ from the other three in properties. Because of its small ionic size (hence a large charge density), beryllium forms covalent-ionic bonds with a number of other atoms. (See $BeCl_2$ in Figure 6.6.)

The compounds of beryllium tend to hydrolyze in water, partly because of the formation of the insoluble hydroxide $Be(OH)_2$. The high charge density of the tiny Be^{2+} ion enables it to attack water; in this and in some other ways it resembles the Al^{3+} ion.

The ions of the alkaline earth elements are colorless and fairly inactive. Many of their simple salts—such as $MgSO_4$, $CaCl_2$, $Ba(NO_3)_2$, and $BeSO_4$—are very soluble. However, the sulfates, carbonates, and phosphates of calcium, strontium, and barium are only slightly soluble.

21•4•2   Oxides   The I_A oxides of the M_2O type (Na_2O, K_2O, and so on) are white solids that are extremely sensitive to moisture and carbon dioxide, reacting to form the hydroxides, MOH, and the carbonates, M_2CO_3, respectively.

The monoxides, M_2O, of the alkali metals can be obtained by heating the metals in a limited supply of dry air at relatively low temperatures (below about 180 °C).

The common II_A oxides have the expected formula, MO. Both lime, CaO, and magnesia, MgO, are made by the high-temperature decomposition of naturally occurring carbonate rocks in *lime kilns*. Magnesia is used for firebrick and as insulation for steam pipes. Lime is used to make mortar and plaster and to neutralize acid soils; it is also the cheapest industrial source of hydroxide ions, $Ca(OH)_2$, formed by the reaction of lime with water.

Because of its extreme affinity for water, calcium oxide is used for the dehydration of such liquids as ethyl alcohol and for the drying of gases. It is becoming increasingly important in removing SO_2 from power plant flue gases. Calcium oxide is also used to control the pH of acid wastes from paper plants and sewage plants, and to remove phosphate ions from sewage.

The group II_A oxides are white solids with very high melting points. They tend to react slowly with water and carbon dioxide in the air to form the hydroxide and the carbonate, respectively.

$$BaO + H_2O \longrightarrow Ba(OH)_2$$
$$MgO + H_2O \longrightarrow Mg(OH)_2$$

$$CaO + CO_2 \longrightarrow CaCO_3$$
$$SrO + CO_2 \longrightarrow SrCO_3$$

Key Chemical: CaO
calcium oxide

How Made

$$CaCO_3 \xrightarrow[\sim 900\ °C]{heat}$$
limestone
$$CaO + CO_2\uparrow$$
lime

$CaO + H_2O \longrightarrow Ca(OH)_2$
lime   slaked lime

U.S. Production; Cost

~20 million tons per year; $32–$42 per ton (different grades)

Major Uses

43% steel (slag forming)
11% making other chemicals
8% treating potable water
5% air and water pollution control processes
5% pulp and paper

Other Uses

cement, plaster, firebrick; reducing soil acidity; stabilizing clay soil for highways, dams, and levies

The reaction of an oxide with water is an exothermic process called **slaking.** In the case of barium oxide, the heat of slaking is so great that, if only a little water is used, the mass may become visibly red hot. When $Ca(OH)_2$, *slaked lime,* is used in a mortar in laying bricks, the setting process involves drying and crystallization, followed by the slow conversion of the slaked lime to calcium carbonate by the action of carbon dioxide from the atmosphere.

The more active alkali and alkaline earth metals form peroxides. Sodium peroxide, Na_2O_2, potassium peroxide, K_2O_2, and barium peroxide, BaO_2, are well-known examples. In the peroxide ion, $:\overset{..}{O}—\overset{..}{O}:^{2-}$, the oxidation state of oxygen is -1. Peroxide ion is a strong oxidizing agent. Sodium peroxide is formed when sodium is burned in a stream of dry oxygen; barium peroxide is formed when barium oxide is heated in air. These two peroxides are sometimes used as sources of oxygen and of hydrogen peroxide (see Section 9.6).

Superoxides and ozonides, such as KO_2 and KO_3, have been studied as *air revitalizers* for manned satellites and rockets. Not only can they be decomposed to yield oxygen for breathing, but the oxides produced can be used to remove exhaled carbon dioxide from the air. First,

$$4KO_2(s) \longrightarrow 2K_2O(s) + 3O_2(g) \quad \text{then} \quad K_2O(s) + CO_2(g) \longrightarrow K_2CO_3(s)$$

21•4•3 Hydroxides Two of the most widely used strong bases are sodium hydroxide and calcium hydroxide. Because of its lower solubility, calcium hydroxide does not form a concentrated solution, but it is a typical strong base. The equations for preparing calcium hydroxide were given in the preceding section. About one fifth of the sodium hydroxide (lye) used in this country is made by a process in aqueous solution in which calcium hydroxide is a raw material:

$$Na_2CO_3 + Ca(OH)_2 \longrightarrow CaCO_3{\downarrow} + 2NaOH$$

or

$$2Na^+ + CO_3^{2-} + Ca^{2+} + 2OH^- \longrightarrow CaCO_3{\downarrow} + 2Na^+ + 2OH^-$$

Of the four possible compounds, calcium carbonate is the least soluble, so it precipitates and leaves the sodium hydroxide in solution.

Sodium hydroxide is made in huge quantities by the electrolysis of sodium chloride brine (see Figure 23.8). This process is the cheapest source of *concentrated* hydroxide solutions for the chemical industry.

Magnesium hydroxide is the familiar *milk of magnesia,* the antacid slurry long used as a household remedy. Lithium hydroxide was used in the *Apollo 11* moon flight to remove carbon dioxide exhaled by the astronauts from the capsule atmosphere:

$$2LiOH + CO_2 \longrightarrow Li_2CO_3 + H_2O$$

21•4•4 Halides Several halides of the alkali and alkaline earth metals occur so abundantly in nature that they serve as the raw material for making other compounds of the metals and halogens. Sodium chloride and potassium chloride are mined directly. During the purification of their ores or solutions,

---

Key Chemical: NaOH
sodium hydroxide

How Made

$$2NaCl + 2H_2O \xrightarrow{\text{electrolysis}}$$

$$Cl_2 + 2NaOH + H_2$$

(see Section 23.3.2)

U.S. Production; Cost
~11 million tons per year;
~$150 per ton

Major Uses
50% making other chemicals
15% pulp and paper
10% aluminum
5% textiles
5% soaps, detergents, foods, other consumer products

Other Uses
the first choice when a soluble, very strong base is needed

other halides—such as lithium, rubidium, and cesium chlorides and some bromides and iodides—that are present to a small extent in the ores are sometimes recovered. Most of the potassium chloride thus obtained, about 3 million tons annually in the United States, is used in fertilizers.

Magnesium chloride is produced from salt wells and from seawater as one step in the production of elemental magnesium. Calcium chloride, which also is found in nature, is produced synthetically as a relatively worthless by-product of the Solvay process (see next section) for making sodium carbonate. Used as a drying agent, it is also put on dusty roads because of its tendency to *deliquesce,* that is, to remove moisture from the air and form droplets of saturated solution.

21•4•5 Carbonates The alkali carbonates, M_2CO_3, are much more soluble than the alkaline earth carbonates, MCO_3, a fact that explains why the latter are found more commonly as beds of sedimentary rock. Carbonates are among the most abundant natural IIA compounds. Calcium carbonate is deposited on the ocean floor as the lowly oyster shell, as lacy coral, and in other forms. Geologic metamorphosis then produces great beds of limestone, or marble, or even beautifully transparent, colorless crystals of calcite. Though their appearances differ, all these forms are essentially $CaCO_3$. Some of the chemical reactions of carbonates are discussed in Section 25.3.3.

Of all the alkali metal compounds, sodium carbonate, Na_2CO_3, is second only to sodium chloride in terms of tons used. It is a base and a source of carbonate ions for the chemical industry. At one time, most of the sodium carbonate used in the United States was produced by the Solvay process, a complex series of reactions that accomplishes indirectly an overall reaction that does not occur spontaneously as written:

$$2NaCl + CaCO_3 \longrightarrow Na_2CO_3 + CaCl_2$$

Since 1948, an increasing fraction of the market has been supplied by purifying and calcining (roasting) a natural deposit called *trona,* $Na_2CO_3 \cdot NaHCO_3 \cdot 2H_2O$, which is found in vast quantities in beds of evaporated salt in California and Wyoming. One of the major differences between the two processes is that the insoluble materials recovered in purifying the trona are relatively innocuous and can be used as landfill, but the by-product calcium chloride and the alkaline wastes of the Solvay process have damaging ecological effects. The Solvay process, which can be operated at about 75 percent efficiency, produces 170 tons of by-products and waste for each 100 tons of sodium carbonate. Most plants have recently closed because they could not afford to comply with the required antipollution measures.

21•4•6 Sulfates Sodium sulfate, Na_2SO_4, is used in making glass and in pulping wood. Potassium sulfate, K_2SO_4, is a valuable ingredient in certain types of fertilizer.

The solubility of the sulfates of the alkaline earth metals decreases markedly from the very soluble beryllium sulfate to the practically insoluble barium sulfate. It is common in chemical analysis to add soluble barium chloride to a sulfate-containing solution and then collect and carefully weigh the barium sulfate precipitate. From the weight of the barium sulfate, the

---

Key Chemical: Na_2CO_3
sodium carbonate

How Made

$2Na_2CO_3 \cdot NaHCO_3 \cdot 2H_2O$
      $\downarrow$ heat
$3Na_2CO_3 + CO_2\uparrow + H_2O\uparrow$

U.S. Production; Cost

~6.5 million tons per year from trona, ~\$55 per ton; ~1.5 million tons per year synthetic, ~\$80 per ton

Major Uses

50% glass (90% containers, 10% flat)
25% making other chemicals
5% pulp and paper
5% detergents

Other Uses

water softening; used as a base when it can be substituted for sodium hydroxide

weight of the sulfate in the original solution can be calculated. Magnesium sulfate and calcium sulfate are mentioned in Section 24.8.

Sodium sulfate decahydrate, $Na_2SO_4 \cdot 10H_2O$ (called Glauber's salt), has properties that make it promising as a storage material for solar heating. It has a convenient transition temperature (32.4 °C),[3] a respectable heat of fusion (250 kJ/kg), and low cost (less than 4 cents/kg). A system developed by General Electric for an average home stores heat in a tank of $Na_2SO_4 \cdot 10H_2O$, perhaps in the basement. During the day, air circulated through panels on the roof transfers the sun's heat to the tank and melts the salt. During the night, the salt crystallizes and gives up its heat of solidification to air that is circulated through the house (see Figure 21.2):

$$Na_2SO_4 \cdot 10H_2O(s) \underset{\substack{\text{exothermic} \\ \text{(heat goes to house)}}}{\overset{\substack{\text{endothermic} \\ \text{(heat from the sun)}}}{\rightleftharpoons}} 2Na^+ + SO_4^{2-} + 10H_2O \quad \text{(solution)}$$

21•4•7 Other Compounds The IA and IIA elements form stable compounds with practically all the known negative ions. Their exceptional solubility in water and their low cost make the sodium compounds the most widely used sources of specific anions for chemical reactions.

Because of their low electronegativities, these elements (except beryllium and magnesium) react at high temperatures with hydrogen to form saltlike ionic hydrides. Some hydrides can be melted and electrolyzed.

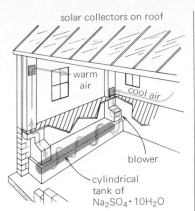

Figure 21•2 Air flow pattern when heat is being drawn from cylindrical tank in which $Na_2SO_4 + 10H_2O$ in solution is changing to solid $Na_2SO_4 \cdot 10H_2O$. Cool air flows around the tank and is warmed. When sun is shining, air flows through solar collectors, then down around tank to melt the solid $Na_2SO_4 \cdot 10H_2O$.
Source: Adapted by permission from a diagram supplied by the General Electric Company, Schenectady, N.Y.

• **Example 21•2** •

Write balanced equations for the following:
(a) lithium oxide + carbon dioxide
(b) barium nitrate (*aq*) + sodium sulfate (*aq*) (the net ionic equation)
(c) electrolysis of molten calcium hydride, showing the cathode and anode half-reactions and the net ionic equation

— • Solution •

(a) $Li_2O + CO_2 \longrightarrow Li_2CO_3$

(b) $Ba^{2+} + 2NO_3^- + 2Na^+ + SO_4^{2-} \longrightarrow BaSO_4\downarrow + 2Na^+ + 2NO_3^-$

(c) cathode: $Ca^{2+} + 2e^- \longrightarrow Ca$

    anode: $\underline{2H^- \longrightarrow H_2 + 2e^-}$

    $Ca^{2+} + 2H^- \longrightarrow Ca + H_2$

See also Exercises 16, 18, and 20.

Transition Metals and Their Neighbors

In the remainder of this chapter we will take up all the transition and inner transition metals and some of their neighbors—about 65 elements in

---

[3] Pure $Na_2SO_4 \cdot 10H_2O$ changes to solid anhydrous Na_2SO_4 and a saturated solution at 32.4 °C (see Figure 10.9). The formation of the anhydrous solid can be avoided by having some additional water present.

all; this includes all the metals not heretofore discussed. We will not be able to devote much attention to each individual element, but we shall describe some of their important physical and chemical properties and point out some important similarities and differences.

21•5   Classification

Based on their positions in the periodic table, the metals under discussion can be grouped into families as follows.

21•5•1   Transition and Inner Transition Metals

The *scandium family:* scandium, Sc, yttrium, Y, lanthanum, La (and the lanthanide series), and actinium, Ac (and the actinide series).

The *titanium family:* titanium, Ti, zirconium, Zr, and hafnium, Hf.

The *vanadium family:* vanadium, V, niobium, Nb, and tantalum, Ta.

The *chromium family:* chromium, Cr, molybdenum, Mo, and tungsten, W.

The *manganese family:* manganese, Mn, technetium, Tc, and rhenium, Re.

The *iron family:* iron, Fe, cobalt, Co, and nickel, Ni.

The *platinum family:* ruthenium, Ru, rhodium, Rh, palladium, Pd, osmium, Os, iridium, Ir, and platinum, Pt.

The *copper family:* copper, Cu, silver, Ag, and gold, Au.

The *zinc family:* zinc, Zn, cadmium, Cd, and mercury, Hg.

21•5•2   Neighbors of Transition Metals

The *aluminum family:* aluminum, Al, gallium, Ga, indium, In, and thallium, Tl.

The *germanium family:* germanium, Ge, tin, Sn, and lead, Pb.

Antimony, Sb, and bismuth, Bi, although metallic in character, are included in the nitrogen family, and polonium, Po, is classified in the sulfur family (see Chapter 24).

21•5•3   Metallo-Acid Elements
The titanium, vanadium, chromium, and manganese families have several common characteristics, one of which is that they form metallic oxides that are acidic, particularly in their higher oxidation states. The term **metallo-acid elements,** a name based on their acid-forming characteristics, refers to the twelve metals in these four families. For example, with chromium(VI) oxide:

$$CrO_3 + H_2O \longrightarrow H_2CrO_4 \quad \text{or} \quad 2CrO_3 + H_2O \longrightarrow H_2Cr_2O_7$$
$$\text{chromic acid} \qquad\qquad\qquad \text{dichromic acid}$$

The salts of such acids are encountered more commonly in chemical reactions than are the acids themselves. For example, potassium chromate, K_2CrO_4, and sodium dichromate, $Na_2Cr_2O_7$, are used as oxidizing agents.

21•6   Physical Properties

The transition elements and their neighbors exhibit typical metallic properties. In general, they are malleable, ductile, and good conductors of heat and electricity, and they have a metallic luster. Silver is the best electric conductor at normal temperatures. Copper is the element most commonly used for electric conductors. Aluminum is used for heavy-duty transmission lines because it compares favorably with copper in cost and conductivity per weight. For home use, aluminum wire has been criticized. It may become corroded at contact points, leading to overheating and possibly to fires.

The structures of these metals are generally of the close-packed types: body-centered cubic (bcc), face-centered cubic (fcc), or hexagonal close-packed (hex), as shown in Figure 8.29. Other structures exist also, such as simple cubic (sc), rhombic (rmb), tetragonal (tet), and diamond (dia), as listed in Tables 21.5, 21.6, 21.7, and 21.8. Except for copper and gold, which have characteristic colors, the metals are similar in appearance, resembling tin or iron to some extent.

The properties of the elements in the first transition series are summarized in Table 21.5. A number of trends that are obvious in this table are typical of all transition elements:

1. The number of electrons in the outside energy level remains constant, except for chromium and copper.

2. The melting and boiling points are uniformly high but do not vary in a regular way.

3. The atomic radii are remarkably constant. In the second and third series these values increase slightly, starting at rhodium, Rh, and iridium, Ir, respectively. But even in these cases, the radii are very nearly the same (see Figure 5.14).

4. The ionization energy tends to increase slightly as the positive charge on the nucleus (atomic number) increases.

These four characteristics reveal strong horizontal similarities among adjacent transition elements.

There are even greater similarities among the members of the first inner transition series, the lanthanides. Some of the lanthanides, especially the adjacent ones, are so nearly identical in chemical and physical properties that their compounds are difficult to distinguish. In nature, compounds of several of them tend to occur together. Undoubtedly, this similarity is due to the presence of similar numbers of electrons in both the outside and the next to the outside energy levels.

The separation of mixtures of lanthanide compounds was very tedious before the development of *ion-exchange chromatography*. This process depends on the differences in attraction between ions in solution and ionized sites of opposite charge on insoluble "exchangers." The exchangers can be either certain natural inorganic materials called *zeolites* (see Section 25.4.3) or certain synthetic organic materials called *resins*. Small particles of these exchange materials are typically packed in a column similar to that shown in Figure 12.16. Figure 21.3 is a schematic diagram of a few individual cation-exchange particles. As a solution of cations flows past the particles in the

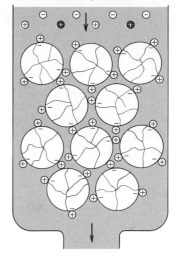

solution containing
two kinds of cations
to be separated

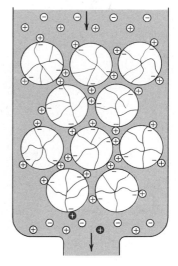

solution containing
a third type of cation
to saturate the column

Figure 21•3 In the top figure, a solution containing two kinds of added cations is starting to flow down through the packed spherical exchange particles. The bottom figure shows the situation some time later. The less strongly held kind of added cation is about to flow out the bottom. Fresh ionic solution coming in at the top will move the more strongly held kind of added cation on through the column.

exchange column, the cations in solution are continually exchanging places with cations that are held momentarily on the stationary, insoluble exchanger. The cations that are held less strongly by the exchanger move through the column more quickly. As the solution flows out the bottom of the column, the portions containing the separated cations are collected in different vessels.

Table 21•5 Physical properties of elements of first transition series

	Sc	Ti	V	Cr	Mn	Fe	Co	Ni	Cu	Zn
melting point, °C	1,539	1,660	1,917	1,857	1,244	1,537	1,491	1,455	1,084	420
boiling point, °C	2,730	3,318	3,421	2,682	2,120	2,872	2,897	2,920	2,582	911
density, g/cm^3	2.99	4.51	6.1	7.27	7.30	7.86	8.9	8.90	8.92	7.1
electron distribution	2,8,9,2	2,8,10,2	2,8,11,2	2,8,13,1	2,8,13,2	2,8,14,2	2,8,15,2	2,8,16,2	2,8,18,1	2,8,18,2
ionization energy, eV	6.5	6.8	6.7	6.8	7.4	7.9	7.9	7.6	7.7	9.4
atomic radius, Å	1.61	1.45	1.32	1.25	1.24	1.24	1.25	1.25	1.28	1.33
electronegativity	1.3	1.5	1.6	1.6	1.5	1.8	1.8	1.8	1.9	1.6
crystal structure	hex	hex	bcc	bcc	sc	bcc	hex	fcc	fcc	hex

Table 21•6 Physical properties of elements of second transition series

	Y	Zr	Nb	Mo	Tc	Ru	Rh	Pd	Ag	Cd
melting point, °C	1,530	1,852	2,477	2,610	2,250	2,427	1,963	1,554	962	321
boiling point, °C	3,304	4,504	4,863	4,646	4,567	4,119	3,727	2,940	2,164	767
density, g/cm^3	4.5	6.5	8.6	10.2	11.5	12.4	12.4	12.0	10.5	8.6
crystal structure	hex	hex	bcc	bcc	hex	hex	fcc	fcc	fcc	hex

Table 21•7 Physical properties of elements of third transition series

	La	Hf	Ta	W	Re	Os	Ir	Pt	Au	Hg
melting point, °C	920	2,222	2,985	3,407	3,180	~2,727	2,454	1,772	1,064	−39
boiling point, °C	3,470	4,450	5,513	5,663	5,687	~5,500	4,389	3,824	2,808	357
density, g/cm^3	6.2	13.3	16.6	19.4	21.0	22.6	22.6	21.4	19.3	13.6
crystal structure	hex	hex	bcc	bcc	hex	hex	fcc	fcc	fcc	rmb

Table 21•8 Physical properties of neighbors of transition elements

	Family IIIA				Family IVA		
	Al	Ga	In	Tl	Ge	Sn	Pb
melting point, °C	660	30[a]	157	303	940	232	328
boiling point, °C	2,447	1,980	2,070	1,487	2,852	2,623	1,751
density, g/cm^3	2.7	5.9	7.3	11.8	5.3	7.3	11.3
electron distribution	2,8,3	2,8,18,3	2,8,18,18,3	2,8,18,32,18,3	2,8,18,4	2,8,18,18,4	2,8,18,32,18,4
crystal structure	fcc	sc	tet	hex	dia	tet	fcc

[a]See Figure 21.4.

Table 21.6 and 21.7 show the melting points, boiling points, densities, and crystal structures for the second and third transition series. A comparison of Tables 21.5, 21.6, and 21.7 reveals the same general trends for the second and third series as those noted above for the first series. However, the melting points and densities of the second transition series are generally higher than those for corresponding members of the first series; those for the third series are the highest of all. For example, osmium, iridium, and platinum in the third series have the greatest densities of all elements—or, for that matter, of all known substances. Tungsten and rhenium in the same series have the highest boiling points of all the elements. Their melting points are exceeded only by that of carbon.

Table 21.8 shows certain properties of the neighbors of the transition metals.

Figure 21•4 As suggested by this picture, the melting point of the metal gallium is between normal room temperature and body temperature.
Source: Courtesy of the Aluminum Company of America.

21•7   Chemical Properties

21•7•1   Binary Compounds and Simple Salts   Practically all the metals form compounds containing well-known anions, such as halide, nitrate, acetate, sulfate, oxide, carbonate, hydroxide, phosphate, sulfide, and silicate. The first four anions tend to form water-soluble compounds and the latter six, insoluble compounds. In all these compounds the metal is in a positive oxidation state. Because most transition metals have more than one oxidation state, they form more than one salt with a given anion. Many of these salts are found in nature.

Desired compounds of metals can be prepared in a variety of ways. Examples of preparations that begin with the elements are

$$Fe + H_2SO_4 \longrightarrow FeSO_4 + H_2\uparrow$$
$$Ag + 2HNO_3 \longrightarrow AgNO_3 + NO_2\uparrow + H_2O$$

The oxidation of iron by hydrogen ion is typical of metals with negative values of $\mathscr{E}^\circ_{red}$. Silver, which has a positive value of $\mathscr{E}^\circ_{red}$, is not oxidized by hydrogen ion, but it is oxidized by concentrated nitric acid.

Salts are commonly formed by precipitation reactions, as described in Section 10.8.2. Examples are

$$CoCl_2 + H_2S \longrightarrow CoS\downarrow + 2HCl$$
$$Al_2(SO_4)_3 + 3Pb(NO_3)_2 \longrightarrow 3PbSO_4\downarrow + 2Al(NO_3)_3$$
$$ZnBr_2 + 2NaOH \longrightarrow Zn(OH)_2\downarrow + 2NaBr$$

Precipitation reactions similar to these are often used to prepare a compound, to remove a certain ion from a solution, or to analyze materials.

21•7•2   Oxidation States   A noteworthy characteristic of the transition metals is that most of them tend to show several oxidation states. This is in contrast to the alkali and alkaline earth metals (groups IA and IIA), which form cations of only 1+ and 2+ charge, respectively. The most common oxidation states for the members of the first transition series (Sc through Cu) are indicated in Figure 5.17. However, there are many other possible oxida-

Key Chemical: TiO_2
titanium dioxide

How Made
$TiO_2 + 2Cl_2 + C \longrightarrow$
crude
ore $TiCl_4 + CO_2$
$TiCl_4 + 2H_2O \longrightarrow$
            $TiO_2 + 4HCl$

U.S. Production; Cost
~700,000 tons per year; $800–$1,050 per ton (different grades)

Major Uses
50% paints, varnishes, etc.
25% paper coatings, fillers
10% plastics

Other Uses
inks, ceramics, special rubber

tion states, for example, the $+6$ state of iron and the $+2$, $+3$, and $+4$ states of vanadium.

Some important generalizations applicable to the transition metals can be made. Reference to Figure 5.17 will show how these apply to the first series.

1. Most of these elements do not have a common oxidation state of less than $+2$.

2. Each of the elements in groups IIIB to VIIB can exhibit the maximum oxidation state for its group. Examples of compounds include titanium(IV) oxide, TiO_2, chromium(VI) oxide, CrO_3, and rhenium(VII) oxide, Re_2O_7.

3. Most of the elements in group VIIIB do not exhibit the maximum oxidation state for this group. Only ruthenium and osmium have an oxidation state of $+8$, the highest shown by any element. Examples of compounds are ruthenium(VIII) oxide, RuO_4, and osmium(VIII) oxide, OsO_4.

By a suitable choice of oxidizing agent, concentration, and temperature, an element can be made to assume any of its known oxidation states. We must emphasize that for oxidation numbers of $+4$ and greater, the ions are not simple. Indeed, no discrete ions may be formed. Rather, the small highly charged particle will form covalent bonds with other molecules or ions if they are available. Examples include osmium tetroxide, OsO_4, in which osmium has an oxidation number of $+8$, even though the Os^{8+} ion does not exist as a separate particle. A similar case is potassium dichromate, $K_2Cr_2O_7$, in which chromium has an oxidation number of $+6$.

21.7.3   General Chemical Activity The general tendency of the metals in the middle of the periodic table to act as reducing agents varies greatly, but it is usually less than that of the alkali and alkaline earth metals. For example, aluminum and zinc are quite active, and iron and lead are moderately so. However, silver, gold, and the platinum family metals are inactive to the point of semi-inertness and so are referred to as *noble metals*. The relative activity of metals can be estimated by comparing their standard electrode potentials. Representative $\mathscr{E}^\circ_{red}$ values are listed in Table 21.9, following the conventions introduced in Table 18.1. The more negative the value of $\mathscr{E}^\circ_{red}$, the more active is the metal as a reducing agent.

1. The group IIIB elements and the members of the lanthanide and actinide series are very active, with reduction potentials usually more negative than that of aluminum, -1.66 V, and ranging down to about -2.5 V.

2. The metallo-acid elements are moderately active, with electrode potentials usually between those of aluminum, -1.66 V, and hydrogen, 0.00 V.

3. The three members of the iron family are also moderately active. Comparable values of electrode potentials are those of iron, -0.44 V, cobalt, -0.28 V, and nickel, -0.25 V.

4. The platinum metals are notable chiefly for their chemical inactivity, and their electrode potentials are in line with this behavior. All electrode

Table 21.9 Typical standard reduction potentials[a]

Couple (ox/red)	$\mathscr{E}^\circ_{red}$, V
Li^+/Li	-3.04
Al^{3+}/Al	-1.66
Fe^{2+}/Fe	-0.44
H^+/H_2	0.00
Cu^{2+}/Cu	$+0.34$
Au^{3+}/Au	$+1.50$

[a]See Table 18.1 for the equations for the reduction reactions.

potentials are more positive than that of copper, ranging from $+0.45$ V for ruthenium up to $+1.15$ V for iridium.

5. The neighbors of the transition metals range widely in activity, with electrode potentials ranging from aluminum, -1.66 V, to tin, -0.13 V.

21•7•4   Corrosion   Closely related to the chemical activity of the metals is the phenomenon of **corrosion,** the chemical attack on a metal by its environment. The most common type of corrosion is due to the action of the atmosphere in conjunction with water and various substances dissolved in the water. Essentially, corrosion is a reaction in which a metal is oxidized. In most cases, the reaction is definitely electrochemical in nature and involves the type of electron transfer that is characteristic of voltaic cells (batteries).

A metal in contact with a solution will corrode (be oxidized) if there are chemical species in solution that can be reduced at a potential more

---

★ Special Topic ★ Special Topic ★ Special Topic ★ Special Topic ★ Special Topic ★ Special Topic ★ Special Topic ★ Special Topic   ★

When Your Car Rusts Out   When your car begins to "rust out"—that insidious process that forecasts an expensive series of visits to car dealers and banks and probably a restriction on your budget for months to come—take a careful look at the corroded area. You will find that the major problem is corrosion of the metal under the paint finish.

Doesn't paint protect iron and steel from corrosion? Yes and no. Yes, if the metal is completely sealed; but no, if the protective coating is broken to expose even a microscopic area of bare metal. If this happens, the metal *under* the paint rusts.

This unexpected result is readily understood by considering the electrochemical corrosion reaction. When iron is in contact with oxygen and water in an acidic environment, the following oxidation–reduction reaction occurs:

$$Fe + \tfrac{1}{2}O_2 + 2H^+ \longrightarrow Fe^{2+} + H_2O$$

This initial corrosion reaction may be divided into two electrochemical half-reactions:

$$Fe \longrightarrow Fe^{2+} + 2e^- \text{ (anodic)} \quad +0.44 \text{ V}$$
$$\tfrac{1}{2}O_2 + 2H^+ + 2e^- \longrightarrow H_2O \quad \text{(cathodic)} \ \underline{+1.23 \text{ V}}$$
$$+1.67 \text{ V}$$

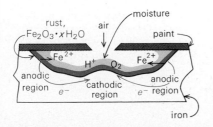

rust,
$Fe_2O_3 \cdot xH_2O$ air   moisture   paint
Fe^{2+} Fe^{2+}
H^+ O_2
anodic region   cathodic region   anodic region
e^- e^-
iron

Rusting of iron near a break in a film of protective paint.

The positive sign of the cell voltage indicates that the reaction is spontaneous. The moderately high voltage shows that the driving force is greater than that for many commercial voltaic cells (see Section 18.9).

The iron(II) is further oxidized by oxygen to form rust, hydrated iron(III) oxide:

$$2Fe^{2+} \longrightarrow 2Fe^{3+} + 2e^- \text{ (anodic)} \quad -0.77 \text{ V}$$
$$\tfrac{1}{2}O_2 + 2H^+ + 2e^- \longrightarrow H_2O \quad \text{(cathodic)} \ \underline{+1.23 \text{ V}}$$
$$+0.46 \text{ V}$$

$$2Fe^{3+} + (x + 3)H_2O \longrightarrow Fe_2O_3 \cdot xH_2O + 6H^+$$

In the film of moisture in the vicinity of a break in the paint finish, there can be substantial variations in the concentrations of the chemical species involved in the two half-cells. In the area most exposed to the air, there is a relatively high concentration of dissolved O_2, and the concentration of H^+ ions is high due to a solution of acidic oxides, such as those of carbon, nitrogen, and sulfur. These conditions favor the cathodic reaction, as we would predict qualitatively from Le Chatelier's principle (see Section 10.4.1) or from the Nernst equation (see Section 18.3.4).

The anodic reaction takes place in the area where the cathodic reaction is not favored, that is, back under the paint. It is there that the metal is most rapidly corroded as iron atoms become iron(II) ions. The concentration of Fe^{2+} ions is kept low because of the reaction to form the iron(III) oxide precipitate, and this favors the anodic reaction. The cathodic and anodic reactions are shown schematically in the accompanying figure.

Source: From Ward Knockemus, "When Your Car Rusts Out," *J. Chem. Educ., 49,* 29 (1972); adapted by permission.

positive than the electrode potential of the metal (compare Figure 18.5). The metal then becomes anodic, usually at local spots due to irregularities on the surface or in the solution. The reaction is

$$M \longrightarrow M^{x+} + xe^- \qquad \text{anode (oxidation, corrison)}$$

Typical cathodic reactions that occur on other regions of the metal or on a less active metal attached to M are the following:

$$2H^+ + 2e^- \longrightarrow H_2$$
$$O_2 + 4H^+ + 4e^- \longrightarrow 2H_2O$$
$$Fe^{3+} + e^- \longrightarrow Fe^{2+}$$

A local corrosion cell is shown in Figure 21.5(a). Some local irregularity, such as a surface scratch or lower hydrogen ion concentration in solution, may result in a local anodic area. The metal corrodes by M^{x+} ions passing into solution. The electrons liberated by this oxidation pass from the anodic to a cathodic region where hydrogen is liberated at the surface.

Moisture is necessary for the atmospheric corrosion of all but the most active metals. If salts are dissolved in the moisture, the corrosion is speeded up, partly because in a solution of electrolytes, charge flow via the movement of ions is made easier. The presence of any substance that dissolves and forms an acid solution—for example, sulfur dioxide or carbon dioxide—usually increases the rate of corrosion. Oxygen that is dissolved in water is an important corrosion agent. Finally, as pointed out in Example 18.4, in pure water an element with an electrode potential more negative than -0.414 V can be spontaneously oxidized (corroded) by H_2O to yield H_2.

The position of a metal in a table of reduction potentials and the cathodic reaction are not the only factors that determine the extent and rate of corrosion. Just as important may be crystal structure (see Figure 21.6 on Plate 4) or the type of film or coating formed on the surface of the metal. Aluminum and magnesium actually react quickly when exposed to air or water, but the thin, closely packed film of oxide that forms on the surface protects the underlying metal from further corrosion. These two metals corrode less completely in air than does the less active metal iron. The rust formed on the surface of iron is so flaky and porous that the corroding chemicals can pass through it easily and attack the underlying metal.

The formation of a protective film explains the paradox of galvanized iron—iron protected from corrosion by a coating of the more active metal zinc. Zinc reacts readily with moisture, oxygen, and carbon dioxide to form a tough film of basic carbonate, $Zn(OH)_2 \cdot ZnCO_3$, that resists further attack. Even if a hole is made in the zinc coating, the underlying iron is protected by the anodic action of the more active zinc. This tends to keep the iron charged negatively, thus preventing the reaction $Fe \rightarrow Fe^{2+} + 2e^-$ as long as any zinc remains; see Figure 21.5(b).

The corrosion protection offered by the tin-plating of steel is inferior to that of zinc in one respect. If the tinplate is broken, the underlying iron corrodes rapidly. In this case the iron acts as the oxidizable anodic material, see Figure 21.5(c), because it has a more negative reduction potential than has tin. (Compare Table 18.1 and Figure 18.5.)

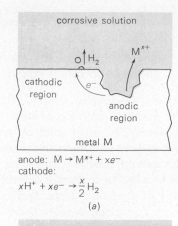

anode: $M \rightarrow M^{x+} + xe^-$
cathode:
$$xH^+ + xe^- \rightarrow \frac{x}{2} H_2$$
(a)

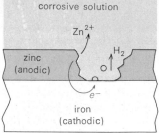

anode: $Zn \rightarrow Zn^{2+} + 2e^-$
(slow, due to protective coating of basic zinc carbonate)
cathode:
$$2H^+ + 2e^- \rightarrow H_2$$
(b)

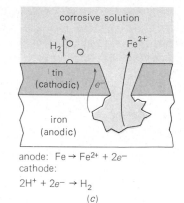

anode: $Fe \rightarrow Fe^{2+} + 2e^-$
cathode:
$$2H^+ + 2e^- \rightarrow H_2$$
(c)

Figure 21.5 In corrosion, charge can flow by direct electron migration in the metal phases and by way of movement of ions in the corrosive solution.

The rusting of underground pipes may necessitate costly repairs or even replacement. One ingenious method of preventing the corrosion of iron pipes is by cathodic protection. Pieces of an active metal, such as magnesium, are buried in the ground near the pipe and connected to it by a wire. Instead of the iron losing its own electrons directly to the oxidizing agents (corrosion agents) that attack it, it merely relays, via the wire, electrons from the more active metal. The slug of active metal corrodes away, but the costly pipeline is protected. This type of protection is also used on steel structures exposed to seawater, such as oil-drilling platforms, barges, and even ships. The cost of protecting and replacing corroded metals is estimated to account for about 4 percent of the total U.S. economy. In 1978 this amounted to about $70 billion.

Our success in producing corrosion-resistant metals makes the reuse of resources more convenient. An example of this is the omnipresent aluminum beverage can. To reclaim and recycle such materials saves both raw materials and energy.

Production of Metals

Control of the environment began when humans learned how to produce metals. Whether used in relatively pure form or mixed with one another to form *alloys,* the elemental metals are the materials that form the tools and machines of modern civilization. Although most metals are rare, a few are among the most common and useful substances known: elements such as nickel, chromium, tungsten, copper, iron, silver, tin, and lead. Knowledge and use of these metals made possible the Bronze Age, the Iron Age, and the Industrial Revolution, and they continue to play leading roles joined by some rarer metals as the Atomic Age joins the Space Age.

21•8   Metallurgy

The subject of **metallurgy** concerns the processing of natural raw materials to obtain elemental metals and with subsequent treatments to produce metals with desired properties. Such treatments include mixing metals to make alloys, careful heating and cooling to influence physical and chemical properties, and shaping by casting and mechanical working. These processes are part empirical and part theoretical. Many of the most useful techniques have been developed over the centuries by trial and error, and many more have been discovered more recently by applying advanced theories and the latest knowledge. With few exceptions, such as magnesium from seawater and some manganese from the floor of the oceans, metals are produced commercially from subsurface deposits. The natural inorganic materials found in the earth's crust are called **minerals.** Minerals that can be used as a source for the commercial production of materials are called **ores.**

The most common ores of the metals are oxides, sulfides, halides, silicates, carbonates, and sulfates. The silicates, although the most abundant minerals, are of relatively little value as ores of metals, because they are so difficult to decompose chemically and because there are simpler ores available that are easier and cheaper to process. For most ores, the general problem is to decompose a compound in which a metal exists in a positive oxidation state, often as positive ions, and to reduce the positive ions of the metal to atoms of the element. A general simplified reduction equation, with x being 1, 2, 3, and so on, is

$$M^{x+} + xe^- \longrightarrow M$$

The metal in its positive oxidation state is not necessarily a simple ion. Indeed, it is often found in an anion with oxygen, for example, $MO_4{}^{x-}$, or in a silicate.

In rare cases, a metal is found in the elemental state and needs only to be separated from impurities.

The initial stages of the metallurgy of most metals involve three steps: (1) concentration of the ore, (2) chemical reduction to the element, and (3) removal of impurities from the metal or adjustment of the composition by various refining processes. For some cases the functions of these stages overlap. For example, removal of impurities may be a part of the ore concentration process so that subsequent refining is not necessary.

21•8•1   Concentration of Ore   The ore that is mined usually contains some worthless rock called **gangue.** If the gangue is objectionable at a later stage, the first step in ore concentration is to remove it. The ore is usually crushed and ground until the particles of the mineral are broken apart from the gangue. If possible, these particles are separated by physical means, such as washing, flotation, or magnetic attraction.

Washing with a turbulent stream of water often washes the lighter gangue away from the desired mineral.

Flotation involves agitating the ore in a vessel with a detergent or foaming agent. The more valuable, denser mineral may stick to the bubbles of foam and float off with it, leaving the gangue behind, or the gangue may be attracted to the foamy layer and float off with it (see Figure 21.7).

With an *electromagnet* some minerals can be drawn out of their crushed ores. An example is magnetite, Fe_3O_4. Also, certain minerals can be charged electrically and then attracted to a charged plate, leaving the gangue behind. (This technique is similar to that used in the Cottrell precipitator, which was described in Figure 12.15.)

If the ore cannot be sufficiently concentrated by physical means, chemical processes are used, such as those described in the following paragraphs.

In many cases the ore is *roasted* to drive off volatile impurities, to burn off organic matter, and to form compounds that are more easily smelted. Roasting in air usually converts sulfides and carbonates to oxides. For example,

$$2ZnS + 3O_2 \longrightarrow 2ZnO + 2SO_2$$

Ores generally contain considerable gangue even after concentration. Often, to remove the last of the gangue, a flux is added during the smelting step. A **flux** is a substance that combines with the gangue and makes a molten material called **slag** as the mixture is heated in a furnace. At

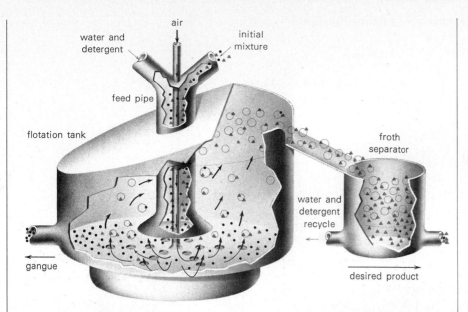

air
water and
detergent
initial
mixture
feed pipe
flotation tank
froth
separator
water and
detergent
recycle
gangue
desired product

Figure 21•7 Diagram of a flotation tank showing the separation of a mixture containing a desired material (colored dots) from gangue (black dots). A rotating paddle at the bottom of the cell distributes the mixture and also sweeps the bubbles around the central pipe. As indicated by the dots in the separated streams, the separation is not perfect.

high temperatures, the slag is a liquid that is insoluble in the molten metal and forms a separate layer. If the gangue is an acidic oxide such as silica, SiO_2, a cheap basic oxide like lime, CaO, will be used for the flux. These two react in a furnance to form the low-melting compound calcium silicate, the slag:

$$SiO_2 + CaO \longrightarrow CaSiO_3$$

If the gangue is basic, for example, calcium or magnesium carbonate, the flux will be a cheap acidic oxide, probably silica.

Because roasting ores commonly produces air pollutants, wet chemical methods probably will be used increasingly in the future. Acids or bases may be used to dissolve part or all of the ore. Sometimes a compound of the desired metal is precipitated from the solution; at other times impurities are precipitated.

21•8•2 **Smelting** The industrial reduction processes are called **smelting**. There are several chemical methods that can be used to reduce a given metal from its oxidation state in the ore to the elemental state. If reduction is difficult for a particular metal, powerful reducing processes will be necessary.

• *Reduction by Heat in Air* • The precious metals in groups VIIIв and Iв are produced easily. Platinum, gold, and sometimes silver are found in their elemental form and have only to be heated to melt them out of the gangue. Because many of the oxides of the less active metals are decomposed by extreme heat, roasting in air is all that is needed for reduction. For example, roasting the sulfide ore of mercury forms the metal rather than the metallic oxide:

$$HgS + O_2 \longrightarrow Hg + SO_2$$

Molten copper(I) sulfide is reduced by blowing air through it:

$$Cu_2S + O_2 \longrightarrow 2Cu + SO_2$$

• *Reduction with Carbon* • Oxides of many moderately active metals can be reduced by carbon. The reactions for cobalt oxide are

$$CoO + C \xrightarrow{\text{heat}} Co + CO$$

$$CoO + CO \xrightarrow{\text{heat}} Co + CO_2$$

This reduction method is suitable for metals of the iron family and for some others such as lead, tin, and zinc. Note that the carbon may be oxidized to carbon monoxide, CO, or carbon dioxide, CO_2. In the presence of carbon (usually as coke) and at high temperatures, CO is the dominant gas and is the effective reducing agent in most smelting processes using carbon.

Carbon tends to form carbides with certain metals such as chromium and manganese; hence it cannot be used for the reduction of all oxide ores of moderately active metals. But it is used when possible because it is both cheap and convenient.

• *Reduction with Hydrogen* • Reduction with hydrogen may be used when carbon is not suitable. Tungsten oxide is reduced in this way, because with carbon as the reducing agent the metal produced is mixed with carbides. The hydrogen reduction reaction is

$$WO_3 + 3H_2 \longrightarrow W + 3H_2O$$

• *Reduction with an Active Metal* • If compounds are not satisfactorily reduced with carbon or hydrogen, an active metal can be used as the reducing agent. Aluminum, magnesium, sodium, and calcium are active enough to be good reducing agents. Uranium(IV) fluoride is reduced with calcium or magnesium. Titanium chloride is reduced with magnesium or sodium:

$$TiCl_4 + 2Mg \longrightarrow 2MgCl_2 + Ti$$

• *Reduction by Electrolysis* • Very active metals, such as the alkali metals and the alkaline earth elements, are most efficiently produced by the electrolysis of anhydrous fused salts:

$$2NaCl \xrightarrow{\text{electrolysis}} 2Na + Cl_2\uparrow$$
$$\text{molten} \qquad\qquad\quad \text{molten} \quad \text{gas}$$

Aluminum is produced by the electrolytic reduction of aluminum oxide or aluminum chloride (see Section 21.9.3). Also, the elements of group IIIB and the lanthanide series are usually prepared by the electrolysis of their molten chlorides.

In addition to the foregoing examples of very active metals, there are some others that are produced electrolytically. Because manganese forms a carbide when it is reduced with carbon monoxide, it is reduced to the metal by electrolysis in water solution.

21•8•3   Refining   The purification or adjustment of the composition of impurities in crude metals is called **refining.** Metals with low boiling points—for example, mercury, bismuth, zinc, and magnesium—can be separated from most impurities by simple distillation. Like salts, metals also can be refined by fractional crystallization.

Probably the most widely used refining process is the electrolytic process. Copper is typical of the metals refined by this method.

A modern purification method that is used with great success on metals is *zone refining* (Figure 21.8). A heating coil is passed slowly along a rod of metal, melting a narrow zone. As the band of molten metal moves along, the relatively impure metal melts on the leading edge and the higher-purity metal crystallizes on the trailing edge of the melted zone. The impurities collect in and move along with the liquid zone to the end section, which is cut off and discarded. The metal spheres shown in Figure 21.6 on Plate 4 were machined from single crystals purified by zone refining.

21•9   Industrial and Environmental Aspects of Metal Production

The study of the impact of human activities on the environment is essentially a study of chemical relationships. This section will focus on industrial activities and industrial products in the belief that industry is the proper concern of every educated citizen. Industry's huge chemical operations have possible dramatic and long-lasting influences on society. Well-informed citizens should be able to work with and influence industry so that benefits can be increased and harmful effects can be minimized.

Our review of major industrial processes begins with the production of three of the major metals. In later chapters we will look at some of the most useful inorganic chemicals and some typical organic chemicals made from petroleum. Each of these processes produces substances or energy that is vital to the standard of living that is enjoyed in developed countries and that is a goal for other less developed countries.

21•9•1   Copper   We can illustrate some of the effects of large-scale technology on a limited geographic area by looking at the production of copper. Copper was one of the first metals to be produced by primitive cultures. Mixed with tin, it formed bronze, a strong, hard alloy from which tools and weapons were fashioned. The Bronze Age began sometime before 3500 B.C. The reason copper was discovered early is because its compounds are easily decomposed. It is likely that crude samples of the metal were formed by the decomposition of rocks in ancient campfires.

Sulfide ores containing copper(I) sulfide, Cu_2S, account for the majority of the copper production in the United States. It is common to find iron(II) sulfide, FeS, in the same mineral deposit, as well as a large proportion of worthless gangue. Usually the gangue is composed mainly of acidic oxides, such as SiO_2, but it may be a basic mineral, such as $CaCO_3$. The crude ore is

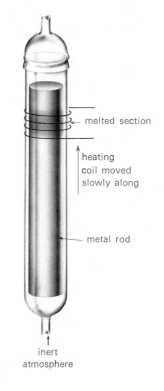

melted section

heating coil moved slowly along

metal rod

inert atmosphere

Figure 21•8 Purification of a sample of metal by zone refining.

concentrated by crushing it and then separating the sulfides from most of the gangue by flotation (see Figure 21.7).

In a modern smelter, the concentrated copper and iron sulfides are heated in a furnace by hot gases produced by burning natural gas or coal dust. The two compounds, Cu_2S and FeS, form a mixture called *matte*. The extreme heat of the furnace melts the matte and the gangue. With the proper flux, the gangue forms a slag, which floats to the top. The principal reactions are

$$2FeS + 3O_2 \longrightarrow 2FeO + 2SO_2$$

The iron oxide combines with added silicon dioxide to form slag:

$$FeO + SiO_2 \longrightarrow FeSiO_3$$

After this slag is drawn off, more air is blown through the molten copper sulfide to form molten copper and gaseous sulfur dioxide:

$$Cu_2S + O_2 \longrightarrow 2Cu + SO_2$$

• *The Copperhill Experience* • We can illustrate the scale of a truly large industrial operation and show how by-products can be handled in different ways by describing a long-time metallurgical operation at Copperhill, Tennessee, near the borders of Georgia and North Carolina. A rich copper deposit there attracted miners and investors before the Civil War. For many years, the sulfide ore was roasted over wood fires in huge open piles. The sulfur dioxide formed went into the atmosphere. The cutting of the trees in the area for the roasting operation plus the action of sulfur dioxide on the rest of the vegetation resulted in the denuding of about 50 square miles of mountain land. Agriculture was adversely affected over a much larger area, but the mining operators were able to carry on their open burning despite continued lawsuits by small farmers in Tennessee, North Carolina, and Georgia.

Although the legal damage suits led eventually to an injunction against the mining companies by the U.S. Supreme Court, the main relief came in the early 1900s when the sulfur dioxide became a valuable product. Plants were built to make sulfuric acid from the sulfur dioxide. One of these was the largest acid plant in the world when it was built, and its production of H_2SO_4 accounted for about 10 percent of the sulfuric acid made at that time in this country. The main product at Copperhill became and remains today sulfuric acid. Some iron, copper, zinc, and even small amounts of noble metals are all recovered, but the principal profit comes from what was once a useless, destructive by-product.

Because erosion of the denuded land led to the loss of the fertile topsoil, it was estimated that it would take 500 to 1,000 years for nature to restore vegetation to the Copperhill area. In the 1930s, the federal government began a reforestation project that today is being pursued by the main private company in the region. Of about 15 million trees planted, possibly half have survived, and the red clay hills are slowly being recovered. Figure 21.9(*a*) and (*b*) on Plate 8 shows the approximately 20-square-mile area that is still bare. This is probably the most obvious unwanted, single-cause environmental effect of human actions on the face of the earth. Only artificial lakes, large cities, and large agricultural areas are as evident in satellite photographs. Figure 21.9(c) and (d) shows recent ground-level photographs.

It is unlikely that such wanton disregard for the environmental consequences of a chemical by-product would be permitted in many countries today. However, the cumulative effects of more subtle actions and substances can be just as serious. The cutting of trees for fuel is causing an alarming increase in the size of desert regions around the world. The living space for perhaps 50 million people, who have a poor standard of living anyway, is being lost. And, as we will see in the special topic on acid rain in Chapter 24, small concentrations of sulfur dioxide distributed over a large area cause major environmental problems.

21•9•2   **Iron**   Because iron is more active than copper, its ores cannot be reduced to the metal simply by heating. One of the most important technical advances in history was the discovery of methods for making elemental iron. The Iron Age began some 2,000 years after the Bronze Age did, and it was another 3,000 years before industrial cast iron was developed in about A.D. 1500.

Iron is a good example of the moderately active metals that are reduced by carbon monoxide. The common oxide of iron, Fe_2O_3, is reduced to the element in a series of steps:

$$3Fe_2O_3 + CO \rightleftharpoons 2Fe_3O_4 + CO_2$$
$$Fe_3O_4 + CO \rightleftharpoons 3FeO + CO_2$$
$$FeO + CO \rightleftharpoons Fe + CO_2$$

• *The Blast Furnace* • The primary reduction of the iron oxide ore is carried out in a mammoth chimney called a **blast furnace.** In this fiery reaction chamber molten iron is formed, and SiO_2, the chief impurity, is largely removed in the form of a slag of calcium silicate, $CaSiO_3$. As shown in Figure 21.10, the molten slag and iron form separate layers at the bottom of the furnace.

A mixture of crushed iron ore, coke, and limestone is added by means of a hopper at the top of the furnace. A blast of hot air is blown up through this mixture from the bottom of the furnace. The complex series of reactions can be summarized as follows:

1.   Near the bottom of the furnace where the blast of hot air enters, coke burns furiously to produce carbon dioxide:

$$C + O_2 \longrightarrow CO_2$$

2.   As the carbon dioxide rises in the chimney, it is reduced almost immediately by hot coke to carbon monoxide:

$$CO_2 + C \longrightarrow 2CO$$

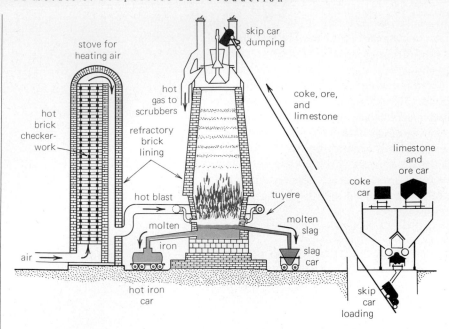

Figure 21•10 A schematic diagram of a blast furnace.
Source: Courtesy of Bethlehem Steel Company.

3. In the middle and upper portions of the furnace, the carbon monoxide reacts with the iron oxide to form metallic iron. The overall equation is

$$Fe_2O_3 + 3CO \longrightarrow 2Fe + 3CO_2$$

4. Near the middle of the furnace, the limestone decomposes to form lime and carbon dioxide:

$$CaCO_3 \longrightarrow CaO + CO_2$$

5. Farther down, the lime and silica react to form the slag:

$$CaO + SiO_2 \longrightarrow CaSiO_3$$

The operation of the furnace is continuous. The mixture of reactants is fed into the top at regular intervals to begin its journey toward the white-hot lower levels. The furnace must be "tapped" about every six hours to drain off the molten iron.

For each ton of **blast furnace iron** or **pig iron** produced,[4] about 2 tons of iron ore, 1 ton of coke, 0.3 ton of limestone, and 4 tons of air are required. The main by-products are 0.6 ton of slag and 5.7 tons of **flue gas,** which is mainly nitrogen and carbon dioxide, but contains about 12 percent CO and 1 percent H_2. Both of the latter, when burned, yield heat energy for a variety of uses: to heat the incoming air; to move the raw materials to the furnace and blow the air; and to fire the furnaces in which coal is changed to

---

[4]The molten blast furnace iron is sometimes run into molds where it hardens into small ingots called *pigs.* However, much of this iron is converted directly into steel, instead of into ingots or pigs. The terms *pig iron* and *blast furnace iron* are synonymous.

coke, and those in which the pig iron is converted to steel. The transformation of brittle, weak pig iron into malleable, tough steel is a complex process, which we will not describe.

The chief environmental effects of steel production result from the open-pit mining of the ore, the mining of the coal, and the waste products of the blast furnace. There are mountains of ash and clinkers to dispose of, and the carbon dioxide and sulfur dioxide in the gases are added to those from other combustion reactions. Also, there is the need to dispose of tremendous quantities of slag from the blast furnace and from the steel furnace. Fortunately, this slag is relatively inert, rocklike calcium silicate, which can be used in building roads and making concrete and as landfill.

21•9•3   **Aluminum**   The third production we will describe is that of aluminum. It is representative of those metals that are so active that it is not practical to reduce their compounds with chemical reducing agents. Although aluminum is the most common metallic element in the earth's crust, it was not until about 100 years ago that an industrial process was invented for producing it. At that time aluminum was as valuable as silver and was used in jewelry, ornaments, and luxury items.

Like other active metals, aluminum is reduced by electrolysis in an anhydrous medium. In 1886, the molten salt cryolite, Na_3AlF_6, was used as a solvent for Al_2O_3 by two inventors working independently. Charles M. Hall in the United States and Paul Héroult in France, each 22 years old, announced practically simultaneously that they had solved a chemical puzzle that had baffled many outstanding chemists.

The main process used for the production of aluminum today, the *Hall process,* is virtually identical chemically with the original process. Anhydrous aluminum oxide (alumina), Al_2O_3, is dissolved in cryolite, where it is thought to be in the form of Al^{3+} and O^{2-} ions. Electrolysis of the solution yields elemental aluminum at the cathode and oxygen at the anode.

A number of electrolytic cells for the production of aluminum are shown in Figure 21.11. An iron tank lined with carbon is the cathode of the cell, and large blocks of carbon serve as anodes. The cryolite is melted in the tank, and purified anhydrous aluminum oxide is added to it.

When current is passed through the cell, molten aluminum forms at the walls and bottom of the tank (the cathode). Essentially all of the oxygen liberated at the anode attacks the carbon and forms carbon dioxide. The electrode half-reactions that are commonly written are[5]

$$\text{at the cathode:} \quad Al^{3+} + 3e^- \longrightarrow Al$$
$$\text{at the anode:} \quad C + 2O^{2-} \longrightarrow CO_2 + 4e^-$$

The carbon anodes are continually used up and hence must be

---

[5] The electrode reactions are more complex than those commonly written. Perhaps the main reactions are

$$\text{at the cathode:} \quad 3Na^+ + 3e^- \longrightarrow 3Na$$
$$3Na + AlF_3 \longrightarrow Al + 3NaF$$
$$\text{at the anode:} \quad 2Al_2OF_6{}^{2-} + C \longrightarrow 4AlF_3 + CO_2 + 4e^-$$

Figure 21•11 A series of electrolytic cells is called a *pot line* in an aluminum plant. Curved hoods trap gaseous emissions. Electric connectors through the hoods carry the current from the carbon anodes.
Source: Courtesy of the Aluminum Company of America, Pittsburgh, Pa.

replaced from time to time. Under normal operating conditions, about 0.44 kg of the anode is consumed per kilogram of aluminum produced. The cost of the carefully made anodes is a major expense.

The raw material for the Hall process is *bauxite*, a mineral that consists mainly of hydrated aluminum oxide, for example, $Al_2O_3 \cdot H_2O$ and $Al_2O_3 \cdot 3H_2O$. There are valuable deposits of it in Arkansas, but we import large quantities from South America and Jamaica.

In 1976 Alcoa began operating a plant that uses the first new commercial method of producing aluminum since the invention of the Hall process. The Alcoa smelting process is based on the electrolysis of molten aluminum chloride, made by chlorinating alumina (see Figure 21.12).

• *Mining and Production* • The scale of production of aluminum—approaching 4.5 million metric tons a year in this country—dictates inevitably that there will be effects on the environment. Some of these are typical of all industries and a few are particularly associated with the aluminum industry.

• *Mining* • Most bauxite is mined from huge open pits where bulldozers and steam shovels scoop up the ore and load it directly into trucks. The richest mines are located in tropical countries where heavy rainfall has leached away the more soluble minerals. In many cases, there have been no effective reclamation programs by the absentee owners of the mines. Reclamation, first to grassy pastures, has been successful on the sites of exhausted bauxite mines, and this action will probably expand in those developing countries that are exercising increased control over the use of their mineral resources.

• *Power Production* • Many of the hydroelectric dams throughout the world supply power for electrochemical industries, particularly the alumi-

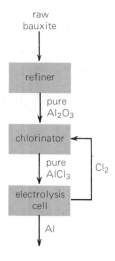

Figure 21•12 Alcoa smelting process.
Source: Adapted by permission from a flow diagram of the Aluminum Company of America, Pittsburgh, Pa.

num industry. The artificial lakes behind these dams result in major modifications of the environment. For the most part these changes have been viewed as desirable, but some have not. Mexican farmers have complained about damage to their agriculture from reduced water flow in the Colorado River and increased salt concentration downstream from dams and irrigation projects in the United States. In Tennessee, completion of a nearly finished dam was long prevented because the lake to be formed would destroy the habitat of a rare small fish. The loss of food-producing land because of flooding by artificial lakes is becoming a greater cost as population and food needs increase.

• *By-products of Electrolysis* • The chief by-product of the electrolysis of Al_2O_3 is the carbon dioxide made at the anode. However, there are also small amounts of fluorides in the gases that are produced in the Hall electrolysis cell. Traces of fluorides emitted by early aluminum plants led to serious agricultural problems. Cattle grazing on vegetation near the plants became lame, and some died. Lawsuits and publicity spurred the companies to develop methods to remove fluorides from waste gases. It was found that the Al_2O_3 that is ready to go into the electrolysis cell is an effective adsorber for gaseous fluorides. As its surface becomes saturated with the fluorides, this Al_2O_3 is simply added to the oxide on its way to be electrolyzed. The new Alcoa smelting process for electrolyzing $AlCl_3$ is designed so that emissions are minimized. Unfortunately, existing Hall process plants cannot be converted to the new process.

While studying the effect of fluorides on plant and animal life, chemists in the aluminum industry discovered a positive, healthful role of fluoride ion. These chemists, working with dentists, found that persons living in areas where the natural water supply contains more than the average amount of fluoride ion have fewer than normal dental caries. This research led to improved medication by dentists, to the formulation of toothpastes containing fluoride, and to the regular addition of fluoride ions to many municipal water systems.

• *Recycling* • In its chemical activity, aluminum is a paradox. It has so negative a reduction potential that it is never found in nature as the element, and it can be reduced by electrolysis only under anhydrous conditions. Yet once metallic aluminum is made, a thin tough film of aluminum oxide protects the metal so efficiently that it appears not to corrode. The result is a shiny metal that refuses to self-destruct but glints at us from ditches and streams in the form of cans, pop-top rings, and pieces of many other discarded items.

Recycling used aluminum is a worthwhile endeavor for the public and the aluminum industry. Almost a million Americans are collecting cans and earning money for themselves or their organizations. The increased attractiveness of the environment is an obvious benefit. The reuse of aluminum in the smelters benefits the environment in several other ways as well. It requires only about 5 percent as much energy to produce a ton of recycled aluminum as it does to produce the element from the ore (see Figure 21.13). Also for every ton that is recycled, there is that much less use of the raw materials and all the resources necessary to process them.

Figure 21•13 Shredded and pressed into blocks, these aluminum cans are on the way to the melting furnace at Alcoa, Tennessee.
Source: Courtesy of the Aluminum Company of America, Pittsburgh, Pa.

Chapter Review

Summary

Most of the chemical elements are metals. The B families of the periodic table consist exclusively of metals classified as **transition elements** and **inner transition elements.** The latter comprise the **lanthanide series** and the **actinide series.** Among the most important metals are the alkalis and alkaline earths, which are main-group elements. They have some unusual physical properties and are chemically highly reactive because of their strong tendency to lose electrons and become cations. Most of them react readily with air, water, and many other substances to form ionic compounds. Many of these are commercially important. Some of the hydroxides (bases) are produced by **slaking** the metallic oxides with water. Other important compounds include the halides, carbonates, and sulfates.

The transition and inner transition metals and their neighbor metals to the right in the periodic table are grouped into families with distinct (though not markedly different) physical and chemical properties. Within each such family the properties are often strikingly similar. (A useful technique for separating compounds of the extremely similar lanthanide elements is **ion-exchange chromatography.**) The chemical properties of these metals are determined largely by their tendency to lose electrons and, for most of them, to show several oxidation states. Most of them form many kinds of salts, but a few, the **noble metals,** are fairly unreactive. Four families of metals are called **metallo-acid elements** because they form oxides that are acidic. Most metals are subject to **corrosion** by substances in the environment; facilitated by moisture, these substances attack and oxidize the metal.

Metallurgy is concerned with producing elemental metals and **alloys** and treating them in special ways to obtain desired properties. Of the many **minerals** in the earth's crust, those that can be used as sources of useful materials are called **ores.** To concentrate a metal ore, most of the **gangue** is removed by some physical means. The ore is then chemically reduced in a **smelting** process. Often a suitable **flux** is added at this stage to combine with the residual gangue and form a molten **slag** that separates from the metal. The crude metal is then purified by **refining.**

If not properly controlled, the large-scale industrial production of metals can have profound adverse effects on the environment and on human health. Adequate pollution-control measures are necessary for such installations as **blast furnaces,** in which **pig iron** is produced; the principal pollutants are found in the **flue gas** from the combustion chamber. Severe pollution problems can also be overcome in the manufacture of copper, which is produced by the smelting of ores, and aluminum, which is produced by the electrolysis of ores.

Exercises

Periodic Relationships

1. Name the three major classes of elements and locate the position of each class in the periodic table.

2. Consider the following twelve elements: Na, K, and Rb in IA; Mg, Ca, and Sr in IIA; P, As, and Sb in VA; and S, Se, and Te in VIA. Looking only at the periodic table inside the front cover, predict which element has the following:
   (a) smallest atomic radius
   (b) greatest tendency to form positive ions
   (c) greatest tendency to gain electrons
   (d) smallest electronegativity
   (e) largest electronegativity
   Check your predictions against Figures 5.8 and 5.14.

3. Consider the six IA and IIA elements listed in Exercise 2 and the six elements of the IB and IIB families. Looking only at the periodic table inside the front cover, make the same predictions listed in Exercise 2 for these twelve elements.

4. With respect to electronic configuration, what distinguishes the transition and inner transition elements from other elements?

5. In the theoretical buildup of atoms by the aufbau principle, what series of elements is terminated when the $4f$ orbitals are filled? When the $5f$ orbitals are filled?

6. Show with a diagram how a periodic table would look if in each period the spaces assigned to each element were uniform in size and immediately adjacent to one another.

Physical Properties of Alkali and Alkaline Earth Metals

7. Just by studying the periodic table inside the front cover, can you determine whether or not the change in the boiling point down the IA family is similar to the change down the VIIA family? Explain.

8. Cite several physical properties of the alkali metals that are quite different from those of common metals.

9. Describe how one would go about identifying a IA or IIA metal by causing the element or its compound to emit radiant energy.

*10. From the first ionization energy of rubidium, 4.2 eV, calculate the frequency of the radiant energy necessary to knock an electron off a rubidium atom.

Chemical Properties of IA and IIA Metals

11. Based on standard reduction potentials (see Table A.9 in the appendix), which alkali metal is most susceptible to oxidation? Does this conclusion agree with that reached

when you consider the electronegativities given in Figure 5.8? Which is the better criterion for making the decision on the oxidation of IA metals? Why?

12. Complete and balance the following equations. State the conditions necessary for each reaction to take place.

(a) $Sr + HCl \longrightarrow$ (b) $Cs + Br_2 \longrightarrow$

(c) $K + H_2O \longrightarrow$ (d) $Sr + H_2O \longrightarrow$

(e) $Na + NH_3 \longrightarrow$ (f) $Mg + H_2O \longrightarrow$

(g) $Ba + O_2 \longrightarrow$ (h) $K + O_2 \longrightarrow$

13. The positive ions formed by the reactions of Exercise 12 are isoelectronic with the atoms of what family?

14. To extinguish burning magnesium, using water may be not only ineffective but also dangerous. Why? Covering the fire with sand or dirt is advised. Explain.

Compounds of IA and IIA Metals

15. Which of the following compounds are classed as water insoluble: $RbCl$, $Sr(NO_3)_2$, $Ca_3(PO_4)_2$, $MgCO_3$, $BaSO_4$, $LiCl$, $Mg(OH)_2$, KOH, $Be(OH)_2$?

16. (a) Explain with equations how sodium carbonate acts as a base. What limits its use as a substitute for sodium hydroxide?

(b) Explain with equations how calcium oxide can affect the pH of sewage and how it can be used to remove phosphate ions from sewage.

*17. Calculate the cost of neutralizing 1,000 kg of hydrochloric acid with sodium hydroxide as compared with sodium carbonate.

18. Show with equations the reactions that occur in making mortar when starting with quick lime and sand and for the subsequent setting of the mortar when used in laying bricks (ignore any reaction of the sand). Would you expect the mortar to set to its final condition in a day, a week, or a longer period of time? Why?

19. About one fifth of the sodium hydroxide used in this country is made by the reaction of calcium hydroxide and sodium carbonate. What weight of sodium carbonate is used to produce the sodium hydroxide if we assume a 95 percent yield?

20. Explain how the following could be prepared in a fairly pure state. Write equations for the chemical reactions necessary for the preparations.

(a) milk of magnesia from magnesium chloride

(b) potassium hydroxide from potassium carbonate

(c) barium sulfate from barium carbonate

21. The early settlers saved their wood ashes for leaching with water to recover a "lye" for soap making. What do you think are the main compound(s) dissolved in the leach water?

22. If more states pass laws requiring the use of returnable bottles, what major chemical will have its production affected most?

*23. In the Solvay process, the following steps lead to the production of sodium carbonate:

$$CO_2 + H_2O \rightleftharpoons H_2CO_3$$

$$NH_3 + H_2CO_3 \longrightarrow NH_4HCO_3$$

$$NaCl + NH_4HCO_3 \longrightarrow NaHCO_3\downarrow + NH_4Cl$$

$$2NaHCO_3 \xrightarrow{heat} Na_2CO_3 + H_2O + CO_2$$

$$CaCO_3 \xrightarrow{heat} CaO + CO_2$$

$$CaO + H_2O \longrightarrow Ca(OH)_2$$

$$Ca(OH)_2 + 2NH_4Cl \longrightarrow 2NH_3\uparrow + CaCl_2 + 2H_2O$$

Note that the overall process reaction is $2NaCl + CaCO_3 \rightarrow CaCl_2 + Na_2CO_3$, although the reaction will not take place directly. Note that the calcium carbonate, in addition to producing carbon dioxide for the reaction, produces lime for the recovery of ammonia so that the ammonia can be used over and over. Calculate the weight of sodium carbonate produced per ton of calcium carbonate, assuming an overall yield of 85 percent. Assuming the same percentage yield, what weight of waste calcium chloride is produced?

24. Consider the data on the Solvay process in Exercise 23. Which would have the greatest impact on the environment: the ammonia, the carbon dioxide, the calcium chloride, or the excess sodium chloride that accumulates in the process?

25. Can the transition temperature for $Na_2SO_4 \cdot 10H_2O$, solid $\rightleftharpoons$ liquid, be reached easily in a panel heated by the sun? Is the transition temperature high enough to heat air for a home? Discuss in terms of Celsius and Fahrenheit temperatures.

*26. To heat an average house requires about 20,000 kJ per day for each 1 °C difference between inside and outside temperatures.

(a) How much heat is required per day for the house to maintain a temperature of 70 °F if the average outside temperature is 0 °F?

(b) How many kilograms of $Na_2SO_4 \cdot 10H_2O$ is required to store the heat calculated in part (a)?

(c) Calculate the volume in cubic meters and cubic feet of a salt–liquid mixture that is required to store the heat. The density of the mixture is $6,180 \, kg/m^3$.

*Physical Properties
of Transition Elements
and Their Neighbors*

27. Other than good conductivity, what physical properties contribute to the wide use of copper as an electric conductor?

28. (a) Ladders are often made of magnesium alloys. Why is magnesium preferred to less expensive iron alloys?

(b) If 0.20 ft³ of metal is needed to make a ladder, how much will an aluminum ladder weigh? A magnesium ladder?

29. Gold atoms are heavier than osmium atoms, yet the density of osmium is greater than that of gold. Give some possible reasons for the difference.

30. (a) A rod of titanium is about twice as strong as a similar size aluminum rod. Would it make sense to use titanium instead of aluminum in some structural rods and beams in airplane construction?

(b) Suppose a titanium rod is substituted for a 5-lb aluminum rod. If the rods are of equal strength, approximately how much will the titanium rod weigh?

31. In what way are the electronic configurations of atoms of the Ib and IIb elements similar to those of the Ia and IIa atoms? In what way are they dissimilar?

*Chemical Properties
of Transition Elements and Neighbors*

32. How do we explain the fact that members of the lanthanide series have such similar properties?

33. Write equations to illustrate the statement: The oxides of Ia and IIa metals react with water to form bases, while the oxides of certain transition metals react with water to form acids. This property has given rise to what names for these transition elements?

34. In rare cases a metal is found in nature in the elemental state.

(a) What would you predict that the standard reduction potential must be for such a metal?

(b) Name nine metals that you think might usually or occasionally be found in nature in the elemental state.

(c) If nine or more metals might be found in nature as elements, is the phrase "in rare cases" justified?

35. Based on atomic structure, account for the difference in the tendencies of Ia, IIa, Ib, and IIb atoms to form positive ions.

36. A house roof of sheet aluminum will normally last longer than one made of galvanized steel. If a house or barn burns down, however, there may be scarcely any trace left of an aluminum roof, but there usually will be pieces of a steel roof left. Account for these observations.

37. (a) What are the minimum and maximum oxidation states exhibited by elements of the Vb group?

(b) Give an example of a known compound in which the metal has an oxidation number of +8.

*38. (a) Calculate the free energy change per mole of Fe corroding to Fe^{3+} when your car rusts out. Use data for standard conditions.

(b) Write equations for the reactions of the dioxides of carbon, nitrogen, and sulfur with water.

(c) Using appropriate algebraic relations, show how the

presence of the dioxides in part (b) can affect the voltage of the corrosion cell reaction and the free energy change of the corrosion reaction.

*39. The corrosion of iron that is fastened to a piece of zinc is slower than the corrosion of iron fastened to a piece of copper. Explain.

40. Titanium dioxide occurs naturally, yet it costs about six times more than synthetic sodium hydroxide. Why?

Production of Metals

41. Predict by means of balanced equations the principal chemical steps for producing the following:

(a) copper from copper(I) sulfide

(b) zinc from zinc carbonate ore

(c) iron from iron(III) oxide

(d) uranium from uranium(IV) oxide

(e) aluminum from aluminum oxide

(f) manganese from manganese(IV) oxide

(g) magnesium from magnesium chloride

42. Explain the fact that aluminum cannot be produced by electrolysis of a water solution of $Al_2(SO_4)_3$, but manganese can be produced by electrolysis of a water solution of $MnSO_4$.

43. Recommend a flux for slag formation during the smelting of a metallic ore that contains limestone as an impurity. Write equations for the reactions.

44. Give a reason for each of the following.

(a) Silver can be obtained by heating its sulfide ore in air, but zinc cannot be obtained from its sulfide ore by this method.

(b) Iron can be obtained by heating its oxide ore with coke, but calcium cannot be obtained this way from calcium oxide.

(c) The cathode product for the electrolysis of a water solution of $CuCl_2$ is Cu, but for the electrolysis of a solution of $BaCl_2$ it is H_2.

45. In the past, the recovery of metals from their sulfide ores, for example, ZnS, NiS, CoS, CdS, Cu_2S, and HgS, resulted in considerable atmospheric pollution.

(a) Account for the formation of the pollutant(s).

(b) Suggest methods for the removal of the pollutant(s) from the wastes.

46. Blast furnace operation to produce iron gives rise to a flue gas that is mainly N_2, CO_2, and CO. Account for the presence of each. Write equations where appropriate.

47. To help relieve our dependency on oil for energy, it has been suggested that more electric energy be produced by use of water power. How might a large move in this direction affect the environment?

48. Give arguments pro and con for the premise "Governmental Environmental Regulations Have Gone Too Far."

Metals II: More About Transition Metals

22

Redox Chemistry
of the Transition Metals

22•1 **Emf Diagrams**

22•1•1 Emf's Between Separated
Oxidation States

22•2 **Stability of Redox Species
in Solution**

22•3 **Effect of pH on Stabilities
of Redox Species in Solution**

22•4 **Effect of Disproportionation
on Stabilities of Redox
Species in Solution**

22•5 **Effect of Atmosphere on
Stabilities of Redox
Species in Solution**

Coordination Chemistry
of the Transition Metals

22•6 **Coordination Compounds**

22•6•1 Examples of Ligands

22•6•2 Coordination Numbers

22•7 **Ligand-Field Theory**

22•7•1 Magnetic Properties

22•7•2 Effect of Ligands on Energies
of *d* Orbitals

22•7•3 Color and Electron Structure

22•8 **Uses of Coordination
Compounds**

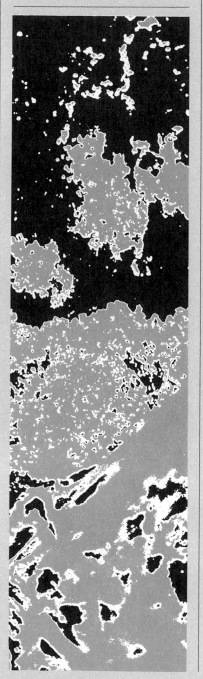

We will begin this chapter with a discussion of the oxidation–reduction solution chemistry of transition metals, continue with a treatment of the structure of coordination compounds, and conclude with the study of some important theoretical aspects of coordination chemistry.

To the chemists of the 1880s and 1890s, the growing accumulation of data on transition metal compounds was quite bewildering. For example, iron in its common +3 oxidation state formed compounds with the molecular formulas $FeCl_3$, $FeCl_2NO_3$, $NaFeCl_4$, Na_3FeCl_6, and $Fe(H_2O)_6Cl_3$. Then in 1892, Alfred Werner, a 26-year-old Swiss lecturer at Zurich, proposed a theory of primary and secondary valences to account for the behavior of certain transition metal compounds that are known today as coordination compounds. He explained the great variety of transition metal compounds with two generalizations:

1. Most transition elements readily exist in two or more oxidation states. Werner called these *primary valences*.

2. Ions of transition metals, particularly of those in higher oxidation states, do not exist as simple particles. Rather the metal is the central atom and is bound to ligands (molecules or anions) to form a complex particle. The ligands act as electron-pair donors, or Lewis bases, toward the central atom. Werner called the number of these bound particles the *secondary valence*.

Today we recognize that the iron compounds mentioned above contain the ions Fe^{3+}, $FeCl_2^+$, $FeCl_4^-$, $FeCl_6^{3-}$, and $[Fe(H_2O)_6]^{3+}$. It is difficult for us to appreciate fully the insight of chemists who worked out such formulas even before the electron was discovered and long before the electronic structure of atoms was developed. Werner's overall ideas about coordination compounds, which have been verified by modern instrumental methods, earned him the Nobel Prize in 1913.

Redox Chemistry of the Transition Metals

The number and stability of its oxidation states are among the most important characteristics of a transition metal (see Section 21.7.2). In addition to the elemental state, most transition metals exhibit two or more oxidation states. The relative stability of any two oxidation states in solution usually can be determined from values of the electrode potentials for oxidation–reduction couples. This difference in stabilities is often the key factor in predicting the course of a chemical reaction.[1] When considering the possible reactions of a given element, it is convenient to collect the standard reduction potentials in the form of an emf (electromotive force) diagram.

---

[1] In making predictions of redox reactions, we will refer first to values of standard reduction potentials. But we must keep in mind the effects of any changes from standard conditions, as pointed out in the last paragraph of Section 18.3.3. Also, see Examples 18.2, 18.3, and 18.4.

22•1 Emf Diagrams

In Section 18.5.2, we saw that the standard free energy for a reaction is directly proportional to the voltage of a cell with standard electrodes:

$$\Delta G^\circ = -nF\mathscr{E}^\circ_{cell}$$

From a thermodynamic point of view, the voltage is the **electromotive force** (emf) that drives the reaction to completion. For an individual element, we can summarize data taken from Table A.9 (or Table 18.1) rather conveniently in an emf diagram.

Consider the element iron. In Table 18.1, there are two cell reactions given for iron:

Fe^{2+}/Fe: $Fe^{2+} + 2e^- \longrightarrow Fe$ -0.44 V
Fe^{3+}/Fe^{2+}: $Fe^{3+} + e^- \longrightarrow Fe^{2+}$ $+0.77$ V

The emf diagram that summarizes these two expressions is

$$Fe^{3+} \xrightarrow{\ +0.77\ } Fe^{2+} \xrightarrow{\ -0.44\ } Fe$$

The convention followed in this type of diagram is to list the species in order of decreasing oxidation states, with the highest state at the left of the diagram.

• **Example 22•1** •

In Table A.9, there are couples involving various oxidation states of manganese at $\mathscr{E}^\circ_{red}$ values of -1.18, $+1.23$, and $+1.70$ V. Construct an emf diagram to summarize these data.

• **Solution** • For -1.18 V, the couple is Mn^{2+}/Mn; for $+1.23$ V, the couple is MnO_2/Mn^{2+}; and for $+1.70$ V, the couple is MnO_4^-/MnO_2. Listing the manganese-containing species in order, beginning with the highest oxidation state, gives

$$MnO_4^- \xrightarrow{\ +1.70\ } MnO_2 \xrightarrow{\ +1.23\ } Mn^{2+} \xrightarrow{\ -1.18\ } Mn$$

See also Exercises 1 and 3 at the end of the chapter.

22•1•1 Emf's Between Separated Oxidation States

In many reactions in solution, there is a direct change from one oxidation state to another that may be two or three steps away, with no experimental evidence that any intermediate ions or compounds are actually formed. For example, under certain conditions, the change from MnO_4^- ions to Mn^{2+} ions can occur with no detectable formation of the intermediate state MnO_2. The electromotive force between widely separated oxidation states of an element can be calculated from the emf values of the intermediate steps.

In calculating the emf between two separated oxidation states, the emf values for the individual steps are not additive. However, *the free energies for the steps are additive.* We can determine ΔG for the overall change and use the relation $\Delta G^\circ = -nF\mathscr{E}^\circ_{cell}$ to determine the emf for the overall change.

Referring to the emf diagram for iron, we find the overall value of $\mathscr{E}^\circ_{\text{cell}}$ for Fe^{3+}/Fe as follows:

$$\text{for } Fe^{3+} \longrightarrow Fe^{2+}: \quad \Delta G = -(1)(+0.77)F = -0.77F$$

$$\underline{\text{for } Fe^{2+} \longrightarrow Fe: \quad\quad \Delta G = -(2)(-0.44)F = +0.88F}$$

$$\text{for } Fe^{3+} \longrightarrow Fe: \quad\quad \Delta G = -n\mathscr{E}^\circ_{\text{cell}}F \quad\quad = +0.11F$$

$$-(3)\mathscr{E}^\circ_{\text{cell}} = +0.11$$

$$\mathscr{E}^\circ_{\text{overall}} = -\frac{0.11}{3} = -0.04 \text{ V}$$

Calculations of this sort can be treated by the relation

$$\mathscr{E}^\circ_{\text{overall}} = \frac{n_1\mathscr{E}_1{}^\circ + n_2\mathscr{E}_2{}^\circ + n_3\mathscr{E}_3{}^\circ + \cdots}{n_1 + n_2 + n_3 + \cdots}$$

where n is the number of electrons involved in each step. For $Fe^{3+} \to Fe$,

$$\mathscr{E}^\circ_{\text{overall}} = \frac{1(+0.77 \text{ V}) + 2(-0.44 \text{ V})}{1 + 2}$$

$$= \frac{-0.11 \text{ V}}{3} = -0.04 \text{ V}$$

We can make the emf diagram for iron more complete by adding this information:

$$\overset{\displaystyle -0.04}{\overline{Fe^{3+} \xrightarrow{\ +0.77\ } Fe^{2+} \xrightarrow{\ -0.44\ } Fe}}$$

— • Example 22.2 • —

Use the data given in Example 22.1 to calculate the emf for the change from MnO_4^- to Mn^{2+}.

— • Solution • For the MnO_4^-/MnO_2 couple, $\mathscr{E}^\circ$ is $+1.70$ V. From the oxidation states assigned, the number n of electrons changed is $7 - 4 = 3$. For the MnO_2/Mn^{2+} couple, $\mathscr{E}^\circ$ is $+1.23$ V and $n = 4 - 2 = 2$. Therefore,

$$\mathscr{E}^\circ_{\text{overall}} = \frac{3(+1.70 \text{ V}) + 2(1.23 \text{ V})}{3 + 2}$$

$$= \frac{+7.56 \text{ V}}{5} = +1.51 \text{ V}$$

See also Exercises 3, 8, 14, and 15.

22.2   Stability of Redox Species in Solution

From an emf diagram of standard reduction potentials, we can get a good idea of the species that are likely to be stable in solutions that are about $1m$ in hydrogen ion. If the values of $\mathscr{E}^\circ$ to the left of the diagram are highly positive, the higher oxidation states are strong oxidizing agents. If the values of $\mathscr{E}°$ to the right of the diagram are highly negative, the lower oxidation states are strong reducing agents.

The emf diagram for manganese drawn in Example 22.1 is characteristic of diagrams that include both strong oxidizing agents (MnO_4^- and MnO_2) and a strong reducing agent (Mn). We would predict that Mn^{2+} is the stable species in solution. To check this prediction for a standard $1m$ solution, we ask:

1. Is the reducing agent Mn (to the right of Mn^{2+}) strong enough to reduce H^+ and liberate hydrogen from the solution?

2. Are the oxidizing agents MnO_4^- and MnO_2 (to the left of Mn^{2+}) strong enough to oxidize H_2O and liberate oxygen from the solution?

We can answer these questions by using the method illustrated in Figure 18.5 for predicting the spontaneity of a pair of redox half-reactions. The value of -1.18 V for the Mn^{2+}/Mn couple shows that Mn will reduce H^+ in standard solutions to H_2:

$$
\begin{array}{ll}
Mn^{2+} + 2e^- \longrightarrow Mn & \mathscr{E}°_{red} = -1.18 \text{ V} \\
Mn \longrightarrow Mn^{2+} + 2e^- & \mathscr{E}°_{ox} = +1.18 \text{ V} \\
2H^+ + 2e^- \longrightarrow H_2 & \mathscr{E}°_{red} = 0.00 \text{ V} \\
\hline
2H^+ + Mn \longrightarrow Mn^{2+} + H_2 & \mathscr{E}°_{cell} = +1.18 \text{ V}
\end{array}
$$

This positive value of $\mathscr{E}°_{cell}$ shows that Mn^{2+} is stable relative to Mn.

The value of $+1.70$ V for the MnO_4^-/MnO_2 couple is greater than the $+1.23$ V for the O_2/H_2O couple. This shows that the conversion of MnO_4^- to MnO_2 will oxidize H_2O and liberate O_2 from standard solutions:[2]

$$
\begin{array}{ll}
O_2 + 4H^+ + 4e^- \longrightarrow 2H_2O & \mathscr{E}°_{red} = +1.23 \text{ V} \\
3(2H_2O \longrightarrow O_2 + 4H^+ + 4e^-) & \mathscr{E}°_{ox} = -1.23 \text{ V} \\
4(MnO_4^- + 4H^+ + 3e^- \longrightarrow MnO_2 + 2H_2O) & \mathscr{E}°_{red} = +1.70 \text{ V} \\
\hline
4MnO_4^- + 4H^+ \longrightarrow 3O_2 + 4MnO_2 + 2H_2O & \mathscr{E}°_{cell} = +0.47 \text{ V}
\end{array}
$$

The case of the MnO_2/Mn^{2+} couple is borderline, because its $\mathscr{E}°$ value just balances the O_2/H_2O couple:

$$
\begin{array}{ll}
2H_2O \longrightarrow O_2 + 4H^+ + 4e^- & \mathscr{E}°_{ox} = -1.23 \text{ V} \\
2(MnO_2 + 4H^+ + 2e^- \longrightarrow Mn^{2+} + 2H_2O) & \mathscr{E}°_{red} = +1.23 \text{ V} \\
\hline
2MnO_2 + 4H^+ \longrightarrow O_2 + 2Mn^{2+} + 2H_2O & \mathscr{E}°_{cell} = 0.00 \text{ V}
\end{array}
$$

Recalling the effect of concentration on potential (see Section 18.3.4), we see that in solutions with the H^+ concentration greater than $1m$, Mn^{2+} would tend to form. In solutions with the H^+ concentration less than $1m$, MnO_2 would not liberate O_2 from solution, so MnO_2 would be the stable species.

---

[2] The liberation of O_2 by MnO_4^- is so slow that permanganate solutions can maintain their concentrations for a long time.

The answers to the two questions originally asked lead us to conclude that Mn^{2+} is stable in solutions more acidic than $1m$ H^+, but in other solutions oxygen from the air would tend to change Mn^{2+} to MnO_2.

To summarize, the lower oxidation state of any couple that has an $\mathscr{E}^\circ_{red}$ value less than 0.00 V will tend to liberate H_2 from standard solutions. The higher oxidation state of any couple that has an $\mathscr{E}^\circ_{red}$ value greater than 1.23 V will tend to liberate O_2 from standard solutions.

In borderline situations, an added margin of driving force usually is needed for a reaction to take place at a measurable rate. This behavior is illustrated by a solution of $Cr_2O_7^{2-}$, which is stable at a pH of 0 even though the $Cr_2O_7^{2-}/Cr^{3+}$ couple has an $\mathscr{E}^\circ_{red}$ of 1.33 V, slightly greater than the theoretical 1.23 V necessary to oxidize water. That is, based on the $\mathscr{E}^\circ$ values, in a water solution that is $1m$ in $K_2Cr_2O_7$ and $1m$ in H^+ (pH $\simeq$ 0), the $Cr_2O_7^{2-}$ ion should be reduced to Cr^{3+} by H_2O:

$$3(2H_2O \longrightarrow O_2 + 4H^+ + 4e^-) \qquad \mathscr{E}^\circ_{ox} = -1.23 \text{ V}$$
$$\underline{2(Cr_2O_7^{2-} + 14H^+ + 6e^- \longrightarrow 2Cr^{3+} + 7H_2O) \qquad \mathscr{E}^\circ_{red} = +1.33 \text{ V}}$$
$$2Cr_2O_7^{2-} + 16H^+ \longrightarrow 4Cr^{3+} + 8H_2O + 3O_2 \qquad \mathscr{E}^\circ_{cell} = +0.10 \text{ V}$$

The fact that acidic dichromate solutions do not change to Cr^{3+} solutions, even though the $\mathscr{E}^\circ_{cell}$ for the above reaction is +0.10 V, is attributed to the need for an added margin of driving force.

Two of the most common strong oxidizing agents used in laboratory work are the dichromate ion, $Cr_2O_7^{2-}$, and the permanganate ion, MnO_4^-. The oxidizing power of the dichromate ion is not as great as that of permanganate, but the solutions are more convenient to prepare and store. $K_2Cr_2O_7$ is available as a very pure compound that is easily dried to a constant weight and composition for making solutions of known concentration.

Permanganate solutions are more easily affected by traces of impurities than dichromate solutions are. Traces of organic matter in the form of dust or grease can reduce MnO_4^- to form small amounts of MnO_2 that catalyze the decomposition reaction. In neutral or basic solutions, the following reactions occur. The emf values are for couples in standard base solutions (see Table A.10):

$$3(4OH^- \longrightarrow O_2 + 2H_2O + 4e^-) \qquad \mathscr{E}^\circ_{ox} = -0.40 \text{ V}$$
$$\underline{4(3e^- + 2H_2O + MnO_4^- \longrightarrow MnO_2 + 4OH^-) \qquad \mathscr{E}^\circ_{red} = +0.59 \text{ V}}$$
$$4MnO_4^- + 2H_2O \xrightarrow{MnO_2} 4MnO_2 + 3O_2 + 4OH^- \qquad \mathscr{E}^\circ_{cell} = +0.19 \text{ V}$$

Such a reaction is called **autocatalytic,** because a product of the reaction acts as a catalyst for further reaction. In laboratory storage bottles of potassium permanganate, a film or sediment of manganese dioxide usually can be seen.

When carefully protected from impurities, permanganate solutions are sufficiently stable to be stored for some time and used for precise analytical titrations. They are reduced to the practically colorless Mn^{2+} in acid solution or to the insoluble MnO_2 in neutral solution. In both cases, the appearance of a trace of deep purple conveniently indicates the presence of excess MnO_4^-, signaling the end of the titration.

22•3 Effect of pH on Stabilities of Redox Species in Solution

The acidity of a solution has an important effect on the ease with which hydrogen or oxygen is liberated. As the acidity decreases, the liberation of hydrogen becomes more difficult, whereas the liberation of oxygen becomes less difficult.

In Example 18.4, it was calculated that a couple must have an $\mathscr{E}^\circ_{red}$ value less than -0.414 V in order to liberate hydrogen from a solution with a pH of 7 (a neutral solution). For a pH of 14, a similar calculation shows that the necessary $\mathscr{E}^\circ_{red}$ value for the couple must be less than -0.828 V.

In standard solutions, with H^+ at $1m$, the pH is 0. As we have already seen, a couple must have an $\mathscr{E}^\circ_{red}$ value greater than 1.23 V to liberate oxygen from such a solution. But at a pH of 7, only 0.815 V is required; and at a pH of 14, only 0.401 V. These data are listed in Table 22.1.

Table 22•1 $\mathscr{E}^\circ_{red}$ **Values required to liberate H_2 or O_2 from solutions**

	to reduce $\overset{+1}{H}$ to $\overset{0}{H_2}$	$\mathscr{E}^\circ_{red}$
pH = 0	$2H^+ + 2e^- \longrightarrow H_2$	< 0.000 V
pH = 7	$2H_2O + 2e^- \longrightarrow 2OH^- + H_2$	< -0.414 V
pH = 14	$2H_2O + 2e^- \longrightarrow 2OH^- + H_2$	< -0.828 V
	to oxidize $\overset{-2}{O}$ to $\overset{0}{O_2}$	
pH = 0	$2H_2O \longrightarrow O_2 + 4H^+ + 4e^-$	$> +1.229$ V
pH = 7	$2H_2O \longrightarrow O_2 + 4H^+ + 4e^-$	$> +0.815$ V
pH = 14	$4OH^- \longrightarrow O_2 + 2H_2O + 4e^-$	$> +0.401$ V

To describe reactions in basic solutions, potentials are determined for electrodes compared with a standard hydrogen electrode in a solution that is $1m$ in OH^- ions. The half-reaction and emf for the standard basic hydrogen electrode are

$$2H_2O + 2e^- \longrightarrow H_2 + 2OH^- \qquad \mathscr{E}_B^\circ = -0.828 \text{ V}$$

The potential of a couple relative to the standard basic hydrogen electrode is defined as the **standard basic reduction potential** $\mathscr{E}_B^\circ$. Values for several couples are given in Table A.10.

22•4 Effect of Disproportionation on Stabilities of Redox Species in Solution

A **disproportionation reaction** is one in which part of a substance is reduced as another part is oxidized. A familiar example is the $+1$ oxidation state of copper:

$$2Cu^+ \longrightarrow Cu + Cu^{2+}$$

The emf diagram for copper is

$$
\begin{array}{c}
\overset{+0.337}{\overbrace{\phantom{Cu^{2+} \xrightarrow{+0.153} Cu^+ \xrightarrow{+0.521} Cu}}} \\
Cu^{2+} \xrightarrow{+0.153} Cu^+ \xrightarrow{+0.521} Cu
\end{array}
$$

When the values for $\mathscr{E}^{\circ}_{red}$ do not decrease from left to right in an emf diagram, we can identify a species that is unstable (that tends to disproportionate). At Cu^+, for example, the emf value increases ($+0.521$ is greater than $+0.153$) instead of decreasing. Therefore, Cu^+ is unstable and disproportionates to form Cu and Cu^{2+}.

We can illustrate additional disproportionation reactions with an emf diagram for manganese that is more complete than the one drawn in Example 22.1:

$$
\begin{array}{c}
\overset{+1.51}{\overbrace{\phantom{MnO_4^- \xrightarrow{+0.56} MnO_4^{2-} \xrightarrow{+2.26} MnO_2 \xrightarrow{+0.95} Mn^{3+} \xrightarrow{+1.51} Mn^{2+}}}} \\
MnO_4^- \xrightarrow{+0.56} MnO_4^{2-} \xrightarrow{+2.26} MnO_2 \xrightarrow{+0.95} Mn^{3+} \xrightarrow{+1.51} Mn^{2+} \xrightarrow{-1.18} Mn \\
\underset{+1.70}{\underbrace{\phantom{MnO_4^- \xrightarrow{} MnO_4^{2-} \xrightarrow{} MnO_2}}} \quad \underset{+1.23}{\underbrace{\phantom{MnO_2 \xrightarrow{} Mn^{3+} \xrightarrow{} Mn^{2+}}}}
\end{array}
$$

At both MnO_4^{2-} and Mn^{3+}, the emf of the step to the right is greater than that of the step to the left. We predict that both of these species are unstable and tend to disproportionate.

On the other hand, consider the emf diagram for chromium:

$$
\begin{array}{c}
\overset{-0.74}{\overbrace{\phantom{Cr^{3+} \xrightarrow{-0.41} Cr^{2+} \xrightarrow{-0.91} Cr}}} \\
Cr_2O_7^{2-} \xrightarrow{+1.33} Cr^{3+} \xrightarrow{-0.41} Cr^{2+} \xrightarrow{-0.91} Cr
\end{array}
$$

The regular decrease in the $\mathscr{E}^{\circ}_{red}$ values from left to right indicates that neither Cr^{3+} nor Cr^{2+} tends to disproportionate.

—— • Example 22.3 • ——

Show that MnO_4^{2-} is unstable (tends to disproportionate) in standard acid solution by calculating the $\mathscr{E}^{\circ}$ for the expected reaction.

—— • Solution • From the emf diagram in Section 22.4, we see that if MnO_4^{2-} disproportionates, it will form both MnO_2 and MnO_4^-. In a disproportionation reaction, the amount of reduction must equal the amount of oxidation, so we write

$$MnO_4^{2-} \xrightarrow{\overset{\text{decrease of}}{\text{2 in oxid. no.}}} MnO_2 \qquad +2.26 \text{ V}$$

$$\underline{2(MnO_4^{2-} \xrightarrow{\overset{\text{increase of}}{\text{1 in oxid. no.}}} MnO_4^-)} \qquad \underline{-0.56 \text{ V}}$$

$$3MnO_4^{2-} \longrightarrow 2MnO_4^- + MnO_2 \qquad +1.70 \text{ V}$$

Because the voltage for the overall reaction is positive, we see that the disproportionation reaction is spontaneous; that is, MnO_4^{2-} is unstable. (Note that the equation does not have to be balanced to calculate $\mathscr{E}°$.)

See also Exercises 11 and 14–16.

22•5 Effect of Atmosphere on Stabilities of Redox Species in Solution

In most laboratory work, solutions are exposed to the atmosphere while scientists are working with them. Three components of ordinary air that may affect redox reagents are dust, carbon dioxide, and oxygen. Dust may contain organic matter that can be oxidized by strong oxidizing agents. Dissolved carbon dioxide will affect the pH of nearly neutral solutions, because carbon dioxide and water form the weak acid carbonic acid. And dissolved oxygen is a strong enough oxidizing agent to react with many reduced species.

From the emf diagram for iron, we would not predict that Fe^{3+} is more stable than Fe^{2+}. The regular decrease in emf values for $Fe^{3+} \rightarrow Fe^{2+} \rightarrow Fe$ shows that Fe^{2+} is stable toward disproportionation. The -0.77 V value for $Fe^{2+} \rightarrow Fe^{3+}$ shows that this reaction is not spontaneous. However, a solution of Fe^{2+} is oxidized to Fe^{3+} by oxygen in the air. The 1.23 V value of the O_2/H_2O couple is more than sufficient to oxidize the Fe^{3+}/Fe^{2+} couple, which has a 0.77-V value in standard acid solution. Therefore, the pale green color associated with Fe^{2+} ions may change to the yellowish-brown of Fe^{3+} ions, unless the solutions are protected from the air. This explains why Fe^{3+} solutions are apparently more stable than Fe^{2+} solutions in the laboratory.

A solution containing Fe^{2+} ions can be stabilized by putting a clean iron nail in it. As long as elemental iron is present, Fe^{3+} will not form, because the $\mathscr{E}°$ value is positive for the reaction $2Fe^{3+} + Fe \rightarrow 3Fe^{2+}$.

Coordination Chemistry of the Transition Metals

22•6 Coordination Compounds

In Section 17.6.2, the formation of a complex ion was described in terms of a metal ion bonded to other particles called its *ligands*. Each ligand, which can be either a molecule or an anion, donates a pair of electrons to form the bond. When both electrons are supplied by one of the bound species, the bond is called a *coordinate covalent bond* (see Section 5.4.3). Compounds in which negative groups or molecules are bonded to metal ions or atoms are called **coordination compounds**.[3]

---

[3]There are coordination compounds in which the central particle is an atom instead of an ion. If the ligands are molecules, the complex itself is a molecule. Nickel combines with carbon monoxide to form such a molecule, $Ni(CO)_4$, nickel carbonyl.

A major part of the chemistry of the transition metals is concerned with their coordination compounds. These compounds are important in laboratory, industrial, and environmental chemistry. Traces of metals that are essential to the health of living organisms are often present as coordination compounds.

22•6•1   Examples of Ligands   Some molecules and anions that are frequently encountered as ligands in coordination compounds are

$$:\overset{..}{\underset{H}{O}}-H \qquad H-\overset{..}{\underset{H}{N}}-H \qquad :C\equiv O: \qquad :\overset{..}{\underset{..}{Cl}}:^- \qquad \left[:C\equiv N:\right]^- \qquad \left[:\overset{..}{N}-\overset{..}{\underset{:\overset{..}{O}:}{O}}:\right]^- $$

If a ligand is large enough to have widely separated atoms with lone pairs, it may be able to substitute for two or more simple ligands. A molecule of ethylenediamine, $H_2NCH_2CH_2NH_2$ or

$$:\overset{H}{\underset{H}{N}}-\overset{H}{\underset{H}{C}}-\overset{H}{\underset{H}{C}}-\overset{H}{\underset{H}{N}}: $$

has two widely separated atoms with lone electron pairs and can often substitute for two simple ligands.

There are many other examples of molecules or ions that can substitute for two or more simple ligands. A ligand that has two or more points of attachment to a central atom is called a **chelate** (Greek *chele*, claw). Some examples are shown in Table 22.2. In the case of $H_2NCH_2CH_2NH_2$, each of the nitrogens is a bonding atom and the ligand is called a *bidentate* (two teeth) chelate, because it forms two bonds with the central atom. There are also *tri-*, *tetra-*, and *hexadentate* and other polydentate chelates. A chelating ligand is held by a central atom more strongly than are monodentate ligands that contain similar atoms.

22•6•2   Coordination Numbers   The secondary valence forces of Werner are now called coordination numbers. In coordination compounds, the **coordination number** (C.N.) is defined as the number of simple ligands bonded to a central atom or ion. Many of the compounds first studied by Werner had a coordination number of 6, and it was one of his great insights to deduce that many metals held six constituents (today called ligands) to form a stable complex ion. He further proposed that such a complex ion would combine with as many other ions as necessary to give a neutral compound.

Although 6 is the most common coordination number, there are examples of stable structures involving two to nine ligands. There are assignments of even higher coordination numbers, but these species tend to be unstable. Typical geometries associated with coordination numbers of 2,

3, 4, 5, and 6 are shown in Table 22.3. In that table, examples are given of complex ions, coordination compounds, and other ions and compounds that do not involve transition metals.

• *Coordination Number 2* • One of the familiar instances of a coordination number of 2 is $[Ag(NH_3)_2]^+$, the ion formed when silver compounds are treated with ammonia (see Section 17.6.2). The characteristic geometry for C.N. = 2 is linear.

• *Coordination Number 3* • Instances of a coordination number of 3 are extremely rare. The only simple one for a transition metal that is known is the $[HgI_3]^-$ anion. The characteristic geometry for C.N. = 3 is trigonal planar.

• *Coordination Number 4* • Four is a common coordination number of several transition metal atoms and ions. On the basis of the VSEPR model, there should be only one geometry for each coordination number (see Figure 6.18 in Section 6.5). For C.N. = 4, we predict tetrahedral geometry by the

Table 22•2 **Examples of chelates (polydentate ligands)**

Name	Abbreviation	Formula	Type of chelate
ethylenediamine	en	$H_2NCH_2CH_2NH_2$	bidentate
acetylsalicylic acid[a]	aspirin	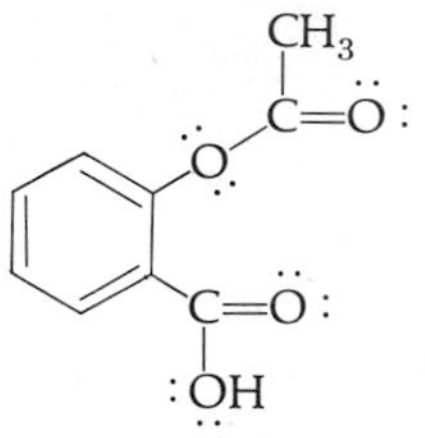	bidentate
dimethylglyoxime	DMG	$CH_3C{=}NOH$ $CH_3C{=}NOH$	bidentate
diethylenetriamine	dien	$H_2NCH_2CH_2NHCH_2CH_2NH_2$	tridentate
triethylenetetramine	trien	$H_2NCH_2CH_2NHCH_2$ $H_2NCH_2CH_2NHCH_2$	tetradentate
ethylenediamine-tetraacetate	EDTA	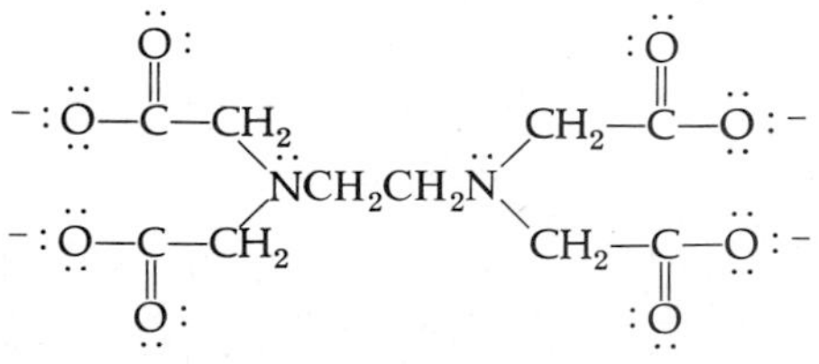	hexadentate

[a]The ligand is usually in the form of the anion of this acid.

theory. However, as seen in Table 22.3, there are two symmetrical geometries, tetrahedral and square planar, for C.N. = 4.

Werner deduced the structures of a pair of interesting square-planar platinum(II) compounds. Both compounds had the formula $Pt(NH_3)_2Cl_2$. Molecules or ions that have the same overall composition but different structural formulas are called **isomers** (Greek *isos*, equal, and *meros*, part). The phenomenon is called **isomerism.** The existence of isomers seemed to be the

Table 22•3 Geometries and hybrid orbital arrangements

Examples	Experimentally determined	No. of equivalent bonds	Proposed hybrids
$BeCl_2$ $HgCl_2$, $[Ag(NH_3)_2]^+$, $[AgI_2]^-$	linear 180° (ideal)	2	*sp*
BCl_3, NO_3^- $[HgI_3]^-$	trigonal planar 120° (ideal)	3	sp^2
CH_4, SiF_4, NH_4^+, SO_4^{2-} $[NiCl_4]^{2-}$, $[Zn(NH_3)_4]^{2-}$	tetrahedral 109°28' (ideal)	4	sp^3
$[Ni(CN)_4]^{2-}$, $[PtCl_4]^{2-}$	square planar 90° (ideal)	4	dsp^2
PCl_5 $Fe(CO)_5$	trigonal bipyramidal 90° (ideal) 120° (ideal)	5	sp^3d dsp^3
SF_6, $[FeF_6]^{3-}$ $[Fe(CN)_6]^{3-}$, $[Fe(CN)_6]^{4-}$ $[Co(NH_3)_6]^{2+}$, $[PtCl_6]^{2-}$	octahedral 90° (ideal)	6	sp^3d^2 d^2sp^3

only way to explain the fact that there were two compounds with the formula $Pt(NH_3)_2Cl_2$.

For a formula of the type MX_2Y_2, Werner recognized that if the shape is square planar, two isomeric arrangements are possible. In $Pt(NH_3)_2Cl_2$, the two chloride ligands (and the two ammonia ligands) can be arranged so they are at adjacent positions, designated cis (Latin, on this side), or at opposite positions, designated trans (Latin, across):

$$
\begin{array}{cc}
\text{Cl} \diagdown \quad \diagup \text{NH}_3 & \text{H}_3\text{N} \diagdown \quad \diagup \text{Cl} \\
\text{Pt} & \text{Pt} \\
\text{Cl} \diagup \quad \diagdown \text{NH}_3 & \text{Cl} \diagup \quad \diagdown \text{NH}_3 \\
\text{cis form} & \text{trans form}
\end{array}
$$

For a tetrahedral shape, only one arrangement is possible. (Making molecular models will help to show why this argument is valid.)

The square-planar isomers can be distinguished from one another because ethylenediamine will react with the cis isomer to replace the two chlorides, but it will not react with the trans isomer. Evidently the $H_2NCH_2CH_2NH_2$ molecule can form two bonds at 90° angles but cannot reach around the Pt to form bonds at 180°. Using en as an abbreviation for ethylenediamine, we can write

$$
\begin{array}{ccc}
\text{Cl} \diagdown \quad \diagup \text{NH}_3 & & \left[\text{en} \; \text{Pt} \diagup \text{NH}_3 \diagdown \text{NH}_3 \right]^{2+} \\
\text{Pt} & + \text{en} \longrightarrow & + 2\text{Cl}^- \\
\text{Cl} \diagup \quad \diagdown \text{NH}_3 & & \\
\text{cis form} & &
\end{array}
$$

$$
\begin{array}{c}
\text{H}_3\text{N} \diagdown \quad \diagup \text{Cl} \\
\text{Pt} \quad + \text{en} \longrightarrow \text{no reaction} \\
\text{Cl} \diagup \quad \diagdown \text{NH}_3 \\
\text{trans form}
\end{array}
$$

Theoretical explanations of why one geometric arrangement for C.N. = 4 is preferred over the other will be discussed in Section 22.7.2.

• *Coordination Number 5* • Instances of coordination number 5 are rare, but they are not as uncommon as coordination number 3. Simple examples are the $[CuCl_5]^{3-}$ ion and iron pentacarbonyl, $Fe(CO)_5$. The characteristic geometry for C.N. = 5 is trigonal bipyramidal, but distorted structures are also known.

• *Coordination Number 6* • As mentioned previously, 6 is the most common coordination number. If a central atom is surrounded by six ligands, there is a limited number of ways of arranging these ligands symmetrically. For a general formula MX_6, some symmetrical possibilities are a planar hexagon, two trigonal pyramids sharing one apex, and an octahedron. The octahedron is favored by VSEPR theory because it enables the six X groups to be as far from one another as possible. Also, the octahedron was the shape proposed by Werner as the probable one for an ion or molecule with C.N. = 6.

Werner used his theory of octahedral symmetry to deduce the structure of a number of interesting cobalt compounds. In addition to square-planar compounds with the formula MX_2Y_2, octahedral compounds with the general formula MX_4Y_2 should give rise to isomers. The two Y ligands can be positioned either adjacent to or opposite one another. Green compounds containing the ion $[Co(NH_3)_4Cl_2]^+$ were known, and Werner proposed that because the two Cl ligands must be either cis or trans, the other isomer could probably be prepared. After several years a violet form was synthesized in his laboratories, which later work showed to be the cis form (see Figure 22.1).

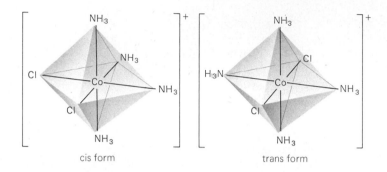

Figure 22•1 Isomeric forms of $[Co(NH_3)_4Cl_2]^+$ ions. The cis form is violet, the trans form is green.

Suppose the four NH_3 ligands in a $[Co(NH_3)_4Cl_2]^+$ ion are replaced by two ethylenediamine (en) ligands to give a $[Co(en)_2Cl_2]^+$ ion. Again cis and trans isomers are possible (see Figure 22.2). Two isomers were known in Werner's time, one green and the other violet. But which was cis and which was trans was a mystery. Werner realized that cis octahedral complexes containing two bidentate ligands could exist as optical isomers. **Optical isomers** are related structurally as mirror images of one another (see Section 27.1.2). In 1911 Werner prepared an optical isomer of the violet form, thus proving the violet compound was the cis isomer.

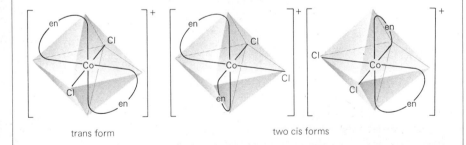

Figure 22•2 Isomers of dichlorobis (ethylenediamine)cobalt(III) ions, $[Co(en)_2Cl_2]^+$. The two cis forms are optical isomers.

For C.N. = 6, in addition to the symmetrical octahedron, the trigonal prism and distorted octahedra are possible structures. Some octahedral complex ions are somewhat distorted in shape because two trans ligands are closer or farther from the central atom than are the other four ligands. One example of a distorted octahedral shape is the hydrated Cu^{2+} ion in aqueous solution shown in Figure 22.3.

22•7   Ligand-Field Theory

As with simple ions and molecules, we seek to understand the properties of complex ions and molecules in terms of their electronic structures. Our aim is to show how the electrons of the ligands and the central ion interact. First we will consider the electronic structure of a simple ion (say, an Fe^{2+} ion), and then we will show how the electronic structure is changed when that ion forms bonds with its ligands.

Descriptions of electronic structure have been developed in terms of crystal-field theory and ligand-field theory. In the original **crystal-field theory,** the net effect of each ligand is treated as a negative charge that repels the electrons of the central ion or atom. The **ligand-field theory** not only takes into account the charge repulsions, but also considers the covalent character of the bond between the ligand and the central ion or atom. In our applications, we will make no distinction between the two theories but will use the term ligand-field theory to refer to all the effects we discuss. Our examples and generalizations will be limited mainly to octahedral and tetrahedral complexes of elements in the first transition series.

The descriptions of electronic structure begin with the known geometric arrangements of the ligands around the central ion or atom. Each ligand, whether a neutral molecule or a negative ion, donates a pair of electrons to form a bond with the central ion or atom. The force exerted on the central ion or atom by the electrons and by the net charge of the ligands is called the **ligand field.**

To show how ligands affect the electronic structure of the central ion or atom, we will discuss first the magnetism of complexes. Second, we will see how ligand-field theory accounts for these magnetic properties. Finally, we will look at the correlation between magnetic properties and color to see how ligand-field theory helps explain many of the colors that transition metal compounds exhibit.

22•7•1   Magnetic Properties

On the basis of behavior in a magnetic field, substances can be classified as **diamagnetic** if they are slightly repelled by or pushed out of the field, **paramagnetic** if they are slightly attracted by or pulled into the field, or **ferromagnetic** if they are very strongly pulled into the field. Ferromagnetism, which is just an extreme manifestation of paramagnetism, is exhibited by only a few substances, such as iron, cobalt, nickel, and certain alloys. Common magnets are made of ferromagnetic substances. Most substances are in one of the two other classes.

The magnetic character of a substance can be measured precisely with a modified balance, as shown in Figure 22.4. When the electromagnet is turned on, a paramagnetic or ferromagnetic substance is attracted by the field and appears to weigh more; a diamagnetic substance appears to weigh less.

The magnetic properties of substances can be explained on the basis of their electronic structures. If they are diamagnetic, the presumption is that all electrons are paired, so the electron spins balance. Paramagnetism indicates that one or more electrons per atom, ion, or molecule are not paired. The total spin contributed by the unpaired electrons gives rise to the overall paramagnetism.

In Table 22.4 the experimentally determined properties of several complex ions are interpreted in terms of the assignment of electron pairs to

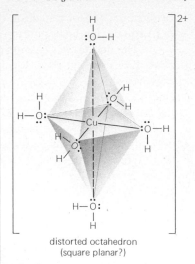

Figure 22•3 The hydrated copper(II) ion holds four H_2O molecules closely and two at greater distances. Considered as $[Cu(H_2O)_4]^{2+}$, the ion is square planar. Considered as $[Cu(H_2O)_6]^{2+}$, the ion is a distorted octahedron.

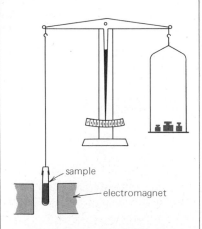

Figure 22•4 The Gouy magnetic balance.

hybrid orbitals. Most transition elements use d orbitals in hybridizing with s and p orbitals. (In heavy atoms, f orbitals also can be involved.)

The first three columns in Table 22.4 record experimental data, and the last two give the theoretical descriptions. Column 1: the formula and charge of a species are determined by elemental analysis and electric measurements. Column 2: magnetic measurements, similar to those made with the Gouy balance shown in Figure 22.4, reveal the number of unpaired electrons per unit. Column 3: the geometry of an ion or molecule is determined commonly by X-ray measurements.

The experimental data are accounted for in the theoretical assignment of electrons to hybridized orbitals as shown in the diagrams in the fourth column of Table 22.4. This hybridization is summarized in the fifth

Table 22.4 Hybrid orbitals in coordination compounds

Ion	Magnetic information	Geometry	Electrons of ion or central atom (black) and electrons of ligands (color)[a]			Hybrid- ization
			$3d$	$4s$	$4p$	
Fe^{2+}	paramagnetic (four unpaired electrons)					
$[Fe(CN)_6]^{4-}$	diamagnetic (no unpaired electrons)	octahedral				d^2sp^3
Fe^{3+}	paramagnetic (five unpaired electrons)				$4d$[b]	
$[FeF_6]^{3-}$	paramagnetic (five unpaired electrons)	octahedral				sp^3d^2
$[Fe(CN)_6]^{3-}$	paramagnetic (one unpaired electron)	octahedral				d^2sp^3
Co^{3+}	paramagnetic (four unpaired electrons)					
$[Co(CN)_6]^{3-}$	diamagnetic (no unpaired electrons)	octahedral				d^2sp^3
Ni^{2+}	paramagnetic (two unpaired electrons)					
$[NiCl_4]^{2-}$	paramagnetic (two unpaired electrons)	tetrahedral				sp^3
$[Ni(CN)_4]^{2-}$	diamagnetic (no unpaired electrons)	square planar				dsp^2

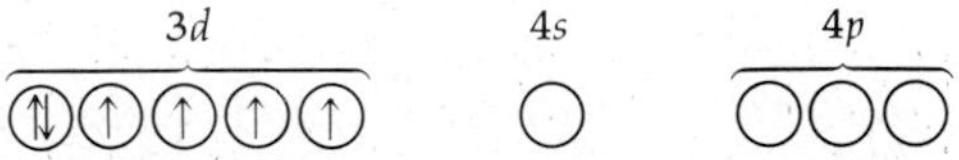

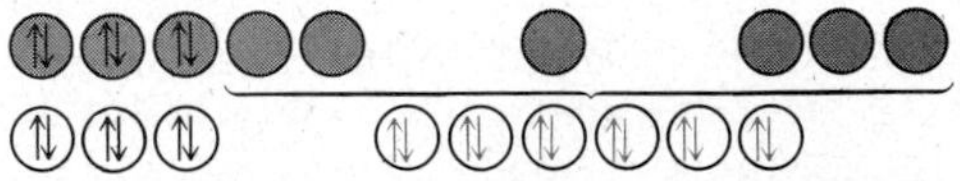

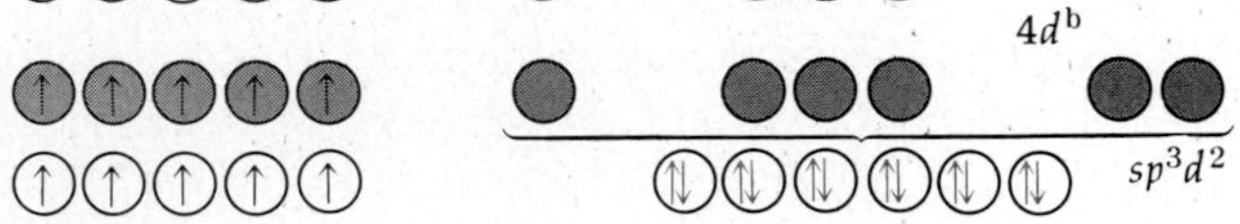

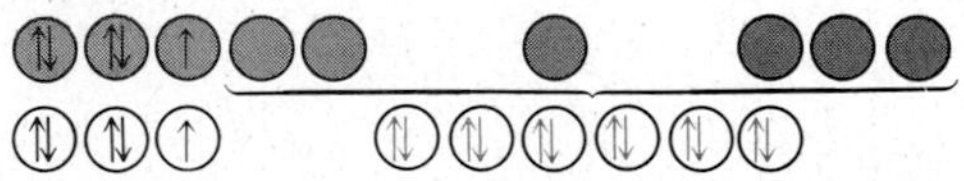

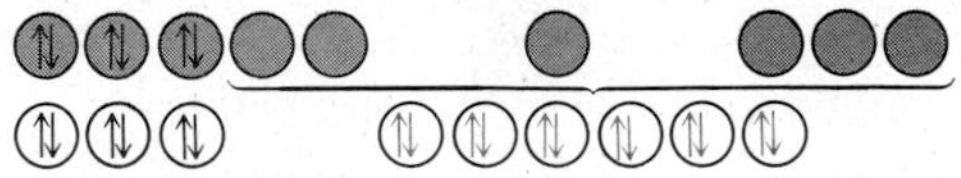

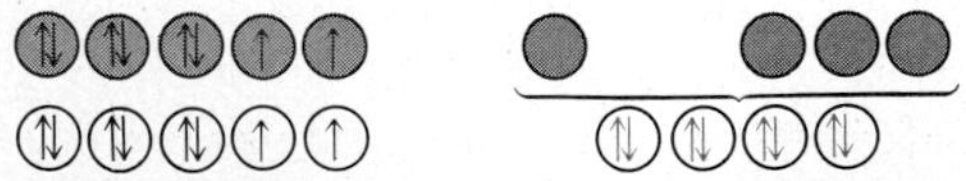

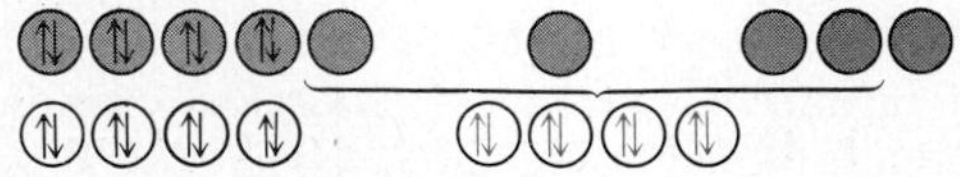

[a]The diagrams that are printed lightly merely show the orbitals of the central atom that are considered in forming the hybrid orbitals.
[b]For $[FeF_6]^{3-}$, only two of the five $4d$ orbitals are shown due to lack of space.

column. We can discuss the first two entries in the table to illustrate how the experimental data are related to the theoretical assignment of electrons to orbitals.

In the case of Fe^{2+}, the electronic assignment is arrived at by loss of the two $4s$ electrons from an Fe atom (see table inside front cover). The assignment of four electrons to separate orbitals is required by the magnetic data and is consistent with Hund's rule (see Section 4.5.3). In the case of $[Fe(CN)_6]^{4-}$, the iron is Fe(II); so first reference is made to the Fe^{2+} structure, which has six electrons in its outside level. The electrons that must be assigned in $[Fe(CN)_6]^{4-}$ are the six electrons belonging to Fe^{2+} and the six pairs of electrons donated to the coordinate bonds by the six ligands.

In column 4, the lightly shaded circles show schematically two imagined steps: first, the pairing of the six electrons associated with Fe^{2+}, because the magnetic data show that no electrons are unpaired in $[Fe(CN)_6]^{4-}$; and second, the choosing of the required number of lowest energy orbitals available to bond the six ligands. The brackets enclose the orbitals that are to be hybridized. The final step shown in column 4 is to draw the dark circles for all the orbitals used and to show with arrows the electrons that occupy them. The six orbitals used in bonding the ligands are drawn close together to indicate that they are hybridized. The hybridization is summarized in the last column: d^2sp^3.

• Example 22•4 •

The magnitude of the paramagnetism of the complex ion $[Cr(CN)_6]^{3-}$ indicates that it has three unpaired electrons. Draw a circle-and-arrow diagram for the electronic structure around the chromium and describe the hybridization involved.

• Solution • In the absence of contrary information, we will assume the ion is octahedral and is bound by six equivalent bonds. Because the oxidation number for the CN^- ion is -1, the chromium must be $-3 - (-6) = +3$.

The electronic configuration for Cr (see inside front cover) is $(Ar)3d^54s^1$. Therefore, the configuration for the Cr^{3+} ion is $(Ar)3d^3$. To represent Cr^{3+} with three unpaired electrons and to show some of its higher energy orbitals that may be used to form bonds with six ligands, we draw

For the complex, we use the six lowest energy orbitals available:

See also Exercises 28 and 29.

22•7•2 Effect of Ligands on Energies of d Orbitals
Some ligands are much more effective than others in changing the magnetic character of a central ion

or atom. Complexes that have relatively great paramagnetism are said to have a high spin, and those with relatively small paramagnetism are said to have a low spin. In Table 22.4, $[FeF_6]^{3-}$ has a high spin like Fe^{3+}, but $[Fe(CN)_6]^{3-}$ has a low spin. Similarly, $[NiCl_4]^{2-}$ has a high spin like Ni^{2+}, but $[Ni(CN)_4]^{2-}$ has a low spin. In both cases, the complex ion with higher spin has the same number of unpaired electrons as the simple ion. The study of many ligands has revealed that, in general, they can be ranked as follows in the order of their effectiveness in reducing the spin of the central ion or atom:

$$CN^- > NO_2^- > NH_3 > H_2O \geq F^- > Cl^- > Br^- > I^-$$

Ligand-field theory assumes that ligands have field effects that change the relative energies of d orbitals. If the effect is strong enough, it will force the d electrons to pair as they occupy the lower energy orbitals. In $[NiCl_4]^{2-}$, the weak effect of the Cl^- ligands leaves two $3d$ electrons unpaired. Therefore, the four ligands must use four sp^3 hybrid orbitals and form a tetrahedral complex. In $[Ni(CN)_4]^{2-}$, the strong effect of CN^- forces the two $3d$ electrons to pair. Therefore, the four ligands must use four dsp^2 hybrid orbitals, which are planar.

The equal energy of the five d orbitals in an isolated ion or atom is illustrated in Figure 22.5. The five d orbitals that are together in this complex figure are shown separately in Figure 4.16. When a transition metal ion or atom forms an octahedral complex, it is surrounded by six ligands, as illustrated in Figure 22.6. Two of the d orbitals, $d_{x^2-y^2}$ and d_{z^2}, lie along the x, y, z coordinates and interact most directly with the ligands as they approach. Three of the d orbitals, d_{xy}, d_{xz}, and d_{yz}, are at $45°$ angles from the x, y, z coordinates, so they do not interact so strongly with the ligands. In an octahedral complex, the ligand field raises the energy of two of the d orbitals relative to the energy of the other three. This *splitting of energy levels* can be diagrammed as in Figure 22.7. With the strong field of six CN^- ligands, electrons can be forced to pair and occupy the d_{xy}, d_{xz}, and d_{yz} orbitals. See $[Fe(CN)_6]^{4-}$, $[Fe(CN)_6]^{3-}$, and $[Co(CN)_6]^{3-}$ in Table 22.4.

When a transition metal forms a tetrahedral complex, it is surrounded by four ligands, as illustrated in Figure 22.8. As the four ligands approach the central ion or atom, they interact more strongly with the three d_{xy}, d_{xz}, and d_{yz} orbitals than with the two $d_{x^2-y^2}$ and d_{z^2} orbitals. The relative energies of the two sets of orbitals are reversed from those diagrammed in Figure 22.7. The splitting of energies by a tetrahedral ligand field is not so strong as in the octahedral case, and electron pairing rarely results. However, the difference in energy of the two sets of orbitals can result in colored compounds, as will be discussed in the next section.

22•7•3   **Color and Electron Structure**   One of the interesting and useful properties of transition metal compounds is the variety of colors they exhibit. A substance is colored when it absorbs some wavelengths of visible light and transmits or reflects the rest. The color we see is that which is not absorbed by the substance.

Radiation in the visible region has only enough energy to excite electrons between energy levels that are rather close together. The levels of d orbitals that have different energies because of ligand-field splitting often make possible such low-energy d–d transitions. One of the intriguing facts

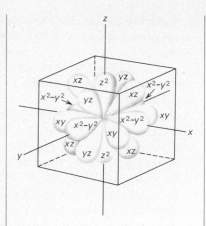

Figure 22•5 Complete set of five d orbitals around a central atom (compare Figure 4.16). The ring region of the d_{z^2} is omitted for clarity. In an isolated atom or ion, the five orbitals have equal energies (they are *degenerate*). The $d_{x^2-y^2}$ and d_{z^2} orbitals are directed toward the middles of the six faces of the cube. The d_{xy}, d_{xz}, and d_{yz} orbitals are directed toward the middles of the twelve edges of the cube.

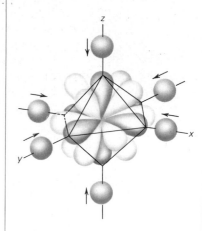

Figure 22•6 With six ligands situated octahedrally, the $d_{x^2-y^2}$ and d_{z^2} orbitals (darkly shaded) have a higher energy of interaction than the d_{xy}, d_{xy}, and d_{yz} orbitals. (Compare Figure 22.8.)

that must be explained is that complexes of low magnetic spin tend to absorb radiation of higher energy than complexes of high spin. This is shown clearly in many cases in which a given central ion or atom forms colored complexes with different ligands. In Figure 22.9, we see that the energy of the radiation absorbed by three complex ions of Cr^{3+} decreases in the order $CN^- > NH_3 > H_2O$. This is part of a long *spectrochemical series*, which is very similar to the series based on the effects of ligands on magnetic properties that was listed in the previous section.

Our theory is that the splitting of the *d* orbitals (see Figures 22.6 and 22.7) is greater by a ligand with a strong field. The ligand CN^-, for example, causes relatively great splitting, so the energy necessary to excite an electron from a lower level *d* orbital to a higher level one is relatively great.

A few examples will illustrate the application of these principles. Compounds of Sc^{3+} and Ti^{4+} (which have no *d* electrons) and Zn^{2+} (which has ten *d* electrons) do not show *d–d* transitions. Most of their compounds are not colored. On the other hand, compounds of Fe^{2+} and Fe^{3+} are colored. In the mineral peridot, $(Fe,Mg)_2SiO_4$, the *d* orbitals in Fe^{2+} are split by a nearly octahedral arrangement of oxygen ions (see Figure 22.10 on Plate 5). The absorbance by peridot at the red end of the spectrum (see Figure 22.11) is due to the *d–d* transition; the yellow-green of the light transmitted by peridot is characteristic of many Fe^{2+} minerals.

The mineral almandine is a garnet and it has the formula $(Fe,Mg)_3Al_2(SiO_4)_3$. It owes its color to Fe^{2+} also, but it is red (see Figure 22.10 on Plate 5). The oxygen-to-iron distance is greater in almandine than it is in peridot, so the energy splitting of the *d* orbitals is less in almandine. The result is that the absorbance curve for almandine (see Figure 22.11) is shifted to lower energies than is the curve for peridot. A greater percentage of the light transmitted by almandine (garnet) is red.

Yellow chrysoberyl (see Figure 22.10 on Plate 5) is Al_2BeO_4 with Fe^{3+} ions substituted for Al^{3+} at many places. Its absorbance curve (Figure 22.11) shows absorption only at the high-energy end of the visible spectrum.

The red color of the ruby and the green of the emerald are both due to the Cr^{3+} ion (see Figure 22.10 on Plate 5). Crystalline corundum, Al_2O_3, is colorless, but with some Cr^{3+} ions substituted for Al^{3+} it becomes ruby. A similar substitution in beryl, $Be_3Al_2Si_6O_{18}$, produces emerald. The absorbance curves are shown in Figure 22.12. The shift in the absorbance for ruby to higher energies is attributed to the fact that the oxygen ligands are more ionic in corundum and more covalent in beryl. The localization of the negative charge on the ions causes a greater repulsion (greater splitting) for the *d* orbitals.

In conclusion, let us note that ligand-field effects due to *d–d* transitions are not the only source of color in transition metal compounds. With their complex electronic structures, these compounds have other ways of absorbing radiation in the visible spectrum.

22•8 Uses of Coordination Compounds

Some of the oldest practical uses of coordination compounds have been due to their colors. Based on art and practice going back to the days of

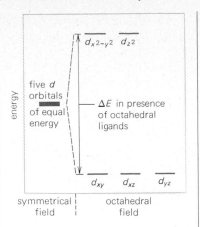

Figure 22•7 In the presence of six ligands at octahedral positions, the *d* orbitals of the central ion or atom are split in energy. Two orbitals have high energy relative to the other three.

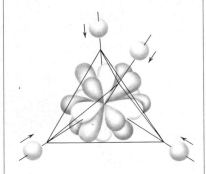

Figure 22•8 With four ligands situated tetrahedrally, the d_{xy}, d_{xz}, and d_{yz} orbitals (darkly shaded) have a higher energy of interaction than the $d_{x^2-y^2}$ and d_{z^2} orbitals. However, because none of the orbitals is precisely along the lines through the apexes of the tetrahedron, the differences in energy are not so great as with octahedral splitting. (Compare Figure 22.6.)

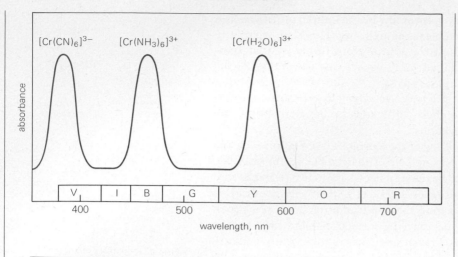

Figure 22•9 Positions of the lowest energy, highly probable *d–d* transitions for three complex ions of Cr³⁺.
Source: Adapted by permission from C. S. G. Phillips and R. J. P. Williams, *Inorganic Chemistry*, Vol. II, Oxford University Press, New York, 1966.

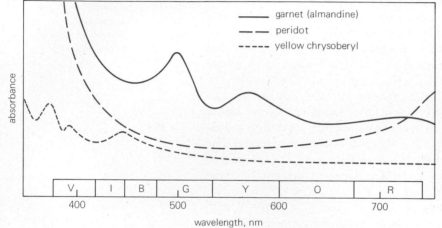

Figure 22•11 Absorbance curves for garnet (almandine), peridot, and yellow chrysoberyl. See also Figure 22.10 on Plate 5.
Source: Adapted by permission from B. M. Loeffler and R. G. Burns, "Shedding Light on the Color of Gems and Minerals," *American Scientist*, 64, 636 (1976).

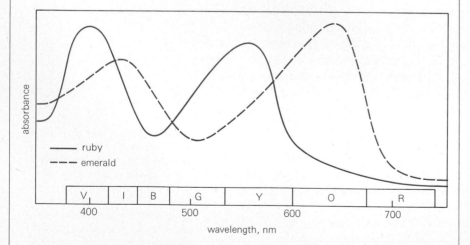

Figure 22•12 Absorbance curves for ruby and emerald. See also Figure 22.10 on Plate 5.
Source: Adapted by permission from B. M. Loeffler and R. G. Burns, "Shedding Light on the Color of Gems and Minerals," *American Scientist*, 64, 636 (1976).

antiquity, chemists and artisans formulated dyes, colored glasses, and glazes for ceramics from substances that we now describe in terms of transition metal coordination chemistry. Complexes of iron(II) and iron(III) cyanides are still known by such names as Turnbull's blue, Prussian blue, and Berlin green. Blueprints are based on an iron cyanide complex. Recently, new dyes, especially those used on some of the synthetic fabrics that are difficult to tint, have been made from coordination compounds.

As illustrated in Section 17.6.2, coordinating ligands can influence greatly the available concentration of simple ions in solution. Phosphate chelating agents are used in water softeners to keep calcium ions in solution; chelated iron is used in fertilizers to provide iron that will not react with soils to form insoluble iron compounds; and chelated metal ions can be formed that are more soluble in oil than in water—the reverse of the usual ionic behavior. Chelating agents are used to remove unwanted impurities from drinking water. They also have been used in separating the mixtures of metal ions found in the fission products of nuclear reactors. A complex containing aluminum and titanium is used as a catalyst in the making of polyethylene, one of the most important commercial processes discovered in this century.

The chelating agent ethylenediaminetetraacetate (EDTA, see Table 22.2) is said to be the most important modern analytical reagent. During the past 30 years, when most new analytical techniques have been based on electric and optical instruments, more than 100 research papers a year have appeared on the use of EDTA in "wet" chemical methods. One of its many other uses is in complexing traces of heavy metal ions in food. A small amount of EDTA prevents such ions from acting as catalysts for the oxidation of fats and oils that occurs as these foods become spoiled or rancid. The effects of lead poisoning in children, a critical problem in run-down urban housing where old lead paint is still exposed, are relieved by treating patients with EDTA, which complexes the lead ion for elimination in the urine.

The cis isomer of $Pt(NH_3)_2Cl_2$ has been used to control tumors and cancers. Its promise in this area was discovered during a study of the effect of an electric field on bacterial growth. Unexpectedly, it was found that minute amounts of the "inert" platinum electrodes form traces of coordination compounds that inhibit cell division in bacteria.

Not only recent discoveries but also many classical chemical reactions are interpreted today in terms of coordination chemistry. The long-known ability of *aqua regia,* a mixture of concentrated HNO_3 and HCl, to dissolve gold and platinum depends in large part on the formation of the complex ions $AuCl_4^-$ and $PtCl_6^{2-}$, respectively.

The precipitation of the Ni^{2+} ion as the brilliant red dimethylglyoxime (DMG) salt is a classical reaction used in the laboratory both as a qualitative test for the presence of nickel ion and to remove nickel from solution quantitatively. The fact that DMG combines with nickel ion but not with ions of iron, cobalt, or most other similar substances was unexplained for many years. Today it is known that two units of DMG, joined by hydrogen bonds, form a planar tetradentate complex with nickel. Thus, DMG nicely satisfies the tendency of nickel to form four dsp^2 hybrid bonds in square-planar complexes. (Compare Tables 22.2 and 22.4 and see Figure 22.13.) There are many such special compounds formed by metals. Heme in red blood cells and chlorophyll are two especially important examples that will be discussed in Section 27.4.

Figure 22•13 The four-coordinate chelate compound of nickel(II) with dimethylglyoxime, $[Ni(DMG)_2]$. Dotted lines indicate hydrogen bonds; arrows indicate coordinate covalent bonds.

Chapter Review

Summary

Most transition metals exist in two or more positive oxidation states. From the known standard reduction potentials between adjacent oxidation states one can calculate reduction potentials between widely separated states. These potentials represent the **electromotive force** (emf) that drives the reaction. For a given transition metal, it is useful to construct an **emf diagram,** showing the principal oxidation states and standard reduction potentials. The relative stability of the various oxidation states at different pH values can be estimated from these potentials. To describe redox reactions in basic solutions, values of the **standard basic reduction potential** are used. Among the factors affecting the stability of some transition metal ions in solution are: (1) **autocatalysis,** in which the reaction is catalyzed by one of its own products, (2) **disproportionation reactions,** in which some of the ions in question are reduced while others are oxidized, and (3) various constituents of the normal atmosphere.

Transition metals often exist as **coordination compounds,** in which the central metal ion or atom has two or more ligands bonded to it by coordinate covalent bonds. Such a compound may be a complex ion with associated ions of the opposite charge, or it may be a neutral complex. A ligand with more than one point of attachment to the central ion or atom is called a **chelating agent.** The geometry of a coordination compound depends in part on the coordination number of the central metal ion or atom. Certain coordination numbers permit **isomerism,** which exists when two compounds have the same empirical formula but different structural formulas. The two compounds are called **isomers.** A special kind of isomerism gives rise to **optical isomers,** which are mirror images of each other.

The electronic structures of transition metal complexes are explained in terms of **crystal field theory** and the more modern **ligand field theory.** The interactions between the central metal ion or atom and the surrounding ligands are determined in part by the nature of the **ligand field** arising from the electrons and net charges of the ligands. The ligand field alters the relative energies of the d orbitals of the transition metal; this is called the **splitting of energy levels.** The resulting electron configuration of the complex underlies its structure and magnetic and optical properties. Almost all substances are classified as either **paramagnetic** or **diamagnetic** (a few substances are **ferromagnetic**), depending on whether they have or do not have any unpaired electrons. Ligands can be ranked in the order of their ability to affect magnetic and optical properties; the ranking for optical properties (color) is called the **spectrochemical series.**

Exercises

Emf Diagrams and Reactions

1. Draw emf diagrams for the following elements using data from Table A.9.
   (a) cobalt (two couples)
   (b) silver (two couples)
   (c) gold (two couples)
   (d) vanadium (five couples)

2. (a) Consider the diagram

$$\overset{a}{\overbrace{M^{3+} \xrightarrow{\ b\ } M^+ \xrightarrow{\ c\ } M}}$$

   If you can look up values for a and c, show how you can calculate b.
   (b) Apply the method you worked out in part (a) to calculate the potential of the Au^{3+}/Au^+ couple. See Table A.9 for data.

3. (a) Complete the following equation and then balance it by the ion-electron method described in Section 18.6.2. See Table A.9 for possible oxidation–reduction products.

$$H_2S_2O_8 + SnSO_4 \longrightarrow$$

   (b) Write the emf diagrams for the half-reactions in part (a).
   (c) Is the reaction described in part (a) spontaneous? Why?

*4. Show how the expression

$$\mathscr{E}^{\circ}_{overall} = \frac{n_1 \mathscr{E}_1^{\circ} + n_2 \mathscr{E}_2^{\circ} + n_3 \mathscr{E}_3^{\circ} + \cdots}{n_1 + n_2 + n_3 + \cdots}$$

   is derived from the relationship $\Delta G^{\circ} = -nF\mathscr{E}^{\circ}_{cell}$.

*5. To test deposits on automobile exhaust pipes for the presence of manganese, a small sample is treated with nitric acid and then sodium bismuthate, $NaBiO_3$ (a strong oxidizing agent), is added. A purple color reveals the presence of manganese.
   (a) Assuming the manganese is present initially as an oxide, write a reasonable equation for the reaction with nitric acid.
   (b) Write an overall equation for the oxidation by sodium bismuthate.

*6. A 1.521-g sample of an iron ore containing Fe_2O_3 was dissolved and reduced to Fe^{2+} with zinc. The iron(II) ion was oxidized to iron(III) with 24.80 mL of a 0.100M solution of $KMnO_4$.
   (a) Write the ionic equation for the reduction to Fe^{2+}.
   (b) Write the ionic equation for the oxidation of Fe^{2+}, assuming the manganese is reduced to Mn^{2+}.

(c) Calculate the percentage of iron in the sample.

*7. Answer the following questions by using the emf diagram you drew for vanadium in Exercise 1(d) and the emf diagrams for copper, manganese, and chromium in Section 22.4.

    (a) To reduce V^{3+} to V^{2+} but not to V, what might you use?

    (b) To reduce Cr^{2+} to Cr, could you use vanadium or one of its compounds? Manganese or one of its compounds? Copper or one of its compounds?

    (c) To oxidize V^{3+} to VO^{2+} but not to $V(OH)_4^+$, what would you use?

    (d) To oxidize V^{3+} to $V(OH)_4^+$, what would you use?

Stability in Solution; Disproportionation

8. Referring to the emf diagrams drawn in this chapter, tell whether the following reactions are spontaneous relative to the standard hydrogen ion–hydrogen comparison reaction.

    (a) $Fe^{3+} \longrightarrow Fe$

    (b) $MnO_4^- \longrightarrow Mn$

    (c) $Cr_2O_7^{2-} \longrightarrow Cr$

9. How can we explain the fact that a solution of $KMnO_4$ tends to decompose on long standing, precipitating a film of MnO_2 on the bottom of the bottle?

*10. Is a water solution of Ag^{2+} stable at a pH of 0? Write equations for possible reactions and then calculate $\mathscr{E}°_{cell}$ (see Table A.9) to decide whether the reactions will occur.

11. (a) Determine by calculation whether Mn^{3+} is an unstable species in standard $1m$ acid solutions.

    (b) Repeat for Fe^{2+}.

    (c) Repeat for Cu^+.

12. How will dissolved CO_2 from the air affect the stability of the following?

    (a) MnO_2 exposed to the solution

    (b) Mn^{2+} in solution

    (c) Cr^{2+} in solution

13. A standard solution of permanganate ion can be used to determine the amount of iron(II) ion in highly acid solution. Assume the most stable form of manganese is formed. Write the overall ionic equation for the probable reaction. (Note that the oxygen from the permanganate ion can react with hydrogen ions to form water.)

*14. Refer to the four emf diagrams you drew in Exercise 1.

    (a) Pick the strongest oxidizing agent.

    (b) Pick the strongest reducing agent.

    (c) Are there any species that are unstable with respect to evolving oxygen or hydrogen from standard acid solutions?

    (d) Identify any species that are unstable with respect to disproportionation.

*15. (a) Draw both acidic and basic emf diagrams for MnO_4^{2-} and its adjacent species with higher and lower oxidation states. Is MnO_4^{2-} less stable in acidic or basic solution? Justify your answer by calculating $\mathscr{E}°$ for the disproportionation reaction.

    (b) Repeat for MnO_2.

*16. Calculate the change in free energy per mole of MnO_4^{2-} that disproportionates according to the reactions summarized in Example 22.3.

17. A solution of $Pt(NH_3)_3Cl_4$ and a solution of $Pt(NH_3)_4Cl_4$ are each treated with silver nitrate solution. From the first solution, 1 mole of AgCl is precipitated per mole of platinum compound, whereas from the second solution, twice as much AgCl is precipitated.

    (a) Write formulas for the ions present in each original solution.

    (b) What is the coordination number of Pt in each complex ion?

    (c) What is the probable geometry of each complex ion?

    (d) What is the oxidation number of Pt in each compound?

18. In diagramming structures of complexes for ourselves, it is convenient to use drawings that are less complicated than those done by an artist. Useful ways to indicate the geometry of tetrahedral and octahedral structures are

tetrahedron   octahedron

To such simple figures, add symbols for central ions and ligands to represent the ions and molecules in Figures 22.1 and 22.2 and in Exercises 20 and 21.

19. Distinguish between oxidation number and coordination number; between normal covalent bonds and coordinate covalent bonds.

20. What are the coordination number and oxidation number of nickel in $Ni(CO)_4$?

21. Is the compound cis-dichlorodiammineplatinum(II) tetrahedral, square planar, or octahedral? Explain.

22. Does ammonia function as a Lewis acid, Lewis base, or neither when forming a complex ion with copper(II) ion? Why?

23. (a) How can the formation of a hydrate be regarded as an acid–base reaction? Give an example.

    (b) Write the balanced equation for the complete conversion of the hydrate $CoCl_2 \cdot 6H_2O$ to the ammoniate. Can this be regarded as an acid–base reaction?

24. Does ethylenediamine function as a Lewis acid or as a Lewis base when forming a complex ion chelate?

*25. In $K_3Fe(CN)_6$, six monodentate ligand ions are held by the Fe^{3+} ion to give a complex ion, $Fe(CN)_6^{3-}$. Show how Fe^{3+} might react with monodentate and one or more bidentate ligands; repeat for tri- and tetradentate ligands.

26. (a) Is cis-trans isomerism possible for ammonium hexanitratocerate(IV), $(NH_4)_2Ce(NO_3)_6$, assuming that the complex ion is octahedral? If so, represent with structural formulas the two isomers.
    (b) Is cis-trans isomerism possible in an octahedral complex ion where Ce^{4+} is coordinated with two ligands of one kind and four of another, such as in $(NH_4)_2Ce(NO_3)_4Cl_2$? If so, represent with structural formulas the two isomers.

27. It is reported that freezing point depression measurments show that $Co(NH_3)_4Cl_3$ exists in solution in the form of $[CoCl_2(NH_3)_4]^+$ and Cl^- ions. Explain in terms of a hypothetical numerical example the sort of data that might have been obtained.

Ligand-Field Theory

28. The electronic structure of the Sn atom is given as $(Kr)4d^{10}5s^25p^2$. Assume that $SnCl_6^{2-}$ is a diamagnetic complex. Using a circle-and-arrow diagram, describe the orbitals that are occupied in this complex ion. State which orbitals are involved in the bonding and predict the overall geometry of the ion.

29. The electronic structure of the Pd atom is given as $(Kr)4d^{10}$.
    (a) Draw a circle-and-arrow diagram for the Pd atom $4d$ electrons.
    (b) Diagram the Pd^{2+} ion, assuming that it is paramagnetic.
    (c) Diagram the $Pd(NH_3)_2Cl_2$ molecule, assuming that it is diamagnetic.
    (d) On the basis of your answer to part (c), what sort of hybridization is involved? What shape is the molecule?

30. The complex ion $Co(CN)_6^{3-}$ is diamagnetic. Would you expect $Co(Cl)_6^{3-}$ to be diamagnetic also? Explain your reasoning.

*31. Draw, in the plane of the paper, the following approximate angles. It is convenient to do parts (b) and (c) together to show the relationship.

(a) angle between the line of approach of an octahedrally positioned ligand and one lobe of the nearest d_{xy}, d_{xz}, or d_{yz} orbital
(b) angle between the line of approach of a tetrahedrally positioned ligand and one lobe of the nearest d_{xy}, d_{xz}, or d_{yz} orbital. How many of these lobes are "nearest" the line of approach?
(c) angle between the line of approach of a tetrahedrally positioned ligand and one lobe of the nearest $d_{x^2-y^2}$ or d_{z^2} orbital. How many of these lobes are "nearest" the line of approach?

32. Draw a diagram of the same type as Figure 22.7 to show the splitting of energies of d orbitals by a tetrahedral ligand field.

33. In garnet, Fe^{2+} is surrounded by eight oxygen atoms. Show, with diagrams, why the d–d split of energy should be less than that for Fe^{2+} surrounded by six oxygen atoms.

34. Why should the splitting of d orbital energies be different for oxygen ligands that have greater ionic character as compared with those that are more covalent? Which type should lead to the greatest difference in energy between $d_{x^2-y^2}$, d_{z^2}, and d_{xy}, d_{xz}, d_{yz}?

Applications of Coordination Chemistry

*35. Which of the following chlorides do you think will form colored solutions in water solution: Co^{3+}, Cr^{3+}, Cu^{2+}, Fe^{2+}, Fe^{3+}, Mn^{2+}, Ni^{2+}, Sc^{3+}, V^{3+}, V^{4+}, Zn^{2+}? Explain your choices.

36. Anhydrous copper sulfate is colorless (commonly a mass of white crystals). Copper sulfate pentahydrate is blue. Suggest in general terms an explanation for this difference.

37. Predict the colors of the three complex ions of Cr^{3+} described in Figure 22.9.

*38. (a) Write equations for the reactions of tin(IV) sulfide and antimony(III) sulfide with hydrochloric acid to form hexachloro complexes.
    (b) Write an equation for the reaction of gold with aqua regia. Repeat for platinum. The oxidizing action of aqua regia is

$$NO_3^- + Cl^- + 4H^+ + 2e^- \longrightarrow NOCl + 2H_2O$$

Groups VIIA and VIIIA: The Halogens and Noble Gases

23

The Halogens

23•1 **Properties of the Halogen Family**

23•1•1 Physical Properties

23•1•2 Chemical Properties

23•1•3 Oxidation–Reduction Chemistry of the Halogens

23•2 **Characteristic Reactions**

23•2•1 Halogens with Halogens

23•2•2 With Metals

23•2•3 With Water (Hydrolysis)

23•2•4 With Hydrogen

23•2•5 With Hydrocarbons

23•2•6 With Certain Nonmetals and Metalloids

23•2•7 With Compounds of Other Halogens

23•3 **Production of the Halogens**

23•3•1 Fluorine

23•3•2 Chlorine

23•3•3 Bromine

23•3•4 Iodine

23•4 **Important Uses of the Halogens**

Special Topic: Mercury and Chlorine in the Environment

23•5 **Metal Halides**

23•6 **Nonmetal and Metalloid Halides**

23•6•1 Hydrogen Halides

23•7 **Halogen Oxy-Acids and Oxy-Salts**

The Noble Gases

23•8 **Isolation of the VIIIA Elements**

23•9 **Physical Properties**

23•10 **Chemical Properties**

Special Topic: Facts, Theories, and Scientific Discovery

23•11 **Compounds of the Noble Gases**

23•11•1 Xenon Compounds

23•11•2 Krypton and Radon Compounds

23•11•3 Analogies with Halogen Compounds

23•11•4 Structures of Xenon Compounds

23•12 **Uses of the Noble Gas**

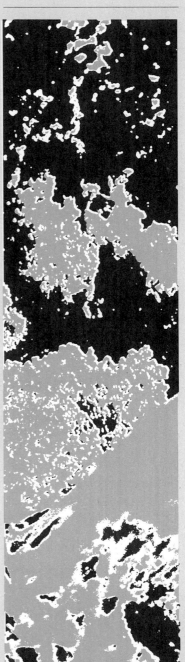

We began our study of families of elements at the left of the periodic table with groups IA and IIA and then had a brief look at a great number of other metals. In this chapter, we will move to the right of the table to our first families of nonmetals, groups VIIA and VIIIA.

Four of the elements in group VIIA, fluorine, chlorine, bromine, and iodine, were known as the **halogen family** of elements even before the formulation of the theory that groups them together in the periodic table. In addition to these four common VIIA elements, there is a rare halogen, astatine, which was made in 1940 by nuclear bombardment experiments. Since then, astatine has been found in nature but only in extremely minute quantities.

In contrast with group VIIA, none of the elements in group VIIIA, the **noble gases,** had been identified at the time Mendeleev constructed the periodic table. This is not surprising, because these elements occur only in small concentrations, they are invisible gases, and they form no natural compounds. After argon (Greek *argos*, lazy) was identified in 1894, the other family members were discovered within just six years.

Family resemblances are strongest in the groups at the far left and far right of the periodic table. In groups VIIA and VIIIA, not only are the members of each group quite similar to one another, but also their differences in properties vary in a regular way from one element to the next. This makes their study somewhat easier than that of other families of nonmetals.

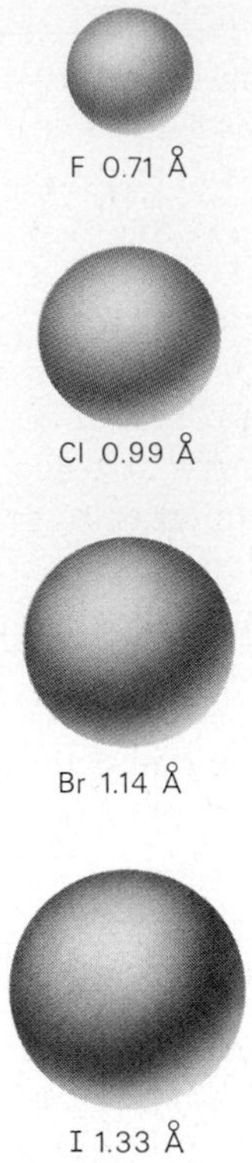

The Halogens

- ### 23•1   Properties of the Halogen Family

- #### 23•1•1   Physical Properties   Table 23.1 lists some of the important physical properties of the halogens; some others are listed in Table 13.5. The striking generality evident from the tabulated data is that any given property changes in a regular way from one element to the next.

   The increase in melting and boiling points with increasing atomic number is explained by the fact that larger molecules have greater van der Waals attraction forces than do smaller ones.

   Except for the noble gases, the halogens have the highest ionization energies and electronegativities of any family of elements. Of the group VIIA elements, fluorine holds most tightly to its electrons, and iodine the least.[1] This trend may be correlated with the sizes of the halogen atoms shown in Figure 23.1.

- #### 23•1•2   Chemical Properties   There is a regular decrease in chemical activity from fluorine to iodine, as shown by the trend in oxidizing strengths. The

F 0.71 Å

Cl 0.99 Å

Br 1.14 Å

I 1.33 Å

Figure 23•1 Relative sizes of halogen atoms, with covalent radii in angstroms.

---

[1]In our discussion of group VIIA elements, astatine is not included; owing to its relative rarity and radioactivity, many of its properties have been studied only slightly.

diatomic fluorine molecule, F_2, is a stronger oxidizing agent than any other element in its normal state.

Both fluorine and chlorine support combustion reactions in the same manner as does oxygen. Hydrogen and the active metals burn in either gas with the liberation of heat and light. The greater reactivity of fluorine compared with that of chlorine is revealed by the fact that ordinary materials, including wood and some plastics, ignite in an atmosphere of fluorine. Even "fireproof" asbestos burns in fluorine. Some of the noble gases combine with fluorine to form stable covalent compounds (see Section 23.10).

All four group VIIA elements are extremely irritating to the nose and throat. Bromine, a deep red liquid at room temperature, has a high vapor pressure, as can be seen by the red vapors of bromine that escape from an open bottle of the liquid. Liquid bromine is one of the most dangerous of the common laboratory reagents because of the effect of the vapor on the eyes and nasal passages. Only 0.1 part bromine per 1 million parts of air (1 ppm) can be tolerated without adverse effects. The liquid also causes severe burns on contact with the skin. Chlorine and fluorine, usually handled as gases, should be used only in hoods and in rooms with good ventilation. Exposure to a concentration of chlorine of greater than 1 ppm in air is hazardous to one's health. A few breaths of chlorine at 1,000 ppm are lethal. All the halogens must be stored out of contact with substances that can be oxidized.

23•1•3 Oxidation–Reduction Chemistry of the Halogens   The emf diagrams for the halogens, halide ions, oxy-acids, and oxy-ions in standard acid and base solutions are shown in Figure 23.2. The data for acid solutions show how strong the family similarities are for the halogens in their variable oxidation states. For any couple with $\mathscr{E}^{\circ}_{red}$ greater than $+1.23$ V, the species listed at the left can oxidize water spontaneously and evolve oxygen. The dissimilarity between the chemistry of fluorine and the other halogens is revealed by the fact that, aside from its elemental form, fluorine is shown only in the -1 oxidation state.

The values of $+3.06$ V for F_2/HF and $+1.36$ V for Cl_2/Cl^- show that acidic fluoride and chloride solutions resist oxidation by oxygen from the air

Table 23•1 **Physical properties of the halogen family**

	Fluorine	Chlorine	Bromine	Iodine
appearance at room temperature	yellowish gas	greenish gas	deep red liquid	purple, almost black, solid
molecular formula	F_2	Cl_2	Br_2	I_2
melting point, °C	-220	-101	-7	114
boiling point, °C	-188	-34	59	184
ionization energy, eV/atom and kJ/mol	17.4 1,680	13.0 1,250	11.8 1,140	10.4 1,000
covalent radius, Å	0.71	0.99	1.14	1.33
ionic radius (X^-), Å	1.36	1.81	1.95	2.16
electronic structure	2,7	2,8,7	2,8,18,7	2,8,18,18,7
electronegativity	4.0	3.0	2.8	2.5

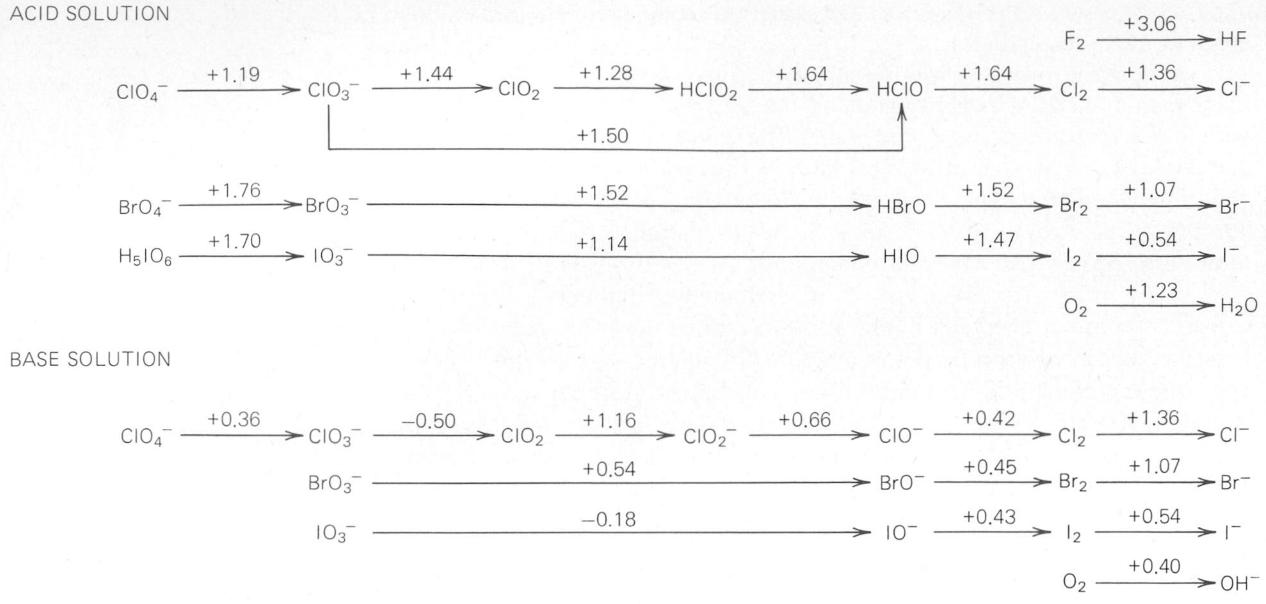

ACID SOLUTION

$$F_2 \xrightarrow{+3.06} HF$$

$$ClO_4^- \xrightarrow{+1.19} ClO_3^- \xrightarrow{+1.44} ClO_2 \xrightarrow{+1.28} HClO_2 \xrightarrow{+1.64} HClO \xrightarrow{+1.64} Cl_2 \xrightarrow{+1.36} Cl^-$$

$$ClO_3^- \xrightarrow{+1.50} HClO$$

$$BrO_4^- \xrightarrow{+1.76} BrO_3^- \xrightarrow{+1.52} HBrO \xrightarrow{+1.52} Br_2 \xrightarrow{+1.07} Br^-$$

$$H_5IO_6 \xrightarrow{+1.70} IO_3^- \xrightarrow{+1.14} HIO \xrightarrow{+1.47} I_2 \xrightarrow{+0.54} I^-$$

$$O_2 \xrightarrow{+1.23} H_2O$$

BASE SOLUTION

$$ClO_4^- \xrightarrow{+0.36} ClO_3^- \xrightarrow{-0.50} ClO_2 \xrightarrow{+1.16} ClO_2^- \xrightarrow{+0.66} ClO^- \xrightarrow{+0.42} Cl_2 \xrightarrow{+1.36} Cl^-$$

$$BrO_3^- \xrightarrow{+0.54} BrO^- \xrightarrow{+0.45} Br_2 \xrightarrow{+1.07} Br^-$$

$$IO_3^- \xrightarrow{-0.18} IO^- \xrightarrow{+0.43} I_2 \xrightarrow{+0.54} I^-$$

$$O_2 \xrightarrow{+0.40} OH^-$$

(O_2/H_2O is only $+1.23$ V). Bromide ion is barely susceptible to such oxidation, but iodide ion is much more so. Actually, a solution of HI, which is colorless, turns brown on exposure to air, because of the formation of some I_2.

In basic solution, however, the halogen in the -1 oxidation state resists oxidation by oxygen from the air. For example, I_2 can react to liberate oxygen and form I^-, which is essentially the reverse of its behavior in acidic solution.

Figure 23.2 Emf diagrams for oxidation–reduction couples of the halogens in $1m$ acid and $1m$ base solutions. If any negative ion exists in solution largely in the form of a weak acid, the protonated form is shown for the acid solutions. Examples are HClO and HF. If the negative ion forms a strong acid, the negative ion formula is shown. Examples are Cl$^-$ (not HCl) and ClO$_3^-$ (not HClO$_3$).

—— • Example 23.1 • ——

Using the data in Figure 23.2, show that chlorine is capable of oxidizing water to oxygen in acid solution. Write a net ionic equation for the reaction.

—— • Solution • We compare the $\mathscr{E}^\circ_{red}$ values for the Cl_2/Cl^- couple and the O_2/H_2O couple, and use their half-reaction equations to arrive at an ionic equation with Cl_2 and H_2O as reactants:

$$
\begin{aligned}
2H_2O &\longrightarrow 4H^+ + O_2 + 4e^- & \mathscr{E}^\circ_{ox} &= -1.23\ V \\
2(Cl_2 + 2e^- &\longrightarrow 2Cl^-) & \mathscr{E}^\circ_{red} &= +1.36\ V \\
\hline
2Cl_2 + 2H_2O &\longrightarrow 4H^+ + 4Cl^- + O_2 & \mathscr{E}^\circ_{cell} &= +0.13\ V
\end{aligned}
$$

The positive emf for the net ionic equation shows that Cl_2 spontaneously oxidizes H_2O.

See also Exercises 7–10 at the end of the chapter.

—— • Example 23.2 • ——

Using the data in Figure 23.2, show that I_2 is capable of reacting with a standard basic solution to liberate oxygen and form I^- ions. Write a net ionic equation for the reaction.

• Solution • As in Example 23.1, we arrive at the net ionic equation by choosing two appropriate half-reactions. We choose two that we can add to give an equation with I_2 and OH^- as reactants.

$$4OH^- \longrightarrow O_2 + 2H_2O + 4e^- \qquad \mathscr{E}^\circ_{ox} = -0.40 \text{ V}$$
$$\underline{2(I_2 + 2e^- \longrightarrow 2I^-) \qquad\qquad\quad \mathscr{E}^\circ_{red} = +0.54 \text{ V}}$$
$$2I_2 + 4OH^- \longrightarrow 4I^- + O_2 + 2H_2O \qquad \mathscr{E}^\circ_{cell} = +0.14 \text{ V}$$

The positive emf for the net ionic equation shows that I_2 can liberate O_2 from standard base solutions.

See also Exercises 7 and 9.

23•2 Characteristic Reactions

23•2•1 Halogens with Halogens

A compound of two halogens is called an **interhalogen compound.** In a reaction between two halogens, the more electronegative element is the oxidizing agent and is assigned a negative oxidation number in the compound. Consider the reaction of bromine with chlorine to produce diatomic molecules of bromine(I) chloride, BrCl:

$$Br_2 + Cl_2 \longrightarrow 2\overset{+1\ -1}{BrCl}$$
$$\text{bromine(I)}$$
$$\text{chloride}$$

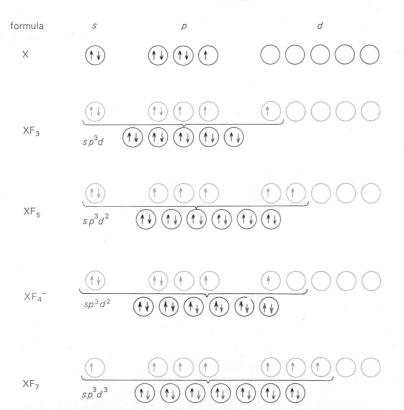

Figure 23•3 Electron assignments in orbitals of atom X in interhalogen compounds of fluorine. Electrons contributed by X are shown by black arrows and those of F by colored ones. The lightly shaded circles represent the imaginary promotion of electrons of the central atom to provide the proper number of bonding orbitals. The bracket denotes the choice of orbitals that are to be hybridized. The final hybridized structures are shown by the dark circles. Orbitals used in bonding have a black and a colored arrow; orbitals with lone pairs have two black arrows. (See also, as examples, ClF_3 in Figure 23.4, IF_5 and ClF_4^- in Figure 23.5, and IF_7 in Figure 23.6.)

In addition to diatomic interhalogens, a number of polyatomic interhalogen compounds and ions are known; the fluorides are the most common. Examples are chlorine(III) fluoride (chlorine trifluoride), ClF_3; iodine(V) fluoride (iodine pentafluoride), IF_5; iodine(VII) fluoride (iodine heptafluoride), IF_7; and the chlorine(III) fluoride ion (chlorine tetrafluoride ion), ClF_4^-.

The structures of interhalogen molecules and ions can often be explained in terms of hybrid orbitals. The hybrid orbitals used in bonds in a halogen X combined with fluorine in compounds XF_n are shown in Figure 23.3. When X is Cl, the orbitals involved are the $3s$, $3p$, and $3d$; when X is Br, the orbitals are the $4s$, $4p$, and $4d$; and when X is I, the orbitals are the $5s$, $5p$, and $5d$. As discussed in Section 6.4, the imaginary steps in bond formation are: (1) promotion of electrons of the central atom to higher energy orbitals, (2) hybridization of the proper number of orbitals to account for the formula and the geometry of the species, and (3) bond formation with the proper number of other atoms.

For ClF_3, sp^3d hybridization is proposed, with the five orbitals directed toward the corners of a trigonal bipyramid. Figure 23.4 shows three possible arrangements of three F atoms and two lone pairs around a Cl atom. Two experimental facts enable us to choose the structure of this compound: (1) the molecule is polar with a dipole moment of 0.59 D; and (2) X-ray studies show that the four atoms are in the same plane. The first fact rules out structure (c), which would be nonpolar. The second fact rules out structure (b). The experimental F—Cl—F bond angles of 87.5° shown in Figure 23.4(a) indicate a modified T-shape for the four atoms in ClF_3. Structure (a) is consistent with VSEPR theory (see Section 6.5 and Example 6.3). Lone-pair–bonding-pair (LP-BP) repulsions are presumably responsible for the bond angles being 87.5° instead of 90°.

For IF_5 and ClF_4^-, sp^3d^2 hybridization is proposed, with the six orbitals directed toward the corners of an octahedron. Their structures are shown in Figure 23.5, where they are compared with that of SF_6. In IF_5, there is only one possible arrangement of the five bonding pairs and one lone pair of electrons. The molecular shape is that of a square pyramid, with the I atom slightly below the plane of the bottom four F atoms. This slight distortion is presumably due to LP–BP repulsions. The molecule is quite polar, with a dipole moment of 2.21 D. The ClF_4^- ion is nonpolar. Therefore, the central Cl atom and the four bonded F atoms are arranged in a square-planar structure.

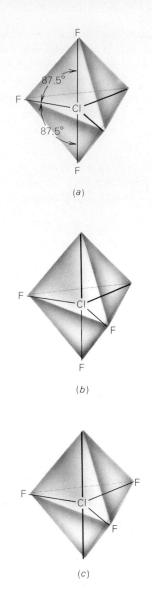

Figure 23.4 Three possible arrangements of three Cl—F bonds and two lone pairs of electrons in a molecule of ClF_3.

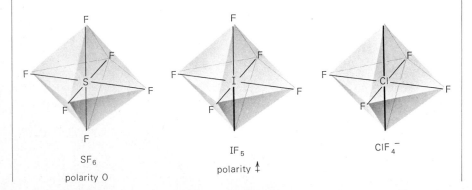

SF₆
polarity 0

IF₅
polarity ↕

ClF₄⁻

Figure 23.5 Typical fluorides that have octahedral, hybrid-orbital symmetry. In the two interhalogen structures, positions in the imaginary octahedra are taken by lone pairs of electrons (indicated by heavier lines). (See also Figure 23.3.)

In this structure, the two lone pairs of electrons are as far apart as possible, a fact that is in accord with VSEPR theory.

The IF_7 molecule has the structure of a pentagonal bipyramid, as shown in Figure 23.6, suggesting sp^3d^3 hybridization.

─── • Example 23•3 • ───

What type of hybridization could be proposed for iodine in ICl_2^-? Based on VSEPR theory, what is the most likely shape of ICl_2^-? What physical measurement might be used to determine whether the proposed shape is reasonable?

─── • Solution • We may assume that ICl_2^- is formed by the reaction of ICl with Cl^-:

$$: \overset{..}{\underset{..}{I}} - Cl + Cl^- \longrightarrow \left[Cl - \overset{..}{\underset{..}{I}} - Cl \right]^-$$

There are five electron pairs in the valence shell of iodine. This suggests sp^3d hybridization and trigonal-bipyramidal orbital geometry, as in the case of ClF_3. Three arrangements of the lone pairs and bonding pairs are possible:

$$\left[\overset{Cl}{\underset{Cl}{-I}} \right]^- \qquad \left[-I \overset{Cl}{\underset{Cl}{<}} \right]^- \qquad \left[Cl - I \overset{}{\underset{Cl}{<}} \right]^-$$

(a) (b) (c)

Structures (b) and (c) are ruled out because the 90° interactions between the lone pairs would give large repulsive energies. Therefore, the proposed shape of the ion is linear, as shown in (a). This proposal could be tested experimentally by determining whether the dipole moment is zero. [Actually, the dipole moment is zero, which is strong evidence that the symmetrical structure in (a) is correct.]

See also Exercises 21–23 and 39.

The interhalogens are very active oxidizing agents; they behave chemically much like mixtures of halogens. For example, ClF_3 is a fluorination agent. It decomposes water:

$$4ClF_3 + 6H_2O \longrightarrow 3O_2 + 2Cl_2 + 12HF$$

and causes spontaneous combustion of most materials that it contacts. ClF_5 is used as an oxidizer in rocket engines. BrCl is an effective alternative to Cl_2 as a disinfectant in waste water treatment.

23•2•2 With Metals The halogens react readily with most metals. Some examples are

$$3Br_2 + 2Al \longrightarrow 2AlBr_3$$
$$F_2 + Cu \longrightarrow CuF_2$$
$$3Cl_2 + 2Fe \longrightarrow 2FeCl_3$$

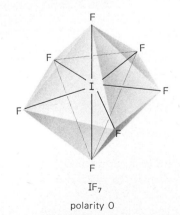

IF_7

polarity 0

Figure 23•6 Hybrid bond structure for IF_7 with sp^3d^3 hybridization.

Bromine and iodine do not react with gold, platinum, or some of the other noble metals, but fluorine and chlorine attack even these inactive elements. The reaction of gaseous chlorine with gold should be contrasted with the fact that no reaction takes place in standard acid solutions:

$$2(Au \longrightarrow Au^{3+} + 3e^-) \qquad \mathscr{E}^\circ_{ox} = -1.50 \text{ V}$$
$$\underline{3(Cl_2 + 2e^- \longrightarrow 2Cl^-)} \qquad \mathscr{E}^\circ_{red} = +1.36 \text{ V}$$
$$2Au + 3Cl_2 \longrightarrow 2Au^{3+} + 6Cl^- \qquad \mathscr{E}^\circ_{cell} = -0.14 \text{ V}$$

23•2•3   With Water (Hydrolysis)   All the halogens except fluorine disproportionate in water solution. For example,

$$Cl_2 + HOH \rightleftharpoons \underset{\substack{\text{hypochlorous} \\ \text{acid}}}{HClO} + HCl$$

These reactions do not take place spontaneously in acid solutions. Using data from Figure 23.2,

$$HClO \xleftarrow{-1.64} Cl_2 \xrightarrow{+1.36} Cl^-$$

we can calculate that the emf is negative for the disproportionation of Cl_2 into HClO and Cl^- in standard acid solution: $-1.64 + 1.36 = -0.28$ V.

However, in standard basic solution the disproportionation reaction has a positive emf of $+1.36 - 0.42 = +0.94$ V:

$$ClO^- \xleftarrow{-0.42} Cl_2 \xrightarrow{+1.36} Cl^-$$

Basic solutions of chlorine or bromine in water are strong oxidizing agents. The bleaching action of chlorine that is dissolved in dilute sodium hydroxide is accounted for by assuming that chlorine first reacts to form HClO, which is then converted to the ClO^- ion:

$$Cl_2 + OH^- \longrightarrow HClO + Cl^-$$
$$\underline{HClO + OH^- \longrightarrow ClO^- + H_2O}$$
$$Cl_2 + 2OH^- \longrightarrow ClO^- + Cl^- + H_2O$$

The ClO^- ion acts as the bleach, oxidizing colored compounds to colorless ones. It is the active ingredient in Clorox and similar household products.

In contrast with the other halogens, fluorine produces hydrofluoric acid and oxygen on reaction with water at room temperature:

$$2F_2 + 2H_2O \longrightarrow 4HF + O_2$$

The compound HOF can be produced at low temperatures. However, on warming to 0 °C and hotter, it decomposes to HF and O_2.

23•2•4   With Hydrogen   The halogens react with hydrogen to form hydrogen halides:

$$X_2 + H_2 \longrightarrow 2HX$$

The reactions of fluorine and chlorine with hydrogen take place with explosive violence, but bromine and iodine react slowly. (See the discussion of hydrogen halides in Section 23.6.1). These reactions proceed by a free radical chain-type mechanism, as discussed in Section 14.7.

• 23•2•5 **With Hydrocarbons** Halogens generally react with hydrocarbons by substituting for hydrogen atoms. The reaction with methane is typical:

$$X_2 + CH_4 \longrightarrow CH_3X + HX$$

• 23•2•6 **With Certain Nonmetals and Metalloids** The halogens react directly with a number of nonmetals and metalloids. Letting Z stand for the nonmetal phosphorus and the metalloids boron, arsenic, and antimony, the general equation is

$$3X_2 + 2Z \longrightarrow 2ZX_3$$

With phosphorus and the metalloids arsenic and antimony, the pentahalides are produced when excess halogen is used:

$$5X_2 + 2Z \longrightarrow 2ZX_5$$

These reactions occur most readily with fluorine and least readily with iodine. Nitrogen does not unite directly with the halogens due to its inactivity (see Chapter 24). Silicon, another metalloid, reacts to give the tetrahalide:

$$Si + 2X_2 \longrightarrow SiX_4$$

• 23•2•7 **With Compounds of Other Halogens** This type of reaction can be thought of simply as a displacement reaction, with the more active halogen displacing a less active one from its compounds. Some examples are

$$F_2 + 2NaBr \longrightarrow Br_2 + 2NaF$$
$$Br_2 + CaI_2 \longrightarrow I_2 + CaBr_2$$
$$Cl_2 + CaF_2 \longrightarrow \text{no reaction}$$

If the reaction occurs in dilute solution, we can calculate the emf from the data in Table 18.1 or Figure 23.2. For example, consider the reaction of chlorine with bromide ion:

$$
\begin{aligned}
2Br^- &\longrightarrow Br_2 + 2e^- & \mathscr{E}^{\circ}_{ox} &= -1.07 \text{ V} \\
Cl_2 + 2e^- &\longrightarrow 2Cl^- & \mathscr{E}^{\circ}_{red} &= +1.36 \text{ V} \\
\hline
Cl_2 + 2Br^- &\longrightarrow 2Cl^- + Br_2 & \mathscr{E}^{\circ}_{cell} &= +0.29 \text{ V}
\end{aligned}
$$

• 23•3 **Production
of the Halogens**

The halogens can be prepared by suitable chemical or electrochemical procedures from naturally occurring compounds, examples of which are listed in Table 23.2. In most cases the chemical change that is involved is the oxidation of a halide ion. Chloride, bromide, and iodide ions can be

Table 23•2 Abundant halogen-containing minerals

Formula	Chemical name	Mineral name
CaF_2	calcium fluoride	fluorspar
Na_3AlF_6	sodium aluminum fluoride	cryolite
$NaCl$	sodium chloride	halite
KCl	potassium chloride	sylvite
KCl and $NaCl$	potassium and sodium chloride	sylvinite
$MgBr_2 \cdot KBr \cdot 6H_2O$	magnesium potassium bromide	bromo-carnallite

Key Chemical: Cl_2
chlorine

How Made

$$2NaCl + 2H_2O \xrightarrow{\text{electrolysis}} Cl_2 + 2NaOH + H_2$$

U.S. Production; Cost

~22 billion lb per year
~5¢-6¢ per pound

Major Derivatives

20% ethylene dichloride, $ClCH_2CH_2Cl$
15% other chlorinated hydrocarbons

Major End Uses

40% diverse chemicals
20% plastics, mostly polyvinyl chloride
15% bleaching pulp and paper

oxidized by strong chemical oxidizers, but the electronegativity of fluorine is so high that chemical oxidizing agents are ineffective. The common way to produce fluorine is by electrolysis of anhydrous molten salts. Note that the $\mathscr{E}^{\circ}_{red}$ of F_2/F^-, $+2.87$ V, is so high that attempted electrolysis in aqueous solution results in the decomposition of water to produce O_2, not F_2. The $\mathscr{E}^{\circ}_{red}$ for O_2/H_2O is only $+1.23$ V (see Figure 23.2).

23•3•1 Fluorine The commercial source of fluorine is the mineral fluorspar, CaF_2. By treating the calcium fluoride with sulfuric acid, hydrogen fluoride is produced (see Section 23.6.1). Elemental fluorine is produced by the electrolytic decomposition of hydrogen fluoride (see Figure 23.7):

$$2HF \xrightarrow[\text{molten } KF \cdot 2HF]{\text{electrolysis in}} H_2 + F_2$$

The electrode reactions are shown in the figure caption.

The direct electrolysis of liquid HF is not practical, because this compound is a nonelectrolyte. A solution of HF in molten $KF \cdot 2HF$ at 80 to 100 °C forms a conducting solution that can be electrolyzed to yield hydrogen and fluorine. HF must be added continually to the molten salt bath to replace the HF that is being decomposed. The electrolysis is carried out under strictly anhydrous conditions.

23•3•2 Chlorine

• *Commercial Production* • Chlorine, a major industrial chemical, is produced commercially in several ways. The two main processes involve the

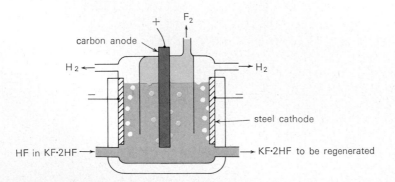

Figure 23•7 Fluorine is produced by the electrolysis of HF in molten $KF \cdot 2HF$. Hydrogen is the by-product. Oxidation at the anode:

$$2F^- \longrightarrow F_2 + 2e^-$$

Reduction at the cathode:

$$2H^+ + 2e^- \longrightarrow H_2$$

electrolysis of concentrated sodium chloride solutions, called *brines*. It is necessary that concentrated solutions be used, because in dilute chloride solutions, O_2 instead of Cl_2 is produced at the anode. In $1m$ solutions, for example, we can calculate from data in Figure 23.2 that water or OH^- ions are oxidized more readily than Cl^- ions.

The major electrolytic method is the **diaphragm process,** which is carried out in equipment like that shown in Figure 23.8. It accounts for more than three fourths of the chlorine/sodium hydroxide produced electrolytically. The reactions at the anode and cathode are shown in the figure. The overall reaction is

$$2NaCl + 2H_2O \xrightarrow[\substack{\text{concentrated} \\ \text{solution}}]{\text{electrolysis}} Cl_2 + H_2 + 2NaOH$$

The process is continuous, with fresh NaCl brine flowing into the cells and the three products, Cl_2, NaOH, and H_2, flowing out. At one time, chlorine was referred to as the product and sodium hydroxide and hydrogen were the by-products of this process. Today, the sodium hydroxide is so valuable that it and the chlorine are called *coproducts*.

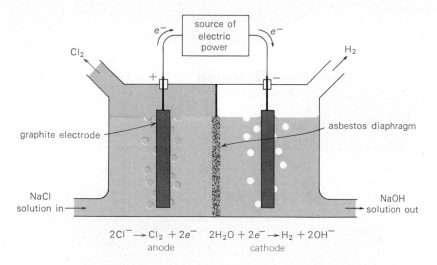

$$2Cl^- \rightarrow Cl_2 + 2e^- \qquad 2H_2O + 2e^- \rightarrow H_2 + 2OH^-$$
anode \qquad\qquad cathode

Figure 23•8 Schematic representation of the electrolysis of concentrated aqueous sodium chloride. (Compare Figure 18.6 on Plate 4.)

The method that accounts for about one fourth of the chlorine/sodium hydroxide produced electrolytically is the **mercury-cell** process. Liquid mercury is used as the anode, and the sodium combines with it to make sodium amalgam, $NaHg_x$. The liquid amalgam is pumped to a decomposing chamber where it reacts with water to regenerate the mercury and form hydrogen and sodium hydroxide:

$$NaHg_x + H_2O \longrightarrow xHg + \tfrac{1}{2}H_2 + NaOH$$

The trend toward building mercury-cell electrolysis plants has practically halted in this country because of the threat of mercury pollution. Ideally, no mercury waste would be produced, because the aim is to recycle the mercury with 100 percent efficiency. Inevitably, however, some mercury is carried out of the decomposer in the product streams, and some is spilled in handling. The net effect has been the pollution of plant sites, with tons of elemental mercury widely and intimately dispersed in the soil.

• **Example 23•4** •

At the site of an abandoned mercury-cell electrolysis plant on a river bank in Appalachia, sediments in old settling ponds contain several thousand kilograms of mercury. If the average river flow is 8.4 m³/s, and 50 mg/s of soluble mercury seeps into the river, what is the average concentration of mercury in the river in parts per billion? What is the mercury concentration in fish in parts per million, if it is 2,000 times that in the water?

• **Solution** • Taking the density of the water as 1 g/cm³, the total mass flow of the river water is

$$\left(\frac{8.4 \text{ m}^3}{s}\right)\left(\frac{100 \text{ cm}}{1 \text{ m}}\right)^3\left(\frac{1 \text{ g}}{1 \text{ cm}^3}\right)\left(\frac{1 \text{ kg}}{1,000 \text{ g}}\right) = 8.4 \times 10^3 \text{ kg/s}$$

The number of milligrams of mercury per 1 billion mg of river water is

$$\frac{\text{mass Hg}}{\text{mass water}} = \frac{50 \text{ mg Hg}}{(8.4 \times 10^3 \text{ kg water})\left(\frac{10^6 \text{ mg}}{1 \text{ kg}}\right)\left(\frac{1 \text{ billion mg}}{10^9 \text{ mg}}\right)}$$

$$= \frac{6.0 \text{ mg Hg}}{1 \text{ billion mg water}} = 6.0 \text{ ppb Hg}$$

The concentration of mercury in the fish is

$$(2,000)(6.0 \text{ ppb Hg})\left(\frac{1 \text{ ppm}}{1,000 \text{ ppb}}\right) = 12 \text{ ppm Hg}$$

See also Exercise 33.

• *Laboratory Preparation* • The laboratory preparation of chlorine can be carried out in a variety of ways; practically all involve the oxidation of the chloride ion. One method uses MnO_2 and concentrated HCl as reactants:

$$MnO_2 + 4HCl \xrightarrow[\text{in water}]{\text{heat}} Cl_2 + MnCl_2 + 2H_2O$$

Other oxidizing agents can be used, such as $KMnO_4$ and $Na_2Cr_2O_7$:

$$2KMnO_4 + 16HCl \longrightarrow 5Cl_2 + 2MnCl_2 + 2KCl + 8H_2O$$
$$Na_2Cr_2O_7 + 14HCl \longrightarrow 3Cl_2 + 2CrCl_3 + 2NaCl + 7H_2O$$

It is instructive to consider these actual laboratory reactions in light of the $\mathscr{E}°_{\text{red}}$ values for the following couples:

MnO_2/Mn^{2+}	$+1.23$ V	Cl_2/Cl^-	$+1.36$ V
$Cr_2O_7^{2-}/Cr^{3+}$	$+1.33$ V	MnO_4^-/Mn^{2+}	$+1.51$ V

In standard solutions only MnO_4^- is strong enough to oxidize Cl^- to Cl_2. The fact that all three oxidizing agents can be used illustrates the influence that changes in concentration and temperature can have on relative emf values.

23•3•3   **Bromine**   Our most abundant sources of bromine compounds are brines from salt wells and from the ocean. Almost all U.S. production of bromine comes from the naturally occurring brines of Arkansas (3,800 to 5,000 ppm Br^- ion) and Michigan (1,300 to 2,100 ppm Br^- ion). Seawater, which was once a U.S. commercial source, contains only about 65 to 75 ppm Br^- ion. The richest known brine is from the Dead Sea, and Israel produces bromine there.

Because the brines usually contain large proportions of chloride ions as well, the production of bromine must involve an oxidizing agent that oxidizes bromide ion but not chloride ion. Chlorine itself is ideal for this:

$$2Br^- + Cl_2 \xrightarrow[\text{solution}]{\text{water}} Br_2 + 2Cl^-$$

The reaction is carried out in a large tower. The brine solution is sprayed in at the top, and steam and chlorine gas are pumped in at the bottom. As the steam and chlorine move up the tower, the chlorine reacts with the Br^- ion in the falling brine droplets. The gases that come out of the top of the tower contain Br_2, Cl_2, and H_2O. They are cooled and condensed, and the crude bromine is separated and distilled to produce pure bromine. The chlorine is recovered and sent back through the tower. About 95 percent of the bromide ion in the original brine is recovered as bromine.

Any of the reactions used for the laboratory preparation of chlorine can be modified for preparing bromine; for example,

$$MnO_2 + 4HBr \longrightarrow Br_2 + MnBr_2 + 2H_2O$$
$$2KMnO_4 + 16HBr \longrightarrow 5Br_2 + 2MnBr_2 + 2KBr + 8H_2O$$

23•3•4   **Iodine**   Methods used in preparing chlorine are used to prepare iodine from brines in which the element is present as the iodide ion. An older source is the compound sodium iodate, $NaIO_3$, found in Chile. Sodium bisulfite, $NaHSO_3$, is the reducing agent:

$$
\begin{array}{ll}
5(HSO_3^- + H_2O \longrightarrow SO_4^{2-} + 3H^+ + 2e^-) & \text{(oxidation)} \\
2IO_3^- + 12H^+ + 10e^- \longrightarrow I_2 + 6H_2O & \text{(reduction)} \\
\hline
2IO_3^- + 5HSO_3^- \longrightarrow I_2 + 5SO_4^{2-} + H_2O + 3H^+ &
\end{array}
$$

or

$$2IO_3^- + 5HSO_3^- \longrightarrow I_2 + 3HSO_4^- + 2SO_4^{2-} + H_2O$$

The mixture of products is evaporated to dryness, and the iodine is separated from the residue by sublimation.

23•4   **Important Uses
of the Halogens**

Chlorine, the halogen produced in the largest amount by far, is used for killing bacteria in drinking water, in power plant water, in swimming pools, and in sewage waste water; for bleaching paper pulp and certain textiles; and in making dyes, drugs, plastics, solvents, and cleansers.

Elemental fluorine, as well as chlorine pentafluoride, has been used as the oxidizing agent in some rockets. Fluorine is used to make a variety of organic fluorine compounds. These include the Freon-type refrigerant gases, such as CCl_2F_2, and the heat-resistant plastic Teflon (Section 27.8).[2]

Both bromine and iodine are used in making medicinal compounds. For years, about half of the thousands of tons of bromine produced annually has been used to make ethylene dibromide, $C_2H_4Br_2$, a component of ethyl gasoline. Lead compounds are added to ethyl gasoline as antiknock agents, but the lead formed during their combustion tends to foul the spark plugs. With ethylene dibromide present, the lead forms volatile lead bromide, which escapes with the exhaust gases to pollute the atmosphere instead of the engine. The change to lead-free gasolines is now prescribed by law in this country for certain types of engines.

Large amounts of bromine are used to make silver bromide, the light-sensitive compound on most photographic films and papers for both

---

[2]Teflon can decompose at a temperature of about 250 °C (482 °F), which is reached in some self-cleaning ovens. The decomposition products are toxic.

★   Special Topic ★ Special Topic ★ Special Topic ★ Special Topic ★ Special Topic ★ Special Topic ★ Special Topic ★ Special Topic   ★

Mercury and Chlorine in the Environment It is characteristic of all human affairs that the most carefully planned activities have unexpected consequences. "The best laid schemes o' mice and men gang aft a-gley." Chemical activities are no exception. Some of our experiences with mercury and chlorine illustrate why the unexpected consequences of large-scale chemical operations are of increasing concern.

Until about 25 years ago, elemental liquid mercury was not considered a serious threat in the environment. Because of its density, mercury tends to settle into the pores of the soil, and the element itself is relatively harmless. Small concentrations of mercury have long been used in medicines and ointments, and in the United States we put about 150 tons annually into our mouths as silver amalgam dental fillings.

In the early 1950s, a mysterious new "disease" was traced to mercury pollution. Residents of a Japanese village on the shore of Minamata Bay were afflicted with severe nervous disorders, skin lesions, and blindness. More than 50 persons died, and twice that number were permanently crippled. The cause was traced to mercury contamination as high as 20 ppm in the fish that constituted a major part of the victims' diet. The mercury in the waters of Minamata Bay, approximately 5 ppb, was traced to the wastes from the mercury-containing catalysts of a nearby chemical plastics plant. The concentration of substances by animals in the biological food chain is a well-documented phenomenon (see Figure 10.8).

Research has shown how mercury changes from the relatively innocuous liquid element into one of the most toxic substances known, methyl mercury, CH_3Hg^+. Mercury has a slight tendency to be oxidized to Hg_2^{2+} or Hg^{2+}, and microorganisms convert these ions to methyl mercury. Incorporated in food, the methyl mercury passes easily from the digestive tract, through the blood, and into the central nervous system. Methyl mercury is so effective in damaging genes that it is the worst known chemical cause of mutations.

The chlorine/sodium hydroxide industry is only one of many sources of extra mercury in the environment. Certain agricultural insecticides deliver their mercury directly to the soil. Many paints contain mercury fungicides, and mercury is released when certain electric and thermal devices are worn out and discarded. It is estimated that the traces of mercury in fossil fuels that are released in smokestack gases worldwide may be equal to the total mercury used in this country, about 3,000 tons per year. The biggest source is thought to be due simply to the disturbance of soil in highway construction, building, mining, and other such projects. Exposing traces of elemental mercury in the soil allows them to evaporate more readily and may account for 30 times more atmospheric pollution than that originating from fossil fuel combustion.

The major environmental effects of chlorine are due to chlorine compounds produced either purposely or inadvertently. In the first category are a variety of pesticides, solvents, and industrial waste products. In the

color and black-and-white photography. Iodine is used as an antiseptic, in the preparation of other medicines, in dye compounds, and to make silver iodide for photographic film.

23•5   Metal Halides

One or more of the elemental halogens react with all metals, even the noble metals. The compounds formed with metals of the lowest electronegativity are predominantly ionic. The metals of intermediate electronegativity may form halides that are more covalent than ionic. The smaller and more positive the metal ion, the greater is its attraction for electrons and the more covalent will be the bond.

Table 23.3 lists a number of metal halides.[3] Note the difference in the melting points of compounds that are mainly ionic as compared with sub-

[3] As you study the metal halides, it will be well to keep in mind that there are some compounds of the so-called *halogenoid ions* that have properties similar to those of halide compounds. Examples of these ions are the cyanide, CN^-, and the thiocyanate, SCN^-, ions. By analogy with well-known chlorides, we would expect sodium cyanide, NaCN, to be ionic and soluble in water, and silver cyanide, AgCN, to be ionic-covalent and insoluble in water.

Special Topic ★ Special Topic ★ Special Topic ★ Special Topic ★ Special Topic ★ Special Topic ★ Special Topic ★ Special Topic

second category are the combustion products of certain plastics, leakages from industrial processes, and, ironically, chlorine compounds formed by the chlorine used to purify drinking water.

In 1908, the population of the United States was 80 million, and 100,000 persons died from gastrointestinal diseases such as cholera. In 1970, in a population of over 200 million, only 1,700 persons died from such ailments. The chlorination of drinking water, along with more sanitary methods of sewage disposal, is given most of the credit for this great improvement.

Recently, with increased attention to the environment and the development of sensitive analytical tools, traces of many undesirable chlorine compounds have been found in chlorinated water. Put into the water to kill infectious microorganisms, the Cl_2 also reacts with many nitrogen-containing organic compounds to form chloramine, NH_2Cl, or substituted chloramines. These substances are extremely toxic to shellfish and other aquatic animals, and it is feared they may harm humans. Chloroform, $CHCl_3$, is another of the some 100 organo-chloro compounds found in chlorinated water that are considered mutagenic (mutation causing), teratogenic (birth defect causing), or carcinogenic (cancer causing).

Sixty percent of the cities in this country take their drinking water from rivers into which other cities have discharged their waste water. The cities upstream may chlorinate their water twice, first to make it fit to drink and again to make it fit to discharge back into the river. Power plants are another large source of chlorinated

water, because they treat their cooling water with chlorine to prevent the growth of microorganisms in cooling towers. Such plants process huge amounts of water. The Quad Cities Power Station near Moline, Illinois, passes one tenth of the flow of the Mississippi River through its coolers, and the Indian Point plant in New York uses half the flow of the Hudson River. The Three Mile Island Plant, on the Susquehanna River in Pennsylvania, has used 2.7 metric tons of chlorine a day to treat the water for its cooling towers.

Most of the objectionable trace compounds in water supplies can be removed in a municipal treatment plant, although the process increases the cost of the water. Filtering the water through deep beds of sand or through beds of activated charcoal is effective, but the material in the beds must be replaced or reconditioned periodically.

Removal of most of the organic substances prior to chlorination is one way of reducing the amounts of chlorine compounds formed in drinking water. Another method is to use ozone rather than chlorine to kill the harmful microorganisms. The challenge is to develop methods that eliminate the recently identified harmful impurities without sacrificing the bacteria-free water that is so healthful in other ways.

Mercury and chlorine are typical of substances that have desirable and even essential uses but also serious ill effects. It is the job of the chemist to minimize and eliminate the latter, while preserving as many as possible of the former.

Table 23.3 Properties of metal halides

Name	Formula	Appearance	Bond type[a]	Melting point, °C	$\Delta H_f°$ per equiv wt, kJ[b]
tin(IV) chloride	$SnCl_4$	colorless liquid	c-i	−34	−127.8
tin(IV) iodide	SnI_4	yellow crystals	c-i	144	?
aluminum chloride	$AlCl_3$	white powder	c-i	194	−235.2
iron(III) chloride	$FeCl_3$	dark brown crystals	c-i	304	−133.2
silver fluoride	AgF	yellow powder	i-c	435	−204.6
silver chloride	$AgCl$	white powder	i-c	455	−127.1
lithium iodide	LiI	white crystals	?	467	−297.9
potassium iodide	KI	colorless crystals	i	681	−327.6
tin(IV) fluoride	SnF_4	white powder	i	705 (subl.)	?
sodium chloride	$NaCl$	colorless crystals	i	801	−411.0
sodium fluoride	NaF	colorless crystals	i	996	−569.0
aluminum fluoride	AlF_3	colorless crystals	i	1,040	−503.3
magnesium fluoride	MgF_2	colorless or violet crystals	i	1,263	−562.1

[a] Very largely ionic, i; more ionic than covalent, i-c; more covalent than ionic, c-i.

[b] In the formation of one equivalent of each of these substances, the same number of halogen atoms will be involved: 6.02×10^{23} atoms.

stances that are ionic-covalent. Ionic substances tend to have high melting points, and covalent substances, low melting points (although there are many exceptions, such as the covalent SiO_2 and SiC, which have extremely high melting points).

Covalent-ionic compounds often react with water (hydrolyze) instead of just dissolving. The covalent character of some metal halides is revealed by the fact that they are soluble in relatively nonpolar liquids. Examples include $AlCl_3$ in carbon tetrachloride, $FeCl_3$ in ether, and SnI_4 in carbon disulfide. The ionic halides are practically insoluble in such liquids.

The electric conductivity of a pure liquid compound is an indication of its ionic character. A mole of sodium chloride has a conductivity about 3 million times that of a mole of aluminum chloride.

23.6   **Nonmetal and Metalloid Halides**

The compounds formed by the halogens with other nonmetals and with the borderline elements (metalloids) are covalent. Examples of some important nonmetal and metalloid halides are boron trifluoride, BF_3, silicon tetrafluoride, SiF_4, carbon tetrachloride, CCl_4, phosphorus trichloride, PCl_3, phosphorus pentachloride, PCl_5, antimony trichloride, $SbCl_3$, sulfur monochloride, S_2Cl_2, nitrogen trichloride, NCl_3 (explosive), oxygen difluoride, OF_2

(explosive), chlorine monoxide, Cl_2O (explosive), and some of the fluorides formed by the noble gases (see Section 23.11).

Many nonmetal and metalloid halides are extremely reactive and must be kept out of contact with water or even moist air. They may undergo hydrolysis readily to yield acidic solutions. Examples of such reactions include

$$BF_3 + 3H_2O \longrightarrow H_3BO_3 + 3HF$$
$$PBr_3 + 3H_2O \longrightarrow H_3PO_3 + 3HBr$$
$$SiCl_4 + 4H_2O \longrightarrow H_4SiO_4 + 4HCl$$

23•6•1 **Hydrogen Halides** The hydrogen halides are a group of nonmetal halides that are so important to the study and practice of chemistry that they are given special attention. Their properties, summarized in Table 23.4, have two important characteristics: (1) the trends from chloride to bromide to iodide are as predicted from their location in the periodic table, and (2) the fluoride compound does not conform to the boiling and melting point trends because of the strong hydrogen bonds fluorine forms (see Section 8.2.3). The melting and boiling points of the four compounds are plotted in Figure 8.7.

The trend in the heats of formation is the same as the trend in the free energies of formation (see Table 13.6). The trend in stability is from the very stable HF to the somewhat unstable HI. Hence, the more negative the $\Delta H_f°$, the more stable is the compound. The percentage dissociation at 1,000 °C is also a measure of the strength of the various H—X bonds; the H—I bond is the weakest of the four.

Table 23•4 Properties of hydrogen halides

	HF	HCl	HBr	HI
boiling point, °C	20	−85	−67	−36
melting point, °C	−83	−114	−87	−51
% dissociation at 1,000 °C for $HX \rightarrow \frac{1}{2}H_2 + \frac{1}{2}X_2$	too slight to measure	0.0014	0.50	33
$\Delta H_f°$ per mole for $\frac{1}{2}H_2 + \frac{1}{2}X_2 \rightarrow HX$, kJ	−271.1	−92.3	−36.4	+26.5
color	none	none	none	none
odor	all very irritating; attack delicate nasal tissues			

All the hydrogen halides are shown to be covalent compounds by the fact that the pure liquid substances do not conduct an electric current. However, when dissolved in water, HCl, HBr, and HI form strong acid solutions; HF forms a weak acid solution, $K_a = 6.6 \times 10^{-4}$, because the strong HF bond prevents a complete reaction with the H_2O molecules to form H_3O^+ and F^-.

Hydrochloric acid, often called *muriatic acid* commercially, is by far the most important of the four acids. It is used to remove rust from scrap iron, to clean mortar from masonry, to clean metal surfaces for electroplating, and to neutralize basic solutions. Hydrobromic and hydriodic acids are not widely used because, although as acids they are almost identical with hydrochloric, they are much more expensive. All three of these compounds act as typical strong acids.

Hydrofluoric acid, unlike the other three, attacks glass and hence is used to etch and frost glass. Very precise markings such as thermometer graduations and even pictures can be etched out of the glass surface. The glass is first completely covered with a thin film of melted paraffin; the wax is then carefully cut away so as to expose the parts that are to be etched. Fumes of HF do not attack the paraffin but they do attack the silicon-containing compounds in the glass. The chemical reaction destroys the glass as the result of the formation of the gaseous silicon compound, SiF_4. The reactions with silicon dioxide and sodium silicate, two compounds that react in a way similar to glass, are as follows:

$$SiO_2 + 4HF \longrightarrow SiF_4\uparrow + 2H_2O$$
$$Na_2SiO_3 + 6HF \longrightarrow SiF_4\uparrow + 2NaF + 3H_2O$$

Concentrated solutions of HCl, HBr, and HI can be shipped in steel or glass containers, but HF solutions must be kept in wax-lined vessels or special plastic tanks.

• *Production of Hydrofluoric and Hydrochloric Acids* • Hydrogen fluoride is usually prepared commercially by the action of hot concentrated sulfuric acid on calcium fluoride:

$$CaF_2 + H_2SO_4 \longrightarrow CaSO_4 + 2HF\uparrow$$

Hydrogen chloride can be prepared in the same way, using NaCl as the raw material, but today very little is produced this way commercially. More than 90 percent of HCl is made as a by-product of the chlorination of organic compounds (see Section 23.2.5).

Hydrogen chloride is also produced by the direct combination of hydrogen and chlorine. In regions where electric power is not too expensive, these gases are produced by the electrolysis of sodium chloride brine for the purpose of making hydrogen chloride.

• *Production of Hydrobromic and Hydriodic Acids* • Hydrogen bromide and hydrogen iodide cannot be prepared by the action of hot concentrated sulfuric acid on a halide, because Br^- and I^- ions are oxidized by this reagent to the elements. Hydrogen bromide can be obtained by direct union of the elements at 200 °C with a platinum catalyst. Both hydrogen bromide and hydrogen iodide can be obtained by the hydrolysis of a nonmetal or metalloid bromide or iodide. An example of such a hydrolysis reaction is that of PBr_3, shown on the previous page.

23•7 Halogen Oxy-Acids and Oxy-Salts

Compounds containing the ions ClO_4^-, ClO_3^-, ClO_2^-, and ClO^- are well known. Bromine and iodine form some similar oxy-ions. In addition to being acids, the compounds HXO_4, HXO_3, HXO_2, and HXO are oxidizing agents, as indicated by Figure 23.2. The salts of these acids, especially the sodium and potassium salts, also serve as oxidizing agents.

The formulas for the hypohalous acids are usually written as HXO, although H—O—X: is the correct Lewis structural formula. The structures of three oxy-halogen species are shown in Figure 23.9. In water solution, the oxy-acid of iodine, in which iodine is in the +7 oxidation state, has the formula H_5IO_6 and is called *paraperiodic acid*. (Iodine is also +7 in periodic acid, HIO_4.) Although bonds are shown by single lines in Figure 23.9, the bonding in these oxy-halogen species also involves pi orbitals, which are not indicated.

Some oxy-acids are formed in disproportionation reactions of the elements. The formation of hypochlorous acid is shown in the first reaction in Section 23.2.3. Chloric acid, $HClO_3$, is obtained by a disproportionation reaction followed by a precipitation reaction:

$$3Cl_2 + 6Ba(OH)_2 \longrightarrow Ba(ClO_3)_2 + 5BaCl_2 + 6H_2O$$
$$Ba(ClO_3)_2 + H_2SO_4 \longrightarrow 2HClO_3 + BaSO_4\downarrow$$

When $KClO_3$ is carefully heated to between 400 and 500 °C, it disproportionates to yield potassium perchlorate and potassium chloride. When a solution of $KClO_4$ and concentrated H_2SO_4 is boiled, $HClO_4$ distills:

$$4KClO_3 \xrightarrow{\text{heat}} 3KClO_4 + KCl$$
$$2KClO_4 + H_2SO_4 \longrightarrow 2HClO_4\uparrow + K_2SO_4$$

Extreme caution must be used when working with concentrated $HClO_4$. Terrific explosions have resulted when solutions have been heated and have become too concentrated.

Potassium chlorate, $KClO_3$, is included with the combustible material in the head of a match. The initial combustion of the match does not depend on atmospheric oxygen; instead, the oxygen comes from the potassium chlorate. This compound is extremely dangerous when mixed with such combustible materials as paper, cloth, and sulfur. Potassium chlorate must be properly disposed of in the laboratory to avoid dangerous fires in waste containers.

The Noble Gases

23•8   **Isolation of the VIIIA Elements**

In 1894, the English chemist William Ramsay identified a new element, argon, as the unreactive gas remaining after all the nitrogen, oxygen, and other substances in a sample of air were removed by chemical reactions. Four years later he prepared another unreactive gas by heating the mineral cleveite.[4] The spectrum of this new gas was similar to certain lines in the

ClO_3^-, pyramidal

ClO_4^-, tetrahedral

H_5IO_6, octahedral

Figure 23•9 Representatives of three oxy-halogen structures. The top two have sp^3 hybridization and the bottom has sp^3d^2. The bonding in these species also involves pi orbitals, which are not indicated here. (See also Figures 23.3 and 6.18.)

---

[4] Alpha emission by radioactive nuclides in certain minerals yields helium atoms that may be trapped in the solid.

spectrum of the sun. In 1868 these lines had been attributed to an element named helium by astronomers (Greek *helios*, sun).

After his discovery of helium on earth, and finding that it was unreactive like argon, Ramsay pursued the idea that a group of elements in the periodic table must be located between the halogens and the alkali metals. By the fractional distillation of liquid argon prepared from liquid air, Ramsay separated and identified three more members of the family, neon, krypton, and xenon. Later he also studied the properties of the radioactive member of the family, radon.

23•9   Physical Properties

Table 23.5 lists some of the important physical properties of the noble gases. As with the VIIₐ elements, any given property changes in a regular way from one element to the next. The elements are liquids over very narrow temperature ranges, less than 4 °C for all except radon. Helium has the lowest melting point and boiling point of any element.

23•10   Chemical Properties

For many years the group VIIIₐ elements were referred to as the *inert gases*, because each seemed to be completely unreactive chemically. It was a shock and a thrill to the chemical world when, in 1962, the Canadian chemist

---

★   Special Topic ★ Special Topic ★ Special Topic ★ Special Topic ★ Special Topic ★ Special Topic ★ Special Topic ★ Special Topic   ★

Facts, Theories, and Scientific Discovery Although the direct combination of xenon and fluorine had been attempted in the 1930s, it had not been successful, probably because of the reaction of the fluorine with the quartz containing vessel. Many attempts were made to form special types of compounds, for example, hydrates of the noble gases at extremely low temperatures. Also, transitory compounds appeared to be formed in gaseous mixtures subjected to high-energy electric discharge. However, repeated failures to prepare any ordinary compounds led to the general acceptance of, first, the idea that these gaseous elements were inert to reaction, and, second, the theory that an atom with eight electrons in the outside main level was incapable of chemical reaction.

The downfall of these views in 1962 has three general lessons for the student of science. First, Dr. Bartlett's test of the accepted theory was not as direct a challenge as the attempt, say, to make a simple binary compound. He had prepared the compound O_2PtF_6, in which a molecule of O_2 loses an electron to form the PtF_6^- ion; but the O_2^+ unit retains its diatomic "molecular" character. By analogy, he decided to see whether the Xe molecule might react similarly to the O_2 molecule, because the ionization energy of the former (12.1 eV) is less than that of the latter (12.2 eV). Second, able chemists had shown

that it was a "fact" that xenon and fluorine would not react; this fact was later shown to be in error. Third, after xenon was shown to react in a direct, simple way with fluorine, the theory of the "inert eight electrons" was easily dropped. The reactivity of xenon was explained in terms of promoting outer electrons to higher energy orbitals, a well-developed theory that had already been applied to many other elements. In summary, our three lessons are these: a well-entrenched, useful theory is rarely challenged directly; facts are the basis of all science, but, like theories, they are open to question; and, when the facts change, the theory must change accordingly.

As a postscript to this case history in scientific method, it must be mentioned that there was at least one seriously considered theoretical prediction that xenon and other members of the family should react chemically. Dr. Linus Pauling, later to win a Nobel Prize in chemistry, held in 1933 that the noble gases possibly could react with fluorine. When laboratory tests of his predictions failed, attempts at direct combination were essentially abandoned for 30 years. His theory, which did not lead to a fruitful test of the facts, is seen now as a suggestion that should have been hammered at longer. But who is to say when a theory has been tested thoroughly enough?

Table 23•5 Physical properties of the noble gases

	He	Ne	Ar	Kr	Xe	Rn
melting point, °C	-272.2^a	-248.6	-189.4	-157.2	-111.8	-71
boiling point, °C	-268.9	-246.0	-185.9	-153.4	-108.1	-62
density, g/L	0.178	0.900	1.78	3.73	5.89	9.73
ionization energy, eV	24.6	21.6	15.8	14.0	12.1	10.7
atomic radius, Å	0.50^b	0.65^b	0.95^b	1.10	1.30	1.45^b
electronegativity^c	2.7	4.4	3.5	3.0	2.6	2.4
amount in air, in ppm (or percent in parentheses)	5.2	18	(0.93%)	1.1	0.087	10^{-15}

^{a} Value is for 26 atm. Helium does not form a solid at 0 K at atmospheric pressure.

^{b} Values are extrapolated from neighboring nonmetals.

^{c} Values are based on indirect calculations and are taken from Bing-Man Fung, *J. Phys. Chem.*, **69**, 596 (1965). Because so few of the VIIIA elements form compounds, and because their electronegativity values are not obtained from bond strengths, chemists tend to ignore the VIIIA elements in discussions and applications of electronegativities.

Neil Bartlett prepared a stable compound thought to have the formula $XePtF_6$. (See Figure 23.10 on Plate 7.) The spell of the inert elements was broken. Very shortly other researchers showed that xenon could react directly with fluorine to form simple binary compounds such as XeF_2, XeF_4, and XeF_6. The term *inert* was no longer appropriate; most chemists began to refer to the family as the *noble gases*, just as the rather unreactive and chemically aloof elements such as gold and platinum are referred to as the *noble metals*.

No compounds of the three lightest noble gases are known. However, the three heaviest elements do combine directly with fluorine. Radon reacts spontaneously with fluorine at room temperature, whereas xenon requires heating or photochemical initiation. Krypton reacts with fluorine only when both are subjected to radiation or an electric discharge.

23•11 Compounds of the Noble Gases

23•11•1 Xenon Compounds
About 200 xenon compounds are known, including halides (mostly fluorides), oxides, oxy-fluorides, fluorosulfates, xenate and perxenate salts, and addition compounds with Lewis acids and bases.

The fluorides XeF_2, XeF_4, and XeF_6 are obtained by reacting xenon with increasing amounts of fluorine. In these compounds, xenon has the even oxidation numbers $+2$, $+4$, and $+6$, which are typical of most xenon compounds. These fluorides are the starting materials for the synthesis of the other xenon compounds. The compounds in which an oxygen atom replaces two fluorine atoms in the hexafluoride are known: $XeOF_4$, XeO_2F_2, and XeO_3. The trioxide XeO_3 is stable in water solution, but the dry white solid is a violent explosive and dangerously sensitive to detonation, as many accidental laboratory explosions have demonstrated.

Although xenon octafluoride is not known, xenon can have a $+8$ oxidation state. Two oxyfluoride derivatives, XeO_2F_4 and XeO_3F_2, as well as the tetroxide XeO_4 are known. Important salts are sodium xenate, $NaH\overset{+6}{Xe}O_4$, and sodium perxenate, $Na_4\overset{+8}{Xe}O_6$. The perxenates are among the most powerful oxidizing agents known. The $\mathscr{E}^\circ_{red}$ for the H_4XeO_6/XeO_3 couple is $+3.0$ V, second only to that of 3.06 V for F_2/HF.

Two of the most interesting xenon compounds are the fluoroxenates $CsXeF_7$ and Cs_2XeF_8. These compounds have eight and nine electron pairs, respectively, in the valence shell of xenon. The octafluoroxenates Cs_2XeF_8 and Rb_2XeF_8 are the most stable xenon compounds known. They can be heated to 400 °C without decomposition.

23•11•2 Krypton and Radon Compounds The only product obtained when krypton reacts with fluorine is the difluoride KrF_2. No oxidation states other than $+2$ are known. Of about a dozen known krypton compounds, all are complex salts derived from KrF_2. One example of the formation of such a salt is

$$KrF_2 + SbF_5 \longrightarrow KrF^+ + SbF_6^-$$

Because radon is radioactive and has a short half-life of four days, its chemistry is difficult to study. However, the existence of both volatile and nonvolatile radon fluorides has been demonstrated.

23•11•3 Analogies with Halogen Compounds Many xenon compounds are similar in their properties to halogen compounds in which a halogen atom exhibits a positive oxidation state. The xenon fluorides XeF_2 and XeF_4 are similar to ClF_3 and IF_5. These compounds are very reactive fluorinating reagents. The xenate ion, XeO_4^{2-}, and the perxenate ion, XeO_6^{4-}, are similar to the chlorate, ClO_3^-, and perchlorate, ClO_4^-, ions. Finally, XeO_3, Cl_2O, and ClO_2 are explosive oxides, although the xenon compound is more sensitive to detonation than the two oxides of chlorine.

23•11•4 Structures of Xenon Compounds The experimentally determined molecular shapes of XeF_2 and XeF_4 shown in Figure 23.11 are predictable from VSEPR theory. In XeF_2, there are two shared pairs and three lone pairs of electrons in a bipyramidal arrangement. The molecular shape is linear, with the three lone pairs directed toward the corners of an equilateral triangle. (This structure is analogous to that of ICl_2^-, described in Example 23.3.) The hybridization proposed for xenon in XeF_2 is sp^3d (see Figure 23.12).

In XeF_4, there are four shared pairs and two lone pairs of electrons. The two lone pairs are located as far apart as possible to minimize repulsions. This requires the four fluorine atoms to be at the corners of a square, giving a square-planar molecular shape. (This shape is analogous to that of ClF_4^- in Figure 23.5.) The hybridization proposed for xenon in the tetrafluoride is sp^3d^2 (see Figure 23.11).

In XeF_6, there are six shared pairs and one lone pair of electrons. The exact position of the shared pairs and lone pair is uncertain, but it is known that the structure is not octahedral. A distorted octahedron and a pentagonal bipyramid have been proposed, with the latter structure based on sp^3d^3 hybridization (see Figure 23.12) and the structure of IF_7.

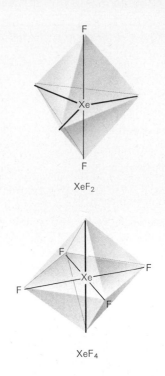

Figure 23•11 Hybridized structures for two xenon fluorides. Heavy lines indicate lone pairs of electrons.

■ **Example 23.5** ■

Consider the noble gas compound xenon oxytetrafluoride, $XeOF_4$. If it is assumed that the repulsive power of an oxygen atom is equal to that of a lone pair of electrons, what is the expected molecular shape of $XeOF_4$? What type of hybridization of xenon would you propose for this compound?

■ **Solution** ■ If we assume that $XeOF_4$ is derived from XeF_6 by replacing two F atoms with one O atom, we have one lone pair, one bonding pair to O, and four shared pairs to distribute around the corners of an octahedron. By analogy with XeF_4, we would expect the four fluorines to assume the corners of the square, and the lone pair and the bonding pair to oxygen to be as far apart as possible. The molecular shape expected is a square pyramid; the orbital geometry is octahedral.

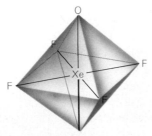

Because six orbitals of xenon are involved in the molecule, we should expect sp^3d^2 hybridization.

See also Exercises 43, 45, and 46.

● **23.12   Uses of
the Noble Gases**

The foregoing discussion of the chemical reactions of the noble gases should not blind us to the fact that the most important property of these

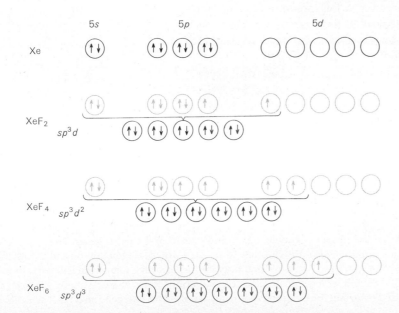

Figure 23.12 Electron assignments in xenon and xenon fluoride compounds. Electrons contributed by fluorine are shown by colored arrows. (See also Figure 23.3 and accompanying caption.)

elements is their extreme resistance to any chemical combination. Most of the present uses of these gases are related to their chemical inactivity.

Helium is found in fairly high concentrations in certain natural gases, largely from wells in Kansas and Texas. Large amounts of helium are used in the fuel pressure systems for rockets, in unreactive atmospheres for welding, and in heat transfer atmospheres for nuclear reactors. Helium is widely used in very low temperature research. With evaporating liquid helium as a coolant, temperatures as low as a few microdegrees Kelvin have been reported. Helium, the least dense of all elements except hydrogen, is preferred for filling meteorological balloons or dirigibles because, unlike hydrogen, it is not combustible. It is also used as a gas for supersonic wind tunnels.

Another interesting use of helium is in an 80 percent helium–20 percent oxygen mixture that is substituted for air for breathing by divers and others who work under high pressures. Too long an exposure to compressed air produces nitrogen narcosis, one effect of which is hallucinations known to divers as "rapture of the depths." Divers so enraptured fail to observe precautions and may drown. Helium was originally proposed as a substitute for nitrogen because it is less soluble in blood than is nitrogen. As divers return to atmospheric pressure, the dissolution of nitrogen gas forms tiny bubbles of gas in the blood, which cause the painful and dangerous "bends."

Argon, which makes up about 1 percent of the atmosphere, has been steadily increasing in industrial importance. It can be used interchangeably with helium in many processes. Welding of titanium and other exotic metals in aircraft and rocket construction requires an inert atmosphere, and argon serves this purpose. It is also used in filling incandescent light bulbs because it does not react with the white-hot tungsten wire as oxygen or nitrogen does.

Neon is used to make high-voltage indicators, lightning arresters, and television tubes. It is also used in making gas lasers. Liquid neon is an economical low-temperature refrigerant. Its most common use is in neon advertising signs.

Krypton is used along with argon as a low-pressure filler for fluorescent lights. It is also used in photographic flash lamps for high-speed photography. One of its sharp spectral lines is used as the standard length for the meter (see Table A.1 in the appendix).

Xenon is used in the manufacture of electron tubes, stroboscopic lamps, and bactericidal lamps. It is used in the atomic energy field in bubble chambers and probes. One of xenon's synthetically produced isotopes, xenon-133, has useful applications as a radioisotope.

Chapter Review

Summary

The elements of the **halogen family** have similar properties that vary in a regular way from one element to the next. All are highly reactive diatomic molecules that are strong oxidizing agents. They react with virtually all the other elements, including (in the case of fluorine and chlorine) the noble metals and several of the noble gases. They react with each other to produce **interhalogen compounds,** both diatomic and polyatomic, in which one of the halogen atoms is in a positive oxidation state; these compounds too are strong oxidizing agents. The halogens also react with many compounds, both inorganic and organic.

The halogens have a wide variety of commercial, industrial, and medicinal uses. They are produced by chemical and electrochemical processes, most of which entail the oxidation of a halide ion. Chlorine, the most important of the halogens, is produced electrolytically from **brines** by the **diaphragm process** and the **mercury cell process,** which also yield sodium hydroxide and hydrogen.

The compounds of the halogens exhibit a wide variety of properties. Their bonding ranges from strongly ionic to strongly covalent, and their stability ranges from inert to explosively reactive. Among the most important compounds are the **hydrogen halides,** which are acids; many of the properties of hydrogen fluoride are anomalous. The **halogen oxy-acids** and their salts, in which the halogen atoms are in positive oxidation states, are oxidizing agents.

The **noble gases,** formerly called inert gases, are as unreactive as the halogens are reactive. Like the halogens, they have similar properties that vary in a regular way. Most of their uses are related to their extreme chemical inactivity. Of the relatively few compounds that *are* formed by the three heaviest noble gases, many are powerful oxidizing agents and many behave similarly to halogen compounds in which the halogen atom is in a positive oxidation state. The discovery of compounds of the noble gases overturned one of the most firmly believed principles of chemistry.

Exercises

Physical Properties of Halogens

1. What similarities in atomic structure exist among the halogens?
2. How many electrons are there in the fourth main energy level of an atom of iodine?
3. Mendeleev placed the four VIIA elements known at that time (1869) into a single family. Was he guided by the similarities in the structures of their atoms? Explain.
4. Which has the greater density at 300 K and 1.2 atm, elemental chlorine or fluorine? How much greater? Calculate the ideal density of fluorine at STP.
5. With respect to chlorine and astatine, which should have the greater (a) atomic radius, (b) electronegativity, (c) ionic radius, (d) number of f electrons, (e) metallic nature, (f) ionization energy?

Chemical Properties of Halogens; emf Calculations

6. The bond dissociation energies of chlorine and fluorine are 242.6 and 157 kJ · mole^{-1}, respectively. Can these data be related to the reactivities of the elements?
7. Using data in Figure 23.2, answer the following:
   (a) Is bromine capable of oxidizing water to oxygen in acid solution?
   (b) Is bromine capable of being reduced to bromide ion in standard base solution?
8. A water solution of HI, which is colorless, turns brown on exposure to air because of the formation of some I_2. Calculate the emf for this reaction in standard acid solution. What is the standard free energy change, $\Delta G_r°$, for this reaction?
*9. Calculate the emf for the BrO_3^-/Br^- couple in both standard acid and standard base solutions from the data in Figure 23.2. Can you predict which would be the stronger oxidation couple without doing the calculations?
*10. For the MnO_2/Mn^{2+} couple, $\mathscr{E}_{red}° = +1.23$ V. Is this couple sufficiently strong to oxidize HBr to Br_2 in standard acid solution? (See Figure 23.2 for needed data.)
11. Will chlorine dioxide, ClO_2, disproportionate in either standard acid or standard base solution? (Consult Figure 23.2.) Calculate the emf for the disproportionation that will occur, assuming that the Cl atom undergoes a change in oxidation number of one unit for each half-reaction.
*12. Can dilute solutions of hypochlorous acid, HClO, undergo disproportionation to chloric acid, $HClO_3$, and hydrochloric acid, HCl, according to data in Figure 23.2?
*13. Repeat Exercise 12 for standard base solutions in which only anions are involved. Write a net ionic equation for the reaction.
14. If you wanted to prepare $HClO_3$ from HClO or NaClO, what would be a good method based on your answers to Exercises 12 and 13?
15. Calculate the change in free energy, $\Delta G_r°$, per mole of Br_2 produced by the reaction between chlorine and bromide ion in standard acid solution.
16. Indicate which oxidizing agents in Figure 23.2 cannot spontaneously oxidize water to evolve oxygen in acid solutions.

Characteristic Reactions of Halogens; Interhalogens

17. What is the oxidation state of each element in the following? (a) ClF_3, (b) IF_5, (c) ClBr, (d) ClF_4^-.
18. Give two acceptable names for each of the following: (a) IF_3, (b) ICl, (c) $BrCl_3$, (d) IBr_5, (e) IF_6^-.
19. Which two of the following formulas most likely represent nonexistent compounds? (a) BrF_3, (b) $ClBr_3$, (c) IF_5, (d) ClF_5, (e) BrI_5.
20. Based on the order of repulsive energies for LP–LP, LP–BP, and BP–BP at approximately 90°, show that structure (*a*) in Figure 23.4 is the best structure based on VSEPR theory.
*21. Iodine is known to react with a water solution of KI. Write a balanced equation for the probable reaction. Propose a possible hybridization state for iodine in the

resulting anion. What is a likely shape for the complex ion?

*22. The IF_4^+ ion is known to be present in certain crystalline salts. Propose a possible hybridization state for iodine and a likely geometry for this complex ion.

23. Sketch a molecule of ClF_5 based on square-pyramidal geometry. What would be the overall geometry of the orbitals, including any lone pairs, about the Cl atom? Would the molecule be polar or nonpolar? Predict the sizes of the F—Cl—F angles.

24. Methane, CH_4, reacts in sunlight with excess chlorine to produce 4 moles of HCl per mole of CH_4. What are the formula and name of the organic product?

25. Write balanced equations for each of the following reactions:
    (a) fluorine (excess) + sulfur
    (b) chlorine + bismuth
    (c) fluorine + silicon
    (d) fluorine + hydrogen
    (e) chlorine + fluorine (excess)
    (f) iodine + fluorine (excess)

Production of Halogen Elements

26. Starting with the mineral fluorspar, CaF_2, show with equations a method of obtaining fluorine. Indicate all necessary conditions.

27. (a) Show with a diagram how you could use laboratory apparatus and chemicals to prepare and collect a container of chlorine gas filled at atmospheric pressure. Write equations for the reactions.
    (b) Could fluorine be prepared by the method outlined in part (a)? Explain.

28. Write balanced equations for three different laboratory preparations of iodine. Which of these three reactions should take place most readily if standard acid solutions are used?

29. In one step in the production of bromine from seawater, the bromine that forms in the water is blown out by a blast of air. This mixture of air and bromine is allowed to react with sodium carbonate to produce NaBr, $NaBrO_3$, and CO_2. The mixture is then acidified with H_2SO_4. Elemental bromine forms along with other products. Write equations for both these reactions.

30. Explain how you could separate iodine from the reaction mixture that results when it is prepared by one of the reactions you wrote in Exercise 28.

31. What is formed at the anode during the electrolysis of a dilute aqueous NaCl solution? During the electrolysis of a concentrated aqueous NaCl solution? Account for the difference.

32. Write the two half-reactions for the preparation of chlorine by the reaction of MnO_2 with HCl. If this reaction were attempted with two standard electrochemical cells, would the reaction be spontaneous? How

do the conditions for the preparation differ from standard cell conditions?

33. Suppose a power plant daily discharges 2.7 metric tons of chlorine (in the form of various chlorine compounds) into a river with a flow of 50 metric tons per second. How much chlorine is in the water (in parts per million) due to the power plant?

Metallic and Nonmetallic Halides

34. Which one of the following compounds should have the highest and which one the lowest melting points: $AlBr_3$, KBr, KF, KI, $LiBr$? Upon what do you base your prediction?

35. When sodium bromide reacts with sulfuric acid, very little hydrogen bromide is obtained. Instead, the major products are H_2O, Br_2, and SO_2. When sodium iodide reacts with sulfuric acid, the major products are H_2O, I_2, and H_2S.
    (a) Write the oxidation and reduction half-reactions and the net ionic equations for both of these reactions.
    (b) Why is H_2S produced instead of SO_2 in the second reaction?

36. Without consulting sources of information, predict which hydrogen halide has the (a) highest boiling point, (b) greatest percentage of decomposition at 600 K, (c) largest K_a in water solution, (d) largest standard heat of formation. Explain in each case how you arrived at your answers.

37. Write balanced equations for each of the following hydrolysis reactions:
    (a) $PCl_5 + H_2O \rightarrow$ (b) $PBr_3 + H_2O \rightarrow$
    (c) $SF_4 + H_2O \rightarrow$ (d) $POCl_3 + H_2O \rightarrow$
    (e) $SOCl_2 + H_2O \rightarrow$ (f) $SO_2Cl_2 + H_2O \rightarrow$

38. Hydrochloric acid is used to remove rust (Fe_2O_3) prior to plating iron with zinc, nickel, or other metals. Write an equation for the rust-removing reaction. Assuming that hydrobromic acid could be bought at the same price per pound as hydrochloric acid, would it be just as good a choice for the job? Why? What happens to the iron compound that forms when either acid is used to remove rust?

Oxy-Acids and Oxy-Salts

39. Propose a likely structure for the ClO_2^- ion. What is a possible hybridization state for the Cl in this ion?

40. Write a balanced equation, using whole-number coefficients, for the reaction that takes place when potassium chlorate and paraffin, $C_{20}H_{42}$, are heated. Assume that the reaction takes place in a way such that all oxidizable material is completely oxidized and all reducible material is completely reduced.

41. The label on a bottle of Clorox indicates that it contains 5 percent sodium hypochlorite. Starting with common salt as the raw material, describe with equations how the Clorox might have been prepared.

The Noble Gases

42. Although argon is much more abundant in the atmosphere than carbon dioxide, it was not discovered until 1894. Why?

43. If XeF_2Cl_2 were prepared, would you expect it to exhibit cis-trans isomerism? Why? Would CCl_2F_2 (Freon)? Why?

44. The first synthesis of the perbromate ion, BrO_4^-, was accomplished by the reaction of xenon difluoride with bromate ion, BrO_3^-, in water. Write a balanced equation for the reaction. Identify the oxidizing and reducing agents.

*45. Write a structural formula for a xenon compound in which Xe has each of these oxidation states: (a) +2, (b) +4, (c) +6, (d) +8.

*46. At present, xenon dioxide, XeO_2, is unknown. If it were obtained, what would be a logical structure for the compound? What would be the hybridization of Xe in the proposed structure?

Groups VIA and VA: The Sulfur Family and the Nitrogen Family

24

The Sulfur Family

24•1 **Properties of the Sulfur Family**

24•1•1 Physical Properties

24•1•2 Chemical Properties

24•1•3 Oxidation–Reduction Chemistry of the Sulfur Family

24•2 **Characteristic Reactions**

24•2•1 With Metals

24•2•2 With Certain Nonmetals

24•3 **Production of the Sulfur Family Elements**

24•4 **Properties of the Binary Hydrogen Compounds**

24•5 **Sulfides, Selenides, and Tellurides**

24•6 **Metal Polysulfides**

24•7 **Oxides and Oxy-Acids of Sulfur**

24•7•1 Sulfur Dioxide

24•7•2 Sulfurous Acid

24•7•3 Sulfur Trioxide

Special Topic: Acid Rain

24•7•4 Sulfuric Acid

24•8 **Sulfates, Selenates, and Tellurates**

24•8•1 Alums

24•8•2 Thiosulfates

The Nitrogen Family

24•9 **Properties of the Nitrogen Family**

24•9•1 Physical Properties

24•9•2 Chemical Properties

24•9•3 Oxidation–Reduction Chemistry of the Nitrogen Family

24•10 **Characteristic Reactions**

24•10•1 With Nonmetals

24•10•2 With Metals

24•11 **Production of the VA Elements**

24•11•1 Nitrogen

24•11•2 Phosphorus

24•11•3 Other VA Elements

24•12 **Properties of the Binary Hydrogen Compounds**

24•12•1 Ammonia

24•13 **Oxy-Acids and Bases of the VA Elements**

24•13•1 Tendency to Act As Acid or Base

24•13•2 Nitric Acid

Special Topic: Protection of Teeth by Fluoride Ions

24•13•3 Phosphoric Acid and Phosphates

24•14 **Miscellaneous Compounds of the VA Elements**

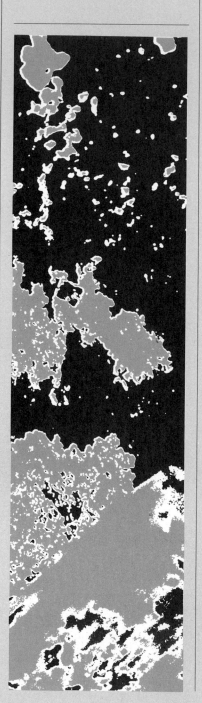

Moving from the noble gases and halogens in groups VIIIᴀ and VIIᴀ to the vertical columns of elements to the left of them in the periodic table brings us to groups VIᴀ and Vᴀ, which we will discuss in this chapter.

The elements in group VIᴀ are oxygen, sulfur, selenium, tellurium, and polonium. Each of these elements has six electrons (s^2p^4) in its outside main energy level. Oxygen is so important in the study of chemistry that it was discussed in Chapter 9. Oxygen is typical of most period 2 elements in that its physical and chemical properties are quite different from those of the other members of its group. We shall consider sulfur, selenium, and tellurium as the typical members of the **sulfur family,** with oxygen mentioned at times for comparison.

The elements in group Vᴀ, nitrogen, phosphorus, arsenic, antimony, and bismuth, are called the **nitrogen family.** These elements, each of which has five electrons (s^2p^3) in its outside main energy level, are similar to one another in some respects but are noted more for their differences. As with oxygen in group VIᴀ, nitrogen is quite different from the other members of group Vᴀ.

These two families of elements illustrate a general trend: *as the atomic number increases in an ᴀ family, the elements become more metallic in character.* In group VIᴀ, oxygen and sulfur at the top are typical nonmetals, and tellurium at the bottom is a metalloid. The stepwise change from nonmetallic to metallic character within an ᴀ family is most evident in the nitrogen family. Nitrogen and phosphorus are nonmetals, arsenic and antimony are metalloids, and bismuth is a metal.

The Sulfur Family

24•1   Properties of the Sulfur Family

24•1•1   Physical Properties Some of the physical properties of elements in the sulfur family are listed in Table 24.1.[1] As the atomic number increases, the important trends are (1) an increase in melting and boiling points, (2) an increase in atomic radius (see Figure 24.1), and (3) a decrease in ionization energy and electronegativity.

Whereas oxygen and sulfur are typical nonmetals with low electric and heat conductivities, tellurium approaches some metals in its electric conductivity. Also tellurium and one form of selenium look like metals. This metallic form of selenium has a rare property: its electric conductivity, though low, is greatly increased when light shines on it. Hence, selenium is used in instruments designed for measuring the intensity of light (even that from stars) and in automatic switches that turn lights on when the sun sets and turn them off again at daybreak.

One of the most popular photocopying devices depends on the photoconductivity of selenium (see Figure 24.2). A rotating aluminum drum

---

[1]Because of its relative rarity and radioactivity, polonium is not included here.

Table 24•1 Physical properties of the sulfur family

	Oxygen	Sulfur	Selenium	Tellurium
appearance at room temperature	colorless gas	yellow, brittle solid	red or gray solid	silver-white solid
common molecular formula	O_2	S_8	Se_8	$(Te_8?)$
melting point, °C	−218.8	115.2	221	450
boiling point, °C	−183.0	444.6	685	1,009
ionization energy, eV/atom and kJ/mole	13.6 1,310	10.4 1,000	9.8 946	9.0 866
covalent radius, Å	0.73	1.02	1.17	1.37
ionic radius (E^{2-}), Å	1.40	1.82	1.98	2.21
electronic structure	2,6	2,8,6	2,8,18,6	2,8,18,18,6
electronegativity	3.5	2.5	2.4	2.1

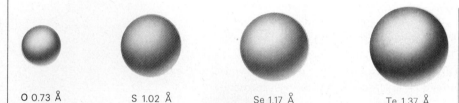

O 0.73 Å   S 1.02 Å   Se 1.17 Å   Te 1.37 Å

Figure 24•1 Relative sizes of group VIᴀ atoms, with covalent radii in angstroms.

coated with amorphous selenium is given a positive charge while protected from light. Next, the drum is exposed to light reflected from the document to be copied. Where irradiated by light, the selenium drum conducts electricity from the supporting aluminum; where in shadow, the drum remains charged positive. Negatively charged powdered pigment is sprinkled on the drum and adheres to the positively charged areas; then the drum is rolled over a positively charged sheet of paper, which attracts the (negative) powdered pigment to give an image of the original document. The last step involves heating the paper and pigment to fuse the latter to the paper.

As noted in Table 24.1, selenium may be either red or gray. This suggests that this element exists in more than one crystalline form. Similar

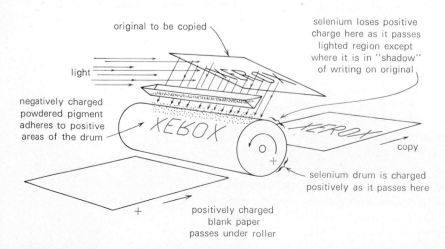

Figure 24•2 Schematic diagram of a copying machine that takes advantage of the photoconductivity of elemental selenium.

polymorphism is exhibited by sulfur, but both common crystalline forms are yellow. In addition to existing in one or more crystalline forms, sulfur, selenium, and tellurium can be prepared in the plastic or amorphous condition by suddenly freezing the hot liquid elements. When solid sulfur is formed very rapidly (for example, by pouring the boiling liquid into cold water), the sulfur molecules do not have time to orient themselves to form a well-developed crystal. As a result, the solid is a mass of tiny crystallites that have no overall pattern; that is, they are amorphous.

The crystalline form of sulfur depends on the temperature. When a solution of sulfur in toluene is evaporated to dryness, the sulfur will crystallize in a rhombic lattice if the temperature is below 95.5 °C, but it will crystallize in a monoclinic lattice if the temperature is above this. When solid sulfur in the rhombic form is raised to a temperature above 95.5 °C, the crystalline pattern slowly changes to monoclinic:

$$S_8 \text{ (rhombic)} \longrightarrow S_8 \text{ (monoclinic)} \qquad \Delta H \text{ is positive}$$

The temperature at which one crystalline form changes to another is called a **transition temperature.** The complete transformation may take days, because the molecules in the solid cannot move easily enough to reorient themselves quickly into the new crystal pattern.

The sulfur molecule, consisting of a ring of eight atoms, is pictured in Figure 24.3. This is the unit particle commonly present in the solid and liquid states, although the rings tend to break up and form chains when the molten sulfur is heated.

The existence of these chains explains the peculiar changes in the viscosity of liquid sulfur. Instead of decreasing regularly between melting and boiling temperatures, the viscosity of sulfur decreases from 115.2 °C to a minimum at about 160 °C, after which the viscosity begins to increase drastically, reaching a maximum at about 190 °C. Above 190 °C the viscosity decreases until the boiling point of 444.6 °C is reached. (See Figure 24.4.) When sulfur first melts, it can be poured easily from a beaker or test tube, but at about 190 °C it has the consistency of thick tar. Evidently, the sulfur molecules still exist as S_8 rings just above the melting point, and these rings slip and roll over one another easily (not viscous). But when the liquid sulfur is heated, the rings are broken into chains that then join, forming very long molecules (see Figure 24.5) that become entangled with one another (very viscous).[2] Above about 200 °C, these chains break up more and more, and sulfur begins to act again like a typical liquid; that is, the viscosity begins to decrease with further rise in temperature.

24•1•2   Chemical Properties A distinctive feature of the elements in group VIA is that their atoms need only two more electrons to attain the electronic configuration s^2p^6 of a noble gas. They therefore frequently react as oxidizing

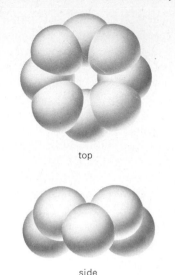

top

side

Figure 24•3 Top and side views of a model of the sulfur molecule S_8.

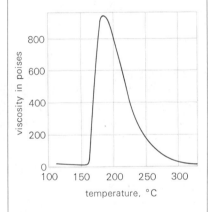

Figure 24•4 The change of the viscosity of liquid sulfur with temperature. (For reference, the viscosity of water at 20 °C is 0.01 poise.) *Source:* Graph adapted from R. F. Bacon and R. Fanelli, *J. Amer. Chem. Soc.,* 65, 642 (1943).

---

[2]This modern explanation is not unlike that given centuries ago by Lucretius (as translated by R. E. Latham): "We see that wine flows through a strainer as fast as it is poured in; but sluggish oil loiters. This, no doubt, is either because oil consists of larger atoms, or because these are more hooked and intertangled and, therefore, cannot separate as rapidly, so as to trickle through the holes one by one."

agents, achieving an oxidation state of -2. Oxygen is the strongest oxidizing agent and tellurium is the weakest. Sulfur, selenium, and tellurium can be oxidized by strong oxidizing agents—for example, oxygen or some of the halogens. When oxidized, the elements tend to be in the $+2$, $+4$, or $+6$ oxidation state; examples are $+2$, SCl_2, $SeCl_2$, $TeCl_2$; $+4$, SO_2, SeO_2, TeO_2; and $+6$, SO_3, SeO_3, TeO_3. However, other oxidation states are also known.

To metals, sulfur acts as an electron acceptor; to most nonmetals, as a donor. The fact that it is both an oxidizing and a reducing agent explains why it combines with all the elements except gold, platinum, and the noble gases. Yet sulfur is not very reactive unless heated above its melting point. Selenium and tellurium have similar chemical properties but are somewhat less reactive than sulfur.

24•1•3   Oxidation–Reduction Chemistry of the Sulfur Family The emf diagrams for the sulfur family elements, the binary acids, oxy-acids, and oxy-ions in standard acid solutions are shown in Figure 24.6. The data for sulfur and some of its oxidation states in standard base solution are also included in the figure.

The following important reactions are suggested by these diagrams:

1. Unlike some of the halogens in their positive oxidation states, none of the species listed to the left of the sulfur family diagrams can oxidize water spontaneously and evolve oxygen.

2. In acid solutions, all oxidation states of sulfur greater than -2 are easily reduced; that is, all are possible oxidizing agents. The oxidizing strength of the $+6$ state increases roughly with atomic number in the sulfur family, but the positions of selenium and tellurium are reversed. The same general trend was observed for the halogen family. However, the weakness of SO_4^{2-} as an oxidizing agent is to be contrasted with ClO_4^- in Figure 23.2. The $+6$ states of selenium and tellurium have greater oxidizing strengths than the $+4$ state, but for sulfur the opposite is true. Note that H_2SO_3 is a stronger oxidizing agent than SO_4^{2-}.

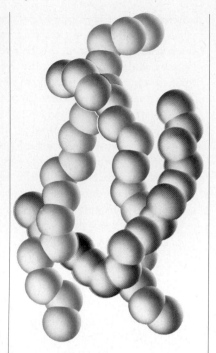

Figure 24•5 Models representing the intertwining chains of sulfur atoms in viscous, molten sulfur; the chains may contain as many as 10,000 atoms each.

Figure 24•6 Emf diagrams for oxidation-reduction couples of the sulfur family elements in $1m$ acid solutions and for sulfur in $1m$ base solutions. See caption to Figure 23.2 for an explanation of protonated versus ionic forms.

ACID SOLUTION

BASE SOLUTION

3. Solutions of H_2S, H_2Se, and H_2Te are oxidized by oxygen from the air. Although this reaction is so slow that H_2S solutions can be used as laboratory reagents, solutions exposed to air are oxidized and become turbid due to the precipitation of finely divided sulfur.

4. In acid solution, H_2S will reduce any positive oxidation state of sulfur to the element.

5. In acid solution, elemental sulfur is stable toward disproportionation. However, in basic solution sulfur does disproportionate.

Examples 24.1 and 24.2 illustrate some of these trends in behavior.

━━━━━ • Example 24•1 • ━━━━━━━━━━━━━━━━━━━━━━━━━━━━━

Using the data in Figure 24.6, show that the following reaction is expected to take place in standard acid solution:

$$2H_2S + S_2O_3{}^{2-} + 2H^+ \longrightarrow 4S + 3H_2O$$

━━━━━ • Solution • We compare the $\mathcal{E}_{red}^{\circ}$ values for the S/H_2S couple and the $S_2O_3{}^{2-}/S$ couple, and use their half-reaction equations to arrive at an ionic equation with H_2S and $S_2O_3{}^{2-}$ as reactants.

$$
\begin{array}{ll}
2(H_2S \longrightarrow 2H^+ + S + 2e^-) & \mathcal{E}_{ox}^{\circ} = -0.14\ V \\
\underline{4e^- + S_2O_3{}^{2-} + 6H^+ \longrightarrow 3H_2O + 2S} & \underline{\mathcal{E}_{red}^{\circ} = +0.50\ V} \\
2H_2S + S_2O_3{}^{2-} + 2H^+ \longrightarrow 4S + 3H_2O & \mathcal{E}_{cell}^{\circ} = +0.36\ V
\end{array}
$$

The positive emf for the net ionic equation shows that the reaction is spontaneous.

See also Exercises 7, 10, and 13 at the end of the chapter.

━━━━━ • Example 24•2 • ━━━━━━━━━━━━━━━━━━━━━━━━━━━━━

Using the data in Figure 24.6, show that oxygen is capable of oxidizing sulfur to sulfate ion in standard basic solution.

━━━━━ • Solution • As in Example 24.1, we arrive at the net ionic equation by choosing two appropriate half-reactions. We choose two that we can add to give an equation with O_2 and S as reactants and $SO_4{}^{2-}$ as a product.

The $\mathcal{E}_{red}^{\circ}$ for the $SO_4{}^{2-}/S$ couple is not given in Figure 24.6, but it can be calculated from the $SO_4{}^{2-}/SO_3{}^{2-}$ and $SO_3{}^{2-}/S$ couples:

$$\mathcal{E}_{overall}^{\circ} = \frac{2(-0.93) + 4(-0.61)}{2 + 4} = -0.72\ V$$

$$
\begin{array}{ll}
2(S + 8OH^- \longrightarrow SO_4{}^{2-} + 4H_2O + 6e^-) & \mathcal{E}_{ox}^{\circ} = +0.72\ V \\
\underline{3(O_2 + 2H_2O + 4e^- \longrightarrow 4OH^-)} & \underline{\mathcal{E}_{red}^{\circ} = +0.40\ V} \\
2S + 3O_2 + 4OH^- \longrightarrow 2H_2O + 2SO_4{}^{2-} & \mathcal{E}_{cell}^{\circ} = +1.12\ V
\end{array}
$$

The positive emf for the net ionic equation shows that oxygen can oxidize sulfur to sulfate ion in basic solution.

See also Exercises 7, 10, 11, and 13.

24•2   Characteristic Reactions

24•2•1   With Metals   The following equations represent the type of reaction that occurs between a metal and members of the sulfur family:

$$Fe + S \longrightarrow FeS$$
$$Fe + Se \longrightarrow FeSe$$
$$Fe + Te \longrightarrow FeTe$$

As a rule, sulfur reacts more energetically than selenium and tellurium and less energetically than oxygen. The heat evolved when a powdered metal is oxidized by powdered sulfur may be great enough to make the products red hot. In fact, some metals—among them copper, silver, and mercury—show a greater chemical affinity for sulfur than for oxygen.

24•2•2   With Certain Nonmetals   Sulfur reacts with hot carbon to form colorless liquid carbon disulfide, CS_2, and with hot boron to form solid boron trisulfide, B_2S_3.

Sulfur burns in oxygen and in fluorine to form gaseous oxides, SO_2 and SO_3, and a gaseous fluoride, SF_6, respectively. It reacts less violently with chlorine to form the liquid sulfur monochloride, S_2Cl_2. Sulfur hexafluoride, SF_6, has a symmetrical octahedral shape (see Figure 6.16), which is consistent with its dipole moment of 0 D. The hybridization of sulfur is presumed to be sp^3d^2. Along with the hexafluoride, a small amount of the tetrafluoride, SF_4, is also obtained. The dipole moment of SF_4 is 0.63 D. As shown in Example 6.3, in SF_4 there are five pairs of electrons, four bonding pairs and one lone pair, in the valence shell of sulfur. This indicates that the orbital geometry is trigonal bipyramidal and that the hybridization of sulfur is probably sp^3d. VSEPR theory predicts a "seesaw" shape for SF_4 [see configuration (1) in Example 6.3].

When hydrogen gas is bubbled through molten sulfur, the two elements react to form the gas hydrogen sulfide:

$$H_2 + S \longrightarrow H_2S$$

The compounds H_2Se and H_2Te can also be prepared by direct combination of the elements.

24•3   Production of the Sulfur Family Elements

The amount of sulfur present in the earth's crust is approximately 0.1 percent by weight. The relative abundance of selenium and tellurium, taken together, is less than a millionth that of sulfur.

All three elements occur naturally both in the free form and in compounds. Elemental selenium is often found mixed with sulfur. When sulfur occurs as the element, it is usually mixed with rocks and earth; such a mixture is easily separated by heating it until the sulfur melts and runs out. The element was known to the ancients as *brimstone*, the stone that burns.

Sulfur is one of the most useful of nature's raw materials. The United States is fortunate in having some of the world's richest deposits of elemental sulfur. Located on the Gulf Coast in Texas and Louisiana are huge beds of rock with veins of sulfur.

Until early in this century, most sulfur was obtained from volcanic sources. Then an American engineer, Herman Frasch, devised the ingenious method now used for mining sulfur. A hole or well is drilled down into the sulfur-containing stratum and three concentric pipes are lowered into it. The sulfur is melted in the rock bed by superheated water (the water is kept under sufficient pressure so that it can be heated to about 180 °C). Compressed air forces the water–molten sulfur mixture up to the surface of the ground, where it is separated. The sulfur is then pumped into a storage vat where it cools and freezes, becoming part of a huge block of sulfur that may be 120 m long, 60 m wide, and 30 m high.

Elemental sulfur is also a by-product in the refining of certain sulfur-containing crude oils. Sulfur can be prepared by the oxidation of sulfides or the reduction of sulfites and sulfates. Selenium and tellurium can be prepared by similar reactions. In the case of sulfur, for many years these reactions were not of industrial importance in this country because the element itself could be mined so cheaply. But today a growing amount of sulfur is being recovered from mineral sulfides and sulfates and from by-products formerly lost to the atmosphere in smokes and gases. This change is spurred by increased energy costs and the increasing cost of the Frasch process as the richest deposits of sulfur are being depleted. Greater emphasis on limiting pollution of the atmosphere by such noxious industrial products as sulfur dioxide gas (see Section 24.7.1) is also a motivating factor.

24•4 Properties of the Binary Hydrogen Compounds

Table 24.2 lists some of the important properties of the binary hydrogen compounds of the first four group VIA elements. The trend in the heats of formation is the same as the trend in the free energies of formation, so the $\Delta H_f°$ values indicate the stabilities of the compounds. Hydrogen oxide is by far the most stable, and hydrogen telluride is the least stable. Hydrogen selenide and hydrogen telluride are in fact endothermic compounds.

Key Chemical: S
sulfur

How Made
melted from underground ores; recovered from H_2S in natural gas and oil and from organic sulfides by oxidation; reduction of SO_2 recovered in smelting ores

U.S. Production; Cost
~12.5 million tons per year
~$52 to $54 per ton

Major Derivatives
89% sulfuric acid, H_2SO_4
5% chemicals

Major End Uses
60% fertilizers
20% chemicals

Other Uses
making gunpowder, insecticides, and tires

Table 24•2 Properties of hydrogen compounds of elements (E) of the sulfur family[a]

	H_2O	H_2S	H_2Se	H_2Te
melting point, °C	0	−85.5	−65.7	−49
boiling point, °C	100	−60.3	−42	−2
$\Delta H_f°$ per mole for $H_2 + E \rightarrow H_2E$, kJ	−286	−20.2	+29.7	+99.6
first ionization constant, K_{a1} $H_2E \rightleftharpoons H^+ + HE^-$	1.0×10^{-14}	1.1×10^{-7}	1.3×10^{-4}	2.3×10^{-3}

[a]The melting and boiling points were plotted in Figure 8.7.

Note the trend in acid ionization constants, K_a, for the solutions of these compounds. In contrast with the hydrogen halides, all these compounds are weak acids. The acid strength increases with increasing atomic number.

Hydrogen sulfide is an important laboratory chemical because it is used widely in qualitative analysis. It can be prepared readily by the action of acids on metallic sulfides or by the hydrolysis of thioacetamide:

$$FeS + 2HCl \longrightarrow H_2S + FeCl_2$$

$$CH_3CSNH_2 + H_2O \longrightarrow H_2S + CH_3CONH_2$$

Hydrogen sulfide is a poisonous gas. Several persons were killed recently when NaHS was inadvertently mixed with acid at a leather factory. The offensive smell of H_2S (like rotten eggs) is well remembered by all who encounter it in the laboratory. It is present in minute amounts in the fumes given off during the burning of sulfur-containing fuel oil and coal.

24•5 Sulfides, Selenides, and Tellurides

The metallic sulfides are important, well-known compounds because they occur widely as minerals and are often encountered in laboratory and industrial chemistry. The selenides and tellurides have similar formulas and structures but are relatively rare.

Aside from the metals in groups Iᴀ and IIᴀ, the metallic sulfides are characterized by their extremely slight solubility. Sulfide ions can be added in analytical or commercial separations to precipitate metal ions that form slightly soluble sulfides.

Sodium sulfide, Na_2S, is important in the organic chemical industry. It is a reducing agent and is used to prepare dyes and remove hair from hides. Cadmium sulfide, CdS, is called *cadmium yellow* and is a valuable commercial pigment. Sodium selenide, Na_2Se, is an active ingredient in certain anti-dandruff shampoos.

Nonmetallic sulfides are also well known, although they hydrolyze in contact with moist air. An exception is carbon disulfide, CS_2, which neither reacts with nor mixes with water. Carbon disulfide is used chiefly in the manufacture of rayon and in the preparation of carbon tetrachloride, CCl_4, from sulfur monochloride, S_2Cl_2:

$$2S_2Cl_2 + CS_2 \longrightarrow CCl_4 + 6S$$

24•6 Metal Polysulfides

Not only do sulfur atoms tend to bond together in elemental sulfur (as rings and chains), but they also may attach themselves to the sulfide ion to form polysulfide ions. For example,

$$\underset{\substack{\text{barium} \\ \text{sulfide}}}{BaS} + 2S \longrightarrow \underset{\substack{\text{barium} \\ \text{polysulfide}}}{BaS_3}$$

The polysulfide ions range in size from S_2^{2-} to S_6^{3-}. Crystals of the most famous polysulfide, the common iron ore *pyrite*, FeS_2, look so much like gold that they are called *fool's gold* (see Figure 24.7).

A common laboratory reagent is ammonium polysulfide, $(NH_4)_2S_x$, where the x indicates a number variable between 2 and 5. Sodium polysulfide is used to make certain synthetic rubber and is important in rayon manufacturing, metallurgy, photography, and engraving.

24•7 Oxides and Oxy-Acids of Sulfur

Of all the sulfur family oxides, only SO_2 and SO_3 have any great use. Both dissolve in water, the former yielding sulfurous acid, H_2SO_3, and the latter yielding sulfuric acid, H_2SO_4.

24•7•1 Sulfur Dioxide

Sulfur dioxide is the colorless, choking gas formed when sulfur burns in air:

$$S + O_2 \longrightarrow SO_2$$

Large quantities are produced during the roasting of sulfide ores, which is a key step in many metallurgical processes (see Section 21.8.2). Iron pyrite is an example:

$$4FeS_2 + 11O_2 \longrightarrow 2Fe_2O_3 + 8SO_2$$

The gas is used as a fungicide, a fumigant (in the form of sulfur candles), and a food preservative.

24•7•2 Sulfurous Acid

Sulfur dioxide is soluble in water, and most of the SO_2 that dissolves is solvated by water molecules. Presumably some SO_2 reacts with water to produce sulfurous acid, H_2SO_3,

$$SO_2 + H_2O \rightleftharpoons H_2SO_3 \tag{1}$$

but no compound with this molecular formula has been identified. The small amount that is assumed to be present acts as a weak acid and reacts with water to produce H_3O^+ and HSO_3^- ions:

$$H_2SO_3 + H_2O \rightleftharpoons H_3O^+ + HSO_3^- \tag{2}$$

Summing the two equilibria, (1) and (2), gives the net ionic equation:

$$\begin{aligned} SO_2 + H_2O &\rightleftharpoons H_2SO_3 \\ \underline{H_2SO_3 + H_2O} &\rightleftharpoons \underline{H_3O^+ + HSO_3^-} \\ SO_2 + 2H_2O &\rightleftharpoons H_3O^+ + HSO_3^- \end{aligned}$$

In the simplified form, using H^+ instead of H_3O^+, the net ionic equation is

$$SO_2 + H_2O \rightleftharpoons H^+ + HSO_3^-$$

Figure 24•7 A crystal of fool's gold, FeS_2, can have a color and luster so like gold that it is easy to indulge in wishful thinking on first seeing it. (Sample is by the courtesy of the Geology Department, The University of Tennessee, Knoxville.)

The concentrations of H_2SO_3, H_3O^+ (or H^+), and HSO_3^- are all quite small. That is, sulfur dioxide is the principal solute in the solution that we call sulfurous acid.

● **24•7•3 Sulfur Trioxide** Sulfur trioxide, SO_3, is formed when sulfur dioxide is heated with oxygen in the presence of a catalyst. At room temperature,

★   Special Topic ★ Special Topic ★ Special Topic ★ Special Topic ★ Special Topic ★ Special Topic ★ Special Topic ★ Special Topic   ★

Acid Rain Rainwater in equilibrium with normal amounts of carbon dioxide has a pH of 5.6. Rain or snow with a pH less than this is called *acid precipitation*.

Between 1930 and 1950 the precipitation in the eastern part of the United States became more acidic. Since the 1950s the pH has continued to fall (see Figure A) and is now near 4 in the area just east of the region bordering the Great Lakes. Similar observations have been made elsewhere. Along the southwest coast of Sweden, which is downwind from industrial and heavily populated areas in England and Germany, the average pH of rain and snow decreased from 5.0 to 4.5 in the thirteen years ending in 1970.

Of approximately 200 mountain lakes in the state of New York, only 4 percent either had a pH of less than 5.0

or had no fish during the period 1929–1937. More recently, 51 percent of these lakes have been found to have a pH below 5.0, and 90 percent of these contain no fish. In Sweden and Norway a similar change in lake pH and a similar severe depletion of fish have occurred (see Figure B). An increase in acidity that is not fatal to adult fish may be sufficient to keep eggs from hatching or to kill the newborn next generation.

There are also deleterious effects on plant life. Cells in leaves of some plants die when rainwater approaches a pH of 3. Values of pH of 2.0 to 3.1 have been measured during individual rainstorms. This rain falls directly on plants, with no possibility of partial neutralization by the soil or dilution by groundwater. The degrading of materials—metal, wood, and stone—is accelerated (see

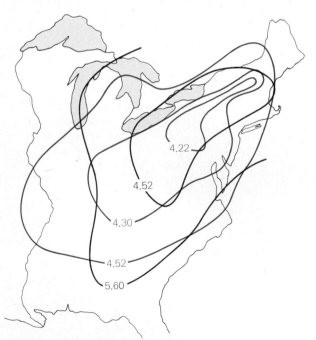

Figure A. The average pH of precipitation east of the Mississippi River. Values along the black lines are for 1955–1956, and values along the colored lines are for 1972–1973.
Source: G. E. Likens, "Acid Precipitation," *Chem. Eng. News*, November 22, 1976, p. 31.

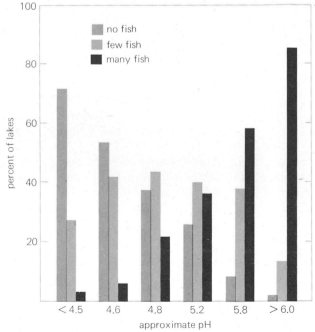

Figure B. Fish population in 1,679 Norwegian lakes as a function of the pH of the water.
Source: Impact of Acid Precipitation on Forest and Freshwater Ecosystems in Norway, SNSF Project, Research Report FR 6/76, from G. E. Likens; see Figure A.

SO_3 is a colorless liquid. Its important use is in the manufacture of sulfuric acid, as will be discussed in the next section.

24•7•4   Sulfuric Acid   Sulfuric acid is most commonly available in the laboratory in a highly concentrated solution that contains about 98 percent H_2SO_4 and 2 percent H_2O. This concentrated acid is a colorless, oily liquid with a density of 1.8 g/mL.

★ Special Topic ★ Special Topic ★ Special Topic ★ Special Topic ★ Special Topic ★ Special Topic ★ Special Topic ★ Special Topic ★

Figure C). Priceless structures such as the Parthenon and the Colosseum have already been affected.

The sources of the increased acidity of rain in recent years are not completely known. Sulfuric and nitric acids from oxides of sulfur and nitrogen are the main culprits. Waste gases from the smelting of metal sulfides are a large and concentrated source of possible air pollution. An even greater source, though dilute, is the combustion of coal, fuel oil, and gasoline by industries, automobiles, and households. Sulfur compounds are found in most coal and oil, sometimes in large amounts. During combustion, the SO_2 formed is emitted as one of the flue gases. There have been reported in this country and elsewhere "killer fogs" produced by an unfortunate combination of weather conditions and large-scale fuel use. In these fogs, sulfur dioxide is the obvious villain that is deadly for persons with respiratory weaknesses.

The amount of nitric acid in rainwater has increased in recent years more than that of sulfuric acid. Much of this is attributed to automobile exhausts. It is ironic that the first catalytic devices installed in large numbers on cars to decrease the amounts of nitrogen oxides formed were found to catalyze the conversion of the weakly acidic SO_2 to the strongly acidic SO_3. The most effective way to prevent pollution by oxides of sulfur is to remove sulfur compounds from petroleum or coal before combustion.

A somewhat exotic process for removing sulfur from coal starts with the solution of the coal in molten iron. Gases are given off that can be used for fuel; the sulfur reacts first with the iron and then passes into a lime slag that floats on top of the iron and is removed.

To remove sulfur dioxide from coal and petroleum furnace flue gases, several processes have been developed, including the following. (1) When flue gases are passed through magnesium oxide, magnesium sulfite is formed. This compound can be decomposed by heat to give magnesium oxide again and concentrated sulfur dioxide gas: $MgSO_3 \longrightarrow MgO + SO_2$. (2) At least three different processes are being used that trap the sulfur in the form of calcium sulfite or calcium sulfate. The gases react with either calcium carbonate or calcium hydroxide. In two of the processes, $CaSO_4 \cdot 2H_2O$ is produced, which is a raw material for making plaster; and in the third, a final reduction produces sulfur. In the most widely used process, a wet limestone–sulfate sludge is produced in quantities so huge that they are used as landfill for highway construction. About 6 million tons of dry sludge are produced a year, and this may increase sixfold by 1987.

The fact that a variety of methods are being developed is a result of both the differences in the waste gases produced under different conditions and the different materials available in different locations that can be economically used.

Acid rain may be the single most serious environmental effect of present human activities, because it is associated with the production of energy for so many purposes. The solutions to such problems must be sought on the broadest scale. One result of Environmental Protection Agency regulations has been a great increase in efforts to identify and control pollutants. The costs of pollution control have added appreciably to the prices of both energy production and manufactured articles.

Figure C. Exterior stone sculpture at Herten Castle, built in 1702 in Westfalia, Germany. The left photograph, taken in 1908, shows the statue after two centuries of exposure. The right photograph, taken in 1969, shows the effect of only 60 more years of exposure to an atmosphere increasingly affected by human activities.
Source: From E. M. Winkler, *Stone: Properties, Durability in Man's Environment,* Springer-Verlag, New York, 1973. Reproduced with permission.

• *Important Properties of Sulfuric Acid* • Sulfuric acid is a strong acid, because it readily donates a proton to water to form the hydronium ion, H_3O^+, and the bisulfate ion, HSO_4^-. The HSO_4^- ion has only a moderate tendency to donate a proton to water molecules (see Table 16.1).

Sulfuric acid is the cheapest acid available for dissolving metals and metal oxides, neutralizing bases, and cleaning corroded metal surfaces, although for the last use, hydrochloric acid may be less expensive overall because it is more easily reconcentrated and used again.

Although SO_4^{2-} is only a weak oxidizing agent in dilute acid solution (see Figure 24.6), hot and concentrated H_2SO_4 is a strong oxidizing agent. Even such inactive substances as copper and carbon are oxidized:

$$Cu + 2H_2SO_4 \xrightarrow{heat} CuSO_4 + SO_2 + 2H_2O$$

$$C + 2H_2SO_4 \xrightarrow{heat} CO_2 + 2SO_2 + 2H_2O$$

The reaction of sulfuric acid with water is extremely exothermic. Because of this affinity for water, concentrated sulfuric acid can be used to remove water from other substances and even to remove hydrogen and oxygen atoms from molecules that do not contain H_2O as such. It removes most of the water vapor from a wet gas, such as humid air, and decomposes some molecules that contain firmly bound hydrogen and oxygen atoms. In the case of ordinary sugar (sucrose), the residue appears charred, with black carbon being formed:

$$C_{12}H_{22}O_{11} + 11H_2SO_4 \longrightarrow 12C + 11H_2SO_4 \cdot H_2O$$

Although it is decomposed at very high temperatures, the stability of H_2SO_4 is in sharp contrast with the instability of H_2SO_3. The high boiling point of sulfuric acid is one of its very important properties, because it makes possible the formation of more volatile acids when sulfuric acid is heated with certain salts:

$$2NaCl + H_2SO_4 \xrightarrow{heat} Na_2SO_4 + 2HCl\uparrow$$

The largest commercial use of sulfuric acid is in the making of phosphoric acid, H_3PO_4, from phosphate rock (see Section 24.13.3).

• *Industrial Production of Sulfuric Acid* • Sulfuric acid is the most important of all synthetic chemicals. Like steel, it is manufactured on such a huge scale and is essential to so many industries that its production is taken as an index of the economic vitality of industrial countries. Even prior to 1800 it was sold by the ton, and today its production is 60 percent higher than that of the next most prevalent chemical, lime (calcium oxide).

Almost all sulfuric acid is made by the **contact process,** so called because the key step occurs when two reactants are in contact with a solid catalyst. This reaction is between sulfur dioxide and oxygen in the presence of platinum or vanadium pentoxide, V_2O_5, to yield sulfur trioxide. Vanadium

---

Key Chemical: H_2SO_4
sulfuric acid

How Made

$$S + O_2 \rightarrow SO_2$$

$$2SO_2 + O_2 \xrightarrow{Pt\ or\ V_2O_5} 2SO_3$$

$$SO_3 + H_2O \rightarrow H_2SO_4$$

U.S. Production; Cost
~40 million tons per year
~$40 per ton, depending on source

Major End Uses
65% fertilizers
5% oil refining
5% metals refining
5% making other chemicals

Other Uses
manufacture of explosives, dyestuffs, parchment paper, and detergents; the electrolyte in automobile batteries (33.5% H_2SO_4, specific gravity 1.250, see Figure 1.5)

pentoxide is commonly used, because platinum is easily poisoned by impurities in the sulfur dioxide.[3]

About 88 percent of the SO_2 used is produced by burning elemental sulfur, and the rest is a by-product of the smelting of metal sulfide ores. Both of these SO_2-producing reactions are followed by the reactions

$$2SO_2 + O_2 \longrightarrow 2SO_3 \qquad \text{(does not occur readily)} \qquad (3)$$
$$SO_3 + H_2O \longrightarrow H_2SO_4 \qquad \text{(occurs readily)} \qquad (4)$$

Reaction (3) occurs readily in the presence of a catalyst, and the control of this reaction is essential for the efficiency of the overall process. In order to obtain the maximum yield, the effects of both temperature and pressure on the following equilibrium have been investigated:

$$2SO_2(g) + O_2(g) \rightleftharpoons 2SO_3(g) \qquad \Delta H = -196 \text{ kJ}$$

An increased yield of sulfur trioxide is favored by a low temperature and a high pressure.

The equilibrium constant for the reaction is

$$K_c = \frac{[SO_3]^2}{[SO_2]^2[O_2]} \quad \text{or} \quad K_p = \frac{(p_{SO_3})^2}{(p_{SO_2})^2(p_{O_2})}$$

The effect of temperature on this equilibrium is shown in Figure 24.8. As indicated by this graph, the conversion of SO_2 to SO_3 is practically complete at equilibrium at temperatures of 700 K and below. However, at these low temperatures the reaction is so slow that SO_3 is not formed within a reasonable time. Compromise conditions are used. In the first part of the catalytic reactor the temperature is allowed to rise to around 875 K, at which temperature the mixture rapidly attains about 70 percent conversion to SO_3. Then the mixture is cooled to around 700 K, at which the final conversion approaches 98 percent.

After a single pass through the catalytic chamber, the exit gases contain up to about 2,000 ppm (0.2 percent) of SO_2. To reduce the SO_2 to less than this concentration costs more than the recovered gas is worth. However, to meet EPA standards for waste gas, SO_3 is removed and the exit gases are sent through another catalytic chamber before they are discharged to the atmosphere with only 100–350 ppm of SO_2.

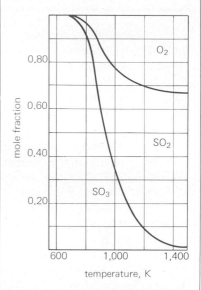

Figure 24•8 Equilibrium composition for a stoichiometric mixture for the reaction $2SO_2(g) + O_2(g) \rightleftharpoons 2SO_3(g)$ **at 1.00 atm pressure.**
Source: W. H. Evans and D. D. Wagman, *J. Research National Bureau of Standards,* 49 (3), 141 (1952).

• Example 24•3 •

From the data in Figure 24.8, calculate the equilibrium constant, K_p, for the system $2SO_2(g) + O_2(g) \rightleftharpoons 2SO_3(g)$ at 1,000 K.

• **Solution** • Reading the graph, the mole fraction of SO_3 is 0.35 at 1,000 K; the mole fraction of O_2 is $1.00 - 0.78 = 0.22$; the mole fraction of

---

[3]In an older process, the *lead chamber process,* the reaction between the same two reactants is catalyzed by the gas nitric oxide. The main steps were given in the discussion of intermediate compounds in Section 14.2.2. At one time this process was the leading method, but it was abandoned as the old plants wore out, and it now accounts for very little of the U.S. production.

SO_2 is $0.78 - 0.35 = 0.43$. (Because of the stoichiometry, we know that the mole fraction of SO_2 must be twice that of O_2, but this graph cannot be read very precisely.) The pressure of SO_3 is $(0.35)(1.00 \text{ atm}) = 0.35$ atm, the pressure of O_2 is $(0.22)(1.00 \text{ atm}) = 0.22$ atm, and the pressure of SO_2 is $(0.43)(1.00 \text{ atm}) = 0.43$ atm.

$$K_p = \frac{(p_{SO_3})^2}{(p_{SO_2})^2(p_{O_2})} = \frac{(0.35)^2}{(0.43)^2(0.22)} = 3.0$$

See also Exercise 24.

Because the reaction to the right involves the combining of three molecules to make two, the formation of SO_3 is favored by high pressure. However, the slight increase in the yield of SO_3 does not justify the increased cost of high-pressure equipment.

Sulfur trioxide is absorbed much more efficiently by concentrated sulfuric acid than by water or by dilute solutions. Therefore, the trioxide gas is passed through concentrated 98 percent H_2SO_4. As the gas is absorbed, it reacts with the small amount of water present to form 100 percent H_2SO_4. The 100 percent acid combines with more sulfur trioxide to form pyrosulfuric acid, $H_2S_2O_7$, or *oleum* as it is called commercially:

$$H_2SO_4 + SO_3 \longrightarrow H_2S_2O_7$$
$$\text{pyrosulfuric}$$
$$\text{acid}$$

This highly concentrated product, which can be diluted to any desired strength by adding it to water, is the form in which much commercial acid is shipped.

Direct pollution by a sulfuric acid plant should be minimal. Rather than polluting the environment, sulfuric acid manufacture is becoming more and more a way of preventing the harmful environmental effects of SO_2 that were described in the special topic on acid rain. Ore smelters, fuel gas works, and oil refineries that are forced to trap their SO_2 and H_2S emissions have constructed plants to produce H_2SO_4.

24•8 Sulfates, Selenates, and Tellurates

In compounds with an oxidation state of $+6$, only the sulfates are common, but selenates and tellurates are known. Because H_2SO_4 and H_2SeO_4 are diprotic acids, there are normal salts—for example, K_2SO_4 and K_2SeO_4—and acid salts—for example, $NaHSO_4$ and $NaHSeO_4$.

Although most sulfates are soluble in water, those of Ca, Sr, and Ba in group IIA are only slightly soluble. Barium sulfate, $BaSO_4$, is used as a filler to give glazed paper a brilliant whiteness and the desired body. Calcium sulfate forms two familiar hydrates: gypsum, $CaSO_4 \cdot 2H_2O$, and plaster of Paris, $(CaSO_4)_2 \cdot H_2O$. Plaster of Paris, the fundamental ingredient of plaster and stucco, is made by heating gypsum:

$$2CaSO_4 \cdot 2H_2O \xrightarrow{\text{heat}} (CaSO_4)_2 \cdot H_2O + 3H_2O$$
$$\text{gypsum} \qquad\qquad \text{plaster of Paris}$$

When the plaster of Paris is mixed with water at room temperature, it slowly picks up enough water of crystallization to turn it back into gypsum. This is the process involved in the setting of plaster.

Among the other useful sulfates are $Al_2(SO_4)_3$ (used for water purification, sizing paper, mordant in dyeing), K_2SO_4 (fertilizer, mild purgative), Na_2SO_4 and $MgSO_4$ (drying agents), and $(NH_4)_2SO_4$ (fertilizer). Some other important hydrates are $MgSO_4 \cdot 7H_2O$ (epsom salts, medicine, dyeing, finishing cotton goods), $CuSO_4 \cdot 5H_2O$ (dyeing, printing, germicides and insecticides, electric batteries), and $Na_2SO_4 \cdot 10H_2O$ (purgative, diuretic, textiles).

24•8•1 **Alums** **Alums** are hydrated *double salts* in which two $SO_4{}^{2-}$ anions are combined with one singly charged cation, one triply charged cation, and twelve molecules of water of crystallization. The general formula can be written as follows:

$$M^+M^{3+}(SO_4)_2 \cdot 12H_2O$$

M^+ can be Na^+, K^+, Tl^+, $NH_4{}^+$, Ag^+, or certain other $+1$ ions, and M^{3+} can be Al^{3+}, Fe^{3+}, Cr^{3+}, Mn^{3+}, or certain others. The crystals of one alum are usually isomorphous with the crystals of the others.

Among the most common alums are the aluminum salts (hence the name *alum*). $NaAl(SO_4)_2 \cdot 12H_2O$ is used in a certain type of baking powder, and $KAl(SO_4)_2 \cdot 12H_2O$ is used in water purification and in fire extinguishers. The latter compound is the common alum of commerce.

24•8•2 **Thiosulfates** In a number of compounds a sulfur atom is considered to be substituted for an oxygen atom. In naming these compounds, the prefix *thio-* (Greek *theion*, brimstone) is often used.

Sulfur reacts with the sulfite ion to form a thiosulfate ion:

$$S + SO_3{}^{2-} \longrightarrow S_2O_3{}^{2-}$$

Sodium thiosulfate pentahydrate, $Na_2S_2O_3 \cdot 5H_2O$, is the well-known "hypo" used by photographers to dissolve silver bromide in developing film.

The Nitrogen Family

24•9 **Properties of
the Nitrogen Family**

24•9•1 **Physical Properties** Some of the important physical properties of the nitrogen family are listed in Table 24.3. As the atomic number increases, the important trends to note are (1) an increase in atomic radius (see Figure 24.9) and (2) decreases in ionization energy and electronegativity. The trend from metal to nonmetal is quite evident. Measurements of the electric conductivity (not included in the table) show that it increases from nitrogen to bismuth.

Table 24.3 Physical properties of the nitrogen family

	Nitrogen	Phosphorus	Arsenic	Antimony	Bismuth
appearance at room temperature	colorless gas	waxy-white, red (violet), or black solid	steel-gray solid	blue-white solid, metallic luster	pinkish-white solid, metallic luster
common molecular formula	N_2	P_4	As_4	Sb	Bi
melting point, °C	−210	44[a] 597[b]	817 (28 atm)	630	272
boiling point, °C	−196	280 416[c]	612 (sublimes)	1,635	1,579
ionization energy, eV/atom and kJ/mole	14.5 1,400	10.5 1,010	9.8 944	8.6 832	7.3 703
covalent radius, Å	0.75	1.10	1.22	1.43	1.52
ionic radius (E^{3-}), Å	1.71	2.12	2.22	2.45	
ionic radius (E^{5+}), Å	0.11	0.34	0.47	0.62	0.74
electronic structure	2,5	2,8,5	2,8,18,5	2,8,18,18,5	2,8,18,32,18,5
electronegativity	3.0	2.1	2.0	1.9	1.9

[a] White phosphorus at 1 atm.

[b] Red phosphorus at 43 atm.

[c] Red phosphorus sublimes.

Both antimony and bismuth have the luster of metals on freshly broken surfaces.

All the elements except nitrogen are solids at room temperature. The elements phosphorus, arsenic, and antimony each have at least two polymorphic modifications—a nonmetallic form of low density and a dense, more closely packed metallic form.

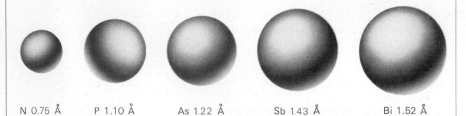

N 0.75 Å   P 1.10 Å   As 1.22 Å   Sb 1.43 Å   Bi 1.52 Å

Figure 24.9 Relative sizes of group VA atoms, with covalent radii in angstroms.

Note in Table 24.3 the difference in the melting points for the white and red forms of phosphorus. Even though these temperatures are not determined at the same pressure, most of the difference is due to the structures of the two. White phosphorus is made up of discrete P_4 molecules attracted to one another by weak van der Waals forces, whereas the red (or violet form) is crystallized in layers of tightly bound atoms (see Figure 24.10). White phosphorus ignites spontaneously when exposed to air at room temperature, is soluble in carbon disulfide, and is poisonous; red phosphorus has none of these properties.

The behavior of arsenic is typical of substances (for example, CO_2) that cannot exist as liquids at atmospheric pressure. When heated, arsenic does not melt; instead, it passes directly from the solid state into the vapor state (sublimes). If heated under pressure, it can exist in the liquid state; the melting point of 817 °C given in Table 24.3 is not comparable with the other data there because it is determined under quite a high pressure, 28 atm.

The formulas N_2, P_4, As_4, Sb, and Bi are informative. Only the metalloids and the nonmetals tend to form discrete polyatomic molecules.[4] The symbols Sb and Bi indicate that these elements crystallize like metals, with the single atom acting as an independent unit. Except for nitrogen, for simplicity's sake we will write only the monatomic symbols in our chemical descriptions of group VA elements.

24•9•2 Chemical Properties The group VA elements may act as oxidizing and reducing agents. When they react as oxidizing agents, they achieve oxidation states of -1, -2, and -3. When they react as reducing agents, oxidation states of $+1$, $+2$, $+3$, $+4$, and $+5$ are obtained. The most common oxidation states of the family members are -3, $+3$, and $+5$.

Perhaps the most striking chemical property of the nitrogen family is the inactivity of elemental nitrogen. Presumably it resists combination with other atoms because of the great affinity that one nitrogen atom has for another. In molecules of elemental nitrogen, the two nitrogen atoms share three pairs of electrons in a triple bond, $N \equiv N$. As shown in Table 13.2, the $\Delta H^{\circ}_{\text{dis}}$ of 945.3 kJ/mole of nitrogen is very large relative to that of other molecules.

The inactivity of nitrogen is apparent in many common processes. In the changes involved in combustion, fermentation, decay, and the respiration of animals, it is the oxygen of the air, not the nitrogen, that participates.

In contrast with the inactivity of nitrogen, phosphorus is very active. The other family members, As, Sb, and Bi, are somewhat less active than P, but they are still considerably more active than N_2.

24•9•3 Oxidation–Reduction Chemistry of the Nitrogen Family In Figure 24.11 the emf diagrams for various couples of nitrogen family elements in standard acid solutions reveal some of the characteristics of their oxidation–reduction chemistry:

1. Several of the species—NO, N_2O, and Bi_2O_5, for example—can oxidize water spontaneously to evolve oxygen.

2. Any compound in which nitrogen is in a positive oxidation state is a relatively strong oxidizing agent.

3. The elemental form is stable for all the elements except phosphorus.

4. Any phosphorus compound (except for oxidation number $+5$, of course) is a relatively good reducing agent.

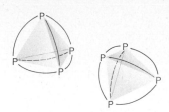

(a) white phosphorus, P_4

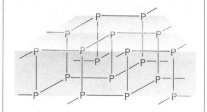

(b) red phosphorus, P

Figure 24•10 Structures of white and red phosphorus. (a) White phosphorus is composed of individual P_4 molecular units. The six covalent bonds arise from the interaction of $3p$ orbitals. The bonding orbitals are bent outward from the lines joining the centers of each pair of atoms. (b) Red phosphorus crystallizes in layers. As in white phosphorus, an individual atom forms three bonds with neighbors. The portion of a layer shown is two atoms thick.

---

[4] There are three solid nonmetals that do not form simple polyatomic molecules under ordinary conditions but rather crystallize in indefinitely large polyatomic crystals. Their symbols indicate independent atoms—boron, B, carbon, C, and silicon, Si.

ACID SOLUTION

$$NO_3^- \xrightarrow{+0.80} N_2O_4 \xrightarrow{+1.07} HNO_2 \xrightarrow{+1.00} NO \xrightarrow{+1.58} N_2O \xrightarrow{+1.89} N_2 \xrightarrow{-1.86} NH_3OH^+ \xrightarrow{+1.42} N_2H_5^+ \xrightarrow{+1.28} NH_4^+$$

$$HN_3 \quad -3.09 \nearrow \quad \searrow -1.25$$

$$N_2O_4 \updownarrow \; 2NO_2$$

$$H_3PO_4 \xrightarrow{-0.96} H_4P_2O_6 \xrightarrow{+0.40} H_3PO_3 \xrightarrow{-0.50} H_3PO_2 \xrightarrow{-0.51} P \xrightarrow{-0.063} PH_3$$

$$H_3AsO_4 \xrightarrow{+0.56} HAsO_2 \xrightarrow{+0.25} As \xrightarrow{-0.61} AsH_3$$

$$Sb_2O_5 \xrightarrow{+0.58} SbO^+ \xrightarrow{+0.21} Sb \xrightarrow{-0.51} SbH_3$$

$$Bi_2O_5 \xrightarrow{+1.59} BiO^+ \xrightarrow{+0.32} Bi \xrightarrow{-0.80} BiH_3$$

$$O_2 \xrightarrow{+1.23} H_2O$$

5.   The +3 oxidation state for arsenic, antimony, and bismuth is relatively stable.

Examples 24.4 and 24.5 illustrate some reactions that can be predicted from Figure 24.11.

Figure 24•11 Emf diagrams for oxidation–reduction couples of the nitrogen family in $1m$ acid solutions. See caption to Figure 23.2 for an explanation of protonated versus ionic forms.

━━━ • **Example 24•4** • ━━━

Use the data in Figure 24.11 to show which one of the following acids tends to disproportionate: H_3PO_4, $H_4P_2O_6$, H_3PO_3, or H_3PO_2. Calculate $\mathscr{E}^\circ_{cell}$ for a possible disproportionation reaction, and write a balanced equation.

━━━ • **Solution** • Because the $H_3PO_4/H_4P_2O_6$ couple has a negative potential whereas that for $H_4P_2O_6/H_3PO_3$ is positive, $H_4P_2O_6$ tends to disproportionate. An emf diagram for a possible disproportionation reaction is

$$H_3PO_4 \xleftarrow{+0.96} H_4P_2O_6 \xrightarrow{+0.40} H_3PO_3$$

The cell potential for these two couples is

$$+0.96 + 0.40 = +1.36 \text{ V}$$

The balanced equation for the disproportionation is

$$H_2O + H_4P_2O_6 \longrightarrow H_3PO_4 + H_3PO_3$$

See also Exercises 33, 34, 37, and 38.

━━━ • **Example 24•5** • ━━━

Nitrogen tetroxide, N_2O_4, was used as the oxidizer in the liquid propulsion system of the first lunar landing craft. One of the fuel components was hydrazine, N_2H_4. Use the data in Figure 24.11 to show that the following reaction is expected to take place in standard acid solution:

$$N_2O_4 + 2N_2H_5^+ \longrightarrow 3N_2 + 2H_3O^+ + 2H_2O$$

• Solution • We wish to compare the $\mathscr{E}_{red}^\circ$ values for the N_2O_4/N_2 couple and the $N_2/N_2H_5^+$ couple, and use their half-reactions to arrive at $\mathscr{E}_{cell}^\circ$. The $\mathscr{E}_{red}^\circ$ values for these couples are not shown in Figure 24.11, but they are easily calculated:

$$\mathscr{E}^\circ N_2O_4/N_2 = \frac{1.07 + 1.00 + 1.58 + 1.89}{4} = +1.38 \text{ V}$$

$$\mathscr{E}^\circ N_2/N_2H_5^+ = \frac{-1.86 + 1.42}{2} = \frac{-0.44}{2} = -0.22 \text{ V}$$

The emf for the diagram may be written as

$$N_2O_4 \xrightarrow{+1.38} N_2 \xleftarrow{+0.22} N_2H_5^+$$

The $\mathscr{E}_{cell}^\circ$ of $+1.38 + 0.22 = 1.60$ V indicates that the reaction is spontaneous.

See also Exercises 38 and 39.

24•10   Characteristic Reactions

24•10•1 With Nonmetals Although nitrogen is relatively inactive, under extreme temperatures and pressures or in the presence of catalysts, nitrogen does react with other elements. For example, nitrogen and oxygen combine when a high-voltage spark (or a bolt of lightning; see Figure 24.12) passes through a mixture of the two gases:

$$N_2 + O_2 \longrightarrow 2NO$$

The nitric oxide then reacts with more oxygen from the air to form nitrogen dioxide, NO_2:

$$2NO + O_2 \longrightarrow 2NO_2$$

The nitrogen dioxide produced during lightning storms dissolves in rainwater, forming a very dilute solution of nitric and nitrous acids:

$$2NO_2 + H_2O \longrightarrow HNO_3 + HNO_2$$

It is estimated that a region with moderate rainfall receives 2 to 3 kg of nitrogen (as HNO_3 and HNO_2) per acre per year. In this way a considerable amount of the inactive elemental nitrogen of the air is converted to nitrogen compounds and is deposited in the soil for use by plants as food (see Figure 24.12). This is one of nature's *nitrogen fixation processes*. **Nitrogen fixation** is any process in which elemental nitrogen reacts to form a compound. The direct union of nitrogen and hydrogen to form ammonia is the most widely used method of artificial nitrogen fixation (see Section 24.12.1).

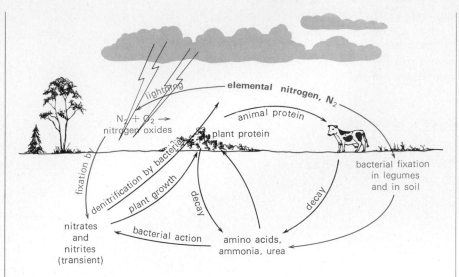

Figure 24•12 The nitrogen cycle in nature. Fixation steps are shown in color.

In contrast with nitrogen, phosphorus burns readily in air to form either phosphorus(III) or phosphorus(V) oxide, depending on the availability of oxygen:

$$4P + 3O_2 \longrightarrow \quad P_4O_6 \qquad\qquad 4P + 5O_2 \longrightarrow \quad P_4O_{10}$$

phosphorus(III)   phosphorus(V)

oxide   oxide

The structures of the two oxides are shown in Figure 24.13.

Arsenic, antimony, and bismuth are not affected by oxygen at ordinary temperatures. At elevated temperatures, however, each burns to an oxide with the empirical formula M_2O_3 (M = As, Sb, Bi):

$$4M + 3O_2 \longrightarrow 2M_2O_3$$

Among the halogens, only fluorine reacts directly with nitrogen to produce the trifluoride:

$$N_2 + 3F_2 \longrightarrow 2NF_3$$

All the halogens react directly with the other group VA elements to produce either the trihalides or the pentahalides. Examples are

$$2As + 5F_2 \longrightarrow 2AsF_5$$
$$2P + 3I_2 \longrightarrow 2PI_3 \qquad (PI_5 \text{ is unknown})$$

The molecular shapes of the trihalides and pentahalides are trigonal pyramidal and trigonal bipyramidal, respectively (see PCl_5 in Figure 6.15).

24•10•2 With Metals The group VA elements do not react appreciably with metals. However, with some of the very active metals, nitrogen does form ionic nitrides. For example,

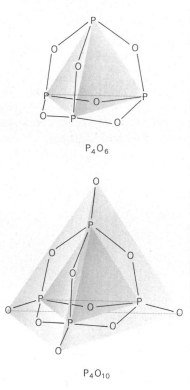

P_4O_6

P_4O_{10}

Figure 24•13 In P_4O_6, phosphorus(III) oxide, each phosphorus atom is bonded to three oxygen atoms; in P_4O_{10}, phosphorus(V) oxide, to four. The four additional oxygen atoms bonded to phosphorus in P_4O_{10} are held by coordinate covalent bonds.

$$3Ca + N_2 \longrightarrow Ca_3N_2$$
calcium
nitride

24•11 Production of the VA Elements

Elemental nitrogen, N_2, makes up 78 percent by volume of the atmosphere, and nitrogen compounds (especially proteins) are constituents of all living things. However, the most abundant element in the nitrogen family is phosphorus. It ranks tenth among the elements in the earth's crust—0.12 percent by weight. The other elements are not abundant on a weight basis, but they are not rare by any means. Only phosphorus is so active that it is not found in nature in the elemental form. Arsenic, antimony, and bismuth are found in nature in both elemental and combined forms and are easily prepared from their compounds.

24•11•1 Nitrogen Nitrogen, like oxygen, is produced most easily by separating it from the atmosphere (see Section 9.2.3).

Ammonium nitrite can be used as a laboratory source of nitrogen gas. This compound is so unstable that it is best made as needed from two more stable compounds by the following process:

- *Step 1*

$$NaNO_2 + NH_4Cl \xrightarrow[\text{water}]{\text{in}} NH_4NO_2 + NaCl$$

- *Step 2*

$$NH_4NO_2 \xrightarrow[\text{solution}]{\text{gently heat}} 2H_2O + N_2$$

The largest use for nitrogen is in the production of ammonia, NH_3. Smaller amounts are used in the hardening of steel and in chambers where an inactive atmosphere is desired. Such chambers range from large rooms, to laboratory hoods, to electric light bulbs. Liquid nitrogen, which boils at $-196\ °C$, is often used in a Dewar vessel in the laboratory to maintain a low temperature.

24•11•2 Phosphorus Phosphorus is produced on a large scale by the reaction of a phosphate mineral, sand, and coke. The process can be summarized with the following equations:

$$2Ca_3(PO_4)_2 + 6SiO_2 \xrightarrow[\text{hot}]{\text{very}} 6CaSiO_3 + P_4O_{10}$$
calcium   silica   calcium   phosphorus(V)
phosphate (sand)   silicate   oxide

then

$$P_4O_{10} + 10C \xrightarrow[\text{hot}]{\text{very}} 10CO + 4P$$

Key Chemical: P
phosphorus

How Made

$$Ca_3(PO_4)_2 + 3SiO_2 + 5C$$
$$\xrightarrow[\text{furnace}]{\text{electric}}$$
$$2P + 3CaSiO_3 + 5CO$$

U.S. Production; Cost
~800 million lb per year
~50¢ to 60¢ per pound

Major Derivatives
85% phosphoric acid and derivatives
10% P_2S_5, PCl_3, $POCl_3$, and their derivatives

Major End Uses
45% detergents
15% foods and beverages
10% metal treatment

As the P_4O_{10} forms, its vapor rises through the mixture and reacts with hot carbon to produce carbon monoxide and phosphorus. The phosphorus is separated from the gaseous CO by running the mixture through cool pipes in which liquid phosphorus condenses.

Elemental phosphorus is used in the manufacture of certain widely used bronze alloys, in the production of phosphoric acid, H_3PO_4, and in the manufacture of matches (see Figure 24.14), smoke bomb mixes, tracer bullets, and pest poisons.

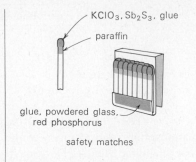

safety matches

24•11•3 Other VA Elements

The three nitrogen family elements arsenic, antimony, and bismuth are used mainly mixed with metals to make alloys with certain desirable properties. A trace of arsenic is added to lead shot to make the lead harder and its surface tension higher.[5] Antimony is used in alloys to dilute more expensive metals, such as lead and tin. Antimony alloys do not have great tensile strength, but they can be used for cheap castings such as white metal (small toys and decorations). Babbitt, an alloy of antimony, is used in bearings.

strike anywhere matches

Figure 24•14 Composition of two types of matches.

Bismuth and antimony alloys tend to expand when they solidify.[6] Hence, they are useful in making good castings. The expanding metal fills even the finest details of the mold. Bismuth is a component of some alloys with very low melting points that are used as safety plugs in steam boilers and in automatic fire-control sprinkler systems.

24•12 Properties of the Binary Hydrogen Compounds

We observed in Table 24.3 that the physical properties of the nitrogen family elements change from those characteristic of nonmetals to those characteristic of metals as one proceeds down the family. The chemical properties of the group VA compounds also show this important trend.

The tendency to be in a negative oxidation state is characteristic of nonmetals. In the nitrogen family, nitrogen has the greatest tendency to have an oxidation state of -3, but bismuth does not form stable compounds in which its oxidation state is -3. The binary hydrogen compounds illustrate this trend very well, as shown by the data in Table 24.4. Only the compound ammonia, NH_3, is stable enough to be made by the direct union of the elements. All of the other four substances are endothermic compounds. They are unstable, flammable, and extremely poisonous. Phosphine, PH_3, has the putrid odor of a mixture of rotten fish and garlic.

In Section 24.4 it was noted that, in contrast with the hydrogen halides, the binary hydrogen compounds of the group VIA elements are all weak acids. Moving from group VIA to VA, a further decrease in acidity is observed. In fact, as we learned in earlier chapters, NH_3 is a weak base in water solution. As with the group VIA binary hydrogen compounds, there is

---

[5] Lead shots are formed by allowing tiny drops of molten lead to fall in air; the greater the surface tension of the liquid, the more nearly spherical the droplet becomes before it solidifies.

[6] The water–ice transformation is the most familiar example of this rare phenomenon.

Table 24.4 Hydrogen compounds of nitrogen family[a]

Name	Formula	mp, °C	bp, °C	$\Delta H_f°$ per mole for $E + \frac{3}{2}H_2 \rightarrow EH_3$, kJ	Remarks
ammonia	NH_3	−77.8	−33.4	−46.1	stable in air; decomposes when strongly heated
phosphine	PH_3	−138.8	−87.8	+23.0	common impure form bursts into flame in air; easily decomposed by heat
arsine	AsH_3	−116.9	−62.5	+66.4	easily decomposed by heat
stibine	SbH_3	−91.5	−18.4	+145.1	decomposes explosively when heated
bismuthine	BiH_3	(?)	16.8	+277.8	decomposes spontaneously at room temperature

[a]The melting and boiling points were plotted in Figure 8.7.

an increase in acid strength (or decrease in base strength) with increasing atomic number in the nitrogen family. Phosphine does not exhibit appreciable basic (or acidic) properties in water solution, although it will form phosphonium salts, PH_4X, when combined with all the hydrogen halides except HF.

24.12.1 Ammonia Ammonia, NH_3, is a common, extremely useful compound that is produced commercially in huge quantities by the direct union of nitrogen and hydrogen. It is made in nature by the decay of proteins in the bodies of dead plants and animals (see the nitrogen cycle in Figure 24.12).

The great solubility of ammonia in water is unusual; about 700 L of the gas dissolves in 1 L of water in room conditions. The attraction of water for ammonia is so great that if a little water is introduced into a chamber filled with the gas, a partial vacuum will result. Much of the attraction between NH_3 and H_2O molecules is attributed to hydrogen bonding of the following type:

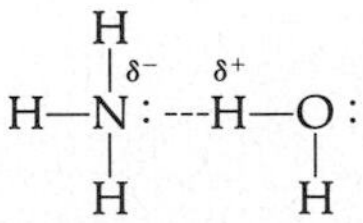

$$
\begin{array}{c}
\text{H} \\
| \quad \delta^- \quad \delta^+ \\
\text{H}-\text{N}:\text{---H}-\text{O}: \\
| \qquad\qquad | \\
\text{H} \qquad\quad \text{H}
\end{array}
$$

Ammonia is more like water than is any other compound. The effect of hydrogen bonding on the melting and boiling points of each liquid was indicated in Figure 8.7. The physical properties of the two liquids are compared in Table 24.5. Below −33.4 °C, ammonia is a colorless liquid. It is an excellent solvent for most of the compounds that dissolve in water. Both ionic and covalent compounds dissolve in ammonia, and chemical reactions involving precipitations, gas evolution, electrolysis, oxidation–reduction, and acids and bases can be carried out.

Just as water ionizes slightly,

$$2H_2O \rightleftharpoons H_3O^+ + OH^- \qquad K_w = 1 \times 10^{-14}$$

hydronium   hydroxide
ion (acid)   ion (base)

so liquid ammonia does likewise:

Table 24.5 Comparison of physical properties of ammonia and water

	NH_3	H_2O
melting point, °C	−77.8	0
boiling point, °C	−33.4	100
critical temperature, °C	132.4	374.2
heat capacity, $J \cdot g^{-1} \cdot K^{-1}$	4.60	4.18
heat of fusion, $J \cdot g^{-1}$	338.9	333.5
heat of vaporization, $J \cdot g^{-1}$	1,366	2,257

$$2NH_3 \rightleftharpoons \underset{\substack{\text{ammonium} \\ \text{ion (acid)}}}{NH_4^+} + \underset{\substack{\text{amide} \\ \text{ion (base)}}}{NH_2^-} \qquad K_a = 1 \times 10^{-33}$$

In water, a typical reaction between a strong acid and a strong base is

$$\underset{\text{acid}}{HCl} + \underset{\text{base}}{KOH} \longrightarrow \underset{\text{salt}}{KCl} + \underset{\text{water}}{H_2O}$$

or

$$\underset{\text{acid}}{H_3O^+ + Cl^-} + \underset{\text{base}}{K^+ + OH^-} \longrightarrow \underset{\text{salt}}{K^+ + Cl^-} + \underset{\text{water}}{2H_2O}$$

In ammonia, the analogous reactions are

$$\underset{\text{acid}}{NH_4Cl} + \underset{\text{base}}{KNH_2} \longrightarrow \underset{\text{salt}}{KCl} + \underset{\text{ammonia}}{2NH_3}$$

or

$$\underset{\text{acid}}{NH_4^+ + Cl^-} + \underset{\text{base}}{K^+ + NH_2^-} \longrightarrow \underset{\text{salt}}{K^+ + Cl^-} + \underset{\text{ammonia}}{2NH_3}$$

A property of liquid ammonia (in addition to its odor!) that is unlike water is that it dissolves alkali metals to form the strongest reducing solutions known. For example,

$$Na + xNH_3 \longrightarrow Na^+ + e^-(NH_3)_x \qquad (5)$$

The ammoniated electron acts as a free ion in moving from place to place. The intense blue color of the solution is due to these ions (see Figure 24.15 on Plate 7). On long standing, the dissolved electron reacts with the ammonia:

$$e^-(NH_3)_2 \longrightarrow \tfrac{1}{2}H_2\uparrow + NH_2^- + (x-1)NH_3 \qquad (6)$$

By adding Equations (5) and (6), an overall equation is obtained for a reaction similar to the one with water:

$$Na + NH_3 \longrightarrow Na^+ + NH_2^- + \tfrac{1}{2}H_2\uparrow$$

$$Na + H_2O \longrightarrow Na^+ + OH^- + \tfrac{1}{2}H_2\uparrow$$

• *Commercial Production of Ammonia* • Shortly before World War I broke out in 1914, the German chemist Fritz Haber invented a process for making ammonia directly from nitrogen and hydrogen. The Haber artificial nitrogen fixation is one of the most important of all industrial processes. Ammonia is a key chemical required for the production of nitric acid, HNO_3, which reacts with NH_3 to produce ammonium nitrate, NH_4NO_3. Ammonium nitrate commonly is used as a fertilizer, but it has also been used as an explosive in bombs and shells. A mixture of ammonium nitrate and TNT (trinitrotoluene) is almost as powerful as the same weight of pure TNT, and it is cheaper.[7]

Some historians have pointed out that Haber's work had a serious impact on military history. In Germany in 1914, there was not a sufficient stockpile of imported nitrates to provide explosives for a long war. When it became evident that the war was not going to be won quickly, the Germans turned to the newly developed Haber process. After months of feverish activity, the process was developed to the point where it provided enough ammonia to supply the munitions industry. Thus it is possible that the Haber process enabled the Germans to prolong the war by two or three years.

The key to the problem of how nitrogen could be made to react directly with hydrogen was to find a catalyst for the reaction. Haber found that iron oxide containing traces of other metal oxides would catalyze the reaction. Although other catalysts have been developed, iron or iron oxide is usually the main ingredient in the modern mixed catalysts.

The equilibrium reaction

$$N_2 + 3H_2 \rightleftharpoons 2NH_3 \qquad \Delta H = -92.2 \text{ kJ}$$

was discussed from several standpoints in Section 15.1. A detailed record of experimentally determined yields of ammonia under various conditions was given in Table 15.2. From that table we see that the amount of ammonia formed at equilibrium is at a maximum when the pressure is high and the temperature is low. On the basis of those data we would pick a temperature of 200 °C and a pressure of 1,000 atm as ideal. But those conditions are not as "ideal" as we might expect, because at a temperature of 200 °C, the reaction

$$N_2 + 3H_2 \longrightarrow 2NH_3$$

is too slow. To make the reaction occur at a reasonably rapid rate, the temperature must be raised to 500 °C or more, even though the percentage of ammonia produced is lowered.

To reach a pressure of 1,000 atm requires a heavy investment in pumps and high-pressure equipment. Moreover, the equipment must be made of a special steel alloy, because, at a high temperature and pressure,

---

[7]Although normally very difficult to detonate, fertilizer-grade ammonium nitrate has been responsible for some disastrous peacetime explosions.

Key Chemical: NH_3
ammonia

How Made
Haber process

$$N_2 + 3H_2 \xrightarrow{\text{catalyst}} 2NH_3$$

U.S. Production; Cost
~17 million tons per year
~$80 per ton

Major Derivatives
25% nitric acid
15% urea
15% ammonium phosphates

Major End Uses
80% fertilizers
10% plastics, resins, and fibers
5% explosives

Other Uses
cleaning agents, refrigeration liquid

hydrogen gas will leak through ordinary steel. Although plants have been constructed that operate at pressures near 1,000 atm, production in the United States is generally more economical at lower pressures. To produce ammonia at the lowest price per ton, the synthesis is carried on at pressures of 200 to 350 atm and at temperatures between 500 and 600 °C.

Considerable ammonia and ammonium compounds are by-products in the production of coke from coal, but more than 90 percent of the U.S. production is made by the Haber process.

24•13   Oxy-Acids and Bases of the VA Elements

24•13•1   Tendency to Act As Acid or Base   The elements in group VA form series of compounds that are analogues of nitric acid, HNO_3 (oxidation number of $N = +5$), and nitrous acid, HNO_2 (oxidation number of $N = +3$). Examples are phosphoric acid, H_3PO_4 ($P = +5$ oxidation number), and phosphorous acid, H_3PO_3 ($P = +3$ oxidation number).

As we proceed down either series, there is a decreasing tendency for the compound to act as an acid and an increasing tendency for it to act as a base. This behavior can be attributed to the gradual change down the family from the nonmetallic element nitrogen, which has a considerable attraction for electrons, to the metallic element bismuth, with a much smaller attraction for electrons (review Section 11.4).

We can write the general formula for the HNO_2 series as HEO_2, where E stands for any of the group VA elements.[8] The Lewis structural formula of HEO_2 is

$$H-\overset{..}{\underset{..}{O}}-\overset{..}{E}=\overset{..}{O}:$$

When E is an element that has a strong attraction for electrons (is nonmetallic), electrons are attracted away from the proton, so the proton is readily donated to proton-attracting groups; that is, HEO_2 tends to act as an acid:

$$HOEO \rightleftharpoons H^+ + OEO^-$$

When E is weakly electronegative (is metallic), HEO_2 tends to act as a base:

$$HOEO \rightleftharpoons HO^- + EO^+$$

Experiments show that, among the HEO_2 compounds of the nitrogen family, nitrous acid is the strongest acid (although it is not very strong); phosphorous acid is next in strength; arsenious acid is weakly acidic and can even act as a base; antimonious acid has no pronounced acidic or basic properties, although it can act as either under the proper conditions; and bismuthyl hydroxide is a weak base. The following equations illustrate some acid–base reactions of the HEO_2 compounds:

---

[8] The actual oxy-acid compound in solution may be H_3EO_3, that is, $HEO_2 + H_2O \rightarrow H_3EO_3$.

As an acid:

$$HONO + NaOH \longrightarrow NaONO + H_2O$$

(written as
HNO_2)
nitrous acid

(written as
$NaNO_2$)
sodium nitrite

As a base:

$$HOBiO + HCl \longrightarrow ClBiO + H_2O$$

(written as
$BiOOH$)
bismuthyl hydroxide

(written as
$BiOCl$)
bismuthyl chloride

As either an acid or a base:

$$HOSbO + NaOH \longrightarrow NaOSbO + H_2O$$

(written as
$HSbO_2$)
antimonious acid

(written as
$NaSbO_2$)
sodium antimonite

$$HOSbO + HCl \longrightarrow ClSbO + H_2O$$

(written as
$SbOOH$)
antimonyl hydroxide

(written as
$SbOCl$)
antimonyl chloride

Hydroxides that can act as either acids or bases are called *amphoteric hydroxides*, as was pointed out in Section 11.2.1. $HAsO_2$ and $HSbO_2$ are examples.

24•13•2 Nitric Acid Hydrogen nitrate, or nitric acid, HNO_3, is a covalent compound that is a colorless liquid at room temperature. The concentrated solution is somewhat unstable and is decomposed slowly by heat and light. The solution becomes yellow due to the presence of dissolved NO_2:

$$4HNO_3 \longrightarrow 4NO_2 + O_2 + 2H_2O$$

Along with sulfuric and hydrochloric acids, nitric acid is one of the most commonly used strong acids.

Nitric acid is also a strong oxidizing agent. The oxidation state of nitrogen actually attained as nitric acid is reduced depends on several factors, including concentration of the nitric acid, activity of the reducing agent, and temperature. Recall that metals that are more active than hydrogen (see Table 18.1) liberate hydrogen gas when they react with hydrochloric acid. With nitric acid, however, hydrogen does not make up an appreciable amount of the final reduction products. Instead, they are largely those that arise from the reduction of the nitrogen with $+5$ oxidation number. The most common products evolved as gases are nitrogen dioxide, NO_2, and nitric oxide, NO. Because NO_2 is brown and NO is colorless, they are easily distinguished.[9] The more concentrated the nitric acid and the less active the metal, the greater is the tendency for NO_2 to be formed.

Consider copper, a metal that is less active than hydrogen. (The value

---

[9]However, NO reacts with oxygen in the air to yield NO_2.

Key Chemical: HNO_3
nitric acid

How Made
Ostwald process (three steps)

$$4NH_3 + 5O_2 \longrightarrow 4NO + 6H_2O$$
$$2NO + O_2 \longrightarrow 2NO_2$$
$$3NO_2 + H_2O \longrightarrow 2HNO_3 + NO$$

U.S. Production; Cost
~8 million tons per year
~$200 per ton

Major Derivatives
85% ammonium nitrate, NH_4NO_3
15% other chemicals

Major End Uses
70% fertilizers
15% explosives

Other Uses
plastics, films, dyes, drugs, metallic nitrates

of $\mathscr{E}^\circ_{red}$ for Cu^{2+}/Cu is $+0.34$ V.) With concentrated HNO_3, the products are $Cu(NO_3)_2$, NO_2, and H_2O:

$$Cu + 4HNO_3 \longrightarrow Cu(NO_3)_2 + 2NO_2\uparrow + 2H_2O$$

When dilute HNO_3 is used, the products are $Cu(NO_3)_2$, NO, and H_2O:

$$3Cu + 8HNO_3 \longrightarrow 3Cu(NO_3)_2 + 2NO\uparrow + 4H_2O$$

Now consider magnesium, a metal that is more active than hydrogen. (The value of $\mathscr{E}^\circ_{red}$ for Mg^{2+}/Mg is -2.37 V.) Depending on the nitric acid concentration, it is possible for the reduction of nitrogen with $+5$ oxidation number to proceed to the formation of $+4$ nitrogen in NO_2, $+2$ nitrogen in NO, $+1$ nitrogen in N_2O, 0 nitrogen in N_2, or -3 in NH_4^+. With concentrated HNO_3, the nitrogen reduction products are NO_2 and NO; with HNO_3 of moderate concentration, the product is largely nitrous oxide, N_2O; but with dilute HNO_3, the product is largely NH_4^+. The equation for the case of NH_4^+ formation is

$$4Mg + 10HNO_3 \longrightarrow 4Mg(NO_3)_2 + NH_4NO_3 + 3H_2O$$

Aqua regia, a mixture of concentrated nitric acid (one volume) and concentrated hydrochloric acid (three volumes), is often more effective in dissolving metals and minerals than nitric acid alone. The mixture dissolves gold and platinum, whereas neither of the acids will do so when used

---

★ Special Topic ★ Special Topic ★ Special Topic ★ Special Topic ★ Special Topic ★ Special Topic ★ Special Topic ★ **Special Topic** ★

Protection of Teeth by Fluoride Ions Dental enamel is mainly hydroxyapatite, $Ca_{10}(PO_4)_6(OH)_2$. The attack on enamel by acids normally present in the mouth, such as lactic or acetic acid, tends to dissolve the enamel and produce caries. In the presence of fluoride ion, insoluble calcium fluoride, CaF_2, forms, and also isomorphic substitution occurs to produce a layer of fluorapatite, $Ca_{10}(PO_4)_6F_2$. The fluoride-treated enamel is more resistant to acid attack.

Tin(II) fluoride, which is particularly effective in treating teeth to reduce the rate of dissolution of enamel in acid solution, is a constituent of some toothpastes. The increased protection in the presence of Sn^{2+} ions is attributed to the adsorption by the fluoride-treated enamel of a very thin layer of basic tin(II) phosphates.

In one laboratory test, extracted teeth were exposed to a buffered lactic acid solution, and the rate of dissolution of enamel was measured. Different samples of teeth were then treated with one of two fluoride solutions of different concentrations, and the rates of dissolution again measured. The results plotted in the accompanying figure show the percent reductions in rate of dissolution of tooth enamel caused by exposure to two fluoride solutions, containing either Sn^{2+} or Na^+. A value of 100 percent reduction would correspond to a dissolution rate of zero, that is, complete protection of the teeth during the standard exposure time of the test.

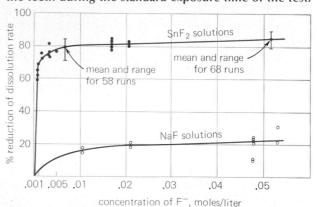

Source: From W. E. Cooley, "Applied Research in the Development of Anticaries Dentifrices," *J. Chem. Educ.*, **47**, 177 (1970). Reproduced with permission.

separately. The efficiency of this mixture of acids is thought to depend on, first, the strong oxidizing action of nitric acid and, second, the tendency of chloride ions to form complex ions. For platinum, the equation is

$$Pt + 8HCl + 2HNO_3 \longrightarrow H_2PtCl_6 + 2NOCl + 4H_2O$$

24•13•3 Phosphoric Acid and Phosphates Anhydrous phosphoric acid, H_3PO_4, is a viscous, colorless liquid at room temperature. A water solution of any concentration can be made easily by dissolving phosphorus(V) oxide, P_4O_{10}, in water:

$$P_4O_{10} + 6H_2O \longrightarrow 4H_3PO_4$$

Commercially, phosphoric acid usually is made from calcium phosphate by an older method:

$$Ca_3(PO_4)_2 + 3H_2SO_4 + 6H_2O \longrightarrow 2H_3PO_4 + 3CaSO_4 \cdot 2H_2O$$

Salts of phosphoric acid are known in which one, two, or three protons are replaced by other cations. Examples are sodium dihydrogen phosphate, NaH_2PO_4, sodium monohydrogen phosphate, Na_2HPO_4, and sodium phosphate, Na_3PO_4.

Soluble phosphates such as Na_3PO_4 hydrolyze in water to make basic solutions. The principal hydrolysis reaction is

$$PO_4^{3-} + H_2O \rightleftharpoons HPO_4^{2-} + OH^-$$

For this reason, solutions of the soluble phosphates lower the surface tension of water and have a soapy feel like sodium hydroxide or potassium hydroxide solutions. Sodium phosphate is used as an additive in many soap and detergent powders. One result of this has been an increase in the phosphate ion pollution in lakes and streams, which has overfertilized aquatic plant growth. When the masses of plants die, the decay reactions use so much of the dissolved oxygen that aquatic animal life is killed. The overall process whereby excess plant food leads to the death of both plant and animal life in a body of water is called **eutrophication.**

Large amounts of phosphate are used in fertilizers. Calcium phosphate, $Ca_3(PO_4)_2$, is not only one of the most common compounds in phosphate rocks; it is also one of the most common components of kidney stones.

24•14 Miscellaneous Compounds of the VA Elements

Some well-known compounds of the nitrogen family not mentioned previously in this chapter are described in Table 24.6.

Key Chemical: H_3PO_4
phosphoric acid

How Made
$Ca_3(PO_4)_2 + 3H_2SO_4 +$
$6H_2O \rightarrow 2H_3PO_4 +$
$3CaSO_4 \cdot 2H_2O$
or $P_4O_{10} + 6H_2O \rightarrow 4H_3PO_4$

U.S. Production; Cost
~10 million tons per year
~$250 per ton, depending on grade

Major Derivatives
55% ammonium phosphates
20% superphosphates (water-soluble phosphates used as fertilizers)

Major End Uses
80% fertilizers
5% detergent additives

Other Uses
tile-cleaning preparations, insecticides, animal feed

Key Chemical: $H_2N-\overset{\displaystyle C}{\underset{\displaystyle O}{|}}-NH_2$
urea

How Made
$2NH_3 + CO_2 \xrightarrow[\text{pressure}]{\text{heat}}$

$H_2N-\overset{\displaystyle C}{\underset{\displaystyle O}{|}}-NH_2 + H_2O$

U.S. Production; Cost
~5 million tons per year
~$120 per ton, depending on grade

Major Derivatives
10% urea–formaldehyde resins
5% other resins

Major End Uses
75% fertilizers
10% animal feeds
10% adhesives and plastics

Table 24•6 Compounds of group VA elements

Formula	Chemical name (common name)	Remarks
N_2H_4	hydrazine	fuel for rockets
NH_4Cl	ammonium chloride	used as a "flux" for cleaning iron before galvanizing and for cleaning metals before soldering; used in dry cells and in preparing fabrics for dyeing
NH_4F	ammonium fluoride	used to etch glass
$(NH_4)_2S$	ammonium sulfide	common reagent for qualitative analysis; used in making polysulfides
$(NH_4)_2SO_4$	ammonium sulfate	fertilizer; except for NH_3, cheapest source of NH_4^+ ions
$(NH_4)_2CO_3$	ammonium carbonate	smelling salts
$(NH_2)_2CO$	urea	constituent of human urine; made synthetically for use as a fertilizer
$(NH_4)_2HPO_4$	ammonium monohydrogen phosphate	fertilizer with both N and P in one compound
N_2O	nitrogen(I) or nitrous oxide (laughing gas)	prepared by cautious heating of NH_4NO_3; mild anesthetic; patient may have such vivid dreams that they are remembered as reality
$NaNO_2$	sodium nitrite	used in meat-packing to preserve the red color associated with fresh meat; may react with certain chemicals in the stomach to produce carcinogenic compounds called nitrosamines
KNO_3	potassium nitrate (saltpeter)	mixed with sulfur and charcoal to make black powder
$Ca(H_2PO_4)_2$	calcium dihydrogen phosphate	produced by action of sulfuric acid on phosphate rock; resulting mixture of monocalcium phosphate and calcium sulfate called superphosphate fertilizer
As_4O_6	arsenic(III) oxide (arsenic)	active ingredient in insecticides, weed killers; fatal human dose about 0.1 g
$Pb_3(AsO_4)_2$	lead arsenate	insecticide for certain fruit trees

Chapter Review

Summary

The elements of the **sulfur family** are found in nature as free elements and as compounds. In the family there is an increase in metallic character with increasing atomic number. The elements can exist in both crystalline and amorphous forms. Sulfur changes from one crystalline form to another at a certain temperature, called a **transition temperature.** At high temperatures in molten sulfur, the normal ring molecules break open to form chains. The sulfur family elements and many of their compounds are oxidizing agents. Because elemental sulfur is an effective reducing agent toward most nonmetals as well as an oxidizing agent toward most metals, it reacts (at high temperatures) with almost all the other elements. Sulfur is less reactive than oxygen but more reactive than selenium and tellurium. Many of its compounds have selenium and tellurium ana-

logues, but these are much less abundant. The binary hydrogen compounds of these three elements are weak acids.

Sulfur and many of its compounds are of great industrial and commercial importance. By far the most important is sulfuric acid, a strong acid that is manufactured by the **contact process.** Concentrated sulfuric acid is a strong oxidizing agent and dehydrating agent. It is the most widely used synthetic chemical. Also very useful are many of the metal sulfates. Certain hydrated sulfate **double salts** containing one singly charged cation and one triply charged cation are called **alums.**

The elements of the **nitrogen family** are also found as free elements (except for phosphorus) and as compounds. In this family there is a more marked increase in metallic character with increasing atomic number, and three of the elements have both nonmetallic and metallic crystalline forms. The nonmetal phosphorus and the metalloid arsenic exist as polyatomic molecules. All the nitrogen family elements can act as oxidizing agents and reducing agents. Nitrogen itself is notably unreactive; phosphorus is very reactive, and the other elements are somewhat less reactive.

Under extreme conditions or with suitable catalysts, elemental nitrogen reacts with some other elements. Any such reaction is called **nitrogen fixation.** In nature, nitrogen oxides are produced which hydrolyze in rainwater to form nitric and nitrous acids. The most important product of artificial nitrogen fixation is ammonia, which is used in the manufacture of nitric acid and nitrate fertilizers. Ammonia is chemically more like water than any other compound, owing in part to its strong tendency to form hydrogen bonds. In aqueous solution it is a weak base.

The nitrogen family analogues of nitric and nitrous acids show a gradual shift from acidic to basic properties with increasing atomic number (and hence metallic character) of the central atom. Nitric acid itself is a strong acid and a strong oxidizing agent. Phosphoric acid is used primarily in the manufacture of phosphate fertilizers and detergent additives. The excessive use of the latter has led to the **eutrophication** of many lakes and streams.

Exercises

Physical Properties of the Sulfur Family

1. What is the electron orbital arrangement in the highest main level of each member of group VIA?

2. Although oxygen is in group VIA, it is not thought of as a typical member of the sulfur family. What are some ways in which oxygen does not resemble other members of the sulfur family?

3. Would the last member of the VIA family tend to be slightly more metallic than the last member of the VIIA family? Explain.

4. How does the viscosity of a liquid normally change as the temperature changes? Is this "normal behavior" characteristic of molten sulfur? Explain.

5. In the mining of sulfur, superheated water at about 180 °C is pumped down the wells.
   (a) How is it possible to heat water to such a temperature?
   (b) Why is not the water heated to a much higher temperature, say, 250 °C, in order to melt the sulfur more efficiently?
   (c) Why is 180 °C preferable to, say, 160 °C as a temperature for the water pumped down the wells?

6. Account for the difference in the observed ionization energies of sulfur and tellurium (see Table 24.1) on the basis of their atomic structures.

Oxidation–Reduction Chemistry of the Sulfur Family

7. Hydrosulfuric acid acts as a reducing agent, sulfuric acid acts as an oxidizing agent, but sulfurous acid can act as either. Explain why for all three acids. Write four equations to illustrate the four possibilities.

8. Use the data in Figure 24.6 to show that $S_2O_6^{2-}$ tends to disproportionate in acid solution. Write a balanced equation for a disproportionation reaction that you would expect to take place spontaneously.

9. Use the data in Figure 24.6 to predict which of the following reactions should occur spontaneously in acid solution:

$$SO_4^{2-} + H_2SeO_3 \longrightarrow H_2SO_3 + SeO_4^{2-}$$
or
$$SeO_4^{2-} + H_2SO_3 \longrightarrow H_2SeO_3 + SO_4^{2-}$$

*10. Based on Figure 24.6 predict whether or not an oxidation–reduction reaction will occur in acidic solution at 25 °C for each of the following:
   (a) $H_2Se + H_2SeO_4 \longrightarrow$
   (b) $H_2Te + H_2SO_3 \longrightarrow$
   (c) $H_2SO_3 + H_6TeO_6 \longrightarrow$
   Write a complete equation for each hypothetical reaction that you predict.

11. In Example 24.2 it was shown that oxygen is capable of oxidizing sulfur to sulfate ion in standard basic solution. Is oxygen also capable of oxidizing sulfur to sulfate in standard acid solution? If so, in which of the two solutions would you expect sulfur to be oxidized to SO_4^{2-} more readily, the standard basic or standard acidic solution? Explain.

*12. Statement 5 in Section 24.1.3 says that sulfur is stable toward disproportionation in standard acid solution but is unstable in standard basic solution. Support this statement with calculations. Write a balanced equation for a spontaneous disproportionation reaction in basic solution.

*13. Using data from any of the figures in Chapters 23 and 24, show with calculations whether each reaction below will or will not take place spontaneously:
   (a) $6H^+ + 4I^- + SO_3^{2-} \longrightarrow 2I_2 + S + 3H_2O$
   (b) $2H_2O + I_2 + SO_2 \longrightarrow H_2SO_4 + 2HI$
   (c) $2H_2Se + H_2TeO_3 \longrightarrow 3H_2O + 2Se + Te$

14. If combustion is defined as an oxidation that proceeds sufficiently rapidly to give off heat and light, is it possible that certain substances will undergo combustion with sulfur? Provide reasons or examples to support your answer.

15. Sulfur can exist in a variety of oxidation states in the compounds it forms. Using sodium and oxygen as the other elements, write Lewis structural formulas and name salts for each of the following oxidation numbers of sulfur: -2, average $= +2$, $+4$, $+6$.

*Reactions and Structures
of Sulfur Family Compounds*

16. Write balanced equations for the formation of each of the following compounds: (a) sulfur hexafluoride, (b) hydrogen sulfide (two methods), (c) sodium sulfide, (d) selenium dioxide, (e) calcium polysulfide, (f) boron trisulfide.

17. Carbon tetrafluoride, CF_4, is a nonpolar molecule, but sulfur tetrafluoride, SF_4, is polar. Account for this difference in terms of molecular structures.

18. Although both SF_6 and SF_4 react with water, the former reacts very slowly under normal conditions. Explain this difference in behavior in terms of molecular structures.

19. Why is it desirable to postulate sp^3 hybridization for H_2O but unnecessary for H_2Te?

20. What do the heats of formation of the hydrogen compounds of the group VIA elements indicate about the metallic–nonmetallic characteristics of the elements?

21. What is the oxidation number of iron in iron pyrite, FeS_2? Write an equation for the reaction of iron pyrite with hydrochloric acid. Would you expect to obtain a clear solution when a small fragment of FeS_2 reacts with HCl? Why?

*Oxides, Oxy-Acids,
and Oxy-Anions of Sulfur*

*22. In Section 24.7.2 the statement is made that sulfur dioxide is the principal component of the solution that we call *sulfurous acid*. Based on this statement, write a balanced overall equation for the reaction of excess sodium hydroxide with sulfurous acid. Also write the net ionic equation for the reaction.

*23. A mixture of sulfur dioxide and oxygen is held at 400 °C in a 1.0-L container until equilibrium is attained. If the initial partial pressures of SO_2 and O_2 are 0.50 and 0.25 atm, respectively, what is the percentage by volume and by weight of the SO_3 in the equilibrium mixture?

$$2SO_2 + O_2 \rightleftharpoons 2SO_3 \qquad K_p = 397$$

24. For Figure 24.8, show that only one of the plotted lines is necessary to give all the information that both lines give.

25. Sulfuric acid has long been used in preparing other acids from their salts. Give examples. What physical property is important here?

26. Write balanced equations showing the action of sulfuric acid as (a) an acid; (b) a dehydrating agent; (c) an oxidizing agent.

*27. Baking powders may consist of bicarbonate of soda, $NaHCO_3$, and, as an acid component, $NaAl(SO_4)_2 \cdot 12H_2O$ (an alum). Explain how this substance is able to act as an acid when added to water and how it liberates carbon dioxide from the bicarbonate of soda. Write the equations.

28. According to Section 24.8.2, sulfur reacts with the sulfite ion to form a thiosulfate ion, $S_2O_3^{2-}$. Does this reaction occur in standard acid or standard base solution? (Consult Figure 24.6.)

*Physical Properties
of the Nitrogen Family*

29. The orbital assignment of electrons in the highest main level of VA elements is s^2p^3. Does this assignment include all of the valence electrons? How would the highest main energy level be indicated in the assignment for a specific element, say, arsenic?

30. Reasoning from information in Figure 24.9, explain why the ionization energy decreases down the VA family. Repeat for the electronegativity.

31. Among the nitrogen family elements, which one has the greatest tendency to form positive ions? Account for this in terms of atomic sizes.

32. (a) Although group VA is called the nitrogen family, nitrogen itself is somewhat different from the rest of the group in properties. Cite some examples.
    (b) Is this relation of nitrogen to its group unusual, or are there similar instances in other groups?

*Oxidation–Reduction
Chemistry of the Nitrogen Family*

33. Based on Figure 24.11, which one of each of the following pairs is the stronger oxidizing agent: arsenic acid, H_3AsO_4, or phosphoric acid, H_3PO_4; nitrous oxide, N_2O, or nitrogen dioxide, NO_2?

34. Solid fertilizer-grade ammonium nitrate has been responsible for some of the most devastating accidental explosions on record. Though it does not explode in solution, would you predict its solutions to be stable or unstable on the basis of the data in Figure 24.11? Explain.

*35. Calculate $\Delta G_r°$ to produce N_2 and H_2O by the reaction of nitrogen tetroxide, N_2O_4, and hydrazoic acid, HN_3, for the pure compounds, based on Table 13.6, and compare with the free energy change for this reaction in standard solutions as calculated from Figure 24.11.

36. By comparing Figures 23.2, 24.6, and 24.11, discuss the relative strengths of the following as oxidizing agents: HCl, H_2SO_4, and HNO_3.

37. Based on Figure 24.11, which reaction would be more likely to occur: the oxidation of nitrogen by H_3PO_4 or the oxidation of phosphorus by HNO_3? Justify your answer.

*38. Use the data in Figure 24.11 to show that N_2O can oxidize water spontaneously in standard acid solution. Assume that N_2 is the nitrogen reduction product, and write a balanced equation for the redox reaction.

*39. Hydrazoic acid, HN_3, is so unstable that it tends to explode, as do its salts such as lead azide, $Pb(N_3)_2$. Based on the data in Figure 24.11, show whether HN_3 can be prepared by the reaction of nitrous acid, HNO_2, with hydrazinium chloride, $N_2H_5^+Cl^-$, in standard acid solution.

*Reactions and Structures
of Nitrogen Family Compounds*

40. Although elemental nitrogen has a higher electronegativity than phosphorus, the latter is much more active chemically. Explain.

41. Describe three nitrogen fixation processes. Tell whether they are natural or artificial.

42. Plants are able to utilize both NO_3^- and NH_4^+ ions in synthesizing proteins. Can humans? What is the advantage, if any, of supplying the nitrogen content of commercial fertilizers with NH_4NO_3 instead of $NaNO_3$? What advantage do organic nitrogen compounds (usually urea or urea derivatives) have over sodium nitrate and ammonium nitrate when used on lawns?

43. What are the electron assignments for hybridized orbitals for PCl_3 and PCl_5? Based on the geometries of the molecules that would be consistent with your assignment, would PCl_3 or PCl_5 molecules have dipole moments? Why?

44. Both arsenic and antimony react with excess fluorine to produce the pentafluorides, but such a reaction will not take place with nitrogen. Explain.

45. What is the oxidation number of the VA element in each of the following: N_2O_4, PH_4Cl, $HSbO_2$, $Ca_3(PO_4)_2$, AsH_3, $Bi_2(SO_4)_3$, SbO^+, HPO_3^{2-}? Name each compound or ion.

46. Outline the chemical steps in the production of elemental phosphorus.

Ammonia

47. Write equations for the reaction of ammonia with (a) nitric acid; (b) water; (c) potassium; (d) sulfuric acid.

48. Show how the reaction of ammonium sulfate and potassium amide in liquid ammonia could be considered an acid–base reaction.

49. Explain why it is theoretically desirable to carry out the Haber process at high pressure and low temperature. Then explain why a moderate temperature and pressure are actually used.

*Oxy-Acids and
Bases of the VA Elements*

50. Write equations for the following for which reactions occur:
(a) $PBr_5 + H_2O \longrightarrow$
(b) $N_2O_5 + H_2O \longrightarrow$
(c) $HOSbO(aq) + KOH(aq) \longrightarrow$
(d) $HONO(aq) + HCl(aq) \longrightarrow$

51. List the formulas and names for ten molecules or ions that are amphoteric. Illustrate the amphoteric character of two ions by writing balanced net ionic equations.

52. Concentrated nitric acid will oxidize elemental sulfur to sulfuric acid. The nitrogen reduction product is nitrogen dioxide. Write a balanced equation for this reaction.

53. When copper reacts with dilute nitric acid in a test tube, the gas formed immediately in the test tube is colorless, but a brownish gas issues from the mouth of the test tube. Why? Write equations for the reactions.

54. Balance the following equations in a systematic way, using either system described in Section 18.6.
(a) $Zn + HNO_3 \longrightarrow Zn(NO_3)_2 + N_2 + H_2O$
(b) $Al + HNO_3 \longrightarrow Al(NO_3)_3 + N_2O + H_2O$

55. Write the balanced equations for the action of aqua regia on gold. One of the products is $HAuCl_4$.

56. Phosphorus(V) oxide, P_4O_{10}, is often used in laboratory work to remove water from gases. Write an equation for the reaction that takes place.

*57. Consider the following oxides of nitrogen: N_2O, NO, N_2O_3, NO_2, N_2O_4, and N_2O_5. Of these six compounds, all but two follow the rule of eight. These two compounds we will designate as A and B. For these two compounds the following reactions are assumed to take place:

$$A + A \longrightarrow A_2$$
$$A + B \longrightarrow AB$$

The compounds designated as A_2 and AB correspond to two of the four oxides that follow the rule of eight.
(a) What are the formulas of A, B, A_2, and AB?
(b) Write Lewis structural formulas for the six oxides.
(c) Write all resonance structures for molecules for which you assume resonance is important.

Carbon, Silicon, and Boron

25

25•1 Properties of Carbon, Silicon, and Boron

25•1•1 Physical Properties

25•1•2 Chemical Properties

Special Topic: Transistors and Solar Cells

25•2 Occurrence

25•2•1 Carbon

25•2•2 Silicon and Boron

25•3 Compounds of Carbon

25•3•1 Carbon Monoxide

25•3•2 Carbon Dioxide

25•3•3 Carbonates and Bicarbonates

25•3•4 Other Inorganic Compounds of Carbon

25•4 Compounds of Silicon

25•4•1 Silicon Dioxide

25•4•2 Silicates

25•4•3 Familiar Silicon-Containing Materials

Special Topic: Extraterrestrial Chemistry

25•5 Compounds of Boron

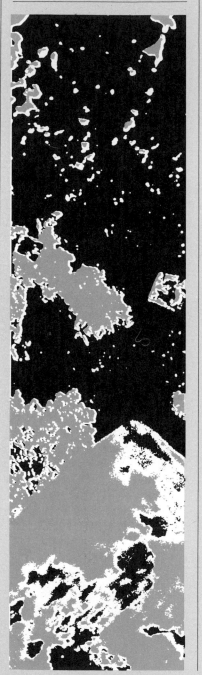

In addition to the noble gases, we have studied three nonmetal families of the periodic table in detail—groups Va, VIa, and VIIa. Of these, the halogens are most like one another. There is a great difference in electronegativity between fluorine and iodine, but all the group VIIa elements are nonmetals. In the next group, the trend from oxygen to tellurium goes from nonmetals to a metalloid element. In group Va, the change is even greater—from the nonmetals nitrogen and phosphorus through the metalloid elements arsenic and antimony to the metal bismuth.

In group IVa, we find that the elements carbon, C, and silicon, Si, differ so much from the other family members—germanium, Ge, tin, Sn, and lead, Pb—that it is not satisfactory to study the five elements collectively. In fact, three of the elements in group IVa (Ge, Sn, and Pb) resemble the elements in group IVb (Ti, Zr, and Hf) more than they resemble the other two IVa elements (C and Si). There is a greater similarity between the a and b families in group IV than in any other group. The trends in these four a families are summarized as follows:

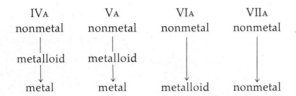

In this chapter, we take up three important elements yet to be discussed—carbon and silicon from group IVa and boron from group IIIa. Boron is physically and chemically similar to silicon. Thus, it is the third of the period 2 elements to show a so-called diagonal relationship to a period 3 element in the adjacent family. The similarities of lithium with magnesium and beryllium with aluminum were mentioned in Section 21.3.3.

25•1   Properties of Carbon, Silicon, and Boron

25•1•1   Physical Properties   The very high melting and boiling points of boron, carbon, and silicon distinguish them from the other nonmetals (see Table 25.1). The atoms of boron, carbon, and silicon are very small compared with other atoms. The calculated ionic radii in crystals of these elements are even smaller, because the atoms are usually in positive oxidation states. Because of their charge densities, the ions do not exist as independent particles in compounds, but are held by covalent bonds.

All three elements are rigid solids that may be thought of as giant molecules consisting of a huge number of atoms. Both boron and silicon have only one crystalline form, whereas carbon occurs in two well-defined crystalline forms. All three elements can be obtained in one or more amorphous modifications. The common amorphous forms of carbon are charcoal, coke, carbon black, and bone-black.[1]

---

[1] Although these forms are often described as amorphous, recent work has shown that some of them are microcrystalline and have the graphite structure.

Table 25.1 Physical properties of boron, carbon, and silicon

	Boron	Carbon	Silicon
melting point, °C	2,177	3,500[a,b]	1,412
boiling point, °C	3,658	3,930[a]	2,680
electron distribution	2,3	2,4	2,8,4
ionization energy, eV/atom	8.3	11.3	8.2
or kJ/mole	800	1,090	790
covalent radius, Å	0.90	0.77	1.18
ionic radius, Å	0.20 (B^{3+})	0.15 (C^{4+})	0.41 (Si^{4+})
electronegativity	2.0	2.5	1.8

[a] Graphite.

[b] Diamond sublimes at 4,000 °C at 64 atm.

The crystalline forms of carbon are famous for their physical differences. One, *graphite,* is a soft black substance that actually feels greasy; as a dry powder it is used as a lubricant, especially for locks. The other, *diamond,* is a colorless solid capable of being cut into brilliant crystals and is the hardest, most abrasive mineral known. Yet both these substances consist of only carbon atoms.

In the case of graphite, the atoms crystallize in a pattern of layers (see Figure 25.1) that slide over one another easily.[2] In contrast, the carbon atoms in the diamond structure have strong bonds with neighbors in three dimensions, each atom being bound by equally strong covalent bonds to atoms on all sides (see Figure 25.2).

Boron and silicon have ionization energies and electronegativities that mark them as borderline elements, but the higher values for carbon show it to be a true nonmetal.

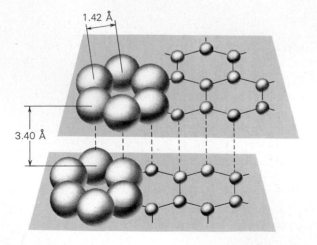

1.42 Å

3.40 Å

Figure 25.1 In graphite, carbon atoms crystallize in layers with hexagonal symmetry. The atoms are much closer to their neighbors in the same layer than to atoms in an adjacent layer.

[2] Layers of clean graphite in a vacuum stick together; but in the atmosphere, a freshly cleaned graphite surface strongly adsorbs molecules of gas and then slips freely.

All three elements are relatively poor conductors of heat and electricity, although the graphite form of carbon conducts electricity better than most other nonmetals.[3] Because it adheres well to many materials and is a conductor, graphite is often brushed over nonconducting surfaces, such as leather or plastic, that are to be electroplated.

Graphite, either natural or synthetic, is used as the black constituent in ordinary pencil leads, as the pigment in black paints, in the manufacture of crucibles and electrodes to be used at extremely high temperatures, and as a dry lubricant. Diamonds, particularly discolored and small ones, are used in industry for making the hardest abrasive powders for grinding wheels and the tips of drills and saws.

The different forms of carbon sublime when heated to high temperatures out of contact with air. On cooling, the vapors condense in the form of graphite. This is the process for the commercial production of graphite from anthracite coal and from coke. At pressures greater than 100,000 atm and temperatures above 2,700 °C, graphite is transformed into diamond. Synthetic diamonds are not so large or so brilliantly transparent as the finest stones found in nature, but in most other characteristics they are identical with natural diamonds.

25●1●2 Chemical Properties All three elements, boron, carbon, and silicon, are quite unreactive at ordinary temperatures. When they do react, there is no tendency for their atoms to lose outer electrons and form simple cations, such as B^{3+}, C^{4+}, and Si^{4+}. Small ions of this type would have such high charge densities that their existence is unlikely. Instead, the atoms usually react by sharing electrons to form covalent bonds.

• *Reactions with Halogens* • Carbon reacts directly with fluorine. Silicon and boron react with the halogens in general, even to the extent of burning in gaseous fluorine (X refers to a halogen atom):

$$C + 2F_2 \longrightarrow CF_4$$
$$Si + 2X_2 \longrightarrow SiX_4$$
$$2B + 3X_2 \longrightarrow 2BX_3$$

• *Common Oxy-Acids* • When heated in air, the elements react with oxygen in highly exothermic combustion reactions to form the oxides B_2O_3, CO_2, and SiO_2. All three of these oxides are *acidic*. The first two react with water to give very weak acid solutions:

$$B_2O_3 + 3H_2O \rightleftharpoons 2H_3BO_3$$
$$\text{boric acid}$$

$$CO_2 + H_2O \rightleftharpoons H_2CO_3$$
$$\text{carbonic acid}$$

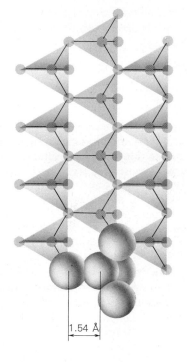

1.54 Å

Figure 25●2 The diamond structure. Carbon atoms crystallize with tetragonal symmetry; each atom has four nearest neighbors.

---

[3]In the diamond structure, the electron pairs are held so strongly that there is very little electric conductivity. In the graphite structure, the electrons in the bonds between the layers are not held so tightly and hence are freer to move through the crystal and conduct electricity.

Carbonic acid is similar to H_2SO_3 (Section 24.7.2) in the following way. Although H_2CO_3 has not been isolated as a compound, we assume some of the compound is present in water solution and write these equations for the equilibria with ions:

$$H_2CO_3 + H_2O \rightleftharpoons H_3O^+ + HCO_3^-$$
$$HCO_3^- + H_2O \rightleftharpoons H_3O^+ + CO_3^{2-}$$

The third oxide, SiO_2, is essentially unreactive with water at ordinary temperatures. However, two simple silicic acids are well known: orthosilicic acid, H_4SiO_4, and metasilicic acid, H_2SiO_3. Both these compounds are practically insoluble in water, but they do react with bases; for example,

---

Special Topic ★ Special Topic ★ Special Topic ★ Special Topic ★ Special Topic ★ Special Topic ★ Special Topic ★ Special Topic

Transistors and Solar Cells One of the most important chemical and physical discoveries of this century is how to use slightly impure silicon (or germanium) in transistors. *Transistors* are devices that control the flow of electricity just as did the old-style radio tubes, but transistors do this more dependably and for a longer time. The consequent development of small calculators, computers, and control devices of various sorts has revolutionized information flow and automation.

Pure silicon is a poor conductor of electricity, but its conductivity can be increased by the presence of very small amounts of certain other atoms. If an atom of phosphorus (or other group VA element) replaces an atom of silicon in a crystal (see Figure A), the extra electron of the phosphorus is called a *negative defect* or an *n*-type defect. The crystal is said to be *doped* with phosphorus. The conductivity of the silicon is increased, and the crystal is called a *semiconductor*.

Another way of increasing the conductivity and making a semiconductor of silicon is to dope the silicon crystal with a few atoms of boron (or other group IIIA element). Because a boron atom has only three valence electrons, its presence produces an electron vacancy or *electron hole* in the crystal (see Figure B). The electron hole attracts electrons, so it is called a *positive defect* or a *p*-type defect.

The number of doping atoms in a semiconductor is a small fraction of the total. Except for the few purposely added doping atoms, the silicon must be extremely pure. Traces of a doping element can be crystallized with the silicon. In the case of phosphorus in silicon, doping can be achieved by neutron bombardment. About 3 percent of silicon atoms are ^{30}Si, and neutron bombardment changes some of these atoms to ^{31}P. This technique provides a precise control of the number of *n*-defects and their locations.

Figure A. When a $\cdot \ddot{P} \cdot$ atom is substituted for an $\cdot \dot{S}i \cdot$ atom, the extra electron is called an *n*-defect.

Figure B. When a $\cdot \dot{B} \cdot$ atom is substituted for an $\cdot \dot{S}i \cdot$ atom, the site of the missing electron is called a *p*-defect.

$$H_4SiO_4 + 4NaOH \longrightarrow Na_4SiO_4 + 4H_2O$$
$$\text{sodium}$$
$$\text{orthosilicate}$$

When partially dried, silicic acid is called *silica gel* (a material that looks something like rock salt). In this form it has great adsorbent capacity for vapors of water, sulfur dioxide, nitric acid, benzene, and other substances. It is widely used as a dehumidifier in small sealed containers.

Boric acid has been a standby in the family medicine chest for years. Available at drugstores in the form of glittering, colorless, thin flat crystals, it dissolves readily in water to yield a mild antiseptic solution.

None of the three acids is of great use commercially as an acid, except that carbonic acid gives carbonated drinks their slightly sharp taste.

★ Special Topic ★ Special Topic ★ Special Topic ★ Special Topic ★ Special Topic ★ Special Topic ★ Special Topic ★ Special Topic ★

A sandwich made of an *n*-type and a *p*-type semiconductor acts as a *rectifier* of electric current, a device that allows current to flow in only one direction (see Figure C). Such a "valve" for electricity plays the key role in the yes–no logic of digital computers.

A dramatic use of semiconductors is in the construction of *solar cells* that convert radiant energy directly into electric energy (see Figure D). In one solar cell design, a thin layer of *n*-type silicon is in contact with a thicker layer of *p*-type. When these two layers are connected by an external circuit, there is a resulting potential that tends to cause electrons to flow from the *n*-layer around through the external wire to the holes in the *p*-layer. However, no current can flow unless the electrons are excited.

When a photon of sunlight of sufficient energy strikes a silicon atom near the interface of the two layers, an excited electron is knocked free and a hole is created. The electron migrates into the *n*-layer, and the hole migrates into the *p*-layer. An electron from the external wire jumps into a hole in the *p*-layer as an electron from the *n*-layer jumps onto the other end of the wire.

Panels of solar cells can convert a theoretical maximum of about 25 percent of solar energy to electric energy, but 10 percent efficiency is a practical expectation. Small units generating 50–100 W of electricity are now used for remote power beacons on harbor buoys and mountain telephone and television relays. A 7 m by 10 m panel of solar cells, generating 5,000 W at midday in the northern United States, could supply the electricity needs of the average house. But present costs make this use uneconomical. One of the largest installations designed to date is a demonstration project at Mississippi County Community College in Blytheville, Arkansas. About 60,000 solar cells will provide between 250 and 350 kW of electric energy. Excess electric energy generated during the day will be stored in batteries for use at night and on cloudy days.

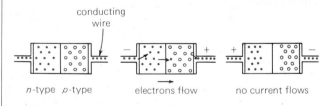

Figure C. A rectifier made of an *n-p* transistor junction will pass current only when a potential is applied (*middle diagram*) that forces electrons into the *n*-type and out of the *p*-type semiconductor. Each time an electron jumps across the junction to fill a hole, another hole is created as an electron jumps onto the wire at the right. Such a rectifier can change alternating current to direct current.

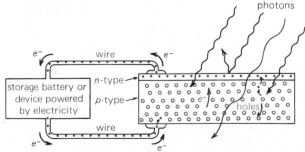

Figure D. A solar cell made of a thin (3×10^{-6} m) film of an *n*-type semiconductor in contact with a thicker (4×10^{-4} m) piece of a *p*-type semiconductor. Photons of sufficient energy knock free electrons and create new holes. Some photons are reflected and some pass through without effect.

• *Salts of Oxy-Acids* • The salts of the acids above are well known. Carbonic acid, as a typical diprotic acid, reacts with bases to give such carbonates and bicarbonates as

K_2CO_3	potassium carbonate
$KHCO_3$	potassium bicarbonate
$MgCO_3$	magnesium carbonate
$Mg(HCO_3)_2$	magnesium bicarbonate

Salts of the two silicic acids include:

Na_2SiO_3	sodium metasilicate
Na_4SiO_4	sodium orthosilicate
Mg_2SiO_4	magnesium orthosilicate
$LiAl(SiO_3)_2$	lithium aluminum metasilicate

All the silicates except those of Na^+, K^+, Rb^+, Cs^+, and NH_4^+ are practically insoluble in water.

A great many simple and complex borates are known, but aside from borax, $Na_2B_4O_7 \cdot 10H_2O$, they are not sufficiently common to be mentioned here.

All the soluble carbonates, silicates, and borates form basic solutions when dissolved in water. The CO_3^{2-}, SiO_3^{2-}, and BO_3^{3-} ions act as bases to remove protons from water (review the discussion of hydrolysis in Section 11.2.3):

$$SiO_3^{2-} + H_2O \rightleftharpoons HSiO_3^- + OH^-$$

• *Formation of Large Molecules* • An important chemical property of carbon, silicon, and boron is their tendency to form huge molecules. We should note two differences between silicon and boron on the one hand and carbon on the other. The first is that carbon atoms join with one another to form a limitless variety of chains or rings of atoms. The second is that carbon atoms tend to form single, double, and triple covalent bonds, whereas silicon and boron tend to form only single bonds (four and three single bonds, respectively). The ability of carbon to form vast numbers of compounds will be discussed in Chapter 26.

Silicon and boron form giant molecules and ions in which oxygen atoms occupy alternate positions. These structures are built up in this fashion:

$$-\overset{|}{\underset{|}{Si}}-O-\overset{|}{\underset{|}{Si}}-O-\overset{|}{\underset{|}{Si}}-O- \qquad -\overset{|}{B}-O-\overset{|}{B}-O-\overset{|}{B}-O-$$

Silicon-containing and boron-containing rocks and minerals are commonly high-melting, hard, brittle solids, each piece of which is a continuous lattice of tightly bound atoms. An example of such a solid is silicon dioxide, which occurs in nature in the form of quartz, agate, sand, and so on. One of the several known SiO_2 crystal structures is shown in Figure 25.3.

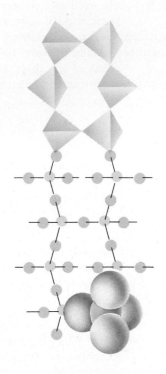

Figure 25.3 The linking of (SiO_4) tetrahedra in one form of silica, SiO_2. Silicon atoms are shown by colored spheres; oxygen atoms, by gray spheres. In each tetrahedron, there is one silicon atom and $4(\frac{1}{2}) = 2$ oxygen atoms.

25•2   Occurrence

25•2•1   Carbon

Carbon occurs in the earth's crust in both the free and combined states. The principal natural compounds of carbon are the organic substances formed in the tissues of living things, both plant and animal, and in materials derived from living things, such as coal and petroleum. Among the common inorganic carbon compounds are carbon dioxide and the carbonate rocks, particularly calcium carbonate, $CaCO_3$. Carbonates of other group IIA elements are well known as minerals: magnesium, strontium, and barium carbonates, $MgCO_3$, $SrCO_3$, and $BaCO_3$, respectively.

25•2•2   Silicon and Boron

In spite of their relative inactivity at moderate temperatures, silicon and boron are not found free in nature. They occur only in oxy-compounds such as silica, silicates, and borates. The principal boron compounds—and they are not abundant—are boric acid, H_3BO_3, and hydrated sodium borate (borax), $Na_2B_4O_7 \cdot 10H_2O$. Borax is found in desert areas where it has been precipitated by the evaporation of salt lakes. It is used in the manufacture of glass, glazes, soap, soldering flux, and artificial gems.

Elemental boron can be prepared as a brown powder by the reduction of boron oxide, B_2O_3, with magnesium or aluminum:

$$B_2O_3 + 3Mg \xrightarrow{\text{heat}} 3MgO + 2B$$

Pure crystalline boron is black and approaches diamond in hardness, but has few uses.

Elemental silicon can be prepared from silica by the same method shown for boron, that is,

$$SiO_2 + 2Mg \xrightarrow{\text{heat}} 2MgO + Si$$

In its crystalline form, silicon is gray or black. It has had few uses; however, it is being used as a component in transistors and in certain of the new solar batteries.

25•3   Compounds of Carbon

25•3•1   Carbon Monoxide

When carbon-containing fuels (for example, wood, coal, gasoline) are burned in the presence of a great deal of air, practically all the carbon combines with oxygen to make carbon dioxide, CO_2, but very little carbon monoxide, CO, is formed. Air pollution by carbon monoxide was discussed in Section 9.2.4. It is present in small amounts in the fumes from practically all fires. The less air (or oxygen) there is available, the greater is the relative amount of carbon monoxide formed. Also, at higher temperatures, carbon dioxide tends to react with hot carbon:

$$CO_2 + C \xrightarrow{700\ ^\circ C} 2CO$$

Carbon monoxide has the strongest bond of any diatomic molecule. It is isoelectronic with the nitrogen molecule and has a triple bond, $:C\equiv O:$, as N_2 has.

Commercially, carbon monoxide has a number of uses. Mixtures of gases containing it have long been used as fuels. Actually, more heat is liberated when carbon monoxide burns to carbon dioxide than when carbon burns to carbon monoxide:

$$C + \tfrac{1}{2}O_2 \longrightarrow CO \qquad \Delta H = -110.5 \text{ kJ}$$
$$CO + \tfrac{1}{2}O_2 \longrightarrow CO_2 \qquad \Delta H = -283.0 \text{ kJ}$$

Four important fuel gases are listed in Table 25.2. Of these, only natural gas contains no significant amount of carbon monoxide.

Table 25.2 Fuel gases

Name	Composition, %		Source	Approximate heat value, BTU[a]/ft^3
water gas	CO	40–50	reaction of steam with hot coal:	300
	H$_2$	45–50	C + H$_2$O → CO + H$_2$	
	CO$_2$	3–7		
	N$_2$	4–5		
producer gas	CO	20–35	coal burned in air and steam	130
	H$_2$	5–10	under conditions that produce	
	N$_2$	55–65	mainly carbon monoxide and hydrogen	
coal gas	H$_2$	45–54	by-product in manufacture of	600
	CH$_4$	28–34	coke by destructive distillation	
	CO	6–7	of coal	
natural gas	CH$_4$	50–92	gas and oil wells	1,000
	C$_2$H$_6$	2–14		

[a]One BTU (British thermal unit) = 1,055 kJ.

As mentioned in Section 21.8.2, carbon monoxide reduces many metal oxides to their elements:

$$MO + CO \longrightarrow M + CO_2$$

25.3.2 Carbon Dioxide Carbon dioxide is produced by the combustion of common fuels, such as coal, petroleum, and wood. It is also a component of the breath exhaled by animals, because it results from the oxidation of food in the body. Present in the atmosphere to the extent of about 0.03 percent, its concentration may rise to 1 percent in a crowded room.

Carbon dioxide is not poisonous, but too high a concentration in the air (10 to 20 percent) is unhealthy because it lowers the oxygen concentration and has harmful physiologic effects (unconsciousness, failure of certain respiratory muscles, and a change in the pH of the blood). There can also be too little carbon dioxide in a person's system (see Sections 10.5 and 16.8.1).

Key Chemical: CO$_2$

carbon dioxide

How Made
by-product of ammonia production and of synthetic fuel production

U.S. Production; Cost
~2,300,000 tons per year
~$45 per ton

Major Uses
30% food refrigeration
10% industrial refrigeration
35% beverages

Other Uses
treating alkaline waste water, pressurizing oil wells, aerosol propellant

Large amounts of the gas are found in the water of some geysers and mineral springs. The atmosphere in a cave or valley where carbon dioxide seeps out of fissures in the ground may be dangerous. Because the gas has such high density (44 g/22.4 L compared to 29 g/22.4 L of air), it tends to stay in low places. It can be poured from one vessel to another like a liquid; it can even remain in an open beaker for a short time as it slowly diffuses out into the air.

Constantly entering the atmosphere in a variety of ways, carbon dioxide is constantly removed by photosynthesis in plants, the formation of carbonate rocks, and the formation of the shells of marine animals (see Figure 25.4). Today the concentration of atmospheric carbon dioxide is increasing because of our intense industrial and agricultural activity. The burning of fuels produces 6×10^9 tons of this gas per year; destruction of forests to create cultivated fields has lessened the capacity of the biosphere to use carbon dioxide by about 2×10^9 tons per year, because cultivated cropland is less efficient in photosynthesis than is forested land.

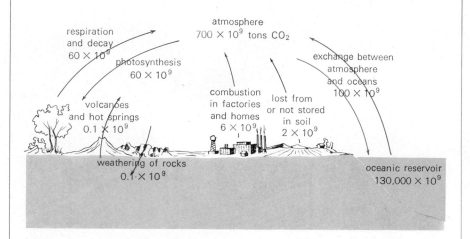

Figure 25•4 Effect of human activities (black arrows) on the carbon dioxide cycle. Quantities are given in tons per year for processes and in tons in place for the atmosphere and oceans.

The huge amounts of CO_2 produced by human activities are sufficient to displace the equilibria involved in natural processes that maintain a rough balance between the earth and its atmosphere. The gradual increase in the amount of CO_2 in the atmosphere in the third quarter of this century is shown in Figure 25.5. This increase may threaten us with a *greenhouse effect*, because CO_2 molecules can absorb infrared radiation from the earth's surface that otherwise would go into outer space. If the CO_2 concentration continues to increase, it is feared that the atmosphere may become so warm that serious changes in climate will occur. Enough polar ice could be melted to raise the ocean levels and flood coastal cities around the world.

Whether or not the continued and increasing burning of fossil fuels will actually produce a serious greenhouse effect is not certain. Smoke particles produced by fires shade the earth's surface and reflect sunlight back into outer space. This tends to cool the earth. We are making great efforts to prevent smoke pollution (see Figure 25.6), and it is difficult at this time to predict the overall effects of combustion on climate.

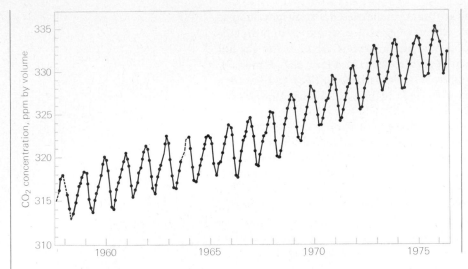

Figure 25•5 Measurements of CO_2 in the atmosphere at the Mauna Loa Observatory in Hawaii. The seasonal oscillations are due to the removal of CO_2 by photosynthesis during the growing season and the release of CO_2 during the fall and winter.
Source: C. D. Keeling et al., "Atmospheric Carbon Dioxide Variations at Mauna Loa Observatory, Hawaii," *Tellus*, 28, 538 (1976); from C. F. Baes, Jr., et al., *The Global Carbon Dioxide Problem*, Oak Ridge National Laboratory, Oak Ridge, Tenn., 1976.

Below a pressure of 5.3 atm (4,030 mm), liquid carbon dioxide does not exist.[4] Solid carbon dioxide, which sublimes at −78.5 °C under a

Figure 25•6 The effect of electrostatic precipitators on power plant smoke production (see also Figure 12.15). (*Left*) Operation with precipitators turned off. (*Right*) Operation with precipitators turned on. *Source:* Courtesy of the Tennessee Valley Authority, John Sevier Steam Plant, Rogersville, Tenn.

---

[4]This behavior can be compared with that of water, which does not exist as a liquid if the pressure is less than 4.6 mm. If some liquid H_2O were thrown into a vacuum chamber in which the pressure was maintained at 4 mm, some of the liquid would evaporate instantly, thereby cooling and freezing the remaining liquid to ice. The ice would then sublime at about 0 °C. Strictly speaking, this would be dry ice.

pressure of 1 atm, is called *dry ice* because it vaporizes without first melting. It is a convenient, clean refrigerant, especially useful when subzero temperatures are necessary. Crushed solid carbon dioxide in a Dewar flask of alcohol or acetone makes an excellent cold bath for laboratory work (see Figure 25.7).

Cylinders of liquid carbon dioxide under about 60 atm pressure are used as fire extinguishers (see Figure 25.8). The cylinder on the left in the figure shows a typical condition inside a liquid carbon dioxide fire extinguisher at room temperature. When the valve is opened, some of the liquid CO_2 escapes to the low pressure of the atmosphere. The liquid rapidly evaporates and the gas formed expands. These two endothermic changes occur so quickly that some of the CO_2 loses enough heat to freeze (an exothermic change). This forms the small CO_2 crystals visible momentarily as the familiar white fog or "snow" that is expelled by an extinguisher.

25·3·3 Carbonates and Bicarbonates

As the most abundant inorganic carbon compounds, the carbonates and bicarbonates are both useful and well-known substances. Most carbonates are only slightly soluble in water—for example, calcium carbonate, $CaCO_3$, barium carbonate, $BaCO_3$, magnesium carbonate, $MgCO_3$, and lead carbonate, $PbCO_3$. Many of the bicarbonates are stable only in water solution. Examples are calcium bicarbonate, $Ca(HCO_3)_2$, and magnesium bicarbonate, $Mg(HCO_3)_2$. All group IA metals, except lithium, form soluble carbonates, the most inexpensive and useful ones being sodium bicarbonate, $NaHCO_3$ (baking soda), and sodium carbonate, Na_2CO_3 (soda ash).

The carbonates and bicarbonates react with most acids to give CO_2. The reaction is quite rapid and the gas is evolved readily. For instance, barium carbonate reacts with hydrobromic acid in this way:

$$BaCO_3 + 2HBr \longrightarrow BaBr_2 + H_2O + CO_2\uparrow$$

A reaction that can be tested in the kitchen occurs when sodium bicarbonate and acetic acid (vinegar) are combined:

$$NaHCO_3 + HC_2H_3O_2 \longrightarrow \underset{\text{sodium acetate}}{NaC_2H_3O_2} + H_2O + CO_2\uparrow$$

The accompanying "fizz" is due to escaping carbon dioxide. Cake rises as a result of being inflated with bubbles of carbon dioxide when the baking soda, $NaHCO_3$, in the baking powder reacts with an acid ingredient in the baking powder. The acid component may be sodium aluminum sulfate, $NaAl(SO_4)_2$, or potassium acid tartrate, $KHC_4H_4O_6$. The sodium bicarbonate and the acid do not react appreciably while in the dry state; but once the baking powder is in the batter, the two dissolve in the water solution and react.

The reactions of carbonates and bicarbonates with acids may be summarized by these ionic equations:

$$CO_3^{2-} + H^+ \longrightarrow HCO_3^-$$

$$HCO_3^- + H^+ \longrightarrow H_2O + CO_2\uparrow$$

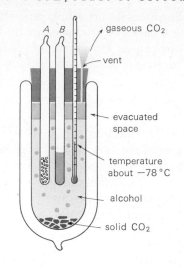

Figure 25·7 A low-temperature bath ($-78.5\,°C$) of isopropyl alcohol cooled with solid carbon dioxide. The sealed glass tube *A* contains solid ammonia, which freezes at $-77.8°C$. Tube *B* contains liquid propane, which condenses at $-42.1\,°C$ but freezes at about $-190\,°C$.

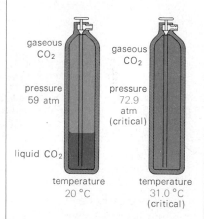

Figure 25·8 Schematic representation of the state of carbon dioxide in two fire extinguishers. The one on the left is at a typical room temperature. The one on the right shows that only gas is present at or above the critical temperature of CO_2.

Bicarbonates are amphoteric substances; that is, they react with both acids and bases. Bicarbonates are unstable; when heated, they decompose to form carbonates. Powdered potassium bicarbonate is used in fire extinguishers because it decomposes readily to yield carbon dioxide:

$$2KHCO_3 \xrightarrow{\text{heat}} K_2CO_3 + H_2O\uparrow + CO_2\uparrow$$

Lithium carbonate is used in the treatment of manic-depressive mental patients. Several lithium compounds are effective, but the carbonate is least likely to upset the stomach.

Calcium carbonate is one of the most widely distributed nonsilicate minerals.[5] As limestone it is the raw material for the useful products lime and slaked lime (see Section 21.4.2). A process designed to produce hydrocarbons by direct reaction under pressure of hydrogen and heated limestone may someday provide huge quantities of hydrocarbons from inorganic source materials.

The solution chemistry of calcium carbonate is both interesting and instructive. If carbon dioxide is passed into a solution of a cation that forms an insoluble carbonate (for example, Ca^{2+}, Ba^{2+}, Mg^{2+}, or Pb^{2+}), a white precipitate will form. Calcium carbonate precipitates when carbon dioxide is bubbled through a solution of lime water, $Ca(OH)_2$:

$$CO_2 + H_2O \rightleftharpoons 2H^+ + CO_3^{2-}$$
$$Ca^{2+} + CO_3^{2-} \longrightarrow CaCO_3\downarrow$$

Surprisingly enough, if we continue to add carbon dioxide, the precipitate will dissolve. This is contrary to the expectation that if a little carbon dioxide causes a precipitate to form, a lot of carbon dioxide should result in even more precipitate. Actually, the solid carbonate dissolves because it is reacting to form the more soluble bicarbonate:

$$CaCO_3 + H_2O + CO_2 \longrightarrow \underset{\substack{\text{calcium} \\ \text{bicarbonate}}}{Ca(HCO_3)_2}$$

If the solution of $Ca(HCO_3)_2$ is heated, the bicarbonate decomposes and the precipitate reappears:

$$Ca(HCO_3)_2 \xrightarrow{\text{heat}} CaCO_3\downarrow + H_2O + CO_2\uparrow$$

Or, if the bicarbonate solution is allowed to simply stand in the open air, the calcium carbonate will reappear as the water evaporates:

---

[5] Although we are not mainly concerned with mineral names, it is of interest that a mineral is named not simply on the basis of its chemical composition, but also on the basis of its general properties and appearance. For example, different forms of $CaCO_3$ are known as calcite, aragonite, limestone, marble, travertine, and chalk. Differences in the mode of formation cause the six to differ in appearance and usefulness.

$$Ca(HCO_3)_2 \xrightarrow{\text{on drying}} CaCO_3 + H_2O\uparrow + CO_2\uparrow$$

• *Cave Formation* • Spectacular limestone caves (such as Mammoth Cave and Carlsbad and Luray Caverns) have been dissolved out of solid rock by the gentle action of the solution formed by carbon dioxide in rainwater. An original fissure or weakness in the rock probably allowed the first trickles of the weakly acidic carbonated water to seep through. As more and more of the carbonate was converted into the soluble bicarbonate, the hole grew in size until there was room for an underground stream or river. Thereafter, erosion by the moving water aided the dissolving action of the carbonated water.

When a cave is open enough to allow a current of air to circulate, the most beautiful changes take place. Groundwater saturated with calcium bicarbonate seeps through from above to the ceiling of the cave and evaporates in the circulating air. As a result of this evaporation, calcium carbonate precipitates on the ceiling (*stalactites*); see Figure 25.9. In the places where the concentrated solution drips to the floor of the cave before evaporating, the carbonate is precipitated on the ground (*stalagmites*).

25•3•4   Other Inorganic Compounds of Carbon

The carbonates are just one of many important classes of inorganic compounds of carbon. Others are the cyanides, the carbonyls, and the carbides.

The cyanides are similar to the halides in many physical and chemical properties. An important difference is that cyanides are deadly poisons. As a base, the CN^- ion forms HCN in the blood, which inhibits the action of an enzyme, cytochrome oxidase, that is essential for energy production in the cells. One effect of the cyanide is to force the hemoglobin to remain oxygenated. Venous blood becomes as red as arterial blood, causing cyanide victims to have a characteristic reddish color. Examples of cyanides:

sodium cyanide, NaCN
hydrogen cyanide, HCN
potassium hexacyanoferrate(III), $K_3Fe(CN)_6$

Examples of carbonyls:

nickel carbonyl, $Ni(CO)_4$
iron carbonyl, $Fe(CO)_5$

Examples of carbides:

calcium carbide, CaC_2
zinc carbide, ZnC_2
aluminum carbide, Al_4C_3
silicon carbide, SiC

Calcium carbide is well known because it is used in the production of calcium cyanamide (see Table 25.3) and of acetylene (see Chapter 26). The gray, solid calcium carbide is obtained in a two-step process from limestone

Figure 25•9 Part of a stalactite with the growth pattern revealed by traces of impurities imbedded in its rings of calcium carbonate that were precipitated slowly over many years.

and coke (obtained from coal), two low-cost and widely distributed natural resources:

$$CaCO_3 \xrightarrow{\text{heat}} CaO + CO_2$$

$$CaO + 3C \longrightarrow CaC_2 + CO$$

A number of other especially useful carbon compounds are listed in Table 25.3.

25·4   Compounds of Silicon

25·4·1   Silicon Dioxide   Silicon dioxide, or silica, is one of the most common chemical compounds. Pure SiO_2 crystals are found in nature in three polymorphic forms, the most common of which is *quartz*. Sand, agate, onyx, opal, amethyst, and flint are silicon dioxide with traces of impurities.

Fused quartz is used to make crucibles and other laboratory vessels that must be heated to extremely high temperatures. Not only does it have a high softening point (about 1,500 °C), but another equally valuable property is its very low coefficient of thermal expansion. A substance with a low coefficient of thermal expansion expands and contracts very little when heated and then cooled. Quartz, therefore, is not likely to crack even if it is heated and cooled rapidly and unevenly.

The forms of silica are some of the truly important crystal structures, not only because silica itself is such an abundant and useful substance, but also because the (SiO_4) structure is the fundamental unit in most minerals. As is evident in Figure 25.3, SiO_2 crystals have two main features: (1) each silicon atom is at the center of a tetrahedron of four oxygen atoms, and (2) each oxygen atom is midway between two silicon atoms.

Attention is often called to the great differences between SiO_2 and CO_2 in physical properties. The former does not soften until heated to about 1,500 °C; the latter sublimes at -78 °C. This and other differences can be correlated with the type of molecule, which in turn depends on the type of bond, Si—O single bonds or C=O double bonds.

Carbon unites with oxygen by forming two covalent bonds with each oxygen atom, O=C=O. Carbon dioxide is made up of tiny triatomic molecules and is a gas at room temperature. In contrast, the seemingly endless structure

$$-O-\underset{|}{\overset{|}{Si}}-O-$$

in silicon dioxide (see Figure 25.3) is a single molecule, whether it is a grain of sand or a magnificent quartz crystal larger than a person's head.

25·4·2   Silicates   Silicon–oxygen compounds are the most abundant of all the compounds in the earth's crust. Most rocks and minerals[6] are silicates

---

[6]The term *mineral* is restricted to naturally occurring, relatively homogeneous inorganic material with characteristic properties and a composition that is within certain limits. *Rocks* may be made of one or more minerals and may be homogeneous or heterogeneous.

with the lattice:

$$-\overset{|}{\underset{|}{Si}}-O-\overset{|}{\underset{|}{Si}}-O-$$

Table 25•3 Some well-known carbon compounds

Formula	Name	Appearance at room conditions	Remarks
SiC	silicon carbide (carborundum)	crystals, ranging from colorless to black	chemically inactive, almost as hard as diamond; used as an abrasive for grinding, cutting, polishing
B_4C	boron carbide	black crystals	next to diamond, boron carbide and boron nitride are hardest substances known; used as an abrasive
HCN	hydrogen cyanide	colorless gas; almond odor	poisonous gas used in pest control and in lethal gas chambers: $H_2SO_4 + 2NaCN \rightarrow 2HCN + Na_2SO_4$ (compare reaction of H_2SO_4 and NaCl); in laboratory keep strong acids away from cyanides
NaCN	sodium cyanide	white solid	solutions of sodium cyanide used to dissolve gold, silver, and platinum for electroplating baths
CS_2	carbon disulfide	colorless liquid; very volatile	solvent for grease, rubber, sulfur; used as solvent in preparation of rayon viscose; very flammable, poisonous
CCl_4	carbon tetrachloride	colorless liquid; volatile	solvent for dry cleaning; used for many years in fire extinguishers, it is now outlawed in many places because it often yields the poisonous gas phosgene, $COCl_2$, upon combustion
$CaCN_2$	calcium cyanamide	white solid	made in one of the few nitrogen fixation processes: $CaC_2 + N_2 \rightarrow CaCN_2 + C$; used for fertilizer or hydrolyzed to yield ammonia

This silicate lattice can be thought of as derived from SiO_2 but with atoms of other elements sometimes attached to the silicon and oxygen atoms and sometimes substituted for these atoms. The formulas and names of some of the more abundant groups of minerals that contain silicon are listed in Table 25.4. The following ideas will help us relate the many silicates to one another:

1. Note how often aluminum is present in silicates. It is the third most abundant element in the earth's crust. Other elements present in the examples in Table 25.4 are iron, calcium, sodium, potassium, and magnesium. Any of them is likely to be found in silicate minerals.

2. The formula of any silicate can be written as a "resolved formula," that is, as a series of simple formulas. For example, the formula of anorthite, $CaAl_2Si_2O_8$, can be resolved into a series of simple oxide formulas:

$$CaO \cdot Al_2O_3 \cdot 2SiO_2$$

Similarly, that for tremolite, $Ca_2Mg_5Si_8O_{22}(OH)_2$, can be resolved into

$$2CaO \cdot 5MgO \cdot 8SiO_2 \cdot H_2O$$

Artificial silicates can be made by melting together mixtures of basic oxides and silicon dioxide. The resolved formulas emphasize the fact that no matter how complicated the formulas are, the total positive oxidation states must equal the total negative oxidation states.

3. Neither the resolved formulas nor the collected formulas represent molecules; they show only the relative number of particles in the huge silicate molecules.

Figure 25•10 Asbestos is a silicate that crystallizes in layers, but there is relatively strong bonding between (SiO_4) tetrahedra in one direction. The lateral binding is weak, so the layers can be pulled apart as long fibers. Widely distributed in nature in small quantities (perhaps even in most "pure" drinking water), asbestos is another of those long-trusted substances that have been revealed as dangerous. Overexposure causes severe lung damage and increases the likelihood of cancer.

Table 25•4 Abundant silicon-containing minerals[a]

Mineral group	Percentage of minerals in earth's crust	Characteristic structure	Representative formulas and common names
feldspars	49	crystal large in three dimensions (boxlike)	$KAlSi_3O_8$, orthoclase $NaAlSi_3O_8$, albite $CaAl_2Si_2O_8$, anorthite $Na_4Al_3Si_3O_{12}Cl$, sodalite
quartz	21	same as above	SiO_2, silica
amphiboles, or pyroxenes[b]	15	crystal large in one dimension (chainlike)	$CaSiO_3$, wollastonite $NaAlSi_2O_6$, jadeite $Ca_2Mg_5Si_8O_{22}(OH)_2$, tremolite (an asbestos)
mica[b]	8	crystal large in two dimensions (layerlike)	$KAl_2Si_3AlO_{10}(OH)_2$, muscovite $K_2Li_3Al_4Si_7O_{21}(OH,F)_3$, lepidolite

[a]The minerals listed here include only silica and the main silicates, yet they account for 93 percent of the total minerals in the earth's crust. The other 7 percent is composed of some of the minor silicates and the myriad of non-silicon-containing minerals, such as the carbonates, sulfates, sulfides, and oxides.

[b]Certain minerals in these classes often have iron ions substituted for other positive ions.

4. The huge silicate molecules tend to link themselves together as chains (amphiboles or pyroxenes, as in Figure 25.10), as layers (micas, as in Figure 25.11), or as a boxlike framework (feldspars). In any of these, the SiO_4 tetrahedron is a principal building unit.

5. Ions other than silicon and oxygen, such as Al^{3+}, Fe^{2+}, Ca^{2+}, Na^+, Cl^-, and OH^-, are often found tucked into holes in the main structure. Also, very small cations (especially Al^{3+}) may be substituted in place of Si^{4+}, and small anions, in place of O^{2-}.

25•4•3 Familiar Silicon-Containing Materials Two of the oldest industries are the making of ceramics and bricks. Much later, cement was developed. The bonds

$$-\overset{|}{\underset{|}{Si}}-O-\overset{|}{\underset{|}{Si}}-O-$$

are chiefly responsible for the strength of both ceramics and cement.

• *Ceramics* • Ceramic products are made of mixtures of various finely divided minerals and rocks that form a strong rocklike mass when heated to a high temperature. The most important ingredient is *clay*, a naturally occurring material that is formed by the action of the weather on certain feldspars. Although clay is not a pure substance, it is chiefly *kaolinite*, $Al_2Si_2O_5(OH)_4$, a soft, easily pulverized mineral. Kaolinite is a hydrated aluminum silicate.

• *Cement* • Cement is a rocklike material that becomes hard as the result of a low-temperature reaction rather than the high temperature used in the ceramic kiln.

Portland cement, first made in the early 1800s in England (and named for its similarity to a natural rock mined on the Isle of Portland), can be made from limestone, clay, and gypsum. Other materials, such as blast furnace slag or iron ore, may be used, depending partly on what is available in the locality of the cement plant.

• *Glass* • The glassy state is between the true crystalline solid and the liquid state. A **glass** is a viscous liquid that becomes rigid without crystallizing as it cools. Unlike a crystalline material, glass does not have a sharp melting point but begins to soften far below the temperatures at which it flows easily.

At temperatures slightly above the softening point, glass can be molded into any desired shape. At somewhat higher temperatures, when it begins to behave like a very viscous liquid, small pieces of molten glass are drawn into spherical drops by surface tension, and two pieces of molten glass will join and become a single piece. In this melted condition, it can be blown, molded, or rolled into sheets.

Glass is so inexpensive and common that it is surprising that the glassy state is really quite rare. Only one element (selenium), a few oxides (such as B_2O_3, SiO_2, GeO_2, and P_4O_{10}), and a few oxy-salts (such as the

Figure 25•11 Micas are silicates in which the molecules form layers of (SiO_4) tetrahedra.

borates, silicates, and phosphates) exist in a glassy form. Neither simple ionic nor simple molecular substances form glasses.

• *Silicones* • Some of the most interesting synthetic compounds, unlike anything found in nature, are the *silicones*. These substances are chainlike molecules of Si, O, C, and H atoms. Methyl silicone is an example:

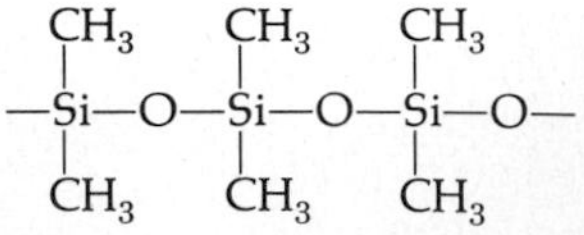

$$
\begin{array}{ccccccc}
 & CH_3 & & CH_3 & & CH_3 & \\
 & | & & | & & | & \\
-Si & -O & -Si & -O & -Si & -O- \\
 & | & & | & & | & \\
 & CH_3 & & CH_3 & & CH_3 &
\end{array}
$$

As the molecular weight changes, the properties change. A silicone made of short-chain molecules is an oily liquid; silicones with medium-length chains behave as viscous oils, jellies, and greases; those with very long chains have a rubber-like consistency.

★   Special Topic ★ Special Topic ★ Special Topic ★ Special Topic ★ Special Topic ★ Special Topic ★ Special Topic ★ Special Topic   ★

Extraterrestrial Chemistry *Extraterrestrial chemistry* is the chemistry of all the universe other than our earth. Our knowledge of such chemistry is growing rapidly today. Instrument-laden rockets have sent back information from four of our neighboring planets, and outer space is revealing its secrets to radioastronomy. One of the chief aims of the study of the chemistry of extraterrestrial matter is to learn more about the origin of the solar system. We also hope to learn more about the origin of life and to look for living organisms outside the earth.

In the late 1960s, the chemistry of the moon's surface was investigated, first remotely by rocket-borne instruments and then by the analysis of actual samples of rocks and soil brought back by astronauts. One of the first analyses for elements was based on the emission from the moon's surface of gamma rays caused by high-energy radiation from the sun. Gamma ray spectrometers on rockets that orbited the moon reported a titanium concentration so high that some geologists felt the data must be incorrect. However, later analyses of rocks brought back from the moon gave similar results. The fact that the analyses made on actual samples agreed generally with the previously reported remote instrumental analyses gives us confidence now in the information returned by radio from rockets sent from as far as Jupiter, 700 million kilometers away.

Of all extraterrestrial bodies, we know most about the moon. Overall, it has relatively more aluminum, titanium, and uranium than the earth, less iron and nickel, and much less of elements with low boiling points, such as sulfur and lead. Not only are there differences in the composition of the crusts of the moon and the earth,

but the fact that the moon's overall density is much less than the earth's indicates that the differences are more than skin deep. These contrasts support the theory that the moon was formed separately from the earth and was later captured by gravitation.

Relatively little is known about Mercury because it has been investigated only by a fly-by rocket and, except for a trace of helium, it has essentially no atmosphere to analyze. But rockets have landed on both Mars and Venus, and analyses of their atmospheres have been made by gas chromatography (see Figure 12.17) and by mass spectrometry (see Figure 20.15). On Venus, an orbiting satellite sent back information for months, but the rocket that landed was put out of operation in about an hour by the conditions of 455 °C surface temperature in an atmosphere featuring sulfuric acid rain clouds and a pressure of 91 atm. It would be more comfortable to be there at night when the temperature is about three degrees lower.

The *Viking Lander,* which went to Mars [see Figure A on Plate 8], found conditions that enabled it not only to analyze the atmosphere, but to land, sample the soil, and take pictures of the surface [Figure B on Plate 8]. With marvelously efficient instrument miniaturization, the *Lander* had the capability of several analytical laboratories, each in a space about the size of a typewriter.

The major elements in the soil were found by X-ray fluorescence to be iron, silicon, calcium, sulfur, and aluminum, with lesser amounts of titanium, rubidium, strontium, and zirconium. Three intensive biology experiments, based on tracing the reactions of radioactive carbon, failed to detect the presence of living organisms.

Other useful properties of the silicones are their resistance to chemical attack and their water-repellent nature (that is, they are *hydrophobic*). Textiles and wood and metal surfaces can be waterproofed with a thin silicone coating; furthermore, silicone is tough and long lasting.

Silicones are among the most promising materials for making tubes, valves, and membranes for medical uses in the human body. The silicone object causes a minimum of irritation to the surrounding tissue and is itself not attacked by body fluids.

• *Zeolites* • The *zeolites* are a class of silicate minerals in which a pair of ions—for example, Al^{3+} and K^+—has been substituted for one Si^{4+}. The trivalent Al^{3+} takes the place of the Si^{4+} in the center of an SiO_4^{4-} tetrahedron, and the monovalent K^+ fits nearby into a hole in the crystal structure. Figure 25.12 shows the structure of *natrolite*, a zeolite in which Na^+ ions fit into the holes; the formula can be written $Na_2(Al_2Si_3O_{10}) \cdot 2H_2O$ to emphasize the freedom and ionic character of the sodium ions. The zeolites are

---

★ Special Topic ★ Special Topic ★ Special Topic ★ Special Topic ★ Special Topic ★ Special Topic ★ Special Topic ★ Special Topic ★

The atmospheric pressure at the surface of Mars is only about 1.3×10^{-3} atm. Gas chromatography–mass spectrometry (GCMS) analyses found mainly carbon dioxide, just as in the case of the very dense atmosphere of Venus:

	Venus	Mars
CO_2	97%	95%
N_2	1–3	2–3
Ar	>0.02	1–2
O_2	0.1	0.3–0.4
H_2O	0.1–0.4	trace

Although there is a much smaller percentage of argon in the atmosphere of Venus, the weight percentage relative to the planet as a whole is greater on Venus than on Mars because the Venusian atmosphere is about 70,000 times more dense than the Martian. Venus has about 300 times the concentration of argon and neon that Earth has, which in turn has 300 times the concentration that Mars has. This is surprising, because Mars is so cold (~0 °C day, ~ −100 °C night) that it would be expected to hold its gases more effectively than the hotter Venus. The ratio of argon:neon is about 3:1 on each of the planets, which contrasts with the 1:30 ratio of the sun. Accounting for these similarities and differences helps theoreticians propose mechanisms for how the planets might have been formed.

One of the most spectacular successes of the United States space program is the 18-month journey of the two *Voyager* rockets to Jupiter. Reaching the great planet in 1979, these fly-by rockets analyzed the atmosphere with infrared and radiofrequency spectrometers and reported the presence of hydrogen, methane, acetylene, ethane, ammonia, phosphine, and water. This massive, cold planet has the hydrogen-rich atmosphere that might be expected in our hydrogen-rich universe.

The planets of our solar system have compositions that are quite different from the average composition of the universe. The major components of the universe, on a percentage basis, are:

$$H: 71\%$$
$$He: 27$$
$$C, N, O, Ne: 1.8$$
$$Na \text{ through } Ti: 0.2$$

Most of the mass of the universe, 90 percent, is in stars, and the rest is in interstellar gas and dust. Even though ultraviolet radiation with bond-breaking energy pervades interstellar space, about 50 molecules or radicals have been detected by radioastronomy. After H and H_2, the most common is CO. Among the others are H_2CO (formaldehyde), HCN, $(CH_3)_2O$ (methyl ether), H_2O, NH_3, SO, SiO, CH_3CH_2CN (ethyl cyanide), and CH_4. About half contain carbon, and about fifteen contain both carbon and nitrogen, which, with hydrogen, are the key elements in the proteins of living organisms. Although there are no known paths by which the scattered molecules in interstellar space interact to form complex biochemicals, it is interesting to find that the building blocks for biochemicals apparently exist throughout the universe.

characterized by a porous structure through which water can circulate rather freely.

← 6.62 Å →

Figure 25•12 A possible arrangement for the naturally occurring zeolite natrolite, $Na_2(Al_2Si_3O_{10}) \cdot 2H_2O$. Each black tetrahedron has an Si atom at the center, whereas each colored one has an Al atom. The order in the natural crystal is not necessarily so regular as here, but the ratio of Si:Al is 3:2. Chains like this are crosslinked with others in three dimensions; Na+ ions and water molecules fit into holes in the lattice.

The great use of the zeolites is in water softening by ion exchange (see Figure 21.3). Certain ions, such as calcium and magnesium, are objectionable in water, because they form insoluble precipitates (curd or scum) with soaps. When water is allowed to flow through a bed of crushed zeolite, the Ca^{2+} and Mg^{2+} ions in solution tend to be attracted to the mineral; K^+ or Na^+ ions leave the zeolite and take the place of the divalent ions in solution. (One Ca^{2+} ion will displace two K^+ ions.) In this way Ca^{2+} and Mg^{2+} ions are exchanged for the unobjectionable Na^+ and K^+ ions.

25•5   Compounds of Boron

Boron compounds are less important than silicon compounds because of the difference in abundance of the two elements. Boric acid and borax were described in Section 25.1.2. On the basis of similarities in structure, the borates resemble the silicates in variety and complexity. Borate anions of several types have been found—discrete ions, chains, rings, and layer structures.

Like carbon and silicon, boron tends to form covalent rather than ionic bonds. Because a boron atom has only three electrons in its outside energy level, it can form three covalent bonds and still not have a complete outside energy level of eight electrons. All the boron halides, BX_3, are Lewis acids (see the discussion of BCl_3 in Section 11.6). They form stable addition compounds with molecules that have lone pairs of electrons. For example, BF_3 reacts with ethyl ether to form a diethyletherate:

$$
\begin{array}{c}
\overset{\displaystyle F}{\underset{\displaystyle F}{|}} \\
F{-}B{-}O{\Big\langle} \begin{array}{l} CH_2CH_3 \\ CH_2CH_3 \end{array}
\end{array}
$$

Boron nitride, BN, exists in two forms; one is similar to graphite and the other to diamond. In the former structure, there are layers in which three B and three N atoms alternate in hexagonal patterns similar to those shown for graphite in Figure 25.1. This form is a soft, white, slippery substance. The form that has properties similar to diamond is the second hardest substance known. It has a cubic lattice and has been made by subjecting the hexagonal form to pressures of about 70,000 atm at about 1,700 °C in equipment similar to that used for making synthetic diamonds. Boron nitride powder is used for industrial grinding and polishing.

The largest class of boron compounds is the boron hydrides or *boranes*. Most have formulas corresponding to B_nH_{n+4} or B_nH_{n+6}. In many of these compounds, hydrogen atoms form bridges between boron atoms. Examples are diborane, B_2H_6, and tetraborane, B_4H_{10}:

$$B_2H_6 \qquad B_4H_{10}$$

Structures of complex compounds such as B_5H_{11}, $B_{10}H_{14}$, $B_{16}H_{20}$, and $B_{20}H_{16}$ are among many worked out by W. N. Lipscomb and co-workers. The unusual types of bonds have required new approaches to bond theory. Some of these compounds, which react explosively with oxygen in the air, are among the most powerful reducing agents known. They have been used as sources of high-temperature flames and for rocket fuel.

Chapter Review

Summary

The three elements carbon, silicon, and boron differ markedly from the other elements in their groups in the periodic table. Carbon is a true nonmetal; silicon and boron are considered as metalloids and are more similar to each other than silicon is to carbon. All three are high-melting, high-boiling solids that exist as giant covalently bonded polyatomic crystals. Carbon exists in two crystalline forms, graphite and diamond, and all three elements can also exist in amorphous forms. Although all three are quite unreactive at ordinary temperatures, only carbon is found as the element in nature. Most carbon, however, exists as carbonate minerals, carbon dioxide, and organic compounds. Silicon and boron are found mainly as oxygen-containing minerals and rocks.

These three elements generally react by the formation of covalent bonds. All of them can form huge molecules. In organic compounds the carbon atoms bond extensively to each other in a great variety of ways, but in complex silicon and boron compounds these elements generally alternate with oxygen atoms to form rigid structures. All three elements react with the halogens, and with oxygen to form acidic oxides. Most of the carbonates, silicates, and borates that are derived from their oxy-acids are insoluble; those that *are* soluble are all mildly basic.

Carbon monoxide is an important constituent of industrial fuel gases. The enormous, steady release of carbon dioxide to the atmosphere due to human activities might cause a **greenhouse effect,** which would raise the average temperature of the atmosphere. The most abundant inorganic carbon compounds are the carbonates and bicarbonates, especially calcium carbonate in its various physical forms. Other important carbon compounds are cyanides, carbonyls, and carbides.

Silicon dioxide and the many silicate minerals are the most abundant chemical compounds in the earth's crust. Various silicates are the main or important constituents of such useful products as ceramics, cement, and **glass.** The latter is an amorphous solid that can technically be regarded as a cold, extremely viscous liquid. Other valuable silicon compounds are **zeolites** (naturally occurring silicate minerals with porous molecular structures) and **silicones** (synthetic polymers with unusual and desirable properties).

The borates resemble the silicates in variety and complexity but are much less abundant. The most important boron compounds are boric acid and borax. The **boranes** are a large class of boron–hydrogen compounds with highly unusual bonding properties.

Exercises

*Physical Properties
of Carbon, Silicon, and Boron*

1. Group IA is made up of metals only; group VIIA is composed of nonmetals only; but group IVA includes

nonmetals, metalloids, and metals. Account for these differences in terms of atomic sizes and the attraction of atomic nuclei for their valence electrons.

2. How many electrons would carbon need to lose in order to attain the electronic structure of a noble gas? How many would it need to gain? What would be the charge of the carbon ion if it lost these electrons? If it gained them? Do carbon atoms normally react by losing or gaining electrons? Explain.

3. Repeat Exercise 2 for boron.

4. Contrast the physical properties of the IIIA and IVA nonmetals with those of the VA, VIA, and VIIA nonmetals.

5. Write a nuclear equation for the bombardment of ^{30}Si with a neutron to produce the nuclide ^{31}P.

Diamond and Graphite

6. Explain in terms of structure why diamond is harder, less ductile, less malleable, and a poorer conductor of electricity than graphite.

7. What is formed when a diamond is heated to a high temperature in the presence of air? When heated to a high temperature in the absence of air and the vapors are allowed to cool and condense?

*8. (a) $\Delta H_r°$ for the reaction C (diamond) $\rightarrow$ C (graphite) is -1.88 kJ. Can we assume that $\Delta G_r°$ for the reaction is also negative? Why?

(b) The standard state of carbon is graphite. Why do not diamonds change spontaneously to graphite? ($\Delta G_f°$ of diamond formed from graphite is $+2.90$ kJ/mole.)

(c) Based on data given in parts (a) and (b), does entropy increase or decrease when graphite is converted to diamond?

9. Once ignited, which would tend to burn more readily, charcoal or diamond? Why?

10. Calculate the molecular weight in atomic mass units of a 1.0-carat diamond, a single, perfect crystal (1 carat = 0.20 g).

Chemical Properties of Carbon, Silicon, and Boron, and Some of Their Compounds

11. Write equations for the following:
(a) Powdered silicon is sifted into a bottle of fluorine.
(b) Boron oxide is dissolved in water and then sodium oxide is added to the solution.
(c) Carbon dioxide is dissolved in water and then some calcium chloride solution is added.
(d) Metasilicic acid is neutralized with sodium hydroxide.

12. Is a water solution of each of the following acidic, basic, or neutral: $NaHCO_3$, K_2CO_3, Na_2SiO_3, K_3BO_3? Write equations to show why you answered as you did.

*13. The acid ionization constants, K_{a1}, for H_3BO_3, H_2SiO_3, and H_2CO_3 are 1.7×10^{-10}, 5.8×10^{-10}, and 4.3×10^{-7}. Do these constants show the trend in acidity expected based on the oxidation states and electronegativities of the central atoms? Explain.

14. The ions BF_4^- and SiF_6^{2-} are known. What would you predict the shapes of these ions to be? What is the basis for your predictions?

15. Both carbon and silicon are in group IVA. Since the SiF_6^{2-} ion is known, would you also expect the CF_6^{2-} ion to exist? Explain.

16. Name the following compounds: K_2SiO_3; $AlBO_3$; K_4SiO_4; $Ti(SiO_3)_2$; H_2CO_3; NH_4HCO_3.

*17. Write Lewis structures for (a) sodium orthosilicate; (b) sodium metasilicate; (c) sodium orthoborate; (d) sodium metaborate.

*18. Consider the formulas $AlBO_3$ and $NaBO_3$. One of these formulas represents a peroxyborate. Which formula must it be? Write a Lewis structural formula for the peroxyborate ion.

Carbon Monoxide and Carbon Dioxide

19. Suggest a method for preparing fairly pure carbon monoxide on a large scale. Write equations for the reactions.

20. The text states that about 93 percent of the carbon monoxide in the earth's atmosphere is due to natural processes over which we have little or no control. Why, then, should we be so concerned about the 7 percent that is contributed by human activities?

21. It is said that a person exposed to a moderate concentration of carbon monoxide over a period of time, for example, in a closed automobile with the motor running and a leaky exhaust system, becomes tired and very sleepy, and shortly falls into a deep sleep from which he or she may never wake. Explain what is happening to produce the tired, sleepy feeling.

*22. Give an example of an equation in which carbon monoxide functions as (a) a reducing agent, (b) an oxidizing agent, and (c) a ligand.

23. Outline a possible series of natural processes by which carbon in atmospheric carbon dioxide could eventually become diamond.

24. Calculate the loss in fuel value when 1.0 ton of carbon is burned to give 10 percent CO and 90 percent CO_2 instead of 100 percent CO_2. Express the loss as a percentage of the maximum value, based on 100 percent conversion to CO_2.

25. Fear or excitement may cause one to breathe so rapidly that the CO_2 concentration in the blood falls. In what way, if at all, will this change the pH of the blood?

26. When liquid carbon dioxide issues from a fire extin-

guisher, some of it turns to gas and some to solid. Explain.

27. Much like sulfurous acid, the water solution we call carbonic acid is largely the dissolved dioxide. Based on this statement, write a balanced overall equation for the reaction of excess potassium hydroxide with carbonic acid. Also write the net ionic equation.

Carbonates and Bicarbonates

28. A common type of fire extinguisher contains a water slurry of potassium bicarbonate and a loosely stoppered bottle of sulfuric acid. When turned upside down, the extinguisher is activated and a stream of liquid is emitted through a hose. Explain the action, using appropriate equations.

29. Buttermilk contains lactic acid, $HC_3H_5O_3$. Write the equation for the reaction that occurs when buttermilk and baking soda are used in baking.

*30. Calculate the weight of $CaCO_3$ left in solution when 500 mL of a $0.10M$ $CaCl_2$ solution is added to 500 mL of a $1.0M$ Na_2CO_3 solution. (See Table 17.1 for the K_{sp} of $CaCO_3$.)

31. A carbonate rock is easily identified by testing it with acid.
    (a) Write a chemical equation for the test and describe the visible results.
    (b) Might not a sulfide rock give similar visible results? Explain with a chemical equation.
    (c) Could the carbonate be distinguished from the sulfide rock by the acid test? Explain.

32. Calcium carbonate is more soluble in rainwater than in absolutely pure water. Explain this action with balanced equations.

33. (a) Show, with at least four specific balanced equations, the amphoteric character of sodium bicarbonate.
    (b) On the basis of the equations that you have just written, explain why a solution of sodium bicarbonate is often kept handy in chemical laboratories as a safety precaution.

34. In localities where the soil tends to be acidic, crushed limestone may be applied by farmers to "sweeten" the soil.
    (a) Why might a soil be acidic?
    (b) Why would limestone lessen this acidity?
    (c) How is the term *sweeten* appropriate for this treatment?

35. If one blows one's breath into lime water through a straw, the liquid becomes milky and turbid. If the blowing is continued for some time, the liquid becomes a clear solution again. Explain with equations.

36. Explain how stalactites and stalagmites are formed. Write equations for the reactions.

*37. In localities where water contains appreciable quantities of calcium and magnesium ions, a hard scale tends to form in boilers and hot water tanks. Write equations for producing two compounds that may be in the scale. How might scale formation be prevented?

Compounds of Silicon and Boron

38. Write resolved formulas for orthoclase, $KAlSi_3O_8$, and tremolite, $Ca_2Mg_5Si_8O_{22}(OH)_2$.

39. Which among the following do not contain an appreciable amount of silicon compounds: clay, sand, granite, limestone, glass, concrete, petrified wood, asbestos, graphite, plaster of Paris, zeolite, gypsum, marble?

40. The formula for natrolite, a zeolite, is given in the caption of Figure 25.12 as $Na_2(Al_2Si_3O_{10}) \cdot 2H_2O$. Does this imply that any sample of natrolite must have this specific composition? Discuss the possibility that the composition does not conform to the law of definite composition.

41. Show with a Lewis structure the compound that is formed when methylamine reacts with boron trifluoride. Which reactant acts as the Lewis base? Why?

42. Boron trifluoride, BF_3, is available commercially as the diethyletherate, $BF_3(C_2H_5)_2O$. Write a Lewis structure for each of the simple compounds and for the addition compound.

43. Write an equation for the combustion of diborane.

Organic Chemistry

26

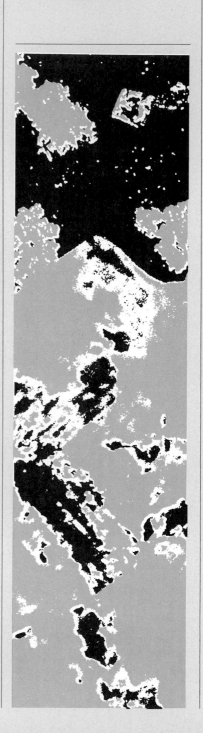

The Aliphatic and Aromatic Hydrocarbons

26•1 **Alkanes**

26•1•1 Structural Formulas of Alkane Molecules

26•1•2 Isomerism in Alkanes

26•1•3 Systematic Nomenclature of the Alkanes

Special Topic: Sources of Natural Gas

26•2 **Alkenes**

26•2•1 Structural Formulas of—and Isomerism in—Alkenes

26•2•2 Systematic Nomenclature of the Alkenes

26•3 **Alkynes**

26•3•1 Structural Formulas and Nomenclature of the Alkynes

26•4 **Benzene Hydrocarbons**

26•4•1 Structural Formulas of Benzene Hydrocarbons

26•4•2 Isomerism in Benzene Hydrocarbons

26•4•3 Systematic Nomenclature of Benzene Hydrocarbons

26•4•4 Fused-Ring Benzenoid Hydrocarbons

26•5 **Cycloalkanes**

26•6 **Sources of Hydrocarbons**

26•6•1 Refining of Petroleum

26•6•2 Processing of Coal

26•6•3 Hydrocarbon Conversions

26•7 **Properties of Hydrocarbons**

26•7•1 Physical Properties

26•7•2 Chemical Properties

Oxygen Derivatives of Hydrocarbons

26•8 **Alcohols**

26•8•1 Classes of Alcohols

26•8•2 Nomenclature

26•8•3 Some Common Alcohols

Special Topic: A Fuel Economy Based on Methanol

26•8•4 Physical Properties

26•8•5 Chemical Properties

26•9 **Phenols**

26•10 **Carboxylic Acids**

26•10•1 Nomenclature

26•10•2 Some Common Carboxylic Acids

26•10•3 Physical Properties

26•10•4 Chemical Properties

26•11 **Esters**

26•11•1 Nomenclature

26•11•2 Chemical Properties

Early chemists drew a sharp distinction between the compounds that originate during the growth of plants and animals and those of which the rocks and soils are composed. In their view, the former came into being only through some *vital force* associated with life processes; their syntheses in the laboratory were assumed to be impossible. Accordingly, those compounds were referred to as *organic* compounds, compounds derived from living organisms.

Friedrich Wöhler, a German chemist, initiated the downfall of the vital force idea when, in 1828, he was able to synthesize urea, a constituent of urine, simply by heating ammonium cyanate, a substance considered to be inorganic:

$$NH_4CNO \xrightarrow{\text{heat}} \quad$$

ammonium
cyanate

urea

From that time on, it became increasingly apparent that the special features of organic compounds lay not in their origin but in their structures.

Organic compounds constitute all or a major portion of petroleum, coal, proteins, fats, carbohydrates, vitamins, hormones, cellulose, anesthetics, antiseptics, antibiotics, enzymes, and a host of other useful products. Of the 50 chemicals produced in the greatest amounts by the U.S. chemical industry, approximately 30 are organic compounds. Eleven of these organic chemicals are **hydrocarbons**—compounds that contain only hydrogen and carbon. Fourteen additional compounds contain only hydrogen, carbon, and oxygen. Each of these 25 chemicals is produced in excess of 1 billion pounds annually.

The Aliphatic and Aromatic Hydrocarbons

We shall begin our coverage of organic chemistry by studying four series of hydrocarbons. Three of these are the **alkanes,** the **alkenes,** and the **alkynes,** frequently referred to collectively as the **aliphatic hydrocarbons.** The fourth series is the **benzene hydrocarbons,** which are part of a larger class of hydrocarbons known as **aromatic hydrocarbons.**

26•1   Alkanes

The atoms of the alkane hydrocarbons are joined to one another only through single bonds. The first ten members of this series are listed in Table 26.1. Note that the molecules of each member differ from those of the preceding and succeeding members by a constant number of atoms (one carbon and two hydrogen atoms). Such a series is called a **homologous series,** and each member is a **homolog** of the series. Furthermore, a general

Table 26•1 Alkanes (C_nH_{2n+2}, general formula)

CH_4	methane
C_2H_6	ethane
C_3H_8	propane
C_4H_{10}	butanes (2)[a]
C_5H_{12}	pentanes (3)
C_6H_{14}	hexanes (5)
C_7H_{16}	heptanes (9)
C_8H_{18}	octanes (18)
C_9H_{20}	nonanes (35)
$C_{10}H_{22}$	decanes (75)

[a] The numbers in parentheses are the calculated numbers of possible isomers for the individual molecular formulas. In the cases of the larger molecules, no attempts have been made to isolate all the isomers.

formula can be assigned to the series that will represent any member. For the alkanes, the general formula is C_nH_{2n+2}, where n is the number of carbon atoms. For example, if the molecule of a homolog contains ten carbon atoms, $n = 10$ and $(2n + 2) = 22$; the formula is $C_{10}H_{22}$.

Because methane, CH_4, is the first member, or parent, of the homologous series of alkanes, this series is also called the **methane series.** Natural gas, which is used extensively in domestic gas ranges and by industry, is made up of 50–94 percent methane. Some ethane, propane, and other alkane hydrocarbons of low molecular weight are also present in natural gas. Methane is the main end product of the anaerobic (without oxygen) decay of plants. The marsh gas that bubbles to the surface of swamps is chiefly methane. It is also the dangerous flammable gas found in coal mines. So-called *bottled gas* or *liquefied petroleum gas* is propane, butane, or a mixture of the two. These gases readily become liquids under pressure and thus are easily transported and stored.

● **26•1•1   Structural Formulas of Alkane Molecules**   Each carbon atom of an alkane molecule is covalently bonded to four other atoms, as is illustrated by the molecular models of ethane, propane, and butane in Figure 26.1. Often *condensed structural formulas* are used instead of Lewis formulas to represent the structures of the molecules. Such formulas provide all the information represented by the Lewis diagrams, but they are simpler to write. Some examples of condensed structural formulas are shown beneath the molecular models in Figure 26.1.

● **26•1•2   Isomerism in Alkanes**   Compounds that have the same molecular formula but different structural formulas were defined in Section 22.6.2 as *isomers.* Isomeric compounds are not possible in the alkane series until we reach the molecular formula C_4H_{10}. For this formula, two arrangements of the carbon chain are possible, and two isomers are known (see Figure 26.2). As shown in Figure 26.2, three arrangements are possible for C_5H_{12}, and three isomers have been found. As the molecular formulas grow larger, the numbers of isomers increase markedly (see Table 26.1). For the molecular formula $C_{20}H_{42}$, there are 366,319 possible isomers.

Isomeric compounds differ both chemically and physically from one another. They can be identified experimentally by differences in their melting points, boiling points, densities, solubilities in certain solvents, chemical activities, and absorption spectra (see Chapter 20).

In the examples of isomerism shown in Figure 26.2, the difference in structure is due to different arrangements of the carbon atoms. For example, in one of the C_4H_{10} alkane isomers, the one called *butane,* there are four carbon atoms joined in a continuous or unbranched chain. In the other C_4H_{10} alkane, the one called *methylpropane* or *isobutane,* three carbon atoms are in a continuous chain arrangement and the fourth is joined as a branch. Such an isomer is called a **branched-chain isomer.** In the condensed structural formulas, the point of attachment of each carbon branch is indicated by a vertical line.

● **26•1•3   Systematic Nomenclature of the Alkanes**   The names shown directly beneath the condensed structural formulas in Figure 26.2 are system-

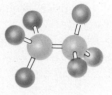

CH₃CH₃

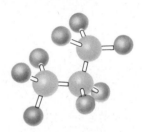

CH₃CH₂CH₃

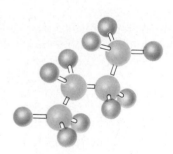

CH₃CH₂CH₂CH₃

Figure 26•1 Ball-and-stick models of molecules of ethane, propane, and butane, with their condensed structural formulas. (See also Figure 6.9.)

atic names. Those shown in parentheses are not systematic and are called common names or trivial names. For alkanes with the molecular formula C_6H_{14} and larger, a systematic system of nomenclature is a necessity.

The discussion of all the rules for systematic nomenclature is beyond the scope of this text. Nevertheless, a few simple rules are easily mastered and enable one to name a great many compounds. Once we have demonstrated how these rules apply to the systematic naming of alkanes, the naming of other organic compounds follows logically.

1. A portion of the molecule is chosen that can serve as the **parent compound.** For alkanes, this portion is the longest continuous chain of carbon atoms.

2. The name of the parent compound is a combination of the proper prefix, which denotes the number of carbon atoms, plus the series ending *-ane* for the alkane series. The names *methane, ethane,* and *propane* are used for alkanes with, respectively, one, two, and three carbon atoms. For the unbranched-chain alkanes with four through ten carbon atoms, the names are the singular forms of those listed in Table 26.1. The prefixes for five or more carbon atoms are derived from the Greek or Latin words for the respective numbers.

3. The carbon atoms of the chain of the parent compound are numbered in order in such a way that the lowest numbers possible locate the positions where hydrogen atoms have been replaced by other atoms or groups of atoms, such as $-CH_3$ (methyl) and $-CH_2CH_3$ (ethyl).

The general term **alkyl group** refers to a monovalent unit derived from an alkane hydrocarbon by removal of a hydrogen atom. If we drop the *-ane* ending from the alkane name and add *-yl*, the corresponding alkyl group name is obtained. Thus, methyl, $-CH_3$, is derived from methane, CH_4, and ethyl, $-CH_2CH_3$, is derived from ethane, CH_3CH_3. There are two possible alkyl groups that can be derived from propane, $CH_3CH_2CH_3$. Removal of a hydrogen atom from a terminal carbon atom gives the normal propyl, or *n*-propyl group. On the other hand, if a hydrogen atom is removed from the central carbon atom, an isomeric propyl group called the isopropyl group is obtained:

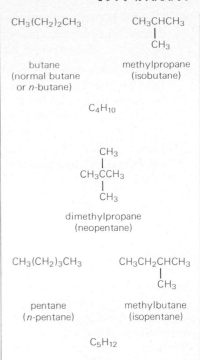

Figure 26•2 Isomers of the C_4H_{10} and C_5H_{12} alkane hydrocarbons. Compare the formula in this figure for normal butane with that in Figure 26.1. Note how $(CH_2)n$ is used in a formula for a carbon chain.

H—C—C—C—	← H—C—C—C—H →	H—C—C—C—H
$CH_3CH_2CH_2-$	$CH_3CH_2CH_3$	CH_3CHCH_3
n-propyl	propane	isopropyl

If a particular alkyl group is substituted more than once for hydrogen atoms in the parent compound, then the prefixes *di-* (two), *tri-* (three), *tetra-* (four), and so on, are attached to the name of the alkyl group. In the case of methyl group attachments, the names are *dimethyl* (two), *trimethyl* (three), *tetramethyl* (four), and so on.

If two different alkyl groups are attached to the parent compound, an alphabetical order (neglecting prefixes, di-, tri-, etc.) is used to decide the placement of the alkyl groups in the name.

4. The complete name of the compound is a combination of (a) the name of the parent portion, (b) the names of the attached groups, and (c) the numbers that show where the groups are attached. The numbers are separated from the alkyl group names by hyphens. If two groups are attached to the same carbon atom, the number is repeated. And if two or more numbers come consecutively, they are separated by commas. Example 26.1 illustrates the rules of systematic nomenclature as applied to the alkanes.

★ Special Topic ★ Special Topic ★ Special Topic ★ Special Topic ★ Special Topic ★ Special Topic ★ Special Topic ★ Special Topic

Sources of Natural Gas Natural gas is one of our most important energy sources, supplying about 30 percent of our fuel demand annually. Unfortunately, our rate of consumption of this valuable fuel far exceeds the rate of discovery of new deposits. A shortage of natural gas has already been felt by homes and industry in this country. Two methods of producing a gas comparable in composition to natural gas are being given serious consideration. These are (1) bioconversion and (2) coal gasification.

Bioconversion is the process by which solid wastes are converted to methane and other gases by anaerobic bacteria. In this process, a large airtight container, called a *digester*, is filled with all sorts of organic waste products, ranging from grass clippings to sewage. Acid-producing bacteria break complex organic molecules down into simpler ones, which are converted by other bacteria into methane. The yield of gas from vegetable wastes is about seven times more than that from animal wastes. The methane and other gases produced are easily separated from the sludge that settles to the bottom of the digester. The gases are passed through limewater, iron filings, and calcium chloride to remove the CO_2, H_2S, and H_2O, respectively.

Although no bioconversion plants are operating presently in the United States, there are more than 2,500 such plants in operation in India. Each of these small plants, known as *gobar* (Hindi, cow dung) plants, can produce as much as $250 \, m^3$ of methane per day. It is estimated that the United States could produce 1 trillion m^3 of natural gas per year from its sewage and garbage. Unfortunately, this amounts to less than 5 percent of the natural gas we consume.

The raw material that is being considered as the main source of a fuel gas is coal. Compared with the reserves of natural gas and oil, the U.S. reserves of coal are tremendous. If gas from coal were developed in a commercially acceptable process, a new source of easily used fuel would be available for centuries to come. In coal gasification, coal that is heated to a high temperature and pressure reacts with steam to produce a mixture of carbon monoxide and hydrogen. The carbon monoxide reacts further with water to generate more hydrogen, and the hydrogen and carbon monoxide react in the presence of a catalyst to produce methane. The following equations are presumed to be involved in the process:

$$coal \longrightarrow CH_4 + C$$
$$C + H_2O \longrightarrow CO + H_2$$
$$C + 2H_2 \longrightarrow CH_4$$
$$CO + H_2O \longrightarrow CO_2 + H_2$$
$$CO + 3H_2 \longrightarrow CH_4 + H_2O$$

It has been estimated that coal gas could furnish more than 60 percent of our energy needs by the year 2000. At present, only pilot plants are operating in our country. The main problem holding back production today appears to be one of economics, because gas from petroleum is still cheaper. A typical gasification plant would cost more than $1 billion.

At the present time, few companies or banks are willing to make such a major investment. As petroleum reserves become more scarce and the cost of producing natural gas increases, however, it is expected that commercial coal gasification plants will be built. In the meantime, the federal government is subsidizing a number of pilot plants in order to help industry test designs for improving the efficiency of coal gas production.

• **Example 26•1** •

Write the systematic name for each of the following alkanes:

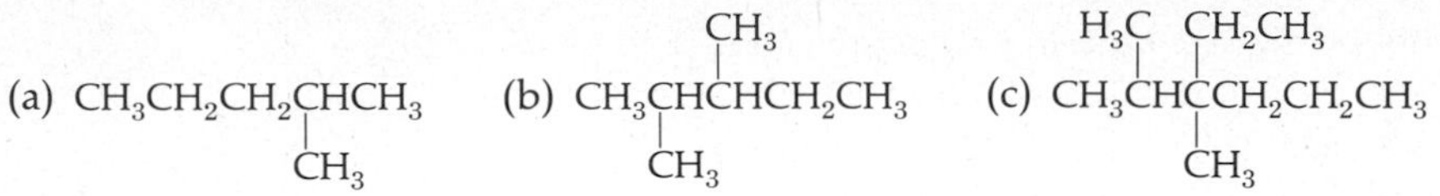

(a) $CH_3CH_2CH_2CHCH_3$
  $\underset{|}{\overset{|}{CH_3}}$

(b) $CH_3CHCHCH_2CH_3$
  with CH_3 above and CH_3 below

(c) $CH_3CHCCH_2CH_2CH_3$
  with H_3C CH_2CH_3 above and CH_3 below

• **Solution** •

(a) There is a methyl group on the second carbon (C-2) from the end of the five-carbon chain. The name is 2-methylpentane. (The name 4-methylpentane is incorrect, because the rule is to use the lowest numbers possible.)
(b) Here there is a methyl group at C-2 and one at C-3. The name is 2,3-dimethylpentane. [Note that to achieve the lowest numbers, this chain is numbered from left to right. In part (a), the chain is numbered from right to left.]
(c) There are methyl groups at C-2 and C-3 and an ethyl group at C-3 of the longest chain of six carbon atoms. Naming ethyl first because it precedes methyl alphabetically, the name is 3-ethyl-2,3-dimethylhexane.

See also Exercises 1 and 3 at the end of the chapter.

26•2 Alkenes

Molecules of the alkene series of hydrocarbons are characterized by having two adjacent carbon atoms joined to one another by a double bond. The first few members of this series are listed in Table 26.2. Any member in this series can be represented by the general formula C_nH_{2n}.

Because ethylene is the parent of the homologous series of alkenes, this series is also called the **ethylene series.** Ethylene constantly is being generated in living plants, from which it escapes as a gas. If the ethylene concentration in a fruit is allowed to accumulate, or if the fruit is sprayed with ethylene, the ripening process is accelerated. Other hydrocarbons with multiple bonds, such as propene, C_3H_6, butene, C_4H_8, and acetylene, C_2H_2, exert the same ripening effect. It is believed that these hydrocarbons stimulate the production of enzymes that control the ripening process.

Industrial production of ethylene exceeds that of any other organic chemical. The ethylene itself has few direct uses, but the compounds produced from ethylene are extremely important to our economy and standard of living. Propene (propylene), C_3H_6, ranks second among organic chemicals in the amount produced in the United States. It too is used primarily to produce a large number of valuable chemical products.

26•2•1 Structural Formulas of—and Isomerism in—Alkenes
A double bond connects one pair of carbon atoms in the molecules of all members of the ethylene series. As shown in Figure 26.3, the other carbon atoms are bonded as in the alkanes. As with the alkanes, condensed structural formulas are frequently used to describe the bonding arrangements in alkenes. Examples for ethylene and propene are shown beneath the molecular models in Figure 26.3.

Key Chemical: $CH_2{=}CH_2$
ethylene

How Made
cracking (thermal decomposition) of petroleum in the presence of a catalyst or steam:

petroleum $\xrightarrow[\text{or steam}]{\text{catalyst}}$
$CH_2{=}CH_2 + H_2$

U.S. Production; Cost
~27 billion lb per year
~12¢ to 13¢ per pound

Major Derivatives
45% polyethylene
20% ethylene oxide and ethylene glycol
15% vinyl chloride
10% styrene

Major End Uses
65% plastics
10% antifreeze
5% fibers
5% solvents

Isomeric compounds are possible in the alkene series for the molecular formula C_4H_8. As illustrated by the following condensed structural formulas, there are three different ways of organizing the four carbons and the double bond into chain-type molecules:

$$CH_3CH_2CH{=}CH_2 \qquad CH_3CH{=}CHCH_3 \qquad CH_3\overset{\displaystyle CH_3}{\underset{|}{C}}{=}CH_2$$

1-butene	2-butene	methylpropene
	(two isomers)	
b.p. −6.3 °C	b.p. 3.7 °C and 0.9 °C	b.p. −6.9 °C

• *Cis–Trans or Geometric Isomerism* • While it is possible for any segment of a carbon chain that is connected by a single bond to rotate essentially freely about the bonds, rotation about a double bond is not possible. Alkyl groups (and other atoms and groups) connected to the two carbon atoms of a double bond are restricted to certain spaces in the molecules. This restriction gives rise to *cis–trans* or *geometric isomerism*. In Figure 26.4, combined Lewis and condensed structural formulas show the two different relative positions of the hydrogen atoms and methyl groups in 2-butene. The designation of cis and trans isomers is analogous to that used for the coordination compounds in Section 22.6.2.

26•2•2   Systematic Nomenclature of the Alkenes Systematic nomenclature for alkenes involves some extension of the rules applied to alkanes:

1. The longest continuous chain of carbon atoms containing the double bond serves as the parent compound.

2. The ending *-ane* of the corresponding alkane hydrocarbon name is replaced by the ending *-ene*.

3. The position of the double bond is indicated by the lower number of the numbers of the carbon atoms to which it is attached. The number that

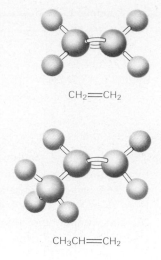

$CH_2{=}CH_2$

$CH_3CH{=}CH_2$

Figure 26•3 Ball-and-stick models of molecules of ethylene and propene, with their condensed structural formulas. (See also Figure 6.13.)

Table 26•2 Alkenes
(C_nH_{2n}, general formula)

C_2H_4	ethene (ethylene)
C_3H_6	propene (propylene)
C_4H_8	butenes (4)
C_5H_{10}	pentenes (6)

Table 26•3 Alkynes
(C_nH_{2n-2}, general formula)

C_2H_2	ethyne (acetylene)
C_3H_4	propyne
C_4H_6	butynes (2)
C_5H_8	pentynes (3)

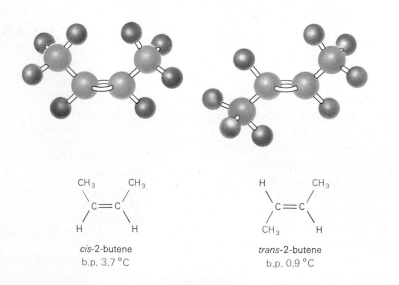

cis-2-butene
b.p. 3.7 °C

trans-2-butene
b.p. 0.9 °C

Figure 26•4 Ball-and-stick models and combined Lewis and condensed structural formulas of molecules of *cis*-2-butene and *trans*-2-butene.

represents this position is placed before the parent compound name. Alkyl groups attached to the parent compound are designated as is done for the alkane hydrocarbons.

4. If a geometric isomer is designated, the name begins with *cis-* or *trans-*.

The following examples serve to illustrate these rules:

$CH_3CH_2CH_2CH_2CH{=}CH_2$ $CH_3CH_2CHCH{=}CH_2$
$\qquad\qquad\qquad\qquad\qquad\qquad\qquad\qquad\quad |$
$\qquad\qquad\qquad\qquad\qquad\qquad\qquad\qquad\;\; CH_3$

$$\begin{array}{c} CH_3 \\ CH_3 \qquad CHCH_3 \\ \diagdown\;\;\;\diagup \\ C{=}C \\ \diagup\;\;\;\diagdown \\ H \qquad\quad H \end{array}$$

  1-hexene   3-methyl-1-pentene   *cis*-4-methyl-2-pentene

26•3   Alkynes

Molecules of the alkyne series of hydrocarbons are characterized by having two adjacent carbon atoms joined to one another by a triple bond. The first few members of this series are listed in Table 26.3. Any member in this series can be represented by the general formula C_nH_{2n-2}. Because acetylene is the parent of the homologous series of alkynes, this series is also called the **acetylene series.**

An old method of preparing acetylene is by the reaction of calcium carbide with water:

$$CaC_2 + 2H_2O \longrightarrow C_2H_2 + Ca(OH)_2$$

Carried out on a small scale, this reaction provides the acetylene flame for carbide lamps. At one time, miners used such lamps, regulating the amount of gas produced by controlling the rate of water dropped into the reaction canister.

The newer commercial method of preparing acetylene is by heating methane and its homologs at a high temperature in the presence of a catalyst. For methane the reaction is represented by the following equation:

$$2CH_4 \xrightarrow[\text{heat}]{\text{catalyst}} C_2H_2 + 3H_2$$

Acetylene is distributed as a bottled gas to burn in oxyacetylene torches used in the welding and cutting of metals. The most important commercial use of acetylene is in the synthesis of complicated organic compounds used to make plastics and synthetic rubber.

26•3•1   Structural Formulas and Nomenclature of the Alkynes   A triple bond connects one pair of carbon atoms in the molecules of all members of the acetylene series. As shown in Figure 26.5, the other carbon atoms are bonded as in the alkanes. Condensed formulas can be used to describe the bonding arrangements in alkynes, as shown beneath the molecular models in Figure 26.5.

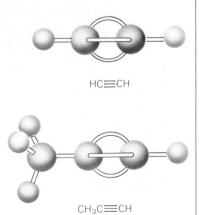

HC≡CH

CH₃C≡CH

Figure 26•5 Ball-and-stick models of molecules of acetylene and propyne, with their condensed structural formulas. (See also Figure 6.14.)

As is true for the alkanes and alkenes, isomeric compounds are not possible among alkynes unless there are at least four carbon atoms. There are two possible arrangements of the carbon atoms and the triple bond in C_4H_6:

$$CH_3CH_2C \equiv CH \qquad CH_3C \equiv CCH_3$$
$$\text{1-butyne} \qquad\qquad \text{2-butyne}$$

Systematic nomenclature for alkynes follows the same rules as for the alkenes except that the ending -yne is used in place of -ene.

26•4   Benzene Hydrocarbons

The most important aromatic hydrocarbons are derived from benzene and have the general formula C_nH_{2n-6}. Some of the members of the benzene series are listed in Table 26.4.

Among organic compounds, benzene ranks third, right after propene, in U.S. production. Close behind benzene in annual production are the aromatic hydrocarbons methylbenzene (toluene) and ethylbenzene. One of the dimethylbenzenes, 1,4-dimethylbenzene (para-xylene), and isopropylbenzene (cumene) also are in the list of the top 50 chemicals. These five aromatic hydrocarbons are the principal raw materials for a large part of the organic chemical industry.

Benzene is a hemotoxin, a substance that damages bone marrow and inhibits the formation of blood cells, much like leukemia. In addition, benzene has been shown to be carcinogenic in animals. Consequently, the Occupational Safety and Health Administration (OSHA) has placed it on a list of chemicals to which exposure should be limited. An exposure limit of an average of 1 ppm in air over eight hours has been proposed. The use of benzene in such consumer products as paint remover, rubber cement, carburetor cleaners, denatured alcohol, and art and craft supplies faces severe restrictions. Safer substitutes, such as toluene, have been proposed for use in these products. What effect the hazards of benzene exposure will have on the industrial importance of benzene is uncertain.

26•4•1   Structural Formulas of Benzene Hydrocarbons

Conventional condensed structural formulas of benzene and substituted benzenes show a hexagon for the six ring carbon atoms. Hydrogen atoms bonded to these carbon atoms are not shown, but groups that replace hydrogen atoms are shown. The additional three pairs of electrons may be represented by double bonds. However, we will use a more modern formula with a circle inside a hexagon. The circle refers to the delocalized pi electron system. For benzene and toluene (methylbenzene), the two types of formulas are drawn as follows:

benzene   toluene
(Kekulé formulas)

benzene   toluene
(modern formulas)

Table 26•4 Benzene-type hydrocarbons (C_nH_{2n-6}, general formula)

C_6H_6	benzene
C_7H_8	methylbenzene
C_8H_{10}	ethylbenzene
	dimethylbenzenes (3)
C_9H_{12}	propylbenzenes (2)
	ethylmethylbenzenes (3)
	trimethylbenzenes (3)

Key Chemical:

benzene

How Made

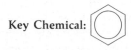

toluene

$+ CH_4$

also obtained from petroleum refining, steam cracking, and distillation of coal tar liquids

U.S. Production; Cost
~11 billion lb (1.5 billion gal) per year
~11¢ per pound (85¢ per gallon)

Major Derivatives
50% ethylbenzene
15% cyclohexane
15% isopropylbenzene

Major End Uses
40% plastics and rubbers
20% nylons

26•4•2 **Isomerism in Benzene Hydrocarbons** As shown in Figure 26.6, four isomeric compounds are possible with benzene hydrocarbons that have the molecular formula C_8H_{10}. One isomer is obtained by substituting an ethyl group for one hydrogen atom of benzene to give a monosubstituted benzene. The other three are obtained by substituting methyl groups for two hydrogen atoms to give disubstituted benzenes. For the molecular formula C_9H_{12}, eight isomers are possible. Two are monosubstituted benzenes, three are disubstituted benzenes, and three are trisubstituted benzenes.

26•4•3 **Systematic Nomenclature of Benzene Hydrocarbons** To facilitate the naming of benzene hydrocarbons, we shall divide them into three classes: monosubstituted, disubstituted, and more highly substituted benzenes.

• *Monosubstituted Benzenes* • Many monosubstituted benzenes are named in a systematic manner by combining the substituent name with the word *benzene*. For example, see ethylbenzene in Figure 26.6. However, a number of monosubstituted benzenes have special names that are an accepted part of systematic nomenclature. For example, the name *toluene* is generally used instead of methylbenzene, and *cumene* instead of isopropylbenzene.

• *Disubstituted Benzenes* • For disubstituted benzenes, the three possible isomers are named using the prefixes *ortho, meta,* and *para* (abbreviated *o, m,* and *p*) to designate the 1,2-, 1,3-, and 1,4- relationships of substituents on the benzene ring:

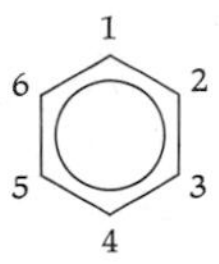

For dimethylbenzene the name *xylene* is used, and the three isomers are called *o*-xylene, *m*-xylene, and *p*-xylene, as indicated in Figure 26.6.

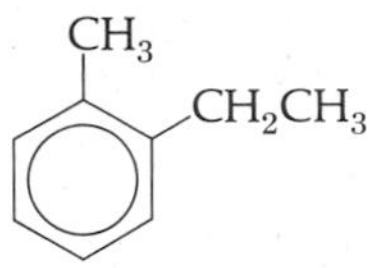

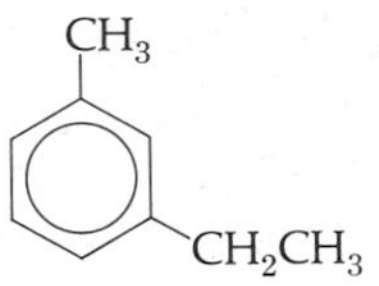

• *More Highly Substituted Benzenes* • For trisubstituted and more extensively substituted benzenes, the numbering system should be used. The following examples are illustrative:

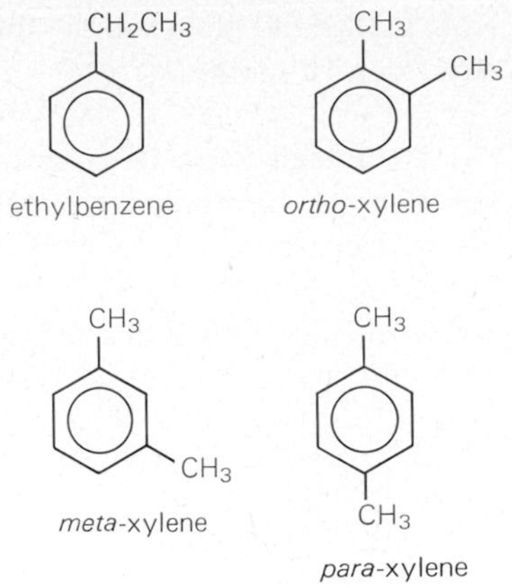

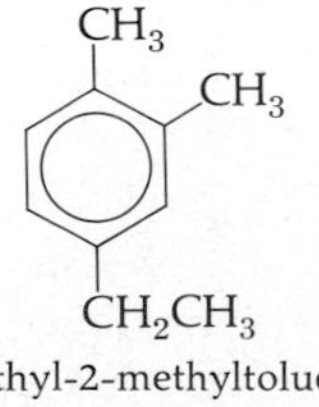

Figure 26•6 Four benzene hydrocarbons with the molecular formula C_8H_{10}. (See also Figure 6.19.)

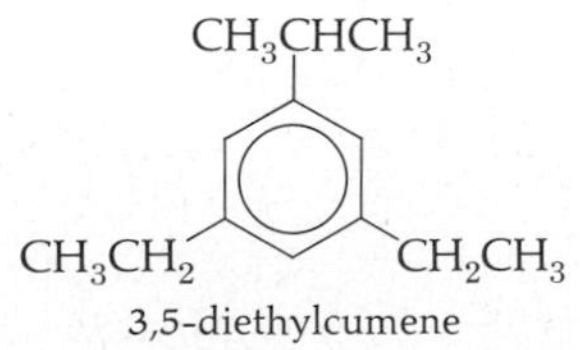

Note that in the first example the substituents are listed in alphabetical order, ethyl before methyl.

26·4·4 Fused-Ring Benzenoid Hydrocarbons

If a hydrocarbon is composed of two or more rings and at least one pair of rings shares two carbons, the hydrocarbon is called a **fused-ring hydrocarbon.** If benzene rings are fused together, the hydrocarbons are called **fused-ring benzenoid hydrocarbons.** Some examples include the following:

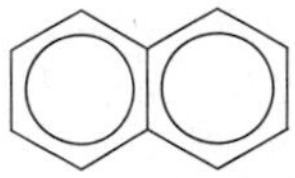

naphthalene, $C_{10}H_8$ 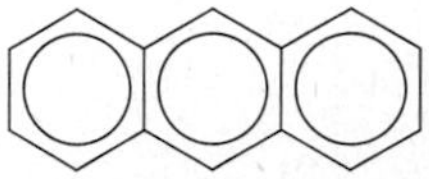 — anthracene, $C_{14}H_{10}$ 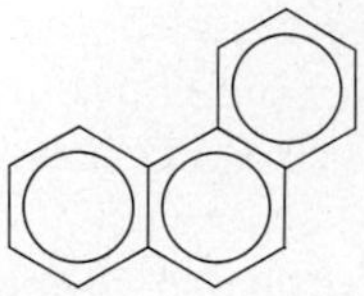 — phenanthrene, $C_{14}H_{10}$ 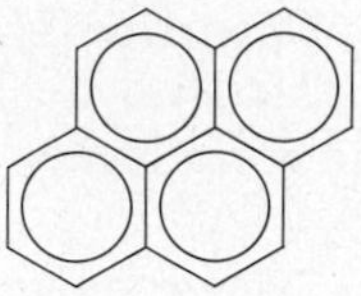 — pyrene, $C_{16}H_{10}$

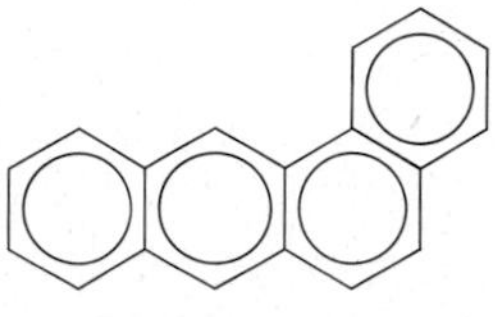

benz[a]anthracene, $C_{18}H_{12}$ 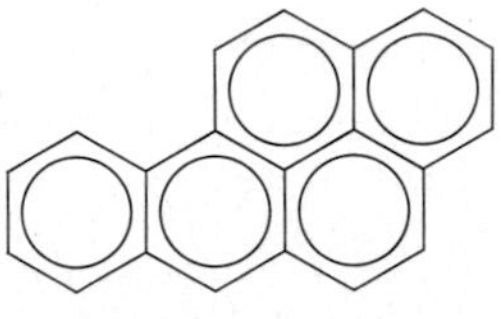 — benzo[a]pyrene, $C_{20}H_{12}$ 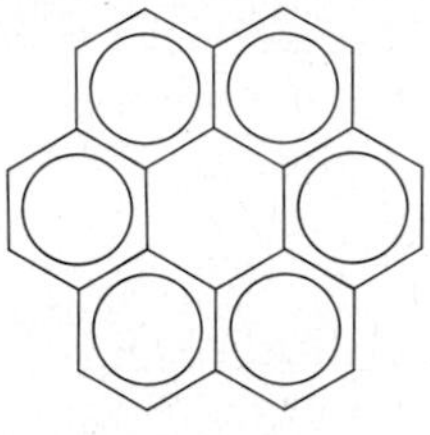 — coronene, $C_{24}H_{12}$

Several of the larger fused-ring hydrocarbons are found in coal tar and soot. Two of the compounds just shown, benz[a]anthracene and benzo[a]pyrene, are known to be carcinogens. Benzo[a]pyrene is found in tobacco smoke, auto exhausts, asphalt streets, and charcoal-broiled steaks. These large fused-ring benzenoid compounds are all quite stable. They tend to be formed when organic molecules are heated to high temperatures in the absence of enough oxygen to undergo complete combustion. For example, in the burning of tobacco in cigarettes, small high-energy organic molecules presumably combine to form the large, stable, fused-ring compounds in the smoke and tar produced. When the number of fused rings becomes very large in two directions, a graphite-like structure results (see Figure 25.1).

26·5 Cycloalkanes

The **cycloalkanes** are members of a homologous series of cyclic hydrocarbons, isomeric with the alkenes (C_nH_{2n}). The first four members without any attached alkyl groups are as follows:

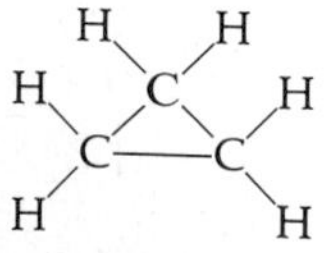

cyclopropane C_3H_6 — cyclobutane C_4H_8 — cyclopentane C_5H_{10} — cyclohexane C_6H_{12}

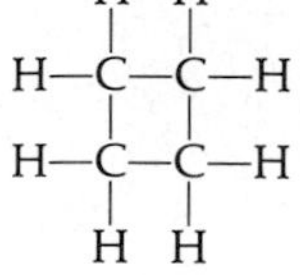

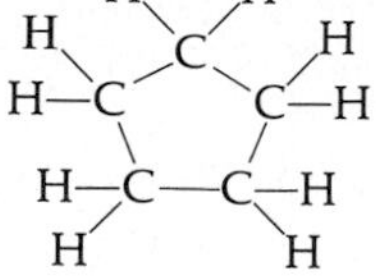

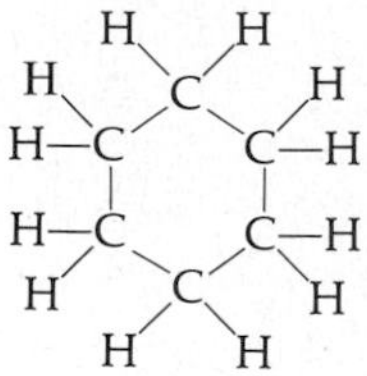

Key Chemical: para-xylene

How Made

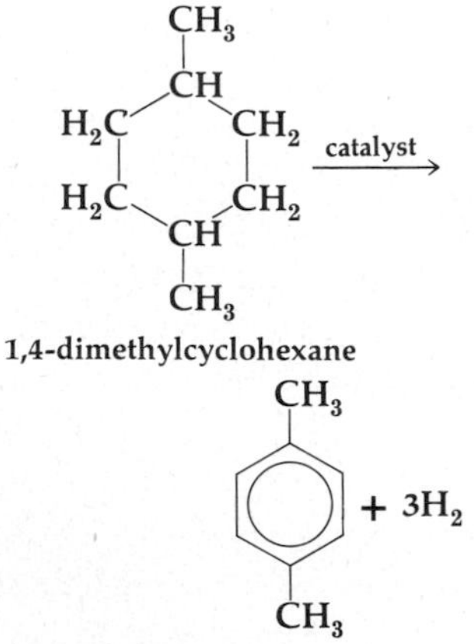

1,4-dimethylcyclohexane $\xrightarrow{\text{catalyst}}$ + $3H_2$

also from distillation of coal tar liquids

U.S. Production; Cost
~3.3 billion lb per year
~12¢ to 14¢ per pound

Major Derivatives
100% terephthalic acid and dimethyl phthalate

Major End Uses
polyester fibers, films, and plastics

Cyclopropane is one of the most potent inhalation anesthetics known, producing unconsciousness within a few seconds after inhaling. (The mechanism of anesthetic action will be discussed in Section 27.7.) The cyclopentane and cyclohexane rings occur in many natural products as fused rings. For example, the important compounds known as *steroids* have one cyclopentane and three cyclohexane rings fused together (see Figure 26.7).

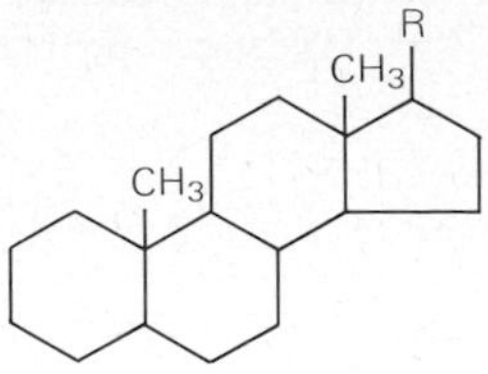

26•6   Sources of Hydrocarbons

Some of the alkane hydrocarbons in nature are the products of living processes. For example, methane is produced by the anaerobic decomposition of vegetable matter (see Section 26.1). Large amounts of alkane hydrocarbons are found in natural gas and petroleum. **Petroleum** is a complex mixture of gaseous, liquid, and solid hydrocarbons that are the end products of the decomposition of animal and vegetable matter that has been buried in the earth's crust for a long time. Small amounts of nitrogen and sulfur compounds are also present.

Another source of hydrocarbons and other organic compounds is **coal,** a solid amorphous substance that is a mixture of very complex molecules; typical of these are fused-ring benzenoid hydrocarbons with —OH and —CO$_2$H groups. One largely untapped source of hydrocarbons is **oil shale,** a porous rock that has petroleum dispersed throughout its structure.

26•6•1   Refining of Petroleum

The refining of petroleum involves the separation of the organic compounds as they exist in nature and the conversion of some of them into other organic compounds. The first step consists of separating the crude oil by fractional distillation (see Section 12.3.2) into batches with different boiling ranges (see Table 26.5).

Crude oil is pumped continuously through a furnace, where it is heated. The hot liquid then passes into a "flash chamber" where, by lowering the pressure, the lower-boiling components are vaporized. The vapors pass into a "bubble tower" (see Figure 26.8). Here, the higher-boiling components are condensed in the lower portion of the tower and flow downward from plate to plate. The ascending vapors bubble through the liquids that have condensed in the plates; hence the more volatile lower-boiling components collect in the upper portion of the tower. These hydrocarbons make up petroleum ether, gasoline, and kerosene and belong mostly to the methane series. The higher-boiling components in the lower portion of the tower make up heating oil and lubricating oil. These hydrocarbons and those in the residue, the nonvolatile hydrocarbons, are composed mainly of alkanes and cycloalkanes of high molecular weights. The separated hydrocarbons may be used as such, or they may be converted into more useful hydrocarbons (see Section 26.6.3).

26•6•2   Processing of Coal

Coal has long been an indirect source of many aromatic compounds. On destructively distilling coal to produce coke, some of the coal is converted into a black, viscous liquid called *coal tar.* This tarlike liquid is separated into pure compounds by fractional distillation, together with crystallization processes, centrifuging, and solvent extraction. One ton of coal yields about 140 lb of tar, along with about 1,500 lb of coke, 10,000 ft^3

Figure 26•7 Fused-ring skeleton of steroids. Double bonds, hydroxyl, and other groups may be present. On the cyclopentane ring, R is a hydrocarbon group or oxygen-type functional group (see Section 26.8).

Key Chemical:

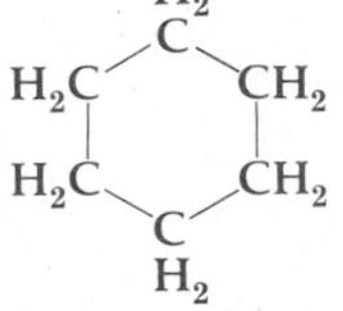

cyclohexane

How Made

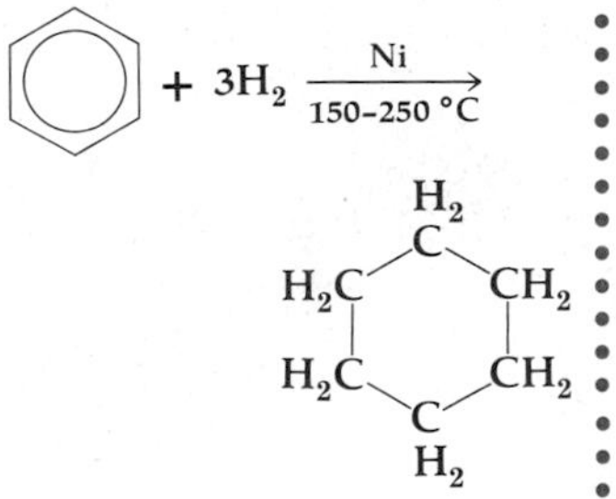

also from petroleum distillate

U.S. Production; Cost
~2.5 billion lb per year
~11¢ per pound (80¢ per gallon)

Major Derivatives
65% adipic acid
30% caprolactam

Major End Uses
60% nylon 66
30% nylon 6

Table 26.5 Hydrocarbon fractions obtained from petroleum

Fraction	Size range of molecules	Boiling point range, °C	Uses
gas	C_1-C_5	−164–30	gaseous fuel; production of carbon black, hydrogen, or gasoline (by polymerization)
petroleum ether (ligroin)	C_5-C_7	30–90	solvent; dry cleaning
gasoline (straight-run)	C_5-C_{12}	30–200	motor fuel
kerosene	$C_{12}-C_{16}$	175–275	illuminant; fuel
gas oil, fuel oil, and diesel oil	$C_{15}-C_{18}$	250–400	furnace fuel; fuel for diesel engines; cracking
lubricating oils, greases, petroleum jelly	C_{16} up	350 up	lubrication
paraffin (wax)	C_{20} up	melts 52–57	candles; waterproofing fabrics; matches; home canning
pitch and tar		residue	artificial asphalt
petroleum coke		residue	fuel; electrodes

of coal gas (mostly hydrogen and methane), and about 25 lb of ammonium salts. The 140 lb of coal tar in turn produces from 8 to 15 lb of phenol and cresols (see Section 26.9), 14 to 18 lb of naphthalene, 1 lb of benzene, 3 lb of toluene and the xylenes, small amounts of many other aromatic compounds, and about 70 lb of pitch.

26.6.3 **Hydrocarbon Conversions** Gasoline produced by the fractional distillation of petroleum is called **straight-run gasoline.** If we had to depend solely on this source for our gasoline supply, the gasoline shortage would have struck many years ago. Fortunately there are other sources of gasoline. These sources involve conversion of one hydrocarbon to another by processes known as cracking and aromatization.

• *Cracking* • Pyrolysis of alkanes at temperatures of around 500 to 700 °C and at pressures as high as 250 psi brings about decomposition into alkenes, hydrogen, and lower molecular weight alkanes. As applied to the petroleum industry, the process is known as **cracking.** Oil refineries crack crude oil that boils at temperatures higher than the gasoline range to produce hydrocarbons that boil within the desired range. To illustrate with a simple example (one that does not produce gasoline-range hydrocarbons), the *thermal cracking* of propane gives some methane and ethene in addition to propene and hydrogen:

$$2CH_3CH_2CH_3 \xrightarrow{500-700 \ °C} CH_3CH{=}CH_2 + H_2 + CH_4 + CH_2{=}CH_2$$

Unbranched-chain hydrocarbons are the main products of thermal cracking of higher molecular weight hydrocarbons.

   Catalytic cracking processes are also used. One widely used procedure involves passage of the hydrocarbon vapors over a solid catalyst of SiO_2 and Al_2O_3 at a temperature of about 450 to 500 °C. Pressures just above atmospheric pressure (50 psi or lower) are used. The alkenes and alkanes formed from catalytic cracking processes tend to be largely of the branched-chain type.

   • *Aromatization* • When alkanes and cycloalkanes are heated at a high temperature (about 500 °C) and a high pressure (about 300 to 700 psi), hydrogen may be evolved and benzene-type compounds may form. This type of chemical change is called **aromatization.** The hydrocarbons hexane, cyclohexane, and methylcyclopentane all react to give benzene. Using hexane, the equation is

$$CH_3(CH_2)_4CH_3 \xrightarrow{\text{catalyst}} \bigcirc + 4H_2$$

The C_7 hydrocarbons heptane, methylcyclohexane, and 1,2-dimethylcyclopentane all react to produce toluene.

   Depending on the hydrocarbon reactant, *cyclization, isomerization,* and *dehydrogenation* may be involved in the aromatization reaction. In the petroleum industry, aromatization is called *reforming.* When platinum is the catalyst, the process is known as *platforming.*

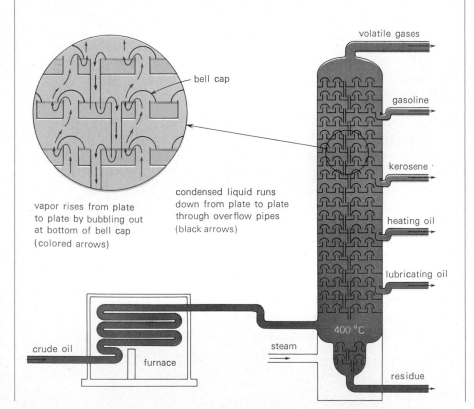

bell cap

vapor rises from plate to plate by bubbling out at bottom of bell cap (colored arrows)

condensed liquid runs down from plate to plate through overflow pipes (black arrows)

volatile gases

gasoline

kerosene

heating oil

lubricating oil

400 °C

crude oil

furnace

steam

residue

Figure 26•8 Diagram of a fractionating tower for petroleum distillation. The cutaway view shows how the vapor and liquid phases are kept in contact with each other so that condensation and distillation occur throughout the column.

Benzene may also be produced from toluene by a process called *hydrodealkylation* (see the Key Chemical section on benzene).

26•7   Properties of Hydrocarbons

26•7•1   Physical Properties   The alkanes, alkenes, alkynes, and benzene hydrocarbons are very much alike physically. All are colorless compounds, insoluble or only slightly soluble in water, but quite soluble in nonpolar solvents. The hydrocarbons with low molecular weights, C_1 to about C_5, are gases; those with intermediate weights are liquids; and those with high molecular weights are solids (see Table 26.5). The actual melting and boiling points vary for molecules with the same number of carbon atoms, depending on the presence or absence of double and triple bonds and the number and kind of chain branches.

26•7•2   Chemical Properties   In many textbooks of organic chemistry, several hundred pages are devoted to describing the chemical reactions of hydrocarbons. In this text we will present only a few typical reactions from this important area.

• *Combustion* •   All hydrocarbons are combustible, burning in oxygen or air to carbon dioxide and water. For example,

$$C_8H_{18} + 12\tfrac{1}{2}O_2 \longrightarrow 8CO_2 + 9H_2O$$
an octane

The combustion of hydrocarbons in a gasoline engine, a diesel engine, an oil furnace, or a gas stove does not take place so completely as the above equation implies. Because of incomplete reaction, some carbon monoxide, carbon (soot), and even hydrocarbons may constitute an appreciable fraction of the combustion products, depending on how the combustion is carried out.

The hydrocarbons in gasoline have boiling points sufficiently low so that gasoline changes to a vapor in the gasoline engine. In the operation of a gasoline engine, the downstroke of the piston draws a mixture of gasoline vapor and air into the cylinder. This mixture is compressed into a small volume at the top of the cylinder by the upstroke. When the piston nears the top of the upstroke, the timing system initiates a spark across the spark plug gap that ignites the gasoline. The hot expanding gases push the piston down, thus turning the crank shaft.

For the engine to run smoothly, burning must start at the proper time at the spark plug and move through the gas mixture at a fast, uniform rate. For straight-run gasoline, the rate of burning tends to be uniform until perhaps three fourths of the gasoline is consumed; then the rate suddenly accelerates up to four times the original. This causes knocking, decreases power, and increases wear on the engine.

High-compression engines, in which there is less space at the top of the cylinder so that the upstroke of the piston puts the fuel mixture under greater pressure, give more power and more efficient operation. Unfortunately, however, as the compression ratio goes up, the tendency to knock also increases. Hence low-compression engines had to be used until gasoline could be produced that had good antiknock properties.

• *Octane Number* • A large percentage of the molecules in straight-run gasoline are of the unbranched-chain variety—the type of hydrocarbon that produces knocking. Hydrocarbons whose molecules are of the branched-chain variety have antiknock characteristics. As a means of measuring the knocking properties of gasoline, two pure hydrocarbons were selected to be used as standards. One, heptane, is of the unbranched-chain type; the other, 2,2,4-trimethylpentane ("isooctane") is of the branched-chain type:

$$CH_3CH_2CH_2CH_2CH_2CH_2CH_3$$

heptane
octane number 0

$$CH_3\overset{\displaystyle CH_3}{\underset{\displaystyle CH_3}{C}}CH_2\overset{}{\underset{\displaystyle CH_3}{C}HCH_3}$$

2,2,4-trimethylpentane
("isooctane")
octane number 100

They were assigned octane numbers of 0 and 100, respectively. When a gasoline is being rated, various mixtures of the two standard hydrocarbons are tried in a test engine until a mixture is found that produces the same knocking as the gasoline being rated. The percentage of "isooctane" in the standard mixture is called the **octane number** of the gasoline being rated. For example, a 70-octane gasoline gives the same knocking in a test engine as a mixture containing 70 percent "isooctane" and 30 percent heptane. Note that whether the gasoline being rated actually contains any "isooctane" and heptane is not determined.

Lead tetraethyl, $Pb(C_2H_5)_4$, inhibits knocking when added to gasoline in concentrations up to 0.01 percent. Some of the gasoline sold in the United States contains lead tetraethyl or similar lead alkyls. Because lead compounds contribute to air pollution, their use in gasolines is being curtailed.

Gasolines produced from alkanes by chemical processes such as cracking and aromatization are made up largely of branched-chain hydrocarbons or of aromatic hydrocarbons. They are high-octane gasolines. By blending them with straight-run gasoline, no-lead gasolines have been made available with octane ratings ranging from about 84 to the low 90s.

• *Substitution Reactions* • The reactions characteristic of alkanes are **substitution reactions,** in which one atom or group of atoms replaces another. For example, at elevated temperatures or in the presence of ultraviolet light, one or more hydrogen atoms in an alkane molecule may be replaced by atoms of chlorine or bromine. If chlorine and ethane are the reactants, the monosubstitution product is chloroethane (ethyl chloride):

$$CH_3CH_3 + Cl_2 \overset{h\nu}{\longrightarrow} CH_3CH_2Cl$$

ethane

chloroethane
(ethyl chloride)

Liquid ethyl chloride, with a boiling point of 12 °C, is often used as a local anesthetic; it absorbs its heat of vaporization and evaporates so rapidly that it freezes tissues and thereby causes partial loss of feeling. It is frequently sprayed on that area of a baseball player's body that has been struck by a pitcher's errant throw.

• *Addition Reactions* • Alkanes are called **saturated hydrocarbons** because their molecules have the maximum number of hydrogen atoms possible. Alkenes and alkynes are called **unsaturated hydrocarbons** because they have bonds available to react with hydrogen or other substances. Whereas alkanes react by substitution reactions, the characteristic reactions of alkenes and alkynes are **addition reactions** at the site of the carbon–carbon multiple bond. Some typical examples of addition reactions are illustrated by the following equations:

$$CH_2{=}CH_2 + Cl_2 \longrightarrow ClCH_2CH_2Cl$$
$$\text{1,2-dichloroethane}$$

$$CH_3CH{=}CH_2 + HBr \longrightarrow CH_3\underset{\underset{Br}{|}}{CH}CH_3$$
$$\text{2-bromopropane}$$
$$\text{(isopropyl bromide)}$$

$$CH_3CH{=}CH_2 + HOH \xrightarrow[\text{catalyst}]{\text{acid}} CH_3\underset{\underset{OH}{|}}{CH}CH_3$$
$$\text{2-propanol}$$
$$\text{(isopropyl alcohol)}$$

$$CH_3CH{=}CH_2 + H_2 \xrightarrow{Pt} CH_3CH_2CH_3$$

$$CH_3C{\equiv}CCH_3 + 2H_2 \xrightarrow{Pt} CH_3CH_2CH_2CH_3$$

In general, most of these reactions are important in the industrial production of valuable organic chemicals. Reactions with hydrogen, called **hydrogenations,** are especially important. One of their important uses is in the conversion of liquid vegetable oils to solid cooking fats (see Section 26.11.2). Other applications are the simultaneous removal of sulfur (as H_2S) and the formation of alkanes from unsaturated hydrocarbons in crude oils, and the conversion of alkenes to alkanes for use as gasoline.

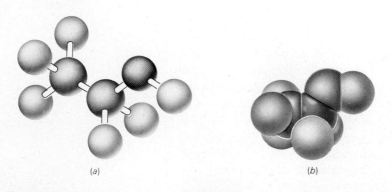

(a) (b)

Figure 26•9 Molecular models of ethanol, CH_3CH_2OH: (*a*) ball-and-stick model, and (*b*) scale model. Carbon atoms, black; hydrogen atoms, light color; oxygen atoms, dark color.

• *Reactions of Benzene* • Benzene and its homologs undergo some of the addition reactions characteristic of the alkene and alkyne hydrocarbons, but usually more drastic conditions such as increased concentrations, higher temperatures, and different catalysts are required. The reactions characteristic of aromatic hydrocarbons are substitution reactions. The following equations illustrate two of the most important substitution reactions:

bromobenzene

nitric
acid

nitrobenzene

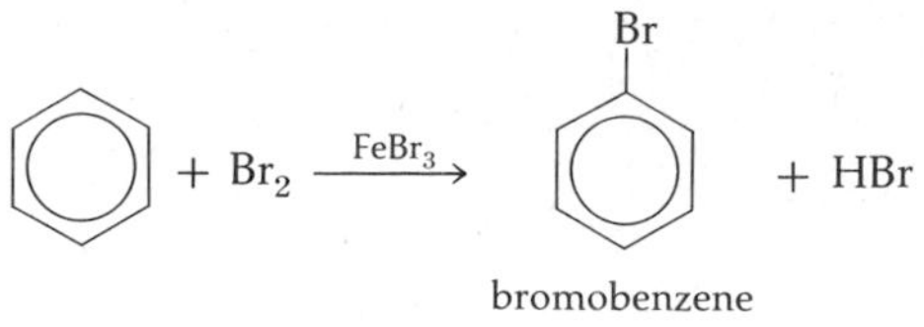

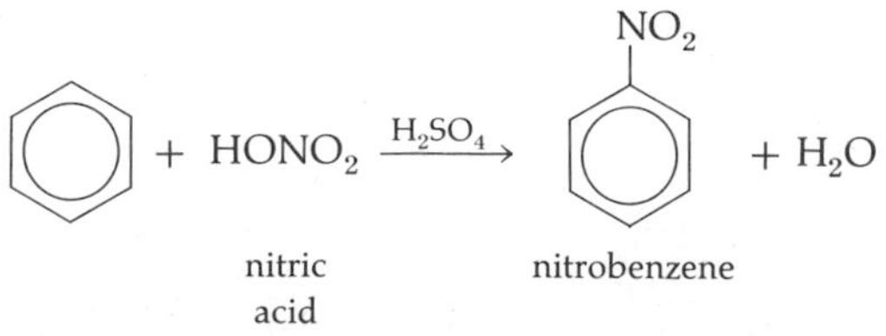

Oxygen Derivatives of Hydrocarbons

The hydrocarbons serve as the basis for the classification of all organic compounds. A nonhydrocarbon compound is considered to be a *derivative* of the hydrocarbon that contains the same carbon chain or ring of carbon atoms. Our study will be confined to simple derivatives that result from replacing one, two, or three hydrogen atoms in hydrocarbon molecules with oxygen atoms or hydroxyl groups. The presence of these atoms or groups of atoms determines to a large extent the physical and chemical properties of the molecules. The atoms or groups of atoms that are most responsible for the properties of a substance are referred to as **functional groups.** Some oxygen derivatives in which hydrogen atoms have been replaced with functional groups are listed in Table 26.6.

26•8   Alcohols

Hydrocarbon derivatives whose molecules contain one or more hydroxyl (—OH) groups in place of hydrogen atoms are known as **alcohols.** The simplest alcohols are derived from the alkanes and contain only one hydroxyl group per molecule. These have the general molecular formula ROH, where R is an alkyl group of the composition C_nH_{2n+1}. Alcohols constituting the first four members of this homologous series are listed in Table 26.7. Molecular models of ethanol are shown in Figure 26.9.

Table 26.6 **Some oxygen derivatives of hydrocarbons**

Derivative	Functional group	General formula[a]	Example	Name[b]
alcohol	—OH	ROH	CH_3OH	methanol
ether	—O—	R—O—R′	$CH_3CH_2OCH_2CH_3$	ethyl ether
aldehyde	$-\underset{\underset{H}{\vert}}{C}=O$	$R-\underset{\underset{H}{\vert}}{C}=O$	$CH_3\underset{\underset{H}{\vert}}{C}=O$	(acetaldehyde)
ketone	$-\underset{\underset{O}{\Vert}}{C}-$	$R-\underset{\underset{O}{\Vert}}{C}-R'$	$CH_3\underset{\underset{O}{\Vert}}{C}CH_3$	(acetone)
carboxylic acid	$-\underset{\underset{OH}{\vert}}{C}=O$	$R-\underset{\underset{OH}{\vert}}{C}=O$	$CH_3\underset{\underset{OH}{\vert}}{C}=O$	(acetic acid)
ester	$-\underset{\underset{O-}{\vert}}{C}=O$	$R-\underset{\underset{OR'}{\vert}}{C}=O$	$CH_3\underset{\underset{OCH_2CH_3}{\vert}}{C}=O$	(ethyl acetate)

[a]R and R′ generally are alkyl groups with the formula C_nH_{2n+1}. However, except for alcohols, R and R′ can also be an aromatic ring or a substituted aromatic ring, for example, C_6H_5— or $CH_3C_6H_4$—. For aldehydes, carboxylic acids, and esters, R can also be H.

[b]Names in parentheses are common names.

26.8.1　Classes of Alcohols　Alcohols, other than methanol, are classified as *primary, secondary,* or *tertiary,* depending on the number of carbons bonded to the carbon atom bearing the —OH group. If one carbon is bonded to this carbon atom, the alcohol is primary; if two carbons are bonded, it is secondary; and if three carbons are bonded, it is tertiary. Primary alcohols among those listed in Table 26.7 are ethanol, 1-propanol, 1-butanol, and methyl-1-propanol. The secondary alcohols are 2-propanol and 2-butanol. The sole tertiary alcohol in the group is methyl-2-propanol.

Table 26.7 **Alcohols derived from alkanes**
($C_nH_{2n+1}OH$, general formula)

Condensed formula	Name
CH_3OH	methanol
CH_3CH_2OH	ethanol
$CH_3CH_2CH_2OH$	1-propanol
$CH_3\underset{\underset{OH}{\vert}}{C}HCH_3$	2-propanol
$CH_3CH_2CH_2CH_2OH$	1-butanol
$CH_3CH_2\underset{\underset{OH}{\vert}}{C}HCH_3$	2-butanol
$CH_3\underset{\underset{CH_3}{\vert}}{C}HCH_2OH$	methyl-1-propanol
$CH_3\underset{\underset{OH}{\vert}}{\overset{\overset{CH_3}{\vert}}{C}}CH_3$	methyl-2-propanol

26•8•2   **Nomenclature**   The name of the alkyl group along with the word *alcohol* forms the common name of an alcohol. Methyl alcohol, ethyl alcohol, *n*-propyl alcohol, and isopropyl alcohol are examples. Unfortunately, people often consider the word *alcohol* as the only significant part of the name; many have lost their lives by drinking alcohol that was not ethyl alcohol. Consequently most manufacturers now use the systematic names instead of the common ones, that is, *methanol* instead of methyl alcohol, *ethanol* instead of ethyl alcohol.

In the systematic naming of alcohols, the following principles are followed:

1. The longest carbon chain that contains the hydroxyl group is considered the parent compound.

2. The -*e* ending of the name of this carbon chain is replaced by -*ol*.

3. The locations of the hydroxyl and any other groups are shown by the smallest possible numbers, with the position of the hydroxyl group having the highest priority.

These principles are illustrated in Table 26.7. Example 26.2 offers additional illustrations.

— • Example 26•2 • —

Name each of the following alcohols by systematic nomenclature. Indicate whether the alcohol is primary, secondary, or tertiary.

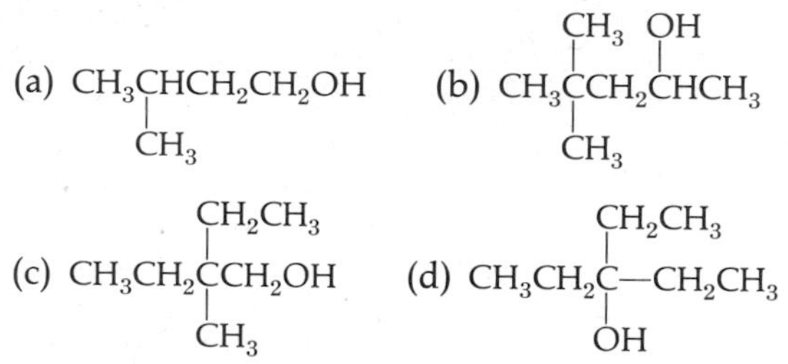

(a) $CH_3CHCH_2CH_2OH$ (with CH_3 below)

(b) $CH_3CCH_2CHCH_3$ (with CH_3, OH above and CH_3 below)

(c) $CH_3CH_2CCH_2OH$ (with CH_2CH_3 above and CH_3 below)

(d) $CH_3CH_2C—CH_2CH_3$ (with CH_2CH_3 above and OH below)

• Solution •

(a) 3-methyl-1-butanol (primary)
(b) 4,4-dimethyl-2-pentanol (secondary)
(c) 2-ethyl-2-methyl-1-butanol (primary)
(d) 3-ethyl-3-pentanol (tertiary)

See also Exercise 34.

26•8•3   **Some Common Alcohols**   In this section we will focus our attention on the preparation of some very important common alcohols and their main uses. The alcohols that will be discussed are methanol, ethanol, ethylene glycol, and glycerol.

• *Methanol* • Methanol (methyl alcohol, wood alcohol) was once made largely by the destructive distillation of hardwoods, such as birch, beech, oak, and maple. One cord of wood yielded about 225 gal of an aqueous distillate containing up to 6 percent methyl alcohol and 10 percent acetic acid. Today, about 99 percent is produced by the catalytic hydrogenation of carbon monoxide. At one time, the carbon monoxide and hydrogen needed for the synthesis were made by passing steam through a bed of red hot coke (see Section 9.4.4). Today it is cheaper to produce most of the CO and H_2 from the methane in natural gas, but the price of gas is rising.

Methanol is toxic and may cause death if taken internally. Blindness can also result from prolonged skin contact or vapor inhalation. Blindness in persons who have ingested methanol has been attributed to the formation of formaldehyde, H_2CO, or formic acid, HCO_2H, which damages cells in the retina. If formaldehyde is the culprit, it is believed to inhibit the formation of ATP, adenosine triphosphate (see Section 27.4.1), which is required for retina cells to function. If formic acid is responsible, it is thought to inactivate iron-containing enzymes responsible for transporting oxygen to the retina. Accompanying the production of formic acid is a condition known as *acidosis*, in which the pH of the blood and body tissues falls.

• *Ethanol* • Ethanol (ethyl alcohol, grain alcohol) is the physiologically active ingredient of beer, wine, and whiskey. It has been produced for centuries by the fermentation of *carbohydrates* (see Section 27.2). In the

---

★ Special Topic ★ Special Topic ★ Special Topic ★ Special Topic ★ Special Topic ★ Special Topic ★ Special Topic ★ Special Topic ★

A Fuel Economy Based on Methanol In Section 9.4.7 the advent of a fuel economy based on hydrogen was discussed. Another candidate as a potential fuel of the future and an especially attractive alternative to gasoline is methanol. Methanol has been described as being superior to hydrogen in many ways. It can be produced from many other fuels such as coal, oil shale, natural gas, petroleum, wood, and farm and municipal wastes. Methanol is easily stored in conventional fuel tanks; not only can it be shipped in tank cars, tank trucks, and tankers, but it can also be transported readily in oil and chemical pipelines.

Of utmost importance is the fact that up to 15 percent of methanol can be added to commercial gasoline in cars now in use without requiring modification of the engines. This methanol–gasoline mixture results in improved economy, lower exhaust emissions, and improved performance, compared with the use of gasoline alone. It has been estimated that existing engines can be converted to use pure methanol at a cost of about $100 per vehicle. Compared with gasoline, the use of methanol in a standard test engine produced one twentieth of the amount of unburned fuel, one tenth of the amount of carbon monoxide, and about the same amount of oxides of nitrogen.

Methanol can be burned cleanly for most of our other energy needs. If handled properly, it is a safe, clean fuel for home heating and can also be burned in power plants to generate electricity with very little pollution of

the atmosphere. In addition, methanol is one of the few known fuels suited for use in fuel cells (see Section 18.9.4).

Unfortunately, the principal drawback to any immediate use of pure methanol as a gasoline substitute and for other fuel applications is that not enough is available. However, it is expected that research will lead to efficient methods of producing methanol from raw materials other than petroleum. Potential future uses of methanol would require many times the amount presently produced. These uses include methanol as a source of synthesis gas (mainly H_2 and CO) or hydrogen and as a nonpetroleum raw material for making ammonia. One oil company has reported a one-step catalytic conversion of methanol to hydrocarbons for gasoline. The refining method involves zeolite (see Section 25.4.3) catalysts with precise pore sizes that permit control of the size of the hydrocarbon molecules. This conversion process could become economically competitive within the next decade. There is no doubt that our supply of fossil fuels is limited. By supplementing our present fuel supply with methanol, we could at least prolong the existence of fossil fuels. An added incentive is that we could simultaneously help clean up our environment.

fermentation processes, organic compounds are broken down into simpler compounds by the action of enzymes. The production of ethanol from *starches* (corn, potatoes, rye, and so on) involves first the enzymatic conversion of the starch into sugar (glucose). The sugar is then converted into ethanol and carbon dioxide by the action of *zymase,* an enzyme produced by living yeast cells. Fermentation must be carried out in dilute water solutions, because the yeast cells cannot live and multiply in concentrated sugar or alcohol solutions. The dilute alcohol solutions thus produced are 12 to 15 percent ethanol. The solutions must be distilled if a more concentrated product is desired. Ethanol produced from starches is used largely in beverages. The reactions are

$$(C_6H_{10}O_5)_x + xH_2O \longrightarrow xC_6H_{12}O_6$$
$$\text{starch} \qquad\qquad\qquad \text{glucose}$$

$$C_6H_{12}O_6 \longrightarrow 2C_2H_5OH + 2CO_2$$
$$\text{glucose} \qquad\qquad \text{ethanol}$$

Industrial ethanol is prepared mostly by two methods: (1) fermentation of black-strap molasses from sugarcane and (2) the addition of water to ethylene in the presence of an acid catalyst (see Section 26.7.2). Although most ethanol is produced by fermentation, about 2 billion pounds are made annually by the second method. Industrial ethanol sells for about $1.50 per gallon.

Denatured alcohol is ethanol that contains an ingredient that renders it unfit for drinking. Methanol and benzene are often used for this purpose. Ethanol is a major therapeutic agent for methanol poisoning, as long as the victim does not have acidosis. However, if sufficient ethanol is taken internally, it can be toxic. When ethanol is consumed, it acts as a *hypnotic;* that is, it tends to produce sleep by depressing the activity of the upper brain.

• *Di- and Trihydroxy Alcohols* • The simplest and most important alcohol that contains two hydroxyl groups is ethylene glycol (the systematic name is 1,2-ethanediol). The most important trihydroxy alcohol is glycerol (the systematic name is 1,2,3-propanetriol):

$$\text{HOCH}_2\text{CH}_2\text{OH} \qquad \text{HOCH}_2\text{CHCH}_2\text{OH}$$
$$\qquad\qquad\qquad\qquad\qquad | $$
$$\qquad\qquad\qquad\qquad\quad \text{OH}$$
$$\text{ethylene glycol} \qquad\qquad \text{glycerol}$$

Ethylene glycol is used extensively as a permanent antifreeze for automobile radiators. Glycerol is a by-product in the production of soap (see Section 26.11.2). Other common names for it are *glycerin* and *glycerine.* It is used in cosmetics and inks and as a sweetening agent, a solvent for medicines, and a lubricant. Glycerol is also a raw material in the production of plastics, fibers, and nitroglycerine.

26•8•4 **Physical Properties** Alcohols boil at considerably higher temperatures than their parent hydrocarbons. This is attributed to the association of

Key Chemical: CH_3OH
methanol

How Made

$$CH_4 + H_2O \xrightarrow[\text{Ni}]{850\,°C}$$
$$CO + 3H_2$$

$$CO + 2H_2 \xrightarrow[\substack{200-300\text{ atm}\\ZnO,\ CrO_3}]{300-400\,°C}$$
$$CH_3OH$$

U.S. Production: Cost
~7 billion lb per year
~7¢ per pound (44¢ per gallon)

Major Derivatives
45% formaldehyde

Major End Uses
60% polymers
10% solvents

Other Uses
as a fuel and to prepare methyl esters of carboxylic acids

Key Chemical: $HOCH_2CH_2OH$
ethylene glycol

How Made

$$2CH_2{=}CH_2 + O_2 \xrightarrow[250\,°C]{Ag}$$
$$2CH_2{-}CH_2$$
$$\qquad\ \ \diagdown\!\diagup$$
$$\qquad\quad O$$
ethylene oxide

$$CH_2{-}CH_2 + H_2O \xrightarrow{H^+}$$
$$\ \diagdown\!\diagup$$
$$\quad O$$
$$HOCH_2CH_2OH$$

U.S. Production; Cost
~4 billion lb per year
~18¢ to 24¢ per pound, depending on grade

Major Derivative
50% polyethylene terephthalate

Major End Uses
50% polyesters
40% antifreeze for automobile radiators

alcohol molecules through hydrogen bonding (the dashed lines indicate the hydrogen bonds):

$$:\overset{R}{\underset{R}{\ddot{O}}}-H---:\overset{}{\underset{}{\ddot{O}}}-H---:\overset{}{\underset{R}{\ddot{O}}}-H$$

Recall that water has a relatively high boiling point for the same reason.

Also, unlike the parent hydrocarbons, the alcohols with low molecular weights are very soluble in water. This, too, is accounted for on the basis of hydrogen bonding between the hydroxyl group of the alcohol and the water molecules. However, as the molecular weight increases, the van der Waals forces between the hydrocarbon portions of the alcohol molecules become more effective in attracting alcohol molecules to each other, thereby offsetting the effect of hydrogen bonding. For this reason, methyl alcohol is soluble in water in all proportions, whereas 1-decanol ($C_{10}H_{21}OH$) is insoluble.

26•8•5   Chemical Properties   Alcohols undergo a number of reactions that involve the hydroxyl group. Two of these are (1) dehydration to form alkenes or ethers, and (2) controlled oxidation to produce aldehydes and ketones.

• *Dehydration* • Reactions involving the loss of H and OH to make H_2O are called **dehydration reactions.** With sulfuric acid as the dehydrating agent, and depending on reaction conditions, alcohols can form alkenes or ethers. Compounds that contain two hydrocarbon groups, R and R', attached to an atom of oxygen are called **ethers** (see Table 26.6).

Consider as an example the dehydration of ethanol with sulfuric acid. Upon heating at 180 °C, dehydration to ethylene results:

$$CH_3CH_2OH \xrightarrow[180\ °C]{H_2SO_4} CH_2{=}CH_2 + H_2O$$

On the other hand, if the reaction mixture is heated at 140 °C, the principal product is ethyl ether:

$$2CH_3CH_2OH \xrightarrow[140\ °C]{H_2SO_4} \underset{\text{ethyl ether}}{CH_3CH_2OCH_2CH_3} + H_2O$$

Ethyl ether, prepared commercially by this process, is the common ether used as a general anesthetic. A scale model is shown in Figure 26.10. In addition to its use as an anesthetic, ethyl ether is widely used as a solvent for fats, waxes, and other substances insoluble in water. It must be used with caution, however, because it is highly flammable.

• *Controlled Oxidation* • The mild, controlled oxidation of primary alcohols produces aldehydes. Derivatives of hydrocarbons that have molecules with a double bond to oxygen in place of two hydrogen atoms at the end of a chain are called **aldehydes** (see Table 26.6).

Figure 26•10 Scale model of a molecule of ethyl ether, $CH_3CH_2OCH_2CH_3$.

Key Chemical: $H{-}\overset{}{\underset{H}{C}}{=}O$

formaldehyde

How Made

$$2CH_3OH + O_2 \xrightarrow[550\ °C]{Cu}$$
$$2H{-}\overset{}{\underset{H}{C}}{=}O + 2H_2O$$

U.S. Production; Cost
~6 billion lb per year
~5.25¢ per pound for 37% solution

Major Derivatives
25% urea–formaldehyde resins
25% phenol–formaldehyde resins

Major End Uses
60% adhesives
10% plastics

Other Uses
general antiseptic, embalming solution, preservative for biological specimens

Common laboratory reagents used for oxidations of alcohols are potassium dichromate ($K_2Cr_2O_7$) in sulfuric acid, chromium(VI) oxide (CrO_3) in acetic acid, and neutral or basic solutions of potassium permanganate ($KMnO_4$). The following general equation illustrates the overall oxidation, in which [O] indicates that the oxygen is derived from some oxidizing agent:

$$RCH_2OH + [O] \longrightarrow \underset{\underset{H}{|}}{RC}\!\!=\!\!O + H_2O$$

primary   aldehyde
alcohol

For large-scale commercial syntheses, oxidation is achieved with atmospheric oxygen whenever possible. One way that acetaldehyde (R is CH_3) is produced commercially is by oxidation of ethanol with oxygen over a copper or silver catalyst:

$$2CH_3CH_2OH + O_2 \xrightarrow{\text{Cu or Ag}} 2CH_3\underset{\underset{H}{|}}{C}\!\!=\!\!O + 2HOH$$

acetaldehyde

Scale models of molecules of formaldehyde and acetaldehyde, the first two members of the homologous series of aldehydes, are shown in Figure 26.11. Some other aldehydes of importance are shown in Figure 26.12.

$CH_2\!\!=\!\!CHC\!\!=\!\!O$
$\quad\quad\quad\quad |$
$\quad\quad\quad\quad H$

acrolein
(in photochemical
smog and overheated fat)

benzaldehyde
(odor and flavor
of almonds)

$-CH\!\!=\!\!CHC\!\!=\!\!O$
$\quad\quad\quad\quad\quad |$
$\quad\quad\quad\quad\quad H$

cinnamaldehyde
(flavor of cinnamon)

Figure 26•11 Scale models of molecules of (a) formaldehyde, $H_2C\!\!=\!\!O$; (b) acetaldehyde, $CH_3\underset{\underset{H}{|}}{C}\!\!=\!\!O$.

Figure 26•12 (*left*) Some aldehydes of interest.

The mild-controlled oxidation of secondary alcohols produces ketones. Derivatives of hydrocarbons that have molecules with a double bond to oxygen in place of two hydrogen atoms at a position other than the end of the carbon chain are called **ketones** (see Table 26.6). The following general equation illustrates the overall oxidation:

$$\underset{\underset{OH}{|}}{RCHR'} + [O] \longrightarrow \underset{\underset{O}{\|}}{RCR'} + H_2O$$

secondary   ketone
alcohol

As is the case with the oxidation of primary alcohols to aldehydes, on a commercial scale, atmospheric oxygen is used whenever possible. For example, acetone (see Figure 26.13), the most widely used ketone, is produced

Figure 26•13 Scale model of a molecule of acetone, $CH_3\underset{\underset{O}{\|}}{C}CH_3$.

commercially mainly by the oxidation of 2-propanol with oxygen over a copper catalyst:

$$2CH_3\underset{\underset{OH}{|}}{C}HCH_3 + O_2 \xrightarrow{Cu} 2CH_3\underset{\underset{O}{\|}}{C}CH_3 + 2HOH$$

acetone

The annual production of acetone is over 2 billion pounds. The greatest use of acetone is as a solvent for waxes, plastics, and lacquers. It is the principal solvent for cellulose acetate in the production of rayon (see Section 27.2.5).

Acetone is found in the body in a concentration of about 1 mg per 100 mL of blood. Its presence is due to the incomplete oxidation of fatty acids during the metabolism of *lipids*—organic compounds that are important constituents of plant and animal tissues. In certain abnormal cases, as with persons who have the disease diabetes mellitus, the acetone concentration in the body may be considerably higher than 1 mg per 100 mL of blood. It is excreted in the urine, where its presence is easily detected. Some other ketones of interest and importance are shown in Figure 26.14.

26•9 Phenols

Compounds that have a hydroxyl group attached to a benzene ring are called **phenols** rather than alcohols. *Phenol,* C_6H_5OH (see Figure 26.15), the simplest member of this class of compounds, is also known as *carbolic acid.* It is extremely destructive in its action on animal tissue.

The methylphenols are commonly known as *cresols.* Three isomers exist, and their condensed structural formulas are

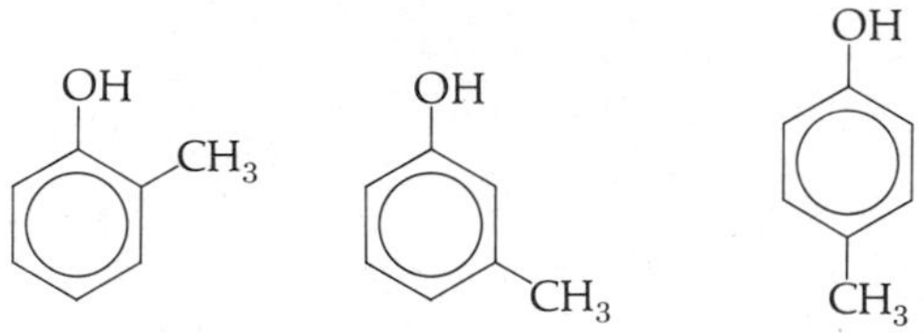

o-methylphenol (*ortho*-cresol) | *m*-methylphenol (*meta*-cresol) | *p*-methylphenol (*para*-cresol)

Creosote oil, an oily liquid obtained by the distillation of coal tar and beechwood tar, contains a large proportion of phenols, including the cresols. It is widely used as a wood preservative.

26•10 Carboxylic Acids

Derivatives of hydrocarbons with a terminal carbon atom that has a double bond to oxygen and a hydroxyl group are called **carboxylic acids.** Those derived from the alkane hydrocarbons have the general molecular formula RCO_2H, which indicates that the *carboxyl group,* $-\underset{\underset{O}{\|}}{C}-OH$, is present

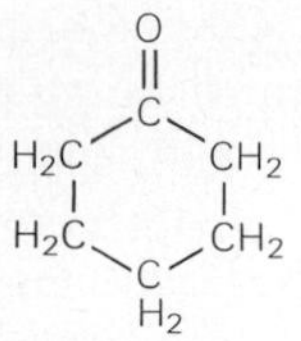

cyclohexanone
(used to make caprolactam, which polymerizes to nylon)

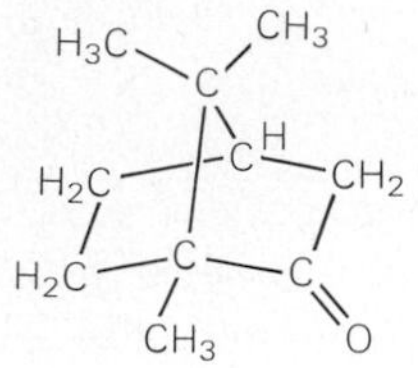

camphor
(from camphor tree, used in medicine and plastic manufacturing)

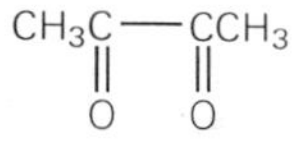

butanedione
(flavoring agent in margarine)

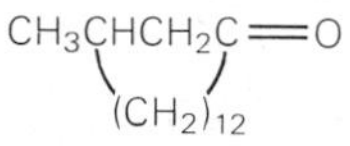

muscone
(from male musk deer, expensive ingredient of perfume)

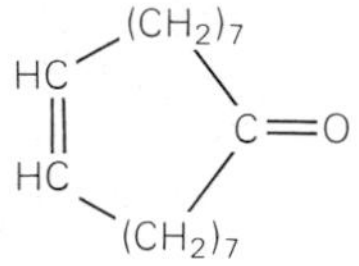

civetone
(from civet cat, expensive ingredient of perfume)

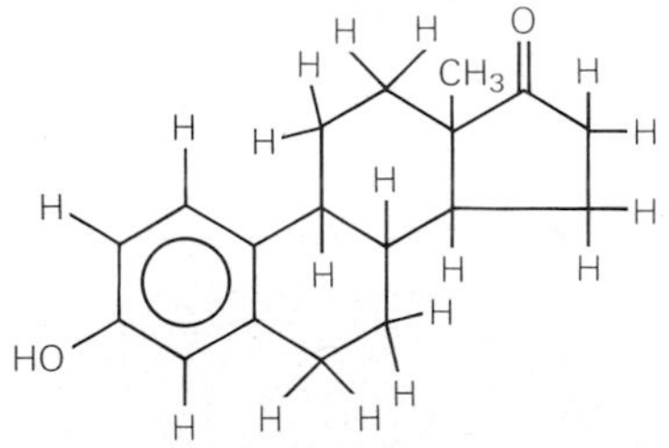

estrone
(from urine of pregnant women)

Figure 26•14 Some ketones of interest.

(see Table 26.6). The first five members of the homologous series of aliphatic carboxylic acids are listed in Table 26.8. A scale model of a molecule of acetic acid is shown in Figure 26.16.

Table 26•8 Some saturated aliphatic carboxylic acids

Condensed formula	Systematic name (common name)
HCO_2H	methanoic acid (formic acid)
CH_3CO_2H	ethanoic acid (acetic acid)
$CH_3CH_2CO_2H$	propanoic acid (propionic acid)
$CH_3(CH_2)_2CO_2H$	butanoic acid (butyric acid)
$CH_3\underset{\underset{\displaystyle CH_3}{\vert}}{C}HCO_2H$	methylpropanoic acid (isobutyric acid)

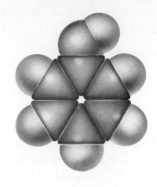

Figure 26•15 Scale model of a molecule of phenol, C_6H_5OH.

Figure 26•16 Scale model of a molecule of acetic acid, $CH_3\underset{\displaystyle O}{\overset{\displaystyle \|}{C}}$—OH.

26•10•1 Nomenclature The systematic naming of carboxylic acids that are derived from alkanes involves replacing the -*e* ending of the corresponding alkane with -*oic* and adding the word *acid*. Many of the carboxylic acids were isolated from natural sources long before systematic naming was introduced. They were given common names indicative of their source, and these names have persisted. For example, methanoic acid was first obtained by the destructive distillation of ants and was called *formic acid* (Latin, *formica*, ant). Similarly, *acetic acid* was first obtained from vinegar (Latin *acetum*, vinegar), and *butyric acid* was found in butter (Latin *butyrum*, butter).

26•10•2 Some Common Carboxylic Acids Acetic acid is the most important commercial carboxylic acid. It is prepared by the air oxidation of acetaldehyde with cobalt or manganese(II) acetate as the catalyst:

$$2CH_3CHO + O_2 \xrightarrow{\text{catalyst}} 2CH_3CO_2H$$

Production in the United States amounts to about 2.6 billion pounds annually, exclusive of that in vinegar. The acid is sold in the pure form as *glacial acetic acid*, so called because it freezes to an icelike solid on cold days. The melting point is 16.6 °C. The acid is used in the manufacture of cellulose acetate, white lead for paint, plastics, perfumes, dyes, and medicines.

Some other common carboxylic acids are shown in Figure 26.17. (Also see the *fatty acids* produced by the hydrolysis of fats listed in Table 26.11.)

26•10•3 Physical Properties The boiling points of carboxylic acids are high relative to those of alcohols, aldehydes, and ketones of approximately the same molecular weight. For example, formic acid boils 23 °C higher than ethanol, although both have the same molecular weight. The high boiling points of carboxylic acids are due to intermolecular hydrogen bonding between two molecules:

$$
R-C
\begin{matrix} \ddot{O}: \text{------} H-\ddot{O}: \\ \ddot{O}-H \text{------} : \ddot{O} \end{matrix}
C-R
$$

The first four members derived from the methane series are completely miscible with water. Thereafter, as the molecular weight increases, the solubility in water decreases. Those that are soluble in water give a sour taste to the solution.

The lower members derived from the alkane series have an acid odor; those with four to eight carbon atoms have disagreeable odors. The odors of rancid butter and strong cheese are due in part to acids in this group.

26·10·4 Chemical Properties The carboxyl functional group, —CO₂H, is made up of a carbonyl group (—C—) and a hydroxyl group (—OH). Most

reactions of carboxylic acids involve only the —OH group. That carboxylic acids are weak acids is revealed by the K_a values for acetic acid and formic acid in Table 16.1.

One of the most common reactions of a carboxylic acid is with an alcohol. For example, an equimolar mixture of acetic acid and ethanol slowly comes to equilibrium to give ethyl acetate and water. The addition of a small amount of strong acid such as H_2SO_4 catalyzes the attainment of the equilibrium concentrations:

$$
\underset{O}{CH_3C}\boxed{OH + H}OCH_2CH_3 \xrightarrow{H_2SO_4} \underset{\underset{ethyl\ acetate}{O}}{CH_3C-OCH_2CH_3} + HOH
$$

$$K_c = 4.0 \text{ at } 25\ °C$$

The equilibrium concentration of ethyl acetate may be increased by using an excess of one of the reactants, usually the alcohol, and by removing the water as rapidly as it forms.

The compounds produced by this type of reaction are called *esters*. The reaction is termed an *esterification*, and the carboxylic acid is then said to be *esterified*.

26·11 Esters

Compounds that may be considered to be derived from carboxylic acids by the replacement of the hydrogen atom of the hydroxyl group with a hydrocarbon group are called **esters** (see RCO_2R' in Table 26.6). Probably the most common ester is ethyl acetate, $CH_3CO_2CH_2CH_3$, a common solvent used in many paint and fingernail polish removers as well as in glue. Ethyl acetate and other esters with ten or fewer carbons are volatile liquids with pleasant fruity odors and are often found in fruits and flowers. Many esters,

Key Chemical:

phenol

How Made

CH_3CHCH_3

cumene

$+ O_2 \longrightarrow$

$CH_3\overset{OOH}{\underset{\ }{C}}CH_3$

cumene hydroperoxide

$CH_3\overset{OOH}{\underset{\ }{C}}CH_3$

$\xrightarrow{H_2SO_4}$

OH

$+ CH_3\overset{O}{\underset{\ }{C}}CH_3$

U.S. Production; Cost
~2.5 billion lb per year
~18¢ to 19¢ per pound

Major Derivatives
45% phenolic resins
20% bisphenol-A
15% caprolactam

Major End Uses
50% adhesives
20% fabricated plastics
20% fibers

Other Uses
antiseptic in dilute solution, disinfectant; used in synthesis of dyes, drugs, oil of wintergreen, and aspirin

both natural and synthetic, are used as flavoring agents. Some esters and their characteristic odors are listed in Table 26.9. The odor and flavor of certain fruits may be due to several esters. For example, ethyl acetate, *n*-butyl acetate, and *n*-pentyl acetate all have the flavor of bananas.

Table 26.9 Some esters and their aromas

Ester	Characteristic aroma
ethyl formate	rum
n-pentyl acetate	bananas
isopentyl acetate	pears
n-octyl acetate	oranges
methyl butyrate	apples
ethyl butyrate	pineapples
n-propyl butyrate	apricots

Naturally occurring esters of long-chain carboxylic acids and long-chain alcohols are called *waxes*. (They are not to be confused with the hydrocarbon type of wax, such as paraffin wax.) Most so-called waxes are usually mixtures of two or more esters and other substances. Such mixtures are solids that melt easily over a wide range of temperature (40–90 °C). When mixed with certain organic solvents, they are applied easily as protective coatings. For example, *carnauba wax* (see Table 26.10) is widely used as a component of automobile and floor waxes.

Table 26·10 Some waxes

Name	Source	Formula[a]	Melting point, °C
spermaceti	head cavity of sperm whale	$C_{15}H_{31}CO_2C_{16}H_{33}$	42–50
carnauba	Brazilian palm leaves	$C_{25,27}H_{51,55}CO_2C_{30,32}H_{61,65}$	83–86
beeswax	honeycomb cells	$C_{23,25}H_{47,51}CO_2C_{32,34}H_{65,69}$	62–66

[a]For carnauba and beeswax, the formulas indicate that more than one ester is present.

26·11·1 Nomenclature The name of an ester is taken from the name of the carboxylic acid and the name of the hydrocarbon group that has replaced the hydrogen of the hydroxyl group. The hydrocarbon group is named first. This is followed by the name of the carboxylic acid, from which both the *-ic* ending and the word *acid* are dropped and to which the ending *-ate* is added. Systematic and common names are formulated in this manner.

26·11·2 Chemical Properties Esters undergo hydrolysis to produce carboxylic acids and alcohols. For example, the hydrolysis of ethyl acetate produces acetic acid and ethanol:

$$CH_3CO_2CH_2CH_3 + H_2O \xrightleftharpoons[]{H_2SO_4} CH_3CO_2H + CH_3CH_2OH \qquad K_c = 0.25 \text{ at } 25 \text{ °C}$$

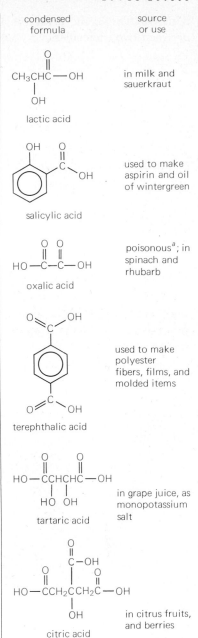

condensed formula	source or use
lactic acid	in milk and sauerkraut
salicylic acid	used to make aspirin and oil of wintergreen
oxalic acid	poisonous[a]; in spinach and rhubarb
terephthalic acid	used to make polyester fibers, films, and molded items
tartaric acid	in grape juice, as monopotassium salt
citric acid	in citrus fruits, and berries

[a]The concentration of oxalic acid in spinach and rhubarb is well below toxic limits.

Figure 26·17 Some common carboxylic acids of interest.

This reaction is the reverse of an esterification.

Esters can also be hydrolyzed by heating them in an aqueous base:

$$CH_3CO_2CH_3 + NaOH \xrightarrow{H_2O} CH_3CO_2Na + CH_3CH_2OH$$

Although the reaction is referred to as hydrolysis, we can see from the balanced equation that the base, NaOH, is the active inorganic ingredient.

• *Hydrolysis of Fats* • When lard, tallow, and other fats are boiled in dilute sodium hydroxide or subjected to the action of specific enzymes (in digestion), the fat molecules are completely hydrolyzed to form simpler molecules. Glycerol, $CH_2OHCHOHCH_2OH$, always constitutes a portion of the product, and acids with high molecular weights make up the remainder. The acids thus produced are called **fatty acids.** Two types are *saturated fatty acids* and *unsaturated fatty acids.* The more common ones are listed in Table 26.11.

Of the unsaturated fatty acids, oleic acid, $CH_3(CH_2)_7CH=CH(CH_2)_7CO_2H$, contains one double bond, linoleic acid contains two, and linolenic contains three. Linoleic and linolenic acids are examples of *polyunsaturated acids.* Table 26.12 shows the relative amounts of fatty acids obtained by the hydrolysis of different fats.

Table 26.11 Some common fatty acids

Saturated	
$C_{11}H_{23}CO_2H$	lauric acid
$C_{13}H_{27}CO_2H$	myristic acid
$C_{15}H_{31}CO_2H$	palmitic acid
$C_{17}H_{35}CO_2H$	stearic acid
Unsaturated	
$C_{17}H_{33}CO_2H$	oleic acid
$C_{17}H_{31}CO_2H$	linoleic acid
$C_{17}H_{29}CO_2H$	linolenic acid

Table 26.12 Fatty acids obtained from hydrolysis of fats and oils (in percentages)[a]

Fat or oil hydrolyzed	Saturated				Unsaturated		
	Lauric	Myristic	Palmitic	Stearic	Oleic	Linoleic	Linolenic
animal fats							
butter	2–5	8–15	25–29	9–12	18–33	2–4	—
lard		1–2	25–30	12–18	48–60	6–12	0–1
tallow		2–5	24–34	15–30	35–45	1–3	0–1
vegetable fats (oils)							
olive		0–1	5–15	1–4	67–84	8–12	—
peanut		—	7–12	2–6	30–60	20–38	—
corn		1–2	7–11	3–4	25–35	50–60	—
cottonseed		1–2	18–25	1–2	17–38	45–55	—
soybean		1–2	6–10	2–4	20–30	50–58	5–10
linseed		—	4–7	2–4	14–30	14–25	45–60
marine oils							
fish		6–8	10–25	1–3	—	—	—
whale		5–10	10–20	2–5	33–40	—	—

[a]Data are adapted from J. R. Holum, *Elements of General and Biological Chemistry*, 5th ed., Wiley, New York, 1979, p. 329. Reprinted by permission of John Wiley & Sons, Inc. The percentages may not total 100% because all the acids are not listed.

Based on the products of the hydrolysis reaction, it is evident that a **fat** may be defined as an ester of glycerol and three long-chain carboxylic acids. A general structural formula is

$$
\begin{array}{l}
\text{three fatty acid parts} \left\{
\begin{array}{l}
\text{R}'\overset{\overset{\displaystyle O}{\|}}{\text{C}}\!-\!\text{OCH}_2 \\[2mm]
\text{R}''\overset{\overset{\displaystyle O}{\|}}{\text{C}}\!-\!\text{OCH} \\[2mm]
\text{R}'''\overset{\overset{\displaystyle O}{\|}}{\text{C}}\!-\!\text{OCH}_2
\end{array}
\right.
\end{array}
\qquad \text{glycerol part} \qquad\qquad
\begin{array}{c}
(\text{RCO}_2)_3\text{C}_3\text{H}_5 \\
\text{abbreviated formula}
\end{array}
$$

R′, R″, and R‴ may or may not be the same in the same molecule; they correspond to $C_{11}H_{23}$, $C_{13}H_{27}$, $C_{15}H_{31}$, $C_{17}H_{35}$, $C_{17}H_{33}$, $C_{17}H_{31}$, and so on. Fats that have melting points below room temperature are called **oils.**

• *Soap* • In soap making, fat is heated in huge iron kettles with aqueous sodium hydroxide until the fat is completely hydrolyzed. Such a reaction is often called *saponification* (Latin *sapo*, soap), because the reaction has been used since the days of the ancient Romans to convert fats and oils to soaps. The equation for the reaction is

$$(\text{RCO}_2)_3\text{C}_3\text{H}_5 + 3\text{NaOH} \longrightarrow 3\text{RCO}_2\text{Na} + \text{C}_3\text{H}_5(\text{OH})_3$$
$$\text{fat} \qquad\qquad\qquad\qquad \text{soap} \qquad \text{glycerol}$$

It is obvious from this equation that common soap may be a mixture of such compounds as sodium stearate, $C_{17}H_{35}CO_2Na$, sodium palmitate, $C_{15}H_{31}CO_2Na$, sodium oleate, $C_{17}H_{33}CO_2Na$, and the sodium salts of other fatty acids. Most natural soap is now made primarily from four fats: beef tallow, palm oil, coconut oil, and olive oil. The soap is precipitated by the addition of salt. It is then removed by filtration, washed, and mixed with dyes, perfumes, or any other special ingredient. After it hardens, it is cut and pressed into cakes. Scouring powders contain soap plus a high percentage of an abrasive, such as volcanic ash or fine sand.

Soap functions in a variety of ways as a cleansing agent. It lowers the surface tension of water, thus enabling the water to moisten more effectively the material being washed; it acts as an emulsifying agent to bring about the dispersion of oil and grease; and it adsorbs dirt.

The so-called *synthetic detergents* are of many types. In all of them a water-soluble saltlike group is attached to a hydrocarbon chain. A typical example is $C_{17}H_{35}-OSO_3Na$. An important difference between ordinary soap and the synthetic detergents is that the latter do not form greasy precipitates with the polyvalent cations (for example, Ca^{2+} and Mg^{2+}) that are normally present in unsoftened water:

$$2C_{17}H_{35}CO_2Na + Ca(HCO_3)_2 \longrightarrow (C_{17}H_{35}CO_2)_2Ca\!\downarrow + 2NaHCO_3$$
$$\text{common soap} \qquad\qquad\qquad\qquad \text{water-insoluble soap curd}$$

$$2C_{17}H_{35}OSO_3Na + Ca(HCO_3)_2 \rightleftharpoons (C_{17}H_{35}OSO_3)_2Ca + 2NaHCO_3$$
$$\text{synthetic detergent} \qquad\qquad\qquad\qquad \text{soluble}$$

• *Hydrogenation of Liquid Fats* • Study of Table 26.12 shows that, when hydrolyzed, most liquid fats from vegetable sources yield a high percentage of unsaturated acids, whereas most solid fats from animal sources yield considerable amounts of saturated acids. This means that the R parts of the molecules of vegetable oils are more unsaturated than is true of the molecules of animal fats. The chemical conversion of liquid oils to solid fats is achieved by the addition of hydrogen to some of the double bonds in the molecules of the oils. Hydrogenation is carried out at elevated temperatures in the presence of powdered nickel or some other catalyst. The following equation shows the complete hydrogenation of glyceryl trioleate to glyceryl tristearate:

$$(C_{17}H_{33}CO_2)_3C_3H_5 + 3H_2 \xrightarrow{\text{catalyst}} (C_{17}H_{35}CO_2)_3C_3H_5$$

For quite some time, most margarine was made in the United States by blending hydrogenated fats (80 percent) with skim milk, salt, lactic acid, flavor-producing bacteria, and carotene (provitamin A). Recently the trend has been toward decreasing the amount of hydrogenated fat in favor of *polyunsaturated fats*, that is, fats with a large percentage of R groups derived from unsaturated fatty acids such as linoleic and linolenic acids (see Table 26.11). There is some evidence that this type of fat is less likely to increase the amount of blood cholesterol to dangerous levels. However, the relationship between one's diet and blood cholesterol level is a topic of considerable controversy at the present time.

Chapter Review

Summary

All organic compounds are either **hydrocarbons** or compounds derived from them. The **aliphatic hydrocarbons** are chain structures that are classified in three **homologous series** of compounds; within each series, each compound is a **homolog** of the others. The **alkanes** (the **methane series**) are **saturated hydrocarbons**: they contain only single bonds. Most alkanes exist as **branched-chain isomers,** which are chemically and physically distinct compounds. Like all other organic compounds, alkanes are given systematic names based on the concept of a **parent compound** with attached groups; groups derived from alkanes are called **alkyl groups.** Many compounds have **common names** that are convenient to use.

The **alkenes** (the **ethylene series**) and the **alkynes** (the **acetylene series**) are **unsaturated hydrocarbons**: they contain one double bond and one triple bond, respectively. Because of the restricted rotation about double bonds, most alkenes exhibit **cis-trans isomerism**. In addition, most alkenes and alkynes exhibit isomerism based on the position of the multiple bond, and, like alkanes, they exhibit branched-chain isomerism.

The **aromatic hydrocarbons** are ring structures that have delocalized pi-electron systems. The most important compounds of this class are the single-ring **benzene hydrocarbons.** Disubstituted benzenes exist as three isomers: *ortho, meta,* and *para*. Most benzenes that are more highly substituted also exist as different isomers. **Fused-ring hydrocarbons** also exist; if the rings in question are benzene rings, the compound is a **fused-ring benzenoid hydrocarbon.** There are also cyclic hydrocarbons that are saturated: these **cycloalkanes** constitute a homologous series that is isomeric with the alkenes.

Most hydrocarbons are obtained from natural gas, **petroleum,** and **coal;** an important potential source is **oil shale.** Petroleum is refined by fractional distillation to separate the components, followed by **cracking** or **aromatization** to convert some of them to other compounds. The most familiar chemical reaction of hydrocarbons is their combustion to carbon dioxide and water. Most other reactions are less drastic. The reactions characteristic of alkanes are **substitution reactions;** those characteristic of alkenes and alkynes are **addition reactions.** Aromatic hydrocarbons undergo both types of reactions, but substitution reactions are much more common.

Most organic compounds are **derivatives** of hydrocarbons: they contain at least one atom of another element, such as oxygen. The properties of the compound are determined largely by its **functional groups.** The hydroxyl group is the functional group of **alcohols,** which may be primary, secondary, or tertiary. Some alcohols contain two or more hydroxyl groups. Alcohols form intermolecular hydrogen bonds. They undergo **dehydration reactions** to form alkenes or **ethers,** and controlled oxidation reactions to form **aldehydes** (from primary alcohols) or **ketones** (from secondary alcohols). Compounds in which a hydroxyl group is attached to a benzene ring are called **phenols.**

The functional group of **carboxylic acids** is the carboxyl group, the hydrogen atom of which is weakly acidic. Carboxylic acids also form intermolecular hydrogen bonds. They undergo **esterification reactions** with alcohols to form **esters.** In the reverse of this type of reaction, esters undergo hydrolysis to form carboxylic acids and alcohols. Esters of long-chain alcohols and long-chain carboxylic acids are called **waxes.** Esters of the trihydroxy alcohol glycerol and three long-chain carboxylic acids (**fatty acids**) are called **fats** or **oils.** The alkaline hydrolysis of fats and oils is called **saponification.** The product, **soap,** consists of the sodium salts of fatty acids.

Exercises

Alkanes

1. Write Lewis formulas, condensed formulas, and names for all the alkanes that have a molecular formula of C_6H_{14}.
2. Do the Lewis formulas written in Exercise 1 correctly portray the shapes of the molecules? Why?
3. Write condensed structural formulas for the following:
   (a) 2,2,4,4-tetramethylhexane
   (b) 3-ethyl-2,2-dimethylpentane
   (c) 4-isopropylheptane
   (d) 4-ethyl-3-methyl-4-n-propyloctane
   (e) 4-ethyl-2,3-dimethylhexane
4. Which of the compounds listed in Exercise 3 are isomers?
5. The alkane with the systematic name 2,4,4-trimethylpentane is known in the petroleum industry as "isooctane." Based on the structural formulas shown in Figure 26.2 for isobutane and isopentane, is the name isooctane a logical common name for this hydrocarbon? If not, write the systematic name and the condensed structural formula for isooctane.

Alkenes

6. (a) Write condensed structural formulas for and name all alkenes with the molecular formula C_5H_{10}.
   (b) Which of these condensed formulas represents two geometric isomers? Draw these geometric isomers in their cis and trans forms by using combined Lewis and condensed formulas similar to those shown in Figure 26.4.

7. Name the following compounds:
   (a) $CH_3(CH_2)_4CH{=}CH_2$
   (b) $CH_3CH{=}CH(CH_2)_3CH_3$
   (c) $CH_3CH_2CH{=}CCH_3$
   $\qquad\qquad\quad | $
   $\qquad\qquad CH_3CHCH_3$

   (d) $CH_2{=}CHCH(CH_2)_3CH_3$
   $\qquad\qquad | $
   $\qquad\quad CH_3CHCH_3$
8. Indicate with an equation how one could prepare ethylene from a hydrocarbon in kerosene, such as $C_{12}H_{26}$. State the conditions necessary for the reaction to occur. What other valuable alkene might form in the process?
*9. Write combined Lewis and condensed formulas for all cis and trans isomers of alkenes that contain six carbon atoms per molecule. Name each isomer.

Alkynes

10. Write Lewis and condensed formulas for all the alkynes with the molecular formula C_5H_8 and C_6H_{10}.
11. Why is the name dimethylbutyne just as satisfactory as 3,3-dimethylbutyne? (See one of your structures in Exercise 10.)
12. The molecular formula C_4H_6 can represent two noncyclic hydrocarbons other than alkynes. Write Lewis and condensed formulas for both.
13. Write the series of equations that illustrate how acetylene can be produced from coal, limestone, and water as the raw materials.

Benzene-Type Hydrocarbons

14. Write condensed structural formulas for and name the eight benzene hydrocarbons with the molecular formula C_9H_{12}.
15. Write condensed structural formulas for the following hydrocarbons: (a) cumene, (b) *para-n*-propyltoluene, (c) *ortho*-xylene, (d) *meta*-xylene.
*16. Write condensed structural formulas and names for the following:
   (a) the disubstituted benzenes with the molecular formula $C_{10}H_{14}$
   (b) the trisubstituted benzenes with the molecular formula $C_{10}H_{14}$
   (c) the tetrasubstituted benzenes with the molecular formula $C_{10}H_{14}$
17. Name the following fused-ring benzenoid hydrocarbons:

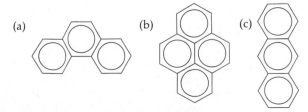

(a)　　　　　　(b)　　　　　　(c)

*18. Write all possible Kekulé formulas for anthracene and phenanthrene. Which compound should have the larger resonance energy? (Review Section 6.6.)

*19. Write condensed structural formulas for all the dimethylnaphthalenes. Are there other substituted naphthalenes that are isomeric with the dimethyl-naphthalenes? If so, write their condensed structural formulas.

Cycloalkanes

20. Write the condensed structural formulas and names for the cycloalkanes that have a molecular formula of C_5H_{10}. These compounds are isomeric with compounds from what other series of hydrocarbons?

21. Cyclobutane and cyclobutene are well-known cyclic hydrocarbons, but cyclobutyne is unknown. Why?

*22. In addition to the two C_4H_6 alkynes shown in Section 26.3.1 and the two C_4H_6 hydrocarbons in Exercise 12, there are other hydrocarbons with this molecular formula. Write Lewis structures for as many of these hydrocarbons as you can.

Sources and Properties of Hydrocarbons

23. List the four principal sources of hydrocarbons. What are the relative costs of obtaining hydrocarbons from these sources on a commercial basis? Do you think the relative costs will be the same in 2080?

24. Differentiate among thermal cracking, catalytic cracking, and aromatization with respect to the type of hydrocarbon obtained by each conversion process.

25. (a) Which of the two cracking processes would give the C_4H_8 alkenes 1-butene and 2-butene among the hydrocarbon products?

(b) Which cracking process would give only 2-methyl-propene as the C_4H_8 alkene among the hydrocarbon products?

26. When heptane and 1,2-dimethylcyclopentane are heated over platinum at about 500 °C and 300–700 psi, toluene and hydrogen are produced. What common intermediate compound presumably is formed during the course of this platforming process?

27. Assuming that the average composition of the hydrocarbons in gasoline equals that of C_8H_{18}, what is the theoretical volume of water formed by the combustion of one gallon of gasoline in an automobile engine if the water vapor is collected, condensed to a liquid, and cooled to 20 °C? (The density of gasoline is 0.70 g/cm³.)

28. Chlorination of 2-methylbutane in the presence of ultraviolet light produces a mixture of chloroalkanes with the molecular formula $C_5H_{11}Cl$. Write condensed structural formulas and systematic names for each of the possible monosubstitution products.

29. How many different chlorine substitution products (monosubstituted, disubstituted, and so on) can be obtained theoretically by the chlorination of ethane? Name each possible product systematically.

30. Write condensed structural formulas for the predominant organic products in each of the following:

(a) ethylene + H_2O $\xrightarrow[\text{catalyst}]{\text{acid}}$

(b) 2-butene + H_2O $\xrightarrow[\text{catalyst}]{\text{acid}}$

(c) cyclopentene + H_2O $\xrightarrow[\text{catalyst}]{\text{acid}}$

(d) acetylene + Br_2 $\longrightarrow$
(1 mol)   (1 mol)

(e) 2-butyne + Br_2 $\longrightarrow$
(1 mol)   (2 mol)

(f) para-xylene + $HONO_2$ $\xrightarrow{H_2SO_4}$

(g) para-dichlorobenzene + Br_2 $\xrightarrow{FeBr_3}$

31. Write the condensed structural formulas and names for the C_7H_7Cl benzene ring substitution products one could obtain from the following reaction:

$$\text{toluene} + Cl_2 \xrightarrow{FeCl_3}$$

32. Nitric acid reacts with naphthalene in the presence of sulfuric acid. In the reaction, a hydrogen atom in naphthalene is replaced by an NO_2 group. Write condensed structural formulas for the substitution products, molecular formula $C_{10}H_7NO_2$, that are possible.

Alcohols

33. Which of the following, if any, could be the molecular formula of an alcohol derived from an alkane hydrocarbon: $C_6H_{14}OH$, $C_7H_{13}OH$, C_4H_7OH, C_6H_5OH, $C_{11}H_{23}OH$?

34. Write condensed structural formulas and systematic names for the eight alcohols with the molecular formula $C_5H_{11}OH$. Classify each alcohol as primary, secondary, or tertiary.

35. Indicate which of the following alcohols are primary, secondary, or tertiary:

36. Ethanol is soluble in water in all proportions, whereas hexanol is only slightly soluble. Account for the high solubility of the one and the low solubility of the other.

37. The alcohol 1-propanol reacts in the presence of sulfuric acid to give two dehydration products. Write equations, using condensed structural formulas for the organic compounds, for the two dehydration reactions. Which reaction should predominate at higher temperatures?

38. Write equations, using condensed structural formulas for organic compounds, that illustrate the preparations of ethanol and 2-propanol from alkenes.

Ethers

39. Write condensed structural formulas and names for all the ethers with the molecular formula $C_4H_{10}O$.

40. When working with ethyl ether in a hospital or elsewhere, one must be careful not to have open flames or electric sparks nearby. Why? Should one be equally cautious when working with *n*-butyl ether? Why?

41. The following equation represents a possible synthesis of ethyl *n*-propyl ether:

$$CH_3CH_2OH + CH_3CH_2CH_2OH \xrightarrow[\text{heat}]{H_2SO_4}$$
$$CH_3CH_2OCH_2CH_2CH_3 + H_2O$$

When the reaction is carried out, considerably less than 50 percent yield is obtained. Why?

Aldehydes and Ketones

42. See Exercises 34 and 35. Which of the alcohols are oxidized to aldehydes? Write condensed structural formulas for the aldehydes produced.

43. The general formula for aldehydes may be written as RCHO. How many isomeric aldehydes are possible when R is (a) C_3H_7, (b) C_4H_9, and (c) C_5H_{11}? Write condensed structural formulas for those in which R is C_5H_{11}.

44. Since alcohols are readily oxidized by atmospheric oxygen at elevated temperatures, why is it necessary to use a catalyst in the commercial method of oxidizing ethanol with atmospheric oxygen to produce acetaldehyde?

45. How many ketones are possible for each of the following molecular formulas: (a) C_4H_8O, (b) $C_5H_{10}O$, (c) $C_6H_{12}O$? Write condensed structural formulas for each ketone.

46. Which of the alcohols in Exercises 34 and 35 are oxidized to ketones? Write condensed structural formulas for the ketones produced.

Phenols

47. The cresols have the molecular formula C_7H_8O. Write their condensed structural formulas. There are two other derivatives of benzene that have this molecular formula.

Write their condensed structural formulas and state to what class of compounds each belongs.

*48. Unlike alcohols, phenols are weak acids in water solutions. One explanation for the acidity of phenol is that the phenoxide ion, $C_6H_5O^-$, is stabilized by resonance. Write Lewis formulas for the resonance contributors to this anion.

49. Phenol reacts rapidly with bromine in water solution to give 2,4,6-tribromophenol. Write the condensed structural formula for this product.

Carboxylic Acids

50. How many carboxylic acids are there for each of the molecular formulas: (a) $C_4H_8O_2$, (b) $C_5H_{10}O_2$?

51. Write condensed structural formulas and names for all the carboxylic acids with the molecular formula $C_5H_{10}O_2$.

52. Benzoic acid is an aromatic carboxylic acid with the molecular formula $C_7H_6O_2$. Write its condensed structural formula.

53. What is the conjugate base of propanoic acid? K_a for propanoic acid is 1.4×10^{-5}. Would the pH of a water solution of sodium propanoate be 7, less than 7, or more than 7? Why?

Esters

54. Write equations, using condensed structural formulas for organic compounds, that illustrate the preparation of the following esters: (a) ethyl acetate, (b) methyl butanoate, and (c) ethyl benzoate.

55. Write condensed structural formulas and systematic names for all the esters that have the molecular formula $C_4H_8O_2$.

*56. A certain ester was hydrolyzed with aqueous sodium hydroxide. The products were 1-propanol and sodium methylpropanoate. Write a condensed structural formula and two names for the ester.

*57. When 4-hydroxybutanoic acid is heated, water and a compound with the formula $C_4H_6O_2$ are produced. The $C_4H_6O_2$ compound is a cyclic ester known as a *lactone*. Write its structural formula.

58. When the unsaturated acids oleic acid, linoleic acid, and linolenic acid are reduced with excess hydrogen over a platinum catalyst, how many moles of hydrogen react with 1 mole of each acid?

59. Name several compounds that are possibly present in a natural soap such as Ivory soap. In what way are the molecules of a natural soap similar to those of a synthetic detergent?

Complex Derivatives of Hydrocarbons; Biochemistry

27

27•1 **Stereoisomers and Chirality**

27•1•1 Chiral Carbon Atoms

27•1•2 Plane-Polarized Light and Optical Activity

27•1•3 Number of Optical Isomers

27•1•4 Chemical Properties of Enantiomers

27•2 **Carbohydrates**

27•2•1 Monosaccharides

27•2•2 Disaccharides

27•2•3 Polysaccharides

27•2•4 Hydrolysis of Di- and Polysaccharides

27•2•5 Cellulose and Its Chemical Modifications

27•3 **Proteins**

27•3•1 Amino Acids

27•3•2 Structure of Protein Molecules

27•3•3 Enzymes

27•4 **Oxidation–Reduction in Cells**

27•4•1 In Animal Cells

27•4•2 In Plant Cells

27•5 **Hormones**

27•6 **Viruses**

27•7 **Drugs**

Special Topic: Biochemical Benefits and Risks of Synthetic Organic Chemicals

27•8 **Polymers**

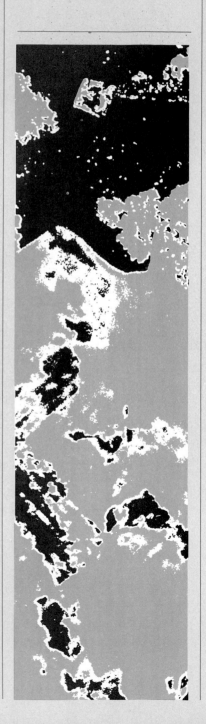

Most organic substances in plants and animals are composed of very complex molecules that, for the purpose of study and classification, may be considered to be derived from hydrocarbons. However, the chemical processes by which plants and animals build complex molecules from simple ones do not involve prior formation of any "parent" hydrocarbons. In this chapter, we shall consider certain complex substances that are of natural origin—carbohydrates and proteins, for example—as well as some that are of synthetic origin. As we discuss these chemical entities, items of biochemical interest will often be mentioned. **Biochemistry** is the branch of chemistry that deals with all the chemical changes occurring in living cells that are responsible for such processes as reproduction, growth, well-being, expenditure of energy, nerve stimulation and response, mental activity, aging, and death. In the 150 years since the overthrow of the vital force theory, great progress has been made, yet much remains to be done to understand some of the fundamental chemistry associated with living and dying.

27•1 Stereoisomers and Chirality

Before discussing carbohydrates and proteins, we must become familiar with certain kinds of isomers known as *stereoisomers*. All isomeric compounds can be placed into two categories, structural isomers and stereoisomers. **Structural isomers** differ from one another in the sequence of atoms bonded together in the molecule, that is, in the structure of the molecule. The alkenes 1-butene and methylpropene are examples of structural isomers. **Stereoisomers** contain the same sequence of atoms but differ in the arrangement of the atoms in space. The alkenes *cis*- and *trans*-2-butene (see Figure 26.4) are examples of stereoisomers. Notice that the sequence of the carbon atoms in these geometric isomers is the same, but the spatial arrangements differ.

A special type of stereoisomerism is shown by two molecules that are mirror images of one another. This relationship is illustrated by our two hands. A left hand has the same arrangement in space as the mirror image of a right hand (see Figure 27.1). Two molecules that are related as a right hand is related to a left hand are identical in all respects except one: their images are not *superimposable*. By this we mean that we cannot visualize one molecule put in the place of the other so that the positions of all parts of the molecules coincide. (An analogy is that a right-handed glove will not fit a left hand.) This general property of objects is called **chirality** (Greek *cheir*, hand) and denotes handedness. An object that is not superimposable on its mirror image is called **chiral.** If it is superimposable, it is called **achiral.**

An example of a chiral compound is lactic acid, $CH_3CHOHCO_2H$, the acid that is formed by the fermentation of sugar in milk and also may be isolated from animal muscle tissue. There are two isomeric lactic acids with molecules that have spatial arrangements that are nonsuperimposable mirror images (see Figure 27.2, top). Two isomers that differ in this way are called **enantiomers.**

27•1•1 Chiral Carbon Atoms A molecule of lactic acid is chiral, because the middle carbon atom has four different atoms or groups attached to it,

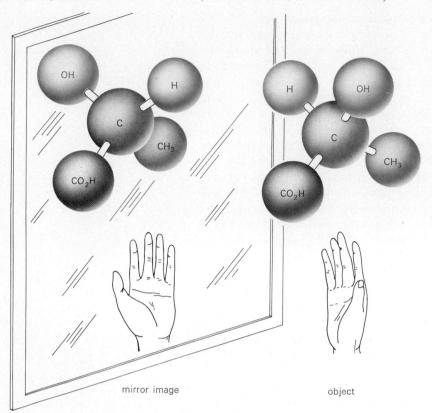

Figure 27.1 Models showing the two possible arrangements in space of the four groups around the chiral carbon atom of lactic acid. Note that the two forms have an object–mirror-image relationship and that one cannot be superimposed on the other for all points to coincide.

namely, —CH$_3$, —H, —OH, and —CO$_2$H. A carbon atom to which four different atoms or groups are attached is called an asymmetric or a **chiral carbon atom.**

One way of depicting three-dimensional formulas that illustrates the nonsuperimposability of enantiomers is to use projection formulas of the type shown in Figure 27.2. If we draw structural formulas in the plane of the paper (two dimensions), we can show the correct spatial relationships by indicating with dotted lines and wedges the groups that lie above and below the plane of the paper. However, when there are a number of chiral carbon atoms in a molecule, projection formulas such as those in Figure 27.2 may be too tedious to draw, and two-dimensional formulas are often used. In order that the two-dimensional formulas convey the three-dimensional aspects of the molecules, certain conventions are followed. For example, if the four different groups attached to chiral carbons are designated a, b, c, and d, the following relationships exist:

$$d—\overset{\displaystyle c}{\underset{\displaystyle a}{C}}—b \equiv \overset{\displaystyle c}{\underset{\displaystyle a}{\overset{d}{\diagdown}C—b}} \quad\Big|\quad b—\overset{\displaystyle c}{\underset{\displaystyle a}{C\diagup^{d}}} \equiv b—\overset{\displaystyle c}{\underset{\displaystyle a}{C}}—d$$

mirror

These relationships may be explained as follows: (1) any two adjacent attached groups of the two-dimensional projection, one group vertically and the other horizontally (a and b, respectively, in the examples shown), are considered to be in the plane of the paper; (2) the other vertically attached group (c in the examples) is projected behind the plane of the paper; and (3) the other horizontally attached group (d in the examples) is projected out from the plane of the paper.

Because the two-dimensional and three-dimensional projection formulas show the relative disposition of attached groups, or the **configuration,** of the chiral carbon atom, they are called **configurational formulas.**

27•1•2 Plane-Polarized Light and Optical Activity

Most of the physical properties of enantiomers are alike. However, enantiomers do differ with respect to the way they interact with plane-polarized light. Polarized light differs from ordinary light in that it vibrates in only one plane, whereas ordinary light vibrates in all directions perpendicular to the path of travel. Ordinary light becomes polarized when it passes through a polarizing material, such as a crystal of calcite, $CaCO_3$, or a sheet of Polaroid. The beam that emerges is said to be **plane-polarized** (see left part of Figure 27.3). When a beam of polarized light is passed through a solution containing one lactic acid enantiomer, the one isolated from animal muscle tissue, the plane of polarization is rotated to the right by a definite number of degrees (see Figure 27.3). However, when the beam is passed through a solution of the other enantiomer at the same concentration, the plane of polarization is rotated by the same amount but to the left. The enantiomer that rotates light to the right (clockwise) is called the *dextrorotatory* form; the one that rotates it to the left (counterclockwise) is called the *levorotatory* form. [The symbols (+) and (−) are used to indicate rotations to the right and left, respectively.] A mixture of the dextrorotatory and levorotatory forms in equal proportions has no rotation.

Optics is the branch of physics that deals with the study of light. The phenomenon of rotation of polarized light is called **optical activity.** Optical activity was noted in 1811 by the French physicist Jean Arago. He found that some quartz crystals rotate the plane of polarized light to the right and others to the left. Somewhat later, the physicist Jean Biot found that certain liquids also rotate polarized light. For example, he found that a glucose solution was dextrorotatory and turpentine was levorotatory. In 1844, the great French chemist Louis Pasteur observed that when a solution of sodium ammonium tartrate, $NaNH_4C_4H_4O_6$, was evaporated, crystals were formed, some of which were mirror images of the others (see Figure 27.4). He separated the two by hand-picking and found that a solution of one type rotated polarized light to the right, and the other, an equal amount to the left. Because enantiomers rotate the plane of polarized light, they are said to be optically active. Consequently, enantiomers are also called **optical isomers.**

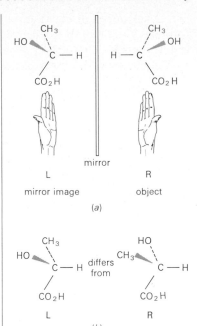

Figure 27•2 Projection formulas showing (*a*) the mirror-image relationship of the optical isomers of lactic acid (L = left; R = right); (*b*) the nonsuperimposability of these isomers. The "right-handed" isomer in (*a*), designated as the object, has been rotated by 180° to give the projection formula in (*b*). In the two projection formulas in (*b*), the —OH and —CH_3 groups are not superimposable and therefore the molecules themselves are also nonsuperimposable.

Figure 27•3 Diagrammatic representation of a polariscope. Light entering the apparatus is shown vibrating in all planes (cluster of arrows) and in one plane after emerging from the polarizing crystal (broken double-headed arrow). After passing through the solution, the light is still vibrating in one plane, but the plane has been rotated. The part containing the analyzing crystal is joined to the rest of the instrument at *B* so as to permit rotation. The angle of rotation, α, required to restore the full intensity of the emergent beam measures the rotation produced by the chiral compound in the solution.

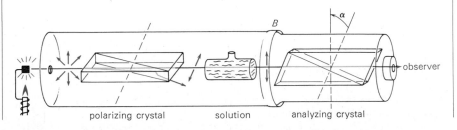

A question may naturally arise as to which of two configurational formulas corresponds to the dextrorotatory optical isomer and which corresponds to the levorotatory form. In the case of the two lactic acids, the configurations of the dextro- and levorotatory forms are as follows:[1]

$$
\begin{array}{cc}
CH_3 & CH_3 \\
HO-C-H & H-C-OH \\
CO_2H & CO_2H \\
(-)\text{-lactic acid} & (+)\text{-lactic acid}
\end{array}
$$

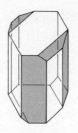

Figure 27•4 Schematic representation of right- and left-handed crystals of sodium ammonium tartrate.

27•1•3 Number of Optical Isomers In a molecule of lactic acid, there is one chiral carbon atom, and two optical isomers may be formed. If a molecule has more than one chiral carbon atom, additional stereoisomers are possible. For example, consider the formula for 2-bromo-3-hydroxybutane, $CH_3\overset{*}{C}HBr\overset{*}{C}HOHCH_3$, with two chiral carbon atoms marked by asterisks. The number of optical isomers corresponding to this condensed structural formula is four; their configurational formulas are shown in Figure 27.5.

By means of molecular models, it can be shown that the number of optical isomers of a given structural formula is 2^n, where n is the number of *unlike* chiral carbon atoms in a molecule.[2] If n is 1, the number of optical isomers is 2^1, or 2; if n is 2, the number is 2^2, or 4; and so on. Note, as in the case of the 2-bromo-3-hydroxybutanes, that the optical isomers for a given structural formula can be grouped into pairs of enantiomers. Each enantiomer rotates polarized light, but only the members of a pair rotate it by the same amount but in opposite directions.

27•1•4 Chemical Properties of Enantiomers Two enantiomers, taken together or separately, behave alike insofar as most chemical reactions are concerned. In laboratory syntheses involving reactions of enantiomers, the difference in the spatial arrangements of the groups around a chiral carbon is of no consequence. For example, both enantiomers of 2-butanol, $CH_3\overset{*}{C}HOHCH_2CH_3$, react with acetic acid to give the acetate enantiomers in the same percentage yield and at the same rate. When the acetic acid approaches either enantiomer, it encounters the same combination of groups in space, except that one arrangement is the mirror image of the other. However, a difference in chemical reactivity between enantiomers is found in their reactions with optically active reagents. Then the interaction between one enantiomer and the reagent will be more favorably disposed toward reaction than will the other enantiomer.

This is especially true in the biosynthesis of many plant and animal materials, proteins and carbohydrates, for example. A molecule with one

Figure 27•5 Configurational formulas of the four stereoisomers of 2-bromo-3-hydroxybutane. Note that they are divided into two pairs of enantiomers.

[1]The methods by which configurations of enantiomers are determined are usually described in comprehensive texts of organic chemistry.

[2]Note that n is the number of *unlike* chiral carbon atoms. If a structure has some *like* chiral atoms, the number of optical isomers is less than 2^n. For example, the two chiral carbons in 2,3-dibromobutane, $CH_3\overset{*}{C}HBr\overset{*}{C}HBrCH_3$, are alike. There are only three stereoisomers of this dibromoalkane, two enantiomers and an achiral compound called a *meso* compound.

spatial arrangement may participate freely in a reaction, whereas its enantiomer will react either not at all or extremely slowly. We will see later that
most of the reactions occurring in plants and animals take place at a measurable rate only if catalytic substances called *enzymes* are present. Both the
molecules undergoing reaction and the enzymes contain one or more chiral
carbon atoms. For the catalytic action to be effective, the molecules involved
must fit snugly at so-called active sites. Just as a right-handed glove fits only
a right hand, an enzyme with one type of shape will fit only a protein or
carbohydrate molecule with a particular shape and will not fit the mirror-
image shape.

27•2   Carbohydrates

The simple sugars and the substances that hydrolyze to yield simple
sugars are called **carbohydrates.** Originally, the name *carbohydrate* was used
because the composition of most sugars, starch, and cellulose corresponds to
that of hypothetical hydrates of carbon, $C_x \cdot (H_2O)_y$. The values of x and y
may range from 3 to many thousands.

A comprehensive classification of the very large number of carbohydrates involves about a half-dozen main classes, with perhaps 40 sub-
classes. In general, all carbohydrates are said to be *saccharides* (Greek
sakcharon, sugar). We will limit our discussion to the following: (1) **monosaccharides,** which do not undergo hydrolysis; (2) **disaccharides,** which may be
hydrolyzed to two monosaccharide molecules; and (3) **polysaccharides,**
which form many monosaccharide molecules upon hydrolysis.

27•2•1   Monosaccharides   Among the most important monosaccharides are
those made up of molecules that contain six carbon atoms, known as the
hexoses, $C_6H_{12}O_6$. When a hexose contains an aldehyde group, it is known
as an **aldohexose;** if it contains a ketone group, it is called a **ketohexose.** The
hexoses are crystalline, water-soluble sweet substances that occur in ripe
fruits and honey. Carbohydrates that hydrolyze to produce hexoses are cane
sugar, malt sugar, milk sugar, starch, and cellulose. The relationship of these
sugars to monosaccharides is discussed in Sections 27.2.2 and 27.2.3.

• *Glucose* • Glucose, $C_6H_{12}O_6$ (dextrose, grape sugar, blood sugar),
one of the isomeric aldohexoses, is an important sugar in nature, both
because of its widespread occurrence and because of the prominent part it
plays in biological processes. It is the sugar into which all other carbohydrates are converted prior to oxidation in the body. Glucose is found in all
ripe fruits; it is especially abundant in grapes. Many other carbohydrates—
for example, maltose, sucrose, starch, and cellulose—yield it on hydrolysis.

Chemical reactions and analyses indicate that glucose molecules
contain five hydroxyl groups and an aldehyde group joined to a six-carbon
chain. Thus glucose can be represented by the following condensed structural formula:

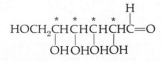

Note that there are four unlike chiral carbon atoms in a molecule of glucose. There are 2^4 or sixteen possible optical isomers; that is, ordinary glucose is one of sixteen aldohexoses, all of which have the same condensed structural formula. All sixteen of these sugars have been isolated and identified.

The spatial arrangement of the four groups about each chiral carbon will not be presented here. However, the configurational formulas for (+)-glucose (common glucose) and its enantiomer (−)-glucose are shown in Figure 27.6.

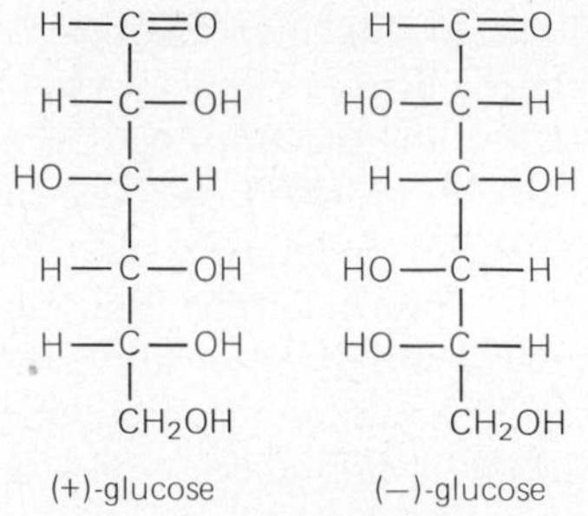

Glucose molecules are now known to exist mostly in closed-chain or cyclic forms. Recall that the valence angles between carbon atoms are about 109.5° rather than the 180° that our two-dimensional structural formulas indicate. Consequently, the aldehyde group on carbon atom 1 can be very close to the hydroxyl group on carbon atom 5 if the chain twists around on itself. This twisting of the chain is represented pictorially as follows: first, using the (+)-glucose as our model, we write the formula horizontally; second, we show the molecule folded around in an arc so that carbon atom 6 is close to the aldehyde group; and finally we rotate about the C_4—C_5 bond axis to obtain an arrangement in which the —OH group of carbon-5 is within bonding distance of the carbonyl group:

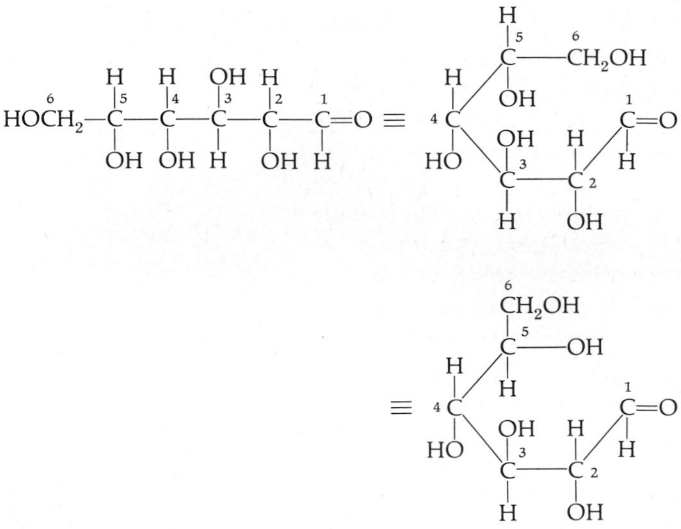

The aldehyde and hydroxyl portions enter into an addition reaction with each other (dotted arrows in formulas below) to form two cyclic molecules, designated as α and β forms:

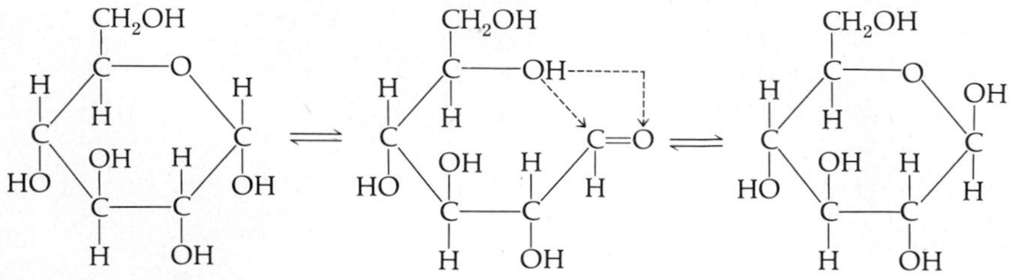

cyclic form (α)   open-chain form   cyclic form (β)

(+)-glucose   (−)-glucose

Figure 27•6 The configurational formulas for (+)-glucose and (−)-glucose.

Formation of these six-membered rings changes the carbon atom of the aldehyde group to an additional chiral carbon atom, which is necessary to account for the pair of optical isomers that can be derived from (+)-glucose. In water solution, the equilibrium concentration of the two forms is 36 percent α and 64 percent β. These isomers have different physical and chemical properties and can be separated from one another. Although the amount of the open-chain form present at equilibrium is very small (0.024 percent), it cannot be completely ignored because glucose acts as an aldehyde in many of its reactions.

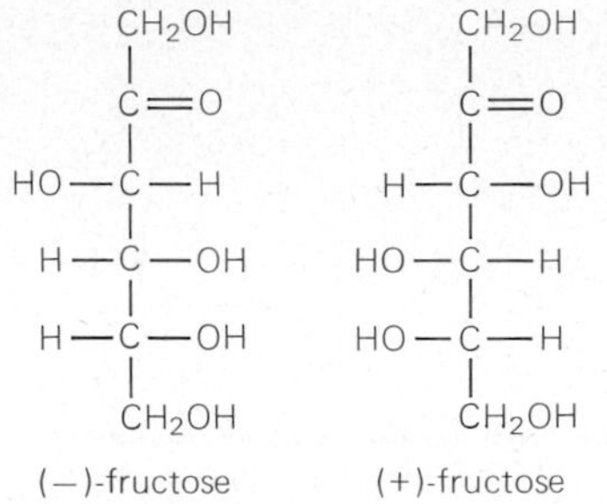

• *Fructose* • Fructose, $C_6H_{12}O_6$ (levulose, fruit sugar), one of the isomeric ketohexoses, is a crystalline sugar that occurs with glucose in honey and in fruits. Chemical reactions and analyses indicate that fructose molecules contain five hydroxyl groups and the ketone carbonyl group at C-2 of the six-carbon chain. Thus fructose can be represented by the following condensed structural formula:

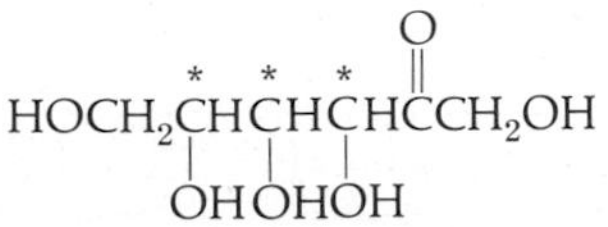

$$\underset{\overset{|}{O}H\overset{|}{O}H\overset{|}{O}H}{HOCH_2\overset{*}{C}H\overset{*}{C}H\overset{*}{C}H\overset{O}{\overset{\|}{C}}CH_2OH}$$

Because there are three chiral carbon atoms, there are 2^3 or eight ketohexoses that have this structure; they differ from one another only in the configurations of these three carbon atoms. Fructose molecules, like glucose molecules, also exist mostly in closed-chain or cyclic forms.[3] The configurational formulas for naturally occurring (−)-fructose in its open-chain form and its enantiomer are shown in Figure 27.7. Note that the configurations at C-3, C-4, and C-5 are the same as those in (+)-glucose.

Figure 27•7 The configurational formulas for (−)-fructose and (+)-fructose.

━━━ • Example 27•1 • ━━━

In addition to the fructose enantiomers shown in Figure 27.7, there are six more 2-ketohexoses. Write their two-dimensional configurational formulas.

━━━ • Solution • First we write formulas for the enantiomers with the three hydroxyl groups at C-3, C-4, and C-5 on the same side; then we write those with the hydroxyls at C-3 and C-4 on the same side; and finally we write those with the hydroxyls at C-3 and C-5 on the same side:

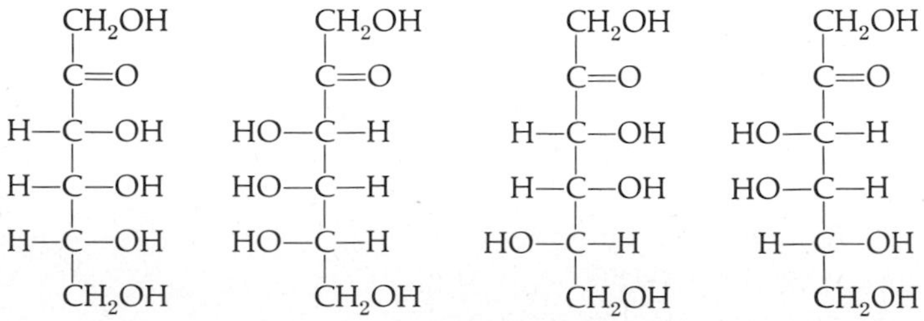

---

[3]In water solution, the equilibrium mixture is complex and apparently involves both five- and six-membered rings.

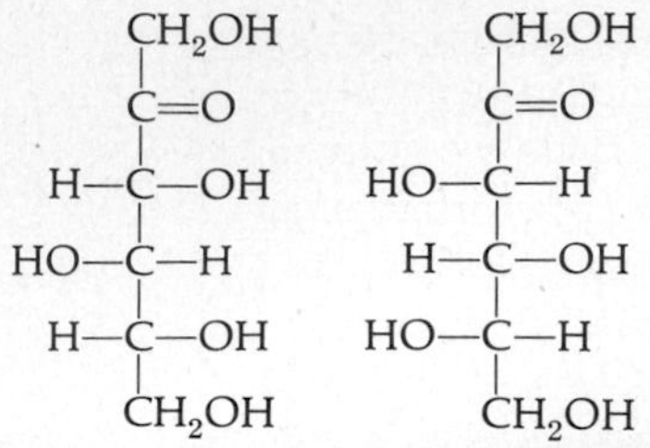

See also Exercise 10 at the end of the chapter.

Both fructose and glucose are energy foods that are ultimately oxidized to carbon dioxide and water in the cells. However, research indicates that the enzymes and hormones that control these oxidations may be different in one or two steps of the overall processes of oxidation.

27·2·2   Disaccharides   Sucrose (table sugar), maltose (malt sugar), and lactose (milk sugar) are important members of the disaccharide group, $C_{12}H_{22}O_{11}$. As the group name indicates, each molecule of these sugars is composed of two monosaccharide units. One can imagine that these units are joined together by bonds that result from the elimination of a molecule of water. For example, a sucrose molecule is composed of a glucose and a fructose unit joined, as shown below:

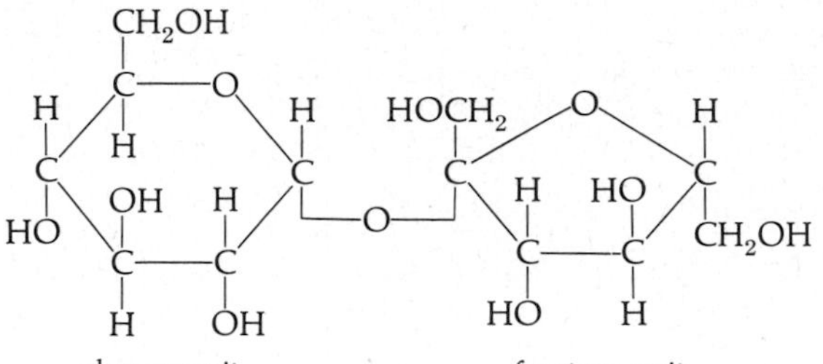

glucose unit   fructose unit

sucrose,
$C_{12}H_{22}O_{11}$

A lactose molecule is also made up of two hexoses: a glucose unit and a galactose unit. Molecules of maltose are built up of only glucose units.

Sucrose is sweeter than glucose but not as sweet as fructose. With sucrose rated as 100, the comparative sweetness of common sugars is:

lactose	16	sucrose	100
maltose	33	invert sugar	130
glucose	74	fructose	173

Several synthetic, noncarbohydrate compounds are known that are considerably sweeter than sucrose. However, most of them have adverse biological activities and are not used in foods. Saccharin, whose structure is shown in Figure 27.8, is perhaps the best-known synthetic sweetener, being about 300 times sweeter than sucrose. It has no food value and has been used in low-calorie foods and drinks. The use of saccharin for this purpose has been challenged in the United States following the discovery that tumors developed in animals that were fed large amounts of the compound.

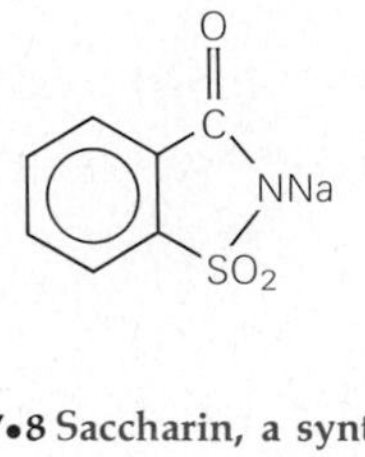

Figure 27·8 Saccharin, a synthetic sweetener.

27•2•3 **Polysaccharides** As the name indicates, molecules of polysaccharides are composed of many monosaccharide units. If the monosaccharide units are pentose sugars, $C_5H_{10}O_5$, the polysaccharide is classed as a **pentosan**, $(C_5H_8O_4)_x$. Pentosans make up a sizable portion of corncobs, oat hulls, and similar woody tissues of plants. If the monosaccharide unit is a hexose sugar, $C_6H_{12}O_6$, the polysaccharide is classed as a **hexosan**, $(C_6H_{10}O_5)_x$.

• *Hexosans* • The most abundant of the hexosan type of polysaccharide, $(C_6H_{10}O_5)_x$, are those in which the hexose unit is glucose. Starch and cellulose are in this group. The manner in which the glucose units are joined in starch molecules is as follows:

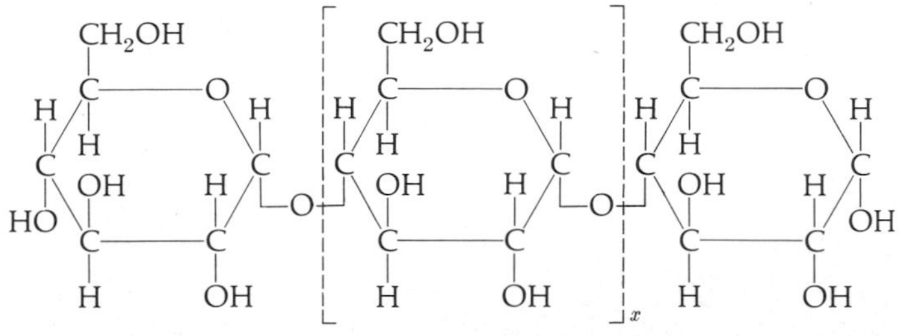

starch

Molecular weight determinations show that starch molecules contain from 200 to 3,000 glucose units per molecule. The weight of cellulose molecules is more difficult to ascertain. However, the best estimates indicate that the number of glucose units per molecule is on the order of several thousand.

There are at least two important differences in the structures of starch and cellulose molecules. In cellulose, the glucose units are attached end to end to form long, filament-like molecules. In starch, these units are generally joined in a branched-chain pattern, although the smaller unbranched molecules are present in varying amounts in most starches. For example, starch from potatoes contains about 20 percent of the unbranched type.

A second difference in the structure of starch and cellulose molecules concerns the spatial arrangement of the groups around the two chiral carbons by which two glucose units are joined. If we think of the arrangement around these carbons in starch as being "right-handed," the arrangement around the same carbons in cellulose is then "left-handed." A comparison of the formula below for cellulose with the one given above for starch will illustrate this relationship:

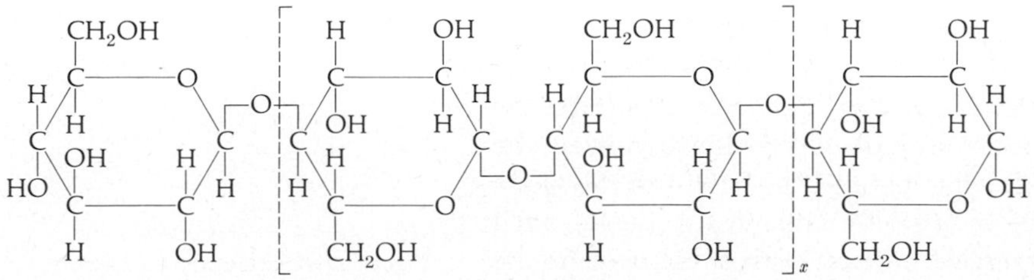

cellulose

Although this difference may appear trivial, it seems to be related to the fact that humans cannot digest cellulose but can digest starch. That is, the enzyme molecules can "fit" the right-handed reactive centers of starch molecules and thereby catalyze the hydrolysis of these large molecules into glucose, but the same enzymes cannot fit properly into the left-handed centers of the cellulose molecules so as to catalyze their hydrolysis.

27•2•4   Hydrolysis of Di- and Polysaccharides   The breaking down (hydrolysis) of complex sugar, starch, and cellulose molecules to monosaccharide molecules is done easily in the laboratory by boiling an aqueous solution or suspension of carbohydrate with dilute mineral acids. In the digestive tract of animals, this hydrolysis is effected at body temperature by enzymes that act as catalysts. Maltose, starch, and cellulose form only glucose on complete hydrolysis:

$$C_{12}H_{22}O_{11} + H_2O \longrightarrow 2C_6H_{12}O_6$$
$$\text{maltose} \qquad\qquad\qquad \text{glucose}$$

$$(C_6H_{10}O_5)_x + xH_2O \longrightarrow xC_6H_{12}O_6$$
$$\text{starch or} \qquad\qquad\qquad \text{glucose}$$
$$\text{cellulose}$$

Maltose, or malt sugar, is made from starch by a hydrolysis that is catalyzed by the enzyme *diastase:*

$$(C_6H_{10}O_5)_x + \frac{x}{2}H_2O \xrightarrow{\text{diastase}} \frac{x}{2}C_{12}H_{22}O_{11}$$
$$\text{starch} \qquad\qquad\qquad\qquad \text{maltose}$$

This enzyme, contained in preparations known as *malt,* is formed during the germination of barley seed.

Sucrose yields equal amounts of glucose and fructose on hydrolysis:

$$C_{12}H_{22}O_{11} + H_2O \longrightarrow C_6H_{12}O_6 + C_6H_{12}O_6$$
$$\text{sucrose} \qquad\qquad\qquad \text{glucose} \qquad \text{fructose}$$

Because sucrose rotates polarized light to the right and the hydrolysis mixture rotates it to the left, the resulting glucose–fructose mixture is called *invert sugar.* This mixture is sweeter than sucrose and also does not crystallize readily, even from concentrated solutions, so it is preferable to sucrose for making some candies and jelly.

In jelly making, the fruit acids catalyze the hydrolysis of sucrose to invert sugar. Consequently, it is important that all the sugar be added at the beginning. In candy making, buttermilk, vinegar, or cream of tartar is sometimes added to serve as an acid catalyst for the hydrolysis.

27•2•5   Cellulose and Its Chemical Modifications   Cellulose constitutes the woody portion of all plants and is present in all plant cells. In wood itself, the long molecules of cellulose are laid down in parallel rows to form the wood fibers; the fibers are bound together by a sticky organic substance called *lignin.* (Wood also contains about 8 percent mineral salts; these become the ash when wood is burned.)

• *Paper* • In making paper, wood is cut into small pieces and cooked in calcium bisulfite or other chemicals to dissolve the lignin. The cellulose is removed by filtration, is bleached with chlorine or hydrogen peroxide, and is then weighted, sized, and rolled into sheets. In weighting and sizing, such materials as starch, glue, casein, rosin, aluminum silicate, and clay are added. *Glazed* paper is heavily weighted with minerals (for example, barium sulfate) to reduce porosity; paper toweling contains little mineral additives. Filter paper is almost pure cellulose.

A large amount of the newsprint made in this country consists of a blend of ground wood (about 90 percent) and cellulose. Because the dry weight of wood is less than 50 percent cellulose, this method of producing paper makes for economical use of wood.

• *Rayon* • Most of the rayon produced in the United States is manufactured by the *viscose* process. Pure cellulose is obtained from wood by the process described above and treated with aqueous sodium hydroxide and carbon disulfide; the syrupy liquid that forms is called viscose. After aging and filtering, the viscose is forced through the tiny holes of a spinneret into a sulfuric acid bath. This precipitates the cellulose as continuous threads that are gathered and twisted into rayon yarn (see Figure 27.9).

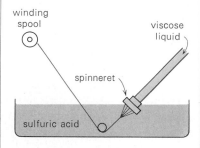

Figure 27•9 Schematic representation of the production of rayon.

• *Cellulose Nitrate and Cellulose Acetate* • Each glucose unit in a cellulose molecule contains three hydroxyl groups. When cellulose reacts with concentrated nitric acid in the presence of concentrated sulfuric acid, one, two, or three of these hydroxyl groups are replaced with nitrate groups, —ONO_2, forming the ester *cellulose nitrate.*

If cellulose is treated with acetic acid and sulfuric acid, or with acetic anhydride, the hydroxyl groups are replaced by acetate groups and *cellulose acetate* is formed. This is used in making acetate rayon and photograph film.

Cellulose nitrate in which a large number of —OH groups have been replaced by —ONO_2 groups is known as *nitrocellulose,* or *guncotton.* This is an important explosive. Not as violent as nitroglycerin, its explosion can take place at a rate suitable for the propulsion of bullets and rockets.

27•3   Proteins

Proteins are present in all living tissue, both plant and animal. The tissues of seeds, lean meat, vital organs, skin, and hair contain greater amounts of proteins than do the fatty tissues.

All protein molecules contain nitrogen in combination with carbon, hydrogen, and oxygen. Many also contain sulfur and phosphorus, and some contain iron, manganese, copper, and iodine. Protein molecules are very large, ranging in molecular weight from about 10,000 to several million amu.

When proteins are boiled in dilute acids or bases or when they are subjected to the action of specific enzymes in digestion, their molecules are hydrolyzed to *amino acids.* Therefore, proteins are like starches and cellulose in the sense that their molecules are built up of repeated units of simpler molecules. The structural units of proteins are amino acids.

27•3•1   Amino Acids   **Amino acids** are compounds with molecules that contain both the amino (—NH_2) and the carboxyl (—CO_2H) functional

groups. Although hundreds of these acids have been synthesized, only 20 have been obtained by the hydrolysis of proteins. In these the amino group is always attached to the carbon atom adjacent to the carboxyl group:

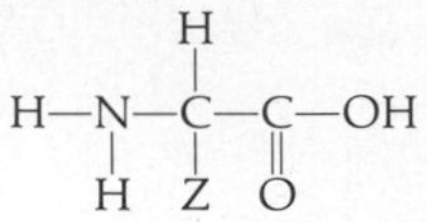

$$H-N-\overset{\overset{\displaystyle H}{|}}{\underset{\underset{\displaystyle H}{|}}{C}}-\overset{\overset{\displaystyle }{}}{\underset{\underset{\displaystyle O}{\|}}{C}}-OH$$

In aminoacetic acid, glycine, the simplest amino acid from proteins, Z in the formula above, is hydrogen. In others, Z may be an alkyl group, a carbon chain that also contains sulfur atoms, a cyclic group, or an extra acidic or basic group. The diverse properties of different proteins are accounted for by these variations both in Z and in the size of the protein molecule.

27.3.2 Structure of Protein Molecules Simple protein molecules are long-chain molecules that are formed by the union of hundreds or even thousands of amino acid molecules. The union is due to bonds, each of which can be considered to originate from the elimination of a molecule of water from an —NH$_2$ and a —CO$_2$H group:

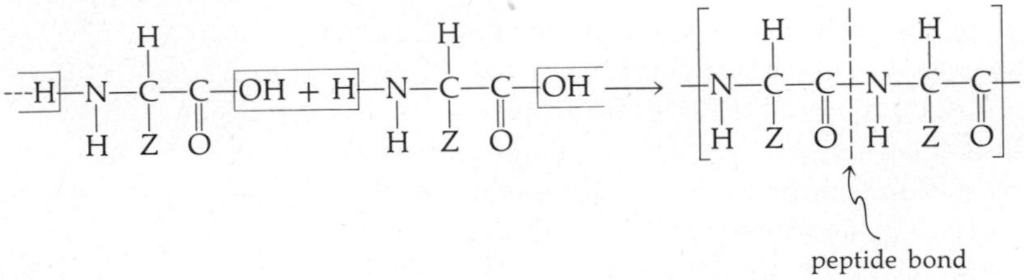

This important linkage is called the **peptide bond.** In Figure 27.10 is shown a portion of a protein molecule formed by joining eight amino acid residues through peptide bonds.

• *Conjugated Proteins* • Conjugated proteins are made up of simple protein molecules linked to nonprotein molecules. The hemoglobin of blood is an example. The protein is combined with *heme*, a complex red compound of iron (see Section 27.4.1). Casein in milk and vitellin in egg yolk are also conjugated proteins in which the protein is combined with phosphoric acid. In other conjugated proteins, the additional group may contain carbohydrates or compounds of nitrogen, magnesium, copper, manganese, cobalt, or other substances.

• *Amino Acid Sequence* • Look again at Figure 27.10. If we number the amino acid residues arbitrarily from top to bottom as 1, 2, 3, and so on, the sequence of amino acids in this portion of the protein molecule is –1–2–3–4–5–6–7–8–. We would expect to find this sequential portion in every molecule of this particular protein. In a different protein, we find a different sequence of amino acid residues. For example, the sequence in a portion of some other protein molecule might be –2–16–4–6–7–13–11–13–. That is, not only is the order different, but some of the amino acids present in the first protein are missing from the second one, and vice versa.

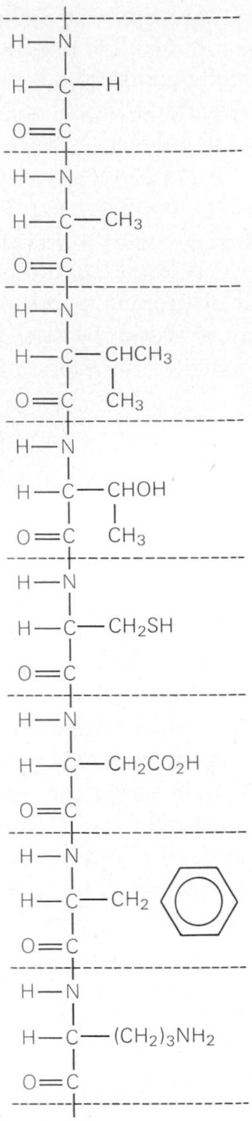

Figure 27.10 A portion of a protein molecule. Each amino acid residue is separated from its neighbors by dotted lines, the portions of the residues that are common to all amino acids are shown in color, and the Z portions are shown in black. The peptide bond joins a carbon of one residue with the nitrogen of the next.

Experimental methods for identifying each amino acid along the protein chain have been developed. F. Sanger and co-workers, during the period 1945–1952, were the first to establish the amino acid sequence in a protein, the hormone insulin (from beef). Insulin is a relatively simple protein, each molecule consisting of only 51 amino acid residues. Since that time, the sequence has been determined for a number of simple proteins. It goes without saying that this type of research is extremely tedious, since the amino acid residues may range from a hundred or so in simple proteins found in certain hormones, viruses, and blood hemoglobin up to several thousand in the more complex proteins comprising muscle, skin, and hair.

A second important aspect of protein structure is concerned with the way that the amino acid residues are organized in space along the polypeptide chain. (Note that Figure 27.10 shows the kinds of atoms in the chain but does not show their relative positions in space.) Linus Pauling was the first to suggest that these units are held in a coiled fashion, called the **alpha helix configuration,** by bonding forces other than those in the chain structure. These bonds are mainly hydrogen bonds (see Figure 27.11). In some proteins—hair, skin, and muscle meat, for example—the alpha helixes are believed to be twisted about one another to form ropelike structures. In other proteins—enzymes and insulin, for example—the chains are folded to give a globular structure.

It will be seen from the general formula of natural amino acids that all these acids, except glycine, are capable of existing as enantiomers, because each contains a chiral carbon atom:

$$H_2N-\overset{\overset{\displaystyle H}{|}}{\underset{\underset{\displaystyle Z}{|}}{C}}-CO_2H \qquad \text{chiral carbon}$$

However, on this planet only those amino acids with a particular configuration are formed and utilized by plants and animals to build proteins.

The proteins of cereals and vegetables are usually lacking one or more of the amino acids essential in the human diet. Accordingly, protein malnutrition frequently occurs among the poor, particularly among the populations of the less developed countries. Considerable research is being carried out on the chemical production of essential amino acids not usually available in cereal proteins. The amino acids so produced could be used to fortify cereal foods such as flour, or they could be used to supplement livestock feed. Lysine, $NH_2CH_2(CH_2)_3CHNH_2CO_2H$, an essential amino acid absent in most cereals, is presently produced at a reasonable price by the fermentation of a number of carbohydrates and by chemical synthesis.

• *DNA and RNA* • The problem of how cells utilize some twenty simple compounds—the amino acids—to build the very complex proteins, each with its unique amino acid sequence, and how this capacity is handed from parent to offspring, is a problem that has fascinated biologists and biochemists for many decades. In the past 25 years, great progress has been made toward its solution. The problem is so complex, involving most of the areas of chemistry and biology, and the amount of research literature that has accumulated is so vast that we can touch only a few points.

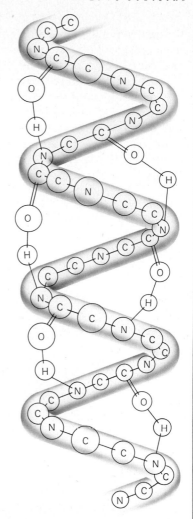

Figure 27•11 The helical structure of proteins. The main strand consists of the —C—C—N— unit of amino acids. Hydrogen bonding, shown by dotted lines from the oxygen atom of one amino acid residue to the hydrogen atom of another, stabilizes the coiled structure. (Only a few hydrogen bonds are shown; Z groups and most hydrogen atoms are omitted for clarity.)

As a beginning, let us remember that many ordinary substances tend to react with each other when brought together under favorable conditions. Iron, oxygen, carbon dioxide, and water react to form iron rust; acetic acid and bicarbonate of soda react to form carbon dioxide; and so on. It is apparent that we have to go a few steps further in analyzing the chemical reactions involving the union of amino acids to form proteins. We must postulate the presence of catalysts and the availability of energy; we must also postulate some sort of guide or coding system to allow for the formation of all the different proteins from the same twenty-odd amino acids. It is now quite clear that the coding systems in all cells, from bacteria to the cells of humans, are embodied in molecules of **deoxyribonucleic acid (DNA)**. (The term *genes* has long been used to refer to the hereditary factors in a fertilized cell. A single gene is now thought to be a part of a molecule of DNA.)

deoxyribose (a sugar)   phosphoric acid   thymine (a nitrogen base)

portion of a polynucleotide chain

The molecular structure of DNA, as proposed by J. D. Watson and F. H. C. Crick in 1953, consists of two parallel **polynucleotide chains** wound around a common axis into a helix and held together by hydrogen bonding between nitrogen bases spaced along the chain. The polynucleotide chain, or **nucleic acid** as it is commonly called, is formed by the repetitious union of the nucleotide unit, a unit formed by joining together a molecule of the pentose sugar *deoxyribose*, a molecule of *phosphoric acid*, and a molecule of a *nitrogen base*. Figure 27.12 shows these structural units and how they are joined to form the polynucleotide chain. Figure 27.13 shows how two polynucleotide chains are wound around a common axis to form a molecule of DNA. The four nitrogen bases that are repeated in this structure project inward and form hydrogen bonds to stabilize the structure.

Sophisticated instruments have helped prove the details of the double helix structure of DNA by direct observation. For example, a portion of a DNA molecule viewed by the electron microscope, more than fifteen years after the theoretical proposal by Watson and Crick, revealed the helical form and the dimensions they had predicted. Also, by the use of X-ray diffraction equipment (see Section 8.6.1), scientists have confirmed that these huge molecules are held in certain configurations by hydrogen bonds.

Figure 27•12 A portion of a DNA molecule, showing how the three structural units are joined to make a huge molecule. Note the use of an outline formula for deoxyribose in the DNA representation.

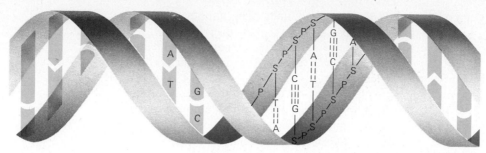

According to the Watson and Crick proposal, the sequence of the four bases along the two polynucleotide strands in the DNA molecule provides the code by which hereditary traits are transmitted. Through its chemical composition, spatial arrangement, and the sequences of the nitrogen bases, DNA controls protein formation, cell metabolism, and its own ability to produce an exact copy of itself from simple chemical substances that come through the cell walls as nutrients. It has been estimated that there are more than 4 million different species of organisms. This means that there must be many times this number of different DNA molecules.

Although DNA governs the amino acid sequence and controls the synthesis of proteins, there is much evidence to indicate that the actual synthesis of most proteins also involves three types of **ribonucleic acid (RNA):** *messenger* RNA, *soluble* or *transfer* RNA, and *ribosomal* RNA. RNA differs from DNA in that the sugar unit in the latter has one less oxygen atom. Transfer RNA is composed of relatively small molecules that contain from 75 to 93 nucleotide units (sugar, phosphoric acid, and nitrogen base). A special feature of a transfer RNA molecule is its coded centers for accepting specific amino acids. Messenger RNA molecules are much larger, consisting of 900 to 1,200 nucleotide units. Ribosomal RNA molecules, which have huge molecular weights, constitute 65 percent of ribosomes, nucleoprotein particles on which proteins are synthesized.

To summarize the protein synthesis process, it is now believed that DNA in the nucleus of the cell serves as the template for the formation of messenger RNA. Messenger RNA then moves into the cytoplasm, where it lines up briefly with the surface of the ribosomal nucleoprotein of the cell. At this point, the transfer RNA molecules with their attached amino acids line up as dictated by the code carried by messenger RNA for the formation of a specific protein. Under these conditions, peptide bonds form rapidly to join the amino acids into a protein molecule, thus freeing the transfer RNA for repetition of the synthesis. The role of ribosomal RNA is not yet clear.

Figure 27•13 The DNA structure is believed to consist of two polynucleotide chains wound around each other in a double helix (P, phosphate, and S, sugar, in the figure). Along each chain, attached to the sugar groups, are the nitrogen bases adenine, thymine, guanine, and cytosine (A, T, G, and C in the pairings in the figure), which form hydrogen bonds with each other to hold the helix together. These features are shown in some detail in the central portion but are simply shown schematically at the left and right portions.

27•3•3   Enzymes   Practically all the chemical reactions that take place in living organisms require the presence of catalytic substances, **enzymes,** in order to proceed at a measurable rate under the mild conditions of temperature and pH that exist in cells. Many enzymes have been obtained in the crystalline state and in a high degree of purity. All are formed in living cells and are proteins. However, the protein structure, including the amino acid sequence, is known for relatively few enzymes. One whose structure has been elucidated is an enzyme from bovine pancreas, a catalyst for the digestion of RNA. This enzyme has been shown to have a molecular weight of 12,600 and to consist of 124 amino acid units. The amino acids are arranged in a specific sequence to give a globular-shaped molecule.

A major difference between the catalytic action of inorganic catalysts and enzymes is that the so-called turnover number is much larger for the latter. The **turnover number,** defined as the number of moles of reactant transformed per mole of enzyme per minute at a definite temperature, varies from about 30 to 3.6×10^7. This means that only a small amount of an enzyme is needed to catalyze a particular reaction.

A second major difference between inorganic catalysts and enzymes is the specificity of the catalytic action of enzymes. For example, platinum will catalyze a great variety of reactions, ranging from the hydrogenation of double bonds to the oxidation of SO_2 in the production of sulfuric acid. On the other hand, urease is a specific enzyme catalyst for splitting urea; it has no effect on other compounds. The enzyme sucrase catalyzes the digestion (hydrolysis) of common sugar but will not catalyze the digestion of maltose. The oxidation of glucose to carbon dioxide and water provides an interesting example of the specificity of enzyme action. The equation for the overall reaction is

$$C_6H_{12}O_6 + 6O_2 \longrightarrow 6CO_2 + 6H_2O$$

In the cell this is a seventeen-step reaction and each step is catalyzed by a different set of enzymes. Because enzyme action is specific, each cell must have a large number of enzymes to serve as catalysts for the many chemical reactions necessary to life and reproduction. The number is estimated to be about 1,500 to 2,000 for many types of cells.

This specificity is due to the folded shape of the protein chain comprising the large enzyme molecule and to the location of active sites on the enzyme surface into which parts of the reactant molecule must fit closely. Only specifically shaped molecules can gain access to these active sites, and in many cases, the fit is due to the proper spatial arrangements of groups around a chiral carbon atom.

Frequently, a small organic molecule called a *coenzyme* is required in the process being catalyzed by an enzyme, and this too must fit properly into the reaction site. Vitamins are changed to act as coenzymes in certain biochemical reactions involving oxidation, reduction, and the removal of $-CO_2H$ groups. Two or three dozen vitamins are now known whose chemical structures have been determined. They are relatively simple compounds as compared with enzymes, and most can now be synthesized in the absence of living matter. However, the body does not synthesize the vitamins it requires; rather, they must be supplied by foods. It is believed that the average person in this country need not fear a vitamin deficiency, providing discrimination is used in the choice of foods.

In the highly competitive field of detergents, manufacturers have been quick to utilize the catalytic activity of enzymes once they became available in large amounts through fermentation processes developed in Europe. Stains of carbohydrate, fat, and protein origin—such as gravies, jams and jellies, chocolates, eggs, blood, grass, coffee, and tea—are removed by reactions similar to those occurring in the digestion of foods in the alimentary canal. That is, the enzymes added to the detergent catalyze the hydrolysis of the large molecules to form smaller, water-soluble molecules that are removed in the wash water.

Current production of enzymes in the United States is more than 10 million pounds annually, exclusive of those used in detergent formulations. Most of this production is for the food industry.

27•4   Oxidation–Reduction in Cells

27•4•1   In Animal Cells   The burning of glucose in a calorimeter with excess oxygen yields carbon dioxide and water. The free energy change, $\Delta G°$, is $-2,880$ kJ/mole. The complete oxidation of glucose in cells releases the same amount of energy; however, the oxidation is achieved at body temperature and in several steps. These spontaneous oxidation steps (as well as the oxidation of other foods) are often utilized by the cell to drive reactions that have positive free energy changes. That is, in the course of the oxidation reaction, new small organic molecules are formed that then react, along with nitrogen compounds or other nutrients, to form large molecules that are necessary to the organism. In general, these synthesis reactions proceed with a positive free energy change and therefore require energy as a driving force.

An interesting coupling of oxidation and synthesis is known to exist in two of the steps in the oxidation of glucose. The energy generated in the oxidation is used in the synthesis of energy-rich molecules of a complex ester of phosphoric acid, ATP (adenosine triphosphate). ATP is used as a convenient source of energy in the cell, for it readily undergoes hydrolysis with a free energy change of about -30 kJ/mole. The energy available from the hydrolysis of ATP is harnessed in some unknown way to produce muscular contraction of all types—in breathing and heartbeat as well as in voluntary movements. ATP is a source of energy for many reactions that have a positive free energy change and are necessary in synthesizing vital components of the organism.

The oxidation of carbohydrates and other nutrients requires the catalytic action of enzymes and coenzymes. Usually, metal ions are a part of these systems; some common ones are iron, copper, zinc, cobalt, manganese, vanadium, and molybdenum. These ions are in many cases bound to ligands, most of which are quite complex, as can be seen from a study of the formula of heme (see Figure 27.14), the iron-containing complex of hemoglobin, and of myoglobin and the cytochrome pigments. These three substances are conjugated proteins; the heme is bound to the protein by coordination bonds between the heme iron and the protein nitrogen and by other bonds.

The initial step in the oxidation process is the adsorption of oxygen by chemical bonding to iron in hemoglobin in humans, or similar iron compounds in other animals (copper replaces iron in certain crabs), for transport to cells. Here, a second heme-containing compound that does not circulate in the blood, myoglobin, receives the oxygen and holds it for release during oxidation.[4]

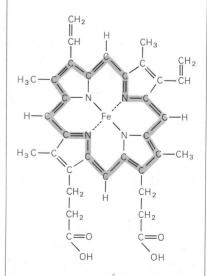

Figure 27•14 Heme. (Compare with Figure 27.15.) The colored areas show bonds involving delocalized pi electrons in the conjugated system.

---

[4] An interesting example of oxygen storage is afforded by the pink color of salmon flesh in the spring. The color is due to relatively large amounts of oxymyoglobin stored to provide for the energy needed to swim long distances to reach spawning grounds.

In one step of glucose oxidation in which two H atoms are removed to form water, oxygen is passed from myoglobin to a set of cytochrome pigments (some containing iron and some containing copper). At the same time, another set of catalysts, the flavoproteins, act as hydrogen acceptors and electron donors, with the net result that hydrogen and electrons are passed along a transport chain of catalysts to oxygen to achieve its reduction to water. As a pair of electrons moves along this transport chain, each carrier is alternately reduced and oxidized. Iron and copper of the heme portion change back and forth from Fe^{3+} and Cu^{2+} to Fe^{2+} and Cu^{+}, respectively. The flavoproteins are not well characterized; however, it appears that some contain iron, some zinc, some copper, and some molybdenum. Many inhibitors are known that block passage of electrons at different points in the chain. And, as has been pointed out in Section 9.2.4, when carbon monoxide is preferentially bonded to iron in heme, the oxidation process cannot proceed for want of oxygen.

The chemical energy released in the oxidation of the two H atoms is utilized in the synthesis of energy-rich ATP. Coupling reactions must exist to allow for the transfer of chemical energy to ATP; however, the coupling mechanism is not yet clear, although a large amount of experimental information about the method has been collected.

27.4.2   **In Plant Cells**   The ultimate source of our common forms of energy is the sun. The chemical energy in the foods we eat and the coal, gasoline, and other petroleum products we burn can be traced to the sun through the **photosynthesis reaction.** This process takes place in green plants; the overall reaction is the conversion of carbon dioxide and water to carbohydrate and oxygen:

$$6CO_2 + 6H_2O \longrightarrow C_6H_{12}O_6 + 6O_2 \qquad \Delta G° = +2,880 \text{ kJ}$$

The use of isotopes and radioactive tracers has shed much light on the mechanism of photosynthesis. For example, using ^{18}O it has been shown that oxygen is liberated from the water and not from the carbon dioxide. When plants are grown in an atmosphere of $^{14}CO_2$, the radioactivity rapidly appears in small molecules containing three and five carbon atoms, and later in molecules of glucose and starch.

It follows that oxygen in water is oxidized from $\overset{-2}{O}$ to $\overset{0}{O_2}$, whereas carbon is reduced from $\overset{+4}{C}$ to some intermediate value in the carbohydrate molecules.[5] However, the flow of electrons from oxygen to carbon will not take place spontaneously; rather, the reverse flow takes place, as was noted for the oxidation of glucose in animal cells.

Since the overall photosynthesis reaction has a positive free energy change, energy must be available for the reaction to occur. Radiant energy from the sun serves this purpose. Many steps are involved in the process, some of which require light energy and hence take place only in daylight.

---

[5] The oxidation states of carbon in CH_4 and CO_2 are -4 and $+4$, respectively, while in formic acid, HCO_2H, it is $+2$, and in methanol, CH_3OH, it is -2. Oxidation states from -3 to $+3$ are characteristic of carbon in oxygenated organic compounds.

These are called *light* reactions. Other steps draw energy from ATP and reduced *ferredoxin* (an iron-containing protein) and do not depend on light. These are called *dark* reactions. The process is initiated by the absorption of light by chlorophyll (see Figure 27.15), a compound very similar to heme except that a magnesium ion instead of an iron ion is coordinated with nitrogen.

The absorption of a quantum of light raises an electron to a higher energy level and produces an excited molecule. Chlorophyll in this condition is an excellent reducing agent and acts through a set of electron transport catalysts to achieve the reduction of carbon in CO_2 to intermediates of glucose. Some of the catalysts in this transport chain are identical with those in the chain for the oxidation of glucose in animal cells, and some are different; heading the chain, that is, the first to receive an electron from the excited chlorophyll, is ferredoxin. In another excited chlorophyll molecule, the electron is passed through a series of catalysts that are coupled to ATP synthesis in the same manner as the coupling of ATP synthesis in animal cells. This electron is eventually used to reduce the chlorophyll in the first transport chain back to the ground state. The chlorophyll in the second chain is at this stage in the oxidized form, since it has lost an electron. As such, its reduction potential is sufficiently high so as to be able to remove an electron from oxygen in water, thereby releasing elementary oxygen.

Thus, two electron transport chains are involved in the synthesis, with the overall effect being the carrying of electrons to carbon by one chain and removing them from oxygen by the other chain. H atoms left from water are passed via the chain to carbon as the catalysts are alternately reduced and oxidized. Chlorophyll molecules pass rapidly back and forth through conditions of excitation (gain of a quantum of radiant energy), oxidation (loss of an electron), and reduction to ground state (gain of electron).

Of course, in each plant cell that contains chlorophyll there are many of these pairs of transport chains operating continuously when radiant energy is available. The small molecules first produced are used in metabolic processes to produce glucose, starch, cellulose, and other plant products, with ATP and ferredoxin supplying energy for reactions that proceed with a positive free energy change. Some of the pathways of these subsequent reactions are well established.

27•5 Hormones

Chemical substances called **hormones** originate in endocrine (ductless) glands and move through the bloodstream to different parts of the body where they exert a regulatory influence on the chemical processes taking place. In humans and other mammals, there is a close connection or relationship among the ductless glands, the nervous system, and the cellular material in order to maintain a proper *hormone balance*. A disturbance of this balance produces abnormal metabolic processes.[6]

Although most of the known hormones are simple chemical compounds, the pathway by which a hormone acts on a cell to regulate one or

Figure 27•15 Chlorophyll *a*. (Compare with Figure 27.14.) The colored areas show bonds involving delocalized pi electrons in the conjugated system.

---

[6]The terms *metabolic process* and *metabolism* refer to all the chemical changes occurring in the cell.

more parts of the cell's metabolism is not known. Some postulated explanations include effects on protein synthesis, RNA formation, and enzyme activity. Table 27.1 lists some of the human hormones along with some of their activities, and Figure 27.16 shows the structure of three hormones.

Table 27•1 Some hormones of humans

Name	Secreted in	Function
epinephrine (adrenaline)	adrenal medulla	increases pulse rate and blood pressure; releases glucose from glycogen and fatty acids from fat tissue
testosterone	testis	maturation and normal functioning of male sex organs
estrone and estradiol	ovary	maturation and normal functioning of female sex organs
insulin	pancreas	metabolism of glucose in cells
cortisol and cortisone	adrenal cortex	metabolic rate of fats, proteins, and carbohydrates
pituitary hormones	pituitary gland	a dozen or so are known and each has a specific function, such as to stimulate the adrenal cortex and the thyroid gland to produce their hormones; to stimulate the activities of testes and ovaries and control the menstrual cycle; to stimulate the mammary gland to produce milk; to control growth

27•6   Viruses

Many common ailments—smallpox, chickenpox, yellow fever, influenza, mumps, poliomyelitis, and common colds, to name a few—are caused by infective agents called **viruses.** In general, all viruses consist of two principal parts: a nucleic acid core and a protein outer shell (See Figure 27.17). The nucleic acid may be either RNA or DNA and, of course, carries the code characteristic of the particular virus. The protein shell serves as a protective covering of the RNA or DNA core and, in some cases, acts to help the core attack and penetrate the walls of cells. Once the RNA or DNA of the virus is inside a cell, it reorganizes the cell's function to manufacture complete virus particles. When a sufficient number of virus particles have been formed, the cell ruptures, spilling out the virus particles to attack other cells.

27•7   Drugs

Many substances not naturally associated with cell metabolism are known that, in relatively small amounts, have a profound effect on cell metabolism including that of the brain and nervous system. These substances are called **drugs.** Included among common types are stimulants, pain killers, tranquilizers, antibiotics, anesthetics, hallucinogens, and nerve poisons. Just how these substances produce their effects is well known for some and completely unknown for others.

epinephrine (adrenalin)

testosterone

estrone

Figure 27•16 Three hormones.

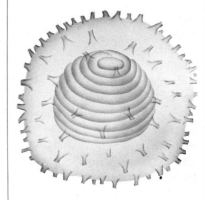

Figure 27•17 The influenza virus has a protein covering that envelops the coiled DNA core. The protein protrusions are believed to aid the DNA in penetrating the host cell. Magnification about 500,000 times. *Source:* Based on electron micrographs and model supplied by Dr. R. W. Horne, University of Cambridge, Cambridge, England. See also L. Hoyle, R. W. Horne, and A. P. Waterson, *Virology,* 13, 448 (1961).

In many cases the action of a drug is due to a specific inhibitory action on a particular enzyme or vitamin. For example, penicillin prevents wall construction in bacteria, presumably by interrupting enzyme and coenzyme reactions responsible for the construction. Sulfa drugs interfere with the uptake of a coenzyme essential to growth and cell division of certain types of organisms. Carbon monoxide and cyanides form compounds with the metal ions that are parts of enzyme systems in humans, thereby blocking their functions. The nerve poison diisopropyl fluorophosphate inhibits acetylcholine esterase, an active site of the enzyme. The effectiveness of insecticides and herbicides is thought to be due to the inhibition of enzyme action in many cases.

There are certain proteins that are necessary for the maintenance of consciousness. In order to function properly, a protein molecule must have a certain shape. It has been proposed that anesthetics cause loss of feeling by affecting the shape of proteins necessary for the functioning of the central nervous system. A logical explanation for the action of an anesthetic is that, as it causes changes in the shapes of essential proteins, it blocks synapses necessary for the maintenance of an upright posture. (*Synapses* are transmitting sites through which nerve impulses pass from one neuron to another.) The blocking concentration depends on both the particular anesthetic and on how quickly the synapses are transmitting the nerve impulses.

Although the exact nature of anesthetic action is not known, it has been suggested that *transmitter release* is the process most sensitive to anesthetic action. For a nerve impulse to pass from one cell to another, a relatively simple chemical called a *transmitter* must be present. When an impulse arrives at a nerve terminal, calcium presumably enters the cell and binds with the protein necessary for transmitter release. Hence, transmitter release is dependent on the calcium concentration. It is theorized that when anesthetics cause changes in the shapes of proteins, charge sites disappear with the result that there is a decrease in the number of sites available for calcium binding. When calcium binding falls below that necessary for sufficient transmitter release, unconsciousness results.

The mechanisms by which many drugs produce their effects are not known. The hallucinogenic drug LSD (see Figure 27.18) has been the subject of extensive study; over 1,000 technical papers have been published about it and its chemical structure and properties are well known, yet the mechanism by which it produces its effect is not known.

The brain has been studied experimentally by a great number of biochemists and physicians, and features such as its oxygen utilization and rate of protein synthesis have been determined. The electric nature of nerve impulses is thought to depend on a concentration gradient of Na^+ and K^+ ions inside and outside nerve cells. As an impulse passes, there is a small change in the voltage of the "electrochemical cells" due to a momentary change in the permeability of the cell walls to the passage of ions. These impulses bring about the release of acetylcholine and other chemicals that stimulate tissue at the nerve endings.

Tranquilizers, stimulants, pain killers, and hallucinogens are examples of drugs that act on the brain. In some cases, the effective dose is a matter of a tiny fraction of a gram—0.000025 to 0.00005 g being an effective LSD dose.

Figure 27•18 LSD (lysergic acid diethylamide).

Biochemical Benefits and Risks of Synthetic Organic Chemicals Ours is the first generation to create synthetic substances in such amounts that our environment is obviously affected on a worldwide scale. Some of these effects are dramatically beneficial. Malaria and typhus, dreadful scourges at one time, have been controlled to such an extent by use of the insecticide DDT that this chemical is said to have saved more lives than were lost in all wars. Synthetic medicines, drugs, and anesthetics have so revolutionized health care that the length and comfort of human life have been greatly enhanced. The impact of plastics on improving our standard of living is difficult to exaggerate. Food, clothing, shelter—the essentials for living have been made more abundantly available for more persons. The proper packaging of food protects its sanitation and makes it less perishable. Clothes made from the new fibers are both cheaper and more durable. And plastic tiles, pipes, siding, screens, insulation, floor coverings, upholstery, and paints make our homes and work places more sanitary and more comfortable.

However, there are also dramatically harmful effects that accompany the synthesis and use of a number of materials that are produced in huge quantities. The most troublesome of these influences are the biochemical ones that affect living organisms, including humans. As pointed out in the special topic in Chapter 23, the worst ones can be mutagenic, teratogenic, and carcinogenic. Most of these influences in the human population can be detected only in studies that involve a large population sample or that require a long time. The latency period for the development of cancer is typically from 15 to 40 years. In experiments to identify potential carcinogens within a much shorter time, test animals are subjected to large doses of substances. However, the results of such tests cannot be applied with certainty to humans.

The range of harmful substances is broad and the specific effects are many. To illustrate some biochemical problems, we will focus on just a few organohalogen compounds. They have some properties typical of chemicals that are produced for one purpose but that have undesirable environmental side effects: (1) they are sufficiently active chemically to react with molecular substances in living organisms, (2) they are carried by air or water over wide areas, and (3) their biochemical effect is to interfere with or to modify some essential cellular function of a living organism.

Vinyl chloride, $CH_2{=}CHCl$, is a key chemical that we have not mentioned previously. Produced at an annual rate of about 7 billion pounds in the United States, and

about equally in both Europe and Japan, it is second in quantity of production only to propylene as a monomer for the manufacture of plastics. The polymerized vinyl chloride, called polyvinyl chloride or PVC, is a versatile material (see Table 27.2). There is no evidence that the polymer is toxic, but the monomer, vinyl chloride, is apparently carcinogenic in humans. The incidence of cancer in workers exposed to the monomer is definitely higher than for the average population, the incidence of birth defects is greater than normal in the babies of women who are exposed, and the chemical causes tumors in test animals. The present strategy of the Environmental Protection Agency and the plastic manufacturers is to take more precautions to protect workers from exposure and to make sure that no traces of the monomer are entrapped in the finished polyvinyl chloride product.

There is some vinyl chloride in the environment that does not arise from manufacturing plastics. For example, minute amounts have been detected in cigarette smoke, probably having been formed during combustion by reaction of traces of chlorine-containing compounds with hydrocarbon radicals from the burning tobacco. Presumably the burning of many other substances produces traces of vinyl chloride.

The herbicide 2,4,5-trichlorophenoxyacetic acid (or 2,4,5-T) is used for agricultural and forest weed control. Among the targets of the 7 million pounds used annually in the United States is the tree-killing kudzu vine. In the production of 2,4,5-T, traces of compounds of the dioxin type are formed as contaminants. The dioxin molecule 2,3,7,8-tetrachlorodibenzo-*p*-dioxin,

is the most toxic teratogen yet identified. In 1979, an unusually high incidence of miscarriages among women in Oregon was blamed on the spraying of nearby forests. In 1976, an explosion in an Italian chemical plant manufacturing a precursor for 2,4,5-T resulted in the dioxin contamination of a large area. Whole communities had to be abandoned. In places the soil was bulldozed and incinerated to destroy the organic chemicals. Subsequently it has been found that dioxin can go unchanged through an incinerator at 1,100 °C, so the result of this effort may have been merely to dilute and spread the dioxin over a wider area. Like DDT, dioxin is now being detected in samples of water and animal tissue from all over the world. The Environmental Protection Agency suspended the use of 2,4,5-T as an emergency in the spring of 1979.

The case of DDT is well known. Savior of millions since the 1940s, it is still being used for mosquito and lice control in many countries. Banned for these purposes, except for emergencies, in the United States, the effect of DDT on birds and other wildlife has been devastating. Although Congress has prohibited its use in this country, its production here and the sale abroad of about 25 million gallons annually is permitted despite clear evidence that the substance persists in the environment as it is carried by wind and water all around the world. The effects of DDT on humans is uncertain, but all of us, thanks to our proud place in the food chain, have some DDT concentrated in our fatty tissues and organs (see Figure 10.8). Two other insecticides, Kepone and Mirex, have produced such obvious ill effects—for example, tremors, sterility, liver damage—in persons exposed to them that their production has been discontinued in this country. Like DDT, they are not easily biodegraded.

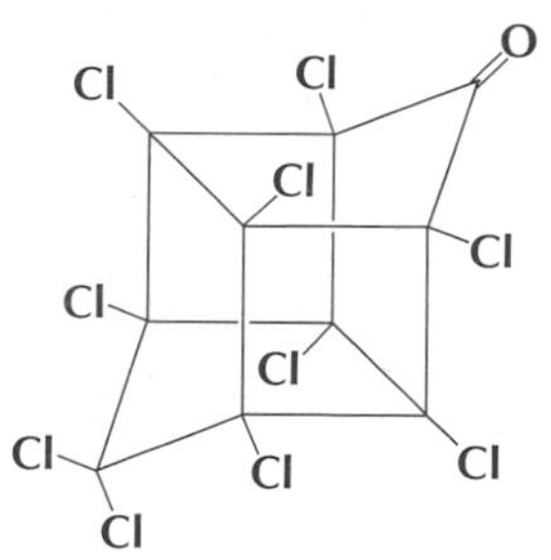

Kepone 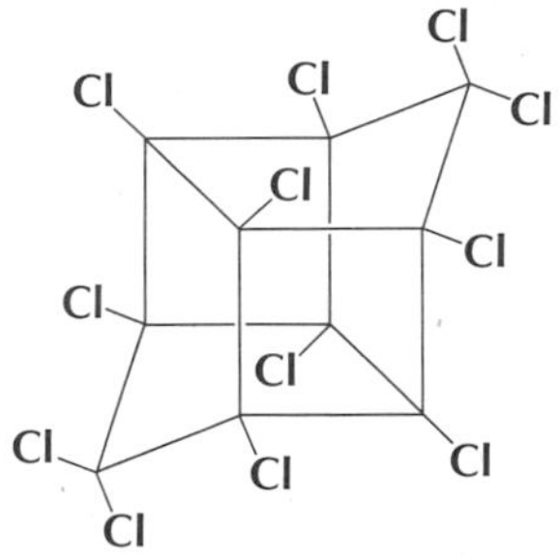   Mirex

A few other examples include the hospital anesthetic halothane (1,1,1-trichloro-2-bromo-2-chloroethane), which apparently causes miscarriage, birth defects, and cancer; the hospital disinfectant hexachlorophene,

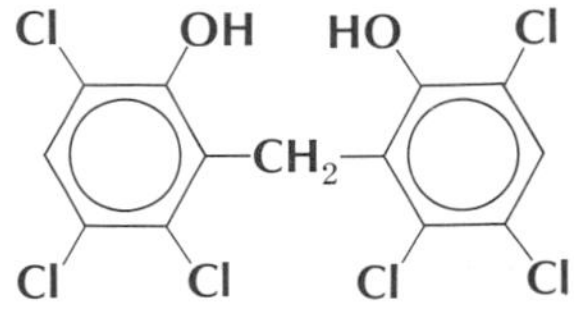

which is associated with miscarriages and birth defects among pregnant nurses and hospital workers; tris-2,3-dibromopropyl phosphate, the flame retardant once used in children's sleepwear, which causes tumors in test animals; and the agricultural insecticide 1,2-dibromo-1-chloropropane, which causes sterility in workers.

With all the compounds described there are questions of benefits versus risks. When the risk is high and substitute materials are available, the obvious action is to discontinue use of the toxic substances. Sometimes this may not solve the problem. Long-used additives in high-octane gasoline are lead tetraethyl plus 1,2-dibromoethane. The latter is known to cause cancer in test animals, so a change to lead-free gasolines would seem to be a solution. However, the combustion of high-octane lead-free gasolines tends to produce more 3,4-benzo[a]-pyrene, a well-known carcinogen (see Section 26.4.4).

A number of very persistent halogen-containing insecticides and herbicides have been replaced by organophosphate substances that are more readily biodegraded. Substitute flame retardants, other anesthetics, and other disinfectants are available. In the case of vinyl chloride, the jobs and consumer products of a huge industry are at stake, so the solution adopted is to limit the number of persons working with the dangerous material and to protect them from exposure.

In addition to direct exposure to synthetic organic chemicals during their manufacture or use, there is also the problem of disposal of waste and excess chemicals. Public and governmental attention was focused on this problem when a huge cache of waste chemicals led to the evacuation and abandonment of the Love Canal section of Niagara Falls in 1979. A small part of a clay-sealed underground dump was 200 tons of 2,4,5-T and its accompanying 60 kg of dioxin. Water from heavy rains forced the chemicals to the surface and into the yards and basements of the homes built over the buried and forgotten wastes. It is estimated that among the 30,000 industrial dumps in the United States, at least 600 could be as dangerous as Love Canal. In one of these, on the flood-threatened banks of a tributary of the Tennessee River, there is buried 4,000 tons of DDT.

The biochemical effects of synthetic organic chemicals are a part of the global picture that includes the greenhouse effect of carbon dioxide and the acid rains of sulfur and nitrogen oxides. The scale of our operations has brought us to the realization that we have to work more in harmony with the great chemical cycles in nature. Although the massive use of technology has brought about most of our pollution diffculties, it is to science and technology that we are looking first for solutions to these problems. The intellect that has made it possible for us to transform parts of nature to our immediate and local advantage must now be used to protect nature and to preserve the web of life of which human life is but one of the fragile and dependent strands.

27·8   Polymers

Compounds composed of very large molecules that are formed by the repetitious union of many small molecules are called **polymers.** The small molecules, called *monomers,* may be of a single kind or several kinds. Cellulose and starch are polymers composed of a single repeating species (glucose units), and proteins are polymers in which the repeating units represent as many as 24 species (amino acid units). As a rule, however, the monomer from which synthetic polymers are formed is of a single type or, at the most, two or three types.

Polymers whose repeating units result from bonds made available by the elimination of simple molecules, such as HOH, HCl, and NH_3, are called **condensation polymers.** Proteins, starch, and cellulose are in this category.

A second kind of polymer is called an **addition polymer,** a polymer in which the repeating units are joined by bonds originally associated with unsaturation, that is, double and triple bonds. Polyethylene is classified as an addition polymer.

Natural polymers are not always satisfactory for specific uses. For example, natural rubber swells and loses its elasticity after prolonged expo-

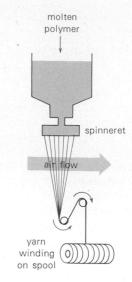

Figure 27·19 Nylon thread is formed by melting, extruding, and cooling.

Table 27·2 Synthetic polymers

Monomers	Polymers	Principal uses
$CH_2{=}CH_2$ ethylene	$-CH_2CH_2CH_2CH_2-$ polyethylene	films and sheets, tubing, molded objects, electrical insulation
$CH_2{=}CHCl$ vinyl chloride	$-CH_2CHCH_2CH-$ Cl Cl polyvinyl chloride	films and sheets, tubing, molded objects, electrical insulation, phonograph records, copolymer with vinyl acetate for floor coverings
$CH_2{=}CHCN$ acrylonitrile	$-CH_2CHCH_2CH-$ CN CN polyacrylonitrile	fibers, for example, Acrilan, Orlon
$CH_2{=}CCO_2CH_3$ $\|$ CH_3 methyl methacrylate	CO_2CH_3 $\|$ $-CH_2C-$ $\|$ CH_3 PMMA, polymethyl methacrylate	Plexiglas, Lucite, paints
$CF_2{=}CF_2$ tetrafluoroethylene	$-CF_2CF_2CF_2CF_2-$ Teflon	objects very resistant to chemical attack, cooking ware coatings
$CH_2{=}CH-CH{=}CH_2$ butadiene $CH{=}CH_2$ ⬡ styrene	$-CH_2CH{=}CHCH_2CHCH_2-$ ⬡ Buna S and GR-S rubber	synthetic rubber

sure to gasoline or motor oil; silk and wool (proteins) are natural foods for certain kinds of bacteria and larvae of insects, as is cellulose; and most natural polymers are hydrophilic, so that they readily absorb water. Moreover, the stability and melting points of natural polymers are such that they cannot be melted and then cast into desired shapes.

During this century, organic chemists have developed hundreds of synthetic polymers that have specific properties; many can be cast into desired shapes, including threads (see Figure 27.19) and sheets. Table 27.2 lists a number of common polymers. Note that the first six are of the addition type, and the remainder are of the condensation type.

An important development in this field is the discovery of specific catalysts that can link the building blocks into ordered structures rather than random structures. As we have seen, in the natural polymers, such as the proteins, starch, and cellulose, the structural units are always arranged in a particular sequence; and in many instances atoms or groups of atoms are arranged around chiral carbon atoms in specific relative positions in space. The fact that variations in positional order can now be achieved with some of the synthetic polymers makes it possible to build, from the same raw materials, polymers that have different properties.

Table 27.2 Synthetic polymers (*continued*)

Monomers	Polymers	Principal uses
HOCH$_2$CH$_2$OH ethylene glycol; terephthalic acid	poly(ethylene terephthalate)	fibers, for example, Dacron, films
NH$_2$CH$_2$CH$_2$CH$_2$CH$_2$CH$_2$CH$_2$NH$_2$ hexamethylene diamine; adipic acid	nylon 66	fibers, molded objects
phenol; H$_2$C=O formaldehyde	Bakelite	molded objects, varnishes, lacquers

Chapter Review

Summary

Biochemistry is the chemistry of life processes. Many natural and synthetic organic molecules exist as **structural isomers** or **stereoisomers**, or both. A special type of stereoisomerism exists in compounds that have **chirality. Chiral** molecules are nonsuperimposable mirror images of each other; they are called **enantiomers**. A **chiral carbon atom** is one to which four different atoms or groups are attached, in one or the other of two enantiomeric **configurations**. These are represented by **configurational formulas**. Enantiomers are **optical isomers** because they exhibit **optical activity**—the rotation of the plane of polarization of **plane-polarized light,** either to the right (**dextrorotatory** form) or the left (**levorotatory** form). Chemically, enantiomers are identical except for their reactivity with other optically active compounds; this difference is vitally important in biochemistry.

All **carbohydrates** are simple or complex sugars, which are called **saccharides. Monosaccharides** are simple molecules that do not hydrolyze. Those that have six carbon atoms are **hexoses.** Glucose, the predominant source of energy in living cells, is an **aldohexose;** fructose is a **ketohexose. Disaccharides,** such as sucrose, hydrolyze to yield two monosaccharides. **Polysaccharides** hydrolyze to yield many monosaccharides. If these are **pentoses,** the polysaccharide is a **pentosan;** if they are hexoses, it is a **hexosan.** The most important hexosans are starch and cellulose, which consist of glucose units linked in different ways.

Proteins consist of long chains of numerous **amino acids,** all having the same enantiomeric configuration. The amino acids are linked by **peptide bonds.** The most important configuration of the resulting chains is the **alpha helix,** whose structure is determined in part by intramolecular hydrogen bonds.

Protein synthesis in living cells is directed by the **nucleic acids. Deoxyribonucleic acid (DNA)** is the master molecule containing all the hereditary information of the organism, in the form of a molecular code. DNA consists of a **double helix** of intertwined **polynucleotide chains** held together by hydrogen bonds. Various segments of the DNA molecule, each with a unique sequence of organic nitrogen bases, constitute the **genes** of the organism. Each gene carries the code for the amino acid sequence of a particular protein. The hereditary information is transmitted from the DNA to the protein being synthesized via two types of **ribonucleic acid (RNA).**

Virtually all chemical reactions of life processes are catalyzed by **enzymes,** all of which are proteins. The high efficiency of enzymes as catalysts is reflected in their typically very large **turnover numbers.** The high specificity of an enzyme in catalyzing only one reaction is due to its unique shape and the presence of **active sites** on its surface. Many enzyme-catalyzed reactions require the presence of a **coenzyme,** such as a vitamin.

The ultimate source of all the chemical energy required for life processes is the sun. In green plants, radiant energy drives the **photosynthesis reaction,** a complex series of redox reactions involving chlorophyll and other compounds, by which carbon dioxide and water are converted to glucose. In animals, much of the energy released in the oxidation of nutrient molecules is used to drive nonspontaneous chemical reactions that are vital for normal development and function. Many life processes are profoundly affected by such potent substances as **hormones, viruses** (which consist of protein and nucleic acid), and **drugs.**

Compounds such as polysaccharides, proteins, and nucleic acids are **polymers**—giant molecules composed of chains of many small molecules, or **monomers.** Depending on the type of linkage of the monomers, the polymers may be **condensation polymers** or **addition polymers.** Many synthetic polymers have been developed that have a great variety of useful properties.

Exercises

Chiral Carbons and Optical Activity

1. Which of the following contain one or more chiral carbon atoms? In each case identify the chiral carbons: CBr_2ClCH_3; $CH_3CHICH_2CH_3$; $CH_3CHClCHClCH_3$; $CH_3CH_2CH_2CHCH_2Cl.$
   |
   CH_3

2. Write condensed structural formulas for those C_4H_9Cl compounds that may exist as enantiomers.

3. Repeat Exercise 2 for the $C_5H_{11}Cl$ compounds.

*4. Draw two- and three-dimensional projection formulas for each pair of enantiomers in Exercises 2 and 3.

5. Tartaric acid, $HO_2CCHOHCHOHCO_2H$, contains two chiral carbon atoms, but there are only two optical isomers. Why are there not four optical isomers?

*6. For tartaric acid (see Exercise 5) there are three stereoisomers. Does this contradict the statement in Exercise 5? Draw three-dimensional configurational formulas for the three isomeric tartaric acids to illustrate why there are only three stereoisomers.

*7. Draw three-dimensional projectional formulas that correspond to the two-dimensional formulas shown in Figure 27.5.

8. Suppose that you have carried out the following synthesis with the purpose of obtaining a sample of lactic acid,

$$CH_3CH_2CO_2H \longrightarrow CH_3CHClCO_2H \longrightarrow$$
$$CH_3CHOHCO_2H$$

and that you have obtained a product that appears to be lactic acid. However, the product is found to be optically inactive when examined by means of a polariscope. Does this mean that the product is not lactic acid? Explain.

Carbohydrates

9. Upon acid hydrolysis, corncobs give about 12 percent of a sugar (xylose) that has the molecular formula $C_5H_{10}O_5$. Chemical tests prove that this sugar has an aldehyde group.
   (a) Write a condensed structural formula for this sugar.
   (b) Classify this sugar in the same manner that glucose is classified.
   (c) What is the total number of optical isomers that have the condensed formula you wrote for part (a)?
10. Write configurational formulas for all the open-chain aldohexoses.
11. (a) Why is it necessary to postulate that most glucose molecules exist in a cyclic form?
    (b) Are cyclic structures also possible for xylose (see Exercise 9)? If so, draw them.
12. Account for the high water solubility of glucose and the very low water solubility of its parent hydrocarbon.
13. The mild oxidation of glucose with $[Ag(NH_3)_2]^+$ gives what compound (condensed formula) as the first oxidation product? This reaction is used in making silver mirrors. In this process, what undergoes reduction as glucose is oxidized?
14. A certain carbohydrate is not appreciably soluble in water, does not have a sweet taste, but does dissolve slowly when heated with dilute HCl. To what class of carbohydrates does it probably belong? Why?
15. Are glucose and fructose isomeric sugars? Why? Are their chemical properties significantly different? Why?
16. Although (+)-glucose and (−)-glucose are isomeric compounds, their ordinary chemical and physical properties are the same. Why?
17. A sucrose solution, when analyzed in a polariscope, was found to rotate plane-polarized light to the right. The analysis was repeated several weeks later, and the same solution was then found to be levorotatory. Propose a theory as to what might have happened.
18. Fructose molecules exist mostly in closed-chain or cyclic form. In water solution, the equilibrium mixture apparently involves both five- and six-membered rings. Draw configurational formulas for the five- and six-membered ring forms of (−)-fructose.
19. When any one of the three isomeric hexoses, (+)-glucose, (−)-fructose, and (+)-mannose, is placed in a slightly alkaline solution, an equilibrium mixture of all three results. This change is thought to involve a proton

shift as shown below (only the C-1 and C-2 atoms are shown):

$$-CHOHCHO \rightleftharpoons [-COH=CHOH] \rightleftharpoons -CHOHCHO$$
(+)-glucose   (+)-mannose

$$-CCH_2OH$$
$$\| \atop O$$
(−)-fructose

(a) If the above representation is correct, what must be true of the configurations about the C-3, C-4, and C-5 atoms in each sugar?
(b) In what way does the structure of (+)-glucose differ from (+)-mannose? From (−)-fructose?
(c) Would you expect the enzyme that catalyzes the fermentation of (+)-glucose to ethanol to catalyze also the fermentation of the other two sugars? Explain.
20. Explain, writing equations where appropriate, the chemistry of each of the following:
    (a) Corn syrup (Karo, for example) is made by cooking cornstarch with dilute HCl solution and then carefully neutralizing the acid with NaOH.
    (b) A sugar is made from wood.
    (c) Buttermilk or vinegar may be added to the recipe when some candies are made with sucrose sugar.
    (d) Cellulose nitrate explodes.
21. One gram of sugar, starch, or other dry carbohydrate requires about 1 L of oxygen for complete oxidation. On the other hand, 1 g of vegetable oil or other dry fat requires about 2 L.
    (a) Explain why, based upon the composition of carbohydrates and fats.
    (b) Assuming for the moment that mineral oil, paraffin wax, and other hydrocarbon materials are foods, would they have fewer calories per unit weight than vegetable oils? Why?
    (c) Discuss the factors that prevent hydrocarbons from being food for humans.

Proteins

22. The complete cleavage by hydrolysis of the peptide bonds in a protein results in the formation of what type of compounds? What substances catalyze the hydrolysis?
23. What is meant by the *amino acid sequence?* How is a given sequence achieved in cells?
24. Show how molecules of cysteine, $HSCH_2CHNH_2CO_2H$, might combine to give a protein molecule. Indicate in your protein structure the peptide bond.
25. Illustrate with electron-dot formulas the portions of a

DNA molecule that are responsible for holding the two strands together.

26. At pH values between 4 and 9, natural amino acids exist in solution largely as a polar ion or zwitterion: $H_3\overset{+}{N}CHZC\overset{-}{O_2}$. At a pH of 12, what would be the predominant ionic type? At a pH of 2? Would the uncharged molecular form, $H_2NCHZCO_2H$, predominate at any pH? Explain.

Oxidation–Reduction in Cells

27. The oxidation of foods in an adult human may produce up to 10–20 moles of CO_2 daily. The CO_2 reacts with water to produce carbonic acid, which ionizes to give an acid condition.
   (a) Show with equations how proteins in blood might react with carbonic acid to help maintain blood in a slightly alkaline condition. (Normally, the pH ranges from 7.3 to 7.5.)
   (b) About 0.1 mole of other acids, such as sulfuric and lactic, is formed in the body daily, but bicarbonate and hydrogen phosphate (HPO_4^{2-}) ions also act as buffers to help maintain the alkalinity of blood. Write equations to illustrate this action.
   (c) Once the blood reaches the lungs, hemoglobin reacts with oxygen to form oxyhemoglobin, and although oxyhemoglobin is a weak acid, it is sufficiently strong to react with bicarbonate ions to effect the release of CO_2 to be exhaled from the lungs. Using HB to represent the formula for hemoglobin, write an equation for this reaction.

*28. Both heme (Figure 27.14) and chlorophyll *a* (Figure 27.15) can be considered derivatives of porphin, $C_{20}H_{14}N_4$. Write the structural formula for porphin.

29. A piece of lean meat is composed mainly of proteins, fats, and water. Describe the chemical changes the meat undergoes during digestion in the alimentary canal.

Hormones, Viruses, and Drugs

30. Given that a certain substance belongs to one of three classes—enzymes, vitamins, and hormones—what determines to which of these three it will be assigned?

31. An injection of synthetic epinephrine can produce the same results—increase in blood pressure and pulse rate, and so on—as natural epinephrine. However, molecules can be synthesized that have the same structural formula shown in Figure 27.16 for epinephrine but that have very little effect when injected into the bloodstream. How can this be accounted for?

32. Why must a virus have a host cell in order to reproduce itself?

Polymers

*33. Which of the following would probably form polymers under normal reaction conditions for the synthesis of polymers: ethane; ethanol; 1-butene; ethylene glycol; propanoic acid; a mixture of ethanol and acetic acid; a mixture of glycerol and oxalic acid? For those that probably form polymers, write equations to show the polymerizations. Classify each polymer as to condensation or addition polymer.

34. A polypropylene molecule might weigh as much as 500,000 amu. How many propene molecules would be needed to form such a molecule?

35. Is a protein an addition polymer or a condensation polymer?

Arithmetic Procedures

In this section we will review several arithmetic procedures that are essential to calculations in different parts of general chemistry. Examples are given with methods of solution and with answers expressed in conventional ways. The use of a hand calculator is not necessary, but it is a great convenience in solving problems. Whether or not you use a calculator, *it is most important to be able to write down the operation or the equation that you are carrying out.*

For many of the following examples, the display on a typical hand calculator (HC) is also given so that you can compare it with the conventional way of writing a number. Please note that different calculators may display answers in different ways. For most of the problems in this text, only the simplest type of calculator is needed. There are a few problems for which the logarithm function on a calculator is very helpful, but these problems can also be solved by use of a table of logarithms (see Table A.8).

1.1 A Few Useful Formulas

For the quadratic equation $ax^2 + bx + c = 0$, $x = \dfrac{-b \pm \sqrt{b^2 - 4ac}}{2a}$

Area of a circle, πr^2

Area of a sphere, $4\pi r^2$

Volume of a sphere, $\frac{4}{3}\pi r^3$

Volume of a cylinder, $\pi r^2 h$

1.2 Significant Figures

In recording a measured or calculated quantity, the number of figures used indicates the precision with which the quantity is known. Measured quantities that have one uncertain figure may be recorded. Suppose we measure the length of a wooden stick and record it as 121.20 cm. This indicates that we have measured the length to the nearest 0.01 cm, although the last figure may be uncertain. It is understood that the last figure is significant even though we may not be able to measure it precisely.

Quantities are often expressed with greater precision than can be justified. The population of a city may be given as 125,762; but 126,000 would probably be sufficiently precise in view of the uncertainty of the counting techniques. Whether zeros at the end of a number are considered as significant or not must often be decided on the basis of common sense. If a friend mentions that he or she has saved $25 in pennies in a piggy bank, one might think the sum of 2,500 pennies had two or three significant figures. But if one purchased $25 in pennies from a bank, the sum of 2,500 pennies would be expected to be precise to four figures.

In the quantity "2,500 pennies," the number of significant figures could be two, three, or four. The common scientific method of expressing a correct number of significant figures is to write the number in exponential form. For example, 2.5×10^3 states the number of significant figures as two, whereas 2.500×10^3 states the number as four.

A little practice will enable you to determine with confidence the number of significant figures in different quantities. The following examples illustrate some typical cases:

Quantity	No. of Significant Figures
1.062 grams	4
751 students	3
0.006110 centimeter	4
1.2×10^8 stars	2
$683,462.02	8
7,685,000 people	4(?)
7.6850×10^6 people	5

In calculating quantities, the available data may vary in precision; that is, different quantities may have different numbers of significant figures. As an example, consider calculating the volume of a rectangular wooden stick from these measurements: length 121.20 cm, width 3.31 cm, thickness 0.19 cm. In recording these measurements, we imply that in each case the last figure is significant but possibly uncertain. Although we have tried to measure to the nearest 0.01 cm, errors of ±0.01 cm may have occurred. By simply multiplying length × width × thickness ($L \times W \times T$) with a calculator, we calculate the volume to be 76.22268 cm³. But to record such a volume would be ridiculous, because it would indicate that we knew the volume precisely to eight significant figures (or to 1 part in over 7.6 million). In this case we round off the volume to 76 cm³. This number has two significant figures, the same number as the least precise of the measurements (the thickness).

An answer to a multiplication or a division problem should have the same number of significant figures as there are in the least precise item of data.

If we were not using a hand calculator, we could have saved time in the volume calculation above by initially rounding the quantities to two significant figures:

$$L \times W \times T = (1.2 \times 10^2 \text{ cm})(3.3 \text{ cm})(0.19 \text{ cm})$$
$$= 75.24 \text{ cm}^3 = 75 \text{ cm}^3$$

This answer is 1 part in 76 less than the first answer, but for the purpose of an answer with two significant figures it is fine. (A change in our measurement of 1 part less in 19 in the thickness would result in an answer that would be even smaller than 75 cm³.)

In calculating the answers to the problems in this text, the practice is to use all the significant figures in the data given or looked up, to enter these figures in a hand calculator, and to round the final answer to the proper number of significant figures.

Some equivalents or conversion factors (see Table A.6) affect significant figures in calculations, and some do not. A relation such as 1 ft³/28.316 L is interpreted as 1.0000 ft³/28.316 L; that is, *the unit is understood to have as many significant figures as its equivalent.* This particular conversion factor would limit a calculation involving it to five significant figures. If a more precise calculation were necessary, a more precise factor would have to be looked up in a reference book, for example, 1 cu ft/28.31605 liters. However, *an exact equivalent does not limit significant figures in a calculation at all.* For example, the ratio 1 in./2.54 cm is exact by definition and is understood to have as many significant figures as needed in any calculation in which it is used, that is, 1.000. . . in./2.540. . . cm. This point is illustrated in Example 2 in Section 1.6 on the factor-units method.

1•3   Expressing Numbers

1•3•1   Use of Exponents   Very large numbers and very small numbers are conveniently expressed as powers of 10. This method of expressing numbers, called *scientific notation*, is so straightforward that a few minutes' study will make you sufficiently expert for most purposes. Consider the following:

Ordinary Number	Exponential Form	Hand Calculator	
1	1×10^0	1.	
10	1×10^1	1.	01
100	1×10^2	1.	02
1,000	1×10^3	1.	03
10,000	1×10^4	1.	04
0.1	1×10^{-1}	1.	−01
0.01	1×10^{-2}	1.	−02
0.001	1×10^{-3}	1.	−03
0.0001	1×10^{-4}	1.	−04

In the exponential form, the exponent is written to the right and above the figure 10. The exponent is shown to the right in a calculator display. A *positive* exponent tells how many times a number must be *multiplied* by 10 to give a certain number. Thus, 1×10^3 means to multiply 1 by 10 three times, and 1×10^3 equals $1 \times 10 \times 10 \times 10$ equals 1,000. Conversely, a *negative* exponent tells how many times a number is to be *divided* by 10. Hence, 1×10^{-5} equals $1 \div 10 \div 10 \div 10 \div 10 \div 10$ equals 0.00001. Other examples:

Ordinary Number	Exponential Form	Hand Calculator	
200	2×10^2	2.	02
340,000	3.4×10^5	3.4	05
0.0000046	4.6×10^{-6}	4.6	−06

In the exponential form, note that the number to be multiplied by 10^a is between 1 and 10. The convention is to write 2×10^2, not 20×10^1; 4.6×10^{-6}, not 0.46×10^{-5}. In some special cases this convention is not followed (for example, see Section 1.5 on taking roots).

In multiplying numbers expressed as powers of 10, exponents are added; in dividing numbers, exponents are subtracted.

usual method:

$$750 \times 2400 = 1,800,000 \qquad (\text{HC, } 1800000)$$

exponential method:

$$(7.5 \times 10^2)(2.4 \times 10^3) = 18 \times 10^5 = 1.8 \times 10^6 \qquad (\text{HC, } 1.8 \quad 06)$$

two other examples:

$$(8 \times 10^3)(5 \times 10^{-7}) = 40 \times 10^{-4} = 4 \times 10^{-3} \qquad (\text{HC, } 4. \quad -03)$$
$$(6 \times 10^4) \div (3 \times 10^5) = 2 \times 10^{-1} \qquad (\text{HC, } 2. \quad -01 \text{ or } .2)$$

1•3•2   Use of Exponents with Units   It is common to use negative exponents with units to indicate division. To set up calculations, it is often clearer to write the relation in the form of a fraction

in place of	write
2.54 cm · in.$^{-1}$	$\dfrac{2.54\ cm}{1\ in.}$
8.3144 J · K^{-1} · mol^{-1}	$\dfrac{8.3144\ J}{1\ K \cdot 1\ mol}$

1·3·3   Abbreviations for Prefixes with Units   Decimal fractions or multiples of units such as grams (g) or meters (m) are frequently expressed by using standard prefixes with units. Standard prefixes and their symbols are given in Table A.3. The symbol of the prefix is combined with the symbol of the unit. For example, kilogram is abbreviated as kg, centimeter as cm, milliliter as mL, nanogram as ng, and so on.

1·4   Logarithms

The expression of numbers in exponential form is closely related to the concept of logarithms. The logarithm to the base 10 of a number x is the exponent a to which 10 must be raised such that $10^a = x$. For example, if x is 1,000, then $\log_{10}$ of 1,000 is 3, because $10^3 = 1,000$.

Because many natural processes can be described in terms of exponential relationships, logarithms are often encountered in chemical and physical equations. In a general chemistry course logarithms are found in the most common equation for describing the rates of a chemical reaction and in equations describing the driving forces of chemical reactions. Also there is the use of the negative logarithm of the hydrogen ion concentration, called the pH, to express the concentration of the hydrogen ion.

There are two frequently used systems for logarithms. The logarithm of a number x to the base 10 is called a common or Briggsian logarithm and is referred to as $\log_{10} x$, or simply $\log x$. In the other system, the logarithm of a number x to the base 2.71828. . . (called e) is called a natural logarithm and is referred to as $\ln x$. The natural logarithm is the form that usually arises in the derivation of equations describing processes in nature. However, common logarithms are easier to relate to our number system, which is based on multiples of ten. Therefore we usually refer to a table of common logarithms to find $\log x$ and use the following relationship if we wish to find $\ln x$:

$$\ln x = 2.303 \log x$$

We can also use a hand calculator to obtain values of either common or natural logarithms.

The logarithm of a number has two parts. A whole number called the *characteristic* tells the location of the decimal point in the number; the *mantissa* tells the order of the digits in the number. Values of mantissas are tabulated in Table A.8. In precise calculations, we must use the same number of significant figures in a mantissa as in the original number. Consider these examples of six numbers, x, for which values of $\log x$ are given:

x	$\log x$, Table A.8	$\log x$, Hand Calculator
2.135	0.3294	0.32939788 (or 0.3294)
21.35	1.3294	1.3293979 (or 1.3294)
213.5	2.3294	2.3293979 (or 2.3294)
2135	3.3294	3.3293979 (or 3.3294)
0.2135	$\bar{1}$.3294	$-.6706021$ (or $\bar{1}$.3294)
0.02135	$\bar{2}$.3294	-1.6706021 (or $\bar{2}$.3294)

For each of the six values of x, the mantissa is 3294, corresponding to the digits 2135. For a value of x between 1 and 10, for example 2.135, the characteristic is 0; for a value of x between 10 and 100, for example 21.35, the characteristic is 1; and so on.

The number corresponding to a given logarithm is called the *antilogarithm*. In the foregoing list of examples, the six values of x are the antilogarithms of the logarithms in the second column. As another example, consider the logarithm 0.8169. Using Table A.8, we find its antilogarithm to be 6.560. Or consider the logarithm -3.12. Following the last two examples of the six tabulated, we write this logarithm with a negative characteristic and a positive mantissa as $\bar{4}.88$. The antilogarithm is $10^{-4} \times 7.6 = 0.00076$.

In certain problems in general chemistry, logarithms may be used in the following operations. An example is given for each, using logarithms from Table A.8. We ignore significant figures in these examples so we can check the answers mentally.

multiplication:
$$\log (x \cdot y) = \log x + \log y$$
$$\log (7 \cdot 5) = 0.8451 + 0.6990$$
$$= 1.5441$$
$$\text{antilog } 1.5441 = 35$$

division:
$$\log (x/y) = \log x - \log y$$
$$\log (18/0.3) = 1.2553 - \bar{1}.4771$$
$$= 1.2553 - (-0.5229)$$
$$= 1.7782$$
$$\text{antilog } 1.7782 = 60$$

raising to a power:
$$\log x^a = a \log x$$
$$\log 5^3 = 3(0.6990)$$
$$= 2.097$$
$$\text{antilog } 2.097 = 125$$

taking a root:
$$\log \sqrt[a]{x} = \frac{1}{a} \log x$$
$$\log \sqrt[5]{243} = \frac{1}{5}(2.3856)$$
$$= 0.4771$$
$$\text{antilog } 0.4771 = 3$$

To practice the use of logarithms, you can make up several problems and solve them with a logarithm table and then either longhand or with a hand calculator. Logarithmic relations are shown in several illustrative examples in the text. Without becoming involved in the particular problems, you can check the relations between logarithms and antilogarithms in Examples 14.5, 14.6, 16.7–16.12, and 18.2–18.6.

1•5   More on Taking Roots

If we do not have a logarithm table or a calculator, the use of exponents is often helpful in taking roots. Suppose we want the square root of 159,000. We state our problem in this way: find x when $x^2 = 159,000$. To find x, we express the number exponentially, taking care to use an exponent that is divisible by two. To do this, we write not 1.59×10^5, but 15.9×10^4:

$$x^2 = 15.9 \times 10^4$$
$$x = \sqrt{15.9} \times \sqrt{10^4}$$
$$= 3.99 \times 10^2 = 399 \qquad \text{(HC, 398.74804)}$$

Or consider taking the square root of 0.0000658:

$$y^2 = 0.0000658$$
$$y = \sqrt{65.8 \times 10^{-6}} = \sqrt{65.8} \times \sqrt{10^{-6}}$$
$$= 8.11 \times 10^{-3} = 0.00811 \qquad \text{(HC, .00811172)}$$

Finally, take the cube root of 16.8:

$$w^3 = 16.8 = 16{,}800 \times 10^{-3}$$
$$w = \sqrt[3]{16{,}800} \times \sqrt[3]{10^{-3}}$$
$$= 25.6 \times 10^{-1} = 2.56$$

1•6   Factor-Units Method

To help master units of measurement, it is desirable to set down the units along with the numbers that they qualify, and then multiply and divide these units as indicated in solving the problem. The following problems illustrate the handling of units in calculations.

• Example 1 •

Calculate the area of a table that measures 30.12 in. by 39.88 in.

• Solution •

$$\text{area} = 30.12 \text{ in.} \times 39.88 \text{ in.}$$
$$= (30.12 \times 39.88)(\text{in.} \times \text{in.})$$
$$= 1201 \text{ in.}^2 \qquad \text{(HC, 1201.1856)}$$
$$= 1.201 \times 10^3 \text{ in.}^2$$

• Example 2 •

Convert 1.201×10^3 in.^{2} to square meters. Given: 1 in. = 2.54 cm; 1 m = 100 cm.

• Solution •

A conversion factor, such as 1 in. = 2.54 cm, is often used in a calculation as a ratio or a factor:

$$\frac{1 \text{ in.}}{2.54 \text{ cm}}, \text{ read as } 1 \text{ inch per } 2.54 \text{ centimeters}$$

or

$$\frac{2.54 \text{ cm}}{1 \text{ in.}}, \text{ read as } 2.54 \text{ centimeters per inch}$$

We set up such ratios or factors in our calculation so that the units cancel to give the desired units in the answer. For our area calculation,

$$\text{area} = 1.201 \times 10^3 \text{ in.}^2 \left(\text{factors that express } \frac{m^2}{\text{in.}^2}\right) = ? \ m^2$$

$$= 1.201 \times 10^3 \text{ in.}^2 \left(\frac{2.54 \text{ cm}}{1 \text{ in.}}\right)^2 \left(\frac{1 \text{ m}}{100 \text{ cm}}\right)^2$$

$$= 1.201 \times 10^3 \text{ in.}^2 \left(\frac{6.4516 \text{ cm}^2}{1 \text{ in.}^2}\right)\left(\frac{1 \text{ m}^2}{10,000 \text{ cm}^2}\right)$$

$$= 0.7748 \ m^2 \quad \text{(HC, 7.7484 }-01\text{)}$$

The five significant figures produced by the calculator, 0.77484 m^2, are rounded to four significant figures because the original area is known to only four significant figures. The two conversion factors are exact and have no effect on the number of significant figures.

• Example 3 •

Calculate in joules the heat required to raise the temperature of 20 g of water by 10 °C. Given: The specific heat of water = 4.184 J per gram per degree Celsius,

$$\frac{4.184 \text{ J}}{1 \text{ g} \times 1 \text{ °C}}$$

That is, 4.184 J is required to raise the temperature of each gram of water by 1 degree Celsius.

• Solution •

$$\text{joules required} = \frac{4.184 \text{ J}}{1\text{g} \times 1 \text{ °C}} \times 20 \text{ g} \times 10 \text{ °C}$$

$$= 836.8 \text{ J} \quad \text{(HC, 836.8)}$$

$$= 8.4 \times 10^2 \text{ J}$$

Solving problems by setting up factors so that units cancel to give an answer containing the desired unit can be quite helpful. However, this method can also become entirely too mechanical. The problems in this text are designed not as exercises in arithmetic, but to develop your understanding of chemical methods and principles. Consequently, problem solving should begin with an analysis of the terms and language used in the problem. This analysis should lead to a definite prediction of what the answer will be before any precise calculations are made. In general, then, solving a problem should involve:

1. analyzing the data and predicting the approximate magnitude and units of the answer
2. setting up mathematical equations and carrying out calculations
3. checking work by canceling units and inspecting the magnitude of the answer

The next two problems and their solutions illustrate this three-step attack.

• Example 4 •

When 12 tons of coke are burned to carbon dioxide, 32 tons of oxygen are required. How much oxygen is needed to burn 100 tons of coke?

• *Analysis* • If 32 tons of oxygen are needed to burn 12 tons of coke, it will take a lot more than 32 tons of oxygen to burn 100 tons of coke. Because 100 is just over 8 times 12, the amount of oxygen used will be between 8 and 9 times 32 tons, that is, between 256 and 288 tons of oxygen.

• **Solution** • For purposes of calculation we set up our equation so that the units cancel to give an answer in tons of oxygen:

$$32 \text{ tons of oxygen} \times \frac{100 \text{ tons of coke}}{12 \text{ tons of coke}} = 267 \text{ tons} \quad (\text{HC, } 266.66667)$$

$$= 270 \text{ tons or } 2.7 \times 10^2 \text{ tons}$$

Because we made a thorough analysis of the problem, we have only to compare this answer with our prediction to see that the answer is reasonable. Such an approach should help in detecting careless mistakes, especially decimal point errors that might lead to such incorrect answers as 26.7 or 2670 tons of oxygen. Note that two of the weights used in the calculation have only two significant figures. Therefore, the answer should be rounded to 270 tons or 2.7×10^2 tons.

• **Example 5** •

A quantity of gas measures 100 mL at a pressure of 15 psi (lb/in.^{2}). What is its volume at 25 psi? Given: As the pressure on a gas increases, the volume decreases proportionately.

• *Analysis* • The volume will be considerably smaller, because the gas will be compressed. However, the volume will not be reduced to one-half its original amount (that is, to 50 mL), because the pressure is not doubled.

• **Solution** •

$$100 \text{ mL} \times \frac{15 \text{ psi}}{25 \text{ psi}} = 60 \text{ mL}$$

Derivation of Integrated Rate Equation

First-Order Reaction

For the reaction

$$A \longrightarrow \text{products}$$

the first-order rate equation for the disappearance of A is

$$\text{rate} = -\frac{d[A]}{dt} = k[A] \tag{1}$$

Rearrangement of Equation (1) gives Equation (2):

$$-\frac{d[A]}{[A]} = k \, dt$$

or

$$\frac{d[A]}{[A]} = -k \, dt \tag{2}$$

Integration of Equation (2) is carried out over the limits of time $t = 0$, when $[A] = [A]_0$, to some later time t, when $[A] = [A]_t$.

$$\int_{[A]_0}^{[A]_t} \frac{d[A]}{[A]} = \int_0^t -k \, dt$$

$$\ln [A]_t - \ln [A]_0 = -kt \tag{3}$$

Multiplying both sides of the equation by -1 gives

$$\ln [A]_0 - \ln [A]_t = kt$$

or

$$\ln \frac{[A]_0}{[A]_t} = kt \tag{4}$$

Conversion of the natural logarithm to the common logarithm gives

$$2.303 \log \frac{[A]_0}{[A]_t} = kt$$

or

$$k = \frac{2.303}{t} \log \frac{[A]_0}{[A]_t} \tag{5}$$

Equation (5) is identical to Equation (11) in Section 14.5.1.

Tables

Table A.1 SI base units

Physical quantity	Name of unit	Symbol
length	meter[a]	m
mass	kilogram	kg
time	second	s
electric current	ampere	A
thermodynamic temperature	kelvin	K
amount of substance	mole	mol
luminous intensity	candela	cd

[a]In most countries the spelling metre is preferred.

Definitions of the SI Base Units[1]

meter: 1,650,763.73 wavelengths in vacuum of the orange-red line of the spectrum of krypton-86

kilogram: the mass of a cylinder of platinum–iridium alloy kept by the International Bureau of Weights and Measures at Paris

second: the duration of 9,192,631,770 cycles of the radiation associated with a specified transition of cesium-133

ampere: the current that, when flowing through each of two long parallel wires separated by 1 meter of free space, results in a force between the wires of 2×10^{-7} newton per meter of length

kelvin: the fraction 1/273.16 of the temperature difference between absolute zero and the triple point of water

mole: the amount of a substance that contains as many entities as there are atoms in exactly 0.012 kilogram of carbon-12

candela: the luminous intensity of 1/600,000 of a square meter of a black body at the temperature of freezing platinum (2045 K)

[1]The definitions and data in Tables A.1–A.6 are taken mainly from the article by the National Bureau of Standards, "Policy for NBS Usage of SI Units," *J. Chem. Educ.,* **48,** 569 (1971), and from a 1975 NBS tabulation.

Table A•2 Certain SI derived units with special names

Physical quantity	Name of SI unit	Symbol for SI unit	Definition of SI unit
force	newton	N	$kg \cdot m \cdot s^{-2}$
pressure	pascal	Pa	$kg \cdot m^{-1} \cdot s^{-2} (= N \cdot m^{-2})$
energy	joule	J	$kg \cdot m^2 \cdot s^{-2}$
power	watt	W	$kg \cdot m^2 \cdot s^{-3} (= J \cdot s^{-1})$
electric charge	coulomb	C	$A \cdot s$
electric potential difference	volt	V	$kg \cdot m^2 \cdot s^{-3} \cdot A^{-1} (= J \cdot A^{-1} \cdot s^{-1})$
electric resistance	ohm	Ω	$kg \cdot m^2 \cdot s^{-3} \cdot A^{-2} (= V \cdot A^{-1})$
frequency	hertz	Hz	s^{-1} (cycle per second)

Table A•3 Prefixes for fractions and multiples of SI units

Fraction	Prefix	Symbol	Multiple	Prefix	Symbol
10^{-1}	deci	d	10	deka	da
10^{-2}	centi	c	10^2	hecto	h
10^{-3}	milli	m	10^3	kilo	k
10^{-6}	micro	μ	10^6	mega	M
10^{-9}	nano	n	10^9	giga	G
10^{-12}	pico	p	10^{12}	tera	T
10^{-15}	femto	f	10^{15}	peta	P
10^{-18}	atto	a	10^{18}	exa	E

Table A•4 General physical constants

Constant	Symbol	Value	SI units	CGS units
speed of light in vacuum	c	2.9979	$\times 10^8$ $m \cdot s^{-1}$	$\times 10^{10}$ $cm \cdot s^{-1}$
elementary charge	e	1.6022	$10^{-19} C$	$10^{-20} cm^{1/2} \cdot g^{1/2}$
Avogadro constant	N_A	6.0220	10^{23} mol^{-1}	10^{23} mol^{-1}
atomic mass unit	amu^a	1.6606	$10^{-27} kg$	$10^{-24} g$
electron rest mass	m_e	9.1095	$10^{-31} kg$	$10^{-28} g$
proton rest mass	m_p	1.6726	$10^{-27} kg$	$10^{-24} g$
neutron rest mass	m_n	1.6750	$10^{-27} kg$	$10^{-24} g$
Planck constant	h	6.6262	$10^{-34} J \cdot s$	$10^{-27} erg \cdot s$
Rydberg constant[b]	R_∞	1.0974	10^7 m^{-1}	10^5 cm^{-1}
gas constant	R	8.3144	10^0 $J \cdot K^{-1} \cdot mol^{-1}$	10^7 $erg \cdot K^{-1} \cdot mol^{-1}$
gas molar volume, STP	V_0	2.2414	10^{-2} $m^3 \cdot mol^{-1}$	10^1 $L \cdot mol^{-1}$

[a] The symbol amu is commonly used in this country, but u is the official symbol.

[b] The symbol R_∞ refers to the Rydberg constant for infinite nuclear mass.

Table A.5 SI equivalents for units to be abandoned eventually

Physical quantity	Name of unit	Symbol for unit	SI equivalents (exact)
length	inch	in.	2.54×10^{-2} m
length	angstrom[a]	Å	1×10^{-10} m
mass	pound (avoirdupois)	lb	0.45359237 kg
pressure	atmosphere[a]	atm	1.01325×10^5 Pa
pressure	torr	Torr	(101,325/760) Pa
pressure	millimeter of mercury	mmHg	$13.5951 \times 980.665 \times 10^{-2}$ Pa
energy	British Thermal Unit	Btu	1.055056×10^3 J
energy	kilowatt-hour	kWh	3.6×10^6 J
energy	calorie	cal	4.184 J
activity (radioactive nuclides)	curie[a]	Ci	3.7×10^{10} s^{-1}
exposure (X or γ rays)	roentgen[a]	R	2.58×10^{-4} C · kg^{-1}

[a]The use of these units is sanctioned for a limited period of time.

Table A.6 Equivalents (conversion factors)

1 m = 39.370 in.

1 in. = 2.54 cm (exact)

1 angstrom (Å) = 1×10^{-10} m (exact)

1 km = 0.62137 mi

1 m^3 = 1,000 L (exact)

1 m^3 = 264.17 gal

1 m^3 = 35.315 ft^3

1 dm^3 = 1 L (exact)

1 L = 1.0567 qt (U.S.)

1 gal (U.S.) = 3.7854 L

1 ft^3 = 28.316 L

1 mL = 1 cm^3 (exact)

1 mL = 0.061024 in.^{3}

1 in.^{3} = 16.387 mL

Δ1 °C = Δ1.8 °F (exact)

1 kg = 2.2046 lb (avoirdupois)

1 lb (avoirdupois) = 453.59 g

1 oz (avoirdupois) = 28.350 g

1 atm = 1.01325×10^5 Pa (exact)

1 atm = 760 mmHg (exact)

1 mmHg = 133.32 Pa

1 atm = 14.696 psi

1 J = 10^7 ergs (exact)

1 J = 0.23901 cal

1 J = 2.7777×10^{-7} kW · hr

1 eV = 1.6022×10^{-19} J

1 eV = 3.8293×10^{-20} cal

1 cal = 4.184 J (exact)

1 Btu = 1054.3 J

1 Btu = 251.98 cal

Table A·7 Vapor pressure of water at different temperatures

Temperature (°C)	Vapor pressure (mmHg)	(kPa)	Temperature (°C)	Vapor pressure (mmHg)	(kPa)
−10	1.95	0.260	33	37.73	5.030
−5	3.01	0.401	34	39.90	5.320
0	4.58	0.611	35	42.18	5.624
5	6.54	0.872	40	55.32	7.375
10	9.21	1.228	45	71.88	9.583
11	9.84	1.312	50	92.51	12.33
12	10.52	1.403	55	118.04	15.74
13	11.23	1.497	60	149.38	19.91
14	11.99	1.599	65	187.54	25.00
15	12.79	1.705	70	233.7	31.16
16	13.63	1.817	75	289.1	38.54
17	14.53	1.937	80	355.1	47.34
18	15.48	2.064	85	433.6	57.81
19	16.48	2.197	90	525.8	70.10
20	17.54	2.338	95	633.9	84.51
21	18.65	2.486	99.0	733.2	97.75
22	19.83	2.644	99.2	738.5	98.46
23	21.07	2.809	99.4	743.9	99.18
24	22.38	2.984	99.6	749.2	99.89
25	23.76	3.168	99.8	754.6	100.6
26	25.21	3.361	100	760.0	101.3
27	26.74	3.565	105	906.1	120.8
28	28.35	3.780	110	1074.6	143.3
29	30.04	4.005	115	1268.0	169.1
30	31.82	4.242	120	1489.1	198.5
31	33.70	4.493	125	1740.9	232.1
32	35.66	4.754	130	2026.2	270.1

Table A•8 Four-place logarithms

Natural numbers	0	1	2	3	4	5	6	7	8	9	Proportional parts								
											1	2	3	4	5	6	7	8	9
10	0000	0043	0086	0128	0170	0212	0253	0294	0334	0374	4	8	12	17	21	25	29	33	37
11	0414	0453	0492	0531	0569	0607	0645	0682	0719	0755	4	8	11	15	19	23	26	30	34
12	0792	0828	0864	0899	0934	0969	1004	1038	1072	1106	3	7	10	14	17	21	24	28	31
13	1139	1173	1206	1239	1271	1303	1335	1367	1399	1430	3	6	10	13	16	19	23	26	29
14	1461	1492	1523	1553	1584	1614	1644	1673	1703	1732	3	6	9	12	15	18	21	24	27
15	1761	1790	1818	1847	1875	1903	1931	1959	1987	2014	3	6	8	11	14	17	20	22	25
16	2041	2068	2095	2122	2148	2175	2201	2227	2253	2279	3	5	8	11	13	16	18	21	24
17	2304	2330	2355	2380	2405	2430	2455	2480	2504	2529	2	5	7	10	12	15	17	20	22
18	2553	2577	2601	2625	2648	2672	2695	2718	2742	2765	2	5	7	9	12	14	16	19	21
19	2788	2810	2833	2856	2878	2900	2923	2945	2967	2989	2	4	7	9	11	13	16	18	20
20	3010	3032	3054	3075	3096	3118	3139	3160	3181	3201	2	4	6	8	11	13	15	17	19
21	3222	3243	3263	3284	3304	3324	3345	3365	3385	3404	2	4	6	8	10	12	14	16	18
22	3424	3444	3464	3483	3502	3522	3541	3560	3579	3598	2	4	6	8	10	12	14	15	17
23	3617	3636	3655	3674	3692	3711	3729	3747	3766	3784	2	4	6	7	9	11	13	15	17
24	3802	3820	3838	3856	3874	3892	3909	3927	3945	3962	2	4	5	7	9	11	12	14	16
25	3979	3997	4014	4031	4048	4065	4082	4099	4116	4133	2	3	5	7	9	10	12	14	15
26	4150	4166	4183	4200	4216	4232	4249	4265	4281	4298	2	3	5	7	8	10	11	13	15
27	4314	4330	4346	4362	4378	4393	4409	4425	4440	4456	2	3	5	6	8	9	11	13	14
28	4472	4487	4502	4518	4533	4548	4564	4579	4594	4609	2	3	5	6	8	9	11	12	14
29	4624	4639	4654	4669	4683	4698	4713	4728	4742	4757	1	3	4	6	7	9	10	12	13
30	4771	4786	4800	4814	4829	4843	4857	4871	4886	4900	1	3	4	6	7	9	10	11	13
31	4914	4928	4942	4955	4969	4983	4997	5011	5024	5038	1	3	4	6	7	8	10	11	12
32	5051	5065	5079	5092	5105	5119	5132	5145	5159	5172	1	3	4	5	7	8	9	11	12
33	5185	5198	5211	5224	5237	5250	5263	5276	5289	5302	1	3	4	5	6	8	9	10	12
34	5315	5328	5340	5353	5366	5378	5391	5403	5416	5428	1	3	4	5	6	8	9	10	11
35	5441	5453	5465	5478	5490	5502	5514	5527	5539	5551	1	2	4	5	6	7	9	10	11
36	5563	5575	5587	5599	5611	5623	5635	5647	5658	5670	1	2	4	5	6	7	8	10	11
37	5682	5694	5705	5717	5729	5740	5752	5763	5775	5786	1	2	3	5	6	7	8	9	10
38	5798	5809	5821	5832	5843	5855	5866	5877	5888	5899	1	2	3	5	6	7	8	9	10
39	5911	5922	5933	5944	5955	5966	5977	5988	5999	6010	1	2	3	4	5	7	8	9	10
40	6021	6031	6042	6053	6064	6075	6085	6096	6107	6117	1	2	3	4	5	6	8	9	10
41	6128	6138	6149	6160	6170	6180	6191	6201	6212	6222	1	2	3	4	5	6	7	8	9
42	6232	6243	6253	6263	6274	6284	6294	6304	6314	6325	1	2	3	4	5	6	7	8	9
43	6335	6345	6355	6365	6375	6385	6395	6405	6415	6425	1	2	3	4	5	6	7	8	9
44	6435	6444	6454	6464	6474	6484	6493	6503	6513	6522	1	2	3	4	5	6	7	8	9
45	6532	6542	6551	6561	6571	6580	6590	6599	6609	6618	1	2	3	4	5	6	7	8	9
46	6628	6637	6646	6656	6665	6675	6684	6693	6702	6712	1	2	3	4	5	6	7	7	8
47	6721	6730	6739	6749	6758	6767	6776	6785	6794	6803	1	2	3	4	5	5	6	7	8
48	6812	6821	6830	6839	6848	6857	6866	6875	6884	6893	1	2	3	4	4	5	6	7	8
49	6902	6911	6920	6928	6937	6946	6955	6964	6972	6981	1	2	3	4	4	5	6	7	8
50	6990	6998	7007	7016	7024	7033	7042	7050	7059	7067	1	2	3	3	4	5	6	7	8
51	7076	7084	7093	7101	7110	7118	7126	7135	7143	7152	1	2	3	3	4	5	6	7	8
52	7160	7168	7177	7185	7193	7202	7210	7218	7226	7235	1	2	2	3	4	5	6	7	7
53	7243	7251	7259	7267	7275	7284	7292	7300	7308	7316	1	2	2	3	4	5	6	6	7
54	7324	7332	7340	7348	7356	7364	7372	7380	7388	7396	1	2	2	3	4	5	6	6	7

Table A.8 Four-place logarithms (*continued*)

Natural numbers	0	1	2	3	4	5	6	7	8	9	Proportional parts								
											1	2	3	4	5	6	7	8	9
55	7404	7412	7419	7427	7435	7443	7451	7459	7466	7474	1	2	2	3	4	5	5	6	7
56	7482	7490	7497	7505	7513	7520	7528	7536	7543	7551	1	2	2	3	4	5	5	6	7
57	7559	7566	7574	7582	7589	7597	7604	7612	7619	7627	1	2	2	3	4	5	5	6	7
58	7634	7642	7649	7657	7664	7672	7679	7686	7694	7701	1	1	2	3	4	4	5	6	7
59	7709	7716	7723	7731	7738	7745	7752	7760	7767	7774	1	1	2	3	4	4	5	6	7
60	7782	7789	7796	7803	7810	7818	7825	7832	7839	7846	1	1	2	3	4	4	5	6	6
61	7853	7860	7868	7875	7882	7889	7896	7903	7910	7917	1	1	2	3	4	4	5	6	6
62	7924	7931	7938	7945	7952	7959	7966	7973	7980	7987	1	1	2	3	3	4	5	6	6
63	7993	8000	8007	8014	8021	8028	8035	8041	8048	8055	1	1	2	3	3	4	5	5	6
64	8062	8069	8075	8082	8089	8096	8102	8109	8116	8122	1	1	2	3	3	4	5	5	6
65	8129	8136	8142	8149	8156	8162	8169	8176	8182	8189	1	1	2	3	3	4	5	5	6
66	8195	8202	8209	8215	8222	8228	8235	8241	8248	8254	1	1	2	3	3	4	5	5	6
67	8261	8267	8274	8280	8287	8293	8299	8306	8312	8319	1	1	2	3	3	4	5	5	6
68	8325	8331	8338	8344	8351	8357	8363	8370	8376	8382	1	1	2	3	3	4	4	5	6
69	8388	8395	8401	8407	8414	8420	8426	8432	8439	8445	1	1	2	2	3	4	4	5	6
70	8451	8457	8463	8470	8476	8482	8488	8494	8500	8506	1	1	2	2	3	4	4	5	6
71	8513	8519	8525	8531	8537	8543	8549	8555	8561	8567	1	1	2	2	3	4	4	5	5
72	8573	8579	8585	8591	8597	8603	8609	8615	8621	8627	1	1	2	2	3	4	4	5	5
73	8633	8639	8645	8651	8657	8663	8669	8675	8681	8686	1	1	2	2	3	4	4	5	5
74	8692	8698	8704	8710	8716	8722	8727	8733	8739	8745	1	1	2	2	3	4	4	5	5
75	8751	8756	8762	8768	8774	8779	8785	8791	8797	8802	1	1	2	2	3	3	4	5	5
76	8808	8814	8820	8825	8831	8837	8842	8848	8854	8859	1	1	2	2	3	3	4	5	5
77	8865	8871	8876	8882	8887	8893	8899	8904	8910	8915	1	1	2	2	3	3	4	4	5
78	8921	8927	8932	8938	8943	8949	8954	8960	8965	8971	1	1	2	2	3	3	4	4	5
79	8976	8982	8987	8993	8998	9004	9009	9015	9020	9025	1	1	2	2	3	3	4	4	5
80	9031	9036	9042	9047	9053	9058	9063	9069	9074	9079	1	1	2	2	3	3	4	4	5
81	9085	9090	9096	9101	9106	9112	9117	9122	9128	9133	1	1	2	2	3	3	4	4	5
82	9138	9143	9149	9154	9159	9165	9170	9175	9180	9186	1	1	2	2	3	3	4	4	5
83	9191	9196	9201	9206	9212	9217	9222	9227	9232	9238	1	1	2	2	3	3	4	4	5
84	9243	9248	9253	9258	9263	9269	9274	9279	9284	9289	1	1	2	2	3	3	4	4	5
85	9294	9299	9304	9309	9315	9320	9325	9330	9335	9340	1	1	2	2	3	3	4	4	5
86	9345	9350	9355	9360	9365	9370	9375	9380	9385	9390	1	1	2	2	3	3	4	4	5
87	9395	9400	9405	9410	9415	9420	9425	9430	9435	9440	0	1	1	2	2	3	3	4	4
88	9445	9450	9455	9460	9465	9469	9474	9479	9484	9489	0	1	1	2	2	3	3	4	4
89	9494	9499	9504	9509	9513	9518	9523	9528	9533	9538	0	1	1	2	2	3	3	4	4
90	9542	9547	9552	9557	9562	9566	9571	9576	9581	9586	0	1	1	2	2	3	3	4	4
91	9590	9595	9600	9605	9609	9614	9619	9624	9628	9633	0	1	1	2	2	3	3	4	4
92	9638	9643	9647	9652	9657	9661	9666	9671	9675	9680	0	1	1	2	2	3	3	4	4
93	9685	9689	9694	9699	9703	9708	9713	9717	9722	9727	0	1	1	2	2	3	3	4	4
94	9731	9736	9741	9745	9750	9754	9759	9763	9768	9773	0	1	1	2	2	3	3	4	4
95	9777	9782	9786	9791	9795	9800	9805	9809	9814	9818	0	1	1	2	2	3	3	4	4
96	9823	9827	9832	9836	9841	9845	9850	9854	9859	9863	0	1	1	2	2	3	3	4	4
97	9868	9872	9877	9881	9886	9890	9894	9899	9903	9908	0	1	1	2	2	3	3	4	4
98	9912	9917	9921	9926	9930	9934	9939	9943	9948	9952	0	1	1	2	2	3	3	4	4
99	9956	9961	9965	9969	9974	9978	9983	9987	9991	9996	0	1	1	2	2	3	3	3	4

Table A.9 Standard reduction potentials, $\mathscr{E}^{\circ}_{red}$, in acid solutions

Couple (ox/red)	Cathode reaction (reduction)	$\mathscr{E}^{\circ}_{red}$, volts
N_2/HN_3	$3N_2 + 2H^+ + 2e^- \rightarrow 2HN_3$	-3.09
Li^+/Li	$Li^+ + e^- \rightarrow Li$	-3.04
K^+/K	$K^+ + e^- \rightarrow K$	-2.92
Rb^+/Rb	$Rb^+ + e^- \rightarrow Rb$	-2.92
Cs^+/Cs	$Cs^+ + e^- \rightarrow Cs$	-2.92
Ra^{2+}/Ra	$Ra^{2+} + 2e^- \rightarrow Ra$	-2.92
Ba^{2+}/Ba	$Ba^{2+} + 2e^- \rightarrow Ba$	-2.91
Sr^{2+}/Sr	$Sr^{2+} + 2e^- \rightarrow Sr$	-2.89
Ca^{2+}/Ca	$Ca^{2+} + 2e^- \rightarrow Ca$	-2.87
Na^+/Na	$Na^+ + e^- \rightarrow Na$	-2.71
Ac^{3+}/Ac	$Ac^{3+} + 3e^- \rightarrow Ac$	-2.6
La^{3+}/La	$La^{3+} + 3e^- \rightarrow La$	-2.52
Ce^{3+}/Ce	$Ce^{3+} + 3e^- \rightarrow Ce$	-2.48
Nd^{3+}/Nd	$Nd^{3+} + 3e^- \rightarrow Nd$	-2.43
Sm^{3+}/Sm	$Sm^{3+} + 3e^- \rightarrow Sm$	-2.41
Eu^{3+}/Eu	$Eu^{3+} + 3e^- \rightarrow Eu$	-2.41
Gd^{3+}/Gd	$Gd^{3+} + 3e^- \rightarrow Gd$	-2.40
Y^{3+}/Y	$Y^{3+} + 3e^- \rightarrow Y$	-2.37
Mg^{2+}/Mg	$Mg^{2+} + 2e^- \rightarrow Mg$	-2.37
Lu^{3+}/Lu	$Lu^{3+} + 3e^- \rightarrow Lu$	-2.26
H_2/H^-	$H_2 + 2e^- \rightarrow 2H^-$	-2.25
H^+/H	$H^+ + e^- \rightarrow H$	-2.11
Sc^{3+}/Sc	$Sc^{3+} + 3e^- \rightarrow Sc$	-2.08
Be^{2+}/Be	$Be^{2+} + 2e^- \rightarrow Be$	-1.85
U^{3+}/U	$U^{3+} + 3e^- \rightarrow U$	-1.79
Al^{3+}/Al	$Al^{3+} + 3e^- \rightarrow Al$	-1.66
Ti^{2+}/Ti	$Ti^{2+} + 2e^- \rightarrow Ti$	-1.63
V^{2+}/V	$V^{2+} + 2e^- \rightarrow V$	-1.19
Mn^{2+}/Mn	$Mn^{2+} + 2e^- \rightarrow Mn$	-1.18
TiO^{2+}/Ti	$TiO^{2+} + 2H^+ + 4e^- \rightarrow Ti + H_2O$	-0.88
SiO_2/Si	$SiO_2 + 4H^+ + 4e^- \rightarrow Si + 2H_2O$	-0.86
Zn^{2+}/Zn	$Zn^{2+} + 2e^- \rightarrow Zn$	-0.76
Cr^{3+}/Cr	$Cr^{3+} + 3e^- \rightarrow Cr$	-0.74
Ga^{3+}/Ga	$Ga^{3+} + 3e^- \rightarrow Ga$	-0.53
Fe^{2+}/Fe	$Fe^{2+} + 2e^- \rightarrow Fe$	-0.44
Eu^{3+}/Eu^{2+}	$Eu^{3+} + e^- \rightarrow Eu^{2+}$	-0.43
Cr^{3+}/Cr^{2+}	$Cr^{3+} + e^- \rightarrow Cr^{2+}$	-0.41
Ti^{3+}/Ti^{2+}	$Ti^{3+} + e^- \rightarrow Ti^{2+}$	-0.37
$PbSO_4/Pb$	$PbSO_4 + 2e^- \rightarrow Pb + SO_4^{2-}$	-0.36
In^{3+}/In	$In^{3+} + 3e^- \rightarrow In$	-0.34
Tl^+/Tl	$Tl^+ + e^- \rightarrow Tl$	-0.34
Co^{2+}/Co	$Co^{2+} + 2e^- \rightarrow Co$	-0.28
V^{3+}/V^{2+}	$V^{3+} + e^- \rightarrow V^{2+}$	-0.26
$V(OH)_4^+/V$	$V(OH)_4^+ + 4H^+ + 5e^- \rightarrow V + 4H_2O$	-0.25
Ni^{2+}/Ni	$Ni^{2+} + 2e^- \rightarrow Ni$	-0.25
$N_2/N_2H_5^+$	$N_2 + 5H^+ + 4e^- \rightarrow N_2H_5^+$	-0.23
AgI/Ag	$AgI + e^- \rightarrow Ag + I^-$	-0.15
Sn^{2+}/Sn	$Sn^{2+} + 2e^- \rightarrow Sn$	-0.14
Pb^{2+}/Pb	$Pb^{2+} + 2e^- \rightarrow Pb$	-0.13
D^+/D_2	$2D^+ + 2e^- \rightarrow D_2$	-0.003
H^+/H_2	$2H^+ + 2e^- \rightarrow H_2$	±0.00

Table A·9 Standard reduction potentials, $\mathscr{E}^\circ_{red}$, in acid solutions (*continued*)

Couple (ox/red)	Cathode reaction (reductions)	$\mathscr{E}^\circ_{red}$, volts
$AgBr/Ag$	$AgBr + e^- \rightarrow Ag + Br^-$	$+0.07$
TiO^{2+}/Ti^{3+}	$TiO^{2+} + 2H^+ + e^- \rightarrow Ti^{3+} + H_2O$	$+0.10$
Sn^{4+}/Sn^{2+}	$Sn^{4+} + 2e^- \rightarrow Sn^{2+}$	$+0.15$
Cu^{2+}/Cu^+	$Cu^{2+} + e^- \rightarrow Cu^+$	$+0.15$
$BiOCl/Bi$	$BiOCl + 2H^+ + 3e^- \rightarrow Bi + H_2O + Cl^-$	$+0.16$
$AgCl/Ag$	$AgCl + e^- \rightarrow Ag + Cl^-$	$+0.22$
Hg_2Cl_2/Hg	$Hg_2Cl_2 + 2e^- \rightarrow 2Hg + 2Cl^-$	$+0.27$
BiO^+/Bi	$BiO^+ + 2H^+ + 3e^- \rightarrow Bi + H_2O$	$+0.32$
Cu^{2+}/Cu	$Cu^{2+} + 2e^- \rightarrow Cu$	$+0.34$
SO_4^{2-}/S	$SO_4^{2-} + 8H^+ + 6e^- \rightarrow S + 4H_2O$	$+0.36$
VO^{2+}/V^{3+}	$VO^{2+} + 2H^+ + e^- \rightarrow V^{3+} + H_2O$	$+0.36$
$Fe(CN)_6^{3-}/Fe(CN)_6^{4-}$	$Fe(CN)_6^{3-} + e^- \rightarrow Fe(CN)_6^{4-}$	$+0.36$
Ag_2CrO_4/Ag	$Ag_2CrO_4 + 2e^- \rightarrow 2Ag + CrO_4^{2-}$	$+0.46$
Cu^+/Cu	$Cu^+ + e^- \rightarrow Cu$	$+0.52$
I_2/I^-	$I_2 + 2e^- \rightarrow 2I^-$	$+0.54$
MnO_4^-/MnO_4^{2-}	$MnO_4^- + e^- \rightarrow MnO_4^{2-}$	$+0.56$
$AgC_2H_3O_2/Ag$	$AgC_2H_3O_2 + e^- \rightarrow Ag + C_2H_3O_2^-$	$+0.64$
Ag_2SO_4/Ag	$Ag_2SO_4 + 2e^- \rightarrow 2Ag + SO_4^{2-}$	$+0.65$
$Au(CNS)_4^-/Au$	$Au(CNS)_4^- + 3e^- \rightarrow Au + 4CNS^-$	$+0.66$
$PtCl_6^{2-}/PtCl_4^{2-}$	$PtCl_6^{2-} + 2e^- \rightarrow PtCl_4^{2-} + 2Cl^-$	$+0.68$
O_2/H_2O_2	$O_2 + 2H^+ + 2e^- \rightarrow H_2O_2$	$+0.68$
HN_3/NH_4^+	$HN_3 + 11H^+ + 8e^- \rightarrow 3NH_4^+$	$+0.70$
$PtCl_4^{2-}/Pt$	$PtCl_4^{2-} + 2e^- \rightarrow Pt + 4Cl^-$	$+0.73$
Fe^{3+}/Fe^{2+}	$Fe^{3+} + e^- \rightarrow Fe^{2+}$	$+0.77$
Hg_2^{2+}/Hg	$Hg_2^{2+} + 2e^- \rightarrow 2Hg$	$+0.79$
Ag^+/Ag	$Ag^+ + e^- \rightarrow Ag$	$+0.80$
NO_3^-/N_2O_4	$2NO_3^- + 4H^+ + 2e^- \rightarrow N_2O_4 + 2H_2O$	$+0.80$
$AuBr_4^-/Au$	$AuBr_4^- + 3e^- \rightarrow Au + 4Br^-$	$+0.87$
Hg^{2+}/Hg_2^{2+}	$2Hg^{2+} + 2e^- \rightarrow Hg_2^{2+}$	$+0.92$
NO_3^-/NO	$NO_3^- + 4H^+ + 3e^- \rightarrow NO + 2H_2O$	$+0.96$
HNO_2/NO	$HNO_2 + H^+ + e^- \rightarrow NO + H_2O$	$+1.00$
$AuCl_4^-/Au$	$AuCl_4^- + 3e^- \rightarrow Au + 4Cl^-$	$+1.00$
$V(OH)_4^+/VO^{2+}$	$V(OH)_4^+ + 2H^+ + e^- \rightarrow VO^{2+} + 3H_2O$	$+1.00$
N_2O_4/NO	$N_2O_4 + 4H^+ + 4e^- \rightarrow 2NO + 2H_2O$	$+1.03$
Br_2/Br^-	$Br_2 + 2e^- \rightarrow 2Br^-$	$+1.07$
ClO_4^-/ClO_3^-	$ClO_4^- + 2H^+ + 2e^- \rightarrow ClO_3^- + H_2O$	$+1.19$
$ClO_3^-/HClO_2$	$ClO_3^- + 3H^+ + 2e^- \rightarrow HClO_2 + H_2O$	$+1.21$
O_2/H_2O	$O_2 + 4H^+ + 4e^- \rightarrow 2H_2O$	$+1.23$
MnO_2/Mn^{2+}	$MnO_2 + 4H^+ + 2e^- \rightarrow Mn^{2+} + 2H_2O$	$+1.23$
Tl^{3+}/Tl^+	$Tl^{3+} + 2e^- \rightarrow Tl^+$	$+1.25$
$N_2H_5^+/NH_4^+$	$N_2H_5^+ + 3H^+ + 2e^- \rightarrow 2NH_4^+$	$+1.28$
$ClO_2/HClO_2$	$ClO_2 + H^+ + e^- \rightarrow HClO_2$	$+1.28$
HNO_2/N_2O	$2HNO_2 + 4H^+ + 4e^- \rightarrow N_2O + 3H_2O$	$+1.29$
$Cr_2O_7^{2-}/Cr^{3+}$	$Cr_2O_7^{2-} + 14H^+ + 6e^- \rightarrow 2Cr^{3+} + 7H_2O$	$+1.33$
NH_3OH^+/NH_4^+	$NH_3OH^+ + 2H^+ + 2e^- \rightarrow NH_4^+ + H_2O$	$+1.35$
Cl_2/Cl^-	$Cl_2 + 2e^- \rightarrow 2Cl^-$	$+1.36$
$Au(OH)_3/Au$	$Au(OH)_3 + 3H^+ + 3e^- \rightarrow Au + 3H_2O$	$+1.45$
PbO_2/Pb^{2+}	$PbO_2 + 4H^+ + 2e^- \rightarrow Pb^{2+} + 2H_2O$	$+1.46$
Au^{3+}/Au	$Au^{3+} + 3e^- \rightarrow Au$	$+1.50$
Mn^{3+}/Mn^{2+}	$Mn^{3+} + e^- \rightarrow Mn^{2+}$	$+1.51$
MnO_4^-/Mn^{2+}	$MnO_4^- + 8H^+ + 5e^- \rightarrow Mn^{2+} + 4H_2O$	$+1.51$

Table A•9 Standard reduction potentials, $\mathscr{E}^\circ_{red}$, in acid solutions (*continued*)

Couple (ox/red)	Cathode reaction (reduction)	$\mathscr{E}^\circ_{red}$, volts
BrO_3^-/Br_2	$2BrO_3^- + 12H^+ + 10e^- \rightarrow Br_2 + 6H_2O$	$+1.52$
Bi_2O_4/BiO^+	$Bi_2O_4 + 4H^+ + 2e^- \rightarrow 2BiO^+ + 2H_2O$	$+1.59$
Ce^{4+}/Ce^{3+}	$Ce^{4+} + e^- \rightarrow Ce^{3+}$	$+1.61$
$HClO/Cl_2$	$2HClO + 2H^+ + 2e^- \rightarrow Cl_2 + 2H_2O$	$+1.63$
$HClO_2/HClO$	$HClO_2 + 2H^+ + 2e^- \rightarrow HClO + H_2O$	$+1.64$
NiO_2/Ni^{2+}	$NiO_2 + 4H^+ + 2e^- \rightarrow Ni^{2+} + 2H_2O$	$+1.68$
$PbO_2/PbSO_4$	$PbO_2 + SO_4^{2-} + 4H^+ + 2e^- \rightarrow PbSO_4 + 2H_2O$	$+1.68$
Au^+/Au	$Au^+ + e^- \rightarrow Au$	$+1.69$
MnO_4^-/MnO_2	$MnO_4^- + 4H^+ + 3e^- \rightarrow MnO_2 + 2H_2O$	$+1.70$
H_2O_2/H_2O	$H_2O_2 + 2H^+ + 2e^- \rightarrow 2H_2O$	$+1.78$
XeO_3/Xe	$XeO_3 + 6H^+ + 6e^- \rightarrow Xe + 3H_2O$	$+1.8$
Co^{3+}/Co^{2+}	$Co^{3+} + e^- \rightarrow Co^{2+}$	$+1.81$
$HN_3/NH_4^+, N_2$	$HN_3 + 3H^+ + 2e^- \rightarrow NH_4^+ + N_2$	$+1.96$
Ag^{2+}/Ag^+	$Ag^{2+} + e^- \rightarrow Ag^+$	$+1.98$
$S_2O_8^{2-}/SO_4^{2-}$	$S_2O_8^{2-} + 2e^- \rightarrow 2SO_4^{2-}$	$+2.01$
F_2/F^-	$F_2 + 2e^- \rightarrow 2F^-$	$+2.87$
H_4XeO_6/XeO_3	$H_4XeO_6 + 2H^+ + 2e^- \rightarrow XeO_3 + 3H_2O$	$+3.0$
F_2/HF	$F_2 + 2H^+ + 2e^- \rightarrow 2HF$	$+3.06$

Source: *Courtesy of Professor André J. de Béthune, Boston College. Entries chosen from the extensive tabulations by A. J. de Béthune and N. A. Swendeman Loud,* Standard Aqueous Electrode Potentials and Temperature Coefficients, *C. A. Hampel, Publisher, Skokie, Ill., 1964. Reproduced with permission.* (*See also Table 18.1 and its footnotes.*)

Table A.10 Standard reduction potentials, $\mathscr{E}_B^\circ$, in basic solutions

Couple (ox/red)	Cathode reaction (reduction)	$\mathscr{E}_B^\circ$, volts
$Ca(OH)_2/Ca$	$Ca(OH)_2 + 2e^- \rightarrow Ca + 2OH^-$	-3.02
$Sr(OH)_2/Sr$	$Sr(OH)_2 + 2e^- \rightarrow Sr + 2OH^-$	-2.88
$Ba(OH)_2/Ba$	$Ba(OH)_2 + 2e^- \rightarrow Ba + 2OH^-$	-2.81
$Mg(OH)_2/Mg$	$Mg(OH)_2 + 2e^- \rightarrow Mg + 2OH^-$	-2.69
BeO/Be	$BeO + H_2O + 2e^- \rightarrow Be + 2OH^-$	-2.61
$Sc(OH)_3/Sc$	$Sc(OH)_3 + 3e^- \rightarrow Sc + 3OH^-$	-2.61
$Al(OH)_3/Al$	$Al(OH)_3 + 3e^- \rightarrow Al + 3OH^-$	-2.30
$H_2BO_3^-/B$	$H_2BO_3^- + H_2O + 3e^- \rightarrow B + 4OH^-$	-1.79
$Mn(OH)_2/Mn$	$Mn(OH)_2 + 2e^- \rightarrow Mn + 2OH^-$	-1.55
$Zn(OH)_2/Zn$	$Zn(OH)_2 + 2e^- \rightarrow Zn + 2OH^-$	-1.24
$SO_3^{2-}/S_2O_4^{2-}$	$2SO_3^{2-} + 2H_2O + 2e^- \rightarrow S_2O_4^{2-} + 4OH^-$	-1.12
SO_4^{2-}/SO_3^{2-}	$SO_4^{2-} + H_2O + 2e^- \rightarrow SO_3^{2-} + 2OH^-$	-0.93
Se/Se^{2-}	$Se + 2e^- \rightarrow Se^{2-}$	-0.92
$Fe(OH)_2/Fe$	$Fe(OH)_2 + 2e^- \rightarrow Fe + 2OH^-$	-0.88
H_2O/H_2	$2H_2O + 2e^- \rightarrow H_2 + 2OH^-$	-0.83
$Cd(OH)_2/Cd$	$Cd(OH)_2 + 2e^- \rightarrow Cd + 2OH^-$	-0.81
$Ni(OH)_2/Ni$	$Ni(OH)_2 + 2e^- \rightarrow Ni + 2OH^-$	-0.72
Ag_2S/Ag	$Ag_2S + 2e^- \rightarrow 2Ag + S^{2-}$	-0.66
$SO_3^{2-}/S_2O_3^{2-}$	$2SO_3^{2-} + 3H_2O + 4e^- \rightarrow S_2O_3^{2-} + 6OH^-$	-0.57
$Fe(OH)_3/Fe(OH)_2$	$Fe(OH)_3 + e^- \rightarrow Fe(OH)_2 + OH^-$	-0.56
$Ni(NH_3)_6^{2+}/Ni$	$Ni(NH_3)_6^{2+} + 2e^- \rightarrow Ni + 6NH_3$	-0.48
S/S^{2-}	$S + 2e^- \rightarrow S^{2-}$	-0.45
$Ag(CN)_2^-/Ag$	$Ag(CN)_2^- + e^- \rightarrow Ag + 2CN^-$	-0.31
$Cu(NH_3)_2^+/Cu$	$Cu(NH_3)_2^+ + e^- \rightarrow Cu + 2NH_3$	-0.12
$Cu(OH)_2/Cu_2O$	$2Cu(OH)_2 + 2e^- \rightarrow Cu_2O + 2OH^- + H_2O$	-0.08
$MnO_2/Mn(OH)_2$	$MnO_2 + 2H_2O + 2e^- \rightarrow Mn(OH)_2 + 2OH^-$	-0.05
$AgCN/Ag$	$AgCN + e^- \rightarrow Ag + CN^-$	-0.02
NO_3^-/NO_2^-	$NO_3^- + H_2O + 2e^- \rightarrow NO_2^- + 2OH^-$	$+0.01$
$S_4O_6^{2-}/S_2O_3^{2-}$	$S_4O_6^{2-} + 2e^- \rightarrow 2S_2O_3^{2-}$	$+0.08$
$Co(NH_3)_6^{3+}/Co(NH_3)_6^{2+}$	$Co(NH_3)_6^{3+} + e^- \rightarrow Co(NH_3)_6^{2+}$	$+0.11$
N_2H_4/NH_3	$N_2H_4 + 2H_2O + 2e^- \rightarrow 2NH_3 + 2OH^-$	$+0.11$
$Mn(OH)_3/Mn(OH)_2$	$Mn(OH)_3 + e^- \rightarrow Mn(OH)_2 + OH^-$	$+0.15$
ClO_3^-/ClO_2^-	$ClO_3^- + H_2O + 2e^- \rightarrow ClO_2^- + 2OH^-$	$+0.33$
Ag_2O/Ag	$Ag_2O + H_2O + 2e^- \rightarrow 2Ag + 2OH^-$	$+0.34$
ClO_4^-/ClO_3^-	$ClO_4^- + H_2O + 2e^- \rightarrow ClO_3^- + 2OH^-$	$+0.36$
$Ag(NH_3)_2^+/Ag$	$Ag(NH_3)_2^+ + e^- \rightarrow Ag + 2NH_3$	$+0.37$
O_2/OH^-	$O_2 + 2H_2O + 4e^- \rightarrow 4OH^-$	$+0.40$
$NiO_2/Ni(OH)_2$	$NiO_2 + 2H_2O + 2e^- \rightarrow Ni(OH)_2 + 2OH^-$	$+0.49$
MnO_4^-/MnO_2	$MnO_4^- + 2H_2O + 3e^- \rightarrow MnO_2 + 4OH^-$	$+0.59$
MnO_4^{2-}/MnO_2	$MnO_4^{2-} + 2H_2O + 2e^- \rightarrow MnO_2 + 4OH^-$	$+0.60$
ClO_2^-/ClO^-	$ClO_2^- + H_2O + 2e^- \rightarrow ClO^- + 2OH^-$	$+0.66$
NH_2OH/N_2H_4	$2NH_2OH + 2e^- \rightarrow N_2H_4 + 2OH^-$	$+0.73$
ClO^-/Cl^-	$ClO^- + H_2O + 2e^- \rightarrow Cl^- + 2OH^-$	$+0.89$
$HXeO_4^-/Xe$	$HXeO_4^- + 3H_2O + 6e^- \rightarrow Xe + 7OH^-$	$+0.9$
$HXeO_6^{3-}/HXeO_4^-$	$HXeO_6^{3-} + 2H_2O + 2e^- \rightarrow HXeO_4^- + 4OH^-$	$+0.9$
ClO_2/ClO_2^-	$ClO_2 + e^- \rightarrow ClO_2^-$	$+1.16$
O_3/O_2	$O_3 + H_2 + 2e^- \rightarrow O_2 + 2OH^-$	$+1.24$

Source: *Courtesy of Professor André J. de Béthune, Boston College. Entries chosen from the extensive tabulations by A. J. de Béthune and N. A. Swendeman Loud,* Standard Aqueous Electrode Potentials and Temperature Coefficients, *C. A. Hampel, Publisher, Skokie, Ill., 1964. Reproduced with permission.* (See also Table 18.1 and its footnotes.)

Glossary

absolute entropy (S) the entropy of a substance relative to its entropy as a perfect crystal at absolute zero [kJ/(mol · K)].

absolute zero 0 K, the lowest possible temperature.

absorption spectrum the pattern of wavelengths produced by the absorption of radiation by a substance. (Compare *emission spectrum*.)

accelerator a machine that increases the velocity of charged particles.

acid see *Arrhenius acid, Brønsted–Lowry acid,* and *Lewis acid*.

acid–base indicator an organic acid or base that changes color within a certain pH range.

acidic solution a solution in which the concentration of H^+ is greater than that of OH^-.

acid ionization constant the equilibrium constant for the ionization of an acid, treating $[H_2O]$ as a constant; $K_a = [H^+][A^-]/[HA]$.

actinide series the series of 14 elements in period 7 of the periodic table in which the $5f$ orbitals are being filled.

activated complex see *transition state*.

activation energy (E_a) the excess energy that reactants must have to form the transition state for a reaction (kJ/mol).

active site a structural feature on the surface of a catalyst or an enzyme molecule where a specific chemical reaction takes place.

addition polymer a polymer whose repeated units are joined by bonds originally associated with unsaturation.

addition reaction an organic reaction in which atoms are added to two atoms joined by a multiple bond.

adhesive forces intermolecular forces between unlike substances.

adsorption the adhesion of atoms, molecules, or ions to the surface of another substance.

alcohol an organic molecule with one or more hydroxyl groups in place of hydrogen atoms.

aldehyde an organic molecule with a double bond to oxygen in place of two hydrogen atoms at the end of a chain.

aldohexose a hexose containing an aldehyde group.

aliphatic hydrocarbon a hydrocarbon containing no aromatic rings.

alkane a chain hydrocarbon containing only single bonds.

alkene a chain hydrocarbon containing one double bond.

alkyl group a monovalent group derived from an alkane by removal of one H atom.

alkyne a chain hydrocarbon containing one triple bond.

allotropes two or more molecular or crystalline forms of a given element.

alloy a homogeneous mixture of two or more metals.

alpha helix the spiral structure of most protein chains.

alpha particle a radioactive emission, identical to a helium nucleus.

alum a hydrated double salt with two $SO_4{}^{2-}$ anions, one singly charged and one triply charged cation, and twelve molecules of water.

amino acids carboxylic acids that contain an amino group; the monomers of proteins.

amphiprotic species an ion or molecule that can either donate or accept a proton.

amphoteric substance a substance that can react as either an acid or a base.

anion a negatively charged ion.

anode in a vacuum tube, the positively charged electrode; in an electrochemical cell, the electrode at which oxidation takes place.

antibonding orbital a molecular orbital in which the electron density is concentrated away from the region between two nuclei.

aromatic hydrocarbon a hydrocarbon containing one or more rings with delocalized π-electron systems.

aromatization the conversion of an aliphatic hydrocarbon to an aromatic hydrocarbon.

Arrhenius acid a substance that dissolves in water to give H^+ ions.

Arrhenius base a substance that dissolves in water to give OH^- ions.

asymmetric carbon atom see *chiral carbon atom*.

atom the smallest particle of an element.

atomic energy   see *nuclear energy*.

atomic number (Z)   the number of protons in the nucleus of an atom.

atomic orbital   a wave function for an electron in an atom; the region in which an electron is likely to be found.

atomic radius   the covalent radius of a nonmetal atom or the metallic radius of a metal atom (Å).

aufbau principle   the filling of energy sublevels and orbitals of atoms in the order of increasing energy.

autocatalysis   the catalysis of a reaction by a product of that reaction.

auxochrome   a substituent group on an organic dye molecule that alters the color due to the chromophore.

average bond energy ($\Delta H_{\text{dis avg}}$)   the average energy per bond required to dissociate 1 mole of a polyatomic molecule to its constituent atoms (kJ/mol).

Avogadro's law   at the same temperature and pressure, equal volumes of different gases contain equal numbers of molecules; V/n = constant.

Avogadro's number   the number of atoms in exactly 12 g of the nuclide ^{12}C (6.0220×10^{23}, to five significant figures).

Avogadro's volume   see *molar gas volume*.

azeotrope   see *constant-boiling mixture*.

background radiation   the natural radiation, due to radioactive decay and cosmic radiation, that is present in the environment.

base   see *Arrhenius base, Brønsted–Lowry base,* and *Lewis base*.

base ionization constant   the equilibrium constant for the ionization of a base, treating [H_2O] as a constant; K_b = [BH^+][OH^-]/[B].

basic solution   a solution in which the concentration of OH^- is greater than that of H^+.

bead test   a test for the presence of certain cations in a solid sample, by the characteristic colors they impart to a borax bead.

benzene hydrocarbon   a hydrocarbon containing a benzene ring.

berthollide   a solid compound of somewhat variable composition.

beta particle   a radioactive emission, identical to an electron.

binary compound   a compound composed of two elements.

binding energy   in atomic structure, the attraction of an atom for one of its electrons (eV); in nuclear structure, the energy equivalent to the mass loss for a given nuclide (MeV).

Bohr theory   the first modern theory of atomic structure based on the quantization of electron energy levels.

boiling point   the temperature at which the vapor pressure of a liquid is equal to the external pressure.

bombardment reaction   a nuclear reaction induced by bombarding a target substance with atomic or subatomic particles.

bond dissociation energy ($\Delta H_{\text{dis}}^{\circ}$)   the amount of energy required to dissociate 1 mole of a given bond in a molecule or radical, with the reactant and products being gases in the standard state (kJ/mol).

bonding orbital   a molecular orbital in which the electron density is concentrated in the region between two nuclei.

bond length   the distance between the nuclei of two atoms that are joined by a chemical bond (Å).

bond moment   the polar moment of an individual chemical bond (D).

bond order   the number of bonds calculated from MO theory to be possible in a given homonuclear diatomic molecule.

borane   a compound containing only B and H atoms.

Born–Haber cycle   a conceptual closed cycle of thermochemical reactions in which the net enthalpy change is zero.

Boyle's law   at constant temperature, the volume of a given mass of gas varies inversely with the pressure. PV = constant.

Bragg equation   the equation for calculating the spacing between parallel planes of atoms or ions in a crystal from X-ray diffraction data; $n\lambda = 2d \sin \theta$.

branched-chain isomer   an isomer of a straight-chain hydrocarbon that has one or more carbon atoms joined as a branch.

Brønsted–Lowry acid   a particle that donates a proton to another particle.

Brønsted–Lowry base   a particle that accepts a proton from another particle.

Brownian movement   the random motion of colloidal particles in a fluid medium.

buffered solution   a solution of a weak acid and a salt of that acid, or of a weak base and a salt of that base; a solution that can react with small amounts of either strong acids or strong bases without much change in its pH.

calorimeter   an instrument used to measure the heat change accompanying a chemical reaction.

capillary rise   the rise of a column of liquid in a narrow tube due to the surface tension of the liquid.

carbohydrate   see *saccharide*.

carboxylic acid   an organic molecule that has a terminal carbon with a double bond to an oxygen atom and a single bond to a hydroxyl group.

catalyst a substance that affects the rate of a reaction without itself being permanently changed.

cathode in a vacuum tube, the negatively charged electrode; in an electrochemical cell, the electrode at which reduction takes place.

cathode ray the electric discharge from the negatively charged electrode in a vacuum tube.

cation a positively charged ion.

cell reaction the algebraic sum of the oxidation and reduction half-reactions occurring at the electrodes of an electrochemical cell.

cell voltage ($\mathscr{E}_{cell}$) the algebraic sum of the oxidation and reduction potentials for the two half-reactions occurring at the electrodes of a voltaic cell (V).

Celsius scale (°C) a temperature scale with 0 °C for the normal freezing point of water and 100 °C for the normal boiling point; formerly called *centigrade scale*.

chain reaction in chemical kinetics, a reaction that is self-sustaining because of the continuous regeneration of a highly reactive species; in nuclear chemistry, a self-sustaining nuclear reaction propagated by neutrons.

charge density the ratio of the charge of an ion to its volume.

Charles's law at constant pressure, the volume of a given mass of gas varies directly with the absolute temperature; V/T = constant.

chelating agent a ligand with two or more points of attachment to a central metal ion or atom.

chemical bond the strong attractive force that holds atoms together in compounds and polyatomic elements.

chemical change any process in which substances are converted to other substances.

chemical energy the energy possessed by a substance by virtue of its chemical state (kJ/mol).

chemical equation a representation of a chemical reaction in which formulas of reactants, an arrow, and formulas of products are shown in order, with equal numbers of atoms of each element to the left and right of the arrow.

chemical equilibrium the state in a reversible reaction in which the rates of the forward and reverse reactions are equal.

chemical kinetics the study of the rates and mechanisms of chemical reactions.

chemical property a property of a substance that causes it to undergo chemical change.

chemical thermodynamics the study of relations of heat, work, and other forms of energy to chemical systems.

chemistry the science of the structure of matter and the changes that matter undergoes.

chiral carbon atom a carbon atom to which four different atoms or groups of atoms are attached (also called *asymmetric carbon atom*).

chirality the property of handedness, possessed by stereoisomers that are nonsuperimposable mirror images of each other.

chromatography a method for separating the components of a fluid mixture by preferential adsorption on a surface.

chromophore that portion of an organic dye molecule that is responsible for its color.

cis-trans isomerism isomerism of molecules or ions due to two parts being either on the same side (cis) or on opposite sides (trans) of the molecule or ion (also called *geometric isomerism*).

coenzyme a small molecule or ion whose presence is required at the active site of an enzyme to facilitate the reaction being catalyzed.

cohesive forces intermolecular forces within a given substance or between like substances.

colligative property a property of a solution that depends on the number of solute particles present but not on their kind.

colligative property law the vapor pressure, freezing point, boiling point, and osmotic pressure of a solution differ from those of the pure solvent by amounts that are directly proportional to the molal concentration of the solute.

colloidal dispersion a dispersion of colloidal matter in another substance.

colloidal state the state of matter that consists of aggregates of particles typically with at least one dimension in the range 10–2000 Å.

combustion a rapid, self-sustaining oxidation reaction that gives off heat and light.

common ion effect the suppression of the ionization of a weak electrolyte in solution by the presence of a compound with which it has one ion in common.

complex ion a metal ion with one or more ligands bonded to it.

compound a pure substance of definite composition (except for the berthollides) that can be decomposed to other substances.

concentration the ratio of the mass or volume of a solute to a specified mass or volume of the solvent or solution (many different units).

condensation polymer a polymer consisting of repeated units linked by bonds made available by the elimination of a simple molecule.

configuration in organic chemistry, the spatial arrangement of the atoms that characterize a given stereoisomer.

conjugate acid the acid that a base becomes when it accepts a proton.

conjugate acid–base pair an acid and its conjugate base, or a base and its conjugate acid.

conjugate base the base that an acid becomes when it donates a proton.

constant-boiling mixture a mixture of liquids that cannot be separated by boiling because the compositions of the vapor and liquid phases are identical.

coordinate covalent bond a covalent bond in which both electrons are donated by one atom.

coordination compound a compound in which one or more ligands are bonded to a central metal ion or atom by coordinate covalent bonds.

coordination number in a crystal, the number of nearest neighbors of a given ion or atom; in a coordination compound, the number of coordinate covalent bonds from ligands to the central metal ion or atom.

corrosion the chemical attack on a metal by substances in the environment.

cosmic radiation high-energy ionizing radiation from deep space.

covalent bond a bond that results from the sharing of one, two, or three pairs of valence electrons between two atoms.

covalent compound a compound consisting of molecules whose atoms are held together in a molecule by covalent bonds.

covalent radius the part of the bond length between two covalently bonded atoms that is attributed to a particular atom (Å).

cracking the thermal or catalytic decomposition of higher-boiling hydrocarbons to lower-boiling hydrocarbons.

critical mass the mass and arrangement of a fissionable material required to make a nuclear chain reaction self-sustaining.

critical pressure the minimum pressure required to liquefy a gas at its critical temperature.

critical temperature the temperature above which a gas cannot be liquefied, no matter how great the pressure.

crystal energy (ΔH_{xtal}) the enthalpy change when widely separated gaseous ions come together to form a crystalline substance (kJ/mol). Also called *lattice energy*.

crystal lattice the regular, three-dimensional array of similar points in a crystalline solid.

cycloalkane a cyclic hydrocarbon containing only single bonds that is isomeric with an alkene.

daltonide a compound of definite composition (the vast majority of all compounds).

Dalton's atomic theory the first modern theory of atoms and molecules as the fundamental particles of all substances.

Dalton's law the total pressure in a mixture of gases is the sum of the individual partial pressures.
$P_{total} = p_1 + p_2 + \ldots + p_n$

decantate the liquid above a precipitate that has settled out of a solution.

decomposition potential the minimum voltage required to bring about the electrolysis of a substance (V).

dehydration reaction an organic reaction in which H_2O or equal amounts of H and OH are removed from a compound or compounds.

deliquescence the absorption of sufficient water vapor from the air by a substance to dissolve it.

delocalization energy see *resonance energy*.

delocalized electrons electrons that participate in more than one chemical bond simultaneously.

density the mass per unit volume of a substance (g/cm^3).

deoxyribonucleic acid (DNA) the nucleic acid containing all the hereditary information of an organism.

derivative an organic compound made from a simpler parent compound by replacement of an atom or group of atoms with a different kind of atom or group of atoms.

descriptive chemistry the record of the properties of substances and the changes that occur in substances.

desiccant a drying agent.

dextrorotatory molecule an optically active molecule that rotates the plane of polarization of plane-polarized light to the right.

dialysis the separation of molecules or ions from colloids by diffusion through a semipermeable membrane.

diamagnetic substance a substance that is slightly repelled by a magnetic field.

dielectric constant the factor by which a given substance reduces the electrostatic force between two charged bodies separated by a vacuum.

digestion a long period of gentle heating of a solution containing a precipitate, to cause either the coagulation or the dissolution of the precipitate.

dipole a body that has opposite charges at two points.

dipole–dipole attraction a relatively weak intermolecular attractive force due to permanent dipole moments.

dipole moment a measure of the degree of polarity of a molecule (*D*).

direct combination reaction a reaction of two elements to produce a compound.

disproportionation reaction a reaction in which one part of a substance is oxidized while another part is reduced.

distillate the condensed vapor from a distillation.

double bond a covalent bond consisting of two pairs of electrons.

double displacement reaction a reaction of two compounds to produce two different compounds by exchanging components.

double helix the structure of the DNA molecule.

double salt a salt containing two different cations.

drug a substance introduced into the body to alter some chemical process.

efflorescence the loss of water of hydration of a crystal to the air at low humidity.

electric energy the energy associated with the flow of an electric current (J).

electrochemistry the study of redox reactions that either produce or consume electric energy.

electrode an electrically charged wire, rod, or plate in a vacuum tube or electrochemical cell.

electrode potential the voltage generated by a given half-cell reaction occurring at an electrode (V).

electrolysis the use of an electric current to bring about a redox reaction.

electrolyte a substance that conducts electricity in its molten state or in solution.

electrolytic cell an electrochemical cell in which an electric current from an external source brings about a redox reaction.

electromotive force the driving force of a voltaic cell (V).

electron a subatomic particle with unit negative charge and relatively light mass ($\sim 5.5 \times 10^{-4}$ amu), which exists outside the nucleus.

electron affinity (ΔH_{ea}) the enthalpy change when a singly negatively charged ion loses one electron (kJ/mol).

electron-dot formula see *Lewis structural formula*.

electronegativity the power of an atom in a molecule to attract the electrons in a covalent bond.

electronic configuration a list of the occupied energy sublevels in an atom, showing the number of electrons in each sublevel.

electrostatic force an attractive or repulsive force due to the interaction of electric charges.

element a substance composed of atoms having the same number of protons in their nuclei.

elementary reaction an individual step in a reaction mechanism.

emf diagram a diagram showing oxidation states and standard reduction potentials for a given element.

emission spectrum the pattern of wavelengths produced by the emission of radiation by a substance. (Compare *absorption spectrum*.)

empirical formula a formula that shows the smallest whole-number ratios of atoms of each kind in a compound.

emulsion a colloidal dispersion of a liquid in another liquid.

enantiomers isomers that are nonsuperimposable mirror images of each other.

endothermic process a process in which heat is absorbed by the system from its surroundings.

endpoint the point in a titration at which an acid-base indicator changes color.

energy the capacity of a body or a system to do work (J).

energy level one of a series of discrete energy states occupied by electrons in an atom.

energy sublevel one of a series of discrete energy states within a given main energy level (except for the first main energy level).

enthalpy (H) the heat content of a substance (kJ/mol).

enthalpy change (ΔH) the heat change for a process carried out at constant pressure (kJ/mol).

enthalpy of ionization (ΔH_{ion}) the ionization energy of a substance (kJ/mol).

entropy (S) the amount of disorder or randomness in a substance or system [J/(mol · K)].

enzyme a protein that acts as a catalyst for a specific biochemical reaction.

equilibrium the state of a system when opposing forces or rates of a process are in balance.

equilibrium constant for a reaction at equilibrium at a specified temperature, the ratio of the product of the concentrations of the reaction products to the product of the concentration of the reactants, with each term raised to a power equal to the stoichiometric coefficient of the species in the reaction. For the reaction $m\text{A} + n\text{B} \rightleftharpoons y\text{C} + z\text{D}$, $K_c = [\text{C}]^y[\text{D}]^z/[\text{A}]^m[\text{B}]^n$.

equivalence point the point in a titration at which chemically equivalent amounts of the two reactants have been mixed.

equivalent a unit of measure, equal to one equivalent weight, for the amount of matter in a sample of a substance.

equivalent weight in a redox reaction, the mass of substance that will gain or lose 1 mole of electrons (g); in an acid–base reaction, the mass of a substance that will liberate or react with 1 mole of protons (g).

ester an organic compound considered to be derived from a carboxylic acid by replacing the hydrogen of the hydroxyl group with a hydrocarbon group.

esterification reaction   an organic reaction in which a carboxylic acid and an alcohol combine to form an ester and water.

ether   an organic compound with two hydrocarbon groups attached to an atom of oxygen.

excited state   any energy condition of an atom above the ground state.

exothermic process   a process in which heat is released by the system to its surroundings.

extrinsic property   a property that is not characteristic of a substance itself, but that depends on its shape, amount, or condition.

Fahrenheit scale (°F)   a temperature scale with 32 °F for the normal freezing point of water and 212 °F for the normal boiling point.

family   one of 16 vertical divisions in the periodic table, in A-and-B pairs constituting the 8 groups.

Faraday's law   during electrolysis, the passage of 1 faraday through a circuit brings about the oxidation of 1 equivalent of one substance and the reduction of 1 equivalent of another substance.

fat   a solid ester of glycerol and three fatty acids.

fatty acid   a long-chain carboxylic acid.

ferromagnetic substance   a substance that is strongly attracted by a magnetic field.

first ionization energy   the ionization energy of the most loosely bound electron in an atom (eV).

first law of thermodynamics   energy can be converted from one form to another but it can be neither created nor destroyed (also called the *law of conservation of energy*).

first-order reaction   a reaction for which the rate is proportional to the first power of the concentration of just one reactant.

flame test   a test for the presence of certain cations in a solution, by the characteristic colors they impart to a flame.

formation constant   see *stability constant*.

formula   a symbolic representation of the kind and number of atoms chemically combined in a unit of a substance.

fourth quantum number (m_s)   the number that specifies the alignment of the electron spin in a given atomic orbital.

fractional distillation   the process by which a mixture of volatile liquids is separated by successive distillations.

fractional precipitation   see *selective precipitation*.

fractionating column   an apparatus used to carry out fractional distillation as a continuous process.

freezing point   see *melting point*.

frequency (ν)   the number of cycles of a wave that pass a given point per unit time (Hz); $\nu = c/\lambda$.

fuel cell   a special kind of voltaic cell that uses a continuous supply of gaseous or liquid reactants for the redox reaction.

functional group   an atom or group of atoms in an organic compound that is largely responsible for the properties of the compound.

fused-ring benzenoid hydrocarbon   a hydrocarbon containing two or more benzene rings that share adjacent carbon atoms.

fused-ring hydrocarbon   a hydrocarbon containing two or more rings, of which at least one pair share two adjacent carbon atoms.

gamma ray   a radioactive emission that is a form of high-energy electromagnetic radiation.

Gay-Lussac's and Amontons's law   at constant volume, the pressure of a given mass of gas varies directly with the absolute temperature. P/T = constant.

Gay-Lussac's law of combining volumes   at the same temperature and pressure, gases react with one another in small whole-number ratios by volume.

gene   a hereditary factor in an organism that is a segment of a DNA molecule.

geometric isomerism   see *cis-trans isomerism*.

Gibbs free energy (G)   the energy of a substance or a system that is available to do useful work (kJ/mol).

glass   an amorphous solid that can be regarded as a cold, extremely viscous liquid.

Graham's law   the rates of effusion of two gases vary inversely with the square roots of their densities; $r_1/r_2 = (d_2/d_1)^{1/2}$.

ground state   the condition of an atom with all its electrons in their lowest energy levels.

group   one of eight vertical divisions in the periodic table, each of which consists of an A and a B family.

half-life   in chemical kinetics, the time required for a given concentration of a reactant to fall to one-half that value; in nuclear chemistry, the time during which half of a given mass of a radioactive nuclide decays to another nuclide.

half-reaction   either the oxidation half or the reduction half of a redox reaction.

halogen family   the elements of group VIIA of the periodic table.

heat   the energy associated with the random vibrations and other motions of the tiny particles that make up matter.

heat capacity   the amount of heat needed to change the temperature of a given mass of a substance by 1 °C or 1 K.

heat of formation of an atom ($\Delta H_{a,m}$ or $\Delta H_{a,n}$)   the amount of energy required to form 1 mole of gaseous atoms from the element in its common state at 25 °C and 1 atm (kJ/mol).

Heisenberg's uncertainty principle it is not possible to measure simultaneously the precise position as well as the momentum of an object.

Henry's law the mass of a gas dissolved by a given amount of a liquid varies directly with the pressure exerted by the gas when it is in equilibrium with the solution.

Hess's law for a given overall reaction, the enthalpy change is always the same, whether the reaction occurs directly or indirectly and in different steps.

heterogeneous equilibrium an equilibrium involving two or more phases of matter.

heterogeneous material a material containing components that are visibly distinct under a microscope.

hexosan a polysaccharide consisting of hexose monomers.

hexose a monosaccharide containing six C atoms.

homogeneous equilibrium an equilibrium involving only one phase.

homogeneous material a material containing no components that are visibly distinct under a microscope.

homolog a member of a homologous series of compounds.

homologous series a series of compounds in which each compound differs from the previous one by the same number of atoms.

homonuclear diatomic molecule a molecule consisting of two atoms of the same element.

hormone an organic compound, produced by the body, that regulates a variety of chemical processes.

Hund's rule within a given energy sublevel each orbital is occupied by one electron before any orbital has two, and the single electrons in different orbitals within a sublevel have parallel spins.

hybridization the promotion of electrons to higher energy levels and the mixing of these levels to form a new set of equivalent atomic orbitals.

hybrid orbital an atomic orbital formed by hybridization.

hydrated ion an ion to which one or more water molecules are attached.

hydrocarbon a compound containing only C and H atoms.

hydrogenation the addition of H atoms to an unsaturated compound.

hydrogen bonding a relatively strong intermolecular or intramolecular attractive force due to the attraction of two strongly electronegative atoms for an H atom covalently bonded to one of them.

hydrolysis any chemical reaction of a substance with water.

hygroscopic substance a substance that attracts water vapor.

ideal gas a gas that obeys the gas laws exactly.

ideal gas constant (R) the proportionality constant in the ideal gas law equation (8.3144 J · K^{-1} · mol^{-1} or 0.082057 L · atm · K^{-1} · mol^{-1}).

ideal gas law a statement of the relations among the pressure, volume, number of moles, and absolute temperature of an ideal gas that combines four basic gas laws; $PV = nRT$.

ideal solution a solution that obeys the colligative property law or Raoult's law exactly.

immiscible liquids liquids that are mutually insoluble.

inner transition elements the lanthanide and actinide series of elements.

inorganic compound any compound that is not an organic compound.

insoluble substance a substance with a very low solubility.

instantaneous rate the rate of a reaction at a given instant during its course (mol · L^{-1} · time^{-1}).

interhalogen compound a compound of two halogens.

intermediate compound a compound that is formed in a reaction but that reacts further under the existing conditions.

internal energy (E) the total energy of a chemical system (kJ/mol).

intrinsic property a property that is characteristic of any sample of a substance, regardless of its size, shape, or condition.

ion a charged particle in which the number of electrons is not equal to the number of protons.

ion exchange a method for the replacement of one kind of ion in a solution by another kind.

ionic bond the electrostatic force that binds oppositely charged ions together.

ionic compound a compound consisting of ions held together by ionic bonds.

ionic equation a chemical equation that shows the individual ions taking part in a reaction.

ionic radius the part of the bond length between two bound ions that is attributed to a particular ion (Å).

ionization energy the energy required to knock an electron off an atom or an ion (eV).

ionization reaction a reaction in which ions are produced.

ionizing radiation radioactive emissions and other high-energy radiation that ionize matter in their paths.

ion product for water (K_w) the equilibrium constant for the ionization of water, $HOH \rightleftharpoons H^+ + OH^-$, treating $[H_2O]$ as a constant; $K_w = [H^+][OH^-]$.

isoelectronic species different species (atoms, ions, or molecules) that have the same number of electrons.

isomerism the existence of two or more molecules or ions with identical composition but different structures.

isomers molecules or ions of identical composition but different structures.

isotopes atoms of the same element that differ in the number of neutrons in their nuclei.

Kelvin scale (K) an absolute temperature scale with 0 K at absolute zero and 273.15 K at the freezing point of water.

ketohexose a hexose containing a ketone group.

ketone a derivative of a hydrocarbon in which a double-bonded oxygen replaces two hydrogens at a position other than at the end of the carbon chain.

kinetic energy (K.E.) the energy possessed by a body by virtue of its motion (J); $K.E. = \frac{1}{2}mv^2$.

kinetic molecular theory the modern theory of the behavior of matter in terms of moving particles.

lanthanide series the series of 14 elements in period 6 of the periodic table in which the $4f$ orbitals are being filled.

lattice energy see *crystal energy*.

law of conservation of energy energy is neither created nor destroyed in any transformation of matter (also called *the first law of thermodynamics*).

law of conservation of mass mass is neither created nor destroyed in any transformation of matter.

law of definite composition a pure compound is always composed of the same elements combined in a definite proportion by weight.

law of multiple proportions when two elements combine to form more than one compound, the different weights of one that combine with a fixed weight of the other are in the ratio of small whole numbers.

law of the equivalence of mass and energy mass and energy are equivalent, interconvertible forms, related by the square of the speed of light; $E = mc^2$.

Le Chatelier's principle a system at equilibrium responds to an applied stress in such a way as to relieve the stress.

leveling effect the reaction of a solvent to reduce a number of different reagents to the same strength.

levorotatory molecule an optically active molecule that rotates the plane of polarization of plane-polarized light to the left.

Lewis acid a substance that acts as an electron-pair acceptor in a chemical reaction.

Lewis base a substance that acts as an electron-pair donor in a chemical reaction.

Lewis structural formula a formula in which all valence electrons are shown. A shared electron pair is shown with two dots or a line; an unshared pair is shown with two dots (also called an *electron-dot formula*).

ligand an ion or molecule that functions as a Lewis base in bonding to a metal ion or atom.

ligand field the electrostatic force exerted by the electrons and the net charge of a ligand on the central metal ion or atom in a transition metal complex.

ligand field theory the modern theory of the electronic structures of transition metal complexes, based on the interaction of the ligand field and the central metal ion or atom.

limiting reactant the substance that reacts completely in a reaction, thereby leaving the other reactants in excess.

London forces weak intermolecular attractive forces due to instantaneous induced dipoles.

lone pair a pair of valence electrons not participating in a chemical bond (also called *unshared pair*).

manometer an instrument used to measure small differences in pressure.

mass a property reflecting the amount of matter in a body (g).

mass loss the difference between the calculated sum of the masses of the nucleons and electrons of an atom and the experimentally determined mass of that atom (amu).

mass loss per nucleon the mass loss of an atom divided by the number of its nucleons (amu/nucleon).

mass number (A) the sum of the numbers of protons and neutrons (the number of nucleons) in the nucleus of an atom.

mass spectrograph an instrument used to measure the masses of atoms and molecules.

melting point the temperature at which the solid and liquid phases of a substance are in equilibrium (also called *freezing point*).

metal an element that is lustrous, malleable, ductile, and a good conductor of heat and electricity.

metallic bond the electrostatic force that binds the positive ions of a metal lattice together by means of a "sea" of valence electrons.

metallic radius half the internuclear distance between two nearest neighbors in the metal lattice (Å).

metallo-acid elements the 12 elements in groups IV_B

through VIIʙ of the periodic table, which form metallic oxides that are acidic.

metalloid   an element that is intermediate between metals and nonmetals in its properties.

metallurgy   the study of the production, properties, and uses of metals and alloys.

mineral   any natural inorganic material found in the earth's crust.

mixture   an intimate combination of two or more substances that retain their chemical identities.

molality (*m*)   a unit of measure for solution concentration: mol solute/kg solvent.

molar gas volume   the volume of 1 mole of any gas at STP (ideally 22.414 L/mol).

molar heat capacity   the amount of heat energy required to change the temperature of 1 mole of a substance by 1 K [J/(mol · K)].

molar heat of condensation   the amount of heat released when 1 mole of a vapor condenses at a given temperature (usually the normal boiling point of the liquid) (kJ/mol).

molar heat of fusion   the amount of heat required to melt 1 mole of a solid at its melting point (kJ/mol).

molar heat of solidification   the amount of heat released when 1 mole of a liquid freezes (kJ/mol).

molar heat of vaporization   the amount of heat required to vaporize 1 mole of a liquid at a given temperature (usually its normal boiling point) (kJ/mol).

molarity (*M*)   a unit of measure for solution concentration: mol solute/L solution.

molar weight   the weight of 1 mole of a substance (g).

mole   the amount of a substance containing the same number of particles as there are in 0.012 kilogram of carbon-12 (that is, 6.0220×10^{23} particles); abbreviated mol.

molecular formula   a formula that shows the actual numbers of atoms of each kind in the molecule.

molecular orbital   a wave function for an electron in a molecule.

molecular orbital (MO) theory   one of the two major theories of the chemical bond, in which bonds arise from the successive filling of molecular orbitals that are characteristic of the molecule as a whole.

molecular weight   the sum of the atomic weights of the atoms constituting a molecule or the smallest unit of a compound.

molecule   the smallest particle of a substance that has the characteristics of the substance.

mole fraction   the ratio of the number of moles of one component of a mixture to the total number of moles of all the components.

mole percent   the mole fraction times 100.

monochromatic radiation   radiation of a single wavelength.

monomer   a small molecule that forms a repeated structural unit in a polymer.

net ionic equation   an ionic equation that shows only the species actually involved in the reaction.

neutralization reaction   a reaction of chemically equivalent amounts of an acid and a base.

neutral solution   a solution in which the concentrations of H^+ and OH^- are equal.

neutron   a subatomic particle with no charge and a mass of about 1 amu.

neutron activation analysis   a method for the identification of a substance by measuring the emissions from nuclides formed by the neutron bombardment of a sample.

90 percent contour   the boundary for a region in an atom within which the probability of finding a given electron is 90%.

nitrogen family   the elements of group Vᴀ of the periodic table.

nitrogen fixation   any process by which N_2 reacts to form a compound.

noble gases   the elements of group VIIIᴀ of the periodic table.

noble metals   silver, gold, and the platinum-family metals; they are relatively inactive chemically.

nonbonding electron pair   see *lone pair*.

nonelectrolyte   a compound that does not conduct electricity to any significant degree in its molten state or in solution.

nonmetal   an element that does not possess the characteristics of a metal to any significant degree.

normal boiling point and **normal freezing point**   the boiling point or freezing point of a substance at a pressure of one atmosphere.

normality (*N*)   a unit of measure for solution concentration: equiv solute/L solution.

nuclear chemistry   the study of natural and artificially induced nuclear reactions and of the chemical reactions of radioactive substances.

nuclear energy   the energy associated with nuclear reactions (also called *atomic energy*).

nuclear fission   the splitting of a heavy nucleus to yield two lighter nuclei.

nuclear fusion   the combining of two light nuclei to form one or more new nuclei.

nuclear magnetic resonance   the low-energy interaction of the magnetic moments of certain nuclei with a strong magnetic field and a radiofrequency field.

nuclear reaction a reaction in which the composition of an atomic nucleus is changed.

nuclear reactor an installation for carrying out a controlled nuclear fission reaction, usually for power generation; also, a fusion reaction device.

nucleic acid a polymer consisting of one or two polynucleotide chains.

nucleon a proton or neutron when it exists in an atomic nucleus.

nucleus the tiny, massive, positively charged center of an atom.

nuclide an atom that is distinguished from others by the numbers of protons and neutrons it contains.

octet rule see *rule of eight*.

oil a liquid ester of glycerol and three fatty acids; or any organic liquid not soluble in water, such as a liquid hydrocarbon.

optical activity the rotation of the plane of polarization of plane-polarized light by chiral molecules.

optical isomers see *enantiomers*.

orbital see *wave function*.

order of a reaction the sum of the exponents of the concentration terms in the rate equation for the reaction.

ore a mineral that serves as a source of useful material.

organic compound any compound of carbon, with a few exceptions that have traditionally been classed as inorganic.

osmosis the flow of a solvent from a solution of lower concentration to one of higher concentration through a semipermeable membrane.

osmotic pressure the pressure required to counteract the flow of solvent (at 1 atm) into a solution through a semipermeable membrane; $\pi = nRT/V$.

oxidation a reaction in which the oxidation state of a substance is increased.

oxidation number a small number related to the degree of positive or negative character of an atom in a molecule or polyatomic ion, or to the charge on a monatomic ion (also called **oxidation state**). These numbers are usually whole numbers, but in rare cases they can be fractions.

oxidizing agent a substance that causes the oxidation of another substance.

paramagnetic substance a substance that is slightly attracted by a magnetic field.

parent compound an organic compound that can be regarded as the base compound for a number of derivatives.

Pauli exclusion principle no two electrons within the same atom can have the same four quantum numbers.

pentosan a polysaccharide consisting of pentose monomers.

pentose a monosaccharide containing five C atoms.

peptide bond the bond that links the amino acid residues in a protein molecule.

percentage yield the percentage of the theoretical yield of a reaction product that is actually obtained.

period one of seven horizontal divisions in the periodic table, each of the first six ending with a noble gas.

periodic law the chemical and physical properties of the elements are periodic functions of their atomic numbers.

periodic table a tabular classification of the elements by atomic number.

pH the negative logarithm of the H^+ concentration in a solution. $pH = -\log[H^+]$.

phase diagram a diagram showing the temperature-pressure relations among the three phases of a substance.

phenol an organic compound in which an —OH group is attached to a benzene ring.

photochemical reaction a reaction initiated by radiant energy.

photoelectric effect the ejection of electrons from a surface by radiant energy.

photon a quantum of radiant energy.

photosynthesis a complex series of redox reactions by which green plants use sunlight to convert carbon dioxide and water to carbohydrate and oxygen.

physical change a process in which the form or certain properties of a substance change, but not its chemical identity.

physical property a property of a substance that distinguishes it from other substances and does not cause chemical change.

pi bond a bond due to electrons in a pi molecular orbital; it is not symmetric about the internuclear axis.

pi molecular orbital a molecular orbital formed by two atomic p orbitals that overlap alongside the internuclear axis.

plane-polarized light light consisting of waves that oscillate in only one plane.

plasma a gas that is partially or completely ionized.

pOH the negative logarithm of the OH^- concentration in a solution. $pOH = -\log[OH^-]$.

polar covalent bond a covalent bond between two unlike atoms, with the shared electrons attracted more toward one of the atoms.

polar molecule a molecule in which one part has a partial positive charge and another part has a partial negative charge.

polyatomic ion an ion containing more than one atom.

polymer a giant molecule composed of one or more chains of small molecules (monomers) that may be identical or of a few similar kinds.

polynucleotide chain a polymer consisting of many nucleotide monomers; the substance of nucleic acids.

polyprotic acid an acid that can donate more than one proton.

polysaccharide a polymer consisting of many saccharide monomers.

potential energy the energy possessed by a body by virtue of its position or its existence in a state other than its normal state of lowest energy (J).

precipitate an insoluble solid that is formed by a chemical reaction in solution.

precipitation reaction a reaction in solution in which an insoluble substance is formed.

principal quantum number (n) the number that specifies the main energy level occupied by an electron in an atom.

probable electron density (Ψ^2) the probability of finding an electron within a certain small volume of space in an atom.

product a substance that is formed in a chemical reaction.

protective colloid a colloid that acts as a stabilizing agent for another colloid.

protein a polymer consisting of many amino acid monomers.

proton (p^+) a subatomic particle with unit positive charge and a mass of about 1 amu.

qualitative analysis the chemical analysis of a material to determine what components it contains.

quantitative analysis the chemical analysis of a material to determine how much of certain components it contains.

quantum a discrete unit of energy, associated with the discontinuous absorption or emission of energy by some form of matter.

quantum mechanics the modern theory of the structure and properties of matter in terms of wave functions for both matter and radiation (also called *wave mechanics*).

radiant energy the energy associated with any form of electromagnetic radiation (J).

radical an atom or group of atoms that has one or more unpaired electrons.

radioactive decay the spontaneous change of an unstable

nuclide to another nuclide by the emission of subatomic particles or gamma rays or both.

radioactive series a group of radioactive nuclides that begins with a parent nuclide and culminates in a stable nuclide.

radius ratio the ratio of the radii of the cation and anion in an ionic crystal. Radius ratio = r_+/r_-.

Raoult's law each component of a solution exerts a vapor pressure equal to its mole fraction in the solution times its vapor pressure as a pure substance. $p_A = \chi_A p_A^\circ$.

rate constant (k) the proportionality constant between the rate of reaction and the concentration term(s) in the rate equation (many different units).

rate-determining step the slowest step in a reaction mechanism.

rate equation the equation that relates the rate of a reaction to the concentration(s) of the reactant(s).

rate of a reaction the change in concentration of a reactant or product in unit time (mol $\cdot$ L^{-1} $\cdot$ time).

reactant a substance that is consumed in a chemical reaction.

reaction intermediate typically an unstable species that is formed in a reaction and that reacts further.

reaction mechanism the stepwise changes (elementary reactions) that take place in the conversion of reactants to products.

recrystallization the process by which a solute is purified by successive crystallizations from a solvent.

redox reaction a reaction in which there is a change in the oxidation state of at least two substances, one of which is oxidized and the other of which is reduced.

reducing agent a substance that causes the reduction of another substance.

reduction a reaction in which the oxidation state of a substance is decreased.

refining the purification or alteration of the composition of crude metals, petroleum, and other raw materials.

relative atomic weight the average weight of the atoms of an element relative to an arbitrary standard (amu).

residue the material left behind when other components of a mixture are removed.

resonance energy the amount by which the energy of a resonance hybrid is less than that of its resonance structures (kJ/mol).

resonance hybrid the assumed actual structure of a molecule or polyatomic ion for which different resonance structures can be written.

resonance structures two or more hypothetical Lewis

structures that can be written for a given molecule or polyatomic ion, differing only in the arrangement of electrons.

reverse osmosis the flow through a semipermeable membrane of a solvent from a solution of higher concentration to one of lower concentration caused by the application of external pressure.

reversible reaction a reaction in which the products can react to form the original reactants.

ribonucleic acid (RNA) any of several nucleic acids, three of which are involved in protein synthesis in cells.

rule of eight the tendency for an atom to attain an electronic structure that is isoelectronic with a noble gas other than helium.

rule of two the tendency for an atom to attain an electronic stucture that is isoelectronic with helium.

saccharide any simple or complex sugar.

salt bridge a device used in some electrochemical cells for completing the internal circuit by providing for the diffusion of ions.

salt an ionic compound.

saponification a reaction that produces soap and glycerol by the basic hydrolysis of a fat or oil.

saturated hydrocarbon a hydrocarbon containing only single bonds.

saturated solution a solution containing the amount of solute required for an equilibrium to exist between dissolved and undissolved solute.

screening effect the reduction of the effective nuclear charge in an atom or ion due to the repulsion between an outer electron and all the other electrons.

second law of thermodynamics the free energy of a system decreases in any spontaneous change; or, the entropy always attains the maximum possible value permitted by the energy of a system.

second-order reaction a reaction for which the sum of the exponents of the concentration terms in the rate equation is two.

second quantum number (l) the number that specifies the energy sublevel occupied by an electron in an atom.

selective dissolution the separation of two substances by dissolving one but not the other.

selective precipitation the separation of two dissolved substances by precipitating one but not the other.

semipermeable membrane a membrane that permits the passage of only certain types of molecules or ions.

sigma bond a bond resulting from electrons in a sigma molecular orbital, symmetric about the internuclear axis.

sigma molecular orbital a molecular orbital formed by two atomic orbitals that overlap end to end along the internuclear axis.

silicone any of a group of chainlike molecules of Si, O, C, and H atoms.

single bond a covalent bond consisting of one pair of electrons.

single displacement reaction a reaction of an element with a compound to produce a different element and compound.

SI units the International System of units of measure.

slaking the reaction of a metal oxide with water.

smelting the chemical reduction of an ore.

solar cell a photoelectric device for converting sunlight directly to electricity.

solubility the amount of a substance that dissolves in a given amount of a solvent to produce a saturated solution (g solute/100 cm^3 solvent).

solubility product constant (K_{sp}) the equilibrium constant for the dissolution of a slightly soluble salt, generally expressed without units; for $A_m B_n \rightleftharpoons m A^{n+} + n B^{m-}$, $K_{sp} = [A^{n+}]^m [B^{m-}]^n$.

solute usually the minor component of a solution, dissolved in the solvent.

solution a homogeneous mixture of two or more substances.

solvation the interaction of solvent molecules with solute particles to form small aggregates.

solvent usually the major component of a solution, in which the solute is dissolved.

solvent extraction the process by which a solute dissolved in one solvent is extracted into another solvent that is immiscible with the first.

specific gravity the ratio of the mass of a substance to the mass of an equal volume of water at a specified temperature.

specific heat the amount of heat energy required to change the temperature of 1 g of a substance by 1 °C or 1 K [J/(g · K)].

spectator ion an ion present in solution that does not take part in a given reaction.

spectrochemical series the ranking of a series of ligands according to their effect on the wavelengths of radiation absorbed by complex ions of a given transition metal.

spectroscope an instrument used to form an optical spectrum of a substance.

spectroscopy the study of the spectra of substances.

spectrum originally the pattern of colors produced when light is dispersed by a prism or a grating; more generally, the pattern produced when a beam of energies or particles is separated into components.

splitting of energy levels in a transition metal complex, the changing of the energies of atomic orbitals of equal energy into different energies by the ligand field.

stability constant (K_f) the equilibrium constant for the formation of a complex ion (also called *formation constant*); $K_f = (K_{f1})(K_{f2}) \cdots (K_{fn})$.

standard basic reduction potential ($\mathscr{E}_B°$) a standard reduction potential relative to a hydrogen electrode immersed in a solution that is $1m$ in OH^- ion (V).

standard electrode an electrode operating at the conventional conditions of 25 °C, 1 atm, and $1m$ with respect to the given ion.

standard enthalpy of formation ($\Delta H_f°$) the enthalpy change for the reaction by which 1 mole of a substance is formed from its elements in their standard states (kJ/mol).

standard free energy of formation ($\Delta G_f°$) the Gibbs free energy change for the reaction by which 1 mole of a substance is formed from its elements in their standard states (kJ/mol).

standard reduction potential ($\mathscr{E}_{red}°$) the electrode potential for a reduction half-reaction occurring at a standard electrode combined with a standard hydrogen electrode to make an electrochemical cell (V).

standard state a specified physical form of a substance; for a solid or liquid it is the pure substance at 1 atm, and for a pure gas it is the hypothetical ideal gas at 1 atm.

standard temperature and pressure (STP) 0 °C and 1 atm.

state function a function whose value depends only on the state of a substance, not on the path by which that state was attained.

stationary state see *energy level.*

stereoisomers isomers that differ in the spatial arrangement of the atoms in the molecule.

stoichiometry the quantitative relations among the reactants and products in a balanced chemical equation.

strong electrolyte a substance that is in the form of ions, completely or nearly completely, in solution.

structural isomers isomers that differ in the sequence of atoms bonded in the molecule.

substitution reaction an organic reaction in which one atom or group of atoms replaces another.

sulfur family the elements of group VIA of the periodic table, excluding oxygen.

supersaturated solution a solution that is more concentrated than a saturated solution.

surface tension the force that tends to make a liquid surface assume a curved shape.

temperature the property of a body that determines the direction of the spontaneous flow of heat.

ternary compound a compound composed of three elements.

theoretical chemistry the fundamental principles underlying all chemical phenomena, and the explanations of chemical changes.

theoretical yield the amount of a product calculated to be obtainable if a reaction goes to completion.

thermochemical equation a chemical equation that shows the heat change for a reaction involving specified physical states of reactants and products.

thermochemistry the measurement and interpretation of the heat changes accompanying chemical processes.

thermodynamics the study of the relations between heat and all other forms of energy.

thermonuclear reaction a nuclear fusion reaction brought about by extremely high temperatures.

third law of thermodynamics a perfect crystal at absolute zero would have zero entropy.

third quantum number (m_l) the number that specifies the particular orbital occupied by an electron in a given energy sublevel of an atom.

threshold frequency (ν_0) the minimum frequency of radiation required to produce the photoelectric effect on a given metal surface (Hz).

titration the method for determining the concentration of a solution by measuring the volume of a solution of known concentration with which it reacts completely.

titration curve in an acid–base reaction, a plot of the pH or pOH versus the volume of reactant added.

tracer a radioactive isotope used to follow the course of a chemical reaction or physical process.

transistor an electronic device made of slightly impure silicon or germanium.

transition elements all the B-family elements in the periodic table, including the actinide and lanthanide elements.

transition state the state that exists at the instant of a potentially effective collision between reacting species (also called *activated complex*).

transition temperature the temperature at which one physical form of a substance changes to another.

triple bond a covalent bond consisting of three pairs of electrons.

triple point the temperature and pressure at which all three phases of a substance are in equilibrium.

turnover number the number of moles of reactant transformed per mole of enzyme per minute at a specified temperature (min^{-1}).

Tyndall effect the scattering of light by a colloidal dispersion.

unit cell the smallest conceptual building block in a crystal lattice.

unsaturated hydrocarbon a hydrocarbon containing one or more multiple bonds.

unshared pair see *lone pair*.

valence the chemical combining capacity of an atom or ion.

valence bond (VB) theory one of the two major theories of the chemical bond, in which bonds arise from the localized overlap of atomic orbitals of the atoms in question.

valence electron an electron in one of the outermost energy levels of an atom, capable of participating in the formation of a chemical bond.

valence shell electron pair repulsion (VSEPR) theory a theoretical model that accounts for the geometries of many simple molecules on the premise that pairs of valence shell electrons tend to be as far from each other as possible.

van der Waals attractive forces the weak intermolecular forces that allow the condensed phases of covalent substances to exist.

van der Waals radius the part of the nonbonded distance between two touching atoms in adjacent molecules that is attributed to the atom in question (Å).

van't Hoff factor (*i*) for a given electrolyte at a given solution concentration, the numerical factor for the amount by which a given colligative property effect is greater than that produced by a nonelectrolyte at the same concentration.

van't Hoff's law an increase in the temperature of a system at equilibrium shifts the equilibrium in the direction that absorbs heat.

vapor the gaseous state of a substance under conditions at which the substance normally exists primarily as a liquid or solid.

vapor pressure the pressure exerted by a vapor when it is in equilibrium with the liquid or solid phase (atm).

virus a self-replicating infectious agent that consists of protein and nucleic acid.

voltaic cell an electrochemical cell in which a spontaneous redox reaction generates an electric current that flows through an external circuit.

volumetric analysis any analytical method based on the measurement of the volumes of reacting substances.

wave function (ψ) a mathematical function that describes a region (orbital) in an atom or molecule within which a given electron is most likely to be found.

wavelength (λ) the distance between identical points on adjacent cycles of a wave (nm); $\lambda = c/\nu$.

wave mechanics see *quantum mechanics*.

wax an ester of a long-chain carboxylic acid and a long-chain alcohol.

weak electrolyte a substance that ionizes only to a small degree in solution.

weight the gravitational force exerted on a body by the earth; this term is often used loosely to denote mass.

work function (W_0) the amount of energy required in the photoelectric effect to eject an electron from an atom at a metal surface (eV).

X ray a form of high-energy electromagnetic radiation.

zeolite any of a group of porous hydrated aluminum silicate minerals.

Note: Units within parentheses, following certain definitions, are commonly used for expressing the quantity.

Answers to Exercises

Chapter 1

1. (a) 3.81×10^8; (b) 1×10^{-4}; (c) 7.63×10^{-4}; (d) 1×10^{-3}; (e) 1×10^{-5}. **2.** (a) 7.2; (b) 5×10^1; (c) 1.0; (d) 6×10^4. **3.** A^xB^y; A^xB^{-y}; 1; A; $\sqrt[n]{A}$. **4.** 503. **5.** 65. **6.** $+\frac{1}{2}$, -2. **7.** (a) 2.9012×10^4; (b) 2.90×10^4. **9.** (a) 1.00×10^4 cm, 109 yd; (b) 77 yd, 7.0×10^{-2} km; (c) 5.0×10^3 m², 6.0×10^3 yd²; Liter Bowl; 10%. **11.** 9.4634×10^{-4} m³. **12.** 1.084×10^{12} km³. **13.** 4.7318×10^3 mL; 4.7318×10^3 cm³; 4.7318 L. **14.** 12.0 km $\cdot$ L^{-1}. **16.** (a) 29 ft $\cdot$ s^{-1}; (b) 32 km $\cdot$ hr^{-1}. **17.** (a) 275 chrons; (b) 2.1×10^3 m $\cdot$ chron^{-1}. **18.** 230 cents/kg; 100 cents/lb. **19.** 3.23%; 96.8%. **20.** 1.0206 kg; 2.06%. **21.** 1.0432 lb; 4.32%. **22.** 2.3 g $\cdot$ cm^{-3}; **23.** 21.5503 g $\cdot$ cm^{-3}. **24.** 110 lb. **25.** m $\cdot$ s^{-2}; kg $\cdot$ m $\cdot$ s^{-2}; kg $\cdot$ m^{-3}. **26.** 32 °C, 305 K; 37.0 °C, 310 K; 22 °C, 295 K; -1 °C, 272 K; -23 °C, 250 K. **29.** 18 °C. **30.** (a) 14 °C; (b) 70 °P. **33.** (a) 4.86 J $\cdot$ g^{-1}; 1.78 J $\cdot$ g^{-1}; 0.90 J $\cdot$ g^{-1}; 2.43 J $\cdot$ g$^{-1} \cdot$ °C^{-1}; 0.89 J $\cdot$ g$^{-1} \cdot$ °C^{-1}; 0.45 J $\cdot$ g$^{-1} \cdot$ °C^{-1}; (b) J $\cdot$ g$^{-1} \cdot$ °C^{-1}. **34.** 105 J; 22.5 °C. **35.** 0.35 J $\cdot$ g$^{-1} \cdot$ °C^{-1}. **36.** 83 g. **37.** 61.4 °C. **38.** 1054.4 J. **39.** 19 g. **40.** (b) 460 kJ; (c) 92 °F. **42.** (a) Compound; (b) yes, no; (c) no. **43.** 2 kg. **44.** (a) 2.2×10^5 J; 4.4×10^5 J; 9.0×10^5 J; (b) 5.2×10^3 J. **46.** (a) 0.21 g H/g N; 0.14 g H/g N; (b) $\frac{2}{3}$. **48.** $\frac{16}{12} = 1.33/1.00$; $2(16)/12 = 2.67/1.00$.

Chapter 2

4. (b) P_4; P. **5.** H_2O_2, hydrogen peroxide. **6.** (a) NH_2; (b) CH; (c) BH_3; (d) C_5H_4; (e) C_5H_{12}; (f) HCO_2; (g) CH_3O; (h) C_3H_2Cl. **8.** (a) 2, 1, 2; (b) 2, 3, 6, 1; (c) 1, 3, 1; (d) 1, 2, 1, 2; (e) 1, 2, 1, 4; (f) 1, 1, 1, 1; (g) 4, 3, 2; (h) 1, 1, 1, 1; (i) 1, 5, 3, 4; (j) 2, 7, 4, 6. **9.** (a) $2Na + Cl_2 \rightarrow 2NaCl$; (b) $2H_2 + O_2 \rightarrow 2H_2O$; (c) $N_2 + 3H_2 \rightarrow 2NH_3$; (d) $H_2 + Cl_2 \rightarrow 2HCl$. **10.** (a) $H_2 + Br_2 \rightarrow 2HBr$; (b) $Si + O_2 \rightarrow SiO_2$; (c) $N_2 + O_2 \rightarrow 2NO$; (d) $4P + 5O_2 \rightarrow P_4O_{10}$; (e) $P_4 + 5O_2 \rightarrow P_4O_{10}$; (f) $2Li + F_2 \rightarrow 2LiF$. **11.** (a) $H_2 + Cl_2 \rightarrow 2HCl$; (b) $C + O_2 \rightarrow CO_2$; (c) $2K + Br_2 \rightarrow 2KBr$; (d) $3Ca + N_2 \rightarrow Ca_3N_2$. **12.** (a) $Ba + H_2 \rightarrow BaH_2$; (b) $3Mg + N_2 \rightarrow Mg_3N_2$; (c) $2NH_4Cl + Pb(NO_3)_2 \rightarrow 2NH_4NO_3 + PbCl_2$; (d) $PBr_3 + 3H_2O \rightarrow 3HBr + H_3PO_3$; (e) $3HCl + Al(OH)_3 \rightarrow AlCl_3 + 3H_2O$. **13.** (a) $2C_8H_{18} + 25O_2 \rightarrow 16CO_2 + 18H_2O$; (b) $C_4H_{10}O + 6O_2 \rightarrow 4CO_2 + 5H_2O$; (c) $C_6H_{12}O_6 + 6O_2 \rightarrow 6CO_2 + 6H_2O$.

14. (a) $H_2 + Ag_2O \rightarrow H_2O + 2Ag$; (b) $Mg + 2HCl \rightarrow MgCl_2 + H_2$; (c) $3Zn + 2Fe(NO_3)_3 \rightarrow 3Zn(NO_3)_2 + 2Fe$; (d) $Cl_2 + 2NaI \rightarrow I_2 + 2NaCl$. **15.** (a) $2NaOH + H_2SO_4 \rightarrow Na_2SO_4 + 2H_2O$; (b) $Zn(NO_3)_2 + H_2S \rightarrow ZnS + 2HNO_3$; (c) $MgCl_2 + Ba(OH)_2 \rightarrow BaCl_2 + Mg(OH)_2$; (d) $ZnSO_4 + H_2S \rightarrow ZnS + H_2SO_4$. **17.** 79.904 amu; 159.81 amu; 342.30 amu. **18.** No, a mole refers to a specific number of particles whose mass does not change; 0.032 kg. **19.** 20.179 g; 70.906 g; 163.9408 g; 92.094 g. **20.** 0.341; 0.166; 0.333; 7.43. **21.** (a) $2Sc + 6HBr \rightarrow 2ScBr_3 + 3H_2$; (b) 0.75 mol; (c) 3.01×10^{23}. **22.** (a) $N_2 + 3H_2 \rightarrow 2NH_3$; (b) 3.00 mol; (c) 51.1 g; (d) 9.07 g. **23.** 123 g. **24.** 270 g. **25.** 10.2 g. **26.** (a) NaOH; (b) 146 g; (c) 2.52 g; (d) 45.0 g. **27.** 10.7 g. **28.** 22 kg. **29.** 1.0 kg. **30.** 1.38 g; N_2, 0.62 g. **31.** 89%. **32.** (a) 7.52 g; (b) 7.45 g Fe. **33.** 88.0 g. **34.** (a) 37.9 g; (b) 60.7%; (c) 100%. **35.** 1.21 g. **37.** (a) C, 52.1%; H, 13.1%; O, 34.7%; (b) 18.50%. **38.** C. **39.** 0.6782 g. **40.** 5.79%. **41.** 4.9%. **42.** Theor., 42.107% C; error, 0.52%; H_2O. **43.** 0.1211 g RbCl; 0.1169 g NaCl. **45.** (a) 91.249% C, 8.751% H; (b) C_7H_8. **46.** (a) CH_2; (b) C_3H_6O. **47.** (a) NH_2; (b) N_3H_6. **48.** N_2O_4. **49.** $C_{12}H_{22}O_{11}$. **50.** $C_3H_7SNO_2$. **51.** (a) P_2O_3; (b) P_4O_6; 219.892 amu.

Chapter 3

4. 9.1096×10^{-28} g; 5.4858×10^{-4} amu; mass H/mass $e^- = 1837.3$. **6.** 1.33×10^{-12} J; 4.47×10^7 mi $\cdot$ hr^{-1}. **7.** 5.3033×10^{-20} C. **16.** (a) No atom has mass 35.453 amu; (b) 2.657×10^{-23} g; 5.857×10^{-26} lb. **17.** (b) 21.0424 amu; 3.4941×10^{-23} g. **18.** (a) $^3_1$H; (b) $^{210}_{82}$Pb; (c) $^{208}_{82}$Pb; (d) $^{34}_{16}$S. **19.** (a) 1, 1, 2; (b) 82, 82, 128; (c) 82, 82, 126; (d) 16, 16, 18. **20.** 10.81 amu. **21.** 62.9296 amu, 69.157%; 64.9278 amu, 30.843%. **22.** $^{51}_{23}$V and $^{50}_{23}$V. **23.** ^{79}Br, $35p^+$, $35e^-$, $44n$; ^{81}Br, $35p^+$, $35e^-$, $46n$. **25.** 780.0 V. **26.** 14.00 amu; nitrogen. **27.** 1 g. **28.** (a) 0.1 nm to 0.5 nm; (b) 10^2 Å, 10 nm; 280 Å, 28 nm; 1 Å, 0.1 nm; 10^{-5} Å, 10^{-6} nm. **29.** 6.6446×10^{-24} g; yes, by the mass of $2e^-$. **30.** 8.5×10^3. **31.** 1×10^{15} g/cm³. **35.** (a) $3e^-$ in outer energy level; IIIA; (b) $1e^-$ in outer energy level; IA. **40.** (a) 18; (b) 10; (c) 2. **41.** (a) K; (b) Kr; (c) Be; (d) S; (e) Fe. **42.** (a) $1s^22s^22p^63s^23p^3$; (b) $1s^22s^22p^63s^23p^64s^23d^5$; (c) $1s^22s^22p^63s^23p^64s^23d^{10}4p^5$; (d) $1s^22s^22p^63s^23p^64s^23d^{10}4p^65s^24d^{10}5p^66s^24f^5$. **43.** (a) [Ne] $3s^23p^3$; (b) [Ar] $3d^54s^2$; (c) [Ar] $3d^{10}4s^24p^5$; (d) [Xe] $4f^56s^2$. **44.** [Kr] $4d^{10}$. **45.** (a) 32; (b) 25; (c) $6s^2$. **46.** 19 sublevels; $8e^-$. **47.** [118] $8s^2$; family IIA. **49.** H_2Te; H_2S. **50.** NaCl, KBr, LiCl, CsF. **51.** (a) 5, 4; (b) 19.

Chapter 4

1. 5.0×10^6 m. **2.** No; 0.122 m. **3.** 3.0×10^{17} Hz. **6.** All will reach earth at the same time; **4.3** years; no. **11.** 4.136×10^{-15} eV · S · photon^{-1}; 1.584×10^{-37} kcal · s · photon^{-1}. **12.** 4.087×10^{-19} J · photon^{-1}. **13.** 4.0×10^{-32} J · photon^{-1}; 1.623×10^{-24} J · photon^{-1}; 3.64×10^{-28} J · photon^{-1} and 1.060×10^{-27} J · photon^{-1}. **15.** (a) 6.6 eV · photon^{-1}; (b) 2.1 eV; (c) 5.1×10^{14} s^{-1}. **16.** Yes; 1.3 eV. **17.** 1.876×10^{-6} m; 1.282×10^{-6} m; 1.094×10^{-6} m; infrared. **18.** Brackett lines: 7.399×10^{13} Hz, 4052 nm; 1.142×10^{14} Hz, 2626 nm; Pfund line: 4.019×10^{13} Hz, 7460 nm; infrared. **21.** 2.410×10^{-19} J · atom^{-1}; 1.451×10^5 J · mol^{-1}. **22.** (a) $(2^2)(13.54$ eV); (b) $(3^2)(13.54$ eV); (c) $(4^2)(13.54$ eV); 338.5 eV. **23.** 1.1×10^{-34} m. **24.** 2.8×10^{-15} m. **25.** 4.85×10^{-12} m. **29.** (a) No; (b) 4.76 Å, 8.46 Å. **30.** (a) 0, 1, 2, 3; (b) 0, ± 1, ± 2, ± 3; $2(l + 1)$. **31.** 16; 3; 5; 7; 2; 2. **40.** 5. **41.** 2.68×10^{17} Hz. **42.** 1s, 3,206 eV; 2s, 326.5 eV; 2p, 249.5 eV; 3s, 29.5 eV; 3p, 16 eV.

Chapter 5

1. Metals: Hf, Pd; metalloids: Sb, Si, As; nonmetals: Rn, S, Br, N. **2.** Bi, 3; S, 2; Ag, 1; O, 2; Sn, 4. **3.** No. **11.** Cations, VO^{2+}, Co^{3+}; anions, NH_2^-, ClO^-. **12.** Sb^{3+}, $1s^2 2s^2 2p^6 3s^2 3p^6 4s^2 3d^{10} 4p^6 5s^2 4d^{10}$; no. **13.** (a) Rb^+, $1s^2 2s^2 2p^6 3s^2 3p^6 4s^2 3d^{10} 4p^6$; Se^{2-}, $1s^2 2s^2 2p^6 3s^2 3p^6 4s^2 3d^{10} 4p^6$; Ga^{3+}, $1s^2 2s^2 2p^6 3s^2 3p^6 3d^{10}$; Zn^{2+}, $1s^2 2s^2 2p^6 3s^2 3p^6 3d^{10}$; (b) Rb^+, Se^{2-}; (c) Ga^{3+}, Zn^{2+}. **14.** Rb^{2+}, S^{3-}, Ba^{3+}. **15.** 54, 10, 42, 36. **18.** SiH_3; $HeCl_2$. **20.** N can apparently have no more than 8 valence e^-. **30.** (a) Se, S, Cl, O; (b) Cs, Rb, Ca, Mg; (c) Ga, Ge, As, S, Cl. **32.** (a) N, same; (b) 0.9, 0; (c) smaller, because predicted bond moment is zero. **33.** More ionic, RbF; others more covalent. **34.** HCl, 1.54×10^{-19} J; HBr, 0.85×10^{-19} J. **35.** (a) N, F; (b) 0.9, 1.0; (c) about same, but opposite in direction; (d) that it contributes significantly. **37.** (a) All except O_2, Cl_2; (c) except SiH_4. **38.** (a) and (d). **39.** All but NCl_3. **40.** ClF. 0.88 D; BrF, 1.29 D; ICl, 0.62 D; BrCl, 0.57 D. **42.** (b) 0.69 D. **43.** (a) ClF, 1.70 Å; IBr, 2.47 Å; (b) ClF, 1.63 Å. **44.** (a) HF, 17%; HCl, 7.1%; HBr, 7.1%; HI, 5.6%. **45.** 2.67 Å, 0.267 nm. **46.** 3.03 Å. **47.** 2.4 Å. **48.** (a) 2.85 Å; (b) 97.0 Å^3, 0.0970 nm^3. **50.** (a) Cl, Br, Ti, Cu, K, Rb; (b) H^+, Na^+, Cl^-, Br^-; (c) Sr^{2+}, Rb^+, Cs^+, Sr, Rb, Cs. **54.** Hydrogen phosphate, hydrogen phosphite, hydrogen hypophosphite, sodium phosphate, sodium phosphite, sodium hypophosphite. **55.** Potassium manganite, potassium permanganate, hydrogen permanganate. **56.** NaBrO, $NaBrO_2$, $NaBrO_3$, $NaBrO_4$; $Ca(ClO)_2$, $Ca(ClO_2)_2$, $Ca(ClO_3)_2$, $Ca(ClO_4)_2$. **57.** Bismuth(III) chloride, bismuth(III) sulfide, silver(I) sulfide, silver(I) oxide, tin(IV) oxide. **58.** Copper(II) sulfate, iron(II) nitrate, nitrogen(I) oxide, nitrogen(II) oxide, nitrogen(IV) oxide, sulfur(III) oxide, manganese(IV) oxide, oxy-

gen(II) fluoride. **59.** Sodium perxenate, sodium xenate; sodium arsenite, sodium arsenate. **60.** No; $+5$ is highest oxidation state for N.

Chapter 6

1. Two. **6.** (a) Si, 2; N, 3; B, 1; (b) Si, 4; N, 1, 2, 3; B, 3. **8.** O, sp^3; C, sp^3. **9.** sp^3; tetrahedral for both. **10.** (a) SiH_4, tetrahedral, tetrahedral; AsH_3, tetrahedral, trigonal pyramidal; H_2Se, tetrahedral, angular. (b) SiH_4, same, same; AsH_3, trigonal pyramidal, trigonal pyramidal; H_2Se, lone pair perpendicular to plane of H—Se—H, angular (AsH_3 and H_2Se use only p orbitals). **11.** 5σ, 1π; 7σ, 0π; 3σ, 2π; 5σ, 0π; 2σ, 2π; 6σ, 1π. **12.** (a) C, sp^2; (b) Si, sp^3; (c) C, sp; (d) N, sp^3; (e) C, sp; (f) C, sp^3 (CH_3); C, sp^2 (C=O). **14.** CO_2; linear; the planes formed by the H—C—H's at each end are mutually perpendicular; C_1 and C_3 are sp^2; C_2 is sp. **18.** (a) sp^3d; (b) sp^3d; (c) sp^3d; (d) sp^3d^2. **19.** H—N—H angle is $\sim104.5°$, analogous to H—O—H in water, which also has two BP and two LP. **20.** H—O—H angle is $\sim107°$, analogous to H—N—H in ammonia, which also has three BP and one LP. **22.** Geometry around both N and B is tetrahedral. **23.** C—O—C; by analogy with angles in H_2O and H_2S. **25.** ClF_3, SF_4, PF_5; no; ClF_3, T-shape; SF_4, seesaw-shape; PF_5, trigonal bipyramidal. **26.** Octahedral; 3s, three 3p, and two 3d. **28.** Linear; sp^3d. **29.** Square planar; sp^3d^2. **30.** sp; sp^2; sp^3; sp^3d; sp^3d^2. **31.** 120°; BF_3 and CO_3^{2-} are examples. **32.** Less, due to more LP–BP repulsion. **34.** All sp^2. **35.** (b) Ion is linear; (c) sp.

Chapter 7

3. (a) Subtracted; (b) 0.4%. **4.** 763.45 mmHg. **5.** (a) 33.9 ft; (b) no [unless atm. pres. $> (35/33.9)$ atm.]. **8.** 101.325 kPa; Miami, 102 kPa; Denver, 84 kPa. **9.** 240 mm. **10.** (a) 9.02 L; (b) 12.3 L; (c) 4.87 L; (d) 2.86 L; (e) 0.987 L. **11.** 0.50 atm. **13.** In (a), (b), and (c), the Hg rises in the right limb; (d) no change. **14.** (a) 263 K; (b) 250 K; (c) 1,000, 273 K; (d) 555,811 K. **15.** (a) 20 L; (b) 231 mL; (c) 523 mL; (d) 6.11 qt. **16.** (b) and (c), 0 °C. **17.** 1.47 L. **18.** -32.8 °C. **19.** 2.6 atm. **20.** 4.5 atm. **21.** 22.4 psi. **22.** 2.2 L. **23.** H_2; 1.11 times. **24.** (a) 319 kg; (b) 675 kg. **25.** No; weight of gas would be greater. **26.** 45.9 amu. **27.** 44.8 amu. **28.** 1.1 L. **29.** 5.10×10^{21}. **30.** 1.292 g · L^{-1}. **32.** (a) 62.36; (b) 8.314; (c) 6.236×10^4. **33.** (a) 6.52 g · L^{-1}, 6.52 kg · m^{-3}; (b) 4.21 g · L^{-1}, 4.21 kg · m^{-3}; (c) 13.0 g · L^{-1}, 13.0 kg · m^{-3}; (d) 8.41 g · L^{-1}, 8.41 kg · m^{-3}. **34.** Dry air; H_2O molecule weighs less than N_2, O_2, or Ar. **35.** 28.9 mL. **36.** 11.2 L. **37.** 2.48×10^{-4} m^3. **38.** 740 g. **39.** (a) 44.6 mol; (b) 4.40 atm; (c) 3×10^{-6} mol; (d) 14,000 K; (e) 0.11 L; (f) 39 atm. **40.** 3.3×10^{15} molecules. **41.** B_2H_5; B_4H_{10}; 53.32 amu. **42.** 100.014 amu. **43.** (a) 16.8 L H_2; (b) 8.4 L O_2; (c) 13.5 g H_2O; (d) 1.5 g H_2. **44.** 6.7 L. **45.** 0.19 L. **46.** 33.6 L. **47.** (a) 50.98 g; (b) 0.0168 m^3; (c) 18.9 kg, 0.76 m^3 O_2 in excess. **48.** 0.00155 mol O_2 remain. **49.** p_{O_2}, 1.1 atm; p_{N_2}, 0.29 atm; p_{SO_2}, 2.0 atm; $p_{\text{total}} = 6.4$ atm.

50. Greater than, assuming negligible O_2 dissolves. **51.** 0.80 L. **52.** 0.386 g. **53.** 16.0 amu. **54.** F_2. **55.** O_3: 22.39 L $\cdot$ mol^{-1}, 0.08197 L $\cdot$ atm $\cdot$ K^{-1} $\cdot$ mol^{-1}, 0.1%; C_2H_6: 22.16 L $\cdot$ mol^{-1}, 0.08113 L $\cdot$ atm $\cdot$ K^{-1} $\cdot$ mol^{-1}, 1.1%. **56.** 2.55 kg $\cdot$ m^{-3}.

Chapter 8

1. At 0 °C, 1.25×10^3 kJ; at 100 °C, 1.13×10^3 kJ. **2.** 12 kJ. **3.** (a) 15 kJ;   (b) H_2O,   54.2 kJ; Br_2,   45.5 kJ. **4.** 31.7 J $\cdot$ mol^{-1} $\cdot$ K^{-1}. **5.** Fusion. 126.0 J $\cdot$ g^{-1} $\cdot$ K^{-1}; vaporization. 394.5 J $\cdot$ g^{-1} $\cdot$ K^{-1}. **7.** Yes for (a), (c), (f), (g), (h), (i). **11.** H_2O, mp ~ -100 °C, bp ~ -80 °C; HF, mp ~ -160 °C, bp ~ -90 °C. **16.** 736 mmHg. **17.** At 20 °C. 10.3 m; at 0 °C. 10.1 m. **18.** 29.5 mm. **19.** (a) Yes; (b) no; (c) no; (d) yes. **20.** (a) 10.0 kg; (b) 52.5%. **21.** 53.8%; 13.8%; 4.42%; 1.68%. **22.** 100%. **23.** (b) $2C_{12}H_{26} + 37O_2 \rightarrow 24CO_2 + 26H_2O$. **25.** (a) 94.7 °C; (b) 83.6 kPa. **27.** Bi fp decreased; other fp's, all bp's increased. **29.** Liquefy, then solidify. **30.** The solid will sublime. **34.** Nanometer, nm; 0.658 nm $\times$ 0.658 nm; K^+, 0.133 nm; Br^-, 0.196 nm. **36.** (a) 2.330 Å. **38.** 0.75. **39.** No. **40.** Exception. **44.** 2.2×10^{16}. **45.** 0.95 Å. **48.** (b) and (c) will float. **50.** (a) C_3H_8; (b) HF; (c) LiF; (d) Kr; (e) H_2O. **51.** 2, bcc; 4, fcc. **52.** 6.04×10^{23}. **53.** (a) CaO, $ScBr_3$; (b) Cl_2, H_2; (c) CH_4, CH_3CH_2OH; (d) NH_4Cl, $NaNO_3$. **54.** $H_2O(l)$; 3 °C; 4 °C. **55.** (a) CH_4; (b) Al, Fe have metallic bonds. **56.** (a) Intrpnt fcc; (b) intrpnt sc.

Chapter 9

4. 2.67; one $\overset{+2}{Fe}$, two $\overset{+3}{Fe}$. **5.** -1, H_2O_2, hydrogen peroxide; $+2$, OF_2, oxygen difluoride. **6.** (a) 19.6 kg O_2; (b) 1.61×10^4 L. **8.** (a) $2Mg + O_2 \rightarrow 2MgO$; (b) using C_8H_{18} for gasoline: $2C_8H_{18} + 25O_2 \rightarrow 16CO_2 + 18H_2O$; (c) $2CO + O_2 \rightarrow 2CO_2$;   (d) $2Au_2O_3 \rightarrow 4Au + 3O_2$; (e) $2H_2 + O_2 \rightarrow 2H_2O$. **9.** $2KNO_3 \rightarrow 2KNO_2 + O_2$;   oxygen;   nitrogen. **10.** $2MnO_2 + 2H_2SO_4 \rightarrow 2MnSO_4 + O_2 + 2H_2O$; oxygen;   manganese. **16.** (a) Fe_3O_4 and   FeO; (b) $Fe_2O_3 + 3CO \rightarrow 2Fe + 3CO_2$. **21.** Tritium, 49.744%. **22.** (b) 2.82 : 1.00. **23.** (a) 2.69×10^{22} in each; H_2, 0.0900 g; D_2, 0.180 g; O_2, 1.43 g; O_3, 2.14 g. (b) All stay constant except O_3; $2O_3 \rightarrow 3O_2$, so number of molecules and pressure increase by factor of 1.5, and temperature increases (reaction is exothermic). **24.** (a) $2K + 2H_2O \rightarrow 2KOH + H_2$;   (b) $2H_2 + O_2 \rightarrow 2H_2O$.   **25.** 27 L.   **26.** (a) $2H^- \rightarrow H_2 + 2e^-$; (b) $2Na^+ + 2e^- \rightarrow 2Na$;   (c) $2NaH \rightarrow 2Na + H_2$. **27.** $CO + H_2 + O_2 \rightarrow CO_2 + H_2O$.   **28.** 6.43×10^4 kJ. **30.** (b) Ethylene, C_2H_4. **31.** (a) $Sr + 2H_2O \rightarrow Sr(OH)_2 + H_2$; (b) $Na_2O + H_2O \rightarrow 2NaOH$; (c) $Ca + 2H_2O \rightarrow Ca(OH)_2 + H_2$;   (d) $N_2O_5 + H_2O \rightarrow 2HNO_3$; (e) $MgCl_2 + 6H_2O \rightarrow MgCl_2 \cdot 6H_2O$. **32.** (b) $\overset{-1}{H}$

oxidized, $\overset{+1}{H}$ reduced. **33.** 0.286 kg;   286 kg;   286 mg. **34.** $CuCl_2 \cdot 4H_2O$. **35.** 50.2 kg. **36.** Examples: (a) NaH, sodium hydride; (b) $Na_2CO_3 \cdot 10H_2O$, sodium carbonate decahydrate; (c) NaOH, sodium hydroxide; (d) Na_2O_2, sodium peroxide; (e) $NaHO_2$. **37.** $H_2S + 4H_2O_2 \rightarrow H_2SO_4 + 4H_2O$. **38.** $H_2S_2O_8 + 2H_2O \rightarrow H_2O_2 + 2H_2SO_4$;   yes. **39.** (a) H—$\ddot{\text{O}}$—$\ddot{\text{O}}$—$\ddot{\text{O}}$—H;   H—$\ddot{\text{O}}$—$\ddot{\text{O}}$—$\ddot{\text{O}}$—$\ddot{\text{O}}$—H; (b) -1, 0, -1; -1, 0, 0, -1; (c) H_2O_4.

Chapter 10

2. (a) 5.64.   **9.** (a) $CuCl_2 + H_2S \rightarrow CuS + 2HCl$. **10.** 1.4×10^{-3} g; 1×10^{-4} g. **12.** Yes; ~ 7.7 g. **14.** Exothermic. **15.** (a) Supersaturated. **18.** (a) 16.0%, 84.0%; (c) 0.217, 0.783. **19.** 0.439. **20.** 0.100; 10.0%. **21.** (a) Cannot be done because of low solubility of $CaCO_3$; (b) 42.9 g $Ca(NO_3)_2$ plus water to make 100 mL. **22.** (a) $0.391M$; (b) $3.12M$; (c) $0.680M$. **23.** 1.3×10^{-5}. **24.** 800 mL. **25.** (a) 5.59%; (b) 3.12%; (c) $1.79m$; (d) $1.70M$. **26.** 11.1; 5.09. **28.** 62.5 mL. **29.** (a) 30 mL; (b) 3 mL; (c) 3 mL. **30.** $SnCl_2$, 94.80 g; $FeCl_3$, 162.21 g.   **31.** (a) 49.03 g;   (b) 0.0625 equiv;   3.06 g. **32.** 1.46 L. **41.** KCl, $CaCl_2$, $NaNO_3$, CsI, RbOH, $Mg(NO_3)_2$, HI, $NaHCO_3$; all of the preceding; H_2S, NH_3, $CH_3CH_2CO_2H$. **43.** (a) $N_2H_4O_3$; (c) ammonium nitrate, NH_4NO_3.

Chapter 11

6. (a) $CHO_2^- + H_2O \rightleftharpoons HCHO_2 + OH^-$; acids, H_2O and $HCHO_2$; bases, CHO_2^- and OH^-; (b) $CH_3NH_3^+ + H_2O \rightleftharpoons CH_3NH_2 + H_3O^+$; acids, $CH_3NH_3^+$ and H_3O^+; bases, H_2O and CH_3NH_2. **7.** F^- is stronger. **8.** NH_4^+ is stronger. **10.** $NH_2^- + H_2O \rightarrow NH_3 + OH^-$. **11.** Most to least acidic: NH_4Cl, $NaHSO_4$; most to least basic: Na_2O, KCN, $NaC_2H_3O_2$, KF. **13.** NH_4F, weakly acidic; $(NH_4)_2SO_3$, weakly basic. **14.** (a) $CO_3^{2-} + H_2O \rightleftharpoons HCO_3^- + OH^-$; (b) $Cu(H_2O)_4^{2+} + H_2O \rightleftharpoons Cu(H_2O)_3(OH)^+ + H_3O^+$; (c) no hydrolysis rxn. **16.** Weakly basic. **20.** HClO, strongest; HIO, weakest. **21.** $CH_3CO_2H < ClCH_2CO_2H < FCH_2CO_2H < CF_3CO_2H$. **22.** NH_3, strongest; $HONH_2$, weakest. **24.** H_3PO_4. **25.** (b) (1) is acidic; (2) is neutral. **26.** (a) Basic, $OH^- + HC_2H_3O_2 \rightarrow H_2O + C_2H_3O_2^-$;   (b) neutral, $OH^- + H^+ \rightarrow H_2O$;   (c) acidic, $NH_3 + HSO_4^- \rightarrow NH_4^+ + SO_4^{2-}$.   **28.** 12.3 g. **29.** (a) 32.7 g; (b) 49.0 g; (c) 98.0 g; (d) 32.7 g. **30.** 40.0 g; 60.0 g. **31.** 40.5 g. **32.** $1.25N$, $0.625M$. **33.** (a) $2.0N$; (b) $2.0N$; (c) $1.5N$; (d) $1.0N$. **34.** 1200 mL. **35.** 107 g. **36.** 59.0 g.   **37.** (b) 28.1 mL;   (c) 18.8 mL;   (d) 1.5 : 1. **38.** 1690 mL.   **39.** 3600 mL.   **40.** (a) 45.0 g;   (b) 90.0 g. **41.** $0.106M$; $0.212N$. **42.** (a) 12.2 g; (b) $MgCl_2$ or $TiCl_4$.

Chapter 12

1. No. **2.** (a) 0 mL; (b) 75 mL, $1.33m$; (c) 225 mL, $1.33m$. **4.** $C_6H_{12}O_6$,   0.00990; H_2O,   0.99010;   54.8 mmHg

(7.31 kPa). **5.** 86.7 mmHg (11.6 kPa). **6.** (a) 1489 mmHg; (b) 1360 mmHg; (c) 1163 mmHg. **7.** (a) 1.6m; (b) 0.028; (c) 100.82 °C. **8.** −0.93 °C. **9.** −9.99 °C; −30.0 °C. **10.** 101.0 °C. **11.** (a) 4.18 mmHg; (b) 4.18 mmHg. **12.** 11.4 °C. **13.** Solution (a). **14.** (a) Solution (a); (b) Solution (b). **15.** 15 kg. **16.** 10 °C/m. **18.** 107 amu. **19.** (a) 0.14m; (b) 160 amu; (c) $C_6H_4N_2O_4$. **20.** (a) 0.456m; (b) 87.6 amu. **21.** $C_4H_{10}O_2$; 90.122 amu. **23.** (a) 0.300; (b) −0.558 °C; (c) 2.60. **24.** (a) −0.264 °C; (b) −0.175 °C; (c) −0.0515 °C; (d) −0.0180 °C. **25.** (a) $i = 1.84$; (b) $i = 1.87$; (c) $i = 2.20$; (d) $i = 1.02$; (e) $i = 1.84$. **26.** $i = 5$; no. **27.** (a) 0.400; (b) −0.372 °C; (c) −0.355 °C. **28.** 10%. **29.** (a) 68 amu; (b) 204 amu; (d) 2.2. **32.** (b) 56.5 °C; (c) 81% F, 19% G; (d) 81% F, 19% G. **34.** 23 m. **36.** 10.3 atm. **37.** (a) 27; (b) 0, 1, 6, 12, 8; (c) 54 cm^2, 108 cm^2. **38.** No; 5×10^3 Å. **39.** 5×10^9 molecules per smoke particle. **40.** (a) 8.90×10^{-21} g; (b) 91; (c) yes. **41.** (a) 82. **42.** (a) 2.1×10^{-19} m^2, 21 Å^2; (b) 2.5×10^{-9} m, 25 Å.

Chapter 13

1. 70 kJ; 7.0×10^2 kJ · mol^{-1}. **4.** −1299.6 kJ. **7.** (a) −36.4 kJ; (b) −239.0 kJ; (c) +588 kJ. **8.** −311.4 kJ. **9.** −3267.6 kJ (for $H_2O(l)$). **10.** $C_2H_4(g)$ liberates 0.38 kJ more than $C_2H_2(g)$. **11.** −1104.28 kJ. **12.** −1077.70 kJ. **13.** −250.27 kJ. **14.** −37.66 kJ. **16.** (a) 631.6 kJ. **17.** (a) 29.9 kJ · mol^{-1}; (b) 61.0 kJ · mol^{-1}. **19.** 209.3 kJ. **21.** 416 kJ. **23.** 468.9 kJ. **24.** (a) 1076.4 kJ; (b) 804.3 kJ. **25.** 489 kJ. **26.** 297 kJ. **27.** 358.8 kJ. **28.** (a) 38.4 kJ; (b) 381.8 kJ. **29.** 410 kJ; 3460 kJ. **31.** (a) 285.0 kJ · mol^{-1}; (b) −971 kJ · mol^{-1}. **33.** −5470.68 kJ. **34.** $H_2O(l)$; (b) NaCl(l); (c) H_2, 1 atm; (d) $2H_2 + O_2$. **36.** −120.40 kJ. **38.** −252.71 kJ. **39.** −429.8 kJ; −395.6 kJ. **40.** (a) +80.44 kJ, no; (b) −1100.40 kJ, yes; (c) −663.30 kJ, yes. **41.** Spontaneous; extremely slow. **42.** 1187 °C.

Chapter 14

3. (a) Two. **4.** (a) Two. **5.** 1.50 L; 1.75 L. **6.** 6.6×10^4 Pa. **7.** (a) Titrate unreacted KOH; (b) titrate HI produced; (c) follow pressure decrease; (d) no simple method, reaction too rapid. **9.** 210 kJ. **16.** (a) 3.5×10^{-5} s^{-1}; (b) 1.9×10^{-10} L · mol^{-1} · s^{-1}. **17.** (b) 2.5×10^{-3} mol · L^{-1} · s^{-1}, 5.0×10^{-3} mol · L^{-1} · s^{-1}, 5.0×10^{-3} mol · L^{-1} · s^{-1}. **18.** Rate = $k[A]^0[B]$. **19.** Rate = $k[H_2][NO]^2$. **20.** 3.5×10^{-4} mol · L^{-1} · min^{-1}; 8.8×10^{-5} mol · L^{-1} · min^{-1}. **21.** Rate = $k[NO]^2[Cl_2]$. **25.** (a) None; (b) none; (c) doubled. **26.** (a) 1.00 h, 2.00 h; (b) 1.00 h, 2.00 h; (c) first order; (d) 0.693 h^{-1}; (e) experiment 1: 1.73×10^{-1} mol · L^{-1} · h^{-1}, 8.66×10^{-2} mol · L^{-1} · h^{-1}, 4.33×10^{-2} mol · L^{-1} · h^{-1}; experiment 2: 3.46×10^{-1} mol · L^{-1} · h^{-1}, 1.73×10^{-1} mol · L^{-1} · h^{-1}, 8.65×10^{-2} mol · L^{-1} · h^{-1};

(f) rate$_2$ = 2(rate$_1$). **27.** (a) 4.00 h, 12.00 h; (b) 2.00 h, 6.00 h; (c) second order; (d) 2.50×10^{-1} L · mol^{-1} · h^{-1}; (e) 6.25×10^{-2} mol · L^{-1} · h^{-1} and 1.56×10^{-2} mol · L^{-1} · h^{-1}, 2.50×10^{-1} mol · L^{-1} · h^{-1} and 6.25×10^{-2} mol · L^{-1} · h^{-1}; (f) rate$_2$ = 4(rate$_1$) at same time interval. **28.** (a) 2.89×10^3 s; (b) 5.78×10^3 s; (c) 8.67×10^3 s; (d) $t_{1/2} = \frac{1}{2}(t_{3/4}) = \frac{1}{3}(t_{7/8})$. **29.** 1.25 g; 0.156; 0.0098 g. **30.** $(1/2^n)100$. **31.** (a) 5.63×10^{-3} min^{-1}; (b) 409 min, 818 min. **32.** (a) 7.92×10^{-2} h^{-1}; (b) 1.97 g. **33.** (a) 7.70×10^{-3} min^{-1}; (b) 180 min. **34.** 3.00×10^8 s. **35.** (a) 1.0×10^6 s; (b) 3.0×10^6 s; (c) $3t_{1/2} = t_{3/4}$. **36.** 2780 s. **37.** L^2 · mol^{-2} · s^{-1}. **38.** First order. **39.** 1.16×10^{-2} min^{-1}. **40.** (a) First order; (b) 0.448 h^{-1}; (c) 1.55 h; (d) 200 mm Hg. **41.** (a) 1.5×10^{-2} s^{-1}; (b) 110 s, 150 s. **42.** (a) $k_{avg} = 0.146$ h^{-1}; (b) $k_{avg} = 0.148$ h^{-1}. **43.** (d) for both. **44.** (a) Step 1: $SO_2Cl_2 \rightarrow SO_2Cl + Cl$; step 2: $SO_2Cl + Cl \rightarrow SO_2 + Cl_2$; (b) step 1. **45.** Initiation: $Cl_2 \rightarrow 2Cl$; propagation: $CH_4 + Cl \rightarrow CH_3 + HCl$; $CH_3 + Cl_2 \rightarrow CH_3Cl + Cl$, and so on; termination: $H + Cl \rightarrow HCl$; $Cl + Cl \rightarrow Cl_2$; $CH_3 + Cl \rightarrow CH_3Cl$; $CH_3 + CH_3 \rightarrow CH_3CH_3$. **46.** As in Exercise 45: $CH_3 + CH_3 \rightarrow CH_3CH_3$; $CH_3CH_3 + Cl \rightarrow CH_3CH_2 + HCl$; $CH_3CH_2 + Cl_2 \rightarrow CH_3CH_2Cl + Cl$. **47.** True: (a), (c), (d), (e), and (i).

Chapter 15

4. (a) 0.45M; (b) 5.0M; (c) 0.59M. **5.** (b) 4.9; (c) unchanged. **6.** 0.0020M. **7.** 0.33. **8.** 0.32. **10.** 1.5×10^{-3}. **11.** 20.9. **12.** 40. **13.** (a) mol · L^{-1}; (b) L^2 · mol^{-2}; (c) mol^2 · L^{-2}; (d) none; (e) L · mol^{-1}. **15.** 1.25M. **16.** 0.67 mol H_2; 0.67 mol I_2; 2.66 mol HI. **17.** Exercise 2, equilibria 1 and 3, and Exercise 13, equilibrium (d); the remaining equilibria; none **18.** $K_c = 100$. **19.** (a) 1.3 mol; (b) 3.0 mol. **20.** (b) 2.9×10^2 g. **21.** 7.5×10^{-4} M. **22.** (a) [H_2O] = [CO] = 0.16M; [H_2] = 0.09M; [CO_2] = 0.34M; (b) [H_2O] = [CO] = 0.19M; [H_2] = 0.06M; [CO_2] = 0.81M. **23.** 5.7×10^{-28} M; 2.62M. **24.** 5.14 g SO_2; 1.28 g O_2; 1.58 g SO_3. **25.** (c) $2CO(g) \rightleftharpoons 2C(s) + O_2(g)$. **26.** 6.1×10^{-2}; yes. **27.** [H_2] = [I_2] = 8.18×10^{-3} M; [HI] = 1.11×10^{-3} M; mole fractions: $H_2 = I_2 = 0.468$; HI = 0.0635. **28.** (a) 72 atm; 63 atm; (b) 6.2%, 21%. **29.** (a) $p_{NH_3} = 52.0$ atm; $p_{N_2} = 12.0$ atm; $p_{H_2} = 36.0$ atm; (b) 4.83×10^{-3}; (c) 10.7. **30.** (a) 1.90×10^{-4}. **31.** (a) 3.39×10^3; (b) 7.35 L. **32.** (a) For 50 atm: $p_{NH_3} = 37.2$ atm, $p_{H_2} = 9.6$ atm, $p_{N_2} = 3.2$ atm; mole fractions = 0.744, 0.192, and 0.064; for 100 atm: $p_{NH_3} = 81.5$ atm, $p_{H_2} = 13.9$ atm, $p_{N_2} = 4.6$ atm; mole fractions = 0.815, 0.139, and 0.046; (b) for 50 atm: [NH_3] = 0.958M, [H_2] = 0.247M, [N_2] = 0.082M; for 100 atm: [NH_3] = 2.10M, [H_2] = 0.358M, [N_2] = 0.12M; (c) no. **33.** 1.33. **34.** (a) 0.250 atm, 0.250 atm, 0.750 atm; (b) no, $(0.250)^2/0.750 = 8.33 \times 10^{-2}$, not 6.0; (d) 0.87 atm.

35. 597 mL. **36.** (a) Atm; (b) atm^{-2}; (c) atm^2; (d) none; (e) atm^{-1}. **37.** 6.4. **38.** 1.08 atm^{-1}. **39.** (a) 5.90; (b) 14.5%. **40.** 89.5%. **41.** (a) 100 atm,   50 atm;   (b) 5.0 $\times$ 10^5. **42.** 0.145. **43.** 6.85; -4.77 kJ. **45.** Endothermic. **46.** Greater. **47.** Greater. **49.** (a) Increase P or decrease T; (b) only ΔT.

Chapter 16

5. 9.3 $\times$ 10^{-6}. **6.** 1.4 $\times$ 10^{-10}. **7.** 2 $\times$ 10^{-5}. **8.** (c), (d), (e). **9.** 3.1 $\times$ 10^{-2} M; 2.1%; no; larger. **10.** (a) 0.60%; 1.3%; 1.9%.   **11.** (a) 1.3 $\times$ 10^{-2} M;   (b) 4.2 $\times$ 10^{-3} M; (c) 1.8 $\times$ 10^{-3} M;   (d) 0.11M;   (e) 1.6 $\times$ 10^{-2} M; (f) 2.3 $\times$ 10^{-4} M.   **12.** 0.100M;   7.9 $\times$ 10^{-6} M. **13.** (a) (i) 6.5 $\times$ 10^{-3} M,   (ii) 6.3 $\times$ 10^{-3} M;   (b) 6.3%. **14.** 0.26M.   **15.** 0.026M; 6.3 $\times$ 10^{-8} M. **16.** [H$^+$] = 0.058M;   [H$_2$PO$_4$$^-$] = 0.058$M$; [HPO$_4$$^{2-}$] = 6.2 $\times$ 10^{-8} M;   [PO$_4$$^{3-}$] = 4.7 $\times$ 10^{-19} M. **17.** [H$^+$] = 0.108M;   [HSeO$_4$$^-$] = 0.092$M$; [SeO$_4$$^{2-}$] = 0.008$M$; K_{a2} = 9 $\times$ 10^{-3}. **18.** (a) Ethylenediamine, 19 times; (c) ethylenediamine, 670; hydrazine, 7.5 $\times$ 10^8.   **19.** (b) K_a = K_b = 1.0 $\times$ 10^{-14};   (c) for H$_2$O $\rightleftharpoons$ H$^+$ + OH$^-$, K_c = 1.8 $\times$ 10^{-16};   for H$_2$O + H$_2$O $\rightleftharpoons$ H$_3$O$^+$ + OH$^-$, K_c = 3.3 $\times$ 10^{-18}. **20.** (a) 5.0 $\times$ 10^{-12} M;   (b) 5.0 $\times$ 10^{-12} M. **21.** 3.3 $\times$ 10^{-3} M; 3.3 $\times$ 10^{-3} M; 3.0 $\times$ 10^{-12} M; 0.60M. **22.** 4.0 $\times$ 10^{-9} = K_a; 7.7 $\times$ 10^{-7} = K_b; 6.2 $\times$ 10^{-11} = K_b; 6.4 $\times$ 10^{-10} = K_a.   **23.** [OH$^-$] = 0.012M; [H$^+$] = 8.3 $\times$ 10^{-13} M.   **24.** (a) 2NH$_3$ $\rightleftharpoons$ NH$_4$$^+$ + NH$_2$$^-$; (b) 3.2 $\times$ 10^{-17} M;   (c) 1 $\times$ 10^{-30} M;   (d) acidic, [NH$_4$$^+$] > [NH$_2$$^-$]. **25.** No; some might be salts that hydrolyze.   **26.** 3.5 $\times$ 10^{-6} M.   **27.** [S^{2-}] = 0.38M; [HS$^-$] = 0.62M;   [H$_2$S] = 9.1 $\times$ 10^{-8} M. **29.** (c) 1.5 $\times$ 10^{-11}; (d) 3.7 $\times$ 10^{-9} M. **30.** (a) Acidic, K_a of NH$_4$$^+$ > K_b of F$^-$; (b) acidic, K_a of Cu(H$_2$O)$_4$$^{2+}$ > K_b of C$_2$H$_3$O$_2$$^-$. **31.** (a) 3.00; (b) 11.00; (c) 2.30; (d) 12.00. **32.** [H$^+$] = 4.9 $\times$ 10^{-9} M, pH = 8.31; [H$^+$] = 2.2 $\times$ 10^{-5} M,   pH = 4.66. **33.** 2.72. **34.** 3.2%. **36.** 8.95; (0.0068)([H$^+$] in NH$_3$–NH$_4$Cl solution). **37.** 7.21. **38.** 1, (g); 2, (a); 3, (f); 4, (b); 5, (d); 6, (c); 7, (e). **39.** 2.70. **40.** (a) 3.83;   (b) 3.74. **42.** 12.70.   **43.** 10.00.   **44.** 4.63. **45.** (a) 9.29; (b) 9.22. **46.** 2:1. **47.** 2:1. **48.** 1.0 $\times$ 10^{-6}.

Chapter 17

2. 7.9 $\times$ 10^{-5}; 7.87 $\times$ 10^{-8}. **3.** 9.1 $\times$ 10^{-12} g; 2.9 $\times$ $^{-7}$ g. **4.** Bi$_2$S$_3$.   **5.** (a) 3.0 $\times$ 10^{-3} M,   0.41 g $\cdot$ L^{-1}; (b) 1.8 $\times$ 10^{-4} M.   **6.** (a) 2.2 $\times$ 10^{-2};   (b) 5.5 $\times$ 10^{-4}; (c) 3.7 $\times$ 10^{-8}. **7.** 7.5 $\times$ 10^{-14}. **8.** 8.5 $\times$ 10^{-12}. **9.** No; [Ca^{2+}][CrO$_4$$^{2-}$] = 2.5 $\times$ 10^{-4} < K_{sp} of   7.1 $\times$ 10^{-4}. **10.** (a) 2.0 $\times$ 10^{-3};   (b) 2.0 $\times$ 10^{-6}.   **11.** (a) 2.0 $\times$ 10^{-3}; (b) moles   Ba^{2+} = 2.4 $\times$ 10^{-10};   moles   SO$_4$$^{2-}$ = 0.018. **12.** (a) 2.0 $\times$ 10^{-3};   (b) 0.018 mol Ba^{2+}, 2.4 $\times$ 10^{-10} mol SO$_4$$^{2-}$. **13.** (a) (5.0 $\times$ 10^{-4})^3(2.5 $\times$ 10^{-2})^2 = 7.8 $\times$ 10^{-14} > K_{sp} of 3.4 $\times$ 10^{-23}; (b) more; (c) less; (d) yes. **14.** 9.3 $\times$ 10^{-13} M.

15. (a) 1.0 g CaCO$_3$;   (b) 2.4 $\times$ 10^{-8} M.   **16.** (a) Yes; 0.070 g;   (b) 3.5 $\times$ 10^{-5} M;   (c) 0.70%. **17.** (a) [H$^+$] = 2.1 $\times$ 10^{-4} M,   [CO$_3$$^{2-}$] = 5.6 $\times$ 10^{-11} M; (b) no,   [Ag$^+$]^2[CO$_3$$^{2-}$] < K_{sp} of   Ag$_2CO_3$. **18.** Yes; [Hg$_2$$^{2+}$][CO$_3$$^{2-}$] > K_{sp} of Hg$_2CO_3$. **19.** No, both ions precipitate. **20.** Ag$_2$CrO$_4$ precipitates but Hg$_2$CrO$_4$ does not; [Ag$^+$] = 8.3 $\times$ 10^{-3} M. **21.** Hg^{2+}, Cd^{2+}, Co$_3$$^{2+}$, Cu^{2+}, and Pb^{2+} will precipitate as sulfides. **23.** (a) MnS; (b) no; make H$_2$S solution 0.010M in H$^+$ by adding HCl. **24.** (a) 0.020; (b) 9.43. **25.** Adjust pH to give [S^{2-}] so only ZnS precipitates. **26.** All of both dissolve. **27.** 1.3 $\times$ 10^{-6} mol. **28.** (a) 7.9 $\times$ 10^{-9} M;   (b) yes,   [Al^{3+}][OH$^-$]^3 > K_{sp} of Al(OH)$_3$; (c) no, Al(H$_2$O)$_6$$^{3+}$ hydrolyzes to give acidic solution.   **29.** FeS, MnS, and ZnS. **32.** (a) Cu^{2+} + 4CN$^-$ $\rightarrow$ Cu(CN)$_4$$^{2-}$; (b) Zn^{2+} + 4OH$^-$ $\rightarrow$ Zn(OH)$_4$$^{2-}$; (c) Cd^{2+} + 4NH$_3$ $\rightarrow$ Cd(NH$_3$)$_4$$^{2+}$. **34.** Make 8.6$M$ in NH$_3$ to dissolve AgBr; make acidic with HNO$_3$. **35.** pH = 13.0: [Zn^{2+}] = 1 $\times$ 10^{-15} M,   [Zn(OH)$_4$$^{2-}$] = 1 $\times$ 10^{-2} M; pH = 14.0: [Zn^{2+}] = 1 $\times$ 10^{-17} M, [Zn(OH)$_4$$^{2-}$] = 1 $\times$ 10^{-2} M.   **36.** 1.8M. **37.** (a) [Ag$^+$] = 9.5 $\times$ 10^{-10} M,   [Ag(NH$_3$)$_2$$^+$] = 0.010$M$; (b) no.   **40.** For   MnS,   [Mn^{2+}] = 2.5 $\times$ 10^{-9} M;   for Mn(OH)$_2$, [Mn^{2+}] = 1.9 $\times$ 10^{-3} M; MnS.

Chapter 18

8. 5.5 cents. **9.** (a) 3.87 V; (b) 2.37 V, 1.50 V. **13.** (a) 0.51 V; (b) 3.02 V;   (c) 0.80 V;   (d) 0.13 V.   **14.** (a) -0.14 V; (b) 0.44 V;   (c) -0.01 V;   only   (b).   **15.** (a) -0.28 V. **17.** 1.01 V. **18.** (a) 0.44 V; (b) 0.38 V; (c) 0.80 V; (d) 0.86 V. **20.** (a) Sn + Pb^{2+} $\rightarrow$ Sn^{2+} + Pb; (b) 0.01 V;   (c) 0.10 V; (d) -0.08 V; (f) 2.2. **21.** (e). **22.** (a) 1.36 V; (b) 1.33 V; (c) 1.39 V; (d) 1.52 V; (e) 1.27 V. **24.** (a) R^{2+} + 2e^- $\rightarrow$ R, -0.45 V;   T^{2+} + 2e^- $\rightarrow$ T, -0.30 V;   (b) 1.2 $\times$ 10^5. **25.** (a) 10, 3.7 $\times$ 10^{15}; (b) reciprocally. **26.** (a) 1 $\times$ 10^{178}; (b) K = [Ni^{2+}]^3/[Au^{3+}]^2;   (c) [Au^{3+}] = 4 $\times$ 10^{-90}. **28.** (a) 0.355 V;   (b) 0.177 V,   0.118 V.   **29.** K_1 = $\sqrt{K_2}$. **30.** (a) 1.1 $\times$ 10^2 kJ,   27 kcal;   (b) no, ΔG is   positive. **33.** (b) 6.09 g;   (c) 0.601 mol;   (d) 7.2 g. **34.** (a) -0.03 V,

$$K = 9.1 \times 10^{-4}; \quad \text{(b) no;} \quad \text{(c) } K = \frac{(p_{Cl_2})^3[Cr^{3+}]^2}{[Cr_2O_7{}^{2-}][H^+]^{14}[Cl^-]^6}.$$

36. (b) 0.66 V;

$$2NO_3{}^- + H_2S + 2H^+ \rightarrow N_2O_4 + S + 2H_2O.$$

37. (a) -0.06 V; (b) $+0.11$ V. **38.** p_{O_2}/[H$_2$O$_2$]^2 = 2 $\times$ 10^{37}; no. **39.** (a) Fe$_2$O$_3$, 26.62 g; S, 8.02 g; (b) NH$_3$, 3.41 g; O$_2$, 8.00 g; (c) KMnO$_4$, 31.61 g; HCl, 36.46 g; (d) N$_2$O, 5.50 g; H$_2$, 1.01 g.   **40.** (a) 2O^{2-} $\rightarrow$ O$_2$ + 4e^-; (b) Al^{3+} + 3e^- $\rightarrow$ Al;   (c) 4Al^{3+} + 6O^{2-} $\rightarrow$ 4Al + 3O$_2$. **41.** 2.25 g Al; 2.00 g O$_2$. **42.** Hg^{2+}; Li$^+$. **43.** 5.58 $\times$ 10^{20} e^-; 89.5 C; 60 s. **44.** 3.47 $\times$ 10^{21} e^-;   557 C; 370 s. **45.** PtCl$_4$. **46.** (a) 6.10 $\times$ 10^{-2} faradays, 1.64 A. **47.** (a) CuBr$_2$, Cu and Br$_2$; NaI, H$_2$ and I$_2$; Mg(NO$_3$)$_2$, H$_2$ and O$_2$; HgSO$_4$, Hg and O$_2$;   HCl(conc.),   H$_2$ and   Cl$_2$;   KOH,   H$_2$ and   O$_2$. **50.** (b) 0.47 g; (c) 0.47 g; (d) 0.32 g.

Chapter 19

2. (a) $^{124}_{54}Xe$, $^{126}_{54}Xe$; (b) all. **7.** $^{227}_{89}Ac$. **8.** $^{215}_{85}At$. **9.** (a) $^{207}_{82}Pb$;
(b) 3. **10.** (a) 100; (b) 1.8×10^{-4} millicurie.
12. (a) 60.5 min; (b) $^{212}_{83}Bi$. **13.** (a) 4.33×10^5 s, 5.01 d;
(b) 20.0 d; (c) 4.59×10^{14}. **14.** 7.8×10^9 y. **15.** 6,400 y.
16. 18,500 y. **20.** (a) ^2_1H; (b) $_{+1}^0e$. **22.** (a) 9.375×10^{-3} amu;
(b) 9.069×10^{-3} amu; (d) 3.589×10^{-2} amu,
3.23×10^9 kJ; (e) 3.04×10^5 kg. **23.** ^{58}Ni, 8.74 MeV; ^{28}Si,
8.45 MeV. **25.** (a) 1.333; (b) $^{14}_6C \rightarrow _{-1}^0e + ^{14}_7N$. **27.** (a) 90,
91, 92, 94; (b) 96. **28.** Yes; 50 p^+. **31.** (b) and (e) are true.
32. ^{135}Xe: (a) 1.4×10^{-6} g; (b), (c), and (d) $<1 \times 10^{-6}$ g;
^{131}I: (a) 0.27 g; (b), (c), and (d) $<1 \times 10^{-6}$ g; ^3H: (a) 0.50 g;
(b) 0.47 g; (c) 1.8×10^{-3} g; (d) $<1 \times 10^{-6}$ g; ^{90}Sr:
(a) 0.50 g; (b) 0.49 g; (c) 0.041 g; (d) $<1 \times 10^{-6}$ g; ^{239}Pu:
(a) 0.50 g; (b) 0.50 g; (c) 0.50 g; (d) 0.38 g. **33.** 8.1×10^4 y.
40. (a) 0.2246 g; (b) 2.019×10^{10} kJ $\cdot$ mol^{-1}.
42. (a) 1.7028 g; (b) 1.530×10^{11} kJ.

Chapter 20

3. 1.5 eV = 8.3×10^{-7} m; 8.0 eV = 1.6×10^{-7} m. **4.** O_2;
Ar. **5.** Yes. **6.** 6.2×10^{-7} m. **7.** 157 kJ; 150 kJ.
9. (a) 2.74×10^{-7} m; (b) 4.93×10^{-7} m; (c) Cl_2.
10. 2.62×10^{-8} m; ultraviolet. **18.** About 25 kJ $\cdot$ mol^{-1}.
19. (a) C—^1H; (b) C=C; (c) C—H. **21.** ^1H; ^2H; ^3He; ^{13}C.
22. Three, 5:2:3. **23.** Four, 2:1:1:3. **24.** Four,
$CH_3CH_2CH_2OH$; three, CH_3CHCH_3. **25.** (a) Examples:
$\overset{|}{OH}$
46 amu, + ion of either; 45 amu, $CH_3OCH_2^+$ and
$CH_3CH_2O^+$; 31 amu, CH_3O^+ and CH_2OH^+; 15 amu, CH_3^+
for either; (b) OH^+ or $CH_3CH_2^+$; (c) CH_3O^+; (d) ether, one
peak; alcohol, three peaks. **26.** ^{35}Cl, ^{37}Cl, $^{12}C^{35}Cl$,
$^{12}C^1H^{35}Cl$, $^{12}C^1H^{37}Cl$, $^{12}C^{35}Cl_2$, $^{12}C^1H^{35}Cl_2$, $^{12}C^1H^{35}Cl^{37}Cl$,
$^{12}C^1H^{37}Cl_2$. **27.** ^{14}N; $^{12}C^1H_2$. **28.** ^{79}Br; ^{81}Br; ^{127}I. **32.** ^{35}Cl,
75.6%; ^{37}Cl, 24.4%.

Chapter 21

2. (a) S; (b) Rb; (c) S; (d) Rb; (e) S. **3.** (a) Zn; (b) Rb;
(c) Zn; (d) Rb; (e) Zn. **10.** 1.0×10^{15} Hz. **11.** Li; no, Cs;
electronegativity. **15.** $Ca_3(PO_4)_2$, $MgCO_3$, $BaSO_4$,
$Mg(OH)_2$, $Be(OH)_2$.
16. (a) $CO_3^{2-} + H_2O \rightleftharpoons HCO_3^- + OH^-$;
$HCO_3^- + H_2O \rightleftharpoons H_2CO_3 + OH^-$;
(b) $CaO + H_2O \rightarrow Ca(OH)_2 \rightleftharpoons Ca^{2+} + 2OH^-$,
$3Ca^{2+} + 2PO_4^{3-} \rightarrow Ca_3(PO_4)_2$. **17.** Na_2CO_3, $88 or $130;
NaOH, $180. **19.** 3.1×10^6 tons. **22.** Na_2CO_3. **23.** 0.90 ton
Na_2CO_3; 0.94 ton $CaCl_2$. **24.** $CaCl_2$. **26.** (a) 7.8×10^5 kJ;
(b) 3.1×10^3 kg; (c) 0.50 m^3; 18 ft^3. **28.** (b) 34 lb; 22 lb.
30. (b) ~4 lb. **34.** (a) $\mathscr{E}^\circ_{red}$ is +; (b) Cu, Ag, Au, Hg, Ru,
Rh, Pd, Os, Ir, and Pt. **38.** (a) -366 kJ $\cdot$ mol^{-1}.

Chapter 22

2. (a) $b = (3a - c)/2$; (b) +1.40 V.
3. (a) $H_2S_2O_8 + SnSO_4 \rightarrow Sn(SO_4)_2 + H_2SO_4$;

(b) $S_2O_8^{2-} \xrightarrow{+2.01} 2SO_4^{2-}$, $Sn^{2+} \xrightarrow{-0.15} Sn^{4+}$; (c) yes,
$\mathscr{E}^\circ_{cell} = +1.86$ V.
5. (a) $MnO + 2HNO_3 \rightarrow Mn(NO_3)_2 + H_2O$;
(b) $2Mn(NO_3)_2 + 5NaBiO_3 + 16HNO_3 \rightarrow$
$2HMnO_4 + 5Bi(NO_3)_3 + 5NaNO_3 + 7H_2O$.
6. (a) $Zn + 2Fe^{3+} \rightarrow Zn^{2+} + 2Fe^{2+}$;
(b) $5Fe^{2+} + MnO_4^- + 8H^+ \rightarrow 5Fe^{3+} + Mn^{2+} + 4H_2O$;
(c) 45.5%. **7.** (a) Cu, Cr, or Cr^{2+}; (b) V, Mn, no; (c) Cu$^+$;
(d) MnO_4^-, MnO_2, or $Cr_2O_7^{2-}$. **8.** (a) No; (b) yes; (c) yes.
10. $Ag^{2+} + e^- \rightarrow Ag^+$, $\mathscr{E}^\circ_{red} = +1.98$ V;
$2H_2O \rightarrow 4H^+ + O_2 + 4e^-$, $\mathscr{E}^\circ_{ox} = -1.23$ V;
$\mathscr{E}^\circ_{cell} = +0.75$ V, reaction is spontaneous, Ag^{2+} is unstable.
11. (a) $2Mn^{3+} \rightarrow Mn^{2+} + MnO_2$, $\mathscr{E}^\circ_{cell} = +0.56$ V, unstable; (b) $3Fe^{2+} \rightarrow 2Fe^{3+} + Fe$, $\mathscr{E}^\circ_{cell} = -1.21$ V, stable;
(c) $2Cu^+ \rightarrow Cu^{2+} + Cu$, $\mathscr{E}^\circ_{cell} = +0.37$ V, unstable.
12. (a) Slightly less stable, but still stable; (b) slightly less
unstable, but still unstable; (c) no effect.
13. $5Fe^{2+} + MnO_4^- + 8H^+ \rightarrow 5Fe^{3+} + Mn^{2+} + 4H_2O$.
14. (a) Ag^{2+}; (b) V; (c) H_2: V and Co; O_2: Co^{3+}, Ag^{2+},
Au^{3+}, and Au$^+$; (d) Au$^+$. **15.** (a) Acid, +1.70 V; base,
+0.04 V; less stable in acid; (b) acid, -1.31 V; base,
-0.85 V; less stable in base. **16.** -109 kJ $\cdot$ mol^{-1}.
17. (a) $[Pt(NH_3)_3Cl_3]^+$, Cl$^-$; $[Pt(NH_3)_4Cl_2]^{2+}$, 2Cl$^-$; (b) 6;
(c) octahedral; (d) +4. **20.** 4; 0. **21.** Square planar.
23. (b) $Co(H_2O)_6Cl_2 + 6NH_3 \rightarrow Co(NH_3)_6Cl_2 + 6H_2O$;
yes. **26.** (a) No; (b) yes. **28.** sp^3d^2 hybridization; octahedral. **29.** (d) dsp^2; square planar. **30.** No. **35.** All but Sc^{3+}
and Zn^{2+}. **37.** $[Cr(H_2O)_6]^{3+}$, purple (compare $KMnO_4$ or I_2
in Figure 20.6); $[Cr(NH_3)_6]^{3+}$, orange; $[Cr(CN)_6]^{3-}$, light
yellow (but very little color). **38.** (a) $SnS_2 + 6HCl \rightarrow$
$H_2SnCl_6 + 2H_2S$; $Sb_2S_3 + 12HCl \rightarrow 2H_3SbCl_6 + 3H_2S$;
(b) $2Au + 3HNO_3 + 11HCl \rightarrow$
$2HAuCl_4 + 3NOCl + 6H_2O$; $Pt + 2HNO_3 + 8HCl \rightarrow$
$H_2PtCl_6 + 2NOCl + 4H_2O$.

Chapter 23

2. 18. **4.** Cl_2, 1.9 times; 1.6952 g $\cdot$ L^{-1}. **5.** (a) At; (b) Cl;
(c) At; (d) At; (e) At; (f) Cl. **7.** (a) No; (b) yes.
8. +0.69 V; -130 kJ $\cdot$ mol^{-1} I_2. **9.** Acid, +1.44 V; base,
+0.61 V. **10.** Yes, $\mathscr{E}^\circ_{cell} = +0.16$ V. **11.** In base,
$\mathscr{E}^\circ_{cell} = +1.66$ V. **12.** Borderline, $\mathscr{E}^\circ_{cell} = 0.00$ V. **13.** Yes,
$\mathscr{E}^\circ_{cell} = +0.39$ V; $3ClO^- \rightarrow 2Cl^- + ClO_3^-$. **14.** Allow
NaClO to disproportionate in 1M base; then add HCl.
15. -56 kJ. **16.** ClO_4^-, Br_2, IO_3^-, and I_2. **19.** (b) and (e).
20. (a) 2BP–BP, 4LP–BP preferred over (b) 2BP–BP, 3LP–BP,
LP–LP or (c) 6LP–BP. **21.** $I_2 + KI \rightarrow KI_3$; sp^3d; linear.
22. sp^3d; seesaw (like (1) in Example 6.3). **23.** Octahedral;
polar; slightly less than 90°. **24.** CCl_4, carbon tetrachloride.
28. (1) $MnO_2 + HI$, (2) $KMnO_4 + HI$, (3) $Cl_2 + HI$;
(2) most readily.
29. $3Br_2 + 3Na_2CO_3 \rightarrow 5NaBr + NaBrO_3 + 3CO_2$;
$5NaBr + NaBrO_3 + 3H_2SO_4 \rightarrow 3Br_2 + 3Na_2SO_4 + 3H_2O$.
31. O_2; Cl_2. **32.** $MnO_2 + 4H^+ + 2e^- \rightarrow Mn^{2+} + 2H_2O$;
$2Cl^- \rightarrow Cl_2 + 2e^-$; no; Cl$^-$ and H$^+$ larger than 1m, T above

298 K. **33.** 0.62 ppm. **34.** Highest, KF, most ionic; lowest, $AlBr_3$, most covalent. **35.** (a) $2Br^- \rightarrow Br_2 + 2e^-$, $SO_4^{2-} + 4H^+ + 2e^- \rightarrow SO_2 + 2H_2O$, $2Br^- + SO_4^{2-} + 4H^+ \rightarrow Br_2 + SO_2 + 2H_2O$; $2I^- \rightarrow I_2 + 2e^-$, $SO_4^{2-} + 10H^+ + 8e^- \rightarrow H_2S + 4H_2O$, $8I^- + SO_4^{2-} + 10H^+ \rightarrow 4I_2 + H_2S + 4H_2O$; (b) I^- is stronger reducing agent than Br^-.
38. $Fe_2O_3 + 6HCl \rightarrow 2FeCl_3 + 3H_2O$; no, mole ratio HCl:HBr = 2.2; dissolves in H_2O. **39.** Angular; sp^3.
40. $3C_{20}H_{42} + 61KClO_3 \rightarrow 61KCl + 6OCO_2 + 63H_2O$.

41. $2NaCl + 2H_2O \xrightarrow{elec} 2NaOH + H_2 + Cl_2$; $Cl_2 + 2NaOH \rightarrow NaCl + NaClO + H_2O$. **43.** Yes, square planar geometry; no, tetrahedral geometry. **44.** $XeF_2 + BrO_3^- + H_2O \rightarrow Xe + BrO_4^- + 2HF$; $XeF_2(ox)$, $BrO_3^-(red)$. **46.** Angular; sp^3.

Chapter 24

8. $S_2O_6^{2-} + H_2O \rightarrow H_2SO_3 + SO_4^{2-}$, $\mathscr{E}°_{cell} = +0.79$ V.
9. Second reaction is spontaneous, $\mathscr{E}°_{cell} = +0.98$ V.
10. (a) $3H_2Se + H_2SeO_4 \rightarrow 4Se + 4H_2O$ will occur ($\mathscr{E}°$ is +1.28 V); (b) $2H_2Te + H_2SO_3 \rightarrow 2Te + S + 3H_2O$ will occur ($\mathscr{E}°$ is +1.19 V);
(c) $H_2SO_3 + H_6TeO_6 \rightarrow H_2SO_4 + TeO_2 + 3H_2O$ will occur ($\mathscr{E}°$ is +0.85 V). For each part there are other possible reactions. **11.** Yes; basic solution. **12.** Acid: $S \rightarrow H_2S + S_2O_3^{2-}$, $\mathscr{E}°_{cell} = -0.36$ V; base: $4S + 6OH^- \rightarrow S_2O_3^{2-} + 2S^{2-} + 3H_2O$, $\mathscr{E}°_{cell} = +0.20$ V. **13.** (a) No, $\mathscr{E}°_{cell} = -0.09$ V; (b) yes, $\mathscr{E}°_{cell} = +0.37$ V; (c) yes, $\mathscr{E}°_{cell} = +0.93$ V. **22.** $2NaOH + SO_2 \rightarrow Na_2SO_3 + H_2O$; $2OH^- + SO_2 \rightarrow SO_3^{2-} + H_2O$. **23.** 74%; 81%. **28.** Base.
33. H_3AsO_4; N_2O. **34.** Unstable; $\mathscr{E}°_{cell}$ for $NH_4^+ + NO_3^- \rightarrow N_2$ is 1.27 V − 0.28 V = +0.99 V.
35. 3670 kJ versus 3450 kJ. **36.** $HNO_3 > H_2SO_4 > HCl$.
37. $HNO_3 + P$. **38.** $\mathscr{E}°_{cell} = +0.66$ V; $2N_2O \rightarrow 2N_2 + O_2$.
39. Yes, $\mathscr{E}°_{cell} = +0.68$ V. **43.** PCl_3, sp^3, yes; PCl_5, sp^3d, no.
45. +4, −3, +3, +5, +3, +3, +3, +3.
52. $6HNO_3 + S \rightarrow 6NO_2 + H_2SO_4 + 2H_2O$.
55. $3HNO_3 + 11HCl + 2Au \rightarrow$ $3NOCl + 2HAuCl_4 + 6H_2O$. **57.** NO_2, NO, N_2O_4, N_2O_3.

Chapter 25

5. $^{30}_{14}Si + ^{1}_{0}n \rightarrow ^{31}_{15}P + ^{0}_{-1}e$. **8.** (a) No; (b) E_a too high; (c) decreases, −0.00342 kJ · mol^{-1} · K^{-1}.
10. 1.2×10^{23} amu. **11.** (d) $H_2SiO_3 + 2NaOH \rightarrow Na_2SiO_3 + 2H_2O$. **12.** All basic, for example, $BO_3^{3-} + H_2O \rightleftharpoons HBO_3^{2-} + OH^-$. **14.** BF_4^-, tetrahedral; SiF_6^{2-}, octahedral. **18.** $NaBO_3$; [:Ö—Ö—B＝Ö:]$^-$.
24. 7.2%. **27.** $2KOH + CO_2 \xrightarrow{} K_2CO_3 + H_2O$; $2OH^- + CO_2 \rightarrow CO_3^{2-} + H_2O$.

29. $NaHCO_3 + HC_3H_5O_3 \rightarrow NaC_3H_5O_3 + H_2O + CO_2\uparrow$.
30. 1.1×10^{-6} g. **31.** (a) Bubbles ($CO_2\uparrow$) form; (b) bubbles ($H_2S\uparrow$) form; (c) H_2S stinks. **38.** $K_2O · Al_2O_3 · 6SiO_2$;

$2CaO · 5MgO · 8SiO_2 · H_2O$. **43.** $B_2H_6 + 3O_2 \rightarrow B_2O_3 + 3H_2O$.

Chapter 26

(Because of space limitations, Lewis formulas and some large condensed formulas are not included in the following answers.)
1. $CH_3(CH_2)_4CH_3$, hexane; $(CH_3)_2CHCH_2CH_2CH_3$, 2-methylpentane; $CH_3CH_2CH(CH_3)CH_2CH_3$, 3-methylpentane; $CH_3CH(CH_3)CH(CH_3)CH_3$, 2,3-dimethylbutane; $CH_3C(CH_3)_2CH_2CH_3$, 2,2-dimethylbutane.
3. (a) $(CH_3)_3CCH_2C(CH_3)_2CH_2CH_3$;
(b) $(CH_3)_3CCH(CH_2CH_3)_2$;
(c) $CH_3CH_2CH_2CH(CH_3CHCH_3)CH_2CH_2CH_3$;
(d) $CH_3CH_2CH(CH_3)C(CH_2CH_3)-$
$(CH_2CH_2CH_3)CH_2CH_2CH_2CH_3$;
(e) $(CH_3)_2CHCH(CH_3)CH(CH_2CH_3)_2$. **4.** (a), (c), and (e) are $C_{10}H_{22}$ isomers. **5.** No; 2-methylheptane, $(CH_3)_2CHCH_2CH_2CH_2CH_2CH_3$. **6.** $CH_3CH_2CH_2CH＝CH_2$, 1-pentene; $CH_3CH_2CH＝CHCH_3$, 2-pentene; $CH_3CH_2C(CH_3)＝CH_2$, 2-methyl-1-butene; $(CH_3)_2CHCH＝CH_2$, 3-methyl-1-butene; $CH_3CH＝C(CH_3)_2$, 2-methyl-2-butene; (b) 2-pentene.
7. (a) 1-heptene; (b) 2-heptene; (c) 2,3-dimethyl-3-hexene; (d) 3-isopropyl-1-heptene.

8. $C_{12}H_{26} \xrightarrow{500-700\,°C} CH_2＝CH_2 + H_2 + CH_4$ + other hydrocarbons; propene, $CH_3CH＝CH_2$. **9.** *cis*-2-hexene, *trans*-2-hexene, *cis*-3-hexene, *trans*-3-hexene, *cis*-3-methyl-2-pentene, *trans*-3-methyl-2-pentene, *cis*-4-methyl-2-pentene, *trans*-4-methyl-2-pentene. **10.** C_5H_8: $CH_3CH_2CH_2C≡CH$, $CH_3CH_2C≡CCH_3$, $(CH_3)_2CHC≡CH$; C_6H_{10}: $CH_3CH_2CH_2CH_2C≡CH$, $CH_3CH_2CH_2C≡CCH_3$, $CH_3CH_2C≡CCH_2CH_3$, $(CH_3)_2CHCH_2C≡CH$, $CH_3CH_2CH(CH_3)C≡CH$, $(CH_3)_3CC≡CH$, $(CH_3)_2CHC≡CCH_3$. **11.** There is only one dimethylbutyne.
12. $CH_2＝CHCH＝CH_2$, $CH_2＝C＝CHCH_3$. **14.** Names: *n*-propylbenzene, isopropylbenzene (cumene), *o*-, *m*-, and *p*-ethyltoluene, and 2,3-, 2,4-, and 3,5-dimethyltoluene.
16. Names: (a) *o*-, *m*-, and *p*-diethylbenzene, *o*-, *m*-, and *p*-*n*-propyltoluene, *o*-, *m*-, and *p*-isopropyltoluene; (b) 3- and 4-ethyl-2-methyltoluene; 2-ethyl-3-methyl-, 2-ethyl-4-methyl-, and 2-ethyl-5-methyltoluene; and 3-ethyl-5-methyltoluene; (c) 2,3,4-, 2,3,5-, and 2,4,5-trimethyltoluene.
18. Anthracene, four formulas; phenanthrene, five formulas; phenanthrene. **19.** Ten dimethylnaphthalenes: 1,2-, 1,3-, 1,4-, 1,5-, 1,6-, 1,7-, 1,8-, 2,3-, 2,6-, and 2,7-; also two ethylnaphthalenes: 1- and 2-. **20.** Cyclopentane, methylcyclobutane, ethylcyclopropane, 1,1-dimethylcyclopropane, 1,2-dimethylcyclopropane (cis and trans isomers); alkenes.
21. Ring is too strained with a triple bond (sp hybridization is associated with linear or near-linear geometry).
22. Cyclobutene, 1- and 3-methylcyclopropene, and bicyclobutane. **25.** (a) Thermal; (b) catalytic. **26.** Methylcy-

clohexane.   **27.** 3.8 L.   **28.** 1-chloro-2-methylbutane, 2-chloro-2-methylbutane, 2-chloro-3-methylbutane, 1-chloro-3-methylbutane. **29.** Nine; chloroethane, 1,1- and 1,2-dichloroethane, 1,1,1- and 1,1,2-trichloroethane, 1,1,1,2- and 1,1,2,2-tetrachloroethane, pentachloroethane, and hexachloroethane.   **30.** (a) CH_3CH_2OH; (b) $CH_3CH_2CHOHCH_3$;   (c) cyclopentanol; (d) $HCBr=CHBr$; (e) $CH_3CBr_2CBr_2CH_3$; (f) 4-methyl-2-nitrotoluene; 2-bromo-1,4-dichlorobenzene. **31.** o-, m-, and p-chlorotoluene.   **32.** 1-   and   2-nitronaphthalene. **33.** $C_{11}H_{23}OH$. **34.** Primary: $CH_3CH_2CH_2CH_2CH_2OH$, 1-pentanol; $(CH_3)_2CHCH_2CH_2OH$,   3-methyl-1-butanol; $CH_3CH_2CH(CH_3)CH_2OH$, 2-methyl-1-butanol; $(CH_3)_3CCH_2OH$, dimethylpropanol; secondary: $CH_3CH_2CH_2CHOHCH_3$, 2-pentanol; $CH_3CH_2CHOHCH_2CH_3$, 3-pentanol; $(CH_3)_2CHCHOHCH_3$, 3-methyl-2-butanol;   tertiary: $(CH_3)_2COHCH_2CH_3$, 2-methyl-2-butanol. **35.** Left to right: primary, secondary, secondary, and tertiary.

37. $CH_3CH_2CH_2OH \xrightarrow{H_2SO_4} CH_3CH=CH_2 + H_2O$;

$2CH_3CH_2CH_2OH \xrightarrow{H_2SO_4}$
$$CH_3CH_2CH_2OCH_2CH_2CH_3 + H_2O;$$
formation of propene. **38.** $CH_2=CH_2 + H_2O \xrightarrow[\text{catalyst}]{acid}$

CH_3CH_2OH; $CH_3CH=CH_2 + H_2O \xrightarrow[\text{catalyst}]{acid} CH_3CHOHCH_3$.

39. $CH_3CH_2OCH_2CH_3$, ethyl ether; $CH_3OCH_2CH_2CH_3$, methyl n-propyl ether; $CH_3OCH(CH_3)_2$, methyl isopropyl ether. **41.** Ethyl ether, n-propyl ether, and some ethylene and propene are also produced. **42.** The primary alcohols. **43.** (a)Two;(b) four;(c) eight; $CH_3CH_2CH_2CH_2CH_2CH=O$, $(CH_3)_2CHCH_2CH_2CH=O$, $CH_3CH_2CH(CH_3)CH_2CH=O$, $CH_3CH_2CH_2CH(CH_3)CH=O$, $(CH_3)_3CCH_2CH=O$, $CH_3CH_2C(CH_3)_2CH=O$, $(CH_3)_2CHCH(CH_3)CH=O$, $(CH_3CH_2)_2CHCH=O$. **44.** To prevent combustion to CO_2 and H_2O. **45.** (a) One, $CH_3CH_2C=OCH_3$;   (b) three, $CH_3CH_2CH_2C=OCH_3$, $CH_3CH_2C=OCH_2CH_3$, $(CH_3)_2CHC=OCH_3$; (c) six, $CH_3CH_2CH_2CH_2C=OCH_3$, $CH_3CH_2CH_2C=OCH_2CH_3$, $(CH_3)_2CHCH_2C=OCH_3$, $CH_3CH_2CH(CH_3)C=OCH_3$, $(CH_3)_3CC=OCH_3$, $(CH_3)_2CHC=OCH_2CH_3$. **46.** The secondary alcohols; for Exercise 34, see part (b) of Exercise 45. **47.** See Section 26.9; $C_6H_5CH_2OH$, alcohol; $C_6H_5OCH_3$, ether. **50.** (a) Two;

(b) four.   **51.** $CH_3CH_2CH_2CH_2CO_2H$,   pentanoic   acid; $(CH_3)_2CHCH_2CO_2H$, 3-methylbutanoic acid; $CH_3CH_2CH(CH_3)CO_2H$, 2-methylbutanoic acid; $(CH_3)_3CCO_2H$, dimethylpropanoic acid.

54. (a) $CH_3CO_2H + HOCH_2CH_3 \xrightarrow{H_2SO_4}$
$$CH_3CO_2CH_2CH_3 + H_2O;$$

(b) $CH_3CH_2CH_2CO_2H + HOCH_3 \xrightarrow{H_2SO_4}$
$$CH_3CH_2CH_2CO_2CH_3 + H_2O;$$

(c) $C_6H_5CO_2H + HOCH_2CH_3 \xrightarrow{H_2SO_4}$
$$C_6H_5CO_2CH_2CH_3 + H_2O.$$

55. $CH_3CH_2CO_2CH_3$, methyl propanoate; $CH_3CO_2CH_2CH_3$, ethyl ethanoate (ethyl acetate); $HCO_2CH_2CH_2CH_3$, n-propyl methanoate (n-propyl formate); $HCO_2CH(CH_3)_2$, isopropyl   methanoate   (isopropyl   formate). **56.** $(CH_3)_2CHCO_2CH_2CH_2CH_3$, n-propyl methyl propanoate (n-propyl isobutyrate). **57.** $CH_2CH_2CH_2C=O$. **58.** One, two, and three, respectively.

$\underset{\rule[1ex]{4em}{0.4pt}\;O\;\rule[1ex]{4em}{0.4pt}}{}$

Chapter 27

1. $CH_3CH_2\overset{*}{C}HICH_3$, $CH_3\overset{*}{C}HCl\overset{*}{C}HClCH_3$, $CH_3CH_2CH_2\overset{*}{C}H(CH_3)CH_2Cl$.   **2.** $CH_3CH_2CHClCH_3$. **3.** $CH_3CH_2CH_2CHClCH_3$, $(CH_3)_2CHCHClCH_3$, $CH_3CH_2CH(CH_3)CH_2Cl$. **6.** One of the stereoisomers is achiral. **8.** No; a 50:50 mixture of enantiomers formed. **9.** (a) $HOCH_2CHOHCHOHCHOHCHO$; (b) aldopentose; (c) 8. **13.** $HOCH_2CHOHCHOHCHOHCHOHCO_2H$; Ag. **14.** Polysaccharide. **19.** (a) They are the same; (b) configurations at C-2 differ.   **26.** $H_2NCHZCO_2^-$;

$\overset{+}{H_3N}CHZCO_2H$; no.

27. (a) $[-NHCH(R)\overset{\displaystyle\parallel}{\underset{\displaystyle O}{C}}-] + H_2CO_3 \rightleftharpoons$

$$[-NHCH(R)\overset{+1}{C}-]^+ + HCO_3^-;$$
$$\underset{OH}{|}$$

(b) $HCO_3^- + H^+ \rightleftharpoons H_2CO_3$; $HPO_4^{2-} + H^+ \rightleftharpoons H_2PO_4^-$; (c) $HBO_2 + HCO_3^- \rightarrow CO_2 + H_2O + BO_2^-$, where HBO_2 stands for oxyhemoglobin. **31.** There are two optical isomers; only one is adrenalin. **33.** 1-butene; ethylene glycol (with terephthalic acid); glycerol and oxalic acid. **34.** About 12,000. **35.** Condensation polymer.

Index

A, mass number, 70
Abelson, P. H., 591
Absolute temperature, 206
Absolute zero, 10, 206
Accelerators, 580–583
Acetaldehyde, 782, 787
Acetic acid, 322, 475, 782, 789
  glacial, 789
Acetone, 782, 787–788
Acetylene, 135, 175, 753, 771
Acetylsalicylic acid
    (Aspirin), 665
Achiral, 799
Acid anhydrides (Acidic
    oxides), 286–287
Acidic oxides, 286–287
Acidic solution, 327
Acidity (pH), 489, 498–502, 661
Acid rain, 716
Acids
  Arrhenius, 322
  Brønsted-Lowry definition,
    322–329
  carboxylic, 782, 788–790
  conjugate, 323
  diprotic, 324, 474
  fatty, 789, 792
  ionization constants, 474
  ionization of, 322, 473–475
  leveling effect, 326
  Lewis definition, 337–338
  monoprotic, 323, 473–474
  naming inorganic, 329
  naming organic, 789
  polyprotic, 474–475,
    479–482
  properties, 322
  relative strengths, 325–327
  triprotic, 324, 475
Acrilan, 822
Acrolein, 787
Acrylonitrile, 822
Actinide series, 625, 634
Actinium, 634
Activated complex, 405
Activation energy, 411
Activity, 535
Activity series
Addition polymers, 822

Addition reaction, 780
Adenosine triphosphate,
    815–817
Adhesive forces, 238
Adiabatic system, 392
Adrenaline, 818
Adsorption, 363–364, 414–415
Aerosol, 362, 364
Agate, 754
Age
  of earth, 576–577
  of organic material, 577–578
Air pollution, 268, 275–278
  and carbon monoxide, 268
  and Ethyl gasoline, 692
  and oxides of nitrogen,
    276–278
  and ozone, 275–278
  and sulfur dioxide, 277
Alcohols, 781–788
Aldehydes, 782, 786–787
Aldohexoses, 803
Aliphatic hydrocarbons,
    765–772, 789
Alkali family, 626–633
  atomic radii, 626
  chemical properties, 627–629
  compounds, 629–633
  flame spectra, 94, 627
  physical properties, 626–627
  preparation, 644
Alkaline earth family, 626–633
  chemical properties, 627–629
  compounds, 629–633
  flame spectra, 94, 627
  physical properties, 626–627
  preparation, 644
Alkanes, 765–769
  isomerism in, 766
  nomenclature, 766–767
  structure, 766
  See also Hydrocarbons
Alkenes
  nomenclature, 770
  structure, 769–770
  See also Hydrocarbons
Alkyl group, 767
Alkynes
  nomenclature, 771–772

  structure, 771–772
  See also Hydrocarbons
Allotropic forms, 264
Alloys, 641, 728
Alpha emission, 567, 569–571
Alpha helix, 811
Alpha particle, 63–65, 564, 569
Alternating current, 308
Aluminum, 634–636
  metallurgy, 649–651
  recycling, 651
Aluminum chloride, 694
Aluminum family, 634–641
Aluminum fluoride, 694
Alums, 721
Americium, 591
Amethyst, 754
Amide ion, 326, 730
Aminoacetic acid, 810
Amino acids, 809–813
Ammonia, 325, 460, 729–732
  comparison with water,
    729–730
  key chemical, 731
  sp^3 hybridization, 173
Ammonia complexes, 520–521
Ammoniated electron, 730
Ammonium carbonate, 736
Ammonium chloride, 138, 736
Ammonium compounds, 736
Ammonium fluoride, 736
Ammonium hydroxide, 312
Ammonium ion, 730
  as an acid, 327, 484–485
Ammonium nitrate, 731
Ammonium nitrite, 727
Ammonium polysulfide, 715
Ammonium sulfate, 721, 736
Ammonium sulfide, 736
Amontons, G., 208
Amontons's law, 208–209
Amorphous substances, 245
Ampere, 5, 535, 836
Amphiboles, 756
Amphiprotic, 324
Amphoteric hydroxides, 733
Amphoteric substance, 324
Amu, 40, 70, 837
Analytical methods, 50

Anesthetics, 775
Angstrom unit, Å, 73, 838
Angular molecule, 181
Anhydrous compound, 287
Anions (Negative ions), 122,
    327–329
Anode, 60–61, 308, 534
Anodic oxidation, 639–640
Anorthite, 756
Antibonding orbitals, 190–191
Antimonious acid, 733
Antimony (Nitrogen family),
    707, 721–736
Antimonyl chloride, 733
Antineutrino, 566
Antimony (III) oxide, 736
Aqua regia, 675, 734–735
Aquated ion, 310
Aquation, 294
Arago, Jean, 801
Aragonite, 752
Argon (Noble gas family), 29,
    680, 697–702
Arithmetical procedures,
    827–834
Aromatic hydrocarbons,
    772–774
  See also Hydrocarbons
Aromatization, 777
Arrhenius theory, 322
Arsenic (Nitrogen family), 707,
    721–736
Arsenic(III) oxide, 736
Arsenious acid, 732
Arsine, 729
Asbestos, 756
Aspirin, 665
Astatine, 680
Aston, F. W., 67
Asymmetric carbon atom, 800
Atmospheric pressure,
    199–201, 838
Atom, 24, 30, 70
  See also Atoms
Atomic bomb, 591–594
Atomic energy, 16, 565–602
Atomic mass (Atomic weight),
    25, 30, 40, 70–71
Atomic mass unit, amu, 40,
    70, 837

Atomic number, 65–66, 79
Atomic orbitals, 106–107
Atomic radius, 149
Atomic structure, 73–87
Atomic theory of Dalton, 24
Atomic volume, 144–147
  table of relative sizes,
    146–147
Atomic weight, 25, 40, 70–71
  determination, 25
  standard of, 70
Atoms, 24, 30, 70
  diagrams, 80
  nuclei, 62–70
  size, 73, 144–150, 722
  structure, 60–80
  wave mechanics model, 104
  weights, 25, 40, 70–71
ATP, 315–317
Attraction, intermolecular,
    235–239
Aufbau principle, 84–85,
    89, 113
Autocatalytic, 660
Auxochromes, 611
Avogadro, Amadeo, 209, 218
Avogadro's law, 209–211
Avogadro's number, 41, 55, 837
  from X-ray measurements,
    248
Avogadro's volume, 210
Azeotropes, 355

Background radiation, 573
Bakelite, 823
Baking powder, 751
Baking soda, 751
Ball-and-stick models, 188
Balmer, J. J., 99
Balmer series, 99
Barium (Alkaline earth family),
    626–633
Barium oxide, 631
Barium peroxide, 631
Barium sulfate, 632, 720
Barometers, 199
Bartlett, Neil, 699
Bases
  Arrhenius, 322
  Brønsted-Lowry definition,
    322–329
  conjugate, 323
  dihydroxy, 324
  ionization constants,
    476–482
  Lewis definition, 337–338
  monohydroxy, 324
Basic oxides, 286–287
Basic solution, 327
Batch process, 355

Battery, 531, 556–559
  alkaline, 557
  mercury(II) oxide–zinc, 557
  nickel–cadmium, 558
  silver oxide–zinc, 557
Bauxite, 650
Becker, H., 380
Becquerel, Henri, 62
Beeswax, 791
Bending of bonds, 612
Benzaldehyde, 787
Benzene, 772–774
  absorption spectrum, 614
  addition reactions of, 781
  isomerism in, 773
  key chemical, 772
  nmr spectrum, 616
  nomenclature, 773
  structure, 184–187, 772
  substitution reactions, 781
Benzoic acid, 797
Benzyl acetate
  absorption spectrum, 614
  nmr spectrum, 616
Berkelium, 591
Berthollet, Claude, 21
Berthollides, 21
Beryllium (Alkaline earth
    family), 626–633
Beryllium chloride, 168–169
Berzelius, J. J., 26
Beta emission, 63, 567, 571–572
Beta particle, 64, 571–572
Bicarbonate ion, 186
Bicarbonates, 751–752
Binary compounds, 155–156,
    329, 637
  of hydrogen 713–714,
    728–732
Binding energy
  electrons, 115–116
  nucleus, 583–589
Biochemistry, 799–823
Biodegradable, 821
Biosphere, 3
Biot, Jean, 801
Bismuth (Nitrogen family),
    707, 721–736
Bismuthine, 729
Bismuthyl ion, 733
Blackett, P. M. S., 579
Blast furnace, 647
Bleaching, 686, 691
Blood sugar (Glucose), 803–806
Body-centered cubic, 254
Bohr, N., 74, 99
Bohr atom, 99–104
  equation, 100
  radius, 108
  theory, 99–104

Boiling point, 28, 242–243
  effect of structure, 237
  normal, 243
  of solutions, 345–346
  variation with pressure, 243
Bombardment reactions,
    579–580
Bond angles, 172, 181, 186
Bond dissociation energy,
    383–385, 609
Bond energy, 384, 608–611
Bonding electrons, 164–166
Bonding orbital, 190–191
Bond length, 144–148
Bond moment, 142
Bond order, 192
Bonds
  chemical, 121
  coordinate covalent,
    136–138
  covalent, 129
  covalent-ionic, 694
  distance, 145–148
  double, 134–137, 174
  electrovalent, 126
  hydrogen, 236
  ionic, 126
  ionic-covalent, 694
  metallic, 148–149
  peptide, 810
  pi, 167, 194
  polar, 142
  polar covalent, 138–144
  sigma, 165–167
  triple, 134–136, 174
Boneblack, 741
Boranes, 761
Borax, 747
Borderline elements
    (Metalloids), 121,
    624–625
Boric acid, 745, 747
Born, Max, 386
Born-Haber cycle, 386–389
Boron, 741–761
Boron nitride, 760
Boron trichloride, 168–169, 338
Boron trifluoride, 694, 760
Bothe, W., 580
Bottled gas, 766
Boyle, Robert, 201
Boyle's law, 201–202
Bragg, W. H. and W. L., 246
Bragg equation, 247
Branched-chain isomer, 766
Breeder reactors, 597–598
Brimstone, 712
British thermal unit, 748, 838
de Broglie, Louis, 104
Bromic acid, 696

Bromine (Halogen family),
    680–697
Bromobenzene, 781
Bromo-carnallite, 688
Brønsted, J. N., 322
Brown, Robert, 363
Brownian movement, 363
Btu, 748, 838
Buffered solution, 493–498
Bumping, 243
Buna rubber, 822
Buret, 335
Butadiene, 822
Butane, 765–766
Butanedione, 788
Butanols, 782
Butene, 770
Butyne, 770
Butyric acid, 789

Cadmium, 634, 636
Calcite, 632, 752
Calcium (Alkaline earth
    family), 626–633
Calcium bicarbonate, 752–753
Calcium carbide, 753, 771
Calcium carbonate, 632, 747,
    751–753
Calcium chloride, 632
Calcium cyanamide, 753, 755
Calcium fluoride, 688
Calcium hydroxide, 631
Calcium oxide, key
    chemical, 630
Calcium phosphate, 727, 735
Calcium silicate, 727
Calcium sulfate, 720
Californium, 591
Calorie, 11, 838
Calorimeter, 374–375
Calvin, Melvin, 578
Camphor, 346, 788
Candela, 5, 836
Capillary rise, 238
Carbides, 753–754
Carbohydrates, 803–809
Carbolic acid, 788
Carbon, 741–761
  chiral carbon atom, 799–781
  determination of, 50
Carbon-14, 577–578
Carbonate ion, 186
Carbonates, 632, 751–753
Carbon black, key chemical,
    742
Carbon dioxide, 267, 748–751
  key chemical, 748
  photosynthesis, 578, 749
  structure, 135, 176
Carbon dioxide cycle, 271, 749

Carbon dioxide fire
    extinguishers, 751
Carbon disulfide, 136, 712,
    714, 755
Carbonic acid, 744–745
    ionization, 474–475
Carbon monoxide, 267,
    747–748
    and air pollution, 268
    heating value, 748
    poisonous action, 267–268
    reduction with, 747–748
Carbon tetrachloride, 130,
    694, 755
Carbonyl group, 790
Carbonyls, 753–754
Carborundum, 755
Carboxyl group, 788, 790
Carboxylic acids, 782, 788–790
    association of, 790
    nomenclature of, 789
Carnauba wax, 791
Catalysts, 265, 277–278,
    413–415
    mechanism of, 413–415
Cathode, 60–61, 308, 534
Cathode rays, 61–62
Cathode ray tube, 61
Cathodic protection, 639
Cations (Positive ions), 122,
    327–329
Caustic soda (Sodium
    hydroxide), 631, 689
Cave formation, 753
Cavendish, H., 262
Cellulose, 807–809
Cellulose acetate, 788, 809
Cellulose nitrate, 809
Celsius, A., 10
Celsius scale, 10
Cement, 757
Centigrade scale, 10
Centimeter, 6
Ceramics, 757
Cesium (Alkali family),
    626–633
Chadwick, J., 69
Chain mechanism, 433
Chain reaction, 432–434,
    592–594
Chalk, 752
Changes
    chemical, 15, 22–23
    in energy, 16
    in matter, 15
    physical, 15
    of state, 230–235
Charcoal, 741
    as adsorbent, 364
Charge density, 250

Charles, J., 206
Charles's law, 206–207
Chelates, 664–665
Chemical bonds, 121
Chemical change, 15, 22–23
Chemical combination
    (Compounds), 18, 123
Chemical energy, 16
Chemical equations, 35,
    37–39, 42
Chemical equilibria, 441–467
Chemical formulas, 26, 30,
    35–37
Chemical kinetics, 404
Chemical reactions, 22–23
    atomic structure and, 122
    endothermic, 17–18, 380
    exothermic, 17–18, 380
    neutralization, 331–335
    oxidation–reduction,
        269–271, 312–315
    volume–volume
        relationships, 218
    weight–volume
        relationships, 217
Chemistry, definition, 2
Chiral carbon atom, 799–801
Chirality, 799–803, 811
Chloric acid, 330–331, 697
Chlorine, 680–697
    in the environment, 692
    key chemical, 688
Chlorine monoxide, 695
Chloroform, 130, 693
Chlorophyll, 816–817
Chlorous acid, 330–331, 475
Chromatography, 364–365
Chromic acid, 634
Chromium, 634–636
Chromium family, 634–641
Chromophores, 611
Cinnamaldehyde, 787
Cis isomer, 667, 770
Citric acid, 791
Civetone, 788
Classification of elements, 121
Clay, 757
Clorox, 686
Cloud chamber, 567–568
Cloud tracks, 567–568
Coal, 768, 775–776
Coal gas, 748
Coal tar, 775
Cobalt, 634, 636
Cohesive forces, 238
Coke, 281, 741, 775
Colligative properties, 343, 477
Colligative properties law, 343
Colloidal dispersions, 362
    color, 361

    properties, 362–364
    types, 362
Colloidal particles, 360–361
Colloidal state, 360–367
Color
    and chemical
        constitution, 611
    of colloids, 361
    of compounds, 611
    and electron structure,
        672–674
Combining weight (Equivalent
    weight), 304–307,
    333–335
Combustion, 20, 267, 778–779
    analysis, 50
Common ion effect, 491–493,
    512–518
Common names, 767
Complexes (Coordination
    compounds), 663–675
Complex ions, 507, 519–523
    See also Coordination
        compounds
Compounds, 18, 123
    composition, 48–51
    intermediate, 267, 413
    ionic, 126
    polar-covalent, 138–144
    systematic naming, 155, 158
Concentration
    effect on reaction rate, 415
    of solutions, 297, 302
Condensation polymers, 822
Conjugate acids and bases, 323
Conjugated proteins, 810
Conservation of energy, 19
    of mass, 19–20
Constants, table of, 837
Contact process, 718–720
Continuous process, 355
Contour, 90-percent, 107
Contour diagrams, 107–108,
    166–167
Control rods, 596
Conversion factors, 838
Converter reactor, 598
Coordinate covalent bond,
    136–138, 663
Coordination
    linear, 665
    octahedral, 667
    planar, 666
    square-planar, 666
    tetrahedral, 665–667
    trigonal bipyramidal, 667
    trigonal planar, 665
Coordination compounds,
    663–675
    color, 672–674

    metal-ligand shapes, 666
    orbitals in, 670–671
    uses, 673
Coordination number,
    249–250, 664–668
Coordination theory, 249
Copper, 634–636, 645–647
Copper(I), -(II), 158
Copper family, 634–641
Copper sulfate
    pentahydrate, 287
Corrosion, 639–641
Cortisone, 818
Cosmic rays, 573
Cottrell precipitator, 364
Coulomb, 74, 535, 553, 837
Coulomb's law, 74
Couples, 539
Covalent bond, 129, 133–134
Covalent compounds, 129, 317
    solid state, 251–252
Covalent-ionic bond, 694
Covalent molecules, 129,
    133–134
Covalent radius, 144–148
Cracking, 776
    catalytic, 777
    thermal, 776
Creosote oil, 788
Cresols, 788
Crick, F. H. C., 812–813
Critical mass, 592–594
Critical pressure, 239
Critical temperature, 239
Cryolite, 649, 688
Crystal defects, 256
Crystal energy, 387
Crystal lattice, 246
Crystalline solids, 245–248
    cubic, 245
    defects in, 256
    hexagonal, 247, 254
    ionic substances, 245
    isomorphous, 245
    polymorphous, 245
    structure determination, 245
    structure of metals, 253
    types, 247, 254
Cupric, 158
Cuprous, 158
Curie, 573, 838
Curie, Marie and Pierre, 573
Curium, 591
Cyanides, 753–754
Cycloalkanes, 774–775
Cyclobutane, 774
Cyclohexane, 774
    key chemical, 775
Cyclohexanone, 788
Cyclopentane, 774

Cyclopropane, 774–775
Cyclotron, 582
Cytoplasm, 813

Dacron, 823
Dalton, John, 24, 60
Daltonides, 21
Dalton's atomic theory, 24–25
Dalton's law, 218–220
Dalton's symbols, 24
Daniell cell, 533
Dating by radioactivity,
    576–578
DDT, 298, 820
de Broglie, Louis, 104
Debye, Peter, 141
Debyes, 141
Decanes, 765
Decomposition potentials, 554
Dees, 582
Defect structure, 256, 744
Definite composition
    isotopes and, 72–73
    law of, 20–21
    limitations of law of, 21
Degenerate orbitals, 116, 191
Degree of rotation (Optical
        isomer), 668
Dehydrating agent, 718
Deliquescence, 295, 632
Delta, Δ, 11
Delocalization energy, 187
Delocalized electrons, 187
Dempster, A. J., 67–68
Denatured alcohol, 785
Density, 8
    of common substances, 9
    of gases, 215–216
    of metals, 626–636
    of water, 253
Dentate, bi-, tri-, et cetera, 664
Deoxyribonucleic acid,
        811–813
Deoxyribose, 812
Derivatives of hydrocarbons,
        781–794, 799–823
    complex, 799–823
    oxygen, 781–794
Desalination of seawater,
        358–359
Descriptive chemistry, 2
Desiccants, 295
Detergents, 793, 814–815
Deuterium, 278
    isotopic weight, 71
Dewar flask, 751
Dextrorotatory
        compounds, 801
Dextrose, 803
Dialysis, 365–366

Diamagnetic, 669
Diamond, 742
Diaphragm process, 689
Dichromate ion, 660
Dichromic acid, 634
Dielectric constant, 141, 295
Diesel oil, 776
Diffusion
    of gases, 220–221, 594
    Graham's law of, 220
    of liquids, 294
Dimethylglyoxime, 665, 675
Dinitrophenol, 498–499
Dioxin, 820
Dipole, 140–144
Dipole–dipole attraction,
        235–236
Dipole moment, 324
Diprotic acid, 324
Direct current, 308
Disaccharides, 803, 806
Dispersion, colloidal, 362
Disproportionation, 661–662
Dissociation energy,
        383–385, 609
Distillate, 354
Distillation, 18, 353–355
    fractional, 354–355
    of petroleum, 777
DNA, 811–813
Döbereiner, J., 26
d orbitals. *See* Orbitals
Double bond, 174
Double salts, 721
Dreiding models, 189
Drugs, 818–821
Dry cell, 556
Dry ice, 751

$E, \Delta E$, 411–412
E_a, 411
$\mathscr{E}_B^\circ$, 345
$\mathscr{E}_{red}^\circ$, 537–539, 842–844
EDTA, 675
Efflorescence, 295
Effusion, 220
Einstein, Albert, 98–99,
        363, 585
Einsteinium, 591
Einstein's mass-energy law,
        585–589
Electrochemical series,
        538–540, 842–845
Electrochemistry, 531, 552–559
Electrode, 60–61
    glass, 546
    hydrogen, 534–535
    potential, 544
    products, 554–555
    standard, 535

Electrodialysis, 358–359
Electrolysis, 281, 552–555
Electrolyte, 307–312
    solutions of, 350–353
    strong, 311–312
    weak, 311–312, 518–519
Electromagnetic radiation,
        93–94
Electron affinity, 386
Electron arrangements, 79–80
Electron cloud, 107
Electronic configuration, 83–85
Electron delocalization, 187
Electron density, 107
Electron dot formulas (Lewis
        formulas), 130–138
Electronegativity, 138–140
    of IA elements, 626
    of IIA elements, 626
    of VA elements, 722
    of VIA elements, 708
    of VIIA elements, 681
    transition elements, 636
Electron jumps, 100
Electron levels (Energy levels),
        77–78
Electrons, 62
    antibonding, 190–191
    arrangement in atoms, 73–74
    arrangement in ions, 75–76
    binding energy, 583–589
    bonding, 190–191
    delocalized, 187
    evidence for, 62
    lone pair, 131
    in metals, 255–256
    ratio of charge to mass, 62
    sharing, 128–132
    transfer, 123–126
    weight, 62
Electron volt, 74–75
Electroplating, 555
Electrostatic force, 74
Electrovalent compounds, 126
Elementary reaction, 404
Elements, 18, 30
    classification, 121
    natural abundance, 264
    synthesis, 583
Emf diagrams, 657–658, 662,
        682, 710, 724
Emissions, nuclear, 64,
        567–573
Emission spectra, 95–96
Empirical formula, 55
Emulsion, 362, 367
Enantiomers, 799–803, 811
Endothermic compounds, 380
Endothermic reactions, 18,
        376, 380

Endpoint of titration, 499
Energy, 16, 600–601
    activation, 411
    atomic, 16, 565–602
    binding, 115–116, 583–589
    bond, 383–385, 609
    chemical, 16
    in chemical changes,
        372–389
    common types, 16
    conservation, 19
    conversions, 16
    crystal, 386
    delocalization, 187
    electric, 16
    electromagnetic radiation,
        93–94
    and a hydrogen economy,
        284–286
    internal, 389
    kinetic, 16
    and mass, 585–587
    of matter, 585
    nuclear, 16, 595–598
    nuclear production, 595–602
    potential, 16
    radiant, 16
    resonance, 185–186
    rotational, 95, 224
    stellar, 599
    vibrational, 611–615
Energy levels, 77–78
Energy sublevels, 81–87
Enthalpy, 374–383, 387
    absolute, 379
    of atomization, 387
    of crystal energy, 387
    of dissociation,
        383–385, 609
    of electron affinity, 386–387
    of formation, 378–380,
        382, 387
    of fusion, 232
    of ionization, 385–387
    of neutralization, 331–335
    of reactions, 381–383
    standard, 376–383
    of sublimation, 387
    of vaporization, 232
Entropy, 391–392
    absolute, 392
Enzymes, 414, 803, 813–815
Epsom salts (Magnesium
        sulfate), 721
Equations
    balancing, 35, 37–39
    chemical, 35, 42
    ionic, 312–315
Equilibrium, 240
    chemical, 441–467

heterogeneous vs.
   homogeneous, 462
ionic, 311
Equilibrium constant, 444–456
   and catalyst, 466–467
   from electrode potentials,
      551–552
   and free energy, 464, 465,
      544–547
   K_c, 445–456
   K_p, 460–465
   and pressure, 456–465
   and temperature, 465–466
Equivalence point, 499–502
Equivalents, 335, 838
Equivalent weight, 304–307,
      333–335
Erg, 838
Esterification, 409, 790
Esters, 782, 790–794
Estrone, 788, 818
Ethane, 133, 172, 765, 767
Ethanoic acid (Acetic acid), 475
Ethanol, 134, 780–785
   absorption spectrum, 614
   nmr spectrum, 616
   vapor pressure, 241
Ethene, 770
Ethers, 782, 786
Ethyl acetate. 782, 790
Ethyl alcohol. See Ethanol
Ethylene, 135, 174, 822
   key chemical, 769
Ethylene bromide, 692
Ethylenediamine, 665
Ethylene glycol, 785, 823
   key chemical, 785
   vapor pressure, 241
Ethyl chloride, 780
Ethyl ether, 782, 786
   vapor pressure, 241
Ethyl gasoline, 692
Ethyl group, 767
Ethyne, 770
Eutrophication, 735
eV, electron volt, 74–75
Excess reactant, 46–48
Excited states, 117
Exothermic compounds, 380
Exothermic reactions, 17,
      376, 380
Exponents, 829–830
Extraterrestrial chemistry,
      758–759

Face-centered cubic, 254
Fact, theory, 698
Factor-units method, 832–834
Fahrenheit scale, 11
Families of elements, 80, 89, 634

Faraday, 552–553
Faraday, Michael, 553
Faraday's law, 552–553
Fats, 792–794
Fatty acids, 789, 792
Feldspars, 756
Fermi, Enrico, 591
Fermi gas model, 591
Fermium, 591
Ferredoxin, 817
Ferric, 158
Ferromagnetic, 669
Ferrous, 158
Field ion microscope, 255
First-order reaction,
      421–428, 835
Fission, nuclear, 588, 591–595
Fission bombs, 591–594
Fixation of nitrogen, 725–726
Flame spectra, 627
Flint, 754
Flotation of ores, 642–643
Fluorescence, 61, 567
Fluorides, teeth protection,
      734
Fluorine (Halogen family),
      680–697
Fluorspar, 688
Flux, 642–643
Foam, 362
Fool's gold, 715
f orbitals. See Orbitals
Formaldehyde, 136, 759, 784,
      786–787, 823
   key chemical, 786
Formic acid, 136, 475, 784, 789
Formulas, 26, 30, 35–37
   arithmetical, 827
   configurational, 801
   empirical, 36, 51–53
   from experimental data,
      49–54
   Lewis, 130–138
   molecular, 36, 53–54
   and molecular weights, 54
   from oxidation numbers,
      154–155
   percentage composition, 50
   projection, 801
   structural, 128–138
Formula weight, 40
Fossil fuel cycle, 284
Fractional distillation,
      355, 777
Fractional precipitation, 515
Fractionating columns,
      355–356, 777
Framework models, 189
Francium (Alkali family),
      626–633

Frasch, Herman, 713
Frasch process, 713
Fraunhofer lines, 97
Free energy, 392–398
   of activation, 411
   from electrode potentials,
      544–547
   and equilibrium, 464–465
   and equilibrium constant,
      464–465
   of formation, 393–396
   influence of temperature,
      396–398
   of reaction, 396
Free radical, 405, 433
Free radical mechanism (Chain
      mechanism), 433
Freezing point, 244, 346–347
   effect of solute, 347
Freon, 276
Frequency of light, 93, 99
Fructose, 805–806
Fuel cells, 556–559
Fuel gases, 748, 768
Fuel oil, 776
Functional groups, 781
Fused-ring hydrocarbons, 774
Fusion, nuclear, 588, 598–599

G, ΔG, 463–464
Gallium, 636–637
Galvanic cells (Voltaic cells),
      531–535, 539–544
Galvanized iron, 640
Gamma rays, 64, 567, 572
Gangue, 642
Gas chromatography, 365
Gas constant, R, 212–213, 222,
      463, 541, 837
Gaseous diffusion, 594
Gaseous state, 198–224
Gases, 198–224
   densities, 215–216
   and the kinetic theory,
      222–224
   particles in, 199
   pressure of, 199–201
   temperature effects on,
      204–209
   volume–volume
      relationship, 218
   volume–weight relationship,
      217–218
Gas ionization counters,
      568–569
Gas laws, 201–224
   deviations from, 221–222
   general equation, 211–216
   ideal gas constant, 212
Gas oil, 776

Gasoline, 692, 775–776
Gay-Lussac, J. L., 208, 218
Gay-Lussac's laws,
      208–209, 218
Geiger, H., 64
Geiger tube, 569
Geometric isomerism, 770
Germanium, 634, 636
Germanium family, 634–641
Ghiorso, A., 591
Gibbs free energy, 393
Glacial acetic acid, 789
Glass, 696, 757–758
Glass electrode, 546
Glauber's salt, 633
Glucose, 803–806
Glycerin (glycerol), 785
Gold, 634, 636
Goldstein, E., 66
Gouy balance, 669
Graham, Thomas, 220
Graham's law, 220–221, 594
Grain alcohol. See Ethanol
Gram, defined, 8
Gram-equivalent weight
      (Equivalent weight),
      304–307, 333–335
Gram-molecular volume
      (Molar gas volume),
      210, 222
Gram-molecular weight (Molar
      weight), 41–43, 55
Graphite, 742–743
Grating spectroscope, 94
Greenhouse effect, 749, 821
Ground states, 100
Group IA (Alkali family),
      626–633
Group IB (Copper family),
      634–641
Group IIA (Alkaline earth
      family), 626–633
Group IIB (Zinc family),
      634–641
Group IIIA (Aluminum family),
      634–641
Group IIIB (Scandium family),
      634–641
Group IVA (Germanium
      family), 634–641
Group IVB (Titanium family),
      634–641
Group VA (Nitrogen family),
      721–736
Group VB (Vanadium family),
      634–641
Group VIA (Sulfur family),
      707–721
Group VIB (Chromium family),
      634–641

Group VIIA (Halogen family), 680–697
Group VIIB (Manganese family), 634–641
Group VIIIA (Noble gas family), 697–702
Group VIIIB (Iron family; Platinum family), 634–641
Groups, 80, 89
Guncotton, 809
Gypsum, 720

h (Planck's constant), 98–105
H, ΔH, 375–376
$\Delta H_{a,m}$, 387
$\Delta H_{a,n}$, 387
ΔH_{dis}, 387
ΔH_{ea}, 386
ΔH_f°, 386
ΔH_{ion}, 386
ΔH_r°, 386
ΔH_{sub}, 386
ΔH_{xtal}, 386
Haber, Fritz, 386, 731
Haber process, 731
Hafnium, 634, 636
Half-life, 426–427, 430, 570, 574–576
Half-reaction, 150, 540
Halides, 693–696
    hydrogen, 695–696
    ionic size, 681
    metal, 631–632, 693–694
    nonmetal, 694–695
Halite, 688
Hall, Charles M., 649
Hall process, 649
Halogen family, 680–697
    biological activity, 681
    chemical properties, 680–681
    compounds of, 683–687
    electronegativity, 681
    occurrence, 688
    oxidation–reduction of, 681–683
    oxyhalogens, 696–697
    physical properties, 680
    production, 688
    reactions, 683–687
    uses, 691–693
Halogenoid ions, 693
Halogen oxy-ions, 696–697
    naming, 157
Halothane, 821
Heat
    of condensation, 232
    of formation of an atom, 383–384

of formation of a radical, 384–385
of fusion, 232
of neutralization, 331–335
of reaction, 373–374
of solidification, 232
of vaporization, 232–233
See also Enthalpy
Heat capacity, 231–232
Heat death of the universe, 392
Heat energy, 16
Heavy hydrogen, 278
Heisenberg, W., 105
Heisenberg uncertainty principle, 105–106
Helium (Noble gas family), 697–702
    discovery, 96, 698
Heme, 810, 815
Hemoglobin, 271, 815
Hemotoxin, 772
Henry, William, 301
Henry's law, 301–302
Heptane, 294, 765, 779
Herbicides, 820–821
Hertz, 93, 837
Hess, G. H., 377
Hess's law, 377–378
Heterogeneous equilibria, 462
Heterogeneous mixture, 18–19, 293
Hexagonal close-packed, 254
Hexagonal crystals, 254
Hexane, 294, 765
Hexosans, 807–808
Hexoses, 803
Hindenburg syndrome, 285–286
Homogeneous equilibria, 461
Homogeneous mixtures, 18, 293
Homolog, 765
Homologous series, 765
Homonuclear diatomic molecules, 191
Hormones, 817–818
Hund's rule, 113
    and ionization energies, 114
Hybrid orbitals, 167–178, 184–185, 666, 670
Hydrated ions, 287
Hydrates, 287
Hydration, 287, 294
Hydrazine, 724, 736
Hydrazoic acid (Hydrogen azide), 381, 739
Hydrides, 282, 633
Hydriodic acid, 696
Hydrobromic acid, 696

Hydrocarbons, 765, 794
    aliphatic, 765–772
    alkanes, 765–769
    alkenes, 769–771
    alkynes, 771–772
    aromatic, 772–774
    aromatization of, 777
    chemical properties, 778–781
    conversions, 776
    cracking, 776
    cycloalkanes, 774–775
    derivatives, 781–794
    fused-ring, 774
    nomenclature, IUPAC, 766–767, 770–773
    physical properties, 778
    saturated, 780
    source, 775–776
    unsaturated, 780
Hydrochloric acid, 495, 695–696
    production, 696
Hydrocyanic acid, 475
Hydrodealkylation, 778
Hydrofluoric acid, 475, 696
Hydrogen, 278–286
    atomic, 278
    compounds of, 279
    determination of, 51
    discovery of, 50
    elemental, 278
    an energy source, 284–286
    isotopes, 278
    occurrence, 278
    preparation, 279–282
    production of, 280–282
    reactions of, 282–283
    uses of, 283
Hydrogen acetate (Acetic acid), 475
Hydrogenation, 283, 780
Hydrogen azide, 381, 739
Hydrogen bomb, 599
Hydrogen bond, 236–237
Hydrogen bromide, 695–696
Hydrogen chloride, 695–696
Hydrogen cyanide, 136, 753, 755
Hydrogen electrode, 536
Hydrogen fluoride, 237, 695–696
Hydrogen halides, 695–696
Hydrogen iodide, 695–696
Hydrogen ion, 310
Hydrogen ion concentration (pH), 489–491
Hydrogen nitrate (Nitric acid), 137, 733

Hydrogen nitrite (Nitrous acid), 137
Hydrogen peroxide, 133, 288, 631
Hydrogen phosphate (Phosphoric acid), 735
Hydrogen selenide, 713
Hydrogen sulfide, 714
Hydrogen telluride, 713
Hydrolysis, 327–329
    constants, 484–487
    of fats, 792–793
    of polysaccharides, 808
    of proteins, 809–810
Hydrometer, 10
Hydronium ion, 729
Hydronium ion concentration (pH), 489–491
Hydrosulfuric acid, 475
Hydroxide bases, 322
Hydroxide ion, 322
Hydroxy compounds, structures of, 330
Hydroxylamine, ionization constant, 476
Hygroscopic, 295
Hypochlorous acid, 330, 475, 697
Hypohalous acids, 697
Hypothesis, 4

i, van't Hoff factor, 351, 465
Ice structure, 252–253
Ideal gas constant, 212, 463, 541, 837
    deviation from, 221–222
Ideal gas law, 212
Ideal solutions, 343, 349–350
Immiscible liquids, 296
Indicators, 498–502
    and pH, 498–502
Indium, 636
Inert gases (Noble gases), 697–702
Infrared radiation, 611
    spectrum, 611–615
Initiation reaction, 433
Inner transition series, 625, 634
Inorganic chemistry, 132
Inorganic compounds, 132, 329
Insecticides, 820–821
Insoluble substances, 295–296
Instantaneous induced dipoles, 235
Insulin, 811, 818
Interatomic distance, 144
Interhalogen compounds, 683–684
Intermediate compound, 267, 413

Intermolecular attractive forces, 235–239
Internal energy, 389
Invert sugar, 806, 808
Iodine (Halogen family), 680–697
Ion, 67–68
  aquated, 310
  electronic structure, 123–128
  polarization of, 250–251
  product for water, 483
  size, 144–150, 249–250
  in solution, 309–312
  See also Complex ions; Negative ions; Positive ions
Ion exchange, 359
Ion exchange resins, 359, 635
Ionic bond, 126
Ionic compounds, 126, 248–251, 309–310
Ionic-covalent bonds, 694
Ionic equation, 312–315
Ionic equilibria, 311
  redox and electrochemistry, 531–559
  solubility and the solubility product, 506–529
  solutions of acids and bases, 472–505
Ionic radius, 149
Ionic solids, 249
Ionic structures, 249
Ionization, 287, 473–502
  of acids and bases, 322–327
  of polar covalent compounds, 310–311
  of water, 482–484
  of weak acids, 473–475
  of weak bases, 475–476
Ionization constants, 473–502
Ionization energies, 74–78
  and atomic numbers, 79
  calculation of, 103
  I_A elements, 626
  II_A elements, 626
  V_A elements, 722
  VI_A elements, 708
  VII_A elements, 680
  of molecules, 608
  transition elements, 636
Ionizing radiations, 567
Ion product of water, 483–484
Iridium, 634, 636
Iron, 636, 647–649
Iron(II), -(III), 158
Iron(III) chloride, 694
Iron family, 634–641

Iron pyrite, 715
Isobutane, 766
Isoelectronic, 127
Isomers, 666–668
  cis, trans, (geometric), 667, 770
  in coordination compounds, 666–668
  optical, 668
  in organic compounds, 766, 769–770, 773, 799–803
  ortho, meta, para, 773
  structural, 799
Isomorphous crystals, 245
Isooctane, 779
Isopiestic system, 392
Isopropyl alcohol, 780
  See also 2-propanol, 782
Isothermal system, 392
Isotopes, 69, 72–73, 88, 278, 565, 579
  table of, 71, 72

Janssen, P., 96
Joule, 11, 837
Joule, J. P., 11

k, 419
k_b, 346
k_f, 347
K_a, 474, 476–482
K_b, 476–482
K_c, 445–456, 482–483
K_f, 520–523
K_p, 460–465
K_{sp}, 510–526
K_w, 483–484
Kaolinite, 757
Kekulé, F. A., 184
Kekulé formula, 184, 772
Kelvin, Lord, 206
Kelvin temperature, 5, 10, 206, 836
Kepone, 820
Kerosene, 775–776
Ketohexoses, 803
Ketone, 782, 787–788
Kilogram, 5, 8, 836
Kilowatt-hour, 838
Kinetic energy, 16
Kinetic molecular theory
  of gases, 222–224
  of liquids, 230
  of solids, 230
Kinetics, 404–434
K level (Energy level), 77–78
Kodachrome process, 612
Krypton (Noble gas family), 697–702

Lactic acid, 791, 799–803
Lactose, 806
Lambda (λ), wavelength, 93
Lanthanide series, 625, 635
Lanthanum, 634
Lattice defects, 256
Lattice energy, 386
Laue, M. von, 246
Laughing gas, 736
Lauric acid, 792
Lavoisier, Antoine L., 20, 262–263
Law, 4
Lawrencium, 591
Laws
  Amontons's, 208–209
  Avogadro's, 209–211
  Boyle's, 201–202
  Charles's, 206–207
  colligative properties, 343
  combining volumes, 208–209, 218
  conservation of energy, 19
  conservation of mass, 19–20
  Coulomb's, 74
  Dalton's, 218–220
  definite composition, 20, 72–73
  definite proportions, 20
  equivalence of mass and energy, 585–589
  Faraday's, 552–553
  first law of thermodynamics, 389–390
  Gay-Lussac's, 208–209, 218
  Graham's, 220–221, 594
  Henry's, 301–302
  Hess's, 377–378
  ideal gas law, 212
  multiple proportions, 21–22
  periodic, 79, 89
  Raoult's, 344
  second law of thermodynamics, 391–392
  third law of thermodynamics, 392
  Van't Hoff's, 465
Lead, 634, 636
  as end product of radioactivity, 570
Lead azide, 739
Lead chamber process, 719
Lead storage cell, 558
Lead tetraethyl, 779
Le Chatelier, Henri L., 299
Le Chatelier's principle, 299–300
  and chemical equilibria, 459–460, 465–466

and reaction rates, 459–460, 465–466
  and solutions, 299–300
Leveling effect, 325–327
Levorotatory compounds, 801–802
Levulose, 805
Lewis, G. N., 130
Lewis acids and bases, 337–338
Lewis formulas, 130–138
Ligands, 519–523, 663–673
Light
  diffusion by colloidal system, 362–364
  plane polarized, 801–802
  See also Radiant energy
Lignin, 808
Lime, 630–631
Limestone, 632, 752
Limiting reactant, 46–48
Linacs, 581–582
Linear, 143, 666
Linear accelerator, 581–582
Linear molecules, 168, 666
Linoleic acid, 792
Linolenic acid, 792
Lipids, 788
Liquefaction of gases, 239–240
Liquid air, 264
Liquids, 235–244
  kinetic molecular theory, 230
Liquid state, 235–244
Liter, 7
Lithium, 626–633
Lithium carbonate, 752
Lithium hydroxide, 631
Lithium iodide, 694
Litmus, 499
Lockyer, J. N., 96
Logarithms, 830–831, 840–841
Logarithm table, 840–841
Lomonosov, M. V., 20
London forces, 235–236
Lone pairs, 131
Lowry, J. M., 322–323
LSD, 819
Lubricating oils, 776
Lucretius, 709
Lye, 631
Lyman, T., 101
  lines, 101–102
Lysergic acid diethylamide, 819

McMillan, E. M., 591
Magic numbers, 590
Magnesia, 630–631
Magnesium (Alkaline earth family), 626–633
Magnesium carbonate, 746, 747, 751

Magnesium chloride, 632
Magnesium chloride
        hexahydrate, 287
Magnesium fluoride, 694
Magnesium sulfate, 721
Maltose, 806, 808
Manganese, 634, 636
Manganese family, 634–641
Manometer, 201
Marble, 632, 752
Margarine, 794
Marsden, E., 64–65
Mass, 8
    conservation of, 19–20
    and energy, 585
    mass and weight, 8
    relativistic, 585
    rest, 585
Mass loss, 583–589
Mass number, 70
Mass spectra, 617–618
Mass spectrograph, 67–69
Matches, 728
Matter
    changes in, 15
    classes of, 18
    properties of, 14
    states, 198–199
    wave nature, 104–105
Mechanism of reaction,
        404–407, 423
    free radical, 433
    and order of reaction,
        431–432
Megaton, 594
Melting point, 244, 251
    effect of structure, 250–251
Mendeleev, D. I., 26
Mendelevium, 591
Meniscus, 238
Mercury, 636
    in the environment, 692–693
Mercury-cell process, 689
Mercury oxide, 264
Messenger RNA, 813
Metabolism, 817
Meta isomers, 773
Metallic bond, 255–256
Metallic hydrides, 282
Metallic properties, 255, 636
Metallic radius, 148–149
Metallic solids, 253–256
Metallo-acid elements, 634
Metalloids, 121, 624–625
Metallurgy, 641–645
Metals, 121, 623–651, 655–675
    corrosion of, 639–641
    oxidation-reduction of,
        656–663
    production, 641–651

properties, 255, 627, 636
reactions of, 628
structure of, 253–255
Metasilicic acid, 744–745
Metastable system, 242
Meter, 5, 836
Methanal. See Formaldehyde
Methane, 765–767
    hybridization in, 168–171
Methanoic acid (Formic acid),
        475, 789
Methanol, 133, 782–785
    as gasoline alternate, 784
    key chemical, 785
Methyl alcohol. See Methanol
Methylamine, 136, 311, 396, 476
Methyl ether, 134, 759
Methyl group, 767
Methyl ketone (Acetone), 782,
        787–788
Methyl mercury, 692
Methyl methacrylate, 822
Methyl orange, 499
Methyl red, 499
Methyl yellow, 499
Metric system, 5–7
    table of equivalents, 836–838
Meyer, Lothar, 26
Mica, 756–757
Milk of magnesia, 631
Milliequivalents, 336
Milligram, 8
Millikan, R. A., 62
Milliliter, 7
Minerals, 641–645, 754
Mirex, 820
Miscible liquids, 294
Mixtures, 18
    constant-boiling, 355
mmHg, 838
Models, molecular, 188
Moderator, 596
Molal boiling point
        constant, 346
Molal freezing point
        constant, 347
Molality, 303
Molar gas volume, 210, 222
Molar heat capacity, 231–232
Molar heats, 232
Molarity, 303–304
Molar weight, 41–43, 55
Mole, 5, 41–45, 55, 71, 836
Molecular attraction, 235–239
Molecular bombardment
        (Brownian
        movement), 363
Molecular geometry, 142
    VSEPR model, 178–182
Molecular models, 188

Molecular motion (Kinetic
        molecular theory),
        222–224, 230
Molecular orbitals, 188–190
Molecular orbital theory,
        188–190
Molecular weight, 40, 55,
        347–349
    from bp and fp change,
        347–349
    from gas densities, 211
    precise, 54
Molecules
    defined, 24, 30
    diatomic, 35, 199
    large, 746. See also Polymers,
        822–823
    monatomic, 199
    nonpolar, 138, 143
    polar, 138
    polyatomic, 36
    shape, 178–182
    sizes, 144–150
    structure. See Chapters 5, 6
    triatomic, 35, 199
    vibrations in, 611–615
Mole fraction, 302–303
Mole percent, 302
Molybdenum, 634, 636
Monochromatic radiation, 246
Monoclinic crystals, 709
Monomer, 822
Monoprotic acid, 323, 473–474
Monosaccharides, 803–806
Moseley, H. G. J., 65–66
MO theory, 188–190
Muriatic acid, 695–696
Muscone, 788
Muscovite, 756
Myristic acid, 792

Naming compounds, 155, 158
Naming inorganic acids,
        329–330
Naphthalene, 774
Natrolite, 759–760
Natural gas, 748, 766, 768
Natural radioactive series, 570
Negative ions, 122
Neighbors of transition
        elements, 634
Neon (Noble gas family),
        697–702
Neptunium, 591
Nernst, Walter, 541
Nernst equation, 541–542
Neutralization, 331–335
Neutral solution (pH), 327, 489,
        498–502, 661
Neutrino, 566

Neutron activation cell, 579
Neutron bombardment, 580
Neutron–proton ratio, 572, 589
Neutron rest mass, 837
Neutrons, 69–70, 580
    discovery, 69
Newlands, J., 26
Newton, 837
Nickel, 634, 636
    as catalyst, 415
Ninety percent contour,
        107–108
Niobium, 634–636
Nitrate ion, 182–184, 187
Nitric acid, 133, 733–735
    key chemical, 733
Nitrides, 726
Nitrobenzene, 781
Nitrocellulose, 809
Nitrogen (Nitrogen family),
        721–736
    structure, 135, 176
Nitrogen cycle, 726
Nitrogen dioxide, 725
Nitrogen family, 721–736
    chemical properties, 723
    compounds of, 725–726
    emf diagrams, 724
    occurrence, 727
    oxidation–reduction
        chemistry of, 723–725
    physical properties, 721–723
    preparation of, 727
    production of, 727–728
    reactions of, 725–726
    uses of, 727
Nitrogen fixation, 725–726
Nitrogen oxides, 276–277, 725
Nitrogen tetroxide, 724
Nitrogen trichloride, 694
Nitroglycerin, 785, 809
Nitrosyl chloride, 735
Nitrous acid, 137, 475, 732
Nitrous oxide, 736
Nmr, 615–617
Nobelium, 591
Noble gas core, 83, 123
Noble gas family, 78–79,
        697–702
    compounds of, 699–700
    electron structure, 700
    properties, 698–699
    uses, 701–702
Noble metals, 638–639
Nomenclature
    inorganic acids, 329–330
    inorganic compounds, 155–158
    of organic compounds,
        766–768, 770–774, 783,
        789, 791

Nonanes, 765
Nonelectrolyte, 307–312, 343–350
Nonmetals, 121, 624–625
Nonpolar molecules, 138, 143
Normal alkanes, 767
Normality, 304–307, 334
Nu (ν), frequency, 93
Nuclear bomb, 591–594
Nuclear chain reaction, 432–434, 592–594
Nuclear chemical equations, 571
Nuclear energy, 16, 595–598
Nuclear explosions, 591–595
Nuclear fission, 588, 591–595
Nuclear fissionable materials, 594–595
Nuclear fission bombs, 591–594
Nuclear fusion, 588, 598–599
Nuclear fusion bombs, 599
Nuclear magnetic resonance, 615–617
Nuclear reactions, 565–602
    fission bombs, 591–594
    fusion bombs, 599
    production of energy, 595–598
Nuclear reactors, 595–598
Nucleic acid, 811–813
Nucleon, 583
Nucleus of the atom, 62–70
    stability, 589–590
    structure, 590
Nuclide, 565, 590
Nylon, 823

Octahedral, 179, 666
Octane rating of gasoline, 779
Octanes, 765, 779
Octet rule (Rule of eight), 126–128
Oil reserves, 285
Oil shale, 775
Olefins (Alkenes), 769–771
Oleic acid, 792
Oleum, 720
Onyx, 754
Opal, 754
Optical activity, 801
Optical isomers, 668, 801–803
Optics, 801
Orbitals, 81–87
    atomic, 106–107
        order of filling, 82–86
        s, p, d, f, 81–87
    molecular, 188–190
        antibonding, 190–191
        bonding, 190–191
        in coordination compounds, 669–673

degenerate, 116
dsp^2, 666
dsp^3, 666
d^2sp^3, 666
sp^3d, 177, 666
sp^3d^2, 177, 666
sp^3d^3, 685, 701
geometries, 666, 684
hybrid, 167–178, 184–185, 666, 670
MO theory, 188–190
order of filling, 113–114
pi, 167, 194
sigma, 166–167
sp, 169, 666
sp^2, 170, 666
sp^3, 170, 666
VB theory, 164–187
Order of reaction, 421–431
Ores, 641–645
    concentration, 642–643
    reduction, 643–645
    refining, 645
    roasting, 642–643
    smelting, 643–645
Organic compounds, 132–133, 765–794
Orlon, 822
Orthoclase, 756
Ortho isomers, 773
Orthosilicic acid, 744
Osmium, 634, 636
Osmosis, 355–360
    reverse, 357
Osmotic pressure, 357–360
Oxalic acid, 791
Oxidation, 150, 268–271
    of alcohols, 786–788
    of hydrocarbons, 778–779
    in living cells, 815–816
Oxidation numbers, 150–158
Oxidation–reduction equations, 270–271, 547–551
Oxidation–reduction reactions, 269–271, 312–315, 656–663
Oxidation state, 150–155
    and acid strengths, 331
    emf diagrams, 657–658, 662, 682, 710, 724
    in naming, 157–158
    from periodic table, 151–153
    rules, 151
    transition elements, 637–638
Oxides
    acidic and basic, 286
    metallic, 286, 630–631
    nonmetallic, 286
Oxidizing agents, 269

Oxy-acids, 329–331
    of halogen family, 696–697
    nomenclature, 329–330
    of sulfur family, 715–720
Oxygen
    atomic, 265
    chemical properties, 265–267
    compounds of, 264
    discovery of, 262–263
    elemental, 263–265
    isotopes of, 263–264
    key chemical, 272
    occurrence, 264
    preparation, 264–265
    reactions of, 265–269
    sources of, 264–265
    (sulfur family), 707–721
    structure, 193
    uses of, 271–272
Oxygen fluoride, 694
Oxy-halogen compounds, 696–697
Oxy-ions, 156
Ozone, 272–278
    in atmosphere, 273–278
    resonance hybrid, 187
Ozonizer, 273

Palladium, 634, 636
Palmitic acid, 792
Paper, 809
para-aminobenzoic acid, 609
Paraffin wax, 776
Para isomers, 773
Paramagnetic, 669
Paraperiodic acid, 697
para-xylene, key chemical, 774
Parent compound, 767
Parent element, 570
Partial ionic bonds (Covalent-ionic bonds), 694
Partial pressures, 218–220
Pascal, 200–201, 837
Paschen, F., 99
    series, 99–104
Pasteur, Louis, 63, 801
Pauli, Wolfgang, 111, 566
Pauli exclusion principle, 111–113
Pauling, Linus, 139, 698, 811
Pentane, 765
Pentene, 770
Pentosan, 807
Pentoses, 807
Pentyne, 770
Peptide bond, 810
Peptization, 367
Percentage composition, 48–51
Period, 80, 89

Periodic acid, 697
Periodic law, 79, 89
Periodic table, 26–30, 80–81
    and boiling points, 28
    and electronegativities, 139
    filling orbitals, 113
    general divisions, 624–625
    and ionization energies, 74–79
    long form, 80–81
    Mendeleev's, 26–29
        and oxidation numbers, 151–153
    trends in, 138–140, 146–149, 624–625, 707, 741
    uses of, 88
Permanganate ion, 659–660
Peroxides, 288, 631
Perrin, Jean, 363
Petroleum, 775–777
Petroleum ether, 775–776
Petroleum jelly, 776
pH, 489, 498–502, 661
    measurement, 491
    of some common solutions, 490
Phase diagrams, 244–245
Phenanthrene, 774
Phenol, 789
    key chemical, 790
Phenolphthalein, 498–499
Phenols, 788, 823
Phlogiston, 262–264
Phosphate ion, 735
Phosphates, 735
Phosphine, 728–729
Phosphoric acid, 475, 732
    key chemical, 735
Phosphorous acid, 732
Phosphorus (Nitrogen family), 721–736
    key chemical, 727
Phosphorus(III) oxide, 726
Phosphorus(V) oxide, 727, 735
Phosphorus pentachloride, 177
Photochemical reaction, 274–275, 277–278
Photoelectric effect, 98–99
Photoelectron spectra
    of atoms, 115–116
    of molecules, 193–194
Photography, color, 612
Photon, 97
Photosynthesis, 272, 816–817
Physical change, 30
Pi bond, 167
Pig iron, 648
Pi orbital, 167, 190–191
Pituitary hormones, 818
pK_a, 496

Planar structure, 168, 179
Planck, Max, 97
Planck's constant, 98–105, 837
Plane-polarized light, 801–802
Plaster of Paris, 720–721
Plastics (Polymers), 822–823
Platforming, 777
Platinum, 634, 636
Platinum family, 634–641
Plücker, J., 61
Plutonium, 591, 594–595
Plutonium-241 series, 572
pOH, 490
Polar bond, 142
Polar-covalent bonds, 138–144, 309–311
Polariscope, 801
Polarized ions, 250–251
Polar molecule, 138
Polar solvent, 293–294
Pollution, 2
  acid rain, 716–717
  asbestos, 756
  carbon dioxide, 749
  and carbon monoxide, 268, 278, 747–748
  and Ethyl gasoline, 692
  eutrophication, 735
  mercury and chlorine, 692–693
  and oxides of nitrogen, 277
  and ozone, 277
  and phosphates, 735
  smog, 277–278
  smoke, 750
  and sulfur dioxide, 277
  synthetic organics, 820–821
Polonium, (Sulfur family), 707–721
Polyacrylonitrile, 822
Polyatomic ions, 154
Polyethylene plastics, 695, 822
Poly(ethylene terphthalate), 725, 823
Polymers, 822–823
Poly(methyl methacrylate), 822
Polymorphous crystals, 245
Polysaccharides, 803, 807–808
Polyunsaturated fats, 794
Poly(vinyl chloride), 822
Population growth, 2
p orbitals. See Orbitals
Positive ions, 122
Positron, 566
Potassium (Alkali family), 622–633
Potassium acid tartrate, 751
Potassium bicarbonate, 746, 752

Potassium chlorate, 697
Potassium chloride, 631, 688, 697
Potassium iodide, 694
Potassium nitrate, 736
Potassium ozonide, 631
Potassium peroxide, 631
Potassium sulfate, 632, 721
Potassium superoxide, 631
Potential energy, 16
Precipitate, 315
Precipitation analysis, 50
Precipitation reaction, 315
Pressure of gases, 199–201, 208–209
Priestley, Joseph, 262–263
Primary alcohols, 782
Primary cell, 556
Principal quantum number, 100
Prism spectroscope, 94
Probable electron density, 107
Producer gas, 748
Product, 55
Propagation reaction, 433
Propane, 765–767
Propanoic acid, 789
Propanols, 782
Propanone (Acetone), 782, 787–788
Propene, 769–770, 772
  key chemical, 772
Properties
  chemical, 15
  extrinsic, 14, 15
  intrinsic, 14
  physical, 15
Propionic acid (Propanoic acid), 789
Propyl alcohol, 782
Propylene. See Propene
Propyl group, 767
Propyne, 770
Protective colloids, 367
Proteins, 809–815
Protium, 278
Proton, 66–67
  rest mass, 837
  size, 73
  weight, 66
Proust, Joseph, 21
Pseudo first-order reaction, 424–425
Pyrite, 715
Pyrolysis, 264
Pyrosulfuric acid, 720
Pyroxenes, 756

Quadratic formula, 827
Qualitative analysis, 507, 523–526

Quanta of energy, 97–98
Quantum mechanics, 106–115
Quantum numbers, 107–111
Quantum theory, 97–98, 99–104
Quartz, 754, 756
Quicklime (Lime), 630–631

R, the ideal gas constant, 212–213, 222, 463, 541, 837
Radial probability, 108
Radiant energy, 16, 607
  effect on substances, 607
  monochromatic, 246
  photon of, 97
Radiation, detection, 567–569
Radical (Free radical), 405, 433
  heat of formation of, 384–385
Radioactive decay, 565–566
Radioactive elements, 565–567
Radioactive emissions, 64, 567
Radioactive nuclides, 565
Radioactive series, 569–573
Radioactivity, 62–64, 565–602
Radium, 567, 570
  See also Alkaline earth family, 626–633
Radius ratio, 249–250
Radon (Noble gas family), 697–702
Ramsay, Sir William, 96, 697
Raoult's law, 344
Rare earth metals (Lanthanide series), 625, 635
Rate
  of diffusion (effusion), 220–221
  of reaction. See Reaction rate
Rate constants, 419
Rate equations (rate laws), 418–420, 423
  derivatives of, 423, 835
Ratio of atoms in compound, 37, 51–53
Rayon, 809
Reactant, 35
  excess, 46–47
  limiting, 46–47
Reaction
  direct combination, 37–39
  double displacement, 37–39
  precipitation, 315
  single displacement, 37–39
Reaction intermediate, 407
Reaction mechanism, 404–407, 423, 431–432
Reaction order, 421–431

Reaction rate, 407–418, 619
  constant, 419
  determining step, 431–432
  factors affecting, 410–418
  instantaneous, 417–418
Reactor, nuclear, 595–598
Recrystallization, 300
Rectifier, 745
Redox equations, 270–271, 547–551
Redox reactions, 269–271, 312–315, 656–663
Reducing agent, 269
Reduction, 150, 269–271
  of ores, 643–645
Reduction potentials, 537–539, 842–845
Refining of metals, 645
Refining of petroleum, 775
Reforming, 777
Relative atomic weights, 25
Relative humidity, 242
Relativistic mass, 585
Rem, 573
Resins, 359, 635
Resonance, 182–187
  benzene, 185
Resonance energy, 185–186
Resonance hybrid, 183–186
Resonance structure, 183–186
Rest mass, 585
Reversible reactions, 441
Rhenium, 634, 636
Rhodium, 634, 636
Rhombic sulfur, 709
Ribonucleic acid, 811–813, 818
Ribosomal RNA, 813
RNA, 811–813, 818
Roasting of ores, 642–643
Rocking motion, 613
Roentgen, 65, 838
Roots, 831–833
Rotational motion, 224
Rotational spectra, 615
Rubidium (Alkaline earth family), 626–633
Rules of eight and two, 126–128
  limitations, 127, 131
Ruthenium, 634, 636
Rutherford, Ernest, 63, 579
Rutherford's experiment, 64–65
Rydberg constant, 99, 102, 104, 837

S, ΔS, 392–393
Saccharides, 803–809
Saccharin, 806

Salicylic acid, 791
Salt bridge, 533
Saltpeter, 736
Salts, 312
  reaction with water, 294,
    311, 484
  solubilities, 296, 299
Sand, 727, 754
Sanger, F., 811
Saponification, 409, 793
Saturated hydrocarbons,
    780, 792
Saturated solutions, 296–297
Scale models, 188
Scandium, 634, 636
Scandium family, 634–641
Scheele, C. W., 263
Schrödinger, E., 106
Schrödinger equation, 106–107
Screening effect, 149
Seaborg, G. T., 591
Secondary alcohols, 782
Second-order reaction, 421,
    428–430
Secondary cell, 556
Selective precipitation, 515
Selenates, 720–721
Selenides, 714
Selenium (Sulfur family),
    707–721
Semiconductor, 744
Semipermeable membranes,
    355–357
Serendipity, 63
Sharing of electrons, 128–132
Shell model, 590
Shells of electrons (Energy
    levels), 77–78
Sigma bond, 165–167, 194
Sigma orbital, 166–167,
    190–191
Significant figures, 827–828
Silica, 727, 754, 756
Silica gel, 745
Silicates, 754–757
Silicic acid, 745
Silicon, 741–761
Silicon carbide, 753, 755
Silicon dioxide (Silica), 727,
    754, 756
Silicones, 758–759
Silicon tetrafluoride, 694
Silver, 634–641
Silver ammonia ion,
    521–522, 665
Silver bromide, 692
Silver chloride, 694
Silver fluoride, 694
Single bond (Covalent bond),
    133–134

SI units, 4–14, 836–838
Slag, 642–643
Slaked lime, 631
Slaking, 631
Smelting, 643
Smog, 277–278
Smoke, 364, 750
Soaps, 793
Soda (Sodium
    bicarbonate), 751
Soda ash (Sodium carbonate),
    632, 751
Sodium (Alkali family),
    626–633
Sodium aluminum
    fluoride, 688
Sodium aluminum sulfate, 751
Sodium ammonium
    tartrate, 802
Sodium bicarbonate, 751
Sodium bisulfite, 691
Sodium borate, 747
Sodium carbonate, 632, 751
  key chemical, 632
Sodium chloride, 249, 631,
    688, 694
Sodium cyanide, 753, 755
Sodium fluoride, 694
Sodium hydroxide, 631, 689
  key chemical, 631
Sodium hypochlorite, 157
Sodium iodate, 688, 691
Sodium nitrate, 691
Sodium peroxide, 631
Sodium phosphate, 735
Sodium silicate, 746
Sodium sulfate, 632, 721
Sodium–sulfur cell, 558
Solar cells, 744–745
Solids, 244–257
  covalent compounds,
    251–253
  crystalline, 245–248
  ionic compounds, 248–251
  kinetic theory of, 230
  in liquids, 299–301
  metallic, 253
Solid solutions, 293
Solid state, 244–257
Sols, 362
Solubility, 296–298
  of certain solids, 296
  factors influencing, 299–301,
    512–514
Solubility product constant,
    507–510
  table of, 510
Solute, 293
Solutions, 293–317
  boiling point, 345–346

buffered, 493–498
concentration, 297–302
defined, 293
of electrolytes, 307–312,
    350–353
factors involved in
    formation of, 294
freezing point, 346–347
gases in, 301
ideal, 349–350
liquid, 293
molality, 303
molarity, 303–304
mole fraction, 302–303
normality, 304–307, 334
properties of, 343–350
saturated, 296–297
seeding of, 297–298
solids in, 299
supersaturated, 297–298
types of, 293
unsaturated, 297–298
vapor pressure, 344–345
Solvated ion, 294
Solvation, 294
Solvay process, 632
Solvent, 293, 298
Solvent extraction, 298
s orbitals. See Orbitals
Specific gravity, 9
Specific heat, 11, 231
  table of, 13
Spectator ion, 313
Spectra, 94, 608
  absorption, 96–97, 607–608
  atomic, 94–104
  band, 607–608
  continuous, 96
  discontinuous, 96
  emission, 95–96
  flame, 627
  infrared, 611–615
  line, 607
  mass, 608, 617–618
  nmr, 615–617
  photoelectron, 115–116,
    193–194
  sun, 96, 698
  ultraviolet, 608–611
  vibrational, 613
  visible, 96, 608–611
  X-ray, 66
Spectrograph, mass, 67–69
Spectroscope, 94
Spectroscopy, 93–94, 607–619
Speed of reaction,
    407–418, 619
Spontaneous reaction, 540–541
Square planar, 143, 179, 666
Stability constant, 520–523

Stalactites, 753
Stalagmites, 753
Standard comparison
    electrode, 536
Standard conditions, 209–210
Standard electrode potentials,
    535–539, 544
Standard enthalpies, 376–383
Standard pressure, 209–210
Standard reduction potentials,
    tables of, 538, 638, 661,
    842–845
Standard states, 376–377
Standard temperature, 209–210
Starch, 785, 807–808
State function, 378–379
States of matter, 198
  changes of, 230–234
Stationary states, 100
Stearic acid, 792
Steel, 649
Stellar energy, 599
Stereoisomers, 799–803
Steric hindrance, 250
Steroid, 775
Stibine, 729
Stoichiometry, 35, 40–54
Storage battery, 531, 556–559
STP, 209–210, 837
Straight-run gasoline, 776
Stretching of bonds, 612–613
Strong acid, 331–333
Strong base, 331–333
Strong electrolyte, 311–312
Strontium (Alkaline earth
    family), 626–633
Structural formula, 128–138,
    766, 771–772
  condensed, 766
Structural isomers, 799
Styrene, 822
Subcritical mass, 593–594
Subgroups, 80, 89, 634
Sublevels (Energy levels), 81–87
Sublimation, 232–233
  enthalpy of, 387
Substitution reactions, 779
Sucrose, 806, 808
Sugars, 803–808
  comparative sweetness, 806
Sulfates, 632–633, 720–721
Sulfides
  metallic, 712, 714
  poly-, 712, 714
Sulfur (Sulfur family), 702–721
Sulfur dioxide, 713, 715, 719
  resonance hybrid, 182
Sulfur family, 707–721
  chemical properties, 709–710
  compounds of, 712–721

Sulfur family (*Continued*)
  emf diagram, 710
  key chemical, 713
  occurrence, 712–713
  oxidation–reduction,
    710–712
  physical properties, 707–709
  polymorphism, 709
  production, 712–713
Sulfur hexafluoride, 177
Sulfuric acid, 330–331,
    717–720
  contact, 718
  ionization, 475
  key chemical, 718
  lead chamber, 719
Sulfur monochloride, 694
Sulfur tetrachloride, 180
Sulfurous acid, 331, 475, 715
Sulfur trioxide, 185–187,
    716–717, 719
Supercritical mass, 593
Superheating, 243
Superphosphate, 736
Supersaturated, 242, 297–298
Surface tension, 238
Sylvinite, 688
Sylvite, 688
Symbols of elements, 25–26
Synapses, 819
Synchrotron, 582
Synthetic organics, 820–821
Synthetic rubber, 771, 822
Synthetic sweeteners, 806
Systematic names, 155,
    766–768
Système International, 4–14,
    836–838

Tantalum, 634, 636
Tartaric acid, 791
Technetium, 583, 634, 636
Teflon, 692, 822
Tellurates, 720–721
Tellurides, 714
Tellurium (Sulfur family),
    707–721
Temperature, 10
  absolute, 10, 206
  and equilibrium
    concentrations, 465–466
  and equilibrium constants,
    465–466
  and pressure, 204–209
  and reaction rates, 412–413
  and solubility, 299–301
  and vapor pressure, 241
  and volume, 206–207
Temperature scales, 10
Terephthalic acid, 791

Termination reaction, 433
Ternary compounds, 156–157
Tertiary alcohols, 782
Testosterone, 818
Tetrahedral, 143, 179, 666
Thallium, 634, 636
Theoretical chemistry, 2
Theoretical yield, 45–46
Theory, 4
Thermal energy, 586
Thermochemical equation,
    374–376
Thermochemistry, 372–389
Thermodynamic properties,
    table, 380, 395
Thermodynamics, 372–399
  first law of, 389–390
  second law of, 391–392
  third law of, 392
Thermonuclear bombs, 599
Thermonuclear power,
    599–602
Thiosulfates, 721
Third-order reactions, 422
Thomson, J. J., 62, 67
Thorium, radioactive, 570–571
Threshold frequency, 98
Thymolphthalein, 499
Tin, 634, 636
Tin(II) fluoride, 734
Tin(IV) chloride, 694
Tin(IV) fluoride, 694
Tin(IV) iodide, 694
Titanium, 634, 636
Titanium dioxide, key
    chemical, 637
Titanium family, 634–641
Titration, 335–337, 499–502
  curves, 500
TNT, 731
Toluene, 772–773
Torr, 200, 838
Torricelli, E., 199–200
Tracers, radioactive, 578
Transfer of electrons, 123–126
Transfer RNA, 813
Trans isomer, 667, 770
Transistors and solar cells,
    744–745
Transition elements, 625,
    633–641, 655–675
Transition state, 405–407
Transition temperature, 709
Translational motion, 224
Transuranium elements, 591
Travertine, 752
Tremolite, 756
Trends in periodic table,
    138–140, 146–149,
    624–625, 707, 741

Trigonal bypyramid, 143,
    179, 666
Trigonal planar, 143, 179, 666
Trinitro benzene, 499
Trinitrobenzene, 499
Triple bond, 134, 174
Triple point, 244–245
Triprotic acid, 324, 475
Trisodium phosphate, 735
Tritium, 278
Trivial names, 155, 767
Trona, 632
Tungsten, 634, 636
Turnover number, 814
Tyndall effect, 362–364

Ultraviolet radiation, 276
  spectrum, 608
Unit cell, 247–248
Units of measure, 5, 8, 10,
    836–838
Universe, heat death of, 392
Unknown, 335
Unsaturated hydrocarbons,
    780, 792
Unshared pairs, 130
Uranium-233, 594–595
Uranium-235, 570–571
Uranium-238, 570–571
Uranium hexafluoride, 594
Uranium radioactivity series,
    569–570
Urea, 735, 736, 765
  key chemical, 735

Valence, 121–122
  primary, 656
  secondary, 654, 664
Valence bond theory, 164–187
Valence electrons, 122
Valence shell electron pair
    repulsion (VSEPR),
    178–182
Vanadium, 634, 636
Vanadium family, 634–641
Vanadium pentoxide, 718–719
van der Waals, J. H., 144
van der Waals forces, 235–239
van der Waals radius, 144–148
van't Hoff factor, 351, 465
Vapor, 240
Vaporization, 240
Vapor pressure, 240–242,
    353–354
  and distillation, 353–354
  effect of structure, 241
  effect of temperature, 241
  measurement of, 241
  of solutions, 344–345
  of water, 241, 839

Vaseline (Petroleum
    jelly), 776
VB theory, 164–187
Verhoek, F. H., 463
Vibrational energy, 611–615
Vibrational motion, 224
Vibrations in molecules,
    611–615
Vinegar, 789
Vinyl chloride, 820, 822
Viruses, 818
Viscose, 809
Visible spectra, 608
Vital force theory, 765, 799
Volatile, 240
Volt, 535, 837
Voltaic cell, 531–535, 539–544
  concentration effects, 541
  potential of, 539–544
  representation of, 539–540
Volume of gases
  and pressure, 389–390
  and temperature, 206–207
Volumetric analysis, 335
VSEPR theory, 178–182

W_0 (Work function), 99
Wagging motion, 613
Wahl, A. C., 114
Wahl diagrams, 115, 166–167
Washing of ores, 642
Washing soda (Sodium
    carbonate), 632
Water
  association of, 253
  chemical behavior, 286–287
  density, 253
  electrolysis, 281
  hydrogen bonds in, 236
  ionic equilibrium, 311
  ionization of, 482
  ion product, 483–484
  pH of, 490
  reaction with salts, 327–329
  softening, 760–793
  sp^3 hybridization in, 173
  structure, 252–253
  triple point, 244–245
  vapor pressure, 241, 839
Water of crystallization, 287
Water gas, 748
Watson, J. D., 812–813
Watt, 837
Wave function, 106–107
Wavelength
  of electromagnetic
    radiation, 93
  table of, 95
Wave mechanics (Quantum
    mechanics), 106–115

Wave nature of matter, 104–105

Waxes, 791
  beeswax, 791
  carnauba, 791
  paraffin, 791
  spermaceti, 791

Weak acid, 332–333, 473–475
Weak base, 332–333, 475–476
Weak electrolyte, 311–312, 518–519
Weight, 8

Werner, Alfred, 664–668
Wilson, C. T. R., 567–568
Wöhler, Frederick, 765
Wolfram (Tungsten), 634–636
Wood (Cellulose), 807–809
Wood alcohol. *See* Methanol
Work function, 99

Xenon (Noble gas family), 697–702
Xenon compounds, 699–701

Xerox principle, 708
X rays, 65–66
  and atomic number, 65–66
  and crystal structure, 246–247
X-ray tube, 65–66
Xylenes, 773–774
  key chemical, 774

Yield
  percentage, 45–46
  theoretical, 45–46

Yttrium, 634, 636

Z, atomic number, 65–66, 79
Zeolite, 359, 635, 759–760
Zero-order reaction, 421
Zinc, 634, 636
Zinc chloride
  hexahydrate, 287
Zinc family, 634–641
Zirconium, 634, 636
Zone refining, 645
Zymase, 785

80 81 82 83   10 9 8 7 6 5 4 3 2 1

Table of Atomic Weights 1977

(Scaled to the relative atomic mass, $A_r(^{12}C) = 12$)

The atomic weights of many elements are not invariant but depend on the origin and treatment of the material. The footnotes to this table elaborate the types of variation to be expected for individual elements. The values of $A_r(E)$ given here apply to elements as they exist naturally on earth and to certain artificial elements. When used with due regard to the footnotes they are considered reliable to ±1 in the last digit or ±3 when followed by an asterisk*. Values in parentheses are used for certain radioactive elements whose atomic weights cannot be quoted precisely without knowledge of origin; the value given is the atomic mass number of the isotope of that element of longest known half-life.

NAME	SYMBOL	ATOMIC NUMBER	ATOMIC WEIGHT	NAME	SYMBOL	ATOMIC NUMBER	ATOMIC WEIGHT	NAME	SYMBOL	ATOMIC NUMBER	ATOMIC WEIGHT
actinium	Ac	89	227.0278^z	helium	He	2	4.00260^x	radium	Ra	88	226.0254^{x,z}
aluminium	Al	13	26.98154	holmium	Ho	67	164.9304	radon	Rn	86	(222)
americium	Am	95	(243)	hydrogen	H	1	1.0079^w	rhenium	Re	75	186.207
antimony	Sb	51	121.75*	indium	In	49	114.82^x	rhodium	Rh	45	102.9055
argon	Ar	18	39.948*^{w,x}	iodine	I	53	126.9045	rubidium	Rb	37	85.4678*^{x}
arsenic	As	33	74.9216	iridium	Ir	77	192.22*	ruthenium	Ru	44	101.07^{xx}
astatine	At	85	(210)	iron	Fe	26	55.847*	samarium	Sm	62	150.4^x
barium	Ba	56	137.33^x	krypton	Kr	36	83.80^{x,y}	scandium	Sc	21	44.9559
berkelium	Bk	97	(247)	lanthanum	La	57	138.9055*^{x}	selenium	Se	34	78.96*
beryllium	Be	4	9.01218	lawrencium	Lr	103	(260)	silicon	Si	14	28.0855*
bismuth	Bi	83	208.9804	lead	Pb	82	207.2^{w,x}	silver	Ag	47	107.868^x
boron	B	5	10.81^{w,y}	lithium	Li	3	6.941*^{w,x,y}	sodium	Na	11	22.98977
bromine	Br	35	79.904	lutetium	Lu	71	174.967*	strontium	Sr	38	87.62^x
cadmium	Cd	48	112.41	magnesium	Mg	12	24.305^x	sulfur	S	16	32.06^w
calcium	Ca	20	40.08^x	manganese	Mn	25	54.9380	tantalum	Ta	73	180.9479*
californium	Cf	98	(251)	mendelevium	Md	101	(258)	technetium	Tc	43	(98)
carbon	C	6	12.011^w	mercury	Hg	80	200.59*	tellurium	Te	52	127.60^{xx}
cerium	Ce	58	140.12^x	molybdenum	Mo	42	95.94	terbium	Tb	65	158.9254
cesium	Cs	55	132.9054	neodymium	Nd	60	144.24*^{x}	thallium	Tl	81	204.37*
chlorine	Cl	17	35.453	neon	Ne	10	20.179^{x,y}	thorium	Th	90	232.0381^{x,z}
chromium	Cr	24	51.996	neptunium	Np	93	237.0482^z	thulium	Tm	69	168.9342
cobalt	Co	27	58.9332	nickel	Ni	28	58.70	tin	Sn	50	118.69*
copper	Cu	29	63.546*^{w}	niobium	Nb	41	92.9064	titanium	Ti	22	47.90*
curium	Cm	96	(247)	nitrogen	N	7	14.0067	tungsten (Wolfram)	W	74	183.85*
dysprosium	Dy	66	162.50*	nobelium	No	102	(259)	unnilhexium	Unh	106	(263)
einsteinium	Es	99	(252)	osmium	Os	76	190.2^x	unnilpentium	Unp	105	(262)
erbium	Er	68	167.26*	oxygen	O	8	15.9994*^{w}	unnilquadium	Unq	104	(261)
europium	Eu	63	151.96^x	palladium	Pd	46	106.4^x	uranium	U	92	238.029^{x,y}
fermium	Fm	100	(257)	phosphorus	P	15	30.97376	vanadium	V	23	50.9415*
fluorine	F	9	18.998403	platinum	Pt	78	195.09*	xenon	Xe	54	131.30^{x,y}
francium	Fr	87	(223)	plutonium	Pu	94	(244)	ytterbium	Yb	70	173.04*
gadolinium	Gd	64	157.25^{xx}	polonium	Po	84	(209)	yttrium	Y	39	88.9059
gallium	Ga	31	69.72	potassium	K	19	39.0983*	zinc	Zn	30	65.38
germanium	Ge	32	72.59*	praseodymium	Pr	59	140.9077	zirconium	Zr	40	91.22^x
gold	Au	79	196.9665	promethium	Pm	61	(145)				
hafnium	Hf	72	178.49*	protactinium	Pa	91	231.0359^z				

Reproduced by permission of the International Union of Pure and Applied Chemistry. *Pure and Applied Chemistry*, **51** (2), 407 (1979).

^{w} Element for which known variations in isotopic composition in normal terrestrial material prevent a more precise atomic weight being given; $A_r(E)$ values should be applicable to any "normal" material.

^{x} Element for which geological specimens are known in which the element has an anomalous isotopic composition, such that the difference between the atomic weight of the element in such specimens and that given in the table may exceed considerably the implied uncertainty.

^{y} Element for which substantial variations in A_r from the value given can occur in commercially available material because of inadvertent or undisclosed change of isotopic composition.

^{z} Element for which the value of A_r is that of the radioisotope of longest half-life.